Studies in Systems, Decision and Control

Volume 487

Series Editor

Janusz Kacprzyk, Systems Research Institute, Polish Academy of Sciences, Warsaw, Poland

The series "Studies in Systems, Decision and Control" (SSDC) covers both new developments and advances, as well as the state of the art, in the various areas of broadly perceived systems, decision making and control–quickly, up to date and with a high quality. The intent is to cover the theory, applications, and perspectives on the state of the art and future developments relevant to systems, decision making, control, complex processes and related areas, as embedded in the fields of engineering, computer science, physics, economics, social and life sciences, as well as the paradigms and methodologies behind them. The series contains monographs, textbooks, lecture notes and edited volumes in systems, decision making and control spanning the areas of Cyber-Physical Systems, Autonomous Systems, Sensor Networks, Control Systems, Energy Systems, Automotive Systems, Biological Systems, Vehicular Networking and Connected Vehicles, Aerospace Systems, Automation, Manufacturing, Smart Grids, Nonlinear Systems, Power Systems, Robotics, Social Systems, Economic Systems and other. Of particular value to both the contributors and the readership are the short publication timeframe and the world-wide distribution and exposure which enable both a wide and rapid dissemination of research output.

Indexed by SCOPUS, DBLP, WTI Frankfurt eG, zbMATH, SCImago.

All books published in the series are submitted for consideration in Web of Science.

Allam Hamdan · Esra Saleh Aldhaen
Editors

Artificial Intelligence and Transforming Digital Marketing

Volume 1

Springer

Editors
Allam Hamdan
College of Business and Finance
Ahlia University
Manama, Bahrain

Esra Saleh Aldhaen
Department of Management and Marketing
Ahlia University
Manama, Bahrain

ISSN 2198-4182   ISSN 2198-4190 (electronic)
Studies in Systems, Decision and Control
ISBN 978-3-031-35827-2   ISBN 978-3-031-35828-9 (eBook)
https://doi.org/10.1007/978-3-031-35828-9

© The Editor(s) (if applicable) and The Author(s), under exclusive license to Springer Nature
Switzerland AG 2024

This work is subject to copyright. All rights are solely and exclusively licensed by the Publisher, whether
the whole or part of the material is concerned, specifically the rights of translation, reprinting, reuse
of illustrations, recitation, broadcasting, reproduction on microfilms or in any other physical way, and
transmission or information storage and retrieval, electronic adaptation, computer software, or by similar
or dissimilar methodology now known or hereafter developed.
The use of general descriptive names, registered names, trademarks, service marks, etc. in this publication
does not imply, even in the absence of a specific statement, that such names are exempt from the relevant
protective laws and regulations and therefore free for general use.
The publisher, the authors, and the editors are safe to assume that the advice and information in this book
are believed to be true and accurate at the date of publication. Neither the publisher nor the authors or
the editors give a warranty, expressed or implied, with respect to the material contained herein or for any
errors or omissions that may have been made. The publisher remains neutral with regard to jurisdictional
claims in published maps and institutional affiliations.

This Springer imprint is published by the registered company Springer Nature Switzerland AG
The registered company address is: Gewerbestrasse 11, 6330 Cham, Switzerland

Foreword

The emergence of digital marketing has changed the way people interact and communicate across the globe using Internet and specialized digital communication supported by technological advancements. One type of such advancements is artificial intelligence (AI), a supportive technological agent that plays an important role in digital marketing. It is used by most of businesses as an effective tool to interact and reach out to stakeholders including customers. Its benefits have gone beyond businesses to include many other fields such as education and health care. Digital marketing is now used more effectively compared with a traditional approach of marketing. Using AI in digital marketing has enhanced marketing communication where many digital tools can be connected to develop an effective communication campaign. Such technological tools include social media, email, and website. Companies that are quickly and wisely adopting such technology enjoy a good level of competitive advantage over their rivals. It is undoubtedly that AI can bring sustainability to companies adopting it where they can manage their current and future resources efficiently and contribute well to the economy and well-being of the society.

This book presents and discusses issues of digital marketing using AI for better marketing penetration, users' satisfaction and business efficiency. It covers important topics such as artificial intelligence, marketing and social media, cultural marketing, artificial intelligence and digital learning, innovation and sustainable operations, AI, banking and financial technology, tech-management in different disciplines and the role of digital marketing and governance and business ethics with AI.

Enjoy reading this book.

Prof. Muneer Al Mubarak
Professor of Management
and Marketing
Ahlia University
Manama, Bahrain

Preface

The use of artificial intelligence has become an essential part of decision-making, the wealth of data collection and data analysis enables drawing targeted audience and setting marketing strategies. The use of artificial intelligence has become vital in all aspects for instance recently AI-based applications including ChatGPT has created a serious fear at different level that AI will be taking over jobs from different fields.

The issue became more acute with the fear of humans that AI will take over jobs in the market and increase the unemployment rate on the other hand, new jobs are being created that requires more advanced skills in technology and digitalization. Several studies highlighted that AI is a potential tool to support humans to perform effectively with clear impact measures. In terms of marketing, the use of AI is proven to be the most effective considering the type of data generated to support market positioning and segmentation. However, considering the booming of AI there are still arising questions in terms of sustainability, would AI be sustainable considering lacking emotional intelligence that impacts social marketing. Other questions and issues are still a challenge for instance there are several ethical standards that need to be considered while using AI specifically while transforming from normal marketing to digital marketing that requires customer preservation of data as per the data protection acts. Therefore, this book provides an insight on future studies that support the integration of AI towards the use of digitalization and transformation of marketing strategies including and restricted to marketing but also managing aspects of AI and the use of digitalization towards transformation. The book aims to attract researchers globally to generate research outcomes that support the digital transformation and the use of AI towards transformation of digital marketing that could make a difference. The outcome of this research outcomes will support researchers and policymakers to have insights on different methods to be adopted to use AI effectively and support sustainability.

This book includes one hundred chapters. All of the chapters have been evaluated by the editorial board and reviewed based on double-blind peer review system by at least two reviewers.

The chapters of the book are divided into five main parts:

I. Artificial Intelligence, Marketing and Social Media.
II. Cultural Marketing, Artificial Intelligence and Digital Learning, Innovation and Sustainable Operations.
III. AI, Banking and Financial Technology.
IV. Tech-Management in Different Disciplines and the Role of Digital Marketing.
V. Governance and Business Ethics with AI.

The chapters of this book present a selection of high-quality research on the theoretical and practical levels, which ground the uses of AI in marketing, business, health care, media, education and other vital areas. We hope that the contribution of this book will be at the academic level and decision-makers in the various economic and executive levels.

Manama, Bahrain   Allam Hamdan
                                                                            Esra Saleh Aldhaen

Contents

Artificial Intelligence, Marketing and Social Media

Digital Marketing and Sustainable Businesses: As Mobile Apps in Tourism 3
Mahmoud Alghizzawi, Mohammad Habes, Abdalla Al Assuli, and Abd Alrahman Ratib Ezmigna

Exploring Small and Medium Enterprises Expectations of Electronic Payment in Kuwait 15
Ayman Hassan, Arezou Harraf, and Wael Abdallah

Content Marketing Strategy for the Social Media Positioning of the Company AD y L Consulting—Peru 27
Húber Rodríguez-Nomura, Edwin Ramirez-Asis, K. P. Jaheer Mukthar, Magdalena Valdivia-Malhaber, María Rodríguez-Kong, Nathaly Zavala-Quispe, and José Rodríguez-Kong

An Exploratory Study on E-commerce in the Sultanate of Oman: Trends, Prospects, and Challenges 37
Dawood Al Hamdani and Aisha Al Wishahi

The Impact of Researchers' Possessed Skills on Marketing Research Success 53
Ahmad M. A. Zamil, Ahmad Yousef Areiqat, Mohammed Nadem Dabaghia, and Jamal M. M. Joudeh

Impact of Online Food Ordering System and Dinning on Consumers Preference with Special Reference to Bangalore 63
J. Chandrakhanthan and C. Dhanapal

Green Economy Mechanisms in the Age of Technology and the Circular Economy 73
CH. Raja Kamal and Surjit Singha

ix

Whether Digital or Not, the Future of Innovation and Entrepreneurship 81
Yacoub Hamdan, Ahmad Yousef Areiqat, and Ahmad Alheet

Customer Buying Intention Towards Smart Watches in Urban Bangalore 87
B. Subha and Jaspreet Kaur

A Fluid-Thermal-Electric Model Based on Performance Analysis of Semi-transparent Building Integrated Photovoltaic Solar Panels 95
Nesrine Gaaliche, Hasan Alsatrawi, and Christina G. Georgantopoulou

The Impact of Social Media Influencer Marketing on Purchase Intention in Bahrain 113
Aysha AlKoheji, Allam Hamdan, and Assma Hakami

Tomorrow's Jobs and Artificial Intelligence 121
Ismail Noori Mseer, Yasser M. Abolelmagd, and Wael F. M. Mobarak

Artificial Intelligence and Security Challenges 129
Ismail Noori Mseer and Syed Muqtar Ahmed

Modern Services Quality and Its Impact on the Satisfaction of the Trade Shows in Jordan 137
Mustafa S. Al-Shaikh and Feras Alfukaha

Artificial Intelligence Application to Reduce Cost and Increase Efficiency in the Medical and Educational Sectors 151
Khaled Delaim, Muneer Al Mubarak, and Ruaa Binsaddig

Data Journalism and Its Applications in Digital Age 163
Abdulsadek Hassan and Mohammed Angawi

Digital Public Relations and Communication Crisis 177
Abdulsadek Hassan and Mohammed Angawi

Employing Applying Big Data Analytics Lifecycle in Uncovering the Factors that Relate to Causing Road Traffic Accidents to Reach Sustainable Smart Cities 193
Mohammad H. Allaymoun, Mohammed Elastal,
Ahmad Yahia Alastal, Tasnim Khaled Elbastawisy, Dana Iqbal,
Amal Yaqoob, and Adnan Sayed Ehsan

The Effects of Leadership Style on Employee Sustainable Behaviour: A Theoretical Perspective 205
Tamer M. Alkadash, Muskan Nagi, Ali Ahmed Ateeq,
Mohammed Alzoraiki, Rawan M. Alkadash, Chayanit Nadam,
Mohammad Allaymoun, and Mohammed Dawwas

Contents

Toward Sustainable Smart Cities: Design and Development of Piezoelectric-Based Footstep Power Generation System 215
M. N. Mohammed, Shahad Al-yousif, M. Alfiras, Majed Rahman, Adnan N. Jameel Al-Tamimi, and Aysha Sharif

Cultural Marketing, Artificial Intelligence and Digital Learning, Innovation and Sustainable Operations

Sustainable Urban Street Design in Jeddah, Saudi Arabia 227
Marwa Abouhassan

Green Discourse Analysis on Twitter: Imperatives to Green Product Management in Sustainable Cities (SDG11) 245
Priya Sachdeva, M. Dileep Kumar, and Archan Mitra

Harnessing Geographic Information System (GIS) by Implementing Building Information Modelling (BIM) to Improve AEC Performance Towards Sustainable Strategic Planning in Setiu, Terengganu, Malaysia 257
Siti Nur Hidayatul Ain Bt. M. Nashruddin, Siti Sarah Herman, Siti Nur Ashakirin Bt. Mohd Nashruddin, Sumarni Ismail, and Siow May Ling

The Role of Contract in Advancing ESG in Construction: A Proposed Framework of 'Green' Standard Form of Contract for Use in Green Building Projects in Malaysia 269
Khariyah Mat Yaman and Zuhairah Ariff Abd Ghadas

Sustainability of Web Application Security Using Obfuscation: Techniques and Effects 281
Raghad A. AlSufaian, Khaireya H. AlQahtani, Rana M. AlAjmi, Roa A. AlMoussa, Rahaf A. AlGhamdi, Nazar A. Saqib, and Asiya Abdus Salam

Heritage Buildings Changes and Sustainable Development 295
Marwa Abouhassan

Global Crucial Risk Factors Associated Stress Among University Students During Post Covid-19 Pandemic: Empirical Evidence from Asian Country 315
Nor Azma Rahlin, Ayu Suriawaty, Siti Aisyah Bahkiar Bahkiar, Suayb Turan, and Siti Nadhirah Ahmad Fauzi

Factors Influencing Purchase Decisions on Online Sales in Indonesia 329
Efa Wakhidatus Solikhah, Indah Fatmawati, Retno Widowati, and M. Suyanto

Re-Addressing the Notion of Patriarchy in Social and Economic
Framework in the Light of Women Empowerment 341
Jenni K. Alex and Dino Mathew

Development of Asset-Based and Asset-Backed Sukuk Issuance:
Case of Malaysia 351
Abdulmajid Obaid Saleh and Mohammed Waleed Alswaidan

Role of Colours in Web Banners: A Systematic Review
and Future Directions for e-Marketing Sustainability 361
Khalid Ali Alshohaib

Relationship of People's Knowledge, Behavior and Governmental
Action Towards Environmental Sustainability 369
Marluna Lim-Urubio and Manolo Anto

Development of Entrepreneurship in the Tourism and Recreation
Sphere: Marketing Research 379
Raisa Kozhukhivska, Olena Sakovska, Svitlana Podzihun,
Valentyna Lementovska, Ruslana Lopatiuk, and Nataliia Valinkevych

The Impact of RUPT on Corporate Environmental Responsibility 391
Acharya Supriya Pavithran, C. Nagadeepa, and Baby Niviya Feston

The Effect of the Characteristics of the Board of Directors
and the Audit Committee on Financial Performance: Evidence
from Palestine 401
Hisham Madi, Ghaidaa Abdel Nabi, Fadi Abdelfattah,
and Ahmed Madi

A Comparative Analysis of Rural and Urban Public Health
Infrastructure and Health Outcomes During Pandemic: With
Special Reference to Karnataka State 415
S. J. G. Preethi and Tinto Tom

Local Community Readiness to Implement Smart Tourism
Destination in Yogyakarta, Indonesia 425
Sri Dwi Ari Ambarwati, Mohamad Irhas Effendi,
and Sri Tuntung Pandangwati

Open Innovation and Governance Models in Public Sector:
A Systematic Literature Review 437
Meshari Abdulhameed Alsafran, Odeh Rashed Al Jayyousi,
Fairouz M. Aldhmour, and Eisa A. Alsafran

Corporate Governance in the Digital Era 453
Mary Yaqoob, Alreem Alromaihi, and Zakeya Sanad

Contents

Environmental, Social, and Governance (ESG) Impact on Firm's Performance 461
Fatema Alhamar, Allam Hamdan, and Mohamad Saif

AI, Banking and Financial Technology

A Stochastic Model for Cryptocurrencies in Illiquid Markets with Extreme Conditions and Structural Changes 479
Youssef El-Khatib and Abdulnasser Hatemi-J

Analysis of the Efficiency, Effectiveness and Productivity of Peruvian Motorcycle Cab Drivers in Times of Covid-19 Pandemic 489
Nelson Cruz-Castillo, Hernan Ramirez-Asis, K. P. Jaheer Mukthar, María Estela Ponce Aruneri, Juan Eleazar Anicama Pescorán, and Wilber Acosta-Ponce

Investment Awareness in Financial Assets—An Exploration Based on the Equity Traders in Bangalore City 501
Aneesha K. Shaji and N. Sivasankar

Cost–Benefit Analysis of Fintech Framework Adoption 509
Zainab Mohammed Baqer Shehab, Muneer Al Mubarak, and Amir Dhia

Financial Technology (Fintech) 525
Adnan Jalal, Muneer Al Mubarak, and Farah Durani

Assessing Opportunities and Challenges of FinTech: A Bahrain's View of Fintech 537
Sabika AlJalal, Muneer Al Mubarak, and Ghada Nasseif

Toward Sustainable Smart Cities: A New Approach of Solar and Wind Renewable Energy in Agriculture Applications 555
Nurhasliza Hashim, Tiffiny Grace Neo, M. N. Mohammed, Hakim S. Sultan, Adnan N. Jameel Al-Tamimi, and M. Alfiras

Digital Transformation Towards Sustainability in Higher Education: A New Approach of Virtual Simulator for Series and Parallel Diodes for a Sustainable Adoption of E-Learning Systems 565
Tan Chor Kuan, Khairul Huda Yusoff, M. N. Mohammed, Adnan N. Jameel Al-Tamimi, Norazliani Md Sapari, Firas Mohammed Ibrahim, and M. Alfiras

Toward Sustainable Smart Cities: Smart Water Quality Monitoring System Based on IoT Technology 577
Lee Mei Teng, Khairul Huda Yusoff, M. N. Mohammed, Adnan N. Jameel Al-Tamimi, Norazliani Md Sapari, and M. Alfiras

2019 Novel Coronavirus Disease (Covid-19): Toward a New Design for All-in-One Smart Disinfection System 595
M. N. Mohammed, M. Alfiras, Hakim S. Sultan, Adnan N. Jameel Al-Tamimi, Rabab Alayham Abbas Helmi, Arshad Jamal, Aysha Sharif, and Nagham Khaled

Modeling of Cutting Forces When End Milling of Ti6Al4V Using Adaptive Neuro-Fuzzy Inference System 605
Salah Al-Zubaidi, Jaharah A. Ghani, Che Hassan Che Haron, Hakim S. Sultan, Adnan N. Jameel Al-Tamimi, Mohammed N. Abdulrazaq Alshekhly, and M. Alfiras

Impact of the Pandemic—Covid-19 on Construction Sector in Bahrain 617
Zuhair Nafea Alani, Mohammed N. Abdulrazaq Alshekhly, Hamza Emad, Adnan N. Jameel Al-Tamimi, and M. Alfiras

The Role of Technology and Market Accessibility on Financial Market Classification 627
Reem Sayed Mansoor, Jasim Al Ajmi, and Asieh Hosseini

A Review of the Recent Developments in the Higher Education Sector Globally and in the GCC Region 635
Elham Ahmed, Amama Shaukat, and Esra AlDhaen

Intra-Industry Trade Trends in India's Manufacturing Sector: A Quantitative Analysis 653
Alan J. Benny, K. P. Jaheer Mukthar, Felix Julca-Guerrero, Norma Ramírez-Asís, Laura Nivin-Vargas, and Sandra Mory-Guarnizo

Risk Management of Civil Liability Resulting from Self-Driving Vehicle Accidents 667
Saad Darwish and Ahmed Rashad Amin Al-Hawari

The Impact of Applying Just-In-Time Production System on the Company Performance in Garment Manufacturing Companies in Jordan 677
Mohammad Abdalkarim Alzuod and Rami Atef Al-Odeh

Evaluation of Large-Scale Solar Photovoltaic Operating Plant by Using Environmental Impact Screening: A Case Study in Penang Island, Malaysia 691
Siti Isma Hani binti Ismail, Loh Yong Seng, Shanker Kumar Sinnakaudan, and Zul-fairul Zakaria

Prefabricated Plastic Pavement for High-Traffic and Extreme Weather Conditions 709
M. E. Al-Atroush, Nura Bala, and Musa Adamu

Contents

Highlighting The Role of UAE's Government Policies in Transition Towards "Circular Economy" 723
Tahira Yasmin, Ghaleb A. El Refae, and Shorouq Eletter

Tech-Management in Different Disciplines and the Role of Digital Marketing

A Systematic Literature Review on Mercantilism 739
Hannah Biju, K. P. Jaheer Mukthar, Norma Ramírez-Asís, Jorge Castillo-Picon, Guillermo Pelaez-Diaz, and Liset Silva-Gonzales

Revisiting the Nexus Between Suicides and Economic Indicators: An Empirical Investigation 751
Keerthana Sannapureddy, Jyoti Shaw, Shahid Bashir, and S. Vijayalakshmi

Dynamics of Sustainable Economic Growth in Emerging Middle Power Economies: Does Institutional Quality Matter? 761
Mithilesh Phadke, Jerold Raj, Sujay Rao, Shahid Bashir, and Jibrael Jos

Land Market in Ukraine: Functioning During the Military State and Its Development Trends 773
Reznik Nadiia, Havryliuk Yuliia, Yakymovska Anna, Halahur Yulia, Klymenko Lidiia, Zhmudenko Viktoriia, and Ischenko Valeriy

A Study on the Tendencies in the Trade Balance of the Indian Economy 783
K. Vinodha Devi and V. Raju

Quality Efficacy Issues in Mangoes: Decoding Retailers Supply Chain 795
Lakshmi Shetty, Kavitha Desai, and Shefali Srivastava

Behavioral Bias as an Instrumental Factor in Investment Decision-An Empirical Analysis 807
Aneesha K. Shaji and V. R. Uma

Creating a Sustainable Electric Vehicle Revolution in India 819
Annie Stephen and M. Vanlalahlimpuii

An Overview of Interest Rate Derivatives in Banking Sector—A Comparasion Between Global and Indian Market 835
Riya Singh and Nidhi Raj Gupta

Adoption of Mobile Banking Among Rural Customers 843
A. J. Excelce and V. G. Jisha

Artificial Intelligence and the Decarbonization Challenge 849
Ismail Noori, Yasser M. Abolelmagd, and Wael F. M. Mobarak

Algorithms Control Contemporary Life 859
Ismail Noori Mseer, W. M. Abd-Elfattah, and A. H. Al-Alawi

**Auditors' Perceptions in Gulf Countries Towards Using
Artificial Intelligence in Audit Process** 867
Ahmad Yahia Mustafa Alastal, Janat Ali Farhan,
and Mohammad H. Allaymoun

**Accounting Students' Perceptions on E-Learning During
Covid-19 Pandemic: Case Study of Accounting and Financial
Students in Gulf University—Bahrain** 879
Ahmad Yahia Mustafa Alastal, Mohamed Abdulla Hasan Salman,
and Mohammad H. Allaymoun

**Factors that Influence the Occupational Safety and Health
for Employees as a Part of Human Resource Management
Practices a Study on Non-government Organization in Palestine
Gaza Strip (UNRWA)** 891
Tamer M. Alkadash

**The Impact of Employee Satisfaction, Emotional Intelligence
and Organizational Commitment on Marketing Service Quality
in Medical Equipment Companies, Bahrain** 903
Tamer M. Alkadash, Mahmoud AlZgool, Ali Ahmed Ateeq,
Mohammed Alzoraiki, Rawan M. Alkadash, Chayanit Nadam,
Qais AlMaamari, Marwan Milhem, and Mohammed Dawwas

**Study of Mechanical Properties of Friction Welded AISI D2
and AISI 304 Steels** 917
Sinta Restuasih, Ade Sunardi, M. N. Mohammed, M. Zaenudin,
Adhes Gamayel, and M. Alfiras

**Assessing the Adoption of Key Principles for a Sustainable
Lean Interior Design in the Construction Industry: The Case
of Bahrain** 925
Aysha Aljawder, Wafi Al-Karaghouli, and Allam Hamdan

**Characterizing the DNN Impact on Multiuser PD-NOMA
System Based Channel Estimation and Power Allocation** 943
Mohamed Gaballa, Maysam Abbod, and Ammar Aldallal

Mobile Wireless Sensor Network: Routing Protocols Overview 967
Maha Al-Sadoon, Ahmed Jedidi, and Hamed Al-Raweshidy

Governance and Business Ethics with AI

**Perception of Young Consumers Towards Electric Two Wheeler
in Bangalore City** 979
C. Surendhranatha Reddy and Ajai Abraham Thomas

Contents

Efficacy of Dividend Announcement on Bluechip Pharma Stocks—An Evidence from the Indian Stock Market 987
R. Madhusudhanan and R. Haribaskar

Examining the Impact of Mentoring on Personal Learning, Job Involvement and Career Satisfaction 997
Roshen Therese Sebastian, L. Sherly Steffi, and Geethu Anna Mathew

Islamic Values Impact on Managerial Autonomy 1007
April Lia Dina Mariyana, Ariq Idris Annaufal, and Muafi

An Empirical Corroboration on Perceived Facilitations for Training and Affective Organisational Commitment 1017
Geethu Anna Mathew, K. Opika, and Roshen Therese Sebastian

An Efficiency Analysis of Private Banks in India—A DEA Approach 1025
Ibha Rani and Arti Singh

A Case Study on the Impact of Tourism on the Tribal Life in Vayalada, Calicut, Kerala 1037
P. T. Retheesh and K. K. Jagadeesh

Pro-Environmental Behavior of Farmers in the Dieng Plateau Indonesia 1047
Dyah Sugandini, Mohamad Irhas Effendi, Yuni Istanto, Bambang Sugiarto, and Muhammad Kundarto

Balanced Economic Development: Barometer and Reflections of Economic Progress Concerning the Economies of India and China 1059
Bijin Philip and Suresh Ganesan

Customer Preference Towards E-banking Services Offered by State Bank of India 1073
Ashwitha Shetty and K. A. Thanuja

The Analysis of the Influence of the January Effect and Weekend Effect Phenomenon on Stock Returns of Companies Listed on the JII Index (2017–2019 Period) 1081
Salsabila Hartono Putri and Nur Fauziah

Integrated Study of Ethical and Economic Efficiency of Society's Perception of Corporate Social Responsibility in India 1093
Mohammad Talha, Marim Alenezi, Syed Mohammad Faisal, and Ahmad Khalid Khan

New Trends in the Banking Sector and the Development of E-Banking 1107
Mohammad Afaneh

The Impact of Economic Globalization Welfare States 1117
Natacha de Jesus Silva, Maria José Palma Lampreia Dos-Santos,
and Nuno Baptista

**The Benefits of Digital Transformation: A Case Study
from Finance and Administrative Departments Within the Oil
and Gas Industry in the Kingdom of Bahrain** 1127
Hamad Aljar and Noor Alsayed

**The Relationship Between IT Governance and Firm
Performance: A Review of Literature** 1141
Noora Ahmed Al Romaihi, Allam Hamdan, and Raef Abdennadher

Carbon Emissions Impact by the Electric-Power Industry 1151
Ismail Noori and Basem A. Abu Izneid

**Employee Engagement Concepts, Constructs and Strategies:
A Systematic Review of Literature** 1159
Hanin Aldoy and Bryan Mcintosh

Board-Level Worker Representation: A Blessing or a Curse? 1175
Fatema Alrawahi, Amama Shaukat,
and Abdalmuttaleb M. A. Musleh Al-Sartawi

The Role of Digital Marketing on Tourism Industry in Bahrain 1181
Yousif Alhawaj and Muneer Al Mubarak

About the Editors

Prof. Allam Hamdan *Dean, College of Business and Finance, Ahlia University, Bahrain,* is a Full Professor, he is listed within the World's top 2% of scientists list by Stanford University, and he is the Dean of College of Business and Finance at Ahlia University, Bahrain. Author of many publications (more than 250 papers, 174 listed in Scopus) in regional and international journals that discussed several accountings, financial and economic issues concerning the Arab world. In addition, he has interests in educational-related issues in the Arab world universities like educational governance, investment in education and economic growth. Awarded the First Prize of Al-Owais Creative Award, UAE, 2019 and 2017; the Second Prize of Rashid bin Humaid Award for Culture and Science, UAE, 2016; the Third Prize of Arab Prize for the Social Sciences and Humanities, 2015, and the First Prize of "Durrat Watan", UAE, 2013. Achieved the highest (1st) scientific research citation among the Arab countries according to Arcif 2018–2022. Appointed an external panel member as part of Bahrain Quality Assurance Authority and National Qualifications Framework NQF as a validator, and appeal committee, General Directorate of NQF, Kingdom of Bahrain. Member of Steering Committee in International Arab Conference of Quality Assurance of Higher Education. Currently leading a mission-driven process for International Accreditation for College of Business and Finance by Association to Advance Collegiate Schools of Business (AACSB).

Dr. Esra Saleh Aldhaen *PFHEA Biography, Executive Director Strategy, Quality and Sustainability, Associate Professor,* holds a Doctorate from Brunel University London, in the area of strategic decisions and quality in the context of higher education, currently an Executive Director Strategy, Quality and Sustainability as well as Associate Professor at Ahlia University Bahrain. Appointed external reviewer and conducted various QA reviews as part of Oman Academic Accreditation Authority (OAAA), Bahrain Education and Training Quality Authority (BQA) for National Qualification Framework validation and National Commission for Academic Accreditation and Assessment (NCAAA) KSA. As HEA Principle fellow and expert in curriculum review, design and mapping to the National Qualification Framework as well as alignment of cross-border qualifications was able to set Assurance of Learning

(AOL) framework to evaluate and assess student direct and indirect learning and chaired the AOL committee at college level.

Highly experienced in HEIs in various countries such as UK, Kingdom of Bahrain, KSA, UAE and Oman, along with the gained diverse background enable to transfer and share knowledge and experience from multiple aspects. In particular, to institutionalize and streamline various adapted QA standards into one quality management system supporting all the HEI targets such as international accreditation and ranking. This area was activated through becoming an active member at governance level which enable revising, implementation, assessing and further planning for the university strategic plan towards achieving the mission and vision of the university that includes setting targets and key performance indicators (KPIs) have also been practised and highly performed focused on teaching and learning excellence, research impact and societal contribution.

An expert in strategic planning, measurement and evaluation for higher education including assessment of internal and external environment. Leading Sustainable Development Initiatives in collaboration with United Nations for promising future and a member of multiple national and international societies tackling sustainable development and quality improvements. Published and awarded best papers by Emerald Publishing in various research areas including strategy, sustainable quality management teaching and learning.

Artificial Intelligence, Marketing and Social Media

Digital Marketing and Sustainable Businesses: As Mobile Apps in Tourism

Mahmoud Alghizzawi, Mohammad Habes, Abdalla Al Assuli, and Abd Alrahman Ratib Ezmigna

Abstract The evolution of digital marketing in tourism from a mobile apps perspective has led to a shift towards personalization and targeting, the use of augmented reality (AR) and virtual reality (VR) technology, and mobile-specific features such as push notifications, and in-app messaging. These developments have allowed tourism businesses to connect with potential customers in a more convenient and personalized way, through easy access to information, recommendations and offers. To achieve this, the study adopted the descriptive analytical approach of previous studies that dealt with mobile and the tourism sector. Accordingly, the results of the study showed that there is a significant and direct impact of mobile applications in developing digital tourism services, in addition to customized advertisements and activating direct communication with tourists in an easy and fast way. As the focus of this study centered on the phenomenon of mobile applications in terms of their impact on tourism and digital marketing. Accordingly, and in light of the global tourism competition, the challenges that were imposed on this sector during the Corona pandemic, and the rapid development of mobile applications and digital services, and accordingly we recommend the need to pursue future studies to keep pace with the development of digital tools and their applications, Employment in digital marketing in the service sectors.

M. Alghizzawi (✉) · A. Al Assuli
Faculty of Business, Marketing Department, Amman Arab University, Amman, Jordan
e-mail: M.alghzawi@aau.edu.jo

A. Al Assuli
e-mail: a.alassuli@aau.edu.jo

M. Habes
Faculty of Mass Communication, Radio and TV Department, Yarmouk University, Irbid, Jordan
e-mail: mohammad.habes@yu.edu.jo

A. A. R. Ezmigna
Azman Hashim International Business School, Universiti Teknologi Malaysia, Johor Bahru, Malaysia
e-mail: Ezmigna@graduate.utm.my

© The Author(s), under exclusive license to Springer Nature Switzerland AG 2024
A. Hamdan and E. S. Aldhaen (eds.), *Artificial Intelligence and Transforming Digital Marketing*, Studies in Systems, Decision and Control 487,
https://doi.org/10.1007/978-3-031-35828-9_1

Keywords Digital marketing · Electronic tourism · Mobile apps · Sustainable businesses · Augmented reality (AR) · Economic growth

1 Introduction

In addition to leveraging rich media, such as text, audio, and video, marketers now have a much enlarged range of pull-based app options as a result of the improved capabilities of mobile applications in the tourism industry [1, 2]. Customers are drawn to applications because they offer a variety of useful and entertaining uses, as well as the ability to customize how they use their devices by adding features and programs that cater to their particular needs [3, 4]. Tourism already employs a number of experience filtering techniques to make selections, and recommendations for travel will spread widely as a result of the availability of richer and more complete personal data. On mobile devices, recommendations (made socially or automatically) can be supplied in real-time alongside mapping services [5, 6] There are a few specific ways that mobile app marketing is different from other web marketing strategies. First of all, ubiquity illustrates the actual features of mobile apps that are eventually used whenever and whenever [7, 8] Specifically, [9]. The widespread use of mobile apps is a multifaceted phenomenon made up of continuity and immediacy. The focus of tourism marketing attention on mobile use has switched from the adoption of mobile technology, which is concerned with attracting new clients, to post-adoption behavior, which emphasizes continuous adoption [10]. Post-adoption behavior has been well explained by stickiness and word-of-mouth WOM in in previous studies [11, 12] The preparation of tourism programs beginning with contracting with visitors up to the conclusion of such programs constitutes the entire effort made to attract local and foreign tourists to pay visits to the tourism locations inside the nation [13, 14]. On this study we will try identify effect phone marketing on Medical tourism in Jordan. Accordingly, the current study sought to try in this study to determine the development of digital marketing in tourism from the perspective of mobile applications through a method that reviews previous studies that dealt with the field of study, relying on secondary sources in an integrated systematic manner, where the theoretical framework of tourism and electronic marketing topics will be reviewed through The mobile phone, in addition to an explanation of the most important features that attract tourists to use these digital services, with an indication of their impact on the development of the tourism sector in light of the high competition, and then the discussions and the conclusion will be dealt with, and finally the future scientific contributions.

2 Literature Review

These apps use data on a user's search history and location to deliver tailored content and offers. This can include recommendations for nearby attractions, hotel deals, and flight promotions. Another feature of mobile apps in tourism is the use of augmented reality (AR) and virtual reality (VR) technology. Mobile apps have become an essential part of the tourism industry, allowing businesses to connect with customers and providing them with a convenient and personalized experience [15]. As more and more people are using their smartphones to research and book travel, it is important for tourism businesses to have a strong presence on mobile platforms. The importance of digital marketing has been recognized by many researchers. Many researches and studies have tried to explain this concept and tried to establish a strong relationship with tourism and digital services as an important and essential part of globalization and economic and health development, so this study will be of importance to marketing companies and stakeholders in the tourism field from private and official bodies [16]. This gives it the importance of marketing via smartphones to tourists through the advantages provided by smartphones and their applications, in order to reach the largest number of tourists and identify the categories covered by advertising and services, as well as to know what affects digital marketing for any specific product or topic and work to reduce those Challenges and high tourism competition [17].

2.1 Mobile Apps in Tourism

The preparation of tourism programs beginning with contracting with visitors up to the conclusion of such programs constitutes the entire effort made to attract local and foreign tourists to pay visits to the tourism locations inside the nation [18, 19] Booking via smartphone apps drives online travel sales. As global sales of digital tourism are growing rapidly, in line with keeping pace with technology. Through Fig. 1, we have the reasons for the adoption of tourism applications through smart phones.

According to [21] Mobile Application in Tourism that provisions customer post buying brand communication and produce use may include measures of devotion effects (attitude as behavioral based loyalty), value of and satisfaction with product use and support. For mobile pull media, the quality of the visitor base may directly affect the expected results of their visit various tourist attractions.

Travelers tend to trust other travelers (friends and review sites) for information about what to see and do during a trip [22]. Figure 2 electronic device use and mobile application In the world using tourism applications in 2015 to search for tourist facilities (Table 1).

The preparation of tourism programs beginning with contracting with visitors up to the conclusion of such programs constitutes the entire effort made to attract

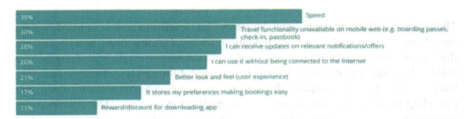

Fig. 1 *Source* [20]

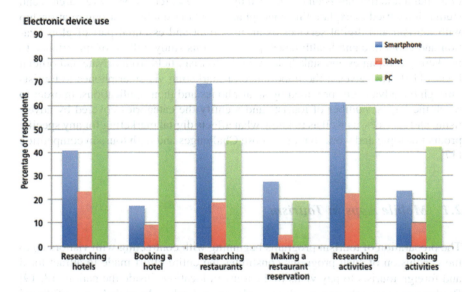

Fig. 2 *Source* Linton [22]

local and foreign tourists to pay visits to the tourism locations inside the nation [14]. Tourism marketing starts by studying the exporting tourism markets, defining their needs for tourism products, identifying available opportunities, studying the tourists' specifics, and meeting their needs. This is different from simply providing tourism services, programs, and offering them both domestically and abroad [27]. And, through developing fresh strategies for marketing the smartphone and other applications, as well as for the growth of contemporary technology.

Digital Marketing and Sustainable Businesses: As Mobile Apps in Tourism

Table 1 Systematic review of importance of mobile apps in tourism

No.	Author	Finding	Society
1	[23]	The study's conclusions 96% of respondents concur that this mobile app helps to promote Indonesian tourism hotspots so that tourists will be interested in visiting. 93.6% of respondents think that Explores offers a new approach to obtain information on tourism in Indonesia	50 Tourist
2	[24]	According to the findings, the perceived ubiquity, informativeness, and personalisation of mobile apps by users are all positively correlated with their usefulness	503 respondents in USA
3	[25]	The findings showed that only PEX and performance expectancy have no bearing on the decision to adopt. Gender was also discovered to moderate the relationship between social influence and intention	474 respondents
4	[26]	The study's findings suggest that timeshare owners' intentions to utilize a mobile app are explained by typical technology adoption antecedents (experience, usefulness, and attitude)	914 respondents in Florida (USA)

2.2 Mobile Application Usage Intention

Mobile app users are more likely to use services in a variety of settings and at all hours of the day, the range of services that are available expands to serve a variety of needs [28, 29]. Commercial services including purchasing tickets, shopping for goods, making reservations, doing financial transactions, and making electronic payments are typical functional apps. Additionally, this category includes informational content services like news, stock quotes, shopping information, maps, and location-based services [30, 31]. The manner that consumers respond to advertising efforts has changed as a result of the widespread usage of mobile phones as a communication tool. Receiving mobile advertising at the ideal time and location can improve perceived value and increase consumers' receptivity to promotions [32]. The spectrum of uses for mobile apps has broadened beyond purely utilitarian goals to mostly include non-utilitarian goals. Mobile apps have become interactive systems that prioritize hedonic demands like socialization, enjoyment, and interaction over utilitarian goals like increased output or performance [33]. The incorporation of intrinsic motivations for using mobile apps and social influences in two forms—subjective norm and social image—reflects social motivations for mobile app users [34]. Table 2 Systematic Review of Importance of mobile apps in tourism.

Table 2 Systematic review of importance of mobile application usage intention

No.	Author	Finding	Society
1	[25]	Only PEX and performance expectancy (PE) are no significant with regard to intention to adopt, according to the findings. Gender also modifies the relationship between social influence (SI) and intention	474 respondents
2	[35]	The study's findings show that users do not always follow through on their positive intentions. While intention is certainly well explained by UTAUT variables, usage is not always predicted by intention. Additionally, we only detect a weak association between self-reported consumption and log data	637 respondents
	[36]	The findings showed that the development of continued use of m-library apps depends significantly on the user's perceptions of utility, expectation confirmation, and contentment. and the evidence supporting the detrimental effects of habit on continuing intention and behavior, as well as the three elements of the information system success model, system quality, information quality, and service quality	396 respondents

2.3 Mobile Apps as Digital Marketing Tools

Tourism is one of the largest sectors, as most countries planned for the success of tourism and diversification of tourism projects. To support the national economy, increase their economic income especially in developing countries [14, 37] These countries diversified the ways of tourism marketing through mobile applications and e-marketing as [4, 38], also found that companies resort to marketing through mobile applications to attract tourists. The use of applications via mobile can change tourists' behavior and their demands, and also provides immediate information, about the problems that tourists can face [29, 39]. Technology plays an important role in the tourism sector so that future technology must meet the needs and aspirations of tourists to a large extent through smartphones.

3 Discussion and Conclusions

The advent of mobile has allowed companies to offer more personalized content, as smartphones are able to closely collect personal information [40]. This does not end with mobile. The Internet of Things increases the possibilities of easy data collection and marketing via these phones. Market smartphones first from a very small base Early on, marketers have already begun to explore a wide range of approaches to take advantage of the smartphone medium, including those developed only to add smartphones and derivatives from other areas such as the Internet [8, 41]. Marketers have made attempts big and small and seen encouraging early successes and useful, informative bombs [41]. Because it is widely recognized as the direct and linear

Digital Marketing and Sustainable Businesses: As Mobile Apps in Tourism 9

descendent of the old advertising messages that were always pushed to customers, direct marketing to tourists has come to be considered as the industry standard for mobile marketing [13]. In addition, the widespread use of mobile devices and the low cost of setting up an advertisement on mobile devices have motivated hotels and tourism stakeholders to invest in mobile advertising [42]. The tourism industry has benefited greatly from digital advertising and marketing via smartphones, including the ability to send tourists personalized messages based on their preferences and interests via SMS, specialized applications, and social media. However, both tourists and marketers must be aware of these benefits, including interactivity, cost, and personalization [23, 43–45]. However, there has been little research on indirect marketing techniques or the many other forms in which smartphone marketing may appear. In conjunction with the development of the industry, a widespread common view of the smartphone advertising trend has emerged. In theory, promoting smartphones can be very beneficial to tourists. From helping people find the moment-important resources they need to helping people find critical information when and where they need it, it's clear that smartphone apps can provide unparalleled convenience and functionality. The evolution of digital marketing in tourism from a mobile apps perspective has seen a significant shift in recent years [46, 47]. Mobile applications have become a crucial tool for tourism organizations to engage with potential clients as more and more people use their smartphones for travel planning and booking. The rising use of personalization and targeting in mobile app marketing has been one of the important trends in this field. This can involve delivering personalized content and offers based on a user's search history and location, as well as utilizing social media and customer reviews to establish credibility and trust. The usage of augmented reality (AR) and virtual reality (VR) technologies in travel-related mobile apps is another development [48, 49]. This can include using AR to provide immersive, interactive experiences for users, such as virtual tours of destinations, or using VR to offer virtual reality experiences like scuba diving or skydiving. In addition, there has been an increase in the use of mobile-specific features, such as push notifications, and in-app messaging, to deliver relevant information and offers to users at the right time and place [50, 51]. Overall, the use of mobile apps in digital marketing for tourism has become an essential part of the industry, and businesses that are able to effectively leverage these tools and technologies will be well-positioned to succeed in the digital age.

3.1 Contributions and Future Research

Despite the progress made in the world in all aspects of life, there is still a weakness in talking about digital tourism and its benefit, whether for individuals or for the economic situation of the country. Knowing that everyone in the family has at least one mobile phone, makes it easy to communicate tourist information to all people and all countries. Therefore, the focus of this study will be on the mobile phone phenomenon in terms of its impact on tourism and digital marketing. Accordingly, in light of the global tourism competition, Future research must be pursued to keep up with the development of digital tools and their employment in digital marketing in the service sectors in light of the challenges placed on this sector during the Corona pandemic as well as the rapid development of smartphone applications and digital services.

3.2 Recommendations

With the development of technology and its impact on the business sector and environment, competition has increased in all sectors, and by examining the tourism sector, which has suffered many crises in the recent past due to the Corona pandemic, it has become necessary to pay attention to decision-makers from international tourism authorities, ministries and all stakeholders in the private sector, including: Investors, to the need to focus on the importance of supporting this sector and ensuring its continuity and growth by employing modern technology through smart phone applications that generated an electronic tourist destination through the advantages it provides for the services provided on these smart applications. In addition to the need for hotels and tourist destinations to take into account the opinions exchanged through these applications, because of their important and influential impact on tourists. Finally, investing in the tourism sector at present must take into account the allocation of sufficient financial resources in terms of technology to stay in touch with tourists, attract them, and see their opinions, which must be taken seriously. Accordingly, there will be sustainable development in the tourism business sector by investing in smart tourism applications because of the advantages it provides, especially since this sector enhances the economy, reduces unemployment, and brings many financial flows.

References

1. Alghizzawi, M.: A survey of the role of social media platforms in viral marketing: the influence of eWOM. Int. J. Inf. Technol. Lang. Stud. **3**(2) (2019)
2. Al-Gasawneh, J.A., Al-Adamat, A.M.: The relationship between perceived destination image, social media interaction and travel intentions relating to Neom city. Acad. Strateg. Manag. J. **19**(2), 1–12 (2020)
3. Alghizzawi, M.: The role of digital marketing in consumer behavior: a survey. Int. J. Inf. Technol. Lang. Stud. **3**(1) (2019)
4. Lončarić, D., Ribarić, I.: The role of electronic word-of-mouth in the tourism market. In: Bijenalni međunarodni znanstveno-stručni kongres Turizam i hotelska industrija 2016: trendovi i izazovi (23; 2016) (2016)
5. Al-Gasawneh, J., Al-Wadi, M., Al-Wadi, B., Alown, B., Nuseirat, N.: The interaction effect of comprehensiveness between social media and online purchasing intention in Jordanian pharmacies (2020)
6. Missaoui, S., Kassem, F., Viviani, M., Agostini, A., Faiz, R., Pasi, G.: LOOKER: a mobile, personalized recommender system in the tourism domain based on social media user-generated content. Pers. Ubiquitous Comput. **23**(2), 181–197 (2019)

Digital Marketing and Sustainable Businesses: As Mobile Apps in Tourism

7. Rahi, S., Mansour, M.M.O., Alghizzawi, M., Alnaser, F.M.: Integration of UTAUT model in internet banking adoption context. J. Res. Interact. Mark. (2019)
8. Alghizzawi, M., et al.: The impact of smartphone adoption on marketing therapeutic tourist sites in Jordan. Int. J. Eng. Technol. 7(4.34), 91–96 (2018)
9. Okazaki, S., Mendez, F.: Perceived ubiquity in mobile services. J. Interact. Mark. 27(2), 98–111 (2013)
10. Kim, Y.B., Joo, H.C., Lee, B.G.: How to forecast behavioral effects on mobile advertising in the smart environment using the technology acceptance model and web advertising effect model. KSII Trans. Internet Inf. Syst. 10(10), 4997–5013 (2016)
11. Alnaser, A.S., Habes, M., Alghizzawi, M., Ali, S.: The relation among marketing ads, via digital media and mitigate (COVID-19) pandemic in Jordan. Dspace.Urbe.University (2020)
12. Racherla, P., Furner, C., Babb, J.: Conceptualizing the implications of mobile app usage and stickiness: a research agenda (2012)
13. Habes, M., Alghizzawi, M., Salloum, S.A., Ahmad, M.F.: The use of mobile technology in the marketing of therapeutic tourist sites: A critical analysis. Int. J. Inf. Technol. 2(2), 48–54 (2018)
14. Magatef, S.G.: The impact of tourism marketing mix elements on the satisfaction of inbound tourists to Jordan. Int. J. Bus. Soc. Sci. 6(7), 41–58 (2015)
15. Gajdošík, T.: Smart tourism: concepts and insights from Central Europe. Czech J. Tour. 7(1), 25–44 (2019). https://doi.org/10.1515/cjot-2018-0002
16. Kalbaska, N., Janowski, T., Estevez, E., Cantoni, L.: When digital government matters for tourism: a stakeholder analysis. Inf. Technol. Tour. (2017). https://doi.org/10.1007/s40558-017-0087-2
17. Al Olaimat, F., Habes, M., Hadeed, A., Yahya, A., Al Jwaniat, M.I.: Reputation management through social networking platforms for PR purposes: a SEM-based study in the Jordan. Front. Commun. 11, 247
18. Rahi, S., Alghizzawi, M., Ngah, A.H.: Factors influence user's intention to continue use of e-banking during COVID-19 pandemic: the nexus between self-determination and expectation confirmation model. EuroMed J. Bus. (2022). https://doi.org/10.1108/EMJB-12-2021-0194 (Ahead-of-p, no. ahead-of-print)
19. Cheng, A., Ren, G., Hong, T., Nam, K., Koo, C.: An exploratory analysis of travel-related WeChat mini program usage: affordance theory perspective. In: Information and Communication Technologies in Tourism 2019, pp. 333–343. Springer (2019)
20. Alternative spaces. Tips on Creating a Mobile Booking App for Travel and Tourism Businesses (2021). Available https://alternative-spaces.com/blog/tips-on-creating-a-mobile-booking-app-for-travel-and-tourism-businesses/
21. Ström, R., Vendel, M., Bredican, J.: Mobile marketing: a literature review on its value for consumers and retailers. J. Retail. Consum. Serv. 21(6), 1001–1012 (2014)
22. Linton, H.: The mobile revolution is here: are you ready? (2015)
23. Safitri, R., Yusra, D.S., Hermawan, D., Ripmiatin, E., Pradani, W.: Mobile tourism application using augmented reality. In: 2017 5th International Conference on Cyber and IT Service Management (CITSM), pp. 1–6 (2017)
24. Kim, S., Baek, T.H., Kim, Y.-K., Yoo, K.: Factors affecting stickiness and word of mouth in mobile applications. J. Res. Interact. Mark. 10(3), 177–192 (2016)
25. Tan, G.W.-H., Lee, V.H., Lin, B., Ooi, K.-B.: Mobile applications in tourism: the future of the tourism industry? Ind. Manag. Data Syst. 117(3), 560–581 (2017)
26. Rivera, M., Gregory, A., Cobos, L.: Mobile application for the timeshare industry: the influence of technology experience, usefulness, and attitude on behavioral intentions. J. Hosp. Tour. Technol. 6(3), 242–257 (2015)
27. Samawi, J.: Medical and wellness tourism challenges in dead sea spas from tourists' point of view. Int. J. Mark. Stud. 9(5), 145 (2017)
28. Habes, M., Alghizzawi, M., Salloum, S.A., Mhamdi, C.: Effects of Facebook Personal News Sharing on Building Social Capital in Jordanian Universities, vol. 295, no. June 2020. Springer (2021). https://doi.org/10.1007/978-3-030-47411-9_35

29. Bazazo, I.K., Alananzeh, O.A., Taani, A.A.A.: Marketing the therapeutic tourist sites in jordan using geographic information system. Marketing **8**(30) (2016)
30. Alghizzawi, M., Habes, M., Salloum, S.A.: The relationship between digital media and marketing medical tourism destinations in Jordan: Facebook perspective. In: International Conference on Advanced Intelligent Systems and Informatics, pp. 438–448 (2019)
31. Cao, Y.Y., Qin, X.H., Li, J.J., Long, Q.Q., Hu, B.: Exploring seniors' continuance intention to use mobile social network sites in China: a cognitive-affective-conative model. Univers. Access Inf. Soc. **0123456789** (2020). https://doi.org/10.1007/s10209-020-00762-3.
32. Erawan, T.: Tourists' intention to give permission via mobile technology in Thailand. J. Hosp. Tour. Technol. **7**(4), 330–346 (2016)
33. Huang, Y.C., Chang, L.L., Yu, C.P., Chen, J.: Examining an extended technology acceptance model with experience construct on hotel consumers' adoption of mobile applications. J. Hosp. Mark. Manag. **28**(8), 957–980 (2019). https://doi.org/10.1080/19368623.2019.1580172
34. Yan, M., Filieri, R., Raguseo, E., Gorton, M.: Mobile apps for healthy living: Factors influencing continuance intention for health apps. Technol. Forecast. Soc. Change **166**, 120644 (2021)
35. Tiefenbeck, V., Kupfer, A., Ableitner, L., Schöb, S., Staake, T.: The uncertain path from good intentions to actual behavior: a field study on mobile app usage. In: DIGIT Pre-ICIS Workshop (2016)
36. Zhao, Y., Deng, S., Zhou, R.: Understanding mobile library apps continuance usage in China: a theoretical framework and empirical study. Libri **65**(3), 161–173 (2015)
37. Alghizzawi, M., Salloum, S.A., Habes, M.: The role of social media in tourism marketing in Jordan. Int. J. Inf. Technol. Lang. Stud. **2**(3) (2018)
38. Gulbahar, M.O., Yildirim, F.: Marketing efforts related to social media channels and mobile application usage in tourism: case study in Istanbul. Procedia-Social Behav. Sci. **195**, 453–462 (2015)
39. Hua, L.Y., Ramayah, T., Ping, T.A., Jun-Hwa, C., Jacky, C.J.H.: Social media as a tool to help select tourism destinations: the case of Malaysia. Inf. Syst. Manag. **34**(3), 265–279 (2017). https://doi.org/10.1080/10580530.2017.1330004
40. Chen, C.C., Tsai, J.L.: Determinants of behavioral intention to use the personalized location-based mobile tourism application: an empirical study by integrating TAM with ISSM. Futur. Gener. Comput. Syst. **96**, 628–638 (2019). https://doi.org/10.1016/j.future.2017.02.028
41. Habes, M., Alghizzawi, M., Elareshi, M., Ziani, A., Qudah, M., Al Hammadi, M.M.: E-marketing and customers' bank loyalty enhancement: Jordanians' perspectives. In: The Implementation of Smart Technologies for Business Success and Sustainability, pp. 37–47. Springer (2023).
42. Kim, H.H., Law, R.: Smartphones in tourism and hospitality marketing: a literature review. J. Travel Tour. Mark. **32**(6), 692–711 (2015). https://doi.org/10.1080/10548408.2014.943458
43. Palos-Sanchez, P., Saura, J.R., Velicia-Martin, F., Cepeda-Carrion, G.: A business model adoption based on tourism innovation: Applying a gratification theory to mobile applications. Eur. Res. Manag. Bus. Econ. **27**(2), 100149 (2021)
44. Law, R., Chan, I.C.C., Wang, L.: A comprehensive review of mobile technology use in hospitality and tourism. J. Hosp. Mark. Manag. **27**(6) (2018). https://doi.org/10.1080/19368623.2018.1423251
45. González-Reverté, F., Díaz-Luque, P., Gomis-López, J.M., Morales-Pérez, S.: Tourists' risk perception and the use of mobile devices in beach tourism destinations. Sustainability **10**(2), 413 (2018)
46. Subramanian, K.R.: Influence of social media in interpersonal communication senior consultant& professor of management. **109**(02), 70–75 (2017)
47. Abuhashesh, M., Al-Khasawneh, M., Al-Dmour, R., Masa'Deh, R.: The impact of Facebook on Jordanian consumers' decision process in the hotel selection. IBIMA Bus. Rev. **2019** (2019). https://doi.org/10.5171/2019.928418
48. Santomier, J.: New media, branding and global sports sponsorship. Int. J. Sport. Mark. Spons. **10**(1), 15–28 (2008). https://doi.org/10.1108/ijsms-10-01-2008-b005

49. Fonseca, L.M., Azevedo, A.L.: COVID-19: outcomes for global supply chains. Manag. Mark. **15**(1), 424–438 (2020). https://doi.org/10.2478/mmcks-2020-0025
50. Erkan, I.: The impacts of electronic word of mouth in social media on consumer's purchase intentions. Int. Conf. Digit. Mark. 11 (2016). https://doi.org/10.5539/ijbm.v9n8p84
51. Habes, M., Pasha, S.A., Ali, S., Elareshi, M., Ziani, A., Bashir, B.A.: Technology-enhanced learning acceptance in pakistani primary education. In: European, Asian, Middle Eastern, North African Conference on Management & Information Systems, pp. 53–61 (2023)

Exploring Small and Medium Enterprises Expectations of Electronic Payment in Kuwait

Ayman Hassan, Arezou Harraf, and Wael Abdallah

Abstract Electronic payments are among the topics on the top of regulators' agendas around the globe. This trend gained further momentum with the lockdown and other measures to front the COVID-19 pandemic. The current study aims to explore Kuwaiti small and medium enterprises' expectations of e-payment services. Semi-structured interviews were conducted with nine participants, either founders or directors of Kuwaiti SMEs. The collected data were analyzed using phenomenological analysis. Findings revealed seven main themes: payment gateway and Infrastructure, clarity, and communications, cost, settlement Period, rigid regulation, perceived benefits, and expected services. The results are meant to give directions that the Kuwaiti Central Bank can build upon to further develop the regulatory framework and infrastructure in a manner that is more inclusive of the SME segment.

Keywords Small and medium enterprises · Electronic payment · Kuwait

1 Introduction

Small and Medium Enterprises SMEs play a significant role in developed and developing economies. The World Bank estimates the contribution of formal SMEs to total employment to be up to 60% and its contribution to the national GDP to be up to 40% in emerging economies. These estimates would be even higher when considering the informal SMEs [17]. The active SME sector facilitates greater utilization of local raw materials, transforming domestic savings into productive economic activities,

A. Hassan
Maastricht School of Management Kuwait, Kuwait, Kuwait
e-mail: AAHassan@CBK.GOV.KW

A. Harraf · W. Abdallah (✉)
Box Hill College Kuwait, Kuwait, Kuwait
e-mail: W.abdallah@bhck.edu.kw

A. Harraf
e-mail: a.harraf@bhck.edu.kw

© The Author(s), under exclusive license to Springer Nature Switzerland AG 2024
A. Hamdan and E. S. Aldhaen (eds.), *Artificial Intelligence and Transforming Digital Marketing*, Studies in Systems, Decision and Control 487,
https://doi.org/10.1007/978-3-031-35828-9_2

enhancing supply chains, and promoting income equality [20]. These benefits are very much needed in resource-dependent economies [16]. In such circumstances, many countries resort to supporting the SME sector in its diversification and job creation endeavors.

In Kuwait, SMEs comprise a significant segment, with estimated 33,000 enterprises that classify as SMEs [1]. This segment has a potential role in job market reforms given its size. The job market in Kuwait needs structural reforms, amongst which is addressing the excessive reliance on the governmental sector for job creation for Kuwaiti citizens. The Central Statistical Bureau's data shows that the Kuwaiti government hires 362,133 citizens as of Mar. 31, 2022, whereas the citizens working in the private sector count for only 72,692 the proportion is roughly 83–17% in favor of governmental jobs. Nevertheless, SMEs in Kuwait, as in the GCC region, suffer from financial exclusion [13].

One of the means to support SMEs is to facilitate their access to efficient payment methods. Payments, settlements, and clearance have witnessed rapid development with the advent of new technologies in recent years. Technologies, such as distributed ledger technology DLT, along with the increasing demand for seamless payments, are revolutionizing how payments are made [11]. Conversely, traditional payment methods are witnessing a negative trend in the GCC region [3].

Adopting electronic payments facilitates accessibility to commodities and services, leading to further consumption and eventually contributing to economic growth. Some researchers found a positive correlation between cashless payment transactions and the level of economic development [6]. In addition, the widespread of cashless transactions has positive implications for reducing shadow operations [19]. Given these benefits, e-payments became one of the regulators' priorities [8].

With the development of e-payments, the role of the regulators is gaining more importance. The role of the regulator expanded to include maintaining the stability and security of the payments system and guaranteeing comprehensive coverage. Regulators are responsible also for setting the standards and overseeing the competition [4]. It is also part of their role to encourage and harmonize schemes among private sector operators and invest in partially or entirely constructing the essential infrastructure. Central banks often directly manage the infrastructure [2]. However, regulators tend to over-regulate e-payments in general. Traditionally, regulation is focused on large financial institutions, and compliance requirements are mainly set accordingly. For smaller fintech, the effort and financial costs of compliance are unaffordable. Add to this the difficulty of classifying financial innovations that fintechs provide under existing financial services categories. This difficulty may invite further tightened regulation to avoid loopholes resulting in an unsupportive regulatory environment. For example, to tackle systemic risks, many regulators require payment service providers to deposit regulatory reserves to cover their risks [9]. According to Porter's five forces model, this will accumulate more systemic importance to that service provider since it reduces the threat of new entrants, lowers competition, allows higher fees, and tolerates inefficiency. A result that is in total contrast to the purpose of the regulation [9].

Therefore, Barr et al. [5], and Frost [12] argue that regulators should design an inclusive payment system. The regulatory environment must offer moderate and balanced regulation and facilitate and ease entry. Providing e-payment services should be feasible. Frost finds a positive correlation between the strictness of law and the level of investment in financial technologies. New entrants to this market are witnessed more in countries with less strict regulatory environments and higher alternative finance volumes [12].

The literature on e-payment in Kuwait has room for growth as, to date, no known literature to authors has discussed the availability and technical adoption of e-payments by SMEs in Kuwait. The current literature on SMEs focuses primarily on lending and access to credit. In addition, the literature on e-payment leans toward the technical aspects in the broader context of financial technologies, including the impact of the adopted information system, the quality of the communication infrastructure, and the introduction of a given function to a specific payment system. This paper aims to address the current gap by including payment services when discussing the financial inclusion of SMEs. The current paper is structured as follows: Sect. 1 introduced the practical and theoretical foundation for the research Sect. 2 outlines the research design and methods. Section 3 presents the findings and the discussion. Section 4 draws conclusions and presents the contributions and limitations of the research.

2 Methods

2.1 Research Design

To further understand and explore the expectations of e-payment services expectations in Kuwaiti small and medium enterprises and to further develop the regulatory framework and infrastructure in a manner that is more inclusive of the SME segment, a qualitative inductive reasoning approach and exploratory research were conducted because of the scarcity of information and data available in literature or secondary data sources on the subject matter. A cross-sectional time frame was adopted for this study.

2.2 Measurement Development

The semi-structured interview offers flexibility, accessibility, intelligibility, and the ability to reveal necessary and often invisible facets of the interviewee's behavior [18]. Therefore, semi-structured interviews were used in the current study to collect the data. Sixteen open-ended questions were prepared based on reviewing the literature to collect data from participants. These questions aimed at obtaining socially constructed meanings of e-payments adoption and usage, the hindrances of optimum use of e-payment, insights into the dynamics of using e-payments on the SME level, and the possible measures to enhance the experience and maximize the benefits of using the e-payments. The main categories of the prepared interview questions were (1) Infrastructure, (2) security and trust, (3) regulations and Policies, (4) perceived benefits, and (5) complexity of the e-payment technology [5, 6, 15].

2.3 Sampling and Population

The number of small and medium enterprises in Kuwait is estimated at 33,000 companies [1]. Kuwaiti laws categorize these SMEs under five main fields of those SMEs in Kuwait: industry, agriculture, crafts, services, and technology. The maximum variation sampling strategy, which aims to Identify critical dimensions of variations and then find cases that vary as much as possible [7], was adopted to collect a vast body of data from different fields, namely industry, agriculture, crafts, services, and technology; our target sample includes 1–2 participants from each field.

2.4 Data Collection Techniques and Procedures

Nine face-to-face semi-structured interviews were conducted with either the founders or the directors of Kuwaiti SMEs. The majority, four out of nine- comes from the service sector, one in the industrial sector, one in the crafts, one in the agriculture sector, and two in the technology sector. The duration of each interview ranged from 50 to 60 min. Two interviews were conducted in Arabic the rest were in English. All of the interviews were conducted between May to July 2022. Five interviewees consented to record the interview, while the rest consented to participate without recording. Researchers took field notes for those who preferred not to record, whereas the recorded interviews were transcribed verbatim.

2.5 Data Analysis

This study adopted phenomenological analysis to analyze the collected data. This method explores the underpinning concepts of a given phenomenon and helps to discover the patterns of human behavior about the studied subject [10] The current research utilized a phenomenological analysis process of four steps: (1) Bracketing – where views about the phenomenon research are determined and outlined. In this way, the researcher brackets out the frames and any assumptions to keep the data pure and isolate the genuine phenomenon, (2) Intuiting – in this step, the researcher sticks around and concentrates on the phenomenon's meaning by the primary research.

Exploring Small and Medium Enterprises Expectations of Electronic … 19

Also, a common understanding of a phenomenon whatsoever being studied is accomplished. Then, the researcher must come up with the variance of the data until a shared understanding is done, (3) Analyzing – In this step, coding is carried out where categorizing and grasping of the primary meanings of the phenomenon is generated, and (4) Describing – In the descriptive step, realization, and definition of the phenomenon are achieved by the researcher. This is aimed at coming up with the final step that offers significant and pivotal descriptions in both written and verbal form [14].

3 Findings and Discussion

All audio recordings and field notes of interviews were transcribed and recorded. Arabic transcriptions and field notes were translated into English for analysis. The phenomenological analysis revealed seven main themes based on the data collected from the interviews with the sample SMEs' representatives (Table 1).

3.1 Theme 1: Payment Gateway and Infrastructure

The online payment gateway provided by K-net company was subject to criticism by many participants. Although two participants expressed satisfaction with the service, the rest were critical and less satisfied. The fact that this is the only competitor that has impacted the competitiveness and efficiency of the payment gateway. The main issues projected by the participants fall under the following sub-themes.

3.1.1 Technical Problems

The participants have reported two main technical problems that reoccur frequently. First, the payer pays the transaction online and the banks deducts the amount from the payer's account. However, the payment gateway does not confirm the completion

	Theme 1	Payment gateway and infrastructure
Table 1 Themes for understanding the expectations of E-payments by the SMEs in Kuwait	Theme 2	Clarity and communications
	Theme 3	Cost
	Theme 4	Settlement period
	Theme 5	Rigid regulation
	Theme 6	Perceived benefits
	Theme 7	Expected services

Source The authors

of the transaction to the payee. In this case, the latter would not take the next step. This issue impacts the payment service providers as they allocate some of their scarce financial and human resources to address such glitches. Allocating human resources or developing automated methods to tackle this issue hints at how frequent it is.

The second technical problem reported by the participants is the frequent interruption of the service. According to one of the participants, when the service is down or if there is any other technical issue, the only communication channel is e-mail.

3.1.2 Lack of Customer Support

To illustrate the magnitude of the problem and how it impacts SMEs, one of the participants gives an excellent example of how difficult it is to communicate with K-net on such technical issues. In their case, it is easier to refund the customer than to go into the hustle of contacting K-net. In most cases, there are no quick fixes when the service is down or when there are overdue payments that the SME cannot afford to refund. A payment service provider describes such a situation as follows:

> We have many issues with the payment gateway. Response time is too slow. For online payments, they don't have 24/7 support; for example, if the payment page is down, the service is completely down, we send them an email. […] K-fast, they [K-net] disabled the feature [K-fast]. It took more than two months for us to get this feature again.

This issue is explained by another participant who stressed the lack of competitiveness of K-net and how it is not a customer-focused entity:

> It [k-net] was designed to serve the banks as a third party. It wasn't designed to be competitive in the market, providing services that help the people.

3.1.3 Underdeveloped Services

Not being designed to provide services for the people may be the reason behind the participants' frequently repeated complaints about the service quality, particularly compared to similar regional entities. Participants noticed that services could be more developed, but the intent and the pace of development could be more robust.

The participants suggested a few areas of enhancement, like the payment mechanism. It is perceived as a complicated process since it involves choosing the bank and entering the account number, PIN code, and OTP. Moreover, a rather specific suggestion was made by one of the participants:

> If you want to pay through a link, it has to go through a bank. Why isn't that done through K-net and universal for everybody?

3.1.4 Lack of Cooperation with Payment Service Providers

According to the participants, the small Fintech firms that provide payment services are swifter in developing new services. They are more capable of understanding

Exploring Small and Medium Enterprises Expectations of Electronic … 21

their customers' needs and catering to them. However, these Fintechs are facing many hindrances, amongst which is the lack of cooperation of the payment gateway. They feel that K-net perceives them as competitors and acts upon that.

3.1.5 Complex Procedures and Expensive Fees of Subscription and Integration

There is a consensus among the participants about this point. Subscribing to the K-net service and integrating the gateway to SMEs' websites is expensive and a long process involving much paperwork and back-and-forth correspondence.

One participant described subscribing with K-net as "fricative; they ask for much paperwork, code from the developer, signature proof, and 1000 KD fees, and it takes 1–2 weeks."

3.2 Theme 2: Regulations Clarity and Communication

Another theme is Regulatory Clarity and Communication. Two of the participants expressed clearly that they expect more clarity in the regulations issued by the central bank. They also expect more effective communication on the part of the regulator. One of the participants expressed a degree of satisfaction with the communication policy adopted by the central bank recently. However, the clarity of the regulation could be more satisfactory. Although the Central Bank of Kuwait has revised it 2018 e-payments regulations, yet, it still has ambiguity for some:

> The mandate was not clear. In the beginning, they said the mandates only covered debit cards. Credit card is not part of it. But the banks were not taking any risks to give us the credit card. Up to this date, when we communicate with CBK and tell them we need a credit card, they say, "go to any bank and take it." When we communicate with the banks, they say, "no, you did not file with us as an agent. We cannot provide it to you". So, the communication it's not very clear. We don't have direct communication with CBK. But later, the banks told us there is communication; it's a one-way correspondence.

This excerpt makes a point of clarity and communication issues so clear. The same article of regulation is understood in different ways by the three parties. The issuer meant something, but the regulated entity (bank) finds it ambiguous, so it prefers to avoid taking any risk, and the SME tries to sort out this miscommunication. Yet, the latter doesn't have any official communication channel with the central bank on that issue.

3.3 Theme 3: Cost

The participants agreed on the cost aspect as one of the essential elements of choosing the service provider. The cost of e-payment depends on the service providers. Starting with K-net, the charges are split per transaction and subscription fees. The participants repeatedly stressed the cost factor in the interviews with variation when comparing K-net to other payment methods. Following is an example:

> If the customer pays using a credit card, we pay 2.5% to the bank on each transaction, deducted from the customer and us. And if he pays using K-net, we pay about 1.5 to the bank.

According to the participants, they also reported high subscription fees ranging between 600 and 1000 KD. Since the payments are primarily processed through K-net, the payment gateway, the other payment providers charge their customer extra fees to cover the cost of transactions paid to K-net. The two-layered price makes it unfeasible for many SMEs to adopt any e-payment service other than the basic service K-net offers. Given the limited services provided, as discussed in theme 1, and the need for more support, this is another pain point for the SMEs in Kuwait.

3.4 Theme 4: Settlement Period

In addition to the high costs of the e-payment, most participants referred to the protracted settlement period with the current e-payment service providers. In some cases, the clearance takes place once a week, and it can be once a month in others.

3.5 Theme 5: Rigid Regulation

Despite some participants commending the enabling regulations for SMEs, particularly licensing home businesses, they still find the rules restricting and rigid. They described e-payments as heavily regulated by the central bank. One participant gave an example of how the regulations prevent saving the subscriber's information, which requires the payer to manually fill in their information every time they need to pay a subscription or make a new payment.

However, there is more criticism of the regulations, whether of the central bank or the other regulators. In the participants' perception, compliance is unaffordable as the rules require capital reserves and highly paid compliance specialists, among other requirements.

The regulator tends to focus on big entities when setting regulations. One example of this issue is given by two of the participants. They stated that CBK requires 10% of the company's capital to be held as a bank guarantee.

In addition, the regulations also require strict measures for anti-money laundering and counter-financing terrorism. These requirements are beyond the capabilities of the SMEs offering payment service.

These requirements will be translated into an additional increase in the service cost, as discussed in theme three above. This is another example of the extensive entity-focused regulations addressed by some participants and the literature.

3.6 Theme 6: Perceived Benefits

Regardless of the critical accounts made by the participants, the e-payment landscape in Kuwait still offers many benefits from their point of view. The demand for cashless transactions by their customers, prospected sales increase, safety, and ease of use, are some of the various benefits reported by the participants.

In general, the participants expressed high trust in the safety and security of e-payments in Kuwait. Reducing cash collection was perceived as essential for eliminating human errors and theft incidents. Adopting e-payments is perceived as secure also from the technical side. For example, one of the participants stated:

> As soon as the payment goes through to the K-net portal, I believe the cyber security is transferred to them. And I've never heard of a problem happening.

The participants were aware of the demand for cashless transactions. They perceived various facets of this topic. This includes the impact of COVID-19 on avoiding direct contact with people for cash payments. In addition, there is an increasing demand for e-payments, even in areas where these transactions are rarely adopted, like the crafts sector. Fulfilling the need for e-payments leads not only to a satisfied customer but also to an increase in purchase will. Ease of payment is also among the perceived benefits expressed by the participants.

3.7 Theme 7: Expected Services

The participant SMEs expressed expectations of the e-payment services in Kuwait to be as frictionless as possible, have a universal payment method, and traceability of the payments. Many participants expressed unease with the long payment process made through the payment gateway. The process starts by requesting the payer's bank information, including the bank, card number, expiry date, and pin code. After that, an OTP is sent to the payer on the phone registered at the bank. Only after entering the OTP is the payment completed.

In addition to that, one of the participants suggested the following:

> K-net can create an app where you can pay by POS, barcode, or QR code, or transfer to the phone number.

Furthermore, two participants mentioned that the payment limits are affecting their businesses. The limit of 5000 KD may be reasonable for personal payments since K-net is not a corporate-oriented service. But compared to the other payment methods adopted by corporates, like cheques and bank transfers, payment over K-net is faster and easier. It settles instantly, unlike bank transfers, which may take one working day to be resolved. This is an area where SMEs are underserved. While 5000 KD may be a fair limit for personal transactions, it is not for SME transactions. Besides, SMEs are more sensitive to the delay of bank transfers and cheques than large-scale corporates.

4 Conclusions

The analysis of the data collected from the participant SMEs resulted in seven basic themes: payment gateway and infrastructure, regulations clarity and communications, cost, settlement period, rigid rules, perceived benefits, and expected services. Figure 1 illustrates these themes in relation to the mandate of the e-payments regulator, i.e. the Central Bank of Kuwait. The inner circle encloses the themes which the central bank can address through internal actions. The middle circle is where the central bank intervenes as a regulator. The outer circle encompasses themes that fall beyond the central bank's mandate.

The objective of this research was to explore the expectations of small and medium enterprises in Kuwait regarding e-payments services. After setting the adopted methodology and reviewing the literature on the topic, data were collected from an SME sample. The data was then analyzed to explore the expectations of the SMEs. The research aims to expand the understanding of the SMEs in Kuwait's expectations

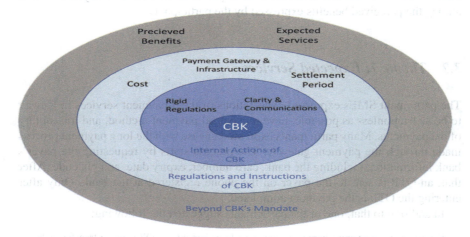

Fig. 1 The SMEs' expectations of E-payments in Kuwait. *Source* The author

from e-payment service providers and how to expand these services for the benefit of SMEs.

The data analysis revealed different expectations. Some of these expectations are related directly to the central bank's mandate. For example, the clarity of communication and enhancing the regulatory environment. Other expectations, such as introducing new services, fall beyond the central bank's mandate. Numerous recommendations were suggested to the central bank to address these expectations that fall under its mandate. On the other hand, the expectations not part of the central bank's mandate were highlighted to be communicated to the concerned parties.

Further research is recommended to complement the findings of this research. For instance, the sample size of this research limits the generalization of the results. Expanding the sample size would enhance the understanding of the SMEs' expectations and eventually develop policymakers' approaches and responses to those expectations. In addition, one must keep in mind that the expectations of SMEs are subject to change along with the development of the service delivered and the technical breakthroughs in e-payments. Quantitative research may be conducted to weigh the factors of each of the various themes found in this research. One of the main limitations is the rapid development in technology and payment solutions which entails continuous changes in customers' demands and expectations. Furthermore, the projectability of the finding is limited due to the qualitative methodology adopted.

References

1. Abdolmonem, H., Talha, A., Ismail, T.: Promoting Micro, Small, and Medium Enterprises in The Arab Countries. Retrieved from Kuwait (2019)
2. Alfonso, V., Boar, C., Frost, J., Gambacorta, L., Liu, J.: E-commerce in the pandemic and beyond. BIS Bullet. **36**(9) (2021)
3. Ali, A., Salameh, A.A.: Payment and settlement system in Saudi Arabia: a multidimensional study (2023)
4. Bank, W.: Governance of Retail Payment Systems: Keeping Pace with Changing Markets. World Bank (2021)
5. Barr, M.S., Harris, A., Menand, L., Xu, W.M.: Building the payment system of the future: how central banks can improve payments to enhance financial inclusion. In: U of Michigan Law & Econ Research Paper (20-038) (2020)
6. Basir, A.A.: Payment systems in Malaysia: recent developments and issues (2009)
7. Benoot, C., Hannes, K., Bilsen, J.: The use of purposeful sampling in a qualitative evidence synthesis: a worked example on sexual adjustment to a cancer trajectory. BMC Med. Res. Methodol. **16**(1), 1–12 (2016)
8. Carstens, A.: Shaping the future of payments. In: BIS Quarterly Review, March (2020)
9. Cong, L.W., Li, Y., Wang, N.: Tokenomics: dynamic adoption and valuation. Rev. Fin. Stud. **34**(3), 1105–1155 (2021)
10. Creswell, J.W., Poth, C.N.: Qualitative Inquiry and Research Design: Choosing Among Five Approaches. Sage Publications (2016)
11. Ehrentraud, J., Prenio, J., Boar, C., Janfils, M., Lawson, A.: Fintech and Payments: Regulating Digital Payment Services and E-money. Bank for International Settlements, Financial Stability Institute (2021)

12. Frost, J.: The economic forces driving Fintech adoption across countries. In: The Technological Revolution in Financial Services: How Banks, Fintechs, and Customers Win Together, p. 70 (2020)
13. Gallego, S., LHotellerie-Fallois, P., López-Vicente, F.: The international monetary fund and its role as a guarantor of global financial stability. Econ. Bull. (DEC), 1–16 (2018)
14. Greening, N.: Phenomenological research methodology. Sci. Res. J. 7(5), 88–92 (2019)
15. Igudia, P.O.: A qualitative evaluation of the factors influencing the adoption of electronic payment systems (SMEs) by SMEs in Nigeria. Eur. Sci. J. (ESJ) 13(31), 472–502 (2017)
16. Movchan, A., Zotin, A., Grigoryev, V.: Managing the resource curse strategies of oil17—Dependent economies in the modern era. Carnegie Moscow Center, Moscow, Russia (2017)
17. Ndiaye, N., Razak, L.A., Nagayev, R., Ng, A.: Demystifying small and medium enterprises'(SMEs) performance in emerging and developing economies. Borsa Istanbul Rev. 18(4), 269–281 (2018)
18. Qu, S.Q., Dumay, J.: The qualitative research interview. Qual. Res. Acc. Manage. (2011)
19. Slozko, O., Pelo, A.: The electronic payments as a major factor for futher economic development. Econ. Soc. 7(3), 130 (2014)
20. Taiwo, M.A., Ayodeji, A.M., Yusuf, B.A.: Impact of small and medium enterprises on economic growth and development. Am. J. Bus. Manage. 1(1), 18–22 (2012)

Content Marketing Strategy for the Social Media Positioning of the Company AD y L Consulting—Peru

Húber Rodríguez-Nomura, **Edwin Ramirez-Asis**,
K. P. Jaheer Mukthar, **Magdalena Valdivia-Malhaber**,
María Rodríguez-Kong, **Nathaly Zavala-Quispe**,
and José Rodríguez-Kong

Abstract The main objective is to propose a content marketing strategy for the SMO positioning of the aforementioned company. The following investigation was applied, since a content marketing strategy was found for the digital marketing and software development company, focused on improving the SMO positioning. Also, descriptive since it is dedicated to detailing the process of content marketing. It was considered mixed, since a quantitative–qualitative analysis will be carried out. To define the study sample, the convenience type non-probabilistic sample design was used. Given the contingency due to COVID 19, it was decided to take as a sample the 10 fixed clients that the company has, as well as 13 potential clients. The techniques used were the survey with the objective of knowing the valuation of the sample. In the case of this research, a questionnaire was used as an instrument. As a result, taking into account the current diagnosis of the company's position in the perception of customers, it was possible to show shortcomings in the content of the publications, since there is no established schedule. A single format has been used in the publications and the response from the audience has not been favorable. In the checklist it was determined that the maximum number of reactions was six. Considering that it is a page with more than 4,200 followers, the scope is not the most appropriate.

H. Rodríguez-Nomura · E. Ramirez-Asis (✉) · M. Valdivia-Malhaber · J. Rodríguez-Kong
Universidad Señor de Sipan, Chiclayo, Peru
e-mail: ramirezas@crece.uss.edu.pe

J. Rodríguez-Kong
e-mail: rodriguezka@crece.uss.edu.pe

K. P. Jaheer Mukthar
Kristu Jayanti College Autonomous, Bengaluru, India

M. Rodríguez-Kong · N. Zavala-Quispe
Universidad Nacional de Trujillo, Trujillo, Peru
e-mail: mprodriguez@unitru.edu.pe

© The Author(s), under exclusive license to Springer Nature Switzerland AG 2024
A. Hamdan and E. S. Aldhaen (eds.), *Artificial Intelligence and Transforming Digital Marketing*, Studies in Systems, Decision and Control 487,
https://doi.org/10.1007/978-3-031-35828-9_3

Keywords Content marketing · Perception · Positioning · Social networks · Strategy

1 Introduction

Providing an appropriate message to the audience, on digital platforms, is important to achieve brand positioning. Attracting the attention of the public with relevant and quality information, drives to establish a communication link between the company and the customer. In this process, the consumer can consider the organization as a benchmark in the industry, placing the brand in a privileged place in the consumer's mind. An organization acquires, through positioning, a differentiating image of its own in the mind of the consumer public. This image is created based on active communication with the audience [1].

According to [2] it is estimated that there are 11.5 million social network users 55% between 8 and 70 years old. Of which 96% have a Facebook account. From the study it is possible to support the importance of obtaining greater positioning in social media, since there is a large influx of public. SMO Positioning or Social Media Optimization is defined as the set of actions through which the search and promotion of content is improved. Optimization will help to achieve greater visibility in terms of related searches in social networks, but also in traditional search engines. It also facilitates interaction, since the content can be shared on other platforms due to its accessibility [3].

AD y L Consulting is located in the city of Chiclayo, founded in 2012. In its beginnings it was only focused on software development. With the passage of time and market demand, it was decided to expand its services. To date, the company is developed based on two business units such as digital marketing and custom software development. Due to the demand, the virtual classroom service for schools was enhanced [4]. AD y L Consulting's vision is to consolidate itself in the national and international market, being the leaders in technology services, always maintaining a level of quality and professionalism. Through its work it provides integral solutions that help companies to be more profitable and make efficient and fast decisions, achieving greater productivity and growth. In terms of web presence, the company has a website, business blog, Facebook page, and Twitter. To date the company has 6 employees, distributed in the areas of administration, design and software development.

In this regard, the following manifestations of the problem have been evidenced: The Facebook page has more than 4,000 followers, the publications on social networks do not have a defined frequency. The content shared on digital platforms is focused only on disseminating the company's services and important civic dates [5]. There is no evidence of a steady increase in the number of "likes" on the Facebook page. Followers rarely interact (comment) on publications. Possible causes of the problem have been identified: Insufficient promotion of services in the content marketing process. Deficiency in the dissemination of varied information in the content marketing process. Insufficient actions in the content marketing process to increase follower interaction. Limited monitoring of social networks in the content marketing process.

These causal manifestations proposed to delve into the study of the Content Marketing process, object of this research. The work done, related to the process of Content Marketing on digital platforms [6], points out that content marketing is a continuous process, in which companies obtain their own means of communication, creating a space to talk to their audience, not with the objective of advertising their service or selling, at first, but with the conviction that, by providing relevant information, the user's loyalty and a subsequent purchase will be won. It is important to emphasize that today's consumer can choose what to watch. A marketing strategy is focused on directing the fulfillment of the area's objectives [7].

The Content Marketing Institute states that 91% of American companies say that they use content marketing and regarding its effectiveness it can be inferred that 77% consider it a very effective technique. Likewise, they point out content quality and strategy as determining factors [8]. "In Peru, content marketing is in an early stage of development, in which it is associated more with an activity to generate branding, and not as a strategy to convert customers" [9] However, based on what has been described, it was considered that there are still not enough practical referents in content marketing focused on improving SMO positioning. Therefore, the field of research is the dynamics of the content marketing process.

According to [10, 11], in their research, they design a model based on natural language; based on this, they make use of a methodology in order to achieve objectives in any organization making use of process marketing with the purpose of capturing as many customers as possible, making use of digital tools such as sitemap, Google cloud, tools among others. Additionally, based on [12] in their publication analyzed 106 Web or Science researches that study the content strategy through social networks with the purpose of analyzing the status, process and design. They concluded that the content strategy has a variety of intellectual and methodologies; also, this strategy focuses on an analysis aimed at customers and organizations, which provides new opportunities for companies that are within a digital framework.

Nowadays, in a digital world there are many ways to exchange information between companies and users; that is why [13], stated that the public has a great inclination to communicate through digital media; thus, the most effective and efficient strategy for the future is content marketing, giving priority to customers seeking information by reaching them directly without being seen as spam, which is a challenge for companies that want to send a personalized marketing communication. Thus, proposing the creation of an application that is easy to use on cell phones to receive relevant and entertaining information about the service provided by the library [14], show in their research an analysis of the use of marketing in organizations and influencers in the tourism sector in order to know if they have used appropriate content to inform and teach about the care against covid-19 and the level of influence it has had on the public. Therefore, these authors concluded that people show a lack of interest in health as a dimension of sustainability in tourism.

Taking into account the performance of the public in social networks allowed the creation of information products; thus, [15] analyzed five native newspapers to identify the impact of the population to create content. Therefore, their results showed that the expected impact of digital media was not obtained and they recommend involving users in the creation of content in a higher percentage. On the other hand, according to [16] argued that, socially responsible labels is the best content marketing strategy that helps to differentiate from competitors and is the most accurate way for buyers to be informed about their social and environmental qualities; although, according to previous research, buyers do not have confidence in product labels. They also concluded that companies should focus on content marketing by investing in the market so that buyers become familiar with the product.

References [17, 18] who in their research aimed to evaluate the essential nutrient characteristics and marketing strategies of wheat and corn in Mexico City. They concluded that, the majority of such processed foods that are sold and advertised through content marketing to the low-income population are hazardous to health. In the paper discussed by [19, 20] they aimed to analyze the impact of content marketing and personal data sharing on young adults and consumption. Concluding that, young adults show a high degree of interest in the news which have greater ease of being able to visualize it through cell phones generating greater attention when these are empathized with the proposed information and taking into account that marketing in the world of virtual video games is no stranger to providing content advertising; [21] in their article mention the most efficient advertising technique for the most recognized brands of video games. Based on this; they determined that, the most interesting marketing strategy is when professional gamers mention the brand of the products to their audience.

2 Methodology

The following research was applied, since a content marketing strategy was found for the digital marketing and software development company, focused on improving SMO positioning. It was also descriptive since it is focused on detailing the content marketing process, describing its characteristics. It was considered mixed, since a quantitative–qualitative analysis will be carried out, regarding the data obtained and its integration in the discussion of results. In order to define the study sample, a non-probabilistic sample design of convenience type was used. Given the contingency due to COVID 19, the number of AD Y L Consulting clients decreased considerably in 2020. Therefore, it was decided to take as a sample the 10 regular clients that the company has, as well as 13 potential clients in order to equal 23, which is the total annual population. The techniques used were the survey, which is the procedure by means of which data will be obtained in a systematic way, with the objective of knowing the valuation of the sample. In the case of this research, a questionnaire was used as an instrument. Observation, this technique was applied in order to obtain data from the exploration by the researcher in the company's social network platforms,

Content Marketing Strategy for the Social Media Positioning … 31

through a checklist. The instrument was a questionnaire that allowed to know the SMO positioning of the company AD Y L Consulting. It was formulated 18 questions, with a Likert measurement scale, which showed to be consistent and reliable with a Cronbach alpha coefficient value of 0.81.

3 Results

In the period from October to December 2020, a low positioning at the moderate level has been evidenced. This is concluded as a result of the application of the checklist in which it was possible to verify that there is no established frequency in the dissemination of content on its page. The company has made 4 publications in the month of October, none in the month of November and finally 4 publications in the month of December.

Likewise, it could be observed that the maximum number of times that the content has been shared has been 3, as for the comments of the users none were found in their publications. On the other hand, the maximum number of reactions obtained is six. Considering that it is a page with more than 4200, it is clearly not achieving its desired maximum reach. All the publications made have the image format, of which almost all (with the exception of one) have the company's symbol, but only in two publications do they present their service. On the other hand, out of the total of 8 publications reviewed, 3 do not contain relevant information for the user.

It is evident that 52% of the total of 12 people are between 21 and 30 years of age, followed by 30% represented by 7 people. From this it can be determined that the highest percentage of companies that have contracted with the company and/ or are within its target audience are young clients. In addition, 95% of the total respondents have a profile on social networks, which highlights the importance of proper management in these digital platforms, since a potential client may visit the page in search of information before contacting. From the results obtained, we know that 69% of the profile is inadequate, so the approach should be redefined, in reference to the profile of the public, in order to have greater reach.

It was determined that the most used social network is Facebook with 56%, followed by Instagram with 21%. From this we verified that there is a greater possibility that the target audience will see the publications made on Facebook, as long as it has an optimal reach. In reference to the format that users prefer, we have 47% who prefer videos, followed by 39% who choose images. The very composition of a short video makes it more attractive to the user, the combination of image and music will help the retention of the message.

The level of brand presence is inadequate, since there is no frequency in publications, the brand is not fully identified by users, limiting the online reputation and interaction on the page. By enhancing the presence, it will be possible to acquire greater visibility and reach, reaching more potential customers. The level of client perception is inadequate at 78%. Knowing that it is the clients who recommend the

Table 1 Test for a sample

Variable	Test value = 11					
	t	gl	Sig. (bilateral)	Difference in averages	95% confidence interval of the difference	
					Lower	Upper
Customer perception	−6.734	22	0.000	−3.4348	−4.493	−2.377

Note Data obtained from the survey application

company and can contribute to the dissemination of its services, special emphasis should be placed on improving this aspect in order to achieve a better positioning.

Table 1 shows that the T-Student test with a significance level of 0.05; its significance or p-value is 0.000, since this value is less than 0.025 (0.05/2 = 0.025 since the contrast is bilateral) we reject the null hypothesis and accept the alternative hypothesis, therefore, "The level of customer perception of the service provided is less than moderate in the company AD Y L Consulting of Chiclayo".

4 Discussion

In the results obtained in terms of the current diagnosis, it was determined that there is no established frequency of publications, therefore, there is no schedule of activities for the dissemination of content on social networks. As concluded by [22], constant updating on the platforms is essential, since it brings users closer to the company's social network, valuing it positively. Likewise, as mentioned by [23] it is necessary to plant a strategy, since companies have the opportunity to generate a communication channel with their audience and it is a growth advantage that should be taken advantage of by creating and sharing content of interest to the audience.

It was also identified that the customer profile is mostly made up of a young population, which has a profile on social networks and 56% use Facebook. It is important to mention that a social network is freely accessible to people, likewise according to [24] this social network has a business approach in which strategies that generate interaction and impact on followers can be raised. As mentioned [25], argued that young adults perceive a great interest in the news content of products offered through social networks that are easier to visualize through their cell phones thus showing empathy with advertising.

In determining what level of brand presence is inadequate, we agree with what [26] points out, stating that brand awareness is fundamental, migrating to digital channels, highlighting that every day the number of users joining these platforms and even the competition is increasing. Likewise [27], indicates that there is a very close relationship between communication in digital media and the positioning obtained by the company. Therefore, it is possible to take advantage of the advantages in order to enhance the presence of the brand by attracting the audience, thus achieving a constant interaction encouraging communication [28]. The relevant content that can be shared is essential to arouse interest in the company, until it is considered a reference by the public.

It can be evidenced that the level of customer perception with the service provided is less than moderate in the company. As [6] states, it is essential to achieve a brand-audience bond in order to generate user trust, allow comments and opinions, but also to attend to them by encouraging dialogue. This is also demonstrated by [11] in their article where they determined that, in order to generate greater perception of video game products, the presence of influencers was necessary, especially professional gamers who mention the brand of the products to their audience.

5 Conclusion

A content marketing strategy proposal was developed for SMO positioning, where the current reality of the company is known, the SWOT analysis, the audiences were identified to later detail the actions to be carried out, as well as those responsible for each one. The implementation of the proposed actions will enhance the growth of the company and optimize the results in terms of service promotion.

The current diagnosis of the company's positioning revealed shortcomings in terms of the content of publications, since there is no established schedule. Only one format has been used in the publications and the audience response has not been favorable. In the checklist it was determined that the maximum number of reactions has been six. Considering that it is a page with more than 4200 followers, the reach is not the most adequate. Also, it was identified that the customer profile is mostly made up of a young population 21–30 years old, who have a profile on social networks and 56% use Facebook. It was also determined that the format they find most attractive in social networks is video, followed by image.

It is also concluded that the level of brand presence is inadequate, since there is no established frequency in the publications and the brand is not fully identified, which is limiting the online reputation and interaction on the page. By optimizing the presence, there will be a greater reach, reaching more potential customers. Finally, regarding the level of customer perception, it was found to be inadequate in 78%. This is a point to work on, since customers can also promote the brand indirectly by recommending the company, thus contributing to the dissemination of services.

References

1. Kotler, P.: Marketing 5.0: Tecnología para la Humanidad. Editorial Almuzara, España (2021)
2. Ramirez GP, Yactayo AC.: Posicionamiento de marca y su relación con el nivel de compra por redes sociales. Revista Arbitrada Interdisciplinaria Koinonia **6**(1), 81–100 (2021). https://doi.org/10.35381/r.k.v6i1.1214
3. Zhang, J., Raza, M., Khalid, R., Parveen, R., Ramírez-Asís, E.H.: Impact of team knowledge management, problem solving competence, interpersonal conflicts, organizational trust on project performance, a mediating role of psychological capital. Ann. Oper. Res. **2021**, 1–21 (2021). https://doi.org/10.1007/s10479-021-04334-3
4. Raza, M., Wisetsri, W., Chansongpol, T., Somtawinpongsai, C., Ramírez, E.H.: Fostering workplace belongingness among employees. Polish J. Manage. Stud. **22**(2), 428–442 (2020). https://doi.org/10.17512/pjms.2020.22.2.28
5. Ramírez, E.H., Colichón, M.E., Barrutia, I.: Rendimiento académico como predictor de la remuneración de egresados en Administración, Perú. Revista Lasallista de Investigación **17**(2), 88–97 (2020). https://doi.org/10.22507/rli.v17n2a7
6. Bahcecik, Y.S., Akay, S.S., Akdemir, A.: A review of digital brand positioning strategies of Internet entrepreneurship in the context of virtual organizations: Facebook, Instagram and YouTube samples. Procedia Comput. Sci. **158**, 513–522 (2019). https://doi.org/10.1016/j.procs.2019.09.083
7. Ramirez-Asis, E.H., Srinivas, K., Sivasubramanian, K., Jaheer, Mukthar, K.P.: Dynamics of inclusive and lifelong learning prospects through massive open online courses (MOOC): a descriptive study. In: Hamdan, A., Hassanien, A.E., Mescon, T., Alareeni, B. (eds.) Technologies, Artificial Intelligence and the Future of Learning Post-COVID-19. Studies in Computational Intelligence, vol. 1019, pp. 679–696. Springer, Cham (2022). https://doi.org/10.1007/978-3-030-93921-2_35
8. Kotler, P., Keller, K.: Dirección de marketing. (Décimo sexta edición). Prentice Hall, Madrid (2017)
9. Velázque-Cornejo, B.I., Hernández-Gracia, J.F.: Marketing de contenidos. Boletín Científico de la Escuela Superior Atotonilco de Tula **6**(11), 51–53 (2019). https://doi.org/10.29057/esat.v6i11.3697
10. Marrón, J.A., Fernández, A.C., Cruz, C., García, A., Pacheco, S., Quezada, A.D., Pérez-Luna, M., Donovan, J.: Perfil nutricional y estrategias de publicidad en el empaque de alimentos procesados de trigo y maíz en la Ciudad de México. Salud Publica de Mexico **63**(1), 79–91 (2020). https://doi.org/10.21149/11252
11. Schoon, A., Mabweazara, H.M., Bosch, T., Dugmore, H.: Decolonising digital media research methods: Positioning African digital experiences as epistemic sites of knowledge production. Afr. J. Stud. **41**(4), 1–15 (2020). https://doi.org/10.1080/23743670.2020.1865645
12. Cuevas, E., Matosas, L., Sánchez, M.: Bibliometric analysis of studies of brand content strategy within social media. Comunicación y Sociedad **e7441**, 1–25 (2019). https://doi.org/10.32870/cys.v2019i0.7441
13. Castillo, N.J., Enciso, L.: Notification of advertising content with bluetooth and beacon technology for library services. Revista Iberica de Sistemas e Tecnologías de Informacao **E46**, 225–238 (2021). https://www.proquest.com/openview/3af57e779f8753593707ab646fbd1038/1?pq-origsite=gscholar&cbl=1006393
14. Wang, C.L.: New frontiers and future directions in interactive marketing: inaugural editorial. J. Res. Interact. Mark. **15**(1), 1–9 (2021). https://doi.org/10.1108/JRIM-03-2021-270
15. Sixto, J., López, X., Toural, C.: Opportunities for content co-creation in digital native newspapers. Profesional de la Información **29**(4), 1–19 (2020). https://doi.org/10.3145/epi.2020.jul.26
16. Carrero, I., Redondo, C.: Decisive factors for purchasing products with social and environmental labels. España Revista de Economía Pública, Social y Cooperativa **83**, 235–250 (2019). https://doi.org/10.7203/CIRIEC-E.83.13425

17. Carvajal, M., Barinagarrementeria, I.: Branded content in Spanish newspapers: Concept, organization and challenges for the journalists involved. Trípodos **44**, 137–152 (2019).. https://doi.org/10.51698/tripodos.2019.44p137-152
18. Martínez, E., Vizcaíno, R.R.: Minors and internet gambling advertising: new formats, advertising content, and challenges in protecting minors. Profesional de la Informacion (2021). https://doi.org/10.3145/epi.2021.jul.20
19. Barrientos, A., Martínez, A.M., Altamirano, V.: COVID-19, a myth in tourism communication. analysis of the contents generated by tourist influencers 2.0 about the pandemic and destinations. Palabra Clave **25**(1), e2518 (2022). https://doi.org/10.5294/pacla.2022.25.1.8
20. Jara, K.S., Miranda, M.D.P., Céspedes, C.P.: Relación entre el neuromarketing y el posicionamiento de marca de una empresa del sector retail. Revista Universidad y Sociedad **14**(1), 554–563 (2022). https://rus.ucf.edu.cu/index.php/rus/article/view/2587
21. Fanjul, C., González, C., Peña, P.: eGamers' influence in brand advertising strategies. a comparative study between Spain and Korea. Comunicar **58**, 105–114 (2019). https://doi.org/10.3916/C58-2019-10
22. Ramírez, E.H., Espinoza, M.R., Esquivel, S.M., Naranjo, M.E.: Emotional Intelligence, competencies and performance of the university professor: using the SEM-PLS partial least squares technique. Revista Electrónica Interuniversitaria de Formación del Profesorado **23**(3), 99–114 (2020). https://doi.org/10.6018/reifop.428261
23. Espinel, G.A., Hernandez, C.A., Rojas, J.P.: Usos, apropiaciones y prácticas comunicativas de los usuarios adolescentes de Facebook. Saber, Ciencia y Libertad **15**(1), 280–296 (2020). https://doi.org/10.18041/2382-3240/saber.2020v15n1.6316
24. Belanche, D., Cenjor, I., Pérez-Rueda, A.: Instagram Stories versus Facebook Wall: an advertising effectiveness analysis. Span. J. Mark.-ESIC **23**(1), 69–94 (2019). https://doi.org/10.1108/SJME-09-2018-0042
25. Martínez, M., Serrano, J., Portilla, I., Sánchez-Blanco, C.: Young adults' interaction with online news and advertising. Comunicar **59**, 19–28 (2019). https://doi.org/10.3916/C59-2019-02
26. Yating, Y., Mughal, N., Wen, J., Ngan, T.T., Ramirez-Asis, E., Maneengam, A.: Economic performance and natural resources commodity prices volatility: Evidence from global data. Resour. Policy **78**, 102879 (2022). https://doi.org/10.1016/j.resourpol.2022.102879
27. Hermoza, R.: El marketing digital y su relación con el posicionamiento de la Empresa Agroindustrias Verdeflor S.A.C, 2018. Master's thesis. Universidad Nacional Federico Villarreal, Perú (2019). http://repositorio.unfv.edu.pe/handle/UNFV/2828
28. Aguirre, M.J.C., Álvarez, J.C., Zurita, C.I.N. Estrategias de Marketing y posicionamiento de marca para el sector artesanal textil. Cienciamatria **5**(1), 245–270 (2019). https://doi.org/10.35381/cm.v5i1.266

An Exploratory Study on E-commerce in the Sultanate of Oman: Trends, Prospects, and Challenges

Dawood Al Hamdani and Aisha Al Wishahi

Abstract Despite being the most understudied and underutilized study area, the Sultanate of Oman has the potential to become a major player in e-commerce due to its strategic location and increasing internet penetration. Compared to other GCC countries, we face a low internet penetration of only 22% limits on e-commerce opportunities. However, as the number of internet users grows, it will open up new opportunities for Omanis and make it easier for them to be introduced to new services. This exploratory used a cross-sectional descriptive survey design to investigate Omani customers' practices and attitudes toward e-commerce development in the Sultanate of Oman. A total of 133 people were chosen from eleven governorates in the country using cluster sampling and an online questionnaire in the Sultanate of Oman. The findings indicate good e-commerce practices and positive attitudes toward e-services/e-shopping. The top challenges identified by the study were: low consumer income, lack of control over internet usage, weak or slow network, lack of a clear legislative framework for regulating this type of trade, and the high cost of Internet access. The study recommends that the Sultanate of Oman consider good practices in e-government, e-services, and e-commerce and address the wide range of challenges.

Keywords E-commerce · Online shopping · Customers · Challenges · UNSDG: Goal#8

1 Introduction

Online shopping is a well-established retail phenomenon that has evolved considerably in recent years. Many consumers now enjoy the convenience of buying everything from books to beauty products and groceries without leaving their homes. Online retailers can profit from new business opportunities by adapting to customer

---

D. Al Hamdani (✉) · A. Al Wishahi
Post-Graduate Center, The State Council, Sohar University, Sohar, Sultanate of Oman
e-mail: alhamdani@su.edu.om

© The Author(s), under exclusive license to Springer Nature Switzerland AG 2024
A. Hamdan and E. S. Aldhaen (eds.), *Artificial Intelligence and Transforming Digital Marketing*, Studies in Systems, Decision and Control 487,
https://doi.org/10.1007/978-3-031-35828-9_4

needs, investing in technology, and developing innovative approaches to marketing and advertising with more and more people turning to the internet for their purchases. In recent years, the expansion of e-commerce has been exponential. When a merchant offers things on a website, users demonstrate acceptance, examine the products' characteristics, prices, and delivery options, purchase products of interest, and then check out [6]. For example, the UK has been at the forefront of this trend for some time now, with China now following suit. Over the past decade, online marketplaces have seen stiff competition from Amazon, with other players like Walmart and eBay being forced to adapt to remain competitive in the market.

Yoon and Occeña [12] identify four types of e-commerce: business-to-business (B2B), business-to-consumer (B2C), consumer-to-consumer (C2C), and business-to-government (B2G). Although mainstream research tends to focus on B2C without making any difference between B2C and C2C, C2C e-commerce is more popular [4].

A range of environmental, organizational, and personal elements not only aid in IT diffusion but also have a significant impact on consumers' opinions and attitudes toward Ecommerce. Environmental considerations, including business-friendly government policies and infrastructure improvements, are vital if Ecommerce is to realize its full potential. Organizational aspects highlight management's role in promoting the usage of Ecommerce within the organization [9]. In general, a consumer's perspective on Ecommerce will influence attitudes and will make managers aware of the potential of Ecommerce if they are to remain competitive and capable of positioning their organizations to meet the requirements of the internationally wired business community. Personal factors influence a person's level of happiness. Simtowe [7] maintains that "technology diffusion (awareness) is an important precondition for adoption to occur". Therefore, it is significant to focus on these two concepts, respectively.

2   Research Questions

This research probes the following four key research questions:

1. What is the current eCommerce trend in the Sultanate of Oman?
2. What are customers' perception and attitude toward the practice of Ecommerce in the Sultanate of Oman?
3. What are the challenges to the practice of Ecommerce in the Sultanate of Oman?
4. Are there any significant differences in the practice of e-commerce in the Sultanate of Oman that can be attributed to the place of residence, age, gender, income, or occupation?

3 Methodology

This exploratory study aims to explore Omani customers' practices and attitudes toward e-commerce development in the Sultanate of Oman using a cross-sectional descriptive survey design. A total of 133 individuals were selected by random cluster sampling via an online questionnaire from eleven governorates in the Sultanate of Oman. Data was collected through a self-administered questionnaire prepared in coherence with the established theoretical framework adopted from a literature review conducted at different levels, including national and global levels as well as the organizational level (governmental sector). However, only 129 questionnaires were found suitable for analysis, as four questionnaires had missing data. Further-more, there were no responses from the Wusta governorate (Table 1). Tables 1, 2, 3, 4, 5, 6 and 7 show the descriptive data of independent variables.

Table 1 The distribution of participants over governorates

Governorates	Frequency	Percent
Muscat	59	45.7
Dhofar	10	7.8
Musandam	8	6.2
Buraymi	4	3.1
Dakhiliyah	11	8.5
North Batinah	13	10.1
South Batinah	5	3.9
South Sharqiyah	4	3.1
North Sharqiyah	10	7.8
Dhahirah	5	3.9
Wusta	0	00
Total	129	100

Table 2 Gender distribution

Gender	Frequency	Percent
Male	80	62
Female	49	38
Total	129	100

Table 3 Age distribution

Age	Frequency	Percent
Less than 24	13	10.1
25–29	18	14
30–34	26	20.2
Over 35 years old	72	55.8
Total	129	100

Table 4 Occupational status

Jobs	Frequency	Percent
Student	11	8.5
Government employee	65	50.4
Private sector employee	25	19.4
Job seeker	6	4.7
Entrepreneur/self-employment	3	2.3
Retired	13	10.1
Housewife	6	4.7
	129	100

Table 5 Average monthly income in Omani rial (as individual)

Incomes	Frequency	Percent
Less than 500 OMR	26	20.2
500 to 999 OMR	37	28.7
1000 to 1500 OMR	34	26.4
More than 1500 OMR	32	24.8
Total	129	100

Table 6 Average monthly expenditure on electronic shopping

Monthly expenditure	Frequency	Percent
Less than 50 OMR	55	42.6
51–99 OMR	27	20.9
100–150 OMR	13	10.1
More than 150 OMR	34	26.4
Total	129	100

Table 7 Devices used for shopping

Device name	No.	Percentage
Mobile phones	98	42.8
Tablets (E.g. iPad)	7	3.1
Laptops	8	3.5
Desktop computers	16	7

An Exploratory Study on E-commerce in the Sultanate of Oman: Trends ...   41

4 Results and Discussion

Because the population sample size was small with respect to the native population in the Sultanate of Oman (2.5 million), and the data is not normally distributed, the researchers considered using nonparametric tests particularly the Mann–Whitney U Test and the Kruskal–Wallis Test in data analysis.

4.1 Question One

What is the current trend of Ecommerce in the Sultanate of Oman in the perception of Omanis, as customers?

Table 8 shows that the most used service is Business to Customer 82% and the least used service is Government to Customer 67%.

Table 9 shows the activities that the precipitants use using online services. 71.4% use buying and selling services and only 11.3% use promotional or marking activities.

Table 10 shows the products and services which are experienced by the participants and as rated from the highest to the lowest. It seems that 45% of users are experienced in buying meals, whilst 10.1% of the users are experienced in buying furniture through Ecommerce.

Table 11 shows the number of times users engage in online shopping, monthly. 64% of the participants (83) have shopped from one time to three times every month whereas only 15% of them used online shopping more than nine times. This indicates a good level of frequency of online shopping.

Table 8 The e-commerce patterns (types) that are used frequently

The pattern	Count
Customer to customer (e.g. buying a product from an individual through an application, or facilitator, such as Open Sooq and OLX)	81
Business to customer (I shop online from companies)	82
Government to customer (I complete my transactions with government entities electronically)	67

Table 9 The activities usually practiced or dealt with electronically

Activities	No.	Percentage
Buying and selling	129	71.4
E-banking services	129	57.1
e-government services	129	50.4
Delivery services	129	43.6
Promotional/marketing activities	129	11.3

Table 10 Experiences of buying products and services as rated by the participant

	No.	Highest	2	3	4	5	6	7	8	Lowest
Meals	129	45	9.3	8.5	11.6	3.9	3.1	3.9	2.3	12.4
Furniture	129	10.1	15.5	17.8	11.6	7.8	5.4	4.7	3.9	23.3
Groceries	129	22.5	14.7	13.2	10.9	8.5	7	5.4	3.1	14.7
Shoes and bags	129	15.5	17.8	18.6	12.4	7.8	4.7	3.9	5.4	14
Beauty/care products	129	22.5	14.7	13.2	10.9	8.5	7	5.4	3.1	14.7
Accessories	129	17.1	17.1	17.8	12.4	12.4	5.4	3.1	3.1	11.6
Perfumes	129	17.8	19.4	14	9.3	10.1	5.4	7.8	5.4	10.9
Clothes	129	28.7	19.4	12.4	14	3.9	4.7	2.3	3.1	11.6
Electronics	129	22.5	17.1	19.4	14	4.7	4.7	4.7	7	6.2

Table 11 Experiences of buying products and services as rated by the participant

Frequency of online shopping	No.	Percentage
1–3 times every month	83	64.3
4–6 times every month	26	20.2
7–9 times every month	5	3.9
More than 9 times per month	15	11.6
Total	129	100

Table 12 shows that most participants use their mobile phones to do online shopping, 76%, and fewer use tablets 5.4%.

Table 13 depicts that 76% of the participants use applications and software when dealing with e-commerce; only 12.4% of them use online shopping website sites such as Amazon, Alibaba, and Talabat.

The results of this section depict the country's readiness for Ecommerce, and they are consistent with Kurnia et al. [4] study, which presented three levels of readiness: organizational readiness (OR), industry readiness (IR), and national readiness (NR) required for successful Ecommerce implementation.

Table 12 Devices used for online shopping

	Frequency	Valid percent
Mobile phones	98	76.0
Tablets (E.g. iPad)	7	5.4
Laptops	8	6.2
Desktop computers	16	12.4
Total	129	100.0

An Exploratory Study on E-commerce in the Sultanate of Oman: Trends ...   43

Table 13 Methods of E-commerce

	Frequency	Valid percent
Apps and software	98	76
Corporate websites	7	5.4
Social media	8	6.2
Online shopping websites (such as Amazon, Alibaba and Talabat)	16	12.4
Total	129	100

4.2 Question Two

What are the customer's perception and attitude toward the practice of E-Commerce in the Sultanate of Oman?

Attitudes Toward E-services Provided by Government Agencies

Electronic government (e-government) aims to "enhance access to and delivery of government services to benefit citizens" [8]. Table 14 shows that the overall mean for the participants' attitude is 3.41 which is above average toward agreeing. Also, this indicates the availability of Ecommerce services in the country. Matbouli and Gao's [5] study emphasizes that in the case of successful adoption of e-commerce, there should be: the availability of the software system and the hardware equipment working properly, and in the case of disaster, the system can recover quickly.

It seems that the participants are concerned about the speed of the internet in all Wilaats, mean = 2.71. Table 15 shows that 23.3% strongly disagree and 26.4%

Table 14 Attitudes toward services provided by government agencies

Items	No.	Means	SD
Government agencies have electronic payment systems in place for the services they provide over the Internet	129	3.84	0.991
Government agencies have websites (e-portals) that they use to provide their services over the Internet	129	3.81	1.006
All Agencies of government agencies have access to the internet	129	3.71	1.071
The application systems used by government agencies are compatible with the services they provide (for example, Tarassud + ROP applications)	129	3.7	1.005
The equipment used by government service agencies is outfitted with the most up-to-date systems and software to facilitate e-commerce practices	129	3.65	1.109
Transactions in e-governance are completed quickly and efficiently	129	2.95	1.227
Government agencies have employees who specialize in e-marketing	129	2.95	1.11
A high-speed internet network connects all Wilayats in the Sultanate of Oman	129	2.71	1.354
Overall means	129	3.41	0.856

disagree, almost 50% of participants are not happy, relatively, and 20.9% are undecided. Kaya and Eyupoglu [2] maintain that the speed of the internet was one of the common issues while using e-government services.

The Importance of E-commerce to Consumers

The participants see that there is an importance for e-commerce for the country, means = 4.02 and SD 0.67, as in Table 16.

It is interesting to notice that the participants see that online shopping provides a variety of products to them. Table 17 shows that 38% agree and 52.7 strongly agree on supporting this online shopping feature. Wang et al. [10] maintain that the growing variety of goods, which encouraged general customers the e-commerce services and using credit systems would promote these services.

Table 15 Percentage of attitude toward a high-speed internet network

Item		Frequency	Percent
A high-speed internet network connects all Wilayats in the Sultanate of Oman	Strongly disagree	30	23.3
	Disagree	34	26.4
	Undecided	27	20.9
	Agree	20	15.5
	Strongly agree	18	14.0
	Total	129	100

Table 16 Attitudes toward the importance of e-commerce to consumers

Descriptive statistics	No.	Mean	Std. D
Providing a variety of products to the consumer	129	4.4	0.755
Taking advantage of special online shopping offers	129	4.3	0.825
Easy access to the product	129	4.29	0.859
Meeting the demand for the purchase of some goods, including goods of privacy	129	4.13	0.823
Meeting the need for reliable and timely delivery services	129	4.02	0.931
Influencing consumer choices as a result of other consumers' opinions/ reviews about products and services on e-commerce platforms	129	3.99	0.906
Purchasing products and services that are similar in quality to those found in traditional shopping	129	3.88	0.997
Price reductions	129	3.85	1.105
Easing the payment process and the possibility of refunding amounts	129	3.75	1.097
Ensuring the safety of the product that you buy online	129	3.62	1.017
Overall means	129	4.02	0.667

An Exploratory Study on E-commerce in the Sultanate of Oman: Trends ...   45

Table 17 Percentage of attitude toward providing a variety of products to the consumer

Item		Frequency	Valid percent
Providing a variety of products to the consumer	Strongly disagree	1	0.8
	Disagree	2	1.6
	Undecided	9	7
	Agree	49	38
	Strongly agree	68	52.7
	Total	129	100

Table 18 Attitude toward E-security, trade laws, and consumer protection

E-commerce transactions	No.	Mean	S.D
Allow the dispensing of paper documents transactions	129	4.08	0.84
Ensure confidentiality for customers	129	3.85	0.84
Take place in a secure electronic environment	129	3.76	0.90
Provide adequate guarantees and clear instructions to maintain the confidentiality of information	129	3.68	0.97
Contribute to the reduction of tax evasion	129	3.66	0.89
Protect consumer rights	129	3.66	0.97
Are carried out according to clear legal systems	129	3.63	0.98
Are carried out in accordance with approved security systems	129	3.61	0.90
Limit financial corruption	129	3.56	0.92
Combat clandestine trade	129	3.46	1.01
Combat money laundering	129	3.46	1.02
Overall means	129	3.67	0.70

E-security, Trade Laws, and Consumer Protection

Table 18 shows that the overall mean for the participants' attitude is 3.67 which is above average. Interestingly enough the item "Allow the dispensing of paper documents transactions" has the highest mean of 4.08. Table 19 shows that almost 78% have a positive attitude toward this item. This result supports the different elements of security that are required to be in place to foster confidence in e-commerce found in Chatterjee's [1] study.

4.3 Question Three

What are the challenges facing the practice of E-Commerce in the Sultanate of Oman in the perception of Omanis, as customers?

Table 19 Percentage of attitude toward the dispensing of paper documents transactions

Item		Frequency	Valid percent
Allow the dispensing of paper documents transactions	Disagree	6	4.7
	Undecided	22	17.1
	Agree	57	44.2
	Strongly agree	44	34.1
	Total	129	100

Table 20 presents that the highest ranged barrier for e-commerce is the income 38% and Lack of control over internet usage/weak or slow network. The absence of a legislative framework regulating this type of trade comes in second, at 34.1%. This means that the most needed intervention is to overcome the challenges. Whereas, Distrust of such websites ranged as lowest 21.7. West [11] argues that in some respects, the e-government revolution has fallen short of its potential to transform service delivery and public trust in government. It does, however, have the possibility of enhancing democratic responsiveness and boosting beliefs that government is effective. Challenges facing e-commerce such as trust, level of readiness and e-commerce security were identified and in order to overcome these challenges, the researchers proposed a number of solutions, which are: designing websites that support privacy through the use of oral passwords to solve the problem of trust, and the researchers also suggested increasing cooperation between industry and government partners.

Table 20 The barriers to e-commerce adoption in the Sultanate

Items	1	2	3	4	5	6	7	8
The low level of Omani consumer income	38	25.6	13.2	10.1	5.4	0.8	4.7	2.3
Lack of control over internet usage/weak or slow network	34.1	27.1	16.3	10.1	3.1	2.3	2.3	4.7
Absence of legislative framework regulating this type of trade	34.1	19.4	23.3	10.1	6.2	1.6	2.3	3.1
High cost of internet access	33.3	22.5	19.4	7.8	7	3.1	3.1	3.9
Factors associated with Omani consumer culture	26.4	32.6	18.6	7.8	7.8	3.9	2.3	0.8
lack of online payment methods	24	19.4	20.9	13.2	7.8	0.8	1.6	12.4
Language factor (lack of Arabic interaction on the Internet)	24	24	20.9	13.2	3.9	4.7	3.9	5.4
Distrust of such websites	21.7	32.6	21.7	14	2.3	0.8	3.9	3.1

An Exploratory Study on E-commerce in the Sultanate of Oman: Trends ...   47

4.4   Question Four

Are there any significant differences in the attitudes and practices of E-Commerce in the Sultanate of Oman attributed to the place of residency, age, gender, income, and occupational status on Shopping?

Influence of Place of Residency in the Practice of E-commerce
As shown in Table 21 the results of the Kruskal–Wallis Test reveal no significant differences in place of residency on practices and attitudes toward services provided by the government agencies. This result goes in line with Kaya et al.'s [3] study which indicates that physical isolation does not create digital isolation.

Influence of Age
As shown in Table 22 the results of the Kruskal–Wallis Test reveal no significant differences in age on practices and attitudes toward services provided by the government agencies.

Influence of Gender
As shown in Table 23 the results of the Kruskal–Wallis Test reveal no significant differences in gender on practices and attitudes toward services provided by government agencies.

Influence of Incomes
The Kruskal–Wallis Test in Table 24 shows there is only significant differences in attitudes toward services provided by the government agencies attributed to incomes.

Table 21 Kruskal–Wallis test for place of residency

Null hypothesis	Sig.	Decision
Attitudes toward services provided by government agencies	0.659	Retain the null hypothesis
Attitudes toward the importance of e-commerce to consumers	0.771	Retain the null hypothesis
Attitude toward E-security, trade laws, and consumer protection	0.955	Retain the null hypothesis

Table 22 Kruskal–Wallis test for age

Null hypothesis	Sig.	Decision
Attitudes toward services provided by government agencies	0.622	Retain the null hypothesis
Attitudes toward the importance of e-commerce to consumers	0.558	Retain the null hypothesis
Attitude toward E-security, trade laws, and consumer protection	0.279	Retain the null hypothesis

Table 23 Kruskal–Wallis test for gender

Null hypothesis	Sig.	Decision
Attitudes toward services provided by government agencies	0.901	Retain the null hypothesis
Attitudes toward the importance of e-commerce to consumers	0.728	Retain the null hypothesis
Attitude toward E-security, trade laws, and consumer protection	0.141	Retain the null hypothesis

Table 24 The Kruskal–Wallis test for incomes

Null hypothesis	Sig.	Decision
Attitudes toward services provided by government agencies	**0.017**	Reject the null hypothesis
Attitudes toward the importance of e-commerce to consumers	0.954	Retain the null hypothesis
Attitude toward E-security, trade laws, and consumer protection	0.624	Retain the null hypothesis

Table 25 shows pairwise comparisons between incomes; indicating that there is a significant difference between participants with more than 1500 and participants with less than 500 OMR incomes. It seems that participants with less than 500 OMR incomes have a positive attitude toward services provided by government agencies, see Fig. 1; this might mean that they trust the agencies more than private ones.

Influence of Occupational Status

The Kruskal–Wallis Test in table 26 shows there is only significant differences of attitudes toward services provided by the government agencies attributed to occupational status.

Table 25 Pairwise comparisons test for incomes

Sample1–Sample2	Test statistic	Std. test statistic	Adj. sig.
More than 1500-From 500 to 999	12.255	1.360	1.000
More than 1500-From 1000 to 1500	22.482	2.445	0.087
More than 1500-Less than 500	28.514	2.893	**0.023**
From 500 to 999-From 1000 to 1500	−10.227	−1.153	1.000
From 500 to 999-Less than 500	16.259	1.702	0.533
From 1000 to 1500-Less than 500	6.033	0.620	1.000

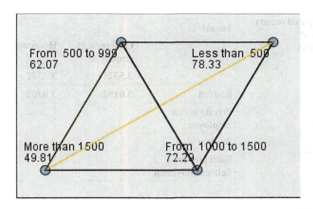

Fig. 1 Pairwise comparisons test

Table 26 The Kruskal–Wallis test for occupational status

Null hypothesis	Sig.	Decision
Attitudes toward services provided by government agencies	0.03	Reject the null hypothesis
Attitudes toward the importance of e-commerce to consumers	0.277	Retain the null hypothesis
Attitude toward E-security, trade laws, and consumer protection	0.823	Retain the null hypothesis

Table 27 Pairwise comparisons test for occupational status

Occupational status	Test statistic	SD, test statistic	Sig
Entrepreneur self-employment-job seeker	61.083	2.314	0.021
Retired-government employee	24.908	2.196	0.028
Retired-housewife	−38.56	−2.093	0.035
Retired-job seeker	49.147	2.667	0.008
Student-job seeker	−40.55	−2.140	0.032
Private sector employee-job seeker	−35.49	−2.092	0.036

Tables 27 and 28 show that retired government employees are more positive toward e-services, mean-3.53 than Entrepreneur/Self-employment; whereas housewives are more positive than Retired. Further, Job seekers hold a more positive attitude, mean = 4.00 than Private sector employees, mean = 3.27, Student, mean = 3.19, and Entrepreneur Self-employment, mean = 2.45.

Table 28 Compared means between groups of occupational status

Means			
	Government employee	Housewife	Job seeker
	3.5327	*3.8542*	*4*
Retired	**3.0192**	*3.0192*	*3.0192*
Private sector employee			*3.275*
Student			*3.1932*
Entrepreneur Self-employment			*2.4583*

5 Conclusion

According to the study, residents of the Sultanate of Oman are eager to adopt e-commerce and e-services practices, as users are willing to use the system despite concerns about trust. These concerns were expressed mainly regarding privacy and cybersecurity, which aligns with previous studies' findings in other populations.

This finding requires tireless effort from policymakers, regulators, and e-businesses to build a high level of trusted e-government, particularly e-services. Achieving increased trust in e-services will lead to a massive surge in the number of e-shoppers in Oman. Building trust comes through implementing well-designed, trustworthy websites with sound third-party policies and security systems. Such an approach will win the confidence of e-shoppers and earn their loyalty towards these trusted e-commerce websites by providing a secure and private reputable e-shopping experience.

The study shed light on challenges such as internet speed, income, and legislation that constitute significant obstacles in building a decent e-environment for both customers and businesses.

Policymakers are advised to address these challenges considering their country's institutional, technological, and cultural aspects. It is important that internet infrastructure is provided at the highest speed and quality and is equally distributed among the country's different geographic and demographic communities. However, the prices of internet services should be in line with the ordinary residents' economic status, enabling them to enjoy the maximum potential of e-services.

Regarding the legislative obstacles, Oman needs to issue an e-commerce law to enhance its legislative framework in this field. The legislative framework should include laws and regulations to protect online intellectual properties, publishing and copyrights, software applications, financial fraud, and electronic signatures on e-transactions, besides consumer protection. Government regulations will enhance trust between e-businesses and e-shoppers and provide security for electronic transactions on e-platforms.

References

1. Chatterjee, S.: Security and privacy issues in e-commerce: a proposed guidelines to mitigate the risk: semantic scholar (1970. Retrieved 2 Nov. 2022 from https://www.semanticscholar.org/paper/Security-and-privacy-issues-in-E-Commerce%3A-A-to-the-Chatterjee/58223b4a0d5fab1fcee6ce4108c25173e640f514
2. Eyupoglu, A., Kaya, T.: E-government awareness and adoption by the residents. Int. J. Public Admin. Digit. Age 7(2), 1–22 (2020). https://doi.org/10.4018/ijpada.2020040101
3. Kaya, T., Sağsan, M., Yıldız, M., Medeni, T., Medeni, T.: Citizen attitudes toward e-government services. Int. J. Public Admin. Digit. Age 7(1), 17–32 (2020). https://doi.org/10.4018/ijpada.2020010102
4. Kurnia, S., Choudrie, J., Mahbubur, R.M., Alzougool, B.: E-commerce technology adoption: a Malaysian grocery SME retail sector study. J. Bus. Res. 68(9), 1906–1918 (2015). https://doi.org/10.1016/j.jbusres.2014.12.010
5. Matbouli, H., Gao, Q.: An overview on web security threats and impact to e-commerce success. In: 2012 International Conference on Information Technology and E-Services (2012). https://doi.org/10.1109/icites.2012.6216645
6. Ribadu, M.B., Ab, W., Rahman, W.N.: An integrated approach toward Sharia compliance e-commerce trust. Appl. Comput. Inf. 15(1), 1–6 (2019). https://doi.org/10.1016/j.aci.2017.09.002
7. Simtowe, F., Muange, E., Munyua, B., Diagne, A.: Technology awareness and adoption: The case of improved pigeon pea varieties in Kenya. In: International Association of Agricultural Economists (IAAE) Triennial Conference, Brazil 18-24 (2012)
8. Solinthone, P., Rumyantseva, T.: E-Government implementation. In: MATEC Web of Conferences, vol. 79, p. 01066 (2016). https://doi.org/10.1051/matecconf/20167901066
9. Stylianou, A.C., Robbins, S.S., Jackson, P.: Perceptions and attitudes about e-commerce development in China. J. Glob. Inf. Manag. 11(2), 31–47 (2003). https://doi.org/10.4018/jgim.2003040102
10. Wang, X., Li. L., Tian, L., Xu, L.: Business to consumer E-commerce enterprises analysis: a case study of Zgpgc.com. In: 2009 International Conference on Computer Technology and Development, pp. 194–197 (2009). https://doi.org/10.1109/ICCTD.2009.33. (n.d.)
11. West, D.M.: E-Government and the transformation of service delivery and citizen attitudes. Public Adm. Rev. 64(1), 15–27 (2004). https://doi.org/10.1111/j.1540-6210.2004.00343.x
12. Yoon, H.S., Occeña, L.G.: Influencing factors of trust in consumer-to-consumer electronic commerce with gender and age. Int. J. Inf. Manage. 35(3), 352–363 (2015). https://doi.org/10.1016/j.ijinfomgt.2015.02.003

The Impact of Researchers' Possessed Skills on Marketing Research Success

Ahmad M. A. Zamil⊚, Ahmad Yousef Areiqat⊚, Mohammed Nadem Dabaghia, and Jamal M. M. Joudeh

Abstract This study aims to determine what specialist thinking will become the most crucial skill for becoming an influential marketing researcher. Some researchers, however, have conflated skills and competencies. Data collection and analysis, for example, have been represented as competencies in several earlier types of research, even though these are abilities that can be learned through training. On the other hand, competencies are natural talents (such as patience, perseverance, and critical thinking) within a person that are displayed differently from person to person. According to the findings, the most crucial abilities required to conduct successful marketing research are critical thinking and tenacity, the capacity to create other people's comprehension of the research problem, as well as crisp, precise, exact, and concise writing, and communication skills, are among the many qualities required to perform successful marketing research. The study also highlights the critical role that universities play in completing successful marketing research initiatives, particularly in the Arab world Universities should devote more attention to marketing research initiatives by investing particular money for this purpose, according to the guidelines. The study also identifies issues that Arab world researchers confront

---

A. M. A. Zamil (✉)
Department of Marketing, College of Business Administration, Prince Sattam bin Abdulaziz University, 165, Al-Kharj 11942, Saudi Arabia
e-mail: am.zamil@psau.edu.sa

A. Y. Areiqat
Department of Business Administration, Business School, Al-Ahliyya Amman University, Amman 19328, Jordan
e-mail: ahmadareiqat@ammanu.edu.jo

M. N. Dabaghia
Department of Accounting, Business School, Al-Ahliyya Amman University, Amman 19328, Jordan
e-mail: mdibageah@ammanu.edu.jo

J. M. M. Joudeh
Department of Marketing, Faculty of Business, Applied Science Private University, Amman 11931, Jordan
e-mail: jamaljoudeh@asu.edu.jo

© The Author(s), under exclusive license to Springer Nature Switzerland AG 2024
A. Hamdan and E. S. Aldhaen (eds.), *Artificial Intelligence and Transforming Digital Marketing*, Studies in Systems, Decision and Control 487,
https://doi.org/10.1007/978-3-031-35828-9_5

regarding English language skills and proposes that university researchers emphasise improving their English language skills.

Keywords Marketing research · Skills · Arab world universities · Academic research · Commercial research

1 Introduction

The concept of "marketing research" employed in this study refers to academic and business-related applied marketing research. The people who benefit from research differ depending on which of these two groups to belong to. Universities, scientific institutions, and undergraduate students are the primary beneficiaries of academic research, whereas corporate sectors, investors, and creditors are the primary beneficiaries of marketing and business research. These advantages are contingent on several aspects, including the research's objective, topic, the efficacy of findings, and the ability to apply those discoveries in the real world. The capacity to successfully appraise these aspects is strongly reliant on the researcher's competencies regarding the research's validity for publication and the beneficiaries' application of its findings. Malhotra [21] discusses four critical topics to concentrate on in marketing research, including focusing on cross disciplines, cross-culture, cross-functional, and customer-centricity.

Academic and commercial research are the two types of international marketing research, with the former undertaken by academics and the latter by businesses. Research is critical because it allows managers to make well-informed decisions [1, 29]. Before conducting marketing research, it is necessary to handle a literature review on recent advances [10, 11]. This research focuses on academic research and determining which researcher skills help the most in generating successful outcomes.

2 Study Problem

The objective of the research, whether inductive or deductive, is to solve a real problem that a community is facing. Social, business, academic, and other issues are such matters. As a result, defining the problem is the most critical step in conducting research. The focus of the paper is to examine if t scientific marketing research's value depends on the researcher's capacity to recognise and comprehend the problem at hand. The following are some of the research's main goals while keeping in mind that the significant context of this study is focused on the Arab region:

1. Determine what abilities a researcher must possess to conduct a successful marketing study.

The Impact of Researchers' Possessed Skills on Marketing Research ... 55

2. Determine the features of scientific study that make marketing research results helpful to others.
3. Determine how a researcher's abilities are related to the rewards derived from search results.

3 Study Objectives

Scientific research reflects a country's and an individual's advancement. Large sums of money have been allocated to scientific studies by private institutions and governments in industrialised countries. Corporate marketing research has been on the diminish for some time, but with the focus on creating a knowledge society, this is an outlier. There is a pressing need to revive marketing research and, as a result, to understand the underlying competencies and skills required to deliver a successful marketing research project [25]. Marketing research is essential because of the diversity and dynamic character of the international marketing environment [5, 29].

According to Madaminovich [18], the United States of America and Japan are the world's top spenders on scientific research. As a result, this research is significant since it focuses on the abilities needed to do successful scientific marketing research. The focus of this paper is to recognise the skills required to conduct good marketing research and rate them in order of importance.

4 Method

The author conducted this qualitative study by evaluating several relevant studies from the literature to discover the researcher competencies that received the most attention. Current researchers, particularly in the Arab World, should demonstrate similar abilities and skills to bring new ideas and inventive solutions to everyday challenges in business, education, or any other field.

5 Literature Review

Scientific research can take many forms, including surveys, case studies, econometric studies, marketing issues, etc. All of these strategies necessitate data gathering from various ways and sources, with the caveat that the data collection method used must be appropriate for the research problem and population [23]. Scientific research is also carried out in steps that should be documented as part of the study design [25]. The structure of the study design, as well as the composition of the research team, are both critical [29, 30]. Regarding international research, there are two schools of thought: "etic" and "emic." The former emphasises cross-cultural distinctions, while the latter believes nation-fair metrics to be globally relevant [7].

However, the majority of the literature agrees that the following are the most significant steps in the research process:

1. Identifying and Comprehending the Research Question

In this step, the researcher should review the literature to determine the problem's risk, promotional, and protective variables [2]. Several researchers [12, 13, 15] suggested that these factors be legal to avoid legal issues. This task necessitates the researcher being well-versed in the law [19, 20]. The research questions should be specific and focused on the objective and statistical technique that will be utilised to get the desired results [16]. In this day of fierce competition and a fast-changing environment, marketing research is more crucial than ever to gain a competitive advantage and profitable growth, as a result, giving priority to diagnosing the problem, speed and efficiency of the technology used, integrative strategy, and broadening the impact area are some of the crucial aspects for marketing research success [3, 17, 28]. In addition, the researcher should be well-versed in information technology to aid in the review of relevant material.

2. Collecting Data

The data-gathering process should be formal and decision-making, sure data-gathering of results and suitable decision-making [19, 20]. The researcher should describe the data-gathering methods and probable sources at this stage. Typically, secondary Data is gathered from the literature, such as books, papers, past studies, and other references. Primary data, on the other hand, should foster collaboration between practitioners and researchers to give alternative solutions to the problem [25]. This necessitates the researcher to identify the data-gathering tools, which could include questionnaires or interviews. As a result, a competent researcher has both qualitative and quantitative data collection abilities and the capacity to frame questions so that sufficient Data is collected [8].

Data collection is more complex than it appears because it may contain flaws such as prejudice, inadequacy, inappropriateness, invalidity, and so on [25]. Some authors have split the literature review procedure from this stage [15]. It was also suggested that technology be used for data collection, such as web-based data collection, emailing, and so on [22].

3. Anticipating Potential Problems

In some cases, the initial problem formulation involves excessive data. As a result, a skilled researcher can tweak the problem formulation to ensure the best level of trust in the findings [25, 26]. This necessitates the ability to reformulate the situation using language realistically. Use of definite terms with clear referents should not have multiple meanings, should be precise and concise, have one clear purpose, state the period and location of the object of study, conditions to research significance, as far as possible, use of the single sentence, and so on are some of the criteria that contribute to the viability of the research question [9].

The Impact of Researchers' Possessed Skills on Marketing Research ...

4. Terms and Ideas Definitions

Many texts imparting specialised knowledge utilise vocabulary misunderstood in other fields by certain persons but are a representation of higher knowledge units. The researcher should now assist readers in comprehending the words and phrases used in the study objectives. For example, if the design is to increase a company's market share, the researcher must first define the word "market share." This step is critical for readers and those who will benefit from the research findings [14].

5. Accurate Population Identification for the Study

The study of a particular phenomenon necessitates determining who will be the study's subjects (called the study community or population). Individuals, products, companies, colleges, and others may make up this population. As a result, to achieve relevant development, the researcher must properly characterise the study population. If a survey aspires to determine job satisfaction at a company, for example, the population should be the employees rather than the company. If the study's goal is to see how sales volume affects profitability over time, the study population will be sales volume vs profits over time. Some researchers have stated that a larger population is a better fit for generalised research on several groups of beneficiaries [22].

6. Analyse Data

The acquired data should be analysed at this step to meet the goals. Answering research questions, testing hypotheses, and delivering final findings that give alternate answers to the study problem are all part of these goals. Assuming the research question is to investigate the difficulties in English language communication among university Arab students (initially believed to be due to a lack of sustainable practice), a questionnaire was used to gather quantitative data from the Arab students that made up the study community during the period —— to ——.

The following is a summary of their responses.

- A total of 120 out of 360 students believe that poor pre-university education is to blame.
- Seventy (70) pupils believe the problem is how English is taught at the beginning.
- Twenty (20) students said they were not interested in learning English.
- Remaining 150 pupils said the reason is due to a lack of conversation quotas in the curriculum.

S. No.	Responses	From No. of students
1	Poor pre-university education	120
2	The problem is the way English is being taught	70
3	Students not interested in learning English	20
4	Not keeping conversational ability practice as a part of the curriculum	150

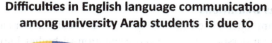

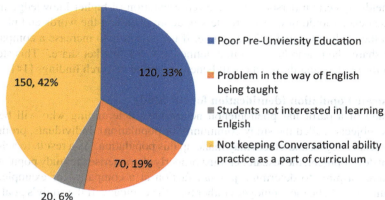

To identify solutions to the problem from the collected perspectives of the respondents, the researcher must have sufficient insight and conscious knowledge.

As a result, generating a review that verifies the researcher relied on the data analysis demands discipline [9]. The researcher may use the findings to suggest solutions to the problem, such as revising the curriculum, shifting the teaching technique from indoctrination to participation, and hiring more qualified teachers.

6 Marketing Researcher Reshaped

Who is a marketing researcher? Who in the organisation does marketing research? In the last century of the past millennium, marketing research was a staff function within an organisation, if there was a place for it. The role of the marketing researcher was well-defined, with specific qualifications and a set location in the organisation. Other than interfacing in particular marketing research projects, marketing researchers needed more interaction with marketing managers and did not participate in marketing decision-making. Likewise, external providers of marketing research required more interaction with marketing managers.

However, as the new millennium has progressed, the border between marketing research and marketing management, and hence the distinction between marketing researchers and marketing managers, has become increasingly blurred. Marketing decision-making is increasingly involving marketing researchers.

Similarly, marketing managers are becoming rising involved in marketing research. In the future, the distinction between marketing researchers and marketing managers will be obliterated. Who is marketing researchers, and what do they do? Who in the company conducts market research? Everyone who works in marketing will be affected. As a result, the response to the question "marketing researcher or

The Impact of Researchers' Possessed Skills on Marketing Research ...

marketing manager?" is "both," and everyone in the marketing department carries both titles. The transition from marketing to research and management has already started. This change is being aided by the availability of more data, better decision tools and decision support systems, and better marketing management and researcher education. Senior managers can now obtain internal and external secondary data from their offices' microcomputers, analyse it, and make choices. As a result, managers are increasingly working as marketing researchers. On the other hand, marketing researchers are becoming more involved in the marketing decision-making process and, as a result, functioning more like marketing managers.

7 Discussions

The purpose of this branch is to explain the literature's current views on the skills and abilities required of researchers. Some authors concentrated on researchers' comprehension abilities, as these abilities aid in summarising contemporary literature [32]. Others point out that while the literature could be more thorough, it might provide insight into how past researchers approached the research subject [6]. Vriens et al. [32] identified six essential researcher talents: humility and openness to criticism, social networking, hard work and intelligent work, having clear goals, going out of one's comfort zone, solid writing skills, and the ability to develop conclusions.

Vriens et al. [32] have conflated competencies and research needs in the following instances: The only skill on his list is good writing ability, whereas the others are prerequisites for conducting research. According to Osorio-Andrade et al. [27], the most significant competency of a competent researcher is the ability to grasp the problem to foster that same knowledge in others.

According to the author, competencies differ from skills in that skills may be learned through training, whereas competencies are akin to talents (primarily innate talents). According to Tzempelikos et al. [31], the primary or tacit competencies required for the various research stages can be learned by training and practice, whereas the knowledge and skills needed for the different research stages can be gained through training and practice [24].

8 Findings and Recommendations

The following are the findings of this research.

1. Researcher competencies differ from researcher skills in that skills can be learned, whereas competencies are inherent.
2. A critical mind and tenacity are the most vital skills.

3. Effective marketing researchers must possess various abilities, the most significant of which is their ability to comprehend an issue, increase others' comprehension of the problem, have practical writing skills, and have clear communication.

The author recommends the following based on these findings.

1. Universities should place a greater emphasis on marketing research.
2. Arab countries should set aside funds specifically for marketing research.
3. To enable marketing researchers to evaluate the literature, Arab universities should emphasise teaching English.

References

1. Babin, B.J., d'Alessandro, S., Winzar, H., Lowe, B., Zikmund, W.G.: Marketing Research. Cengage Learning (2020)
2. Cluley, R., Green, W., Owen, R.: The changing role of the marketing researcher in the age of digital technology: practitioner perspectives on the digitisation of marketing research. Int. J. Mark. Res. **62**(1), 27–42 (2020)
3. da Fonseca, A.C.R.: Marketing research in the digital era: a comparison between adaptive conjoint analysis methods (2018)
4. Dzhalolovna, M.S.: The role of marketing research in the formation of competitive advantages. JournalNX **6**(10), 403–406 (2020)
5. Dzwigol, H.: Innovation in marketing research: quantitative and qualitative analysis (2020)
6. Esteban-Bravo, M., Vidal-Sanz, J.M.: Marketing Research Methods: Quantitative and Qualitative Approaches. Cambridge University Press (2021)
7. Ganaee, M.A., Ahmed, S., Khan, M.A., Gulzar, A.: Bibliometric analysis of researchers' competencies through web of science. Library Philos Practice 1–16 (2021)
8. Grönroos, C.: Service marketing research priorities. J. Serv. Mark. (2020)
9. Grönroos, C.: Service marketing research priorities: service and marketing. J. Serv. Mark. **34**(3), 291–298 (2020)
10. Gui, R.I.: Endogeneity in marketing research. Doctoral dissertation. University of Zurich (2021)
11. Gupta, D.K., Gupta, B.M., Gupta, R.: Global library marketing research: a scientometric assessment of publications output during 2006–2017. Library Manage. (2018)
12. Kadirov, D., Bahiss, I., Bardakcı, A.: Causality in Islamic marketing research: building consistent theories and stating correct hypotheses. J. Islamic Mark. (2020)
13. Klaus, P., Zaichkowsky, J. AI voice bots: services are marketing research agenda. J. Serv. Mark. (2020)
14. Kolb, B.: Marketing research for the tourism, hospitality and events industries. Routledge (2018)
15. Kruja, D.: Destination marketing research. In: The Routledge Handbook of Destination Marketing, pp. 35–48. Routledge (2018)
16. Kumar, V., Leone, R.P., Aaker, D.A., Day, G.S.: Marketing Research. Wiley (2018)
17. Lindgreen, A., Di Benedetto, C.A., Thornton, S.C., Geersbro, J.: Qualitative research in business marketing management (2021)
18. Madaminovich, X.Q.: Marketing Research of 2021 Trends. In: E-Conference Globe, pp. 20–21 (2021)
19. Maison, D.: Qualitative methods: the different tools in the hands of a marketing researcher. In: Qualitative Marketing Research, pp. 48–71. Routledge (2018)

20. Maison, D.: Qualitative Marketing Research: Understanding Consumer Behaviour. Routledge (2018)
21. Malhotra, N.K.: Marketing research: current state and next steps. Braz. J. Mark.-BJMkt Revista Brasileira de Marketing–Re Mark Special Issue **17**, 18–41 (2018)
22. Masud, M.F.: Marketing research and integrated marketing communication plan of Oreo. Mark. Res. Integr. Mark. Commun. Plan Oreo **65**(1), 24–24 (2020)
23. McDaniel, C., Jr., Gates, R.: Marketing Research. Wiley (2018)
24. Meyers E.M., Erickson, I., Small, R.V.: Digital literacy and informal learning environments: an introduction. Learn. Media Technol. (2013). https://doi.org/10.1080/17439884.2013.783597
25. Nunan, D., Malhotra, N.K., Birks, D.F.: Marketing Research: Applied Insight. Pearson, UK (2020)
26. Nuttavuthisit, K.: Qualitative Consumer and Marketing Research: The Asian Perspectives and Practices. Springer (2019)
27. Osorio-Andrade, C.F., Murcia-Zorrilla, C.P., Arango-Espinal, E.: City marketing research: a bibliometric analysis. Revista EAN **89**, 113–130 (2020)
28. Pedko, I., Hordiienko, V.: Marketing research in the marketing information system. ЭкономическийвестникДонбасса **4**(58), 138–143 (2019)
29. Ranfagni, S.: Online marketing research: new trends and challenges. Micro Macro Mark. **28**(2), 227–230 (2019)
30. Steinmetz, J., Posten, A.C.: Guidelines on Acquiescence in Marketing Research. ACR North American Advances (2020)
31. Tzempelikos, N., Kooli, K., Stone, M., Aravopoulou, E., Birn, R., Kosack, E.: Distribution of marketing research material to universities: the case of an archive of the market and social research (AMSR). J. Bus. Bus. Mark. **27**(2), 187–202 (2020)
32. Vriens, M., Rademaker, D., Verhulst, R.: The Business of Marketing Research. Cognella Incorporated (2018)

Impact of Online Food Ordering System and Dinning on Consumers Preference with Special Reference to Bangalore

J. Chandrakhanthan and C. Dhanapal

Abstract In the eatery business, consumer loyalty can be connected straightforwardly to cafe deals; in this manner, it is critical to recognize which seen quality factors all the more emphatically influence consumer loyalty or disappointment. This Exploration paper is to realize about the purchaser's inclination on Feasting in a cafe and requesting food utilizing on the web food conveyance applications. This study assists with distinguishing the variables that influence the client's choice on eating in the cafe and request food on the web. Factors like assortment of food varieties, amount of food, Vibe of the cafe and cleanliness were the elements that influence customer's choice on eating the eatery. What's more, nature of food, quick conveyance, deals and limits, expectation issues were the fundamental causes that influence the customer's choice on requesting food on the web.

Keywords Food delivery apps · Hotel · Quality of food · Customer service · Promotional offer

1 Introduction

Eating delicious cuisine, especially when doing so with loved ones, is one of life's pleasures. We all understand that those who follow healthy eating habits and manage their caloric intake will likely have plenty of energy to work and experience fewer ailments and other illnesses. Today individuals are progressively consuming homemade meals, including eating at entire administration and drive-thru eateries.

Eatery is business environment where individuals can pick a feast to be ready and served to them. One of the many entities that affects the country's financial development is the restaurant industry. To discover the ultimate showcasing system, numerous restaurants have invested extensive energy besides exertion attempting to support their status in a serious eatery market.

J. Chandrakhanthan (✉) · C. Dhanapal
Kristu Jayanti College, Bengaluru, India
e-mail: chandrakhanthanj@gmail.com

© The Author(s), under exclusive license to Springer Nature Switzerland AG 2024
A. Hamdan and E. S. Aldhaen (eds.), *Artificial Intelligence and Transforming Digital Marketing*, Studies in Systems, Decision and Control 487,
https://doi.org/10.1007/978-3-031-35828-9_6

In eateries the consumer loyalty is a critical consider their serious fields. As consumer loyalty could significantly influence foreseeing client post-buy conduct. Hence, in the event that eatery advertisers realize which saw quality variables greatest affect cafe consumer loyalty or disappointment, they could have an effective method for tracking down the main component of progress or disappointment in an eatery's administration.

This study's goal was to identify the customers perceived quality. This analysis exposed the eminence criteria in the food industry had a more pronounced influence on consumers' satisfaction. Concentrate on results can assist the restaurateurs with creating advertising techniques that can be effectively utilized by full-administration cafe administrators.

As many as 50,000 cafes in India give home conveyance, and are frequently simply ready to see minimal benefits from their remove areas. This demonstrates a huge potential in a still-untapped market. The below model indicates the relationship among the key areas engaged with food-conveyance areas. Low-priced Foodstuff the straightforward important point/conveyance area has seen tremendous droplets in edges.

The emergence of advanced innovation is transforming business undertakings. Customers have started placing orders online using applications or websites, with the utmost ease and simplicity, anticipating the exact experience they would receive from the power source itself. Applications are providing clients with additional offices and services to align with their expectations as buyers. Food Panda, Zomato, Swiggy, Box8.

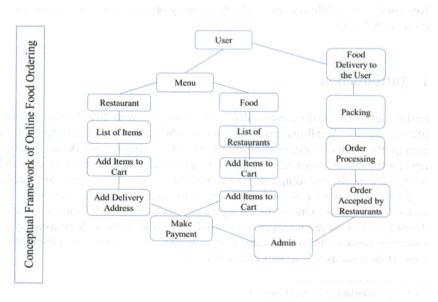

Impact of Online Food Ordering System and Dinning on Consumers ... 65

These sustenance conveyance applications consume not just expanded the benefit portion of pecking orders yet additionally expanded the interest. These applications offer overwhelming offers, limits which specially makes it enticing for the customers food, regardless of whether they weren't pondering requesting in. With these conveyance applications, individuals can have food very close to home lacking conceding the nature of food and their plan for getting work done. Thus, clients favor requesting in, rather than the problem of cooking and trusting that their food will show up.

2 Objectives

1. To investigate consumer preferences for online meal delivery services.
2. To examine the aspects which stimulus the customers to decide on the online order and restaurant.
3. To identify the problems faced by the consumer in online food ordering service.

3 Review of Literature

Pai and Mayya [1] The dawn of technology has expanded the e-businesses in India. This study focused on to assess the consumer preferences and perceptions of online food ordering amenities. This study determine the elements that impact the consumer's decision to order food online. This study helps to understand better customer perceptions and preferences for online food ordering amenities. It demonstrates the swiftness in the consumers to discover the best restaurants or select their favorite dish from the menu as per their want with the feel of dining at home, with hot food on the Table with quick delivery at the door. This study helped to find out the preferred app for online food delivery.

Krishna Kumari [2] argued that using an online ordering system was a convenient and simple option for consumers to purchase meals by not spending time in a restaurant. This technique were advantageous, protected, solid and it is upsetting the present cafe industry. To explore the elements inducing purchasing behaviour and the relationship between online food administration and offices received, organised polling was used.

Jeneefa and Rajalakshmy [3] this study reveals that people are connected with mobile apps, it affected the companies operations. Companies have changed their traditional business strategies into online marketing to suit customer needs and taste at any time. This research paper reveals the consumer's perception towards the online food ordering and delivery services in Pune. The research is focused on the study and analysis of data collected from all those users who are already using the online food delivery services. This study reveals the influencing factors, their perceptions, needs, positioning of various attributes of different online portals in their mind and overall satisfaction towards online food delivery services.

Manju [4], it focus on Client Inclination and Effect of Online Food Administration Applications states around the effect of different food requesting applications in our everyday life and what elements have added to this extraordinary arrangement of food administration application utilization.

Hyde et al. [5], Customers are given an organised platter of goods together with an expanding range of important data that influences their purchasing decisions. Additionally, it demonstrates how segment factors affect consumers' purchase behaviours.

Tribhuvan [6], The review found that the many consumers said they ate one time in a month. Generally speaking, they only went out to eat on trips or special occasions with friends or family. Chi-square analysis has been used to analyse the differences between customers' preferences for family-style restaurants and fast food establishments.

Chavan et al. [7], concentrate on uncovered that client utilizing a Cell phone was considered as a fundamental supposition for the framework. At the point when the client way to deal with the eatery, the saved request can be confirmed by contacting the Cell phone. The rundown of chosen preordered things will be displayed on the kitchen screen, and when confirmed, request slip will be printed for additional request handling. The arrangement gives simple and advantageous method for choosing pre-request exchange structure clients.

Chorneukar [8], He sheds insight on how people choose to buy food through electronic requesting in his focus on Client Impression of Electronic Food Requesting. Electric meal ordering reveals the client's satisfaction. People who work in the IT field tend to favour these computerised requests.

Prabhavathi et al. [9], Cafes have been replaced in recent years by a web-based food delivery system. It has grown significantly in recent years. In the contemporary global economy, the administrative sector contributed 70% of all national output. For their highest level of comfort and simplicity, the number of customers who regularly order for breakfast, midmeal, and supper is speedily growing in the country's major cities.

4 Method of Study

The researcher used descriptive study to conduct this research. The researcher used formal set of questions to conduct this survey. The primary data were collected from the respondents in Bengaluru. The researcher collected the data from 220 respondents. The researcher used statistical tools to analyse the data.

Impact of Online Food Ordering System and Dinning on Consumers ... 67

5 Analysis and Interpretation

5.1 Age Wise Respondents

Particulars	Respondents	Percent
18–25	196	89.0
26–35	18	8.2
Above 35	06	2.7
Total	220	100.0

It is inferred that 89% of respondents are belongs to the age category of 18–25.

5.2 Preference Towards Dine Outside and Online Food Delivery Apps

Particulars	Respondents	Percent
Dine outside	150	68.2
Order online	70	31.8
Total	220	100.0

It is clear that 68.2% of respondents prefers to dine outside.

5.3 Preference Towards Online Food Delivery App

Particulars	Respondents	Percent
Swiggy	134	61.0
Zomato	60	27.0
Food Panda	16	7.0
Others	10	5.0
Total	220	100

It is inferred that 61% of respondents are preferring Swiggy.

5.4 Reason for Prefer the Online Food Delivery App

Particulars	Respondents	Percent
Discounts	106	48.2
Fast delivery	76	34.5
Free delivery	10	4.5
Good customer service	28	12.7
Total	220	100.0

It is inferred that 48.2% of respondents are mentioned discounts is the reason for preferring online food delivery app.

5.5 Reason for not Preferring Online Food Ordering Service

Particulars	Respondents	Percent
Quality of food	52	23.6
Late delivery	14	6.4
Trust issues	54	24.5
Over priced	26	11.8
All the above	74	33.6
Total	220	100.0

It is inferred that 33.6% of the respondents are responded that Quality of food, Late delivery, Trust issues, Overpriced are the reason for not preferring the online food ordering service.

5.6 Age and Customer Preference Towards Dine Outside and Order Food in Online

Customer preference towards ordering food in online or restaurant				
		Dine outside	Order online	Total
Age	18–25	154	42	196
	26–35	12	06	18
	Above 35	04	02	06
Total		170	50	220

From the above table it is clear that 85 respondents are responded that dine outside is the preference.

5.7 Age and Satisfaction Level of Customers on Online Food Ordering System

		Satisfaction level					Total
		Extremely satisfied	Satisfied	Neutral	Not satisfied	Extremely not satisfied	
Age	18–25	8	102	66	12	8	196
	26–35	0	10	6	2	0	18
	Above 35	0	2	4	0	0	6
Total		8	114	76	14	8	220

From the above table it is clear that 57 respondents are responded that they are satisfied with online food ordering system.

5.8 Age and Factors and Customers Looking in the Restaurant

		Customers looking in the restaurant					Total
		Ambience	Service	Hygiene	Variety of deals	Price	
Age	18–25	46	24	76	20	30	196
	26–35	02	2	14	0	0	18
	Above 35	0	4	2	0	0	6
Total		48	30	92	20	30	220

From the above table it is clear that 46 respondents are responded that Hygiene is the major factor customer looking in the restaurant.

5.9 Correlation

		Online	Restaurant	OR1
Online	Pearson correlation	1	0.241*	0.772**
	Sig. (2-tailed)		0.011	0.000
	N	220	220	220
Restaurant	Pearson correlation		1	0.802**

(continued)

(continued)

		Online	Restaurant	OR1
	Sig. (2-tailed)			0.000
	N	220	220	220
OR1	Pearson correlation			1
	Sig. (2-tailed)			
	N	220	220	

*Correlation is significant at the 0.05 level (2-tailed)
**Correlation is significant at the 0.01 level (2-tailed)

Online and restaurant ordering are very closely related.

The Pearson correlation for online and restaurant is 77.2% and 80.2%. Thus, both of the criteria that affect customers' decisions to eat at a restaurant outside or place an online food order.

5.10 Chi-Square Tests

	Value	df	Asymptotic significance (2-sided)
Pearson chi-square	4.280	2	0.118
Likelihood ratio	4.164	2	0.125
Linear-by-linear			
Association	2.258	1	0.133
N of valid cases	220		

H0: There is no significance relation between gender and consumer preference towards selecting online food system and restaurant.

Here the P value is 0.118 which is greater than 0.05.

So accept the hypothesis. Hence there is no significance relation between gender and Customer preference towards selecting online food system and restaurant.

6 Findings

- 89% of the responses are has a place with the age class of 18–25.
- 68.2% of respondents likes to feast outside.
- 91% of respondents are utilizing on the web food conveyance applications.
- 91% of respondents are utilizing these web-based food conveyance applications every day.
- 50.09% of respondents are utilizes online food conveyance application sometimes.

Impact of Online Food Ordering System and Dinning on Consumers ... 71

- 61% of respondents are inclining toward Swiggy.
- 48.2% of respondents are inclining toward these web-based food conveyance applications for limits.
- 34.5% of respondents are incline toward online food conveyance applications for quick conveyance.
- 48.2% of respondents are fulfilled utilizing on the web food conveyance applications.
- 54.5% of respondents are impacted by the incessant pop of commercials and limits notices.
- 44.5% of respondents have trust issues on web-based food conveyance administrations.
- 29.1% of respondents are leaning toward the customary strategy for feasting in an eatery for the explanation of agreeable.
- 41.8% of respondents are answered that first search for cleanliness when they go to a cafe.
- 58.2% of respondents like for family eatery.
- 44.5% of respondents gets draws in for unique proposals on dinners and costs.

7 Recommendations

From the above discoveries the majority of the respondents like to feast in the cafes. Yet, there are not many difficulties for them to eat in the cafes. Cafes need to focus favoring the nature of food sources they serve to the clients. Since buyers are more careful on cleanliness food sources. The cafes proprietors should focus on the feeling of the eateries it ought to be arranged in the legitimate area which eco-accommodating. The cafes should be perfect and clean and the framework of the eateries ought to be alluring. To draw in additional clients subject based cafe is prescribed to the eatery proprietors. Since the greater part of the respondents favors family cafe then the subject based eateries will draw in more family clients and furthermore draws in additional youths to eat in the cafe.

From the above discoveries 90% of the respondents utilizes online food conveyance applications yet the use of online food conveyance administration is extremely less. The fundamental explanation is the trust issue on the internet based food conveyance applications. Online food aggregators ought to acquire the trust of their clients to expand their benefits. The Internet based food aggregator ought to make their administration straightforward to their clients and ought to expand the amount and nature of the food sources.

8 Conclusion

The buyer conduct has turned into an arising area of examination inside the business discipline here different variables impacts shopper conduct. The examination researches the buyer's inclination on eating in a cafe and requesting food on the web. Indeed, even web-based food conveyance administration is a developing business sector in India the purchaser lean towards the customary approach to feasting in the cafe. A large portion of individuals lean towards a cafe with great feel and give cleanliness food. Many customers have online food conveyance applications in their phones. This study reveals that the shoppers not leaning toward online food conveyance administration is absence of faith. The study reveals that the buyers are looking for unique in proposals and cost, so restaurant people may work on that area. Online food app may focus on more discounts and fast delivery to attract customers. Buyers are frightened about the filled food, therefore online food distributors need to win their faith in order to grow their business.

References

1. Pai, S., & Mayya, S.: A study on consumer preferences with reference to online food delivery amenities. Int. J. Manag. Technol. Soc. Sci. 144–166 (2022). https://doi.org/10.47992/IJMTS.2581.6012.0214
2. Krishna Kumari, V.: A impact of customer behavior towards online food services. Int. J. Recent Technol. Eng. 8(1C2), 497–499 (2019)
3. Jeneefa, H., & Rajalakshmy, M.H.: A study on consumer satisfaction and preference towards online food ordering applications. J. Emerg. Technol. Innovative Res. 6(3), 132–138 (2019)
4. Manju, M.S.: A Study on Customer Preference and Impact of Online Food Service Apps.
5. Hyde, A.M., Jain, D., Verma, S.K., Jain, A.: A study of exploratory buying behavior tendencies in FMCG sector. Int. J. Recent Trends Bus. Tour. 1(2), 16–27 (2017)
6. Tribhuvan, A.: A study on consumers perception on food apps. Int. J. Adv. Res. Innovative Ideas Edu. 6, 36 (2020)
7. Chavan, V., Jadhav, P., Korade, S., Teli, P.: Implementing customizable online food ordering system using web based application. Int. J. Innov. Sci. Eng. Technol. 2(4), 722–727 (2015)
8. Chorneukar, M.J.: To study the customer perceptions of electronic food ordering. Thesis submitted to bangalore university, Reg. No.5712360002 (2014). http://sjput.in/pdf/Marketing%20sample%20project.pdf
9. Prabhavathi, Y., Kishore, N.K., Kumar, M.R.: Consumer preference and spending pattern in Indian fast food industry. Int. J. Sci. Res. Publ. 4(2), 16–22 (2014)

Green Economy Mechanisms in the Age of Technology and the Circular Economy

CH. Raja Kamal and Surjit Singha

Abstract Circular economics goes beyond sustainability and green economics. The green economy prioritizes investments, employment, and skills for sustainable economic growth. A green economy is one whose economic operations aim towards sustainability without hurting the environment through lowering environmental hazards and environmental scarcities. Circular economies prevent extracting, producing, and disposing of resources. The circular economy reduces waste by making products durable, reusable, repairable, and refurbishable and using recyclable materials. In recent years, technologies have been used to promote circular economies and resilience, where productive and economic development does not depend entirely on intense natural resource use and waste creation.

Keywords Green economy · Circular economy · Technology

1 Introduction

The circular economy is an economic concept related to sustainability and green economics, but going beyond the latter. The European research and innovation policies on environmental issues are designed to support a shift towards the restoration EU and to identify and promote a transformational agenda for the economy and the whole society towards achieving genuinely sustainable development. Circular development thus supports circular economies, creating new societies, aligned to a new focus on waste management, sustainable development, and meeting citizens' needs. Circular economy practices have been found to have positive correlations with ecological and economic productivity. Industry 4.0 technologies are a significant boost for circular economy practices. Rethinking access to goods and services, fostering sharing and collaboration, and facilitating the sharing and development of knowledge could unlock opportunities for functional circular economies.

CH. Raja Kamal · S. Singha (✉)
Department of Commerce, Kristu Jayanti College (Autonomous), Bengaluru, India
e-mail: surjitsingha@kristujayanti.com

© The Author(s), under exclusive license to Springer Nature Switzerland AG 2024
A. Hamdan and E. S. Aldhaen (eds.), *Artificial Intelligence and Transforming Digital Marketing*, Studies in Systems, Decision and Control 487,
https://doi.org/10.1007/978-3-031-35828-9_7

2 Green Economy

The green economy provides a macroeconomic approach to sustainable economic growth, focusing primarily on investments, jobs, and skills. It has been at the forefront of green economy discussions across the European area since the 2005 Ministerial Conference on Environment and Development in Asia and the Pacific in Seoul, where the Seoul Initiative for Environmentally Sustainable Economic Growth (Green Growth) was established. As part of its green economy initiative [5], UNEP commissioned an early author of Blueprint for a Green Economy to produce a report entitled The Global Green New Deal, published in April 2009, which proposed a combination of policy actions which would GGND spur an economic recovery while improving the sustainability of the global economy. The UN Environment Programme (UNEP) defines a Green Economy as one that leads to improvements in human welfare and social equality while substantially reducing environmental risks and ecological scarcity.

A green economy is one whose economic activities are focused on reducing environmental risks and ecological scarcities, aiming for sustainability while not degrading the environment. In green economies, growth in employment and income is driven by government and private investments in economic activities, infrastructure, and assets, such as those allowing reductions in carbon emissions and pollution, improvements in energy and resource efficiency, and the prevention of loss of biodiversity and ecosystem services [6]. Suppose a green economy is defined as interconnected economic activities promoting global-scale sustainable development, poverty alleviation, environmental protection, and ecologically efficient, low-carbon development. In that case, the circular economy is a regenerative economic growth strategy focusing on recovery, renewable use, and eliminating toxic chemicals and waste by better designing materials, products, systems, and business models. The International Chamber of Commerce (ICC), representing global businesses, defines a green economy as one where economic growth and environmental responsibility work together in mutually reinforcing ways, supporting advances in social development. Instead, green growth provides a practical, flexible approach to achieving concrete, measurable progress across its economic and environmental pillars whilst fully considering the social consequences of green economies' growth dynamics [7].

When we combine traditional green professions, such as renewable energy, energy efficiency, and environmental management, with sustainability-related green employment, we may obtain a more comprehensive picture of a sustainable workforce and economy. Therefore, it will contribute to the creation of a sustainable economy. This includes the Green Jobs Guarantee, but it is not limited to just that. Accelerating investments and innovations is necessary if we are going to foster green growth effectively. Both of these things would help growth continue and provide new economic possibilities.

3 Circular Economy

A circular economy avoids extracting, manufacturing, and discarding resources. The circular economy aims to eliminate waste by designing goods to be durable, reusable, repairable, and refurbishable and utilizing materials that can be recovered and repurposed. A circular economy shares, rents, reuses, repairs, renovates and recycles resources and goods as long as feasible. Instead of wasting and exploiting new resources, circular economies encourage product reuse. A circular economy reduces materials, reworks them to consume less, and reuses trash to make new goods. An economy optimizes resource output by transferring goods, components, and materials that are most helpful in the technological and biological cycles at all times.

In a circular economy, renewable energy sources are utilized, raw materials, components, and products are wasted as little as possible. In contrast, the circular economy is a regenerative industrial paradigm that aims to boost resource productivity and mitigate climate change and economic unpredictability. The circular economy might reduce greenhouse gas emissions, raw material use, agricultural productivity, and linear model externalities. The current economic paradigm of Take, Make, throw—where resources are harvested, processed into goods, utilized, and discarded—is wasteful. The circular economy is a different way to do business, forcing companies to rethink everything from designing and producing products to how they relate with customers. The transition to the circular business model is a radical, restorative, regenerative way to approach business that requires a fundamentally different mindset. Many new models, materials, and products will emerge from entrepreneurs, but existing organizations are also playing a crucial role by rethinking their existing strategies. Transition is challenging but also brings new infrastructure, energy, and manufacturing opportunities, with infrastructure adaption to fit a circular economy model. Circle Economy should be implemented at every level (national, regional, industry, enterprise, product, process or material) to create conditions for and eliminate barriers to the circular economy shift.

4 Correlation of Green Economy and Circular Economy

The Circular Economy is a framework for tackling the challenges on a global scale, such as climate change, biodiversity loss, waste, pollution, and the biocapacity gap. The circular economy supports transformations and developments in industries and infrastructure toward sustainable consumption and production (SCP). The drive for a sustainable bioeconomy could also be seen to promote SCP. The Green Economy is a system of economic activity that produces, distributes, and consumes products and services to improve long-term human well-being and avoid exposing future generations to significant ecological hazards and ecological shortages. In contrast, a Circular Economy represents a strategy for development which allows for economic growth without increasing the consumption of resources and which also lowers environmental impacts.

The UN Environment Program (UNEP) defines the green economy as one which improves human well-being and social equality, whilst substantially reducing environmental risks and deficits. Suppose a green economy is defined as interconnected economic activities promoting global-scale sustainable development, poverty elimination, environmentalism, and ecologically efficient, low-carbon development. In that case, the circular economy is a regenerative economic growth development strategy focusing on recovery, renewable energy usage, and eliminating toxic chemicals and waste by better designing materials, products, systems, and business models. Technological innovation, ecological sustainability, energy efficiency, and the utilization of renewable resources all work together to give the circular economy all the characteristics of a new, virtuous system. The green economy is a driving force that will drive investments and innovations to achieve sustained growth and new economic opportunities.

There is a shift away from the anthropocentrism that is, in some ways, at the heart of a green economy, built around implementing environmentally friendly technical solutions, toward an understanding of unity and interdependence in resource movements within the natural and economic systems in the context of the circular economy. Circularity is focused on cycles of resources, whereas resilience is related to humans, the planet, and the economy in more general terms.

In this circular economy, products are recycled with no waste, and all products must be easy to disassemble into various components which can be used to make new products or systems. If new input materials are needed, these should be obtained sustainably to avoid harming natural and human environments. To ensure there are sufficient commodities in the future to provide food, housing, heat, and other necessities, our economies need to be circular.

The results may impact efforts to terminate the material cycle and promote a circular economy where goods, materials, and resources retain their worth. Thus, increased consumption elsewhere in the economy may counterbalance efficiency improvements. The bounceback effect occurs when the economy produces the same number of products and services with less material and energy.

5 Usage of Technology to Promote Green Economy and Circular Economy

Various global IT companies are helping companies efficiently manage their finite resources; the concept of embedding the principles of a circular economy into all major business processes began to gain serious traction in 2018 with a design challenge called 'The Plastics Challenge', launched to challenge employees, customers, and citizens to eliminate disposable plastics through the use of innovative technologies. These processes allow organizations to address unanticipated disruptions, lower trucking, improve operational efficiency, and increase the sustainability of smart technologies. Similarly, data captured from sensors helps to drive waste management, automate and enhance recycling, and promote reuse strategies to support a circular economy. Let us take a look at three of the circular economy principles that IoT powers. This circular economy principle calls on industries to overhaul supply chain processes to avoid waste and emissions, from plastics to electronics and automotive tires. Specifically, the green economy is a system of economic activity linked to producing, distributing, and consuming goods and services that improves long-term human welfare and protects future generations from significant environmental risks and ecological scarcity. In contrast, the circular economy is a development strategy that permits economic growth without increasing resource consumption and reduces environmental impact.

Managing the lifecycle of natural resources, from their extraction, through product design and manufacture, through what is considered waste, is fundamental for Green Growth and is part of developing a resource-efficient, circular economy in which nothing is wasted. In recent years, technologies have been put to work in service of the Earth, constituting the tools needed to shift to circular economies and resilience, whereby productive and economic development does not solely rely on intensive natural resource usage and waste production. Technological innovation, ecological sustainability, energy efficiency, and renewable resources all go hand-in-hand, giving a circular economy all the characteristics of a new, virtuous system. There is no comparison with helping preserve the environment and enforcing a circular economy through the application of technologies and innovations in resource management, waste management, and pollution within the value chain.

The Commission adopted a Circular Economy Package designed to spur European businesses and consumers towards the circular economy, in which resources are used more sustainably. The Circular Economy Model increases resource efficiency and lowers environmental impacts on natural capital by designing more recyclable products, deploying effective technologies, and turning waste into a resource.

As part of BSH Hausgerate—which includes leading appliances brands like Bosch, Siemens, Neff, and Gaggenau—BlueMovement is an obvious example of how major companies can be responsible for and transition to the circular economy through its products-as-a-service business model. JEPLAN is committed to creating a circular ecosystem for the contemporary fashion industry, developing recyclability technologies and building a circular supply chain, focusing on changing consumer behaviour. Business-to-business customers and consumers alike have an opportunity to purchase more sustainable products, as well as the responsibility to figure out how to avoid wasteful materials and to package as much as possible–for example, by using reusability models on products or recycling them into useful reprocess.

6 Industry 4.0 and Circular Economy

By using innovations, including robotics, to promote a Circular Economy sector, the waste management industry would be the core of a more sustainable world. It could secure their own, yet far-beyond, "business-as-usual" future. CE can be transformed by using I4.0 technologies on the operational, tactical, and strategic staffs in organizations and on the potential utilization strategies within particular research areas such as supply chain management and manufacturing of products in general.

Recent studies extend SSMEs in alignment with other emerging practices, including circular economies (CEs) and supply chain innovations in Industry 4.0 [4] For practitioners, the concept of CE is embedded within a Take-Make-Disposal economic model that is guided by three principles, namely, eliminating waste and pollution, keeping products and materials in use, and rebuilding natural systems [3]. In contrast to a linear economy, which generally involves manufacturers using input materials to produce products, selling to consumers, and then disposing of waste, the CE takes a more circular approach. Circular economies (CEs) assist organizations with a supply chain cycle to keep the cycle flowing, optimizing their usage of resources, such as energy and materials, to enhance their triple-bottom-line gains [1].

In this context, CE plays a significant role in industrial production, contributing characteristics like recycling resources and minimizing material and energy use. That is, CEA is committed to promoting a just and inclusive circular economy, separating economic growth from resource usage, eliminating waste through design, and combating climate change. The CE is perceived as a new economic model where it is expected to achieve equilibrium and harmony between the economy and society [9]. The concept has been broadly appreciated worldwide to address the challenges in the implementation of a green economy and the most efficient use of ecological resources [2].

In that sense, experts agreed that manufacturing industries must focus on the digital skills of workers, financing the enterprise itself, and the environmental quality regulations, which are the most likely and the highest impact factors to implement a CE by 2030. 4.0 changes operational management in areas like automation and manufacturing, supply chain management, lean manufacturing, and total quality management [8]. These technological features can enhance materials reuse, upcycling, circularity, and recycling programs and the performance management of the circular supply chain.

7 Conclusion

Circular economics transcends sustainability and green economics. European environmental research and innovation policies help restore the EU by establishing and pushing a transformational economic and social agenda for sustainable development. Circular economy methods enhance ecological and economic productivity. Industry 4.0 promotes a circular economy. Rethinking access to goods and services, fostering sharing and collaboration, and exchanging knowledge may allow circular economies.

References

1. Bressanelli, G., Saccani, N., Pigosso, D.C., Perona, M.: Circular economy in the WEEE industry: a systematic literature review and a research agenda. Sustain. Prod. Consump. **23**, 174–188 (2020). https://doi.org/10.1016/j.spc.2020.05.007
2. Chauhan, A., Jakhar, S.K., Chauhan, C.: The interplay of circular economy with industry 4.0 enabled smart city drivers of healthcare waste disposal. J. Clean. Prod. **279**, 123854 (2021). https://doi.org/10.1016/j.jclepro.2020.123854
3. Circular economy introduction: Ellen MacArthur Foundation (n.d.). https://ellenmacarthurfoun dation.org/topics/circular-economy-introduction/overview
4. Gao, D., Xu, Z., Ruan, Y.Z., Lu, H.: From a systematic literature review to integrated definition for sustainable supply chain innovation (SSCI). J. Clean. Prod. **142**, 1518–1538 (2017). https://doi.org/10.1016/j.jclepro.2016.11.153
5. Green economy: European Environment Agency (2016). https://www.eea.europa.eu/publicati ons/europes-environment-aoa/chapter3.xhtml
6. Green Economy: UNEP—UN Environment Programme (n.d.). https://www.unep.org/regions/ asia-and-pacific/regional-initiatives/supporting-resource-efficiency/green-economy
7. International Chamber of Commerce (ICC): Green Economy Roadmap and Ten Conditions for a Transition toward a Green Economy. UN SDG (n.d.). https://sustainabledevelopment.un.org/ index.php?page=view&type=400&nr=674&menu=1515
8. Kristoffersen, E., Blomsma, F., Mikalef, P., Li, J.: The smart circular economy: a digital-enabled circular strategies framework for manufacturing companies. J. Bus. Res. **120**, 241–261 (2020). https://doi.org/10.1016/j.jbusres.2020.07.044
9. Ma, S., Zhang, Y., Yang, H., Lv, J., Ren, S., Liu, Y.: Data-driven sustainable intelligent manufacturing based on demand response for energy-intensive industries. J. Clean. Prod. **274**, 123155 (2020). https://doi.org/10.1016/j.jclepro.2020.123155

Whether Digital or Not, the Future of Innovation and Entrepreneurship

Yacoub Hamdan, Ahmad Yousef Areiqat, and Ahmad Alheet

Abstract This paper aims to investigate recent aspects of entrepreneurship and innovation in Jordan. Furthermore, this paper differs from other work on the subject in that it examines entrepreneurship and innovation from a more global perspective, with the permissible international outcome.

Keywords Jordan · Innovation · Entrepreneurship · Emerging economies · Economic growth

1 Introduction

The emergence of novel and powerful digital platforms, infrastructures, and technologies has significantly altered entrepreneurship and innovation. Digital technologies have wider repercussions for value creation and capture than providing new opportunities for entrepreneurs and innovators. The role of digital technologies in transforming organisations and social relationships must be explicitly acknowledged in research that aims to comprehend the digital transformation of the economy, embrace ideas and concepts from various fields and disciplines, and incorporate multiple and cross-levels of analysis. Openness, affordances, and generativity are three major digitisation-related themes that we outline in terms of broad research questions to help realise this research agenda. Such articles inborn to computerised innovations could be a typically calculated stage for interfacing issues at various levels and coordinating thoughts from different disciplines/regions.

Y. Hamdan · A. Y. Areiqat (✉) · A. Alheet
Department of Business Administration, Business School, Al-Ahliyya Amman University, Amman 19328, Jordan
e-mail: ahmadareiqat@ammanu.edu.jo

Y. Hamdan
e-mail: y.hamdan@ammanu.edu.jo

A. Alheet
e-mail: a.alheet@ammanu.edu.jo

© The Author(s), under exclusive license to Springer Nature Switzerland AG 2024
A. Hamdan and E. S. Aldhaen (eds.), *Artificial Intelligence and Transforming Digital Marketing*, Studies in Systems, Decision and Control 487,
https://doi.org/10.1007/978-3-031-35828-9_8

Over the past few years, many novel and potent digital technologies, platforms, and infrastructures have significantly altered innovation and entrepreneurship, with broad organisational and policy repercussions. Indeed, the term "digital transformation" has become widely used in contemporary business media to refer to the transformational or disruptive effects of digital technologies on businesses, such as the creation of new business models, product/service types, and customer profiles [1, 2], also generally, to point out how established businesses may need to change to succeed in the rapidly developing digital world fundamentally [3].

These repercussions have been the subject of recent research in innovation and entrepreneurship. Digital technologies, for instance, have been shown to accelerate the outset, scaling, and evolution of new ventures, embrace networks, ecosystems, and communities, integrate digital and non-digital assets, and fuel new arrangements of innovation and entrepreneurial dynamism that cross traditional industry/sectoral boundaries (e.g., [4, 5]). In a similar vein, research has also documented the methods by which established large companies like GE, Volvo, Johnson Controls, and Caterpillar, along with Boeing, have attempted to radically restructure their innovation strategies and practices in response to digitisation (e.g., [6]). In general, research have observed that the introduction of new digital technologies alters the nature of the uncertainty that is inherent in innovation and entrepreneurship, both in terms of the processes and the outcomes, encouraging a radical rethink of how individuals, groups, and organisations can pursue creative endeavours [7].

2 Design, Methodology and Path

The paper postulates an overview of the literature on entrepreneurship, innovation, and critical related topics such as content and economic growth, as well as links to relevant international works.

Notably, the digitisation of innovation and entrepreneurship has repercussions on a broader scale at the regional, national, and societal levels. These repercussions have the potential to provide stakeholders and decision-makers with information. Studies, for instance, have shown that digitisation can lead to improvements in innovation productivity, increased regional entrepreneurial activity, and overall progress in the economy and society [8, 9]. Similarly, digital infrastructures and platforms have made it possible the development of novel work structures that redraw industry/sectoral boundaries and influence economic health locally and regionally [10].

In addition, digitisation has forced government agencies and other public institutions to reevaluate a variety of laws, regulations, and policies about a variety of topics, such as intellectual property rights, data privacy and security, consumer rights, worker experience and training, entrepreneurial financing and guarantees, incubator/ accelerator programs, and regional/local economic expansion (e.g., [11–16]).

The ongoing discussion regarding the digitisation of entrepreneurship and innovation reveals three significant issues. First, digitisation has implications not only at various levels of analysis (individual, organisation, ecosystem/community, regional/

societal) but also significantly across levels—a feature that has arguably received insufficient attention. Take, for instance, data security and privacy. Social media, mobile computing, and cloud computing are just a few new digital infrastructures and platforms that raise essential privacy and security issues for individual users or customers. However, and this is important, these concerns also produce ripple effects that extend to problems at the firm level (such as relationships between businesses and consumers, reputation), as well as societal issues (such as the misuse of social media as a surveillance tool, a lack of trust in the media, and democratic institutions) [15, 17, 18]. According to Aguinis and Edwards [19], conducting such studies that consider phenomena across levels of analysis continues to be difficult and complicated. Eckardt and others [20]; According to Zhang and Gable, the underlying themes and concepts of digitisation may provide new conceptual bridges and empirical opportunities.

Second, research on the digitisation of innovation and entrepreneurship has primarily been confined to specific fields or disciplines (such as marketing, economics, information systems, operations, and strategy), and it is conceivable that little effort has been put into adopting a more interdisciplinary perspective of the underlying issues up to this point. For instance, I suggest that you consider crowdfunding. Entrepreneurial finance researchers (e.g., Drover et al. are particularly interested in crowdfunding because it represents a different method of obtaining venture capital [21]; But it is important to note that the context of crowdfunding also has sociological underpinnings because crowd behaviour can influence both the processes and the results. Similarly, the crowdfunding platform, a digital platform, possesses distinctive technological features that can affect how participants interact with one another and the outcomes. Despite this, most crowdfunding research appears to be divided into functional silos. Only some studies have attempted to combine concepts and ideas from venture financing, sociology, information systems, and other related fields.

Thirdly, digitisation is more than just a setting for entrepreneurship and innovation. Digital technologies are increasingly able to play the role of operant resource, which means they can actively contribute to developing novel projects [22]. To theorise about the nature and process of innovation and entrepreneurship [23], it becomes crucial that studies take into account characteristics inherent to digital technologies.

However, such a comprehensive approach must still be incorporated into digital innovation and entrepreneurship scholarship outside information systems. This suggests that non-digital factors are given too much weight and that opportunities to gain a more nuanced understanding of how digital technologies encourage innovation and entrepreneurship have yet to be noticed.

Research that incorporates issues at various or cross levels of analysis clasp ideas and concepts from multiple fields/disciplines, and explicitly endorses the role of digital technologies, as well as contribute to a broader understanding of the implications of digitisation for innovation and entrepreneurship, is the goal of this particular issue. The 11 articles chosen for this special issue achieve the above to changing degrees and assist with outlining the commitment and potential for such an interdisciplinary and cross-level exploration plan on the digitisation of development and business venture. By focusing on three broad digitisation-related themes—openness, affordances, and generativity—we present such a research agenda in more concrete terms in the remainder of this essay. We propose that identifying such digitisation-related themes will help realise the promise of multidisciplinary and cross-level research to inform innovation and entrepreneurial conducting in an increasingly digital world and improve the coherence of future research efforts.

Snippets from the section "Themes in the Digital Revolution of Innovation and Entrepreneurship" We propose that a more comprehensive research agenda could be based on themes that are dwellers to digital technologies and, at the same time, are open to broader interpretations. This would help ensure that issues are connected across all levels of analysis because of the central role those digital technologies play. It examines those issues by integrating ideas from other fields and areas. Openness, flexibility, and creativity are the three main themes we will be focusing on here.

Our primary goal in developing our framework, which is centred on the three concepts of openness, affordances, and generativity, was to illustrate the potential for future research to take a more comprehensive approach to take into account the effects of digitisation on innovation and entrepreneurship on a variety of levels and from a variety of disciplinary perspectives. Policy and practice can benefit significantly from such analysis.

3 Findings

According to the author's analysis, the study of entrepreneurship and innovation shall be placed in the context of a country's economic development, institutional environment, and other factors.

4 Originality and Value

The study by the authors investigates the co-evolution of firm strategies, the institutional habitat in entrepreneurship and innovation in emerging and transition economies, and their local–global and parent-subsidiary connectivity. The papers in this event are summarised and summed up by the authors to provide findings and recommendations for future research on entrepreneurship, innovation, and the creation of new businesses, which is thought to be a significant driver of economic growth in the coming years.

References

1. Boulton, C.: What is Digital Transformation? A Necessary Disruption (2018)
2. Boutetiere, H., et al.: Unlocking Success in Digital Transformations (2018)
3. Kosack, E., et al.: Information management in the early stages of the COVID-19 pandemic. Bottom Line **34**(1), 20–44 (2021). https://doi.org/10.1108/BL-09-2020-0062
4. Fischer, E., et al.: Social interaction via new social media: (How) can interactions on Twitter affect effectual thinking and behavior? J. Bus. Ventur. (2011)
5. Rayna, T., et al.: Co-creation and user innovation: the role of online 3D printing platforms. J. Eng. Technol. Manage. (2015)
6. Fitzgerald, M., et al.: Embracing digital technology: a new strategic imperative MIT Sloan. Manage. Rev. (2014)
7. Balawi, A., Ayoub, A.: Assessing the entrepreneurial ecosystem of Sweden: a comparative study with Finland and Norway using Global Entrepreneurship Index. J. Bus. Socio-econ. Dev. **2**(2), 165–180 (2022). https://doi.org/10.1108/JBSED-12-2021-0165
8. Brynjolfsson, E.: ICT, innovation and the e-economy EIB Papers (2011)
9. Burtch, G, et al.: Can you gig it? An empirical examination of the gig economy and entrepreneurship. Manage. Sci. (2018)
10. Abusaq, Z., et al.: A flexible robust possibilistic programming approach toward wood pellets supply chain network design. Mathematics **10**(19), 3657 (2022). https://doi.org/10.3390/math10193657
11. Agrawal, A., et al.: Some simple economics of crowdfunding. Innov. Policy Econ. (2014)
12. Goldfarb, A., et al.: Introduction to "economic analysis of the digital economy". In: Economic Analysis of the Digital Economy (2014)
13. Goldfarb, A., et al.: Privacy and innovation. Innov. Policy Econ. (2012)
14. Bruton, G., et al.: New financial alternatives in seeding entrepreneurship: microfinance, crowdfunding, and peer-to-peer innovations. Entrep. Theory Pract. (2015)
15. Martin, K.: The penalty for privacy violations: how privacy violations impact trust online. J. Bus. Res. (2018)
16. Greenstein, S., et al.: Digitization, innovation, and copyright: what is the agenda? Strateg. Org. (2013)
17. Allcott, H., et al.: Social media and fake news in the 2016 election. J. Econ. Perspect. (2017)
18. Grinberg, N., et al.: Fake news on Twitter during the 2016 US presidential election Science (2019)
19. Aguinis, H., et al.: Methodological wishes for the next decade and how to make wishes come true. J. Manag. Stud. (2014)
20. Eckardt, R., et al.: Reflections on the Micro–macro Divide: Ideas from the Trenches and Moving Forward (2018)
21. Drover, W., et al.: A review and road map of entrepreneurial equity financing research: venture capital, corporate venture capital, angel investment, crowdfunding, and accelerators. J. Manage. (2017)
22. Kumar, S., et al.: Challenges and opportunities for mutual fund investment and the role of industry 4.0 to recommend the individual for speculation. In: Nayyar, A., Naved, M., Rameshwar, R. (eds.) New Horizons for Industry 4.0 in Modern Business. Contributions to Environmental Sciences & Innovative Business Technology. Springer, Cham (2023). https://doi.org/10.1007/978-3-031-20443-2_4
23. Al Dhaen, E.S.: The use of information management towards strategic decision effectiveness in higher education institutions in the context of Bahrain. Bottom Line **34**(2), 143–169 (2021). https://doi.org/10.1108/BL-11-2020-0072

Customer Buying Intention Towards Smart Watches in Urban Bangalore

B. Subha and Jaspreet Kaur

Abstract A market for wearable technology has developed as the marketing industry works to connect with mobile consumers. This industry would require businesses to create and enhance marketing strategies in the sector, as well as significant innovation initiatives in the field of digital marketing. The wearable tech business has made significant strides in its development, which enables researchers to do more studies on the topic. This article aimed to determine the factors that influence customer buying intention towards smartwatches among the most prominent wearables in urban Bangalore. A sample of 163 respondents was selected using the convenience sampling method among the students and young professionals residing in Urban Bangalore. Multiple regressions were applied to determine the relationship between the independent variables and the intention to purchase a smartwatch. The results show that there is a significant and positive relationship between independent variables' perceived usefulness, ease of use, fashion product, product feature, health and fitness tracker, and dependent variable customer buying intention towards smartwatches.

Keywords Customer buying intention · Smartwatches · Digital technology · Internet of things · UNSDG17

1 Introduction

With the exponential growth of the Internet of Things (IoT) in recent years, people can now receive information from anywhere at any time. Smartwatches and other wearable tech gadgets are examples of IoT applications. These gadgets give consumers access to health-related information and feedback on a variety of physical activities [17]. Smartwatches are currently being developed and sold by businesses, and sales are growing quickly. Smartwatches' distinctive qualities, such as their ability to send

B. Subha (✉) · J. Kaur
Department of Management, Kristu Jayanti College, Autonomous, Bengaluru 560077, India
e-mail: subhaashok3@gmail.com

© The Author(s), under exclusive license to Springer Nature Switzerland AG 2024
A. Hamdan and E. S. Aldhaen (eds.), *Artificial Intelligence and Transforming Digital Marketing*, Studies in Systems, Decision and Control 487,
https://doi.org/10.1007/978-3-031-35828-9_9

notifications, pair with smartphones to access numerous applications, keep time, and have many watch faces, have drawn many consumers to adopt this technology [16]. Wearable computers that can conduct a variety of daily functions to assist users in managing their daily job are known as smartwatches [11]. The global smartwatch market was estimated to be worth $20.64 billion in 2019 and is expected to grow to $96.31 billion by 2027, representing a CAGR of 19.6% from 2020 to 2027 (www. alliedmarketresearch.com). Today's technical advancements and the widespread use of mobile devices like smartphones and smartwatches around the world have made information accessible in real time whenever and wherever. The idea of "mobility" is also evolving, moving from simple carriable equipment to flawlessly wearable devices. Increasing the prevalence of human communication as a result. In terms of technology, wearables are sophisticated electronic gadgets and computers that may be incorporated into a variety of daily-use clothing accessories and worn on or connected to the body. The structure of this study is as follows. The introduction and relevance of smartwatches are covered in the first section, which is then followed by a review of the literature and the study's goal. Additionally, the research methodology, which includes the method of data collecting, the demographic profile, and data analysis, In the final section, the results have been presented, followed by limitations, suggestions for further research, a summary, and a conclusion.

2 Review of Literature

Smartwatch adoption is still in its initial stages, and elements like perceived usefulness and visibility are crucial [6]. Wearable devices, notably smart watches, are receiving considerable attention and massive investments from major smartphone companies (Apple, Samsung, LG, Google, and others), demonstrating the evolving era of smartwatches [8] The ease of use has a greater impact on consumers' perceptions and intentions to embrace smartwatches. Furthermore, the intention to purchase a smartwatch does not depend on gender [20].

According to [5], smartwatches are fashion accessories, and customers who value originality are more likely to use them. Consumer value smartwatch functionality more than brand recognition and product cost. Because smartwatches record a lot of physical activity data while being worn, privacy risks have a detrimental impact on users' behavioural intentions. The literature has a lot of studies on the adoption and use of smartwatches [2, 19]. Numerous factors affect how smartwatches are used. According to Scott [19], privacy concerns are one of the factors affecting the uptake of smartwatches in the USA.

According to a comparable study [19] conducted in four developed countries, including the USA, the UK, France, and Germany, the adoption of smartwatches is influenced by age and gender. It was also found that adoption attitudes were unaffected by a product's perceived ease of use. However, a South Korean study indicated that perceptions of ease of use significantly influenced people's intentions to adopt smartwatches [2]. Other elements, like prior knowledge, emotional quality, and anxiety related to technology all, all impacted persons' adoption of smartwatches [4]. The adoption of smartwatches was shown to be highly influenced by four factors, including perceived benefits, healthology, IT innovation, and smartwatches as luxury products, according to another study [14] carried out in Malaysia. Hokoh et al. in their study concluded that design is more important for younger users than for older ones. Graphical user interfaces (GUI), health information support (HIS), weight control, and sleep improvement are other factors that are significant.

Smartwatches have been the subject of previous research known as "fashionology," with an emphasis on its aesthetic and fashion characteristics [3, 6]. Another strategy is to examine the innovative features of smartwatches to see if they have an effect on the uptake of wearable technology [15, 18]. According to some experts, smartwatches can even be considered a kind of entertainment [10]. Additionally, a user-friendly application and fashionable appearance will boost consumers' adoption. The smartwatch is a consumer device that straddles the line between cutting-edge IT and regular fashion products. A particular form of accessory considered to be fine jewellery is a "watch." The smartwatch is more likely to be seen as a type of status symbol for fashion and wealth, similar to luxury goods. The 'Apple Watch Edition,' a luxury version of the Apple Watch that can cost up to $20,000, was soon sold out in the Chinese market, indicating the potential for the smartwatch to become a status or wealth symbol. Therefore, a smartwatch would have distinctive qualities that are akin to what customers would anticipate from high-end fashion items [1].

Chun et al. [7] investigated the usability and usage of smartwatches from a qualitative point of view. Based on a thorough interview, the goal of this study was to learn how smartwatch owners used them. They discovered that many users find mobile devices convenient for checking pushed alerts. However, it offers inadequate responses when individuals wish to interact with graphically rich content. According to the report, individuals enjoy using smartwatches to play music and check the weather. Additionally, users were assisted by the smartwatch through health and medical-related applications [12]. On the basis of the technology acceptance model (TAM), the unified theory of acceptance and use of technology (UTAUT), the theory of planned behaviour (TPB), the innovation diffusion theory (IDT), and other theories, several academics have built their own research models. In the modern day, a smartwatch is frequently described as both a wearable technology device and a high-technology product. Users prefer smartwatches with circular screens over those with square ones, according to [13]. On square-screen smartwatches, however, managing features are more practical [13].

3 Objectives of the Study

The major goal of this study is to establish the significance of the relationship between smartwatch demand among students and young professionals residing in urban Bangalore and perceived usefulness, ease of use, fashion product, product feature, and health and fitness tracker.

4 Materials and Methods

Residents of urban Bangalore who had previously used smartwatches and were considering buying a new one served as the study's respondents. A convenience sample of the researchers' community of contacts, from undergraduate to graduate level students, along with peers from professional circles, was used to collect data using an online questionnaire. For data analysis, SPSS software version 20.0 was employed. Out of 163 respondents, 56% are male and 44% are female. Table 1 displays a thorough demographic profile. The age range of most respondents is 20 to 29. A majority of those who responded have bachelor's degree. 56% of respondents are students and 44% of respondents are working professionals. A statistical method known as multiple regression analysis can be used to examine the relationship between a single dependent variable and a number of independent (predictor) variables [9]. Hence multiple regression analysis is used in this study to determine the relationship between the smartwatch buying intention of customers and other independent variables such as perceived usefulness, ease of use, fashion product, product feature, and health and fitness tracker. In order to investigate the relationship between the dependent variable and the independent variables, the following hypotheses have been constructed.

H1: Perceived Usefulness has a positive influence on customer buying intention of smartwatches.

Table 1 Demographic characteristics of study participants

Variable name	Category	Frequency	Percentage of the sample
Gender	Male	92	56
	Female	71	44
Age (in Years)	Less than 20	8	5
	20 to 29	94	58
	30 to 39	61	37
Education	Higher secondary	21	13
	Bachelor degree	95	58
	Master degree	47	29
Family income (in Rupees)	Below 20,000	11	7
	20,000–40,000	43	26
	40,000–60,000	52	32
	>60,000	57	35
Occupation	Students	91	56
	Working professionals	72	44

H2: Ease of Use has a positive influence on customer buying intention of smartwatches.

H3: Fashion Product has a positive influence on customer buying intention of smartwatches.

H4: Product Feature has a positive influence on customer buying intention of smartwatches.

H5: Health and Fitness Tracker has a positive influence on customer buying intention of smartwatches.

5 Results

Out of 163 respondents, 56% are male and 44% are female. Table 1 displays a thorough demographic profile. The age range of most respondents is 20 to 29. A majority of those who responded have bachelor's degree. 56% of respondents are students and 44% of respondents are working professional.

5.1 Multiple Regression Analysis

Regression lines' capability to represent for all of the variation in the dependent variable is shown in Table 2. The dependent variable's variance can be used to calculate its variation. By taking into account perceived usefulness, ease of use, fashion product, product feature, health and fitness trackers, the multiple correlation coefficient (R) of the model is 0.66 (Table 2). The R2 determination coefficient is 0.48 (Table 2). This demonstrates that the independent variable is responsible for 48% of the change in the dependent variable, leaving 55% of the dependent variable unexplained. With an adjusted R2 of 0.42, the model was able to predict 42% of changes in customers' intentions to purchase smartwatches using a combination of five independent factors.

The regression equation's capacity to explain the variability of the response variable is shown in Table 3's analysis of variance (ANOVA) table. The outcome suggests that there is a strong correlation between the independent and dependent variables. As a result, the regression model statistically predicts the outcome variables. The

Table 2 Model summary

Model	R	R square	Adjusted R square	Std. error of the estimate	Durbin-Watson
1	0.662[a]	0.483	0.423	0.41079	1.351

[a]Predictors: (Constant), Perceived usefulness, ease of use, fashion product, product feature, health and fitness tracker
[b]Dependent Variable: Customers Buying Intention towards smartwatches

Table 3 ANOVA[a]

Model		Sum of squares	df	Mean square	F	Sig.
1	Regression	31.872	3	10.628	60.324	0.000[b]
	Residual	41.431	246	0.169		
	Total	73.364	249			

[a]Dependent Variable: Customers Buying Intention towards smartwatches
[b]Predictors: (Constant), Perceived usefulness, ease of use, fashion product, product feature, health and fitness tracker

Table 4 Coefficient[a]

Model	Unstandardized coefficients	Sig.
Constant	1.102	0.000
Perceived usefulness	0.275	0.000
Ease of use	0.160	0.013
Fashion product	0.280	0.000
Product feature	0.295	0.000
Health and fitness tracker	0.292	0.000

model's regression coefficients and statistics for collinearity are displayed in Table 3. The model's p-value (Sig.), which is 0.000 and smaller than the alpha value of 0.05, as indicated in Table 3, is considered significant. As a result, the independent variables in this study can account for the variation in customers' intentions to purchase smartwatches. Table 4 shows that with a sig value less than 0.05, the coefficient for all independent variables is statistically significant. Hence H1, H2, H3, H4, and H5 are accepted in this study. The value used to predict the dependent variable by the independent factors is known as the unstandardized coefficient. The regression equation of this study is as follows.

Customer Smartphone Buying Intention $= 1.102 + 0.2759$ (perceived usefulness) $+ 0.160$ (Ease of Use) $+ 0.280$ (fashion product) $+ 0.295$ (product feature) $+ 0.292$ (Health and Fitness Tracker).

6 Limitations and Future Research Directions

The current study's shortcomings provide a foundation for future research. Only 163 surveys were gathered and taken into account due to time restrictions. In the future, it would be wise to use a larger sample size for the research. Respondents from a variety of demographic groups, such as adults, seniors, or respondents from different Indian states, could make up a sample from a potential study. Future research may take into account additional factors that are hypothesized to be influencing consumers' propensity to buy smartwatches.

7 Summary and Conclusion

With its utility and usability qualities, wearable technology has firmly established itself in our daily lives. Sports, health, and social wearable technologies offer a variety of tasks, from reminding users of their meeting schedules to measuring heart rate, or from accessing the most recent notifications to counting steps. Smartwatches, one of the most well-liked and commonly utilized wearable technology devices, are the subject of this study. It has been noted that very few studies and study on the subject have been carried out in Bangalore. According to the researcher, analyzing the sector's developments and current state will be crucial to the literature's contribution. The purpose of this survey is to ascertain consumer attitudes and intent to purchase smartwatches. The purpose of this survey is to ascertain consumer attitudes and intent to purchase smartwatches. Five variables have been discovered in this regard such as perceived usefulness, ease of use, fashion product, product feature, and health and fitness tracker. Demand for wireless accessories and Smartwatches has grown significantly in the Indian market. Customers are becoming more and more interested in using these devices, and they value high-quality goods. The customer wants everything to be digitalized and portable, whether it be a smartwatch, earbuds, or any other accessory, since the adoption of Digital India is on the rise. Hence Smartwatch manufacturers should offer offers top-notch comfort, simple use, and stylish design smartwatch device so that users may use them for extended periods of time .

References

1. Barrie, J.: The $20,000 gold Apple Watch Edition sold out in China in less than an hour. Business Insider (2015)
2. Baudier, P., Ammi, C., Wamba, S.: Differing perceptions of the smartwatch by users within developed countries. J. Glob. Inf. Manag. **28**, 1–20 (2020)
3. Blazquez, M., Alexander, B., Fung, K.: Exploring millennial's perceptions towards luxury fashion wearable technology. J. Fashion Market. Manage. Int. J. **24**(3), 343–359 (2020)
4. Choe, M.J., Noh, G.Y.: Combined model of technology acceptance and innovation diffusion theory for adoption of smartwatch. Int. J. Contents **14**, 32–38 (2018)
5. Choi, J., Kim, S.: Is the smartwatch an IT product or a fashion product? A study on factors affecting the intention to use smartwatches. Comput. Hum. Behav. **63**, 777–786 (2016)
6. Chuah, S.H.-W., Rauschnabel, P.A., Krey, N., Nguyen, B., Ramayah, T., Lade, S.: Wearable technologies: the role of usefulness and visibility in smartwatch adoption. Comput. Hum. Behav. **65**, 276–284 (2016)
7. Chun, J., Dey, A., Lee, K., Kim, S.: A qualitative study of smartwatch usage and its usability. Human Factors Ergon. Manuf. Serv. Ind. **28**(4), 186–199 (2018)
8. Dehghani, M., Kim, K., Dangelico, R.: Will smartwatches last? Factors contributing to intention to keep using smart wearable technology. Telematics Inform. **2018**(35), 480–490 (2018)
9. Hair, J.F., Black, W.C., Babin, B.J., Anderson, R.E., Tatham, R.L.: Multivariate Data Analysis, 7th edn. Pearson, New York (2010)
10. Herweijer, C., Combes, B., Johnson, L., McCargow, R., Bhardwaj, S., Jackson, B. Ramchandani, P.: Enabling a sustainable fourth industrial revolution: how G20 countries can create the

conditions for emerging technologies to benefit people and the planet. Economics Discussion Papers, No. 2018-32 (2018)

11. Hsiao, K.-L., Chen, C.: What drives smartwatch purchase intention? Perspect. Hardware Softw. Des. Value **35**, 103–113 (2018)

12. Kalantarian, H., Sarrafzadeh, M.: Audio-based detection and evaluation of eating behavior using the smartwatch platform. Comput. Biol. Med. **65**, 1–9 (2015)

13. Kim, J.K.: Round or square? How screen shape affects utilitarian and hedonic motivations for smartwatch adoption. Cyberpsychol. Behav. Soc. Netw. **19**(12), 733–739 (2016)

14. Lazaro, M., Lim, J., Kim, S., Yun, M.: Wearable technologies: acceptance model for smartwatch adoption among older adults. In: Human Aspects of I. T. for the Aged Population, Technologies Design and User Experience (2020)

15. Li, H., Wu, J., Gao, Y., Shi, Y.: Examining individuals' adoption of healthcare wearable devices: an empirical study from privacy calculus perspective. Int. J. Med. Inform. **88**, 8–17 (2016)

16. Lyons, K.: What can a dumb watch teach a smartwatch? Informing the design of smartwatches. In: Proceedings of the ACM International Symposium on Wearable Computers, pp. 3–10 (2015)

17. Niknejad, N., Ismail, W.B., Mardani, A., Liao, H., Ghani, I.: A comprehensive overview of smart wearables: the state-of-the-art literature, recent advances, and future challenges. Eng. Appl. Artif. Intell. **90**, 103529 (2020)

18. Saheb, T.: An empirical investigation of the adoption of mobile health applications: integrating big data and social media services. Heal. Technol. **10**(5), 1063–1077 (2020)

19. Scott D.A.: A Correlation Study of Smartwatch Adoption and Privacy Concerns with U.S. Consumers Using the UTAUT2. Dissertation, Colorado Technical University (2020)

20. Wu, L.-H., Wu, L.-C., Chang, S.-C.: Exploring consumers' intention to accept smartwatch. Comput. Hum. Behav. **64**, 383–392 (2016)

A Fluid-Thermal-Electric Model Based on Performance Analysis of Semi-transparent Building Integrated Photovoltaic Solar Panels

Nesrine Gaaliche, Hasan Alsatrawi, and Christina G. Georgantopoulou

Abstract The combination of solar cells with semi-transparent windows allows for multifunctional usage, such as power generation and natural light penetration, resulting in enhanced total energy efficiency. The effectiveness of semi-transparent structures with integrated photovoltaic windows on office building facades in Bahrain's subtropical desert environment was investigated in this study. The thermal performance and electrical capacity of semi-transparent integrated photovoltaic solar windows were investigated using parametric research. The COMSOL coupled solver was used to execute 3D fluid-thermal-electric multiphysics simulations for five design window geometries. The computational findings were compared to data from the manufacturer. The effects of glazing type, insulation gas type, and photovoltaic cell insertion on the thermal performance of the window were quantitatively investigated under various weather conditions, including solar irradiance and air temperature. The findings showed that krypton-filled triple-glazing windows with integrated PV might provide better thermal insulation. Furthermore, they demonstrated improved thermal performance. At 1000 W/m^2, the highest electrical efficiency of the PV module is 17.61%, and the maximum produced power is 248.42 W. Furthermore, the study demonstrates that semi-transparent window glazing is effective and saves 12% on energy.

Keywords Windows · Semi-transparent · PV · Efficiency · Energy performance · Heat transfer · Optimal design

Nomenclature

A	Area
a_0	Modified diode ideality factor
A	Window surface area, m^2

N. Gaaliche (✉) · H. Alsatrawi · C. G. Georgantopoulou
School of Engineering, Bahrain Polytechnic, PO Box 33349, Isa Town, Bahrain
e-mail: nessrine.gaaliche1@gmail.com

© The Author(s), under exclusive license to Springer Nature Switzerland AG 2024
A. Hamdan and E. S. Aldhaen (eds.), *Artificial Intelligence and Transforming Digital Marketing*, Studies in Systems, Decision and Control 487,
https://doi.org/10.1007/978-3-031-35828-9_10

a_{ref}	Modified diode ideality factor at STC
E	Total energy
E_g	Band-gap energy of PV cell material (eV)
$E_{g,ref}$	Band-gap energy of PV cell material at STC (eV)
G	Incident irradiance, W/m^2
H	Enthalpy
I	Unit vector
I_L	Light generated current (A)
$I_{L,ref}$	Light generated current at STC (A)
I_o	Diode reverse saturation current (A)
$I_{o,ref}$	Diode reverse saturation current at STC (A)
m	Temperature dependence parameter for a_0
n	Surface normal
NCS	Number of PV cell in series in PV panel
p	Pressure
Q	Heat flow rate per unit area [w.m^{-2}]
$q"$	Heat flux vector
R_{sh}	Series resistance at STC (Ω)
$R_{sh,ref}$	Shunt resistance (Ω)
S	Plane-of-array absorbed solar radiation at operating conditions (W/m^2)
S_E	Radiation energy source
S_{ref}	Absorbed solar radiation at STC (W/m^2)
T_{cell}	PV cells temperature (K)
$T_{cell,ref}$	PV cells temperature at STC (K)
U	Overall heat transfer coefficient [W.m^{-2}.K],
μ_{isc}	Temperature Coefficient of Short circuit current
v	Continuum velocity
V	Volume
v_r	Relative velocity

1 Introduction

Energy is a key element and is regarded as the principal factor that impacts practically all aspects of human existence, as well as the demand for country progress in order to achieve long-term objectives [1, 2]. Solar (PV) technologies, being the primary source of photovoltaic power generation, are a renewable resource that aid in the reduction of global warming [3]. It is among the most important ways to obtain clean electricity for structures, which could help reduce the energy shortage [4–6]. Photovoltaic minerals such as monocrystalline silicon, polycrystalline silicon, and amorphous silicon convert sunlight into electrical energy, as stated by Green et al. [7]. Building-integrated photovoltaic systems are being investigated as a way which meets users' requirements effectively and consistently [8, 9].

These are PV technologies that are employed when photovoltaic modules are incorporated into building construction [10, 11]. The generation of power from building-integrated PV results in consumption that is very similar, and the building envelope offers decorative, practical, and technological solutions [12]. Nevertheless, the electricity created by these systems is affected by a variety of parameters, including sunlight, temperatures, humidity, position, and winds [13–15].

According to Yang et al. [16], office towers are ideal for building integrated PV since they use electricity mainly during the day, during which the photovoltaic cells gather and transform sunlight into electrical energy; consequently, the use of energy storage may be eliminated. The semi-transparent integrated photovoltaic system is very commonly utilized in office buildings because it produces a comfortable atmosphere by permitting sunlight to penetrate the office, preventing heating during the summer, and generating power. Semi-transparent building-integrated PV systems provide for strong illuminance, whereas opaque PV systems do not enable sunlight entry; therefore, they cannot be utilized in windows [17, 18]. Manufacturer-reported efficiency data are typically obtained using research lab circumstances that include a solar irradiation of 1000 W/m^2, a temperature of 25 °C, and an air mass value of 1.5 [19, 20]. In most situations, data generated for research lab circumstances doesn't really reflect the real situation obtained in particular locations, resulting in considerable discrepancies in test data [21]. Generally, cost-effectiveness analysis and payback period are not taken into account for the building of integrated PV based on real variables [22]. However, research comparing the functional efficiency of semi-transparent buildings integrated with PV to photovoltaic solutions is currently lacking in the Gulf region. Most of the studies on the functional impact of semi-transparent, building integrated PV systems have sought to use theoretical models of semi-transparent cells that may not accurately represent commercial systems [23, 24]. The kingdom of Bahrain, located in the Gulf region, is often characterized by hot and humid weather conditions mostly all year, leading to the use of air conditioning throughout the year.

This study aims to develop different 3D finite element models to investigate the thermal and electrical performance of semi-transparent building integrated PV systems for office applications, primarily in Gulf climates. The coupled thermal, electric, and optical multiphysics model is used to calculate photovoltaic panel thermal and electrical performance. First, the impacts of the type of glazing, the type of insulation gas, and the insertion of photovoltaic cells on the thermal performance of the window is numerically investigated. The thermal conductivity values of Argon, Krypton, and Xenon insulation gases are listed in Table 1. Second, the performance metrics include electrical power output, photovoltaic cell temperatures, and the outer and inner surface temperatures of the window. Lastly, the overall energy performance of the semi-transparent building-integrated photovoltaic systems and its energy saving potential has been studied with respect to weather and operating parameters.

Table 1 Thermal conductivity values of Argon, Krypton, and Xenon [25]

Gases	Argon	Krypton	Xenon
Thermal conductivity W/(m·K)	0.01772	0.00949	0.00565

2 Methodologies

2.1 Thermal 3D Model and Simulation Setup

Numerical analysis is carried out using fluid-thermal-electric multiphysics. The software Comsol Multiphysics 5.6 is used for the Simulation results. The three-dimensional building-integrated PV unit models with various configurations are tested. Conduction and convection heat transfers are taken into consideration in the model. Different types of insulated gases including argon, krypton and xenon having various thermal conductivity values were used in this simulation, in order to compare their heat flux (Table 1). As a result, the goal of this paper is to include the governing equations into COMSOL Multiphysics using the COMSOL coupled solver to perform a more reasonable and precise fluid-thermal-electric multiphysics analysis on the building-integrated photovoltaic module. The COMSOL coupled solver takes into account all of the physical features, which is the most appropriate technique for solving the fluid-thermal-electric multiphysics model.

Nieti et al. [26] created a 3D CFD model of the photovoltaic panels resting on the frame at a particular inclination. The simulation findings are supported by experiments. The photovoltaic panel's dimensions are similar to those of a real-world panel that measures 1600×1000 mm ($L_p \times W_p$).

A photovoltaic panel made from crystalline silicon is composed of sheets of various materials. The PV cell is made up of a glass panel of 3 mm thickness on the front surface, a PV layer of 0.5 mm thickness with Ethylene Vinyl Acetate (EVA) inserted between the EVA sheets [27], and a 0.5 mm Tedlar in the back surface. The Photo voltaic sheet and the EVA sheets are isolated surfaces, and the Photo voltaic width may range from 180 to 350 m [27]. Nevertheless, since the EVA and Photovoltaic layers are thin (especially relative to other computational regions), the temperature differential between the EVA and Photovoltaic sheets is ignored in the current study's heat transfer model. As a result, for heat transfer modeling, the EVA sheets and Photovoltaic sheets are combined into a single domain. Nizetic et al. [26] followed a similar method, in which tests supported their findings. Following the previous assumptions, the glass surface, Photovoltaic cells and EVA layers, and back surface are adhered to each other [28]. The model specifies the normal values for each material's physical parameters, including density, thermal conductivity, and specific heat capacity Cp. By establishing these parameters, the simulation can correctly replicate heat transfer. The simulation recreates a building-integrated Photovoltaic system model in the Kingdom of Bahrain during peak hours. The input section of the outside space was designated as a velocity inlet border, with specifications determined by local meteorological parameters. The sun irradiation is 1000 W/m^2 and the air outside temperature is 41.8 °C [29].

After creating the three-dimensional model, it is meshing using fine tetrahedron elements with a minimum size of 4.5 × 10^{-3} m and a growth rate of 1.188. The contact area of the cell and the air contain even smaller prismatic meshing elements.

The finite element analysis used the shear stress transfer (SST) k-omega turbulent model, which has demonstrated that the modeling results are correct [30]. The SST model is a combination model that includes the benefits of both the k-epsilon and k-omega models. As previously stated, the current investigation used a 3D model in a steady-state condition. The SST k-omega model, which is derived from the Reynolds-averaged Navier–Stokes (RANS) and the energy equations, was computed to determine the temperature and velocity results. The following partial differential Eqs. (1) continuity equation, (2) momentum equation, and (3) energy equation were used as a governing equation derived from the mass conservation law and momentum conservation law [31].

$$\frac{\partial}{\partial t}\left(\int_V \rho dV\right) + \oint_A V_r.da = 0 \tag{1}$$

$$\frac{\partial}{\partial t}\left(\int_V \rho dV\right) + \oint_A \rho v_r \otimes v.da = -\oint_A pI.da + \oint_V f_b dV \tag{2}$$

$$\frac{\partial}{\partial t}\int_V \rho E dV \cdot \oint_A [\rho H V_r + v_g p].da = -\oint_A q''.da + \int_A T.vdA$$

$$+ \oint_V f_b.vdV + \oint_V S_E dV \tag{3}$$

where v presents continuum velocity, v_r is the relative velocity; V is volume; a is the area; p is the pressure; I is the unit vector; T presents the viscous stress tensor; E is the total energy; H is the enthalpy; q'' is the heat flux vector; $\otimes$ is a Kronecker product; and S_E radiation energy source.

2.2 Heat Transmittance: U-factor

The heat transmittance for each window can be calculated from the heat flux density using the following equation [32]:

$$U = \frac{Q}{\Delta T} \tag{4}$$

where, U is the overall heat transfer coefficient [W.m^{-2}.K], Q is the heat flow rate per unit area [w.m^{-2}] and ΔT is the temperature difference [K].

2.3 Electrical Model

In this study, the electrical model represented a photovoltaic device [33] by an equivalent electric circuit of Fig. 2 [34], and the current–voltage expression for a photovoltaic cell is given by the following equation:

$$I = I_L - I_o\left(\exp\left(\frac{V + I.R_s}{R_{sh}}\right) - 1\right) - \frac{V + I.R_s}{R_{sh}} \tag{5}$$

To implement the model, first determine the parameters I_L, I_o, a, R_s, and R_{sh} at a reference state and then translate them to the operational state given in Eqs. (5)-(9). Both maximum power values calculated at a higher temperature and lower irradiation are used to determine the variables m and n.

$$a_0 = a_{ref}\left(\frac{T_{cell}}{T_{cell,ref}}\right)^n \tag{6}$$

$$I_L = \left(\frac{S}{S_{ref}}\right)^m \left(I_{L,ref} + \mu_{isc}\left(T_{cell} - T_{cell,ref}\right)\right) \tag{7}$$

$$I_o = I_{o,ref}\left(\frac{T_{cell}}{T_{cell,ref}}\right)^3 e^{\left(\frac{NCS, T_{cell,ref}}{a_{ref}}\left(\frac{E_{g,ref}}{T_{cell,ref}} - \frac{E_g}{T_{cell}}\right)\right)} \tag{8}$$

$$R_{sh} = \frac{S}{S_{ref}} R_{sh,ref} \tag{9}$$

$$R_{sh} = R_{sh,ref} \tag{10}$$

2.4 Estimating of Electrical Performances

Electrical performance was determined by implementing the electrical model of an equivalent circuit and estimating the power generation capacities of the semi-transparent photovoltaic systems, which were also examined by other authors [35, 36]. The electric characteristics, including the maximum rated power P_{max} and the power conversion efficiency of the photovoltaic modules, were calculated from the formula (11):

$$\eta = \frac{P_{max}}{GA} \tag{11}$$

where: G is the incident irradiance, W/m², and A is a window surface area, m².

3 Results and Discussion

3.1 Thermal Performance of Window Configurations Designs

The boundary conditions met the assumptions, specifically 22 °C and 41.8 °C for both outside and inside temperatures, respectively (see Table 2). The solver settings were imposed to ensure adequate homogeneity with the real-world situations.

3.1.1 The Effect of the Type of the Glazing

Figure 4 presents the heat flux contours for the aforementioned simulated cases. The window has a surface temperature of 38.850 °C on the outside and 27.664 °C on the room side (Fig. 3a).

The heat flux was found to be 20.229 W/m², as shown in Fig. 4a. The triple-glazed window was designed without the use of a PV panel first to compare it to double-glazed windows (Fig. 3b). In this case, the window has three glass insulated gas (see Fig. 1c). The window has a surface temperature of 39.761 °C on the outside and 25.361 °C on the room panes instead of two in the double-glazing window, and each glass pane is separated by argon as the side.

The heat flux was also simulated and found to be 12.838 W/m², as shown in Fig. 4b. The thermal simulation for the window is performed, and the results show that the surface temperatures of the window were 39.395 °C on the outside and 25.574 °C on the room side (Fig. 3c).

Table 2 Surface temperatures obtained from CFD simulation together with heat flux and heat transmittance showing the effect of the type of the glazing

Simulated scenarios	Outer suface temperature (°C)	Inside surface temperature (°C)	Heat flux (W/m²)	Heat transmittance U-value (W/m².K)
Case A	38.850	27.664	20.229	1.756
Case B	39.761	25.361	12.838	0.882
Case C	39.395	25.574	12.838	0.882

Fig. 1 Schematic illustration of building-integrated PV unit model: **a** Double-glazing window, **b** Triple-glazing window, **c** Double-glazing window with PV and argon, **d** Triple-glazing window with PV and Krypton, and **e** Triple-glazing window with PV and Xenon

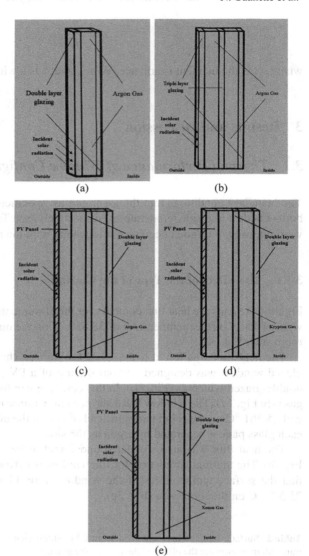

Fig. 2 Equivalent circuit of an actual PV cell [34]

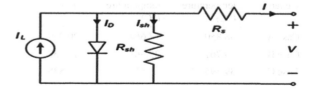

A Fluid-Thermal-Electric Model Based on Performance Analysis ...

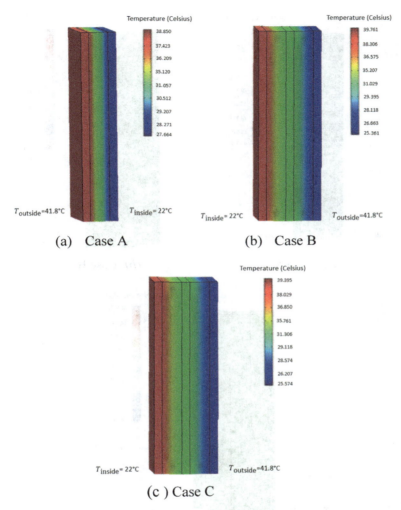

Fig. 3 Simulated temperature distribution across the window: **a** Case A: Double Glazing, **b** Case B: Triple Glazing, and **c** Case C: Triple Glazing with PV Panel Windows

The heat flux density was found in the PV integrated window to be 12.838 W/m^2, as shown in Fig. 4c. It is clearly shown that the inside temperatures and heat flux of the building-integrated PV and the triple-glazed window were extremely close, but the outer temperature of the building-integrated PV was higher than that of the triple-glazed window by approximately 1 °C.

As shown in Table 2, the U-factor of the double-glazing window is significantly larger than that of the building's integrated PV. The primary explanation for the considerable disparity in U-factor between the three kinds of windows was ascribed

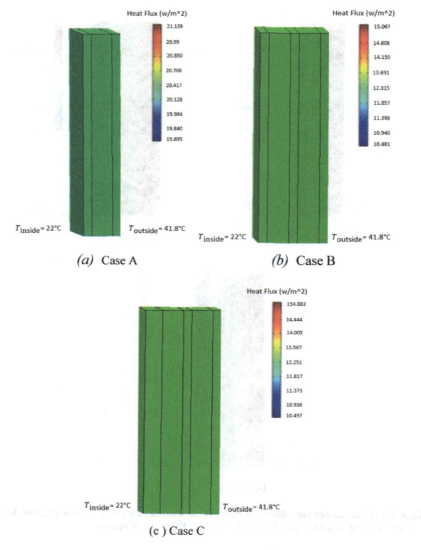

Fig. 4 Simulated heat flux distribution across the window: **a** Case A: Double glazing; **b** Case B: Triple glazing; and **c** Case C: Triple glazing with PV panel windows

to the varying installation positions and number of glass panes. For the double-glazed window, the double layer of glazing was separated by argon gas. As a result, it is not possible for longwave infrared light to be completely reflected back into interior spaces. The U-factor of the building-integrated PV is greatly decreased, and its performance of heat insulation is enhanced because the PV cell is made up of a glass panel of 3 mm thickness on the front surface.

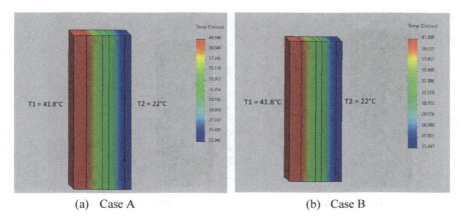

Fig. 5 Simulated temperature distribution across the window: **a** Case A: Insulated gas Krypton with PV, and **b** Case B: Insulated gas Xenon with P V windows

3.1.2 The Effect of the Insulating Gas

Figure 5 depicts different surface temperature distributions of both building-integrated PV systems using Krypton (see Fig. 5a) and Xenon (see Fig. 5b) as insulating gases. It is shown that the inside temperatures using Krypton and Xenon insulated gases were extremely close, 23.942 °C and 23.247 °C, respectively, but the outer surface temperature of the building-integrated PV system with Xenon insulated gases was higher than that with Xenon insulated gas by 0.5 °C (see Fig. 5). The heat flux of both building-integrated PV systems using Krypton and Xenon was measured and compared. As shown in Fig. 6, the building-integrated PV systems with Xenon have lower heat flux and U-factors than those with krypton because of the lower thermal conductivity of Xenon isolated gas (Table 1). The average heat flux of the building-integrated PV systems using Xenon and Krypton was 7.778 W/m^2 and 4.995 W/m^2, respectively. Consequently, the average U-factors of the Xenon and Krypton insulating gases were 0.468 W/m^2.K and 0.281 W/m^2.K, respectively. Thus, Xenon can be fully effective to prevent direct radiation into a building (Table 3).

3.2 *Electrical Performance*

The electric features of the power generated by the argon-filled triple-glazing windows with integrated PV panels (see Fig. 1e) are simulated. Figure 7 depicts the effects of varying irradiance conditions on the semi-transparent window pane. From the curve, the efficiency and all factors were calculated using the formula (10). Table 4 summarizes the electric performance parameters. The greatest efficiency attained was 17.61% at 1000 W/m^2 for a PV temperature of 39.76 °C, which was similar to the manufacturer's efficiency, whereas at 1100 W/m^2, the efficiency declined to

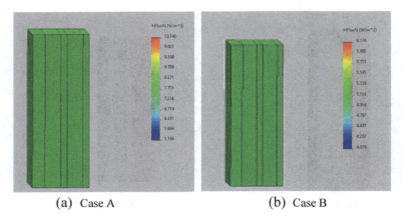

Fig. 6 Simulated heat flux distribution across the window: **a** Case A: Insulated gas Krypton with PV, and **b** Case B: Insulated gas Xenon with PV windows

Table 3 Surface temperatures obtained from CFD simulation together with heat flux and heat transmittance showing the effect of the insulating gas

Simulated scenarios	Outer surface temperature (°C)	Inside surface temperature (°C)	Heat flux (W/m^2)	Heat transmittance U-value (W/m^2.K)
Case A	40.566	23.942	7.778	0.468
Case B	41.009	23.247	4.995	0.281

11.49% for a PV temperature rise of 52.18 °C. The greatest efficiency attained was 17.61% at 1000 W/m^2 for a PV temperature of 39.76 °C, which was similar to the manufacturer's efficiency, whereas at 1100 W/m^2, the efficiency declined to 11.49% for a PV temperature rise of 52.18 °C. Incident irradiances and ambient temperatures have an effect on the electricity generated by the PV-integrated windows. The greater the sun irradiation, the greater the electricity. For example, at G = 650 W/m^2 and module temperature = 49.42 °C, the power seemed to be less than that at G = 810 W/m^2 and module temperature of 49.23 °C (Table 4). Other investigations [37, 38] conducted in the tropics of Spain and China, respectively, found that by raising the PV temperature, the power generated decreased. Figure 4 depicts a simulation of the change in power generated vs. solar radiation from morning to night. The power produced appears to be dependent on solar irradiation, with more radiation resulting in higher output power and vice versa. Morning irradiances increase with increasing power output, whereas midday irradiances change somewhat before decreasing in the night according to other parameters, including PV temperature. Other researchers [39, 40] further demonstrated that irradiance has a direct proportionate effect on power production as temperature changes. The coefficient R2 was determined to be 0.9616 in the regression analysis, which is between two sets of data; this shows a direct reciprocal link between the power produced and sun irradiation.

Fig. 7 Power generated by the building-integrated PV system versus solar intensity

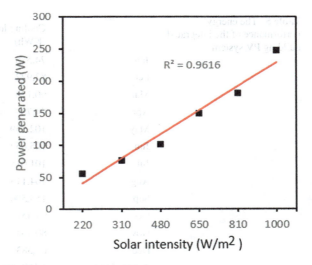

Table 4 Electrical power: investigating the impact of solar irradiance and module temperature

Irradiance, G (W/m^2)	Maximum power, P$_{max}$ (w)	Efficiency, η (%)	Module temperature (°C)
220	56.59	3.18	30.59
310	75.78	6.72	34.87
480	100.58	12.54	40.74
650	150.09	16.29	49.42
810	180.37	16.89	49.23
1000	248.42	17.61	39.76
1100	177.81	11.49	52.18

3.3 Energy Performance of the Building-Integrated PV System

The electricity generated by the semi-transparent solar cell accounts for about 12% of the total electricity consumption to fulfill the cooling demand of the building's occupancy. The computational investigations of solar power in Bahrain indicate that the highest solar power yield is approximately 14,629 kWh/month in the summer months (June to September) and closely 11,094 kWh/month in the winter months (December–February), with an annual monthly average of 12,980 kWh. The expected annual solar power generated by the building facade using 500 panels is nearly 140,755 kWh (Table 5). The electricity consumption for cooling supply during the winter and summer seasons increased from 75,624.kWh to 116,071.kWh, or about 34%. As shown in Table 5, we could infer from the above that the cooling load is

Table 5 The energy performance of the integrated building PV system

	Cooling load (KWh)	Solar power (KWh)
Jan	74,578	10,442
Feb	83,922	11,747
Mar	90,128	12,617
Apr	94,789	13,270
May	102,559	14,358
Jun	105,657	14,793
Jul	101,015	14,140
Aug	104,113	14,575
Sep	153,500	15,011
Oct	99,451	13,923
Nov	80,804	11,312
Dec	68,383	9572
Annual load	1,158,885	140,755
Annual electricity consumption	1,018,140.128	
Annual electricity % saving	12	

greater in the summer than in the winter. And it is easy to conclude that 12.6% of the electricity consumption for cooling supply is saved since the integrated-building PV system produces more solar energy in the summer.

4 Conclusions

In this paper, a numerical study of the thermal and electrical performance of the semi-transparent building-integrated PV system used in office buildings appropriate for the Gulf climate was implemented. The present findings were evaluated under several test circumstances that included different irradiances during daytime and air temperature. The current study's results are consistent with common assumptions about the performance of building-integrated PV systems. The thermal and electrical performances of various models of solar window configurations were computationally investigated using a coupled thermal-electric model by Comsol Multiphysics under Kingdom of Bahrain weather conditions.

Among the simulated cases, examining the different types of glazing showed that the triple-glazed window and building-integrated PV can provide lower U-factors compared to a double-glazed window.

Evaluating the type of insulate gases indicated that a building-integrated PV system with Xenon showed greater results in terms of a lower U-factor and outer surface temperature, and reducing the need for electrical temperature control. Despite greater irradiation, the electrical study demonstrates that performance falls as module temperature rises. The greatest efficiency attained was 17.61% at 1000 W/m^2 with a PV temperature of 39.76 °C, whereas at 1100 W/m^2 and a temperature of 52.18 °C, the efficiency declined to 11.49%. Furthermore, this research shows that semi-transparent window glazing saves between 11 and 19% on energy. As a result, testing its electrical performances in real-world operating situations is required to encourage system integration in office buildings and aid in more efficient energy consumption in accordance with sustainable energy targets. The computational investigation shows that the yearly energy savings potential due to the use of integrated-building PV windows is about 12%. These investigations addressed the solar energy produced under different boundary conditions, including the summer season and various local climate conditions. It is a very remarkable accomplishment. Further research and development of solar cell energy conversion efficiency, optimum transmittance, and thermal insulation qualities would make this technology more energy-efficient and inexpensive. It should be emphasized that the above energy performance data was simulated only from numerical models. In the future, we will concentrate on experimental data to assess the yearly total energy performance of integrated-building PV windows in various climate zones.

References

1. Shukla, A.K., Sudhakar, K., Baredar, P.: Exergetic analysis of building integrated semitransparent photovoltaic module in clear sky condition at Bhopal India. Case Stud. Thermal Eng. **8**, 142–151 (2016)
2. Matungwa, B.J.: An Analysis of PV Solar Electrification on Rural Livelihood Transformation: A Case of Kisiju-Pwani in Mkuranga District, Tanzania. University of Oslo, Blindern, Norway (2014)
3. Tripathy, M., Kumar, M., Sadhu, P.: Photovoltaic system using Lambert W function-based technique. Sol. Energy **158**, 432–439 (2017)
4. Cao, X., Dai, X., Liu, J.: Building energy-consumption status worldwide and the state-of-the-art technologies for zero-energy buildings during the past decade. Energy Build. **128**, 198–213 (2016)
5. Ban-Weiss, G., Wray, C., Delp, W., Ly, P., Akbari, H., Levinson, R.: Electricity production and cooling energy savings from installation of a building-integrated photovoltaic roof on an office building. Energy Build. **56**, 210–220 (2013)
6. Tripathy, M., Sadhu, P., Panda, S.: A critical review on building integrated photovoltaic products and their applications. Renew. Sustain. Energy Rev. **61**, 451–465 (2016)
7. Green, M.A., Dunlop, E.D., Levi, D.H., Hohl-Ebinger, J., Yoshita, M., Ho-Baillie, A.W.: Solar cell efficiency tables (version 54). In: Progress in Photovoltaics: Research and Applications, p 27 (2019)
8. Gonçalves, J.E., Hooff, T.V., Saelens, D.: Understanding the behaviour of naturally-ventilated BIPV modules: a sensitivity analysis. Renewable Energy **161**, 133–148 (2020)
9. Gholami, H., Nil, H., Steemer, K.: The contribution of building-integrated photovoltaics (BIPV) to the concept of nearly zero-energy cities in Europe: potential and challenges ahead. Energies **14**(19), 6015 (2021). https://doi.org/10.3390/en14196015
10. Yin, H., Pao, F.: Building Integrated Photovoltaic Thermal Systems (2022)

11. Angèle, H.M.E., Reinders, Müller, M.F.: Comprehensive Renewable Energy, 2nd ed. (2022)
12. Aghaei, M., Castañón, J.M.: Photovoltaic Solar Energy Conversion (2020)
13. Kang, J., Reiner, D.M.: What is the effect of weather on household electricity consumption? Empirical evidence from Ireland. Energy Econ. 111, 106023 (2022). ISSN 0140-9883. https://doi.org/10.1016/j.eneco.2022.106023
14. Razavi, R., Gharipour, A., Fleury, M., Akpan, I.J.: Occupancy detection of residential buildings using smart meter data: a large-scale study. Energy Build. 183, 195–208 (2019). https://doi.org/10.1016/J.ENBUILD.2018.11.025
15. Sim, T., Choi, S., Kim, Y., Youn, H., Jang, D.-J., Lee, S., Chun, C.-J.: eXplainable AI (XAI)-based input variable selection methodology for forecasting energy consumption. Electronics 11(18), 2947 (2022). https://doi.org/10.3390/electronics11182947
16. Yang, L., Liu, X., Qian, F.: Optimal configurations of highrise buildings to maximize solar energy generation efficiency of building-integrated photovoltaic systems. Indoor Built Environ. 28(8), 1104–1125 (2019)
17. Ghosh, A., Sundaram, S., Mallick, T.K.: Colour properties and glazing factors evaluation of multicrystalline based semitransparent photovoltaic-vacuum glazing for BIPV application. Renewable Energy 131, 730–736 (2019)
18. Sarkar, D., Kumar, A., Sadhu, P. K.: A survey on development and recent trends of renewable energy generation from BIPV systems. IETE Tech. Rev. 1–23 (2019)
19. Shukla, A.K., Sudhakar, K., Baredar, P.: Recent advancement in BIPV product technologies: a review. Energy Build. 140, 188–195 (2017)
20. El Tayyan, A.A.: PV system behavior based on datasheet. J. Electron Dev. 9, 335–341 (2011)
21. Femia, N., Petrone, G., Spagnuolo, G., Vitelli, M.: Power Electronics and Control Techniques for Maximum Energy Harvesting in Photovoltaic Systems. CRC (2017)
22. Tripathy, M., Joshi, H., Panda, S.K.: Energy payback time and life-cycle cost analysis of building integrated photovoltaic thermal system influenced by adverse effect of shadow. Appl. Energy 208, 376–389 (2017)
23. Halme, J., Mäkinen, P.: Theoretical efficiency limits of ideal coloured opaque photovoltaics". Energy Environ. Sci. 12(4), 1274–1285 (2019)
24. Ahmed, O.K., Hamada, K.I., Salih, A.M.: Performance analysis of PV/Trombe with water and air heating system: an experimental and theoretical study. Energy Sources, Part A 1–21 (2019)
25. Vargaftik, N.B., Vasilevskaya, Y.D.: Thermal conductivity of krypton and xenon at temperatures up to 5000°K. J. Eng. Phys. 39, 1217–1222 (1980). https://doi.org/10.1007/BF00824745
26. Nižetić, S., Grubišić-Čabo, F., Marinić-Kragić, I., Papadopoulos, A.M.: Experimental and numerical investigation of a backside convective cooling mechanism on photovoltaic panels. Energy 111, 211–225 (2016)
27. Tiwari, G.N., Tiwari, A., Shyam: Handbook of Solar Energy: Theory, Analysis and Applications. Springer Singapore, Singapore (2016)
28. The structure of a photovoltaic module | ECOPROGETTI—Specialist in photovoltaic production process (2012).
29. Alnaser, W., Alnaser, N.W., Batarseh, I.: Bahrain's Bapco 5MW PV grid-connected solar project. Int. J. Power Renew. Energy Syst. 1, 72–84 (2014)
30. Menter, Z.F.R.: Two Equation Kappa-Omega Turbulence Models for Aerodynamic Flows, NASA Technical Memorandum (1993)
31. Cho, K.-J., Cho, D.-W.: Solar heat gain coefficient analysis of a slim-type double skin window system: using an experimental and a simulation method. Energies 11, 115 (2018)
32. Huertas, J.D., Rodríguez, R., Moyano, J., Delgado, F., Marín, D.: Determining the U-value of façades using the thermometric method: potentials and limitations. Energies 11(2), 1–17 (2018). https://doi.org/10.3390/en11020360
33. Siddiqui, M.U.: Multiphysics modeling of Photovoltaic panels and Arrays with auxiliary thermal collectors. MS Thesis, King Fahd University of Petroleum & Minerals, Saudi Arabia (2011).
34. Duffie, J.A., Beckman, W.A.: Solar Engineering of Thermal Processes, 2nd edn. Wiley, New York (1991)

35. Kha, N.P.: Semi-Transparent Building-Integrated Photovoltiac (BIPV) Windows for the Tropics (2013)
36. Shukla, A.K., Sudhakar, K., Baredar, P.: A comprehensive review on design of building integrated photovoltaic system. Energy Build. **128**, 99–110 (2016)
37. Qiu, C., Yang, H., Zhang, W.: Investigation on the energy performance of a novel semi-transparent BIPV system integrated with vacuum glazing. Build. Simul. **12**(1), 29–39 (2019)
38. Chemisana, D., Moreno, A., Polo, M. et al.: Performance and stability of semitransparent OPVs for building integration: a benchmarking analysis. Renewable Energy **137**, 177–188 (2019)
39. Yang, S., Cannavale, A., Prasad, D., Sproul, A., Fiorito, F.: Numerical simulation study of BIPV/T double-skin facade for various climate zones in Australia: effects on indoor thermal comfort. Build. Simul. **12**(1), 51–67 (2019)
40. El Desoky Faggal, A., Mohamed Shams Eldin, A., El Sayed Ali, R.: The impact of using semi-transparent photovoltaic in office building facades on improving indoor thermal performance in Egypt. J. Al-Azhar Univ. Eng. Sector **14**(51), 613–625 (2019)

The Impact of Social Media Influencer Marketing on Purchase Intention in Bahrain

Aysha AlKoheji, Allam Hamdan, and Assma Hakami

Abstract The marketing field is rapidly developing with the advancement of internet and social media platforms. As a result, new digital marketing tools are being used such as social media influencer marketing. Influencers are individuals who are shaping the customer's preferences and opinions through sharing photos and videos related to a specific product. This paper is focusing on social media influencers and how they are impacting purchase intention. The study evaluates how credibility factors such as attractiveness, expertise and trustworthiness are having an effect on the purchase decision. The research methodology of this study is based on literature review. It concludes that the effect of these factors is positive to the purchase intention.

Keywords Influencer marketing · Social media influencers · Source credibility · Source attractiveness · Purchase intention

1 Introduction

Social media platforms are digitalized interactive websites or applications that allow users to interact and communicate with each other in a virtual network by sending and receiving digital content and information while sharing thoughts and ideas as well [3, 30]. YouTube, Twitter, Instagram, Facebook, TikTok and snapchat are examples of applications and websites that are considered to be social media platforms where users can create their virtual profiles with a personalized profile picture that represents them to other users. It is also where users start sharing videos, photos, or texts. Moreover, users can interact with each other by several basic functions such as: reply and comment on other users' posts, and like or share posts as well [21].

A. AlKoheji · A. Hamdan (✉)
Ahlia University, Manama, Bahrain
e-mail: allamh3@hotmail.com

A. Hakami
University of Business and Technology, Jeddah, Saudi Arabia

© The Author(s), under exclusive license to Springer Nature Switzerland AG 2024
A. Hamdan and E. S. Aldhaen (eds.), *Artificial Intelligence and Transforming Digital Marketing*, Studies in Systems, Decision and Control 487,
https://doi.org/10.1007/978-3-031-35828-9_11

As a result of accessibility to internet in the world and the spread of smartphones, social media platforms have become extremely popular around the globe. In 2015, the number of users in all social media platforms were around 2 billion users [9], where in 2022 the total number of users around the world jumped to approximately 4.7 billion making it more than half of the world's population [26]. By comparing the 4.7 social media platform users to the internet users around the world which are 5.03 billion, we can clearly see that 93.4% of them are using social media platforms which is an indicator to the magnitude of usage for these applications.

These numbers are a representation of what is clearly happening in Bahrain. Based on a report conducted in 2021, the number of internet users in Bahrain were around 1.71 million making it 99% of the population. Moreover, active social media users reached 1.50 million which is considered 87% of the population [12]. These statistics clearly give an indication of the high usage of social media platforms in Bahrain.

The beginning of the social media concept was in 2003 when the website My Space was launched. My Space allowed users to create personal pages sharing pictures, interests, and hobbies. Users could also link their pages with other user pages to connect with their friends and to create new ones [22]. In 2004, My Space became the most popular website as it was the first social media platform to reach 1 million monthly active users at the time. Between 2005 and 2008, My Space was the most visited social media network website in the world. In addition, in 2006, it became the most visited website in The United States surpassing Google and Yahoo. In 2008, Facebook which was previously established in 2004, had begun having high numbers of unique users equal to My Space. These numbers revealed around 115 million unique users. Since then, My Space visitors started to rapidly decline as users started to shift to Facebook. Facebook surpassed My Space in the number of unique users by 130 million users [22, 28]. As of January 2022, Facebook became the most popular website worldwide with more than 1 billion registered accounts and 2910 million unique active users [10].

After 19 years of social media growth, many popular platforms such as Hi5, My Space and Friendster had disappeared as they were competitors to Facebook in the social media market. Eventually, these disappearing platforms lost the little shares they had left and exited the market [22].

It has been noticed that social media platforms that have survived the last decade and continued to operate have applied significant updates and changes to their platforms. This is done to make sure current customers keep using the applications and new customers start adopting them [13]. Twitter is a great example to illustrate this point. The mega application started as a social platform that allows users to register and share their ideas and thoughts in short messages called tweets using 280 words or less. In 2011, Twitter had changed their strategy allowing users to upload videos and pictures which currently makes up more than 50% of their viewed tweets. Another great example for the evolution of social media applications is Instagram. The platform started as a simple application for sharing photos. Later on, more features were added such as Direct Messages which allowed users to exchange information via the application [22]. The application did not stop there, Instagram kept evolving itself with the times and adding new features such as hashtag and location browsing options. It also added its story feature which gave users the ability to post photos and video that can only be viewed for 24 h after the original posting [7]. These changes and updates in the applications, allowed them to survive and thrive as they were adapting with the market needs and changes [11].

2 Digital Marketing

Digital marketing or e-marketing is a term that has appeared in 2005. It is mainly marketing through digital channels using the internet to allow a vast number of receivers [8]. The two most important factors that allowed for the transition of traditional marketing to digital marketing is the acceleration of globalization and the wide use of the internet [18].

Nowadays, digital marketing has become one of the most essential aspects of marketing. Any company, brand, small to medium sized business or home-based business uses at least one social media platform to promote their products. These businesses also use social media platforms to create awareness for their brands and drive visitors to their websites and pages.

Digital Marketing has taken many forms using advanced technologies. Search Engine Marketing (SEM), Social media marketing, content marketing and influencer marketing are just some of the digital marketing tactics used by many companies [31].

2.1 Social Media Influencers

While active users on social media platforms are constantly increasing, a new type of digital marketing has become a common theme on these applications. Social media influencer marketing has been the wave of change for social media marketing in the last ten years [23].

A social-media influencer "is first and foremost a content generator: one who has a status of expertise in a specific area, who has cultivated a sizable number of captive followers—who are of marketing value to brands—by regularly producing valuable content via social media" [16].

As one of the famous advertisement strategies, it was common that a majority of organizations were using celebrity advertisement as a marketing strategy [15]. It consists of mainly having an agreement with one of the celebrities or public figures, such as: actors, singers, sport players and models to recommend and promote their products or services through traditional channels, like: TV, Radio, Newspapers, etc.

The same concept applies for social media influencer marketing, but the difference is that the influencers are non-celebrity individuals who start creating content on their accounts in fields, such as: health, sport, food, photography, fashion, or technology.

After regularly posting on any social media platform, an influencer starts to gain followers and increases their popularity and brand image.

Brands usually invite influencers who have a huge following to try their products or services and share their own experiences [27]. This involvement helps brands promote and endorse their products by creating a positive impact to their image. The mentioned impact allows followers to gain valuable feedback and product reviews which in turn affects their purchase intention regarding specific products and brands [32, 16].

An apparent advantage of advertising through social media influencers is that companies can get insights and real interaction figures from influencers' pages on social media platforms. Examples of these insights can be: number of likes, comments, shares and saves of specific photos related to the promoted product [32]. Insights such as these can draw a clear image on the influencers' effectiveness in shifting the consumers behavior.

Social media influencer marketing depends mainly on opinion leaders and electronic word of mouth (eMOW) [16, 24].

The concept of opinion leaders refers to the ability of individuals to shape opinions and attitudes of a certain group of people due to knowledge, experience, or expertise in any field. Opinion leaders, such as social media influencers, have shown to have a great impact on the purchase intention of followers who have reviewed certain products from their profiles [29]. Moreover, electronic word of mouth has been viewed to affect social media users in a positive or sometimes negative way. When social media influencers post a product review, this is considered as a type of electronic word of mouth where followers can share these posts between them and send them through any social media platforms.

2.2 Companies

Many companies and organizations utilize social media platforms as digital marketing tools that aid their progress due to their success and popularity with the consumers [2]. They represent important keys to socialize companies' and business as well as to achieve certain goals, such as building relationships with targeted audiences and enriching the brand's image and values [5].

Moreover, Companies take advantage of social media platforms and use them in posting, advertising or creating content with extremely low costs compared to traditional advertisements, such as TV, fliers, and billboards [25].

Companies use social media marketing to promote products and services through their accounts in order to provide their targeted audience with the required information as well as enhance their image and increase their sales [17].

With the enhancement of social media platforms, 200 million businesses are now marketing their products and creating Instagram business accounts to promote their brands on this application alone [20]. Companies are trying to promote their brands by posting pictures with messages reflecting the needs and personalities of their targeted audience. They are using social media platforms to interact with their followers and build a positive brand image. However, individuals keep skipping posts from brands and business accounts as it is obvious to be an advertisement [14].

When social media platforms started presenting a new type of marketing tool which was independent third-party endorser, businesses started shifting to influencers marketing as people were facing a massive amount of data on their social media platforms. On Instagram to be precise, users upload more than 1000 post every second, which means, around 100 million photos daily [1]. The amount of time that users give to each post and story is only a few seconds before scrolling down the feed.

2.3 Purchase Intention

Purchase intention is a dependent variable that depends on "An individual's conscious plan to make an effort to purchase a brand" [4]. The experiences customers go through can greatly affect their purchase intentions. If the customer is satisfied, his purchase intention will increase, while having a positive brand image, which will make his purchasing decision easier in future.

Companies are always trying to increase the purchase intention of their customers by providing and focusing on their needs. This is done in order to satisfy the customer and increase his purchase intention as well. Purchase intention is a very essential element in marketing, as marketers can analyze and study those results to evaluate the advertisements and create more tailored advertising campaigns which will attract more customers and increase marketers' sales and profit [6, 19].

3 Conclusion

The literature review concludes that social media influencers are positively influencing the purchase intention and customer behaviors. The framework of this study shows how social media influencers are affecting purchase intention through factors such as: trustworthiness, expertise, and attractiveness. The results of previous papers and literature reviews specifically showcase how consumers are affected by those factors positively or negatively. This study will contribute to the marketing and advertising field in GCC and Bahrain as GCC countries share a very similar culture. Moreover, this study is considered new evidence in Bahrain whereas no studies related to this area was conducted before. Companies, influencers, and marketing agencies depend on the results of studies such as these to set their marketing strategies and advertisements and evaluate their results.

References

1. Ahlgren, M.: 40+ Instagram Statistics & Facts for 2022. https://www.websiterating.com/research/instagram-statistics/ (4 Nov 2022)
2. Annisa, L., Er, M.: Impact of alignment between social media and business processes on SMEs' business process performance: a conceptual model. Procedia Comput. Sci. **161**, 1106–1113 (2019). https://doi.org/10.1016/j.procs.2019.11.222
3. Appel, G., Grewal, L., Hadi, R., Stephen, A.T.: The future of social media in marketing. J. Acad. Mark. Sci. **48**(1), 79–95 (2020). https://doi.org/10.1007/s11747-019-00695-1
4. Arachchige, E., Karunarathne, C.P., Thilini, W.A.: Advertising value constructs' implication on purchase intention: social media advertising. Manage. Dyn. Knowl. Econ. **10**(3), 287–303 (2022). https://doi.org/10.2478/mdke-2022-0019
5. Arora, A.S., Sanni, S.A.: Ten years of 'social media marketing' research in the journal of promotion management: research synthesis, emerging themes, and new directions. J. Promot. Manag. **25**(4), 476–499 (2019). https://doi.org/10.1080/10496491.2018.1448322
6. Barber, N., Kuo, P.-J., Bishop, M., Goodman Jr, R.: Measuring psychographics to assess purchase intention and willingness to pay. J. Consum. Mark. (2012)
7. Blystone, D.: Instagram: What It Is, Its History, and How the Popular App Works. https://www.investopedia.com/articles/investing/102615/story-instagram-rise-1-photo0sharing-app.asp#citation-21 (2022)
8. Chomiak-Orsa, I., Liszczyk, L.: Digital marketing as a digital revolution in marketing communication. Bus. Inf. (2022)
9. Dean, B.: Social Network Usage & Growth Statistics: How Many People Use Social Media in 2022? https://backlinko.com/social-media-users#how-many-people-use-social-media (2022)
10. Dixon, S.: Most popular social networks worldwide as of January 2022, ranked by number of monthly active users. https://www.statista.com/statistics/272014/global-social-networks-ranked-by-number-of-users/ (26 Jul 2022)
11. Dubey, P.: The effect of entrepreneurial characteristics on attitude and intention: an empirical study among technical undergraduates. J. Bus. Socio-economic Dev. (2022). https://doi.org/10.1108/JBSED-09-2021-0117
12. Kemp, S.: Digital 2021: Bahrain. https://datareportal.com/reports/digital-2021-bahrain (2021)
13. Kosack, E., Stone, M., Sanders, K., Aravopoulou, E., Biron, D., Brodsky, S., Al Dhaen, E.S., Mahmoud, M., Usacheva, A.: Information management in the early stages of the COVID-19 pandemic, The Bottom Line, **34**(1), 20–44 (2021). https://doi.org/10.1108/BL-09-2020-0062
14. Kumar, S., Srivastava, M., Prakash, V.: Challenges and opportunities for mutual fund investment and the role of industry 4.0 to recommend the individual for speculation. In: Nayyar, A., Naved, M., Rameshwar, R. (eds) New Horizons for Industry 4.0 in Modern Business. Contrib. Environ. Sci. Innovative Bus. Technol. Springer, Cham. (2023). https://doi.org/10.1007/978-3-031-20443-2_4
15. Lim, X.J., Mohd Radzol, A.R. bt Cheah, J.-H., Wong, M.W.: The impact of social media influencers on purchase intention and the mediation effect of customer attitude. Asian J. Bus. Res. **7**(2) (2017a). https://doi.org/10.14707/ajbr.170035
16. Lou, C., Yuan, S.: Influencer marketing: how message value and credibility affect consumer trust of branded content on social media. J. Interact. Advert. **19**(1), 58–73 (2019). https://doi.org/10.1080/15252019.2018.1533501
17. Marzouk, W.: Usage and effectiveness of social media marketing in Egypt: an organization perspective. Jordan J. Bus. Adm. **12**(1), 209–238 (2016)
18. Melović, B., Jocović, M., Dabić, M., Backović Vulić, T., Dudic, B.: The impact of digital transformation and digital marketing on the brand promotion, positioning and electronic business in Montenegro. Technol. Soc. **63** (2020). https://doi.org/10.1016/j.techsoc.2020.101425
19. Mohammed, U., Karagöl, E.T.: Remittances, institutional quality and investment in Sub-Saharan Africa. J. Bus. Socio-Econ. Dev. (2023). https://doi.org/10.1108/JBSED-07-2022-0077

20. Mohsin, M.: 10 Social media statistics you need to know in 2022. https://www.oberlo.com/blog/social-media-marketing-statistics#:~:text=to%20research%20products.-,71%25%20of%20consumers%20who%20have%20had%20a%20positive%20experience%20wi (2022)
21. Nations, D.: What is Social Media? Take a closer look at what social media is really all about. https://www.lifewire.com/what-is-social-media-explaining-the-big-trend-3486616#:~:text=Social%20media%20are%20web-based%20communication%20tools%20that%20enable,on%20a%20more%20specific%20subcategory%20of%20social%20media (2021)
22. Ortiz-Ospina, E.: Social media sites are used by more than two-thirds of all internet users. When did the rise of social media start and what are the largest sites today? https://ourworldindata.org/rise-of-social-media (2019)
23. Phua, J., Jin, S., Kim, J.: Gratifications of using Facebook, Twitter, Instagram, or Snapchat to follow brands: the moderating effect of social comparison, trust, tie strength, and network homophily on brand identification, brand engagement, brand commitment, and membership intention. Telematics Inform. **34**(1), 412–424 (2017). https://doi.org/10.1016/j.tele.2016.06.004
24. Saima, Khan, M.A.: Effect of social media influencer marketing on consumers' purchase intention and the mediating role of credibility. J. Promot. Manage. **27**(4), 503–523 (2020). https://doi.org/10.1080/10496491.2020.1851847
25. Sipior, J.C., Ward, B.T., Volonino, L.: Benefits and risks of social business: are companies considering e-discovery? Inf. Syst. Manag. **31**(4), 328–339 (2014). https://doi.org/10.1080/10580530.2014.958031
26. Statista: Most popular social networks worldwide as of January 2022, ranked by number of monthly active users. https://www.statista.com/statistics/272014/global-social-networks-ranked-by-number-of-users/ (2022)
27. Tapinfluence: What is influencer marketing? https://www.tapinfluence.com/blog-what-is-influencer-marketing/ (2017)
28. The Editors of Encyclopaedia Britannica. (n.d.). My Space—Website
29. Tobon, S., García-Madariaga, J.: The influence of opinion leaders' eWOM on online consumer decisions: a study on social influence (2021). https://doi.org/10.3390/jtaer
30. Torres, P., Augusto, M., Matos, M.: Antecedents and outcomes of digital influencer endorsement: an exploratory study. Psychol. Mark. **36**(12), 1267–1276 (2019). https://doi.org/10.1002/mar.21274
31. Vaibhava, D.: Digital marketing: a review. Int. J. Trend Sci. Res. Dev. 196–200 (2019). 2456-6470. https://www.ijtsrd.com/papers/ijtsrd23100.pdf
32. De Veirman, M., Cauberghe, V., Hudders, L.: Marketing through Instagram influencers: the impact of number of followers and product divergence on brand attitude. Int. J. Advert. **36**(5), 798–828 (2017). https://doi.org/10.1080/02650487.2017.1348035

Tomorrow's Jobs and Artificial Intelligence

Ismail Noori Mseer, Yasser M. Abolelmagd, and Wael F. M. Mobarak

Abstract According to James Brown, the executive director of the STEM Education Coalition in Washington, D.C., "STEM is the future of the economy. "There will be work there in the future." That claim is supported by data from the U.S. Bureau of Labor Statistics (BLS). It is anticipated that between 2012 and 2022, employment in professions related to STEM (science, technology, engineering, and mathematics) will increase to more than 9 million. That represents an increase in employment of roughly 1 million jobs from 2012 levels. This article examines roughly 100 professions from a list compiled by a group made up of representatives from various federal agencies, giving readers an overview of STEM work. The article's introduction provides a succinct overview of the subjects of computer science, engineering, mathematics, and the life and physical sciences. The second half of the report contains information on the STEM professions with the highest employment, as well as predicted job vacancies and growth. The benefits and difficulties of STEM job are covered in the third part. How to get ready for a profession in a STEM field is covered in the fourth section. The article's conclusion includes a list of references for additional information.

Keywords Artificial intelligence · Future jobs · Machine learning · Decentralization · Digitization

---

I. N. Mseer (✉)
Ahlia University, Manama, Bahrain

Y. M. Abolelmagd · W. F. M. Mobarak
Civil Engineering Department, College of Engineering, University of Business and Technology (UBT), Jeddah, Saudi Arabia
e-mail: Yasser@ubt.edu.sa

W. F. M. Mobarak
e-mail: W.fawzy@ubt.edu.sa

© The Author(s), under exclusive license to Springer Nature Switzerland AG 2024
A. Hamdan and E. S. Aldhaen (eds.), *Artificial Intelligence and Transforming Digital Marketing*, Studies in Systems, Decision and Control 487,
https://doi.org/10.1007/978-3-031-35828-9_12

1 Introduction

There will always be a need for workers who can synthesize information, think critically, and be adaptable in how they behave in various settings, according to Andrew Ng, head of the Stanford Artificial Intelligence Laboratory in Palo Alto, California. The jobs of today won't be the same as those of tomorrow though [1]. "Workers will probably need to pursue professions involving more cognitively challenging tasks that machines can't match," says Ito in his conclusion. According to Frey, those jobs also often call for higher education. "Education and technology are in a race," As increased productivity and efficiency enable exponential growth in some areas of the global economy, more people, companies, and governments are embracing artificial intelligence (AI) and machine learning (ML) [2]. The efficiency and productivity gap between industries and enterprises that have benefited from AI and ML and those that have not, however, is also expanding tremendously. As a result, there may be less chance for those at the bottom to catch up to the top and fall more behind.

Most of the world's largest economies only revealed their own AI efforts in 2017 and 2018, and most nations have only recently started to seriously consider their own AI futures. The others must imagine a world in which the few nations with the largest economies, the greatest AI-focused expertise, and a sizable pool of state resources that can be used to achieve AI dominance will control all aspects of technology [3], the economy, and the military. Future ramifications of a small number of nations possessing cutting-edge AI are significant. On the one hand, these highly developed nations may end up serving as de facto guardians of AI, ensuring that substantial resources be invested in its long-term growth. Furthermore, it is certain that these nations' top businesses will establish and keep a sizable lead in the global economy, giving them a significant competitive advantage. Future AI developments would almost probably be most advantageous to these nations' forces [4, 5], igniting a race for more advanced autonomous weapons systems and leading to perilous new ways of fighting wars.

2 Economic Paradigm

Multi-polarity is gradually displacing the well-known economic paradigm of a single dominant economic pole, a major technology, and a leading style of governance. Companies are being forced to deal with an array of paradigms, technology, and governance principles. The new shipping lanes are the information superhighways [6]. Storage facilities like warehouses and shipping containers are being replaced by cloud storage. The traditional methods of communication and commerce are being replaced by decentralization and digitization [7]. How will we move from a world dominated by things that cannot always be seen or felt to one where we are all familiar with and at ease with actual, tactile goods? The exponential growth of silicon is already generating significant disruptions in economies throughout the world through new global economic flows [8]. Physical products are losing ground to online platforms in importance. For instance, the largest money depository in the world, which operates entirely on cryptocurrency, has any physical structures or safes. Software, which is based on knowledge and procedures recorded by automation, powers all of them. Value chains that combine digital technology with more traditional, low-cost technologies [9], enable greater integration across products and services and take advantage of the emergence of independent global platforms for the exchange of goods and services will replace the globally optimized value chain, a familiar feature of the current phase of globalization. The most pressing short-term challenge in the era of ML is how to go from the existing economic paradigm [10], which is based on traditional manufacturing and fossil fuels, to a new model that is based on technology advancement that, until recently, was only possible in science fiction. How will we move from a world dominated by things that cannot always be seen or felt to one where we are all familiar with and at ease with actual, tactile goods? Virtual reality is already a reality for us in the cyber world, where it is also much desired by many of us. We are lured to this daring new world because its possibilities entice us. The AI universe that awaiting us will operate similarly [11].

However, it is possible—and even likely—that AI-powered robots could increasingly replace highly educated and skilled professionals, such as doctors, architects, and even computer programmers. Conventional wisdom suggests that AI will continue to benefit higher-skilled workers with a greater degree of flexibility, creativity, and strong problem-solving skills [12]. The connections between the technological revolution and other significant global developments, such as demographic shifts like aging and migration, climate change, and sustainable development, require much more consideration and study. Many of these subjects have either not even been mentioned or have just recently become the focus of substantial debate in international fora. An AI-dominated future might lead to the biggest concentration of resources and power the world has ever seen, rather than serving to reduce the level of global equality. Although it is apparent that the increasing capacity of AI to solve difficult problems on its own might radically alter our economies and societies [13], it will be many years before it is known how AI will affect a wide range of concerns. AI is like an amoeba that is constantly undergoing metamorphosis, changing its shape and adapting to its environment, even when solutions seem to be taking shape [14]. It is not difficult to see a future in which power, wealth, and technology become much more concentrated than they already are, even if the effects of the AI revolution on the world order have only just been given thought.

Future conflicts could affect more than just territory, resources, and populations; they could also affect how the human race develops in the future [15]. An AI-dominated future might lead to the biggest concentration of resources and power the world has ever seen, rather than serving to reduce the level of global equality. The world cannot afford to let nature run its course or wait for governments and companies to solve important concerns related to AI governance [16, 17], regulation, and rule of law when it may be more convenient for them to do so. The future direction of the new global economy can be significantly influenced by the multilateral system [18]. Therefore, it is imperative that multilateral organizations address the best way to shape and manage our global AI future through improved communication, resource allocation, and action.

How can we ensure that artificial intelligence doesn't go rogue and acts as a positive factor instead? International agreements and governance frameworks on the use of artificial intelligence (AI) are required in a variety of domains. For instance, the necessity for globally agreed-upon regulations [19], governing autonomous weaponry and the use of facial recognition technology to target minorities and stifle dissent is vital [20]. Eliminating prejudice in algorithms used in social media curation [21], credit allocation, criminal sentencing, and many other sectors should be a key subject of research and best practice dissemination. Unfortunately, we have already let the genie out of the bottle when it comes to the more general question of whether we will dominate our artificial creations or if they will rule us [22]. Through a straightforward thought experiment, Nick Bostrom suggested that the future development of AI could pose an existential threat to humans in his book [23] Superintelligence: Paths, Dangers, Strategies. A self-improving AI has been tasked with managing a paper clip plant. This AI is able to learn from experience and automatically enhance its performance. Making as many paper clips as it can is its task. As it develops superintelligence, it concludes that humans are a hindrance to achieving their goal and ends up annihilating us all. In a more lyrical take on that story, Elon Musk had a robot that picked strawberries declare that humans stand in the way of "strawberry fields forever [24].

We don't realize that we've already built such systems, though. Even while they are not yet totally independent from their human creators or super intelligent, they are already misbehaving in the manner that Bostrom and Musk predicted [25]. And most of our efforts to control them are failing. It is crucial to comprehend how such systems operate in order to explain why such is the case. I'll begin with a straightforward example. I once owned a piggy bank that sorted coins when I was younger. I enjoyed dropping in a clutch of spare change and witnessing the coins magically sort themselves into columns according to size after sliding through transparent tubes. When I was a little older, I understood that vending machines operated similarly and that you could trick one by inserting a foreign coin of the proper size or even a slug of metal that had been punched out of an electrical junction box [26]. Actually, the machine knew nothing about the worth of money. It was merely a device designed to trip a counter when the proper size and weight disk fell through a slot [27].

You may learn quite a deal about systems like Google search, social media newsfeed algorithms, email spam filtering, fraud detection, facial recognition, and the most recent developments in cybersecurity if you know how that piggy bank or coin-operated vending machine operates. These devices are sorting devices [28]. A system is created to identify characteristics in an incoming data set or stream and to sort the data in some way. (Coins come in a variety of weights and sizes. In addition to having sources, click through rates, and countless other attributes, emails, tweets, and news articles all contain keywords. A picture can be categorized as Tim O'Reilly or not Tim O'Reilly, or a cat or not a cat [29]. People attempt to spoof these systems, just like my teenage friends and I did with vending machines, and the mechanism designers are incorporating more and more data properties to reduce errors [30].

3 Conclusion

Vending machines are easy to use. There are only so many ways to fake currency [31], and changes to it happen seldom. But because content is infinitely changeable, creating new techniques to account for every fresh subject, fresh material source, and fresh attack is a Sisyphean process. introducing machine learning [32]. In a conventional method, the programmer specifies the attributes to be inspected, the permissible values, and the action to be performed when creating an algorithmic system for recognizing and sorting data. (A feature of the data is frequently referred to as the union of an attribute and its value.) To develop a model of what good and bad look like, a system is exposed to a huge number of samples of both excellent and terrible data using a machine-learning approach. The programmer simply knows that the machine-learning model produces results that seem to match or outperform human judgment when applied to a test data set; the programmer may not always be fully aware of the aspects of the data the machine-learning model is relying on. The system is then given access to data from the real world. The system can be made to keep learning beyond the initial training. You've taken part in a portion of that training process if you've ever used the face recognition tools in Apple or Google's photo programs to locate images of you, your friends, or your family. After giving names to a few faces, you are then presented with a collection of images that the algorithmic system is reasonably certain are of the same face, as well as some images with a lower confidence level, and asked to affirm or deny their identity. The application grows better the more times you correct its assumptions. As a result of my efforts, my photo application has become increasingly accurate at telling me apart from my brothers and occasionally, even from my daughters. It can identify the same person from their early years to their old age.

References

1. Robinson, G.L.: 6. How the future of work impacts the workforce of technical organizations. In: Wingard, J., Farrugia, C. (eds.) The Great Skills Gap: Optimizing Talent for the Future of Work, pp. 59–66. Stanford University Press, Redwood City (2021). https://doi.org/10.1515/9781503628076-010
2. Butcher, J., Rose-Adams, J.: Part-time learners in open and distance learning: revisiting the critical importance of choice, flexibility and employability. Open Learning: J. Open Distance e-Learn. 30(2), 127–137 (2015). https://doi.org/10.1080/02680513.2015.1055719
3. Davenport, T., Guha, A., Grewal, D., et al.: How artificial intelligence will change the future of marketing. J. Acad. Mark. Sci. 48, 24–42 (2020). https://doi.org/10.1007/s11747-019-00696-0
4. Brynjolfsson, E., Rock, D., Syverson, C.: Artificial intelligence and the modern productivity paradox: a clash of expectations and statistics. In: NBER Working Paper 22401. National Bureau of Economic Research, Cambridge, MA (2017)
5. Kurdy, D.M., Al-Malkawi, H.-A.N., Rizwan, S.: The impact of remote working on employee productivity during COVID-19 in the UAE: the moderating role of job level. J. Bus. Socio-Econ. Dev. (2023). https://doi.org/10.1108/JBSED-09-2022-0104

6. Tennant, J.P., Waldner, F., Jacques, D.C., Masuzzo, P., Collister, L.B., Hartgerink, C.H.:. The academic, economic, and societal impacts of Open Access: an evidence-based review. F1000Research **5**, 632 (2016). https://doi.org/10.12688/f1000research.8460.3
7. The World Bank: Centralized platforms to decentralized networks, enabling equitable market access and trust. https://www.worldbank.org/en/events/2022/04/20/-centralized-platforms-to-decentralized-networks-enabling-equitable-market-access-and-trust (May 4, 2022)
8. McKinsey Global Institute: Disruptive technologies: advances that will transform life, business, and the global economy. https://www.mckinsey.com/~/media/mckinsey/business%20f unctions/mckinsey%20digital/our%20insights/disruptive%20technologies/mgi_disruptive_t echnologies_full_report (May 2013)
9. Bárcia de Mattos, F., Eisenbraun, J., Kucera, D., Rossi, A.: Disruption in the apparel industry? Automation, employment, and reshoring. Int. Labour Rev. **160**(4), 519–536 (2021)
10. Bosch, X.: An open challenge. Open access and the challenges for scientific publishing. EMBO Rep. **9**(5), 404–408 (2008). https://doi.org/10.1038/embor.2008.60
11. Sapoval, N., Aghazadeh, A., Nute, M.G., et al.: Current progress and open challenges for applying deep learning across the biosciences. Nat. Commun. **13**, 1728 (2022). https://doi.org/10.1038/s41467-022-29268-7
12. Bloom, N., Garicano, L., Sadun, R., Van Reenen, J.: The distinct effects of information technology and communication technology on firm organization. Manage. Sci. **60**(12), 2859–2885 (2014)
13. Bickley, S.J., Chan, H.F., Torgler, B.: Artificial intelligence in the field of economics. Scientometrics **127**, 2055–2084 (2022). https://doi.org/10.1007/s11192-022-04294-w
14. Hassani, H., Silva, E.S., Unger, S., Mazinani, M.T., Mac Feely, S.: Artificial intelligence (AI) or intelligence augmentation (IA): what is the future?" AI **1**(2), 143–155 (2020). https://doi.org/10.3390/ai1020008
15. Stewart, F.: Root causes of violent conflict in developing countries. BMJ (Clin. Res. Ed.) **324**(7333), 342–345 (2002). https://doi.org/10.1136/bmj.324.7333.342
16. Erman, E., Furendal, M.: The global governance of artificial intelligence: some normative concerns. Moral Philos. Politics **9**(2), 267–291 (2022). https://doi.org/10.1515/mopp-2020-0046
17. Alshaikh, H.: Exploratory study on judicial rulings issued by Bahraini Courts in litigations related to the lease ending with ownership (Ijarah Muntahiya Beltamleek) mode of financing. Int. J. Bus. Ethics Gov. **5**(2), 11–43 (2022). Available at: https://www.ijbeg.com/index.php/1/article/view/104
18. Piper, N., Foley, L.: Global partnerships in governing labour migration: the uneasy relationship between the ILO and IOM in the promotion of decent work for migrants. Global Public Policy and Governance, 1 (2021). https://doi.org/10.1007/s43508-021-00022-x
19. Yudkowsky, E.: Artificial intelligence as a positive and negative factor in global risk. In: Bostrom, N., Ćirković, M.M. (eds.) Global Catastrophic Risks, pp. 308–345. Oxford University Press, New York (2008)
20. Renda, A., Arroyo, J., Fanni, R., Laurer, M., Sipiczki, A., Yeung, T., Maridis, G., Fernandes, M., Endrodi, G., Milio, S., Devenyi, V., Georgiev, S., de Pierrefeu, G.: Study to Support an Impact Assessment of Regulatory Requirements for Artificial Intelligence in Europe, Final Report (D5) (2021)
21. Harini, V.: A.I. 'bias' could create disastrous results, experts are working out how to fight it. https://www.cnbc.com/2018/12/14/ai-bias-how-to-fight-prejudice-in-artificial-intelligence.html (14 Dec 2018)
22. Tai, M.C.: The impact of artificial intelligence on human society and bioethics. Tzu Chi Med. J. **32**(4), 339–343 (2020). https://doi.org/10.4103/tcmj.tcmj_71_20
23. Oxford University Press's Academic Insights for the Thinking World: Nick Bostrom on Artificial Intelligence Superintelligence: Paths, Dangers, Strategies (Sept 8, 2014)
24. McLay, R.: Managing the Rise of Artificial Intelligence. https://tech.humanrights.gov.au/sites/default/files/inline-files/100%20-%20Ron%20McLay.pdf (2017)

25. Khatchadourian, R.: Will artificial intelligence bring us utopia or destruction? THE DOOMSDAY INVENTION A REPORTER AT LARGE, The New Yorker. https://www.newyorker.com/magazine/2015/11/23/doomsday-invention-artificial-intelligence-nick-bostrom (23 Nov 2015)
26. Kubik, C., Knauer, S.M., Groche, P.: Smart sheet metal forming: importance of data acquisition, preprocessing and transformation on the performance of a multiclass support vector machine for predicting wear states during blanking. J. Intell. Manuf. **33**, 259–282 (2022). https://doi.org/10.1007/s10845-021-01789-w
27. Adeosun, O.T., Owolabi, T.: Owner-manager businesses and youth employee perceptions. J. Bus. Socio-economic Dev. **3**(2), 97–117 (2023). https://doi.org/10.1108/JBSED-03-2021-0032
28. Bharadwaj, R.:, Artificial intelligence for ATMs—6 Current Applications Avatar. https://emerj.com/ai-sector-overviews/artificial-intelligence-for-atms-6-current-applications/ (November 22, 2019)
29. Spark, D., O'Reilly, T.: A Machine Can Learn Just Like a Child. https://insights.dice.com/2011/08/15/tim-oreilly-a-machine-can-learn-just-like-a-child/ (15 Aug 2011)
30. Mathew, D., Brintha, N.C., Jappes, J.T.W.: Artificial intelligence powered automation for industry 4.0. In: Nayyar, A., Naved, M., Rameshwar, R. (eds) New Horizons for Industry 4.0 in Modern Business. Contrib. Environ. Sci. Innovative Bus. Technol. Springer, Cham (2023). https://doi.org/10.1007/978-3-031-20443-2_1
31. Petropoulos, G.: The dark side of artificial intelligence: manipulation of human behavior, Transparency over systems and algorithms, rules and public awareness are needed to address potential danger of manipulation by artificial intelligence. https://www.bruegel.org/blog-post/dark-side-artificial-intelligence-manipulation-human-behaviour (02 Feb 2022)
32. Hyvärinen, N.: Business Security, Videos, Artificial Intelligence, and Machine Learning in Cyber Security. https://blog.f-secure.com/artificial-intelligence-and-machine-learning-in-cyber-security/ (12 Apr 2018)

Artificial Intelligence and Security Challenges

Ismail Noori Mseer and Syed Muqtar Ahmed

Abstract Big data processing, vast computing power, information technology, improved machine learning (ML) and deep learning (DL) algorithms are driving the recent growth in AI technologies. With more conventional methods, Google would not have been able to reduce its field device management costs by 40% as much as it has by deploying deep-mind AI technologies. The energy sector can benefit from AI technology by utilizing the expanding opportunities that result from the use of the Internet of Things (IoT) and the incorporation of renewable energy sources. Super-computers, power electronics, cyber technologies, information, and bi-directional connectivity between the control center and equipment are only a few of the sophisticated infrastructures available to the smart energy sector. The infrastructures of the current electricity system are too old, ineffective, outdated, unreliable, and do not offer enough protection from fault circumstances. But energy production, distribution strategy, and financial sustainability are crucial for the world economy. The integration of renewable energy sources was not intended to be managed by the conventional power system (RES). Meeting the fluctuating demands of the power system is made more difficult by changes in the characteristics of RES (such as wind, solar, geothermal, and hydrogen). The energy sector is undergoing a change thanks to recent developments in AI technologies, such as machine learning, deep learning, IoT, big data, etc. Many nations have implemented AI technology to carry out many types of jobs, including managing, predicting, and effective power system operations. Photovoltaic (PV) systems may be controlled effectively by inverters thanks to, which also improves the ability to track power points. Artificial maximum power point tracking (MPPT) techniques are efficient and can improve performance compared to conventional MPPT techniques. Due of its simplicity and speed of calculation, particle swarm optimization for MPPT is preferred by swarm intelligence classes Predictive technologies are frequently used to anticipate load demand, electricity costs, generation from RES (such as wind, hydro, solar, and geothermal energy), as

I. N. Mseer (✉)
Ahlia University, Manama, Bahrain
e-mail: imseer@ahlia.edu.bh

S. M. Ahmed
Jeddah College of Engineering, University of Business and Technology, Jeddah, Saudi Arabia

© The Author(s), under exclusive license to Springer Nature Switzerland AG 2024
A. Hamdan and E. S. Aldhaen (eds.), *Artificial Intelligence and Transforming Digital Marketing*, Studies in Systems, Decision and Control 487,
https://doi.org/10.1007/978-3-031-35828-9_13

well as fossil fuels (such as oil, natural gas, and coal). Probabilistic forecasting (forecasting future events, for example) and non-probabilistic forecasting (forecasting fuel purchase management, generation planning, distribution scheduling, various forms of investment programs, maintenance schedules, and security purposes) are both possible.

Keywords Maximum power point tracking · Machine learning · Deep learning · Internet of Things

1 Introduction

Russell stated that "there are several key breakthroughs that have to occur, and those could come very swiftly." He continued, "It's very, very hard to forecast when these conceptual breakthroughs are going to happen," citing British physicist Ernest Rutherford's description of the quick transformational effect of nuclear fission (atom splitting) in 1917 [1]. But if they do, he highlighted how important preparation is. This necessitates starting or continuous conversations regarding the moral application of AGI and if regulation is necessary. That entails attempting to eradicate data bias, which now poses a serious threat to AI by distorting algorithms. That requires creating and improving security systems that can control technology. Having the humility to understand that just because we can, doesn't mean we should, is also a requirement. "Most AGI researchers predict the development of AGI within a few decades; if we enter this situation unprepared, it will likely be the largest error in human history. In his TED Talk, Tegmark warned that this might lead to a ruthless global dictatorship marked by never-before-seen levels of inequality, surveillance, pain, and possibly even human extinction. But if we take care, we may arrive at a great future where everyone is better off, both the poor and the wealthy as well as everyone being healthy and free to pursue their ambitions.

Over the past few decades, technology has advanced rapidly before our eyes, and artificial intelligence is one of the things that has everyone either fascinated or even a little afraid. For those who don't know, one of the numerous subfields of computer science is artificial intelligence. In other words, it is a simulation of human intelligence, and these machines are trained to think and act like people. It is used to create intelligent computers capable of carrying out activities that would ordinarily need human intelligence. Some characteristics that an "AI" would have include the ability to recognize speech, make decisions, and see the world. The ability of these computers to learn, reason, and perceive like humans has long been the core objective of artificial intelligence. AI is being employed more frequently than one might imagine in a variety of global businesses, including finance and health care. One might interact with an AI on a regular basis, as evidenced by intelligent assistants like Alexa and Siri, self-driving cars, conversational bots, spam filters for email, and Netflix suggestions.

Artificial Intelligence and Security Challenges

There are numerous AI-powered language learning platforms and applications that enable individuals to work and learn at their own pace, [2] focusing on the topics they find the most challenging, involving them in various activities, appealing to their interests to motivate them to keep learning, and taking into account elements like cultural background. Platforms for AI language learning can immediately point out mistakes [3] and provide various solutions to avoid doing them in the future. It can be stressful to have to wait weeks for the results of a test you worked hard on. It becomes more difficult to recall how or why you made the inevitable blunders after that point, but this kind of technology offers immediate feedback. AI enables students to study at their own pace, from any location in the globe, with their own goals, and a tailored study plan that adapts to their needs and availability. However, keep in mind that the greatest method to learn a language is through practical experiences, such as spending time in a Spanish-speaking nation to learn about its language as well as its culture and history. Madrid, Spain's capital city, is a great place to start learning Spanish. Online learning and artificial intelligence shouldn't ever replace traditional learning; as previously stated, they are tools, [4] not replacements, and students gain just as much from attending in-person classes as they do when they use AI language learning platforms.

To reinforce each new concept you learn, I therefore advise you to incorporate a language learning app into your Spanish courses in Madrid [5]. You only need to have your phone with you in order to use it, so it is both quick and somewhat practical. GPT-3: A modern natural language system has acquired more knowledge than only coding. GPT-3 [6], or Generative Pre-trained Transformer 3, is the third iteration of the GPT-n series and was developed by OpenAI. It employs deep learning to produce text with a language structure resembling that of a human. The system, which took months to develop, studied tens of thousands of digital books, Wikipedia, social media, and the internet as a whole to learn how to produce new natural language on its own. This system can now, among other things, produce tweets, email summaries, respond to trivia questions, and translate languages [7]. Contrary to popular belief, artificial intelligence is not coming to replace people, take our jobs, or otherwise eliminate us. With the help of this tool, content producers may design more effective content strategies and increase the caliber of their output.

2 Wide Variety of Content

Because of technology that can produce a wide variety of content on a particular subject, authors can save a significant amount of time and money that would otherwise be used to research the subject thoroughly before considering writing an article on it. Like in any other occupation, there are technologies that make people more productive, and GPT-3 is the one that content producers should start utilizing [8]. This system, like many others in the field of artificial intelligence, has a long way to go before it can fully replace humans. Can you imagine how much time you would save by utilizing this program to generate merely the bare bones of your text?

Yes, you will need to spend some time curating, refining, and generally making the information GPT-3 has just made more human-like. What we refer to as "writer's block" is another issue that the AI may profit from. It is well known that the reason for this creative block is not a lack of writing ability or commitment issues, but rather the sheer volume of information we are required to produce in a short period of time. No matter how long an essay is, commencing is typically the most difficult portion of writing it. Therefore, why not give GPT-3 a title or primary theme and explore all the alternatives it provides? I can guarantee that you'll be delighted and find it useful.

Once more, one of the biggest benefits of employing this technology is the time you may save when performing research because GPT-3 can gather data about what we're writing. Making ensuring the topic you're providing [9] is as clear and specific as possible is crucial if you use this option. "Machine Learning: Bane or Blessing for Mankind?" is the title of a blog entry. I pointed out that the eminent theoretical physicist Stephen Hawking, together with Stuart Russell, Max Tegmark, and Frank Wilczek, advise proceeding slowly in the development of artificial intelligence (AI), [10] particularly in the field of autonomous weapon systems. But Hawking and his colleagues are aware that there is no turning back the AI genie now that it has been let out of the bottle. [11] Ron Neale replies in response to Hawking's worries, Such a caution regarding the deployment of AI and its derivative intelligent machines (IMs), especially in the domain of military application, would be relevant. But what if instant messaging is actually just a new branch on the evolutionary tree that has brought us from the first Protists to where we are now? Artificial and machine intelligence: Reality or Science Fiction? The EE Times, May 13, 2014 Neale is doubtful that a system like Skynet would ever emerge, despite finding the idea of such a system terrifying (Skynet is the AI system in "Terminator" that takes control and begins eradicating humanity). He clarifies: "Do not be alarmed, for in my opinion, IMs cannot exist without a special evolutionary key [—Synergistic Evolution (SE)—] and it is that element of evolution that reveals why it might never happen. A species must get assistance from another species in order for synergistic evolution (SE) to take place. This is distinct from serving as a food source, in which the existence of a lower-ranking species serves as the food or fuel that enables others higher up the food chain to live and evolve. A species can more easily find food to exist and evolve in situations where animals like dogs or horses that exist at the same time, on a separate branch, are present. The closest analog to SE would be a species variation like selective breeding (unnatural selection), where human intervention is utilized to produce a trait, such more meat or milk in cattle or in hunting animals like dogs or horses. In any wild speculation, I believe that three possibilities must be taken into account: the first is the evolution of some incredibly clever tools, weapons, and body parts that become a crucial part of the human species tree; the second is the creation of a new branch on the evolutionary tree; and the third is the extension of the human branch.

Sincere to say, I'm not sure the parallel of biological evolution is a valid one. Biological evolution essentially modifies a species' organic properties such that advantageous features can be passed on to succeeding generations [12]. Because

Artificial Intelligence and Security Challenges

they aren't organic, two of Neale's possibilities aren't genuinely biological evolution (they fit much neater in the transhumanist framework). The only choice that could lead to evolution was his "new branch of the tree"; however, it wouldn't be a new branch; rather, it would be a completely new tree. No matter how artificial intelligence evolves in the coming years, almost all commentators concur that the world will change drastically as a result [13]. In spite of the fact that technology is exploding this decade into a world of the Internet of Things and the propulsion into artificial intelligence, Bill Gates insisted during a speech at the American Enterprise Institute that "the mindset[s] of the government and people have not adjusted to view the future." Victoria Wagner Ross, San Diego Technology Examiner, 14 March 2014 "Bill Gates defines the mindset of A.I. for occupations in the future" [14].

The quantity of jobs that AI systems are likely to replace is the biggest concern for many observers, including Gates. According to Mark van Rijmenam, "Artificial intelligence has immense potential [15]. In fact, Oxford University research from 2013 predicted that in the near future, AI might replace roughly half of all occupations in the United States." Is AI on the verge of forever altering how we conduct business? Smart Data Collective, March 8, 2014 Ross also uses the research from Oxford University. She says: According to Carl Benedikt, there are 702 jobs that will be impacted by automation in the future of artificial intelligence [16] and robots. Frey, an economist with a Ph. D., was taken aback by how quickly the computer replaced the loan officer. A 98% likelihood of replacement was anticipated for the loan officer. Journalists were a safer occupation with only an 11% probability. The probability was lowest for surgeons and elementary school instructors. Gates allegedly said that "there are a few of decades to re-set the thinking and prepare for the next occupation." Ross quotes Gates on this. What specific jobs should students be preparing themselves to fill? Whatever new jobs are created, according to Gates, the highest-paying and most secure ones will necessitate a solid background in science, technology, engineering, and mathematics (STEM). Ross says, "In the future, opportunity will meet preparation, and Gates educates the public to be ready to accept the exciting new world of artificial intelligence" [17]. The majority of the finest occupations that will be created will necessiate close computer and human interaction [18]. Thankfully, we are raising a generation that has already experienced this kind of cooperation.

3 Conclusion

According to Zachary John, "AI has allowed us humans to create robots that are perfectly suited to fit into our daily lives, [19] suggesting speedier routes to work, offering TV series we might like, even telling us jokes when we're feeling bad." Guardian Liberty Voice, "Robots: The Possibilities of Artificial Intelligence," 22 March 2014 "AI is only getting better, as computational intelligence approaches keep on improving, becoming more precise and faster due to significant advances in processing speeds," Kevin Curran, a technical specialist at the Institute of Electrical and Electronics Engineers (IEEE), told Lee Bell. The Inquirer, 14 March 2014) AI will play an important part in our future, [20] just don't anticipate robot butlers [21]. Like many others, Curran thinks AI systems will continue to replace humans in positions that are currently held by machines, especially "humans executing monotonous automated chores." According to Aki Ito, "Artificial intelligence has entered the American workplace, giving rise to technologies that mimic human judgements that were too subtle and complex [22] to be reduced into instructions for a machine. Engineering teams no longer have to explicitly type out every command thanks to algorithms that "learn" from prior examples. According to Bloomberg Business-Week on March 12, 2014, "Your Job Taught to Machines Puts Half U.S. Work at Risk." Like the authors previously mentioned, Ito cites the Oxford University study as evidence that the workforce of the future need a significant makeover [23]. The worldwide workforce would need to change, according to Frey, the study's co-author, who told Ito. These shifts have occurred in the past, said Frey. This time, technological transformation is occurring considerably faster and might have a bigger impact on a wider range of jobs. Will there be enough excellent employment to keep the global economy expanding? is arguably the biggest unsolved question. Because customers are essential to economic progress, AI systems cannot function as consumers.

References

1. The University of Manchester, 2 November 2017, Rutherford's Legacy—the birth of nuclear physics in Manchester. https://www.manchester.ac.uk/discover/news/rutherfords-legacy--the-birth-of-nuclear-physics-in-manchester/
2. Chen, X., Zou, D., Xie, H., Cheng, G.: Twenty years of personalized language learning: topic modeling and knowledge mapping. Educ. Technol. Soc. **24**(1), 205–222. https://www.jstor.org/stable/26977868
3. Manser Payne, E.H., Dahl, A.J., Peltier, J., Peltier, J.: Digital servitization value co-creation framework for AI services: a research agenda for digital transformation in financial service ecosystems. J. Res. Indian Med. **15**(2), 200–222 (2021)
4. Fernández-Martínez, C., Hernán-Losada, I., Fernández, A.: Early introduction of AI in Spanish middle schools a motivational study. Künstl. Intell. **35**, 163–170 (2021). https://doi.org/10.1007/s13218-021-00735-5
5. Salas-Pilco, S.Z., Yang, Y.: Artificial intelligence applications in Latin American higher education: a systematic review. Int. J. Educ. Technol. High. Educ. **19**, 21 (2022). https://doi.org/10.1186/s41239-022-00326-w
6. Yamins, D.L., DiCarlo, J.J.: Using goal-driven deep learning models to understand sensory cortex. Nat. Neurosci. **19**(3), 356 (2016)
7. Johnson, S.: A.I. Is Mastering Language. Should We Trust What It Says? https://www.nytimes.com/2022/04/15/magazine/ai-language.html (2022)
8. Korngiebel, D.M., Mooney, S.D.: Considering the possibilities and pitfalls of generative pre-trained transformer 3 (GPT-3) in healthcare delivery. NPJ Digit. Med. **4**, 93 (2021). https://doi.org/10.1038/s41746-021-00464-x
9. Gpt Generative Pretrained Transformer, Almira Osmanovic Thunström, Steinn Steingrimsson. Can
10. GPT-3 write an academic paper on itself, with minimal human input?. 2022. ffhal-03701250f
11. Hawking, S., Tegmark, M., Russell, S.: Transcending complacency on super intelligent machines. https://www.huffpost.com/entry/artificial-intelligence_b_5174265.(2014)

12. O'Reilly, T.: We have already let the genie out of the bottle. https://www.rockefellerfoundation.org/blog/we-have-already-let-the-genie-out-of-the-bottle/
13. Gess, N.: The "Tropological Nature" of the Poet in Müller and Benn. In: Primitive Thinking: Figuring Alterity in German Modernity, Chap. 7, pp. 205–235. De Gruyter, Berlin (2022). https://doi.org/10.1515/9783110695090-007
14. Roman, M., ch3n81, Alice.: Play I: unfolding of a concept: information. In: Hovestadt, L., Bühlmann, V. (eds.) Play Among Books: A Symposium on Architecture and Information Spelt in Atom-Letters, pp. 55–192. Birkhäuser, Berlin (2021). https://doi.org/10.1515/9783035624052-004
15. Fleming, P.: Robots and organization studies: why robots might not want to steal your job. Organ. Stud. **40**(1), 23–38 (2019). https://doi.org/10.1177/0170840618765568
16. Bhardwaj, A., Kishore, S., Pandey, D.K.: Artificial intelligence in biological sciences. Life **12**(9), 1430. https://doi.org/10.3390/life12091430
17. Illéssy, M., Huszár, Á., Makó, C.: Technological development and the labour market: how susceptible are jobs to automation in Hungary in the international comparison? Societies **11**(3), 93 (2021). https://doi.org/10.3390/soc11030093
18. Ernst, E., Merola, R., Samaan, D.: Economics of artificial intelligence: implications for the future of work. IZA J. Labor Pol. **9**(1). https://doi.org/10.2478/izajolp-2019-0004
19. Kanade, V.: What Is HCI (Human-Computer Interaction)? meaning, importance, examples, and goals. https://www.spiceworks.com/tech/artificial-intelligence/articles/what-is-hci (2022)
20. Tai, M.C.: The impact of artificial intelligence on human society and bioethics. Tzu Chi Med. J. **32**(4), 339–343 (2020). https://doi.org/10.4103/tcmj.tcmj_71_20
21. DeAngelis, S.: Artificial intelligence: ascendant but not transcendent. https://enterrasolutions.com/artificial-intelligence-ascendant-transcendent/ (2014)
22. Torres, É.P.: Opinion How AI could accidentally extinguish humankind. https://www.washingtonpost.com/opinions/2022/08/31/artificial-intelligence-worst-case-scenario-extinction/ (2022)
23. Brynjolfsson, E.: The turing trap: the promise & peril of human-like artificial intelligence. https://digitaleconomy.stanford.edu/news/the-turing-trap-the-promise-peril-of-human-like-artificial-intelligence/ (2022)

Modern Services Quality and Its Impact on the Satisfaction of the Trade Shows in Jordan

Mustafa S. Al-Shaikh and Feras Alfukaha

Abstract This study objective is to ascertain the service quality provided by Conex for exhibitions through dimension of trust, reliability, responsiveness and efficiency, design and price. These variables represented the independent, and these variables were examined with how affected the dependent variable, which a measure of trade shows participants satisfaction. The size of the population was (350) companies. A total of 240 questionnaire were distributed, the size of the study sample analyzed was (199) companies, in this study we used simple random sampling method to communicate with the respondents. The results showed that the impact of the is statistically significant service quality of all the dimensions covered in this study, which is provided by Conex Company, in addition to the price on their customer's satisfaction, and it has a strong positive connection to the dependent variable, with the contribution of (61.5%). Study recommends that Conex Company should develop the capabilities of its employees to offer the best services possible satisfy their customers, as well as the continuity of innovation and development in the designs of the participant's booth to meet their satisfaction.

Keywords Modern quality of services · Trade shows · Jordan

1 Introduction

The study of [31] mention that the satisfaction of professionalism and convenience of the visitors to the corporate exhibition hall led to a positive corporate image in the future.s A trade show's success is dependent on the close working together between organizers together with exhibitors potential visitors [4, 8]. Conex seeks to develop its services through achieving the highest standards of accuracy, efficiency

---

M. S. Al-Shaikh (✉)
Zarqa University, Zarqa, Jordan
e-mail: allamh3@hotmail.com

F. Alfukaha
Conex, Amman, Jordan

© The Author(s), under exclusive license to Springer Nature Switzerland AG 2024
A. Hamdan and E. S. Aldhaen (eds.), *Artificial Intelligence and Transforming Digital Marketing*, Studies in Systems, Decision and Control 487,
https://doi.org/10.1007/978-3-031-35828-9_14

and credibility of treatments and focusing on its qualities through its designs and costs within the insights of the concept of meeting and exceeding customer expectations, encouraging them to submit their feedback about the provided services on one hand and taking into consideration their notes on the other.

2 The Problem of Research

The proportion between service quality dimensions could provide appositive perception of trade shows participants satisfaction in the light of perceptions that service companies adopt in organizing and preparing tradeshow exhibitions in drawing their marketing policies that ultimately have profit maximization as their goal, which is pushing the trade shows companies to apply modern methods to guarantee their service quality and achieve their overarching goals. Therefore, the issue with the study clearly demonstrates the existence of weakness of the Conex for Exhibitions during presenting services, also the lack of adoption of Conex for Exhibition management of an integrated program of quality dimensions. According to the problem statement the following main question is designed: How does this affect Conex provided service excellence on the satisfaction of exhibitors participating in trade shows? The subsequent questions are derived from this basic query are expected: What is the impact of Conex trust, reliability, responsiveness and efficiency, prices, and designs factors on the satisfaction of exhibitors participating in trade shows?

3 Research Importance

The practical significance of the article arises based on the significance of the quality of services provided by Conex for exhibitions to improve its performance in a way that allows it to fulfill its delegations to achieve the participating companies' satisfaction in trade shows. And the scientific importance arises from the vast knowledge development on all basis which have led to the importance of economical and organizational necessities that push Conex Company to think of modern strategies related to services qualities.

4 Research Objectives

The study's objectives specify the impact of trust, reliability, prices, and design on the satisfaction of participating exhibitors in trade shows.

5 Model

The research design developed according to the study variables and after returning to the study of [30, 32].

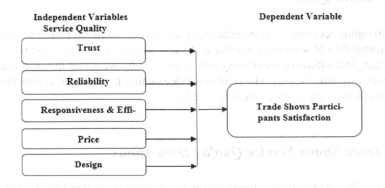

6 Research Hypotheses

The main hypothesis: H0: service Conex's quality has no statistically significant impact (p > 0.05) on trade show attendees' satisfaction. The following supporting hypotheses are deduced from this core hypothesis:

- **H01**: The trust component has no statistically significant impact (a > 0.05).
- **H02**: The dependability factor has no statistically significant impact (a > 0.05).
- **H03**: There is no statistically significant impact ($\alpha \leq 0.05$) of responsiveness and efficiency factor.
- **H04**: There is no statistically significant impact ($\alpha \leq 0.05$) of prices factor.
- **H05**: The designs element has no statistically significant impact (p > 0.05).

7 Theoretical Framework

A trade show is one of several marketing strategies that businesses can use to advertise to attendees and potential customers. Recent trade fairs have established themselves as one of the most effective marketing and promotional tools, and most industrialized nations include them in their corporate strategies. The marketing management operations are connected to the consumer distribution of goods and services. According to Jerome McCarthy's classification, the marketing mix has been classified into four elements: product, place, price, and promotion known as 4 Ps [1, 34]. The promotion activities have been categorized in the literature as personal selling, advertising, sales promotion, and public relations. Trade exhibitions are seen as a form of sales promotion.

7.1 Service Quality

Any intangible act or performance that one party renders to another without resulting in the ownership of something is referred to as a service [15]. Due to the fact that it is an intangible offer, the other party will evaluate it by weighing the pleasure they will receive against the cost. The level of service quality is attained if clients feel that the services have met their expectations [13].

7.2 Trade Shows Service Quality Dimensions

- **Trust**: The level of benevolence is how much the service provider is thought to care about the consumer beyond just making a profit. Integrity is the perception by the client that the service provider honors commitments, upholds moral standards, and gives accurate information [29].
- **Reliability**: Is "the ability of an organization to accurately achieve its services in the proper time and according to the promises it has made to its clients" [10, 22]. In case of trade show field, reliability means that the exhibition company has to be able to perform the promised service dependably and precisely.
- **Responsiveness and Efficiency**: Is "the ability to respond to customer requirements timely and flexibly" [20]. In the exhibits sector, responsiveness is defined as the capacity and willingness to assist attendees and provide rapid attention.
- **Price**: As an explanation of a concept connected to customer satisfaction is "assessment of the possible outcomes compared to the costs paid by the consumers" [9, 24].
- **Design**: The design of the exhibition booth can be seen from three sides, location, the comfortable layout, and the contents.

7.3 Trade Shows Participants Satisfaction

Customer satisfaction can be considered as among the most important aspect that may be helpful the companies to intensify their sales, and in the trade show market it's a good indicator of the company competition position [5]. Customer satisfaction is "the consumer's fulfillment response, it is a judgment that a product or service feature, or the product or service itself, provides a pleasurable level of consumption-related fulfillment" [33]. We can infer from the aforementioned definitions that the success of future events heavily depends on how satisfied exhibitors are and whether they plan to attend future events [31]. In addition, from the researcher's point of view the exhibitor's satisfaction can defined as Participant's sense of pleasure caused by the exhibition service provider through covering all their needs from all aspects as perfectly as promised.

7.4 Trade Shows Exhibitions

Trade fairs are "events held on particular days where several companies display their newest items and offer them to current customers and/or future customers" [12, 26] has viewed trade fairs as planned market events that take place at particular times and locations and at which several businesses display their products so that consumers can learn more or make purchases. According to the above information, the researchers can define the trade shows exhibitions as: it is a worthy promotional opportunity for every organization that seeks to market its products or services effectively by meeting its target market in a particular place and time.

8 Previous Studies

Study of [30, 31] demonstrates that when delivering services, tangibles, assurance, dependability, empathy, and responsiveness all help to increase client satisfaction [18] showed that four key components (product quality, environment quality, delivery quality, and social quality) of the exhibition service quality of medical tourism were identified [34] revealed that the exhibition service quality of medical tourism included four main components: product quality, environment quality, delivery quality, and social quality [2, 16]. The future of the exhibition depends on how satisfied the exhibitor is with the level of service provided [3, 14] analyzes the interactions between exhibition service quality, perceived value, emotions, satisfaction, and behavioral intentions for exhibition visitors with the goal of determining the characteristics of exhibition service quality. The results show that there are four main aspects and thirteen supporting dimensions to how exhibition visitors judge the quality of the services they receive [17, 23] lists the aspects of ATM service quality that have an impact on client satisfaction. Price, dependability, responsiveness, convenience, security, and service quality all had good and significant results when utilized as statistical tests for the investigation of correlation and regression analysis, according to SPSS 20 [7, 11] provide an empirical analysis of the hotel industry's service quality. Although service quality is thought to play a significant role in determining customer happiness and repurchase intention, little research and analysis has been done in the Albanian context. The statistical analysis of the data obtained reveals the significance of visitors' perceptions of service quality and the significant influence of perceived service quality on hotel guests' levels of customer satisfaction [21, 27] the banking industry is becoming more interested in customer service as a result of the changing economic climate. Customer satisfaction was found to be a key predictor of future actions and financial success in this industry [14, 19] through a pilot test, scale purification, and validation, a measuring scale that took into account the significant responsibilities of three main stakeholders—trade show attendees, exhibitors, and organizers—in a trade show environment was developed.

9 Methodology and Procedures

9.1 Methodology

The researchers relied on the questionnaire to collect data in order to analyze and test the hypothesis, and they used a literature study for the theoretical portion. The researchers employed descriptive statistical analytical methods to describe the phenomena of the sample or population. The research sample's personal and functional characteristics were covered in the first portion of the questionnaire, which also measured five variables.

9.2 Research Population

Conex has been working with (350) different type of companies from different sectors, so all companies that participated in the shows are the population of research. Therefore, the research population size is (350) companies, some are domestic and others are foreign.

9.3 Research Sample

In accordance with [28] and considering the size of the population (350), the representative sample shouldn't be less than (196), for more accuracy, the researchers distributed (240) questionnaires for the population. About (199) of the questionnaires were valid to be analyzed, that represent (57%) out of the distributed questionnaires.

- **Reliability** a test is carried out Coefficient Internal consistency or reliability are both measured by Cronbach's Alpha. The recommended minimum values are compatible with a Cronbach's Alpha score of (0.7) or higher

The table shows that all research variables have greater than (0.7).Therefore, the research results can be accepted according (Table 1).

Modern Services Quality and Its Impact on the Satisfaction of the Trade ...   143

Table 1 Variable reliability (cronbach alpha)

Number	Variables	Cronbach's alpha
1	Trust	0.755
2	Reliability	0.791
3	Responsiveness and efficiency	0.795
4	Price	0.865
5	Design	0.733
6	Consumer satisfaction	0.747
	All items	**0.901**

10   Data Analysis and Findings

10.1   Description Analysis

Three levels have been assigned to the importance of the respondents' responses, and the following table lists them.

The study evaluated five independent variables and one dependent variable. The discussion that follows reveals the "Mean" and "Standard Deviation" for the study variables (Tables 2 and 3).

All of the independent variables are at a high level, but the trust variable's mean value is the highest at (4.32), which highlights how crucial the variable is to customer satisfaction. Reliability, on the other hand, has the lowest variable with the lowest mean value despite being classified at a high level (4.02). The mean value of all independent variables is high level, and (4.15).

Table 2 Statistical criteria for interpreting the study's arithmetic mean variables

Level	Means
High	3.67–5
Medium	2.33–3.66
Low	1–Less than 2.32

Table 3 The independent variables' means and standard deviations

Variables	Means	Standard deviation	Importance	Level
Trust	4.32	0.654	1	High
Reliability	4.02	0.771	5	High
Responsiveness and Efficiency	4.09	0.697	3	High
Price	4.06	0.772	4	High
Design	4.27	0.626	2	High
Average	**4.15**	**0.704**		**High**

Table 4 Independent variable correlations

Independent variables	Trust	Reliability	Responsiveness and efficiency	Price	Design
Trust	1	0.571^{**}	0.615^{**}	0.564^{**}	0.527**
Reliability		1	0.755^{**}	0.428^{**}	0.580^{**}
Responsiveness and efficiency			1	0.644^{**}	0.639**
Price				1	0.448^{**}
Design					1

** The 0.01 level of significance for correlation (2-tailed)

10.2 Test of Data Validity

The Multicollinearity test for data validity. We utilize the correlation indicator to assess multicollinearity; the correlation between the variables must be (0.9) or lower [25]. Table 4 demonstrates that there is no multicollinearity between the independent variables because all coefficient relations are less than (0.9).

11 Hypothesis Testing

Using multiple, two types of linear regression to assess hypotheses and simple and following table show the results.

11.1 Main Hypothesis

Table 5 demonstrates the multiple regression's main hypothesis outcome.

Table 5 The main hypothesis's multiple regression results

Dependent variable	R	R^2	F	DF	SIG	Independent variable	B	T	Sig
Customers satisfaction	0.784	0.615	61.637		0.000	Trust	0.048	0.876	0.382
				5		Reliability	0.165	3.084	0.002
				193		Responsiveness and efficiency	0.307	4.158	0.000
				198		Price	0.122	2.888	0.004
						Design	0.134	2.000	0.047

Modern Services Quality and Its Impact on the Satisfaction of the Trade ... 145

Table 6 First sub-hypothesis simple regression results

SIG	t–calculated	t-table	Independent variable	R^2	R	Dependent variable
0.00	9.530	1.962	Trust	0.316	0.562	Customers satisfaction

Table 5 because F significant (0.00) is smaller than (0.05), indicating that the dependent and independent variables are significant, we reject the null hypothesis and accept the alternative one. The dependent and independent variables have a solid link with one another. It exceeds (0.5) [6], R = 0.784. Additionally, $R^2 = 0.615$ indicates that the contribution of the independent variable to the dependent variable is 61.5%.

11.2   First Sub Hypothesis

The Simple Regression Results for the First Sub Hypotheses are shown in Table 6.

Table 6 shows that the estimated t-value (9.530) is larger than the t-table value (1.96), indicating that the independent variable has a statistically significant effect on the dependent variable. The researcher rejects the null hypothesis and adopts the alternative one since the significant value of t is smaller than (0.05). Table 6 further demonstrates that the independent and dependent variables have a positive association, as evidenced by the R value (R = 0.562), which is more than 0 [6]. Additionally, the independent variable adds to customer satisfaction by 31.6% according to the (R^2 = 0.316) statistic.

11.3   Second Sub Hypothesis

The simple regression result for the second sub-hypothesis is shown in Table 7.

The fact that the estimated t-value (12.491) is higher than the t-table value (1.96), as shown in Table 7, suggests that the independent variable has a statistically significant effect on the dependent variable. The researcher rejects the null hypothesis and adopts the alternative one since the significant value of t is smaller than (0.05). Table 7 clearly demonstrates that the independent and dependent variables exhibit a positive correlation, as seen by the Rvalue (R = 0.665), which is greater than 0.5 [6]. Additionally, the reliability component adds to customer satisfaction by 44.2% according to the (R^2 = 0.442) statistic.

Table 7 Second sub-hypothesis simple regression results

SIG	t–calculated	t-table	Independent variable	R^2	R	Dependent variable
0.00	12.491	1.962	Reliability	0.442	0.665	Customers satisfaction

146   M. S. Al-Shaikh and F. Alfukaha

Table 8 Third sub-hypothesis simple regression results

SIG	t–calculated	t-table	Independent variable	R^2	R	Dependent variable
0.00	15.674	1.962	Responsiveness and efficiency	0.555	0.745	Customers Satisfaction
0.00	15.674	1.962	Responsiveness And efficiency	0.555	0.745	Customer satisfaction

11.4 Third Sub Hypothesis

The outcome of simple regression for the third sub hypothesis is shown in Table 8.

Table 8 shows that the estimated t-value (15.674) is larger than the t-table value (1.96), indicating that the independent variable has a statistically significant effect on the dependent variable. The researchers reject the null hypothesis and embrace the alternative one since the significant value of t is smaller than (0.05). Table 8 further demonstrates that the independent and dependent variables have a positive correlation, as shown by the R value (R = 0.745), which is more than 0.5 [6]. Additionally, the ($R^2 = 0.555$) indicates that the variable relating to responsiveness and efficiency adds to customer satisfaction by 55.5%.

11.5 Fourth Sub Hypothesis

The Fourth sub-hypothesis' simple regression result is shown in Table 9.

The estimated t-value, which is larger than the t-table value (1.96) in the table, indicates that the independent variable has a statistically significant effect on the dependent variable. The researcher rejects the null hypothesis and adopts the alternative one since the significant value of t is smaller than (0.05). The independent and dependent variables are positively correlated, as shown by the R value (R = 0.585), which is greater than 0.5, in Table 9 [6]. Additionally, the price variable adds 34.2% to consumer satisfaction ($R^2 = 0.342$).

Table 9 Fourth sub-hypothesis simple regression results

SIG	t–calculated	t-table	Independent variable	R^2	R		Dependent variable
0.00	10.120	1.962	Price	0.342	0.585		Customers satisfaction

Modern Services Quality and Its Impact on the Satisfaction of the Trade ...   147

Table 10 Fifth sub-hypothesis simple regression results

SIG	t–calculated	t-table	Independent variable	R^2	R	Dependent variable
0.00	10.064	1.962	Design	0.340	0.583	Customers Satisfaction

11.6 Fifth Sub Hypothesis

The outcome of simple regression for the Fifth sub hypothesis is shown in Table 10.

12 Conclusions

- The research included the following variables: trust, dependability, responsiveness and efficiency, price, and design. The study demonstrated that there is statistically significant impact of Conex supplied service quality on the satisfaction of trade show participants. Additionally, the fact that all factors that affect customer satisfaction contribute roughly (61.5%) shows that Conex is focused on integrating marketing communications to reflect a distinctive great image among its customers in exhibitions.
- Conex's responsiveness and efficiency have a statistically significant impact on trade show attendees' satisfaction, and they have a strong relationship with and have the highest contribution to customer satisfaction (55.5%). This means that the company closely monitors the work of its employees and gives them training to set themselves apart from rivals.
- Conex dependability has a statistically significant impact on trade show attendees' satisfaction, and it has a strong relationship and the second-highest contribution to customer satisfaction (44.2%), indicating that the company is constantly working to maintain excellent relations with its clients.
- Conex's price has a statistically significant impact on how satisfied trade show attendees are, and it has a strong relationship with customer satisfaction, contributing the third-highest percentage (34.2%), indicating that Conex strives to offer the best price in the market.
- Conex design has a statistically significant impact on trade show attendees' contentment, and it has a strong relationship with and a 34% contribution to customer satisfaction. This finding suggests that the corporation should pay more attention to the design of the conferences it delivers.
- Conex trust has a statistically significant impact on trade show participants' satisfaction, and it has a relationship with and a low impact on customer satisfaction (31.6%), which could occasionally lead customers to avoid doing business with the company.
- Most Conex clients are satisfied with the company's services, and their level of confidence is at its highest (4.32).

- Customers were quite delighted with the Conex-provided booth design, which also has the second-highest mean among the variables (4.27).
- Conex's clients may have confidence in the company's personnel and services, which results in consumer agreement with Conex's dependability, and it has the lowest mean (4.02).

13 Recommendations

- Conex should keep developing services in the trade shows, and get customers' feedback about the services. They should start.
- Building of Conex staff capacity on how to improve service quality and speak different languages to understand customer's needs.
- Analyzing competitors' prices and reformulating the company's price policy to match with the market prices and competitors' pricing policy.
- Enhancing customers' trust in Conex services and employees by conducting diagnostic studies to explore Conex weaknesses and to increase customer confidence.
- Developing the designs of the exhibitors' booths in the trade show, to be attractive on one hand and serving the displaying of their products on the other hand.

References

1. Al Jariri, Saleh Amr: Impact of internal marketing and service quality in customer satisfaction (Practical Study of a Sample of the Yemeni Banks), PhD Thesis. Institute of Economic Sciences, Baghdad University, Iraq (2006)
2. Aditya, B.: Exhibition service quality and its influence to exhibitor satisfaction. Semantic Scholar (2019)
3. Wu, H.C., Cheng, C.C., Ai, C.H.: A study of exhibition service quality, perceived value, emotion, satisfaction, and behavioral intentions. Event. Manag. **20**, 565–591 (2016)
4. Chidambaram, V., Ramachandran, A.: A review of customer satisfaction towards service quality of banking sector. J. Soc. Manage. Sci. **2**(2), 71–79 (2012)
5. Chowdhuri, K.: Service quality dimensionality: a study of the Indian banking sector. J. Asia Pac. Bus. **8**(4), 21–38 (2007)
6. Cohen, J.W.: Statistical Power Analysis for the Behavioral Science, 2nd edn (1988)
7. Tabaku, E., Cerri, S.: Assessment of service quality and customer satisfaction in the hotel sector. J. Tourism Hospitality Ind., Congr. Proc., 480–489 (2016)
8. Guerrero, J., Jimenez, J., Tarifa-Fernandez, J.: Measurement of service quality in trade fair organization. Sustainability **12**(22) (2020)
9. Hirunsupachot, P.: Exhibitors Satisfaction with Service Quality: Acomparative Study of Three Main Exhibition Centers in Thailand. Thesis (MBA). Assumption University, Thailand (2005)
10. Kabir, M., Carlsson, T.: Expectations, Perceptions and Satisfaction about Service Quality at Destination Gotland-Master Thesis. Gotland University (2010)
11. Kim, J.: The effect of the utilization and level of exhibition marketing by small and medium companies on the improvement of the companies: productivity-focused on business performance. Prod. Rev. **19**(1), 135–157 (2005)

Modern Services Quality and Its Impact on the Satisfaction of the Trade ...

12. Kirchgeorg, M., Springer, C., Kastner, E.: Objectives for Successfully participating in trade shows. J. Bus. Ind. Mark. **25**(1), 63–72 (2010)
13. Kitapci, O., Akdogan, C., Dortyol, I.T.: The impact of service quality dimensions on patient satisfaction, repurchase intentions and word-of-mouth communication in the public healthcare industry. Procedia. Soc. Behav. Sci. **148**, 161–169 (2014)
14. Kotler, P., Armstrong, G.: Principles of Marketing, p. 290. Printice Hall, USA (2012)
15. Kotler, P., Keller, K.L., Jha, M.: Marketing management: a South Asian perspective. Pearson Prentice Hall, New Delhi (2009)
16. Kotler, P., Keller, L.: Marketing Management, p. 789. Pearson Education Inc., New Jersey (2009)
17. Landau, S., Everilt, B.: A Handbook of Statistical Analysis Using SPSS. Chapman & Hall press, NY (2004)
18. Lee, D.H., Yun, EJ.: The effect of exhibition service quality of medical tourism in attendance satisfaction and behavioral intention. Event Manag. **25**(5), 535–548 (2021)
19. Lin, Y., Kerstetter, D., Hickerson, B.: Developing a trade show exhibitor's overall satisfaction measurement scale. In: Tourism Travel and Research Association: Advancing Tourism Research Globally.28, TTRA International Conference (2015)
20. Mariappan, V.: Changing the way of banking in India. J. Econ. Behav. Stud. **26**(2), 26–34 (2006)
21. McCole, P.,Williams, J., Ramsey, E.: Trust considerations on attitudes towards online purchasing: the moderating effect of privacy and security concerns. J. Bus. Res. **63**(9–10), 1018–1024 (2010)
22. Mohammad, A., Alhamadani, S.: Service quality perspectives and customer satisfaction in commercial banks working in Jordan. Middle Eastern Financ. Econ. (14), 61 (2011)
23. Naeem, Raza, A., Siddiqi, U., Umair, S., Ijaz, M.: Impact of A.T.M service quality on customer satisfaction: an empirical study in Kasur Pakistan banking sector. Int. Rev. Manag. Bus. Res. **5**(2) (2016)
24. Oliver, R.: Customer satisfaction research, the handbook of marketing research: uses, misuses, and future advances. Thousand Oaks, pp. 569–587. Sage Publications, Calif (2006)
25. Pallant, J.: SPSS survival manual, Philadelphia, Open University Press (2003)
26. Pantano, E.: Cultural factors affecting consumer behavior: a new perception model. Euro. Med. J. Bus. **6**(1), 117–136 (2011)
27. Revathi S., Saranya, A.S.: Dimensions of service quality and customer satisfaction: banking sector. Int. J. Adv. Sci. Res. Develop. **3**(3), 55–66 (2016)
28. Sekaran, U.: Research Methods for Business, pp. 294. NY Willey & Sons (2003)
29. Lee, S., Sedigheh, M.: The dimension of service quality and its impact on customer satisfaction trust, and loyalty: a case of Malaysian banks, Asian. J. Bus. Acc. **8**(2), 98 (2015)
30. Zygiaris, S., Alsubaie, M., Ur Rehman, S.: Service quality and customer satisfaction in the post pandemic world: a study of Saudi auto care industry. Front. Psychl. **13** (2022)
31. .Seung-Wan, J.: A Study on the effect of service quality on satisfaction and corporate image in enterprise exhibition hall. J. Syst. Manag. Sci. **12**(2), 443–461 (2022)
32. Water, W., Bulte C.: The 4P classification of the marketing mix revisited. J. Mark. **56**(4), 83–93 (1992)
33. Arthur, Y., Sekyere, F., Kantanubah Marlle, E., Banuenumah, W.: Impact of service quality on customer satisfaction in Obuasi electricity company of Ghana (ECG) the customers perspective. Int. J. Contemp. Appl. Sci. **3**(3) (2016)
34. Zeithaml, V.A., Bitner, M.J., Gremler, D.: Services Marketing: Integrating Customer Focus Across the Firm. McGraw-Hill, NY (2013)

Artificial Intelligence Application to Reduce Cost and Increase Efficiency in the Medical and Educational Sectors

Khaled Delaim, Muneer Al Mubarak, and Ruaa Binsaddig

Abstract Machine-enabled functionalities are being pushed to the limit by artificial intelligence (AI). This cutting-edge technology enables machines to do iterative tasks with a degree of autonomy, resulting in a more efficient use of resources. Next-generation workplaces rely on seamless communication between the enterprise system and employees, thanks to the help of artificial intelligence (AI). This means that human resources are not going away; they are being enhanced by new technology. When it comes to freeing up resources, AI is a boon to organizations. This study is focusing on the use of AI to reduce costs and be more effective and make people focus on more important tasks instead, it will reveal the future of applying AI in medical and education sectors. Thus, it will give more insight for people interested on the AI about how they can know its applications at the present and the future. The study went through main concepts of AI to make it easier for the reader to understand how it works. Then, the study revealed that education has a lot of AI applications coming in the next future in the teaching, and assessment methods. Moreover, the study revealed that medicine have a promising AI application in the coming future in the cardiology, pulmonology, and all other fields.

Keywords Artificial intelligence · Predictive algorithm · Natural language processing · Education · Healthcare

---

K. Delaim
The George Washington University, 2121 I St NW, Washington, DC 20052, USA

M. Al Mubarak (✉)
Ahlia University, Manama, Bahrain
e-mail: malmubarak@ahlia.edu.bh

R. Binsaddig
College of Business Administration, University of Business and Technology, Jeddah, Saudi Arabia

© The Author(s), under exclusive license to Springer Nature Switzerland AG 2024
A. Hamdan and E. S. Aldhaen (eds.), *Artificial Intelligence and Transforming Digital Marketing*, Studies in Systems, Decision and Control 487,
https://doi.org/10.1007/978-3-031-35828-9_15

1 Introduction

Machines and systems that resemble human intellect are known as artificial intelligence (AI). These systems or machines are capable of continuously improving themselves depending on the information they gather. Artificial Intelligence (AI) presents itself in a variety of ways. Few examples are as the following: Chatbots employ artificial intelligence (AI) to better comprehend client concerns and give more efficient responses. Recommendations in Netflix given based on analyzing the behavior of the user [11]. The AI is not about some specific format or function, rather AI is focusing more about process and the potential of tremendous and super thinking and analyzing capacities. It is not with the target of replacing humans by AI machines, but to make this AI machines helps humans to focus on major issues while AI machines do the other easier things on behalf of humans. AI will assist humanity and the business for the best future.

John McCarthy's concept of artificial intelligence (AI) in a 2004 study is one of several formulations that have appeared over the last few decades: "It is the science and engineering of making intelligent machines, especially intelligent computer programs. It is related to the similar task of using computers to understand human intelligence, but AI does not have to confine itself to methods that are biologically observable" [8]. One example of how powerful AI is, is that just hours after the first instances of a fresh coronavirus epidemic were discovered in Wuhan, China, Canadian business BlueDot employed artificial intelligence to identify the outbreak in China. Data from local news stories, social media accounts, and governmental papers were gathered by the infectious disease analytics business, which foresaw the catastrophe a week before the WHO did.

Predictive algorithms might help us avoid pandemics and other global hazards, but AI's long-term effect is hard to foresee [9]. Artificial Intelligence (AI) is a hot subject these days, but not everyone is aware of how much it may help their company. A 40% boost in productivity can be achieved with the current AI technology [7]. Netflix was able to save $1 billion in 2017 by using machine learning. 34 percent of consumers spend more money on a company's goods and services when they use AI. In addition, 49% of respondents said they would purchase from the firm more often if it used AI. The research problem to be discussed and analyzed is the use of AI to reduce costs and be more effective and make people focus on more important tasks instead, so in general turning human to focus on important causes and leave easy jobs done automatically by AI machines and tools. This will help businesses to get more profits by using AI in their marketing approach. By using AI, the businesses will create specific algorithms for doing jobs by those AI standalone technologies, all manufacturing and banking and all industries will replace humans by AI robots to do all tasks. Human may have less interventions just in very specific issues like failures of machines and major maintenance. Companies might benefit from looking at AI from a commercial perspective rather than a technology one. Automating corporate operations, getting insight via data analysis, and interacting with customers and workers are three areas in which artificial intelligence (AI) may help businesses.

2 Literature Review

As mentioned in the introduction the scope of this study is to find out how to use AI to reduce costs and be more effective and make people focus on more important tasks instead, so in general turning human to focus on important causes and leave easy jobs done automatically by AI machines and tools. The following are resources that were used to obtain literature review with the knowledge gained from each that is part of the scope of the study.

2.1 Artificial Intelligence for the Real World

In this article reference the authors mentioned that it is useful for industries to use AI to help them in the automation of the process of their business. "At the same time, the cancer center's IT group was experimenting with using cognitive technologies to do much less ambitious jobs, such as making hotel and restaurant recommendations for patients' families, determining which patients needed help paying bills, and addressing staff IT problems. The results of these projects have been much more promising: The new systems have contributed to increased patient satisfaction, improved financial performance, and a decline in time spent on tedious data entry by the hospital's care managers. Despite the setback on the moon shot, MD Anderson remains committed to using cognitive technology—that is, next-generation artificial intelligence—to enhance cancer treatment, and is currently developing a variety of new projects at its center of competency for cognitive computing" [4, p. 4]. Other thing is to use AI to get deep understanding of data analysis, last thing is cognitive engagements to support the dealing between customers and employees [4, p. 6].

The process automation can get into the administrative and financial functions of the business. The authors recommended using RPA as it will be the most efficient where it almost replaces humans in the cognitive AI skills for the tasks "RPA is the least expensive and easiest to implement of the cognitive technologies we'll discuss here, and typically brings a quick and high return on investment. (It's also the least "smart" in the sense that these applications aren't programmed to learn and improve, though developers are slowly adding more intelligence and learning capability.) It is particularly well suited to working across multiple back-end systems" [4, p. 4]. The more cognitive insight into data for the business can be achieved by using AI, where algorithms are created and used to analyze volume of data and keep learning patterns and trends of behaviors. This will help industries in many ways, such as to know which customers is more likely to buy, to identify the fraudulent activities instantly, to target the right customers in on the net advertisements. The machine learning will give industry competitive edge in competing with other competitors "Cognitive insights provided by machine learning differ from those available from traditional analytics in three ways: They are usually much more data-intensive and detailed, the models typically are trained on some part of the data set, and the models get better—that is, their ability to use new data to make predictions or put things into categories improves over time" [4, p. 4].

The last thing is the cognitive engagement, where AI can give benefits such as 24/7 customer service intelligent agents that are deep learned all databases of the company and frequently asked questions to mimic human agents. It will answer all customers, also it will answer employees' questions and provide training for HR policies, employee benefits and IT. It will give product and service recommendations systems for retailer to increase sales and personalization, thus it will help companies to deal with the increases number of interactions between customers and employees without the need to increase number of employees [4, p. 6].

The author advises a four-step plan for confronting the obstacles of implementing AI in their companies. First, understanding the technologies that will help them from AI solutions "Before embarking on an AI initiative, companies must understand which technologies perform what types of tasks, and the strengths and limitations of each. Rule based expert systems and robotic process automation, for example, are transparent in how they do their work, but neither is capable of learning and improving" [4, p. 7]. Second, creating a portfolio of projects based on the needs and capabilities. Third is to launch pilots for design and redesign. Fourth is scaling up with full base business integrations and adaptation. "In time, cognitive technologies will transform how companies do business. Today, however, it's wiser to take incremental steps with the currently available technology while planning for transformational change in the not-too-distant future. You may ultimately want to turn customer interactions over to bots, for example, but for now it's probably more feasible—and sensible—to automate your internal IT help desk as a step toward the ultimate goal" [4, p. 8].

2.2 Artificial Intelligence

The authors started the research with statement about the excitement of professional bodies about the use of AI in the world as assistive tool to mimic cognitive human behaviors is some applications in the business sector. This is based on the advancement of mathematics and physics that model human cognitive behavior by machines "Perhaps most notably, human intelligence was the central exemplar around which early automation attempts were oriented. The goal was to reproduce intelligent human behavior in machines by uncovering the processes at work in our own intelligence such that they could be automated. Today, however, most researchers want to design automated systems that perform well in complex problem domains by any means, rather than by human-like means" [5, p. 2]. But after years of research and application the researchers and developers have debated about how the human cognitive skills and behaviors are interpreted by science. AI today works by suitable algorithms for simulating specific human cognitive power for specific tasks. Machine learning are made to work for specific task where they try to do prediction after taking huge inputs to some form of neural network and outputs prediction rule related to the domain in polynomial time [5].

2.3 What to Expect from Artificial Intelligence

The authors started with the fact that the AI is sought by many IT companies (google, Apple) to reduce cost by replacing human cognitive skills through AI machines and technologies. Comparing the rise of computers which made reduced cost of equivalent work done by labor, the AI can be another breakthrough to do something similar to reduce cost and replace cognitive human behaviors by AI machines "AI presents a similar opportunity: to make something that has been comparatively expensive abundant and cheap. The task that AI makes abundant and inexpensive is prediction—in other words, the ability to take information you have and generate information you didn't previously have" [1, p. 3]. The power of AI recently shown in the machine learning where predications are created after continuous learning of existing and earlier data "Environments with a high degree of complexity are where machine learning is most useful. In one type of training, the machine is shown a set of pictures with names attached. It is then shown millions of pictures that each contain named objects, only some of which are apples. As a result, the machine notices correlations—for example, apples are often red. Using correlates such as color, shape, texture, and, most important, context, the machine references information from past images of apples to predict whether an unidentified new image it's viewing contains an apple" [1, p. 3]. What makes AI more cost effective and applicable nowadays are the advancement of computational speed, data storage, data retrieval, sensors and algorithms and their good cost [1].

The best way to describe how those AI are made is to show the anatomy of a task and the predications as shown in the following figure [1] (Fig. 1).

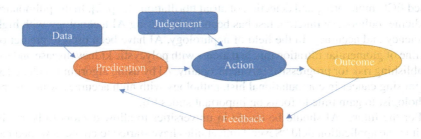

Fig. 1 Task anatomy [1]

Today many factors are playing a role that makes AI is cost-effective and practical. The advances in the AI made the cost of prediction lower. The huge availability of data in the databases and easy accessibility makes which are main inputs for the predictions of AI. The authors continued discussing the issue of AI making judgment and the related effect of different application fields of business. The clearer decisions logics the easier algorithms can be created for AI to replace human judgment and decision making. The main concerns to develop AI applications in the real world are to work harder on making scientific solutions for limitations of the AI predictions [1].

2.4 Artificial Intelligence in Medicine: Today and Tomorrow

The authors have discussed the use of medical technology specifically AI enabled technologies in the early diagnosis, also in reducing complications, and optimizing treatment by providing less invasive treatments "The expression Medical Technology is widely used to address a range of tools that can enable health professionals to provide patients and society with a better quality of life by performing early diagnosis, reducing complications, optimizing treatment and/or providing less invasive options, and reducing the length of hospitalization" [3, p. 1]. This intelligent medical technology with the power of AI have met professional and people interest because it covers the 4P model of medicine (Predictive, Preventive, Personalized, and Participatory). Smartphones and smartwatch with other biosensors can be used by people to assist this technology and give good information for healthcare, this will help the development of the augmented medicine. In the present time several AI based algorithms FDA approved several AI based technologies, also many medicines technology approved for use such as surgical navigation systems for computer assisted surgery [3].

The current application of AI in medicine allowed for early detection of atrial fibrillation and risk of cardiovascular disease by mobile remote technology using smartphones and smartwatch, this is in the field of cardiology "AliveCor received FDA approval in 2014 for their mobile application Kardia allowing for a smartphone-based ECG monitoring and detection of atrial fibrillation" [3, p. 2]. In the pulmonary medicine, pulmonary function test has been applied using AI technology with high efficiency and accuracy. In the field of nephrology, AI have been used to predict of decline of glomerular filtration rate in patients with polycystic kidney disease and for establishing risk for progressive IgA nephropathy. AI by using algorithms helped for diagnosing cancer in computational histopathology with high accuracy which helps pathologist to gain time to focus on important slides [3].

For the future, AI should be taught in universities to allow doctors to be ready for it in the application field "Several universities have started to create new medical curriculum, including a doctor-engineering (18), to answer the need of educating future medical leaders to the challenges of artificial intelligence in medicine (53). Such curricula see a stronger approach to the hard sciences (such as physics and

Artificial Intelligence Application to Reduce Cost and Increase ...   157

mathematics), and the addition of computational sciences, coding, algorithmics, and mechatronic engineering" [3, p. 4]. AI can be used to get perfect medical records of people by having devices all the time with patients to monitor and notify healthcare, but this have some kind of privacy preach which need to be normalized agreed legally, Continuous monitoring and privacy violations, for example, have the potential to increase stigma surrounding chronically ill or more disadvantaged citizens, as well as penalize those citizens who are unable to adopt new healthy lifestyle standards, such as by reducing access to health insurance and care; little to no debate has been focused on these potential and critical pitfalls in health policy making. Doctors will not be replaced by AI but rather supported by AI technologies power [3].

2.5   Becker, Brett. *Artificial Intelligence in Education: What is It, Where is It Now, Where is It Going?*

Authors mentioned about the use of AI is mainly to perform complex tasks that is similar to tasks solved by human cognitive intelligence. The use of AI in education needs a firm decision by decision makers since the educational system didn't seek any huge technological advancement to its conventional system [2]. The future of AI education will be impressive, it is expected that AI takes some roles of the teachers but not the total inspirational role of teachers because still the general AI is not strong. It will help in teaching, assessing, embody new insights from the learning sciences, and be learning partners.

The learner's full educational status and advancement and required development targets will be analyzed continuously by AI to help the learning process and give recommendation for teaching approach for teachers to develop students. Also, AI will use some techniques for learning analytics by providing just in time information about learners' success, challenges and needs that can be used to create learning plans. The AI will help in assessment to let teacher know the level of memory of students and patterns of answers logic of the student to get more insight about. Also, it will analyze emotional factors of students with analyzing behaviors in the answers to get more idea about emotional readiness of the students. The assessment of students will change from normal test of small sample of what have been studied to assessments involving meaningful learning activities, such as games and collaborative projects and will assess all of the learning taking place as it happens "Artificial Intelligence Education will help us do away with the stop-and-test approach that pervades assessment today. As described by Luckin et al., instead of traditional assessments which rely upon testing small samples of what students have been taught, AIEd-driven assessments will be built into meaningful learning activities, such as games and collaborative projects, and will assess all of the learning taking place, as it happens" [2, p. 46]. AI will also help the science by creating better model that help influence learner progress and motivation. The best approach is to combine human power with AI power to make education better for the future of mankind [2].

2.6 China Has Started a Grand Experiment in AI Education. It Could Reshape How the World Learns

Authors mentioned how china used AI in education in the last few years and the success of the trial. The tech giants in china took the lead over the world tech companies in this pilot attempt to use advanced AI with the cooperation of the educational institutes. In an experiment in one school has shown students school gets higher for students by using AI assistive teaching. Using best teachers and IT experts many centers are working to enhance this AI experience in china schools. This AI advancement in china was supported by three main things. The tax breaks to help this AI efforts, the academic competition between students which led the parents to seek whatever high tech to enable their children receive best scores and better future in the profession. Last is the businessmen have a lot of data as the china privacy system is much relaxed and many data are collected from people which makes it easier to create better AI algorithms for education purpose. Many studies have shown that face to face teaching have highly increase the students' scores, the same idea is made through AI online teaching.

The AI online teacher will ask students many questions and by algorithm identify the gap needed to be bridged in the knowledge of student and tailor a study plan for the student. The validation of this was done comparing it with normal teaching and it showed that student using AI has better scores in math. But still in more efforts demanding teaching in areas such as creativity, collaboration, communication, and problem solving, the human teacher did better. So, it is better to free the human teachers to do the more unique teaching while AI support in the easy teaching and by this make educational system much less cost and more efficient "online learning platform is meant to supplement a traditional classroom. Knowledge that can be exercised through adaptive learning, like vocabulary words, is practiced at home through the app. So are skills like pronunciation, which can be refined through speech-recognition algorithms. But anything requiring creativity, like writing and conversation, is learned in the classroom. The teacher's contribution is vital" [6, p. 7]. On the other hand the problem of privacy is a concern, Using cameras and other sensors to determine a student's emotions or mental condition is an application that has little scientific footing and might lead to over-surveillance, but some professionals argues the data gathered in an intelligent classroom could be valuable, That's why the data mostly shared for teachers and not everybody, because we haven't yet done scientific studies [6].

2.7 AI in Healthcare: Medical and Socio-Economic Benefits and Challenges

It was the goal of this study to examine the advantages and disadvantages of artificial intelligence in health care. Medical and economic and social advantages were separated out into their own subcategories, and those benefits were then subdivided again. Smart data inclusion contributes significantly and helps improve the quality of decision-making, according to the findings. Robotic surgery has led to a significant increase in accuracy and predictability. Furthermore, intraoperative guiding via video images has proven to be effective in cases when clinics are difficult to access, travel restrictions, or pandemics occur. Emotional expressions in language are analyzed, interpreted, and responded to via sentiment analysis. Incorporating natural language processing (NLP) and sentiment analysis, data scientists were able to develop algorithms that can decipher human emotions from written text. Rebalancing a doctor's workload using AI would provide them more time with patients and thereby improve patient care. One of the biggest problems is that the data often reflects the systemic inequalities and biases that exist within it.

In order to meet the demand for large datasets, developers are encouraged to collect data from a wide range of patients When it comes to privacy, there is another set of dangers. Developers are motivated to collect data from a large number of patients because of the need for large datasets. Data sharing between large health institutions and AI startups has sparked complaints from concerned patients who fear their privacy will be violated. Patients may be harmed, or other health-care difficulties may arise as a result of AI systems being inaccurate at times. There is a risk of discrimination and inequality in health-care AI. An artificial intelligence system can learn from the data it has access to, as well as the biases it has picked up. If AI data is acquired largely in academic medical institutions, the future AI systems will learn less about—and hence treat patients from populations who do not frequently visit academic medical centers—and thus will treat them less effectively. AI systems that use speech-recognition algorithms to transcribe interactions between providers and customers can fail to work properly if the provider is of a race or gender that is underrepresented in the training data.

New technology is not considered to always be beneficial; it has the potential to be harmful. There are both positive and negative aspects to the current situation, and additional research is needed to address these issues.

It is possible to employ artificial intelligence to increase the efficacy of existing medications, as well as to assist in the development of new ones. Even though AI systems in healthcare are now restricted, the medical and financial benefits are too significant to ignore. Advances in healthcare technology will open new avenues for forward-thinking companies to stay ahead of their competitors in high-cost areas. Only a fraction of what is possible when AI is applied to healthcare delivery may be outlined here. Collaboration between government and private sector industrial participants is essential to realize this potential, which cannot be exaggerated or underestimated. As people around the world live longer and the prevalence of chronic disease increases, the cost of healthcare will continue to be a major concern for healthcare stakeholders. There may be a need for machines to assist us [10].

3 Conclusion

All business can benefit from AI to reduce cost and increase efficiency it; AI is useful for industries to help them in the automation of the process of their business and to get to get deep understanding of data analysis, along with using it for cognitive engagements to support the dealing between customers and employees. The business shall use the four-step plan for confronting the obstacles of implementing AI in their companies. Where understanding the technologies that will help them from AI solutions. Second, creating a portfolio of projects based on the needs and capabilities. Third is to launch pilots for design and redesign. Fourth is scaling up with full base business integrations and adaptation. The power of AI recently shown in the machine learning where predications are created after continuous learning of existing and earlier data. Which makes AI more cost effective and applicable nowadays are the advancement of computational speed, data storage, data retrieval, sensors and algorithms and their good cost.

This study has gone through several resources of information and discussed the different aspects of the use of AI in the real life to help mankind. The healthcare systems can benefit in many ways to reduce the cost and work more efficiently, by using many AI technologies that incorporates the smartphones and smartwatches along with other bio sensory devices which will make the healthcare providers more engaged with live information to speed up the diagnosing of health issues. Also, it will enable better guidance for best treatment to be used and give better medications. Better operations done by AI robotics with better decision making and accuracy. An important aspect is the power of AI imaging where more information and details about the condition of the patient are discovered. The access to data engines for hospitals in the advanced AI ways promotes the quality of medical services to higher extent.

When it comes to developing life-saving drugs, artificial intelligence (AI) can cut costs by billions of dollars, which can be passed on to taxpayers. Artificial Intelligence (AI)-powered algorithm that predicts the efficacy of Ebola therapies based on millions of simulations and analyses. This saves money and, most importantly, lives. Biomarker monitoring frameworks, which enable gene-level sickness diagnosis, and a huge number of patient data points, which can be evaluated in seconds using at-home devices, can be used to optimize medicine development in clinical trials through the use of artificial intelligence.

In the field of education, the AI will make big impact in reducing cost and improving efficiency when it is incorporated more into the educational systems. The learning process and the assessment methods will be in different level. The learner's educational status and even emotional status will be carefully monitored analyzed and action taken more effective. The assessment methods will be much more sophisticated and more insight into the better identification of the best learning progress and required development of the learners. People will need to focus on the talents that remain unique: creativity, teamwork, communication and problem-solving while computers become better at monotonous activities. Automating more and more jobs will require them to adjust fast. Instead of imparting classical knowledge that was more appropriate for an industrial age classroom, 21st century classrooms should encourage students to discover their own talents and interests. n principle, AI may make this easier. Teachers might devote more time to one-on-one interactions with students if it took over some of the more routine responsibilities in the classroom. As to what that would look like, there are a variety of theories. A variety of possibilities exist for how artificial intelligence (AI) could be used in the classroom, from helping teachers keep track of student progress to giving students more influence over their own education. Regardless, the ultimate goal is to provide students with a personalized education. The assisting AI will focus on each student separately and measure level of knowledge and create plans to develop them in the best ways which will elevate their scores and performance to the level that will make countries reduce cost of education and let the teacher concentrated more in elevating the quality of knowledge and work better for research and development fields.

3.1 Implications

AI in Health and education sectors is very helpful to reduce cost and increase productivity. In educational Teachers might devote more time on the more unique skills that only human can transmit to other humans and AI cant be successful, at the same time human teachers use AI teacher to do most of the other tasks and analysis to help them get students to learn better. In the Health sector AI is useful in creating better environment for patients and healthcare staff where it will help in making AI databases for early diagnosis and fast medical intervention, it will also support better surgical quality. The governments, people, private sector, and technology developers shall work closely in this transformation to be very effective.

3.2 Suggestions for Future Research

It is suggested that next researchers on this topic shall focus on the benefits of AI on other sectors such as construction, governments services, national security. Also, more attention to the comprehensive integration of different sectors that AI systems can contributed in.

References

1. Agrawal, A., Gans, J., Goldfarb, A. What to expect from artificial intelligence (2017)
2. Becker, B.: Artificial intelligence in education: what is it, where is it now, where is it going. Ireland's Yearb. Educ. **2018**, 42–46 (2017)

3. Briganti, G., le Moine, O.: Artificial intelligence in medicine: today and tomorrow. Front. Med. **7**. https://doi.org/10.3389/fmed.2020.00027
4. Davenport, T.H., Ronanki, R.: Artificial intelligence for the real world. Harv. Bus. Rev. **96**(1), 108–116 (2018)
5. Dick, S.: Artificial intelligence. (1) (2019). https://doi.org/10.1162/99608f92.92fe150c
6. Hao, K.: China has started a grand experiment in AI education. It could reshape how the world learns. MIT Technol. Rev. **123**(1) (2019)
7. Haponik, A: Reduce Operating Costs and Improve Efficiency Using AI. Addepto. https://addepto.com/reduce-operating-costs-and-improve-efficiency-using-ai/ (2019)
8. IBM Cloud Education: Artificial Intelligence (AI). IBM. https://www.ibm.com/cloud/learn/what-is-artificial-intelligence?lnk=fle (2020)
9. Koenig, S.: What does the future of artificial intelligence mean for humans? Tech Xplore. https://techxplore.com/news/2020-07-future-artificial-intelligence-humans.html (2020)
10. Shaheen, M.Y.: AI in Healthcare: medical and socio-economic benefits and challenges. ScienceOpen Preprints (2021). https://doi.org/10.14293/s2199-1006.1.sor-.pprqni1.v1
11. What is Artificial Intelligence (AI)? (n.d.). Oracle. https://www.oracle.com/artificial-intelligence/what-is-ai/

Data Journalism and Its Applications in Digital Age

Abdulsadek Hassan and Mohammed Angawi

Abstract The study aims to identify data journalism, its types and importance, the role of the Corona pandemic in increasing the importance of data journalism, and the developments in data journalism in the Arab world. The greatest impact in changing many journalistic practices, and "data journalism" is one of these new phenomena that modern communication technology has mainly helped in developing and presenting in different, more interactive and attractive forms. Data journalism relies mainly on analyzing available data and numbers and embodying them in tables, curves, maps and other visual forms, to allow new facts and news that would not have been easy to obtain without those numbers. The study also revealed that data journalism is one of the tools that facilitated the interpretation and presentation of health topics. In addition to presenting news stories by processing large sets of data.

Keywords Data journalism · Corona virus · Electronic journalism

1 Introduction

The tremendous development in the means of communication and the abundance of data on the Internet and social networking sites, and the revolution we are currently experiencing and the flow of large and complex data, posed a great challenge in the ways of displaying and disseminating data in traditional press templates, which led to the emergence of new methods and concepts related to journalism, including data journalism [1].

Data journalism has spread due to the development of Internet uses and its implications for the context of press production in the era of information technology integration [2]. This new press seeks to facilitate the assimilation of information

A. Hassan (✉)
Ahlia University, Manama, Bahrain
e-mail: aelshaker@ahlia.edu.bh

M. Angawi
University of Business and Technology, Jeddah, Saudi Arabia

© The Author(s), under exclusive license to Springer Nature Switzerland AG 2024
A. Hamdan and E. S. Aldhaen (eds.), *Artificial Intelligence and Transforming Digital Marketing*, Studies in Systems, Decision and Control 487,
https://doi.org/10.1007/978-3-031-35828-9_16

163

by the public in light of the rapid flow of data through contemporary channels of communication [3].

Any journalist can now create data-driven press stories through a set of steps that start with extracting data from huge databases or websites of government institutions, then verifying, cleaning, and analyzing the data, and then using graphics programs or free applications specialized in designing data, which will help in the graphic representation process in order to show the story in an attractive and interesting way [4].

The Journalism was not born fully developed and in a definitive form, for press direction developed according to the different aesthetic directorial schools and was not embodied in one go [5]. The journalistic genres, in turn, were not born at once, but took different paths with different characteristics [6]. With time, electronic platforms have developed, and the means of communication have multiplied, so we are talking about blogs. During that time, there were many forms of journalism, so let's first talk about transportation journalism, which is based on re-publishing in another media [2]. Then we started talking about opinion journalism during the nineteenth century, and newspapers turned into platforms for political battles [7]. Between the end of the nineteenth century and the beginning of the twentieth century, news journalism emerged that sought to regulate the profession of journalism [3], which later paved the way for communication journalism and the media moved from focusing on the event to focusing on building a relationship with the audience [8]. Data journalism, which coincided with the Internet revolution, is an extension of the forms that journalism has known during the past centuries [9].

This type of specialized journalism highlights the increasing role of digital and graphic data in the production of news stories that include a simplified visual presentation of the data that is easy for the audience to understand and assimilate, which takes many forms such as static, moving, interactive, or even video graphs, all of these types of express methods of data adaptation [10].

Interest in data journalism has increased locally and globally to become part of press institutions, for which specialized websites are established on the Internet, and there are departments and journalists based on the design, development and production of data journalism, whether in print or electronic newspapers [11].

2 Data Journalism Definition

Data journalism is a type of journalism that reflects the increasing role that digital data uses in the production and distribution of information in the digital age [11]. It reflects the increased interaction between content producers (journalists) and many other fields such as design, computer science and statistics [6]. From the journalists' point of view, it represents "an overlapping set of competencies drawn from disparate fields" [12].

Data journalism is style of journalism based on four main elements, which are data as sources, the context that organizes this data and de-clutters it, telling the story to be attractive, and finally the technology that has become available through many tools that do not need programmers, pointing out that the journalist must act as a "link" between these four elements [13], where he raises questions about the data, organizes it, chooses the appropriate medium for displaying it, whether in the press, television or radio, and finally uses technology to display it [14].

And that obtaining data is no longer a problem even in the absence of laws for freedom of information, as many agencies in the world issue statistical reports, with the journalist having to master the use of some available and free tools to manage, process, analyze and visually display data, the most important of which is Microsoft Excel, Florish, and others [15].

The concept of data journalism intersects with multiple concepts, including structural journalism and solutions journalism, and its practice requires traditional journalism skills such as creating ideas and simplified writing, dealing with numbers [16], employing computer programs, social networking sites, and the Internet in general, in addition to scientific research skills such as formulating hypotheses and analysis, and using statistics software such as SPSS [17].

Accordingly, data journalism can be defined as specialized journalism or a new specialization in journalism to highlight the increasing role of digital and graphic data in the production and dissemination of information in the era of the digital revolution. It reflects the increasing interaction between journalistic content producers [18], and specialists in a number of other fields such as technical design and computer science and statistics [19]. From the point of view of journalists, data journalism represents an overlapping set of skills used in different fields to present data in a simplified form in an attractive graphic template for the readership [11].

3 The History of Data Journalism

The term data-driven journalism has been in use since 2009, and one of the earliest examples of computers being used with journalism dates back to CBS's 1952 attempt to use a mainframe computer to predict the outcome of a presidential election, but it wasn't until 1967 that the use of computers to analyze data was more widely adopted [8].

Philip Meyer worked for the Detroit Free Press at the time and used a mainframe computer to better report on the riots spreading throughout the city [4]. With a new former group analyzing data in journalism, Mayer teamed up with Donald Barlett and James Steele to look at patterns with conviction sentencings in Philadelphia during the 1970s [11]. Mayer later wrote a book called Precision Journalism that advocated the use of these techniques to incorporate data analysis into journalism [8].

Towards the end of the 1980s, important events that helped to organize the field of computer-aided reporting began formally taking place [18]. Investigative reporter Bill Deadman of The Atlanta Journal-Constitution won a Pulitzer Prize in 1989 for The Color of Money, his 1988 series of stories using CAR techniques to analyze racial discrimination by banks and other mortgage lenders in middle-income black neighborhoods [2]. The National Institute of Computer Aided Reporting (NICAR) [20] was formed at the Missouri School of Journalism in collaboration with Investigative Reporters and Editors (IRE) [4]. The first conference dedicated to CAR was jointly organized by NICAR with James Brown at Indiana University and held in 1990 [21]. NICAR conferences have been held annually since then and are now the largest single gathering of data journalists [22].

Data journalism constitutes one of the modern episodes in international journalistic thought, seeking to achieve depth and comprehensiveness, to confront television and social networking sites, and to solve problems of urgent journalistic practices in dealing with events [23]. Precision journalism, to bridge the gap between journalistic work and social scientific research, using its tools and assumptions, which marked the beginning of data journalism, until the modern term spread in 2006–2007 [24].

Although data journalism has been used unofficially by practitioners of computer-aided reporting for decades, the first recorded use by a major news organization is The Guardian, which launched its data blog in March 2009 [13]. The paternity of the term is disputed, and it has been in widespread use since the leak of WikiLeaks documents on the Afghan war in July 2010 [14].

The Guardian's coverage of War Records took advantage of free data imaging tools such as Google Fusion spreadsheets, another common aspect of data journalism. Facts Sacred by The Guardian's data blog Editor Simon Rogers describes data journalism like this [14].

Investigative data journalism combines the field of data journalism with investigative reporting. An example of investigative data journalism is the search for large amounts of textual or financial data. Investigative data journalism can also relate to the field of big data analytics to process large data sets [25].

Since the concept was introduced, a number of media companies have created "data teams" that develop visualizations for newsrooms [8]. Most notably teams such as Reuters, [22] Pro Publica, [21] and La Nacion (Argentina) [26]. In Europe, both The Guardian [1] and Berliner Morgenpost [27] have highly productive teams, as well as public broadcasters [23].

Data-driven journalism can play an investigative role, dealing with "unopened" data also known as confidential data at times [7].

4 Importance of Data Journalism

From the inception of journalism until now, data has been and still is a major component of journalistic work, but data journalism has gained great importance in the era of information flow [2]. Data journalism is not a substitute for traditional journalism, but in addition to it, it serves two important purposes [13]. It works to create unique news materials away from the traditional method, and it helps journalism to fulfill its role as an instrument of public control over government policies [28].

Data Journalism and Its Applications in Digital Age

Stream data filtering

After the journalist was suffering from a lack of information, information became abundant and flowing on the Internet due to the development of social networks, which contributed to the rapid flow of data, so the journalist had to select from this information and carry out a process of filtering and auditing [5].

Process data to extract attractive press reports

Data journalism helps to tell richer journalistic reports. The journalist's job is no longer just to gather information and write a news report, but rather to process and analyze data, taking advantage of digital technology to tell a distinctive journalistic story [21].

Increasing the credibility and reliability of press reports

Data journalism is based on analyzing real data from specific sources mentioned by the journalist, which increases the credibility of the information contained in his press report [6]. International, such as the World Bank, the World Health Organization, the state's national statistics agency, the Ministry of Finance, and others, which gives its report accuracy and reliability, and the skeptical reader can refer to these data to verify its authenticity [29].

Arousing public opinion about societal issues

Data journalism helps to explore the unclear relationships between data and helps the reader to obtain more information that is difficult for him to access or understand and extract information that is useful to him [4]. On analyzing the data of Hurricane Andrew, which struck southern Florida in 1992, and mapping the extent of damage that occurred in the vicinity of the hurricane, and another map showing wind speed, which revealed that new homes were more vulnerable to damage from the storm [29], which means that building laws in the state and the absence of supervision and inspection of the new housing units caused an increase in the extent of the damage [16]. Doig won the prestigious Pulitzer Prize in 1993 for this report, which is a huge inspiration for what can be done with data [14].

Connect the reader to data on a personal level

Data journalism contributes to strengthening the link between the press report and the individual [4]. For example, the BBC and the British Financial Times make the budget in an interactive manner, where one can see the impact of the budget on personally and not in general [29].

Data journalism also addresses the problem of the inability to absorb the flow of information and deal with it at the same speed and quantity that it reaches, as the information taken from the media, whether printed, visual or audio, affects the choices and behavior of citizens, so data journalism is considered an aid to the individual in making decisions [24].

Attracting a wider audience of readers

Data journalism deals with the reality of changing the readership; In the era of technology and high speed, the reader no longer has the time or perseverance to read an entire newspaper or a lengthy press report that depends on the verbal narration only, but many are browsing the headlines of newspapers on their mobile phone, and press reports flow to them through news applications, so data journalism has emerged. To keep pace with this development, the press narrative was changed to attractive images, graphics, interactive maps, and headlines [17].

5 The Relationship Between Electronic Journalism and Data Journalism

Although some believe that the electronic press is the pioneer in using data journalism, the archives of Arab and international newspapers have a remarkable history of using infographics and attention-grabbing informational illustrations to communicate information to the reader without using many words [30].

The relationship between data and the press has witnessed more strength and interdependence all over the world, especially in light of what the world is experiencing now from the height of the digital age in which data accumulates in real time more than ever before, to the extent that some call data the term "new oil", due to its ability to Attract a large number of readers by simplifying information and presenting it in interactive visual forms [4].

Thus, the momentum of information has had a great impact on the tasks assigned to the journalist, as his main task has moved from going to the field to scanning piles of data with the aim of extracting the relationships between different documents, and the links between different patterns, to tell an effective news based on providing new information based on real data [17]. To reach results that are abstract from personal opinions that prevent the public from taking decisions without political, financial or legal restrictions [8].

6 Types of Data Journalism

There are three main categories of data journalism: quantitative, qualitative, and visual [24]. Quantitative data journalism focuses on numbers and statistics, qualitative data journalism uses interviews, surveys and other methods to gather people, and visual data journalism uses images and graphics to convey information, and each type of data journalism has its own set of advantages and disadvantages [11].

Quantitative data journalism

Quantitative data journalism is commonly found in digital publications such as The Guardian, Fast Company, and Quartz [25]. An example is this British Academy publication Stand out and be count, a guide that shows concrete steps you can take to become proficient with numbers and statistics [3].

Advantages of quantitative data journalism

All forms of quantitative data journalism rely on numbers and statistics [13].

Complex statistical models can be created that highlight actual world trends [18].

Quantitative data journalism can help organizations identify consumer preferences.

Easy to read, analyze and explain [19].

Even if you've never taken a math class in your life, quantitative data journalism is easy to work with [4].

Qualitative data journalism

Qualitative data journalism is often written by journalists who interview experts, review documents and records, and conduct extensive research [16].

Advantages of qualitative data journalism

Qualitative data journalism is unlikely to be biased because it relies on humans rather than computers [10].

Not facing the challenge of creating complex mathematical formulas [17].

Possibility to practice telling stories [30].

People love to hear stories [10].

Visual data journalism

Visual data journalism is the latest form of data journalism, combining text and images to tell a story [17]. Visual data journalism includes: The Atlantic, Business Insider, and the BBC [28]. Charts are becoming increasingly popular because they are fast and easy to produce [4].

Advantages of visual data journalism

Pictures and graphics make it easy to understand graphs [12].

It gives readers a better look at new products and services [27].

It highlights current events [30].

It's fun to watch and share [2].

Charts help people learn things faster [9].

As for the most important types of data sources, data sources are usually also divided into three main types:

The data is open source: anyone can access, reuse, and produce it, including but not limited to national statistics, legislation, election results in chambers of commerce and municipal councils [28].

- **Closed-source data**: data that is protected within the specialized institutions, and can only be accessed by permission or penetration, such as: security data, banks and the private sector [18].
- **Data is shared**: it is between open source and closed source together, where it can be shared with a small group of people for a specific purpose such as modification and statistics [11].

Technical tools used in the investigation (maps, satellites, search engines, databases, verification tools, storytelling tools)

Investigation is one of the important features of journalism [28]. Journalists are not satisfied with sitting and waiting for events; Rather, they take action through investigation [4]. Investigative reporting requires different skills than just gathering facts [5].

The good news is that technology helps facilitate investigative work. Technology used in investigative journalism includes maps, satellites, search engines, databases, verification tools, storytelling tools, and more [17].

7 Maps

When it comes to reporting investigations, maps play an important role [23]. The investigator needs to locate and track the locations mentioned in the story, and Maps assists the investigators in doing so, and Google Maps is a good example of this [30]. In addition, using maps allows reporters to share their findings with readers and help others find similar stories [10].

Satellite images

Using satellite imagery provides another way to track locations. Satellite images are useful because, instead of relying on traditional mapping techniques, they capture entire areas within seconds [18]. And because it captures the entire image rather than just a small section, satellite images can reveal details missed on maps [12]. Examples of satellite imagery tools include: Landviewer, EO Browser, Earthdata Search, Remote Pixel, Google Earth and many more [12]. For example: NASA Earth Observatory uses satellite imaging to produce unique 3D models of the world. These maps show us what our planet looks like at night and during meteor showers [11].

Search engines

Search engines are very important tools for investigative journalists, by entering keywords in Google, Bing, etc., an investigator can quickly find sources of information within minutes [28]. Once the investigator has identified relevant information, he or she can use the search engine results to validate certain claims made in a published story [4].

Data Journalism and Its Applications in Digital Age 171

8 Databases

A database is a set of related information stored in a computer system [18]. Databases are often used in preparing investigation reports [8]. For example, the International Atomic Energy Agency maintains a database that contains an index of thousands of nuclear sites around the world [9]. Similar to a traditional library, the database contains information on many topics [25]. Examples of public databases available to journalists include ProPublica, US News and World Report, and many others [20].

However, databases are organized differently, while libraries organize things according to the category of subject matter, databases organize things according to individual records containing information about a particular subject [17].

Verification Tools

To write true stories; Investigative journalists have to do the check [5]. Checks allow them to confirm whether something is right or wrong [12]. Online websites allow users to verify claims made on social media platforms such as Facebook [26].

Storytelling tools

Because storytelling is so central to the journalism profession, there are a range of tools that can help a journalist build a digital story [8]. Examples of these tools are: Piktochart, Timeline, Adobe Spark, and others [29].

Video News Release (VNR)

One of the most significant developments in recent years has been the emergence of Video News Releases (VNRs), which are short films primarily used to promote stories [21]. Unlike traditional press files, which tend to be rather long and complex, VNR files are often between three and ten minutes in length and do not necessarily depend on graphics [20]. Instead, it contains text and audio clips accompanied by various stills [4]. An important advantage of video news data is its ability to reach a large audience quickly and at minimal cost [27].

Data collection

Another area in which technology has made a difference is data collection [4]. Not long ago, journalists relied for information gathering on newspapers, magazines, and TV stations, and even today bloggers are still doing the same [13]. In the future, we will see an increase in the amount of data available to journalists and the Internet will play an even more important role in helping journalists navigate the sea of data [19].

9 What Skills Does a Journalist Need to Be a Data Journalist?

Let's start with the skills you don't need: You don't need to be a tech genius, good at math, a seasoned journalist, or any kind of journalist at all [11]. Data journalism has been poisoned by technical chauvinism, and the belief that technology is always the answer [18]. Data journalism is not about technology; Rather, it is about stories that explain the human experience [24]. You must be passionate about the topic you are covering and then you must be able to follow a proven step-by-step process of research that takes you from idea to completed investigation and validation that gets to the root of the problem [3]. You must adhere to transparency and the highest standards of documentation while developing your data skills [14]. If you are passionate and punctual, you already have what it takes to become a data journalist. The rest is learning the tools and processes to help you find and tell your story [12].

10 The Reality of Data Journalism in the Arab World

The status of data journalism is still in its early stages, and what is noticed is the idea of the attraction towards presenting perceptions of data by highlighting the value primarily [24], while other contexts that require more extended investigations have been absent [22]. The second: that data journalism is a continuous interaction between two existing models: The first sees that data journalism refers to a style of journalism that relies on reporting facts based on quantifiable evidence [29]. The second model sees that data journalism refers to the intermingling and overlap between the logic of computing and journalism in order to search beyond information and data structures and to come up with a new angle for the journalistic story [30].

According to a survey conducted by the Arab Data Journalists Network in 2017, on data journalism in the Arab world, which included 60 journalists from 8 Arab countries, 71.9% of the respondents confirmed the difficulty of obtaining data in their countries, while 22.8% described it as very difficult, while 5.3% considered it easy to obtain data in their country [5]. The survey indicated that there are 100 countries around the world that have laws that allow the free circulation of information and access to data; Jordan passed a law on freedom of information in 2007, and similar laws were passed in Tunisia and Morocco in 2011, and in Yemen in 2012 [1]. However, despite the existence of freedom of information laws in some Arab countries, 83.1% of those surveyed said that they did not use their right to obtain information, compared to 16.9% who did [4].

Arab experiences in data journalism are still limited due to the challenges and problems it faces [23]. Data journalism is still developing and spreading in Tunisia, Egypt, Lebanon and the rest of the Arab countries, due to the efforts of trained journalists working to spread knowledge [17].

11 Data Journalism and the Coronavirus

After the blow to traditional journalism and more digital journalism, data journalism's role in telling fuller stories about the new virus has strengthened [18].

In the first months of the year 2020, press stories dealing with the numbers of corona infection and related deaths dominated, but data journalism is not only related to publishing data or presenting it in innovative graphic forms [1]. Daily reports about Corona numbers in themselves fail to provide real value to the public regarding other aspects related to the pandemic, such as the level of severity of disease outbreaks among people, identifying age groups most at risk of infection, in addition to evaluating the efficiency of public health institutions [13].

Here comes the role of data journalism in contextualizing this data and explaining it to the public [11]. The journalist must analyze the data provided by the official authorities, as they are the main source for the numbers related to the virus, and he must be keen on understanding and interpreting them accurately [6].

Although these numbers can be presented to the public, they may not mean much to them in and of themselves, especially with regard to their relationship to other aspects of the pandemic, such as the extent of infection spread in a city [4], its percentage of the total population, determining the number of beds available in hospitals, and knowledge of recovery rates or stable cases, and assessing the efficiency of public health facilities in dealing with them and the quality of services they provide [14]. The analysis of this data also helps to uncover misinformation and misinformation [22].

In the early stage of the outbreak of the Corona epidemic, data specialists collaborated with scientists in order to understand what this virus was and to get clear facts about it [3]. There are noteworthy projects that helped illustrate the impact of the pandemic, through the use of maps and graphs, in which the of analysis obtained on the data collected becomes clear [9]. For example, the owners of these projects were interested in explaining how the authorities dealt with the virus and comparing the results to determine which methods were most effective in controlling the spread of infection with it [12]. Some data-based reports also attempted to determine the feasibility of complete closures between a country and a country, while other reports tracked the feasibility of adherence to protective masks, and the possibility of transmission of infection through the air in enclosed spaces: at home, cafe or classroom [19]. This topic alone, which was published by the Spanish newspaper "El Pais", related to the transmission of infection through the air, generated more than 40 million views of the pages it deals with [7].

Data teams have achieved a great deal of success during the pandemic due to excellent reporting that relied on data to illustrate the severity of the impact of the pandemic [22], and it has become a staple of most press reporting during the crisis and has mainly helped to overcome the obstacles imposed by working from home [21]. The crisis we are living in today has proven that it is no longer possible to dispense with data, and that it is not just an alternative, show-stopping way to talk about numbers [26]. It also stressed the necessity of applying and relying on them in the Arab world [25].

Data journalism will continue to strive to enhance its position in covering the ongoing epidemiological crisis that has reached the stage of dealing with and distributing vaccines today [8], coinciding with the emergence of new mutations from the epidemic in the United Kingdom and others [9], which created new areas of probing these developments through the available data [28]. Vaccine development has taken nearly a year since the pandemic began, and with news of this new mutation on the virus, data will remain essential in following up, understanding and estimating its effects [17].

12 The Future of Data Journalism

The past years have witnessed the rise of data as a primary source of news rather than traditional news in itself, and the future is seen that there will be multiple sources of data, those that organizations and governments disclose as part of transparency, data that individuals provide voluntarily, data related to electronic platforms, and leaked data [8].

Accordingly, there is a tendency for data journalism to replace traditional news journalism [15], and it is expected to be the most preferred field for journalists, in the sense that it has become strongly imposing itself, and also represents a new direction for the future of journalism related to technology and visual illustrations [11].

It seems from all this that the field of journalism and media is no longer theoretical anymore [22]. Through these great challenges, media institutions are waiting for years of development, as well as for independent journalists, colleges and departments of journalism across the world [3].

The future is the king of data-driven journalism, so every journalist should be proficient in dealing with data technology [12]. In the past, a journalist would get material by chatting with people in bars and maybe still is sometimes [14], but now the situation has changed and It is necessary for a journalist to study the data while equipping himself with the tools to analyze it and pick out what is interesting while keeping it all in an illustration, to help people see all angles of the subject and really understand what is going on around them [28].

Data journalism fills the gap between statisticians and wordsmiths, by finding outliers and searching for new trends, which often have not only statistical significance but are also related to the question of fragmentation and simplification of the nature of our current complex world [16].

13 Conclusion

The results of the study revealed that data journalism is, in fact, a quick automated process that aims to transfer a huge and scattered amount of data through a mechanism to deal with it automatically in a smart way away from the traditional method that the classicists call the press kitchen (the editor's kitchen), which has been automatically converted and thus became journalistic topics visual information that reduces information and then designs it in forms, graphs and graphs until it is presented to the reader in an attractive image with the help of intelligence applications and computer programs by collecting, organizing or analyzing that data and then broadcasting or publishing it, and the audience interacts with it more to take advantage of the data-rich world, and technology is used to extract information that is of interest by the audience and not only the skill of using modern technology, while traditional journalism relies on people's narration and conveying the news in an interesting and interesting way to the public.

The results also revealed that the press moved from focusing on the event to focusing on building a relationship with the public and extending bridges of communication with the masses until it reached, today to data journalism that coincided with the Internet revolution and the use of smart means that imposed changes on the stages of the publishing or news production process.

Keeping abreast of the news of the Corona pandemic gave a new breath in the journalism, and it was a factor that accelerated opportunities for cooperation between journalists and digital experts to crystallize news based on a journalistic vision and technical mechanisms that allow the provision of news material in interesting and more attractive formats with sound and image, and this momentum may be in resorting to data journalism, an opportunity to develop the production of the concept of story and journalistic narration, as well as encourage the return of journalists to learning benches to develop their digital capabilities, under the supervision of technical experts, as the pandemic has taught media professionals a lesson in exploiting the heaps of numbers and information widely available and treating them using news templates that rely on data journalism to deliver information that is reduced and simplified to opinion.

References

1. Kosterich, A.: Reengineering journalism: product manager as news industry institutional entrepreneur. Digit. Journalism (2021). https://doi.org/10.1080/21670811.2021.1903959
2. De-Lima-Santos, M.-F., Salaverría, R.: From data journalism to artificial intelligence: challenges faced by La Nación in implementing computer vision in news reporting. Palabra Clave 24(3), 1–40 (2021)
3. Stalph, F., Borges Rey, E.: Data journalism sustainability: an outlook on the future of data-driven reporting. Digit. Journalism 6(8), 1078–1089 (2018)
4. Hepp, A., Loosen, W.: Pioneer journalism: conceptualizing the role of pioneer journalists and pioneer communities in the organizational re-figuration of journalism. Journalism 22(3), 577–595 (2021)
5. Chan-Olmsted, S.M.: A review of artificial intelligence adoptions in the media industry. Int. J. Media Manag. 21, 193–215 (2019)
6. Hong, H., Oh, H.J.: Utilizing bots for sustainable news business: understanding users' perspectives of news bots in the age of social media. Sustainability 12, 6515 (2020)
7. Heravi, B.R., Lorenz, M.: Data journalism practices globally: skills, education, opportunities, and values. J. Media 1, 26–40 (2020). https://doi.org/10.3390/journalmedia1010003

8. de-Lima-Santos, M.-F., Mesquita, L.: Data journalism beyond technological determinism. Journalism Stud. **22**(11), 1416–1435 (2021)
9. Palomo, B., Teruel, L., Blanco-Castilla, E.: Data journalism projects based on user-generated content. How la Nación data transforms active audience into staff. Digit. Journalism **7**, 1270–1288 (2019)
10. Appelgren, E., Lindén, C.-G., van Dalen, A.: Data journalism research: studying a maturing field across journalistic cultures, media markets and political environments. Digit. Journalism **7**, 1191–1199 (2019)
11. Jamil, S.: Artificial intelligence and journalistic practice: the crossroads of obstacles and opportunities for the Pakistani journalists. Journalism Pract. 1–23 (2020)
12. Heft, A., Alfter, B., Pfetsch, B.: Transnational journalism networks as drivers of Europeanisation. Journalism **20**(9), 1183–1202 (2019)
13. Graefe, A., Bohlken, N.: Automated journalism: a meta-analysis of readers' perceptions of human-written in comparison to automated news. Media Commun. **8**, 50–59 (2020)
14. Fahmy, N., Attia, M.A.M.: A field study of Arab data journalism practices in the digital era. Journalism Pract. **15**(2), 170–191 (2021). https://doi.org/10.1080/17512786.2019.1709532
15. Splendore, S., Brambilla, M.: The hybrid journalism that we do not recognize (anymore). Journalism Media **2**(1), 51–61 (2021)
16. Ali, W., Hassoun, M.: Artificial intelligence and automated journalism: contemporary challenges and new opportunities. Int. J. Media Journalism Mass Commun. **5**, 40–49 (2019)
17. Bastos, M., Mercea, D.: The public accountability of social platforms: lessons from a study on bots and trolls in the Brexit campaign. Philos. Trans. Roy. Soc. A: Math. Phys. Eng. Sci. **376** (2018)
18. Davenport, L., Fico, F., Detwiler, M.: Computer–assisted reporting in Michigan daily newspapers: more than a decade of adoption. In Newspaper Division, AEJMC National Convention. Phoenix: AZ. Available online: http://online.sfsu.edu/jjohnson/Courses&Syllabi/BU-JO807/Bibli&Articles/Davenport(2000).pdf. Accessed on 19 May 2022
19. Mutsvairo, B., Bebawi, S., Borges-Rey.: Data Journalism in the Global South: Comparative Perspectives. Palgrave, London, UK (2019)
20. de-Lima-Santos, M.-F., Ceron, W.: Artificial intelligence in news media: current perceptions and future outlook. Journalism Media **3**, 13–26 (2022). https://doi.org/10.3390/journalmedia 3010002
21. Heravi, B.: 3Ws of data journalism education: what, where and who? Journalism Pract. **13**, 349–366 (2018)
22. de-Lima-Santos, M.F.: ProPublica's data journalism: how multidisciplinary teams and hybrid profiles create impactful data stories. Media Commun. **10**(1), 5–15 (2022)
23. Wright, K., Zamith, R., Bebawi, S.: Data journalism beyond majority world countries: challenges and opportunities. Digit. Journalism **7**, 1295–1302 (2019)
24. Young, M.L., Hermida, A., Fulda, J.: What makes for great data journalism? A content analysis of data journalism awards finalists 2012–2015. Journalism Pract. **12**, 115–135 (2018)
25. Kotenidis, E., Veglis, A.: Algorithmic journalism—current applications and future perspectives. Journalism Media **2**, 244–257 (2021). https://doi.org/10.3390/journalmedia2020014
26. Wright, S., Doyle, K.: The evolution of data journalism: a case study of Australia. Journalism Stud. **20**, 1811–1827 (2019)
27. Ojo, A., Heravi, B.: Patterns in award winning data storytelling: story types, enabling tools and competences. Digit. Journalism **6**, 693–718 (2018)
28. Mutsvairo, B.: Data Journalism: International Perspectives, Oxford Research Encyclopedia. Communication (Oxfordre.Com/Communication). Oxford University Press USA, pp. 1–16 (2022)
29. Leppänen, L., Munezero, M., Granroth-Wilding, M., Toivonen, H.: Data-driven news generation for automated journalism. Paper presented at the 10th International Conference on Natural Language Generation. Santiago de Compostela, Spain, September 4–7, pp. 188–197 (2017)
30. Appelgren, E., Lindén, C.G.: Data journalism as a service: digital native data journalism expertise and product development. Media Commun. **8**(2), 62–72 (2020)

Digital Public Relations and Communication Crisis

Abdulsadek Hassan and Mohammed Angawi

Abstract The study aims to identify the nature of digital public relations (PR) and the different means of communication in addressing the crises facing different institutions, and the study reached several results, the most important of which are: Digital public relations plays an important role in crisis management, as it is the link between the masses of the crisis and the crisis management team, administratively and in the media. Institutions that have prior plans for crisis management view the media as a good way to solve crises through the masses' knowledge of all information. Public relations is the art and science of communicating with the public benefiting from the services of an organization at a time of prosperity or crises in order to gain their confidence and build a mental image and the good reputation of the enterprise because it is the first line of defense that protects institutions when exposed to crises.

Keyword Digital public relations · Communication crisis · Strategies

1 Introduction

Digital public relations is the use of various media on the Internet and social media, to create and share content about the institution or person, and to reach the target audience and communicate with it through electronic media [1], that is, it is building and managing the reputation of the institution, and this is because forming the correct image of the institution in the minds of people is a matter very important in conveying the brand or organization to a distinct and pioneering identity [2].

In light of the development of social media and its communication networks, and the increasing use of them by the business sector, the media, with modern technologies, have thus contributed to highlighting the work of public relations in various

---

A. Hassan (✉)
Ahlia University, Manama, Bahrain
e-mail: aelshaker@ahlia.edu.bh

M. Angawi
University of Business and Technology, Jeddah, Saudi Arabia

© The Author(s), under exclusive license to Springer Nature Switzerland AG 2024
A. Hamdan and E. S. Aldhaen (eds.), *Artificial Intelligence and Transforming Digital Marketing*, Studies in Systems, Decision and Control 487,
https://doi.org/10.1007/978-3-031-35828-9_17

sectors and institutions on both the private and public levels [3]. In light of this development, the development of digital media represented in the professional use of social networks and their employment in the service of public relations, the digital public relations crystallized in the transition of relationship management and communication with the public from the traditional stage to the modern stage emerged. That keeps pace with the era of technology and its users [4]. It is no longer limited to media public relations and the sending of press releases only, but today digital public relations has become an art of integrating what was limited to working with traditional media and channels with new digital-marketing content [5], which serves the goals of the institution through what it publishes and communicates today [6]. On social media platforms and networks, which was reflected positively on the institution itself in search engines across the Internet [3].

With the development of the field of media and communication, and the increasing speed of the Internet's dominance in all fields and sectors, digital public relations is more comprehensive and more diverse, giving public relations and communication officials tremendous possibilities in conveying their message to the target audience in an effective manner [7]. This strategy is not limited to mastering the methods of speaking and rhetoric, in order to improve the image of the institution, but goes beyond it to work on knowledge collection and analysis, keeping pace with the requirements of the times, and focusing on value, speed and creativity by relying on modern technologies [8].

2 Types of Digital Public Relations

Digital PR professionals have experience communicating with a large variety of people on behalf of organizations in all sectors [9].

Given this scope, there are many different types of PR and different areas of focus [9].

1. Digital Strategic Communication

Every action a public relations professional takes should fall under strategic communication [10]. This basically means that all public relations efforts are coordinated to help the institution achieve its business goals [11].

Understanding the organization's priorities from the outset is a must, while defining communications objectives and subsequent activities to support these priorities [12].

2. Digital Media Relations

A good working relationship with the media is needed in order to disseminate key messages to the target audience [12]. Public relations professionals can shine the spotlight on clients by sending out press releases and giving interviews in the media that enable these companies to reach the most desirable and important audiences [13].

Digital Public Relations and Communication Crisis

Journalists need a consistent flow of news to fill websites, so creating compelling news stories for media that include organizations looking for a media offering is a win–win [14].

3. Digital Community Relations

While websites are an important outlet for PR professionals, sometimes the most effective means of communication is direct engagement with the community or audience [15].

Engagement with the community in which the organization operates must work in two ways. For example, when a institution opens a new facility, getting feedback from the local community via social media is just as important as highlighting the benefits to the local economy [1]. Good listening skills and the ability to coordinate events are essential for this [11].

4. Internal Communication

Internal communications quickly became a major focus area in public relations [1]. Employees can be either the institution's biggest advocates or its harshest critics, so keeping them satisfied, motivated, and loyal is critical to the institution's overall success [4].

Developing ongoing programs to keep employees engaged and informed, while understanding their needs and concerns, is a challenge for companies and internal communications professionals now play a critical role in helping them [11].

5. Crisis Communication

Contrary to popular opinion, the communications team should not be brought in only when a crisis occurs; That's too late [9]. Organizations must take a planned and consistent approach to crisis management, with a clear crisis communications plan in place, strong relationships with both stakeholders and the digital media created over time that they can rely on in such times [5]. This makes crisis communication as rewarding and valuable as it is challenging [16].

6. Public Affairs

Public affairs is one of the most important types of public relations [7]. Those who work in public affairs or lobbying participate in building and developing relationships between an organization and politicians, governments and other decision-makers [17].

It is a relatively distinct subgroup within public relations and those who work in this field have a keen interest in the political system and the process of enacting legislative change [18]. They can also add amazing value by providing assistance to organizations in areas such as regulatory compliance, corporate communications, and trade associations [16].

7. Online and Social Media Communication

In today's world of instant communication, it is imperative for businesses to have a strong online presence to stand out from their competitors.

Customers are increasingly turning to the web to do their own research before making a purchase, so digital PR has become of great importance in generating leads, building business relationships with blogs and social media, and attracting new talent [9].

Both organizations and PR professionals today must be adept at choosing the best social media platforms and other digital channels to achieve their communication goals [11].

3 Advantages of Digital Public Relations

3.1 Brand Awareness

One of the main advantages of digital PR is brand awareness [12]. The simple truth of brand awareness is that no business [16], no matter the industry, can succeed without it [18].

People don't buy from brands they don't know or invest in, and they can't learn about business if they aren't exposed to your brand and content [19].

This applies to specialized publications, as well as general industry publications [15]. Digital PR enables you to position your brand in front of the movers and shakers of your chosen industry [20].

3.2 Thought Leadership

Similarly, digital PR also provides an opportunity for thought leadership [14]. When you or an influencer issues a by line or provides an expert comment, it in turn helps establish your brand as an industry expert [7].

It is also important to note here that individuals who are consistently featured in the media are more likely to be considered industry experts and thought leaders for doing so [43].

3.3 Unpaid Visits

The next benefit of using digital PR is that it brings in unpaid (membership) traffic. Getting high-quality links through media placements are "votes" in Google's eyes, which Google then uses as a sign of trust and can boost your ranking [16].

When you bring more traffic to your website and actively work to improve the credibility of brand, people inevitably respond [2].

Digital Public Relations and Communication Crisis

This means more attention to brand and content, more people choosing service or product, and more revenue going into your institution's coffers [7].

3.4 Appearance and Ranking in Search Engine

The ultimate benefit goes hand in hand with unpaid visits [8]. When your content is constantly viewed and clicked on by untapped audiences, it plays a huge role in your ability to rank higher in search engines [14].

This is because Google looks at backlink cookies and the number of times people click on your links [4].

4 Digital Public Relations and Crises

Public relations originally originated and developed during crises, as during turmoil, strikes and social problems, the media becomes difficult, so it is necessary for the institution to explain and communicate with the public [13]. In this time full of media, it may be possible for an official to know that there are disturbances in his institution through the media and in this atmosphere [12], the interventions that complicate the situation and overlap the factors are increasing, which makes the work of the public relations official sensitive towards the external audience, as for the internal audience, the situation is unenviable [14].

Public turmoil weakens trust in the institution and creates an atmosphere of instability in the face of the overwhelming chaos and contradictory requirements of the situation [18]. To satisfy the public, the institution must be able to carry out a positive reaction [11]. This positive reaction cannot be successful unless it is prepared in advance and its application is thoughtful [18]. From here, the public relations official must master crisis management, because good crisis management may lead to giving a good image of the institution later, by exploiting what is happening [21]. Crisis management is in other words: knowing the identification of the problem or possible crises and devising appropriate solutions to them while maintaining calm during the storm [8]. So, for better crisis management, it is necessary to know what to do before, during and after the crisis [11].

5 Strategies for Facing Digital Public Relations Crises

1. The strategy of violence in dealing with the crisis

This strategy is used with the unknown crisis for which there is insufficient information [13]. And hitting the igniting fuel for the crisis or stopping feeding the crisis with the fuel necessary for its continuation [16].

2. The strategy of stopping growth

This strategy aims to focus on accepting the fait accompli and making efforts to prevent its deterioration and at the same time striving to reduce the degree of impact of the crisis and not reach the degree of explosion [12]. This strategy is used in the case of dealing with issues of public opinion and strikes [7]. Some concessions and meeting some requirements in order to create the conditions for direct negotiation and resolution of the crisis [4].

3. The fragmentation strategy

This strategy depends on the study and analysis of the constituent factors and the influencing forces, especially in large and strong crises, where they can be transformed into small crises, which facilitates dealing with them [14]. Here, a conflict of interest can be created between the large parts of the crisis and the struggle to lead and co-opt the parts and present temptations to strike alliances [2].

4. The strategy of aborting the thought that creates the crisis

The thought behind the crisis in the form of certain trends represents a severe impact on the strength of the crisis [1]. This strategy focuses on influencing this thought and weakening the foundations on which it is based, as some forces depart from it and weaken the crisis [16]. Here, questioning the constituent elements of thought can be used [18]. Solidarity with this thought and then abandoning it and causing division [3].

5. The strategy of pushing the crisis forward

This strategy aims to speed up pushing the forces participating in the crisis industry to an advanced stage, showing their differences and accelerating the existence of conflict between them [6]. This strategy is used to leak false information and make tactical concessions to be a source of conflict and then benefit from it [8].

6. Lane change strategy

It aims to deal with sweeping and severe crises that are difficult to stand in front of and focuses on riding the crisis vehicle and walking with it for the shortest possible distance, then changing its natural path and turning it into paths far from the direction of the top of the crisis [20].

6 Traditional Public Relations Methods and Digital Public Relations Methods in Crisis Management

6.1 First, the Traditional Methods

1. **Denial of the crisis**: where a media blackout is practiced on the crisis and the denial of its occurrence, and to show the solidity of the situation and that the conditions are fine, in order to destroy the crisis and control it [22]. This method is often used under dictatorial regimes, which refuse to acknowledge the existence of any defect in their administrative entity [11]. The best example is the denial of exposure to the epidemic or any health disease and so on [7].
2. **Suppressing the crisis**: It means postponing the emergence of the crisis, and it is a kind of direct dealing with the crisis with the intent of destroying it [8].
3. **Extinguishing the crisis**: It is a very violent method based on a public and violent clash with the forces of the Azmawi current, regardless of human feelings and values [18].
4. **Underestimating the crisis**: that is, underestimating the crisis (from its impact and results) [7]. Here the existence of the crisis is recognized, but as an insignificant crisis [11].
5. **Venting the crisis**: It is called the method of venting the volcano, where the manager resorts to venting the pressures inside the volcano to relieve the state of boiling and anger and prevent the explosion [15].
6. **Unloading the crisis**: According to this method, multiple and alternative paths are found in front of the main and secondary impetus that generates the current of the crisis to turn into many alternative paths that absorb his effort and reduce his danger [19].
7. **Isolating the forces of the crisis**: the crisis manager monitors and identifies the forces behind the crisis and isolates them from the course of the crisis and its supporters, in order to prevent its spread and expansion, and thus ease of dealing with it, and then solve it or eliminate it [1].

6.2 Second: Digital Methods

They are appropriate methods for the spirit of the age and compatible with its variables [11]. The most important of these methods are the following:

1. **Teamwork method**: It is one of the most widely used methods at the present time, as it requires the presence of more than one expert and specialist in different fields in order to calculate each of the factors and determine the required behavior with each worker [13]. These methods are either temporary or permanent work methods of specialized cadres that are formed and prepared to face crises and times of emergency [11].

2. **Tactical reserve method to deal with crises**: where weaknesses and sources of crises are identified, so a tactical and preventive reserve is formed that can be used if the crisis occurs [2]. This method is often used in industrial organizations when there is a crisis in raw materials or a shortage of liquidity [11].
3. **The method of democratic participation to deal with crises**: It is the most effective method and is used when the crisis is related to individuals or is centered on a human element [6]. It means this method discloses the crisis and its seriousness and how to deal with it between the boss and subordinates in a transparent and democratic manner [9].
4. **Containment method**: In trapping the crisis in a narrow and limited scope [3]. Examples of this are labor crises, where the method of dialogue and understanding is used with the leaders of those crises [13].
5. **The method of escalation of the crisis**: It is used when the crisis is not clear-cut and when there is a cluster at the stage of the formation of the crisis, so the person dealing with the situation deliberately escalates the crisis to break this cluster and reduce the pressure of the crisis [19].
6. **The method of emptying the crisis of its content**: It is one of the most successful methods used, where each crisis has a specific content, which may be political, social, religious, economic, cultural, administrative, and others [3].
7. **The method of breaking up crises**: It is the best if the crises are severe and dangerous [5]. This method depends on studying all aspects of the crisis to know the forces that constitute the crisis alliances and to determine the framework of conflicting interests and potential benefits for the members of these alliances [5], and then multiply them by creating artificial leaders and finding gains for these trends that are in conflict with the continuation of Aesthetic alliances [16]. Thus, the major crisis turns into small, fragmented crises [14].
8. **The method of self-destructing the crisis and detonating it from the inside**: It is one of the most difficult non-traditional ways to deal with crises and it is called the method violent confrontation or direct confrontation and is often used in the absence of information, and this is its danger and is used in the case of certainty that there is no alternative [15].
9. **Containing and diverting the course of the crisis**: It is used with very violent crises whose escalation cannot be stopped [4]. Here, the crisis is transferred to alternative paths. The crisis is contained by absorbing its consequences [9], acquiescing to it, recognizing its causes, then overcoming them, and treating its secretions and consequences, in a manner that reduces its risks [19].

Digital Public Relations and Communication Crisis

7 The Foundations of Digital Public Relations Dealing with Crises and Its Principles

Confronting the crisis since its inception through the stage of reducing its danger and even overcoming it requires commitment to several basic principles that are the beginning of its success, and confronting it is as follows:

1. Setting goals and priorities

This factor is one of the most important factors for success in facing the crisis, especially the main goal, which is often not clear. Knowing the main cause represents 50% of treating it and confronting it [2]. It is necessary to coordinate the goals and determine their priority, as the main goal of facing the entire crisis may not be possible or outside the available capabilities so it will be fragmented, and setting the goal does not mean selecting the risk factor that may involve some failures or successes [18].

2. Freedom of movement and speed of initiative

This step is the first step to achieving the goal, as it distances decision-makers from being affected by shocks and allows them to take the initiative that subject the crisis to the factor of backlash, so it can be controlled, and its danger reduced [7].

3. Surprise

Surprise almost achieves complete control over the crisis and for appropriate periods, as declaring the method of confronting it may result in the failure of the efforts made to solve it, while the results of the surprise allow reducing its danger and eliminating it, as close to the target as possible [16].

4. Mobilizing and organizing forces

Possession of force is one of the factors of success in facing the crisis and having the desired effect in the local and international environment according to its scope [16]. The organization of forces aims to mobilize all material and human capabilities and mobilize them morally to enable them to face the crisis and eliminate it [11]. The force organizes multiple components, some of which are related to the place of the crisis [13]. The other is related to the time of the crisis and the stage it reached [4]. The mobilization of power includes five basic aspects, which are the geographical power resulting from human interaction with the place, the environmental resources [1], the economic forces that are represented in the available resources, the military power in terms of its size, the type of its formations, and its morale, and the influence aspect, which means the organized influence effort [14]. In public opinion at home and abroad, which limits the ability and effectiveness of the other party and weakens its forces [12]. The crowd should not be illusory, it must take into account the techniques and human expertise that can be activated to face the crisis [4].

5. Cooperation and effective participation

The available capabilities may be unable to confront the emerging crisis, whether it is local or international, so it is imperative to seek help from external support that doubles the energies to confront it, but rather helps to broaden the vision, comprehensiveness, specialization and integration of the confrontation [14], in addition to the speed and accuracy resulting from the diversity of expertise, skills and capabilities [12].

6. Constant control of events

The rapid and growing succession of the events of the crisis intensifies its negative effects resulting from the polarization of external factors supporting it [5]. Therefore, dealing with it requires superiority in controlling its events through full knowledge of its developments [4]. This process also requires dealing with the factors causing the crisis and the forces supporting it [7].

7. Comprehensive insurance for people, property and information

Physical insurance of people and property is an absolute necessity to face the crisis, as the minimum natural insurance must be provided for each of the people, property and information before it occurs, and means of prevention must be provided [5]. Likewise, the additional vital insurance when the crisis actually occurs and its forces must be confronted with stronger forces than them to stop its growth and limit the extension of its fields [7]. Its strength is the formation of effective precautions that the entity may need to overcome the crisis [16].

The existence of an insurance system is an absolute necessity to confront crises, and this system prevents penetration of the hostile side, withholds information from it and isolates it internally and externally, which is the beginning of the end of the crisis [11].

8. Quick confrontation to the events of the crisis

The scientific progress that the world has witnessed had a profound impact on the nature of crises, which have become rapidly evolving and necessitated a rapid response to them, which necessitates the presence of scientific cadres trained to confront crises [18]. If a quick confrontation is vital, it must be precise in order not to exacerbate the crisis [13].

9. Economy in the use of force

Determining the capabilities designated to confront the crisis must be subject to accurate calculations [8]. Excessive use of force is a waste of potential in terms of its expenditure compared to the rate of safety that it provided [14], and the adverse reaction resulting from the crisis [6]. In addition, the excessive and exaggerated use of force have a reaction and transform the public manifestations of the crisis into a hidden pressure that is difficult to follow or to notice its development accurately [7].

8 For Professional Overlap Between Digital Public Relations and Crisis Management

The follower of the history of public relations sees that it is a product of the events, problems and chronic issues that have afflicted the past two centuries, which has strengthened the position of digital public relations, which has become looking for ways and means to address these problems [12].

The institution is in the humanitarian fields, and this confirms the close relationship between problem management, and considering this management an important task that characterizes digital public relations [3].

The tasks and duties performed by digital public relations are increasing day by day, at a time when these tasks seem clear between the role of public relations and the crisis if it occurs [18]. Public relations and crisis management are two different sciences on the surface, but they are similar in many foundations and rules [14]. The good, which is followed by honest effective media and crisis management, and it depends on the same rule, and in order for this to succeed [8], it is necessary to plan for all the various activities and tasks that it undertakes in order to achieve the desired goal, which is to build a good image of the institution [7]. The biggest loss for the institution is usually its image in front of its audience, and here lies the role of public relations in improving this image [18].

Achieving this goal pushes digital public relations towards building bridges of mutual trust between its different audiences, which reduces the opportunities available for the occurrence of crises [1]. In addition, the research and studies conducted by the institution play a role in predicting them and taking the necessary measures to avoid their occurrence [15]. It is worth mentioning that the great transformation that occurred in administrative thought and the modernization movement witnessed in administrative studies led to the increasing recognition of the role and importance of digital public relations in contemporary institutions [7]. The public relations manager is often exposed to many problems during his work, which require urgent solutions to these problems [7]. As for the crisis, it is considered a real test for it. Either the institution succeeds or fails [4].

9 Case Study

9.1 The Role Digital Public Relations to Confront the Corona Crisis

Digital public relations have benefited greatly from the technical development and mechanization of the work cycle within the institution, which has been positively reflected on the conduct of its business during the Corona crisis.

Attracting major international and local brands operating in the market in various economic sectors [14]. It has also succeeded in organizing many major economic events, and with the Corona crisis, the world has been in a state of severe confusion that affected the conduct of business in all sectors [19].

Taking serious steps in anticipation of the world coming to a close, so we had only technology to support the business cycle [2].

Digital public relations have benefited greatly from the digital transformation that has occurred in the world in recent years [19], and the infrastructure and that telecommunications companies have contributed greatly to the business cycle of companies during the Corona crisis and enabled them to easily accomplish their work with the same efficiency [8].

Supporting employees with various communication tools and devices that help them accomplish their work, including (communication lines, internet, printers, scanners), so technology and its tools were the hero that helped us overcome the previous stage [12].

About the aspect related to news monitoring in newspapers that have PDF issues, and those newspapers were addressed to send their issues periodically to us, and with regard to the accounts sector, a very large part of the invoicing was transferred to electronic, and partners were supported with direct bank transfers instead of checks [14].

Holding periodic meetings on a daily basis via video through various applications, including Zoom, and Webinar, and that they succeeded in maintaining the efficiency of the work cycle, which allowed presence in various events and conferences as one of the pillars of the normal work cycle in this field [5], and with Corona, digital public relations thinking in thinking about the means of technology to hold conferences remotely, and great positives were discovered for the new form, including saving time for the entire system's members (clients, media, and employees) [13]. The crisis, but it does not give the same effect in communication in its natural form, the side discussions that were contributing to the development of business have been lost, and no matter how technology is able to play a certain role, the idea of social and human convergence during conferences cannot be compensated by technology [16].

Journalists responded to the new form of communication with sources through digital public relations through modern technologies, because there were no other alternatives, so modern methods became a convenient and accurate means, pointing out that the media industry had a precedent in managing editorial galleries with technology before the Corona crisis, so it was natural Journalists accept the new form of communication [18].

The matter is divided into two parts, the first is if the epidemic continues and there is a state of coexistence, technology will contribute to maintaining the new caveats, but if we reach a complete closure, it is impossible for any means to be able to achieve something, due to the interruption of work clients [16].

The work cycle will not return in the same traditional way as before, and that whoever returns in the same way to before the Corona crisis represents an injustice to institution and takes steps back [3], so we have some departments that will perform their role some days of the week from home, and this leads to supporting the country economically [7, 23], and reduces the bill of life daily for the employee and the institution, and instead of the employee spending 25% of his income because of his presence at the institution 's headquarters, transportation and food, he will provide them with additional income, in addition to giving him the opportunity to perform other duties that increase the social dimension positively, and the institution can reduce its area [4]. It leads to saving expenses that will reflect positively on customers by reducing the pricing of services [11].

The media and public relations play an essential role in managing crises and linking the bodies working to address crises with each other so that they can bear fruit and so that they can exchange information and coordinate among themselves to work in order to confront any ordeal the state is going through, as well as informing the people of the latest developments in the work of government agencies [16], and reassuring the people that the state and its institutions [13]. It harnesses all its capabilities to deal with crises [8].

The media, with its smart tool "public relations", can eliminate rumors and false information and confront them by obtaining the correct information from the source and publishing it in the official newspapers and the state TV channel [2]. The cases infected with the Corona virus and their history on these cases or not, as these rumors would affect the security of the country and its public order [18].

The ministries of health in the world have resorted to the media platform through public relations in several statements explain in a diligent, diligent and daily manner about all developments [19], and public relations take its modern form in social media, where it is inform the public about cases that continue to recover directly, as well as any update on critical cases and quarantine [14].

The role of public relations today is not hidden from anyone in the crisis stage and is confirmed by the masses by addressing the problems and taking responsibility in all statements and striving to avoid mistakes in pursuit of the optimal use of technological resources [7].

However, a major problem that institutions face today is the formation of an integrated system in their various activities that makes the Public Relations a real platform based on improving the good image of the institution in front of the public and clarifying the role of the importance of public relations in addressing and treating crises of all kinds [4], in proportion to the tendencies and capabilities of creativity, and this requires Providing an ideal work environment managed in a way that develops a sense of belonging, enhances sincerity and the desire to contribute, and pushes employees to take initiatives without fear or failure in front of the public [15].

10   Conclusion

The results indicated that digital public relations in light of the Corona pandemic is concerned with conveying and presenting the distinguished point of view supported by evidence, their appearance on the search engines, and the readiness to develop a long-term strategy based on planning for the future, while continuing to follow up on competitors in the sector and monitoring what they are talking about; to outperform them.

The results revealed that public relations manages and solves crises and problems, due to the dimensions that these crises contain that affect the social, economic, environmental and even political aspects.

The results also revealed that the function of the administration, which helps in establishing a line of mutual communication, acceptable understanding and cooperation between the institution and its public, especially in times of crisis, and includes managing problems and issues, and helps the administration to communicate and deal with public opinion, stresses the responsibility of the administration in serving the community and taking into account its interest, and maintaining survival along with management, and to benefit as much as possible from the changes in an effective way, it is also considered as an early warning system that precedes events and its way is the use of research, honest and specialized communication, and public relations is based on completing and enhancing the work of the institution, through specialized consultants, seeking to strengthen the institutional identity of the institution, they also aim to promote a modern image of the society, in which the institution is active, taking into account the general loyalty to the provision of traditional services to the institution.

References

1. Kutoglu Kuruç, U., Opiyo, B.: Social media usage and activism by non-western budding PR professionals during crisis communication. Corp. Commun. Int. J. **25**, 98–112 (2020)
2. Bachmann, P.: Public relations in liquid modernity: how big data and automation cause moral blindness. Public Relat. Inquiry **8**(3), 319–331 (2019)
3. Dozier, D.M., Shen, H., Sweetser, K.D., Barker, V.: Demographics and internet behaviors as predictors of active public. Public Relat. Rev. **42**(1), 82–90 (2016)
4. Davies, C., Hobbs, M.: Irresistible possibilities: examining the uses and consequences of social media influencers for contemporary public relations. Public Relat. Rev. **46**(5), 101983 (2020)
5. Ayman, U., Kaya, A.K., Kuruç, U.K.: The impact of digital communication and PR models on the sustainability of higher education during crises. Sustainability **12**, 8295 (2020). https://doi.org/10.3390/su12208295
6. Sommerfeldt, E.J., Yang, A.: Notes on a dialogue: twenty years of digital dialogic communication research in public relations. J. Public Relat. Res. **30**(3), 59–64 (2018).
7. Hagelstein, J., Einwille, S., Zerfass, A.: The ethical dimension of public relations in Europe: digital channels, moral challenges, resources, and training. Public Relat. Rev. **47**(4), 102063 (2020)
8. Arief, N.N.: Public Relations in the Era of Artificial Intelligence. Simbiosa Rekatama Media, Bandung (2019)
9. Nguyen, M.H., Gruber, J., Fuchs, J., Marler, W., Hunsaker, A., Hargitta, E.: Changes in digital communication during the COVID-19 global pandemic: implications for digital inequality and future research. Soc. Media + Soc. 1–6 (2020)
10. Suciu, L.: Methods for increasing the efficiency of the interdisciplinary learning in the master's program communication, public relations and digital media. Proc. Soc. Behav. Sci. **46**, 3588–3592 (2012)
11. Salma, A.N.: Analyzing online public sentiment toward corporate crisis in the age of big data and automation. J. Soc. Media **6**(1), 188–206 (2022)

12. Timothy Coombs, W., Holladay, S.J.: Strategic intent and crisis communication: the emergence of a field. In: Zerfass, A., Holtzhausen, D. (eds.) The Routledge Handbook of Strategic Communication. New York (2015)
13. Verčič, D., Verčič, A.T., Sriramesh, K.: Looking for digital in public relations. Public Relat. Rev. **41**(2), 142–152 (2015)
14. Husaina, K., Abdullaha, A.N., Ishakb, M., Kamarudina, M.F., Robania, A., Mohinc, M., Hassana, M.N.S.: A preliminary study on effects of social media in crisis communication from public relations practitioners' views. In: The International Conference on Communication and Media 2014 (i-COME'14), 18–20 October 2014, Langkawi, Malaysia. Proc. Soc. Behav. Sci. **155**, 223–227 (2014)
15. Alexander, D.M.: What digital skills are required by future public relations practitioners, and can the academy deliver them? PRism **13**(1), 1–13 (2016)
16. Almansa-Martínez, A., Fernández-Souto, A.-B.: Professional public relations (PR) trends and challenges. El profesional de la información **29**(3), e290203 (2020)
17. Vercic, D., Vercic, A.T., Sriramesh, K.: Looking for digital in public relations. Public Relat. Rev. **41**(2), 142–152 (2015)
18. Macnamara, J., Zerfass, A.: Evaluation stasis continues in PR and corporate communication: Asia Pacific insights into causes. Commun. Res. Pract. **3**(4) (2017)
19. Camilleri, M.A.: Strategic dialogic communication through digital media during COVID-19 crisis. In: Camilleri, M.A. (ed.) Strategic Corporate Communication in the Digital Age. Emerald, Bingley, UK, pp. 1–22 (2020)
20. Gulerman, N.I., Apaydin, F.: Effectiveness of digital public relations tools on various customer segments. J. Manage. Market. Logistics (JMML) **4**(3), 259–270 (2017)
21. Wirtz, J.G., Zimbres, T.M.: A systematic analysis of research applying 'principles of dialogic communication' to organizational websites, blogs, and social media: implications for theory and practice. J. Public Relat. Res. **30**(1–2), 5–34 (2018)
22. Tankosic, M., Ivetic, P., Vucurevic, V.: Features of interactive public relations: using web 2.0 to establish a two-way communication with the consumers. Int. J. Econ. Manage. Syst. **1**, 290–295 (2016)
23. Ben Ayed, W.: The Tunisian stock market before invoking article 80 of the constitution: the (in)direct impact of government interventions during the sanitary crisis. J. Bus. Socio-economic Dev. Vol. ahead-of-print No. ahead-of-print (2022). https://doi.org/10.1108/JBSED-02-2022-0022
24. Cheng, Y., Wang, Y., Kong, Y.: The state of social-mediated crisis communication research through the lens of global scholars: an updated assessment. Public Relat. Rev. **48**(2), 102172 (2022)
25. Huang, Y.H.C., Wu, F., Huang, Q.: Does research on digital public relations indicate a paradigm shift? An analysis and critique of recent trends. Telematics Inform. **34**(7), 1364–1376 (2017)

Employing Applying Big Data Analytics Lifecycle in Uncovering the Factors that Relate to Causing Road Traffic Accidents to Reach Sustainable Smart Cities

Mohammad H. Allaymoun, Mohammed Elastal, Ahmad Yahia Alastal, Tasnim Khaled Elbastawisy, Dana Iqbal, Amal Yaqoob, and Adnan Sayed Ehsan

Abstract This paper aims to identify the factors that relate to causing road traffic accidents to reach sustainable smart cities—this issue is significant to society and people. To that end, the paper pro-poses a model for uncovering such factors using big data analytics lifecycle: discovery, data preparation, model planning, model building, communication of results, and operationalization. In other words, the paper presents a simplified methodology taking advantage of the big data analysis life cycle and the velocity of analyzing data to reduce traffic accidents—velocity is the most important characteristic of big data. The velocity in analyzing data and identifying the factors

M. H. Allaymoun (✉)
Administrative Science Department, College of Administrative and Financial Science, Gulf University, Sanad 26489, Kingdom of Bahrain
e-mail: alkarak1@yahoo.com

M. Elastal
Mass Communication and Public Relations, College of Communication and Media Technologies, Gulf University, Sanad 26489, Kingdom of Bahrain
e-mail: dr.mohammed.elastal@gulfuniversity.edu.bh

A. Y. Alastal · T. K. Elbastawisy · D. Iqbal · A. Yaqoob · A. S. Ehsan
Accounting and Financial Science, College of Administrative and Financial Science, Gulf University, Sanad 26489, Kingdom of Bahrain
e-mail: dr.ahmed.alastal@gulfuniversity.edu.bh

T. K. Elbastawisy
e-mail: 190101114975@gulfuniversity.edu.bh

D. Iqbal
e-mail: 190101114925@gulfuniversity.edu.bh

A. Yaqoob
e-mail: 190101115029@gulfuniversity.edu.bh

A. S. Ehsan
e-mail: 200101115241@gulfuniversity.edu.bh

© The Author(s), under exclusive license to Springer Nature Switzerland AG 2024
A. Hamdan and E. S. Aldhaen (eds.), *Artificial Intelligence and Transforming Digital Marketing*, Studies in Systems, Decision and Control 487,
https://doi.org/10.1007/978-3-031-35828-9_18

that relate to causing car accidents helps decision-makers quickly make appropriate decisions. Worthy of mentioning that Google Data Studio (GDS) was used here to produce the visualizations and reports needed. Hopefully, researchers will build on this methodology in the future to achieve sustainability, to propose a more inclusive and effective methodology to minimize the negative effects of traffic accidents in the smart cities.

Keywords Big data analysis lifecycle · Big data · Decision making · Google data studio · Road traffic accidents · SDG11

1 Introduction

Information technology is one of the key elements of the modern era and has recently attracted greater attention [1]. As a result of the advancement of these technologies and the significance of having analytical tools capable of swiftly and accurately extracting knowledge, the volume of data has continued to grow and increase. What best describes the huge amounts of data we have today is the term "big data" [2]. Big data refers to complex data that is unstructured and whose quantity exceeds the capacity of traditional analytical tools to store, process, and distribute it [3]. According to Zhang et al. [4] big data is a dataset that cannot be recognized, acquired, managed, processed, or studied using the available methods. Big data is a term used to describe a combination of recent and antiquated technologies that help businesses gather useful data. To enable real-time analysis and feedback, big data is defined as the capacity to manage a large volume of diverse data promptly, Furthermore, the three categories that are typically used to categorize big data are the volume of the data, velocity (how quickly data can be processed), and variety of data [5, 6].

Prescriptive, predictive, diagnostic, and descriptive analytics are the four main subfields of big data analytics. One of the most well-known technologies for analyzing large amounts of data is Hadoop, which was created in 2006. Hadoop is entirely open source and may be used with cloud infrastructure to handle data on a bigger scale. Hadoop can also use network computers to handle a wide range of issues [7]. Another useful tool is Qlik, created in 1993 as a business intelligence tool. It is a user-friendly tool that enables reliable report development [8]. The ability to swiftly evaluate data from websites, sensors, and social networking sites—three of the most significant big data sources—is one of the most crucial characteristics of big data. Since the examination of this data enables correlations between a collection of separate data to exist, various characteristics can be revealed [9]. For instance, predicting commercial business trends, tackling crime in the security industry, and other things. These forecasts also give decision-makers cutting-edge tools to better comprehend the circumstances and make wise judgments that accomplish the desired aims [4].

Nowadays, the number of cars has increased and, as a result, the number of accidents has increased. In this paper, traffic records will be analysed using Google

Data Studio to produce graphical results and statistics, which decision-makers can use to make appropriate decisions to deal with this issue. In order to reach a proper decision, there is a specific cycle to the data analysis that must be followed, It consists of the following six phases: discovery, data preparation, model planning, model building, communicating results, and operationalization. Each of these phases has its importance and characteristics to come up with the intended results [6, 9, 10].

This research paper aims to examine the factors that relate to causing traffic accidents such as age, gender, and years of driving experience. The paper also helps to establish a pattern based on a number of hypotheses that must be applied to uncover the factors that relate to causing traffic accidents. Additionally, it discusses all stages of the big data analytics lifecycle, starting from the discovery phase and ending up with the operationalization phase. The research will use big data analytics lifecycle to produce useful and appropriate solutions to the problem. In other words, this paper aims to discover whether there is a relationship between the independent variables such as age, gender, and years of driving experience and traffic accidents (the dependent variable). To that end, Google Data Studio will be used to analyze data and produce graphical results to investigate the problem and provide decision-makers with solutions to that problem.

2 Literature Review

Big data is a term used to describe massive and complicated data sets that are challenging to store, analyze, and visualize for use in following operations or outcomes. Big data analytics is the process of analysing massive amounts of data to find undiscovered patterns and connections [11]. This crucial knowledge can give businesses and organizations deeper and richer insights that will give them a competitive advantage.

The three key components of big data are volume, velocity, and variety. Its sources include social media, portable devices, and the internet of things (IoT) [12]. Its forms include structured, semi-structured, and unstructured data. Such data come from social networking connections, science data, sensors, mobile phones, and the apps that go along with them, as well as online transactions, emails, videos, audio, photos, click streams, logs, postings, and search queries [13].

Big data analytics helps businesses access a variety of various data sources, obtain fresh insights, and take the right action. This leads to quicker and better decision-making. Additionally, identifying the trends and insights that aid in more precisely predicting corporate performance lowers costs and operational efficiency. It also aids in locating and identifying hazards by developing patterns for traffic accidents [14]. There are several advantages to big data analysis. Still, two in particular stand out: developing effective algorithms that can accurately predict future observations and gaining an understanding of the relationship between traits and response for scientific purposes [15]. Big data also helps with two additional objectives: comprehending heterogeneity and commonality across various subpopulations. These are made possible by the enormous sample size. Alternatively, big data promises to: (a) explore the hidden structures of each data subpopulation, which was previously impossible to do and can even be treated as "outliers" when the sample size is small, and (b) extract significant common traits across many subpopulations, despite significant individual differences [16].

The typical purpose of the data analytics lifecycles is to address issues big data-using enterprises face. With the use of the big data analytics lifecycle, which consists of six stages, analysts may determine priorities, specify requirements, pinpoint needs, and provide a general sense of expected results. This methodical approach aids businesses in redefining their goals and organizing operations and processes including data collection, processing, and analysis [17]. The big data analytics lifecycle is a circular process that includes discovering the problem, data preparation, model planning, model building, communicating the results, and operationalizing the results. A big data-based application, like the one used by insurance firms, produces large amounts of data about traffic, which can be grouped into a model for predicting what will happen and generating reports to benefit individuals and insurance firms. The complete data process in the system is represented by the data life cycle [18].

In order to successfully control traffic, a city in Zhejiang connected over 100 intelligent monitoring checkpoint systems, over 300 electronic police checkpoints, and over 500 video monitoring systems, according to a case study from Intel [19]. The city traffic management division's data center houses all structured data from collection devices such as time, location, and vehicle information, while a centralized data center houses semi-structured data such as images and videos. The traffic management division analyses the real-time data to monitor traffic conditions, traffic accident statistics, traffic infractions, and driver histories, among other things. The size of traffic monitoring operations has increased along with the city's continued growth. Reliable storage is required for the data gathered by monitoring equipment. Additionally, the size of the photographs is larger than before due to the advancement of new technologies and the update of the electronic police checkpoint system to high-definition video images, which calls for improved storage performance [20]. The local administration collaborated with Hangzhou Trustway Technology Co., Ltd. to address its big data dilemma (Trustway). Trustway developed a dynamic monitoring system for essential vehicles based on Apache Hadoop and Intel Xeon processor E5 series in accordance with the demands. The solution integrates with the city's checkpoints, video surveillance, traffic flow sensing, signal systems, and other devices for traffic monitoring. Effective decision-making can be achieved by using data analysis. In the future, Trustway will keep collaborating with Intel on huge data analysis, utilizing newer technology to deliver more intelligent traffic management solutions [21].

A new data visualization program called Google Data Studio seems to be an easy-to-use tool for displaying enormous data sets in a way that is aesthetically pleasing and accessible and with graphical outputs appropriate for a variety of analytical studies [22]. Data Studio provides capabilities including the ability to merge several sources into a single report, dynamic data updates, and interactive visualizations in addition to using the conventional combination of charts and graphs to convey information. Though its main function is to help users understand data and web analytics, Data Studio also supports a broad variety of data sources, such as MySQL and spreadsheets, suggesting that academics can use it to quickly analyse their data. However, if Data Studio's support for outside data sources was enhanced, it might be a useful tool for aspiring data photographers [23].

3 Research Methodology

The idea of this research paper is to review the phases of the big data analytics lifecycle to discover the factors that relate to causing traffic accidents. In other words, it aims to discover whether there is a relationship between the independent variables such as age, gender, and years of driving experience and traffic accidents (the dependent variable). To that end, the paper proposes a model for uncovering such factors using big data analytics lifecycle: discovery, data preparation, model planning, model building, communication of results, and operationalization. In addition, relevant previous studies will be surveyed, and primary data will be analyzed using Google Data Studio.

The primary semi-structured data analyzed here were collected from Kaggle. After that, a model was built based on two hypotheses developed to help identify the factors that relate to causing traffic accidents [24].

4 Big Data Analytics Lifecycle

In this part, all phases of the big data analysis lifecycle will be reviewed to reach a methodology to identify the factors that relate to causing road traffic accidents this issue is significant to society and people. In other words, the methodology proposed here was developed to help decision makers to uncovering the factors that relate to causing traffic accidents. To that end, the paper used hypothetical data and models to discuss the results.

4.1 Phase 1: Discovery

Car accidents have been increasing through the past period involving all genders and ages of drivers. The expert analyst found connections between the rate of accidents and the age of drivers and the type of car they drive. The analyst collected data during the discovery phase as they had different opinions and data about a variety of difficulties.

- **Learning the Business Domain**

The analyst will be using multiple tools in order to gain an understanding of the information. The analyst will use google data studio, which will produce the charts and graphs for the information presented. The data is also taken into consideration the impact of car accidents on drivers around the region.

- **Resources**

Data is a collection of structured and unstructured data such as numbers, concepts, dates, locations, and various angles. A documented financial plan, figure, and analysis are also provided by the information, and these will be used to confirm the group's suspicions. Both structured and unstructured data is used in this part, but research groups frequently share unstructured data.

- **Framing Problem**

Data on drivers and the impact of the percentage of the accident were used in this study; this has been done to check the team's assumption; both structured and unstructured information are presented in this analysis, it also showed how the experience of the drivers and type of vehicles used affect the rate of accidents occurred, there the lower the experience, the higher percentage of accidents, and if the car is smaller and private the accident rate is lower, The analysts have to perform research, gather data, evaluate it, and simplify it using graphs so that management and the board of directors can make informed decisions. The model's key goals are to provide a complete report using professional methods that demonstrates the presence of a problem for drivers regarding car accidents.

- **Developing Initial Hypotheses**

    H1: The relation between years of driving experience and car accidents [25].
    H2: The connection between the type of car driven and car accidents [26].

4.2 Phase 2: Data Preparation

At this stage, information regarding the drivers has been collected, and the big data available for the drivers has been analyzed. The analysts did the work on an extensive database, and there was a clear connection between the number of years of driving experience and the type of car on the one hand and the number of accidents on the other hand. After analyzing the data, we found the main issue.

- **Preparing Analytic Sandbox**

An analytical sandbox allows professionals to do discovery and situational analytics. Sandboxes examine the behavior while operating, making them useful against

Employing Applying Big Data Analytics Lifecycle in Uncovering ... 199

malware that avoids static analysis. A sandbox is safer than other behavior analysis tools since it avoids executing a suspect item in the entire business infrastructure. This method is aimed at business analysts.

- **ETLT**

We have taken the data from the data source used, then the data was turned into CVS and put into google data studio to create the report and dashboard used in this report that is shown clearly and easily.

- **Learning About the Data**

Results are obtained regarding the drivers excellently, and the problem at hand is illustrated and shown in the graph as the impact and issue are clear.

- **Data Conditioning**

The data management and optimization techniques used to efficiently route, optimize, and safeguard data for storage or data flow in the system are known as data conditioning. In the cleaning process, the data is segregated, and important information is separated from non-important ones.

4.3 Phase 3: Model Planning

The data taken and shown is on the drivers' accident rate and the causes of these accidents, showing how logical judgment has been taken to talk about the problem at hand. Establishing efficient hypothesis testing for the data to be tested takes much work.

H1: The relation between years of driving experience and car accidents.

H2: The connection between the type of car driven and car accidents.

In this paper, objective criteria developed were used in order to generate ideas that will help with the analysis; the following has been included in the analysis done: -

- Years of Experience Driving
- Gender of Driver
- Type of Vehicle
- Service Year of the Vehicle
- The area of the accident occurred.

Figure 1 shows how the Age of the driver, place of accident, type of car, and time of the accident has affected how the accident happened and how did these factors make the accident happen more for these categories; 28% of the accidents occurred because of the age of the driver on a rate of 33 accidents and 32 vehicles were involved, 31% of the accidents occurred because of the place of the accident on a rate of 25 accidents and 30 vehicles were involved, 23% of the accidents occurred because of the type of car on a rate of 22 accidents and 31 vehicles were involved. Finally, 18% of the

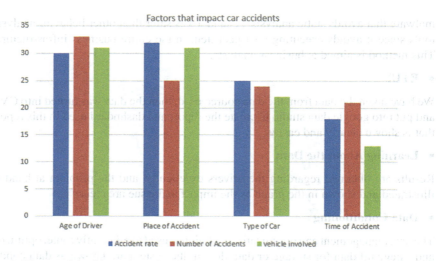

Fig. 1 The age of the driver, place of accident, type of car, and time of the accident

accidents occurred because of the type of car driven in 19 accidents involving 16 vehicles.

- **Explanation**

It can be concluded that many factors can increase the rate of accidents. When these factors are there, the accident rate is higher.

This shows as the younger the driver is, the higher the percentage of accidents that happens as he will have less time driving and will be vulnerable to getting into an accident; you can see also the type of car is also essential as the smaller the car, the more accidents will happen as the car will be light and affected by the wind and road. The timing of the accident also is very critical as it shows that in the night-time, the accident rates are higher due to lack of vision from the drivers.

4.4 Phase 4: Model Building

Methods for analysis of the causes of car accidents between drivers due to various factors through multiple analytic approaches. This contains the effort of data analysts who used different methodologies to find the issue at hand and solve it. In addition, analysts used modern and new technologies to perform the analysis in the best way possible.

Google Data Studio

In this part, a detailed graphical report, extracted by Google Data Studio, reviews the relationships arising from linking variables, which were built based on theories that

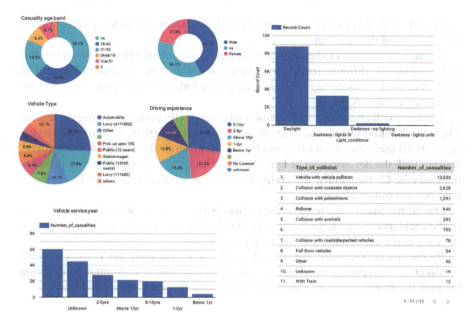

Fig. 2 The age of the driver, place of accident, type of car, and time of the accident

were imposed in the discovery phase. In addition, with the use of Google Studio, it can read the vast amounts of data available in Excel, which displayed the driver's age, location of the accident, type of car, and time of the accident using graphs. This made it easier for readers to under-stand that the driver's age is a significant factor in traffic collisions, in addition to the location of the accident, as it helped us identify the best and worst contributing factors.

Figure 2 show that the highest number of casualties was between 31 and 50 years; interestingly, most drivers fall into this group. Regarding the type of car driven, automobiles were the most involved in traffic accidents, followed by lorries and pick-up cars. The Charts also show that males were involved in traffic accidents more than females (42.7% versus 21.3% consecutively). Regarding the years of driving experience, the Charts tell that 28% of drivers involved in accidents had 5–10 years of driving experience, and 21.2% had less than five years of experience. Finally, the Charts reveal that most accidents took place during daylight time.

4.5 Phase 5: Communicate Results

The team came to several conclusions regarding the traffic collision. Using the tools we mentioned previously, we could compare the conclusions to those drawn from factors like the driver's age, the location of the accident, the kind of car involved, and the time of the accident.

The results will need to be evaluated, and the government will need to decide how to present them to different parties, for instance, the team from the General Directorate of Traffic, while considering several variables, including driving experience and the various types of cars driven.

The team will also need to use caution when correctly communicating the results. The team should select the most robust findings in addition to highlighting and trying to establish their investigation's conclusion. In general, the techniques employed, starting with the development of hypotheses, resulted in a satisfactory conclusion to the analysis. Consequently, the ability to devise a practical and helpful solution to help prevent traffic accidents.

One of the many outcomes of the project is the ability to identify the causes of traffic accidents by identifying elements like the driver's age, the accident's location, the car's make and model, and the accident's period.

By the end of this phase, the team should be able to analyze key project research results and record preliminary findings. In order to evaluate its performance regarding the predetermined goals, the team must also solicit feedback from other parties.

4.6 Phase 6: Operationalization

The cycle's operationalize phase, which deals with concluding, entails programming, briefings, presenting documents to stakeholders, and creating an extensive report highlighting the initial findings. In order to compare the information to the targets, it is now directly observed in the same way it was recognized in the sandbox. If input data are inconsistent or ineffective, the team could go back to any stage and change them, producing new outputs.

In this case, the General Directorate of Traffic team will be able to assess the benefits of their project from a broader perspective and reach a decision to submit the work in a controlled manner initially.

While introducing stage six to implement the new analytics technique for the first time, going back to stage 4 when the framework was sandboxed for analytics will minimize the risk as well as allows the team to handle it effectively by implementing it on a lesser level at first rather than immediately deploying the model to a larger scale. This issue's causes should be determined to create a system that can effectively prevent traffic accidents.

Along with the implementation of the big data analytics lifecycle. The driver's experience and the type of vehicle being driven must both be carefully inspected by the compliance officer. The analysis will involve comparing documents, processes, and data to data in the system. The data gathered over time helps identify the causes of auto accidents.

5 Conclusion and Future Work

In this paper, a proposed model for uncovering the factors that relate to causing traffic accidents is presented, using the six phases of the data analytics life cycle: discovery, data preparation, model planning, model building, communicating results, and operationalization. The proposed model was built on these two hypotheses: (a) The relation between years of driving experience and car accidents and (b) the connection between the type of car driven and car accidents.

Besides using big data analytics lifecycle to identify the factors that relate to causing car accidents, previous relevant studies were also surveyed for this study. This paper contains a complete study and analysis of the problem using similar research papers. In addition to helping to improve sustainability in smart cities.

It is worthwhile to mention that besides the analytical results this paper provides for uncovering the factors that relate to causing traffic accidents, it allows for building a more complex model, using more hypotheses to help figure out what factors relate to causing accidents in smart cities.

References

1. Xu, Y., Chong, H.-Y., Chi, M.: Blockchain in the AECO industry: current status, key topics, and future research agenda. Autom. Constr. **134**, 104101 (2022)
2. Satija, S.: A Review Study on Applications of Big Data in Business Organizations (2022)
3. Naeem, M., et al.: Trends and future perspective challenges in big data. In: Advances in Intelligent Data Analysis and Applications, pp. 309–325. Springer (2022)
4. Zhang, H., Zang, Z., Zhu, H., Uddin, M.I., Amin, M.A.: Big data-assisted social media analytics for business model for business decision making system competitive analysis. Inf. Process. Manage. **59**(1), 102762 (2022)
5. Allaymoun, M.H., Qaradh, S., Salman, M., Hasan, M.: Big data analysis and data visualization to help make a decision-Islamic banks case study. In: International Conference on Business and Technology, pp. 54–63. Springer (2023)
6. Allaymoun, M.H., Hamid, O.A.H.: Business intelligence model to analyze social network advertising. In: 2021 International Conference on Information Technology (ICIT), pp. 326–330. IEEE (2021)
7. Bhathal, G.S., Singh, A.: Big data: Hadoop framework vulnerabilities, security issues and attacks. Array **1**, 100002 (2019)
8. Goundar, S., Bhardwaj, A., Nur, S.S., Kumar, S.S., Harish, R.: Industrial internet of things: benefit, applications, and challenges. In: Innovations in the industrial internet of things (IIoT) and smart factory, pp. 133–148 (2021)
9. Ali, F., et al.: An intelligent healthcare monitoring framework using wearable sensors and social networking data. Futur. Gener. Comput. Syst. **114**, 23–43 (2021)
10. Arooj, A., Farooq, M.S., Akram, A., Iqbal, R., Sharma, A., Dhiman, G.: Big data processing and analysis in internet of vehicles: architecture, taxonomy, and open research challenges. Arch. Comput. Methods Eng. **29**(2), 793–829 (2022)
11. Anshari, M., Almunawar, M.N., Lim, S.A., Al-Mudimigh, A.: Customer relationship management and big data enabled: personalization and customization of services. Appl. Comput. Inform. **15**(2), 94–101 (2019)
12. Aceto, G., Persico, V., Pescapé, A.: Industry 4.0 and health: internet of things, big data, and cloud computing for healthcare 4.0. J. Ind. Inf. Integr. **18**, 100129 (2020)

13. Kibe, L.W., Kwanya, T., Owano, A.: Characteristics of big data produced by the Technical University of Kenya and Strathmore University. In: Proceedings of 20th Annual IS Conference, vol. 18, p. 279 (2019)
14. Ghasemaghaei, M., Calic, G.: Does big data enhance firm innovation competency? The mediating role of data-driven insights. J. Bus. Res. **104**, 69–84 (2019)
15. Allaymoun, M.H., Al Saad, L.H., Majed, Z.M., Hashem, S.M.A.: Big data analysis and data visualization to facilitate decision-making-mega start case study. In: International Conference on Business and Technology, pp. 370–379. Springer (2023)
16. Triguero, I., García-Gil, D., Maillo, J., Luengo, J., García, S., Herrera, F.: Transforming big data into smart data: an insight on the use of the k-nearest neighbors algorithm to obtain quality data. Wiley Interdisc. Rev. Data Min. Knowl. Disc. **9**(2), e1289 (2019)
17. Ren, S., Zhang, Y., Liu, Y., Sakao, T., Huisingh, D., Almeida, C.M.: A comprehensive review of big data analytics throughout product lifecycle to support sustainable smart manufacturing: a framework, challenges and future research directions. J. Clean. Prod. **210**, 1343–1365 (2019)
18. Liu, Y., Zhang, Y., Ren, S., Yang, M., Wang, Y., Huisingh, D.: How can smart technologies contribute to sustainable product lifecycle management? J. Clean. Prod. **249**, 119423 (2020)
19. Chu, D., Cao, Y.: Typical intelligent transportation applications. In: Intelligent Road Transport Systems, pp. 545–608. Springer (2022)
20. Cui, Y.: AI applications in the rule of law. In: Blue Book on AI and Rule of Law in the World, pp. 77–114. Springer (2020)
21. Dong, X., Yan, M., Hu, Y.: Management transformation and system building in line with international standards. In: Huawei, pp. 165–199. Springer (2023)
22. Allaymoun, M.H., Khaled, M., Saleh, F., Merza, F.: Data visualization and statistical graphics in big data analysis by google data studio–sales case study. In: 2022 IEEE Technology and Engineering Management Conference (TEMSCON EUROPE), pp. 228–234. IEEE (2022)
23. Kemp, G., White, G.: Starting your data studio journey. In: Google Data Studio for Beginners, pp. 1–13. Springer (2021)
24. Road Traffic Accidents: Kaggle (2022). https://www.kaggle.com/datasets/saurabhshahane/road-traffic-accidents
25. Tseng, C.M.: Social-demographics, driving experience and yearly driving distance in relation to a tour bus driver's at-fault accident risk. Tour. Manage. **33**(4), 910–915 (2012)
26. Karlberg, L., Undén, A.L., Elofsson, S., Krakau, I.: Is there a connection between car accidents, near accidents, and type A drivers? Behav. Med. **24**(3), 99–106 (1998)

The Effects of Leadership Style on Employee Sustainable Behaviour: A Theoretical Perspective

Tamer M. Alkadash⬚, Muskan Nagi, Ali Ahmed Ateeq,
Mohammed Alzoraiki, Rawan M. Alkadash, Chayanit Nadam,
Mohammad Allaymoun, and Mohammed Dawwas

Abstract The main objective of this study is to examine the effects of leadership style on employee sustainable bahaviour of employees from a theoretical perspective. The literature evaluation served as the foundation for the predicted relationship between transformational leadership, transactional leadership, and e employee sustainable behaviour. The study contributes to the existing literature review on adding information to the theories of leadership. More study is needed to offer a more complete picture of these interactions. Much of the future research that has to be done has been determined because of the study's limitations. Further research is needed to understand the link between transactional leadership style, psychological empowerment, job satisfaction, and employee sustainable behaviour.

Keywords Leadership style · Employee sustainable behaviour · Transformational leadership · Transactional leadership · Theoretical perspective

1 Introduction

With the world becoming more aware of sustainability, businesses need to understand their role in the global economy. Organizations all over the world need to make sure that their operations are sustainable for long-term success and to contribute positively to the country's economy. Due to the importance of employee sustainable behavior [1]. In today's world, employee sustainable behaviour is no longer an

---

T. M. Alkadash (✉) · M. Nagi · A. A. Ateeq · M. Alzoraiki · M. Allaymoun · M. Dawwas
Administrative Science Department, College of Administrative and Financial Science
Department, Gulf University, Sanad 26489, Kingdom of Bahrain
e-mail: tamer.alkadash@gmail.com

R. M. Alkadash
Al Azhar University, Gaza, Palestine

C. Nadam
Global Business Center of British Petroleum, Kuala Lumpur, Malaysia

© The Author(s), under exclusive license to Springer Nature Switzerland AG 2024
A. Hamdan and E. S. Aldhaen (eds.), *Artificial Intelligence and Transforming Digital Marketing*, Studies in Systems, Decision and Control 487,
https://doi.org/10.1007/978-3-031-35828-9_19

option—it is a necessity. Organizations must develop and implement policies that prioritize employee behaviour and create an environment that encourages, acknowledges, and rewards sustainable initiatives. This will ensure that sustainable practices are ingrained in the organization's culture and that employees become strong advocates for sustainability [2].

Leaders in companies should think about encouraging employees to carry out their responsibilities and functions as effectively and efficiently as feasible to increase sustainable development in the organizations. As a result, leaders are involved in motivation in the workplace is essential and crucial since it may alter employee behavior for the better. Effective leadership is required in this highly competitive climate to lower the attrition rate. Only via successful leadership styles is it feasible for a company to realize its production goals. Leadership styles have an impact on employee productivity and performance; management practices and their effectiveness are mostly based on different styles that exist in different cultures because methods, approaches, and management styles change [3].

The effect of leadership style on employee sustainable behaviour is significant. According to [4–7] Leaders who encourage and support sustainable behaviours among their employees are more successful at creating a culture of sustainability within the organisation. A transformational or participatory leadership style, for example, can empower employees to develop and implement effective strategies that will result in environmentally friendly practices. In addition, when leaders communicate the importance of sustainability in terms of personal values and outcomes, employees are more likely to adopt sustainable behaviours. By emphasising recognition, motivation, and reward for eco-friendly initiatives and engaging with staff through regular consultation regarding sustainability efforts and results, leaders can create an atmosphere conducive to fostering sustainable behaviour among their team members.

Employee sustainable behavior is important for organizations as it directly affects the effectiveness of their environmental protection efforts [1, 20]. It is also a key factor in promoting sustainability [2] and has been linked to economic, social, and environmental variables [3]. Research has shown that employees' sense of working for a purpose-driven company can have an effect on their workplace sustainability behaviors [4], while other studies have explored how employee green behavior impacts organizational performance [5, 17]. Leadership and employee sustainable behaviour are two interdependent facets within an effective and thriving business. An effective leader cultivates the behaviours required to create a sustainable organisation, while employees embrace and nurture the behaviours that enable the leadership team to successfully implement their vision. Sustainable behaviours go beyond temporary shifts in behaviour in reaction to current events; instead, employees and leaders attempt to achieve long-term, positive social and ecological outcomes [18].

Leadership shapes employee behaviours in many ways. Setting a culture of transparency can help employees understand the company's values and understand how their work is contributing to the company's strategy [21]. Through communication and clear vision, leaders can outline how their actions impact the organisation and demonstrate how sustainable behaviours contribute to business success. Additionally, they must continually invest in the development of their people to identify and cultivate the skills and knowledge necessary to develop and promote the behaviours the organisation values. The study served to close the gap between the theoretical, conceptual, and empirical arguments surrounding the influence of leadership style (transformational and transactional) on employee sustainable behaviour.

2 Literature Review

2.1 Leadership Style (Transformational Leadership, and Transactional Leadership)

Transformational Leadership and Transactional Leadership are two distinct leadership styles used to motivate and inspire teams. Transformational Leadership is a highly participatory approach that encourages creativity, innovation, and problem-solving among team members [8, 9]. This style of leadership focuses on setting a clear vision for the organization, inspiring enthusiasm through shared goals, and leading by example. It relies heavily on communication between all levels of the organization to ensure employees understand their value within the team. Transactional Leadership, on the other hand, focuses more on setting expectations and then rewarding or punishing results accordingly. This style can often lead to higher productivity and consistency of performance but can also create an environment that ignores individual employee needs and intrinsic motivation in favor of maintaining rules-based transactional processes. Ultimately, both styles have their place within professional organizations—Transformational Leaders foster engagement and innovation while Transactional Leaders provide clear tasks with feedback loops that help identify areas for improvement quickly and efficiently. In addition to that, Transformational and transactional leadership styles are two of the most popular leadership approaches used today. Transformational Leadership is a style of leading that focuses on getting employees to think outside the box while emphasizing the importance of cultural values, vision, organization development, and engaging followers. This method empowers leaders to build relationships with subordinates by inspiring them to contribute beyond expectations. Transaction Leadership, on the other hand, concentrates more on a reward and punishment system to motivate its staff. It uses a top-down approach where managers set goals, provide incentives if results are achieved, and punish people if they fall short of expectations. Both these styles have proven their effectiveness in different organizational contexts and may be combined together in order to achieve success efficiently in dynamic business environments [10–12].

It is crucial to highlight that leaders are not limited to a particular leadership style. Combining diverse techniques can improve organizational outcomes in many circumstances. Transformational and transactional leadership are two well-studied leadership styles, and a particular leader may demonstrate variable degrees of both.

2.2 Effect of Leadership Transformational Leadership and Employee Sustainable Behaviour

Transformational leadership has been found to have a positive influence on employee sustainable performance [13], with green transformational leadership (GTFL) having a direct influence on environmental performance [15]. GTFL helps employees understand green work styles and values, fostering commitments to environmentally responsible behavior [13]. The nexuses between transformational leadership and employee green organizational citizenship behavior (GOCB) are further strengthened by environmental attitude and perceived organizational support [14]. However, transformational Leadership can help foster employee sustainable behaviour through its ability to motivate, inspire and provoke deep thinking. Through empowering dialogue, transformational leaders can engage their teams in creating meaningfully sustainable solutions that align with the company's mission and values. This method of leadership is known to amplify motivation levels amongst employees as they are able to feel more personally responsible for the success of their behaviour and initiatives. Additionally, Transformational Leadership encourages staff collaboration which helps build trust and transparency within a team dynamic whilst promoting innovative ideas which are beneficial for meeting sustainability objectives [13, 14]. Therefore, Transformational Leaderships capability to provide meaningful feedback; challenge current systems; lead creative collaborations; and act as role models demonstrates its invaluable contribution to driving Employee Sustainable Behaviour.

The link between transformational Leadership on employee sustainable behaviour, is the key to transformational Leadership to enabling employees to be more engaged in the mission and vision of their organisation. Leaders must possess qualities like charisma and motivation, while communicating a compelling vision that sparks change. These leaders must ensure that their teams share in this elevated level of engagement by inspiring them to put forth maximum effort into achieving common objectives as they work collaboratively towards organisational success. Through providing an effective environment with strong support, these leaders will serve as catalysts for positive behaviours within the workplace; creating a ripple effect that leads to improved morale which can result in a greater commitment from employees to sustain the organisation's development for the future. Transformational leaders can encourage employee engagement by fostering their subordinates' positive behaviors and attitudes towards work, supporting their self-efficacy, and modeling enthusiasm and high power [1, 4].

Transformational leadership has been found to be strongly associated with employee sustainable behaviour. In particular, transformational leadership encourages employees to think and act in a way that serves both the long-term interests of the organisation as well as their own. From an employer perspective, this means enabling and inspiring employees to take an active role in developing meaningful and sustainable organisational initiatives and behaviours that benefit their team, their department or even their entire organisation. Moreover, through a process of differentiated recognition and rewards offered for successful behaviour those who achieve desired outcomes could be rewarded or recognized publicly. Consequently, organisations through adopting transformational leaderships are more likely to create cultures of employee engagement that foster greater innovation and performance amongst staff members which are necessary for promoting sustainability in organisations [16]. Hence, on the basis of the previously mentioned argument, it is believable that transformational leadership plays a vital role in employee sustainable behaviour. Therefore, we posit that:

Hypothesis 1 (H1). Transformational Leadership Has a Positive Relationship on Employee Sustainable Behaviour

2.3 Effect of Transactional Leadership Style on Employee Sustainable Behaviour

Employees have the power to positively affect the sustainability of the organization [19]. With their daily decisions, they can set the tone for the rest of the organization, by forming habits and mindsets that respect the environment and promote its conservation. These behaviours are especially important for organizations that have any type of negative impact on the environment. Companies with strong sustainability behaviour among their employees can minimize the effects of their operations by employing these behaviour techniques. Having transactional leadership actively engaged in promoting and recognizing employee sustainable behaviour is key—they must model desired sustainability behaviours and practice what they preach. Thus, organizations that prioritize sustainability in their operations and culture also have an advantage in the recruitment and retention of top-tier talent. They can attract individuals that share their values and build a team of employees who are passionate about their mission to reduce environmental impact. This can lead to having informed and engaged employees, who remain with the organization for a longer period of time. This is in line with Transactional leadership style encourages leaders to adjust their behaviour and to understand subordinates' expectations. So, it is considered transactional leadership is a leadership style where leaders rely on rewards and punishments to achieve optimal job performance [1]. It focuses on results, conforms to the existing structure of an organization, and relies on rigorous checks and balances throughout a company [22, 23]. It is a more structured approach to management that can effectively motivate team members to maximize productivity [1, 23], create achievable goals for individuals at all levels [1], and provide a simple and low-cost approach with clear instructions that are easy to follow [4]. transactional leadership include motivating team members to maximize productivity [1, 22], creating achievable goals for individuals at all levels [22], boosting productivity [23], providing a simple and low-cost approach [4], being a clear and easy to follow style [24], being a powerful motivator [24], and getting results fast [10, 24]. It is based on the idea that workers need structure, instruction, and monitoring in order to complete tasks correctly and on time [3]. This style of leadership also utilizes control, organization, and short-term planning [4], as well as rigorous checks and balances throughout a company [5].

Transactional leadership should also promote green behavior through the person-organization fit (POF) model, which emphasizes the importance of aligning individual values with organizational values [20]. To create behavior change in an organization, transactional leadership should encourage and reward positive behavior [21]. This can be done through recognition programs or other incentives. Thus, transactional leadership should engage employees to create a sustainable business by finding ways to get all employees personally engaged in sustainability initiatives [24]. Leaders can also inspire employees to challenge old ideas and use new methods to solve environmental problems through green intellectual stimulation [10]. Based on the above literature, it is recommended that transactional leadership also acts as an essential part in determining employee sustainable behaviour actions. Therefore, the following hypothesis is offered.

Hypothesis 2 (H2). Transactional Leadership Has a Positive Relationship on Employee Sustainable Behaviour.

3   Discussion and Future Research

The study set out to explore the relationship between transformational leadership, transactional leadership and employee sustainable behaviour. The study found there is a significant relationship between the transformational and transactional leadership and employee sustainability behaviour. In turn, having sustainable employees can be beneficial to both the organisation and their own personal development [18]. Employees are more likely to be engaged in their work if they understand how their job is contributing to the company's overall mission. It is also beneficial for employees to develop their own sustainable behaviours, as it can positively enhance their commitment, efficiency and productivity [19]. Furthermore, exercising sustainable behaviours can have positive external impacts, such as improved public image and a reduced ecological footprint. When it comes to creating long-term change in sustainability behaviour, both leaders and their employees need to work collaboratively. Leaders need to collaborate with their employees and empower them to recognise and report any questionable behaviour or practices [17]. Working together to identify potential barriers to sustainable behaviour enables leaders and their team to develop the strategies and technologies needed to overcome those barriers. Additionally, energy-saving practices, waste reduction initiatives and other organisational programmes can be developed in partnership between leaders and employees, strengthening the bond between the two and fostering a culture of sustainability [20]. For an organisation to remain competitive and viable in the longer term, it is essential that both leaders and employees embrace sustainable behaviour and adopt a sustainable mindset within the workplace. Leaders must demonstrate to their employees the importance of sustainable behaviour and create policies, processes and rewards that help and encourage employees to adopt those behaviours. At the same time, employees must recognise sustainability as part of their job and take personal responsibility in developing and promoting sustainable behaviours in the workplace. When done together, both groups have the potential to develop an ethical and resilient organisation that can help shape a sustainable future. As the discussion above shows, employee sustainable behaviour is increasingly important for organizations as the demand for corporate social responsibility and environmental goals steadily grows. This type of behaviour involves how employees make decisions that support the company's sustainability objectives and how they live a sustainable lifestyle both inside and outside of work. As such, promoting employee sustainable behaviour is a critical component of any long-term sustainability strategy. Leaders in organizations needs to build several strategies for developing employee sustainable behavior in collaboration with HR department. Leaders should run employee awareness campaigns to educate employees on sustainability and the importance of green behavior [1]. Additionally, organizations can lower energy consumption by providing screensaver messages on office computers with tips on how to reduce energy consumption [1].

Future research should focus on the effects of leadership style on employee sustainable behaviour in different sectors. Leadership styles have been shown to have a significant impact on sustainable performance [1], and can influence employees in terms of both cognition and behaviour [2]. Sustainable leadership practices such as valuing employees, shared vision, social responsibility, and amicable labor needs to be consider in future research and it's effects on employee sustainability behaviour. Inclusive leadership has also been found to improve sustainability outcomes by promoting connectedness and belongingness among employees [4]. Additionally, further research is needed to understand the link between transactional leadership style, psychological empowerment, job satisfaction and employee sustainable behaviour. Moreover, Future research into the effects of leadership style on employee sustainable behaviour should consider a theoretical perspective that takes into account the cognitive-affective system theory [5]. This theory suggests that leadership can contribute to employee green behaviour by influencing their attitudes and beliefs [5].

Previous studies have found that sustainable leadership can improve employees' behaviour, making them more responsible [1]. Additionally, research has shown that different leadership styles can have an impact on sustainable performance [2], as well as various outcomes such as employees' performance [3] and motivation [4].

Therefore, future research should focus on how different leadership styles can influence employee sustainable behaviour from a cognitive-affective system perspective.

References

1. McQuaid, R., Webb, A.: The importance of SMEs as innovators of sustainable inclusive employment: some issues resulting from shocks to the economy imposed by the COVID-19 pandemic. In: The Importance of SMEs as Innovators of Sustainable Inclusive Employment, pp. 33–46. Rainer Hampp Verlag (2020)
2. Wadström, P.: Advancing Strategy Through Behavioural Psychology: Create Competitive Advantage in Relentlessly Changing Markets. Kogan Page Publishers (2022)
3. Rezaei, F., Khalilzadeh, M., Soleimani, P.: Factors affecting knowledge management and its effect on organizational performance: mediating the role of human capital. Adv. Human-Comput. Interact. **2021**, 1–16 (2021)
4. Iqbal, Q., Ahmad, N.H.: Sustainable development: the colors of sustainable leadership in learning organization. Sustain. Dev. **29**(1), 108–119 (2021)
5. Islam, T., Khan, M.M., Ahmed, I., Mahmood, K.: Promoting in-role and extra-role green behavior through ethical leadership: mediating role of green HRM and moderating role of individual green values. Int. J. Manpow. **42**(6), 1102–1123 (2021)
6. Çop, S., Olorunsola, V.O., Alola, U.V.: Achieving environmental sustainability through green transformational leadership policy: can green team resilience help? Bus. Strateg. Environ. **30**(1), 671–682 (2021)
7. Ahmad, S., Islam, T., Sadiq, M., Kaleem, A.: Promoting green behavior through ethical leadership: a model of green human resource management and environmental knowledge. Leadersh. Organ. Dev. J. (2021)
8. Freihat, S.: The role of transformational leadership in reengineering of marketing strategies within organizations. Probl. Perspect. Manag. **18**(4), 364–375 (2020)
9. Iriqat, R.A., Abu-Zaid, M.K.: The impact of gender leadership styles on the individual creativity in Palestinian non-governmental organisations. Int. J. Bus. Innov. Res. **29**(2), 194–220 (2022)
10. Aboramadan, M., Kundi, Y.M.: Does transformational leadership better predict work-related outcomes than transactional leadership in the NPO context? Evidence from Italy. Voluntas: Int. J. Voluntary Nonprofit Organ. **31**(6), 1254–1267 (2020)
11. Alrowwad, A.A., Abualoush, S.H., Masa'deh, R.E.: Innovation and intellectual capital as intermediary variables among transformational leadership, transactional leadership, and organizational performance. J. Manage. Dev. **39**(2), 196–222 (2020)
12. Cho, Y., Shin, M., Billing, T.K., Bhagat, R.S.: Transformational leadership, transactional leadership, and affective organizational commitment: a closer look at their relationships in two distinct national contexts. Asian Bus. Manag. **18**, 187–210 (2019)
13. Kotamena, F., Senjaya, P., Prasetya, A.B.: A literature review: is transformational leadership elitist and antidemocratic? Int. J. Soc. Policy Law **1**(1), 36–43 (2020)
14. Busari, A.H., Khan, S.N., Abdullah, S.M., Mughal, Y.H.: Transformational leadership style, followership, and factors of employees' reactions towards organizational change. J. Asia Bus. Stud. **14**(2), 181–209 (2020)
15. Atan, J.B., Mahmood, N.: The role of transformational leadership style in enhancing employees' competency for organization performance. Manage. Sci. Lett. **9**(13), 2191–2200 (2019)
16. Milhem, M., Muda, H., Ahmed, K.: The effect of perceived transformational leadership style on employee engagement: the mediating effect of Leader's emotional intelligence. Found. Manage. **11**(1), 33–42 (2019)
17. Alzgool, M.: Nexus between green HRM and green management towards fostering green values. Manage. Sci. Lett. **9**(12), 2073–2082 (2019)
18. Mujtaba, M., Mubarik, M.S.: Talent management and organizational sustainability: role of sustainable behaviour. Int. J. Organ. Anal. **30**(2), 389–407 (2022)
19. Fatoki, O.: Hotel employees' pro-environmental behaviour: effect of leadership behaviour, institutional support and workplace spirituality. Sustainability **11**(15), 4135 (2019)

20. Ojo, A.O., Tan, C.N.L., Alias, M.: Linking green HRM practices to environmental performance through pro-environment behaviour in the information technology sector. Soc. Responsib. J. **18**(1), 1–18 (2022)
21. Ansari, N.Y., Farrukh, M., Raza, A.: Green human resource management and employees pro-environmental behaviours: examining the underlying mechanism. Corp. Soc. Responsib. Environ. Manag. **28**(1), 229–238 (2021)
22. Nurlina, N.: Examining linkage between transactional leadership, organizational culture, commitment and compensation on work satisfaction and performance. Golden Ratio Human Resour. Manage. **2**(2), 108–122 (2022)
23. Purwanto, A., Bernarto, I., Asbari, M., Wijayanti, L.M., Hyun, C.C.: Effect of transformational and transactional leadership style on public health centre performance. J. Res. Bus. Econ. Educ. **2**(1) (2020)
24. Young, H.R., Glerum, D.R., Joseph, D.L., McCord, M.A.: A meta-analysis of transactional leadership and follower performance: double-edged effects of LMX and empowerment. J. Manag. **47**(5), 1255–1280 (2021)

Toward Sustainable Smart Cities: Design and Development of Piezoelectric-Based Footstep Power Generation System

M. N. Mohammed, Shahad Al-yousif, M. Alfiras, Majed Rahman, Adnan N. Jameel Al-Tamimi, and Aysha Sharif

Abstract Smart cities involve information and communication technology (ICT) as the main features. The Internet of Things and 5G will aid smart sensing gadgets by allowing them to connect quickly and securely. Nevertheless, a significant problem in this industry will always be the requirement for active sensing devices that are self-powered and sustainable sources. Piezoelectric energy-harvesting systems are intriguing prospects for creating smart and active self-powered sensors with a variety of applications in addition to powering wireless sensor nodes sustainably. This paper presents a study on sustainable power generation based on the footstep utilizing a piezoelectric sensor. The harvester is made up of a piezoelectric oscillator that is self-biased to control the transistor's switching frequency and amplitude in the boost converter. This device enables the controlling of the output DC voltage in a simple manner that does not require any biasing external source. Piezoelectric-Based Footstep Power Generation Systems can play an important role in providing energy to smart cities. This research highlights their contribution towards sustainable smart cities while offering a blueprint for integrating footstep power generation systems into urban infrastructure.

Keywords Piezoelectric · Power generation · Sustainability · Smart cities · Energy harvesting · Smart cities · Internet of Things · IoT · SDG 7 · SDG 11

M. N. Mohammed (✉) · A. Sharif
Mechanical Engineering Department, College of Engineering, Gulf University, Sanad, Kingdom of Bahrain
e-mail: dr.mohammed.alshekhly@gulfuniversity.edu.bh

S. Al-yousif · M. Alfiras
Electrical and Electronic Engineering Department, College of Engineering, Gulf University, Sanad, Kingdom of Bahrain

M. Rahman
Department of Engineering and Technology, Faculty of Information Sciences and Engineering, Management and Science University, Shah Alam, Selangor, Malaysia

A. N. J. Al-Tamimi
College of Technical Engineering, Al-Farahidi University, Baghdad, Iraq

© The Author(s), under exclusive license to Springer Nature Switzerland AG 2024
A. Hamdan and E. S. Aldhaen (eds.), *Artificial Intelligence and Transforming Digital Marketing*, Studies in Systems, Decision and Control 487,
https://doi.org/10.1007/978-3-031-35828-9_20

1 Introduction

Smart cities perform a significant task as innovation triggers for enterprises in various sectors, including health, environment, and information and communication technology. Future smart cities will use smart innovation ecosystems to raise residents' standard of living in general [1–3] in the situation of smart cities, it is suitable to address issues such as the Internet of Things (IoT), smart homes, sensors, and controllers are pertinent practically everywhere. The benefit of IoT is that it enables quick and wireless connections between objects, improving communication and monitoring primarily [4–7]. With respect to a Smart City, sustainability is an essential concern. Researchers are challenged by the complexity of replacing fossil fuels to investigate new renewable, sustainable, and green energy generation techniques. Green energy is conceptualized as a sustainable and alternative source at the same time, being one that comes from renewable sources with no or very little environmental effect. With expanding daily demand, electricity has currently turned into a lifeline for the populace. Modern technology demands enormous amounts of electrical power for its many functions. The production of power is the leading cause of emissions worldwide [8, 9]. As a result, the development and responsible use of alternative energy sources have been perceived as increasing worry over the gap between the supply and demand for electricity. On the other side, as the world's population continues to grow, the demand for energy is rising linearly every day. In addition, the gap between supply and demand for energy is rising exponentially. For this reason, researchers and innovators in the field of energy collecting are investigating viable applications for renewable energy sources [10]. In recent decades, power generation and distribution have become integral components of daily living. As the demand for petroleum products skyrockets, it will eventually become impossible to produce energy using non-sustainable power sources. The condition and economy are also being forced by this overconsumption and its associated risks. This level of CO emissions and greenhouse gas absorption into the atmosphere has led to concerns about increasing ocean levels, rising average temperatures, and unfavorable weather conditions. Unexpectedly, as modern technology and devices become more prevalent, there is an increase in the demand for power supply [11]. Unfortunately, oil fuel generators, which are outmoded and unreliable due to the significant environmental damage they inflict, are the primary source of electricity today. So there is a huge need for an oil fuel generator replacement that is intelligent, clean, dependable, and less polluting. For future sustainable cities, a smart grid that can manage erratic power supplies will be required. Utilizing as many locally available renewable energy sources as is practical is one of the main objectives. Numerous researchers have shown interest in small-scale energy-harvesting devices like vibration-based harvesters over the years. There are several motion and vibration sources in smart cities that could be used for energy harvesting [12]. The direct conversion of kinetic energy from human motion, wind and air pressure, vehicles, ocean waves, water pressure, buildings and structures, bridges, and roads into electrical energy is made possible by piezoelectric energy-harvesting systems. One of the potential sources of energy in smart cities is human movement. Humans execute walking more frequently than any other activity because it produces a significant amount of displacement. To capture the energy generated by walking, piezoelectric materials have been developed. These materials can either be incorporated into shoes to harvest energy from human weight, or they can be attached to the leg or foot so that the energy harvester can be excited by a walking or swing motion. Piezoelectric energy, which produces consistent energy by transforming mechanical energy (applied load) into electrical energy and voltage, is one of the most effective and clean energy sources [13]. When a mechanical stress is applied, such as when the substance is squeezed or stretched, a piezoelectric substance produces an electric charge. Piezoelectric energy harvesters hold great promise for generating power from ambient vibration sources. They offer the benefit of transforming mechanical strain into electrical energy without drawing any more power, have a huge power density, are simple to use, and can be manufactured at various scales. If the resonance frequencies of piezoelectric energy-harvesting systems coincide with the excitation vibration frequencies, they can output their maximum amount of power. Microelectronic devices may use this electrical energy or batteries may store it. The transition of vibrational energy from the environment—generated by human motion, air flow, and water current/wave—has been the subject of numerous studies. Footstep Power Generation Using Piezoelectric Sensor is presented in this study. The harvester is made up of a self-biased piezoelectric oscillator that will be used to regulate the boost converter transistor's switching frequency and amplitude. This gadget makes it simpler and unnecessary to use an external source for biasing to control the converter's DC output voltage.

2 Methodology

A piezoelectric layer, or beam, is attached to a vibrating mechanical component in a piezoelectric energy harvester's basic construction. In this piezoelectric layer, the vibration creates bending (stress), and the stress in turn induces electric charges. It is common practice to add a proof mass to the end of a piezoelectric beam in order to convert the vibration—or, more accurately, acceleration—into an effective inertial force that will increase the beam's bending. The mass can also be changed to adjust the structure's effective resonant frequency to the frequency of vibrations, hence increasing output power. The developed system comprises of three major elements which are: input, process, and output. Figure 1 depicts developed system's block diagram.

Piezoelectric sensor, LCD, ARDINO UNO, battery, LED, rubber pad, bridge rectifier, and connecting wires are all parts of the constructed project. In any area where we have tiles with piezoelectric sensors, the energy or force is supplied by human stepping. The hardware and software components of this project are combined during the implementation phase. The usage of a foam cushion in this instance, as can be seen in the Figure above, is done to prevent the Piezo sensors from disintegrating under load. As illustrated in Fig. 2, the boot loader has been pre-programmed into the Arduino

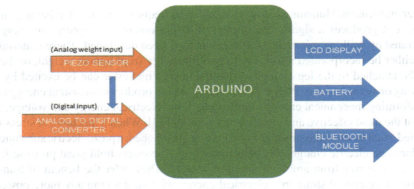

Fig. 1 Block diagram

microcontroller to make it easier to upload these programs to the chip's memory. The UNO Arduino, an optiboot type device, serves as the boot loader by default. The boards have program code loaded to another computer via a serial link. Some serial Arduino boards employ a shift registers logic circuit to convert between RS-232 logic and transistor-transistor logic. The modern Arduino boards were constructed utilizing a USB-to-serial FIDI FT232 chip and were programmed using a universal serial bus. The Arduino board's structure comprises of numerous controllers and microprocessors, as well as a collection of analog/digital input output pins for connecting to other development panels like shields and other circuits in the system. Furthermore, a number of models' serial communication interfaces, including USB, are utilized to load software from computer devices. The microcontrollers have traditionally been programmed utilizing the dialect features of the C and C ++ programming languages.

By using the concept of mechanical vibration, the piezo sensor turns the applied force into voltage. This sensor essentially transforms kinetic energy into electrical energy types. sensors array should be linked serially mode in order to produce a sufficient amount of electrical power. The two types of sensors that are readily accessible on the market are PZT and PVDF, and they are both coupled to filters to regulate the output voltage of the sensors, as illustrated in Fig. 3.

Fig. 2 Platform of Arduino type UNO

Fig. 3 Piezo sensor

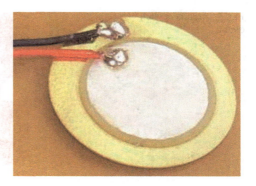

A booster type utilized as a DC/DC power converter is the step up converter. With this converter, the input voltage is increased while the output load's current is decreased. This converter's switched mode power supply class includes transistors, diodes, and energy storage components like inductors and capacitors. To reduce voltage ripple, an inductor or capacitor combination circuit is included as a filtration component. This kind of converter is depicted in the system shown in Fig. 4. Here, a boost is being employed to increase the piezoelectric transducer's erratic voltages.

The switching circuit and Bluetooth module are shown in Figs. 5 and 6, respectively. The Switching circuit enables the Arduino to manage the system's power flow. When to permit and not permit a given level of electricity to pass through can be determined by the Arduino's programming. The permissible voltage in this situation is 3.0 V. The Bluetooth module type HC-05 has been utilized in this system, and once the system reaches a charge point of 3.0 V, it permits the switching circuit to let electricity flow through and recharge the battery at its normal voltage level of 3–5 V.

The compiler type IDE is a cross-platform tool created in the Java programming language for Processing with the aim of introducing artists to computer programming and software development. This required the use of the code editor, syntax highlighting that matched, and automatic indentation that can upload and build the program to the panel with just one click. The electronics type that can be employed as an inaccessible organization for a particular remote entity is the universal android

Fig. 4 DC–DC buck converter

Fig. 5 Arduino switch circuit

Fig. 6 HC-05 module

software used to show and control the data transfer via Bluetooth. Comparing this component's interface design to that of other software applications, it is easily adaptable.

The Proteus framework is under constant development, turning into efficient, affordable, and marketable EDA packages. Its embedded co-simulation program, called VSM Proteus, combines animated components, a microprocessor model, and mix mode SPICE simulation to make it easier to simulate an entire microcontroller. After several attempts, it was discovered that the Piezo Sensors must be placed in order to absorb the most kinetic energy possible from the weight of the footstep, as illustrated in Figs. 7 and 8. This was discovered during the design phases of the stepping mechanism. Thus, after numerous experiments, the star-shaped 14 piezo design was selected to replicate the movements of the heel and forefoot as people walk, as seen in Fig. 9.

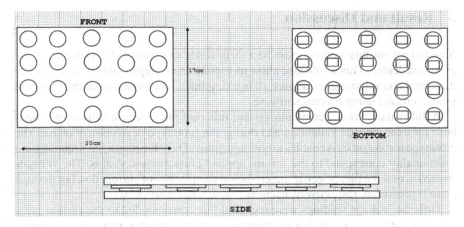

Fig. 7 Illustrating final design for the stepping mechanism

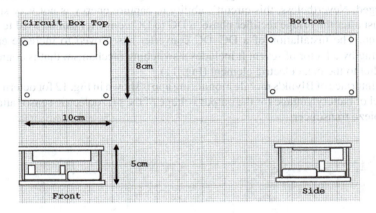

Fig. 8 Case design for circuitry

Fig. 9 Piezo placement on foam pad

3 Result and Discussion

Simulations on this system must be run in order to confirm the viability of the proposed design. This would also allow us to confirm the dependability of the above-discussed control measures. Figure 10 is a simplified representation of the proposed system created using Proteus software.

A full smart power generation system utilizing piezoelectric sensors has been built for experimental validation and during the prototype development phase. The prototype testing process enables a thorough grasp of how the prototype's features and functionality correspond to the various applied pressures and strains. The voltages created across the piezoelectric materials and the quantity of current flowing through them is measured, respectively, using voltmeters and ammeters.

Different voltage readings were recorded that corresponded to the various pressure and strain readings as they were tested on the piezoelectric material. The capacitor may be charged to store energy, and depending on the situation, it can also be discharged. Nevertheless, this circuit's ability to capture energy is not very significant. Just after the bridge rectifier phase, a DC to DC converter can be used to solve this issue. The installation of a DC–DC converter has proved to increase energy harvesting by a factor of seven. It includes a switching mechanism that is connected in parallel to the piezoelectric element (Fig. 11).

The interface of Blynk's mobile monitoring app is shown in Fig. 12 for determining the level of battery voltage for the output voltage of the piezoelectric sensor attached to the piezo transducer.

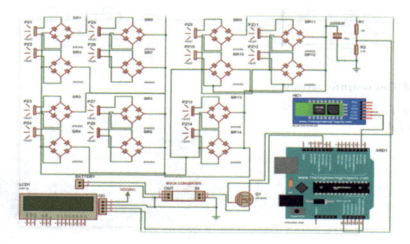

Fig. 10 Circuit diagram of pizoelectric power system

Toward Sustainable Smart Cities: Design and Development ... 223

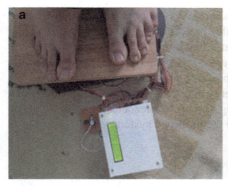

Fig. 11 System under. **a** Minimum load and **b** Full load

Fig. 12 Output voltage of the battery connected from the Piezo transducer

4 Conclusions

By utilizing technology to manage the available resources, smart cities seek to improve the life quality of residents. Applications that have been implemented in such cities and have a lower negative environmental impact include renewable energy harvesters. Because they can power tiny micro-electronic devices and real-time IoT sensors, ambient energy-harvesting sources in smart cities have drawn a lot of interest. Materials that are piezoelectric are excellent choices for this use. This paper presents a proposal for energy harvesting based on the footstep by using piezoelectric materials. It is an additional strategy for encouraging the implementation environmentally friendly innovations that protect the planet from environmental contamination. Although this study indicates that piezoelectricity in a footstep has good potential for energy harvesting, there are still numerous ways to broaden the scope of this study. According to the proposed technology, piezoelectric materials might convert an independent potential energy into electrical power. The sensors and microcontrollers-based voltage monitoring system of piezoelectric sensor energy generation were effectively integrated into the proposed system.

References

1. Mohammed, M.N., Al-Rawi, O.Y.M., Al-Zubaidi, S., Mustapha, S., Abdulrazaq, M.: Toward sustainable smart cities in the Kingdom of Bahrain: a new approach of smart restaurant management and ordering system during Covid-19. In: 2022 IEEE International Conference on Automatic Control and Intelligent Systems (I2CACIS), pp. 7–11. IEEE (2022)
2. Izadgoshasb, I.: Piezoelectric energy harvesting towards self-powered internet of things (IoT) sensors in smart cities. Sensors 21, 8332 (2021). https://doi.org/10.3390/s21248332
3. Choiri, A., Mohammed, M.N., Al-Zubaidi, S., Al-Sanjary, O.I., Yusuf, E.: Real time monitoring approach for underground mine air quality pollution monitoring system based on IoT technology. In: 2021 IEEE International Conference on Automatic Control and Intelligent Systems (I2CACIS), pp. 364–368. IEEE (2021)
4. Alfiras, M., Yassin, A.A., Bojiah, J.: Present and the future role of the internet of things in higher education institutions. J. Positive Psychol. Wellbeing 6(1), 167–175 (2022)
5. Mohammed, M.N., Alfiras, M., Al-Zubaidi, S., Al-Sanjary, O.I., Yusuf, E., Abdulrazaq, M.: 2019 novel coronavirus disease (Covid-19): toward a novel design for smart waste management robot. In: 2022 IEEE 18th International Colloquium on Signal Processing & Applications (CSPA), pp. 74–78. IEEE (2022)
6. Pangestu, A., Mohammed, M.N., Al-Zubaidi, S., Bahrain, S.H.K., Jaenul, A.: An internet of things toward a novel smart helmet for motorcycle. AIP Conf. Proc. 2320(1), 050026 (2021)
7. Mohammed, M.N., Dionova, B.W., Al-Zubaidi, S., Bahrain, S.H.K., Yusuf, E.: An IoT-based smart environment for sustainable healthcare management systems. In: Healthcare Systems and Health Informatics, pp. 51–74. CRC Press
8. Jettanasen, C., Songsukthawan, P., Ngaopitakkul, A.: Development of micro-mobility based on piezoelectric energy harvesting for smart city applications. Sustainability 12, 2933 (2020)
9. Belli, L., Cilfone, A., Davoli, L., Ferrari, G., Adorni, P., Di Nocera, F., Dall'Olio, A., Pellegrini, C., Mordacci, M., Bertolotti, E.: IoT-enabled smart sustainable cities: challenges and approaches. Smart Cities 3, 52 (2020)
10. Calvillo, C.F., Sánchez-Miralles, A., Villar, J.: Energy management and planning in smart cities. Renew. Sustain. Energy Rev. 55, 273–287 (2016)
11. Clerici Maestosi, P.: Smart cities and positive energy districts: Urban. Perspectives in 2020. Energies 14, 2351 (2021)
12. Wei, C., Jing, X.: A comprehensive review on vibration energy harvesting: modelling and realization. Renew. Sustain. Energy Rev. 74, 1–18 (2017)
13. Alotaibi, I., Abido, M.A., Khalid, M., Savkin, A.V.: A comprehensive review of recent advances in smart grids: a sustainable future with renewable energy resources. Energies 13, 6269 (2020)

Cultural Marketing, Artificial Intelligence and Digital Learning, Innovation and Sustainable Operations

Sustainable Urban Street Design in Jeddah, Saudi Arabia

Marwa Abouhassan

Abstract The streets in a city are important because they determine how people get about. Structures, people, vehicles, utilities, greenery, signs, furniture, and lighting are all accommodated. The emergence of compact and sustainable cities is largely attributable to the close relationship between urban design and sustainability, with streets constituting most of the public realm in most communities. Both cities and streets exist. Introducing sustainable urban streets can make cities more hospitable for residents. Sustainable streets are multimodal rights of way that are planned, built, and maintained to provide positive outcomes in the areas of transportation, ecology, and community as part of a larger sustainability agenda that prioritizes environmental protection, social justice, and economic vitality. This paper's focus is on determining methods for incorporating sustainable design requirements into city roadways. This paper will argue for the improved sustainability of urban streets by providing a more general definition of sustainability. Taking into account and evaluating sustainable principles can help urban streets serve people, communities, the economy, and the environment in a more positive way. Using these guidelines for eco-friendly street planning, we'll examine few streets as case studies in Jeddah, Saudi Arabia. Therefore, a complete street upgrade is unquestionably necessary to guarantee a good sustainable street design. The paper concludes with some suggestions for developing the main components of sustainable streets that are derived from the theoretical study in order to design a sustainable street for a livable community.

Keywords Sustainability · Urban street · Street design · Walkability · SDG11

M. Abouhassan (✉)
Architecture Department, University of Business and Technology, Jeddah, Saudi Arabia
e-mail: m.abouhassan@ubt.edu.sa

© The Author(s), under exclusive license to Springer Nature Switzerland AG 2024
A. Hamdan and E. S. Aldhaen (eds.), *Artificial Intelligence and Transforming Digital Marketing*, Studies in Systems, Decision and Control 487,
https://doi.org/10.1007/978-3-031-35828-9_21

1 Introduction

Cities may adapt to the challenges posed by a warming planet by designing streets that respond to their surroundings. The UN Sustainable Development Goals (UNSDGs) and other international organizations and agendas have brought greater attention to environmental sustainability, greenhouse gas emissions, and global warming. The time has come to emphasize the advantages of wonderful streets for the environment. By emphasizing enhanced environmental consequences and increasing contribution to attaining a city's environmental goals, investment in sustainable streets can be recruited [1].

A change in the quality of streets is required as an increasing number of towns rediscover the value of their streets as significant public places for many elements of everyday life. People require streets that are secure for walking or crossing, that provide locations for gathering, and that have a diverse range of retail. As cities strive towards more sustainable development and lifestyle patterns, more people must be able to walk and ride bicycles in their neighborhoods. As a result, more and more communities are considering changing the way their roadways are laid up. In addition, there is a need for creativity in design and flexibility in how the current design rules are applied to site-specific project requirements [2]. The current idea is to accommodate all users of the transportation system, including drivers and passengers of motor vehicles, bicycles, pedestrians, and users of public transportation [3, 4]. This idea has also been referred to as "complete streets" and "walkable thoroughfares".

By establishing a transition toward sustainable mobility, the objective is to increase accessibility and quality of life [5]. The establishment of a community that fosters safety, connection, and beauty through a transportation network that accommodates all modes, all ages, and all abilities is supported and encouraged by recent standards and best practices developed by a few cities.

Communities are re-discovering their streets as public spaces and complete streets design principles at the same time as they are also becoming aware of the importance of street design to the sustainability of the community as a whole. It has long been recognized that how we plan our neighborhoods affects both the demand for travel and the method of transportation [6, 7]. The increased travel demand has an effect on other community-impacting issues such as obesity, traffic congestion, crashes, and emissions of pollutants and greenhouse gases [8].

Promoting active ways of transportation has positive effects on road safety as well. All roadway designs and reconstruction projects must therefore strictly adhere to the full streets design principles. With the COVID-19 pandemic outbreak, the need to encourage the healthy and safer growth of cities has grown even more essential, underscoring the necessity for policymakers and urban planners to refocus their efforts on more user-oriented planning and urban redevelopment [9].

Fig. 1 Sustainability pillars. *Source* author

2 Definition of Sustainability

Sustainability means meeting our own needs without compromising the ability of future generations to meet their own needs. In addition to natural resources, we also need social and economic resources. Sustainability is not just environmentalism. Embedded in most definitions of sustainability we also find concerns for social equity and economic development. The motivations behind sustainability are often complex, personal and diverse. It is unrealistic to create a list of reasons why so many individuals, groups and communities are working towards this goal. Yet, for most people, sustainability comes down to the kind of future we are leaving for the next generation. Sustainability as a value is shared by many individuals and organizations who demonstrate this value in their policies, everyday activities, and behavior. Individuals have played a major role in developing our current environmental and social circumstances. The people of today along with future generations must create solutions and adapt [5] (Fig. 1).

3 Sustainable Development Goals (UNSDGs) and Sustainable Urban Street Design

Sustainable cities mentioned in the 2030 Agenda for Sustainable Development (UNSDGs). This is appeared in goal 11 (SDG11), which is about cities, specifically about the need to make human settlements and cities inclusive, resilient, safe, and sustainable through planning, sustainable urbanization, and management. Goal 11 targets by 2030, provide access to safe, affordable, accessible and sustainable transport systems for all, improving road safety, notably by expanding public transport, with special attention to the needs of those in vulnerable situations, women, children, persons with disabilities and older persons [10].

Sustainable urban streets are "multimodal rights of way designed and operated to create benefits related to movement, ecology, and community that together support a broad sustainability agenda that includes the three E's: environment, equity, and economy" [11], and putting them into place can make communities easier to live in.

There is a strong link between urban design and sustainability, which is why cities and streets have become more compact. Sustainability in architecture tries to make buildings less harmful to the environment by using raw materials, energy, and the development environment in a way that is efficient. This is called studying sustainable development for the future. It aims to improve the high level of economic growth, make social progress better, protect the environment well, and use natural resources wisely. When designing a sustainable street, there are several goals: to use less energy, to use less material resources, to affect the environment less, to support healthy urban communities, and to support sustainability while the street is being built [12].

4 The Problems of Urban Street

Modern cities are inherently environmentally unsustainable because they require the importation of food, energy, and raw materials; they generate more waste than they can handle within their boundaries (Fig. 2).

- Sewerage and rainwater drainage: The expansion of major cities, combined with an increase in per capita water consumption, has resulted in a crisis in capacity to meet demand.
- Pollution of the air: Factories near residential areas make many processes produce particulates and gases that are offensive, toxic, or both, even if they are not greenhouse gases.
- Noise pollution: The faster the development, the more likely it is that traffic, police sirens, fire engines, and trains will be reflected.

Fig. 2 The problems of urban street. *Source* author

Sustainable Urban Street Design in Jeddah, Saudi Arabia

- Lack of greenery: Many streets do not have a suitable amount of greenery for pedestrian to increase the environmental aspect.
- Waste disposal: According to an analysis of the domestic waste stream, recycling and composting can reduce the amount of waste going to landfill sites [13].

5 Objectives of Sustainable Urban Street

Cities are finally rediscovering the benefits of designing safe and livable streets that balance the needs of all users after decades of designing streets to move large numbers of vehicles as efficiently as possible. It is time to change practices and redefine what makes a street successful. Streets should not be evaluated solely as transportation projects. Instead, each design presents an opportunity to consider the overall benefits [1].

5.1 Public Health and Safety

Millions of people die needlessly each year from avoidable causes like traffic accidents, chronic illnesses brought on by poor air quality, and a lack of exercise. In order to encourage safe surroundings for all users and provide wholesome options that support active transportation like walking, cycling, and taking public transportation. Streets should provide accessibility to wholesome food options, reduce noise levels, and include landscaping and trees that enhance the quality of the air and water.

5.2 Quality of Life

Cities from all over the world are vying for the title of "most livable city," recognizing the importance of quality-of-life initiatives in luring and keeping businesses and inhabitants. The livability of a city is heavily reliant on its streets since people experience cities through their public places. A city's livability and sense of community will be influenced by how safe, cozy, effective, and vibrant its streets are. Streets may promote social contact, have designs that provide natural surveillance, and help communities become stronger and safer.

5.3 Environmental Sustainability

Street initiatives offer a chance for local activities to enhance a city's environmental sustainability and resilience in the face of extraordinary climate challenges.

Promoting environmentally friendly modes of mobility through thoughtfully planned streets can reduce carbon emissions and enhance overall air quality. The addition of trees and landscaping can enhance water management, promote biodiversity, and expand access to the outdoors.

5.4 Economic Sustainability

Excellent streets draw customers and companies. Positive economic consequences of street projects include better retail sales and higher property prices. These projects also raise safety, enhance the public realm, and invite multimodal use. Long-term economic gains come from investing in roadways.

5.5 Social Equity

In a time when inequality is getting worse, cities must make sure that their most valuable public spaces can be used safely and fairly by everyone, regardless of ability, age, or income. This gives the most vulnerable users more mobility options that are safe and reliable. The design of a city's streets can make it easier for people to get to jobs and schools, improve people's health, make the city cleaner, and build strong communities (Fig. 3).

6 Sustainable Urban Street for People

Urban streets are used by people for movement or stationary activities, for recreation or employment, out of necessity or desire. Streets are experienced and needed in many ways by people of all ages and capacities. The numerous activities that streets accommodate and facilitate, whether they be sitting, strolling, cycling, utilizing collective or personal transport, transporting goods, offering city services, or conducting business, shape the accessibility and livability of the city. Numerous factors, including the time of day, the street's size, the urban environment, and the local weather, influence the different user groups and the total number of persons on a given street. Within the constrained geometry of the street, each user walks at a different speed and occupies a varied amount of space. The variety of transportation modes that the street design supports will therefore be what determines the roadway's overall capacity.

In order to provide a welcoming environment that ensures access, safety, comfort, and enjoyment for everyone, streets should be designed to balance the demands of a variety of users [1] (Figs. 4 and 5).

Sustainable Urban Street Design in Jeddah, Saudi Arabia 233

Fig. 3 Objectives of sustainable urban street. *Source* author

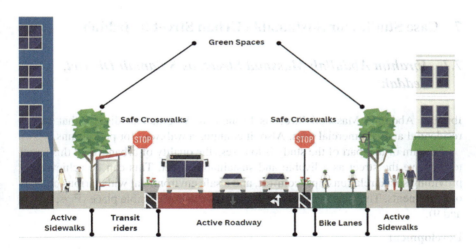

Fig. 4 The different transportation modes that the street design supports. *Source* author

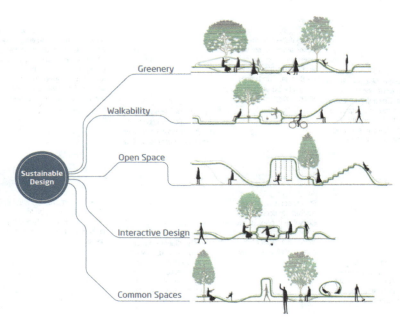

Fig. 5 Designing streets for people. *Source* author

7 Case Studies for Sustainable Urban Street in Jeddah

7.1 Ibrahim Abdallah Massoud Street, as Salamah District, Jeddah

Ibrahim Abdallah Massoud Street is located in As Salamah district that serves residential and commercial uses. Also, it contains a walkway for pedestrians.

The aim of this part of the study is to assess the quality of urban life in this street to improve the street as a livable and sustainable street. This can be achieved by providing social relationships, comfort, and a safe environment with various choices of movements. That sets the street as an attractive and accessible place (Figs. 6, 7, 8 and 9).

Development

- The Visual interception improved after reorganizing the site by replacing the huge greenery and scattering it in a rhythm which also made space for pedestrian walkways and bike paths.
- Improved the site by adding speed humps at all intersecting points to ensure safety.
- A shading system is developed by depending on natural trees with specific heights without affecting the visual interception.

Fig. 6 Ibrahim Abdallah Massoud street location. *Source* author

Fig. 7 Ibrahim Abdallah Massoud street problems. *Source* author

7.2 Abdulmaqsoud Khojah Street, Ar Rawdah District, Jeddah

Located in Jeddah off Sari Road in Ar Rawdah district, Abdulmaqsoud Khojah is a street that serves both residential and commercial locations. Also, cafes and restaurants.

All these activities interfere with traffic and pedestrians. So, this street needs a sustainable development (Fig. 10).

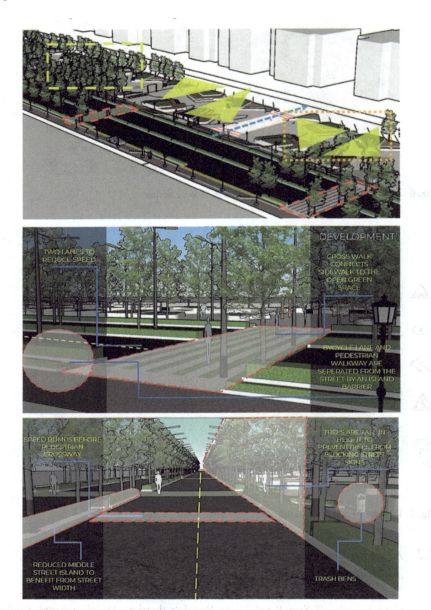

Fig. 8 Ibrahim Abdallah Massoud street development. *Source* author

Sustainable Urban Street Design in Jeddah, Saudi Arabia 237

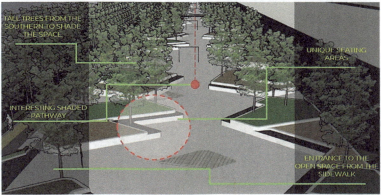

Fig. 9 Ibrahim Abdallah Massoud street sustainability. *Source* author

Fig. 10 Abdulmaqsoud Khojah street location. *Source* author

Problem Statement

The street is surrounded by cafes, restaurants and residential homes, The area is crowded and there is no parking to accommodate visitors. The residential area would benefit from improving and preparing the area for activities such as cycling, and walking improves the quality of life on this street (Fig. 11).

Development

The main goal of this design is to enhance the urban factors of this street by adding the following (Figs. 12, 13 and 14).

- Pathways to increase walkability.
- Creating a comfortable environment.

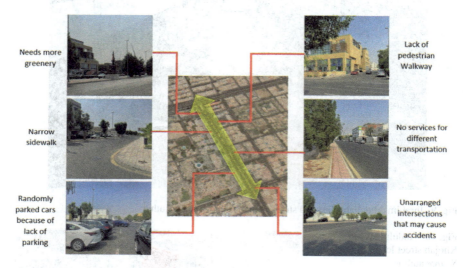

Fig. 11 Abdulmaqsoud Khojah street problems. *Source* author

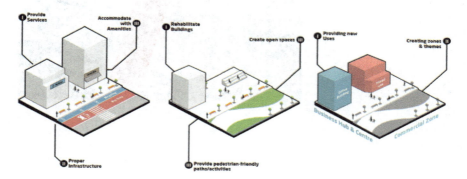

Fig. 12 Abdulmaqsoud Khojah street goals. *Source* author

Sustainable Urban Street Design in Jeddah, Saudi Arabia 239

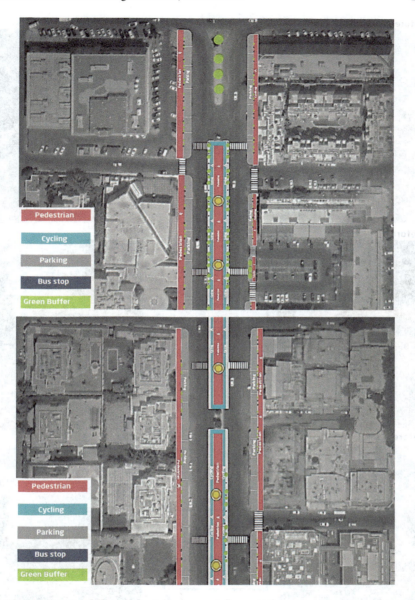

Fig. 13 Abdulmaqsoud Khojah street development. *Source* author

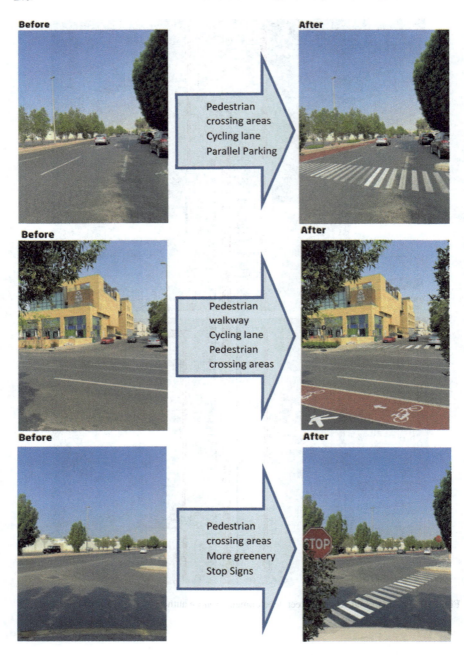

Fig. 14 Abdulmaqsoud Khojah street sustainability. *Source* author

Sustainable Urban Street Design in Jeddah, Saudi Arabia

- Seating areas to provide rest stops.
- Greenery to provide cleaner air.
- Bus stops to support public transport.
- Parking to help regulate traffic flow.
- Cycling lanes to increase environmentally friendly transport.

8 Recommendations

8.1 Sidewalks

Sidewalks paving materials should be chosen for their durability and permeability, and to reduce life-cycle environmental impacts use locally sourced paving materials manufactured using sustainable practices. Also, reduce the urban heat island effect by using cool pavements wherever possible.

Using long-lasting materials will help you save money on maintenance. The material and texture should reinforce the area's distinctiveness and improve its appearance.

All damaged stone, metal, and concrete elements should be replaced as soon as possible. Landscaping, trash cans, seating, and lighting should be provided on street corners.

Establishing barriers between sidewalks and street channels. There should be a barrier-free ramp at every street crossing.

8.2 Planting

Trees planted along city streets should be both large enough to shield people and parked vehicles from the sun and resilient enough to survive the city's pollution, heat, and glare.

Strips should be designed with clearly demarcated borders for pedestrians and cars.

It is very important to plan the planting of trees in conjunction with the installation of streetlights and utility lines.

8.3 Street Furnishing

The width of the sidewalk must be considered while designing street furniture and use recyclable and renewable materials and energy efficient items whenever possible.

Seating on benches should be pleasant and require little upkeep. It ought to be long-lasting and made from materials that won't wear down easily and can't be easily cracked.

Benches in public spaces should be easily accessible from the surrounding sidewalks. Also, should be placed where they won't be in the way of foot traffic but will still be easily reached. Be sure to place them in areas that get plenty of light and have plenty of plants nearby.

8.4 Lighting

The location and design of lighting elements should take into account the effects of light pollution, energy efficiency, and any potential negative impacts.

To achieve sustainability, solar lighting should be used. The concept of "dark skies" should be considered because it reduces extraneous light and directs it to areas and surfaces that need to be illuminated.

Lighting should serve an architectural purpose and draw attention to the street's unique architectural details to give the street its identity.

8.5 Bus Shelter

Bus entrance ramp should be free of any obstacles for passengers with mobility impairments. When we select the material it should be made from recycled sources for environmental friendliness.

Standardized signs ought to be installed at each bus stop. The signs themselves, as well as the accompanying benches, light fixtures, and garbage cans, should be well-kept and aesthetically pleasing.

8.6 Bicycle Facilities

Provide bike lanes along the street to contribute to a sustainable streetscape. These lanes must be designed not blocking entrances or pedestrian walkways.

Lanes should be designated plainly and simply.

8.7 Accessibility and Walkability

When designing streets, other modes of mobility should be considered in addition to automobiles. To ensure that people may utilize the streets for walking and cycling, the roadways and sidewalks must be devoid of obstructions such as plants, trash cans, and electricity generators. Moreover, the concept of walkability and connected streets should be considered at all design levels to guarantee that streets are connected to neighborhood, district, and municipal activities and facilities.

8.8 Security and Safety

Crosswalks should be governed by laws and rules, and there should be public awareness campaigns to teach drivers how to keep pedestrians safe. To do this, the street layout will need to be carefully planned to find good crossing points, bike lanes, and sidewalks for people to walk on. Also, the right barriers need to be made to keep people from walking into traffic. People shouldn't be able to use the central median as a public space because it makes people less safe and secure.

8.9 Activities

People need to realize that streets are more than just ways to get around. They are also important public spaces in a city. For this change to happen, people will need to know what activities and different uses streets can support based on where they are in a city.

References

1. Global Designing Cities Initiative Rockefeller Philanthropy Advisors: Global Street Design Guide, 2nd edn. Island Press, Washington, DC (2016)
2. Transportation Research Board: International Perspectives of Urban Street Design. Transportation Research Circular E-C097. Washington, DC, USA (2006)
3. Gregg, K., Hess, P.: Complete streets at the municipal level: a review of American municipal complete street policy. Int. J. Sustain. Transp. **13**, 407–418 (2019)
4. McCann, B.: Completing Our Streets: The Transition to Safe and Inclusive Transportation Networks. Island Press, Washington, DC (2013)
5. Rupprecht, S., Brand, L., Böhler-Baedeker, S., Brunner, L.M.: Guidelines for Developing and Implementing Sustainable Urban Mobility Plan, 2nd ed. Rupprecht Consult Editor, Cologne (2019)
6. Department for Transport: Manual for Streets. Thomas Telford Publishing, London (2007)
7. Transport for London: Streetscape Guidance, 4th edn. Transport for London, London (2019)
8. Wegman, F., Aarts, L.: Advancing Sustainable Safety: National Road Safety Outlook for 2005–2020. SWOV Institute for Road Safety Research, Leidschendam (2006)
9. Song, X., Cao, M., Zhai, K., Gao, X., Wu, M., Yang, T.: The effects of spatial planning, well-being, and behavioural changes during and after the COVID-19 pandemic. Front. Sustain. Cities. **3**, 686706 (2021)

244

10. United nations UN Homepage. https://www.un.org/sustainabledevelopment/cities/. Accessed 24 Jan 2023
11. Ellen, J.: Greenberg: Sustainable Streets. Transportation Research Board, New York (2009)
12. Timothy, A., Ondrej, S., John, A.: Sustainable urban street design and assessment. Third Urban Street Symposium, Washington, DC (2007)
13. New York City Department of Transportation: Measuring the Street. New Metrics for 21st Century Streets, New York, NY (2012)

Green Discourse Analysis on Twitter: Imperatives to Green Product Management in Sustainable Cities (SDG11)

Priya Sachdeva⊙, **M. Dileep Kumar**⊙, **and Archan Mitra**⊙

Abstract Pollution threatens the planet. The report discusses how green rhetoric dominates social media sites like Twitter. We don't worry about environmental impacts when buying a product since we have an egoistic value that doesn't allow us to alter. Thus, social media raises awareness of green items that are ecologically friendly if they are marketed as carbon-free and sustainable. Objective: To study how green rhetoric can affect consumer attitudes to promote ecologically friendly products. The study examines how people talk about "green" products. This study helps designers of environmentally friendly items create better products that give more sustainable long-term pollution solutions. The study must be relevant to the environment because it supports sustainable development. Analysis: Topic analysis using Twitter API hashtag data scraping is the study's analysis. NVIVO coding references helped analyse. Findings: People's attitudes toward green products are mostly positive with some neutral aspects that cannot be analysed. Conclusion: Social media influences green product management. The findings suggest individuals like green items. Implications: This research helps brands to understand consumer behaviour for green products. This research helps the researchers, policy makers to understand the people's sentiment to process critical information regarding big data in regards to product sustainability.

Keywords Discourse analysis · Green product management · Green discourse · Twitter analysis · SDG11 · Sustainability

P. Sachdeva (✉)
Amity School of Communication, Amity University, Sec-125, Noida, India
e-mail: sachdevapriya35@yahoo.in

M. D. Kumar
Research and Institution Area Nigeria, Nile University of Nigeria (Honoris United Universities Network), Abuja, Nigeria
e-mail: Dileep.KM@nileuniversity.edu.ng

A. Mitra
School of Media Studies, Presidency University, Bangalore, India

© The Author(s), under exclusive license to Springer Nature Switzerland AG 2024
A. Hamdan and E. S. Aldhaen (eds.), *Artificial Intelligence and Transforming Digital Marketing*, Studies in Systems, Decision and Control 487,
https://doi.org/10.1007/978-3-031-35828-9_22

1 Introduction

A consumer is unaware of how much carbon or pollution they produce when buying a product. So, the research identifies these attitudes during purchasing decisions and promotes sustainability. Since the 1972 UN Conference on the Human Environment in Stockholm, most governments have prioritised environmental sustainability and global climate change. Businesses may be significant. Businesses can promote environmental sustainability by making eco-friendly products [1]. Green products—those that "use fewer re-sources, have fewer impacts and risks to the environment, and prevent waste generation already at the conception stage" [2]—have been credited with creating a "new growth paradigm and a higher quality of life through wealth creation and competitiveness." Green product innovation (GPI) research has grown rapidly in recent years, despite environmental issues' limited position in innovation research. People now realise that sustainability is vital to creative solutions [3].

As environmental issues like resource scarcity, high energy consumption, rising atmospheric CO_2 levels, rising global temperatures causing ozone depletion, population explosion, declining natural resources, and other effects of global industrial growth gained attention in the late 1980s and early 1990s, "Green Product Development (GPD)" became popular. The Oslo Roundtable defined "sustainable" in 1994 [4]. According to Vinodh and Rajanayagam [5], a green manufacturing system combines product and process design with process planning and control to identify, measure, assess, and manage environmental waste. Green product creation serves customers, stakeholders, and businesses while reversing environmental damage [6]. Eco-design—combining environmental concerns with commercial design goals—is also used in green products [7]. According to Howarth and Hadfield [8], firms and designers should prioritise the selection of "environmentally friendly" raw materials, appropriate manufacturing and distribution networks, and product usage and disposal that minimises social and environmental effect [9]. Florida and Davidson [10] recommend an EMS to meet "lean and green" aims. Green management uses a systematic approach to improve an organization's financial and environmental performance. Green management is quantified by environmental impact assessment, ISO 14001 or EMAS certification, process eco-innovations, and clean technology [11]. Green product management considers items from conception to disposal. The Green Product Management strategy promotes technology-driven product innovation and sustainable production to meet customer expectations.

Today, many readers respect social media sites like Twitter and Facebook for their ease of use and ability to share ideas and opinions on a variety of topics [12]. Twitter, a microblogging and social networking platform, receives millions of "tweets" daily. Twitter's 140-character limit and hashtags to facilitate processing and search have encouraged researchers to study the data for a number of purposes. Many people are discussing green products and management on social media. This survey seeks Twitter users' green product management views.

1.1 Objectives

To gain an understanding of how and what kind of green discourse can help achieve sustainability in the marketing of environmentally friendly products through changes in consumer attitude.

2 Theoretical Foundation

An abundance of studies has been conducted on many aspects of Twitter networks, including its structure, the interactions and types of messages between users, the diffusion of information in the Twitter network, the veracity of the information, and its potential as a sentiment predictor. Differences in Twitter's impact due to a hashtag's meaning have also been examined [13]. Sentiment analysis, of which Twitter is a part, is a major topic of research in the field of computer linguistics. Opinions expressed in text can be detected and evaluated with the use of sentiment analysis automation methods. News pieces in traditional media [14], product reviews on social media [15–17], online discussion boards [18, 19], have all been subjected to sentiment analysis [20, 21]. Research into sentiment analysis has expanded alongside the social media market because of the growing interest among corporations and academics in the insightful opinions of sizable user populations.

3 Research Methodology

The study makes use of a research method known as sentiment analysis, which determines the tone of people's conversations by analysing the words they use [22, 23]. Because of this, the researcher will have an easier time comparing the success of sales of environmentally friendly products with how people react to them on social media sites like Twitter, which is where the vast majority of the debate takes place. This helps us comprehend the idea that using such products can help save the environment and that there is a sense of positivity surrounding the use, but even in this case, people might not be aware of the usefulness in the long run. This helps us comprehend the idea that using such products can help save the environment and that there is a sense of positivity surrounding the use [24–26]. The mentality that individuals have about the things and the language that they use to talk about them are likely the primary contributors to this phenomenon [27–29]. The research method is qualitative in nature and uses the topic analysis approach called sentiment analysis which helps in finding out the attitude of the buyers.

3.1 Sampling

The researchers utilised a #hashtag topic analysis approach to identify the sample, and twitter was specifically chosen as the social media platform to use for the study [30–32]. Because of the substantial volumes of data utilised, there is not a predetermined size for the study's sample population [33, 34], therefore a random sampling technique computer generated was used for the study. The significance of the word was highlighted by the researcher by using the hashtag #greenproduct [35–37]. During the month of December 2022, the research team collected data from more than 18,000 tweets, and a total of 1800 tweets that had been filtered and then coded were included in the study which was mentioned in the coding reference. The month of December was specifically chosen for this analysis of tweets since, on average, the air quality index in India is lower during that month [38, 39] which impacts peoples attitude towards environment and purchasing decisions. Because of the nature of massive data scraping, it was not possible to properly geotag the data sample; nevertheless, this is a field of research that can be specified in the future [40, 41]. As a result, the vast majority of the tweets that were collected were from an audience in India. The twitter API was used to collect data from twitter which is a legitimate and provides data authenticity.

3.2 Coding Reference

The sentiment data acquired by the researchers ($n = 1800$) tweets from the hashtag #greenmarketing are coded into items for analysis. The software NVIVO codes the data accordingly as shown in Table 1. The table reflects the number of tweets coded for polarity against mixed, neutral, negative and positive coding. Later the analysis is done on the basis of polarity extremely negative to extremely positive based on Likert scaling in order.

3.3 Using NVIVO for Qualitative Research Analysis

NVivo is one of the computer-assisted qualitative data analysis softwires (CAQDAS). Through the use of this application, qualitative research that goes beyond the coding, sorting, and retrieval of data is made possible [42]. In addition to that, the integration of coding together with qualitative linking, shaping, and modelling was expected to take place [43]. The analysis of the study was carried out with NVivo software (version 12.0) where text analysis, theme analysis and word cloud analysis is made possible through the software's coding, which demonstrates one of the primary functions of NVivo that assists qualitative researchers in managing their data [44]. The study serves as an example. When doing a thematic analysis of the data obtained from

Green Discourse Analysis on Twitter: Imperatives to Green Product ...

Table 1 Coded tweets

Codes	Number of coding references	Number of items coded
(19) #greenproducts Twitter	1800	20
(19) #greenproducts Twitter—mixed	200	2
(19) #greenproducts Twitter—negative	100	1
(19) #greenproducts Twitter—neutral	1200	12
(19) #greenproducts Twitter—positive	500	5

Source Nvivo

transcripts with NVIVO, the codes are saved as nodes inside the NVivo database. Using nodes created in NVivo, which are comparable to post-it notes stuck to the page, the researcher may indicate that a certain portion pertains to a particular topic or problem by marking it with a node. NVivo nodes, in contrast to sticky notes, are easily retrievable, may be easily organised, and give the researcher the freedom to add new nodes, delete existing nodes, alter existing nodes, or combine existing nodes at any moment [45].

3.4 Research Analysis Overview

The key method that the study utilised in order to get a variety of perspectives on the research concerns that were related to the area of green product marketing and sentiment. A fresh perspective on a social phenomenon may be gained via the use of qualitative transcripts, which enable respondents to think and reason about a variety of issues in a manner that is entirely their own. As a consequence of this, the production of discourse to identify the relationship between sentiments, word use about green product and product sales. Because 'such persons not only provide with insights into a matter but also can suggest sources of corroborating or contrary evidence,' were seen as an option within many that gives insight into the phenomenon that is being investigated, "key informants" were the sampling strategy for the study, and they are frequently essential to the success of a contextual study. Due to the fact that transcripts were the primary form of data collection, it was essential to pay close attention to the type of data analysis carried out in the early phases of the project. What is beneficial to macro marketers about the expansion of the conceptualization of all things green (green marketing, green consumption, green product design. policy discourse through a case study on the Toronto Green Standard (TGS).

4 Research Findings

Any thread of tweets that goes around are reinforced by the influencers who help push the data thread of information up into the higher brackets of search. The influencers play a critical role when it comes to understanding the source of the data. In regards to this research where the researchers are trying to find out how green product sentiment on social media tires to influence the sales of green products in actuality or are there any loopholes. From the Fig. 1 given below we can find bambooz99, green products, green marketing etc.

The diagram (Fig. 2) helps to understand the usage of words thread in green product usage. The tweet discourse is from network of users who use green products or who are ought to use green products even from companies and organizations who wish to create a positive thread of discourse towards his/her product chain which is environmentally friendly and sustainable.

In looking at Fig. 3, we can see that the total number of data that are extremely negative is 50, the total number of data that are moderately negative is 250, the total number of data that are moderately positive is 900, and the total number of data that are very positive is 600. If we combine data that is very negative with data that is moderately positive in the negative column, we get a total of 300 for the negative data; on the other hand, if we combine data that is very positive with data that is moderately positive in the positive column, we get a total of 1500 for the positive data. The difference between the two figures spans 1200 digits of numerical space in total. It is not difficult to see that the number of positive data points far exceeds the number of negative data by a significant margin. In situations like this one, the

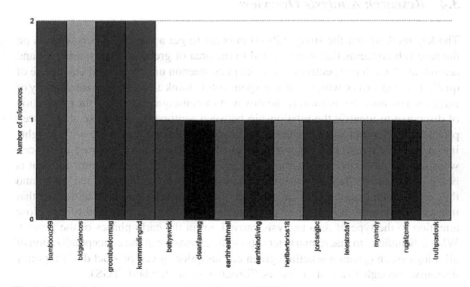

Fig. 1 Twitter influencer on green product. *Source* NVIVO

Green Discourse Analysis on Twitter: Imperatives to Green Product ...

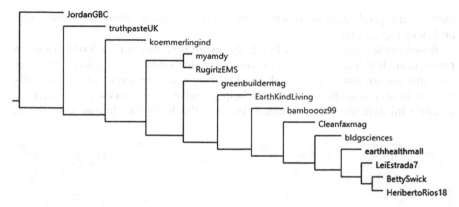

Fig. 2 Network diagram on green product. *Source* NVIVO

optimistic approach will almost always emerge victorious over the other possible courses of action (Table 2).

The researchers can see that there is a large space of neutral data on the graph in the figure titled "Sentiment Polarity with Neutral on Green Products" (Fig. 4). Additionally, researchers can see that there is a high degree of positive data points and a considerably low number of negative data points. Thus, we are able to clearly

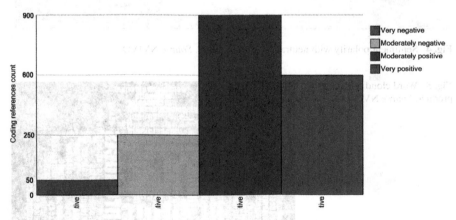

Fig. 3 Sentiment polarity on green product. *Source* NVIVO

Table 2 Sentiment polarity on green product

Hashtags	A: very negative	B: moderately negative	C: moderately positive	D: very positive
#greenproducts	50	250	900	600

Source Nvivo

evaluate the good response of environmentally friendly products on Twitter after analysing Figs. 3 and 4.

Based on the data presented in Fig. 5, the researchers were able to derive that green products are both the present and the future. Products that are considered green are ones that are not just natural but also produce no trash. The use of environmentally friendly items like sanitary napkins can help spread awareness of the need of leading a healthy lifestyle while simultaneously boosting the demand for Indian-made goods.

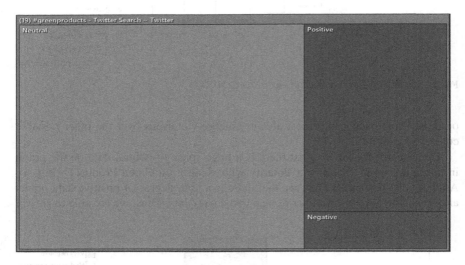

Fig. 4 Sentiment polarity with neutral on green product. *Source* NVIVO

Fig. 5 Word cloud on green product. *Source* NVIVO

Green Discourse Analysis on Twitter: Imperatives to Green Product ...

Table 3 Word references

Word	Number		Percentage	Hashtags
Sustainable	12	9	2.45	#sustainability, #sustainable, sustainability
Green	5	5	1.36	#green, green
Cleaning	9	4	1.09	#cleaning, clean, cleaning
Natural	8	4	1.09	#natural, natural, nature

Source Nvivo

The usage of words accordingly is given in Table 3 above depicts the usage of words which are associated with green products as used by the consumers and non-consumers to depict the attitude of consumers. The words used like green, clean and nature are three most important and used words. The green discourse is a much-researched word where consumer uses the word in discourse gives us clear indication of the fact that green related to the product sustainability and hence, we can understand the attitude of the consumer.

4.1 Implication of the Study

The study helps us to identify the sentiment of the consumers who use or plan to use green products to create low carbon footprint. The sentiment helps the manufactures to understand the people's sentiment who are customers or prospective target market. This will help to process critical information regarding big data in regards to product sustainability.

5 Conclusion

The objective of the research demands to underhand the sentiment so from the findings we can conclude with the fact that with over 900 coded references for green product management the sentiment of positivity in respect to the data was the majority.

The researchers have inferred from first analysis of influencers where we can find how the influencers play a crucial role in green product marketing sustainability. We find that influencer posts are shared and commented which creates a buzz about the product. The first finding led to the second where we can find that majority of the green discourse is positive in nature. When we see the box diagram Fig. 4 of the research, we see that many of the data is neutral in nature. Neutral information is mainly extracted from companies who neither go against or for the product and lets the consumers choose their interests. Sometimes the researchers also noticed the fact that neutral information comprises of consumers who are unsure of the fact that how the green product can bring positive or negative change in their life.

Focusing on the aspect of word analysis we can see that the use of word sustainability which is in discussion with the green products. The addition of the word in the discourse changes the dynamics of the total research as we can find that the word suitability has greatly affected the green product market and hope to affect more in the future green cities or sustainable smart cities.

Recommendations: for *policy makers* -the research creates a blueprint on how to manage discourse on social media and use it to amend regulations which can make create a rich and green marketing interface. Lot of carbon footprint is created in many sectors of marketing using such green practices which lot of the users of the social media depict online. Policy makers can listen to them analyze them and use them for their own benefit and for the benefit of the environment. If we talk about *authors* the researchers can say that the research sheds light on to the need to assess the online discourse and priority to not only quantitative sales of green products but rather what the sellers and buyers have to say about the product and its qualitative aspect. Judging from the quantitative side can be hazardous for the producers because as the producers may know how much sales are increasing per financial transaction timeline but they cannot fathom the change in human attitude without qualitative understanding of the same. Therefore, for future authors this research will definitely lead the path into not only quantitative aspects but the importance of qualitative aspects as well.

5.1 Limitations of the Research

In terms of the analysis of big data, the analysis has to rely on a predetermined dictionary of lexical coding and the library functions of predetermined programmers, each of which has a predetermined grammatical function and structure. As a result, it is challenging to depict the data in its entirety as it actually exists. Because of this, the researchers are able to say that there is a minimal degree of error that we have to consider when conducting big data sentiment analysis. This degree of error, however, cannot be reduced to numbers in the same way that it is possible to do for statistical analysis, nor can the degree of confidence be assigned to it.

5.2 Future Scope of Study

There is a lot of potential for advancement in this area of research, as we can conduct in-depth analyses of each and every sentiment, and we can try to find out separately what the themes of the data are. Additionally, there is room for growth in terms of understanding the neutral data, and there is room for growth in terms of analysing how we can use it.

References

1. Zhao, L., Gu, J., Abbas, J., Kirikkaleli, D., Yue, X.G.: Does quality management system help organizations in achieving environmental innovation and sustainability goals? A structural analysis. Econ. Res. Ekonomska Istraživanja **15**, 1–24 (2022)
2. Commission of the European Communities: Green Paper on Integrated Product Policy (2001)
3. Nidumolu, R., Prahalad, C.K., Rangaswami, M.R.: Why sustainability is now the key driver of innovation. IEEE Eng. Manag. Rev. **2**(41), 457 (2013)
4. Lertzman, D.A., Vredenburg, H.: Indigenous peoples, resource extraction and sustainable development: an ethical approach. J. Bus. Ethics **56**(3), 239–254 (2005)
5. Vinodh, S., Rajanayagam, D.: CAD and DFM: enablers of sustainable product design. Int. J. Sustain. Eng. **3**(4), 292–298 (2010)
6. Fuller, D.A., Ottman, J.A.: Moderating unintended pollution: the role of sustainable product design. J. Bus. Res. **57**(11), 1231–1238 (2004)
7. Karlsson, R., Luttropp, C.: Eco design: what's happening? An overview of the subject area of eco design and of the papers in this special issue. J. Clean. Prod. **14**(15–16), 1291–1298 (2006)
8. Howarth, G., Hadfield, M.: A sustainable product design model. Mater. Des. **27**(10), 1128–1133 (2006)
9. Jasti, N.V.K., Sharma, A., Karinka, S.: Development of a framework for green product development. Benchmark. Int. J. **22**, 426–445 (2015)
10. Florida, R., Davison, D.: Gaining from green management: environmental management systems inside and outside the factory. Calif. Manag. Rev. **43**(3), 64–84 (2001)
11. Albino, V., Balice, A., Dangelico, R.M.: Environmental strategies and green product development: an overview on sustainability-driven companies. Bus. Strat. Environ. **18**(2), 83–96 (2009)
12. Alayba, A.M., Palade, V., England, M., Iqbal, R.: Arabic language sentiment analysis on health services. In: Proceedings of the 2017 1st International Workshop on Arabic Script Analysis and Recognition (ASAR), pp. 114–118. IEEE (2017)
13. Chong, M.: Sentiment analysis and topic extraction of the twitter network of# prayforparis. Proc. Assoc. Inform. Sci. Technol. **53**(1), 1–4 (2016)
14. Tetlock, P.C.: Giving content to investor sentiment: the role of media in the stock market. J. Finan. **62**(3), 1139–1168 (2007)
15. Dave, K., Lawrence, S., Pennock, D. M.: Mining the peanut gallery: opinion extraction and semantic classification of product reviews. In: Proceedings of the 12th International Conference on World Wide Web, pp. 519–528 (2003)
16. Gamon, M.: Sentiment classification on customer feedback data: noisy data, large feature vectors, and the role of linguistic analysis. In: COLING 2004: Proceedings of the 20th International Conference on Computational Linguistics, pp. 841–847 (2004)
17. Pang, B., Lee, L., Vaithyanathan, S.: Thumbs up? Sentiment classification using machine learning techniques. arXiv preprint cs/0205070 (2002)
18. Abbasi, A., Chen, H., Salem, A.: Sentiment analysis in multiple languages: Feature selection for opinion classification in web forums. ACM Trans. Inform. Syst. **26**(3), 1–34 (2008)
19. Das, S.R., Chen, M.Y.: Yahoo! for Amazon: sentiment extraction from small talk on the web. Manag. Sci. **53**(9), 1375–1388 (2007)
20. Ortigosa, A., Martín, J.M., Carro, R.M.: Sentiment analysis in Facebook and its application to e-learning. Comput. Hum. Behav. **31**, 527–541 (2014)
21. Troussas, C., Virvou, M., Espinosa, K.J., Llaguno, K., Caro, J.: Sentiment analysis of Facebook statuses using Naive Bayes classifier for language learning. In: IISA 2013, pp. 1–6. IEEE (2013)
22. Ravi, K., Ravi, V.: A survey on opinion mining and sentiment analysis: tasks, approaches and applications. Knowl. Based Syst. **89**, 14–46 (2015)
23. Tausczik, Y.R., Pennebaker, J.W.: The psychological meaning of words: LIWC and computerized text analysis methods. J. Lang. Soc. Psychol. **29**(1), 24–54 (2010)
24. Buckingham, M., Clifton, D.O.: Now, discover your strengths. Simon and Schuster (2001)

25. Ghodeswar, B.M.: Building brand identity in competitive markets: a conceptual model. J. Prod. Brand Manag. **17**(1), 4–12 (2008)
26. Knight, K.E.: A descriptive model of the intra-firm innovation process. J. Bus. **40**(4), 478–496 (1967)
27. Lantolf, J.P., Pavlenko, A.: (S)econd (L)anguage (A)ctivity theory: understanding second language learners as people. In: Learner Contributions to Language Learning, pp. 141–158. Routledge, New York (2014)
28. Moscovici, S.: Notes towards a description of social representations. Eur. J. Soc. Psychol. **18**(3), 211–250 (1988)
29. Wells, C.G.: Dialogic Inquiry, pp. 137–141. Cambridge University Press, Cambridge (1999)
30. Bruns, A., Stieglitz, S.: Towards more systematic Twitter analysis: metrics for tweeting activities. Int. J. Soc. Res. Methodol. **16**(2), 91–108 (2013)
31. Felt, M.: Social media and the social sciences: how researchers employ big data analytics. Big Data Soc. **3**(1), 828 (2016)
32. Zhan, Y., Liu, R., Li, Q., Leischow, S.J., Zeng, D.D.: Identifying topics for e-cigarette user-generated contents: a case study from multiple social media platforms. J. Med. Internet Res. **19**(1), e5780 (2017)
33. Bender, D.J., Contreras, T.A., Fahrig, L.: Habitat loss and population decline: a meta-analysis of the patch size effect. Ecology **79**(2), 517–533 (1998)
34. Stanfield, A.C., McIntosh, A.M., Spencer, M.D., Philip, R., Gaur, S., Lawrie, S.M.: Towards a neuroanatomy of autism: a systematic review and meta-analysis of structural magnetic resonance imaging studies. Eur. Psych. **23**(4), 289–299 (2008)
35. Fakhira, K., Mismiwati, M., Africano, F., Riduwansah, R., Riski, O.S.: The effect of green product, halal label and Safi cosmetic brand image on purchase decisions moderated by word of mouth in the Muslim community of Palembang city. J. Bus. Stud. Manag. Rev. **5**(2), 266–274 (2022)
36. Lee, I., Mangalaraj, G.: Big data analytics in supply chain management: a systematic literature review and research directions. Big Data Cognit. Comput. **6**(1), 17 (2022)
37. Sekhon, T.S., Armstrong Soule, C.A.: Conspicuous anticonsumption: When green demarketing brands restore symbolic benefits to anticonsumers. Psychol. Market. **37**(2), 278–290 (2020)
38. Lasko, K., Vadrevu, K.P., Nguyen, T.T.N.: Analysis of air pollution over Hanoi, Vietnam using multi-satellite and MERRA reanalysis datasets. PLoS ONE **13**(5), e0196629 (2018)
39. Mahato, S., Ghosh, K.G.: Short-term exposure to ambient air quality of the most polluted Indian cities due to lockdown amid SARS-CoV-2. Environ. Res. **188**, 109835 (2020)
40. Hogenboom, F., Frasincar, F., Kaymak, U., De Jong, F., Caron, E.: A survey of event extraction methods from text for decision support systems. Decis. Support Syst. **85**, 12–22 (2016)
41. Lovelace, R., Birkin, M., Cross, P., Clarke, M.: From big noise to big data: toward the verification of large data sets for understanding regional retail flows. Geogr. Anal. **48**(1), 59–81 (2016)
42. Richards, T., Richards, L.: Using hierarchical categories in qualitative data analysis. Comput. Aided Qual. Data Anal. Theory Methods Pract. **157**, 80–95 (1995)
43. Richards, L., Bazeley, P.: The Nvivo Qualitative Project Book. SAGE Publications. https://www.google.co.in/books/edition/The_Nvivo_Qualitative_Project_Book/90XvXCwnqusC?hl=en&gbpv=0. Accessed 13 Oct 2000
44. Roberts, L.D., Breen, L.J., Symes, M.: Teaching computer-assisted qualitative data analysis to a large cohort of undergraduate students. Int. J. Res. Method Educ. **36**(3), 279–294 (2013)
45. Bonello, M., Meehan, B.: Transparency and coherence in a doctoral study case analysis: reflecting on the use of NVivo within a 'Framework' approach. Qualit. Rep. **24**(3), 483–498 (2019)

Harnessing Geographic Information System (GIS) by Implementing Building Information Modelling (BIM) to Improve AEC Performance Towards Sustainable Strategic Planning in Setiu, Terengganu, Malaysia

Siti Nur Hidayatul Ain Bt. M. Nashruddin, Siti Sarah Herman, Siti Nur Ashakirin Bt. Mohd Nashruddin, Sumarni Ismail, and Siow May Ling

Abstract The spread of Covid-19, which began in early 2020, has an impact on Malaysian tourism development. Destination planners and managers are fully aware of the problem with the extremely fragmented tourism industry's lack of coordination and cohesion [1]. Traditional methods of engaging people in participatory planning exercises are limited in their reach and scope. Simultaneously, socio-cultural trends and technological innovation provide opportunities to rethink the status quo in master planning [2, 3]. This advanced tool is used as an instrument to monitor and coordinate the entire project planning process through all stakeholders, including the community, via the internet. Currently, there is the potential to embark on online master planning in the digital world by integrating BIM into GIS to design, monitor, and coordinate at any time. Building information modelling (BIM) is frequently used in the tourism industry as a result of the industry's use of information and communication technologies. The systematic application of BIM in the tourism sector, on the other hand, is fraught with difficulties. In the context of the tourism industry, this project uses BIM to create strategic planning in Malaysia's Setiu Wetland, Terengganu. The initial strategies are based on a thorough review of the relevant literature. A BIM modeler, town planner, architect, engineer, contractor (AEC), project manager,

---

S. N. H. A. Bt. M. Nashruddin (✉) · S. S. Herman · S. Ismail · S. M. Ling
Faculty of Design and Architecture, University Putra Malaysia, 43400 Serdang, Selangor, Malaysia
e-mail: dr.ainnashruddin@gmail.com

S. S. Herman
e-mail: h_sitisarah@upm.edu.my

S. N. A. Bt. M. Nashruddin
Institute of Microengineering and Nanoelectronics (IMEN), Level 4, Research Complex, The National University of Malaysia, 43600 Bangi, Selangor, Malaysia

© The Author(s), under exclusive license to Springer Nature Switzerland AG 2024
A. Hamdan and E. S. Aldhaen (eds.), *Artificial Intelligence and Transforming Digital Marketing*, Studies in Systems, Decision and Control 487,
https://doi.org/10.1007/978-3-031-35828-9_23

programme manager, and the Setiu Wetland in-charge were all interviewed, yielding a total of 30 effective experts with diverse backgrounds. The seven essential methods identified were financial support, clearly defined plans and objectives for implementing BIM competencies and BIM-related skills, collaborative BIM execution, managing project modifications and risks, organizational and delivery mechanisms to ensure BIM implementation, and government policy and incentives to encourage BIM implementation. The additional main component analysis categories the strategies and identifies latent characteristics like institutional governance, change adaptation, technological environment, cooperation, and resources. This conclusion demonstrates that the tourism industry's context is inextricably linked to systematic BIM deployment, and that improving the way BIM is currently used in business requires adopting managerial, institutional, and technical techniques to adapt to change. The findings have implications for BIM adoption in the tourism industry and may increase productivity. The findings of this investigation, which is still ongoing. This investigation's findings are based on the most recent information available. Building information modelling (BIM) will be used to validate data in the future to ensure the sustainability of wetland tourism.

Keywords Sustainable strategic planning · Building information modelling (BIM) · Geographic information system (GIS) · AEC performance · Wetland

1 Introduction

Wetland ecosystems provide distinct roles and benefits as one of the most diverse ecological viewpoints and significant living habitats for people. However, because it is a major population center, the city's natural condition is undoubtedly concerning [4]. As a result of severe phenomena and conditions such as rapidly expanding urban scope, rapidly growing population, and rapidly changing economic position, there are several ecological security challenges, including declining wetland areas and overexploitation of wetland resources. Urban expansion has the greatest impact on the rate of extinction of local species and frequently causes the extinction of the majority of local species among human activities that cause habitat loss [5]. The distinction between urban and other natural wetlands is that the former is found in or near cities and are heavily influenced by the urban environment, economic growth, and social and cultural forms [6]. People's perceptions of urban wetland parks are skewed, which not only keeps them from being fully appreciated but also contributes to the low quality of their landscape architecture.

The most visible outward manifestation of urbanization and social and economic development is the significant transformation of urban landscapes. Simultaneously, it results in a significant reduction in urban river wetland and the loss of wetland landscape characteristics. In fact, many cities have lost a significant portion of their wetlands. China has the fourth-largest wetland area in the world, but wetlands are one of the aspects of urbanization that are disappearing the fastest [7]. As a result, it is critical to protect and study urban wetland parks [8].

BIM implementation has increased significantly on a global scale in recent years. According to McGraw-Hill Construction, BIM adoption is now required for some projects in Denmark, Finland, Norway, Singapore, Sweden, the United Kingdom, and the United States (2014). Furthermore, the Australian government has a three-year strategy for implementing BIM in public AEC projects and promoting BIM use across the country's industries [9]. BIM was also identified as an important effort in China's National Economic and Social Development Plan.

Development plans to promote its use in the tourism industry [10]. In the industry, BIM has a couple of observed values. According to a recent Dodge Data and Analytics (2017) survey, 25, 48, 37, and 35% of participants from the United States, the United Kingdom, France, and Germany claimed that their BIM-adopted projects had a 25% higher return on investment. Furthermore, a recent case study revealed that using BIM in a building project could resolve and clarify problems that cost 15.92% of the total cost of the project [11].

However, strategic planning is required for BIM adoption in the tourism industry in order to accommodate its systematic deployment and realize its potential benefits. There have been no studies on this topic. However, little strategic planning has been devoted to improving BIM implementation in wetland-focused tourist practice initiatives.

Furthermore, much of the available BIM research indicates that the implementation of BIM in practise is a methodical and stage-by-stage process. According to a few studies [12–14], there are three steps to implementing BIM, with integrated collaboration being the final level. Jin et al. [15] also clarified the relationships between the benefits of BIM, value-realized variables, and barriers to BIM practice. It is critical to incorporate BIM into IPD in order to improve project performance [16, 17]. However, due to the complexities of the construction process, systematic BIM application in AEC projects fails [18].

Despite the fact that industry deployment and organizational methods are part of BIM practice, cultural resistance in tourism projects has a significant impact on BIM implementation [19–21]. Certain studies, such as Cao et al. [22] and Liao and Teo [23], have linked BIM to projects, but more needs to be learned about how to advance its use in projects. As a result, this research analyses the strategies for advancing BIM toward integrated collaboration and frames them in the context of a tourism project, thereby encouraging BIM in the sector. The strategies are in-depth and cover various BIM implementation aspects in tourism projects due to the advanced features of BIM. This study helps to better plan strategies by adopting BIM in Malaysian tourism master planning in Setiu Wetland practices and increasing levels of BIM use in the tourism industry.

2 Study Area

Setiu, Terengganu Coastal Wetland area located about 55 kms from Kuala Terengganu, Setiu also offers many interesting places to visit (Fig. 1). Setiu is situated along the coast with crystal clear water and white sandy beaches which it is sure to offer memorable memories for those who visit it. Setiu is rich in flora and fauna, cultural art and beautiful beach scenery consist of: (1) Pantai Penarik, (2) Setiu Boardwalk, (3) Terrapuri Heritage Village.

This study demonstrates that BIM has started to conduct in-depth research into the efforts of the tourism industry in conjunction with the integration of BIM and nature. The study identifies the starting point and foothold, and summaries and applies them to the tourism industry, as well as the strategic planning to comply with ecological design methodologies. The design of an urban wetland environment that takes into account function and other factors also maintains the viability of developing tourism. Through a review of the literature and interviews with people with different levels of competence, this article will also undertake in-depth research and analysis to support the development of wetland tourism using BIM.

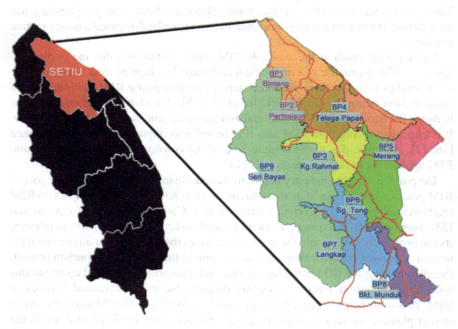

Fig. 1 Location of the study area

Harnessing Geographic Information System (GIS) by Implementing ... 261

Table 1 The previous study of the strategies for BIM implementation

Code	Strategies	Sources
S1	Clearly defined plans and objective	Dainty et al. [28]; Ding et al. [29]; Liu et al. [27]
S2	Financial support	Dainty et al. [28]; Jin et al. [25]
S3	Capabilities and skills	Ding et al. [29]
S4	Collaborative working to execute	Eadie et al. [20], Liu et al. [27]
S5	Managing changes and risks in projects	Jin et al. [15]; Ding et al. [29]
S6	Organizational and delivery measures	Liu et al. [27]
S7	Government policy and incentives	Ding et al. [29]

3 Research Method

Factor analysis is frequently used in a variety of academic fields. The study of key success factors (CSFs), for instance, is widely used in project management and strategic management.

3.1 Identification of the Strategies Through Literature Review

A comprehensive review of BIM implementation studies aids in the identification of strategies. To begin, the initial literature review referred to factor studies such as Eadie et al. [20], Rogers et al. [24], and Jin et al. [25] to provide an overview of BIM implementation. The investigation that followed focused on qualitative research about BIM practice, such as that of Ahn et al. [26], Liu et al. [27], and Dainty et al. [28]. These two types of studies were used to create strategies. However, in order for strategies to be practical and contribute to improving BIM implementation in projects, they are based on qualitative studies that directly research BIM practice rather than empirical factor studies, and they include at least one source of qualitative BIM practice studies. Through the review, 7 strategies related to BIM implementation in projects were identified as presented in Table 1.

3.2 Identification of the Strategies Through Interview

3.2.1 Data Collection

Based on qualitative data collected from primary sources, the study used an exploratory qualitative research approach [30]. In-depth interviews with eight experts from the BIM modeler, programme and project manager, AEC, town planners, and Setiu wetland expertise that related to tourism and BIM implementation were conducted recently. In-depth interviews began with open-ended questions to explore themes.

The interviews were recorded using a recording device for both face-to-face interviews, video calls, and online interviews (zoom). The researcher manually transcribed the audio interview into a text write up, which was then analyzed using Nvivo software. The interview transcripts were categorized and analyzed using a pattern coding technique based on the simple themes.

3.2.2 Data Analysis

Thematic analysis was performed using NVivo software during the research study. Because the literature for BIM adoption for sustainable wetland tourism is still emerging, we use an inductive approach to identify and learn from the research expertise's views, opinions, knowledge, experiences, and values. As a result, we were able to easily sort out broad themes using the NVivo software.

4 Findings and Analysis

The research was analyzed by gathering and transcribing the interviews document and audio into NVivo software. The result of the analysis indicated different level of stakeholders' perceptions towards adoption of BIM for sustainability of wetland buildings for tourism attraction. The analyses indicate that the expertise adoption decisions would be based on their knowledge of BIM and their accessibility to functionality of BIM applications. Figure 2 shows the various perceptions of expertise and several managers towards adoption of BIM for complex wetland building projects. The positive response reaction indicate that theses expertise believed adoption of BIM is possible for the wetland building case study project, mixed response indicates expertise who are not quite sure if there are any BIM benefits for adoption on wetland building project case study while negative response expertise perceptions indicate they are not ready to adopt BIM yet.

From Fig. 2, most of the architect has enough knowledge about BIM, and hence, have greater positive response or influence. While, engineers and town planners not have enough knowledge about BIM, and hence, some have mixed response. The manager representative organizations are completely neutral and are not sure about BIM even though they fully know the benefits of BIM for such projects. The BIM modelers and Setiu wetland expertise are fully informed and are ready to accept BIM while some of the engineer, town planner and contractor seem to depend on the clients for their decision to implement BIM. Other expertise such as BIM managers, project managers and contractors' attitudes are unbiased to implement BIM for the adopted multi-purpose buildings in wetland.

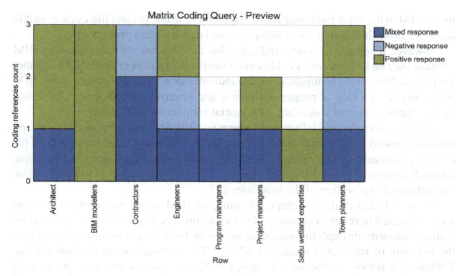

Fig. 2 Perception of interviewed expertise towards the needs to adopt BIM for strategic planning in wetland

5 Discussion

The purpose of this study was to investigate the strategies planning awareness and adoption of better strategies planning for enhancing BIM implementation in wetland Malaysian tourism practices and increasing levels of BIM use in the tourism industry.

The findings emphasize the importance of using BIM for strategic planning in wetlands, as well as the latent factors influencing BIM strategy adoption. As a result, the discussion covers the implications of the criticality analysis and strategy planning. Setiu wetland in Terengganu is used as a case study in this study. Previous studies have primarily focused on the technical aspects of BIM [31] for building construction.

The interpretative research in this study was analyzed by gathering and transcribing the audio into Nvivo software. The findings revealed a disparity in expertise perceptions of BIM adoption and its flexibility in construction and planning at Setiu Wetland. Figure 2 depicts the various perceptions of various levels of BIM expertise in construction and development, as well as the Person in Charge in Setiu Wetland (expertise) towards future tourism development and planning. The positive response reaction indicates that expertise agreed to adopt BIM in the wetland planning for the case study projects as future strategic planning, while the mixed response indicates those who are unsure about BIM adoption, and the negative response specifies they are not ready to adopt BIM.

Although BIM benefits include increased productivity, better decision making, more information, competitive advantages, and better risk management by reducing project errors during construction [32, 33]. Other advantages include cost savings, reduced redundant work, and ease of monitoring and coordination. Furthermore, by using BIM in the Built Environment throughout the whole project life cycle to tackle risks that arise throughout the construction and development process [34].

However, a negative response indicates that they are not yet ready to adopt BIM because they lack the necessary skills and understanding to operate BIM throughout the project lifecycle. The mixed response indicates those who are unsure about BIM adoption due to a lack of proper instruction and government policy to refer to and follow. Another reason was a lack of financial support for BIM adoption.

Table 2 shows the critical strategic planning of BIM implementation in wetland based on various insights of professional workers towards implementation of BIM for complex wetland building projects. Since the analysis for exploratory interpretative research is semi-structured, the interview begins with open-ended questions to allow for additional exploration of the responses [35].

Seven (S1–S7) critical strategic planning of BIM implementation in wetland were identified in order to improve BIM execution. The results, we summarized by coding similarity through the nodes, as shown in Table 2 critical strategic, because the majority of interviews identified S2, S3, S7 as saturation intersection results. Following a review of the need to adopt BIM for strategic planning in wetland projects suggested by participants, solutions for existing or new development of multi-purpose building in wetland were offered (Table 3).

As a result, by referring to code S2, financial assistance in the form of financial resources would be provided by the government to each level of the construction team or consultants in order to maximise the values of project delivery. Using the code S3, capabilities and skills will be developed by providing a BIM training workshop for the construction team or consultants, which will be organised by a professional BIM trainer from a legal company such as Autodesk to develop a set of skills using BIM tools. Finally, using code S7, developing government policy to implement BIM at the early stages of projects will raise awareness among stakeholders. In order to encourage wetland development, incentives should be provided to attract interest among public and private industry players, at once increase values in tourism industry.

Table 2 The overview of the participant demographic information

Code	Role type	No. of expert	Years of experience
Architect	Design	5	5–20
BIM modeller	Design and coordinate	5	5–10
Contractor	Off-site works	5	5–20
Engineer	Civil and mechanical and electrical	5	5–20
Program manager	Overall management	1	5–12
Project manager	Facility management	3	5–12
Town planner	Planning	1	5–20
Setiu wetland expert	Off-site coordination	5	5–10
Total		30	

Table 3 The critical strategic planning of BIM implementation in wetland

Code	Nodes
S1	• Clearly defined plans and objectives on the given project time-line basis
S2	• Financial support to adopt BIM software for each level of construction team or consultants in order to maximize the values of project delivery
S3	• Capabilities and skills will be developed by provide the BIM training workshop for the construction team or consultants that will be organize by professional BIM trainer
S4	• Collaboration prior to project execution through VA/VE review in BIM adoption
S5	• Using a BIM integration solution to manage project modifications and risks
S6	• Organizational and execution department should be setup for each consultant's department consists of AEC team and planning team
S7	• Formulate government policy to implement BIM at the very beginning stage and incentives should be given for wetland development project in order to increase values in tourism industry

6 Conclusion

Although BIM has many applications in the industry and get an attention in academia, its systematic implementation in tourism projects is still left behind. The strategies for BIM planning were summarized in this case study.

Acknowledgements This research was funded by UPM. The authors would like to prompt their deepest gratitude to the experts who assisted us with data collection. The authors would also like to thank the anonymous reviewers for their understanding comments.

References

1. Gunn, C. A.: Vacationscape: Designing tourist regions. Van Nostrand Reinhold (1988)
2. Pearce, D., Turner, R. K.: Packaging waste and the polluter pays principle: a taxation solution. J. Environ. Planning Manage. 35(1), 5–15 (1992). https://doi.org/10.1080/09640569208711905
3. Sieber, R.: Public participation geographic information systems: A literature review and framework. Ann. Assoc. Am. Geogr. 96(3), 491–507 (2006). https://doi.org/10.1111/j.1467-8306.2006.00702.x
4. Wang, Y.: Research on plant landscape construction technology and construction strategy of urban wetland park-taking Huzhou city wetland park landscape construction as an example. Abstracts Chin. Horticult. 32(4), 3 (2016)
5. Wang, S.: Research on landscape design in wetland park construction. Smart City 12(12), 1 (2016)
6. Yan, L.: The application of humanized landscape design in urban theme parks-Taking Liuzhou City Wetland Park as an example. Mod. Horticult. 10(10), 2 (2018)
7. He, J.: Landscape Design Method of Urban Wetland Park Using the 2022 (2022)
8. Qiao, L., Li, M., Zhang, Y., Zhao, X.: Research on the construction of the urban wetland park environment based on resource saving and environment friendliness. Nat. Environ. Pollut. Technol. 15(1), 221–226 (2016)

9. Building SMART Australasia.: National building information modelling initiative (2012). https://ipweaq.intersearch.com.au/ipweaqjspui/handle/1/2836
10. Bernstein, H.M., Jones, S.A., Gudgel, J.E.: Business Value of BIM in China. McGraw-Hill Construction Smart Market Report. Dodge Data&Analytics, Bedford, MA (2015)
11. Kim, S., Chin, S., Han, J., Choi, C.-H.: Measurement of construction BIM value based on a case study of a large-scale building project. J. Manage. Eng. 33(6), 05017005 (2017)
12. Khosrowshahi, F., Arayici, F.: Roadmap for implementation of BIM in the UK construction Industry. Eng. Constr. Architectural Manage. 19(6), 610–635 (2012)
13. Porwal, A., Hewage, K. N.: Building Information Modeling (BIM) partnering framework for public construction projects. Autom Constr. 31, 204–214 (2013)
14. Succar, B.: Building information modelling framework: A research and delivery foundation for industry stakeholders. Autom. Constr. 18(3), 357–375 (2009)
15. Jin, R., Hancock, C., Tang, L., Chen, C., Wanatowski, D., Yang, L.: An empirical study of BIM-implementation-based perceptions among Chinese practitioners. J. Manage. Eng. 33(5), 04017025 (2017). https://doi.org/10.1061/(ASCE)ME.1943-5479.0000538
16. Miettinen, R., Paavola, S.: Beyond the BIM utopia: approaches to the development and implementation of building information modeling. Autom. Constr. 43(7), 84–91 (2014). https://doi.org/10.1016/j.autcon.2014.03.009
17. Rowlinson, S.: Building information modelling, integrated project delivery and all that. Constr. Innov. 17(1), 45–49 (2017). https://doi.org/10.1108/CI-05-2016-0025
18. Mancini, M., Wang, X., Skitmore, M., Issa, R.: Editorial for IJPM special issue on advances in building information modeling (BIM) for construction projects. Int. J. Project Manage. 35(4), 656–657 (2017). https://doi.org/10.1016/j.ijproman.2016.12.008
19. Davies, K., McMeel, D.J., Wilkinson, S.: Making friends with Frankenstein: hybrid practice in BIM. Eng. Constr. Archit. Manage. 24(1), 78–93 (2017). https://doi.org/10.1108/ECAM-04-2015-0061
20. Eadie, R., Browne, M., Odeyinka, H., McKeown, C., McNiff, S.: A survey of current status of and perceived changes required for BIM adoption in the UK. Built. Environ. Project Asset. Manage. 5(1), 4–21 (2015). https://doi.org/10.1108/BEPAM-07-2013-0023
21. Ma, X., Xiong, F., Olawumi, T.O., Dong, N., Chan, A.P.C.: Conceptual framework and roadmap approach for integrating BIM into lifecycle project management. J. Manage. Eng. 34(6), 05018011 (2018). https://doi.org/10.1061/(ASCE)ME.1943-5479.0000647
22. Cao, D., Li, H., Wang, G., Zhang, W.: Linking the motivations and practices of design organizations to implement building information modeling in construction projects: empirical study in China. J. Manage. Eng. 32(6), 04016013 (2016). https://doi.org/10.1061/(ASCE)ME.1943-5479.0000453
23. Liao, L., Teo, E.A.L.: Organizational change perspective on people management in BIM implementation in building projects. J. Manage. Eng. 34(3), 04018008 (2018). https://doi.org/10.1061/(ASCE)ME.1943-5479.0000604
24. Rogers, J., Chong, H.-Y., Preece, C.: Adoption of building information modelling technology (BIM) perspectives from Malaysian engineering consulting services firms. Eng. Constr. Archit. Manage. 22(4), 424–445 (2015). https://doi.org/10.1108/ECAM-05-2014-0067
25. Jin, R., Hancock, C.M., Tang, L., Wanatowski, D.: BIM investment, returns, and risks in China's AEC industries. J. Constr. Eng. Manage. 143(12), 04017089 (2017). https://doi.org/10.1061/(ASCE)CO.1943-7862.0001408
26. Ahn, Y.H., Kwak, Y.H., Suk, S.J.: Contractors' transformation strategies for adopting building information modeling. J. Manage. Eng. 32(1), 05015005 (2015). https://doi.org/10.1061/(ASCE)ME.1943-5479.0000390
27. Liu, Y., Van Nederveen, S., Hertogh, M.: Understanding effects of BIM on collaborative design and construction: an empirical study in China. Int. J. Project Manage. 35(4), 686–698 (2017). https://doi.org/10.1016/j.ijproman.2016.06.007
28. Dainty, A., Leiringer, R., Fernie, S., Harty, C.: BIM and the small construction firm: a critical perspective. Build. Res. Inf. 45(6), 696–709 (2017). https://doi.org/10.1080/09613218.2017.1293940

29. Ding, Z., Zuo, J., Wu, J., Wang, J.: Key factors for the BIM adoption by architects: a China study. Eng. Constr. Archit. Manage. **22**(6), 732–748 (2015). https://doi.org/10.1108/ECAM-04-2015-0053
30. Binder, P., Holgersen, H., Moltu, C.: Staying close and reflexive: an explorative and reflexive approach to qualitative research on psychotherapy. Nordic Psychol. **64**, 103–117 (2016). https://doi.org/10.1080/19012276.2012.726815
31. Chong, H. Y., Lee, C. Y., Wang, X.: A mixed review of the adoption of Building Information Modelling (BIM) for sustainability. J. cleaner Prod. **142**, 4114–4126 (2017)
32. Bryde, D., Broquetas, M., Volm, J.M.: The project benefits of building information modelling (BIM). Int. J. Proj. Manage. **31**(7), 971–980 (2013)
33. Chen, Y., Kamara, J. M.: A framework for using mobile computing for information management on construction sites. Autom. Constr. **20**(7), 776–788 (2011)
34. Arayici, Y.: Towards building information modelling for existing structures. Struct. Surv. **26**(3), 210–222 (2008). https://doi.org/10.1108/02630800810887108
35. Oates, J., McMahon, A., Karsgaard, P., Al Quntar, S., Ur, J.: Early Mesopotamian urbanism: a new view from the north. antiquity, **81**(313), 585–600 (2007). https://doi.org/10.1017/S0003598X00095600

The Role of Contract in Advancing ESG in Construction: A Proposed Framework of 'Green' Standard Form of Contract for Use in Green Building Projects in Malaysia

Khariyah Mat Yaman and Zuhairah Ariff Abd Ghadas

Abstract The Malaysian construction industry is working to create a more sustainable built environment in response to the worldwide agendas for 'Sustainable Development' (SD) and 'Sustainable Development Goals' (SDGs). In tandem with SDG 17—Partnerships for the Goals, a mechanism or an instrument must be in place to bring industry participants together to cultivate a sustainable construction ecosystem in both processes and the end result, i.e., 'green' or sustainable buildings. This paper focuses on the role of contracts in meeting these objectives; particularly the construction industry's prevalently used standard contract forms. This study aims to determine how Malaysia's most used standard contract forms could be improved to meet the specific needs, approaches, and risks of green building construction. In Malaysia, no standard form has yet been developed specifically for green building projects. This study employed doctrinal legal research methodology to highlight the gaps in the standard contract instruments and recommend a solution using descriptive analysis, content analysis, and analytical analysis approaches. The relevant Malaysian rating tools were examined and critically evaluated within the common structure of the standard contract forms. These analyses developed a model framework of a 'green' standard form of contract for use in green building construction in Malaysia. The model framework allows green-specific terms and conditions to be added to the standard forms as the governing contract. It recommends areas and strategies for a well-drafted green standard form of contract for green building projects in Malaysia.

Keywords ESG in construction · Green building standard form · SDG 17

K. M. Yaman (✉) · Z. A. A. Ghadas
Faculty of Law and International Relations, Universiti Sultan Zainal Abidin, Gong Badak Campus, Gong Badak, 21300 Kuala Nerus, Terengganu, Malaysia
e-mail: khariyah.matyaman@gmail.com

K. M. Yaman
Messrs Hisham, Sobri & Kadir, Wisma HSK, 48G, Jalan PJU 5/22, Encorp Strand, Kota Damansara, 47810 Petaling Jaya, Selangor, Malaysia

© The Author(s), under exclusive license to Springer Nature Switzerland AG 2024
A. Hamdan and E. S. Aldhaen (eds.), *Artificial Intelligence and Transforming Digital Marketing*, Studies in Systems, Decision and Control 487,
https://doi.org/10.1007/978-3-031-35828-9_24

1 Introduction

The construction industry is grappling with the issue of fostering sustainability, as its activities consume significant amounts of earth's resources (particularly energy), generate polluting toxins and waste, create conditions conducive to soil and biodiversity loss, interfere with life support systems (such as the water cycle, soil systems, and air quality), and exacerbate urban sprawl, traffic pollution, social inequalities, and alienation. If economic and social development is to continue without harming the environment that sustains us, every member of the construction industry must contribute to the discovery of new sustainable future pathways [1, 2].

Malaysia's construction industry transformation programmes focus on green building and 'sustainable design and construction' (SDC) practices. The 11th Malaysia Plan's Green Technology Master Plan 2017–2030 (GTMP) targets six sectors, including the building sector [3]. The Construction Industry Transformation Programme (CITP) supported GTMP as a national construction industry transformation initiative [4]. CITP identified, among other things, the prevalence of inefficient construction practices that endanger the environment and compromise quality, safety, and productivity, necessitating a paradigm shift in construction methods and processes [4].

A significant milestone towards meeting its Sustainable Development Goals (SDGs), causally related to green buildings and SDC, is the introduction of sustainability rating tools into the Malaysian construction market. The first voluntary sustainability rating tool was developed by the Green Building Index Sdn. Bhd. (GBI) and introduced in 2009, known as the GBI rating tool [5]. For the government sector, the Construction Industry Development Board Malaysia (CIDB) produced a rating tool called the Malaysian Carbon Reduction and Environmental Sustainability Tool (MyCREST) [6], and the Department of Works Malaysia (PWD) came up with the *Skim Penarafan Hijau JKR* (pHJKR) [7].

Despite these measures, GTMP recognised a critical concern that could derail the national agenda: the need for a holistic approach that brings together all building and construction sector players to agree on future green building goals [3]. Currently, green building construction projects in Malaysia are carried out on bespoke terms. This approach hinders the development of the law on the subject matter and participation from a bigger pool of industry players for want of knowledge, awareness, and requisite skills. This study recommends a standard contract tailored to green building requirements to achieve the desired green building ecosystem. The study, in essence, posits a bottom-up approach by focusing on engaging the construction industry players to work towards SDG 17 to attain broader goals of SDGs [8].

This study examines the provisions of the prevalently used standard contract forms in both public and private projects in Malaysia. It identifies areas that require modifications in terms of approaches, strategies, and legal safeguards to accommodate the requirements of SDC vis-à-vis green building construction.

The Role of Contract in Advancing ESG in Construction: A Proposed ... 271

2 Problem Statement

Green buildings have distinct criteria and risks from conventional buildings [9]. Green building projects are exposed to "additional risks and uncertainties resulting from the many objectives it carries and associated conflicts arising" [10]. Risks include meeting building performance goals, funding and timing for green materials, products, and technologies, and finding suitable substitutes [10], and risks associated with "innovative design that does not work or proves impossible to construct" [9].

From the literature, the standard contract forms generally fail to address the following matters concerning SDC and green building requirements:

- The lack of 'green' definitions and related terms makes the green expectation ambiguous [11, 12].
- The responsible party/parties to achieve green certification or sustainability goals may not be specified [10, 13].
- The party responsible for project registration and green certification may not be designated [12, 14].
- Green building-related insurance policies may not be required, provided, or available [12, 13].
- Decertification consequences may not be covered in the contract [15].
- Due diligence for green products and technologies may be overlooked [15, 16].
- Consequential damages associated with green buildings may not be addressed [15, 17].
- The impact of the long lead time to achieve a green certification may not be addressed or addressed in a detrimental manner [12].

The standard contract forms in Malaysia also do not address any of these issues. Consequently, there is no contractual guidance and coverage regarding SDC and green building principles and procedures, including the standards and requirements of the respective rating tools for obtaining green certification and green building status and the associated inherent risks. These lacunae, if left unaddressed, may render the 'green' expectation ambiguous and ultimately lead to dispute.

3 Materials and Methods

This study employed doctrinal legal research methodology to identify gaps in the standard contract instrument in addressing the new development of SDC vis-a-vis green buildings and propose improvements [18]. This methodology systematically describes the laws governing a legal category, analyses their relationship, explains problems, and may forecast future developments [18].

For the analysis, this study examined the standard forms of contract written and commissioned by the Public Works Department Malaysia (PWD) for government projects and the standard forms of contract commissioned by Pertubuhan Akitek

Malaysia (PAM), which are mainly adopted in private sector projects in Malaysia [19]. For completeness, the analysis included the standard forms produced by the Malaysian Construction Industry Development Board (CIDB) and the Asian International Arbitration Centre (AIAC) to appraise the perspectives of statutory authority and an arbitral institution, respectively. The forms shall be collectively referred to as the 'Standard Forms' in the discussion.

The Standard Forms and the rating tools, in particular, MyCREST [20], pHJKR [7], GBI [21], Melaka Green Seal [22] and GreenRE [23], are central to the whole analyses. Reference was also made to other standard forms of contract published in other jurisdictions, mainly those designed for sustainable projects in the United Kingdom (UK) and the United States (US). The analyses were further supported by published guidelines, official reports and statistics by the relevant government ministries, departments, and agencies of Malaysia and other local and international industry standards. To facilitate data analysis, the study adopted descriptive analysis [18], directed content analysis method [24] and analytical analysis [18, 25] to establish the SDC parameters, identify the gaps in the Standard Forms and correspondingly propose the improvements.

4 Results and Discussion

The study found that SDC-specific terms and conditions can and should be developed. Standardising these SDC terms and conditions would help the industry players understand and execute them. An 'Addendum' to the Standard Forms can include these SDC-specific terms and conditions (the SDC Addendum). The study found that introducing the SDC standard terms and conditions via the SDC Addendum does not require a complete reorganisation of the Standard Forms, except for minor parallel modifications needed to be compatible with it. This approach would ensure a smooth transition to embedding and embracing sustainability within the contract structure familiar to Malaysian construction industry players.

By employing a directed content analysis method, the study appraised the components and subcomponents of each primary sustainability theme that comprise the baseline model of the SDC parameters. This baseline model of the SDC parameters consists of the common submittal requirements under MyCREST, pHJKR, GBI, GreenRE, and Melaka Green seal. The sustainability themes identified as characterising green buildings and SDC practices in Malaysia include (i) sustainable project planning and management, (ii) sustainable site planning and management, (iii) transportation and urban/community connectivity, (iv) water efficiency, (v) energy efficiency, (vi) indoor environmental quality, health and well-being, (vii) sustainable materials and resources, (viii) waste management, and (ix) innovation with a view of optimising all the sustainability measures. These baseline model features were then analysed within the framework of SDC-specific approaches and delivery systems to

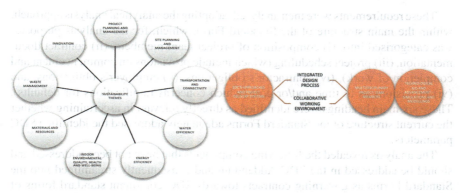

Fig. 1 A simplified depiction of the main components of the SDC parameters (own work)

determine the SDC's complete parameters and, accordingly, appraise the legal risks inherent within these parameters (see Fig. 1).

From these analyses, the study found that the SDC features complemented a concept known as the whole-systems approach. This approach adopted strategies that culminated in what is known as the IDP or the 'integrated design process' [26]. The IDP recognises the involvement of an integrated multidisciplinary project team [27] and the application of technological advancements [28]. From the analysis, the SDC parameters could be classified into four general taxonomies of (i) planning, management, and execution of the project using the IDP methodology and the allocation of associated risks, (ii) SDC-compliant technical documents (i.e., drawings and specifications) to facilitate the execution and implementation of the SDC parameters, (iii) technical requirements of the nine primary sustainability themes and risks allocation, and (iv) Integration of green rating certification procedures into the terms and conditions of the contract.

The implementation of item no. (i), i.e., the project planning, management, and execution via the IDP approach would further entail:

- A project needs assessment must include market conditions, physical needs, the environmental goal, a rating instrument and certification level, and the amount of capital allocated to green initiatives.
- The appointment of multidisciplinary project team members with clear roles and responsibilities.
- Engaging a green building consultant (GBC) or project manager familiar with the green building market, the rating tool, the required degree of certification, product category, and other aspects of sustainable or green construction.
- Establishing a collaborative working environment with decisions based on broad stakeholder feedback and a green-savvy workforce.
- Tech aids this collaborative work atmosphere. The SDC parameters include using computer simulations in some areas and building information modelling (BIM) for integration throughout the construction process.

These requirements were then analysed, adopting the analytical analysis approach, within the main structure of the Standard Form, which, for the analysis purposes, was categorised into: (i) composition of project team members, (ii) contract documentation, (iii) project scheduling (which include provisions on commencement and completion of work), (iv) contractor's obligations, (v) employer's obligations, and (vi) provisions as to time and/or cost (which concerns ramifications for defaults). These six main headings were identified as directly relevant in ascertaining whether the current structure of the Standard Forms adequately addressed the identified SDC parameters.

The analysis revealed the following components that have not been addressed and should be addressed in the SDC Addendum and consequently streamlined into the Standard Forms as governing contracts towards SDC-compliant standard forms of contract.

4.1 Green Project Goals and Objectives

The first component that must be expressly specified in the SDC Addendum, as gathered from the analysis, is the 'green project goals and objectives'. The contract must assign and clarify special definitions to reflect these goals and objectives. In the context of the proposed SDC Addendum, this component entails the drawing up of a 'sustainability plan' or any equivalent document that should address, among others: (i) the sustainable objective and targeted sustainable measures; (ii) implementation strategies selected to achieve the sustainable measures; (iii) the project team's roles and responsibilities associated with achieving the targeted sustainable measures; (iv) the specific details about design reviews, testing or metrics to verify achievement of the targeted sustainable measures; and (v) the sustainability documentation required for the project.

4.2 SDC Requirements, the IDP Team, and Allocation of Duties/Risks

The SDC requirements affect the fundamental duties of the respective parties involved within the main structure of the Standard Forms. The analysis found that the fundamental obligations of the employer, employer's representatives (ERs), and contractor typically structured into the Standard Forms will remain relevant and be maintained. For SDC implementation, these fundamental obligations must be augmented by green project goals and objectives.

For the employer, the identified augmented obligations that the SDC Addendum should address include the obligations to (i) elect the rating tool, identify the targeted green status to be achieved under the elected rating tool and the green measures required to be implemented for certification to achieve the green status; (ii) prepare or cause to be prepared the specification, drawings and the contract bills which conform to the green measures identified, showing and describing the works to be carried out by the contractor; (iii) provide and supply the necessary instructions as to the carrying out of the works, which include the green measures; (iv) giving possession of the site that conforms with the identified green measures; (v) nomination of specialist sub-contractors and suppliers; and (vii) appointment and nominations of a qualified person that conforms with the SDC requirements. GBC and commissioning specialists are among the qualified person identified for appointment under the SDC requirements.

The appointment of these qualified persons, in particular GBC, necessitates the inclusion of relevant contractual provisions to cater to potential collaboration and interfacing between them. The potential interfacing identified from the analysis includes any change in the approved green measures or sustainability plan to attain the elected green status. Instructions for any change that results in a variation to the contractor's works (variation orders or VOs) can only be issued by ERs under the authority given to them by the relevant provisions of the Standard Forms [29]. In this situation, the GBC, albeit concerning his/her scope of work, does not have the authority to issue the VOs. Similarly, inspection and testing of materials or goods to be incorporated into the works are under the purview of ERs under the Standard Forms.

On the other hand, the requirements under the SDC parameters dictate the involvement of GBC in verifying the adequacy of the materials and goods to be incorporated into the works following the relevant submittal requirements. In this situation, verification from GBC may be needed by ERs before any approval can be made. Another aspect concerns the authority conferred on ERs as certifiers. It is found that there is a need to tie the issuance of the 'certificate of practical completion', 'certificate of making good defects' and 'final certificate' contingent upon the contractor's compliance with the green measures identified for their execution and submission of all the relevant documentation required from the contractor. This mechanism would necessitate an interface whereby GBC will have to verify compliance on the contractor's part for ERs, acting as certifiers, to issue the relevant certificates to the contractor.

Similar to the employer and ERs, the contractor's fundamental obligations under the main structure of the Standard Forms will also remain applicable. However, as far as the SDC implementation is concerned, these fundamental obligations will now have to be augmented with the green project goals and objectives and the related requirements. Therefore, all green measures within the contractor's purview must be identified in the sustainability plan (or its equivalent) and contract documents with the introduction of a contractual term imposing the contractor to perform those green measures identified to be within the contractor's responsibility. Identifying these obligations in the sustainability plan and contract documents will resolve the issues concerning the contractor's obligations in implementing the SDC parameters. Provisions should also be made in the SDC Addendum concerning (i) contractor's submittals, (ii) waste management, and (iii) time for completion and green certification, affecting the contractor's obligations.

4.3 Design Charrette Process

The Standard Forms do not cater to this activity as it is to be carried out prior to the construction stage or pre-construction activity. If the design charrette process is to be implemented, the Standard Forms must reflect the process, timelines, and activities. Accordingly, provisions should be made under the SDC Addendum to allow for the design charrette process to be carried out before the construction works commence. However, finality must be afforded if the contractor's involvement during the workshop(s) concerns the project's sustainability plan or its equivalent. The fact that the inputs given in the workshop(s) serve to assist and facilitate the employer, GBC, and other project team members to finalise the document must be made clear under the SDC Addendum. This strategy ensures finality to the sustainability plan or its equivalent, which, in any event, contains the employer's project's objectives and requirements. Apart from that, the SDC Addendum should also discuss the contractual status of the design charrette report.

4.4 SDC-Compliant Technical Working Documents: Drawings and Specifications

The analysis pointed to the importance of drawings and specifications in implementing the SDC requirements. The functions of these technical working documents, under the SDC parameters, are augmented to serve as documentary evidence for the intended certification. Therefore, these technical working documents should incorporate and integrate the requirements of the SDC. The timeline that ties the preparation and production of the 'contract drawings' by the employer and the 'as-built' drawings by the contractor should also be streamlined to the certification timeline.

4.5 Technological Aid and Advancement

Two aspects of the SDC requirements for technological aid and advancement are identified under the SDC parameters. Firstly, the requirements for computer simulations and calculations as specified in certain credit requirements under the rating tools. Secondly, the use of BIM software as a medium of integration throughout the construction process. Regarding the required computer simulations and calculations, the main issue identified is the allocation of risks. Therefore, the requirements to carry out the simulations and calculations are very much related to under whose obligations the credit requirements that necessitate the carrying out of the simulations and calculations will be allotted.

In comparison, the use of BIM software would render a more significant impact as it would affect the structure of the construction process and require familiarity with a specific system throughout the construction process by all the stakeholders. One possible option is to incorporate a BIM protocol into the Standard Forms. BIM protocol is a contractual document that 'enables contractual incorporation of BIM terms, allowing for legal implementation of the processes and procedures required to produce a model that adheres to the employer's information requirements and BIM execution plan' [30]. Necessary modifications should be made to the Standard Forms' terms and conditions to streamline the BIM protocol requirements and other related BIM-specific documents.

5 Conclusion

This study established that the components of contractual terms that need to be devised for an SDC-compliant standard form of contract could be identified, developed and standardised without having to revamp the Standard Forms' structures, save for the parallel modifications that may be required to streamline the provisions, as shown in Fig. 2. This strategy would ease the transition from not having the requirements (related to SDC implementation) to understanding the requirements within the contract procurement structure of the Standard Forms familiar to the Malaysian construction industry players.

Upon identification of this framework, the next step is for the relevant stakeholders to draft the relevant terms and conditions, catering to all aspects highlighted, to be incorporated into the SDC Addendum and introduced into the Standard Forms as the governing contracts. This measure can be taken up by the respective producers of the Standard Forms to have the SDC Addendum introduced into the governing standard form for use in green projects.

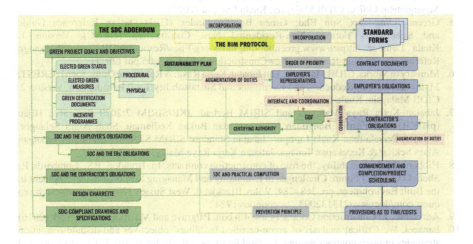

Fig. 2 The proposed framework for SDC-compliant standard forms (own work)

Similarly, the producers of the Standard Forms should also consider the development of the BIM Protocol as proposed in this study. This study has shown that the BIM implementation can also be incorporated into the Standard Forms, subject to the introduction of the BIM Protocol. For the BIM Protocol, it is recognised that further studies must be carried out to ascertain, among others, the best approach to suit the needs and capacity of the Malaysian construction industry players, particularly in dealing with the collaborative nature of the BIM processes and deliverables.

These findings are significant to the Malaysian construction industry to fill in the lacuna in the Malaysian construction industry's standard practice. This study's proposed SDC-compliant standard form of contract would provide a platform to familiarise the industry players with the requirements of the SDC and allow for a bigger pool of participation. Ultimately the goal is to bring the Malaysian construction industry players together in realising the national sustainability agendas.

References

1. Ayarkwa, J., Joe Opoku, D.G., Antwi-Afari, P., Man Li, R.Y.: Sustainable building processes' challenges and strategies: the relative important index approach. Clean. Eng. Technol. **7**, 455 (2022). https://doi.org/10.1016/j.clet.2022.100455
2. Mazhorina, M.V.: ESG principles in international business and sustainable contracts. Actual. Probl. Russ. Law **16**(12), 185–198 (2022). https://doi.org/10.17803/1994-1471.2021.133.12.185-198
3. Ministry of Energy Green Technology and Water Malaysia (KeTTHA): Green Technology Master Plan Malaysia 2017–2030. Ministry of Energy Green Technology and Water Malaysia (KeTTHA), Putrajaya. https://www.pmo.gov.my/wp-content/uploads/2019/07/Green-Technology-Master-Plan-Malaysia-2017-2030.pdf (2017)
4. Ministry of Works (MOW) Malaysia and Construction Industry Development Board (CIDB) Malaysia: Construction Industry Transformation Programme (CITP) 2016–2020. Programme Management Office, CIDB Malaysia, Kuala Lumpur (2015)
5. Greenbuildingindex Sdn Bhd: Green Building Index (GBI) Design Reference Guide and Submission Format: Non-Residential New Construction (NRNC), Version 1. Kuala Lumpur. https://www.greenbuildingindex.org/Files/Resources/GBITools/GBIDesignReferenceGuide-Non-ResidentialNewConstruction(NRNC)V1.05.pdf (2011)
6. Ministry of Works, CIDB Malaysia, and Jabatan Kerja Raya: A Reference Guide for MyCREST: Malaysian Carbon Reduction and Environmental Sustainability Tool: Introduction, Version 2. CIDB Malaysia, Kuala Lumpur (2020)
7. Jabatan Kerja Raya Malaysia and SIRIM Berhad: JKR/SIRIM 2:2020 Penarafan Hijau (pHJKR) bagi Fasiliti Bangunan Kediaman dan Bukan Kediaman. Shah Alam Selangor, Wilayah Persekutuan Kuala Lumpur: Jabatan Standard SIRIM STS Sdn Bhd dan Cawangan Alam Sekitar & Kecekapan Tenaga, Jabatan Kerja Raya Malaysia (2020)
8. Hibberd, P.: Sustainability: the role of construction contracts. In: Brandon, P.S., Lombardi, P., Shen, G.Q. (eds.) Future Challenges in Evaluating and Managing Sustainable Development in the Built Environment, pp. 268–284. Wiley Blackwell, West Sussex (2017). https://search.proquest.com/docview/2131320932?accountid=17242
9. Adriaanse, J.: Construction Contract Law, 4th edn. Palgrave and Macmillan, New York (2016)
10. Ismaeel, W.: Critical analysis of green-certified buildings' objectives and dispute-resolution strategies through contract writing. In: Building Innovatively Interactive Cities Horizons and Prospects, pp. 437–456 (2017)

The Role of Contract in Advancing ESG in Construction: A Proposed ... 279

11. Shaw Development v. Southern Builders, No.: 19-C-07-011405
12. Perkins, M.L.: Identifying and managing the risks unique to 'green' construction: what sureties should know. In: Proceedings of the Northeast Surety and Fidelity Claims Conference, pp. 1–26 (2009)
13. Ghazaleh, S.N.A., Alabady, H.S.: Contractual suggestions for the contractor in green buildings. J. Law Policy Global. **61**, 32–40 (2017)
14. Abdul-Malak, M.A.U., Khalife, F.G.: Managing the risks of third-party sustainability certification failures. J. Legal Affairs Dispute Resolut. Eng. Constr. **12**(3), 1–13 (2020). https://doi.org/10.1061/(asce)la.1943-4170.0000407
15. Maura, A., James, B., Heady, E.J.: Hidden legal risks of green building. Florida Bar J. **84**(3), 35–41 (2010)
16. Polat, G., Turkoglu, H., Gurgun, A.P.: Identification of material-related risks in green buildings. Proced. Eng. **196**, 956–963 (2017). https://doi.org/10.1016/j.proeng.2017.08.036
17. Prum, D.A., del Percio, S.: Green building contracts: considering the roles of consequential damages and limitation of liability provisions. Loyola Consum. Law Rev. **23**(2), 113–146 (2010)
18. Nuraisyah, C.A.: Legal Research Methodology. Thomson Reuters Asia Sdn Bhd, Subang Jaya (2018)
19. Rajoo, S.: Standard Form of Building Contracts Compared, vol. 2. LexisNexis Malaysia Sdn Bhd, Kuala Lumpur (2022)
20. Ministry of Works, CIDB Malaysia, and Jabatan Kerja Raya: A Reference Guide for MyCREST: Malaysian Carbon Reduction and Environmental Sustainability Tool—Design and Construction Stage, Version 2. Kuala Lumpur (2020). https://www.cidb.gov.my/en/construction-info/sustainability/construction-sustainability/mycrest
21. Greenbuildingindex Sdn Bhd: GBI Assessment Criteria for Non-Residential New Construction (NRNC), 1st Edn. Greenbuildingindex Sdn Bhd, Kuala Lumpur (2009). https://www.greenbuildingindex.org/gbi-tools/
22. Corporation, M.G.T.: Guidelines Melaka Green Seal. Melaka Green Technology Corporation, Melaka (2019)
23. GreenRE Sdn Bhd: GreenRE Design Reference Guide Non-Residential Building, Version 3. GreenRE Sdn Bhd, Petaling Jaya (2021)
24. Hsieh, H.F., Shannon, S.E.: Three approaches to qualitative content analysis. Qual. Health Res. **15**(9), 1277–1288 (2005). https://doi.org/10.1177/1049732305276687
25. Yaqin, A.: Legal Research and Writing. Malayan Law Journal Sdn Bhd, Bangsar (2007)
26. Kubba, S.: Handbook of Green Building Design and Construction: LEED, BREEAM, and Green Globes, 2nd edn. Butterworth-Heinemann, Kidlington, Oxford and Cambridge (2017)
27. Busby Perkins+Will and Stantec Consulting and BC Green Building Roundtable: Roadmap for the Integrated Design Process. Busby Perkins+Will and Stantec Consulting, Vancouver (2007)
28. Lawrence, T., Darwich, A.K., Means, J.K.: ASHRAE GreenGuide Design, Construction, and Operation of Sustainable Buildings, 5th edn. ASHRAE Atlanta, Atlanta (2018)
29. Furst, S., Ramsey, V.: Keating on Construction Contracts, 11th edn. Sweet & Maxwell, London (2021)
30. Rock, S., May, W.: The winfield rock report: overcoming the legal and contractual barriers of BIM. BIM-Legal **34**(6), 426–477 (2018)

Sustainability of Web Application Security Using Obfuscation: Techniques and Effects

Raghad A. AlSufaian, Khaireya H. AlQahtani, Rana M. AlAjmi, Roa A. AlMoussa, Rahaf A. AlGhamdi, Nazar A. Saqib, and Asiya Abdus Salam

Abstract In software development, obfuscation is one of the most straightforward approaches developers use to protect the source code and the property. This paper discusses obfuscation techniques in web applications from a security perspective. Various obfuscation methods can be grouped into three main obfuscation technique types. We analysed the effect of the obfuscation technique on web technology in terms of intellectual property, malicious code, and attacks. Obfuscation techniques are exposed to reverse engineering and the use of malicious code, which results in multiple defence solutions with the use of obfuscation for firmer resistance that will be proposed in this paper.

Keywords Obfuscation · Web technology · Intellectual property · Malicious code · Resistance to attacks · Security enhancement · Secure code development

---

R. A. AlSufaian (✉) · K. H. AlQahtani · R. M. AlAjmi · R. A. AlMoussa · R. A. AlGhamdi · N. A. Saqib
Networks and Communications Department, Imam Abdulrahman Bin Faisal University, Dammam 31441, Saudi Arabia
e-mail: 2190004263@iau.edu.sa

K. H. AlQahtani
e-mail: 2190001593@iau.edu.sa

R. M. AlAjmi
e-mail: 2190001000@iau.edu.sa

R. A. AlMoussa
e-mail: 2190003001@iau.edu.sa

R. A. AlGhamdi
e-mail: 2190001867@iau.edu.sa

N. A. Saqib
e-mail: nasaqib@iau.edu.sasss

A. Abdus Salam
Computer Information Systems Department, Imam Abdulrahman Bin Faisal University, Dammam 31441, Saudi Arabia
e-mail: aasalam@iau.edu.sa

© The Author(s), under exclusive license to Springer Nature Switzerland AG 2024
A. Hamdan and E. S. Aldhaen (eds.), *Artificial Intelligence and Transforming Digital Marketing*, Studies in Systems, Decision and Control 487,
https://doi.org/10.1007/978-3-031-35828-9_25

1 Introduction

With the increasing reliance on software technology, the opportunity to steal source code has increased due to disseminating code in an architecturally neutral format. This behaviour has alarmed software companies that want to preserve their intellectual property. Although copyright laws prohibit direct piracy of software, developers are concerned about the theft of proprietary data structures and algorithmic designs. Though there are various methods for securing software, including encryption, server-side execution, and native code, obfuscation is the most cost-effective and straightforward approach [1]. On the other hand, hiding malicious codes is the prevalent use of obfuscation techniques that threaten web security and, as a result, obfuscated malicious scripts bypass detection [2].

Moreover, cyber attackers have access to a wide variety of weapons, from malware to sophisticated reverse engineering tools. Disassemblers, decompilers, and other tools give them access to and analyse the application source code. In particular, applications that suffer from a lack of protection against code injection and code obfuscation are most vulnerable to hacking attacks [3]. In fact, Computer coding is obfuscated to distract the reader, making it challenging to figure out the meaning of the information they are reading by using complicated syntax. Moreover, Obfuscation does not change the content of a program's original code but instead makes the delivery and presentation of that code more confusing. In addition, Obfuscation does not alter the program's functionality. Obfuscation can also be used to trick antivirus tools and other software [4].

2 Background

The web application is software installed and operated on a remote web server through the internet using the HTTP protocol. Web applications have two types: Static and Dynamic. Static web apps avoid modifying web content based on user input, whereas dynamic allows web content to change based on user input. The back end of a web application is typically tied to the server-side implemented by PHP, Java, Python etc. In contrast, the front-end is related to the client-side, implemented using HTML, CSS, JavaScript etc. [5] (Fig. 1).

When malicious code is executed in a script from the website on the victim's PC, it is called an XSS attack. In SQL injection attacks, the attacker inserts their code into a pre-existing query. In order to protect against these two types of attacks, a novel obfuscation technique has been developed [5].

Obfuscation has been recommended by OWASP for years, but it is still not so popular among developers. Where overusing Obfuscation can hinder software performance, which is why many developers stay away from it. Apart from this, some believe that Obfuscation is easier than protecting data, but indeed hiding information is entirely different from protecting it [3]. The importance of code obfuscation is to

Fig. 1 Web application [5]

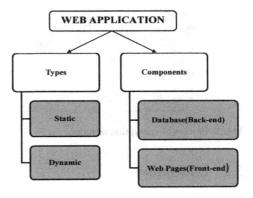

make the resulting code unintelligible for humans and difficult to reverse engineer or be tamper with by automated tools. Code obfuscation transforms the source code to make it more challenging to analyse by attackers. Its primary goal is to make attacking complicated enough to repulse attackers rather than formally proving the strength of algorithms. The measures to evaluate the effectiveness of code obfuscation techniques are generally measured from the application's potency, resilience, cost, and stealth [6].

This paper will discuss the obfuscation technique in Sect. 3, the effect of obfuscation techniques on web security in Sect. 4 with three classifications Sect. 4.1 intellectual property, Sect. 4.2 malicious code, and Sect. 4.3 resistance to attacks. Also, in Sect. 5, possible existed solutions and defence techniques using obfuscation will be presented, a discussion and analysis in Sect. 6, proposed future work in Sect. 7, and the conclusion in Sect. 8.

3 Obfuscation Technique

In code obfuscation, implementation details are hidden from the adversary, for instance, converting the original program into a semantic equivalent that is much more difficult to understand [7]. It involves using transformations to alter the physical appearance of a code while maintaining its black-box specifications [1]. Data obfuscation has multiple techniques grouped into three main categories: encryption, data masking, and tokenization [8].

Fig. 2 Process of encryption technique [8]

3.1 Encryption

To make the security level of obfuscation techniques stronger, NIST recommends using AER algorithms for encryption to safeguard the intellectual property of the US national information security system. Obfuscation is a method to hide the code while preserving its content with the aim of keeping it from unauthorised access. However, this technology was more vulnerable to disintegration. A key difference between encryption and obfuscation is that obfuscation does not require a key, whereas encryption does. Despite this, obfuscation without encryption is still seen as insecure in confidentiality. Therefore, obfuscation with encryption is the ideal method for increasing security. Without a doubt, Programmers are advised to implement obfuscation in PHP applications in order to protect sensitive and confidential information such as usernames and passwords and processes used to log into the application. Since obfuscation in PHP extension code uses the AES algorithm, the information is rendered unreadable, thus ensuring a high level of security [3]. Data encryption is Performing specific mathematical calculations on data based on a cryptography technique. And cannot decrypt it until they know the secret key [8]. Data encryption is considered the least effort compared to data masking [9]. Figure 2 shows how to implement the encryption technique [8].

Data Encryption benefits [9]:

- Remain encrypted even if it's on an external disk or backed up.
- The size of the database is unchanged.
- Effective risk mitigation for other attack vectors.

3.2 Data Masking

Data masking secures sensitive data while keeping its confidentiality, and even if the values are modified, the controlling party can retrieve them by inverting the operation. It will change the data values, so the actual values will not know. Developers use masking when developing the code or testing it. Data masking tools use for users who want to publish their work on the internet using GitHub or any other web to download,

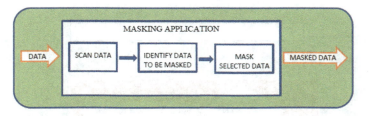

Fig. 3 Process of masking technique [8]

read, or execute for hosting source code but preserve the proprietary information. It masks the variable name, functions, or even classes—an example of the data masking tool JumbleDB [10]. Figure 3 show the process of masking [8].

1. Data scanning.
2. Identify sensitive data by the system.

This process uses a one-way process, so there is no way to know the original masked data [8]. Data Masking benefits [9]:

1. The application will work as expected with data masking.
2. There's no need to be worried regarding unauthorised access or data extraction.

3.3 Tokenization

Tokenization is the process of replacing a portion of sensitive data with a surrogate value. Usually used token in a payment card. This transition performs in a one-way function, so it won't be easy to reassemble the original information without using the tokenization system's resources [8]. Figure 4 will show the process of tokenization [8]:

1. The application will pass data that wants to generate tokenization and authentication data to the tokenization system.
2. If authentication doesn't match, it will stop the process.
3. If authentication matches, the token will be generated using a one-way algorithm.
4. Token data is passed to the application to use.

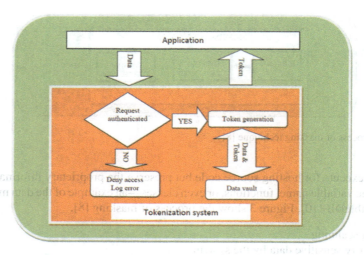

Fig. 4 Process of tokenization technique [8]

4 Obfuscation Techniques Effect on Web Security

4.1 Intellectual-Property

Generally, a web application's distributed source code should be protected on the client-side. A solution was proposed to secure web application source code. It demonstrates how the intellectual property of developed code can be protected when distributed as source code by adopting a multi-layered protection solution. The three layers of protection are the following [11]:

- **The basic level**: A watermark (logo and licensee name) is added to the GUI of the application.
- **The intermediate level**: It uses code obfuscation techniques to make source code unreadable and unmodifiable by programmers.
- **The server level**:
  1. Monitors clients and determines action plans with respect to their licenses.
  2. Provides the client with the missing source code.
  3. Verifies that the client's code hasn't been tampered with and that the logo and licensee's name conform to the license.

Obfuscating provides a method for preventing unauthorized reuse of HTML content. A big obstacle for digital content is finding the right balance between accessibility and security. HTML Obfuscation with Text Distribution to Overlapping Layers (HOTDOL) has been proposed to obfuscate the text of HTML contents to discourage unpermitted parties from overusing and redistributing HTML content. HOTDOL is used to divide the text of an HTML document into several translucent layers. To

Sustainability of Web Application Security Using Obfuscation ... 287

maintain the exact visual representation, these layers on display overlap each other. "Hotdolized content" refers to the entire or even a portion of web content that has been processed with HOTDOL. One of the main advantages of HOTDOL is that when viewing web content in a web browser, it is hard for consumers to pick and replicate the source text, either in portions or in entire. Even when hotdolized content is loaded into other programs, it isn't easy for users to modify it [12].

4.2 Malicious Code

JavaScript can be run on almost every platform in the Internet-distributed environment, and programs written in Java are used anywhere on the network. Therefore, Java can be useful for web programming. Thus, remarkable work is performed to protect the output of Java, which is used widely and frequently using "obfuscation". Using JavaScript obfuscation methods can help developers preserve the program or ensure security when transmitting the program via communication [13].

Malicious websites commonly use obfuscated JavaScript. By using obfuscated JavaScript, The analysis of JavaScript takes a considerable time, and the program that blocks malicious websites can be temporarily bypassed. Virus writers often employ obfuscation to change the signature of the code, for instance, the sequence of instructions, and as a result, a virus scanner won't be able to detect the existence of the virus using search strings by using various obfuscation techniques that change the code's layout [1]. The static analysis method, in this instance, is inefficient as it requires too much time and is difficult to set up the analysis environment with the dynamic analysis method. Therefore, detecting and controlling malicious websites has become a crucial concern [13].

Malware obfuscation technologies include dead-code insertion, register reassignment, subroutine reordering, instruction substitution, code transposition, and code integration. The primary use of these techniques is to avoid antivirus scanners by polymorphic and metamorphic malware. Despite the fact that this technology was designed to protect software developers' intellectual property, malware makers have widely used it to avoid detection. These obfuscation tactics will get more advanced and complex in the future. While focusing on these obfuscation techniques customised for web and mobile malware [14]. Malicious users infiltrate the client's system by embedding malicious JavaScript code in a regular web page. So, they have the ability to thieve confidential information, install malware on client systems, etc. Security systems use signatures to detect JavaScript codes in malicious web pages to defend against attacks. As a result, detecting obfuscated strings in malicious web pages that contain JavaScript code has been proposed by using a novel methodology [15]. Suggested solutions using obfuscation will be presented.

4.3 Resistance to Attacks

Obfuscation doesn't always assure that the program will not be interfered with or reverse engineered, but it does add an auxiliary level of protection by making it costly and more difficult for an intruder to understand the underlying features of the secured program [16]. Software security in the context of reverse engineering is a challenging job that has been studied over a number of years and is still extremely energetic. Simultaneously, this field is essentially workable and thus has applications in the industry. It is challenging to protect a portion of software from interfering, malicious adjustments, or reverse engineering. By employing Obfuscator-LLVM (ollvm), a set of obfuscating code modifications is applied as middle-end routes in the LLVM compilation suite. The vast ample amounts of protection mechanisms that have been deployed are open-source and publicly available. A solution proposed is to implement numerous code evolutions to augment software opposition to reverse engineering and tampering [17].

Also in a SQL injection attack, the attacker tries to exploit the database during the run time of the website by providing a query to the web application. An example of queries that an attacker may enter to bypass the check: "OR '1' = '1' –". And to protect these applications or websites from SQL attacks, they use obfuscation [18]. Using tokenization can provide detestation of spaces, double quotes, single quotes, and check all strings in the input. If the suspect symbols or strings founded, it will detect the injection. Before sending a query to a DBMS, use an obfuscation-based approach to identify the existence of probable SQL Injection Attacks (SQLIA). These approaches combine static and dynamic phases as follows:

- **Static phase**: replace queries in the application with queries in obfuscated form.
- **Dynamic phase**: Before submitting the query to the DBMS, a de-obfuscation procedure is done to retrieve the original query. This phase checked for the existence of probable SQLIA performed on those atomic formulae that have been flagged as vulnerable.

Many problems have been solved by using the obfuscation method, Making reverse engineering of the program logic tougher, preventing common flaws, and keeping software from being altered without permission. Finally, hiding data makes programs more difficult to understand and implement [16].

5 Defence Techniques Using Obfuscation

As the Internet's popularity grew, it became an attractive target for criminals. An investigated deceptive defence strategies for web servers often exposed on the Internet will be discussed. Additionally, the usual security features like firewalls do not adequately protect web servers. Because of this, multiple ways are described to make web servers more secure. By conducting experiments on the Internet, it has

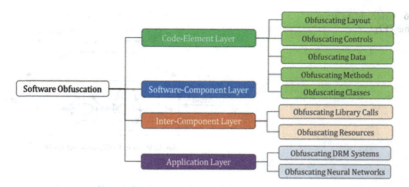

Fig. 5 Software obfuscation technique taxonomy for layered security [20]

been demonstrated that honey tokens could mislead attackers from their targets. Furthermore, information defence strategies based on deception techniques such as feints, defence its, delays, misdirection, fake responses, and obfuscation are used to reduce the effectiveness of intrusions [19]. So, it is recommended to combine more than one obfuscation technique for more security complexity [7]. And this idea is inspired by the classical layered security principle for risk management. A promising method of obfuscating software is proposed where layered obfuscation is employed.

Layered obfuscation combines various obfuscation approaches into a single solution to reduce the risk of reverse engineering and attacks [20].

Based on the differences in the obfuscation target, the taxonomy classifies the obfuscation techniques into four layers as shown in Fig. 5 [20]. The four layers are the code-element layer, software-component layer, inter-component layer, and application layer. If obfuscation targets can be classified further, each layer can be divided into multiple subcategories. A taxonomy of obfuscation techniques can be used to help developers choose the most appropriate obfuscation techniques and design layered solutions that match their specific needs [20].

Moreover, an obfuscation scheme for protecting susceptible software code segments was presented. Given such sensitive software parts, it proposes a four-step approach to obfuscate them. To begin, divide the delicate areas into logical code segments. Secondly, for each segment, create non-trivial code clones. Third, using dynamic predicate variables, to originate a proper control flow track connect the clones to the corresponding original fragments, one of which at run time is picked randomly. This results in an exponential number of semantically similar paths relating to a single way present in the original code necessitating an attacker to execute and analyse numerous execution traces to recognise the functionality of the sensitive code. Fourth, use identification renaming and any other user-specified deobfuscation technique(s) to thwart the assailant's efforts to decipher the obfuscated code. This technique is resisted static, dynamic and clone detection attacks [17].

Figure 6 depicts this process, in which ellipses represent logical code clones created for obfuscation.

Fig. 6 Code obfuscation scheme [17]

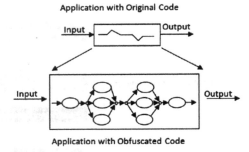

Additionally, the privacy-aware obfuscation method handles weaknesses associated with existing methods. It is a generic method that can apply to various applications of Web data. Developing this method aims to predict the privacy risk of Web data that includes all key aspects of privacy, namely uniqueness, uniformity, and link ability of Web data. And the data that entails high privacy risks will be obfuscated using probabilistic methods. Experiments on two web databases have shown that privacy risk in Web data can be predicted and obfuscated for data with high privacy risk using this method [21]. Another approach by using control flow diversification to obfuscate software, making it quite challenging for an attacker to correlate structural information obtained through running a program several times and logging its trace. As a result of splitting the code into small chunks (gadgets) before diversification, a complex control flow graph is achieved, and static analysis can only disclose limited local information about the program. Evaluating the strength of this approach against automated deobfuscation demonstrates its efficacy in increasing an attacker's effort [22].

Figure 7 shows a more complex control flow graph in order to increase obfuscation security. Thus, static reverse engineering is prevented, and dynamic analysis is limited [22].

In terms of malware detection, instead of static and dynamic analysis that is less affected against obfuscated malware, one solution is a JavaScript obfuscation-based system that analyses JavaScript density, frequency, JavaScript entropy in its entirety and entropy of each characteristic. The ratio of non-detected malicious websites is smaller when the analysis time is short, and therefore more websites are detected [13]. Similar to the previous technique, detecting obfuscated strings in malicious web pages that contain JavaScript code has been proposed by using a novel methodology. Extracted three metrics as regulations for detecting obfuscated strings by analysing standard and malicious JavaScript code patterns. They are Entropy, N-gram, and Word Size. Counts how many times each byte code appears in a string by N-gram. Checks on used byte codes are distributed by Entropy. Determines whether a very long string has been used by Word size. This method detects obfuscated strings well. Although, it has boundaries for the reason that it only discovers a few patterns of real malicious web pages, this methodology can detect a limited number of patterns [15].

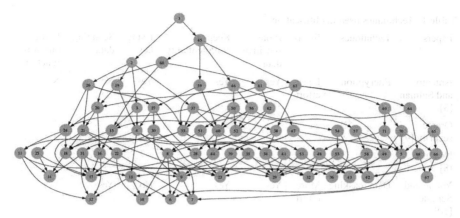

Fig. 7 Diversified control flow graph [22]

6 Discussion and Analysis

In this section, we compare the previous techniques used in obfuscation based on the effort to protect sensitive data, keep confidentiality, use key, scan data, and use a one-way function.

According to Table 1, these papers discussed the three main obfuscation techniques: encryption, Data masking, and Tokenization. Encryption is the least effort compared to data masking. Also, it uses a key to encrypt and decrypt the data, whereas the other techniques do not. They use a one-way process to protect the data. A one-way function is difficult to disassemble to original information from masked data or even Token data. The last comparison is scanning data with high privacy to preserve the proprietary host information in data masking only. And all of these techniques protect sensitive data from unauthorised users while keeping their confidentiality.

Moreover, after analysing the various obfuscation techniques, we found that standard existing techniques do not meet all criteria for obfuscation effectiveness, such as resistance to attacks and reverse engineering. Also, it could be a helpful tool to hide malicious code in the web application that leads to security threats. Due to that, we discovered various schemes and approaches that help resist the above threats. First, we can specify the high-risk and most vulnerable portions of the web data that will be obfuscated using the privacy-aware obfuscation method making it more difficult for an attacker to correlate the structural information obtained. Combining multiple layers and techniques for obfuscating can reduce the risk of reverse engineering and attacks and protect sensitive software code fragments. Layered obfuscation, Obfuscation scheme and control flow diversification are proposed to increase the attacker's effort. And when it's come to web technology security, the most common threat that might be faced is malware. To enhance web security in this aspect, Javascript obfuscation-based system and a novel methodology for detecting obfuscated strings might be used to analyse malware. When malicious code is obfuscated, it's hard to

Table 1 Techniques used in obfuscation

Papers	Techniques	Effort	Protect sensitive data	Keep confidentiality	Using key	Scanning data	Using one-way function
Ismanto and Salman [3]	Encryption	Least effort	Yes	Yes	Yes	No	No
Oracle security [9]							
Neamtiu [8]							
Young and Semple [10]	Data masking	More effort	Yes	Yes	No	Yes	Yes
Oracle security [9]							
Neamtiu [8]							
Neamtiu [8]	Tokenization	–	Yes	Yes	No	No	Yes

be detected using static and dynamic analysis, which results in analysing it using obfuscation criteria. In addition, we have noticed that the most efficient solutions are based on combining more than one obfuscation technique or beside it. It should be implemented thoughtfully. Otherwise, it will negatively affect performance and functionality [17].

7 Future Work

It would be beneficial to conduct studies on developing hybrid malware analysis systems. Also, the need for alternative methods of website analysis besides JavaScript must be studied. Finally, a research is required in order to determine which techniques yield the best results.

8 Conclusion

In conclusion, the primary goal of this work was to explore the use of obfuscation as a means to improve the security of web applications. We discussed various strategies for enhancing web server security, as well as the use of obfuscation technologies to protect intellectual property. Additionally, we examined how obfuscation can be

Sustainability of Web Application Security Using Obfuscation ... 293

used to conceal malware, such as using malicious JavaScript code that employs obfuscation to evade detection by virus scanners. While obfuscation can be used maliciously, it can also be used benignly to preserve code privacy or intellectual property. However, it is essential to consider that the obfuscation code should be unreadable by humans to ensure its effectiveness. Using code obfuscation tools makes it difficult for attackers to reverse-engineer and understand the inner workings of your web application. This can help to protect sensitive information, such as encryption keys and business logic. There are several practical recommendations that authors and policymakers can consider when implementing obfuscation techniques in web applications, such as:

1. Use minification techniques to reduce the size of the code, making it harder to read and understand.
2. Use code obfuscation tools to change the structure of the code, making it harder to reverse engineer.
3. Use code encryption to protect sensitive data and algorithms from being stolen or compromised.
4. Use anti-debugging and anti-tampering techniques to prevent attackers from using debuggers or other tools to reverse engineer the code.
5. Regularly update the obfuscation techniques used in the application to stay ahead of new attack methods.
6. Monitor and log application activity to detect and respond to potential security breaches.
7. Conduct regular penetration testing to identify and remediate vulnerabilities in the application.

It's important to note that while obfuscation techniques can make it more difficult for attackers to reverse engineer web applications, they do not provide complete protection. Thus, they should be used in conjunction with other security measures for reliable security.

References

1. Balakrishnan, A., Schulze, C.: Code Obfuscation Literature Survey. Computer Sciences Department University of Wisconsin, Madison (2022)
2. Al-Taharwa, I., Lee, H., Jeng, A., Wu, K., Ho, C., Chen, S.: JSOD: JavaScript obfuscation detector. Sec. Commun. Netw. 8(6), 1092–1107 (2014)
3. Ismanto, R., Salman, M.: Improving security level through obfuscation technique for sources code protection using AES Algorithm. Scholar.google.com (2022)
4. ACM Computing Surveys: Protecting Software Through Obfuscation: Can It Keep Pace with Progress in Code Analysis? ACM Computing Surveys, vol 49. ACM Computing Surveys (2022)
5. Kumar, D., Kumar, A., Singhc, L.: Enhance web application security using obfuscation. Turk. J. Comput. Math. Educ. 12(12), 168 (2021)
6. Sebastian, S., Malgaonkar, S., Shah, P., Kapoor, M., Parekhji, T.: A study and review on code obfuscation. In: Proceedings of the 2016 World Conference on Futuristic Trends in Research and Innovation for Social Welfare (Startup Conclave) (2016)

7. Behera, C., Bhaskari, D.: Different obfuscation techniques for code protection. Proced. Comput. Sci. **70**, 757–763 (2015)
8. Neamtiu, O.: Tokenization as a Data Security Technique. Zeszyty Naukowe AON, pp. 124–135 (2016)
9. Oracle Security Software Data Masking and Encryption (2013)
10. Young, G., Semple, N.: Software Data Masking for Showcasing Programming Skills. Csis.pace.edu (2022)
11. Simonetta, A., Rinaldi, F.: Code Protection Techniques When Distributed in Source Format: An Adobe Connect Pod Written in JavaScript (2021)
12. Han, S., et al.: HOTDOL: HTML Obfuscation with Text Distribution to Overlapping Layers (2014). https://doi.org/10.1109/CIT.2014.104
13. Kim, B.I., Im, C.T., Jung, H.C.: Suspicious Malicious Web Site Detection with Strength Analysis of a JavaScript Obfuscation. Korea Internet & Security Agency (2018)
14. You, I., Yim, K.: Malware Obfuscation Techniques: A Brief Survey. Citeseerx.ist.psu.edu (2022)
15. Choi, Y., et al.: Automatic detection for JavaScript obfuscation attacks in web pages through string pattern analysis. In: Future Generation Information Technology Anonymous. Springer, Berlin, Heidelberg, pp. 160–172 (2018)
16. Hosseinzadeh, S., et al.: Diversification and obfuscation techniques for software security: a systematic literature review. Inform. Softw. Technol. **104**, 72–93 (2018)
17. Junod, P., et al.: Obfuscator-LLVM: Software Protection for the Masses (2015). https://doi.org/10.1109/SPRO.2015.10
18. Halder, R., Cortesi, A.: Obfuscation-based analysis of SQL injection attacks. In: The IEEE Symposium on Computers and Communications (2010)
19. Fraunholz, D., Schotten, H.D.: Defending Web Servers with Feints, Distraction and Obfuscation (2022)
20. Xu, H., Zhou, Y., Ming, J., Lyu, M.: Layered obfuscation: a taxonomy of software obfuscation techniques for layered security. Cybersecurity **3**(1), 157 (2020)
21. Masood, R., Vatsalan, D., Ikram, M., Kaafar, M.: Incognito. In: Proceedings of the 2018 World Wide Web Conference on World Wide Web—WWW '18 (2018)
22. Schrittwieser, S., Katzenbeisser, S.: Code Obfuscation against Static and Dynamic Reverse Engineering. Information Hiding, pp. 270–284 (2011)
23. Xu, W., Zhang, F., Zhu, S.: The power of obfuscation techniques in malicious JavaScript code: a measurement study. In: Proceedings of the 2012 7th International Conference on Malicious and Unwanted Software (2012)
24. Kulkarni, A., Metta, R.: A new code obfuscation scheme for software protection (2014). https://doi.org/10.1109/SOSE.2014.57
25. Kushwaha1, S., Soni, S.: A survey on malware and session hijack attack over web environments. IOSR J. Comput. Eng. **20**, 30–35 (2018)

Heritage Buildings Changes and Sustainable Development

Marwa Abouhassan

Abstract Heritage is seen as a major component of quality of life, features that give a city its unique character and provide the sense of belonging that lies at the core of cultural identity. In other words, heritage by providing important social and psychological benefits enrich human life with meanings and emotions and raise quality of life as a key component of sustainability. This paper deals with the phenomena of the changes in the facades of heritage buildings which are subjected to distortion and loss, to conserve them because they are the important facet of our culture and identity. First, the paper presents the planning studies for the case study which is in the Northwest part of the old city center of Alexandria, Egypt. Next, deduce some analytical studies which will shed light on the current state of the case study. Then, conducting an architectural study for the heritage buildings in the case study in relation to the elements of the architecture in the external facade. Through these studies, we would be able to deduce that a lot of changes have occurred in these heritage buildings. Finally, the paper is ended by general results and recommendations which help us to control these changes and to stress the existing links between cultural heritage and sustainable development.

Keywords Heritage buildings · Sustainable development · Architectural elements · Identity · SDG11

1 Introduction

Since the late 1970s, the idea of sustainability has been an important part of development work. At a United Nations conference on the environment in Stockholm in 1972, the idea of "Sustainable Development" was put forward. It was looked at from different angles to reduce the bad effects of human activities and protect the quality of life for both the current and future generations [1].

M. Abouhassan (✉)
University of Business and Technology, Jeddah, Kingdom of Saudi Arabia
e-mail: m.abouhassan@ubt.edu.sa

© The Author(s), under exclusive license to Springer Nature Switzerland AG 2024
A. Hamdan and E. S. Aldhaen (eds.), *Artificial Intelligence and Transforming Digital Marketing*, Studies in Systems, Decision and Control 487,
https://doi.org/10.1007/978-3-031-35828-9_26

295

The UN's 2030 Agenda for Sustainable Development (UNSDGs), especially Goal 11: "Make cities and human settlements inclusive, safe, resilient, and sustainable," includes links between cultural heritage and sustainable development. Even though several global and regional networks, like UCLG, and other stakeholders are working to improve the SDGs, there aren't many clear references to cultural aspects in the goals. UCLG, on the other hand, thinks that cultural aspects are key to achieving goals in all areas of sustainable development [2].

The World Commission on Culture and Development (WCCD) brought the concept of sustainability to the field of cultural development. A new multidisciplinary approach, cultural sustainability, has evolved to elevate the importance of culture and its factors in local, regional, and global sustainable development in response to the report's advocacy for the long-term requirements of future generations for access to cultural resources [3].

Statistics, indicators, and data on the cultural sector, as well as operational actions, have consistently emphasized that culture can be a potent driver for development, with far-reaching social, economic, and environmental implications for communities throughout the past decade. Planning communities with a focus on preserving their cultural and natural heritage while also meeting other sustainability goals can have a positive "triple bottom line" effect, leading to environmental, economic, social, and cultural benefits that are important for building sustainable cities [4].

Socially, heritage helps build social capital between people by giving them a sense of identity and pride in their community. It has the power to make communities stronger when people connect the historic environment to their shared identity, their connection to a place, and their everyday lives. In this way, the memories and values that come with heritage often give people a sense of belonging and connection to a community. This helps build social cohesion and inclusion, which are key parts of sustainable cities, and also improves cultural well-being and quality of life [5].

Unfortunately, many of the cities facing problems with its heritage buildings. One of these cities is Alexandria [6]. It was an important economic port after the establishment of the state of Mohamed Ali in 1805. It was the refuge to foreigners especially from the countries of the Mediterranean Sea [7]. The buildings were affected by architectural styles prevalent in Europe during the nineteenth century and early twentieth century [8]. But after the beginning of twentieth century, urban changes happened in these areas and have accumulated a lot of problems associated with indiscriminate growth, also the number of demolished villas increased and were replaced by high buildings that did not fit in with the character of the area [9].

Hence, the paper shall set controls and laws to keep these areas and buildings that are representing an important part of the history and civilization of Alexandria, which gives it good and irreplaceable value and increase the sustainable development.

2 Sustainable Development Goals (UNSDGs) and Cultural Heritage

Cultural heritage mentioned in the 2030 Agenda for Sustainable Development (UNSDGs). This is appeared in goal 11 (SDG11), which is about cities, specifically about the need to make human settlements and cities inclusive, resilient, safe, and sustainable through planning, sustainable urbanization, and management. Heritage is an important factor for urban sustainable development in the New Urban Agenda (NUA). There are many points that emphasize the importance of cultural heritage (both intangible and tangible) in urban sustainable development [10].

Culture should be considered when promoting and implementing sustainable production and consumption patterns. It is known as a key element in the human settlements and humanization of cities, playing an important role in rehabilitating and revitalizing urban areas, and in strengthening the exercise of citizenship and social participation.

Furthermore, NUA emphasizes the cultural heritage in the development of sustainable, vibrant, and inclusive urban economies, as well as in the maintenance and support of urban economies in their progressive transition to higher productivity [11].

In the adoption of planning instruments such as master plans, zoning guidelines, coastal management, building codes policies, and strategic development policies that protect a diverse range of intangible and tangible cultural heritage. Culture is an important element of strategies and urban plans, so it is required to protect them from potential disruptive impacts of urban development.

Furthermore, some international organizations, such as UNESCO and the International Council on Monuments and Sites (ICOMOS), emphasize the importance of culture heritage in achieving sustainable development [12].

3 Study Area

The study area is located in the old city center of Alexandria, which is a linear axis that extends along the tram line from Mansheya area (Unknown Soldier), called the Commerce Chamber Street, Saad Zaghloul Street, Alexander street and Omar Lotfi street. The total length of this axis is about 4 km. This study area had been divided into four zones defined by three main streets intersect with the axis: Safia Zaghloul street, Suez Canal street, and Mohamed Shafiq street, so as to facilitate the study process (Fig. 1).

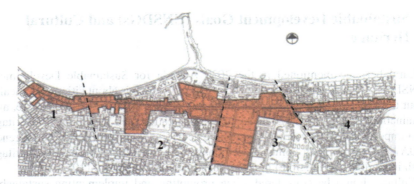

Fig. 1 Study area. *Source* author

4 Reasons for Selecting the Study Area

1. The area is overlooking the main traffic artery in the city of Alexandria.
2. It contains many of European-style buildings, in spite of the current situation, it is still expressing the architectural character of the city of Alexandria.
3. The axis is characterized by clear architectural changes at frequent intervals which facilitate to monitor the trends of change, its features and its causes.

5 Methodology of Work in the Study Area

- Planning studies

Works in the first phase depended on collecting the necessary planning data such as heights, conditions and uses of the buildings, with photography to document the current situation, especially for heritage buildings.

- Analytical studies

It is a compilation of various data for areas and compare them to each other, to make analyzes and studies on the current status, and to make a comprehensive planning study of the buildings that overlook the spatial axis.

- Architectural Studies

A detailed architectural study has been made for the facades of heritage buildings because it contains a lot of architectural vocabularies that characterize this era, which must be preserved.

6 Planning Studies

A survey of all buildings that overlooking the axis had been made and putted in tables such as Table 1.

7 Analytical Studies

From the previous analytical studies, we note the following (Figs. 2, 3 and 4):

- The ratio of heritage buildings that overlook the axis is 65% of the total buildings and therefore occupy the largest number of buildings and facades, and this is evidence of the importance of the study area.
- The largest area of modern buildings is located in the second zone, although the fourth zone was the highest in terms of number of modern buildings, and less areas in modern buildings is in the first zone and therefore it is most of the areas that contain the architectural character yet.
- The area of non-heritage buildings is a very small area up to 1.2% of the total buildings that overlook the axis, and therefore it is not under consideration in this axis.
- Most of the commercial residential buildings in the first, second and fourth zone. The third zone is a public building area.
- 4–6 storey buildings occupy a large percentage of the buildings and give a distinctive skyline to the heritage buildings.
- Most of 7–10 storey buildings found in the second and fourth zone and these are the areas that contain many changes.

8 Architectural Studies for the Exterior Facades of Heritage Buildings

8.1 The Domes

Very few buildings that overlook the axis used domes as one of the architectural vocabularies, either to confirm one of the pillars of the building, or to confirm a central element in the building (Fig. 5).

Table 1 Planning studies. *Source* author

Code	Nom	Street name	Heights	Uses			Style			Condition			
				Residential	Commercial	Public	Heritage	Non heritage	Modern	Good	Moderate	Bad	Notes
39	109	Omar Lotfy	1 + 1	•			•				•		
40	111	Omar Lotfy	1 + 3		•				•	•			
41	113	Omar Lotfy	1 + 5	•			•					•	
42	115	Omar Lotfy	1 + 4	•				•			•		

Heritage Buildings Changes and Sustainable Development 301

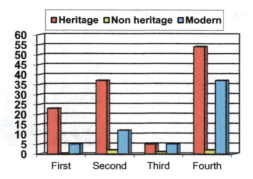

Fig. 2 Buildings style. *Source* author

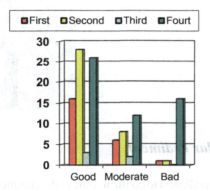

Fig. 3 Buildings conditions. *Source* author

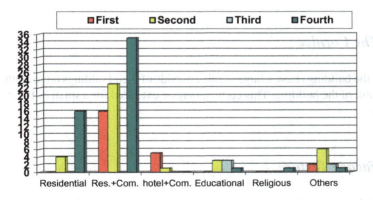

Fig. 4 Buildings uses. *Source* author

Fig. 5 The domes. *Source* author

Fig. 6 The pediment. *Source* author

8.2 The Triangular Pediment

They use the triangular pediment to confirm some of the entrances of public buildings. And it was not used in residential buildings (Fig. 6).

8.3 The Cornice

Most of the buildings have a large cornice outside of their buildings on to confirm the upper limit of the building. This cornice may be extended for a small or big distance (Fig. 7).

8.4 Facade Texture

Some of them are soft texture by using paints without any divisions, and others with a coarse texture using horizontal lines or large stone (Fig. 8).

Fig. 7 The cornice. *Source* author

Fig. 8 Facade texture. *Source* author

8.5 Sornaga Brick in Facades

Used for cladding some parts of the facades for emphasizing them and give a sense of balance between Sornaga bricks with small division and other smooth areas, and between the dark and light color of the rest of the facade (Fig. 9).

Fig. 9 Sornaga brick in facades. *Source* author

Fig. 10 The skyline. *Source* author

Fig. 11 The rhythm. *Source* author

8.6 Skyline of the Facades

Some buildings have horizontal skyline confirmed by a huge cornice. Others with variable heights in certain places, so as to give the movement and dynamic to the top of the building (Fig. 10).

8.7 Rhythm in the Facades

The treatment of the terraces and the openings with a regular rhythm, with the use of Cornice to give a sense of horizontal direction. Sometimes vertical towers are used to feel the vertical direction (Fig. 11).

8.8 The Columns

Columns were used to confirm the entrances of public buildings or the terraces as they were used to decorate the openings. It is observed that most of the columns follow the classic models such as: Ionic, Corinthian and Doric (Fig. 12).

Heritage Buildings Changes and Sustainable Development 305

Fig. 12 The columns. *Source* author

Fig. 13 The arches. *Source* author

8.9 The Arches

Used to decorate and confirm the terraces and the openings. They were with multiple formats including half circular—pointed—triangular—ribbed (Fig. 13).

8.10 The Terraces

Sometimes there were separate, and sometimes connected especially in the last floor. It sometimes confirmed by arches and columns. It sometimes solid or with opening and decorations (Fig. 14).

Fig. 14 The terraces. *Source* author

Fig. 15 The windows.
Source author

Fig. 16 The ornaments.
Source author

8.11 The Windows

It is main important element in facades, give a sense of balance between voids and Solid walls. It sometimes rectangular and decorated with columns or arches half circular, pointed and other (Fig. 15).

8.12 The Ornaments

It had been chosen to decorate and highlight some of the important elements can be divided into the following types: Plant—Engineering—Striping—human—Animal and symbolic decorations (Fig. 16).

8.13 The Doors

We find that most of them are wooden doors with slots with motifs of iron, which are either rectangular doors or with arches, and other doors made of iron with geometric and floral motifs, very few of the doors were made of solid wood only (Fig. 17).

Fig. 17 The doors. *Source* author

9 Problematic of Change and Its Causes in the Case Study

9.1 Functional or Economic Reasons

- Change the facades by occupants as a result of increasing population and the inability of individuals to move to a larger housing unit, the phenomenon began to change in the facades and plans to enlarge a room area or add a new space.
- Change of the facades by Owners: Old buildings owners who suffer from a lack of rate of return, as a result of the depreciation of rents, are building new floors not matching with the existing style, leading to distortion of exterior facades of the old buildings.

9.2 Effects of Chronological Development of the Architectural Output for External Facades

In the current reality, many people do not follow the spirit of the architectural heritage, and cognitive impairment sense of beauty as a value and demand, appeared architectural buildings patterns result of missing identity. Also selling private unfinished units, made the owner of the unit depend on the personal perception and behaviors, to finish units in a manner commensurate with his requirements, needs and concept.

9.3 Environmental Reasons

Due to increasing population density rate and rising public buildings density, and therefore represented by the emergence of high-rise towers have led to the decline of privacy, and lack of green spaces, so a large proportion of the occupants skipped the terrace element and replaced with glass windows, especially with the crowed street with cars and pollution, noise and exhaust, making it a source of pollution, and not a place for recreation and rest.

9.4 Legislative Reasons

- The fluctuation of urban laws—between continuing and cancellation—with regard to heights and population density of each region.
- There is no respect for the laws and legislation governing the construction process which imposes a specific style of architectural design and planning for all residential communities.
- The occupants are not committed to the signed contract with the owner of the building, which states not to change anything in the external facade.
- Many people do not know the importance of the architect role, leads to the use of professionals to do the same role, which leads to that most of the residential buildings have problems in the design.

10 Study the Changes in the Facades of Heritage Buildings

10.1 Advertisements Boards

Some are setting on the terraces covering the beautiful decorations on them, or at the entrances of the buildings and some took the entire side of the building (Fig. 18).

Fig. 18 Advertisements boards. *Source* author

Fig. 19 Adding new floors. *Source* author

Fig. 20 Commercial changes. *Source* author

10.2 Adding New Floors

Some owners added new modern floors without any restrictions or requirements make this floors follow the original facade, not only the architectural vocabulary, also changed the original skyline (Fig. 19).

10.3 Commercial Changes

Most of the shops when designing their elevation did not care about the original building facades. Some have had to remove architectural elements or changing the colors and the proportions of openings (Fig. 20).

10.4 Additions in the Original Floors

Some terraces had been closed, changing the proportions of the openings and using different cladding. All of these additions resulting in a distortion of the original facades (Fig. 21).

Fig. 21 Additions in the floors. *Source* author

Fig. 22 The absence of maintenance. *Source* author

10.5 The Absence of Maintenance

A large number of heritage buildings have been missed by a regular maintenance, leading to the decay of the architectural vocabulary that characterizes the facades and this affects the safety factor of the building (Fig. 22).

10.6 Air-Conditioning and Dishes

Technological advances signs like air-conditioning, dishes and antenna were placed in positions not matching with the heritage buildings (Fig. 23).

10.7 Modern Buildings Do not Fit with the Style

As a result of the social and technological developments in various fields, it featured high-rise towers missing the sense of beauty and distort the aesthetic system of heritage buildings (Fig. 24).

Fig. 23 Air-conditioning and dishes. *Source* author

Fig. 24 Modern buildings do not fit with the style. *Source* author

11 Recommendations

11.1 Administrative and Legal Recommendations

Administrative and legal rules have to be set for modern buildings in the heritage areas to respect the style, condition and uses. Also, it is not important to pass 100 years on the building to be registered so that there will be exceptions for buildings of special architectural and historical value.

Update the laws regulating construction work to include a new article to regulate the decoration of facades of shops, banks and… etc. to suit the historical buildings and registered districts.

11.2 Technical Recommendations

It is very important to create restoration centers and research laboratories to provide technical and scientific expertise specialized in dealing with heritage areas. Making a survey of all heritage buildings to be easy access for them and encourage studies and research for heritage buildings.

11.3 Urban Planning Recommendations

Set some urban planning rules to control the building densities in the heritage areas to reduce overcrowding and set specific rules for the other planning processes in the heritage areas.

Also, the maintenance of heritage buildings has to be under the supervision of the authorities of planning to eliminate infringements.

11.4 Political Recommendations

The commitment of workers in urban management to laws and legislation to preserve the heritage. Exempt the heritage buildings from property taxes.

Also, the government agencies which leased or owned heritage buildings have to preserve them and not to make any modifications only after the approval of the responsible authorities.

11.5 Economic Recommendations

We need to increase the financial allocations to maintain the heritage buildings and to set up some projects that are contributing to the revival of the heritage areas, to increase the fund's resources.

11.6 Financing and Motivational Side

Provide some grants like a direct grant to the owners of heritage buildings to preserve the buildings. And regarding the taxes, we can make an exemption of the owners of heritage buildings from the taxes. On the other side we can make an imposition of a part of the sales tax is earmarked for maintenance operations to be deducted from the airlines, tourism, hotels, shops, and services benefiting from the development and upgrading of the heritage area.

Also, raising the rental values of residences and shops in the heritage district for the owners to help them to participate in the preservation of the heritage buildings.

References

1. UNESCO.: Culture urban future: global report on culture for sustainable urban development. UNESCO, Paris (2016)
2. UCLG.: Culture in the sustainable development goals: a guide for local action. UCLG, Barcelona (2018)
3. Throsby, D.: Cultural Sustainability. Edward Elgar Publishing Limited, Cheltenham (2003)
4. Fitch, J.M.: Historic Preservation. McGraw Hill Inc., New York (1992)
5. Cheshmehzangi, A., Heat, T.: Urban identities: influences on socio-environmental values and spatial inter-relations. Procedia Soc. Behav. Sci. **36**, 253–264 (2012)
6. GOPP.: General strategic urban plan Alexandria 2032. GOPP, Cairo (2015)
7. Mckenzie, J.: The Architecture of Alexandria and Egypt. Yale University Press, New Haven (2007)
8. Nassar, D.: Heritage conservation management in Egypt the balance between heritage conservation and real-estate development in Alexandria. Environ. Behav. Proc. J. **1**(4) (2016)
9. Soliman, A., Soliman, Y.: Exposing urban sustainability transitions: urban expansion in Alexandria, Egypt. Int. J. Urban Sustain. Dev. **14**, 33–55 (2022)
10. United Nations.: Draft outcome document of the United Nations Conference on Housing and Sustainable Urban Development (Habitat III). United Nations, New York (2016)
11. Hosagrahar, J., Soule, J., Fusco Girard, L., Potts, A.: Cultural heritage, the UN sustainable development goals, and the new urban agenda. ICOMOS, Charenton-le-Pont (2016)
12. Potts, A.: The Position of cultural heritage in the new urban agenda a preliminary analysis prepared for ICOMOS. ICOMOS, Charenton-le-Pont (2016)

Global Crucial Risk Factors Associated Stress Among University Students During Post Covid-19 Pandemic: Empirical Evidence from Asian Country

Nor Azma Rahlin, Ayu Suriawaty, Siti Aisyah Bahkiar Bahkiar, Suayb Turan, and Siti Nadhirah Ahmad Fauzi

Abstract The pandemic Covid-19 is about to alter the planet entirely, negatively impact the global economy, decrease mental health, and increase stress as well as reduce quality of education. The study focuses on two crucial risk factors in stress by examining financial issues and online learning as the major factors behind university students' stress during pandemic covid-19. Two hundred self-administered questionnaires were distributed to obtain quantitative data. Google Form was used to build the questionnaire. The questionnaire was then circulated via messaging apps like WhatsApp, Telegram, Facebook and Instagram. The partial least squares structural equation modelling (PLS-SEM) software version 3.6 was used for hypothesis testing. A total of 120 questionnaires represented of 60% response rate and useable data for further analysis. The survey result shows that financial issues have a more significant effect on stress than online learning. Path regression coefficient indicated that financial issues were the most important predictors of stress among university students. These results allow us to understand crucial risk factors in students stress to better and drive us to substantially consider them in time-oriented policy and plan for the management of the education system may decrease the stress of university students. Future research and practice implications and recommendations will be presented.

N. A. Rahlin (✉)
Faculty of Business, Economic and Accountancy, Universiti Malaysia Sabah (UMS), 88400 Kota Kinabalu, Sabah, Malaysia
e-mail: norazma.rahlin@ums.edu.my

A. Suriawaty · S. A. Bahkiar Bahkiar
Indah Water Konsortium Ltd, No 44, Jalan Dungun, Damansara Heights, 50490 Kuala Lumpur, Malaysia

S. Turan
Faculty of Economics and Administration Science, Cankiri Karatekin University, 18100 Çankırı Merkez/Çankırı, Turkey

S. N. A. Fauzi
Faculty of Business and Management, Universiti Sultan Zainal Abidin (UNISZA), 21300 Kuala Nerus, Terengganu, Malaysia

© The Author(s), under exclusive license to Springer Nature Switzerland AG 2024
A. Hamdan and E. S. Aldhaen (eds.), *Artificial Intelligence and Transforming Digital Marketing*, Studies in Systems, Decision and Control 487,
https://doi.org/10.1007/978-3-031-35828-9_27

315

Keywords Financial issues · Online learning · Quality education · Stress · Structural equation modelling

1 Background of the Study

The pandemic Covid-19 is about to alter the planet entirely. While it is impossible to predict the exact economic impact of the global COVID-19 coronavirus pandemic, experts generally agree that it has a significant negative impact on the global economy. Preliminary projections predicted that if the virus became a worldwide pandemic, most major economies would lose nearly 2.9% of their GDP by 2020. This prediction has already been revised to a 3.4% drop in GDP. Global GDP is expected to be around 84.54 trillion US dollars in 2020, implying that a 4.5% decline in economic growth leads to nearly 2.96 trillion US dollars of lost economic output [1]. Besides, diverse afflicted individuals were focused on the COVID-19 pandemic. As we all know, epidemics may aggravate or generate new sources of stress, including anxieties and concerns for yourself or loved ones, physical and social activity limits due to isolation and changes in your abrupt and radical lifestyle.

Additionally, key findings from empirical research revealed that covid-19 impacts students' mental health problems, including an overwhelming majority of college students (85%) experiencing increased stress and anxiety [2]. In addition, there is little evidence that student financial issues are related to mental health among higher education students [3]. Moreover, the impacts of high financial issues from university student stress have less likely been identified by the researchers in the Malaysian context. The challenges of COVID-19 have made online and blended learning the new normal and sought to recognise online learning to prevent further disruption to students [4]. However, few studies have addressed the stress of implementing online learning among university students in Malaysia. Therefore, this current study focuses on two crucial risk factors in stress by examining financial issues and online learning as the major factors behind university students' stress.

2 Literature Review

2.1 Stress Among University Students

University is not only a place to gain knowledge. Students or individuals, as well as the community and society, received huge benefits from the university. Students experience significant stress levels connected to their education, which can harm their health, quality of life, and academic performance [5–7]. Therefore, understanding stress from the perspective of university students, identifying crucial risk factors and the significant effect of stress among university students are a way to reduce stress at the higher education level. The prevalence of epidemics of Covid-19 creates new sources of stress, including fears and worries about yourself or your loved ones, restrictions on physical and social activities due to isolation, and sudden and radical lifestyle changes [8]. All this has led to an experience of stress for both students and educators. Low physical activity and spending all-time at home are creating a negative impact, which all of us try to overcome by using many other activities to cope with this condition [9]. The researcher noted that the results of a previous study highlighted that students used to illustrate academic pressure as the biggest stress in their lives [9]. Therefore, understanding critical factors of stress could help to reduce stress and eliminate the negative impacts of stress faced by the university student. In this study, the researcher focuses on two critical risk factors of stress: financial issues and online learning.

2.2 The Relationship Between Financial Issues and Student Stress

Financial issues related to university students became the attention of many parties during Pandemic Covid-19. Financial stress is part of financial issues and has been widely characterised as difficulties meeting one's financial responsibilities, with particular economic behaviours best understood in light of people's attitudes and beliefs about resource availability and management [10]. Several studies suggest that students with increasing financial issues may relate to trends in reducing student mental health [3]. Financial expenditures and income inequalities split, and teachers and students are struggling with new and sudden laws of teaching and learning that are fully implemented in the technical field [11]. A previous study before the pandemic revealed that financial factors rank quite high as stress factors for college students [10]. A study conducted by [12] found that students' financial status and performance have a positive relationship. Indeed, researchers explained that students with stronger financial support positively associated with their academic performance. Another group of researchers found that male students have more financial problems interfere with than female students [13]. Furthermore, results indicated that young students are more likely to be satisfied with their financial position than old students and seem to think that their parents worry more about disappointing them for materialistic reasons [13]. There is no comprehensive understanding of the causal mechanisms related to financial issues on stress among university students. Given the serious need on examines the causal relationship between financial issues and stress, this study assumes that:

Hypothesis 1: *There is a significant effect of financial issues on student stress among university students.*

2.3 The Relationship Between Online Learning and Student Stress

The challenges of COVID-19 have made online and blended learning the new normal and sought to recognise online learning to prevent further disruption to students [4]. Many problems in online teaching include targeting specific remote locations where network connectivity is lacking, instructors less experience with new technologies, lags in terms of digital learning information gap, officials have lately started to provide schools and teachers with web services and email accounts, and the complex home environment remains unsolved [14, 15]. Besides that, [16] indicated that, students at private institutions (80.6%) were found to be more psychologically burdened than students at public universities (7%). On the other hand, an empirical study from other Asian country proves that 63% of them are suffering from financial problems which affect their mental and psychological wellbeing [17]. Similar to the previous study, low socioeconomic status and poor self-perceived health are also perceived as determinants of perceived stress [18]. Collectively, these studies outline the critical role of financial factors in determining stress from various communities at the international level. Thus, the researcher postulated that.

Hypothesis 3: *There is a significant effect of online learning on student stress among university student.*

3 Methodology

The research was carried out quantitatively. The study's target population included college and university students from the East of Malaysia. A self-administered questionnaire was used to obtain quantitative data. The first section of the questionnaire collected demographic information from respondents such as gender, age, year of study, graduation, family income, and scholarship. The second section of the survey required respondents to rate a set of questionnaire questions on a five-point Likert scale, with one indicating strongly disagree and five indicating strongly agree. The questionnaire was adopted based on established measures from the past study on stress [19], financial issues adapted from [20] and online learning adapted from [17].

Convenient sampling is in the form of a non-probability sampling technique adopted in this study since difficult to reach respondents physically during the Covid-19 pandemic, as suggested by [21] and data was gathered utilising online forms. Google Form was used to build the questionnaire. The questionnaire was then circulated via messaging apps like WhatsApp, Telegram, Facebook and Instagram. Thereby the sample size falls within 200–500, as recommended by [22, 23]. Two hundred (200) questionnaires were distributed to respondents. The response rate is 60% accounting for 120 useable responses for further hypothesis testing.

A pilot study involving 30 respondents reached Cronbach alpha value ranging from 0.812 to 0.923, which is above the cutting value of internal consistency of the questionnaire [22, 24]. Following the data screening procedures, the software partial least squares structural equation modelling (PLS-SEM) version 3.6 was used for further analysis. The two-step strategy of evaluating the model is proposed by [24]. The common method bias and measurement model validity were tested in the first phase to find common variance issues. Next, the suggested model's reliability and validity are assessed to verify that it measures what the survey meant to measure. The second step examined the structural model to test the study's hypotheses. Finally, the results of hypothesis testing are presented in the following section.

4 Results and Discussion

4.1 Demographic Profile of the Respondents

The respondents in this study comprised of 72 male and 48 female responses (59.5% and 40.5%, respectively) with more than half of them possess an undergraduate education level (81.8%). A breakdown of age group is as follows: 51.2% were between 20 and 25 years old, 36.4% were between 25 and 30 years old, 9.1% were between 30 and 35 years old, and 3.3% were age 40 years old and above. Regarding years of study, 37.2% of respondents are in their first year of study, 35.5% in their second year of study, 25.6% in their third year, and the remainder in their fourth or above year (1.7%). The proportion of respondents from families earning between RM7000 and RM10000 and above RM10000 shares the same result, which is 47 respondents (38.8%). There are 16 respondents (13.2%) from families that earn from RM2000 to RM5000, and 11 respondents (9.1%) are from families that earn from RM5000 to RM7000, respectively. Looking at the findings of the demographics, it seems that over 90% of respondents do not receive any scholarship, and just 20.7% do receive a scholarship.

4.2 Descriptive Analysis for Stress Level

The result of a descriptive analysis of stress levels shows a mean of 2.24–2.52 on a scale of 1–5, while the standard deviation for each item stretched from 1.149 to 1.293. The maximum value of the measurement is five, divided into three groups; high (3.34–5), medium (1.68–3.33), and low (0–1.67). It is evident from the results in Table 1 that stress levels among university students are at a medium level.

Table 1 Validity and reliability of the measurement model

Construct	Item	Indicator reliability	Convergent validity	Internal consistency reliability	
		Outer loadings	AVE	Composite reliability	Cronbach's alpha
Student stress	S1	0.776	0.563	0.885	0.845
	S2	0.770			
	S3	0.755			
	S4	0.722			
	S5	0.670			
	S6	0.802			
Financial issues	FI1	0.796	0.651	0.914	0.816
	FI2	0.880			
	FI3	0.779			
	FI4	0.816			
	FI5	0.752			
	FI6	0.813			
Online learning	OL1	0.883	0.727	0.885	0.845
	OL2	0.877			
	OL3	0.840			
	OL4	0.810			

4.3 Assessment of Common Method Variance

This study uses Harman's single factor test and full collinearity techniques to test the common method variance. Harman's single factor test is conducted to determine whether a single factor surfaces from the principal component analysis or if a distinct factor explains the majority of covariance among the variables in the un-rotated factor analysis. From the result, it can be seen that Harman's Single Factor techniques estimate the variance of the common method at 32.718%, which is below the widely accepted threshold of 50% [25, 26]. Therefore, this study concluded that there is no issue of common method variance [25]. The variance inflation factor (VIF) should be below 5.0 to avoid any common method variance issues, as suggested by [22]. Hence, by employing [27], it is confirming again that there is no issue on common method variance in this study since the value of pathological VIFs for all constructs are 1.162 [28].

4.4 Assessment of the Measurement Model and Path Coefficients

Table 1 presents the result of the reflective measurement model consisting of average variance extracted (AVE), Composite Reliability, and Cronbach's Alpha to assess the measurement model's reliability and validity. The AVE for each construct is more than 0.5, and the composite reliability and Cronbach alpha are above 0.7 for each construct. This threshold is recommended by [22]. All of the items remained since every item had a factor loading greater than 0.5, ranging from 0.67 to 0.88. Overall, the results in Table 1 indicate the validity and reliability of the collected data for further analysis [24].

The results of discrimination based on the Fornell-Larcker criterion are shown in Table 2. Based on result of Cross-Loadings method in Table 2, it is proved that all indicators scored highly on their respective constructs. Thus, it is confirming the validity of the measurement as a discriminant [29]. Under the Fornell-Larcker Criteria, bold numbers indicate the square root of the AVE, and any unbolded values indicate the correlation between the structures. According to Mustafa and Noorhidawati [30], the value of the bold numbers should be higher than other constructs in its row. Based on the findings, the Fornell-Larcker criterion's minimal condition is met [30].

The variable attitude value of R^2 to the stress level is 0.263, which indicates that this endogenous latent construct is explained by up to 26.3% [27, 31]. The R^2 values of these two variables are deemed optimal for the structural model since they are higher.

The effect sizes are evaluated to examine the strength of the relationship between latent variables [32]. The proportionate impact on endogenous buildings is calculated by a predictor building. It is resulted that financial Issues (0.156) has medium effect on stress level management, while online learning (0.073) has small effect [28].

4.5 Structural Model: Hypotheses Testing

The evaluation of the path coefficient and the t-value are used for assessing the hypotheses in this study on the basis from the previous structural model evaluation.

Hypothesis 1 predicts that there is a significant effect between financial issues and student stress among university student. The result reveals an indication of a significant value exists ($\beta = 0.365$, t $= 2.911$, p < 0.05). While, Hypothesis 2 predicts that there is a significant effect between online learning and student stress among university student. The result reveals an indication of a significant value exists ($\beta = 0.249$, t $= 2.150$, p < 0.05). Hence, these results support the hypothesized for both relationships.

Table 2 Discriminant validity using Fornell-Larcker criterion and cross-loading method

Fornell-Larcker criterion			
	FI	OL	S
FI	**0.807**		
OL	0.374	**0.853**	
S	0.458	0.386	**0.750**
Cross loadings			
FI1	**0.796**	0.277	0.434
FI2	**0.880**	0.393	0.422
FI3	**0.779**	0.297	0.346
FI4	**0.816**	0.223	0.275
FI5	**0.752**	0.230	0.345
FI6	**0.813**	0.360	0.352
OL1	0.304	**0.883**	0.383
OL2	0.416	**0.877**	0.341
OL3	0.281	**0.840**	0.314
OL4	0.264	**0.810**	0.257
S1	0.354	0.403	**0.776**
S2	0.321	0.258	**0.770**
S3	0.317	0.279	**0.755**
S4	0.413	0.240	**0.722**
S5	0.267	0.213	**0.670**
S6	0.367	0.311	**0.802**

Note: *FI* Financial Issues, *OL* Online Learning, *S* Stress Level

It is shown in Table 2 that the value of beta for financial issues is higher than the value of beta for online learning. In this study, financial issues are more relevant to stress levels than online learning.

5 Discussion

This paper studied stress level management among university students during the Covid-19 pandemic. It is found that the level of stress management among university students is at a medium level. This result seems to be consistent with other research, which found almost half of the respondents had moderate stress [33]. The result is not surprising as, during the Covid-19 pandemic, Malaysian authorities enforced movement control orders to break the chain of Covid-19. Despite this action, it affects the entire Malaysian population, including university students. In accordance with the previous study, authors believed the lockdown-related issues are therefore alleged to be causing students to suffer from mental health issues [34]. Graduation time will be delayed due to the closure of educational institutions, which ultimately affects the future careers of the university students [35]. Having insecurity about their future journey can increase their stress level. The consequence of these issues is related to financial issues and online learning issues. Lastly, this will make a difference in the lives of university students psychologically, physiologically, financially, and academically [36, 37].

The result from this study revealed that financial issues have a significant relationship with student stress among university students. It has been shown that students agree that issues have an impact on their level of stress management. The results are consistent with the previous studies which found that financial crisis give a higher impact on mental stress [16, 35, 38]. They found that some of reasons that cause financial crisis such as high tuition fees and job loss of the students or family members. It is possible that they are concerned about managing their expenses for education. Aside from this, students who depend on jobs for support may be more likely to suffer from depression and worry due to economic hardship [39]. Another previous study, [40], found finance to be a source of stress, which is in line with current study.

In addition, this study found that online learning also affects university students' stress levels. This result is consistent with the study by [16]. Due to financial constraints, the researcher believed students might be unable to buy internet data or appropriate devices. Rural areas may also have slow or no internet connection, with the result that students cannot attend their online classes [16, 41] which will affect their learning performance. Consequently, stress levels can rise [42]. Moreover, this finding supports evidence from previous observations [43, 44]. Researchers found online learning contributed to enormous stress for university students since most lecturers or instructors used the same curriculums and learning outcomes as face-to-face classes [44]. Furthermore, online learning does not provide them the same amount of input as face-to-face instruction [43]. Thus, the stress management levels of students will differ between face-to-face and online learning due to their different vibes. It was found by [45] that students might lack motivation during online learning compared to face-to-face learning. Hence, it is proved that stress levels among students might be influenced by school-related and pandemic psychological issues [46].

One interesting finding is that financial issues impacted university students' stress levels more than online learning. In accordance with the present result, a previous study by [16] demonstrated that the stress level for university students in public universities is more influenced by financial issues than online courses. However, these findings cannot be extrapolated to all university's students since the same study also found that stress levels for private university students received more affected by online courses than the financial crisis [16]. So, we can conclude that the results may seem different depending on which type of university the students attend. Stress levels among students are more impacted by financial issues than online learning since they tend to focus less on academics when they are struggling financially. A study by previous researchers has proved this fact [39]. In another study, finding employment during the Covid-19 pandemic was critical to their ability to support their learning needs, especially the internet data for online education [47]. Thus, it shows that financial issues can be a crucial factor in online learning. If university students have financial stability, they do not face difficulty in online learning and finally reduce their stress level.

6 Implication of the Study

This study proved that university students' stress levels had affected during Covid-19 because of financial issues and online learning. Thus, this study can be one of the supportive resources for the Malaysian government and university administration to find solutions to help their students and support their well-being. For example, they could improve online learning methods and provide financial assistance. It can therefore reduce stress levels and enhance academic performance.

7 Limitation and Recommendation for Future Study

Despite the validation of measurement model and several other interesting findings, this study has some limitations. First, this study was limited to students of University of Malaysia Sabah. Thus, there is a small sample size used in this study. Selection of limited respondents was due to limitation of time and financial issues. Besides, this study was conducted during Covid-19 pandemic. Due to the increasing number of positive Covid-19 cases, the Malaysian authorities issued a Movement Control Order (MCO), which also influenced university policy. Consequently, this will affect students' emotions and increase their stress levels, thereby interfering with data collection. Future studies could expand the research area to include other universities in Malaysia and increase the number of respondents. In addition, the researcher suggests using mixed methods (qualitative and quantitative approaches) to improve the robustness and credibility of the results.

8 Conclusion

This research identifies and examines the relationship between the dependent variable, Stress Level, and the independent variables, Financial Issues and Online Learning. The data collected from 119 respondents revealed that students' stress levels were in the medium range. Furthermore, stress levels are significantly influenced by financial issues and online learning. In addition, financial issues were more influential on stress levels during the pandemic of COVID-19 than online education. The results of this study may be useful to higher education authorities in Asian countries in improving higher education policies and regulatory authorities in the higher education sector. Future research on stress among university students should include other independent factors for comparative analysis. In addition, it is recommended that future research on stress and its consequences consider the multicultural context of Asian university students.

References

1. Szmigiera, M.: Impact of the coronavirus pandemic on the global economy—statistics and facts. Coronavarius: impact on the global economy (2022)
2. Neal, K.: College Students' mental health continues to suffer from COVID-19. New survey by TimelyMD Finds—TimelyMD. https://timely.md/college-students-mental-health-continues-to-suffer-from-covid-19-new-survey-by-timelymd-finds/ (2020). Accessed 25 July 2022
3. McCloud, T., Bann, D.: Financial stress and mental health among higher education students in the UK up to 2018: rapid review of evidence. J. Epidemiol. Community Health 1978, 977–984 (2019). https://doi.org/10.1136/jech-2019-212154
4. McMillan, M.: COVID19—the catalyst for educational change: applying/using E-PBL for accelerated change. J. Probl. Based Learn. 7(2), 51–52 (2020). https://doi.org/10.24313/jpbl.2020.00297
5. Baker, A.R., Montalto, C.P.: Student loan debt and financial stress: implications for academic performance. J. Coll. Stud. Dev. 60(1), 115–120 (2019). https://doi.org/10.1353/csd.2019.0008
6. Pascoe, M., et al.: Physical activity and exercise in youth mental health promotion: a scoping review. BMJ Open Sport Exerc. Med. 6(1), 1–11 (2020). https://doi.org/10.1136/bmjsem-2019-000677
7. Pascoe, M.C., Hetrick, S.E., Parker, A.G.: The impact of stress on students in secondary school and higher education. Int. J. Adolesc. Youth 25(1), 104–112 (2020). https://doi.org/10.1080/02673843.2019.1596823
8. Brooks, S.K., et al.: The psychological impact of quarantine and how to reduce it: rapid review of the evidence. Lancet 395(10227), 912–920 (2020). https://doi.org/10.1016/S0140-6736(20)30460-8
9. Chandra, Y.: Online education during COVID-19: perception of academic stress and emotional intelligence coping strategies among college students. Asian Educ. Dev. Stud. 10(2), 229–238 (2021). https://doi.org/10.1108/AEDS-05-2020-0097
10. Robb, C.A.: College student financial stress: are the kids alright? J. Fam. Econ. Issues 38(4), 514–527 (2017). https://doi.org/10.1007/s10834-017-9527-6
11. Brooks, S.K. et al.: The psychological impact of quarantine and how to reduce it: rapid review of the evidence. Lancet 395(10227), 912–920. Lancet Publishing Group (2020). https://doi.org/10.1016/S0140-6736(20)30460-8
12. Benson-Egglenton, J.: The financial circumstances associated with high and low wellbeing in undergraduate students: a case study of an English Russell Group institution. J. Further High. Educ. 43(7), 901–913 (2018). https://doi.org/10.1080/0309877X.2017.1421621
13. Utkarsh, A.P., Ashta, A., Spiegelman, E., Sutan, A.: Exploration of financial stress indicators in a developing economy. Strateg. Change 29(3), 285–292 (2020). https://doi.org/10.1002/JSC.2327
14. Fawaz, M., Samaha, A.: E-learning: depression, anxiety, and stress symptomatology among Lebanese university students during COVID-19 quarantine. Nurs. Forum (Auckl) 56(1), 52–57 (2021). https://doi.org/10.1111/nuf.12521
15. Malik, M., Javed, S.: Perceived stress among university students in Oman during COVID-19-induced e-learning. Middle East Curr. Psychiatry 28(1) (2021). https://doi.org/10.1186/s43045-021-00131-7

16. Shafiq, S., Nipa, S.N., Sultana, S., Rahman, M.R.U., Rahman, M.M.: Exploring the triggering factors for mental stress of university students amid COVID-19 in Bangladesh: a perception-based study. Child. Youth Serv. Rev. **120** (2021). https://doi.org/10.1016/J.CHILDYOUTH.2020.105789
17. Hossain, S.Z., Alam, M.S., Lemon, A., Omar, N.B., Begum, F., Islam, U.N.: Combating the impact of COVID-19 on Public University students through subsidized online class: evidence from Bangladesh. J. Educ. Pract. (2020). https://doi.org/10.7176/jep/11-27-17
18. Kocalevent, R.D., Hinz, A., Brähler, E., Klapp, B.F.: Determinants of fatigue and stress. BMC Res. Notes **4** (2011). https://doi.org/10.1186/1756-0500-4-238
19. Cohen, S., Kamarck, T., Mermelstein, R.: A global measure of perceived stress. J. Health Soc. Behav. **24**(4), 385–396 (1983)
20. Bennett, D.: The impact of financial stress on academic performance in college economics courses (2015)
21. Saunders, M., Lewis, P., Thornhill, A.: Research Methods for Business Students, vol. 5 (2009)
22. Hair, J.F., Babin, B.J., Krey, N.: Covariance-based structural equation modeling in the journal of advertising: review and recommendations. J. Advert. **46**(1), 163–177 (2017). https://doi.org/10.1080/00913367.2017.1281777
23. Kline, R.B.: Principles and Practice of Structural Equation Modeling, 3rd edn. PsycNET, Guilford Press (2011)
24. Hair, J.F., Risher, J.J., Sarstedt, M., Ringle, C.M.: When to use and how to report the results of PLS-SEM. Eur. Bus. Rev. **31**(1), 2–24 (2019). https://doi.org/10.1108/EBR-11-2018-0203
25. Lee, J., Song, J.H.: Developing a measurement of employee learning agility. Eur. J. Training Dev. **46**(5–6), 450–467 (2022). https://doi.org/10.1108/EJTD-01-2021-0018
26. Podsakoff, P.M., MacKenzie, S.B., Lee, J.Y., Podsakoff, N.P.: Common method biases in behavioural research: a critical review of the literature and recommended remedies. J. Appl. Psychol. **88**(5), 879–903 (2003). https://doi.org/10.1037/0021-9010.88.5.879
27. Awang, Z.: Analyzing the SEM structural model. In: A Handbook on SEM, pp. 71–86 (2015)
28. Hair, J., Hollingsworth, C.L., Randolph, A.B., Chong, A.Y.L.: An updated and expanded assessment of PLS-SEM in information systems research. Ind. Manag. Data Syst. **117**(3), 442–458 (2017). https://doi.org/10.1108/IMDS-04-2016-0130
29. Gill, A.A., Shahzad, A., Ramalu, S.S.: An examination of post implementation success determinants of enterprise resource planning: Insights from industrial sector of Pakistan. Int. J. Supply Chain Manag. **8**(3), 1101–1106 (2019)
30. Mustafa, A., Noorhidawati, A.: Adoption and implementation of evidence-based library acquisition of electronic resources. Malays. J. Libr. Inf. Sci. **25**(1), 1–29 (2020). https://doi.org/10.22452/mjlis.vol25no1.1
31. Rahlin, N.A., Awang, Z., Abd Rahim, M.Z., Bahkia, A.S.: The direct effect of climate emergency and safety climate on intention to safety behavior: a study. Humanit. Soc. Sci. Rev. **8**(3), 178–189 (2020). https://doi.org/10.18510/hssr.2020.8319
32. Chin, W.W.: How to write up and report PLS analyses. Springer, Berlin, Heidelberg (2010). https://doi.org/10.1007/978-3-540-32827-8_29
33. AlAteeq, D.A., Aljhani, S., AlEesa, D.: Perceived stress among students in virtual classrooms during the COVID-19 outbreak in KSA. J. Taibah Univ. Med. Sci. **15**(5), 398–403 (2020). https://doi.org/10.1016/j.jtumed.2020.07.004
34. Al Mamun, F., Hosen, I., Misti, J.M., Kaggwa, M.M., Mamun, M.A.: Mental disorders of Bangladeshi students during the covid-19 pandemic: a systematic review. Psychol. Res. Behav. Manag. **14**(April), 645–654 (2021). https://doi.org/10.2147/PRBM.S315961
35. Islam, S.M.D.U., Bodrud-Doza, M., Khan, R.M., Haque, M.A., Mamun, M.A.: Exploring COVID-19 stress and its factors in Bangladesh: a perception-based study. Heliyon **6**(7), 1–10 (2020). https://doi.org/10.1016/j.heliyon.2020.e04399
36. Birmingham, W.C., Wadsworth, L.L., Lassetter, J.H., Graff, T.C., Lauren, E., Hung, M.: COVID-19 lockdown: impact on college students' lives. J. Am. Coll. Health 1–15 (2021). https://doi.org/10.1080/07448481.2021.1909041

37. Kokkinos, C.M., Tsouloupas, C.N., Voulgaridou, I.: The effects of perceived psychological, educational, and financial impact of COVID-19 pandemic on Greek university students' satisfaction with life through Mental Health. J. Affect. Disord. **300**(November 2021), 289–295 (2022). https://doi.org/10.1016/j.jad.2021.12.114
38. Kwaah, C.Y., Essilfie, G.: Stress and coping strategies among distance education students at the University of Cape Coast, Ghana. Turk. Online J. Distance Educ. **18**(3), 120–134 (2017). https://doi.org/10.17718/tojde.328942
39. Kecojevic, A., Basch, C.H., Sullivan, M., Davi, N.K.: The impact of the COVID-19 epidemic on mental health of undergraduate students in New Jersey, cross-sectional study. PLoS One **15**(9 September), 1–16 (2020). https://doi.org/10.1371/journal.pone.0239696
40. Patten, E.V., Vaterlaus, J.M.: Prevalence of depression, anxiety, and stress in undergraduate dietetics students. J. Nutr. Educ. Behav. **53**(1), 67–74 (2021). https://doi.org/10.1016/j.jneb.2020.10.005
41. Şahin, M.: Opinions of university students on effects of distance learning in Turkey during the Covid-19 pandemic. Afr. Educ. Res. J. **9**(2), 526–543 (2021). https://doi.org/10.30918/aerj.92.21.082
42. Cornine, A.: Reducing nursing student anxiety in the clinical setting: an integrative review. Nurs. Educ. Perspect. **41**(4), 229–234 (2020). https://doi.org/10.1097/01.NEP.0000000000000633
43. Deepika, N.: The impact of online learning during covid-19: students' and teachers' perspective. Int. J. Indian Psychol. **8**(2), 783–793 (2020). https://doi.org/10.22214/ijraset.2020.32277
44. Sundarasen, S., et al.: Psychological impact of covid-19 and lockdown among university students in Malaysia: implications and policy recommendations. Int. J. Environ. Res. Public Health **17**(17), 1–13 (2020). https://doi.org/10.3390/ijerph17176206
45. Segbenya, M., Bervell, B., Minadzi, V.M., Somuah, B.A.: Modelling the perspectives of distance education students towards online learning during COVID-19 pandemic. Smart Learn. Environ. **9**(1) (2022). https://doi.org/10.1186/s40561-022-00193-y
46. Ahmed, O., Ahmed, M.Z., Alim, S.M.A.H.M., Khan, M.D.A.U., Jobe, M.C.: COVID-19 outbreak in Bangladesh and associated psychological problems: an online survey. Death Stud. **46**(5), 1080–1089 (2022). https://doi.org/10.1080/07481187.2020.1818884
47. Rotas, E.E., Cahapay, M.B.: Difficulties in remote learning: voices of Philippine University students in the wake of COVID-19 crisis. Asian J. Distance Educ. **15**(2), 147–158 (2020)

Factors Influencing Purchase Decisions on Online Sales in Indonesia

Efa Wakhidatus Solikhah, Indah Fatmawati, Retno Widowati, and M. Suyanto

Abstract Along with the growth of times like today, where communication and information technology continues to develop, the internet has become a trend for many consumers before making a purchase process. Every company wants the level of product sales to increase every year. There are many things that can influence consumer purchasing decisions. However, in this study will only examine several variables that can influence consumer buying decisions. Several things that can be done by a company to increase product sales are product innovation, web interestingness, electronic word of mouth, and electronic trust (e-trust). This variable was chosen because it is widely used as a reference by consumers in the online purchasing decision process. The purpose of this study was to determine the effect of product innovation, web attractiveness, electronic word of mouth, and e-trust on purchasing decisions. The sample of this study were 142 customers. This explanatory quantitative research with data collection using a questionnaire and analysis techniques using SPSS. The results showed that all variables had a significant positive effect on purchasing decisions.

Keywords E-WoM · E-Trust · Web attractiveness · Innovation · Purchase decision · Digital marketing

---

E. W. Solikhah (✉) · I. Fatmawati · R. Widowati
Universitas Muhammadiyah Yogyakarta, Yogyakarta, Indonesia
e-mail: efawakhidatus@gmail.com

M. Suyanto
Universitas Amikom Yogyakarta, Yogyakarta, Indonesia

E. W. Solikhah
Universitas Ahmad Dahlan, Yogyakarta, Indonesia

© The Author(s), under exclusive license to Springer Nature Switzerland AG 2024
A. Hamdan and E. S. Aldhaen (eds.), *Artificial Intelligence and Transforming Digital Marketing*, Studies in Systems, Decision and Control 487,
https://doi.org/10.1007/978-3-031-35828-9_28

1 Introduction

The internet has become a trend for many consumers before making a purchase process. This is supported by the continued increase in internet users from year to year around the world [1]. Websites that provide online transaction services are called e-commerce. Online purchases are developing and becoming an alternative choice for people's shopping because in terms of effectiveness, service, security and popularity they are considered good. According to Harris and Prideaux [2] shopping online is definitely different from shopping in traditional markets because it is able to offer a more varied product and service and the access and convenience provided are not limited by space and time. Nguyen et al. [3] also said that shopping online has a risk, that is, there is uncertainty about quality that is not necessarily the same as consumer expectations. However, to reduce the risk of quality uncertainty Voramontri and Klieb [4] said that consumers tend to seek information in advance about the product or service to be purchased on social media.

The internet offers a variety of ways to get information about goods or services from other customers, claim [5]. Social networking sites are characterized as public media by Ahmadinejad and Asli [6] where users can write, store, and publish content online. Internet users have altered how customers communicate and share their ideas or reviews of consumed goods and services, according to Babić Rosario et al. [7]. Electronic Word of Mouth (EWoM) refers to the method of online consumer communication (e-WOM). Ismagilova et al. [8] define e-WOM as a communication channel for the exchange of knowledge about a good or service that has been used by customers who have never met before. E-WoM is a claim based on good, neutral, or negative experiences made by prospective customers, current customers, or past customers on a product, service, brand, or company that is developed and shared via Internet media, according to Prasad et al. [9]. One definition of electronic word of mouth (E-WoM) is the online dissemination of review information from past users of a product. E-WoM as a representation of the product's real and potential strengths or flaws. You can find information about this product review on social networking websites [10].

Innovation will raise a product's added value and result in a new product that can provide consumers better answers to their problems. This study aims to examine the influence of the role of e-wom on purchase decisions, in addition to the effect of innovation on purchase decisions, the influence of e-trust on purchase decisions, and the effect of website attractiveness on purchase decisions.

2 Literature Review

Purchase decisions are defined as a person's choice of a certain good or service made available by the business [11]. Making a purchase choice is the last step before a consumer actually makes a purchase [12]. Additionally, cultural, social, emotional, and psychological aspects all have an impact on purchase decisions [12].

Since the beginning of human society, word-of-mouth (WOM) has been acknowledged as one of the most significant sources of information transmission. Before the advent of the internet, consumers communicated their product experiences to one another through conventional WoM (such as conversations with friends and family) [13]. Due of the internet's global reach, consumers who have never met can now communicate with e-WOM [14]. Customers can leave reviews on online marketplaces that affect prospective buyers. E-WOM, which is accessible to many individuals and organizations via the internet, is a remark made by a prospective, present, or past customer regarding a product or company. According to research [15], e-WOM has a favorable impact on a consumer's decision to make a purchase. According to the study's findings [16], e-WOM significantly influences a consumer's decision to buy.

H0: E-WoM has a significant positive effect on Purchase Decision
H1: E-WoM has a significant negative effect on Purchase Decision

Likewise, e-WOM is measurable because comments on a product are written and available on the Website. The World Wide Web is a service provided by the internet, which uses internationally accepted standards for storing, re-accessing, formatting, and displaying information found on the internet [17]. Website is a site that contains information that can be used to create an online store (online). Website design is one of the many visualizations to convey information from the company and can provide interest in placing advertisements in its product offerings online [18]. Other people's thoughts provide a plan in analyzing and studying the visualization of a product. Attractive customer experience design or product design can be one of the physical attractions [19]. Like advertisements, the attractiveness of website design can also be an attraction for customers.

According to Mandal et al. [20] website attractiveness has been defined as a series of interactive features that attract the attention of customers and ensure easier interaction, navigation and use to maintain long-term relationships. These websites consist of almost all the text and color page elements, formatted text, images, animations, videos, and sounds needed to build a commercial marketplace. The goal of a website [21] is to deliver content to customers and sales to complete transactions.

It is known that the attractiveness of website design [22] has an influence on the purchasing decision level of a consumer when visiting certain e-commerce websites. The behavior of customers in shopping online is greatly influenced by the information available, the attractiveness of website design, and the easy way to make purchasing decisions [23]. According to earlier research findings [19], website design influences purchasing decisions in a good and significant way. This means that a user-friendly website layout will promote customer convenience and have an impact on consumers' decisions to buy the products or services being offered. Customers will form a bad opinion of the online retailer and believe that it is a less-than-professional company if the website's appearance is merely modest. When a customer visits an online store, the first thing they see determines the seller's reputation in the eyes of the customer. A positive consumer response to goods and services will demonstrate a strong first impression, and it is not unlikely that they will be more likely to make a purchase.

H2: Web Attractiveness has a significant positive effect on Purchase Decision
H3: Web Attractiveness has a significant negative effect on Purchase Decision

Online consumer reviews, a subset of e-WOM, are statements made by customers about goods purchased via online stores. Contrarily, customers who read evaluations that are generally unfavorable to a product could disregard or detest it since they find it uncomfortable to disagree with others [13].

Consumer trust, according to Dam [24], is a company's readiness to rely on business partners. It depends on a variety of inter-organizational elements, including how competent, honest, and ethical the corporation is seen to be, as well as its policies. Tran and Vu [25] assert that consumer confidence in online purchasing websites depends on the popularity of the website itself. The better the website, the more assured consumers will be in the website's dependability. According to Vongurai et al. [26], consumers have a favorable expectation of manufacturers that they will be able to produce goods that will satisfy buyers. According to these experts' perspectives, trust is a situation that arises when a customer has faith in the reliability and honesty of an internet website. E-trust is a belief that a product or service provider can be relied upon to behave in such a way that the long-term interests of consumers can be fulfilled [27].

The results of previous research [28] found that there was a direct and indirect relationship between E-Trust and Purchase Decisions so that E-Trust had a significant positive impact on Purchase Decisions. The results of the study [29] also show that e-trust has a significant effect on repurchase intention.

H4: E-Trust has a significant positive effect on Purchase Decision
H5: E-Trust has a significant negative effect on Purchase Decision

Innovation can come from anything, like a product or marketing strategy. Customers' purchase intentions may also be impacted by this, resulting in an increase in sales. Rayi and Aras [30] define innovation as the process of introducing a new product as a result of an increase in a product's added value, which distinguishes it from competing items. It has been documented that product innovation in the service sector can significantly affect consumer satisfaction [31]. Product innovation, novelty, distinctiveness, and origin of items were proven to have a statistically significant favorable impact on purchase decisions in earlier research [32]. In the context of developing innovative and sustainable new products, research [33] discovered a favorable association between buying attitudes and purchase intentions (Picture 1).

H6: E-WoM has a significant positive effect on Purchase Decision
H7: E-WoM has a significant negative effect on Purchase Decision

Picture 1 Research model

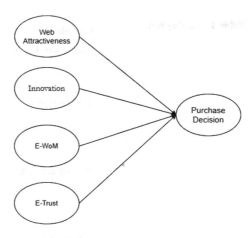

3 Research Methods

A local shoe brand company that sells online using its website account is the subject of this study. Purposive sampling was used to choose the sample for this study. According to Hair et al. [34], the sample for this study consisted of 159 respondents. The types and sources of data used in this study are primary data. Primary data obtained from the results of questionnaires that have been distributed to respondents which is the sample in this study. The data analysis used was IBM SPSS Statistics 25 application.

4 Results and Discussion

Before being analyzed, validity and reliability tests were first carried out on the distribution of questionnaires (Table 1).

In the validation test, the results of the loading factor of Innovation (INV) with an indicator denoted by INV1 item is 0.795, INV2 item is 0.738, and INV3 is 0.842. Furthermore, the Electronic Word of Mouth (EW) variable gets the results of the loading factor of each indicator denoted by item EW1 which is equal to 0.637, item EW2 which is equal to 0.676, item EW3 which is equal to 0.727 and EW4 which is equal to 0.811. The next variable is the Web Attractiveness (WA) variable. The Web Attractiveness variable gets the results of the loading factor of each indicator denoted by the WA1 item which is equal to 0.720, the WA2 item which is equal to 0.737, the WA3 item which is equal to 0.810 and the WA4 which is equal to 0.645. Next, the Electronic Trust (ET) variable gets the loading factor results of each indicator denoted by ET1 item which is equal to 0.689, ET2 item which is equal to 0.694, ET3 item which is equal to 0.779 and ET4 which is equal to 0.752. In addition, the Purchase Decision (PD) variable gets the results of the loading factor of each indicator denoted by the PD1 item which is

Table 1 Validity test

Variable	Item	*Loading factor*	Description
Innovation (INV)	INV1	0.795	Valid
	INV2	0.738	Valid
	INV3	0.842	Valid
E-Wom (EW)	EW1	0.637	Valid
	EW2	0.676	Valid
	EW3	0.727	Valid
	EW4	0.811	Valid
Web attractiveness (WA)	WA1	0.720	Valid
	WA2	0.737	Valid
	WA3	0.810	Valid
	WA4	0.645	Valid
E-Trust (ET)	ET1	0.689	Valid
	ET2	0.694	Valid
	ET3	0.779	Valid
	ET4	0.752	Valid
Purchase decision (PD)	PD1	0.830	Valid
	PD2	0.879	Valid
	PD3	0.892	Valid
	PD4	0.856	Valid
Limit		> 0.5	Accepted

equal to 0.830, the PD2 item which is equal to 0.879, the PD3 item which is equal to 0.892 and the PD4 which is equal to 0.856.

The validation test showed valid results because the loading factor values were obtained from the INV, EW, WA variables. ET and PD have an overall value of > 0.5 so that the data can be accepted, then all items can proceed to the next stage of analysis (Table 2).

In the reliability test, the results obtained from the reliability construct of the Innovation variable (INV) with items denoted by INV1, INV2, and INV3 were 0.8350 and the value of Average Variance Extracted (AVE) was 0.6285. Furthermore, the E-Wom (EW) variable with items EW1, EW2, EW3, EW4 has a construct reliability value of 0.8064 and AVE of 0.5122. Next, we also obtained values from construct reliability and average variance extracted (AVE) from Web Attractiveness (WA) with WA1, WA2, WA4, and WA3 items, namely 0.8196 and 0.5334. Furthermore, the construct reliability and AVE values respectively from the Electronic Trust (ET) variable with the items ET1, ET2, ET3, and ET4 are 0.8194 and 0.5321. Finally, the variable Purchase Decision (PD) with items PD1, PD2, PD3, and PD4 has a construct reliability value of 0.9220 and the value of average variance extracted (AVE) is 0.7474.

Factors Influencing Purchase Decisions on Online Sales in Indonesia 335

Table 2 Reliability test

Variable	Item	Construct reliability	AVE
Innovation (INV)	INV1	0.8350	0.6285
	INV2		
	INV3		
E-Wom (EW)	EW1	0.8064	0.5122
	EW2		
	EW3		
	EW4		
Web attractiveness (WA)	WA1	0.8196	0.5334
	WA2		
	WA3		
	WA4		
E-Trust (ET)	ET1	0.8194	0.5321
	ET2		
	ET3		
	ET4		
Purchase decision (PD)	PD1	0.9220	0.7474
	PD2		
	PD3		
	PD4		

The results of reliability testing on the research questionnaire showed the value of construct reliability of all items ≥ 0.70. In addition, the research also obtained the value of Average Variance Extracted (AVE) for each item ≥ 0.50. So the results of reliability testing on the research questionnaire show reliable results. Therefore, the data obtained can be further processed for model testing.

Testing the Path Analysis of the Research Model

The decision to determine whether to accept or reject the hypothesis is based on the resulting significance probability, the value of the significance level is $\alpha = 5\%$ (0.05). The research results are presented in Table 3.

The results of the significance test in Table 4 show that Electonic Word of Mouth has a significant effect on the Purchase Decision (0.000 < 0.05), so H0 is accepted.

Table 3 Path model significance test

Variable	Regression coefficient	P	Description
PD ← INV	1.296	0.000	Significant
PD ← EW	0.580	0.000	Significant
PD ← WA	0.711	0.000	Significant
PD ← ET	0.954	0.000	Significant

Furthermore, Web Attractiveness has a significant effect on Purchase Decision (0.000 < 0.05) then H2 is accepted, E-trust has a significant effect on Purchase Decision (0.000 < 0.05) then H4 is accepted. Finally, there is a significant direct effect of Innovation on Purchase Decision (0.000 < 0.05), so H6 is accepted.

Discussion

E-wom has a positive and significant effect on Purchase Decision

The findings demonstrated that hypothesis 0, which held that electronic word of mouth (E-WOM) had a favorable and significant impact on purchase decision, was accepted. E-WOM is a claim about a product, service, brand, or business that is produced and disseminated via Internet media based on the positive, neutral, or negative experiences made by prospective customers, current customers, or past customers. Because customers can engage in internet-based social communication where users can transmit and receive product-related information online, e-wom can have an impact on purchasing decisions. A choice to buy the product will be made if the reviews offered match the parameters that the potential buyer wishes to satisfy. Reviews provided by past customers or other users can be seen by potential customers. Customers often like online shops with a lot of positive customer evaluations that include details about the products.

Web Attractiveness has a significant positive effect on Purchase Decision

The results showed that hypothesis 2 was accepted, namely Web Attractiveness had a significant effect on Purchase Decision (0.000 < 0.05). Web Attractive Web site attractiveness has been defined as a set of interactive features that grab the attention of customers and ensure easier interaction, navigation and use to maintain long-term relationships. This explains that the attractiveness of a website that is fun, entertaining, can encourage users to have a purchasing decision. An increase in purchasing decisions can be due to users looking at the display on the website so that they are interested in the content and information provided related to the product. This can stimulate consumers to decide to buy the products offered on the website.

E-Trust has a significant positive effect on Purchase Decision

The study's findings support hypothesis 4, which states that electronic trust significantly influences purchase decisions. Electronic Trust (E-Trust) is the conviction that a provider of goods or services can be counted on to act in a way that serves consumers' long-term interests. This demonstrates that respondents believe the website provides good service in meeting customer needs so that customers can have faith in it. In order for customers to pay closer attention and tailor their purchases to their needs, the website also offers comprehensive information and specs about the things being sold. After customers have favorable interactions with merchants, trust in online transactions will grow, resulting in customers buying these things without hesitation.

Innovation has a positive and significant effect on Purchase Decision

The results showed that hypothesis 6 was accepted, namely innovation had a positive and significant effect on purchase decisions. Innovation is the process of introducing a new product due to an increase in the added value of a product, which makes it different from other products. Consumers will be more interested in products that have their own uniqueness or have innovations that can be very useful to support activities. A product that has an innovation that is able to create a sense of interest and has functional value for its users will make consumers curious and also increase purchase intention which influences consumer purchasing decisions.

In this study, researchers only measure purchasing decisions based on the variables Innovation, E-wom, Web Attractiveness and E-trust. Therefore, this research still cannot describe anything that can influence the purchase decision of consumers as a whole. This is because there are still many other variables that can influence purchase decisions. In this study, researchers only used the website as a reference in the questionnaire, for further research they can add variables that can influence purchase decisions and use social media which is more frequently used in Indonesia, so that it is expected to obtain even better results.

Based on the research results, it can be concluded that there is a significant influence of innovation on purchase decisions. Furthermore, Electronic Word of Mouth has a significant effect on Purchase Decision. Next, there is a significant effect of Web Attractiveness on Purchase Decision. Electronic Trust also has a significant effect on Purchase Decision.

References

1. Pandita, R.: Internet a change agent: an overview of internet penetration and growth across the world. Int. J. Inf. Dissem. Technol. **7**(2), 83 (2017). https://doi.org/10.5958/2249-5576.2017. 00001.2
2. Harris, A., Prideaux, B.: The potential for eWOM to affect consumer behaviour in tourism. In: Dixit, S.K. (ed.) The Routledge Handbook of Consumer Behaviour in Hospitality and Tourism, 1st edn, pp. 366–376. Routledge (2017). https://doi.org/10.4324/9781315659657-41
3. Nguyen, D.H., de Leeuw, S., Dullaert, W.E.H.: Consumer behaviour and order fulfilment in online retailing: a systematic review: order fulfilment in online retailing. Int. J. Manag. Rev. **20**(2), 255–276 (2018). https://doi.org/10.1111/ijmr.12129
4. Voramontri, D., Klieb, L.: Impact of social media on consumer behaviour. Int. J. Inf. Decis. Sci. **11**(3), 209 (2019). https://doi.org/10.1504/IJIDS.2019.101994
5. Felix, R., Rauschnabel, P.A., Hinsch, C.: Elements of strategic social media marketing: a holistic framework. J. Bus. Res. **70**, 118–126 (2017). https://doi.org/10.1016/j.jbusres.2016.05.001
6. Ahmadinejad, B., Asli, H.N.: E-business through social media: a quantitative survey (Case Study: Instagram). Int. J. Manag. Acc. Econ. **4**(1) (2017)
7. Babić Rosario, A., de Valck, K., Sotgiu, F.: Conceptualizing the electronic word-of-mouth process: what we know and need to know about eWOM creation, exposure, and evaluation. J. Acad. Mark. Sci. **48**(3), 422–448 (2020). https://doi.org/10.1007/s11747-019-00706-1
8. Ismagilova, E., Dwivedi, Y.K., Slade, E., Williams, M.D.: Electronic Word of Mouth (eWOM) in the Marketing Context: A State of the Art Analysis and Future Directions. Springer (2017)

9. Prasad, S., Garg, A., Prasad, S.: Purchase decision of generation Y in an online environment. Mark. Intell. Plan. **37**(4), 372–385 (2019). https://doi.org/10.1108/MIP-02-2018-0070
10. Kudeshia, C., Kumar, A.: Social eWOM: does it affect the brand attitude and purchase intention of brands? Manag. Res. Rev. **40**(3), 310–330 (2017). https://doi.org/10.1108/MRR-07-2015-0161
11. Hanaysha, J.R.: Impact of social media marketing features on consumer's purchase decision in the fast-food industry: brand trust as a mediator. Int. J. Inf. Manag. Data Insights **2**(2), 100102 (2022). https://doi.org/10.1016/j.jjimei.2022.100102
12. Kamil, N.A.I., Albert, A.: The effect of e-wom and brand image towards Sushi Masa consumer purchasing decision. J. Soc. Stud. (JSS) **16**(1), 19–34 (2020). https://doi.org/10.21831/jss.v16i1.31020
13. Harahap, D.A., Hurriyati, R., Gaffar, V., Amanah, D.: The impact of word of mouth and university reputation on student decision to study at university. Manag. Sci. Lett. 649–658 (2018). https://doi.org/10.5267/j.msl.2018.4.027
14. Srivastava, M., Sivaramakrishnan, S.: The impact of eWOM on consumer brand engagement. Mark. Intell. Plan. **39**(3), 469–484 (2021). https://doi.org/10.1108/MIP-06-2020-0263
15. Prasad, S., Gupta, I.C., Totala, N.K.: Social media usage, electronic word of mouth and purchase-decision involvement. Asia Pac. J. Bus. Adm. **9**(2), 134–145 (2017). https://doi.org/10.1108/APJBA-06-2016-0063
16. Perkasa, D.H., Suhendar, I.A., Randyantini, V.: The effect of electronic word of mouth (eWOM), product quality and price on purchase decisions. Int. J. Educ. Manag. Soc. Sci. **1**(5) (2020)
17. Angelova, N., Kiryakova, G., Yordanova, L.: The great impact of internet of things on business. Trakia J. Sci. **15**(Suppl.1), 406–412 (2017). https://doi.org/10.15547/tjs.2017.s.01.068
18. Moreno, F.M., Lafuente, J.G., Carreón, F.Á., Moreno, S.M.: The characterization of the millennials and their buying behavior. Int. J. Mark. Stud. **9**(5), 135 (2017). https://doi.org/10.5539/ijms.v9n5p135
19. Khalil, H., Umapathy, K., Goel, L.C., Reddivari, S.: Exploring relationships between e-tailing website quality and purchase intention. In: Nah, F.F.-H., Siau, K. (eds.) HCI in Business, Government and Organizations. ECommerce and Consumer Behavior, vol. 11588, pp. 238–256. Springer International Publishing (2019). https://doi.org/10.1007/978-3-030-22335-9_16
20. Mandal, S., Roy, S., Raju, A.G.: Exploring the role of website attractiveness in travel and tourism: empirical evidence from the tourism industry in India. Tour. Plan. Dev. **14**(1), 110–134 (2017). https://doi.org/10.1080/21568316.2016.1192058
21. Qalati, S.A., Vela, E.G., Li, W., Dakhan, S.A., Hong Thuy, T.T., Merani, S.H.: Effects of perceived service quality, website quality, and reputation on purchase intention: the mediating and moderating roles of trust and perceived risk in online shopping. Cogent Bus. Manag. **8**(1), 1869363 (2021). https://doi.org/10.1080/23311975.2020.1869363
22. Baeshen, Y., Al-Karaghouli, W., Ghoneim, A.: Investigating the effect of website quality on eWOM and customer purchase decision: third parties hotel websites (2017)
23. Tang, M., Zhu, J.: Research of O2O website based consumer purchase decision-making model. J. Ind. Prod. Eng. **36**(6), 371–384 (2019). https://doi.org/10.1080/21681015.2019.1655490
24. Dam, T.C.: Influence of brand trust, perceived value on brand preference and purchase intention. J. Asian Financ. Econ. Bus. **7**(10), 939–947 (2020). https://doi.org/10.13106/JAFEB.2020.VOL7.NO10.939
25. Tran, V.-D., Vu, Q.H.: Inspecting the relationship among e-service quality, e-trust, e-customer satisfaction and behavioral intentions of online shopping customers. Glob. Bus. Financ. Rev. **24**(3), 29–42 (2019). https://doi.org/10.17549/gbfr.2019.24.3.29
26. Vongurai, R., Elango, D., Phothikitti, K., Dhanasomboon, U.: Social media usage, electronic word of mouth and trust influence purchase-decision involvement in using traveling services. Asia Pac. J. Multidisc. Res. **6**(4) (2018)
27. Al-Khayyal, A., Alshurideh, M., Kurdi, B.A.: The impact of electronic service quality dimensions on customers' e-shopping and e-loyalty via the impact of e-satisfaction and e-trust: a qualitative approach. Int. J. Innov **14**(9) (2020)

28. Pop, R.-A., Săplăcan, Z., Dabija, D.-C., Alt, M.-A.: The impact of social media influencers on travel decisions: the role of trust in consumer decision journey. Curr. Issues Tour. **25**(5), 823–843 (2022). https://doi.org/10.1080/13683500.2021.1895729
29. Miao, M., Jalees, T., Zaman, S.I., Khan, S., Hanif, N.-A., Javed, M.K.: The influence of e-customer satisfaction, e-trust and perceived value on consumer's repurchase intention in B2C e-commerce segment. Asia Pac. J. Mark. Logist. **34**(10), 2184–2206 (2022). https://doi.org/10.1108/APJML-03-2021-0221
30. Rayi, G., Aras, M.: How product innovation and motivation drive purchase decision as consumer buying behavior. J. Distrib. Sci. **19**(1), 49–60 (2021). https://doi.org/10.15722/JDS.19.1.202101.49
31. Ayodele, M.S., Oluwayemi, B.: Effect of product innovation on customer satisfaction: an overview of insight into Nigerian service market (2019)
32. Timtong, J., Lalaeng, C.: Product innovation and digital marketing affecting decision to purchase OTOP products. Int. J. Crime Law Soc. Issues **8**(1) (2021)
33. An, D., Ji, S., Jan, I.U.: Investigating the determinants and barriers of purchase intention of innovative new products. Sustainability **13**(2), 740 (2021). https://doi.org/10.3390/su13020740
34. Hair, J.F., Black, W.C., Babin, B.J., Anderson, R.E.: Multivariate Data Analysis, 8th edn. Cengage Learning (2018)

Re-Addressing the Notion of Patriarchy in Social and Economic Framework in the Light of Women Empowerment

Jenni K. Alex and **Dino Mathew**

Abstract Patriarchy—male dominance in all aspects—a powerful and dominant concept that existed from the beginning of humanity, has started diminishing its sharpness and intensity over the years. Even though, this very notion still prevails in society in different forms it makes an explicit hindrance to the development and advancement of women in the public and private sector of society. For many decades, women were behind the shadow of men, and begin to explore the world where they excel through their talents, efficiency, and merit. Hence, it's high time to evaluate and understand the nuances of the patriarchal dimension in the progress and empowerment of women who has a significant and prolific role to play in the process of nation-building from a wider perspective. This review paper attempts to describe and analyze how patriarchy makes obstacles to the growth and progress of women in society. In addition to that, this paper encapsulates the impact of patriarchy on the economic and social sectors of society as economic and social features are the inevitable features and pillars of sustainability. Finally, this article will conclude with some of the recommendations and insights which lend a hand to the promotion and furtherance of women who can be molded as the champion of change in society and add value and uniqueness to their own lives in a large sense.

Keywords Patriarchy · Women empowerment · Sustainability · Nation building

1 Introduction

When did the subservience of women begin? Does it go back to the beginning of the history of human beings? Why are women subjugated? Does it mean once women were subordinated, they could still be under suppression and exploitation? These are

---

J. K. Alex
Department of Economics, Newman College, Thodupuzha, Kerala, India

D. Mathew (✉)
Mahatma Gandhi University, Kottayam, Kerala, India
e-mail: dinokallikattu@gmail.com

© The Author(s), under exclusive license to Springer Nature Switzerland AG 2024
A. Hamdan and E. S. Aldhaen (eds.), *Artificial Intelligence and Transforming Digital Marketing*, Studies in Systems, Decision and Control 487,
https://doi.org/10.1007/978-3-031-35828-9_29

the questions to which specific answers are to be sought and addressed and that sets off the subject matter of heated discussion in the modern world. To substantiate views and opinions about the above-mentioned questions, one must have to put more light on the concept of Patriarchy which kept women subjugated for many centuries. The subjugation to their husband and delineation from society due to patriarchy was a stumbling block to the growth and development of women. Women, one of the core concepts of the development of our nation, cannot be thrown to the corner; rather they must be strengthened and motivated to make tremendous changes and impact in the society and nation through their active and innovative participation. These revolutionary changes should be a *'mantra'* for the progress and prosperity of our nation.

2 The Notion of Patriarchy

Patriarchy—an ideology and a social system—was derived from the Greek word *'patriarkhēs'* (πατριάρχης) meaning "ruler of the family". The term patriarch originated from the old testament where it is perceived as the paternal ruler of a family [1]. Though there are numerous definitions of patriarchy available, a more appealing and broad sense, relating to the social aspect is as follows. "Patriarchy is a system of social structure and practices, in which, men dominate, oppress and exploit women" [2].

Patriarchy presumes men be superior to women and dominate them in all aspects of life. It is a social system that assets the father or male child as the head/superior of the family who controls everything according to their discernment [3]. This societal system promotes all privileges to males. When we consider the domination of males in the family, domination, in most cases, can take the form of violence. Henceforth patriarchal domination is a threat to women in the family and all the organized and unorganized sectors of society. Violence and suppression are not only the aftermath of patriarchy but also the exploitation of women and their femininity which is always perceived as a lower position in social structure. Hence as a result of this suppression, all sorts of inequalities originated to devalue and defame the dignity of women [4]. In this context, a redefining and readdressing of the concept of patriarchy becomes a subject matter of the modern times where a topic like gender equity and women empowerment has its significance. This study aims to highlight the impact of patriarchy made on women in relation to social and economic dimensions. It also encompasses how women can be empowered and lifted from the dim light of patriarchy to society-building and nation-building.

3 Evolution of Patriarchy in Indian Society

During the Vedic and Epic periods (Ancient times) women had prominence in society and family in terms of autonomy and power [5]. In this period women largely occupied an equal position in society equivalent to men [6]. But after the Vedic period, we see a polarization of sexes where patriarchy and its manifold forms started to restrict and control women and their mobility. People overpraised the purity and sanctity of women who are ready to subjugate themselves through their sacrifice and surrender wholeheartedly to their husbands and the family they belong. A scarifying mother and faithful wife were highly acknowledged and lauded in society [7]. As the Mugul invasion (Medieval Period) took place the status of women begins to deteriorate to a lower position and declined to the control and governance of the husband or father of the family. The senseless beliefs and rituals of this time gave birth to brutal practices like female infanticide, child marriage, enforced widowhood, and Sati [8, 9]. The era of colonization and post-colonization called attention to the supreme right of the man who was controlling, dominating, and asserting power in family and society [10]. Even after many decades of colonization women had to fight hard to rewrite the ill effect of patriarchy and its norms and conditions. Even though the shadows of patriarchs are still prevalent in our society, the presence and involvement of women in the family and society are felt. Readdressing the notion of patriarchy to gender equality and women empowerment enhances women to bring out their innate talents and potential to the maximum for the betterment of the family and society.

4 Women in Patriarchy

Patriarchy is a blind belief [11], social institution, and ideology in which men are in control and women are in subjugation [12]. As Stephan Goldberg implies patriarch may be viewed as an organization that includes—social, economic, cultural, and political—where the chief position in its functionalities and activities is vested in the hands of men [13]. Even though the subordination of women is not the centrality of patriarchy, it rallies on the masculinity of men. In a patriarchal context, women are less considered human and seen as sexual objects, always under the superiority of men who are perceived as sexual subjects [14]. Jessica McCallister asserted that women who live in a society that is patriarchal in nature have the following characteristics namely, male dominance, male identification, male centeredness, and obsession with control, that confine them to demonstrate their independence and suggest opinions to change the social order [15].

1. Male Dominance—Patriarchy is viewed as the first structure and system of domination which results in the violence of men over women [16].
2. Male Identification—Masculinity is used to expand male identification. The very expression of masculinity identifies men to sit in the driving seat and control women to fulfill their roles as housekeepers and caretakers [17].

3. Male Centeredness—The central attraction in family and society belongs to men who look after, take responsibility for, and administer all the household and societal matters. Hence, the patriarchal society designates him the *'alpha position'* in all spheres of family and society which emphasizes the centrality of man [18].
4. Obsession with Control—The obsession with control can be originated from the concept of domination of man. This obsession can pave the way to violence which is a consequence of patriarchal social arrangements and ideologies, that become the vulnerability of women who are at the constant menace of being oppressed [19].

5 Patriarchy—Social and Economic Dimension

We analyze the influence and impact of patriarchy that restrict the mobility of women in family and society from the social and economic point of view. Because, the sustainable development and empowerment of women has to take place especially in economic and social realm. Social and economic sustainability paves the way for transforming women into leaders of tomorrow and hence changing the world around them to an abode of peace and love. The social dimension includes education, decision-making, and the role and status of women whereas the economic dimension mainly discusses employability and workplace wage distribution.

1. **Social Dimension**

   A. Education

      Looking at Indian history, patriarchal and religious practices adversely affected the promotion of the education of women. The dreams and opportunities of acquiring knowledge for women were reduced and shattered by meaningless misogynistic practices and ideologies [20]. In ancient times, education for men was encouraged and it was limited only to the upper caste sections of society. Meanwhile, the education of women was not important and necessary. The advent of colonization promoted the education of women but not in its fullest sense. The first woman who spearheaded the importance of education for the change of the stature of women was Savitribai Phule, a social reformer and the first female teacher [21]. From there the transformation of women's education took place and still continuing through various organizations and institutions.

   B. Decision Making

      Johnson and Leones argued that in a patriarchal context, every activity of women is controlled and governed by men including reproductive activities and the size of the family [3, 22]. In the earlier days, women who were forced into early marriage, are totally at the feet of their husbands, requesting permission for numerous matters of household activities. Women's say which is a component of their well-being, is not taken into account when decisions are made relating to the prosperity and progress of the family. Visit their own home, parents, siblings, and friends are always at the mercy of their husbands [23]. Even when they stepped into society for holding various positions, they are bounded to take decisions regarding prime matters of society, and it should be in consultation with male authorities. In recent days, women's decision-making power has improved a lot and they are free to make their choices for the family and society [24].

C. Role and Status of Women

The role and status of women in India under patriarchal societies were a complex subject of discussion. In the olden days, women were considered and lauded like goddesses. Once the caste system burst into society, the status and role of women declined to the second position behind the man and in society. The lowered secondary and submissive position of women restricted them from taking a major role in societal activities [25]. The status of women in a patriarchal society was shrunk to good daughters, wives, and mothers [26]. This status of women is refined when they begin to acquire knowledge through proper education and hence, they become a decisive factor in the inevitable function of the family and society.

2. Economic Dimension

A. Employability

During the patriarchal time, women were limited to going beyond the front door of the house for a job or any paid activities. Quoted from the Tamil classical literature (Sangam), "soul of man is in working outside and that of woman is caring her husband and family". The socio-cultural tradition of India at the time of patriarchy considers a woman who works for payment outside the home, less valued and less respected. The employability of women at that time is baring and raring children and attending to household chores [17].

B. Workplace Wage Distribution

It was during the contemporary period, the mobility of women for jobs was well established and from that time onwards gender disparities were prevalent in whatever occupation women engaged in. Women were always at the receiving end of a—'double burden'—preferring paid and unpaid labors like cleaning, caring for their husbands and children, cooking, and fetching water [27]. Women are paid less fair regardless of their effort and hard work, especially in unorganized sectors of the market. Women are seldom preferred for higher jobs and are forced to work in unhealthy conditions due to discrimination and that forces biased wage imbalances in the realm of occupational concentration [28].

6 Glitches Made by Patriarchy in Women's Empowerment

1. Gender Discrimination

   It is said that gender discrimination begins from the birth of a child. Men prefer a boy child rather than a girl for a reason that a male child adds a number to the family and preserves the name of the family [3] whereas a female child is a financial burden [5]. Moving to employment, most of the appointments and opportunities are reserved according to gender rather than merit. Looking from the educational point of view, a boy child would be preferred to continue their education while the path of education of a girl child would be shut forever midst of the financial crisis of the family [29]. The labor and economic contribution of women were not given proper recognition and acknowledgment which could have been a boost to their economic productivity and input to the family [26].

2. Exploitation of Women

   A society governed by patriarchal norms exploited women for the advantage of men. Women are always delineated from the main streams of social structures and family setup [5]. Men being the head of the family and society controls women's sexuality, labor, reproduction, and mobility [12]. The concept of women being the trophy given to men for signaling and symbolizing man's success and at the time of his failure, women will be at the receiving end of blame, pain, and torture [4].

3. Violence

   Patriarchy is viewed as the first structure and system of dominance which results in the violence of men over women [16]. Patriarchal social arrangements and ideologies amplified the superiority of man which becomes the root cause of all sorts of violence—domestic and societal [19]. Being the *'alpha position'* of society and family, anything that goes against the will of men by women leads to torture and violence.

7 Areas Where Empowerment is Required (Recommendation)

1. Educational Empowerment

   Education a revolutionary concept for women has a significant role to play in the advancement of women in family and society. Education is the entry point where all the other doors are opened for women to explore the world and showcase their talents and potential for active social participation [30]. Education is an effective weapon to irradicate the notion of gender discrimination and disparities [31]. Furthermore, education is the surest means to escalate the capacity of women to reach out to the world and deal with its current affairs to transform the world inside and the world outside [32].

2. Economic Empowerment

The effective and successful mean for the eradication of poverty, gender inequality, and economic and educational disparities is investing in the economic empowerment of women [33]. Women's economic empowerment has a wide range of implications for their own life and society and that accelerate and bolster the improvement of the self-image of women, decision-making capacity, better education, and quality living conditions [34].

3. Social Empowerment

Social empowerment includes magnifying the role and status of women in society, promoting gender equity, enhancing work-labor justice, and eradicating exploitation at home and in society. The challenge of the time is improving the image of women and promoting the holistic development of women not in peripheral but in core depth. Therefore, the prerequisites for the sustainable development of women can be achieved through a society that is supportive, enriching, and empowering economically and socially [35].

4. Psychological Empowerment

The patriarchal beliefs and practices made women powerless and devalue their dignity of women. Hence psychological empowerment to motivate and inspire the attitude and social demands of women becomes an essential and inevitable theme in the modern world. To make women powerful and skillful, a smart strategy or psychological strengthening for self-determination and commitment to their work is highly demanding [6].

5. Gender Empowerment

Gender empowerment can be phrased as gender equity—an action of being fair to both men and women. Equity leads to equality. Gender equality doesn't indicate men and women are on the same page, but it refers to the equal sharing of socially valued goods, opportunities, resources, and rewards irrespective of gender discrimination. Gender equality—a way of empowering women—ensures the active participation of women in social activities especially in decision-making regarding familial and social matters and opens access to acquire all the resources of society like men.

8 Projects Focused on Women's Empowerment

The Indian government has prime concern for the empowerment of women. Hence, the country has established numerous projects and schemes for empowering and uplifting the educational, economic, social, and political sectors of women in India. The ultimate aim of those projects is to render a helping hand to women who are potential and capable but remain in utter chaos and darkness due to their worse situations and circumstances. A Few projects are (Table 1).

Table 1 Projects organized for the women empowerment in India

No.	Project/scheme	Purpose of the project/scheme
1	One Stop Center Scheme	It is a national mission for the empowerment and advancement of women. It is also known as 'Sakhi'. The main objective is to support and encourage women who are physically, sexually, economically, emotionally, and psychologically tortured and oppress
2	Mahila Shakti Kendra Scheme	This scheme aims to provide an environment to realize the full potential of women, quality education, health and safety, employability, etc. Through proper training and capacity-building programs, this scheme assists women to attain entitlement and empowerment
3	Nirbhaya	It focuses on the safety and security of women in society and the country. This project makes sure that the mobility of women can be done without fear and anxiety in any corner of the country
4	Women Helpline Scheme	Women helpline scheme aims to provide 24-h toll-free telecom service to women facing violence domestic and societal helpline
5	Nari Shakti Puraskar	The motive behind 'Nari Shakti Puraskar' is to acknowledge and highlight the contribution given by an eminent woman or institution or organization who magnificently contributes to the rejuvenation of vulnerable and marginalized women in society
6	Beti Bachao Beti Padhavo Scheme	This scheme ensures the protection of all women child and the promotion and implementation of their education

Though there are several projects and schemes by the central government, state government, and other institutions, the biggest challenge before us is to address the following questions. Do the women in our society and country still safe, secure, and empowered through the activities of various projects set for the empowerment of women? Do they live with dignity and pride? Do these projects provide a means for the recalibration of women to contribute like men?

9 Conclusion

Moving towards a less patriarchal society where patriarchal norms, beliefs, and practices are diminishing to the end, should be the target of today's society, and that becomes the need of the time. The empowerment of women strategies and policies should incorporate the basic requirement such as gender equality to cater to the needs and wants of women, providing quality education for the advancement of their own life and society, economic stability to contribute to the income of the family and progress of the nation, psychological and physical motivation to succeed the complex and difficult situation of life, and open the door for the free mobility and exploration of the world in a wider perspective. A change in attitude and approach towards the status of women in society is highly recommended in this fast-growing technological world. Systematic and updated transformation should be reflected in the economic, social, educational, and psychological spheres of development and empowerment to upgrade the role and responsibilities of women in the family and society. Backing up their rights and formulating policies for economic securities, just wages, domestic violence, educational opportunities, and finally, an entry to the modern technical world must be the agenda for the advancement, upliftment, and empowerment of our women's society.

References

1. Wilson, A.: Patriarchy: feminist theory. In: International Encyclopedia of Women: Global Women's Issues and Knowledge, pp. 1493–1497 (2000)
2. Walby, S.: Theorising Patriarchy. Basil Blackwell Ltd. (1991)
3. Mudau, T.J., Obadire, O.S.: The role of patriarchy in family settings and its implications to girls and women in South Africa. J. Hum. Ecol. **58**(1–2), 67–72 (2017). https://doi.org/10.1080/097 09274.2017.1305614
4. Artanti, S.K., Wedati, M.T.: Subalternity in Amitav Ghosh's sea of poppies: representation of Indian women's struggle against patriarchy. Prosodi J. Ilmu Bahasa Dan Sastra Program Studi Stastra Inggris Universitas Trunojoyo **14**(1), 1–14 (2020)
5. Habib, A.: Patriarchy and prejudice: Indian women and their cinematic representation. Int. J. Lang. Lit. Linguist. **3**(3), 69–72 (2017). https://doi.org/10.18178/ijlll.2017.3.3.113
6. Rawat, P.S.: Patriarchal beliefs, women's empowerment, and general well-being. Vikalpa **39**(2), 43–55 (2014). https://doi.org/10.1177/0256090920140206
7. Desai, N., Maitreyi, K.: Women and Society in India, 2nd edn. Ajantha Book International (1987)
8. Naaz, I.: Women status in India. Indian Streams Res. J. **2**(I). http://www.isrj.net/PublishArtic les/502.pdf (2012)
9. Thilagam, G.P., Santhi, S., Ratha, R., Paranthaman, G.: Indian women status: a historical perspective. MJSSH **3**(2), 258–266 (2019)
10. George, M.: Religious patriarchy and the subjugation of women in India. Int. J. Interdisc. Soc. Sci. **3**(3), 21–29 (2008). https://doi.org/10.18848/1833-1882/cgp/v03i03/52558
11. Hossen, M.S.: Patriarchy practice and women's subordination in the society of Bangladesh: an analytical review. J. Soc. Sci. Humanit. **2**(3), 53 (2020)
12. Tarkeshwari, N.: Patriarchy and the women. CASIRJ **8**(4), 119–133 (2017)
13. Goldberg, S.: Why Men Rule: A Theory of Male Dominance, 2nd edn. Open Court, Chicago (1993)
14. Becker, M.: Patriarchy and inequality: towards a substantive feminism. Univ. Chic. Leg. Forum **1999**(1), 3 (2015)
15. McCallister, J.: Patriarchal system: definition and overview. https://study.com/academy/lesson/ patriarchal-system-definition-lesson-quiz.html (2015)
16. Facio, A.: What is patriarchy? In: Patriarchy in East Asia, vol. 1988, pp. 5–26 (2013). https:// doi.org/10.1163/9789004247772_003

17. Sivakumar, I., Manimekalai, K.: Masculinity and challenges for women in Indian culture. J. Int. Women's Stud. **22**(5), 427–436 (2021)
18. Gandhi, P., Tilak, S.: Masculinity and trapped in patriarchal: looking at gender stereotypes and male domination using popular Indian media and scholarly discourse, p. 10 (n.d.)
19. De Lima, G.T.: Definitions of patriarchy: a literature review (2015)
20. Vallon, S.: Women's education in India: what you need to know. The Borgen Project. https://borgenproject.org/womens-education-in-india/ (2021)
21. Lohia, A.: A short history of education for women in India, pp. 1–6 (2020)
22. Johnson, M.P., Leone, J.M.: The differential effects of intimate terrorism and situational couple violence: Findings from the national violence against women survey. J. Fam. Issues **26**(3), 322–349 (2005). https://doi.org/10.1177/0192513X04270345
23. Jayachandran, S.: The roots of gender inequality in developing countries (2015). https://doi.org/10.1146/annurev-economics-080614-115404
24. Mawa, B.: Challenging patriarchy: the changing definition of women's empowerment. Soc. Sci. Rev. **37**(2), 239–264 (2021). https://doi.org/10.3329/ssr.v37i2.56510
25. Ritumbara, T., Rekha, T.: Patriarchal system and the condition of women. Notions **7**(4), 1–6 (2016)
26. UKEssays.: The status of women in patriarchal India. https://www.ukessays.com/essays/sociology/the-status-of-women-in-patriarchal-india-sociology-essay.php?vref=1 (2018)
27. Avni, A.: Trickle-down wage: analysing Indian inequality from a gender lens. https://www.orfonline.org/expert-speak/trickle-down-wage-analysing-indian-inequality-from-a-gender-lens/ (2022)
28. Rustagi, P.: Understanding gender inequalities in wages and incomes in India. Indian J. Labour Econ. **48**(2), 319–334 (2005)
29. Fred, K.: Negative effects of patriarchy on the wellbeing of women in the society today, p. 3. https://www.academia.edu/33600191/NEGATIVE_EFFECTS_OF_PATRIARCHY_ON_THE_WELLBEING_OF_WOMEN_IN_THE_SOCIETY_TODAY (2017)
30. Deka, P.P.: Education and women empowerment: co-partners in social development. Int. J. Appl. Res. **1**(9), 149–152 (2015)
31. Chandio, A.R., Ali, M.: A critical analysis of women empowerment and social issues in Sindh: a case study. Int. J. Eng. Inf. Syst. (IJEAIS) **5**(4), 86–91 (2021)
32. Law, J., Kirkpatrick, A.E.: Films, screens and cassettes for mammography. Br. J. Radiol. **62**(734), 163–167 (1989). https://doi.org/10.1259/0007-1285-62-734-163
33. Gumede, V.: Economic empowerment of women. In: Political Economy of Post-Apartheid South Africa, pp. 55–62 (2019). https://doi.org/10.2307/j.ctvh8r1rm.12
34. Hunt, A., Samman, E.: Women's economic empowerment: navigating enablers and constraints, Issue September (2016)
35. DAC Network on Gender Equality.: Women's economic empowerment, vol. 9789264168, Issue April (2012). https://doi.org/10.1787/9789264168350-6-en

Development of Asset-Based and Asset-Backed Sukuk Issuance: Case of Malaysia

Abdulmajid Obaid Saleh and Mohammed Waleed Alswaidan

Abstract The purpose of this study is to explain the concept of sukuk, show the benefits of sukuk issuance in the economy, and highlight some key features of sukuk markets as opposed to bond markets. The study also explored two types of sukuk: asset-based sukuk and asset-backed sukuk. The sukuk market is one of the alternative tools for investment and fundraising activities around the world due to its flexibility and the Sharia compliance mechanisms associated with its application.

Keywords Issuance of sukuk · Case of Malaysia · Asset-based sukuk and asset-backed sukuk · Corporate and government sukuk

1 Introduction

Sukuk have increasingly expanded their share of the global market over the last few decades. The world sukuk market, which was first established mainly in countries with Muslim-majority populations, has seen significant growth in the last ten years, with several issuances from leading companies and several government issuances.

A sukuk is a security structured in a Shariah-compliant manner that aims to achieve returns that are comparable or superior to traditional fixed income products such as bonds. Sukuk is one of the most important funding tools in modern Islamic banking and finance. However, not only Islamic countries but also many other countries such as Japan, UK etc. are interested in issuing sukuk. However, the study analyzes the volume of sukuk worldwide and the world's largest sukuk market, Malaysia. Another objective of this study is to understand the viability of the sukuk market and how it performs differently from government and corporate bonds.

A. O. Saleh (✉)
IIUM Institute of Islamic Banking and Finance, Kuala Lumpur, Malaysia
e-mail: alamri@iium.edu.my

M. W. Alswaidan
ETF Bureau, New York, USA
e-mail: mswaidan@etfbureau.com

© The Author(s), under exclusive license to Springer Nature Switzerland AG 2024
A. Hamdan and E. S. Aldhaen (eds.), *Artificial Intelligence and Transforming Digital Marketing*, Studies in Systems, Decision and Control 487,
https://doi.org/10.1007/978-3-031-35828-9_30

351

In addition to presenting the volume of sukuk issuances, especially in 2021, and how the Sukuk industry has continued to grow during the COVID-19 Pandemic. The research method used is qualitative by analyzing the issuance of corporate sukuk and bonds outstanding from 2016 to 2020. The study also looks at key players in the sukuk market, analyzing Malaysian Securities Commission data to compare the feasibility of domestic sukuk and bond issuance.

2 Literature Review

2.1 Sukuk and Its Differences from Conventional Bond

According to the Securities Commission of Malaysia (SCM), a sukuk is an equivalent denoting an undivided ownership or investment in assets using Shariah principles and ideas endorsed by Malaysia's Shariah Advisory Council (SAC) is a certificate of value. Sukuk is purposely built to provide investors with a benefit compared to asset-backed securities. They are issued and sold in accordance with Sharia rules prohibiting the payment of liba or interest. Unlike bondholders, sukukholders receive ownership of an asset or corporate finance and the rate of return depends on the income generated by the underlying asset [1].

According to Abubakar [2], the use of sukuk is a major corporate strategy, such as multinational corporations, global development companies, and government agencies, to raise funds that would be impossible to meet through individual investors alone. The establishment of a Sukuk Fund is a powerful tool to mobilize resources in accordance with the principles of Islamic Sharia and Maqasid Syariah. Also, a big issue for many investors is whether there are diversified income streams available from multiple investment channels that can also be liquidated quickly to meet needs as they arise. This fear is overcome if sukuk is used because there are opportunities in the secondary market for sukuk trading wherein investors can sell their sukuk in whole or in part to get cash, including profits if the investment is profitable.

Additionally, Sukuk is a great risk management option. Islamic securities can be issued with varying levels of risk and reward, allowing investors to customize their portfolios to suit their risk management needs. Sukuk risk is much more difficult to manipulate artificially than other forms of asset. This is because the value and risk of sukuk are always tied to a real asset that proves its value through its physical or virtual existence, rather than debt and interest payments like traditional bond markets [3].

There are many aspects that differentiate Sukuk from conventional bonds, according to Zolfaghari [4], Sukuk represents ownership of a particular asset, but bonds denote a financial obligation, whereby the bond issuer and buyer have a relationship like that of a lender and borrower. Bonds are issued using cash as collateral, whereas Sukuk is issued with physical and real assets. Moreover, according to the definition of Sharia, the assets subject to sukuk must be approved by the Sharia

Supervisory Committee and must be an asset that can be bought and sold, while conventional bonds may be backed by assets, those assets that are not recognized by Sharia.

The major players in the global sukuk market are the Gulf Cooperation Council countries and Southeast Asian countries. The United Arab Emirates, Saudi Arabia, Bahrain, Qatar, and Kuwait are among his GCC countries that display sukuk most commonly. However, Malaysia is a major player in Southeast Asia. Besides Malaysia, Indonesia and Singapore also contribute to the global sukuk market from this region. Other Asian countries such as Pakistan, Japan, USA, UK, Germany, Turkey, Egypt, Gambia also contribute to the market.

Furthermore, according to the International Islamic Financial Markets [5] report, the improvement in the quality of the sukuk market has led to the emergence of the sukuk market as an alternative source of finance and the growth of the market from Europe to Asia, to the Arab Gulf states, the British Commonwealth, Africa, North America.

2.2 Understanding Different Types of Sukuk

Sukuk can be classified into four different categories based on their structure: asset-based, asset-backed, exchangeable and hybrid sukuk. However, in this section, the discussion will deal with asset-based and asset-backed instruments.

Asset-based sukuk require the issuer to own the underlying assets and use the funds from the issued securities (sukuk) to invest, trade or lease them on behalf of investors (sukuk holders). there is. This arrangement is commonly known as a developer sale and lease, in which the developer makes an enforceable promise to buy back the underlying asset when it expires. Only sukuk holders can force the originator to acquire the underlying in this form. As a result, Sukuk owners have an unsecured claim against the author contained in the purchase price to be paid upon performance of binding purchase obligations.

This means that sukuk holders cannot fully rely on the underlying asset and cannot be used as collateral. Asset-based sukuk give ownership to sukuk holders only to beneficiaries, so investors have no claim to assets in the event of default. The originator typically transfers beneficial ownership of the intended issuer to the investor. Meanwhile, sukuk owners have the right to transfer their assets under Islamic law [4].

In an asset-backed sukuk, sukuk holders receive an interest in a physical asset or business venture in exchange for a portion of the overall risk. In this structure, there is a true sale transaction where the originator sells the underlying assets to a special purpose vehicle (SPV), the SPV holds the assets and issues sukuk. Sukuk purchasers have no right to claim against authors if payments are less than expected. In the event of default or liquidation, a genuine sale means that the issuer's assets are not added to the issuer's assets. If the sukuk asset is damaged, the sukuk holder must bear the loss. As a result, asset-backed sukuk are less popular in the sukuk market because they are more like stocks than liabilities [4].

3 Result and Analysis

3.1 Global Sukuk Issuances and Outstanding

According to Bashar [6], the total value of sukuk issued in 2021 increased by 36% annually to $252 billion. The issuance was controlled by central banks, governments, and international entities. Despite hurdles with AAOIFI regulations, Covid-19 disruptions, and increased oil costs, this has been achieved. Sukuk issued in local currencies accounted for 80% of the total issuance. In 2021, the major GCC jurisdictions Malaysia, Indonesia, Turkey, and Pakistan issued US$230 billion in Sukuk issuances. Sukuk has also been issued by non-core sovereigns in the market such as the United Kingdom, the Maldives, and Nigeria. In 2021, Fitch Ratings rated the Sukuk at a total of $132 billion, of which 80% is investment grade. Negative prospects declined to 8.8% in the fourth quarter of 2021 from 23.4% in the fourth quarter of 2020, while the percentage of sukuk issuers increased.

Furthermore, sukuk issuance is expected to increase by 10% over the next five years, reaching $290 billion by 2026. Malaysia, Saudi Arabia, Indonesia, Turkey, and Kuwait are in the top five with 90% of sukuk emissions. Total emissions in Q3 2021 [7].

In 2021, the value of global sukuk outstanding amounted to 711 billion US dollars, an increase of 12.7% over the previous year. Green and sustainable sukuk volume increased by 17.2% to $15 billion in 2021, with the trend expected to continue in 2022 [6]. Moreover, in the third quarter of 2021, the value of global sukuk outstanding was $7775.4 billion, an increase of 2.8% over the second quarter of 2021. The outstanding amount of Fitch-rated sukuk was $132.2 billion, with a ratio of 79.2 in cent of the issue as an investment [8].

3.2 Size of Islamic Capital Market in (Malaysia)

First, Islamic Capital Markets (ICM) won his 59.19% of the total Malaysian market. His market capitalization in 2017 was RM1.893 trillion, up from RM1.692 billion in the previous year. This includes a market capitalization of RM1134 billion in Shariah-compliant shares and RM760 billion of sukuk outstanding (Fig. 1).

Additionally, in 2018, the total volume of Islamic Capital Markets (ICM) in Malaysia was RM1.881 billion. However, the sukuk market was about 60.55% of the total volume with a volume of RM844 billion. It is worth noting that the share of the sukuk market in 2019 witnessed a significant increase, as it constituted 63.57% of the

Fig. 1 Size of the Islamic capital market

total volume of the Islamic capital market, amounting to about RM939 billion of the market value, compared to the market value of other instruments that are compliant with the provisions of Islamic law. Which was about RM1097 billion. In 2020, the sukuk market recorded the highest percentage of the total Islamic capital market at 65.085% with a value of RM 1017 billion. However, the size of the Sharia-compliant securities market was RM1238 billion, and the total value of the Islamic capital market was RM2256 billion.

3.3 Number of Shariah-Compliant Securities and Share of Market Capitalization

Figure 2 shows the percentage of the total Islamic capital market and securities compared to the securities listed in Malaysia and the percentage of Islamic capital market capitalization compared to the total market capitalization.

In 2017, these shares increased to 688 Shariah-compliant securities, accounting for 76.19% of the 903 securities listed on Bursa Malaysia. The market capitalization of Shariah-compliant securities reached RM1134 billion, accounting for 59.46% of the total market capitalization, up 10.02% from the end of 2016. However, the number of Shariah-compliant securities increased by one in 2018, and the number of Islamic Shariah-compliant securities also increased. The total number of listed securities decreased to 75.3% of the 915 securities listed on the Malaysian Stock Exchange. Market capitalization decreased to RM103.7 billion, representing 60.96% of total market capitalization. The number of Shariah-compliant securities increased significantly in 2019, reaching 714, accounting for 76.86% of the total of 929 securities listed on Bursa Malaysia. Moreover, the market capitalization of Shariah-compliant securities reached RM1.97 billion, accounting for 64.06% of the total market capitalization, up 5.80% from the end of 2018.

Fig. 2 Number of Shariah-compliant securities and share of market capitalization

In 2020, Shariah compliant securities increased from 714 to 742 at the end of 2019, accounting for 79.27% of the 936 securities listed on Bursa Malaysia. The market capitalization of Shariah-compliant securities reached RM1.239 billion, representing 68.15% of the total market capitalization, up 12.94% from the end of 2019.

4 Corporate Sukuk Issuances and Outstanding

Table 1 shows the volume of corporate Sukuk issuance in Malaysia between 2016 and 2020 and compares it to corporate bonds.

In 2016, corporate bond sukuk accounted for 75.68% of corporate bond issuance and sukuk value. The total issue of corporate bonds and sukuk was RM85.65 billion. However, corporate sukuk issuance amounted to about RM64.82 billion of total issuance. Even though the value of corporate sukuk rose to MYR87.65 billion in 2017, the ratio of sukuk to total bond and sukuk issuance fell sharply to 70.19%.

Furthermore, the share of sukuk in total bond and sukuk issuance continued to decline in 2018, reaching 68.92%. However, the value of corporate sukuk has fallen

Table 1 Corporate sukuk issuance

Corporate sukuk issuance					
Total issuance (RM billion)	2016	2017	2018	2019	2020
Sukuk issuance	64.82	87.65	72.68	102.39	76.98
Total corporate bonds and Sukuk issuance	85.65	124.88	105.45	132.82	104.58
Percentage of sukuk to total issuance of corporate bonds and sukuk (%)	75.68	70.19	68.92	77.09	73.61

Development of Asset-Based and Asset-Backed Sukuk Issuance: Case ... 357

Table 2 Total sukuk outstanding (RM billion)

Total sukuk outstanding (RM billion)	2016	2017	2018	2019	2020
Sukuk outstanding	393.45	454.49	497.21	555.50	593.43
Total corporate bonds and sukuk outstanding	532.76	604.88	654.11	698.04	732.39
Percentage of sukuk to total issuance of corporate bonds and sukuk (%)	73.85	75.14	76.01	79.58	81.03

to RM72.68 billion. It should be noted that corporate sukuk issuance increased significantly in 2019, reaching RM10.239 billion, and the ratio of sukuk to total sukuk and bonds issued in that particular year rose to 77.09% there is. However, this value fell again in 2020, reaching 73.61%, with the total amount of corporate sukuk issued this year at RM76.98 billion.

Table 2 shows sukuk and bonds issued in Malaysia between 2016 and 2020. In 2016, corporate bond sukuk 73.85% of total issued corporate bonds and sukuk. However, total outstanding of corporate bonds and sukuk in 2016 was RM532.76 billion and total corporate bond issuance in 2016 was RM393.45 billion. In addition, corporate sukuk outstanding increased significantly in 2017, reaching RM45.449 billion with a 75.14% share. Moreover, the value of outstanding Sukuk continued to increase in 2018, 2019 and 2020, reaching RM49.721 billion, RM55.550 billion and RM59.343 billion respectively. The ratio of sukuk to total outstanding corporate bonds and sukuk has continued to increase since then, reaching 81.03% in 2020.

5 Government and Corporate Sukuk Issuances and Outstanding

Table 3 shows the values of government and corporate Sukuk issuance from 2016 to 2020 in Malaysia.

In general, the value of government and corporate sukuk increased during the mentioned years, with the value of sukuk issuances reaching RM129.45 billion in 2016 and ending at RM223.94 billion in 2020. However, the total value of government, corporate and corporate sukuk and bonds increased from RM240.56 RM1 billion in 2016 to RM385.93 billion in 2018, then the total value decreased by about 1

Table 3 Government and corporate sukuk issuance

Government and corporate sukuk					
Total issuance (RM billion)	2016	2017	2018	2019	2020
Sukuk issuance	129.45	168.68	199.90	235.20	223.94
Total bonds and sukuk issuance	240.56	317.94	385.93	384.85	366.67
% of sukuk to total bonds issuances (%)	53.81	53.05	51.80	61.11	61.07

Table 4 Government and corporate sukuk outstanding (RM billion)

Total sukuk outstanding (RM billion)	2016	2017	2018	2019	2020
Sukuk outstanding	661.08	759.64	844.21	938.96	1017.79
Total bonds outstanding	1172.91	1291.91	1405.78	1490.28	1609.01
% of Sukuk to total bonds outstanding (%)	56.36	58.80	60.05	63.01	63.26

billion in 2019 and reached RM366.67 billion in 2020. However, the total percentage of government and corporate sukuk decreased from 53.81 to 51.80% between 2016 and 2018 then it increased to 61.07% in 2020.

Table 4 includes the value of government and corporate sukuk and bonds based in Malaysia from 2016 to 2018. In general, the value of governmental and corporate sukuk has been increasing exponentially over the years. In 2016, the value was RM661.08 billion, however, it has reached RM1017.97 billion in 2020. Moreover, the total value of government bonds and outstanding bonds continued to rise from 2016 to 2020. However, the value increased at around RM400 billion, while, in 2016, the value was RM1172.91 billion and reached RM1609.01 billion in 2020. Similarly, the percentage of government and corporate sukuk to total bonds outstanding which has increased significantly over the years and reached to 63.26% in 2020.

6 Conclusion

The Sukuk market is one of the alternative tools for investment and fundraising activities worldwide, due to its flexibility as well as the Sharia-compliance mechanism associated with its applications. It demonstrated the issuance of asset-backed sukuk and the extent of the dynamics of the sukuk market, however, the sukuk market has certain rules that the issuer must adhere to in other countries to achieve a successful sukuk issuance process. The asset to be used in the sukuk structure must be approved by the Shariah Supervisory Board indicating that not all assets are eligible to be used for fundraising in the sukuk market, and the assets to be used meet all the provisions of Islamic Shariah as well. As with other legal requirements, it is accepted as the underlying asset of the Sukuk. The sukuk asset is considered one of the valuable assets that can be sold, bought, as well as rented in the sukuk market, as the asset will not be accepted as a sukuk asset without a valid usufruct right. Sukuk developed and it was also learned that the Malaysian Sukuk market has become the leading Sukuk market due to the strict Sharia complaint mechanism as well as the legal framework governing Sukuk activities in the country.

References

1. Bixmalaysia.: What is sukuk. Retrieved from https://www.bixmalaysia.com/learning-center/articles-tutorials/what-is-sukuk (2019)
2. Abubakar M.: Sukuk–meaning, benefits and challenges. Retrieved from https://saraycon.com/sukuk-meaning-benefits-and-challenges/ (2019)
3. Naveed M.: The main benefits that result from sukuk. Retrieved from https://www.sukuk.com/article/main-benefits-result-sukuk-29/#/?playlistId=0&videoId=0 (2014)
4. Zolfaghari, P.: An introduction to Islamic securities (Sukuk) (2017)
5. IIFM Newsletter.: Issue 1, p. 7 (2017)
6. Fitchratings.: Promising outlook for global sukuk market following robust 2021. Retrieved from https://www.fitchratings.com/research/islamic-finance/promising-outlook-for-global-sukuk-market-following-robust-2021-11-01-2022 (2022)
7. Refinitiv.: Global sukuk issuance to hit $180 billion by end of 2021. Retrieved from https://www.salaamgateway.com/story/global-sukuk-issuance-to-hit-180-billion-by-end-of-2021 (2021)
8. Fitchratings.: Sukuk momentum to stay intact, short-term headwinds to persist. Retrieved from https://www.fitchratings.com/research/non-bank-financial-institutions/sukuk-momentum-to-stay-intact-short-term-headwinds-to-persist-13-10-2021#:~:text=Global%20outstanding%20sukuk%20reached%20USD775,of%20issues%20being%20investment%20grade (2021)

Role of Colours in Web Banners: A Systematic Review and Future Directions for e-Marketing Sustainability

Khalid Ali Alshohaib

Abstract Empirical studies investigating the impact of colour hue, saturation, and value on consumers' purchase intention and emotional state of mind, are widely addressed in scientific experiments, from advertising to sociology and psychology. However, the evidence base for the web banner advertisement from colour hue, saturation, and value perspective impact enrichment on consumers' pleasure, arousal, and dominance emotional state of mind has not been widely investigated in marketing literature. The present study, which considers the colour combination as one significant factor in designing web banner ads, draws together evidence on the impact of the web banner advertisement with the combination of colour hue, saturation, and value impact on consumer' emotional state of mind and purchase intention under the umbrella of "Ensure sustainable consumption and production patterns" 12th sustainable development goals. With no robust models currently available to evaluate how colour hue, saturation, and value combinations of web banner advertisement impacts consumers' purchase intention and emotional state of mind, this systematic review consolidates the research that has measured correlates of web banner ads in colour hue, saturation, and value measures the consumers' emotional state of mind. This study aims to assess how previously published studies investigated and concluded the impact of colour hue, saturation, and value on consumers' purchase intention and emotional state of mind. The systematic review found five studies met full-text selection criteria. Following the discussion from the reviewed papers, the researcher argues that it is necessary to consider the colour hue, saturation, and value to understand better the isolated colours' impacts on consumers' purchase intention and emotional state of mind. In this regard, to summarise those papers' main findings and future directions, the researcher describes novel approaches and insights to explore further how colour hue, saturation, and value can contribute to developing web banner advertisements in the marketing context. This may contribute to the literature and practice about developing web banner advertisements following the assessed colours.

K. A. Alshohaib (✉)
Jeddah College of Advertising, University of Business and Technology, Jeddah 23435, Saudi Arabia
e-mail: k.alshohaib@ubt.edu.sa

© The Author(s), under exclusive license to Springer Nature Switzerland AG 2024
A. Hamdan and E. S. Aldhaen (eds.), *Artificial Intelligence and Transforming Digital Marketing*, Studies in Systems, Decision and Control 487,
https://doi.org/10.1007/978-3-031-35828-9_31

362 K. A. Alshohaib

Keywords Color hue · Saturation · Value · e-Marketing · Web banners · Ensure sustainable consumption and production patterns goal

1 Introduction

The diverse link between colours and purchase intention has been studied broadly across the most scholarly work and scientific experiments, from advertising to sociology and psychology [1–6]. Thus, the impact of a brand's logo, packing, and store colours (i.e., hue, saturation, and value) on consumers' emotional state of mind (i.e., pleasure, arousal, dominance) and purchase intention is still in its early stages of empirical investigation, largely in the marketing literature [6–8]. Mainly associating colour hue, saturation, and value with consumers' emotional reactions in online shopping [9] and in-store shopping [10]. Accordingly, Williams and Son [11] concluded that the different colours affect the consumers' attitudes and intentions to buy.

First, the colour hue (i.e., red, yellow, blue, and green) is determined by the wavelength of light [5]; yellow and red hues are also categorized as warm colours [12], while blue and green are categorized as cool colours [13]. Furthermore, the red colour hue encourages strength, passion, energy, intolerance, anger, and fear [5, 14]; the yellow colour hue elicits joy, happiness, and fear [15], blue colour hue determines the confusion, trust, confidence, confusion, and sadness [16], and green colour hue predicates faith and greed [17]. Therefore, from a physiological perspective, web banners in warm colours affect users' perception from different points of view, such as; eye blinking, maximising blood pressure, and developing high tension-based emotions [6, 18]. Similarly, Moore et al. [19] concluded that web banners in cool colours positively attract consumers' intentions by developing a pleasurable emotional state.

Second, web banners in colour saturation are further categorized into higher and lower saturation, where high saturation colours indicate vivid colours [20] and lower saturation emphasizes dull colours [21]. Prior studies highlighted that the high saturation colour of web banners positively links with the consumers' emotional state of mind [17, 21–23] consumers' feel excited while experiencing web banners in high saturation colours [22]. Similarly, White et al. [6] state that vivid web banners express a virtual product's reality, which tends to impact consumers' emotional state of mind and also plays an important role in changing consumers' intention to buy.

Third, web banners in colour value are categorized into high and low [23] higher-value colours contain bright colours, i.e., white, which is associated with equality, simplicity, and purity [5]; lower-value colours are relatively darker, i.e., black, which is associated with mystery, sophistication, and strength [6, 24]. Prior studies reported that the online shopping environment in high and low-value colours failed to build positive consumers' perceptions and emotional states of mind [9, 25]. In this regard, De Bortoli and Maroto [21] debate, most probably, that companies are not preferring to develop web banners using high or low-value colours. On the other hand, Huang and Lu [4] argue that designing and determining a real store environment following the value of colour positively impacts consumers' emotions.

Due to the novelty of this area, a lack of studies shed light on concluding how colour combinations (i.e., hue, saturation, and value) impact consumers' emotional state of mind and purchase intention [3, 26]. Previously, retailers and advertisers paid less attention to comprehend the role of colours in designing web banners advertisements [24]; thus, in the context of marketing, studies on colours and consumers' phycological functions are lacking to confirm a particular phenomenon of colours designing the web banners [27]. Accordingly, Lin [28] suggests that the researcher examines and confirms the impact of colours on consumers' emotional state of mind and purchase intention. More recently, White et al. [6] investigated the role of colours in developing web banner advertising and its impact on consumers' perceptions and emotions. For instance, they do not study the colour hue, saturation, and value separately and broadly. Thus, the area is still lacking in addressing the impact of colour hue, saturation, and value on consumers' emotional state of mind and purchase intention.

However, the present study aims to contribute to the literature by systematically reviewing the colour hue, saturation, and value combinations applied to the web banner advertising literature, identifying the contextual gaps, and consequently, addressing the future study direction in this domain. The main goal of the present study is to systematically explore and understand the possible impact of colour hue, saturation, and value on consumers' emotional state of mind and purchase intention. This study is structured as follows. Section 2 discusses the methodology. Section 3 presents the findings extracted by the systematic review process. Lastly, discussion on the findings and future research directions are discussed.

2 Methodology

To determine the aim of the present study, the researcher includes five articles that investigated and concluded the link between colour (hue, saturation, and value), perception and emotional state of mind of consumers [3–6, 27]. First, Wilms and Oberfeld [27] conducted a study in Germany of 62 respondents to examine the impact of colour hue, saturation, and brightness on the respondents' emotions. The authors employed ANOVA test approaches to predicate the impact of colour hue, saturation, and value on the consumers' emotional state and confirmed positive impact. Second, Kim et al. [5] conducted a study in China; the authors aimed to investigate the impact of luxury hotels' colour hue, saturation, and value on consumers' emotional state of mind and aesthetic perception. The authors applied ANCOVA and structural equation modelling to test the proposed hypotheses. Third, Huang and Lu [4] studied 114 students in Canada and USA. The study aimed to investigate the impact of food packing colours and health perception on purchase intention. The researchers applied a linear regression approach to test the hypotheses. Fourth, a quantitative study was carried out by Countryman and Jang [3] in the USA. They aimed to investigate and conclude the impact of light colours combination and styling decoration of hotels' lobbies on the consumers' impression. The authors applied the structural equation modelling approach to obtain the study's findings. Finally, White et al. [6] gathered data from 409 respondents in Portugal. The authors intended to shed light on the effect of colour hue, saturation and value on the consumers' attitude and purchase intention towards web banner advertising. Therefore, the summary of the key studies is presented in Table 1.

Table 1 Summary of the key studies

Authors	Findings	Colours studied	Future direction
Wilms and Oberfeld [27]	Colour hue, saturation, and value positively impact consumers' arousal and emotional state of mind; more importantly, saturation colours have a systematic effect on the emotional state of a person viewing the colour	27 chromatic colours and 3 brightness categorized in hue, saturation, and value	Future studies may contribute by examining the impact of a broader range of other hue colours, such as purple or yellow, and wider ranges of high and low saturation on consumers' emotions
Kim et al. [5]	The results indicate that the bright and muted colour hue play an important role in predicate the emotional state of mind and aesthetic perceptions	Green, red, blue, and yellow	Future studies are encouraged to investigate the colour hue, saturation, and value effect on consumers' pleasure, arousal, and dominance emotional state in a physical environment
Huang and Lu [4]	Food packed in blue-light and red-light colours positively impact consumers' intention to buy	Red, blue, regular/ whole, light/ skim	In the future would be possible to examine the impact of food packing and the food itself on the utilization intention of consumers
Countryman and Jang [3]	Light colours combination and styling decoration enhance the impression of the store in the physical environment	High saturated colours	Scholars are suggested to use different colour combinations in-store environment and investigate their impression
White et al. [6]	Findings reveal that the blue-green colours hue tend to elicit higher levels of consumers' attitude towards web banner ad and also enhance the purchase intention	Orange, red, blue, green, yellow, and purple	Future studies may examine the impact of blue-purple and blue-orange colour hue impact on consumers' attitudes towards banner ads and purchase intention

3 Discussion

The systematic review of previous empirical studies concluded that the colour hue, saturation, and value of online and offline shopping environments play an important role in associating with consumers' emotional state of mind (i.e., pleasure, arousal, and dominance) and purchase intention. Furthermore, the researcher emphasises that this area of colour combinations is still young in web banner advertising design. A lack of scholarly work addresses how different combinations of colours could develop a strong link between observed consumers and the products and/or services [18]. However, Pelet and Papadopoulou [8] conclude that web banner ads' colour hue, saturation, and value play an important role in developing a positive attitude and intention to buy. Similarly, Cheng et al. [9] affirm that the colour combination in web banners widely changes the consumers' online shopping mood, which ultimately links with consumers' emotional state of mind.

Noticeably, White et al. [6] highlight that the blue-green combination of web banners enhances the consumers' confidence level and agents to develop a positive attitude that fundamentally ends with buying. However, authors make wide claims that the still impact of colours on pleasure, arousal, and dominance emotional state is unknown, mainly in the marketing literature. Therefore, an empirical study by Kim et al. [5] confirms that a physical store environment in light colour hue (i.e., blue and red) significantly influences consumers' pleasures state of mind. Thus, still, no study confidently claims the impact of the light colour hue of web banners on consumers' pleasure, arousal, and dominance state of mind. Similarly, Wilms and Oberfeld [27] examined and claimed that colour saturation stunningly impacts consumers' arousal emotional state. However, considering the current competition in the retail sector, retailers need to understand the wide role of different colour hues, saturation, and value on consumers' emotional state of mind and then accordingly design and launch the web banners and/or other online and/or instore marketing campaigns [6, 18, 25].

Moreover, the systematic review process reveals several gaps in the understanding of the mechanisms of colours combination that associate consumers with products and/or services in a virtual shopping environment. However, several studies investigate and claimed the different impacts of colour hue, saturation, and value on the consumers' emotional state of mind. Thus, what remains unclear is how different colour combinations (i.e., hue, saturation, and value) play an important role in encouraging a different emotional state of mind (i.e., pleasure, arousal, and dominance) in the online shopping environment. The scholarly work in the field of colour combinations has focused on psychological studies rather than marketing [29–32]. Considering the current rareness of literature on the mind, the researcher suggests future works and limitations stated below:

4 Future Studies and Limitations

The present study faced a number of constraints and challenges, which may provide guidelines for future research. The following section clarifies limitations that the researcher encountered, besides recommendations for the future. The systematic review analysis highlighted that: (i) to date, limited scholars have focused on conducting studies in this field; (ii) empirical studies have to employ narrow small sample sizes with little diversity in participants and; (iii) there is a wide array of practical and methodological presentation procedures being employed. In conjunction with these limitations, as yet, no study investigated the impact of separate colour hue, saturation, and hue on consumers' emotional state of mind (i.e., pleasure, arousal, dominance) towards web banner advertising. Recently, White et al. [6] revealed that web banners in a blue-green hue stunningly develop consumers' positive attitudes, ultimately enhancing their purchase intention. Thus, no study investigated the impact of web banner ads in different applications of colour hue, saturation, and value on consumers' perception and pleasure, arousal, and dominance emotional state of mind. Accordingly, Kim et al. [5] encourage researchers to investigate and conclude how colour hue, saturation, and value impact consumers' pleasure, arousal, and dominance emotional state of mind. Accordingly, Wilms and Oberfeld [27] suggest that future researchers should encounter the impact of the wide range of colour hues and high and low saturation colours on consumers' emotional state of mind and purchase intention. More importantly, Cheney et al. [33] suggest that researchers should also test the colour blindness of the participants through the "Ishihara test" to avoid the biases of the study's overall findings.

References

1. Kaya, N., Epps, H.H.: Relationship between color and emotion: a study of college students. Coll. Stud. J. **38**(3), 396–405 (2004)
2. Gilbert, A.N., Fridlund, A.J., Lucchina, L.A.: The color of emotion: a metric for implicit color associations. Food Qual. Prefer. **52**, 203–210 (2016)
3. Countryman, C.C., Jang, S.: The effects of atmospheric elements on customer impression: the case of hotel lobbies. Int. J. Contemp. Hosp. Manage. (2006)
4. Huang, L., Lu, J.: The impact of package color and the nutrition content labels on the perception of food healthiness and purchase intention. J. Food Prod. Mark. **22**(2), 191–218 (2016)
5. Kim, D., Hyun, H., Park, J.: The effect of interior color on customers' aesthetic perception, emotion, and behavior in the luxury service. J. Retail. Consum. Serv. **57**, 102252 (2020)
6. White, A.R., Martinez, L.M., Martinez, L.F., Rando, B.: Color in web banner advertising: the influence of analogous and complementary colors on attitude and purchase intention. Electron. Commer. Res. Appl. **50**, 101100 (2021)
7. Koo, D.M., Ju, S.H.: The interactional effects of atmospherics and perceptual curiosity on emotions and online shopping intention. Comput. Hum. Behav. **26**(3), 377–388 (2010)
8. Pelet, J.É., Papadopoulou, P.: The effect of colors of e-commerce websites on consumer mood, memorization and buying intention. Eur. J. Inf. Syst. **21**(4), 438–467 (2012)

9. Cheng, F.F., Wu, C.S., Yen, D.C.: The effect of online store atmosphere on consumer's emotional responses–an experimental study of music and colour. Behav. Inf. Technol. **28**(4), 323–334 (2009)
10. Labrecque, L.I., Patrick, V.M., Milne, G.R.: The marketers' prismatic palette: a review of color research and future directions. Psychol. Mark. **30**(2), 187–202 (2013)
11. Williams, A.S., Son, S.: Sport rebranding: the effect of different degrees of sport logo redesign on brand attitude and purchase intention. Int. J. Sports Mark. Sponsorship (2021)
12. Hardin, C.L.: Red and yellow, green and blue, warm and cool: explaining colour appearance. J. Conscious. Stud. **7**(8–9), 113–122 (2000)
13. Greene, T.C., Bell, P.A.: Additional considerations concerning the effects of 'warm' and 'cool' wall colours on energy conservation. Ergonomics **23**(10), 949–954 (1980)
14. Elliot, A.J., Maier, M.A.: Colour-in-context theory. In: Advances in Experimental Social Psychology, vol. 45, pp. 61–125. Academic Press (2012)
15. Wu, C.S., Cheng, F.F., Yen, D.C.: The atmospheric factors of online storefront environment design: an empirical experiment in Taiwan. Inf. Manage. **45**(7), 493–498 (2008)
16. Jenny, B., Kelso, N.V.: Color design for the color vision impaired. Cartogr. Perspec. **58**, 61–67 (2007)
17. Chu, A., Rahman, O.: Colour, clothing, and the concept of 'green': colour trend analysis and professionals' perspectives. J. Glob. Fash. Market. **3**(4), 147–157 (2012)
18. Panigyrakis, G.G., Kyrousi, A.G.: Color effects in print advertising: a research update (1985–2012). Corp. Commun. Int. J. (2015)
19. Moore, R.S., Stammerjohan, C.A., Coulter, R.A.: Banner advertiser-web site context congruity and color effects on attention and attitudes. J. Advert. **34**(2), 71–84 (2005)
20. Lohtia, R., Donthu, N., Hershberger, E.K.: The impact of content and design elements on banner advertising click-through rates. J. Advert. Res. **43**(4), 410–418 (2003)
21. De Bortoli, M., Maroto, J.: Colours across cultures: Translating colours in interactive marketing communications. In: European Languages and the Implementation of Communication and Information Technologies, pp. 1–27 (2001)
22. Singh, N., Srivastava, S.K.: Impact of colors on the psychology of marketing—a comprehensive over view. Manage. Labour Stud. **36**(2), 199–209 (2011)
23. Pelet, J.E., Conway, C.M., Papadopoulou, P., Limayem, M.: Chromatic scales on our eyes: how user trust in a website can be altered by color via emotion. In: Digital Enterprise Design and Management 2013, pp. 111–121. Springer, Berlin, Heidelberg (2013)
24. Jouttijärvi, S.: The role of creative design in capturing consumer attention with effective banner advertising: an eye tracking approach (2019)
25. White, A.E.R.: Complementary colors and consumer behavior: emotional affect, attitude, and purchase intention in the context of web banner advertisements. Doctoral dissertation (2018)
26. Nitse, P.S., Parker, K.R., Krumwiede, D., Ottaway, T.: The impact of color in the e-commerce marketing of fashions: an exploratory study. Eur. J. Mark. (2004)
27. Wilms, L., Oberfeld, D.: Color and emotion: effects of hue, saturation, and brightness. Psychol. Res. **82**(5), 896–914 (2018)
28. Lin, I.Y.: The interactive effect of Gestalt situations and arousal seeking tendency on customers' emotional responses: matching color and music to specific service \scapes. J. Services Mark. (2010)
29. Aslam, M.M.: Are you selling the right colour? A cross-cultural review of colour as a marketing cue. J. Mark. Commun. **12**(1), 15–30 (2006)
30. Kumar, J.S.: The psychology of colour influences consumers' buying behaviour–a diagnostic study. Ushus J. Bus. Manage. **16**(4), 1–13 (2017)
31. Xi, C., Jingying, L.: The impact of social crowding on color saturation preference: the mediating role of threat to freedom and avoidance behavior. Afr. J. Bus. Manage. **16**(8), 170–181 (2022)
32. Liu, S.Q., Wu, L.L., Yu, X., Huang, H.: Marketing online food images via color saturation: a sensory imagery perspective. J. Bus. Res. **151**, 366–378 (2022)
33. Cheney, K.L., Green, N.F., Vibert, A.P., Vorobyev, M., Marshall, N.J., Osorio, D.C., Endler, J.A.: An Ishihara-style test of animal colour vision. J. Exp. Biol. **222**(1), jeb189787 (2019)

Relationship of People's Knowledge, Behavior and Governmental Action Towards Environmental Sustainability

Marluna Lim-Urubio and Manolo Anto

Abstract This study aims to substantiate that the success of environmental sustainability is related to people's knowledge and behavior towards environment as well as governmental action which of which environmental sustainability is embodied in UNESCO ESD Roadmap 2030. Descriptive research design is used to gather information through survey, guided questions and observations of practices done with the respondents. Data were analyzed quantitatively, where descriptive and linear regression methods were applied. Descriptive statistics measured information such as percentages and means while Pearson correlations and R-Square were employed in testing association between variables. Hypotheses were analyzed through multiple linear regression using Stata Version 17 software. Multiple linear regression is used to understand the strength of the relationship between two or more independent variables and one dependent variable, In this case, the relationship of the independent variables—environmental knowledge, behavior and government action to the dependent variable which is the environmental sustainability. Major findings proved that achieving environmental sustainability will be successful if both government and people will work hand in hand to ensure prevention of environmental degradation. However, on people' part, committing to environmental sustainability and sustainability of action would require knowledge of ecology. Another important factor is people's behavior toward environment which has been supported by other studies as critical for promoting environmental sustainability. With knowledge comes the behavior to choose what is good or detrimental to the environment. People's behavior can be positive or negative where positive attitude are attitudes that lead to environmental preservation thus, improving the quality of environment while negative attitude are actions that harm the environment. Lastly, government actions, are very crucial in promoting environmental sustainability as they are the ones that can impose laws and regulations and further implement them along with sufficient funding in order to protect the environment.

M. Lim-Urubio (✉) · M. Anto
Business Informatics Programme, College of Administrative and Financial Sciences, University of Technology Bahrain, Salmabad, Bahrain
e-mail: mlurubio@utb.edu.bh

© The Author(s), under exclusive license to Springer Nature Switzerland AG 2024
A. Hamdan and E. S. Aldhaen (eds.), *Artificial Intelligence and Transforming Digital Marketing*, Studies in Systems, Decision and Control 487,
https://doi.org/10.1007/978-3-031-35828-9_32

Keywords Environmental sustainability · Single-use plastics · UNESCO SDG · Ecology

1 Introduction

Environmental advocates call for environmental sustainability to address the current state of environment which experiences climate change, pollution, environmental degradation, and resource depletion among others. Since 1970, CO_2 emissions have increased by about 90%, with emissions from fossil fuel combustion and industrial processes contributing about 78% of the total greenhouse gas emissions increase from 1970 to 2011 [1]. Humans are undoubtedly the greatest contributor through negligent actions and behavior, lack of concern, and poor decisions. Humans impact the physical environment in many ways: overpopulation, pollution, burning fossil fuels, and deforestation. Decisions and activities result to changing the earth's atmosphere faster due to burning of fossil fuels in industry and by vehicles that releases carbon dioxide and other greenhouse gases as well as human's personal choices like utilizing single use plastic bags that contribute to pollution and climate change.

One reason why environment deteriorates is pollution. Land pollution takes place due to littering and dumping of household trash which includes a huge volume of plastic bags and inability to do waste segregation. Industrial companies also produce high volume of industrial waste that affect land and water.

Water pollution is caused by spills or leaks from oil and chemical containers, garbage and plastic materials thrown in the bodies of water. Plastic materials have a substantial contribution to pollution as it is one of the most lasting materials that can take hundreds of years to degrade. A study showed that plastic waste makes up 80% of all marine pollution and around 8 to 10 million metric tons of plastic end up in the ocean each year [2].

UNESCO has included goals that are pertaining to environmental sustainability like Sustainable Development Goal 6 (SDG 6)—which is clean water and sanitation, SDG 7—Affordable and Clean Energy, SDG 11—Sustainable Cities and Communities, SDG 13—Climate Action, SDG 14—Life below water and SDG 15—Life on Land [3] is clearly discussed in the Education for Sustainable Development Roadmap 2030.

Countries have responded to this call and have done numerous measures to ensure the sustainability of the environment. The use of sustainable technology which is SDG 7, like solar technology and other renewable energies minimizes burning of fossil fuel [3, 4].

Other countries have built sustainable communities which maintain green spaces while others have been continuously guarding their land, water and air against pollution through careful watch on the disposal of plastic items [5]. Despite all the actions, pollution proved to be a serious issue causing environmental degradation [6].

Air pollution is mainly caused by carbon monoxide emission from vehicles, use of fuel and even natural gas to heat homes, use of power generation to include coal-fueled power plants and also the generation of fumes from chemical production. In addition to this, burning of garbage that include plastic materials also contributes to air pollution. The use and improper disposal of plastic bags results to environmental degradation.

Plastic bags dumped in the garbage bins and landfills release gasses when exposed to heat coming from the sunlight. Worse is, as plastics are exposed to sunlight, these give birth to the so called micro plastic which results to more damaging gasses like methane and ethylene. Plastics stay for many decades. They clog drainage system, rivers, ocean and produce leachate when thrown in the landfills.

Bahrain News Agency reported that the Kingdom of Bahrain intends to stop the utilization of plastic by July 2019. Phasing out of the use of plastic will be done in phases where the initial phase will be on the use of single-use plastic bags. The first phase also include putting a stop the importation of non-biodegradable plastic bags. However, as of this time, stores, malls, groceries and other business establishments are also using plastic bags.

Reports also said that in 2018, Bahrain Airport Company (BAC) which operates the Bahrain International Airport (BIA) has done the recycling of more than 1.2 tons of plastic. Part of the BAC's waste management practice and system is the strengthening of the plastic usage prevention, reusing if possible, recycling and recovering when possible and correct disposal of the plastic waste [7].

In Africa, the Southern African Development Community (SADC) acknowledged that plastic utilization has led to plastic pollution the need an intervention to manage if not totally control its manufacturing and use. One reason for failure or low turn -out of the action is the lack of awareness campaign thus people has no sufficient knowledge on the harm of the use of plastic and the possible alternatives [8].

In Jakarta, Indonesia, it has begun its ban on the utilization of single-use plastic bags in malls, centers, markets in order to reduce the plastic waste. Thirty three cities and regions to include Bali adapted the same measure to reduce the plastic waste equal to 34% of 7702 tons/day waste [9].

Guatemala, Ecuador, Guyana, Haiti, Mexico, Panama and Colombia did the same. They have banned and put levy in the use of disposable plastic bags. While Colombia became a bit successful in the reduction of usage of plastic [10].

In May 18, 2018 as published in UN Environmental Program [11], the "Single-Use Plastics: A Roadmap for Sustainability" newsletter, where members of the European Commission forwarded its proposal to ban the use of single use plastic.

Zimbabwe and Kenya imposed total ban on use, production and selling of single use plastic and a violation of this in Kenya will be penalized with imprisonment, while Taiwan has restricted the use of plastic bags, straws, cups and even utensils. Australia, Tasmania and Queensland imposed a statewide ban on the utilization of single—use plastic while Rwanda has totally banned its use [12].

For Italy who has banned the use of single use plastic in May 1, 2018 and imposed a fine for violation of this with an amount of 500 euros, people became very cautious and careful in ensuring that they will not violate the rule. In 2011, the usage of plastic bag has decreased by 55% [13, 14].

In Greece, after the levy, the plastic bag consumption is reduced by 75 to 80%, and the sale and use of reusable eco bags have also strengthened [15].

In Netherlands, the consumption of plastic bags has been reduced by 40% after a levy has been put on the plastic bags while in Portugal, the reduction was 74%. Also in Portugal, the reusable plastic bags were made exempted from the levy.

The problem on single use plastic directed countries like UK, Spain, India, and France to ban specific categories of single-use plastics. Canada is set to join these countries in implementing the ban on plastics where the sale of plastic cutleries, plastic grocery bags and straws will be stopped after Dec 2023 [16].

In Rio de Janeiro, customers will be offered disposable plastic bags by supermarkets. Rio de Janeiro's state law demands that only recyclable items/bags can be used by the establishments and a campaign will be made to raise the awareness of people about the harmful effects of plastic bags [17].

Morocco adopted a law to reduce plastic pollution through the Zero Mika Law which banned the production, import, sale or distribution of single-use plastic bags [18].

The literatures summarized above confirmed that the utilization of single use plastic bags is destructive to the environment. Daily, people have been exposed to the use of plastic in form of disposable plates, cups, cutleries, coffee cups, plastic carry bag, disposable water bottle and a lot more. Each person uses an average of 350 pcs of plastic bags in a year. This is plastic bags only, what about other plastic materials? Plastic bags are not decomposing. It is not degradable and therefore it stays in the environment for decades.

For plastics to decompose, it takes 20–500 years and it depends on the material itself and its structure. On the other hand, dumping them in landfills exposed to heat will result to emission of methane gas as well as carbon dioxide gases. Moreover, leachate will be produced as garbage in the landfill decompose together with the plastic that will penetrate the soil.

The literatures proved that both people and government have a hand in achieving environmental sustainability. The role of government in environmental sustainability validated the importance of the role in protecting the environment that is beyond simply formulating policies [19]. Furthermore, policies are not enough for the government to address the environmental issues but actions and implementation should go together. Government has the very significant role in environmental sustainability through its policies, policy implementation and collective actions to address the issue [20].

A research validated the crucial role of local governments in policy formulation that aims to protect the environment where it reiterated the role the government has to play in transforming the environment through policies, laws and other actions advocating environmental protection [21, 22].

On the other hand, knowledge on ecology and environmental issues is a prerequisite for a positive environmental attitude. It also confirms that having a comprehensive knowledge on ecology is considered a vital factor for a successful implementation of

Relationship of People's Knowledge, Behavior and Governmental ... 373

ecological programs [8]. Moreover, people's beliefs, values and practices in connection with the surrounding, the nature, the environment or the ecology is related to how they see environmental issues.

The literature served as springboard for the conduct of this study on "Relationship of People's Knowledge, Behavior and Governmental Action Towards Environmental Sustainability" which aims to validate the effect of people's knowledge of ecology and their behavior and the government's action on achieving environmental sustainability.

This study used the Theory of Reasoned Action [23] which indicates that preexisting attitudes of consumers are important in the decision making process where there is a relationship of behaviors and attitude to consumers action. This theory can be linked to the attitude of consumers towards environment. If his desired outcome is to protect the environment, then his choices will definitely show that desire. The Deep Ecology Theory considers humankind as an integral part of its environment therefore, human should be absolutely changing its relationship one that acknowledges the inherent value of nature [24]. It believes that everything and everyone has equal rights to exist and live fully. It also claims that since everything is interrelated, then harming the nature, harming the environment would mean harming ourselves, This theory has naturally linked to aiming for environmental sustainability, avoiding the destruction of environment by the own careless acts of people living in this planet. Now to be able to purse environmental sustainability, people should have environmental sensitivity and knowledge of ecology from which he will get his desire to protect the environment, Human should be made responsible for their behavior and action that can lead to environmental sustainability or environmental destruction.

2 Methods

This is a descriptive study where the researcher collected data in an attempt to describe systematically about a situation, problem, phenomenon, programmed, or provide information towards the issue. This study also used quantitative approach and correlational research in nature to describe the relationship between independent variables towards dependent variable.

In this study, the total population is unknown, therefore, the study follows the criteria given in Krejci and Morgan table, that if the population is maximum of 1,000,000 + or unknown, then 385 sample will be best considered for analysis [25]. The current study used the maximum suggested sample of 385 expatriates who are walk in clients along the Manama area to be the sample of this study but still subject to convenience sampling method.

Data where analyzed quantitatively, where descriptive, and linear regression methods were applied. Descriptive statistics measured information such as percentages and means. Pearson Correlations, and R-Square were employed in testing association between variables. Lastly, the hypotheses were analyzed through multiple linear regression using Stata Version 17 software.

3 Results

This study was conducted to determine the relationship of people's knowledge of ecology and behavior and government's action on environmental sustainability. Multiple regression analysis was used in addition to descriptive statistics such as mean and standard deviation. Correlation Analysis was used to test the strength of relationship between the variables.

Descriptive statistics and correlation. Table 1 presents the mean and standard deviation of all the variables used in this study. Likewise, bivariate correlations are presented. Results of the Pearson correlation indicated that people's actions leading to environmental sustainability ($M = 16.84$, $SD = 3.17$) has a moderate positive relationship with people's knowledge of ecology ($M = 39.18$, $SD = 5.99$, $r = 0.596$, $p < 0.001$), a strong positive relationship with people's behavior ($M = 21.77$, $SD = 3.4$, $r = 0.748$, $p < 0.001$), and a strong positive relationship with government's action ($M = 12.34$, $SD = 2.50$, $r = 0.654$, $p < 0.001$).

The regression model in Table 2 with an F-value of 191.93 and p-value of < 0.001 indicates a significant model that explains the relationships between variables. The r-squared value is 0.6018, which means that 60.18% of the people's actions leading to environmental sustainability can be explained by the model. The remaining 39.82% is due to other factors not captured by the study.

Table 1 Descriptive statistics and Pearson r correlation

	Mean	Standard deviation	Knowledge	Behavior	Government action
Knowledge	39.18	5.99	–	–	–
Behavior	21.77	3.4	0.652**	–	–
Government action	12.34	2.5	617***	0.698***	–
People's actions	16.84	3.17	0.596***	0.748***	0.654***

Note ***$p < 0.001$, **$p < 0.01$, *$p < 0.05$

Table 2 Results of the multiple regression analysis

DV: People's actions	Beta Coeff.	Std. Err.	t	$P > t$
Knowledge	0.1526268	0.05317	2.87	0.004
Behavior	0.596285	0.05714	10.44	< 0.001
Government actions	0.2073355	0.04499	4.61	< 0.001
Constant	0.0954155	0.18432	0.52	0.605

r-squared value = 0.6018, Prob > F (< 0.001), F = 191.93

4 Discussion and Conclusion

Since all the variables have *p*-values less than 0.05, all were considered significant predictors of people's actions. For Knowledge: For every unit increase in the knowledge of respondents about environmental sustainability, their action also increases. If quantified using the same scale, the increase is 0.153 units on the average, holding the other variable constant. In other words, the higher the knowledge about environmental sustainability, the actions leading to environmental sustainability tends to increase as well.

On the other hand, for the Behavior: For every unit increase in the attitude of respondents about environmental sustainability, their action also increases. If quantified using the same scale, the increase is 0.596 units on the average, holding the other variable constant. In other words, the more positive attitude about environmental sustainability, the actions leading to environmental sustainability tends to increase as well.

The data shows that for every unit increase in the government actions about environmental sustainability, the people's actions also increase. If quantified using the same scale, the increase is 0.207 units on the average, holding the other variable constant. In other words, the more actions done by the government about environmental sustainability, the actions of the people leading to environmental sustainability tends to increase as well.

Environmental sustainability is everybody's concern and both the government and people should work hand in hand to ensure that the environment is protected. However, on the part of people, knowledge plays an important part in committing to environmental sustainability and sustainability. This knowledge covers knowledge on environment itself, awareness on how people's action can contribute to either degradation or protection.

For business organizations, government can propose for an income tax reduction scheme that will encourage business organizations to reduce the utilization of plastic materials and shift to sustainable ones. Shopping malls and grocery stores use the biggest amount of single use plastic bags and having them stop the use of plastic bags will be a great amount of help in attaining environmental sustainability.

Another way to get people's cooperation towards this move is to give tax bonus for every use of reusable eco bags or even minimal discount if they will opt to use eco bags. The success of this is a joint effort by the government and people.

Local government can also come up with awareness campaign through having film/movie showing on climate change, environmental degradation and the likes—free of charge to public. Local schools can adapt the same for its students by organizing an Environmental Week where programs and activities will be done to raise the students' awareness on environmental sustainability. The Environmental Week can include film showing, debates and lectures form prominent environmental advocates.

The government can set up a "Help Desk" in strategic areas and locations that can give informative materials and leaflets. Business organizations can also be mandated to have environmental awareness week where employees can be grouped to work on environmental initiatives that can be adapted within the organization. Business organization can use the concept of Quality Control Circles where people are group and working on separate project—the only difference is that projects are about environmental initiatives.

Lectures can also include the information on proper waste management system covering disposal, reduction, reuse, recycle and prevention of waste. Students and/or employees can come up with "products' coming from recycled materials. Making this type of activity as a competition can motivate people to become creative thus promoting using of recycled materials.

National campaigns in national TV, newspaper may play a key role in raising environmental awareness and knowledge which will lead to individual's actions on environment. However, although awareness does not automatically lead to positive action, the knowledge about the impact, for example of pollution caused by plastics may trigger a change in behavior and values which would ultimately lead to behavioral change or promote behavioral change [8].

Government actions is very crucial to environmental sustainability since they have legislative powers and can propose and implement laws and regulations [26, 27]. Government funding is equally significant to put plans into concrete actions that have been planned to promote environmental sustainability therefore a heavier weight is put upon the government if environmental sustainability should be achieved.

Recommendations as far as people's behavior is concerned include ensuring the presence of a support group that can monitor their actions aside from leading by example, supporting good habits and giving the right feedback.

Summarizing the conclusions, environmental sustainability can be achieved by choosing sustainable products which are widely used in the daily lives of people. Single Use Plastic bag is one product that can be replaced with a more sustainable bag for use everywhere, whether in small stores, market, groceries, malls and others. However, the option to drop the utilization of this non-biodegradable bags depends on the people's knowledge of its adverse effect to environment, his behavior and action that support his knowledge and lastly by how the government supports the advocacy to shift to a more sustainable bag through environmental laws and policies that will be implemented. This conclusion is also supported by the findings in other researches/studies on similar topic [28].

References

1. Global Greenhouse Gas Emissions Data, US, Environmental Protection Agency
2. Fava, M.: Ocean plastic pollution an overview: data and statistics (2022). https://oceanliteracy.unesco.org/plastic-pollution-ocean/
3. https://en.unesco.org/sustainabledevelopmentgoals
4. Güney, T.: Solar energy and sustainable development: evidence from 35 countries. Int. J. Sustain. Dev. World Ecol. **29**(2) (2021)
5. Kwon, O.H., Hong, I., Yang, J., et al.: Urban green space and happiness in developed countries. EPJ Data Sci. (2021). https://doi.org/10.1140/epjds/s13688-021-00278-7

Relationship of People's Knowledge, Behavior and Governmental …

6. Trusted science for a changing world. Health Effects Institute Annual Report (2021). https://www.healtheffects.org/system/files/hei-annual-report-2021.pdf
7. https://www.constructionweekonline.com/products-services/261517-bahrain-airport-company-recycled-more-than-12-tonnes-of-plastic-in-2018
8. Bezerra, J.C., et al.: Single-use plastic bag policies in the Southern African Development Community, Elsevier B.V. (2021). https://doi.org/10.1016/j.envc
9. Yulisman, L.: Jakarta steps up battle against plastic waste with ban on single-use plastic bags. Strait Times (2020)
10. Suarez, L.: They ban single-use plastic in Colombia, Sheryl Sandberg leaves meta and more (2022). https://impactotic.co/en/they-ban-the-use-of-single-use-plastic-in-colombia-sheryl-sandberg-leaves-goal-and-more-news.
11. UNEP.: Single-use plastics: a roadmap for sustainability (2018). ISBN: 978-92-807-3705-9
12. Calderwood, I.: 16 Times countries and cities have banned single-use plastics (2018). https://www.globalcitizen.org/en/content/plastic-bans-around-the-world/
13. Kontogianni, B.: Greece bans single-use plastics to protect environment (2021). https://greekreporter.com/2021/07/05/greece-bans-single-use-plastics-protect-environment/
14. Sipsas, E.: Greece places tax on single-use plastic cups and lids in attempt to meet EU goal (2022). https://newseu.cgtn.com/news/2022-02-07/Greece-targets-plastic-waste-by-placing-tax-on-single-use-cups-17o2br9SJKE/index.html
15. Sweden: Parliament votes to adopt tax on plastic bags (2020). https://www.loc.gov/item/global-legal-monitor/2020-01-31/sweden-parliament-votes-to-adopt-tax-on-plastic-bags/
16. Muzdakiz, M.: Canada joins growing list of nations banning single-use plastics (2022). https://mymodernmet.com/canada-bans-single-use-plastic/
17. Nitahara, A.: Plastic bags banned from Rio Markets (2019). https://agenciabrasil.ebc.com.br/en/geral/noticia/2019-06/plastic-bags-banned-rio-markets
18. El Mekaoui, A., et al.: Plastic bags ban and social marginalization: evidence from Morocco. Police J. Environ. Stud. **30** (2021)
19. Kulin, Seva.: A little more action, please: increasing the under-standing about citizens' lack of commitment to protecting the environment in different national contexts. Int. J. Sociol. (2018). ISSN: 0020-7659 (Print) 1557-9336 (Online). https://www.tandfonline.com/loi/mijs20
20. Kulin, Seva.: The role of government in protecting the environment: quality of government and the translation of normative views about government responsibility into spending preferences. Int. J. Sociol. **49**(2) (2019)
21. Salvador, M., Sancho, D.: The role of local government in the drive for sustainable development public policies. An analytical framework based on institutional capacities. Sustainability **13**, 5978 (2021). https://doi.org/10.3390/su13115978
22. Racoma, A.: Environmental attitude and environmental behavior of Catholic Colleges' employees in Ilocos Sur, Philippines (2019). https://hal.archives-ouvertes.fr/hal-02330424
23. Theory of Reasoned Action for Predicting Consumer Behaviour (2021). https://accounting.binus.ac.id/2021/11/15/theory-of-reasoned-action-for-predicting-consumer-behaviour/
24. Biswas, N.G., Prakash, G.: Samkhya philosophy, deep ecology and sustainable development. Probl Sustain Dev 17(1), 288–292 (2022). https://doi.org/10.35784/pe.1.26
25. Bukhari, S.A.R.: Sample Size Determination Using Krejcie and Morgan Table (2021). https://doi.org/10.13140/RG.2.2.11445.19687. https://www.researchgate.net/publication/349118299_Sample_Size_Determination_Using_Krejcie_and_Morgan_Table
26. Onofre, et al.: Impacts of the allocation of governmental resources for improving the environment. An empirical analysis on developing European countries. Int. J. Environ. Res. Public Health (2020). https://doi.org/10.3390/ijerph17082783. PMCID: PMC7216234,PMID: 32316606
27. Oguge, N., Oremo, F., et al.: Investigating the knowledge and attitudes towards plastic pollution among the youth in Nairobi, Kenya. Social Sci. **10**, 408 (2021). https://doi.org/10.3390/socsci10110408
28. Isak, A.A.: Survey on the impact of the waste of plastic-made materials, its disposal and their effects on human health and environment: a case study in Mogadishu City, The capital of Somalia, Horn of Africa (2021)

Development of Entrepreneurship in the Tourism and Recreation Sphere: Marketing Research

Raisa Kozhukhivska, **Olena Sakovska**, **Svitlana Podzihun**, **Valentyna Lementovska**, **Ruslana Lopatiuk**, and **Nataliia Valinkevych**

Abstract The processes of natural restructuring, crisis economic phenomena that take place in Ukraine have a negative impact on the economic situation. Under such conditions, it is relevant to study new approaches, principles, theoretical and methodological provisions regarding the directions and perspectives of entrepreneurship development at the destination level, as well as the development of innovative and investment forms of entrepreneurial activity in the field of tourism and recreation. The article determines that the scientific basis that defines the strategic development of Ukraine in the foreseeable future is the concept of tourism development based on the use of initiative, economic independence and innovative capabilities of business entities of the national market and its territorial and sectoral constituentsIt has been established that entrepreneurship in the field of tourism and recreation creates opportunities for further economic development of regions, formation of investment and

---

R. Kozhukhivska (✉) · O. Sakovska
Department of Tourism, Hotel and Restaurant Business, Uman National University of Horticulture, Uman, Ukraine
e-mail: ray80@ukr.net

O. Sakovska
e-mail: sakovska_lena@ukr.net

S. Podzihun
Department of Marketing, Management and Business Administration, Pavlo Tychyna Uman State Pedagogical University, Uman, Ukraine
e-mail: spodzigun@ukr.net

V. Lementovska
Department of Marketing, Uman National University of Horticulture, Uman, Ukraine

R. Lopatiuk
Department of Management of Foreign Economic Activity, Hotel and Restaurant Business and Tourism, Vinnytsia National Agrarian University, Vinnytsia, Ukraine

N. Valinkevych
Department of Economics, Entrepreneurship and Tourism, Polissia National University, Zhytomyr, Ukraine
e-mail: natali1573@ukr.net

© The Author(s), under exclusive license to Springer Nature Switzerland AG 2024
A. Hamdan and E. S. Aldhaen (eds.), *Artificial Intelligence and Transforming Digital Marketing*, Studies in Systems, Decision and Control 487,
https://doi.org/10.1007/978-3-031-35828-9_33

innovation policy and provision of competitive services in new forms. The singularities of forming the entrepreneurial concept of tourism and recreational activities have been determined. With the purpose of developing the tourism and recreational industries of Ukraine and providing better tourism services, the article proposes an innovative form of methodology for the formation of a competitive tourism environment at the destination level using the marketing principles of customer focus. The author proposes a managerial and economic mechanism for carrying out entrepreneurial activities by the means of recreation and tourism entities at the level of destinations.

Keywords Tourism · Recreation · Entrepreneurial activity · Marketing · Management · Destination · Service · Consumer

1 Introduction

At the present time, Ukraine is facing the issue of developing and creating a new form of national tourism product (service) that can be realized through further sustainable development of business structures in the tourism and recreational sectorConducting entrepreneurial activity in tourism is a complex socio-economic process. Entrepreneurship in tourism is based on the ideology of market proportionality and adequacy of consumer production programs. It provides ways and opportunities for the further development of tourist destinations, the formation of an innovation and investment climate and competitive advantages, etc.

Comprehension of the role and importance of the effective formation, functioning and development of tourism and recreational activities in the context of overcoming the global crisis and economic globalization has necessitated further theoretical comprehension, substantiation and solution of applied character, determined the choice of the topic, formed the purpose and objectives of the scientific research.

2 Literature Review

The research of the issues of enterprise peculiarities in the sphere of tourism and recreation is devoted to the works of Aleynikova [1], Beydyk [2], Kyfyak [3], Lyubitseva [4]. The resolution of regional problems of tourist and recreational complex is devoted to the works of Borushchak [5], Gerasimenko and Nezdojminov [6], Nezdoyminov [6, 7], Stechenko [8] and so forth.

Through the analysis of available scientific and methodological sources, it is worth noting that the number of tourist and recreational services being provided and implemented in Europe is increasing. However, it is worth noting that in Ukraine, sales of tourism and recreation services are less numerous in comparison to other services. However, given the dynamic development of tourism and the understanding of public authorities that tourism is a sector that generates significant profits, and at the same time the government's intention to support this type of economic activity, it can be expected that sales of tourism and recreation services will increase in the future. Many services and websites that sell travel services already use I-commerce elements. I-commerce services are based on the principles of artificial intelligence and can be efficient in customer service. The use of I-commerce services makes it possible to personalize services and identify those that are best suited to a particular customer. This aspect adds advantages to the implementation of the tourist service system [9, p. 1008].

The specificity of investigating the issue of creating an effective investment policy for Ukrainian tourism and recreation enterprises is an important component of business development. Therefore, we can say that the tourism business needs to create a favorable economic environment. And such an environment can be formed on the basis of innovative principles of attracting finances and investments [10, p. 32]. In the tourism and recreation sector the priority areas are: development of tourism infrastructure; reconstruction and modernization of tourism and recreational facilities; development of rural «green» tourism, especially in areas that have preserved folk customs and traditions. Expansion of the range of services and provision of subsidized services; increasing the level of «stardom» of recreation and leisure facilities; preservation and protection of the natural and resource potential of the territory. It should be noted that the lack of proper managerial and financial flexibility is one of the main obstacles for business entities to enter the global market of tourism and recreation services [11, p. 38]. Some aspects of external economic activity (particularly: changes in the economy, instability of the external environment, etc.) have a direct impact on the sphere of tourism and recreation and cause the problem of revising the methodological apparatus in the management of tourism business [12, p. 40].

Under a given set of conditions, the issues of identifying new and improving existing marketing tools, mechanisms for stimulating consumer demand for tourism and recreation services, developing of an innovative means of marketing communications, etc., become acutely relevantю. It is possible to increase the efficiency of a tourism enterprise (recreation or leisure facility) and improve the ability to implement and satisfy consumer demand for tourism services through the following operational measures: invention, development and implementation of innovative approaches in the marketing activities of the enterprise; rejection of such methods of managing the work of tourism enterprises that are obsolete and authoritarian in nature; introduction of innovative marketing technologies that actualize the possibilities of realization and satisfaction of consumer demand for quality tourism services.

The aforementioned operational measures to improve the efficiency of the tourism enterprise require the appropriate use of innovative strategies [13, p. 680; 14, p. 238]. In other words, the use of innovative strategies, with the assistance of which enterprises create innovative products and services and form the latest economic processes that add value to the market, has a significant impact on the stabilization of the economic situation of both the enterprise itself and the external environment in which it operates [15, p. 8]. The conducted empirical studies have made it possible to ascertain the fact that enterprises that use innovative strategies in their activities are forced to constantly evaluate and control their external and internal environment. The purpose of this analysis is to identify new opportunities to strengthen their competitive position [16, p. 206].

Given the retrospective analysis of scientific and methodological material, it can be stated that innovation activity is one of the main criteria for the economic development of tourism and recreation enterprises. Meanwhile, it is worth noting that the study of the issue of introducing innovations, in particular in the marketing activities of tourism and recreation enterprises, is insufficiently researched and therefore requires additional study and critical analysis.

Furthermore, it should be noted that the issue of forming an organizational and economic mechanism for carrying out entrepreneurial activities by recreation and tourism entities at the destination level is poorly understood. The importance of these issues indicates the relevance of the chosen research area and determines its purpose and objectives.

3 Purpose of the Study

The purpose of the article is to is to investigate the key principles of business organization and the formation of components of the business concept in the field of tourism and recreation.

Accomplishment of this goal necessitated solving the following tasks: to analyze the essence and specific aspects of entrepreneurial activity in the tourism and recreation sector; to determine the peculiarities of forming a business concept for tourism and recreation entities; to propose an organizational and economic mechanism for building and implementing entrepreneurial activity by tourism and recreation entities at the destination level.

4 Methodology

The basis of the research methodology is an analysis of the works of domestic scholars in the field of tourism and generalization of foreign practices in the development of tourism and recreation; the personal judgment of the authors of this article was also used.

In the course of the theoretical analysis to determine the specifics of entrepreneurial activity in the tourism and recreation sphere, the author used analytical and abstract-logical research methods. The use of analytical approaches made it possible to determine the peculiarities of the formation of an entrepreneurial concept for the subjects of the tourism and recreation sphere. The application of the methodology of empirical analysis made it possible to develop an organizational and economic mechanism for the implementation of entrepreneurial activity in the field of tourism and recreation at the destination level.

5 Results

With the development of integration processes in the tourism sector and the aggravation of competitive struggle, management of entrepreneurial activity at the regional level is not simplified but becomes more complex, acute and multifaceted problem. Entrepreneurial activity in the tourism sector is a complex socio-economic phenomenon, and at the same time—a certain type of activity of tourist market actors, i.e. a process that requires consistent implementation of logically interrelated stages and phases. This process is based on the idea from which every new entrepreneurial activity—business.

«The tourism business is the only one economic and technological system for the formation and sale of the tourism product in order to meet the demand for tourism services» [17, p. 1592]. The latest research carried out by Ukrainian scientists, experts, and their publications [1, 2, 4] regarding the strategy of economic policy of the state in the sphere of tourism confirm that Ukraine has chosen an economic-balanced model of the national tourism product and which can be practically realized only in the context of further development of entrepreneurship and high-quality tourist service in tourist-recreational regions.

«Structural changes in the economy, instability of the environment require a revision of the forms and methods of managing the market of tourism services. The topical issue is to define new and improving existing marketing tools, mechanisms for shaping the demand of consumers for tourism services» [18, p. 5283].

«Customer orientation has an impact not only on production and consumption, but also on the associated common practices, which are not purely economic, for example, consumption practices» [19, p. 1372].

«The application of innovations in tourism greatly facilitates the process of providing tourist services, which begins with informing about them and ends with their final consumption. For example, the emergence of the World Wide Web has led to the simplification of information exchange, improvement of methods of implementing the marketing cycle in tourism, the development of electronic means of payment with customers and suppliers and more» [20].

The formation of the essence of the concept of entrepreneurship and its socio-economic character is conditioned by the fact that it is directly affected by factors of industrial and commercial activity and the external environment and public consciousness.

By the definitions of scientists [1, p. 52; 7, p. 84] the concept of «entrepreneurship» is revealed from different positions: as commercial, risky activity what is directed on getting profit, as a method of competition, as a general plan reflecting the multiscope and importance of this concept for modern market activity of entrepreneurs.

The research conducted by us on the development of business activity allows us to state qualitative changes in the business paradigm in tourism, namely, new direction—entrepreneurship, that is, activity aimed at achieving the goals on the basis of the opportunities of entrepreneurship. It can be considered as activity on the basis of integration of entrepreneurial opportunities of the employee and tourist enterprise.

Summarizing the presented theoretical approaches [1, 4, 5, 7] to define the essence of the concept of entrepreneurship in the tourism and recreation sector, it should be concluded that entrepreneurship in tourism and recreation has the following characteristic features.

First—entrepreneurship in the tourism and recreation sphere performs the function of combining tourism and recreational resources with the satisfaction of the demand of the population and tourists in the reproduction of physical, spiritual, psychological forces at the expense of their own material resources and entrepreneurial abilities.

The second consits of the fact that the satisfaction of the needs of tourists and recreationists is carried out in the conditions of market competition of all subjects of tourist business in the specialized territories of tourist and recreational complex.

Generalizing the considered theoretical approaches to define the essence of the concept of «entrepreneurial activity» in the tourism and recreation sector, we propose the following interpretation: «entrepreneurial activity in the tourism and recreational sphere» is an economic activity that combines resources, functions of recreation and tourism and is technologically related to the production, sales and consumption of tourism products, concomitant and specific tourist services and products to achieve the goal of business and satisfy the needs of society.

Taking into consideration the peculiarities of entrepreneurial activity in the tourism and recreational sphere as a multidimensional economic and socio-cultural process, it is necessary to determine the types of entrepreneurship by functional and sectoral activities in tourism. According to the efficacious Classifier of Economic Activities, tourism and recreational activities can be defined as a process of combining actions that lead to the receipt of a specified set of products and services based on the use of certain resources: raw materials, equipment, labor, technological processes, etc. Some scientists [4, 6, 7] consider the classification of entrepreneurship in tourism through the prism of economic and civil relations, sectoral and functional activities, by geographical criterion.

The author's approach to the definition of «entrepreneurial activity in the tourism and recreational sphere» allows to expand the classification of entrepreneurship by the following criteria:

- firstly, the regulatory policy of the state in the field of tourism, the mechanism of licensing of tour operator and travel agency activities restricts access to the tourism activities of all enterprise entities;
- secondly, the development of the market of tourist and recreational services leads to the specialization of enterprises and the emergence of the newest forms of territorial and economic entities of the tourism business by type of tourism and geography of Tour Operating;
- thirdly—the intensification of competition in the international and regional markets causes an intensification of entrepreneurs' efforts to identify the manufacturer of tourism products or services, which is provided by the form of positioning in foreign and regional markets;

- fourthly, the objective basis for the formation of the industry is, first of all, the necessity to separate the producer of tourist products from the intermediary in the system of division of labour, which has a specific type of economic activity by the object of entrepreneurial activity.

Our proposed classification of entrepreneurship is the basis for the development of further organizational and economic foundations for the development of tourism and recreation complex of the region and scientific substantiation of the entrepreneurial concept of tourism and recreational business entities.

Further development of regional reforms in Ukraine, associated with the introduction of modern market-oriented methods of management in theory and practice, has made significant changes in the key factors of production: land, labour and capital, added to them a new factor—entrepreneurial potential (the potential ability of tourism entities to efficiently use the totality of human and material resources).

The formation and exploitation of this potential by entrepreneurs of the regional tourism market will be the essence of the notion of «entrepreneurial concept». In our opinion, the feature of our proposed interpretation of the entrepreneurial concept in tourism and recreation, is its comprehensive integration into the system of national and international tourism industry, where the term «tourist product» is used in one definition to combine goods and services, offered by entrepreneurs in the tourism industry.

In the course of analyzing the forms of entrepreneurship in the national and regional markets, for the formation of the entrepreneurial concept of tourism business, we propose to extend the classical and innovative models of entrepreneurship with an alternative combination of their constituent economic elements of the entrepreneurial concept (Fig. 1).

The general mechanism for the formation and implementation of the entrepreneurial concept of tourism and recreation entities is considered by us as a system of organically linked organizational, economic and informational measures that provide the prerequisite conditions for its effective functioning (Fig. 2).

All the aforementioned measures should ensure not only the achievement of the ultimate goal, but also quality of management in tourism enterprises. The key components of the generalized scheme are the:

- management system;
- economic indicators;
- development management;
- cooperation,
- integration,
- clustering and joint venture.

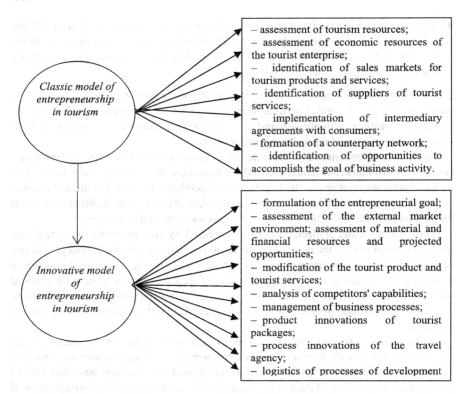

Fig. 1 Typological model of entrepreneurship in the tourism and recreation sector. *Source* Compiled by the authors

6 Conclusion

Thus, the entrepreneurial concept as well as the process of entrepreneurship in the tourism and recreation sector is an open system that depends on the mutual exchange of resources and results of activities with the external world, while simultaneously, influencing it through business relationships and commercial transactions with business partners.

The defined models of entrepreneurship and the proposed organizational and economic mechanism for the implementation of the entrepreneurial concept of recreation and tourism entities can be used in the establishment of innovative tourist and recreational complexes, which have the following features:

(1) vividly highlighted tourism and recreational specialization;
(2) significant number of private entrepreneurs providing their accommodation and nutrition services, organization of leisure and excursions on semi-legal conditions, i.e. the presence of shadow business;

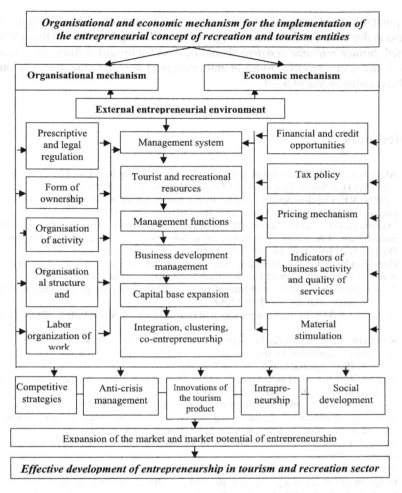

Fig. 2 Diagram of the mechanism of implementation of the entrepreneurial concept in the tourism and recreation sector. *Source* Compiled by the authors

(3) aspiration of each enterprise and state administrative structures to get the maximum income in their territories;
(4) discursiveness and unmanageability of processes that are taking place;
(5) deficiency and costliness of a specific marketing investigations;
(6) lack of information regarding business activities and successes (failures);
(7) indirection of regional strategy of entrepreneurial development in the field of recreation and tourism.

Thuswise, the issue of formation of the mechanism of realization of business activity in the sphere of tourism and recreation is an additional argument on necessity of building the proper entrepreneurial concept and estimation of business from the point of view of its survival and ability to compete, as well as expediency of investments in this sphere at the state and regional level. Taking into account the aforementioned, *further scientific research* should be directed toward defining tendencies of development of innovative forms of entrepreneurial activity in the sphere of tourism and recreation at macro level.

References

1. Aleynikova, H.M.: Organization and Management of Tourism Business. DIT, Donetsk (2012)
2. Beydyk, O.O.: Tourist Resources of Ukraine. Altrpress, Kyiv (2010)
3. Kyfyak, V.F.: Development of tourism and recreation in the context of the region's economic policy strategy. Bull. Donetsk Inst. Tour. Bus. **10**, 112–119 (2010)
4. Lyubitseva, O.O.: Tourist Services Market (Geospatial Aspects). Altrpress, Kyiv (2013)
5. Borushchak, M.A.: Conceptual approaches to defining the definition of «tourist region». Bull. Donetsk Inst. Tour. Bus. **10**, 104–112 (2009)
6. Gerasimenko, V.G., Nezdojminov, S.G.: Strategy for the development of entrepreneurial potential in tourist regions of Ukraine. Bull. Socio-economic Res. **30**, 64–70 (2011)
7. Nezdoyminov, S.G.: To the problems of forming the modern concept of entrepreneurship in tourism. Scien. Bull. **2**(58), 82–95 (2014)
8. Stechenko, D.M.: The scientific paradigm of regional tourism. Scien. Bull. CTEI **IV**(1), 38–43 (2012)
9. Kozhukhivska, R., Sakovska, O., Skurtol, S., Kontseba, S., Zhmudenko, V.: An analysis of use of internet technologies by the consumers of tourism industries in Ukraine. Int. J. Adv. Sci. Technol. **29**(6s, Special Issue), 1007–1013 (2020)
10. Ivanova, Z.: Theoretical aspect functioning of the financial-investment mechanism of tourism and recreation industry. Naukovi zapiski **159**, 32–34 (2014)
11. Shkola, I.M., Korolchuk, O.P.: Tourism of Management. Books-XXI, Chernivci (2011)
12. Gorina, G., Barabanova, V.: Marketing aspects of developing tourism services market in Ukraine & the Baltic countries. Baltic J. Econ. Stud. **5**, 39–47 (2019)
13. West, M.A., Anderson, N.R.: Innovation in top management teams. J. Appl. Psychol. **81**(6), 680–693 (1996)
14. Wong, A., Tjosvold, D., Liu, C.: Innovation by teams in Shanghai, China: cooperative goals for group confidence and persistence. Br. J. Manag. **20**(2), 238–251 (2009)
15. Gundry, L.K., Kickul, J.R., Iakovleva, T., et al.: Women-owned family businesses in transitional economies: key influences on firm innovativeness and sustainability. J. Inno. Entrepreneurship **3**(8), 2–17 (2014)
16. Rogers, P.R., Bamford, C.E.: Information planning process and strategic orientation: the importance of fit in high-performing organizations. J. Bus. Res. **55**(3), 205–215 (2002)
17. Kozhukhivska, R., Chuchmii, I., Harbar, O., Kostiuk, M., Nechytailo, V., Sakovska, O.: Development of the tourist sphere in Ukraine: socio-economic aspect. In: Proceedings of the 35rd Conference, IBIMA 2020: Education Excellence and Innovation Management: A 2025 Vision to Sustain Economic Development during Global Challenges, 1–2 April, 2020. Seville, Spain, pp. 1591–1597 (2020)
18. Kozhukhivska, R., Sakovska, O., Maliuga, L., Maslovata, S.: The formation of a system of investment prospects of Ukrainian tourism and recreation sector enterprises on terms of benchmarking. In: Proceedings of the 33rd Conference, IBIMA 2019: Education Excellence and Innovation Management through: Vision 2020, 10–11 April, 2019. Granada, Spain, pp. 5282–5290 (2019)

19. Kozhukhivska, R., Sakovska, O., Shpykuliak, O., Podzihun, S., Harbar, O.: Social customer-oriented technologies in the tourism industry: an empirical analysis. TEM J. **8**(4), 1371–1383 (2019)
20. Kozhukhivska, R., Sakovska, O., Maslovata, S., Dluhoborska, L., Chuchmii, I.: Managing innovation in tourism and hospitality industry: international experience. Proc. Conf. **2413**(1). 040007. https://aip.scitation.org/doi/pdf/10.1063/5.0089854, https://doi.org/10.1063/5.0089854

The Impact of RUPT on Corporate Environmental Responsibility

Acharya Supriya Pavithran⬤, C. Nagadeepa⬤, and Baby Niviya Feston⬤

Abstract Each and every terminology connected to human aspect has a tendency to connect to its roots—Nature. Terms like 'Global Village' imply the connect to everything. Living in a multiple ecosystem, all its internal and external connections that might be overlooked in an attempt to manage and implement managerial strategy effectively. In short, in enduring times, people always look for satisfying their outmost desire and needs which counterparts their willingness to accept the situation or not. The global environment has taken a setback on the willingness of people especially corporates in adopting a strategy which will help them to endure the present scenario with minimal setback. Corporates as Global leaders look into solutions like RUPT to overcome the challenges with one right solution. The study attempts to unleash the relation that will arise out of the challenges that RUPT has offered corporates to decide on a better and correct environmental sustainability programme for its organisation. The study reveals the challenges and solutions that RUPT has to offer in the context of environmental sustainability. The study attempts to reveal the benefit at the increased level in creating consciousness of both the environmental and professional aspects of the large and small enterprises alike who are under pressure to contribute for the greater good.

Keywords RUPT · Corporate environmental responsibility · Company · Ecosystem · Sustainability

---

A. S. Pavithran (✉) · B. N. Feston
Garden City University, Bangalore, India
e-mail: supriyap23@gmail.com

C. Nagadeepa
Kristu Jayanti College Autonomous, Bangalore, India

© The Author(s), under exclusive license to Springer Nature Switzerland AG 2024
A. Hamdan and E. S. Aldhaen (eds.), *Artificial Intelligence and Transforming Digital Marketing*, Studies in Systems, Decision and Control 487,
https://doi.org/10.1007/978-3-031-35828-9_34

1 Introduction

1.1 RUPT (Rapid, Unpredictable, Paradoxical and Tangled) from VUCA

RUPT is an alternative to VUCA (Volatile, Uncertainty, Complexity and Ambiguity). VUCA was first introduced in 1987 based on the theories of Warren Bennis and Burt Nanus, elucidating the collapse of the USSR retorted by the US Army War College in the early 1990s. As a result, a new approach to identifying the enemy—seeing and reacting, was framed. Innovation being defined as to engage in sincere cooperation, taking on concise responsibilities. This requires liberty, creativity, flexibility, rapidity and corporate culture connecting the people with the organisation. VUCA world promises to anticipate the future by strengthening the cooperation in companies with newer and better modernized solution. VUCA is being used by many organisations to describe their way of complying with the changing world where they dwell.

In the recent times, VUCA does not capture what the corporates are experiencing, or how to lead through it. As a result, a new acronym, which defines the need to navigate the disruption in leadership through—RUPT.

Rapid: Change is rapid. Resembling the crashing of the waves in the ocean, unpredictable of its next impact.

Unpredictable: The managers analyze, strategize and forecast the future, unaware of the unexpected. Thus, bringing them back to challenging their so framed strategies and reframing their thinking from square one.

Paradoxical: There are times, when a leader misses to see the different solutions to the complex problem and concludes on one and only right solution. This reprimands the teams as well as the organisation to conduct their duty for the long term and for the short term too. The temptation arises in choosing between short term and long term problem solving. Being able to tolerate ambiguity and withstand paradoxical is the new rule for the leaders.

Tangled: Everything is connected to everything else. Like a tangled ball of thread—the more struggle is done to detangle the thread, the more tangled it becomes. For this, the leader should know to pull the right thread, to detangle the situation slowly but steadily.

To overcome the disRUPT situation, the leaders in business need to collaborate and communicate to connect their attributes.

Collaborate: Collaboration happens through positive approach. The leader must decide to bring together his teams, departments and consult them on inspiring and connecting the detangled problem. They have to put aside their rivalry and competitiveness to encourage the leader in overcoming the problem with the right solution.

Communicate: Communication is the most important task that every other person working in the organisation should follow. The leader has the responsibility to undertake that the employees are able to contact each other and communicate the right agenda in a proper manner. He should not burden them with unnecessary updates

The Impact of RUPT on Corporate Environmental Responsibility 393

or lengthy paper-works, which ultimately misses the point of the meeting. The employees should burden (in a positive sense) with concrete work, which gives positive and concrete solution. This will optimize the productivity with minimal miscommunication among the employees.

As a leader, in ruptured and disruptive times, it is most important to find the point of calm to speak more loudly and find the courage to be innovative, patient and to care, collaborate and communicate. Even though leadership is a lonely and uncertain place, with the right people to support the decisive action, the leader has very less to worry on improving his organisational betterment.

1.2 Corporate Environmental Responsibility (CER), Sustainability and Corporate Environmental Performance (CEP)

CER, a component of Corporate Social Responsibility (CSR) states the company's duty to protect the natural environment and refrain from destroying it. In the Brundtland Report of 1987 [1] under the guidance of the World Commission on Environment, issues pertaining to sustainable development were discussed. This directs towards the scholars, business owners specifically mangers, studying the reason of why the corporates invariably include environmental aspects in their policies. ISO 14063 defines this aspect of the environment inducted in the management policies.

The Brundtland Report, 1987 defined sustainability as "meeting the needs of the present generations without compromising the ability of future generations to meet their own needs." Elkington [2] developed the Triple Bottom Line, emphasizing on three major aspects—profits, people and planet. The Global Reporting Initiative (GRI) of 2011 has defined "environmental dimension of sustainability concerns an organization's impacts on living and non-living natural systems, including ecosystems, land, air and water". The inputs of GRI includes materials, water and energy; the output includes emissions, waste and effluents; covering the other aspects of biodiversity, environmental certifications and its expenditures.

CEP, as per ISO 14063 is defined as "the results of an organisation's management of its environmental aspects". Judge and Douglas [3] have mentioned the concept of CEP being "an effectiveness of a firm's commitment to reach environmental excellence". In 2007, Xie and Hayase [4] designed a two-dimensional theory defining EMP-Environmental Management Performance and EOP-Environmental Operational Performance.

2 Literature Review

Doheny et al. [5] said that VUCA is changing the nature of competition evolving among the industries and its markets.

Warwick-Ching [6] suggested that volatility provides profit opportunity.

Hemingway and Marquart [7] pointed out that uncertainty is opportunity.

Boston Consulting Group [8] have printed that IT simplification helps to grab key opportunities.

Amerasia Consulting Group [9] suggested that opportunity equals ambiguity.

Brooks [10] suggested that the decision of Southwest Airlines to hedge their jet fuel helped in the company's low-cost operating strategy.

Organisations showing higher sustainability levels in connection with environmental change tend to adapt the change rapidly in comparison to firms neutral to the changing environmental demands are less effective to change their existing structures and processes [11].

3 RUPT on CER, Sustainability and CEP

RUPT describes the global adaptation of the socio-economic system as rapid, unpredictable, paradoxical and tangled. VUCA is present in the various studies across the organisation, which changes recognition and responses of the management and its planning. These findings identify the dependence of the human well being on the functionality of the eco-system on a wide range of threats imposed on them. The contents of RUPT are interconnected to each other, which are examined to understand its relevance on the organizations' environmental sustainability.

Rapid: Rapid means pattern in the dynamics and the rate of change in the eco-system. In the present conditions it can be defined as major fluctuations in financial markets, macroeconomic situations and commodity prices, and also to a great extent on the environmental pressures affecting the nation. According to Svenning and Sandel [12] 'static state' principles has defined various scientific and business models, evidential to the non-linear state of both the natural and cultural system pertaining to climatic changes, tech innovations and rapid fluctuations in the world market [13]. Even though, it is difficult to determine the rapid change in the precise elements with the system. There is a rapid speed in these changes. In order to provide more detailed version of rapidity, requires a detailed method of monitoring for a longer period. However, a short time review of the present situation can put a clear picture of the rapidity. For example, on the release of the lockdown situation that gripped the world, people were rapid to stock their houses with supplies on a large scale, not caring about the necessity of the other individual. In this context, various healthcare organisation and institutions were rapid enough to introduce their version of healthcare products at a higher rate, which would attract the individual to invest in them, regardless of the price that these products carry. In terms of corporate environmental responsibility, many organisations introduced various programmes in the guise of providing healthcare products to the hospitals, healthcare units, and to the community at large. In this event, the individual is more prone to stocking their supplies then worry about the product quality or branding. This sudden act (rapid) means that extreme events are more prone to accentuate the existing scenario even worse.

Unpredictable: Unpredictability is emerged out of uncertainty. Multiple feedback loops with interactions are the resultant of unpredictability which are essential to the complex system [14]. When such surprises occur, the system behaves in a way that is fundamentally different from what was anticipated [15], which can put at danger the assets under conservation and the efficiency of management measures. As argued by Renn [16] that risks are the mental 'constructions' an illusion of the human mind and not the real event. That perception is clouded by the fact that, in contrast to actual objects, they must be made and chosen by human actors rather than being able to be observed and counted in the environment. Tversky and Kahneman [17] suggest that one's risk factor is the Act of God for another and sometimes an opportunity to someone else. In the current situation for example, the people and the organisation are uncertain about the future, which makes them vulnerable to act accordingly. Till recently organisations were adamant on approach only on the socio-economic cause, but the present scenario demands them to reconcile their approach. They have to consider their predicament of the enduring future and reprimand for the benefit of the public.

Moore and McCarthy [18] have pointed that any uncertainty suffered by the environmental managers, encourage them to decide on all the possible solutions and strategies to solve these uncertainties. Technology used that places a greater emphasis on knowledge-based decisions at the expense of no-knowledge-based action, poses greater risks than using already available knowledge and evidence [19].

Paradoxical: Being in a paradoxical situation is having complexity. Complexity is referred to offering an intricate and widespread structural network with dynamic pathways in the contents of the system. This principle is built on chaos and subject to tipping points. There is no close linkage among cause-affect and space-time as a result of the inherent properties in the complex system. The system becomes more paradoxical as there are more parts and linkages between those parts. Paradoxical attempts to cause several challenges for managers. For instance, Australia, America and Europe have all witnessed protests. Numerous migrant labourers in India started going back to their homes. The coronavirus pandemic on the one hand, and economy and politics on the other, are exerting increasing pressure on the authorities. This is a case of the prevention paradox, which happens during pandemics when society or a government enacts a policy that comes at a cost in terms of resources (money, time and efforts) and inconvenience (though relatively minimal on an individual level). The same prevention paradox is being faced by authorities dealing with the new coronavirus outbreak in India and worldwide. Is it advisable to lift the ongoing lockdown and risk of spreading the novel coronavirus? Those who have not been personally affected by the virus transmission may find it difficult to comprehend the effectiveness of lockdown in containing the virus spread. However, the coronavirus lockdown is undoubtedly hurting their freedom, businesses and jobs.

Tangled: Where there is uncertainty concerning the nature of cause-and-effect relationships, the situation is tangled. If enough information is gathered, it is possible to forecast what might happen in a complicated situation. There is limited historical precedent for predicting the results of specific causes or courses of action, making a complicated situation more difficult. The unnecessary accumulation of resources could be beneficial in a rapid situation, but it is a huge waste of time and effort in tangled situation. At its most basic level, there are three ways to look at the issue that the world's government are facing. These are 'storm in a teacup', 'house on fire' and 'holding back the tide' respectively. Both the US's initial downplaying of the situation with the US President's statement that the virus is 'very well under control in the USA' and China's now-famous early attempts to hide news of the pandemic make sense only in the 'teacup' interpretation. This perspective of the scenario states that the key objective is to prevent unduly panicking individuals because the issue with eventually be resolved. First in China, South Korea and Japan, then in Europe and the US, the 'house on fire' interpretation has led to unprecedented lockdowns. According to this viewpoint, combating the virus should be the top concern. Similar to collateral water damage caused by the firefighter's hose, the economic and social effects of shutting down or significantly reducing sports, restaurants, airlines and other activities will be significant. However, they must be accepted. The strategy for this action's finish is to start out strong and move quickly. After the press conference with the Prime Minister, the UK government is focusing on the third narrative. According to the 'holding back the tide' perspective, fighting the virus is merely temporary fortifications. Herd immunity would be the goal in the end, and it is anticipated that this would happen once the virus had infected and healed between 60 and 80% of the population. To control the virus spread through the population as smoothly as possible, tiered strategy is necessary.

4 Implications

Because things change so quickly, the external environment for an organisation can be complicated, making it challenging for leaders to understand it and take the appropriate action. There is someone to disRUPT the world, when it is RUPT. Someone will introduce a brand-new paradigm, a product or service that is so ground-breaking that it overtakes all else. Finding novel paradigms and future-relevant solutions is the only way to escape an RUPT environment. Even more radical and far-reaching changes are being made to the current business climate. The organisations are coming up with fresh and creative suggestions for more refined environmental programmes and procedures to enhance their business. The better and more appealing their environment investment portfolio is the more interested investors are in their company. This should not put the organisation in an RUPT scenario. For this purpose, the model created by Col. John Boyd of the US Airforce, the OODA loop was defined for combat flying. OODA stands for Observe, Orient, Decide and Act (Fig. 1) [20].

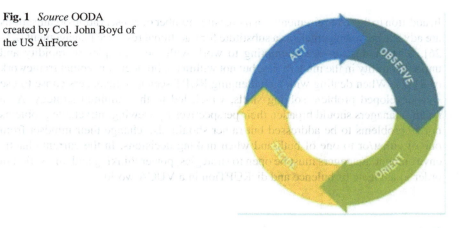

Fig. 1 *Source* OODA created by Col. John Boyd of the US AirForce

Observe: It is important to keep an accurate perception of how the world is evolving, while observing it. However, since the same thing can be perceived in a variety of ways, perception is not a straightforward process. We need to go beyond the apparent, and effective observation is a key component of good entrepreneurial practice.

Orient: There is a need to gather intelligence about several forms of development and modifications in 360 degrees. Integrating this knowledge and getting one's bearings on their options is difficult. This calls for creativity, invention and sometimes even using one's instincts to connect the links.

Decide: It is time to make a choice now that creativity and innovation have been applied to find the options. The best decision of disruption must be chosen, whether it be the creation of a new product, a new service, a new method of managing operations and costs, or a new method of delivery.

Act: Act on the decision made. The unique method of starting is with frugal experimentation. Rapidity, awareness of what has been observed, oriented, decided and acted in today's world.

Those organisations will be able to survive that will be able to continuously deliver value to all the stakeholders, and their valuable investors.

5 Conclusion

Humankind has significantly changed and harmed the biosphere during the past 50 years [21]. The Millenium Ecosystem Assessment (2005) revealed the complexity of both nature and our relationship with it as well as the vulnerability of human survival should the loss and degradation of global ecosystems go unchecked. Building and maintaining ecological resilience as well as the social adaptability required to deal with innovative, and adapt to rapid change are the ways to promote conservation [22].

In addition to these requirements, an increasing number of conservation organisations are advocating management as a substitute for traditional resource management [23–26]. Many industries are attempting to work with the principles of rapidity and unpredictability in the modern day, but not within the broader conceptual framework of RUPT. When dealing with a challenging RUPT scenario, managers chose to use well-developed problem-solving skills, which led to this minimal strategy. As a result, managers should broaden their perspectives by viewing intractable problems not as problems to be addressed but rather should also change their mindset from one of either/or to one of both/and when making decisions. In the current chaotic environment, managers must be open to strategies, power thinking and innovation in order to navigate turbulence and disRUPTion in a VUCA world.

References

1. Brundtland, G.H.: Our common future. United Nations World Commission on Environment and Development (Brundtland Commission). Oxford University Press, Oxford (1987)
2. Elkington, J.: Partnerships from cannibals with forks: The triple bottom line of 21st-century business. Environ. Qual. Manag. **8**(1), 37–51 (1998). https://doi.org/10.1002/tqem.3310080106
3. Judge, W.Q., Douglas, T.J.: Performance implications of incorporating natural environmental issues into the strategic planning process: an empirical assessment. J. Manage. Stud. **35**(2), 241–262 (1998)
4. Xie, S., Hayase, K.: Corporate environmental performance evaluation: a measurement model and a new concept. Bus. Strateg. Environ. **16**(2), 148–168 (2007)
5. Doheny, M., Nagali, V., Weig, F.: Agile operations for volatile times. McKinsey Quarterly (2012, May). Retrieved from http://www.mckinsey.com/insights/operations/agile_operations_for_volatile_times
6. Warwick-Ching, L.: Currency wars: volatility provides profit opportunity. The Financial Times (2013, March 25). Retrieved from http://www.ft.com/intl/cms/s/0/e17e1ab0-8714-11e2-9dd7-00144feabdc0.html#axzz2r3QHfbxW
7. Hemingway, A., Marquart, J.: Uncertainty is opportunity: engage with purpose. Edelman (2013, June 27). Retrieved from http://www.edelman.com/post/uncertainty-is-opportunity-engage-with-purpose/
8. Boston Consulting Group.: Simplifying IT complexity a major opportunity for many companies [Press Release] (2013, March 21). Retrieved from http://www.bcg.com/media/PressReleaseDetails.aspx?id=tcm:12-130333
9. Amerasia Consulting Group.: Ambiguity equals opportunity: the story of the new HBS application (2013, June 3). Retrieved from http://www.amerasiaconsulting.com/blog/2013/6/3/ambiguity-equals-opportunity-the-story-of-the-new-hbs-application
10. Brooks, R.: A life cycle view of enterprise risk management: the case of Southwest Airlines jet fuel hedging. J. Fin. Edu. **38**(3/4), 33–45 (2010)
11. Heugens, P.P.M.A.R., Lander, M.W.: Structure! Agency! (and other quarrels): a meta-analysis of institutional theories of organization. Acad. Manag. J. **52**(1), 61–85 (2009)
12. Svenning, J.-C., Sandel, B.: Disequilibrium vegetation dynamics under future climate change. Am. J. Bot. **100**, 1266–1286 (2013)
13. Hall, P.A.: Aligning ontology and methodology in comparative research. In: Mahoney, J., Rueschemeyer, D. (eds.) Comparative Historical Analysis in the Social Sciences, pp. 373–406. Cambridge University Press, Cambridge, UK (2003)
14. Gunderson, L.H., Holling, C.S.: Panarchy: Under-Standing Transformations in Human and Natural Systems. Island Press, Washington, DC, USA (2002)

The Impact of RUPT on Corporate Environmental Responsibility 399

15. Gunderson, L.H.: Ecological resilience in theory and application. Annu. Rev. Ecol. Syst. **31**, 425–439 (2000)
16. Renn, O.: Precaution and ecological risk. In: Jorgensen, S.E., Fath, B.D. (eds). Encyclopedia of Ecology, pp. 2909–2916. Elsevier B.V., Amsterdam, The Netherlands (2008)
17. Tversky, A., Kahneman, D.: Judgment under uncertainty: heuristics and biases. Science **185**, 1124–1131 (1974)
18. Moore, A.L., McCarthy, M.A.: On valuing information in adaptive-management models. Conserv. Biol. **24**, 984–993 (2010)
19. Sutherland, W.J., Pullin, A.S., Dolman, P.M., Knight, T.M.: The need for evidence-based conservation. Trends Ecol. Evol. **19**, 305–308 (2004)
20. Boyd, J.R.: Destruction and Creation (1976). http://goalsys.com/books/documents/DESTRU CTION_AND_CREATION.pdf. Accessed March 6, 2013
21. Vitousek, P.M., Mooney, H.A., Lubchenco, J., Melillo, J.M., Series, N., Jul, N.: Human domination of Earth's ecosystems. Science **277**, 494–499 (1997)
22. Holling, C.S.: Understanding the complexity of economic, ecological, and social systems. Ecosystems **4**, 390–405 (2001)
23. Holling, C.S.: Adaptive environmental assessment and management. John Wiley & Sons, Ltd., Chichester, New Hampshire, USA (1978)
24. Walters, C.J.: Adaptive management of renewable resources. Macmillan, New York, New York, USA (1986)
25. Gunderson, L.H., Holling, C.S., Light, S.S.: Barriers and bridges to the renewal of ecosystems and institutions. Columbia University Press, New York, New York, USA (1995)
26. Berkes, F., Folke, C., Colding, J.: Linking social and ecological systems: management practices and social mechanisms for building resilience. Cambridge University Press, Cambridge, UK (1998)

The Effect of the Characteristics of the Board of Directors and the Audit Committee on Financial Performance: Evidence from Palestine

Hisham Madi, Ghaidaa Abdel Nabi, Fadi Abdelfattah, and Ahmed Madi

Abstract The objective of this study is to explore the relationship between the board of directors' characteristics, audit committee characteristics, and financial performance of companies listed on the Palestine Exchange (PEX). Return on assets (ROA) and return on equity (ROE) were used to proxy financial performance. The characteristics of the board were indicated by the proportion of non-executive directors and the board's size, whereas the characteristics of the audit committee were depicted by the number of non-executive members in the audit committee, the financial expertise of the audit committee, the frequency of audit committee meetings, and the size of the audit committee. To meet the purpose of the study, all companies listed on the exchange that published annual reports between 2011 and 2015 were analyzed. The analysis revealed that board size is negatively significantly related to ROE; audit committee meetings are significantly associated with ROE; firm size is negatively associated with ROA; a positive relationship between leverage and ROE is observed, but only at the 10% significance level; and other relationships with financial performance are statistically insignificant. This implies that the results are in most of its parts inconsistent with agency theory and resource dependence theory. This might be because corporate governance mechanisms in Palestine are still cosmetic mechanisms.

H. Madi (✉) · F. Abdelfattah
Modern College of Business and Sciences, Al-Khuwair, 133 Muscat, Oman
e-mail: hisham.madi@mcbs.edu.om

F. Abdelfattah
e-mail: fadi.abdelfattah@mcbs.edu.om

G. A. Nabi
The Islamic University of Gaza, 108 Gaza, Palestine

A. Madi
Adjunct Business Faculty, Higher Colleges of Technology, Abu Dhabi, UAE

F. Abdelfattah
College of Business Administartion, A'sharqiyah University, Ibra, Oman

© The Author(s), under exclusive license to Springer Nature Switzerland AG 2024
A. Hamdan and E. S. Aldhaen (eds.), *Artificial Intelligence and Transforming Digital Marketing*, Studies in Systems, Decision and Control 487,
https://doi.org/10.1007/978-3-031-35828-9_35

402 H. Madi et al.

Keywords Board of directors · Audit committee · Financial performance · Palestine exchange

1 Introduction

In very severe political circumstances, unstable economic and uncertain financial environments, such as Palestine, public listed companies become increasingly vulnerable to agency issues, problems arising from two main forms of conflict of interest, one between managers and shareholders and the other between majority and minority shareholders [1]. In the first type of conflict of interest, the growing control of management posed a risk to the interests of the shareholders. Even though they have the skills and important information to run the company for the shareholders' benefit, their growing power puts them in a position where they may put their own interests ahead of the shareholders' [2, 3]. In the second form of conflict of interest, majority shareholders expropriate their power and influence to benefit themselves at the expense of minority shareholders [1, 4, 5]. Hence, to mitigate agency problems, public shareholding companies may use corporate governance mechanisms (board of directors and audit committee) as monitoring tools to reduce agency conflict as indicated by the agency theory [6]. Additionally, the theory of resource dependency places a significant emphasis on the central role that the board of directors plays in the acquisition of vital resources for the firm. These resources are considered crucial for the continued growth and success of the organization, and their acquisition is seen as a key factor in improving the firm's performance [7]. Reguera-Alvarado and Bravo [8] indicate that the board of directors is an essential corporate governance tool and a governance mechanism to protect a firm and its shareholders. Al-Matari [9] argues that among its responsibilities, the board of directors plays a key monitoring role including controlling management actions and operations to mitigate agency problems and overseeing the Chief Executive Officer's decision and management. Moreover, the board of directors is responsible to draw strategic policies of a firm as well as ensure shareholders' value and rights are protected [10]. Under the umbrella of corporate governance, the audit committee (AC) is the board subcommittee, established to oversee the financial reporting process [6, 11]. The AC is charged with monitoring the financial and non-financial reporting activities and hence reduces information asymmetry between management and shareholders [6]. According to Zhou et al. [12], the AC is responsible for overseeing and monitoring the firm's internal control system and reviewing the firm's operations risks. Therefore, the AC is expected to improve the transparency of the financial statements and risk management practices and thus enhance the quality of financial reporting. Along a similar line of argument, the AC plays a significant role in protecting the shareholders' interests in relation to the financial and non-financial reporting process and control [12].

Despite the challenges posed by economic and political instability, which are anticipated to negatively affect the stock market's performance and lead to a loss of confidence among domestic and foreign investors, the Palestine Exchange (PEX)

has managed to attract, retain, and incorporate substantial amounts of foreign and institutional investors [2, 3, 5, 13]. The stock market's performance in an unstable and risky environment raises important questions about market dynamics and factors driving performance. Would the introduction and implementation of a corporate governance code have contributed to this surprising outcome? If so, does this suggest that sound governance practices become increasingly crucial in unstable business environments? If not, are these practices merely superficial in nature?

Prior research has explored the impact of the board of directors and AC characteristics in detail. The majority of this research focused on well-governed corporations, a stable and regulated economy, investor rights protection, and legal enforcement [8, 10, 12, 14–19]. In spite of this, not a lot of research has been done to investigate how the composition of the board of directors and the AC influences the performance of a company, particularly in a country like Palestine, which has a very uncertain economic environment. Therefore, PEX offers an interesting and unique environment to study the relationship between board characteristics, AC characteristics, and financial performance. Moreover, investigating the applicability of the dominant theoretical frameworks, namely agency theory and resource dependence theory, in a Palestinian context will contribute to the existing literature by offering a new perspective and illuminating the validity of these theories. This study aims to highlight the role of the board of directors and the AC as essential corporate governance mechanisms for the effectiveness as well as success and survival of companies listed on the PEX.

2 Background and Hypothesis Development

2.1 Prior Studies and Hypotheses Development

Board of Directors Characteristics

From the point of view of the agency theory, non-executive directors are seen as a mechanism controlling managers' decision-making processes [20]. They can better protect shareholder interests and conduct monitoring and control functions to align corporate resources for improved performance [21]. Therefore, the introduction of non-executive directors is intended to give an objective and impartial assessment of management performance. Consequently, the shareholders' interests are protected and maximized [8]. Employing resource dependence theory, Cohen et al., [7] demonstrated that non-executive directors contribute to the value of a company because their different backgrounds provide expert counsel, function as a safeguard, and bring skills, knowledge, and experience.

Prior empirical research on the relationship between the effectiveness of non-executive directors and firm performance yielded inconsistent results. Some studies [22, 23] revealed that the inclusion of non-executive members on the corporate board has a favorable impact on business performance. However, Costa [24] have shown no relationship between non-executive directors and company performance. Additionally, Guo and Kumara [25] and Arosa et al. [26] revealed that non-executive directors have a negative impact on corporate performance. Based on the above discussions, the following hypothesis is empirically tested:

H1: The proportion of non-executive directors is positively associated with financial performance.

Accounting researchers continue to discuss whether large or small boards are more successful at performing the monitoring job and enhancing business performance. Some research found a negative correlation between the size of a company's board and its financial performance [26–28]. This conclusion is supported by the fact that a large board may lead to communication and coordination problems, hence diminishing its efficacy and monitoring efficiency. Additionally, a larger board size is inefficient regarding higher maintenance expenditures and more planning, issue processing, and decision-making difficulties [23]. The results of these studies were in line with the idea of the agency theory and showed that there was a correlation between a larger board size and lower firm performance. Other scholars, however, observed positive associations between board size and financial performance [22, 26, 29]. These studies show that large corporate boards are more effective because they have more knowledge and expertise between them. Furthermore, the research showed that as the board size grows, it becomes increasingly likely that the perspectives and interests of all stakeholders will be considered in decision-making and less likely that decisions will be made to benefit only a select few board members. The results of these studies supported the resource dependence theory and showed that there is a positive link between the size of the board and how well a company does. Other data, besides agency theory and resource dependency theory, have shown that there is no correlation between business success and board size [29, 30]. The following hypothesis is therefore developed as a result of previous discussion and supporting evidence:

H2: Board size is negatively associated with firm performance.

Audit Committee Characteristics

The AC is considered a key decision-making body with respect to the financial reporting process, as it is responsible to oversee the financial reporting practices and reduce the conflicts of interest and information symmetry between management and shareholders [11, 31]. It is generally agreed upon that the non-executive members of the AC are an important aspect of the effectiveness of the committee, in addition to being a vital component of accomplishing high-quality financial reporting [32, 33]. The agency theory argues that the AC, made up primarily of non-executive directors, performs an effective oversight role that enhances the quality of information and disclosures [6]. Moreover, it is believed that the inclusion of non-executive members strengthens the AC's capability to oversee and restrict the manipulative behavior of corporate management [32]. Appointing non-executive members to the AC permits them to exercise independent judgment in circumstances when internal management and shareholders have conflicts of interest, such as reviewing financial reporting statements and reducing the consumption of benefits by executives [20, 34]. Hence, non-executive directors of AC would improve the firm performance. According to the resource dependence theory, non-executive members may bring resources through their broader ranges of competence, which can lead to better decision-making that leads to better performance [35]. Hence, appointing non-executive members to the AC enable the firm to extract useful resources, which in turn helps the firm achieves its goals and improves performance [7]. Empirically, some research indicates a correlation between the percentage of non-executive members of AC members and financial performance [14, 36]. In contrast, other studies [32] disclosed that audit committees consisting of non-executive members are negatively associated with financial performance. However, other researchers [37, 38] found no association between the proportion of non-executive members in the AC and financial performance. Based on the above discussion, the following hypothesis is formulated:

H3: There is a positive relationship between non-executive members of the audit committee and financial performance.

According to the agency theory, for AC members to effectively oversee the reliability of a company's financial reporting process and the conduct of its management, they must have a solid grounding in accounting, finance, or financial literacy [36]. Consequently, as a result, the financial competence of AC members will strengthen the committee's ability to develop and seek inquiries that require management to investigate matters more completely and conduct more extensive audits [39]. This would improve the quality of corporate reporting and reduce information asymmetry [1]. Hence, AC with financial expertise members is more likely to protect investors' wealth and thus enhance firm performance [40]. From the resource dependency perspective, AC with more financial expertise members is more likely to bring to the firm different skills, experiences, and ideas that may enhance such committee effectiveness and hence improve firm performance [33]. Therefore, according to agency theory and resource dependency theory, it is anticipated that an increase in the number of members with accounting knowledge of ACs will minimize agency costs and thereby boost company performance [11]. Therefore, the following hypothesis is structured:

H4: Audit committee financial expertise is positively associated with financial performance.

The frequency of AC meetings is essential to the committee's audit and control quality performance [32]. According to Abbott et al. [41], committee meetings enable the AC to periodically examine the company and resolve any challenges found by management. Furthermore, an AC that holds frequent and active meetings has ample opportunity to effectively oversee the financial reporting process, efficiently monitor the actions of management, and ensure the proper functioning of internal controls [12]. Therefore, by holding frequent meetings, the AC improves its monitoring function, which could lead to enhanced business performance [41]. Previous studies on the impact of AC meeting frequency on business performance produced contradictory findings [28, 31]. The literature shows that frequent meetings of the AC can lead to an improvement in the financial accounting systems, resulting in enhanced performance [12, 17, 31, 41]. Jensen [28] stated that frequent AC meetings may enhance firm performance because it improves monitoring quality [42]. However, some studies revealed a negative relationship between the two variables (e.g., [43, 44]). Other researchers discovered no correlation between AC sessions and performance, such as [1, 12, 17]. It is, therefore, hypothesized that:

H5: Audit committee meetings are positively associated with financial performance.

Resource dependence theory argues that a larger AC may provide more diverse resources and perspectives, which can help enhance firm performance [7]. This is due to the fact that a larger AC can use its experience and skills to aid the committee in overseeing the process of financial reports. In addition, non-executive directors are recruited due to their access to information, industry expertise, and resources that enable management to effectively communicate and take decisive action in a timely and efficient manner [7]. Therefore, AC members with strong industry experience are more likely to have adequate information to analyze the business operations and risks of a firm as well as the accuracy of its financial reports [7], resulting in improved financial performance. According to Naveen and Singh [21], large audit committees offer greater monitoring resources for senior management. These committees could also help the organization improve its internal governance practices, which would make more resources available for internal monitoring. On the contrary, supporters of the agency theory argue that if the size of a committee is excessive, it is probable to result in low performance [28, 34, 42]. It was reported that the larger AC lost focus and became less interactive than the smaller AC. In comparison to those with a larger number of members, audit committees that have fewer members typically have a higher level of participation [33]. It is anticipated that significant AC may struggle to fulfill their duties effectively due to potential communication and coordination issues, leading to a decline in the efficiency of AC monitoring [42]. Researchers from both developed and developing countries looked into the link between the size of an AC and indicated a negative association between the two variables. Nonetheless, much research in both developed and developing countries have revealed a positive correlation (e.g. [24, 36, 39]. Several other studies, in addition to those undertaken by proponents of the agency theory and the resource dependency theory, revealed no association between AC size and firm success. Included in this research are Ben Barka and Legendre [16], Zhou et al. [12], and Ghabayen [38]. Based on these results, the researcher proposes the following hypothesis:

H6: Audit committee size is negatively associated with financial performance.

3 Research Design

The data used in this study was obtained from the annual reports of companies listed on the Philippine Stock Exchange (PSE) for the period between 2011 and 2015, as well as from the annual guides published by PEX for the companies. Initially, all the listed companies were included in the sample. However, companies that did not have a complete record of data related to variables employed in this study were excluded. This results in a sample of 29. Drawing from the work of some prior researchers, [23, 27, 38], accounting measures used in this study to the proxy financial performance of Palestinian companies are Return on assets (ROA) and return on equity (ROE). The explanatory variables employed in the current study are non-executive directors, the board size, non-executive members of the AC, AC financial expertise, AC meetings, and AC size. The model also includes firm size and leverage, which are two control variables for company characteristics. To investigate the effect of board and AC characteristics on financial performance, the following models are tested:

$$ROA_t = \alpha + \beta_1 NED_t + \beta_2 BS_t + \beta_3 NEMAC_t + \beta_4 ACFINEXP_t$$
$$+ \beta_5 ACMET_t + \beta_6 ACS_t + \beta_7 FIRMS_t + \beta_8 LEVE_t + e$$
$$ROE_t = \alpha + \beta_1 NED_t + \beta_2 BS_t + \beta_3 NEMAC_t + \beta_4 ACFINEXP_t$$
$$+ \beta_5 ACMET_t + \beta_6 ACS_t + \beta_7 FIRMS_t + \beta_8 LEVE_t + e$$

where

ROA_t Return on assets; measured by the ratio between net income and total assets; ROE_t Return on equity; measured by the ratio between net income and total stockholder's equity; α Intercept. NED_t Non-executive directors; are measured by the ratio between the number of non-executive members and the total members on the board. BS_t Board size; is measured by the total number of directors on the board. $NEMAC_t$ Non-executive members of the audit committee; are measured by the ratio between the number of non-executive members of the AC and the total members in the committee. $ACFINEXP_t$ AC financial expertise; measured any accounting or non-accounting professional experience and knowledge the AC members may possess. $ACMET_t$ AC meetings; are measured by the number of meetings held annually. ACS_t AC size; is measured by the total number of members on the committee. $FIRMS_t$ Firm size; is measured by the natural logarithm of total assets. $LEVE_t$ Leverage; is measured by the ratio of total debt to total assets. β_1 To β_8 Parameters of the model. e Random error.

Table 1 The descriptive statistics of the dependent, independent, and control variables

Variables	Observations	Minimum	Maximum	Mean	Std. Dev.
ROA	145	− 0.020800	0.071900	0.046540	0.019220
ROE	145	0.057400	0.205000	0.137480	0.055973
NED	145	0.000000	0.545500	0.0946000	0.156200
BS	145	6.000000	15.000000	9.280000	0.214000
NEMAC	145	0.000000	0.500000	0.100000	0.200693
ACFINEXP	145	0.750000	1.000000	0.900000	0.122899
ACMET	145	6.000000	12.000000	7.200000	2.408319
ACS	145	2.000000	4.000000	3.200000	0.750925
FIRMS	145	7.860000	7.880000	7.868000	0.007509
LEVE	145	0.621000	0.702000	0.652000	0.027988

4 Empirical Results

4.1 Descriptive Statistics

Table 1 provides a summary of the descriptive statistics of the variables used to estimate the regression models. During 2011–2015, the mean financial performance as measured by ROA for the entire sample ranged from − 2.08% to 7.15%. Similarly, the average financial performance as measured by ROE ranged from 5.74 to 20.5%. The proportion of non-executive directors on the board ranged from 0 to 54.55%, with a mean of 9.46% for the entire sample. Regarding the board size variable, the number of directors ranged from six to fifteen, with a sample-wide mean of roughly nine. In terms of AC financial competence, 90% of committee members have financial expertise on average. This demonstrates that over half of the committee is financially literate. In terms of AC meetings, the frequency of meetings ranged from a minimum of 6 to a maximum of 12, with a mean of about 7 meetings in a year. As for the control variables, the total company assets (in logarithms) varied between 7.86 and 7.88, with a mean of 7.86. The range of leverage was between 62.1 and 70.2%, with a mean of 65.2%.

4.2 Correlation

A Pearson correlation coefficient matrix was constructed to determine the nature and magnitude of the association between the independent variables. The matrix was reported in Table 2. The table indicates that the correlation coefficient between independent variables is larger than 60%, indicating that multicollinearity is not viewed as a significant issue.

The Effect of the Characteristics of the Board of Directors and the Audit ... 409

Table 2 Pearson correlation

	NED	BS	NEMAC	ACFINEXP	ACMET	ACS	FIRMS	LEVE
NED	1							
BS	− 0.002	1						
NEMAC	− 0.250	0.250	1					
ACFINEXP	0.408	− 0.408	0.480	1				
ACMET	− 0.250	0.250	0.400	0.408	1			
ACS	− 0.133	0.040	− 0.801	− 0.080	− 0.080	1		
FIRMS	− 0.133	0.010	0.327	0.327	− 0.534	0.070	1	
LEVE	0.143	− 0.143	0.005	0.649	0.060	− 0.018	− 0.312	1

4.3 Regression Analysis

The results of the regressions for model 1 and model 2 are reported in Table 3. The results show that R^2 values are 31% and 29% respectively, indicating that 31% and 29% of the return on assets and return on equity variations are determined by the board of directors and AC characteristics employed in this study. Furthermore, the p-value of models 1 and 2 is 0.000 and hence the models are significant in explaining the performance of Palestinian listed firms.

Table 3 Regression models using ROA and ROE as dependent variables

Independent variables	Model 1 (ROA) coefficient	Model 2 (ROE) coefficient
NED	− 0.015	0.132
BS	− 0.002	− 0.034***
NEMAC	0.037	− 0.063
ACFINEXP	0.004	− 0.091
ACMET	0.0033	0.032***
ACS	− 0.004	0.041
FIRMS	− 0.033***	− 0.041
LEVE	− 0.006	0.129*
SIG	0.000	0.000
R^2	%31	%29
Mean of residual	0.000	0.000
White	0.018	0.009
(Cross-section fixed)	0.000	0.000
(Cross-section random)	0.1889	0.135

***Significant at 1%, *Significant at 10%

As indicated in Table 3, there is no significant correlation between the financial performance (as measured by ROA and ROE) and the non-executive directors and AC on the board of directors. As a result, hypotheses H1 and H3 are not supported. The findings contradict both agency theory and resource dependence theory. The non-executive directors of Palestinian companies lack effective monitoring systems and industry knowledge to influence financial performance. The findings are similar to other studies that reported no association between the variables [15, 17, 24]. Possible reasons for this outcome include the fact that non-executive directors are not aware of all of the activities and businesses that are conducted by the company, as a result, non-executive directors will not be able to oversee the management's activities. Consistent with expectation, board size has a significant and negative impact on financial performance represented by accounting–based measure (ROE). Thus, hypothesis 2 is supported. This result is in the line agency theory perspective, as it suggests that more directors on board may decrease board effectiveness in monitoring management action due to coordination and free rider problems. This finding is consistent with other studies that reported a negative association between the two variables [6, 13, 26, 27, 30]. As the number of directors increases, the costs of the board meetings increase (such as travel expenses, refreshments, and director's remunerations) and the reported profits decrease, therefore, negatively influencing financial performance. However, the result is inconsistent with the perspective of resource dependence theory which posits that the board size considers an essential resource for monitoring management's activities and hence decreases financial reporting fraud. The result also is inconsistent with the findings of Al-Matari [9] and Pucheta-Martínez and Gallego-Álvarez [10] who found that board size is appearing to have a positive impact on firm performance.

Contrary to agency and resource dependence theories, financial competence within the AC framework has little to no effect on financial performance as measured by ROA and ROE. Therefore, H4 is not confirmed. This conclusion is in line with prior research that found no correlation between the variables [11, 14, 31]. The insignificant impact of AC financial expertise may be explained through the fact that AC members may lack sufficient accounting skills, knowledge, and experience, hence such a committee would not be able to evaluate the firm's operation and thus firm performance would not increase [7, 40].

Consistent with hypothesis 5, the result indicates that the frequency of meetings of AC members has a significant and positive influence on the accounting-based measure (ROE). This finding is consistent with resource dependence theory and agency theory. This finding is also consistent with the findings of earlier research [6, 9]. The significant and positive association between AC meetings frequency and corporate performance can be explained through the fact that AC members meeting regularly would be able to discuss and oversee management operations and hence improve financial performance. Moreover, it is expected that the size of the board of directors has a positive and substantial effect on financial performance, however, the given results do not support this hypothesis, indicating that a large number of directors on the board does not contribute to increased business performance. This result is inconsistent with resource dependency theory. This indicates that more directors on

AC would not add value to firm performance this is because such directors are lacking the experience that is needed to handle issues related to management operations. This outcome aligns with findings from previous research efforts by Salehi et al. [31], Al Farooque et al. [15] and Hassan et al. [1].

As for the two control variables, firm size and leverage, they showed a significant association with ROA and ROE, respectively (Leverage is found only significant at a 10% level). However, large firms are more likely to utilize sophisticated accounting and auditing procedures to decrease these monitoring and agency expenses caused by manager and shareholder conflicts. Consequently, any rise in these expenses would have a negative effect on ROA. On the other hand, the positive relationship between leverage and ROE may be explained by the fact that the increase in leverage levels automatically leads to an increase in firm operations and thus increase in reported profits.

5 Conclusion

The study examines the association between the characteristics of the board of directors and audit committee (AC) and financial performance in a sample of Palestine Exchange-listed companies from 2011 to 2015. The differences in financial performance among these publicly traded companies were investigated using agency theory and resource dependence theory. From the analysis conducted, it revealed that board size is negatively associated with ROE; AC meetings are positively related with ROE; firm size is negatively associated with ROA; a positive relation between leverage and ROE is observed, but only significant at a 10% level, while other relations are statistically insignificant with financial performance. This suggests that the findings are, for the most part, contradictory to agency theory and resource dependence theory. Even though the predictions of the agency theory are the ones that are most commonly found in studies of corporate governance, describing other theories will help us learn more about the relationships between the efficiency of the board of directors, the audit committee, and the company's financial performance. In fact, the governance system that has been implemented by Palestinian companies is still in its infancy, and the companies have very limited experience with using the system to improve their overall corporate performance. However, the outcome of the study is expected to assist policymakers together with management and shareholders. As the findings of this research become clearer, policymakers and legislative authorities will be better able to work within the constraints of the current corporate law. Furthermore, the study also suggests that authorities should also enforce the rules governing corporate governance in a more stringent manner and impose additional penalties on those who break the laws. The study's results would further emphasize the significance of various board and AC characteristics in Palestine. The findings revealed that Palestinian non-executive directors do not have sufficient expertise, skills, and knowledge to understand financial reporting details. Therefore, PEX needs to ensure that all directors participate in the continuous education program to enhance their competence and professionalism, and to publish in their annual reports whether their directors have received such training. Companies should make sure that the majority of their board members are non-executives. This means that the directors don't work for the company and don't depend on it for their income. This way, the directors can keep an eye on what's going on and offer their advice in the best interest of the company without fear of retaliation. However, this research has several limitations and caution should be exercised when drawing any conclusions from it. The research only focused on the quantitative aspects of corporate governance, but this ignores the fact that there are qualitative aspects as well. It would be more reliable if interviews were conducted with directors, managers, and shareholders to learn about their perspectives on the procedures that contribute to good governance. Moreover, another drawback of this study is that the financial performance proxies used in the model only considered accounting-based measures, not market-based measures. These limitations leave room for further research.

References

1. Hassan, Y., Hijazi, R., Naser, K.: The influence of corporate governance on corporate performance: evidence from Palestine. Afro-Asian J. Fin. Account. 6(3), 269–284 (2016)
2. Saleh, M.W., Zaid, M.A., Shurafa, R., Maigoshi, Z.S., Mansour, M., Zaid, A.: Does board gender enhance Palestinian firm performance? The moderating role of corporate social responsibility. Corp. Govern. Int. J. Bus. Soc. 21(4), 685–701 (2021)
3. Zaid, M.A., Wang, M., Abuhijleh, S.T.: The effect of corporate governance practices on corporate social responsibility disclosure: evidence from Palestine. J. Glob. Responsib. 10(4), 134–160 (2019)
4. Saleh, M.W., Islam, M.A.: Does board characteristics enhance firm performance? Evidence from Palestinian listed companies. Int. J. Multidiscip. Sci. Adv. Technol. 1(4), 84–95 (2020)
5. Hassan, Y., Hijazi, R.: Determinants of the voluntary formation of a company audit committee: evidence from Palestine. Asian Acad. Manage. J. Account. Fin. 11(1), 27–46 (2015)
6. Musallam, S.R.: Effects of board characteristics, audit committee and risk management on corporate performance: evidence from Palestinian listed companies. Int. J. Islam. Middle East. Financ. Manag. 13(4), 691–706 (2020)
7. Cohen, J., Gaynor, L.M., Krishnamoorthy, G., Wright, A.M.: Academic research on communications among external auditors, the audit committee, and the board: implications and recommendations for practice. Curr. Issues Audit. 2(1), A1–A8 (2008)
8. Reguera-Alvarado, N., Bravo, F.: The effect of independent directors' characteristics on firm performance: tenure and multiple directorships. Res. Int. Bus. Financ. 41, 590–599 (2017)
9. Al-Matari, E.M.: Do characteristics of the board of directors and top executives have an effect on corporate performance among the financial sector? Evidence using stock. Corp. Govern.: Int. J. Bus. Soc. 20(1), 16–43 (2019)
10. Pucheta-Martínez, M.C., Gallego-Álvarez, I.: Do board characteristics drive firm performance? An international perspective. RMS 6(4), 1251–1297 (2020)
11. Appuhami, R., Tashakor, S.: The impact of audit committee characteristics on CSR disclosure: an analysis of Australian firms. Aust. Account. Rev. 27, 400–420 (2017)
12. Zhou, H., Owusu-Ansah, S., Maggina, A.: Board of directors, audit committee, and firm performance: evidence from Greece. J. Int. Account. Audit. Tax. 31, 20–36 (2018)
13. Saleh, M.W., Latif, R.A., Bakar, F.A., Maigoshi, Z.S.: The impact of multiple directorships, board characteristics, and ownership on the performance of Palestinian listed companies. Int. J. Account. Audit. Perform. Eval. 16(1), 63–80 (2020)

14. Kallamu, B.S., Saat, N.A.M.: Audit committee attributes and firm performance: evidence from Malaysian finance companies. Asian Rev. Account. **23**(3), 206–231 (2015)
15. Al Farooque, O., Buachoom, W., Sun, L.: Board, audit committee, ownership and financial performance–emerging trends from Thailand. Pac. Account. Rev. **32**(1), 54–81 (2020)
16. Ben Barka, H., Legendre, F.: Effect of the board of directors and the audit committee on firm performance: a panel data analysis. J. Manage. Governance **21**, 737–755 (2017)
17. Al-Okaily, J., Naueihed, S.: Audit committee effectiveness and family firms: impact on performance. Manag. Decis. **58**(6), 1021–1034 (2019)
18. Merendino, A., Melville, R.: The board of directors and firm performance: empirical evidence from listed companies. Corp. Govern. Int. J. Bus. Soc. **19**(3), 508–551 (2019)
19. Kao, M.F., Hodgkinson, L., Jaafar, A.: Ownership structure, board of directors and firm performance: evidence from Taiwan. Corp. Govern.: Int. J. Bus. Soc. **19**(1), 189–216 (2018)
20. Fama, E., Jensen, M.: Separation of ownership and control. J. Law Econ. **26**(2), 301–325 (1983)
21. Naveen, K., Singh, J.P.: Outside directors, corporate government and firm performance. Empirical evidence from India. Asian J. Fin. Account. **109**(8), 89 (2012)
22. Haniffa, R., Hudaib, M.: Corporate governance structure and performance of Malaysian listed companies. J. Bus. Financ. Acc. **33**(7–8), 1034–1062 (2006)
23. Muller, V.: The impact of board composition on the financial performance of FTSE 100 constituents. EL-Sevier J. **109**, 969–975 (2014)
24. Costa, M.: The impact of CEO duality and board independence on FTSE small cap & Fledgling company performance. World Rev. Bus. Res. **5**, 1–19 (2015)
25. Guo, Z., Kumara, U.: Corporate governance and firm performance of listed firms in Sri Lanka. EL-Sevier J. **40**, 664–667 (2012)
26. Arosa, B., Iturralde, T., Maseda, A.: The board structure and firm performance in SMEs: evidence from Spain. EL-Sevier J. **19**(3), 127–135 (2014)
27. Dogan, M., Yildiz, F.: The impact of the board of directors size on the bank's performance: evidence from Turkey. Eur. J. Bus. Manag. **5**(6), 130–145 (2013)
28. Jensen, M.C.: The modern industrial revolution, exit, and the failure of internal control systems. J. Fin. **48**(3), 831–880 (1993)
29. Bebji, A., Mohammed, A., Tanko, A.: The effect of board size and composition on the financial performance of banks in Nigeria. Afr. J. Bus. Manage. **9**(16), 590–598 (2015)
30. Garg, A.K.: Influence of board size and independence on firm performance: a study of Indian companies. J. Decis. Makers. **32**(3), 39–60 (2007)
31. Salehi, M., Tahervafaei, M., Tarighi, H.: The effect of characteristics of the audit committee and board on corporate profitability in Iran. J. Econ. Admin. Sci. **34**(1), 71–88 (2018)
32. Bedard, J., Gendron, Y.: Strengthening the financial reporting system: can audit committees deliver. Int. J. Audit. **14**(2), 174–210 (2010)
33. Jackling, B., Johl, S.: Board structure and firm performance: evidence from India's top companies. Corp. Govern. Int. Rev. **17**(4), 492–509 (2009)
34. Beasley, M.S., Carcello, J.V., Hermanson, D.R., Neal, T.L.: The audit committee oversight process. Contemp. Account. Res. Spring. **26**(1), 65–122 (2009)
35. Ahmed, K., Sehrish, S., Saleem, F., Yasir, M., Shehzad, F.: Impact of concentrated ownership on firm performance (evidence from Karachi stock exchange). Interdiscip. J. Contemp. Res. Bus. **13**(5), 201–210 (2012)
36. Chan, K., Li, J.: Audit committee and firm value: evidence on outside top executives as expert-independent directors. Corp. Govern. Int. Rev. **16**(1), 16–31 (2008)
37. Hsu, H.E.: Audit committees in U.S. Entrepreneurial firms. J. Am. Acad. Bus. **13**, 121–128 (2008)
38. Ghabayen, M.: Board characteristics and firm performance: case of Saudi Arabia. Int. J. Account. Fin. Report. **2**, 168–200 (2012)
39. Salehi, M., Tarighi, H., Safdari, S.: The relation between corporate governance mechanisms, executive compensation and audit fees: evidence from Iran. Manage. Res. Rev. **41**(8), 939–967(2018)

40. Chaudhry, N.I., Roomi, M.A., Aftab, I.: Impact of expertise of audit committee chair and nomination committee chair on financial performance of firm. Corp. Govern.: Int. J. Bus. Soc. **20**, 621–638 (2020)
41. Abbott, L.J., Parker, S., Peters, G.F., Raghunandan, K.: An empirical investigation of audit fees, nonaudit fees, and audit committees. Contemp. Account. Res. **20**(2), 215–234 (2003)
42. Hassan, Y.M.: Determinants of audit report lag: evidence from Palestine. J. Account. Emerg. Econ. **6**(1), 13–32 (2016)
43. Danoshana, M.S., Ravivathani, M.T.: The impact of the corporate governance on firm performance: a study on financial institutions in Sri Lanka. Merit Res. J. Account. Audit. Econ. Fin. **8**(1), 118–121 (2013)
44. Hsu, W., Petchsakulwong, P.: The impact of corporate governance on the efficiency performance of the Thai non-life insurance industry. Geneva Papers Risk Insurance Issues Pract. **35**(1), S28–S49 (2010)

A Comparative Analysis of Rural and Urban Public Health Infrastructure and Health Outcomes During Pandemic: With Special Reference to Karnataka State

S. J. G. Preethi and Tinto Tom

Abstract Efficiency of human capital determined by Education standard of living and most importantly Health. Health infrastructure is an essential factor to connect health input and health output. Health input has been taken through preventive methods or curative methods. To access these method country needs equipped health infrastructure. Present study analyses the existence of health infrastructure and its association with health Outcome. Study period taken into account for the present study is pandemic period. Physical public Health infrastructure like availability of Sub Centres, primary health Centres and Community health Centres in rural and urban area taken in to account. In the case of health workforce Nurses, Doctors, Lab technician and Pharmacists in position considered for this analysis. Health outcome variable such as Infant Mortality Rate, Birth Rate and Death rate used for analysis to find out the relationships. Study identified that there is a strong correlation between health infrastructure in rural and urban area of Karnataka with Health outcomes. Research also identified that there is close connection between availability of health workforce and health outcome during the study period.

Keywords Health infrastructure · Health outcome · Covid · IMR · DR health manpower

1 Introduction

Health is very essential ingredient to economic growth in every economy. Social medicine, Community medicine and Preventive and Curative medicines are the integral part of Public Health. Good health is very important and a contributing factor to our nation. To achieve good health all the country should promote health care

---

S. J. G. Preethi (✉) · T. Tom
Kristujayanti College, Bangalore, Karnataka 560077, India
e-mail: preethi.s@kristujayanti.com

© The Author(s), under exclusive license to Springer Nature Switzerland AG 2024
A. Hamdan and E. S. Aldhaen (eds.), *Artificial Intelligence and Transforming Digital Marketing*, Studies in Systems, Decision and Control 487,
https://doi.org/10.1007/978-3-031-35828-9_36

services, prevent diseases and help people to make their healthy choices. Health infrastructure are the basic services or social capital of a country. It is very much essential for understanding the healthcare policy and welfare mechanism in a country. Hence, in this paper will be covering about the health infrastructure in rural and urban areas in Karnataka. When we talk about rural and urban, there is a vast disparity in every aspects especially regarding health infrastructure. There is inequality in publicly provided goods such as healthcare facilities and it diverge among urban and rural areas. The government policies and ideologies are different among urban and rural areas in Karnataka. Health infrastructure is lot more advanced in urban areas. It is because the government is less concentrating on the rural areas. Advancement of health infrastructure in both rural and urban areas are necessary for a better economy. The paper we will be looking into the causes for this disparity, negligence of government, measures to be taken to improve the status of health infrastructure.

At present status of Health Infrastructure in India is not in a commendable position. As per the NITI Aayog 2021, report identified that only 35% of population out of 50% can able to access hospital beds. According to Indian Public Health Standards Health Infrastructure gap between rural and urban should be reexamined. The development of infrastructure and health care facilities, the position of the workforce, and the quality of service delivery are important challenges that are confronting healthcare centers in rural India. The importance of the good health of people cannot be minimized, as it has been considered as one of the most important components of the human capital. Health is essential factor to enhance the productivity. Due to its vital importance, the economics of health is attracting researchers and policy makers more rapidly in the recent decades. It is well said in the theory of the human capital that people should invest themselves in terms of education, health and skill development programmes. Health is a major segment of the human capital. If the quality of the human capital is not good, then the physical capital and natural resources cannot be utilized properly and the growth can neither be sustained nor be qualitative.

The covid 19 pandemic brought about significant changes in the health infrastructure all over the world. It exposed the gaps in healthcare for underserved and vulnerable populations. Ensuring health equity in a pandemic requires more resilient public health infrastructure during normal times. The SARS-CoV-2 virus does not differentiate between urban and rural areas. It affects people who do not have the immunity against the virus. The virus was spreading inevitably. The problem is arises from the fact that the rural India is not well equipped to deal with the diseases. According to World Bank Data 65% of people resides in the rural areas. The problem stems out from the fact that the country was least focused on developing the rural health infrastructure. A parliamentary committee report on 'Management Of Covid 19 Pandemic and Related Issues' presented to the chairperson of Rajya Sabha on December 21, 2020 indicated the possibility of diseases reaching rural areas. It had advised the government to ensure testing infrastructure and upgraded health infrastructure in remote and rural areas. The National Health Mission, which predicts achievement of universal access to equitable, affordable and quality healthcare services, receives about 50% of budgetary allocation for health. However, the allocations for rural component has decreased by 3% since last year. Paper analyzing the existence of disparities between rural and urban health care system.

2 Literature Review

Adams et al. [1] examined the impact of health infrastructure with economic growth in 30 subsharan Africa regions during 1990–2014. They identified though capital investment is supportive other resources like labour force and political environment not facilitated the health infrastructure to increase the GDP.

Varkey et al. [2] identified the relationship between health infrastructure, health outcome and economic growth of India for major states. Author found that there is relationship between health infrastructure index with Life expectancy and negative relationship with IMR.

Hati [3] paper examined health infrastructure and health outcome India through composite indices. Results of the study recommending there is strong relationship between health status and economic wellbeing. Study also proved that strong association between primary health infrastructure with preventive and curative health achievements.

Siddu [4] in his article analyzed Karnataka's health infrastructure with Kerala. He identified that Kerala health infrastructure and health indicators are comparatively greater than Karnataka. When we take the case of Gulbarga, we can realize that most of the parts are rural and hence the health infrastructure in those areas is quite low. If the health infrastructure in those areas is low, then the people living in those areas are not productive enough and hence, they will not be able to contribute to the nation in a proper way resulting in poverty.

Present research emphasis health infrastructure association with health outcome during covid period. This research gap has been identified in the present research.

3 Methodology

Present study analyzing the public health Infrastructure and its association with health outcome in Karnataka during covid. To prove that researcher chosen the following variables such as Public physical health infrastructure like number of Sub centres, Primary Health centres and community Health Centres in Rural and Urban areas with Public health manpower such as Nurses, Lab technician, Pharmacists and Doctors in position in PHC's and CHC's taken in to account. To measure the Health outcome Infant Mortality Rate, Birth Rate and Death Rate has been Chosen.

3.1 Objectives of the Study

There are three main objectives of this paper;

First one is to find the association between Public health infrastructure and Health outcome in Karnataka during 2018–19 to 2020–21.
Second Objective helps to identify the differences between rural and urban health infrastructure and health outcome during covid period.
Last objective is addressing the relationship between the existence of health personal and their association with health outcomes during the study period.

3.2 Data Collection and Study Period

Present study only focusing the relationship between Health Infrastructure and Health outcome during pandemic period i.e. 2018–19 to 2020–21. For this data has been collected from rural health statistics report of three years. Other information such as health status, health policy, and health manpower information has been collected from Ministry of health and family welfare and Karnataka health department websites.

4 Results and Discussion

Table 1 and Fig. 1 shows that there is an increasing number of health infrastructure in respective with Sub Centres, Primary Health Centres and Community health Centres in Rural and Urban Areas of Karnataka during pandemic period from 2018–19 to 2020–21.

Table 2 and Fig. 2 shows that health personal number has been progressing during covid period in Karnataka. This growth has been identified in Primary and Community health centres of Karnataka.

The information in Table 3 and Fig. 3 proved that there is significant improvement in health outcomes even in covid period. Its observed that there is drastic reduction in IMR, death Rate. Birth rate has been reduced at moderate level.

Table 1 Status of public health care system during covid

Year	SC		PHC		CHC	
	R	U	R	U	R	U
2018–19	9187	251	1995	364	198	9
2019–20	9188	247	2176	358	189	19
2020–21	8891	550	2141	390	182	30

A Comparative Analysis of Rural and Urban Public Health ...

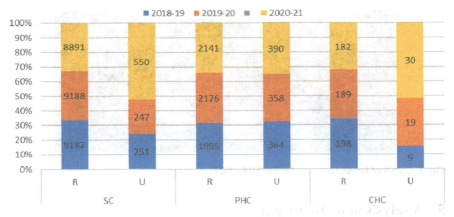

Fig. 1 Status of public health care system during covid

Table 2 Status of health manpower in PHC and CHC during covid

Year	PHC				CHC			
	Nurse	LT	Pharma	Doc	Nurse	LT	Pharma	Doc
2018–19	3334	1653	2149	2492	1110	183	268	475
2019–20	3721	1834	1849	2427	1788	238	224	312
2020–21	3215	1770	1675	2165	1715	306	1420	271

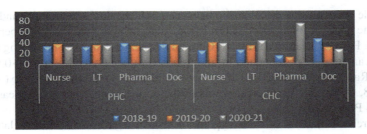

Fig. 2 Status of health manpower in PHC and CHC during covid

Table 3 Status of health outcome during 2018–2021

Year	IMR		BR		DR	
	R	U	R	U	R	U
2018–19	27	19	18.5	16.2	7.9	4.9
2019–20	25	20	18.1	15.9	7.2	4.8
2020–21	23	18	17.8	15.4	7.1	4.6

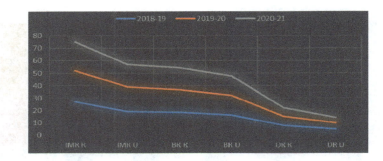

Fig. 3 Status of health outcome during 2018–2021

5 Analysis and Discussion

1. Relationship between Public Health infrastructure and Health Outcome:

Relationship between Public Health infrastructure and Health Outcome:

H_0: There is no relationship between Public health infrastructure Facility and Health Outcome during Covid.
H_1: There is an relationship between Public health infrastructure Facility and Health Outcome during Covid.

Table 4 explains the relationship between public health infrastructure and Health outcome during Covid period. Correlation values proves that when number of Sub Centre, Primary health centres and Community health centres is increasing there is significant amount of reduction in Infant mortality rate (-0.65, -0.74, -0.99) and Death Rate (-0.18, -0.97, -0.78). There is reduction in the Birthrate also (-0.427, -0.898, 0.898) due to efficient norms of the government to adopt the measures to control Population.

Therefor Rejection of Null hypothesis due to significance in the relationship between public health infrastructure and health outcome during Covid period.

2. Relationship between public health infrastructure in rural and urban areas and its health outcome:

H_0: Comparatively Urban areas health outcome and Public health infrastructure relationships are significantly lower than rural areas during Covid period.

Table 4 Relationship between public health infrastructure and health outcome

Public health infrastructure	IMR	BR	DR
SC	−0.655	−0.427	−0.189
PHC	−0.746	−0.898	−0.979
CHC	−0.990	−0.898	−0.786

H_1: Comparatively Urban areas health outcome and Public health infrastructure relationships are significantly greater than rural areas during Covid period.

Table 5 significantly explains during covid time period urban areas public health infrastructure are more positively correlated with health outcomes. Infant Mortality, Birth rate and Death Rate has been reduced when number of Sub Centre, Primary health centres and Community health centres were increasing in urban areas. For instance SC and IMR, BR and DR are in urban area has negative correlation − 0.86, −0.872, −0.815, −0.924, −0.587, −0.941. It shows Subcentres number are increasing Infant Mortality reduced at significant level. But in rural areas though the Public health infrastructure has been strengthen during covid, its relationship with IMR, BR and DR are not significantly connected. It proved its positive connectivity with health outcomes. For instance SC and IMR, BR and DR in rural area has positive correlation when SC in rural areas is increasing IMR, BR and DR also increasing 0.865, 0.867, 0.820, 0.927, 0.594, 0.944. Therefore alternative hypothesis has been accepted and rejecting the null hypothesis.

3. Public health manpower and its association with health outcome during covid period (Table 6).

H_0: There is no association between number of Public Health manpower in position with Health Outcome during covid period.

Table 5 Relationship between public health infrastructure in rural and urban areas and its health outcome

Health outcome	Rural/urban	SC		PHC		CHC	
		R	U	R	U	R	U
IMR	R	0.865	−0.860	−0.760	−0.764	0.997	−1.000
	U	0.867	−0.872	0.182	−0.941	0.436	−0.524
BR	R	0.820	−0.815	−0.811	−0.709	1.000	−0.994
	U	0.927	−0.924	−0.660	−0.849	0.977	−0.993
DR	R	0.594	−0.587	−0.956	−0.445	0.944	−0.906
	U	0.944	−0.941	−0.624	−0.872	0.966	−0.987

Table 6 Public Health manpower and its association with health outcome during covid

Health outcome	PHC—manpower in position				CHC—manpower in position			
	Nurses	LT	Pharma	Doc	Nurses	LT	Pharma	Doc
IMR	−0.405	−0.480	−0.942	−0.970	−0.688	−0.992	−0.934	−0.867
BR	0.144	−0.699	−0.698	0.614	−0.858	−0.990	−0.803	0.969
DR	−0.106	−0.854	−0.884	−0.785	−0.959	−0.923	−0.630	−0.940

H_1: There is an association between number of Public Health manpower in position with Health Outcome during covid period.

Association table proves that there is an relationship between Public health Manpower and its positive connection with health outcomes such as IMR, BR and DR. Health workforce in PHC and CHC are proving that when number of health manpower (i.e. Nurses, lab technician, pharmacists and Doctors) is increasing there is significant level reduction in IMR (-0688 and 0.405, -0.480 and -0.992, -0.942 and -0.867) in PHC and CHC respectively. It shows there is a close association existing between health workforce and health outcome. Therefore alternative hypothesis is accepted.

6 Good Health and Wellbeing (SDG-3) Status in India

As per United nation-Sustainable Development Goal 3 India's Indicators List, we could find out in many parameters India doing better performance and few parameters Its running low. Indicators related with health infrastructure like Percentage of children in the age group of 9–11 Months fully immunized at India 91 and Karnataka its 94 out target 100. In the case of total physicians, Nurses and Midwives at India level 36.84 and in Karnataka 70.2 out of 45. In the same way percentage of Institutional deliveries out of total deliveries reported 94.4 and 99.9 in India and Karnataka out of 100. In India Ministry of Health and family welfare developed broad framework for reproductive, maternal, newborn, child, adolescent health, communicable and Non-communicable dieses with the support of United Nations.

7 Conclusions and Policy Implications

This paper has empirically investigated the relationship between Health infrastructure and health outcome during covid period in Karnataka. The study used correlation analysis to prove the association between the chosen variable. It identified that there is an relationship between health infrastructure facility with health outcome. However, during pandemic period there is an lot of health issues reported in public health care system present study observed that strengthening health infrastructure is an significant action has taken by government of Karnataka in order to progress the health outcome.

Though pandemic, lead the economy in difficult path government consistently investing in health infrastructure. This positively reflected in the health outcome of the people.

An important policy implication based on the general result of the study is that to enhance the economic growth individual productivity is mandatory. This can be achieved through good health. To achieve the good health, Health infrastructure should be strengthen at all circumstances. Therefore, the respective governments should stabilize the economic environment to investing on health infrastructure at any situation. Human capital efficiency significantly connected with health.

Followed by that Government should pay attention to develop and implement innovative technology to improvise the efficiency of Health Infrastructure. Health work force should be trained in such a manner they could serve this technological advancement in rural areas too.

Country like India Health service provision during emergency period like pandemic to all needy people is less practical. Government should spread the awareness of Health education especially preventive health care measure to the public. This could improvise the Indian health care system.

References

1. Adams, S., Klobodu, E.K.M., Lamptey, R.O.: Health infrastructure and economic growth in Sub-Saharan Africa. In Health Economics and Healthcare Reform: Breakthroughs in Research and Practice, pp. 146–163. IGI Global (2018)
2. Varkey, R.S., Joy, J., Panda, P.K.: Health infrastructure, health outcome and economic growth: evidence from Indian major states. Health 7(11), 2020 (2020)
3. Hati, K.K., Majumder, R.: Health Infrastructure, Health Outcome and Economic Wellbeing: A District Level Study in India (2013)
4. Siddu, V.H., KD, K.M., Revankar, R.: Status and infrastructure of the health sector in Karnataka. Artha J. Soc. Sci. 11(3), 15–45 (2012)

Local Community Readiness to Implement Smart Tourism Destination in Yogyakarta, Indonesia

Sri Dwi Ari Ambarwati, Mohamad Irhas Effendi, and Sri Tuntung Pandangwati

Abstract Local economic development for relatively long period of time. Tourism also allows rapid recovery of local economy after COVID-19 outbreak. Amongst different kinds of approaches for tourism development, smart tourism destination (STD) has been widely believed as an excellent approach to rapidly recover tourism sector during post-pandemic. STD concept also allows improvement of adaptive capacity that is critical to anticipate any future disturbances including the unpredicted ones. However, a question remains regarding the readiness of local communities for implementing STD. Therefore, this research aims to address this gap by examining the readiness of managers of community-based tourism in the Special Region of Yogyakarta province, Indonesia. An online survey was conducted to collect data from various community-based tourism destinations located in this region. Data were analyzed using descriptive statistics to measure the overall readiness of community-based tourism to implement STD. For triangulation purpose, interviews were conducted to complement the online survey.

Keywords Smart tourism · Community based tourism · Desa Wisata · Economic recovery

---

S. D. A. Ambarwati (✉) · M. I. Effendi
Universitas Pembangunan Nasional Veteran Yogyakarta, Jl. SWK Jl. Ring Road Utara No. 104, Ngropoh, Condongcatur, Kec. Depok, Kabupaten Sleman, Daerah Istimewa Yogyakarta, 55283 Yogyakarta, Indonesia
e-mail: dwiari.ambarwati@upnyk.ac.id

S. T. Pandangwati
Universitas Gadjah Mada, Bulaksumur, Caturtunggal, Kec. Depok, Kabupaten Sleman, Daerah Istimewa Yogyakarta, 55281 Yogyakarta, Indonesia

© The Author(s), under exclusive license to Springer Nature Switzerland AG 2024
A. Hamdan and E. S. Aldhaen (eds.), *Artificial Intelligence and Transforming Digital Marketing*, Studies in Systems, Decision and Control 487,
https://doi.org/10.1007/978-3-031-35828-9_37

425

1 Introduction

Tourism has been a major sector in Indonesia especially during the last decades. The contribution of tourism sector to Indonesia's total GDP is about 4.1% and it provides roughly 10.5% of total employment in 2017 [1]. Travel restriction during COVID-19 pandemic has significantly impacted tourism sector in Indonesia. Yogyakarta, as one of significant tourist destinations in Indonesia was severely impacted by COVID-19 crisis. Many tourism-dependent sectors either formal or informal were disrupted by COVID-19 restrictions. Consequently, local economic growth was declined significantly during this crisis. Although current local economy in Yogyakarta has slowly recovered since borders were opened early this year, it is still reeling from the pandemic conditions. Smart Tourism Destination (STD) could be a promising solution to foster the eco-nomic recovery. STD aims to create a tourist destination in which information and communication technologies (ICT) were used to improve knowledge and information accessibility, facilitate sustainable innovations amongst stakeholders, enable public-private partnership and maintain relationships with customers [2]. For instance, ICT can be used to analyze the market, map prospective tourists' locations and also maintain communication with customers. ICT also can improve social network amongst stakeholders and this can facilitate social learning as well as collaborations. STD does not simply mean using digital technologies for marketing, but it comprehensively covers every aspect of tourism management [3]. In this era of uncertainty, tourism management needs to be adaptive to changing situations and STD can improve adaptive capacity and adaptive management of tourist destinations [4]. To enable physical distancing during post COVID-19, ICT can provide real time information of the number of visitors. Artificial reality and teleconference technology can be used to allow tourists virtually visit the destinations or enjoy local artists performances during pandemic restrictions. Digital marketing can support local economic recovery during post pandemic through providing user information and behavior tracking which are important to formulate further marketing strategies. However, the readiness of local communities to implement STD has not clearly understood especially in the context of developing countries. Indonesia has been focusing on community-based tourism to foster local economic development [5, 6]. In this case, the role of local communities is significant for tourism development. Therefore, this research aims to measure the readiness of local communities to apply STD. Measuring the readiness of local communities to implement STD is critical because they are the key actor and locomotor of the community-based tourist destination. Furthermore, understanding communities' readiness is critical to inform future STD development strategies and planning policies. This paper starts by a brief review of recent studies on STD. The main objective of this literature review is to identify the research gap addressed in this study. After which the research methods were explained to provide an overview of how data col-lection and analysis were conducted. Then the findings were presented and discussed, before we conclude by some policy recommendation and further research opportunities.

2 Literature Review

The concept of Smart Tourism Destination (STD) is basically the modified version of traditional tourism which make use of smart information and communication technology (ICT) to improve management and innovations within a tourist destinations. Many scholars like Bulchand-Gidumal [6] and Boes et al. [2] adopted Cohen's [7] smart city wheel in their research of smart tourism. STD is rooted from the concept of smart city and it is defined as a concept that aims to create tourist destinations that are innovative, adaptive and sustainable through utilization of advanced smart technology and infrastructure to ameliorate destination competitiveness, enhance tourist experience and ensure local residents' well-being [8, 9]. Coupling the smart concept and tourism destination development requires synergies and trade-offs between tourists and the actors involved in the governance of the tourism destinations. Through a smarter platform for data collection and distribution, tourism actors can track their destination progress and formulate strategies to increase its competitiveness. Moreover, STD is excellent for integrating different tourism actors across various levels. These aspects are critical to ensure that tourism destinations can sustain and thrive over time. According to Shaafiee et al. [10] the existing research on STD covers discussion on causal conditions, context conditions, intervening conditions, actions/interactions and consequences. Causal conditions are things or situations that encourage the growth of STDs such as the emergence of smart city, advanced ICT, big data and socio-economic development. Studies about context conditions examine the factors that are required for successfully coupling smart technologies and tourism destinations. In terms of intervening conditions, the existing studies discuss how government and other agencies support STD development. The rest of studies focus on analyzing various actions responding to STD phenomenon and the consequences of implementing STD strategies. However, there has been a relatively limited research that focuses on the readiness of community-based tourism especially in an Indonesian context. More research is needed to delve into the community dimensions of STD [11]. The existing Indonesian studies on smart tourism mainly analyze the readiness of a city or a tourism zone, but rarely examine local community readiness in great depth. For example, Farania et al. [11] analyzed the readiness of The City of Surakarta to implement smart tourism, and Saputra and Roychansyah [12] quantitatively examined the implementation of STD concept in three tourist destinations in The City of Yogyakarta. If community-based tourism is important to boost local economic development especially during post-pandemic situations, more research focusing on local community e-readiness is needed.

3 Research Methodology

This research applied quantitative approach to measure local community readiness for implementing STD concept. The data were gathered through an online survey. Google form was used to create the online questionnaire which consists of a set of Likert scale questions. This questionnaire was adapted from Bulchand-Gidumal's framework of STD [6] and designed to enable tourist destination managers to self-assess their readiness to implement STD. Besides, interviews with tourist destination managers in Yogyakarta Province were also conducted to complement this online survey. The online survey link was shared to some of tourist destination managers that we interviewed and they shared the link to their peers through an online chat application and social media. By utilizing this online method, we got a total of 51 respondents. The aspects that we used to measure community readiness to implement STD are: economy, sustainability, governance, tourist and residents. These aspects adopted from Bulchand-Gidumal's smart destinations wheel framework [6], but the survey questions of each aspect were modified to make them relevant with the contexts in Yogyakarta. The data were analyzed using descriptive statistics approach. This analysis focuses on identifying how ready the local communities to implement STD. Table 1 shows the statements used in the Likert scale questions and STD aspects they belong to:

Respondents were asked to what extend they agree with above mentioned questions by choosing the relevant score in the *Likert* scale which consists of number 1–5 in which 1 means strongly disagree and 5 means strongly agree. This *Likert* scale is effective for capturing respondents' perceptions or self-assessment. Due to the absence of tourism data at community levels, this method is appropriate because it enables the managers of community-based tourism to self-assess their readiness. This also adds a participatory/democratic element to this research. An assessment done exclusively by an external examiners may be more objective than self-assessment, but a participatory assessment would create more impacts by raising local actors awareness regarding their capacity [13]. This participatory assessment also allows local actors to contemplate and identify their current strengths, weaknesses, challenges and opportunities. This self-evaluation will lead to motivations for making improvement in the future [14]. Therefore, instead of simply undertaking traditional evaluation study, self-assessment aspect was added to the research design of this study.

The *Likert* scale data for each variable were used to calculate the index based on Formula (1) below. The indexes resulted from this calculation then translated based on criteria outlined in Table 2. This data analysis approach were adapted from existing studies using a *Likert* scale for measuring readiness [15, 16]. However, combining a set of *Likert* data into indexes has a limitation in terms of validity [17]. Therefore, this indexes then compared with the mode and median of each variable.

Displayed equations are centered and set on a separate line.

$$\text{Readiness Index} = \frac{(1n_1) + (2n_2) + (3n_3) + (4n_4) + (5n_5)}{AN} \tag{1}$$

Local Community Readiness to Implement Smart Tourism Destination ... 429

Table 1 The Likert scale statements presented in the online survey and their STD aspects

STD aspects	Code	Statements scored by respondents using the *Likert* scale
Smart economy	Y1	Developing new economic sectors other than tourism
	Y2	Creating new businesses using ICT that complement tourism
	Y3	Applying incentive campaigns to increase demand
	Y4	Providing online payment facilities
	Y5	Using social media for marketing and tracking users' information and behaviors
	Y6	Implementing communication campaign
	Y7	Providing real time information of tourism data (visitors, parking, etc.)
	Y8	Using online media for marketing and promotion
Smart sustainability	Y9	Planning to regularly close the destination to allow maintenance, infrastructure development and environmental regeneration
Smart governance	Y10	Using ICT for managing tourism resources
	Y11	Developing collaborations amongst tourism managers, government, NGO, private sectors and universities
	Y12	Creating strategies to anticipate future crisis
	Y13	Receiving economic incentives such as soft loans and tax exemption
	Y14	Collaborating with tour agents
	Y15	Developing tourism models using virtual reality technology to allow virtual visits to the destination
	Y16	Using smart application for managing tourism data
	Y17	Integrating various tourism destinations and attractions within a region
	Y18	Using smart application for contact tracing
Smart tourist	Y19	Developing new tourist attractions or modifying the old ones to accommodate physical distancing amongst visitors
Smart residents	Y20	Applying measures to maintain safety and health of visitors, local residents, and hospitality and tourism workers

Source Adapted from Bulchand-Gidumal's framework of STD [6]

Table 2 The classification criteria for readiness indexes

Range of indexes	Descriptions
0–0.24	Strongly not ready
0.25–0.49	Not ready
0.50–0.74	Ready
0.75–100	Strongly ready

n_1 number of respondents for strongly disagree
n_2 number of respondents for disagree
n_3 number of respondents for neutral
n_4 number of respondents for agree
n_5 number of respondents for strongly agree
A highest weight
N total number of respondents

Complementing the online surveys, we also conducted ten interviews with local leaders of community-based tourism destinations in August 2022. These interviews were recorded and transcribed, then coded using a thematic analysis approach. Themes were self-emerged during this analysis process. The results of this qualitative data analysis are used to complement and tringulate the findings from the online surveys.

4 Results

The respondents of this research came from a range of tourism types: Village tourism, art and craft tourism, eco-tourism and agro-tourism. The occupation of these respondents are farmer, sand miner, laborer, employee, craftsman and civil servant. Before pandemic, the average annual revenue of destinations they manage ranges from 5 to 400 million. After pandemic, it ranges from 5 to 60 million per year. In terms of media, 89% of total sample already used different kinds of social media for marketing, networking, tracking costumer enthusiasm, education, coordination and ticketing. Roughly 61% of respondents indicated that their destination has more than one admin staff that manage social media. The indexes resulted from the calculation of Likert scale data start from 0.60 to 0.69. This means that the destinations managed by respondents are ready to implement STD. Although overall results suggest that these destinations are ready to be a smart destination, the mode of some aspects are low (Table 3). Based on the mode (the value that appears most often), the variables that mostly perceived have very low readiness (scale 1) are: Development of new economic sectors other than tourism (Y1) and development of strategies to anticipate future crisis (Y12). The variables that mostly perceived have low readiness (scale 2) are: Application of incentive campaigns to increase demand (Y3), the collaborations amongst stakeholders (Y11), and provision of economic incentives (Y13). In addition, the use of ICT for managing tourism resources (Y10) is mostly perceived as moderately ready (scale 3). This result suggests that the destinations are perceived strongly ready to implement STD mainly in terms of sustainability, tourists and residents, but improvements need to done in some aspects of smart economy and governance.

These findings accord with the results from our interviews. In terms of application of incentive campaigns, the interviewees also did not talk about this. The current ICT application for marketing is merely using social media and websites to

Table 3 The readiness index for each variable

STD aspects	Code	Index	Mode
Smart economy	Y1	0.60	1
	Y2	0.62	4
	Y3	0.60	2
	Y4	0.66	5
	Y5	0.64	5
	Y6	0.68	4
	Y7	0.65	4
	Y8	0.67	5
Smart sustainability	Y9	0.64	5
Smart governance	Y10	0.63	3
	Y11	0.67	2
	Y12	0.60	1
	Y13	0.65	2
	Y14	0.65	5
	Y15	0.62	4
	Y16	0.68	5
	Y17	0.69	5
	Y18	0.66	5
Smart tourist	Y19	0.67	5
Smart residents	Y20	0.66	5

provide information and recent activities in the destinations. There has no incentive campaigns implemented either embedded in social media or in off-line events. In terms of smart economy, the destinations merely focus on tourism sector and there hasn't been any initiatives for developing a new economic sector other than tourism. Many community-based tourism destinations in Yogyakarta were originated from non-tourism activities such as agriculture and craft industry. So, sectors other than tourism are usually exist and developed before tourism sector emerged as explained by Nanda. This explains why the score of this smart economy aspect is low. "This village tourism offers visitors to experience rural life and involve in local residents' daily activities such as farming and making bamboo crafts. These activities has been practiced in this village for decades." (Nanda, Diro Tourism Village) Most interviewees also mentioned that the existing governance of tourist destinations they manage has not formulated visionary strategies to anticipate any possible future crisis. When being asked about their strategy for future crisis, their responses are varied but the main idea is explaining that they were still focusing on economic recovery after pandemic crisis and infrastructure improvement. This shows that they currently have no specific strategies to anticipate future crisis. "To anticipate future restrictions we focus on providing appropriate facilities and infrastructure to ensure public health." (Nur, Wukirsari Tourism Village) However, Yulianto, one of our interviewees mentioned that a new attraction for domestic/local tourist was developed during pandemic as a strategy to cope with the absence of international tourists due to national border closure. This destination used to depend on international and inter-regional tourists. Due to border closure, the number of visitors were significantly decreased during pandemic. By creating a new tourist attraction, Yulianto and his peers hope that the destination will thrive and no longer solely depend on international tourists. "We developed a new culinary hub near a dam that aims to attract local tourists. In that place, visitors can enjoy the scenery and food provided by local residents. If border closure happens in the future, we can thrive due to this new tourist destination." (Yulianto, Kebonagung Tourism Village) Furthermore, the interviewees also suggest that the collaborations amongst stake-holders were still limited. Based on our interviews, each community has focused on developing its tourism destination, but rarely collaborate with other community-based tourist destinations. Furthermore, there has been reluctance amongst community members for partnering with other organizations or institutions managing the destination as reported by Yulianto as follow: "The members are not willing to collaborate with village-owned enterprises (BUMDes) because they afraid that the destinations that they initiated and developed will be acquired by these enterprises." (Yulianto, Kebonagung Tourism Village) There has also been limited access to economic incentives. As explained by Giyono, there has been no economic incentives provided to residents although the destinations' core product is craft especially the traditional woven fabric and it depends on the local residents' willingness to preserve this traditional craft. This finding suggest that the local community has no long-term strategies. Incentive is one of many ways to motivate local residents to preserve traditional culture marketed by this tourist destination. In this modern era, it is challenging to preserve traditional culture. Hence, local residents need either financial and in-kind support to prevent them from abandoning this traditional practices. If no incentives provided, this traditional weaving practice and its products wont sustain in a long term. "We understand that our tourism destination is depended on residents who still have and use traditional weaving tools, but there has been no economic incentives for them to maintain it." (Giyono, Gamplong Tourism Village) Overall, the community-based tourist destinations in Yogyakarta are already prepared to be developed as smart tourism destinations (STDs). However, improvements need to be done to enhance its readiness. Smart governance aspects that need be focused on are: the strategy to anticipate future disturbances, the collaboration with different stakeholders and the provision of economic incentives for local residents. Furthermore, smart economy aspects that need to be improved are incentive campaigns and development of a new non-tourism sectors.

5 Discussion

The results presented above show that community-based tourism destinations in Yogyakarta generally ready to be STDs especially in terms of ICT application. However, the findings reveal that some non-ICT aspects still need to be improved. Based on our STD framework, the smartness is not only about ICT application, but also covers the capacity of the whole tourism system. This findings add to the existing literature by showing some examples of non-ICT aspect of STD that should be the attention of current smart tourism policy. Yogyakarta is one of developed regions in Indonesia, hence why its tourism destinations are relatively ready in term of ICT application. Internet network, ICT infrastructure and knowledge are available and accessible for local residents. However, to be considered as a smart destination, the tourism governance and economy should also be smart in creating strategies and enhancing adaptive capacity. This finding support the recent conceptualization of smart tourism which includes soft smartness aspects such as innovation, human capital, leadership and social capital [18, 19]. These soft components covers the smartness of human and institutions governing a tourist destination. They are equally important as the ICT application and act as an activator of a smart destination. Boes et al. [17] suggest that coupled hard and soft smartness systems are required to enhance the sustainability and competitiveness of tourism destinations. The findings of our study support existing literature on STD by emphasizing the significance of soft smartness particularly in terms of tourism governance and economic innovations. The result of our study also adds to the existing literature by providing some in-sights from local actors that illustrate some challenges for enhancing the soft smartness of community-based tourism destinations in an Indonesian context. Further works are needed to improve marketing innovations, diversify economic sectors complementing tourism, set strategies to anticipate future stressors, create collaborations between different tourism actors, and provide economic incentives for local residents. The previous studies on community-based tourism also indicate that innovations either social or technical are a crucial component for community-based tourism [20–22]. The diversification of economy within a tourism destination has also been highlighted as an important but also challenging component of tourism development in previous studies [23, 24]. In terms of tourism resilience and adaptive capacity, previous studies also note that muti-actors engagement is crucial for ensuring the capacity of tourism destinations in facing future disturbances [4, 25]. In this era of uncertainty, destinations need to be prepared for any sudden unpredictable changes or crisis [26]. Lastly, existing literature on cultural and ecological tour-ism also recommends economic incentives as an effective strategy to sustain the cultural and ecological practices of local residents [27]. This does not only good for the sustainability of tourism destinations, but it is also critical to ensure that local residents benefit from tourism development. The welfare of local residents and environmental sustainability are the ultimate goals of smart tourism [28]. Advanced ICT application at the destinations would not make the destination completely smart if it does not coupled with these components of soft smartness.

6 Conclusion

Community-based tourism has been the focus of economic development especially during this post-pandemic era. It is believed as a strong booster for local economy which can create positive domino effects to other related sectors. In this era of modern technology, the role of community-based tourism can be accelerated by accommodating STD concept. This relatively new approach in tourism development aims to optimize the use of ICT in every aspects of tourism. However, it is still unclear to what extend local actors who directly manage and operate tourism destinations are ready to implement STD concept. This study addresses this by conducting a research in Yogyakarta province, Indonesia and the findings of this study suggest that generally tourist destinations in Yogyakarta are ready to implement STD especially in terms of economy and sustainability aspects. However, these destinations still need to improve the readiness of governance system as well as smart economy. Accommodating smartness into tourism governance is challenging but it is critical to enhance destinations' competitiveness and sustainability. Governance is the core of tourism management and it can be more effective and powerful if supported by smart technologies. Marketing innovations, economy diversification, anticipation of future unpredicted crisis, collaboration amongst stakeholders, and economic incentive are some examples of soft-smartness aspects that need to be fostered in community-based tourism destinations in Yogyakarta. These soft-smartness aspects also can be enhanced by using hard-smartness (ICT). Therefore, this study recommends that tourism policies and planning should focuses on building the capacity of the governance systems for example by creating a platforms to accommodate collaborations and social learning amongst stakeholders. Some training is also needed to raise awareness of marketing innovations and provision of economic incentives. Community-based tourism destinations in Yogyakarta are already good in terms of hard-smartness, but future improvements and tourism policies are needed to promote the soft-smartness of these destinations. This paper presents some initial findings and further analysis needs to be done to identify the determinants of local communities' readiness to implement STD. The results of this study show that the readiness of community-based tourism in Yogyakarta are relatively good, but questions remains regarding what factors influence this readiness. This is important to understand the conditions required to build community readiness to implement STD. Moreover, this research is merely based on quantitative data and statistical analysis. Although some interviews were done to complement this numeric data, further research that qualitatively examine community readiness to implement STD is needed. A qualitative research can further explore local actors' experiences and community dynamics in great depth. This can capture many detailed things that cannot be recorded in statistical data.

Based on UNSDGs indicator, the goals of this journal was to foster local economic development. It was related to 8th point on UNSDCs indicator; Promote Inclusive and Sustainbale Economic Growth. Having a job does not guarantee the ability to get out of the clutches of poverty. The continuing lack of decent work opportunities, inadequate investment and low consumption lead to the erosion of the fundamental social contract that underpins society's motion: All progress must be shared.

References

1. OECD.: https://www.oecd-ilibrary.org/urban-rural-and-regional-development/oecd-tourism-trends-and-policies-2020_6b47b985-en. Last accessed 15 Aug 2022 [Jovicic, D. Z.: From the traditional understanding of tourism destination to the smart tourism destination. Curr. Issues Tourism **22**(3), 276–282 (2019)]
2. Boes, K., Buhalis, D., Inversini, A.: Conceptualising smart tourism destination dimensions. In: Tussyadiah, I., Inversini, A. (eds.) Information and Communication Technologies in Tourism 2015, pp. 391–403. Springer, Cham (2015)
3. Gretzel, U., Scarpino-Johns, M.: Destination resilience and smart tourism destinations. Tour. Rev. Int. **22**(3–4), 263–276 (2018)
4. Yanti, A.I.E.K.: Community based tourism Dalam Menyongsong new normal Desa Wisata Bali. Jurnal Komunikasi Hukum (JKH) **7**(1), 72–86 (2021)
5. Suryani, A., Soedarso, S., Rahmawati, D., Endarko, E., Muklason, A., Wibawa, B.M., Zahrok, S.: Community-based tourism transformation: what does the local community need? IPTEK J. Proc. Ser. **7**, 1–12 (2021)
6. Bulchand-Gidumal, J.: Post-COVID-19 recovery of island tourism using a smart tourism destination framework. J. Destin. Mark. Manag. **23**, 100689 (2022)
7. Cohen, B.: https://boydcohen.medium.com/blockchain-cities-and-the-smart-cities-wheel-9f6 5c2f32c36. Last accessed 15 Aug 2022
8. Habeeb, N.J., Weli, S.T.: Relationship of smart cities and smart tourism: an overview. HighTech Innovation J. **1**(4), 194–202 (2020)
9. Mandić, A., Praničević, D.G.: Progress on the role of ICTs in establishing destination appeal: implications for smart tourism destination development. J. Hosp. Tour. Technol. **10**(4), 791–813 (2019)
10. Shafiee, S., Ghatari, A.R., Hasanzadeh, A., Jahanyan, S.: Developing a model for sustainable smart tourism destinations: a systematic review. Tourism Manag. Perspect. **31**, 287–300 (2019)
11. Farania, A., Hardiana, A., Putri, R.A.: Kesiapan Kota Surakarta Dalam Mewujudkan Pariwisata Cerdas (Smart Tourism) Ditinjau Dari Aspek Fasilitas dan Sistem Pelayanan. Region: Jurnal Pembangunan Wilayah dan Perencanaan Partisipatif **12**(1), 36–50 (2017)
12. Saputra, A., Roychansyah, M.S.: Penerapan smart tourism destination di Tiga Destinasi Wisata Kota Yogyakarta. Jurnal Sinar Manajemen **9**(1), 122–129 (2022)
13. Chens, C.Y., Sok, P., Sok, K.: Evaluating the competitiveness of the tourism industry in Cambodia: self-assessment from professionals. Asia Pacific J. Tourism Res. **13**(1), 41–66 (2008)
14. Marthasari, G.I., Hayatin, N., Wahyuni, E.D., Kristy, R.D.: Measuring user readiness of web-based encyclopedia for kids based on technology readiness index. Jurnal Media Informatika Budidarma **4**(2), 294–301 (2020)
15. Pratama, Y.: Making of digital forensic readiness index (DiFRI) models to Malware attacks. Cyber Security dan Forensik Digital **3**(2), 1–5 (2020)
16. Allen, I.E., Seaman, C.A.: Likert scales and data analyses. Qual. Prog. **40**(7), 64–65 (2007)
17. Boes, K., Buhalis, D., Inversini, A.: Smart tourism destinations: ecosystems for tourism destination competitiveness. Int. J. Tourism Cities **2**(2), 108–124 (2016)
18. Gretzel, U., Zhong, L., Koo, C.: Application of smart tourism to cities. Int. J. Tourism Cities **2**(2) (2016)
19. Mendoza-Moheno, J., Cruz-Coria, E., González-Cruz, T.F.: Socio-technical innovation in community-based tourism organizations: a proposal for local development. Technol. Forecast. Soc. Chang. **171**, 120949 (2021)

20. Songkhla, R.N., Wanvijit, W., Charoenboon, P., Ninaroon, P.: The marketing efficiency development to create value-added for product and service of community-based tourism. Study Case for Phatthalung Province. J. Environ. Manag. Tourism **12**(1), 266–276 (2021)
21. Ditta-Apichai, M., Kattiyapornpong, U., Gretzel, U.: Platform-mediated tourism micro-entrepreneurship: implications for community-based tourism in Thailand. J. Hosp. Tour. Technol. **11**(2), 223–240 (2020)
22. Benur, A.M., Bramwell, B.: Tourism product development and product diversification in destinations. Tour. Manage. **50**, 213–224 (2015)
23. Weidenfeld, A.: Tourism diversification and its implications for smart specialisation. Sustainability **10**(2), 319 (2018)
24. Pyke, J., Law, A., Jiang, M., de Lacy, T.: Learning from the locals: the role of stakeholder engagement in building tourism and community resilience. J. Ecotour. **17**(3), 206–219 (2018)
25. Bui, P.L., Wickens, E.: Tourism industry resilience issues in urban areas during COVID-19. Int. J. Tourism Cities **7**(3), 861–879 (2021)
26. Loukaitou-Sideris, A., Soureli, K.: Cultural tourism as an economic development strategy for ethnic neighborhoods. Econ. Dev. Q. **26**(1), 50–72 (2012)
27. Wunder, S.: Ecotourism and economic incentives—an empirical approach. Ecol. Econ. **32**(3), 465–479 (2000)
28. Santos-Júnior, A., Almeida-García, F., Morgado, P., Mendes-Filho, L.: Residents' quality of life in smart tourism destinations: A theoretical approach. Sustain. **12**(20), 8445 (2020)
29. Qonita, M., Giyarsih, S.R.: Smart city assessment using the Boyd Cohen smart city wheel in Salatiga, Indonesia. GeoJournal 1–14 (2022)
30. Reid, D.G., Mair, H., George, W.: Community tourism planning: a self-assessment instrument. Ann. Tour. Res. **31**(3), 623–639 (2004)

Open Innovation and Governance Models in Public Sector: A Systematic Literature Review

Meshari Abdulhameed Alsafran ⓘ, Odeh Rashed Al Jayyousi ⓘ, Fairouz M. Aldhmour ⓘ, and Eisa A. Alsafran ⓘ

Abstract Innovation in the public sector coincided with different waves of administrative reforms which can be due to technological change, restructuring, and Open Innovation (OI). The OI process revolves around new forms of knowledge inflows and outflows that accelerate interior innovation and expand the markets for the external use of innovation. OI within the public sector is an umbrella term that describes processes, outcomes, and business models of a new form of innovation creation. The Gulf Cooperation Council Countries (GCC) vision is framed around innovation and transformation of the public sector to ensure competitiveness. The significance of this research stems from its contribution to explore the factors and outcomes of innovation in the public sector within different regions. This review outlined key thematic angles, models, factors, and research methods of OI and innovation outcomes in public sector. The systematic review included 191 articles extracted from Google scholar, Scopus, and ScienceDirect data bases, that were published from 1990 to 2022. The existing research shows there were limited publications in this domain in the Middle East and GCC. This paper attempts to shed light on the enablers for fostering public sector innovation through harnessing OI. The findings of this review contribute to framing PSI and OI in the digital era.

Keywords Public sector · Open innovation · GCC · Innovation models · Innovation management

---

M. A. Alsafran (✉) · O. R. Al Jayyousi · F. M. Aldhmour
Arabian Gulf University, Manama, Bahrain
e-mail: Meshariaes@agu.edu.bh

O. R. Al Jayyousi
e-mail: Odehaj@agu.edu.bh

F. M. Aldhmour
e-mail: Fairouzm@agu.edu.bh

E. A. Alsafran
Algonquin College, Al Jahra, Kuwait
e-mail: Ealsafran@gmail.com

© The Author(s), under exclusive license to Springer Nature Switzerland AG 2024
A. Hamdan and E. S. Aldhaen (eds.), *Artificial Intelligence and Transforming Digital Marketing*, Studies in Systems, Decision and Control 487,
https://doi.org/10.1007/978-3-031-35828-9_38

1 Introduction

Innovation in the public sector coincided with different waves of administrative reforms which were influenced by technological change, innovation, and private sector participation [1]. Scholars such as Karo [2] argue that the networked, and adaptive public sector should be informed by institutional paradigms adapted from the private sector. Saxena [3] examined the potential role that Open Government Data (OGD) could play in assisting the GCC region's economic reform agenda. Durugbo et al. [4] outlined models of innovation management in GCC. Tomer et al. [5] explored the relationship between smart governance and sustainable development. Besides, public innovation labs were explored as arenas where public sector organizations can design innovative policies since policy labs provide a learning space for policy formulation [6]. This paper applied a systematic literature review (SLR) to gain insights into models and enablers for Open Innovation (OI) in the public sector. The following section outlines definitions of OI in the public sector.

The term innovation originated from the Latin word "innovare", meaning to create something new [7]. According to Bloch [8] Oslo Manual is committed to the business sector, it is conceded that innovation may be specular in the public sector. Bloch and Bugge [9] identify the innovation in the public sector. Concerning public-sector administration, practitioners [10] focused on the definitions, understandings, and uses of innovation. Tjønndal [11] anticipated that different forms of innovation take place in sports organizational innovation, social innovation, and community-based innovation. Glor [12], highlighted the value of fostering the innovation process in the public domain.

1.1 Open Innovation in Public Sector

The OI process is defined as an innovative form of inflows and outflows of information [13]. The term OI in the public sector is considered an umbrella term that defines outcomes, business models, and processes of a new and novel innovation creation form. The domains of OI inside the public sector include drivers of adoption, success factors, and innovation outcomes [14–18]. The concept of OI was coined in the private sector on the way to address the usage of inflows knowledge and outflows knowledge that led to accelerated organizational innovation [13, 18] demonstrating that applying OI in the civic domain con-tributes to creates public value.

It is argued that OI requires an operating model that can utilize more ideas and technologies from outside an organization [19]. In the public sector, OI can take place in many different settings and at different levels including urban labs, policy labs, and innovation living labs for ideation and experimentation [20]. A key feature of OI in the public sector is collaborative innovation that is underpinned by internal and external stakeholders' engagement through harnessing open government, and open data (OD) initiatives [1, 18, 21–24]. Recent studies in OI show a paradigm shift in public governance [25] is taking place through adopting policy innovation labs and digital platforms to promote open, transparent, and collaborative schemes. Innovation policy labs, social innovation labs, digital platforms, and toolkits are means to foster OI [26].

There are variations in the context and drivers of innovation in the private and public sectors [9]. Public sector innovation (PSI) encompasses a broad continuum of process, service, governance, conceptual, and innovative products [15] but framing adequate innovation process in the public domain needs a strategy for innovation adoption and diffusion [27] due to variations of public sector services [28, 29]. Besides, conventional innovation models are inadequate in the public sector [30].

1.2 Open Innovation Models

Research in OI revealed that business models that leverage knowledge and openness provide a competitive advantage [31]. A study by Hawi et al. [32] examined the state-of-the-art around models of innovation in the private and public sectors to foster effectiveness and efficiency. They also proposed the "stage-gate" model developed by Cooper (1990) and Chesbrough. This change was described by Chesbrough [13, 33] in his articles as a paradigm shift from closed to OI. According to Fuglsang [34], the development of services in the public sector is becoming more relevant to an OI pattern. Moreover, it is realized that OI is a future area of research since it contributes to address grand societal challenges in public sector [35, 36]. In some studies, inbound OI has been the focus and focal actor of studies on OI in the public sector [36] e-citizen engagement [37], innovation practices [13, 33], Open Government [3]; Data external linkages [38], public–private partnership [39].

Collaborative innovation based on network government is significant for fostering OI [27]. Similarly, the collaborative innovation process within the public sector supported by several characteristics of governance [40], The main characteristic stressed, was the significant tool or intermediation strategy the use of a technological platform, innovation laboratories. Concepts such as OI [13] and user innovation [41] were both initially developed in the private sector but can be extended to public sector. According to Schuurman and Tõnurist [42] specific platforms for open and user innovation in the public sector are provided via innovation labs. A shift in perspective of policymakers and academics is required to encourage a more OI model that takes advantage of the opportunities provided by citizen, entrepreneur, and civil society collaboration as well as new emerging technologies [43].

This paper outlined key enablers for fostering PSI and with the effect of OI. It aims to discover key thematic angles, models, factors, and research methods of OI and innovation outcomes in public sector. To achieve this aim, the study adopts SLR thus seeking to gain insights on the OI in the public sector to map governance models. Public authorities, administrative, as well as local governments are becoming more open and transparent to give value to the public [29], with the assistance of Good Public Governance [44], E-Government [45], and Performance Management [46].

Researchers and practitioners find out other concepts of Open Government [47] include such as "collaborative public management" [48], "citizen engagement" [49, 50], "wiki government" [51], and "coproduction with public sector clients," [52].

2 Method

This SLR method depends on the framework developed by Gough et al. [53] and Jesson et al. [54]. The review intended to identify relevant models, Enablers, PSI outcome and drivers of OI in the public sector. The motivation for adopting a SLR was to gain a useful insight to know the state of the art about OI within PSI in GCC. The process for conducting the SLR includes three phases, i.e., review protocol, evaluation, and synthesis as outlined in Fig. 1.

The study research question is: "explore the linkages between OI and public sector performance", guides the search strategy and the review protocol. The search was based on papers published in English in ScienceDirect, Scopus, and Google Scholar during 1990–2022. The review used the keywords (Query's = "Open Innovation" AND "Public Sector Innovation"). Science Directs (accessed on 21 Nov. 2022), Scopus (accessed on 18 Nov. 2022), and Google scholar (accessed on 12 Dec. 2022) provide the electronic database for the review. This study excludes conference proceeding papers, master's theses, doctoral dissertations, textbooks, and unpublished working papers.

After many trials and testing, a final selection of these keywords was concluded. Keyword searches through journal titles, abstracts, keywords, and topics yielded 122 articles. Figure 2 shows the compiled body of literature from 1990 to 2022.

It is essential to explain how we filtered through the numerous results. We set noticeable inclusion and exclusion criteria. The results were limited to "peer-reviewed" journal articles only. It was also limited to "full text" journal articles that were available or we could get access to it. We included "review articles" and "research articles" while excluding books, book chapters, conference proceeding papers, master's theses, doctoral dissertations, and unpublished working papers and

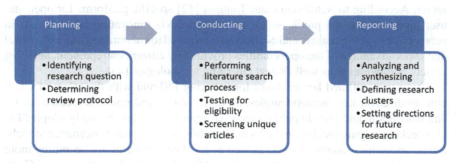

Fig. 1 The process for the SLR. *Source* Adaptive from Al-Jayyousi et al. [55]

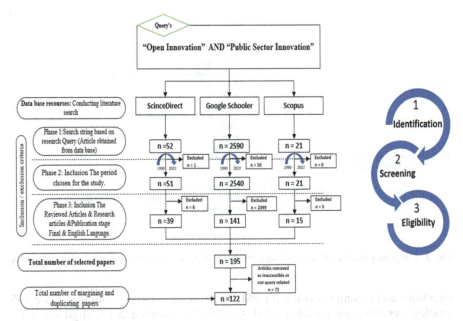

Fig. 2 The flow diagram for screening process in the SLR (PRISMA)

other kinds of articles. Subsequently, the articles in the review were published in a diverse set of journals, including 164 journals. We have filtered the results by the list of journals it was noticeable that most papers were published in Government Information Quarterly, Sustainability, and innovation as shown in Fig. 3, which presents the frequency of articles in published journals. The other lists were also included to keep the high-quality bar for the literature research, but not all of that is open access. This led to a high-quality list of about 195 journal articles.

The figure present summarily the diverse forms and approaches on the current state of literature pertaining to innovation in public sector. The following section outlines descriptive analysis, then concluding with findings and conclusion of the study.

3 Descriptive Analysis: Themes and Concepts

In line with the review scope, themes and clusters emerge from cross-referencing and screening to eliminate duplicates and exclude irrelevant sources process based on assessing research similarities and differences in terms of regions, core issues, methodologies, and sectors. The synthesis phase entailed combining key findings after the evaluation phase. Guided by the research question, this phase analyses the OI models and practices and uses insights from the literature to frame governance

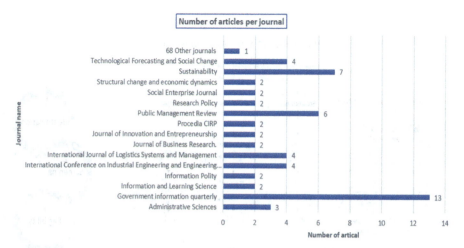

Fig. 3 Frequency number of articles distribution per publication journals that analyzed in the study

directions and recommendation for PSI. The articles obtained from this phase 195 articles, were then analyzed to exclude criteria. This resulted in 122 eligible articles used for this review as depicted in Fig. 2.

In this subsection, we provide the results addressing in Fig. 3 shows the frequency and number of publications in the domine of OI in public sector within various of journals. As well, Fig. 4 Well present the number of publication articles per years.

The research examines conducted articles from 1990 to 2022. The relevant research literature on this subject started appearing in 2007, to the best of our knowledge. Figure 3 shows the journal name and the number of articles published by them. The highest journal in publication appears "Government information quarterly", "sustainability", "Public management review", then it comes three journals equal in

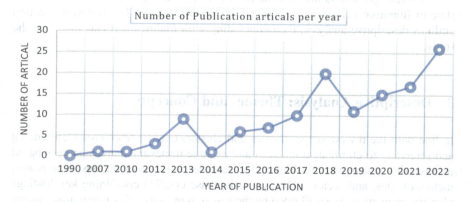

Fig. 4 Numbers of articles publications distribution per year that used in the study

the amount of research papers, then the rest they have very few of publication. 83 journals in totals. The literature review (LR) was concluded in the very last part of 2022, and hence we have a significant number of articles published in 2022. It is observed the high rate of publication in 2018, and 2022, by 20 and 26 publications papers respectively.

As seen in Fig. 4 the distribution of the number of articles per year. To break it down Initially from 1990 to 2006 there was no single research extracted. Then between 2007 and 2014 most of the research papers have a decent distribution below 5, except in 2013 it almost reaches 10 papers per year which that indicate researchers shows some interest in covering the research subject of this paper. Then in the last period between 2014 and 2022, it has a precipitous increase to the final year of the graph (112) papers, except for 2019. Where the increase in the last period after the 2019 drop was healthy, as it occurs as a reaction to the crisis. The pandemic of COVID-19 is a crisis that affects worldwide, which led to a huge focus on solutions to deal with and over-come this crisis. It is noteworthy, however, that concerning the governmental level, the data is somewhat skewed towards innovation adoption and sustainability. There has been a rapid increase in the number of papers in high-quality journals included in the study. It shows the growing interest in OI in the public sector/ governance.

It is illustrated there is a discernible pattern of increased interest in this area that began in 2012. In a similar van [56, 57] conduct a LR that reveals the peak edge indicator for several documents was found trendy in 2015 and 2017. These two topics studied OD and OI in such databases. As a result, this indicates the domine of OI is attracting the scholars to explore it wider. The study expects a continuous increase in the number of research publications in 2023 than in the previous years.

Our systematic review adds some findings, many papers used literature review and adapted qualitative methods. Where the conceptual research got the least amount of attention among all the articles reviewed Fig. 5 and Table 1.

Article keywords provide great indication for the scope of the research, so the paper has outlined this in Figs. 6 and 7. Keywords were automatically clustered in separate groups identified by different colors using the software VOS viewer to map them in various data bases. For instant the keywords for the articles found in three data bases related to the research topic shows that the main cluster words were highlighted. Thus, the clustering identified the significance of public sector concepts such as "open innovation", "public sector innovation", "innovation" and "social innovation". Figure 6 outline, the keywords emerging in Google scholar database (n = 141). It was evident that the main attribute of OI includes drivers, social innovation, lab.

However, certain subfields and applications emerged: Living Labs as well as specific applications were found in the context of PSI by adopting OI; E-government seemed to be related to the citizen sourcing phase to present OI for the PSI. It rose the keywords OI, PSI, OI, innovation, social innovation, and driver in (Scopus articles), (ScinceDirect articles), and (Google scholar articles) respectively. This keywords analysis was done to improve readability; where it was found that it is compatible with [58].

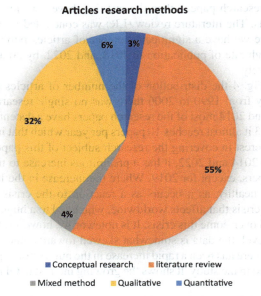

Fig. 5 The methodology accumulated from the review articles. By the authors

Table 1 Show the type of methods with the frequency used in articles

Methods	Articles No.
Conceptual research	3
literature review	57
Mixed method	4
Qualitative	33
Quantitative	6

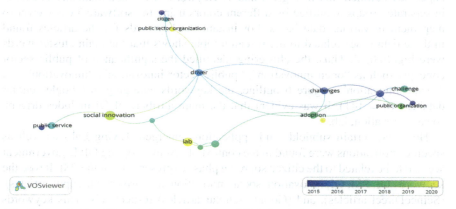

Fig. 6 Visualization of keywords from the reviewed Google scholar articles. *Source* Adapted from VOS viewer software

Fig. 7 Visualization of keywords in reviewed articles. *Source* Adapted from Word Art software

The analysis of the overall reviewed articles keywords related to the Topic research with highest occurrences based on Word Art Software were illustrated in Fig. 7. It shows the OI, innovation, public sector then the government sector is the main keyword with higher occurrences.

Moreover, Articles were reviewed to gain insight on the themes, sector, enablers, and the highest cited articles Top 20 are exemplified and presented through an accessible link further down, and Figures respectively. Whereas the distribution of research articles carried out by country (or region) represent in Fig. 10.

The Extracted themes, concepts, sub-themes, and authors name are synthesized as illustrated in the link:

https://www.dropbox.com/scl/fi/eu32yni6eplf08t0cht2l/Themes-of-OI-in-public-sector.docx?dl=0&rlkey=4lj57usz7843xgt1h8v10pu4y

The themes were captured from the carefully selected analysis papers for the study, which is categorized into 16 different categories. These categories include innovation policies, innovation outcomes, enablers and drivers, barriers, sector, private sector participation, activities, co-creation and co-production, collaboration, crowdsourcing, technology, governance, process, practices, leadership, and innovation models.

Whereas the study review shed light on the top cited papers found in the study period 1990–2022, and it was captured and presented through the link:

https://www.dropbox.com/scl/fi/2l1sy6p0dppbwoz1law4j/The-Top-20-Most-Cited-articles-in-the-review.docx?dl=0&rlkey=5bgwtgszttmftmsh870bnzqyp

By using this link, the top 20 key articles will be found with their number of citations among 122 articles related to the topic research. This section indicates the systematic review article papers of Voorberg et al. [59] and De Vries et al. [15] with a title "co-creation and co-production: Embarking on the social innovation journey"; and "Innovation in the public sector" has the highest citations 1957; 1441 cite respectively, among the analyzed papers in this study using three data sources Scopus, Google Scholar, and ScienceDirect.

Furthermore, as shown in Fig. 8 the sectors covered in the OI articles reviewed. The key sectors include health, education, ICT, government sector, public service, and urban planning. The mean prevalent sector was mainly the public sector as it the focused of the study in the research.

A synthesis of the 122 conducted articles extracts the key enablers that included digital, social, data innovation, tools, network, knowledge, and governance. Which is visualized in the following illustration Fig. 9.

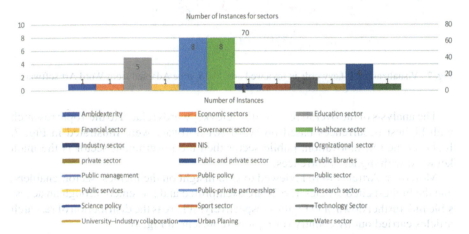

Fig. 8 Sectors that have innovation in the review

Fig. 9 Enablers for innovation in the review. *Source* Adapted from Word Art software

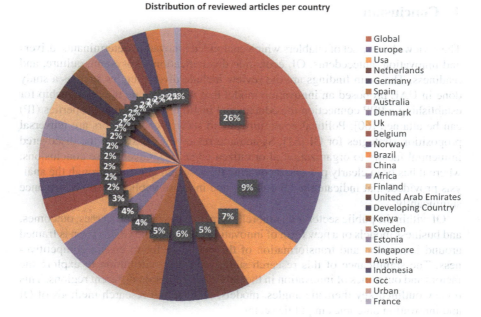

Fig. 10 Distribution of reviewed articles per country or region

Research on OI in the reviewed literature shows its appearance in different fields and countries. Where from Fig. 10 the mainly originate countries that have the most research articles were global with 32 frequency, Europe 11, USA 9, Germany 7, and the Netherlands, Spain with 6 instances as shown in Fig. 10. Along with a systematic review by Torab-Miandoab et al. [57], OI in Healthcare Ecosystem where his study shows the rate of health OI articles based on countries throw his analysis, an indication of the highest countries was Germany 14% then the United States and Sweden and 13%, neither the GCC countries were identified, nor the non-Western contexts were analyzed.

This review guide policy framing for GCC which should form a mission-oriented innovation policy; digital innovation; PPP policy and NIS. Overall, the literature shows coverage of different sectors including public sector, (Urban planning, government, education, sport, ICT, PPP, healthcare, technology, organizational, and service) sectors.

4 Conclusion

The review shows a set of enablers which enclouded innovation determinants, drivers and innovation, antecedents, OI, economic diversification, ecosystem, culture, and readiness. The main findings are the review include diver's angles such as a study done in UAE proposed an innovation model that will facilitate the relationship for establishing a PPP connection the educational institutes Intellectual Properties (IP) can be attained [60]. Political commitment and service of mediators are universal proposition strategies for OI [61]. Systematic review by broken up the discovered influential factors to organization-al or citizen edge [59]. Given these conclusions, where it has been clearly presented covering the research question through the analysis provided, that indicate the OI enhance and increase public sector performance significantly.

OI within the public sector is an umbrella term that describes processes, outcomes, and business models of a new form of innovation creation. The GCC vision is framed around innovation and transformation of the public sector to ensure competitiveness. The significance of this research stems from its contribution to explore the factors and outcomes of innovation in the public sector within different regions. This review outlined key thematic angles, models, factors, and research methods of OI and innovation outcomes in public sector.

References

1. Osborne, S.P., Strokosch, K.: It takes two to tango? Understanding the co-production of public services by integrating the services management and public administration perspectives. Br. J. Manag. **24**(S3) (2013). https://doi.org/10.1111/1467-8551.12010
2. Karo, E.: Modernizing governance of innovation policy through 'decentralization': a new fashion or a threat to state capacities? Innovation Manage Policy Practice, 923–959,(2012). https://doi.org/10.5172/impp.2012.923
3. Saxena, S.: Prospects of open government data (OGD) in facilitating the economic diversification of GCC region. Inf. Learn. Sci. **118**(5–6), 214–234 (2017). https://doi.org/10.1108/ILS-04-2017-0023
4. Durugbo, C.M., Al-Jayyousi, O.R., Almahamid, S.M.: Wisdom from Arabian creatives: systematic review of innovation management literature for the gulf cooperation council (GCC) region. Int. J. Innovation Technol. Manag. **17**(6) (2020). https://doi.org/10.1142/S02198770 20300049
5. Tomor, Z., Meijer, A., Michels, A., Geertman, S.: Smart governance for sustainable cities: findings from a systematic literature review. J. Urban Technol. **26**(4), 3–27 (2019). https://doi.org/10.1080/10630732.2019.1651178
6. Criado, J.I., Dias, T.F., Sano, H., Rojas-Martín, F., Silvan, A., Filho, A.I.: Public innovation and living labs in action: a comparative analysis in post-new public management contexts. Int. J. Public Adm. **44**(6), 451–464 (2021). https://doi.org/10.1080/01900692.2020.1729181
7. Tidd, J., Bessant, J.R.: Managing innovation: integrating technological, market and organizational change. Translated by Arasti et al. Tehran, Rasa (2013)
8. Bloch, C.: Assessing recent development in innovation measurement: the third edition of the Oslo Manual. Sci. Public Policy **34**(1), 23–34 (2007). https://doi.org/10.3152/030234207X19 0487

9. Bloch, C., Bugge, M.M.: Public sector innovation—from theory to measurement. Struct. Change Econ. Dyn. **27**, 133–145 (2013). https://doi.org/10.1016/j.strueco.2013.06.008
10. Hagen, A., Higdem, U.: Calculate, communicate, and innovate: do we need "innovate" as a third position? J. Plan. Lit. **34**(4), 421–433 (2019). https://doi.org/10.1177/0885412219851876
11. Tjønndal, A.: Sport innovation: developing a typology. Eur. J. Sport Soc. **14**(4), 291–310 (2017). https://doi.org/10.1080/16138171.2017.1421504
12. Glor, E.D.: The innovation journal: the public sector innovation journal fellow. In Canada The Innovation Journal: The Public Sector Innovation Journal, vol. 26, issue 2 (n.d.)
13. Chesbrough, H.: Open innovation the new imperative for creating and profiting from technology. Null (2003)
14. Bakici, T., Almirall, E., Wareham, J.: Technology analysis & strategic management the role of public open innovation intermediaries in local government and the public sector. Strateg. Manag. **25**, 311–327 (2013). https://doi.org/10.1080/09537325.2013.764983
15. De Vries, H., Bekkers, V., Tummers, L.: Innovation in the public sector: a systematic review and future research agenda. Public Adm. **94**(1), 146–166 (2016). https://doi.org/10.1111/padm.12209
16. Dias, C., Escoval, A.: The open nature of innovation in the hospital sector: the role of external collaboration networks. Health Policy Technol. **1**(4), 181–186 (2012). https://doi.org/10.1016/j.hlpt.2012.10.002
17. Mergel, I.: 11-Open collaboration in the public sector: the case of social coding on GitHub. Gov. Inf. Q. **32**(4), 464–472 (2015). https://doi.org/10.1016/j.giq.2015.09.004
18. Mergel, I., Desouza, K.C.: Implementing open innovation in the public sector: the case of Challenge.gov. Public Adm. Rev. **73**(6), 882–890 (2013). https://doi.org/10.1111/puar.12141
19. Mulgan, G.: Ready or Not? Taking Innovation in the Public Sector Seriously (2007)
20. De Coninck, B., Gascó-Hernández, M., Viaene, S., Leysen, J.: 10-Determinants of open innovation adoption in public organizations: a systematic review. Public Manag. Rev. (2021). https://doi.org/10.1080/14719037.2021.2003106
21. Chesbrough, H., Bogers, M.: Explicating open innovation: clarifying an emerging paradigm for understanding innovation keywords. New Front. Open Innovation 1–37 (2014)
22. Lakomaa, E., Kallberg, J.: Open data as a foundation for innovation: the enabling effect of free public sector information for entrepreneurs. IEEE Access (2013). https://doi.org/10.1109/access.2013.2279164
23. Lee, M., Almirall, E., Wareham, J.: Open data and civic apps. Commun. ACM **59**(1), 82–89 (2015). https://doi.org/10.1145/2756542
24. Zuiderwijk, A., Helbig, N., Gil-García, J.R., Janssen, M.: Special issue on innovation through open data—a review of the state-of-the-art and an emerging research agenda: Guest editors' introduction. J. Theor. Appl. Electron. Commer. Res. **9**(2) (2014). https://doi.org/10.4067/S0718-18762014000200001
25. Sangiorgi, D.: Designing for public sector innovation in the UK: design strategies for paradigm shifts. Foresight **17**(4), 332–348 (2015). https://doi.org/10.1108/FS-08-2013-0041
26. Carstensen, H.V., Bason, C.: Powering collaborative policy innovation: can innovation labs help? Helle Vibeke Carstensen & Christian Bason. Innovation J. Public Sector Innovation J. **17**(1), 2–27 (2012)
27. Hartley, J.: Innovation in governance and public services: past and present. Public Money Manag. **25**(1), 27–34 (2005). https://doi.org/10.1111/j.1467-9302.2005.00447.x
28. Kattel, R., Cepilovs, A., Drechsler, W., Kalvet, T., Lember, V., Tonurist, P.: Can we measure public sector innovation? A literature review. Lipse Project Paper, WP 6 Socia **2**, 2–38 (2013)
29. Moore, M.H.: Break-through innovations and continuous improvement: two different models of innovative processes in the public sector. Public Mony Manag. **25**(1), 43–50 (2005)
30. Borins, S.F.: The persistence of innovation in government: a guide for innovative public servants. In Innovation Series—the IBM Center for The Business of Government, vol. 8 (2014)
31. Timmers, P.: Business models for electronic markets. Electron. Mark. **8**(2), 3–8 (1998). https://doi.org/10.1080/10196789800000016

32. Hawi, T.A., Alsyouf, I., Gardoni, M.: Innovation models for public and private organizations: a literature review. IEEE Int. Conf. Ind. Eng. Manag. **565–569** (2018). https://doi.org/10.1109/IEEM.2018.8607562
33. Chesbrough, H.: Open Innovation—Open Innovation. Harvard Business School Press **2006**(193), 1–9 (2006)
34. Fuglsang, L.: Capturing the benefits of open innovation in public innovation: a case study. Int. J. Serv. Technol. Manage. **9**(3–4), 234–248 (2008). https://doi.org/10.1504/IJSTM.2008.019705
35. Bekkers, V., Tummers, L.: Innovation in the public sector: towards an open and collaborative approach. Int. Rev. Adm. Sci. **84**(2), 209–213 (2018). https://doi.org/10.1177/002085231876 1797
36. Kankanhalli, A., Kankanhalli, A., Kankanhalli, A., Zuiderwijk, A., Tayi, G.K.: Open innovation in the public sector: a research agenda. Gov. Inf. Q. (2017). https://doi.org/10.1016/j.giq.2016.12.002
37. Hilgers, D., Ihl, C.: Citizensourcing: applying the concept of open innovation to the public sector. Int. J. Public Participation **4**(1), 67–88 (2010)
38. Hameduddin, T., Fernandez, S., Demircioglu, M.A.: Conditions for open innovation in public organizations: evidence from Challenge.gov. Asia Pacific. J. Public Adm. **42**(2), 111–131 (2020). https://doi.org/10.1080/23276665.2020.1754867
39. Osborne, S.P.: The new public governance? Public Manag. Rev. **8**(3), 377–387 (2006). https://doi.org/10.1080/14719030600853022
40. Lopes, A.V., Farias, J.S.: How can governance support collaborative innovation in the public sector? A systematic review of the literature. Int. Rev. Adm. Sci. **88**(1), 114–130 (2022). https://doi.org/10.1177/0020852319893444
41. von Hippel, E.: Democratizing innovation: the evolving phenomenon of user innovation. Int. J. Innovation Sci. **1**(1), 29–40 (2005). https://doi.org/10.1260/175722209787951224
42. Schuurman, D., Tõnurist, P.: Innovation in the public sector: exploring the characteristics and potential of living labs and innovation labs. Technol. Innovation Manag. Rev. **7**(1), 78–90 (2016). https://doi.org/10.22215/timreview1045
43. Gasco-Hernandez, M., Sandoval-Almazan, R., Gil-Garcia, J.R.: Open Innovation and Co-Creation in the Public Sector: Understanding the Role of Intermediaries. Lecture Notes in Computer Science (Including Subseries Lecture Notes in Artificial Intelligence and Lecture Notes in Bioinformatics), 10429 LNCS, pp. 140–148 (2017). https://doi.org/10.1007/978-3-319-64322-9_12
44. Peters, B.G., Pierre, J.: Citizens versus the new public manager: the problem of mutual empowerment. Adm. Soc. **32**(1), 9–28 (2000). https://doi.org/10.1177/00953990022019335
45. Stowers, G.N.L.: Becoming cyberactive: state and local governments on the world wide web. Gov. Inf. Q. **16**(2), 111–127 (1999). https://doi.org/10.1016/S0740-624X(99)80003-3
46. Holzer, M., Kloby, K.: Public performance measurement: an assessment of the state-of-the-art and models for citizen participation. Int. J. Prod. Perform. Manag. (2005)
47. Obama, B.: Memorandum for the heads of executive departments and agencies. Presidential Stud. Q. **39**, 429+ (2009)
48. McGuire, M.: Collaborative public management: assessing what we know and how we know it. Public Adm. Rev. **66**(1), 33–43 (2006). https://doi.org/10.1111/j.1540-6210.2006.00664.x
49. Hickley, M.: The Grounds for Citizen Engagement and the Roles of Planners. VDM Publishing (2008)
50. OECD, O.: The OECD principles of corporate governance. Contaduría y Administración, 216 (2004)
51. Noveck, B.S.: Wiki Government: How Technology Can Make Government Better, Democracy Stronger, and Citizens More Powerful. Brookings Institution Press (2009)
52. Alford, J.: Engaging Public Sector Clients: From Service-Delivery to Co-Production. Springer (2009)
53. Gough, D., Thomas, J., Oliver, S.: Clarifying differences between review designs\rand methods art. System. Rev. 1–9 (2012). https://doi.org/10.1186/2046-4053-1-28

54. Jesson, J., Matheson, L., Lacey, F.M.: Doing Your Literature Review: Traditional and Systematic Techniques (2011)
55. Al-Jayyousi, O., Tok, E., Saniff, S.M., Wan Hasan, W.N., Janahi, N.A., Yesuf, A.J.: Rethinking sustainable development within Islamic worldviews: a systematic literature review. Sustainability **14**(12), 7300 (2022)
56. Corrales-Garay, D., Mora-Valentín, E.M., Ortiz-de-Urbina-Criado, M.: 15-Open data for open innovation: an analysis of literature characteristics. Future Internet, **11**(3) (2019). https://doi.org/10.3390/fi11030077
57. Torab-Miandoab, A., Samad-Soltani, T., Rezaei-Hachesu, P.: Open innovation in healthcare ecosystem—a systematic review. Res. Square (2021).
58. Greve, K., De Vita, R., Leminen, S., Westerlund, M.: 51-living labs: from niche to mainstream innovation management. Sustainability (Switzerland) **13**(2), 1–24 (2021). https://doi.org/10.3390/su13020791
59. Voorberg, W.H., Bekkers, V.J.J.M., Tummers, L.G.: A systematic review of co-creation and co-production: embarking on the social innovation journey. Public Manag. Rev. **17**(9), 1333–1357 (2015). https://doi.org/10.1080/14719037.2014.930505
60. Al Hawi, T., Alsyouf, I.: A proposed innovation model for public organizations: empirical results from federal government innovation experts in the United Arab Emirates. Int. J. Syst. Assur. Eng. Manag. **11**(6), 1362–1379 (2020). https://doi.org/10.1007/s13198-020-00994-9
61. Mu, R., Wang, H.: A systematic literature review of open innovation in the public sector: comparing barriers and governance strategies of digital and non-digital open innovation. Public Manag. Rev. **24**(4), 489–511 (2022). https://doi.org/10.1080/14719037.2020.1838787

Corporate Governance in the Digital Era

Mary Yaqoob, Alreem Alromaihi, and Zakeya Sanad

Abstract Digitalization is the application of technology to promote efficiency and long-term competitive advantage. There is a consensus that corporate governance is all about transparency, accountability, and people. But how has the digital revolution impacted the business world and particularly how the artificial intelligence and big data could influence corporations' corporate governance? The purpose of our study is to examine the influence of digitization on corporate governance. Besides, to check if the advantages outweigh the implications and to identify what are the associated factors and risks.

Keywords Corporate governance · Digital transformation · Digital corporate governance · Artificial intelligence · Blockchain · Cyber risk · Big data · Board of directors · Goal 8: decent work and economic growth · Goal 9: industry · Innovation and infrastructure

1 Introduction

Nearly every aspect of modern life has been profoundly impacted by the process of digitization, from education to work to how we shop and socialize. Digitalization is the application of technology to promote efficiency and long-term competitive advantage. Artificial intelligence (AI), cloud computing, blockchain, and the Internet of Things (IoT), among other emerging forms of digital technology, are thought to have significant impact on businesses [1].

M. Yaqoob (✉) · A. Alromaihi · Z. Sanad
Ahlia University, Manama, Bahrain
e-mail: mmaarryyd24@gmail.com

A. Alromaihi
e-mail: 201910056@ahlia.edu.bh

Z. Sanad
e-mail: zsanad@ahlia.edu.bh

© The Author(s), under exclusive license to Springer Nature Switzerland AG 2024
A. Hamdan and E. S. Aldhaen (eds.), *Artificial Intelligence and Transforming Digital Marketing*, Studies in Systems, Decision and Control 487,
https://doi.org/10.1007/978-3-031-35828-9_39

There is a consensus that corporate governance is all about transparency, accountability and people [2]. Corporate governance is the system by which a company manages itself. In the broadest sense, it's a framework of regulations and rules, procedures and practices, etc., determines and controls how a firm conducts its operations. A company's alignment with its mission, achievement of its goals, and fulfillment of its obligations to all stakeholders are all ensured by good corporate governance [3]. Reference [4] stressed that good governance is a result of a diverse board that is independent and regularly evaluated, the company being transparent, defining shareholder rights and proactively managing risks.

But how has the digital revolution impacted the business world and particularly how the artificial intelligence and big data could influence corporations' corporate governance? Digitalization and corporate governance are of great importance on their own, however, according to the literature, combing them, among other areas aids in increased transparency, process efficiency, better decision making, strong strategic planning and enhanced information flow [5].

Nevertheless, all this works well in theory, but things in practice can be quite different. It is important to observe how digital corporate governance works in real life. Corporate governance has long been a topic of interest and attention and conflict of interest is one of the major problems, but now as digital components are being added new problems are arising contributing to the old. Hence, the purpose of our study is to examine the influence of digitization on corporate governance. Besides, to check if the advantages outweigh the implications and to identify what are the associated factors and risks.

This paper contributes to UNSDGs 8, 9 and 16. For goal 8: Decent Work and Economic Growth the contents of this study may help with the target of achieving higher levels of economic productivity through technological upgrading and promotion of development-oriented policies that support decent job creation [6]. For goal 16: Peace, Justice and Strong Institutions, with targets "Develop effective, accountable and transparent institutions at all levels" and "Ensure responsive, inclusive, participatory and representative decision-making at all levels" [7]. As our research encourages technological progress and investment in the area, this aligns with goal 9: Industry, Innovation and Infrastructure [8].

Having established the definitions, stated the importance, the issue and the purpose of the research, Sect. 2 is organized by publications and the year they were published, from oldest to newest, and presents data from various nations that is in accordance with the study's goals. Each article outlines key elements, pros and cons, some talk about tools and software, how they are used in practice, and give real life examples. The breadth of digital corporate governance procedures at the present time is discussed in [9–11]. Reference [5] also offers recommendations on how to maximize the benefits of digital corporate governance. Finally, Sect. 3 starts with our verdict on the question do advantages of corporate governance digitalization rundown on the inadequacies, next shortcomings of the matter are highlighted followed by practical suggestions for policymakers, a brief analysis of the literature review and pointing out the associated risks and factors, the research gap and lastly, future study and research method recommendations.

2 Literature Review

To conduct the research, we looked through a number of studies to understand the effects of digitalization on corporate governance. The papers used were published in different countries. Reference [12] discussed the use of blockchain, artificial intelligence, and big data analytics in corporate governance to promote more open and honest relationships amongst stakeholders.

One of today's most significant technological advancements is artificial intelligence (AI). According to the book "Artificial Intelligence in Practice," most businesses have begun to investigate how they may leverage AI to enhance consumer experiences and expedite corporate processes [13]. According to the researcher, compared to the resources needed for a human specialist to study the same amount of information, artificial intelligence skills enable us to analyze massive data arrays. As artificial intelligence can go through more information it increases efficiency. The author also mentioned that a Venture Foundation in Hong Kong incorporated artificial intelligence to the board and their activities. One reason is to help and improve firms' decision making. But according to the theory of singularity, it is important to consider that this integration might lead to artificial intelligence becoming the dominant force in the organization and not humans this will lead to serious consequences.

Reference [12] added that blockchain, which [14] lists as one of the technologies to have the biggest impact on business and society, according to several scientists and experts, blockchain technology has the potential to drastically alter company governance by introducing a new type of corporation based on DAO (Decentralized Autonomous Organization) technology. Overall, it has vulnerabilities but through time and model improvements many positive effects are expected.

A year later, an article by Polidi et al. [10] discussed the key elements to gain corporate governance excellence. Only a small percentage of firms currently use integrated management automation systems, according to the official figures. Managing relationships with external counterparties and core business operations are the two areas where information technology is applied in corporate governance. The authors added that it is compulsory to develop management based on a balance of interests and to lay performance indicators in corporate management information systems to measure the level of quality of corporate governance. The authors perceive that this would help to increase the effectiveness of corporate management.

Another 2019 study conducted in Russia by Ivaninskiy [15] discussed the effect of digitalization, more specifically blockchain technology on corporate governance and argues that blockchain technology can be used to solve and mitigate several board member related conflicts, such as the lack of transparency in the voting process and empty voting. Also, the author claimed that certain blockchain technology, such as "smart contracts" which is a blockchain technology that only runs if certain requirements are met, can improve the governance process, and increase its transparency and reliability.

However, [15] admitted that the usage of blockchain technology in corporate governance can pose certain risks and might create new problems. Finally, the author stressed that the utilization of blockchain in several aspects of corporate governance has many benefits, including increasing the role of shareholders in governance, which leads to improving organizational dynamics, as well as increasing the organization's overall transparency.

A 2020 study conducted by Moşneanu [16] in Romania discussed the inevitability of digitalization. The Author claimed that a benefit of digitalization in corporate governance is "digitized recruitment", which is using technological tools, such as DM software technology (decision making software), a decision-making software that categorizes possible members and ranks them based on pre-logged requirements, to better the election of board members process and lower its risk of corruption, bias, and abuse.

The author added that digitized recruitment also offers an interesting perspective as it completely eliminates the human factor in the election process, and instead depended on algorithms and software's, which does have the potential to lower biases. Another technology that the author posed as a potential tool to improve corporate governance in the digital era is called the "RACI matrix method", which is a technology that is dependent on responsibility, as it assigns members tasks and roles they must take on to complete projects. The researcher highlighted that this same technology led to the improvement of the Romanian public administration's framework against abuse.

The digitization process, as discussed in [17], impacts all stakeholders, including shareholders, management, and suppliers, in addition to the internal operating environment of a firm. Reference [17] covers using communication technologies and other developing technologies. Business goals that were formerly profit-focused have been replaced with ones that focus on adding value in the long run. By ensuring that stakeholders have access to the information they require, which is transparent and accurate and on time, the effect of digitalization on the accounting function, one of the most significant supporting pillars of corporate governance, enables a more developed framework of accountability and responsibility.

Reference [17] add that the quantity of data utilized in corporate decision-making has increased due to the new technologies involved in this process. As a result, opportunities for richer material to be included in both non-financial and financial reporting have also evolved. Corporate governance procedures, accounting standards, and other company operations have all been revolutionized thanks to blockchain technology, artificial intelligence, and big data analysis.

A study conducted in Italy by Lepore et al. [18], between the years 2017–2019, hypothesized that the increased moderation role of stakeholder e-engagement on social-media applications such as Facebook, Linked-in, and Twitter can greatly impact corporate social responsibility, an extension of corporate governance [19]. The authors used a sample of 347 Italian firms and utilized an economic panel data dependance technique to conduct their research.

Results validated the authors hypothesis as it showcased a prevalent and strong connection between positive corporate social responsibility and stakeholders having a visible moderating role when it comes to e-engagement on social media. Reference [18] opens up new spaces for future research regarding social media potentially becoming a tool for good governance. Despite positive findings, the study still faced some limitations. For instance, the sample used only included Italian companies, therefore new studies adopting and applying the same approach elsewhere are needed to further refine the findings. Also, the author suggests mapping out stakeholder's social media engagement in different Euro-zones and comparing the findings to provide much needed insight on the reasons behind the differences or similarities between the companies.

Reference [11] examines Bahrain's digital corporate governance practice, particularly on the employment of artificial intelligence, which is one of the main variables of this paper. The study has focused on the ethical use of AI's capabilities for corporate governance which leads to changes in cost, decreased effort, timeliness and precision of disclosure, increased data accessibility for stakeholders and time reduction. According to [11], (1) the legal basis of digital transformation and (2) the legal foundation of disclosure and its electronic activation are two legal bases for applying AI in corporate governance.

In the legal frameworks of Bahrain and UAE, artificial intelligence tools are plainly insufficient for stock market trading, licensing and electronic company registration. The legislature is urged to expand the digital transformation to encompass all of the legally required responsibilities of corporate registrar, to track businesses' performance digitally, to carry out audits and electronic inspections of corporate accounts, and to evaluate the financial standing of these accounts.

In research done by Arachchi et al. [9] with the aim of investigating the application and extent of digital corporate governance, 160 Colombo Stock Exchange listed firms were sampled mainly using quantitative methods. The study observed that although Sri Lanka is not in pace with digital transformation of business, to an extent the country was involved in the application and pondering on the subject. COVID-19 accelerated this process and many businesses begun to digitize every aspect of their operations including corporate governance.

The researchers used 4 main variables namely IT experience of board members, transparency, cyber security management and virtual meetings. The main results of the study were, first, more than 50% of the 160 sample firms were practicing digital corporate governance. Second, mixed results were received regarding the variables more research is required. Third, Cyber security is one of the most vulnerable areas. Companies must develop stringent cyber security laws which are necessary for stronger corporate governance systems.

A recent paper by Zinyuk et al. [5] focused on the unique characteristics of corporate governance in the context of digital transformation and lists the pros and cons of corporate governance digitalization. For instance, [5] mentioned that communication is key for good corporate governance. Technology incorporation in corporate governance practices lead to improved collaboration and increased participation as it enables board members to conduct meetings and work together anywhere at any time. Flow of information is now enhanced as all members can share and review documents and information which also increases transparency. As now it is easy to access and analyze historical records, better decisions and plans can be made. Positive results relating to conflict of interest were also observed. Table 1 provides an

Table 1 Current and potential benefits and drawbacks of corporate governance digitalization

Current and potential benefits	Current and potential drawbacks
• Decision-making is expedited and easier • Improvement and speeding of business processes • Quick report creation • Enhancing the company's reputation with clients and workers • Corporate governance digital solutions are dynamic • Utilizing cloud technology for safe data storage • Production and service activities are fully automated • Gaining a competitive edge • Virtualizing workspaces	• Not all workers are prepared to adapt to and use digital technology • Increased costs • The necessity of enlisting knowledgeable personnel with information systems expertise • Continuous staff training • Frequent updating and upgrading of systems • Cybercrime • Problems with regulators

overview of the existing and implications of digital corporate governance in light of the SWOT analysis performed by [5].

Based on their SWOT analysis, the authors have given recommendations to make the most through digital corporate governance. Using appropriate technology is one of them. The technology must support the company's strategy and address its problems. Presence of an IT director, which goes hand in hand with the first point as a digital director with all his relevant knowledge and experience will be the one to choose the right software among other tasks. Data shows that enterprises are 1.6 times more likely to carry out successful digital transformation when a digital director is present. Next suggestion requires not only a team that is specialized in technology but employees from all departments to have a decent understanding of the matter. Other suggestions include focusing on company culture and defining the vision.

3 Conclusion

We have concluded through our research that digitalized corporate governance is inevitably on the rise, and although the technologies mentioned in our studies are fairly recent and do need further development, it is evident that the beneficial aspects of digital corporate governance practices, which include tools such as blockchain, AI, DM software and integrated management automation systems far outweigh the risks associated with these technologies.

Also, we have found that continuing to utilize outdated, traditional corporate governance practices leaves the organization vulnerable to certain risks, such as board member fraud and lack of cyber-security, forming a weakened corporate governance framework. It was also clear to us that there is a lack of funding and research when it comes to studies discussing digital corporate governance, as most of our sources had concluded stating that further research is needed.

As big part of the improvement responsibility relies on the decisions of the policy makers. Thus, we recommend they shift their focus to their research and development department and increase their budget, as research and development are an essential part of digital governance improvement. With more focus and a bigger budget, the department could predict trends and unexpected conditions that are on the rise and develop the necessary tools and strategies to deal with them. Another recommendation for policymakers is the hiring of experienced and competent IT directors. The directors must have the necessary knowledge and experience to compact the different challenges the corporation and its governance may face in this digital age. Finally, we stress the importance of awareness and training. Digitalization can be unpredictable and ever changing, and because of its vastness and complexity, policymakers as well as employees must be aware and educated on the manner. This can be achieved through recurring training workshops and seminars.

All in all, we observed that the majority of the previous studies shared similarities and no contradictions were found and concurred that digitalization increases transparency and improves role of shareholders/company dynamics. The shortcomings could be limited research in this field and privacy concerns. Also, some aspects need further investigation, for example, factors and risks were identified but the extent of their effectiveness or damage not as much due to the fact companies are in the process of incorporation. We believe with time results in lacking areas will be clearer. The most repeated and significant variables in the previous studies that tackled this issue were: Artificial intelligence, IT experience of board, Blockchain as well as Risk—Cybercrime.

This study suggests conducting more studies on corporate governance and digital technology in general as there are research gaps, this will lead to attainment a definite idea about the topic. For instance, as found in [15] study, blockchain technology might pose certain risks, therefore, further quantitative studies are needed to understand clearly what are the risks associated with this technology in terms of corporate governance.

Furthermore, prior studies highlighted the need of safeguarding corporate governance by using software as there is potential for corporate blackmail in this digital era.

Hence, future studies should focus on this aspect as well. In addition, research on digital corporate governance should be conducted in the GCC as most studies found were from USA, Europe and Asia (excluding GCC). Also, although [16] study suggested that digitized recruitment might be the answer to several board election issues and a potential way to decrease fraudulent corporate governance practices, this suggestion needs further quantitative research and testing. Finally, future studies should use SWOT analysis method as discussed earlier as we found it to a good tool to understand this critical topic.

References

1. Digitalization of Corporate Governance in Emerging Markets.: Frontiers (n.d.). https://www.frontiersin.org/research-topics/39106/digitalization-of-corporate-governance-in-emerging-markets
2. Akgiray, V.: Blockchain Technology and Corporate Governance: Technology, Markets, Regulation and Corporate Governance. Organisation for Economic Cooperation and Development (OECD), 1 (2018)
3. Corporate Governance in the Digital Age.: Lukasa (n.d.). https://lukasa.com/
4. Lagercrantz, M.: How To Ensure Good Corporate Governance in 10 Simple Steps. Boardclic (2021). https://boardclic.com/corporate-governance/how-to-ensure-good-corporate-governance/
5. Zinyuk, M., Deeva, N., Bogatyreva, K., Melnychenko, S., Faivishenko, D., Shevchun, M.: Digital transformation of corporate management. Finan. Credit Act. Probl. Theory Pract. 5(46), 300–310 (2022). https://doi.org/10.55643/fcaptp.5.46.2022.3807
6. #Envision2030 Goal 8: Decent Work and Economic Growth | United Nations Enable (n.d.). https://www.un.org/development/desa/disabilities/envision2030-goal8.html
7. #Envision2030 Goal 16: Peace, Justice and Strong Institutions | United Nations Enable (n.d.). https://www.un.org/development/desa/disabilities/envision2030-goal16.html
8. Sustainable Development Goals | United Nations Development Programme.: UNDP (n.d.). https://www.undp.org/sustainable-development-goals?utm_source=EN
9. Arachchi, S., Dissanayake, H., Deshika, T., Iddagoda, A.: Digital corporate governance practices: evidence from Sri Lankan listed companies. Econ. Insights Trends Chall. 2022(2), 79–92 (2022). https://doi.org/10.51865/eitc.2022.02.06
10. Polidi, A.A., Goukasyan, Z.O., Maslova, I.A., Fedorenko, R.V.: Some aspects of the quality of corporate governance in digital economy. In SHS Web of Conferences, vol. 62, p. 04002. EDP Sciences (2019)
11. Al-Obeidi, M.Y.M.: Legal regulation for adopting artificial intelligence systems in corporate governance according to Bahraini law (comparative study). J. Law 17(2) (2020)
12. Nikishova, M.I.: Prospects of digital technologies application in corporate governance. In 8th International Conference Social Science and Humanity, pp. 86–95 (2018)
13. Marr, B.: The 7 Biggest Technology Trends in 2020 Everyone Must Get Ready For Now. Forbes (2019). https://www.forbes.com/sites/bernardmarr/2019/09/30/the-7-biggest-technology-trends-in-2020-everyone-must-get-ready-for-now/?sh=1d81f44a2261
14. These Eight Technologies Will Have the Greatest Impact on Business and Society (n.d.-b). https://www.knowit.eu/services/experience/digital-strategy-and-analysis/strategy-and-digitalization/knowits-take-on-tomorrow/eight-technologies/
15. Ivaninskiy, I.: The impact of the digital transformation of business on corporate governance. An overview of recent studies. Корпоративные финансы 13(3), 35–47 (2019)
16. Moşneanu, D.: Corporate governance in the digital world. Proc. Int. Conf. Bus. Excellence 14(1), 333–342 (2020). https://doi.org/10.2478/picbe-2020-0032
17. Varoglu, A., Gokten, S., Ozdogan, B.: Digital corporate governance: inevitable transformation. In Financial Ecosystem and Strategy in the Digital Era, pp. 219–236. Springer, Cham (2021)
18. Lepore, L., Landriani, L., Pisano, S., D'Amore, G., Pozzoli, S.: Corporate governance in the digital age: the role of social media and board independence in CSR disclosure. Evidence from Italian listed companies. J. Manag. Gov. 1–37 (2022)
19. Xu, E.G., Graves, C., Shan, Y.G., Yang, J.W.: The mediating role of corporate social responsibility in corporate governance and firm performance. J. Clean. Prod. 375, 134165 (2022)

Environmental, Social, and Governance (ESG) Impact on Firm's Performance

Fatema Alhamar, Allam Hamdan, and Mohamad Saif

Abstract This research article aims to review the relationship between corporate environmental, social, and governance (ESG) disclosure and firm performance. The study looked at three aspects of a company's performance: return on assets (ROA), return on equity (ROE), and Tobin's Q. In addition, the goal of this study is to investigate how ESG and public disclosures of mismanagement to various stakeholders have influenced a company's market valuation, which we use as one of the indicators of shareholder wealth. The findings also imply that governance-related issues lower market valuations more than environmental and social issues, with the latter two remaining value neutral throughout the study period. At the same time, the substantial gap between the minimum and maximum in the examined event windows suggests that the materiality of ESG problems remains a significant consideration.

Keywords Environmental · Social · Governance · (ESG) · Disclosure · Performance · Firms

1 Introduction

The ESG is being used to determine the firms' performance either in a positive or negative way. The ESG is a shortcut to Environmental, Social, and Governance factors that impacts the firms' performance. The United Nations Responsible Investment Principles (UNPRI) aims to identify the impacts of ESG and to give enough support to the investors to integrate all these issues that matter in the investment practices they are having. Moreover, it required the investors to take into their consideration the issues of ESG when they evaluate the firms' performance. The (UNPRI) is a result of a collaboration between the finance effort of the United Nations Environment Program,

---

F. Alhamar · A. Hamdan (✉)
Ahlia University, Manama, Bahrain
e-mail: allamh3@hotmail.com

M. Saif
University of Business and Technology, Jeddah, Saudi Arabia

© The Author(s), under exclusive license to Springer Nature Switzerland AG 2024
A. Hamdan and E. S. Aldhaen (eds.), *Artificial Intelligence and Transforming Digital Marketing*, Studies in Systems, Decision and Control 487,
https://doi.org/10.1007/978-3-031-35828-9_40

the United Nations Global Compact, the investment industry, governmental agencies, and intergovernmental organizations [1].

Also, the ESG gives the firms the information it needs to know how to control the company's finances and where the money is invested. Therefore, the stakeholders and the investors are very worried and interested in the factors of ESG. As it been mentioned in this paper ESG is a shortcut for the Environmental, Corporate Social Responsibility, and Corporate Governance factors. Here are some examples of the factors in each term. First the Environmental contains the changes in the climate, the protection of the natural environment, and the effects that occur to the environment because of the business operation. Second, the social factor considers human rights, equity, diversity, and contributing to the community. Third the Governance factor for example the revelation of the firms' information, translucence, the rights of the minority shareholders, ownership structure, treating all the shareholders in the same ways, and board independence [2].

The stakeholders and the non-government organizations became very careful and accurate when it comes to the ESG factors, and this is a reason for the repeated scandals of the firms in the financial part. Moreover, the companies are trying to give the stakeholders very specific ideas about the firms' effort and practices. As a result of paying attention to the ESG factors and scandals financial firms that made the firms improve in different aspects over the years. Finally, many firms are now playing a big part in the reveal of ESG components and activities [3].

For the socially conscious investor, ESG agencies are valuable sources of information. However, when it comes to explaining the evaluation criteria utilized, particularly those relating to risk management, agencies are tight-lipped. As a result, investors will have a tough time analyzing this data directly. Investors, on the other hand, have another option for incorporating sustainability into their decision-making: sustainability indices. These indices assist investors in making investment decisions, however, they limit the use of an investor's own social responsibility criteria [4].

Some investors, shareholders, decision-makers, and regulators may not have a clear vision of the effect of ESG on a firm's performance. And the relationship between ESG and company performance remains the focus of attention and that is because ESG investing is a developing field. Moreover, the findings of the relationship between ESG and firms' performance are very important for the investors either in positive or negative ways, ESG investing is also a way for some investors to express their views on how the world should be and shareholders, decision-makers, and regulators need to have a clear vision on the effect of the ESG on firms performance.

2 Literature Review

In the last few years, different studies have tried to do the measurement for the performance and valuation impact of ESG factors. A portion of this study looked at the factors that drive a company's ESG disclosure, as well as the potential valuation implications of such disclosure. However, the main concern of how ESG operations and the extent of ESG-related transparency affect a company's value is understudied. According to some theories, the relationship between a company's ESG operations and its valuation is regulated by the company's disclosures regarding such efforts. However, the consequence of disclosure in this circumstance is not immediately apparent. One may expect a positive effect if openness reduces information asymmetries and helps investors better understand the firm's ESG strengths and faults. Alternatively, if investors perceive ESG disclosure as "cheap talk" or "greenwashing," it may detract from a company's value [5].

The ability to raise money at minimal prices is one of the main reasons companies turn to capital markets. This is contingent on corporations' ability to sufficiently compensate investors, as investors prefer companies that deliver a higher return per unit of risk. In this regard, corporations place a premium on the quality of institutional communication with the market via reports that provide accurate, comprehensive, and trustworthy data [6]. In addition, corporate governance practices and social and environmental performance have become increasingly important in the last decade for legislators and the general public and investors. As a result, increased profit margins are likely to coexist with better governance, social, and environmental policies [7].

Some studies have shown that a company's commitment to sustainability and CSR reduces uncertainty, business risk, and, as a result, the firm's cost of capital. If the business risk is defined as the possibility that the firm will not achieve its objectives, companies that pollute or have unfair employee relations may face penalties, fines, or even be compelled to discontinue operations, resulting in financial losses. Furthermore, irresponsible business behavior poses a huge danger of reputational loss [8].

3 Environmental Impact on Firms' Performance

The whole world is facing the same problem which is the changes that are happening in our weather such as global warming, pollution, and the problems that are facing the environment that causing changes in the climate. As a reason for all the changes in the environment may affect the firms' performance. So, to be able to overcome this problem each firm should be aware of the environmental rules and expose their engagement to the environmental issues [9].

In the past, the firms did only pay attention to improving flexibility, progressing quality, decreasing costs, and minimizing led times. Nowadays, the future firms should become more considered about environmental factors. Moreover, firms should look at the environmental factors as very important asset. Unfortunately, some of the firms are struggling in managing their environmental performance and that may lead them to a strategic issue [10].

According to Ann, Zailani, and Wahid, They found that the environmental factor has a significant positive effect on the firms. Also, it impacts the efficiency and quality of products [10].

464 F. Alhamar et al.

In an article written by Junquera and Sánchez, they talked about the advantages and the positive competition between the firms to create a proactivity environment. Moreover, to be able to enhance the firm performance and to enhance the managers understanding of the good performance, the firm needs to have a well understanding of environmental management [11].

In the past, the firms were focusing on enhancing quality, decreasing cost, being more flexible, and decrease lead time. But nowadays firms are more responsive to the environment around them and how it affects the firms because the environment become an asset that is highly considered in deciding the value of the firms and it is a strategic issue to a lot of firms around the world [10].

The international organization of standardization (IOS) set around 14000 standers in 1996. Those standers aim to help the firms combine the environmental aspects in the decision-making process and make it more systemic, harmonic, regular, organized, and methodical. Moreover, it helps the firms to grow their operations based on their consideration of the environmental standard [10].

According to Ann, Zailani, and Wahid, they assume that the environment impacts the firms' qualification, efficiency, validation, and effectiveness positively [10].

James set five important leading points for the firms' performance: financial stakeholders, nonfinancial stakeholders, the biosphere, public, and buyers. After some years Haines add one of the most significant parts that affect the firms' performance which is the environmental impact [12]. According to Turulja and Bajgoric they say that the environment plays a very big role in improving innovation inside the firms and in their opinion it affects innovation more than just the firm performance [13].

(EPI) is a shortcut for the Environmental Performance Indicators which means the numeral measurement, both financial and non-financial aspects that prepare the firms with the full information about the environmental effect, stakeholder relationships, organizational systems, and regulatory compliance. Moreover, it focuses on the size of interaction between the firms and the environment. Also, it spots the light on the quantification of the efficiency and effectiveness of environmental behavior and impact on the firms. Moreover, it measures the internal and external impact of the environment on the firms. Finally, the EPI is an element in environmental management accounting (EMA) that refers to the management of the environmental performance over the improvement of suitable environmental-related accounting systems routines [14].

According to Journeault and Henri, they prefer that firms should make sure that their managers are having the knowledge that using the EPI helps the firms to have more functional environmental strategies and support, discuss, and report this strategy all over the company [14].

All the changes happening all over the world, especially in the environment made the firms more concerned especially when it comes to the green marketing strategy because it made the firms think of the impact on the environment whether it will damage or improve the company. Moreover, the firms are paying lots of attention to the environment because of the very strict regulations, and the stress from the stakeholders investigating the environment aspects in their decision-making [15].

Eneizan et al. [16] says that the green marketing strategy, means that firms consider seven factors in their decision-making: green product, green price, green distribution, green promotion, green people, green process, and green physical evidence, is expected to produce more revenue than firms that do not consider the green marketing strategy. It also has a positive impact on financial and non-financial performance.

Porter and der Linde they are supporting that there is a huge connection between the environment and the financial performance of the firms. Moreover, they assume that reducing pollution will extend future cost savings, lower future liabilities, boost efficiency and decrease compliance costs [17].

Research was written by Sambasivan, Bah, and Ann state four main points of the environmental impact on the firms. First, the environmental proactivity measure supports the firms to enhance their operational performance, financial performance, and organizational learning and satisfy the stakeholders. Second, the relationship between the environment and financial performance is mediated by stakeholder satisfaction. Third, the relationship between the environment and operational performance is unaffected by the technology being used (prevention and control) [18]. Finally, the relationship between the environment and stakeholder satisfaction is not mediated by the environment [19].

4 Social Impact on Firms' Performance

Firms should put in their minds that the public is knowledgeable about the social firms are responsible for. Therefore, the managers and the stakeholders should be more careful with their reactions toward the social issues that matter. Moreover, lots of firms specialized many efforts and resources to discover the information that is related to social responsibility in their reports [20].

According to Buallay, Kukreja, Aldhaen, and Hamdan they think that firms should have a main target which is to have a very positive social responsibility between the firms and society and people. Moreover, they define the social responsibility as the long-term effect of the actions the firms are taking on the well-being of society. Also, firms should have a very strong relationship with the stakeholders and that's by having more charity events and considering that the employees are in a good health [9].

The firms should understand that to low the risk in their business they should high their socially responsible exposure exposure. Moreover, to be able to earn the trust of the stakeholders the firms should have a clear social responsibility that is used accurately and boost the organization's sustainability. Finally, here are some examples of the advantages of social responsibility in the firm. First, enhance the firms' financial performance. Second, create positive competition. The third, is to have an excellent reputation for the firm. Fourth, to raise revenues and decrease costs. Fifth, allow the firms to become more transparent. Finally, to enhance the employees to realize the sustainable development of the firms [9].

The social corporation affects all kinds of firms the little and the large management. The bad social corporation will impact the firms in negative ways such us: it can raise the firm's risk, have a bad impact on firm's reputation, reduce the firm's value, and drive to a bad relationship with the stakeholders and in some cases could be the end of the hole firm [21].

Moreover, as Fiori, Donato, and Izzo said that social responsibility could affect the firms' economic stage positively for example: first, expand into new sections of the market that the consumers are not just paying attention to prices and quality of the market but also put in their consideration the ethical code which can enhance the sales. Second, enhance the loyalty of the suppliers and customers. Third, the ability to keep talented people which will give a good reputation and benefit the managers. Fourth, decrease the interest rate. Fifth, formatting strategic alliances and collaborations. Finally, to have a superior quantity and quality exposure [21].

According to McWilliams and Siegel, they said that corporate social responsibility help to increase the stakeholders and enhance the demand from the consumers, community, investors, and employees. For example, consumers are calming to have organic products and that will need to operate innovation by the farmer, also its need the natural foods retailer to produce new products. On the other hand, the employees ask to advance labor relations policies, proper workplace, and safety and to increase the working staff who are well trained to apply the corporate social responsibility policies in the firms. Finally, the firms are trying to merge their products with the corporate socially responsible attributes [22]. The interaction between corporate social responsibility and the firms performance will impact the service quality, customer satisfaction, increase financial performance, and job satisfaction. Moreover, customer satisfaction and firm satisfaction mediate the link between corporate social responsibility and the firm's financial performance because they complete each other [23]. Jo and Harjoto found that corporate social responsibility relates to the internal and external firm's attributes, monitoring mechanisms, board leadership, firm ownership, board independence, and analysts. Finally, Jo and Harjoto finds that social responsibility impacts the firms positively [24]. Abiodun said that each firm is different from the other based on the amount they paid to the commitment to social responsibilities. Moreover, based on the data Abiodun had to collect says that most of the firms invested around ten percent or less of their annual firm profit social responsibilities [25].

The different types of firms in the market such as: financial services, software, health care, consumer goods, and different types are spending a big amount of money on corporate social responsibility activities. Lately, Financial Times reported that around 500 firms spent more than 15 billion dollars on corporate social responsibility activities this amount of money was spent on donating free medicine like in Johnson and Johnson company, donating free software in Oracle company, in Prudential they operate in educational plans in the developing countries, make more profitable work environment for different minatory groups in the Chicago Fed and different examples in different companies [26].

Berland suggests that the customers put in their minds that the firms are paying so much because of the corporate social responsibilities activates when they purchase from it. Based on the customers' considerations, they may be willing to pay more or also they will be more interested to pay. Nielsen1 did a global survey, they had 29,000 responders and 50% of these responders across 58 different countries responded that they have the desire to pay more or higher prices for services or products that have been developed by the firms that operate in corporate social responsibilities activates [27].

Luo and Bhattacharya shows that in the last few years, social responsibilities has a positive impact on firms and also it will drive the company to a lot of different commercial benefits for the companies. Moreover, the social responsibilities activities will have an affirmative impact on the company's brand evaluation, brand recommendations, brand choices, consumer satisfaction, customer loyalty, customer participation in the products harm crises situation, customers' high willingness to buy more, and customers firm identification [28]. Shen and Benson claims that one of the most important aspects of social responsibilities is to provide employees with the feeling of being safe in the firm, sensations of belonging, and higher self-esteem, making them act with a greater sensitivity to work and for sure all these points will make them more motivated to work and produce more. Moreover, the firms should purposefully bring in skill-enhancing corporate social responsibilities factors that will improve employee performance [54].

Jain, Vyas, and Chalasani claims that corporate social responsibility is defined as the firm's activities in volunteering to merge the company with the environment and social interest into the firms and the business and across stakeholders to obtain sustainability. "People, Planet, and Profit" is a triple bottom line term that takes into consideration the main values to measure the company's success for example in the economic, social, and environmental standards. Moreover, corporate social responsibility procedures are like a guideline and a plan to enhance responsible behavior depending on the business environment and risk [30]. The positive affiliation is found in stakeholders theory, which shows that the firms that boost the social responsibility performance will earn a better financial performance than the other companies in the long term. On the other hand while using accounting-based, market-based, and investor-based to investigate the connection between corporate social responsibilities and firms the result shows that it aims to elevate the firms' performance. Moreover, many studies show that corporate social responsibilities help to develop a positive reputation, have a high degree of accuracy in predicting a firm's value, it has the potential to become a long-term competitive advantage for the firm and the firm's long-term relationships with its stakeholders are being improved [31].

5 Governance Impact on Firms' Performance

All the firm aims to attract the investors' confidence so they will have better investments and that can be achieved by concentrating on corporate governance. The framework for better corporate governance should include the following. First, boost the transparency. Second, to have active markets Third, to become more regular with the rule of law. Finally, to divide the responsibility with supervisor, regulatory, and enforcement authorities [32].

Corporate governance is critical in fostering a culture of awareness, transparency, accountability, and openness inside the firms [33].

Corporate governance can be defined as the combination of the structures and processes that are being used to direct the firm's business. Moreover, The firms should have corporate governance to represent responsibilities and rights with the other parts that are having a stake in the firms. Also, the firms should put in their consideration that the most important thing is to enhance and improve the shareholder wealth. Whenever the firms start to implement this thing they will have an active and successful corporate system that will help include an effective division between the shareholders, the board of directors, and management. Moreover, according to [34]. "Corporate governance is not just corporate management; it is something much broader to include a fair, efficient, and transparent administration to meet certain well-defined objectives. It is a system of structuring, operating, and controlling a company with a view to achieving long-term strategic goals to satisfy shareholders, creditors, employees, customers, and suppliers, and complying with the legal and regulatory requirements, apart from meeting environmental and local community needs. When it is practiced under a well-laid out system, it leads to the building of a legal, commercial and institutional framework and demarcates the boundaries within which these functions are performed." Finally, the firms that are having for both domestic and international businesses, well-functioning corporate governance efforts to boost economies are critical investors that are concerned with the massive opportunities that the economies expand [32].

Finally, Ahmed Diab and Ahmed Aboud think that it is not important for the firms to just be listed in the ESG index but it is also more important to have a high rank. That is because the firms that rank in the higher place are found to have more value than the others [35].

Also, Amina Buallay thinks that each element of the ESG (Environment, Social and Governance) affect the firms in a certain way but overall the ESG affects the firms positively [36].

The research that is talking about the relationship between governance and company performance has taken great consternation. The reason behind all this huge number of literatures is the clash of interest between the managers shareholder and the transmitted-in-principle agent relationship. Moreover, that clash of interest raises the firms costs according to the shortage of controlling devices that are in the ownership of the shareholders. To sums up corporate governance is being used to align the principal agent's interests and the consequent progress in the firms' performance [37].

According to Hassan and Halbouni, the result they found about the effect of corporate governance based on the dependent principles of corporate governance can be also applied to emerging firms. The results are very necessary for organizers, investors, administrators, and researchers to participate in enhancing new policies that set a superior legal and regulatory infrastructure to develop investors' confidence and catch the attention to the new foreign investment and also to improve the firms' development. Nevertheless, the results of the consequence of the impact of corporate governance on the firms explain the extra challenges that emerging firms' economies must deal with to become able to perform the very high standards of corporate governance and promote new regulations of governance [38].

Constantinos says that governance scholars have established that repairing corporate governance drives the firms' victory and success in various countries. Moreover, corporate governance is supposed to raise the controlling tools by the council of directors which will be able to reduce the firms' problems between the management and the shareholders [39].

On the other hand, the United Arab of Emirates is paying so much attention to the ways that will help to preserve and save the investors so, as a reason to that the United Arab of Emirates has legislated comprehensive economic regulations and laws to simplify and organize investments. Lately, the United Arab of Emiratis government force the corporate governance code that applies to all public firms. Moreover, the United Arab of Emiratis Ministry of Economy issued the corporate governance code in 2009 in a Ministerial resolution and imposed to apply it on all firms, especially in the financial markets [40].

In their research, Farhan, Obaid, and Obaid assume that corporate governance develops and increases the firm's performance. Moreover, corporate governance helps firms to enhance their experience and improve their skills [40].

Gompers, Ishii, and Metrick, examined the consequences of corporate governance on the firms' performance and they discovered that the stock returns of firms with powerful shareholders are much more on a risk-adjusted basis returns of the firms with powerless shareholders' rights by around 8.5% for per year. Looking at those outcomes will put forward major concern according to the functional market hypothesis because all those portfolios are able to be built with publicly available data. Regarding the policy domain, corporate governance supporter has prominently bear witnessed this founding as proof that good governance has a positive impact on the firms' performance [41].

6 Environmental, Social, and Governance (ESG) Impact on the Emerging Markets

The impact of the ESG on emerging firms shows in systemic hazards, the role of investors and regulatory agencies is critical. Moreover, if the investors ignore the changes that occur to the firm because of the environment or society for example the changes in the climate or they ignore the equality or working circumstances in the poor communities. The firms will face so many risks [8].

Ali and Jamieson founding's shows a compelling case for the benefit of applying Environmental, social, and Governance based investment strategies through emerging firms, especially in the equity's asset class. Also, they supposed that according to the historical improvement of the Environmental, social, and Governance (ESG) integration the Environmental, social and Governance (ESG) investing opportunities and standards will remain to enhance in precision and clarity as the institutional investors are using the possibility in the Environmental, social and Governance (ESG) investing to become able to invest to boost returns and boost risk diversification. Moreover, the Environmental, social, and Governance ((ESG) opportunities in investing might also let the institutional investors earn the advantages of risk diversification from emerging firms without being subjected to the same degree of turbulence, low liquidity and lack of data on which the traditional emerging firm's investment [42].

Pollard and Sherwood assumed that Environmental, social, and Governance (ESG) investing has a positive impact on the financial performance of emerging firms [43].

Tarmuji, Maelah, and Tarmuji mentioned in their paper that firms nowadays are globally connected with each other. Also, stakeholders realize that the Environmental, social, and Governance (ESG) are accountable for a firm's integration to its performance and to the long-term sustainability. Moreover, the Environmental, social and Governance (ESG) impact on emerging firms is such: first, it posts a business spirt. Second, creates an environment combined with firm's integrity within society and the confidence of its stakeholder. Third, the firms earn a good reputation. Fifth, it helps to enhance investor trust. Sixth, optimum resource use. Finally, maintain a competitive edge [44].

According to Refinitiv these are some definitions related to the stakeholder theory: "The Workforce Score assesses a company's performance in terms of job satisfaction, a healthy and safe workplace, diversity, equal opportunity, and growth possibilities for its employees."; "The Human Rights Score is a metric that assesses a company's performance in adhering to basic human rights conventions."; "The Community Score assesses a company's commitment to being a good citizen, preserving public health, and adhering to ethical business practices."; "A company's commitment and effectiveness in adopting best practice corporate governance principles are measured by the Management Score." And "A company's dedication and efficacy in decreasing environmental emissions in production and operational operations is measured by its Emission Reduction Score [45]."

CSR is defined by the European Commission as "enterprises' responsibility for their social consequences to incorporate social, environmental, ethical, human rights, and consumer issues into their business operations and core strategy in close partnership with their stakeholders [46]."

According to Dyllick and Hockerts, when measuring CSP, maintains that businesses should think about the economic, environmental, and social consequences for society, as well as for specific stakeholders [47].

Corporate governance performance is widely used as an additional component of CSP. Both studies and the capital market agree that ESG variables can be used to collect CSP [48]. Rating agencies and participants of the UN-backed Principles for Responsible Investment (UN PRI) program, for example, employ ESG elements to evaluate socially responsible investments [46].

A stakeholder, according to Freeman, is "any group or individual who can influence or is influenced by the achievement of the firm's objectives." Shareholders, for example, are among the firm's stakeholders, as are creditors, employees, customers, suppliers, public interest groups, and governmental agencies [49].

A case can be made, based on stakeholder theory, that greater levels of CSP are associated with decreased financial risk. Low levels of CSP are more likely to result in law cases and legal costs, whereas high levels of CSP can lead to more solid relationships with the government and financial community [50]. Moreover, market participants are more likely to contribute resources to organizations with greater CSP levels, implying that CSR engagement can help enterprises overcome capital restrictions [51].

Although stakeholder theory shows a negative relationship between a company's CSP and risk, managerial opportunism theory suggests a positive relationship between CSR performance measurements and risk [52].

Business-stakeholder connections are frequently viewed as a management issue in the corporate world. Related stakeholder theories examine how corporations collaborate with stakeholders to obtain critical resources. Stakeholders, or stakeholder management strategies, are viewed as tools that assist businesses in achieving their goals [53].

This viewpoint can be traced all the way back to Freeman's seminal book "Strategic management: A stakeholder approach," the foundation for modern stakeholder theory was laid by this book. The so-called "hub-and-spoke" stakeholder model, which depicts companies as the center of a wheel and stakeholders as the ends of spokes around the wheel, was one of his most prominent stakeholder theories. The corporate perspective of stakeholder theory is well illustrated by researchers focused on how the hub (i.e. the firm) can turn faster (i.e. perform better) with the available spokes (i.e. its stakeholders) [54].

When academics are attempting to better understand stakeholder claims, strategies, and behaviors, they neglect firms and their performance in favor to focus on the business-society interaction from the stakeholders' perspective. Some of these researchers investigate the legitimacy and status of different stakeholder groups (for example, the controversy about whether nature can be considered a stakeholder or not) [55]. Others examine the resources and tactics that stakeholders use to support their claims, as well as their level of success. According to Frooman, stakeholder strategies characterized by a stakeholder's dependent on the firm and the firm's dependency on the stakeholder. He continues his assessment by stating that "knowing how stakeholders may try to influence a firm is crucial knowledge for any manager." Moreover, demonstrates how, in some situations, corporate-centric management reasoning can be adapted to different points of view. Theories that prescribe recipes for stakeholders and explain how to engage effectively with enterprises are an ideal-typical approach from a stakeholder perspective, comparable to a corporate perspective [56].

Provided that Donaldson and Preston examined the normative, descriptive, and instrumental components of stakeholder theory from a business standpoint, the following question remains: how does their work belong to a larger typology in which the corporate perspective is only one of three? Because the three aspects are generic heuristic procedures "rooted in a centuries-old philosophy of science," they supply the answer to this query [57]. In this historical context, the normative part resembles the deductive technique, which tries to apply universal rules to specific circumstances (typically based on ethical considerations). On the other hand, the descriptive aspect, resembles the inductive technique, which attempts to infer general rules and conclusions from examining individual situations. Finally, the instrumental aspect encompasses both the normative/deductive and descriptive/inductive approaches; its distinguishing trait is that it emphasizes causality by connecting means and ends. The three "What should happen?" questions are "What happens?" (descriptive/empirical aspect), and "What would happen if?" (Normative aspect). (instrumental aspect) effectively summarizes the general nature of the three elements [58]. The importance of the three elements of stakeholder theory, when viewed as such heuristic devices, goes far beyond what Donaldson and Preston had in their thought: they are applicable to all three perspectives of stakeholder theory, not just the corporate [59].

7 Conclusion

The gap in this research is the type of ESG disclosure a company utilizes has no bearing on its ESG score. In addition, various types of ESG can have different implications on business performance [60]. Furthermore, when different kinds of ESG disclosure are combined into a single score, the related effects may cancel out, leaving only the true effect to be identified. Moreover, more studies on the effects of ESG disclosure for other publicly traded companies, like those in developing countries or for various industries in the United States, are encouraged, since this will provide helpful and intriguing comparisons. Finally, it would also be interesting to investigate the influence of various factors in modulating the ESG-firm performance link.

References

1. Caplan, L., Griswold, J., Jarvis, W.: From SRI to ESG: the changing world of responsible investing. Inst. Educ. Sci. 3–12 (2013)
2. Atan, R., Alam, M., Said, J., Zamri, M.: The impacts of environmental, social, and governance factors on firm performance: panel study of Malaysian companies. Manag. Environ. Qual. 182–194 (2017)
3. Alareeni, B., Hamdan, A. ESG impact on performance of US S&P 500-listed firms. Emerald Publishing Limited, pp 1409–1428 (2020)
4. Olmedo, E., Torres, M., Angeles, M.: Socially responsible investing: sustainability indices, ESG rating and information provider agencies. Int. J. Sustain. Econ. 442–461 (2010)
5. Fatemi, A., Glaum, M., Kaiser, S.: ESG performance and firm value: the moderating role of disclosure. Global Finan. J. 45–64 (2018)
6. Mendes-Da-Silva, W., Onusic, L.: Corporate e-disclosure determinants: evidence from the Brazilian market. Int. J. Disclosure Governan. Adv. Online Publ. 1–20 (2012)

7. Cheng, B., Ioannou, I., Serafeim, G.: Corporate social responsibility and access to finance. Strat. Manag. J. 1–23 (2014)
8. Garcia, A., Mendes-Da-Silva, W., Orsato, R.: Sensitive industries produce better ESG performance: evidence from emerging markets. J. Clean. Prod. 135–147 (2017)
9. Buallay, A., Al Hawaj, A., Hamdan, A.: Integrated reporting and performance: a cross-country comparison of GCC Islamic and conventional banks. J. Islam. Market. 759–833 (2020)
10. Ann, G., Zailani, S., Wahid, N.: A study on the impact of environmental management system (EMS) certification towards firms' performance in Malaysia. Emerald Insight, 73–80 (2006)
11. Junquera, B., Sánchez, V.: Environmental proactivity and firms' performance: mediation effect of competitive advantages in Spanish wineries. MDPI 1–12 (2018)
12. Tyteca, D.: On the measurement of the environmental performance of firms—a literature review and a productive efficiency perspective. J. Environ. Manag. 281–308 (1995)
13. Turulja, L., Bajgoric, N.: Innovation, firms' performance and environmental turbulence: is there a moderator or mediator? Eur. J. Innov. Manag. 213–232 (2019)
14. Journeault, M., Henri, J.: Environmental performance indicators: an empirical study of Canadian manufacturing firms. J. Environ. Manag. 165–176 (2007)
15. Leonidou, C., Katsikeas, C., Morgan, N.: "Greening" the marketing mix: do firms do it and does it pay off? J. Acad. Market. Sci. 151–170 (2013)
16. Eneizan, B., Abd.Wahab, K., Zainon M. S., Obaid, T.: Effects of green marketing strategy on the financial and non-financial performance of firms: a conceptual paper. Arab. J. Bus. Manag. Rev. 14–27 (2020)
17. Porter, M., der Linde, C.: Green and competitive: ending the stalemate. Harvard Bus. Rev. 1–27 (1995)
18. Patra, G., Roy, R. K. (2023). Business sustainability and growth in Journey of Industry 4.0-A Case Study. In: Nayyar, A., Naved, M., Rameshwar, R. (Eds.) *New Horizons for Industry 4.0 in Modern Business. Contributions to Environmental Sciences & Innovative Business Technology.* Springer, Cham. https://doi.org/10.1007/978-3-031-20443-2_2
19. Sambasivan, M., Bah, S., Ann, H.: Making the case for operating "Green": impact of environmental proactivity on multiple performance outcomes of Malaysian firms. J. Clean. Produc. 69–82 (2013)
20. Nekhili, M., Nagati, H., Chtioui, T., Rebolledo, C.: Corporate social responsibility disclosure and market value: Family versus nonfamily firms. J. Bus. Res. 41–52 (2017)
21. Fiori, G., Donato, F., Izzo, M.: Corporate social responsibility and firms performance—an analysis on italian listed companies. SSRN 1–14 (2007)
22. McWilliams, A., Siegel, D.: Corporate social responsibility: a theory of the firm perspective. Acad. Manag. Rev. 117–127 (2001)
23. Yuen, K., Thaib, V., Wongc, Y., Wangc, X.: Interaction impacts of corporate social responsibility and service quality on shipping firms' performance. Sci. Direct J. 397–409 (2018)
24. Jo, H., Harjoto, M.: Corporate governance and firm value: the impact of corporate social responsibility. J. Bus. Ethics 351–383 (2011)
25. Abiodun, B.: The impact of corporate social responsibility on firms' profitability in Nigeria. Eur. J. Eco. 39–51 (2012)
26. Bhardwaja, P., Chatterjeeb, P., Demirb, K., Turutb, O.: When and how is corporate social responsibility profitable? J. Bus. Res. 206–219 (2018)
27. Berland, P.: Corporate social responsibility branding survey. Burson Marsteller, pp 1–20 (2010)
28. Luo, X., Bhattacharya, C.: Corporate social responsibility, customer satisfaction, and market value. J. Market. 1–36 (2006)
29. Shen, J., Benson, J.: When CSR is a social norm: how socially responsible human resource management affects employee work behavior. J. Manag. 1–36 (2014)
30. Jain, P., Vyas, V., Chalasani, D.: Corporate social responsibility and financial performance in SMEs: a structural equation modelling approach. SAGE J. 630–653 (2016)
31. Nguyen, T., Nguyen, C.: Corporate social responsibility and firm financial performance: a literature review. J. Manag. Dev. 401–409 (2017)

32. Amba, S.: Corporate governance and firms' financial performance. J. Acad. Bus. Ethics 2–8 (2014)
33. Al-ahdala, W., Alsamhib, M., Tabash, M., Farhan, N. (2020). The impact of corporate governance on financial performance of Indian and GCC listed firms: an empirical investigation. Elsevier, pp 1–13
34. Al Amosh, H., Khatib, S.F.A.: Ownership structure and environmental, social and governance performance disclosure: the moderating role of the board independence. J. Bus. Socio-Econ. Dev. 2(1):49–66 (2022). https://doi.org/10.1108/JBSED-07-2021-0094
35. Aboud, A., Diab, A.: The impact of social, environmental and corporate governance disclosures on firm value: evidence from Egypt. J. Account. Emerg. Econ. 1–33 (2018)
36. Buallay, A.: Is sustainability reporting (ESG) associated with performance? Evidence from the European banking sector. Manag. Environ. Qual 98–115 (2019)
37. Ghazali, N.: Ownership structure, corporate governance and corporate performance in Malaysia. Int. J. Comm. Manag. 109–119 (2010)
38. Hassan, M., Halbouni, S.: Corporate governance, economic turbulence and financial performance of UAE listed firms. Stud. Econ. Finance 118–138 (2013)
39. Constantinos, C.: The effect of the mandatory adoption of corporate governance mechanisms on executive compensation. Int. J. Account. 138–174 (2011)
40. Farhan, A., Obaid, S., Obaid, S.: Corporate governance effect on firms' performance—evidence from the UAE. J. Econ. Adm. Sci. 66–80 (2017)
41. Gompers, P., Ishii, J., Metrick, A.: Corporate governance and equity prices. Quart. J. Econ. 107–155 (2003)
42. Ali, U., Jamieson, O.: ESG investing in emerging and frontier markets. J. Appl. Corp. Finance 96–101 (2016)
43. Pollard, J., Sherwood, M.: The risk-adjusted return potential of integrating ESG strategies into emerging market equities. J. Sustain. Finance Invest. 26–44 (2017)
44. Tarmuji, I., Maelah, R., Tarmuji, N.: The impact of environmental, social and governance practices (ESG) on economic performance: evidence from ESG score. Int. J. Trade Econ. Finance 67–74 (2016)
45. Signori, S., San-Jose, L., Retolaza, J., Rusconi, G.: Stakeholder value creation: comparing ESG and value added in European companies. Sustainability 1–16 (2021)
46. Sassen, R., Hinze, A.-K., Hardeck, I.: Impact of ESG factors on firm risk in Europe. J. Bus. Econ. 867–904 (2016)
47. Dyllick, T., Hockerts, K.: Beyond the business case for corporate sustainability. Bus. Strat. Environ. 130–141 (2002)
48. Murphy, D., Mcgrath, D.: Australian class actions as a potential motivator for environmental, social and governance (ESG) reporting. Res. Gate 1–14 (2015)
49. Freeman, E.: Strategic Management: A Stakeholder Approach. Cambridge University Press, Cambridge (2010)
50. McGuire, J., Sundgren, A., Schneeweis, T.: Corporate social responsibility and firm financial performance. Acad. Manag. J. 854–872 (2017)
51. Chang, K., Kim, I., Li, Y.: The heterogeneous impact of corporate social responsibility activities that target different stakeholders. J. Bus. Ethics 211–234 (2014)
52. Bouslah, K., Kryzanows, L., M'Zalic, B.: The impact of the dimensions of social performance on firm risk. J. Banking Finan. 1258–1273 (2013)
53. Figge, F., Schaltegger, S.: What is stakeholder value? Developing a catchphrase into a benchmarking tool. Sustain. Value 1–14 (2000)
54. Andriof, J., Waddock, S., Husted, B., Rahman, S.S.: Unfolding stakeholder thinking 2: relationships, communication, reporting and performance. J. Environ. Assess. Policy Manag. 9–12 (2003)
55. Phillips, R., Reichart, J.: The environment as a stakeholder? A fairness-based approach. J. Bus. Ethics 185–197 (2000)
56. Frooman, J.: Stakeholder influence strategies. Acad. Manag. Rev. 191–205 (1999)

57. Donaldson, T., Preston, L.: The stakeholder theory of the corporation: concepts, evidence, and implications. Acad. Manag. Rev. 65–91 (1995)
58. Jones, T.: Instrumental stakeholder theory: a synthesis of ethics and economics. Acad. Manag. Rev. 404–437 (1998)
59. Steurer, R.: Mapping stakeholder theory anew: from the stakeholder theory of the firm to three perspectives on business society relations. Bus. Strat. Environ. 55–69 (2006)
60. Hillman, A., Keim, G.: Shareholder Value, Stakeholder Management, and Social Issues: What's the Bottom Line? pp. 125–139. Wiley (2001)

AI, Banking and Financial Technology

A Stochastic Model for Cryptocurrencies in Illiquid Markets with Extreme Conditions and Structural Changes

Youssef El-Khatib and Abdulnasser Hatemi-J

Abstract Cryptocurrencies are increasingly utilized by investors and financial institutions worldwide. The current paper proposes a prediction model for a cryptocurrency that encompasses three properties observed in the markets for cryptocurrencies—namely high volatility, illiquidity, and regime shifts. By using Ito calculus, we provide a solution for the suggested stochastic differential equation (SDE) along with a proof. Moreover, numerical simulations are performed and are compared to the real data, which seems to capture the dynamics of the price path of a cryptocurrency in the real markets.

Keywords Stochastic modeling · Cryptocurrencies · Illiquid · High volatility · Regime switching · CTMC

1 Introduction

With the fast development of digital finance over the past decade and with the introduction of blockchain technology, there has been an immense expansion in the trading of cryptocurrencies. According to [32] there exists around 10,000 cryptocurrencies and the most prominent one with the largest market value is bitcoin, which was initiated in the report [28] written by an unknown author under the pseudonym [28]. Researchers are not in total agreement regarding the usefulness of cryptocurrencies. For example, [32] doubts that cryptocurrencies qualify for even being called currencies. In contrast, [20] believes in cryptocurrencies as the newcomers that will reshape

---

Y. El-Khatib
College of Science, Department of Mathematical Sciences, UAE University, P.O. Box 15551,
Al-Ain, United Arab Emirates
e-mail: Youssef_Elkhatib@uaeu.ac.ae

A. Hatemi-J (✉)
College of Business and Economics, Department of Accounting and Finance, UAE University,
P.O. Box 15551, Al-Ain, United Arab Emirates
e-mail: AHatemi@uaeu.ac.ae

© The Author(s), under exclusive license to Springer Nature Switzerland AG 2024   479
A. Hamdan and E. S. Aldhaen (eds.), *Artificial Intelligence and Transforming Digital Marketing*, Studies in Systems, Decision and Control 487,
https://doi.org/10.1007/978-3-031-35828-9_41

the entire financial system, which needs to be reshaped according to him. As of the first week of June 2022, the market capitalization of cryptocurrencies is estimated to exceed 900 billion dollars at [4]. Modeling cryptocurrency prices is then an important issue. Reference [8] provides an interesting presentation of the economics of cryptocurrencies. On the other hand, [3] shows that the market for cryptocurrencies is vulnerable to bubbles and crises. References [6] and [7] demonstrate empirically that a cryptocurrency can function as a very useful hedge. In [33] the author observes that the market of the bitcoin is informationally inefficient. A comprehensive survey on the publications on cryptocurrencies is provided by Ref. [13], which shows that the number of publications on the underlying topic has as a strong positive trend since its start in the year 2013. The work of [29] provides a detailed survey of numerous publications that deal with the security concerns of cryptocurrencies. Citation [25] suggests an index that can be used for measuring uncertainty in cryptocurrencies. The basis for this index is the news coverage from the mass media. Reference [2] investigates empirically the factors that determine the adaptation of cryptocurrencies and blockchains in 137 countries. In [24] a three-factor model for cryptocurrencies is presented. They find that the return of each individual cryptocurrency is explained by the return of the entire market for the cryptocurrencies, the size, and the momentum. Citation [23] finds empirical evidence that the returns of cryptocurrencies are robustly linked to the network factors but not to the production factors.

Since cryptocurrencies are energy intensive, there are naturally environmental effects. This important issue has been investigated by Ref. [12] using asymmetric causality tests developed by Ref. [14]. It is found that there are negative asymmetrical environmental causal impacts of the demand for major cryptocurrencies. The authors suggest that policymakers should introduce environmental taxes imposed on cryptocurrency transactions to dampen the damaging effects of cryptocurrencies on the environment. By applying asymmetric causality tests, the authors of [19] obtain empirical evidence that supports the dynamic interaction and risk transmission between the oil market and bitcoin.

Publication [22] explores the potential portfolio diversification gains between ten main cryptocurrencies, which results in finding that diversifying across these cryptocurrencies leads to better investment outcomes indeed. In [16] the potential portfolio diversification benefits between bitcoin, stocks, bonds, and the US dollar in the global market are explored. The authors find that there are no portfolio diversification benefits that can be obtained from this cryptocurrency if the portfolio is created by the standard approach pioneered by Ref. [27]. However, if the portfolio is created by using the [15] approach, which combines risk and returns in the optimization problem, including bitcoin in the portfolio results in a higher risk-adjusted return. In addition, [18] obtains empirical support for increasing return per unit of risk for investors from the Middle East if they add cryptocurrencies to their portfolios.

Cryptocurrencies are increasingly chosen as financial assets that are included in investment portfolios by individual investors and financial institutions worldwide. Many studies on cryptocurrencies have emerged investigating principally portfolio diversification profits, market effectiveness, hedging, or capturing the data generating process for the volatility of the cryptocurrencies. The main contribution of this paper is to suggest a prediction model for cryptocurrencies. Market observations imply several differences between cryptocurrency and traditional asset prices. Cryptocurrencies have higher volatility compared to regular assets. Besides, cryptocurrencies are widely less liquid than conventional financial instruments. The paper of [26] examines the relationship between cryptocurrency liquidity, herding behavior, and profitability during extreme price movement periods and it reveals that herding behavior variations have a decreasing magnitude. When building a prediction model for cryptocurrency prices we need to keep in mind that these instruments are different from traditional assets. Cryptocurrency modeling has been studied by a great number of authors in the literature. Many of these works investigated prediction models. In [1] the authors suggest a method for predicting price variations in bitcoin and Ethereum using Twitter data and Google Trends data. Reference [17] provides a review of the research works on predicting cryptocurrency prices from 2010 to 2020.

Most of these studies utilize machine learning for price prediction. Other papers investigate the cryptocurrency price volatility and prediction using econometrics, and statistical models on time-series data. As far as we know, no previous research has suggested a prediction model using stochastic differential equations (SDEs) dealing with cryptocurrencies' high fluctuations and illiquidity. In this paper, we propose an SDE to model future trajectories of cryptocurrency values. Our suggested model comprises three of the most significant stylized facts of cryptocurrencies: illiquidity, high volatility, and regime switching (RS).

The paper is organized as the following. Section 2 describes the construction of the model. Section 3 studies the existence, uniqueness, and positivity of the SDE solution. In Sect. 4, numerical simulations are conducted, and several figures are presented to illustrate the performance of the model. The final section expresses the concluding statements.

2 Model Formulation

We follow the works of [21] and more recently [11] to construct our cryptocurrency prediction model. First of all, a filtered probability space and sources of randomness living on it are to be specified. In [11] and [21], only the Brownian motion is generating the randomness. The idea of this paper is to include a second source of randomness that is independent of the Brownian motion namely a continuous time Markov chain-CTMC. Then, the model becomes a regime-switching model which can be seen as an expansion of the [11].

2.1 Probability Space and Sources of Randomness

Consider a filtered probability space $(\Omega, F, (F_t)_{t \in [0,T]}, P)$. Let $(B_t)_{t \in [0,T]}$ be a Brownian motion and denote by $(F_t^B)_{t \in [0,T]}$ the filtration generated by $(B_t)_{t \in [0,T]}$. We assume that there is a second source of randomness living in the probability space. Let $(Z_t)_{t\in[0,T]}$ denote a Markov jump process defined values in a finite state space $S := \{1, 2, ..., N\}$ and denote by $F^Z = (F_t^Z)_{t\in[0,T]} := \sigma(Z_t, 0 \le t \le T)$, which is the natural filtration generated by the Markov process Z under P. The filtration F is then defined as $F := F^B \vee F^Z$.

2.2 The Model

Let $C = (C_t)_{t\in[0,T]}$ be the rate of the cryptocurrency with regard to the US dollar. Assume that μ is the expected return of the underlying cryptocurrency, and σ is its volatility. The price impact factor of the broker is $\lambda(t, C_t)$ and θ_t represents the number of units of the underlying cryptocurrency that the trader owns at time t. Thus, $\lambda(t, C_t)d\theta_t$ represents the price impact of the investor's trading. We assume that the cryptocurrency's value is therefore governed by the following stochastic differential equation:

$$dC_t = \mu(Z_t)C_t dt + (\sigma(Z_t)C_t + g(t))dB_t$$
$$+ \lambda(t, C_t)C_t d\theta_t, \quad t \in [0, T],$$
$$C_0 = x > 0. \tag{1}$$

where $g(t)$ is originally suggested by Ref. [5] and it signifies a function that is deterministic and captures well the impact of increase in the volatility that takes place during a special period. It is also able to embody Sornette's [31] empirical encompassing of the market price index during a crunch via a dissipative harmonic oscillator characterized by a sinusoidal function that is exponentially decreasing. This function is expressed as the following:

$$g(t) = c_1 + c_2 e^{c_3 t} \sin(c_4 t),$$

where $c_i, i = 1, 2, 3, 4$ are real constants. Let θ_t be the number of the underlying cryptocurrency that the trader owns and assume that it satisfies the following process:

$$d\theta_t = \eta_t dt + \zeta_t dB_t, \quad t \in [0, T]. \tag{2}$$

The above proposed model can be seen as generalisation of the models considered in [9] on the valuation of options during crisis and in [10], and [11] where the authors investigated option's pricing in illiquid and high volatile situations. The SDE (1) expands the previous models by adding regime switching which permits to variate the parameters according to different economic situations.

2.3 Solution Existence and Uniqueness Analysis

The below proposition outlines the existence and uniqueness of the solution of the SDE of our model (1). We assume that the price impact factor of the broker is λ is deterministic.

Proposition 1 Let $a_t := \mu(Z_t) + \lambda_t \eta_t$, $b_t := \sigma(Z_t) + \lambda_t \zeta_t$, and

$$\xi_t = \exp\left(\int_0^t \left(a_u - \frac{b_u^2}{2}\right) du + \int_0^t b_u dB_u\right), \quad \xi_0 = 1. \tag{3}$$

Then

$$C_t = \left(C_0 - \int_0^t g(u)[\sigma(Z_u) + \lambda_u \zeta_u]\xi_u^{-1} du + \int_0^t g(u)\xi_u^{-1} dB_u\right)\xi_t. \tag{4}$$

Proof We consider the process $(\xi_t)_{t\in[0,T]}$ defined by the SDE

$$d\xi_t = a_t \xi_t dt + b_t \xi_t dB_t \quad \xi_0 = 1, \tag{5}$$

where a_t and b_t are defined at the begining of the above proposition. The SDE provides a geometric Brownian motion with solution given by (3). We use the variation of constants method. First, assume that

$$C_t = Y_t \xi_{1t}, \quad Y_0 := C_0. \tag{6}$$

Then Y_t can be obtained using the below integration by parts for stochastic processes

$$dY_t = d\big(\xi_t^{-1} C_t\big) = \xi_t^{-1} dC_t + C_t d\xi_t^{-1} + \big[d\xi_t^{-1}, dC_t\big]. \tag{7}$$

The solution of the regime switching model SDE (1) given by (4) can be obtained after embedding (2) into Eq. (1), then calculating $d\xi_t^{-1}$ by using the Ito formula applied to $f(\xi_t) = \frac{1}{\xi_t}$ and then by making use of (7), (3), and (6). $\blacksquare$

Remark 1 The Eq. (4) demonstrates the existence and uniqueness of a solution for the SDE (1) but does not guarantee non-negative values. For the positivity, the below condition is required

$$C_0 + \int_0^t g(u)\xi_u^{-1} dB_u \geq \int_0^t g(u)[\sigma(Z_u) + \lambda_u \zeta_u]\xi_u^{-1} du. \tag{8}$$

The above condition can be satisfied with a careful choice of g and λ. It should be pointed out that choices of g and λ that do not satisfy the condition (8) but produce a rate of cryptocurrencies approaching zero or become negatively with very low probability could be considered as reasonable choices. This can be interpreted in the sense that any rate that reaches zero implies that the underlying cryptocurrency does not survive and disappears from the market, which in fact can happen in the real markets.

3 Numerical Results and Simulations

3.1 Methodology

Consider the one dimensional stochastic process $X := (X_t)_{t \in [0,T]}$ driven by the following SDE:

$$
\begin{aligned}
dX_t &= a(t, Z_t, X_t)dt + b(t, Z_t, X_t)dB_t, \\
X_0 &= x \quad \text{is a given constant.}
\end{aligned}
\tag{9}
$$

To obtain a discretized trajectory of X_t from the SDE (9) using Euler–Maruyama scheme, the following steps need to be implemented:

1. simulate ΔB_k as normally distributed random variable $N(0, \Delta t)$
2. simulate Z_k the Continuous Time Markov Chain
3. set $\tilde{X}_0 := X_0 = x$ and evaluate $\tilde{X}_{k+1}$ using

$$
\tilde{X}_{k+1} = \tilde{X}_k + a\left(k\Delta t, Z_k, \tilde{X}_k\right)\Delta t + b\left(k\Delta t, Z_k, \tilde{X}_k\right)\Delta B_k,
\tag{10}
$$

for $k = 0, \ldots, N - 1$. Notice that $\Delta B_k = B_{k+1} - B_k$. We will not use the symbol "tilde" for a discretized version of a given SDE from now on. The application of (10) to the model (1) gives the system

$$
\begin{aligned}
C_{k+1} &= C_k + \mu(Z_k)C_k\Delta t + (\sigma(Z_k)C_k + g(k))\Delta B_k \\
&\quad + \lambda(k, C_k)C_k\Delta\theta_k, \\
\theta_{k+1} &= \theta_k + \eta_k\Delta t + \zeta_k\Delta B_k,
\end{aligned}
\tag{11}
$$

where we have discretized the time into M time steps t_k with equal sizes $\Delta t = t_{k+1} - t_k = \frac{T}{M}$, for $k = 0, \ldots, M - 1$.

3.2 Algorithm

Our algorithm consists of the following phases:

1. We simulate a trajectory for the Brownian motion: $(B_k)_{k=0,...,M-1}$.
2. We simulate independently a trajectory for the CTMC: $(X_k)_{k=0,...,M-1}$.
3. For $k = 1$, (C_0, θ_0) and all the parameters of the model are given, we use (11) to calculate (C_1, θ_1),
4. for $k = 2$, (C_1, θ_1) and the parameters of the model are all known from the previous step, then apply (11) to calculate (C_2, θ_2),
5. repeat the previous two steps to $k = M - 1$.

3.3 Illustrations

The above algorithm is implemented by creating a code in Python. It is assumed that the CTMC has 3 possible states, $Z \in \{0, 1, 2\}$. Here, for the illiquid with high volatility case the parameters have the values: $C_0 = 10$, $g(t) = \alpha(Z_t) \cos(\pi t/4)$, $\lambda = 1.5, \eta_t = t, \zeta_t = \sin(\pi t/4), T = 40, N = 10{,}000$. State one when $Z_t = 0$ corresponds to a bad economic situation for cryptocurrencies. The following parameters values are used for covering this specific situation:

(a) expected return $\mu(0) = 0.005$
(b) volatility $\sigma(0) = 2.5$,
(c) increase of the volatility factor in function g is $\alpha(0) = 10$.

State two when $Z_t = 1$ represents a normal economic circumstance. This case is dealt with via the following parameters values:

(a) expected return $\mu(0) = 0.045$
(b) volatility $\sigma(0) = 0.5$,
(c) increase of the volatility factor in function g is $\alpha(0) = 1$.

State three when $Z_t = 2$ depicts a good economic condition. The parameters values for this situation are below

(a) expected return $\mu(0) = 0.2$
(b) volatility $\sigma(0) = 0.3$,
(c) increase of the volatility factor in function g is $\alpha(0) = 0.5$.

Figures 1 and 2 provide two simulations of cryptocurrency to dollar values trajectory in an illiquid market under stress. The graphs seem to accord well with the reality since cryptocurrency prices tend to be exceptionally volatile during certain periods. Figure 3 offers the bitcoin to dollar values from November 2021 to August 2022 taken from the website of yahoo finance. We have divided the time period into four parts. These four periods can be seen as four different states for the Bitcoin to dollar values.

Fig. 1 Realizations of the cryptocurrency to dollar value. A first run of the simulations

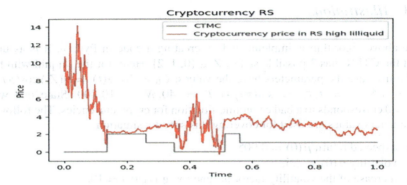

Fig. 2 Realizations of the cryptocurrency to dollar value. A second run of the simulations

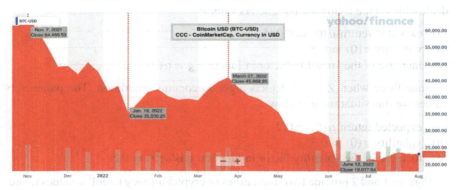

Fig. 3 Bitcoin to the US dollar as of August 7, 2022, divided into 4 states. *Source* https://finance.yahoo.com

4 Conclusions

A cryptocurrency is a digital form of payment that is based on cryptography. Cryptocurrencies are increasingly considered a serious alternative measure of payment versus traditional fiat currencies. Despite having their advantages, cryptocurrencies are extremely volatile and risky. They are also characterized by markets that can suffer from the illiquidity issue. Structural breaks or regime shifts are also taking place in the markets for cryptocurrencies. The current paper provides a model that can be used for predicting the ex-ante path of the exchange rate of cryptocurrencies. The suggested approach contributes to the existing literature on the topic by considering simultaneously three explicit characteristics of cryptocurrencies—namely—(1) illiquidity, (2) high volatility, and (3) regime shifts. A solution for the SDE modeling the exchange rate of the cryptocurrency is provided along with mathematical proof. Numerical simulations are provided, which can capture the situations in which the dynamism of the cryptocurrency rates with regard to the US dollar operates in real markets. There are massive publications on cryptocurrencies. However, these publications are mainly empirically dealing with issues like portfolio diversification benefits, market efficiency, hedging, or capturing the data-generating process for the volatility of the cryptocurrencies. To our best knowledge, this is the first attempt to suggest a stochastic differential equation for modeling the exchange rate (i.e., the pricing) of the cryptocurrencies that covers the three mentioned characteristics.

References

1. Abraham, J., Higdon, D., Nelson, J., Ibarra, J.: Cryptocurrency price prediction using tweet volumes and sentiment analysis. SMU Data Sci. Rev. **1**(3), 1 (2018)
2. Bhimani, A., Hausken, K., Arif, S.: Do national development factors affect cryptocurrency adoption? Technol. Forecast. Soc. Chang. **181**, 121739 (2022)
3. Cheah, E.T., Fry, J.: Speculative bubbles in bitcoin markets? An empirical investigation into the fundamental value of bitcoin. Econ. Lett. **130**, 32–36 (2015)
4. Coinmarketcap. [Online]. Available: https://coinmarketcap.com/charts/. Verified online on 15 July 2022
5. Dibeh, G., Harmanani, H.-M.: Option pricing during post-crash relaxation times. Physica A **380**, 357–365 (2007)
6. Dyhrberg, A.H.: Hedging capabilities of bitcoin. Is it the virtual gold? Finan. Res. Lett. **16**, 139–144 (2016a)
7. Dyhrberg, A.H.: Bitcoin, Gold and the dollar—a GARCH volatility analysis. Financ. Res. Lett. **16**, 85–92 (2016)
8. Dwyer, P.G.: The economics of bitcoin and similar private digital currencies. J. Financ. Stab. **17**, 81–91 (2015)
9. El-Khatib, Y., Hatemi-J, A.: Option valuation and hedging in markets with a crunch. J. Econ. Stud. **44**(5), 801–815 (2017)
10. El-Khatib, Y., Hatemi-J, A.: Option pricing in high volatile markets with illiquidity. AIP Conf. Proc. **2116**, 110007 (2019). https://doi.org/10.1063/1.5114100
11. El-Khatib, Y., Hatemi-J, A.: Option pricing with illiquidity during a high volatile period. Math. Methods App. Sci. **45**(5), 3213–3224 (2022)

12. Erdogan, S., Ahmed, M.Y., Sarkodie, S.A.: Analyzing asymmetric effects of cryptocurrency demand on environmental sustainability. Environ. Sci. Pollut. Res. **29**(21), 31723–31733 (2022)
13. Fang, F., Ventre, C., Basios, M., Kanthan, L., Martinez-Rego, D., Wu, F., Li, L.: Cryptocurrency trading: a comprehensive survey. Finan. Innovation **8**(1), 1–59 (2022)
14. Hatemi-J, A.: Asymmetric causality tests with an application. Empir. Econ. **43**(1), 447–456 (2012)
15. Hatemi-J, A., El-Khatib, Y.: Portfolio selection: an alternative approach. Econ. Lett. **135**, 141–143 (2015)
16. Hatemi-J, A., Hajji, M.A., Bouri, E., Gupta, R.: The benefits of diversification between bitcoin, bonds, equities and the us dollar: a matter of portfolio construction. Asia Pac. J. Oper. Res. 2040024 (2022)
17. Khedr, A.M., Arif, I., El-Bannany, M., Alhashmi, S.M., Sreedharan, M.: Cryptocurrency price prediction using traditional statistical and machine-learning techniques: a survey. Intell. Syst. Acc. Fin. Manage. **28**(1), 3–34 (2021)
18. Kumaran, S.: Portfolio diversification with cryptocurrencies—evidence from Middle Eastern stock markets. Invest. Anal. J. **51**, 14–34 (2022)
19. Li, D., Hong, Y., Wang, L., Xu, P., Pan, Z.: Extreme risk transmission among bitcoin and crude oil markets. Resour. Policy **77**, 102761 (2022)
20. Lipton, A.: Cryptocurrencies change everything. Quant. Financ. **21**, 1257–1262 (2021)
21. Liu, H., Yong, J.: Option pricing with an illiquid underlying asset market. J. Econ. Dyn. Control **29**, 2125–2156 (2005)
22. Liu, W.: Portfolio diversification across cryptocurrencies. Financ. Res. Lett. **29**, 200–205 (2019)
23. Liu, Y., Tsyvinski, A.: Risks and returns of cryptocurrency. Rev. Financ. Stud. **34**(6), 2689–2727 (2021)
24. Liu, Y., Tsyvinski, A., Wu, X.: Common risk factors in cryptocurrency. J. Financ. **77**(2), 1133–1177 (2022)
25. Lucey, B.M., Vigne, S.A., Yarovaya, L., Wang, Y.: The cryptocurrency uncertainty index. Financ. Res. Lett. **45**, 102147 (2022)
26. Manahov, V.: Cryptocurrency liquidity during extreme price movements: is there a problem with virtual money? Quant. Financ. **21**(2), 341–360 (2021)
27. Markowitz, H.: Portfolio selection. J. Financ. **12**(7), 77–91 (1952)
28. Nakamoto, S.: Bitcoin: a peer-to-peer electronic cash system. Technical Report. Available online at: https://bitcoin.org/bitcoin.pdf. Accessed 5 Aug 2022
29. Quamara, S., Singh, A.K.: A systematic survey on security concerns in cryptocurrencies: state-of-the-art and perspectives. Comput. Secur. **113**, 102548 (2022)
30. Sharpe, W.-F., Alexander, G.-J., Bailey, J.-V.: Investments. Prentice Hall, New Jersey (1999)
31. Sornette, D.: Why Stock Markets Crash: Critical Events in Complex Financial Markets. Princeton University Press, Princeton, NJ (2003)
32. Taleb, N.N.: Bitcoin, currencies, and fragility. Quant. Financ. **21**, 1249–1255 (2021)
33. Urquhart, A.: The inefficiency of bitcoin. Econ. Lett. **1148**, 80–82 (2016)

Analysis of the Efficiency, Effectiveness and Productivity of Peruvian Motorcycle Cab Drivers in Times of Covid-19 Pandemic

Nelson Cruz-Castillo[ID], Hernan Ramirez-Asis[ID], K. P. Jaheer Mukthar[ID], María Estela Ponce Aruneri[ID], Juan Eleazar Anicama Pescorán[ID], and Wilber Acosta-Ponce[ID]

Abstract The objective of this study is to analyze the effect of COVID-19 on the efficiency, effectiveness, and productivity of motorcycle taxi drivers in the Paramonga-Lima district. The study applied a face-to-face survey to 166 motorcycle taxi workers over 18 years of age. The efficiency, efficacy and productivity indices were constructed using the multivariate statistical method of Multiple Correspondence Analysis; therefore, to test the hypothesis, the MANCOVA was used. The results show that all ages are productive, the most productive being motorcycle taxi drivers who are between 18 and 29 years old and in good health; It also indicates that those who were vaccinated and used a mask show stable efficacy, efficiency and productivity, with higher productivity. Likewise, having been vaccinated against COVID 19, having had protective measures and being infected with COVID-19 have no effect on efficiency, effectiveness and labor productivity when age is controlled.

N. Cruz-Castillo · M. E. P. Aruneri · J. E. A. Pescorán
Universidad Nacional Mayor San Marcos, Lima, Peru
e-mail: ncruzca@unmsm.edu.pe

M. E. P. Aruneri
e-mail: mponcea@unmsm.edu.pe

J. E. A. Pescorán
e-mail: janicamap@unmsm.edu.pe

H. Ramirez-Asis (✉)
Universidad Nacional Santiago Antunez de Mayolo, Huaraz, Peru
e-mail: ehramireza@unasam.edu.pe

K. P. Jaheer Mukthar
Kristu Jayanti College Autonomous, Bengaluru, India
e-mail: jaheer@kristujayanti.com

W. Acosta-Ponce
Universidad Cesar Vallejo, Lima, Peru
e-mail: wacostapo@ucvvirtual.edu.pe

© The Author(s), under exclusive license to Springer Nature Switzerland AG 2024
A. Hamdan and E. S. Aldhaen (eds.), *Artificial Intelligence and Transforming Digital Marketing*, Studies in Systems, Decision and Control 487,
https://doi.org/10.1007/978-3-031-35828-9_42

Therefore, there is a resilience on the part of motorcycle taxi drivers in the face of the pandemic crisis.

Keywords Covid-19 · Decent employment · Efficiency · Effectiveness · Motorcycle taxi driver · Labor productivity

1 Introduction

Since the COVID-19 pandemic appeared, the world economy has suffered a drastic contraction; likewise, it brought with it the disappearance of many jobs, but it has also brought the strengthening, maintenance, transformation and creation of new jobs; Parallel to this, a growing inequality and the enhancement of labor productivity are expected, primarily due to the trend of digitalization in the world [1]. Unemployment increased, so informal employment increased rapidly during the COVID-19 contagion process [2]. Within this framework in Latin America, which includes Peru, the service sector such as formal and informal transportation has specifically increased the service provided by smaller vehicles or the so-called motorcycle taxis.

These minor vehicles consist of a three-wheeled vehicle model, they are destined for places where public transport in a city does not reach, it emerged in Peru in the eighties, becoming more established in the nineties due to the need for employment; and it is used basically because people want to move with light, fast, low-cost mobility that is adaptable to the geography [3]. In 2018 in the city of Lima, the capital of Peru, there were approximately 600,000 motorcycle taxis, of which 40% were informal [4] 200 km. To the north of Lima, in the province of Barranca, it is estimated that there are currently 5000 formal motorcycle taxi drivers, and in its agro-industrial district of Paramonga there are around 1000 motorcycle taxis that transport people.

The motorcycle taxi service is restricted in large cities, it is generally used by people with lower economic resources; therefore, it is considered as an inferior service; that is, if people's income increases, their use decreases and vice versa [5]. Thus, in a survey carried out by Pilco et al. [6] it was found that the number of minor vehicles or motorcycle taxis that work in the city of Cartagena and the expected profit are the most relevant determinants to violate the measure of peak and license plate, which are restrictions on the movement of this type of vehicle. The need to receive more income has caused some motorcycle taxi drivers to have an additional vehicle [7].

So, it is necessary to ask, how does COVID-19 affect the efficiency, effectiveness and labor productivity of motorcycle taxi drivers in the Paramonga district. Lima-Peru? In this regard we have that, if by increasing the amount of work in one unit an additional amount of production is increased, it is said that the marginal product of labor (PML) has increased, and according to the classics of economics, this is equal to the salary real. In that sense, classical theory predicts the existence of a strong relationship between real wages (W/P) and labor productivity (Y/L), and the Cobb–Douglas production function says that the marginal product of labor is proportional to the average productivity of Labor (Y/L); then it can be concluded that if productivity increases, the quality of life of workers should increase since their real wages increase [8].

Chiavenato and González [9] expresses that, on the one hand, efficiency measures the use of resources in the processes; it is a technical relationship between what goes in and what goes out, which tells us that this relationship is between costs and benefits. Likewise, efficiency as a normative measure is one that is in charge of the means, procedures and methods to achieve the optimal use of resources or productive factors [10]. And on the other hand, effectiveness measures the achievement of results that means reaching the stated objectives, it indicates the ability to satisfy a need through the products it offers [11]. As the two concepts are different, a person or organization can be efficient but not effective, or vice versa.

It is thus deduced that efficiency means achieving the goals with the least number of resources, so it is about saving or reducing resources to a minimum; for example, the Solow growth model considers labor efficiency as a variable, which measures health, education, knowledge of the economically active population, and professional quality [12]. Efficiency focuses on how to produce a good, effectiveness on the ability to produce it. Productivity refers to the product-input relationship in a given period considering quality improvement and control. Productivity, efficiency and effectiveness are the objectives of the organization that are related to the group of actions that reflects job performance; that is, job performance is the relationship between the objectives of the organization such as efficiency, quality and effectiveness criteria with the results of the work done by people [13].

In this sense, the objective of this study is to analyze the effect of Covid-19 on the efficiency, effectiveness and productivity of motorcycle taxi drivers in the Paramonga district, which in this case determined the size of the sample of this population and the respective survey in the final days of the month of November 2021. By this date the second wave of infections had already passed, the number of deaths daily had been considerably reduced and the third dose of vaccination had already begun the previous month [14]. As background we have Fig. 1 where it is observed that, between November 29 and 30, 2021, the date on which the surveys were conducted, the number of people who completed the initial vaccination protocol against COVID-19 reached 19 million. Considering that the Peruvian population reaches around 33 million people [15]; which comes to represent 57.6% of the total population, a fact that means a relative advance considering that Peru has had a delay in the vaccination process against this pandemic.

Likewise, Fig. 2 shows that the daily new deaths confirmed by COVID-19 reduced to a minimum, being around 34 deaths at the end of November 2021; which means that the vaccination process and the sanitary measures gave results.

For this reason, the survey gives us the result that there are 95% of vaccinated, as well as a large percentage of motorcycle taxi drivers who are in good health, mainly due to the fact that they have taken protective measures against COVID-19 and because of the ability to resilience or the ability to have overcome problems and come out stronger benefiting personally [16]. Although the risk still continues, the effects are less and less, but the main question is relevant to know and analyze the

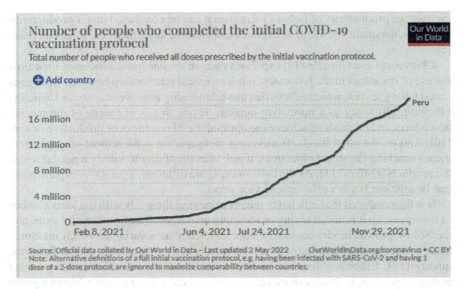

Fig. 1 Number of people who completed the initial vaccination protocol against COVID-19

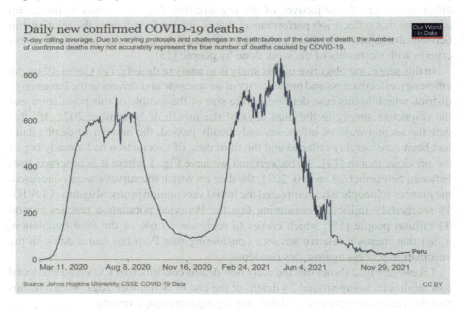

Fig. 2 Daily new deaths confirmed by COVID-19

effect that this pandemic has had on the efficiency, effectiveness and productivity of motorcycle taxi workers who risk their lives every day.

2 Methodology

The study is non-experimental, explanatory and correlational. A face-to-face survey was carried out during the day and at night and in different points of the city; The sample consists of 166 motorcycle taxi drivers from a population of approximately 1000 motorcycle taxi drivers from the Paramonga district. Although minors have been found doing this work and people at risk such as those over 60 years of age, the survey was applied to people over 18 years of age. The questionnaire of questions consisted of 8 questions related to health, COVID-19 and the motorcycle taxi driver's family, and 18 questions related to efficiency, effectiveness and productivity.

With the questionnaire questions referring to efficiency, effectiveness and productivity, the efficiency, effectiveness and productivity indices were constructed using the multivariate statistical method of Multiple Correspondence Analysis [17], to evaluate the effects the MANCOVA (Multivariate Analysis of Covariance, generalization of ANCOVA) to test the hypotheses shown in Fig. 5.

The unrestricted multivariate general linear model has application in multivariate regression analysis and multivariate analysis of variance, where several dependent variables (continuous quantitative variables) with one or more factors (categorical variables) are considered, and if they can also be included. One or more covariates would have multivariate analysis of covariance (MANCOVA). The factors divide the population into groups. Using this multivariate general linear model procedure, it is possible to test null hypotheses about the effects of the factor on the dependent variables when they are controlled by one or more covariates (continuous quantitative variables) [18].

3 Results

The results show that 86.8% of motorcycle cab drivers are between 18 and 59 years of age; this majority group is not at high risk of COVID-19 infection. In addition, we found that 69.3% of those surveyed are in good and/or very good health. On the one hand, it indicates that 96.4% of those surveyed have taken precautions or protections against contagion; on the other hand, it should also be noted that motorcycle taxi drivers are associated or affiliated with an association that has a business character, which allows them to plan and coordinate their actions among themselves and with the municipal administration. However, there are informal operators, as in the Peruvian economy as a whole. Table 4 shows that 94.6% of motorcycle taxi drivers were vaccinated. Finally, they show that motorcycle taxi drivers have the following characteristics: 95.2% are men, 72.1% have a high school education and 65.1% have monthly incomes between 501 and 1000 soles.

It is found, then, that motorcycle taxi drivers have an adequate state of health and that the effect of the COVID 19 pandemic on their health has not been important. Apparently age, protection measures and vaccination have also been the factors that have allowed motorcycle taxi drivers to have a good job performance.

According to Table 2, 23.8% of rural microentrepreneurs have a low level of tax culture, 15.2% have a medium level and 13.5% have a high level of tax obligations, while 12.3% have a medium level.

Figure 3 shows that all ages are productive but the highest percentage of productive motorcycle taxi drivers are between 18 and 29 years old and in good health. While the effectiveness and efficiency is low in all ages, but with motorcycle taxi drivers in good health.

The first row of Fig. 4 indicates that motorcycle taxi drivers who were vaccinated and used a mask show stable efficacy, efficiency and productivity, with productivity being higher; while those who were not vaccinated have productivity, efficacy and efficiency that are not very stable but higher productivity. The second row shows a lot of heterogeneity in efficacy, efficiency, and productivity between motorcycle taxi drivers who used a mask and alcohol, whether they were vaccinated or not, but the greatest variability is shown by those who were not vaccinated, however, both showed the higher productivity. But those who were vaccinated showed greater efficiency and those who were not vaccinated showed greater efficacy.

To test the general hypothesis: COVID-19 has an effect on the efficiency, effectiveness and labor productivity of motorcycle taxi drivers in the Paramonga-Lima

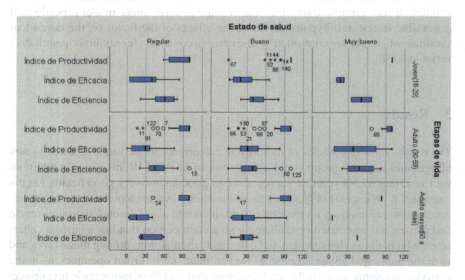

Fig. 3 Motorcycle taxi drivers by index of productivity, effectiveness and efficiency; health status and life stages

Analysis of the Efficiency, Effectiveness and Productivity of Peruvian ... 495

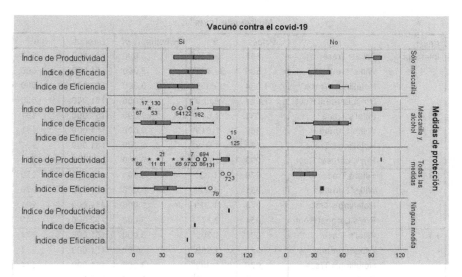

Fig. 4 Motorcycle taxi drivers by index of productivity, effectiveness and efficiency; was vaccinated and protective measures against COVID-19

district. You have to: F3 = Protection measures; F4 = He was vaccinated against COVID-19, F5 = He was infected with COVID-19, as the factors. Then, according to the multivariate tests observed in Table 9, for a level of statistical significance of 5%, the null hypothesis is not rejected, therefore, we have that:

1. The age covariate does not influence efficiency, effectiveness and productivity
2. Protection measures, getting vaccinated and getting COVID-19, each of these factors has no effect on the efficiency, effectiveness and labor productivity of motorcycle taxi drivers in the Paramonga district, when they are controlled for age.
3. Protective measures and getting vaccinated against COVID-19 do not exert an interaction in the effect on the efficiency, effectiveness and work productivity of motorcycle taxi drivers in the Paramonga district, when they are controlled for age.
4. The protection measures and getting COVID-19 do not exert an interaction in the effect on the efficiency, effectiveness and labor productivity of motorcycle taxi drivers in the Paramonga district, when they are controlled by age.
5. Getting vaccinated and getting COVID-19 do not exert an interaction in the effect on the efficiency, effectiveness and labor productivity of motorcycle taxi drivers in the Paramonga district, when they are controlled for age.
6. The protection measures, getting vaccinated and getting COVID-19 do not exert an interaction in the effect on the efficiency, effectiveness and labor productivity of motorcycle taxi drivers in the Paramonga district, when they are controlled by age (Fig. 5).

Multivariate Tests[a]

Effect		Value	F	Hypothesis df	Error df	Sig.
Intercept	Pillai's Trace	,665	99,859[b]	3,000	151,000	,000
	Wilks' Lambda	,335	99,859[b]	3,000	151,000	,000
	Hotelling's Trace	1,984	99,859[b]	3,000	151,000	,000
	Roy's Largest Root	1,984	99,859[b]	3,000	151,000	,000
Edad	Pillai's Trace	,032	1,649[b]	3,000	151,000	,181
	Wilks' Lambda	,968	1,649[b]	3,000	151,000	,181
	Hotelling's Trace	,033	1,649[b]	3,000	151,000	,181
	Roy's Largest Root	,033	1,649[b]	3,000	151,000	,181
F3	Pillai's Trace	,070	1,222	9,000	459,000	,279
	Wilks' Lambda	,931	1,222	9,000	367,645	,280
	Hotelling's Trace	,073	1,219	9,000	449,000	,281
	Roy's Largest Root	,054	2,777[c]	3,000	153,000	,043
F4	Pillai's Trace	,025	1,309[b]	3,000	151,000	,274
	Wilks' Lambda	,975	1,309[b]	3,000	151,000	,274
	Hotelling's Trace	,026	1,309[b]	3,000	151,000	,274
	Roy's Largest Root	,026	1,309[b]	3,000	151,000	,274
F5	Pillai's Trace	,010	,520[b]	3,000	151,000	,669
	Wilks' Lambda	,990	,520[b]	3,000	151,000	,669
	Hotelling's Trace	,010	,520[b]	3,000	151,000	,669
	Roy's Largest Root	,010	,520[b]	3,000	151,000	,669
F3 * F4	Pillai's Trace	,034	,866	6,000	304,000	,520
	Wilks' Lambda	,966	,867[b]	6,000	302,000	,520
	Hotelling's Trace	,035	,867	6,000	300,000	,519
	Roy's Largest Root	,033	1,690[c]	3,000	152,000	,172
F3 * F5	Pillai's Trace	,043	1,106	6,000	304,000	,359
	Wilks' Lambda	,957	1,110[b]	6,000	302,000	,356
	Hotelling's Trace	,045	1,114	6,000	300,000	,354
	Roy's Largest Root	,044	2,218[c]	3,000	152,000	,088
F4 * F5	Pillai's Trace	,017	,864[b]	3,000	151,000	,461
	Wilks' Lambda	,983	,864[b]	3,000	151,000	,461
	Hotelling's Trace	,017	,864[b]	3,000	151,000	,461
	Roy's Largest Root	,017	,864[b]	3,000	151,000	,461
F3 * F4 * F5	Pillai's Trace	,012	,599[b]	3,000	151,000	,617
	Wilks' Lambda	,988	,599[b]	3,000	151,000	,617
	Hotelling's Trace	,012	,599[b]	3,000	151,000	,617
	Roy's Largest Root	,012	,599[b]	3,000	151,000	,617

a. Design: Intercept + Edad + F3 + F4 + F5 + F3 * F4 + F3 * F5 + F4 * F5 + F3 * F4 * F5

b. Exact statistic

c. The statistic is an upper bound on F that yields a lower bound on the significance level.

Fig. 5 Multivariate tests

4 Discussion

There is a significant number of motorcycle taxi drivers who are between 30 and 59 years of age, which represent 63.9%; Likewise, if we evaluate their quality of life, it is noted that 62.7% have a good state of health and due to the characteristics of the work, the vast majority are men (95.2%), which contrasts with the similar study by Refs. [12, 19]; Figs. 3 and 4 show that there is heterogeneity in the results, however, it can be seen that all ages are productive, the most productive being motorcycle taxi drivers between 18 and 29 years of age and in good health; It also indicates that motorcycle taxi drivers who were vaccinated and used a mask show stable efficacy, efficiency and productivity, with higher productivity. This contrasts with [1] who mentions that labor productivity will gradually grow in Latin America after having a "typical" economic crisis, because the gross domestic product grows and therefore employment in production sectors that have medium and high productivity. It should be taken into consideration that current motorcycle taxi drivers use more modern machinery, equipment and means of communication which allow them to have more and better skills and competencies. Table 9 shows that having been vaccinated against COVID 19, protective measures and getting infected with COVID-19 have no effect on the efficiency, effectiveness and labor productivity of motorcycle taxi drivers when age is controlled; These evidences induce us to carry out new studies with other variables, or perhaps apply new methods that contrast the individual interaction of each factor or variable of the model.

5 Conclusion

The objective of this study was to analyze the effect of Covid-19 on the efficiency, effectiveness, and productivity of motorcycle taxi drivers in the Paramonga-Lima district. The socio-business group or the association that the motorcycle taxi drivers represent served him in a transcendental way to continue working in his transport units, since they made agreements such as protection measures, getting vaccinated and being in constant communication. There is resilience on the part of motorcycle taxi drivers, since they have been able to overcome the pandemic crisis. It should also be considered that this sector of workers was not confined, nor in social isolation, unless one of them had the COVID 19 disease; This is not a remote job, so their resilience has been strong, so they have not had frequent symptoms of stress and anxiety since they have not had long periods of confinement [20, 21]. The lack of face-to-face supervision of employees who work remotely has been one of the possible causes of their low productivity; that is, it is concluded that working from home has a negative impact on the productivity of workers [22–24].

References

1. Weller, J. (2020). La pandemia del COVID-19 y su efecto en las tendencias de los mercados laborales. Documentos de Proyectos (LC/TS.2020/67), Santiago, Comisión Económica para América Latina y el Caribe (CEPAL). https://repositorio.cepal.org/bitstream/handle/11362/45759/1/S2000387_es.pdf
2. Raza, M., Wisetsri, W, Chansongpol, T., Somtawinpongsai, C., Ramírez, E.H.: Fostering workplace belongingness among employees. Polish J. Manage. Stud. **22**(2), 428–442 (2020). https://doi.org/10.17512/pjms.2020.22.2.28
3. Chumacero, P., Peralta, J., León, F.: Carga de trabajo, somnolencia y accidentes de tránsito: ¿se potencian en conductores de mototaxi? Rev. Méd. Hered. **27**(4), 268–269 (2016). http://dx.doi.org/https://doi.org/10.20453/rmh.v27i4.3000
4. Finck-Carrales, J.C.: Governance through storytelling and possible futures: motorcycle-cab service planning in Mexico City.J. Environ. Planning Manage. 1–18 (2022).https://doi.org/10.1080/09640568.2022.2113045
5. Hernández, B.B., Osorio, P.F.: El servicio de mototaxis: Una fuente alternativa de trabajo en Puebla. DÍKÊ. Rev. Inv. Derecho, Criminología y Consultoría Jurídica **8**(15). https://doi.org/10.32399/rdk.8.15.168
6. Pilco, J.M., Contreras, A.M., Tito, J.M., Condori, O.: Propuestas legales y participación ciudadana en el ordenamiento del transporte urbano de mototaxis. Cienc. Lat. Rev. Científica Multidisciplinar **6**(6), 999–1018 (2022). https://doi.org/10.37811/cl_rcm.v6i6.3591
7. Ramírez, E.H., Colichón, M.E., Barrutia, I.: Rendimiento académico como predictor de la remuneración de egresados en Administración, Perú. Rev. Lasallista Investig. **17**(2), 88–97 (2020). https://doi.org/10.22507/rli.v17n2a7
8. Mankiw, N.: Macroeconomía. Antoni Bosch editor, S.A, España (2014)
9. Chiavenato, I., González, E.: Comportamiento organizacional: La dinamica del exito en las organizaciones (Tercera ed.). Mc Graw Hill, Mexico D.F. (2017)
10. Rueda, J.C.C., Harris, N.A.V., Vives, J.D.C., Almario, P.A.P., Riascos, I.C.A., Ávila, F.J.M.: Nivel de satisfacción y factores que inciden en la elección del servicio de transporte entre estudiantes de la Universidad de Cartagena, sede Piedra de Bolívar. Rev. Jóvenes Investigadores Ad Valorem **3**(2), 84–99 (2020). https://doi.org/10.32997/RJIA-vol.3-num.2-2020-3241
11. Zhang, J., Raza, M., Khalid, R., Parveen, R., Ramírez-Asís, E.H.: Impact of team knowledge management, problem solving competence, interpersonal conflicts, organizational trust on project performance, a mediating role of psychological capital. Ann. Oper. Res. 1–21 (2021).https://doi.org/10.1007/s10479-021-04334-3
12. Tanadi, M., Sihombing, S.O.: Investigating the impact of brand relationship toward brand evangelism: an empirical study of it-based transportation. In: Proceeding of International Conference of Management Science—ICoMS, vol. 22, pp 127–135 (2017). http://repository.umy.ac.id/handle/123456789/10598
13. Ramírez, E.H., Espinoza, M.R., Esquivel, S.M., Naranjo, M.E.: Emotional intelligence, competencies and performance of the university professor: using the SEM-PLS partial least squares technique. Rev. Electrónica Interuniversitaria Formación Profesorado **23**(3), 99–114 (2020). https://doi.org/10.6018/reifop.428261
14. Ministerio de Salud—MINSA: Dosis de refuerzo: Minsa inició vacunación contra la COVID-19 a colegios profesionales de la salud. Nota de prensa. 21 de noviembre (2021). https://www.gob.pe/institucion/minsa/noticias/553216-dosis-de-refuerzo-minsa-inicio-vacunacion-contra-la-covid-19-a-colegios-profesionales-de-la-salud
15. Instituto Nacional de Estadística e Informática—INEI: Perú: Estado de la Población en el año del Bicentenario (2021). https://www.inei.gob.pe/media/MenuRecursivo/publicaciones_digitales/Est/Lib1803/libro.pdf
16. González-Ospina, L.M., Paredes-Núñez, L.S.: Apego y Resiliencia. CienciAmérica **6**(3), 102–105 (2017). http://201.159.222.118/openjournal/index.php/uti/article/view/102
17. Watkins, M.W.: Exploratory factor analysis: a guide to best practice. J. Black Psychol. **44**(3), 219–246 (2018). https://doi.org/10.1177/0095798418771807

18. McQuitty, S.: The purposes of multivariate data analysis methods: an applied commentary. J. Afr. Bus. **19**(1), 124–142 (2018). https://doi.org/10.1080/15228916.2017.1374816
19. Buchely, L., Castro, M.V.: "Yo me defiendo": entendiendo la informalidad laboral a partir del trabajo de las mujeres mototaxistas en Barranquilla, Colombia. Rev. CS (SPE) 23–47 (2019). https://doi.org/10.18046/recs.iespecial.3223
20. Castagnola Sánchez, C.G., Carlos-Cotrina, J., Aguinaga-Villegas, D.: La resiliencia como factor fundamental en tiempos de Covid-19. Propósitos Representaciones **9**(1) (2021). https://doi.org/10.20511/pyr2021.v9n1.1044
21. Flores-Vargas, N.J.: La importancia de la resiliencia en tiempos de Covid-19. ConcienciaDigital **4**(1.2), 269–285 (2021). https://doi.org/10.33262/concienciadigital.v4i1.2.1593
22. Rayees, F., Almaas, S.: The potential impact of the COVID-19 pandemic on work from home and employee productivity (2021). https://doi.org/10.1108/MBE-12-2020-0173
23. Zhang, X., Husnain, M., Yang, H., Ullah, S., Abbas, J., Zhang, R.: Corporate business strategy and tax avoidance culture: moderating role of gender diversity in an emerging economy. Front. Psychol. **13** (2022).https://doi.org/10.3389/fpsyg.2022.827553
24. Yating, Y., Mughal, N., Wen, J., Ngan, T.T., Ramirez-Asis, E., Maneengam, A.: Economic performance and natural resources commodity prices volatility: evidence from global data. Resour. Policy **78**, 102879 (2022). https://doi.org/10.1016/j.resourpol.2022.102879

Investment Awareness in Financial Assets—An Exploration Based on the Equity Traders in Bangalore City

Aneesha K. Shaji and N. Sivasankar

Abstract Financial assets are belongings obtained through ownership of securities reflecting another company's equity or through contracts that provide for future cash flows. In the present study, an investigation was made among the equity traders in Bangalore to understand their socio-economic background, awareness level, and investment priorities. The current study also examined the dependence of equity investors on financial intermediaries. Primary data collection among 200 respondents was conducted using a structured questionnaire. A cluster sampling technique is used to identify the samples. The results help to understand the influence of awareness level on the equity investment decisions.

Keywords Investment awareness · Financial assets · Equity traders · Financial intermediaries · Investment decision

1 Introduction

Investors have access to a wide range of investment opportunities. Investors frequently, although not always, are aware of it. An individual investor must prioritize capital growth. Suppose people are happy with the financial advantages they get from only one or two investment options. In that case, they are less likely to look for new investment opportunities that might offer a better return. Due to a variety of factors that influence investors' interests and interest rates, it has been noticed that the market for interest on government securities, bank deposits, and other fixed deposits as a whole has been steadily declining year after year.

On the other hand, as time passes, the investor's attention shifts from stocks and bonds to mutual funds and other corporate instruments. Mutual funds provide a better rate of return than bank deposits. The return from the equity market is also

---

A. K. Shaji (✉) · N. Sivasankar
Department of Professional Accounting and Finance, Kristu Jayanti College, (Autonomous), Bangalore, Karnataka, India
e-mail: aneesha@kristujayanti.co

© The Author(s), under exclusive license to Springer Nature Switzerland AG 2024
A. Hamdan and E. S. Aldhaen (eds.), *Artificial Intelligence and Transforming Digital Marketing*, Studies in Systems, Decision and Control 487,
https://doi.org/10.1007/978-3-031-35828-9_43

quite significant compared to mutual funds. There are a lot of risks involved with these securities. Therefore, it is crucial to assess investors' awareness of financial assets and their familiarity with the stock market's workings, protocols, and financial intermediaries. These factors are considered in the current study on investor awareness and investing preferences for financial assets.

2 Literature Review

Chandra and Sharma [1] there are disparities in investors' preferences depending on their gender, educational background, age, occupation, and annual income. Guiso and Jappelli [2] lack of financial knowledge significantly impacts studying the various equity market problems and anticipating investment costs. Kabra et al. [3] their age and gender primarily determine investors' ability to take risks. Joseph Anbarasu and Annette [4] among the uneducated, older age group, and daily wage class, financial literacy needs to be raised because it is low. Equity investors typically perceive risk at a moderate level. Factors contributing to this impression include information screening, investment education, fundamental and technical expertise others [5]. The lessening of information asymmetry made possible by financial literacy and accounting expertise allows investors to invest in riskier instruments. Even when an investor's preference for riskier assets shifts as he gets older and becomes more knowledgeable, it doesn't imply he still likes stock investing; he just does it in the hopes of generating dividend returns rather than capital gains [6].

3 Research Gap

The extensive review of past articles revealed that most of the studies are based on the socio-economic factors of investors rather than their awareness effect on investment decisions. The studies related to the awareness level of equity investors are also limited in the Indian context. The current research focuses on examining the socio-economic factors along with the awareness level of the equity investors in the Indian context with special reference to Bangalore city.

4 Statement of the Problem

A variety of factors influence investors' perceptions. The interest rates have resulted in a consistent decline in the market. But as time goes on, investors become more interested in stocks, mutual funds, and other financial products. Mutual funds offer a better rate of return than bank deposits. The equity market's return is also quite significant compared to mutual funds. Of course, many risks are involved in employing these technologies. It is crucial to comprehend how knowledgeable investors are about financial assets, how at ease they are with the functioning of the stock market, and how comfortable they are with financial intermediaries. The current study examines the investment awareness of equity traders in financial assets. The study also tried to identify how far equity investors depend on financial intermediaries. In short, this study can act as a medium for exploring investment awareness.

5 Objectives

The objectives of the present study include

1. To assess the investment awareness of equity traders in financial assets
2. To identify how far equity investors are dependent on financial intermediaries.

6 Hypotheses

H_0: There is no significant relationship between Education and Awareness of index.

H_1: There is a significant relationship between Education and Awareness of index.

7 Methodology

The study followed a descriptive research design. The sample consists of 200 equity investors residing in Bangalore city. Snowball sampling technique is employed for identifying the samples. A survey method is followed with a structured questionnaire to collect the relevant data. Secondary data sources were also used to support the present study. Analysis tools such as descriptive statistics tools, ranking methods and one-way ANOVA were used to analyze the collected data. The study was carried out only with equity investors in Bangalore city.

8 Results and Discussions

See Table 1.

It is evident that most of the respondents are aware of the investment through friends and relatives (67.0%) and advertisement (31%) and only 9.5% per cent of the respondents get aware through professional advisors and 8.0% of the respondents through company executives. Therefore, it is inferred that most of the respondents get aware through friends and relatives.

504 A. K. Shaji and N. Sivasankar

Table 1 Sources of awareness

Source	No. of the respondents	Percent
Advertising	31	15.5
Company executives	16	8.0
Friends and relatives	134	67.0
Professional advisors	19	9.5
Total	200	100

Source Primary data

Table 2 shows the determinants in selecting investment outlet.

It clearly shows that the investors consider safety, return and liquidity as important factor in section of investment outlet. Among the three factors safety is the first and foremost determinant factor. The next factor considered by the investor is return. The last factor is liquidity. The investor needs safety of their investment and moderate return on their investment.

Table 3 shows the awareness about index of the respondents.

Table 2 Determinants in selecting investment outlet

Determinants	Level of importance			Total
	High	Moderate	Low	
Safety	178	15	7	200
Percent	(89)	(7.5)	(3.5)	(100)
Return	125	59	16	200
Percent	(62.5)	(29.5)	(8)	(100)
Liquidity	44	89	66	200
Percent	(22)	(44.5)	(33)	(100)

Source Primary data

Table 3 Awareness about index

Index	Level of awareness			Total
	High	Moderate	Low	
BSE sensex	185	14	1	200
Percent	(92.5)	(7.0)	(0.5)	(100)
NSE Nifty	175	25	1	200
Percent	(87.0)	(12.5)	(0.5)	(100)
NASDAQ	17	130	53	200
Percent	(8.5)	(65.0)	(26.5)	(100)
OTCEI	30	55	115	200
Percent	(15.0)	(27.5)	(57.5)	(100)

Source Primary data

Investment Awareness in Financial Assets—An Exploration Based … 505

It is clear from the Table that the investors' awareness about various Indian Stock Exchange and index is high. But majority (92.5%) of the investors are aware only BSE.

Table 4 shows the investors awareness about regulation of trading of the respondents.

It is found from the Table that (53.0%) of the investors are aware of the terms price band and (57.5%) of the investors are moderately aware about the price rigging which is used in the security trading.

Table 5 shows the investors awareness about settlement and delivery.

It is found from the table majority of the investors have moderate awareness about the terms of Settlement and delivery which is used in the security trading.

These statements are the important factor in the preference of the financial intermediaries. Hence, the researcher has given the collected information in Table 6.

Table 4 Investors awareness about regulation of trading

Terms	Level of awareness			Total
	High	Moderate	Low	
Price band	106	78	16	200
Percent	(53.0)	(39.0)	(8.0)	(100)
Price rigging	45	115	40	200
Percent	(22.5)	(57.5)	(20)	(100)

Source Primary data

Table 5 Awareness about settlement and delivery

Terms	Level of awareness			Total
	High	Moderate	Low	
Bad delivery	80	21	99	200
Percent	(40)	(10.5)	(49.5)	(100)
Clearing	22	137	41	200
Percent	(11.0)	(68.5)	(20.5)	(100)
Forward trading	63	95	42	200
Percent	(31.5)	(47.5)	(21.0)	(100)
No delivery period	68	95	37	200
Percent	(34.0)	(47.5)	(18.5)	(100)
Pay-in	55	103	42	200
Percent	(27.5)	(51.5)	(21.0)	(100)
Pay-out	46	115	39	200
Percent	(23.0)	(57.5)	(19.5)	(100)
Record date	64	104	32	200
Percent	(32.0)	(52.0)	(16.0)	(100)

Source Primary data

Table 6 Testing of investor preference of the financial intermediaries

Statement	Strongly agree	Some what Agree	Agree	Disagree	Strongly disagree	Percent
An enriched trading experience Percent	33 (16.5)	31 (15.5)	130 (65.0)	5 (2.5)	1 (0.5)	200 (100)
Personalized services Percent	21 (10.5)	100 (50.0)	61 (30.5)	15 (7.5)	3 (1.5)	200 (100)
Customized market watch Percent	17 (8.5)	47 (23.5)	101 (50.5)	32 (16.0)	3 (1.5)	200 (100)
Providing free software to the investors Percent	23 (11.5)	48 (24.0)	76 (38.0)	42 (21.0)	11 (5.5)	200 (100)
Optimistic advice to the investor Percent	16 (8.0)	51 (25.5)	81 (40.5)	39 (19.5)	13 (6.5)	200 (100)
Providing multiple products on a single trading platform Percent	20 (10.0)	41 (20.5)	69 (34.5)	59 (29.5)	11 (5.5)	200 (100)
Equity returns are highly satisfactory Percent	15 (7.5)	43 (21.5)	79 (39.5)	49 (24.5)	14 (7.0)	200 (100)
I have invested in equity because of the ease of investment Percent	12 (6.0)	49 (24.5)	74 (37.0)	50 (25.0)	15 (7.5)	200 (100)
Liquidity in equity is high Percent	17 (8.5)	46 (23.0)	81 (40.5)	41 (20.5)	15 (7.5)	200 (100)
Capital appreciation in equity is high Percent	16 (8.0)	45 (22.5)	90 (45.0)	36 (18.0)	13 (6.5)	200 (100)

Source Primary data

Investment Awareness in Financial Assets—An Exploration Based ... 507

Table 7 Reasons for preference of using the intermediaries

Reasons	1	2	3	4	5	Weighted total	Average	Rank
Time saving	23	48	37	62	30	572	2.86	4
Personal experience in the field	36	45	52	41	26	624	3.12	1
Proper guidance for investment	57	39	18	27	59	608	3.04	3
Market trend/close touch with market	46	31	55	27	41	614	3.07	2
Help to encase the securities whenever needed	27	21	49	57	46	526	2.63	5

Source Primary data

From the above table it conclude majority of the respondents agree with all statements.

Table 7 shows the reasons for preference of using intermediaries of the respondents.

It is observed form Table that personal experience in the field has been identified as the most important reason for the preference of using the intermediaries by the investors with average score of 3.1. Market trend/Close touch with market has been pointed out as the second important reason with a average score of 3.07. The remaining reasons, namely, Time saving, Proper guidance for investment, Help to encash the securities whenever needed have been ranked III, IV and V with a mean score of 3.04, 2.86 and 2.63 respectively.

9 Education and Awareness about index

Null Hypotheses (H0): There is no significant relationship between Education and Awareness about index.

Alternate Hypotheses (H1): There is a significant relationship between Education and Awareness about index Table 8.

The calculated value is more than the 5%. So we reject the null hypothesis, therefore, there is a significant relationship between educational qualification and their awareness about index.

10 Discussion

As the education qualification increases awareness level of the investors also increase. Most of the respondents are highly dependent on their financial intermediaries while making their equity investment decisions. When we consider the previous literatures

Table 8 Education and awareness about index

Awareness about index		Sum of squares	Df	Mean square	F	Sig
Awareness about index- BSE sensex	Between groups	0.088	4	0.022	0.258	0.905
	Within groups	16.632	195	0.085		
	Total	16.720	199			
Awareness about index- NSE nifty	Between groups	0.180	4	0.045	0.349	0.845
	Within groups	25.175	195	0.129		
	Total	25.355	199			
Awareness about index- NASDAQ	Between groups	2.189	4	0.547	1.740	0.143
	Within groups	61.331	195	0.315		
	Total	63.520	199			
Awareness about index- OTCEI	Between groups	1.217	4	0.304	0.551	0.699
	Within groups	107.658	195	0.552		
	Total	108.875	199			

the current study also supports the results obtained from various studies which the researcher conducted in the similar field.

11 Conclusion

It is always important to enhance the awareness level of the investors to achieve more returns from the equity investments. Financial intermediaries are playing avital role in the performance of equity investments of the most of the investor. If the financial intermediaries are able to provide better advisory services to the investors they can excel in the investment portfolios.

References

1. Chandra, A., Sharma, A.: Investor's awareness, preference and pattern of investment in various financial assets. Bus. Manage. Rev. **10**(4), 45–52 (2019)
2. Guiso, L., Jappelli, T.: Awareness and stock market participation. Rev. Finance **9**(4), 537–567 (2005)
3. Kabra, G., Mishra, P.K., Dash, M.K.: Factors influencing investment decision of generations in India: an econometric study. Asian J. Manage. Res. **2010**, 308–326 (2010)
4. Joseph Anbarasu, D., Annette, B.: Study on socio economic status and awareness of Indian investors of insurance (2010)
5. Singh, R., Bhattacharjee, J.: Measuring equity share related risk perception of investors in economically backward regions. Risks **7**(1), 12 (2019)
6. Lodhi, S.: Factors influencing individual investor behaviour: an empirical study of city Karachi. J. Bus. Manag. **16**(2), 68–76 (2014)

Cost–Benefit Analysis of Fintech Framework Adoption

Zainab Mohammed Baqer Shehab, Muneer Al Mubarak, and Amir Dhia

Abstract A rise in the amount of attention dedicated to financial technology (Fintech) services has been seen in recent years. Because of the inherent risks associated with Fintech, several people are skeptical about its widespread acceptance. However, although many financial experts and practitioners believe that Fintech will have a substantial influence. We need to understand why consumers are thrilled or nervous about using financial technology, as well as what factors influence their choice to accept or reject services. A benefit-risk approach is proposed, which takes into account both the good and negative aspects of its implementation. It is based on the net valence framework, which is conceptually anchored in the idea of reasoned action. This study analyzes whether perceived advantage and risk have a major influence on the motivation to utilize Fintech technology using empirical data obtained from Fintech users. Following that, as per our findings, legal risk has the biggest negative influence on Fintech adoption intentions, while convenience has the highest favorable impact. The differences in behavior between early and late adopters might be attributed to a variety of variables, including legal risk and convenience.

Keywords Fintech · Financial Technology · Information Technology

---

Z. M. B. Shehab
The George Washington University, 2121 I St NW, Washington, DC 20052, USA

M. Al Mubarak (✉)
Ahlia University, Manama, Bahrain
e-mail: malmubarak@ahlia.edu.bh

A. Dhia
University of Business and Technology, Jeddah, Saudi Arabia
e-mail: amirdhia@ubt.edu.sa

© The Author(s), under exclusive license to Springer Nature Switzerland AG 2024
A. Hamdan and E. S. Aldhaen (eds.), *Artificial Intelligence and Transforming Digital Marketing*, Studies in Systems, Decision and Control 487,
https://doi.org/10.1007/978-3-031-35828-9_44

1 Introduction

As a consequence of recent advancements in technology of information, flood of new and creative financial services, collectively known as Financial Technology, have risen to prominence (Fintech). The terms "financial technology" and "financial services" are combined in the term "fintech." Technology in the financial sector (fintech) is a rapidly developing and inventive subject that is attracting public interest and increasing investment. According to Skan et al. [1], global investment in Fintech enterprises and start-ups has surged dramatically in only In one year, the amount raised from $4.05 billion to $12 billion. This is a considerable rise from the $4.05 billion invested in 2013 to the $12 billion invested in 2014. Financial technology, Increasing openness, Cutting expenses, removing intermediaries, and improving access to information are all key objectives (Fintech) creates new opportunity for people to gain more power [2]. According to the Financial Technology Association, Fintech firms are transitioning from the net to portable devices, such as mobile payment and remittance, along with non-financial vendors to provide innovative and distinctive financial services beyond typical internet banking by financial institutions.

This study used a net disposition framework based on the idea of perceived behavioral control was employed advantages along with dangers of Fintech implementation goals. In accordance with the study's aims, the following questions will be answered:

1. Is the perceived advantage and danger of Fintech adoption important in deciding whether or not to use it?
2. What particular benefits and risks considerations have an impact on Fintech adoption intention via the perception of the customers?
3. Will the perceived advantage and risk differ between early and late adopters?

Fintech acceptance is still in question, even though several experts and practitioners think it has the potential to change the financial sector in the future. As a result of the significant hazards associated with Fintech, some consumers are wary of implementing it. The most significant adoption barriers are financial concerns (for example, a cash loss plus an additional cost), regulatory concerns Legal and privacy problems, and Fintech companies' insufficient processes or systems are also risks. Customers should examine the benefits and risks of a product at the same time, and only adopt it if the benefits outweigh the risks. Consequently, Fintech businesses are pushed to improve the possible advantages while simultaneously reducing the potential hazards while providing Fintech to clients [3]. As a result, we need to learn more about whether and why clients are keen to join a developing financial service, because this information is crucial for practitioners to make informed decisions.

Furthermore, consumers' perceptions of benefit and risk fluctuate based on user-centric variables, owing to the fact that each group has its own set of benefits and risks. These distinctions assist Fintech organizations in gaining a complete grasp of the uniqueness Fintech adoption is aided by the ability the capacity of each individual user to offer its services successfully while meeting the expectations and wants of their customers as a consequence, we split Fintech users among 2 categories: early users and late users, and examined each group's actions. Despite the fact that much earlier research has identified the major drivers influencing user behavior.

This investigation may aid operators in better understanding consumers' return and possibility view, which can then be utilized to create advantage and disadvantage tactics to boost financial technology adoption. An outcomes for this study also provide advise to Fintech businesses on which factors to highlight and which to avoid while offering services to clients.

2 Literature Review

2.1 Fintech

'It is a financial service (or industry) that has emerged in recent years that combines financial and information technology (IT) (or industries)'. Various authors use different definitions of financial technology (Fintech). Sweeney mentioned and Kuo Chuen and Teo [4] have uncovered a new financial service (or business) that combines financial and information technology (IT) skills in recent years. [It is a financial service (or industry) that has emerged in recent years that combines financial and information technology (IT) capabilities (or industries)]. Various writers utilize various definitions of financial technology to describe what they are talking about in their books (Fintech). Financial technology (Fintech) has been applied by Sweeney, as well as Kuo Chuen and Teo [4] define highly creative and disruptive service technology-based goods or services in financial service firms. Financial technology (Fintech) is described by Freedman as the system development for representing, valuing, and processing financial assets such as connections, supplys, agreements, and cash. According to Ernst and Young, a Fintech development is one which automation plays a significant role (2015). As described by Lee, a Fintech firm is one that offers financial services via the use of technological innovations such as computer hardware and software. Arner et al. described Fintech as financial solutions that are backed by technology, which was previously discussed.

Regardless of the fact that the connection between finance services and information technology is not a new thing, it is becoming increasingly important, Fintech's opportunities, hazards, and regulatory implications are distinct from those of traditional electronic financial services. It is not technology itself that is driving the current concerns among politicians and corporate leaders, but rather the individuals who are adopting financial technology [5] who are producing the concerns. The increasing and growing importance of information technology is also evident in the Fintech business. Instead of serving as a facilitator or enabler in the delivery of successful financial services, information technology in financial technology serves as a true innovator or disruptor in the disruption of the current value chain by circumventing the traditional distribution channel, as defined by the International Organization for

Standardization. "Fintech" is defined as "new and disruptive financial services where information technology (IT) is the fundamental component in nonfinancial firms," according to the findings of this study. Mobile payment and remittance are two examples of Fintech, and they are the ones that are most extensively used throughout the world. Client-facing fintech services include personal financial management, peer-to-peer lending, fundraising, and a fair funding in the form of stock options. According to Kuo Chuen and Teo [4], Teo et al. [6], fintech goods or services in financial service organizations are characterized as those that are based on highly creative and disruptive service technology. The system was created to represent, assess, and handle financial data like bonds, shares, contracts, and money is known as financial technology. According to Ernst and Young, a Fintech innovation is one in which technology is the major facilitator (2015).

Moreover, Although the process of finance and ICT operations is still not recent, Fintech's possibilities, hazards, and regulatory consequences are distinct from traditional electronic finance. Regulators and companies are concerned about who is using financial technology, not so much about the technology itself [5]. Furthermore, Fintech is distinguished by the increasing and strengthening importance of information technology. In Fintech, information technology serves as a genuine innovator, not as a provider or operator for successfully delivering wealth management or disruptor who bypasses the traditional route to disrupt the current value chain. As a result of these factors, Fintech was characterized as "new and disruptive financial services where information technology (IT) is the major feature in nonfinancial firms" in this investigation. Fintech encompasses payment platform as well as settlement, along with these two instances represent Fintech the most over the world. Clients can benefit from fintech services such as cash management, mentoring lending, crowdfunding, and equity investing Barberis [7].

2.2 Impact Analysis for Information Technology Acceptance Models

Existing technology of information adoption theories, as well as the Technology Acceptance Model, are primarily concerned with how the advantages of implementing a new information technology development could persuade users to accept it. For example, If a technology can minimize both physically and mentally effort required by individuals to utilize it while simultaneously improving their work performance, It has a better chance of being approved. The author [8] describes the process of creating a synthesis of two or more sources of information. Statistically, these "plus" effects of adopting a technology appear to be a good motivator for that technology's adoption. Using new technology, on the other hand, does not always result in benefits; for example, when replacing an old solution with a new technology, the outcome is always accompanied with a number of uncertainties, making the decision to embrace the new technology riskier (c.f. Mitchell [9]). Full analysis, that conclude both the benefits along with risks associated with applying information technology, as a result, lead to a more complete knowledge of information technology adoption.

Decision science is concerned with evaluating the over desirability of a certain future action by taking into account both rewards obtained and sacrifices made. Expectancy utility theory [10] and prospect theory [10] are two examples of ideas that have been presented and widely used to aid managers and project leaders in financial and economic decision-making processes (c.f. Conchar et al. [11]). Accordingly, Because a participant's choise or behavior frequently often has societal and financial ramifications that are impossible to forecast with certainty, sacrifices have usually been measured in terms of risk. (i.e. see. Campbell and Brown [12]; Rashid and Hayes [2]; Zinkhan and Karandde [13]). Marketing research on individual consumer behavior may also reveal similar data. (i.e. Dardis and Stremel [14]; Wood and Scheer [15]). Dardis and Stremel [14, p. 554], for example, a conducted advantages and disadvantages analysis to see if customers were prepared to take risks. They defined risk rating as "both the probability of certain occurrences and the implications of such outcomes expressed in dollar terms." They evaluated the usefulness of certain items in marketing, among other applications, using a risk–benefit ratio. Consumer views of the benefits and risks connected with new technologies in the food market, according to Bruhn [16], determine their acceptability.

In contrast to past studies that examined risk objectively, Latest market analysis, such as the perceived risk hypothesis, addresses risk from a subjective viewpoint. The study of information systems has a long and storied history. This idea has been frequently applied (i.e. Lee [17]; Featherman and Pavlou [18]). As in current study, it is argued that, like two sides of a coin, perceived risk and perceived benefit must both be included in order to offer a more perspective of clients' choice methods.

2.3 Explaining Adoption with a Benefit-Risk Framework

In a variety of scenarios, consumers make decisions based on uncompleted or I may say incorrect information [19]. In consequence of these factors, client adoption decisions are frequently fraught with uncertainty and risk. However, in the case of adoption, risk is not the only consideration on which consumers rely, because perceived advantage serves as an additional motivator for consumers to engage in adoption behavior [20]. Peter and Tarpey [21] created a net valence paradigm by combining perceived advantage and perceived danger. This theory believes that consumers perceive both positive and negative aspects in items or services, and that customers make decisions in order to optimize the net valence generated by the decision's negative and positive features. Valence is also congruent with notions given by Lewin [22] and Bilkey [23], both of which serve as the theoretical underpinning for this research.

The use of reasoned action in this study also allowed us to obtain a better comprehension of the net valence framework. A recent study by Transparency Research Associates found that consumers' views about Fintech adoption are determined by their behavioral beliefs, which in turn impact their likelihood to utilize Fintech. Furthermore, the advantages and disadvantages of Fintech adoption can be thought of as a set of behavioral (both good and negative) ideas that shape people's opinions, which in turn drive behavioral intents and behaviors [24].

A good attitude toward Fintech adoption, on the other hand, increases perceived benefits, whereas a negative attitude toward Fintech adoption increases perceived dangers. On the basis of this hypothesis, the outcomes of this study indicated that consumers recognize specific advantages and hazards that may arise as a consequence of Fintech adoption, and then integrate these individual benefits and dangers into an overall perceived benefit and risk. Because of the findings, as a consequence of the findings, there is a broad attitudinal evaluation of Fintech adoption results in a desire to embrace Fintech.

2.4 Hypothesis Building

Consumers may make judgments based on information that is inadequate or misleading [19]. As a result of these considerations, customer adoption decisions are often plagued with risk and uncertainty. Perceived benefit, on the other hand, serves as an incentive for consumers to engage in adoption behavior, therefore risk isn't the sole factor on which consumers base their decisions in the adoption context [20]. As developed by, Peter and Tarpey [21] a framework by combining perceived benefit and risk. Consumers see items or services to include pros and cons features, and they make judgments in order to optimize the overall reactivity produced by the decision's favorable and unfavorable attributes. Furthermore, concept of the project is compatible along with hypothesis advanced by Lewin [22] and Bilkey [23], both who served as the theoretical foundation for the investigation.

Furthermore, by the use of reasoned action in this study, we were able to get a better grasp of the net valence framework. According to TRA, customers' views about Fintech adoption are influenced by their behavioral beliefs, which in turn impact their likelihood to utilize Fintech. Even more explicitly, the advantages and hazards of Fintech adoption may be viewed as behavioral (beneficial and harmful) beliefs that attention and influence, which in turn drive behavioral intentions and actions [24].

As a result, favorable attitudes toward Fintech adoption increase perceived benefits, whereas negative attitudes toward Fintech adoption increase perceived dangers. Consumers notice specific advantages and hazards that may occur as a result of Fintech adoption, and then integrate these individual benefits and dangers into an overall perceived benefit and risk, according to the findings of this study. As a result of the findings, there is an overall attitudinal assessment of Fintech adoption that leads to an intention to embrace Fintech.

H1: Their intention to adopt Fintech is positively related to their perceived benefit.

H2: Their intention to use Fintech services is inversely correlated with their perception of risk.

It is possible to categorize the perceived advantages of Fintech technology into two categories: (1) Convenience, and (2) the transaction procedure are all factors to consider. The most prevalent and persistent motivation for Fintech, has discovered by Kuo Chuen and Teo [4] as being economic advantage. Fintech indicates cheaper transaction and capital costs as compared to conventional financial services, resulting in a benefit to the end user behavior [20]. The ease that comes from mobility and fast accessibility is often mentioned as one of the obvious advantages of Fintech [4, 25]. A convenient time and place is defined as one that is flexible [17, 19, 22]. Because the mobile devices is the main device used by fintech so the benefits connected to the mobile device that effect the adoption. Fintech delivers a number of transaction-related benefits, referred to as transaction process benefits, in the case of financial transactions (e.g., buying, money transferring, lending, and investing). Furthermore, a seamless transaction, a key component of Fintech, eliminates the need for a middleman by allowing consumers to perform and manage their financial transactions on low-cost platforms [24, 26]. In comparison to traditional financial transactions, Fintech users may increase the speed and efficiency of their financial transactions by utilizing a smooth transaction procedure. We feel that the economic advantage, convenience, and transaction process are the three possible benefits of Fintech adoption. As well as the transaction process, will all have an influence on the total perceived benefit associated with positive Fintech adoption. As a result, the hypotheses listed below have been proposed:

H3: The perceived advantage is proportional to the economic benefit.

H4: Its perceived advantage is positively connected to its convenience.

H5: The perceived reward of a transaction is positively connected to the transaction procedure.

This study looked at four types of risks as perceived Fintech hazards: There are a variety of risks to consider, such as financial, operational and most importantly legal. Financial risk in Fintech claim to the probability of financial loss in practically all financial transactions [27]. To be more specific, The most constant predictor of online and mobile user behavior is perceived financial risk [28]. This refers to the legal ambiguity surrounding Fintech and the lack of generally applicable regulation. A slew of financial and other relevant rules prevent new players from entering and growing in the Korean Fintech business. Nonfinancial enterprises undertaking financial operations, in particular, are subject to stringent rules, which makes it difficult to expand the Fintech industry in Korea. Security risk is defined in the Fintech business as the risk of financial loss as a result of fraud or a hacker violating the security of a financial transaction. Not only can both fraud and hacker infiltration result in monetary losses for consumers, but they also breach their privacy, which is a significant source of worry for many online and mobile users [27]. Operational risk in Fintech refers to any potential losses that may arise as a result of weak or failed internal procedures, staff, and systems. Customer adoption of Fintech will be hindered if

Fintech businesses have issues with their finance systems and operations. Due to a lack of operational skills and prompt solutions to system and transaction difficulties, clients are distrustful and dissatisfied, which makes it difficult for them to embrace financial technology. As a result, four types of risks may have a significant influence on risk perception, which has a negative impact on Fintech acceptance and adoption. As a result, we provide the following advice:

H6: The risk perception of liquidity loss is positively connected.

H7: The perceived danger of legal risk is inversely proportional to the actual serious possibility.

H8: The perceived danger of security is inversely connected to the actual risk.

H9: There is a positive link between operational risk and perceived risk.

The pace of adoption of new technology is influenced not only regarding the abilities of the new system, but also by the qualities of the users Karanna and colleagues. Customers tend to adapt with the new system at so many different rates, and they are classified into variety of adopter categories based on their starting adoption journey (i.e., period of adoption) and the degree to which they are in the process of adopting and utilizing the innovation (i.e., degree of adoption) [29]. Clients adopt to this new technology with a variety of rates, and they are divided into distinct adopter groups based on when they start (i.e., early adopters). In this study, the time and behavioral aspects of a new technological service adoption were utilized to divide Fintech consumers split to 2 categories, either early adopter or late adopters. According to researcher findings, previous empirical studies Lewin [22], Bilkey [23], early adopters are separate people who collect different data in order to understand more about the advantages of embracing new technologies, whereas late adopters are those who are more cautious about accepting new services and are suspicious about adoption. In many situations, early adopters act as thought leaders, urging people to accept new technology by sharing their own experiences and views [29]. Early adopters make significant judgments about technologies, even when the advantages and costs of such innovations are not yet fully understood [30]. Late adopters, on the other hand, are not only averse to change, but they are also skeptical of those who bring about change [31]. The majority of late adopters want to be confident that the novelty and utility of innovative goods will not be compromised before committing to them [31]. According to Escobar-Rodrguez and Romero-Alonso, Early adopters are more likely to employ new data technology and have a good attitude toward innovation than late adopters. Late adopters, on the other hand, are resistive to change and have an adverse attitude regards latest technology enabled services uptake. As a result of these findings, we may deduce that the projected benefit and risk of Fintech adoption vary depending on the kind of Fintech user. As a result, the following theories have emerged:

H10: The influence of perceived advantage on Fintech adoption intention is larger among early adopters than late adopters.

H11: Early adopters are more likely than late adopters to be influenced by perceived risk.

As Lee discovered [17], 5 different types of risk while investigating. There are several risks associated with Internet banking adoption, including performance risk, social risk, time risk, financial risk, and security risk, to name a few. In order to forecast account uptake, Featherman and Pavlou [18] classified the essential factors that perceived risk as performance, financial, time, psychological, social, privacy, and overall risk. Using the perceived risk theory of risk, many research have been undertaken to evaluate the adoption of commerce-related information technology advancements. We've also made perceived risk a key component of our study approach, which is founded on the concept of perceived risk. The extent to which an individual's subjectively held thoughts regarding the possible losses caused by uncertainties associated with the usage of mobile payment technology are backed by actions is described as perceived risk in the context of mobile payment services. In light of the above discussion, the following suggestion is made:

H12: Risk perception has a negative relationship with intention to use.

According to previous research, perceived risk is assessed as a multidimensional construct in this study. Lee [17], Cunningham [31], Featherman and Pavlou [18] are only a few examples. The essay delves into three major characteristics of perceived risk connected with mobile payment: financial risk, psychological risk, and privacy risk. They have supplied the following definitions:

Financial risk: The possibility of incurring a disproportionate financial loss as a result of a mobile service transaction, such as exorbitant pricing or malicious billing [32].

Privacy risk: The potential loss of private information of consumers exposed in mobile services [32].

Psychological risk: The probability that customers would experience mental stress as a result of their usage of technology [33]. "Consumers do not desire technology; rather, they seek goods that deliver certain benefits," the author claims [16, p. 555]. In other words, buyers make a risky decision not for the purpose of taking a risk, but rather to obtain benefits or profits. When the benefits of adopting new information technology are significant, users are more likely to overcome the obstacles. Porter and Donthu released an article in 2006 titled Lee [17] observed that there are a range of benefits to implementing Internet banking technology, including financial rewards, faster transaction speeds, and more data transparency, when conducting study on the subject. Lee's [17] research also revealed that perceived benefit had a significant impact on intention to use, with the effect being even bigger than the influence of attitude. According to Dholakia and Uusitalo's [34] research, perceptions of shopping benefits should be included when examining how customers choose between physical and electronic merchants. According to Melenhorst et al. [35], cost–benefit analysis is utilized to analyze communication technology adoption, with the finding that users evaluate the individually perceived advantages and costs when determining whether or not to embrace the technology. Similarly, perceived benefits have been utilized as a direct predictor of specific IS adoption and have been proved to be beneficial [17, 36]. As a result, perceived benefit is seen as a critical construct in the suggested framework, which is defined as the entire benefits that a person perceives as a result of adopting a certain information technology. Kim and Olfman [37] On the basis of the above discussion, the following hypothesis is advanced:

H2: Perceived benefit positively relates to intention to use.

The framework's applicability.

Many information systems researchers have proposed that the predictive capacity of a certain theory or construct may be restricted in its relevance to specific information technology categories under consideration. ... For example, van der Heijden discovered that when used to evaluate utilitarian information systems, perceived usefulness has minimal predictive value, but reported enjoyment is a better predictor when used to evaluate hedonic information systems. Similarly, Liu et al. [38] conducted research on mobile learning acceptance and discovered that perceived long-term usefulness is a more important predictor of educational information system adoption. It is crucial to highlight that similar arguments may be found in a wide range of previous studies. (See, for example, Sun and Zhang [39]; Verhagen et al. [40]; Wen et al. [41]). As a result, it is critical to assess the relevance of the proposed framework in the context of the perceived risk theory, which is described further below. Furthermore, because the framework is based on perception-based risk theory, whether or not perceived risk theory is relevant impacts the framework's applicability. People are exposed to risk when their decision or action has the potential to cause significant economic and social harm (cf. Zinkhan and Karande [13]); on the other hand, in "less complex situations or routine choice situations, individuals are more likely to resort to simpler processes or even ignore risk issues entirely." Mitchell [9], as mentioned in Conchar et al. [11], p. 424. It should be especially useful for investigating information technology improvements in the commercial sector. As previously noted, perceived risk theory appears to be one of the most often employed theories in consumer behavior research, as seen by its broad use in fields like as e-commerce.

3 Scope

This study created a benefit and risk paradigm that considers both the positive and negative factors that impact the decision to adopt Fintech. Based on previous research, perceived benefit and risk were interpreted as multidimensional factors in this study. The researchers observed three major factors of perceived advantage in this study, which are as follows: monetary reward, convenience, and transaction method. Furthermore, the four primary variables utilized to analyze perceived risk are financial, legal, security, and operational threats.

A second goal was to compare the influence of perceived advantage and risk on Fintech adoption intention in two Fintech groups (i.e., early adopters and late adopters). As a result, we projected that perceived benefit and risk had a significant influence on Fintech adoption intent. This research also took into account the fact that different Fintech user segments create a variety of expected benefits and risks.

4 Conclusion

The purpose of this research was to determine why consumers are eager or unwilling to utilize an emerging financial service, as well as to identify distinct benefit and risk perceptions based on the user categories that participated in it. Service in the financial technology industry. According to the results of this research, the most important factors and impediments to Fintech adoption intention are successfully identified.

Many issues are raised by our research in terms of theory. First and foremost, this study shed light on critical difficulties connected to client intentions regarding Fintech that had previously gone unnoticed by earlier research. A few empirical research on perceptions of Fintech adoption have been done, even though Fintech has received increasing attention. Before any major progress can be made, it is necessary to have a better knowledge of Fintech acceptance in the information technology area. As a result, this study used both positive and negative criteria that promote Fintech adoption, which were derived from IS research. Their interaction with one another, or how they function together to impact adoption intention and choice, is shown in this research. As a result of this study, a significant addition to research has been made in the area of the relationship between perceived benefit and perceived risk on Fintech adoption. Second, this study illustrates the decision-making process of consumers, which might be useful for researchers in gaining insight into their behavior. In this study, we discovered the advantages and hazards that influence the creation of adoption intentions, as well as the amount to which they influence the formation of adoption intentions. It is hoped that this research will contribute to a better understanding of the process of weighing multiple salient ideas about benefits and risks before making a choice. As a result, the decision-making process is made more visible and traceable than before. In addition, this research demonstrates that the impacts of perceived benefit and risk on Fintech adoption intention varied depending on the kind of Fintech user. When it comes to new technologies, the speed with which a newly developed service is adopted relies not only on the qualities of the service in question, but also on the characteristics of the users who are adopting it [42]. In order for positive and negative variables to have the desired integrative impact on Fintech adoption intention, Fintech businesses must first evaluate the user types who will be using their services. When it comes to the success of a Fintech firm, user-specific aspects are important.

5 Implications

The study's findings show, first and foremost, that perceived advantage is a stronger influence in financial technology adoption than perceived threat. According to data, customer enthusiasm in adopting Fintech is strong; nevertheless, there are various barriers blocking them from doing so. As a result, managing the risks associated with Fintech is just as critical as maximizing the advantages of this burgeoning industrial area. Consider how much more difficult it is to create a risk-free transaction environment than it is to provide advantages to customers. As a result, Fintech firms should explore risk-mitigation measures that will allow them to inspire high levels of trust in potential clients. Furthermore, our research provides Fintech executives with insights into the components that they should highlight or avoid while providing Fintech services to their clients. Managers must evaluate this conclusion before moving forward in order to determine how to deploy resources in order to sustain and develop their present client base. As a consequence of this research, managers will be able to put key practical advice on how to improve the use of financial technology into action. Third, managers should be well-versed in the distinctions between benefit and risk based on the type of user who will be impacted by the decision. This distinction enables Fintech companies to obtain a deeper knowledge of the characteristics of each Fintech user and to build services that are both effective and meet their customers' expectations and wishes, hence boosting the acceptance of their services among the general public. These elements should be taken into account by the executives of Fintech companies in order to increase the likelihood of their products being adopted.

6 Limitations

Despite the many contributions that have been made so far, there are a number of limits to this study. Before moving on to other topics, this research concentrated on a specific set of perceived advantages and dangers that were comparable to those explored in prior studies. It is probable that in future research, the inclusion of additional advantages and hazards linked with the use of Fintech services may become more significant. As a result, shown that a creative nature of Fintech and the nascent stage of Fintech service implementation, this study primarily examined behavioral intention as the dependent variable in order to assess theory-driven real behavior during the early adoption stage of Fintech service implementation. As a result, future research should aim to improve measurement reliability by incorporating additional methods, such as a field study and/or longitudinal analysis, to allow for more in-depth observation and investigation of differences between adopters and non-adopters in the later stages of Fintech service implementation, especially in the early stages. The third item to emphasize is that all of the components in this study were evaluated at the same time and using the same self-reported instrument, which is unique to this study. As Straub et al. [43] point out, common techniques variance is possible. Future study will consequently necessitate the creation of a comprehensive research strategy that combines quantitative and qualitative methodologies. To be effective, researchers must employ this multi-methodological triangulation method to discover likely variables that may contribute to explaining the lowered variations of the dependent variable. The fourth goal of this study was to provide extensive but cost-effective decision-making model for Fintech service uptake based on multidimensional benefit and risk considerations rather than a single benefit and risk consideration. However, behavioral intentions only account for 39.1% of the variance in behavior, according to the current model. This is our intention. In future study, incorporate new factors into the equations to boost the model's explanatory capacity even further.

Last but not least, the investigation was restricted that fintech comes in four varieties of services: Contactless wallet, cellphone payment, mentoring lending, and crowdfunding. To avoid being generalized, our findings cannot be generalised to other financial technology services (e.g., bitcoin, Ethereum, internet banking, personal financing, equity financing, retain investment, and Fintech tool and software). Fintech services study might be expanded and included into future studies. In terms of the final point, our findings may not be totally generalizable because our sample was limited to Korea, which differs from other Fintech-advanced nations in terms of national characteristics such as the United States, the United Kingdom, China, or Singapore. As a result, exercise with caution when interpreting the findings of this study. Future study will likely explore a variety of national variables in order to better understand the challenge of Fintech service acceptance across different countries.

References

1. Skan, J., Dickerson, J., Masood, S.: The future of fintech and banking (2014)
2. Rashid, M.M., Hayes, D.F.: Needs-based sewerage prioritization: alternative to conventional cost-benefit analysis. J. Environ. Manage. **92**(10), 2427–2440 (2011)
3. Chan, R.: Asian regulator seek fintech balance. In: Finance Asia, Finance Asia, Hong Kong (2015)
4. Kuo Chuen, D.L., Teo, E.G.: Emergence of FinTech and the LASIC principles. J. Financ. Perspect. **3**(3), 24–36 (2015)
5. Arner, D.W., Barberis, J.N., Buckley, R.P.: The evolution of fintech: a new post-crisis paradigm? (2015)
6. Teo, T., Lee, C.B., Chai, C.S.: Understanding pre-service teachers' computer attitudes: applying and extending the technology acceptance model. J. Comput. Assist. Learn. **24**(2), 128–143 (2008)
7. Barberis, J.: The rise of finTech: getting Hong Kong to lead the digital financial transition in APAC. Fintech Report. Fintech HK **13**(4) (2014)
8. Davis, F.D.: Perceived usefulness, perceived ease of use, and user acceptance of information technology. MIS Q. **13**(3), 319–340 (1989)
9. Mitchell, V.-W.: Consumer perceived risk: conceptualizations and models. Eur. J. Market. **33**(1/2), 163–195 (1999)
10. Kahneman, D., Tversky, A.: Prospect theory: an analysis of decision under risk. Econometrica **47**(2), 263–291 (1979)

11. Conchar, M.P., Zinkhan, G.M., Peters, C., Olavarrieta, S.: An integrated framework for the conceptualization of consumers' perceived risk processing. J. Acad. Mark. Sci. **32**(4), 418–436 (2004)
12. Campbell, H.F., Brown, R.P.C.: A multiple account framework for cost-benefit analysis. Eval. Program Plann. **28**(1), 23–32 (2005)
13. Zinkhan, G.M., Karandde, K.W.: Cultural and gender differences in risk-taking behavior among American and Spanish decision-makers. J. Soc. Psychol. **131**(5), 741–742 (1991)
14. Dardis, R., Stremel, J.: Risk-benefit analysis and the determination of acceptable risk. Adv. Consum. Res. **8**(1), 553–558 (1981)
15. Wood, C.M., Scheer, L.K.: Incorporating perceived risk into models of consumer deal assessment and purchase intent. Adv. Consum. Res. **23**(1), 309–404 (1996)
16. Bruhn, C.M.: Enhancing consumer acceptance of new processing technologies. Innov. Food Sci. Emerg. Technol. **8**(4), 555–558 (2007)
17. Lee, M.-C.: Factors influencing the adoption of internet banking: an integration of TAM and TPB with perceived risk and perceived benefit. Electron. Commer. Res. Appl. **8**(3), 130–141 (2009)
18. Featherman, M., Pavlou, P.A.: Predicting e-services adoption: a perceived risk facets perspective. Int. J. Hum Comput. Stud. **59**(4), 451–474 (2003)
19. Kim, D.J., Ferrin, D.L., Rao, H.R.: A trust-based consumer decision-making model in electronic commerce: the role of trust, perceived risk, and their antecedents. Decis. Support Syst. **44**(2), 544–564 (2008)
20. Wilkie, W.L., Pessemier, E.A.: Issues in marketing's use of multi-attribute attitude models. J. Mark. Res. **17**(3), 428–441 (1973)
21. Peter, J.P., Tarpey, L.X.: A comparative analysis of three consumer decision strategies. J. Consum. Res. **2**(1), 29–37 (1975). https://doi.org/10.1086/208613
22. Lewin, K.: Forces behind food habits and methods of change. Bull. Nat. Res. Council **108**(1043), 35–65 (1943)
23. Bilkey, W.J.: A psychological approach to consumer behavior analysis. J. Mark. **18**(1), 18–25 (1953)
24. Chishti, S.: How peer to peer lending and crowdfunding drive the fintech revolution in the UK. In: Banking Beyond Banks and Money, pp. 55–68. Springer (2016)
25. Sharma, S., Gutiérrez, J.A.: An evaluation framework for viable business models for m-commerce in the information technology sector. Electron. Mark. **20**(1), 33–52 (2010)
26. Porter, C.E., Donthu, N.: Using the technology acceptance model to explain how attitudes determine Internet usage: The role of perceived access barriers and demographics. J. Bus. Res. **59**(9), 999–1007 (2006)
27. Forsythe, S., Liu, C., Shannon, D., Gardner, L.C.: Development of a scale to measure the perceived benefits and risks of online shopping. J. Interact. Mark. **20**(2), 55–75 (2006)
28. Melewar, T., Alwi, S., Tingchi Liu, M., Brock, J.L., Cheng Shi, G., Chu, R., Tseng, T.-H.: Perceived benefits, perceived risk, and trust: influences on consumers' group buying behavior. Asia Pacific J. Mark. Logistics **25**(2), 225–248 (2013)
29. Rogers, E.M.: Diffusion of Innovations, 4th edn. The Free Press, New York (1995)
30. Karahanna, E., Straub, D.W., Chervany, N.L.: Information technology adoption across time: a cross-sectional comparison of pre-adoption and post-adoption beliefs. MIS Q. **19**(2), 183–213 (1999)
31. Cunningham, S.: The major dimensions of perceived risk. In: Cox, D. (ed.) Risk Taking and Information Handling in Consumer Behavior. Harvard University Press, Cambridge, MA (1967)
32. Yang, Y., Zhang, J.: Discussion on the dimensions of consumers' perceived risk in mobile service. In: Eighth International Conference on Mobile Business, Dalian. China (2009)
33. Lim, N.: Consumers' perceived risk: sources versus consequences. Electron. Commer. Res. Appl **2**(3), 216–228 (2003)
34. Dholakia, R.R., Uusitalo, O.: Switching to electronic stores: consumer characteristics and the perception of shopping benefits. Int. J. Retail Distrib. Manage. **30**(10), 459–469 (2002)

35. Melenhorst, A.S., Rogers, W.A., Caylor, E.C.: The use of communication technologies by older adults: exploring the benefits from the user's perspective. Proceedings of the Human Factors and Ergonomics Society Annual Meeting, **45**(3), 221–225 (2001). https://doi.org/10.1177/154193120104500305
36. Lacovou, C.L., Benbasat, I., Dexter, A.S.: Electronic data interchange and small organizations: adoption and impact of technology. MIS Quart. **19**(4), 465–485 (1995)
37. Kim, D., Olfman, L.: Determinants of corporate web services adoption: a survey of companies in Korea. Commun. Assoc. Inf. Syst. **29**(1), 1–24 (2011)
38. Research and Applications, 2(3), 216–228. Liu, Y., Li, H., Carlsson, C.: Factors driving the adoption of m-learning: an empirical study (2010)
39. Sun, H., Zhang, P.: Causal relationship between perceived enjoyment and perceived ease of use: an alternative approach. J. Assoc. Inf. Syst. **7**(9), 618–645 (2006)
40. Verhagen, T., Feldberg, F., van den Hooff, B., Meents, S.: Understanding virtual world usage: a multipurpose model and empirical testing. In: ECIS, paper 98 (2009)
41. Wen, C., Prybutock, V.R., Xu, C.: An integrated model for customer online repurchase intention. J. Comput. Inf. Syst. **52**(1), 14–23 (2011)
42. Sweeney, et al.: The role of perceived risk in the quality-value relationship: a study in a retail environment. J. Retail. **75**(1), 77–105 (1999)
43. Straub, D., Limayem, M., Karahanna-Evaristo, E.: Measuring system usage: implications for IS theory testing. Manage. Sci. **41**(8), 1328–1342 (1995)

Financial Technology (Fintech)

Adnan Jalal, Muneer Al Mubarak, and Farah Durani

Abstract Financial technology (FinTech) is a terminology used for applying technology in financial aspects. This technology started in its early days being very vague and unreliable, nowadays almost all financial institutes apply it in their operation. Added to that, it became so reliable that even small and medium enterprises use it in their transactions and asset management. Regulating authorities in certain sectors made the use of FinTech mandatory for daily operation of those sectors. The importance of this technology is growing rapidly to become mandatory in most of the sectors around the world. The investment in financial technology companies raised from 1200 million USD in 2008 up until 12,200 million USD in 2014 within a time frame of six years this increase was around 10 times and analysts claim that this figure will keep increasing every year (Truong in How FinTech industry is changing the world, 2016). Recently, FinTech is the scale by which the development of the financial institutes is measured at. The ease and readily availability of this technology on mobile devices saved not only time but it threatened the existence of face-to-face transactions. This study will cover a literature in which the evolution, types, success factors and models of FinTech in which a thorough look is given to these elements based on existing studies.

Keywords Financial technology · Fintech · Internet of Things

A. Jalal
The George Washington University, 2121 I St NW, Washington, DC 20052, USA

M. Al Mubarak (✉)
Ahlia University, Manama, Bahrain
e-mail: malmubarak@ahlia.edu.bh

F. Durani
University of Business and Technology, Jeddah, Saudi Arabia
e-mail: f.durani@ubt.edu.sa

© The Author(s), under exclusive license to Springer Nature Switzerland AG 2024
A. Hamdan and E. S. Aldhaen (eds.), *Artificial Intelligence and Transforming Digital Marketing*, Studies in Systems, Decision and Control 487,
https://doi.org/10.1007/978-3-031-35828-9_45

1 Introduction

Financial technology or Fintech is a terminology used to define the use of technology in financial services. The benefit of this service is not only to businesses or companies, it also includes the end users and the consumers. It helps in easing the financial operations among businesses and customers. In fact, this revolutionary technology is threatening the paper currency used for trading to be substituted with balance indicated in the account of the user. Added to that, Fintech is used in payments, lending, money transfer, loans and even in asset management.

The history and evolution of Fintech started way back in 1866 and it is categorized in three phases. Fintech 1.0 is the era between 1866 till 1967. In this period technology such as stream ships, telegraphs and railroads used for financial operations. In the second phase so called Fintech 2.0 is between 1967 till 2008 in which with help of technology and introduction of internet the financial institutes became more customer centric by giving solutions to all financial requirements in one place. Finally, the introduction of Fintech 3.0 after 2008 till present more different types of technology is introduced such as e-banking, and it continuous to develop given the demand in the market [13].

With the evolution of internet the need for this technology to develop was necessary to cope up with the financial needs of its users. It can be arguably said that the introduction of Automatic Teller Machines so called ATM in 1967 by Barclays bank is what commenced the evolution of Fintech around the globe [2]. Today with a fingertip touch on a mobile screen can transfer money without the need of bank visit, thanks to the evolution that Fintech has reached.

Moreover, Fintech development is highly desirable in large economies. It changes the market structure of nations. It has been noted that financial services through financial institute branches is expensive and time consuming, to agile this process Fintech developing firms are aware of the shortcomings in the industry [15].

The Fintech like any other technology is continuously facing challenges to continue its existence. One of the major concerns for Fintech is security management, which simply can be achieved by identifying the assets of a certain firm and protecting them. As we are all aware that any technology on the internet cloud is prone to breaches that could make it exposed and put the data and assets of the user in risk of cyber-attack. Also, the challenge of continuous development in Fintech urges developers to emphasize on customer centricity and innovation. In addition to that, the need to comply with regulations when adopting an innovative Fintech or in case of introduction of new regulation, Fintech needs the assistance of so called Regtech companies to do it for them effectively and economically [12].

The outbreak of Covid-19 pandemic forced Fintech developers to introduce services that minimizes the human-to-human interaction. Surely this was beneficial to Fintech developers as regulatory authorities such as governments eased the process of implementation of such into smallest daily life transactions. For example, in Kingdom of Bahrain a widely used mobile application called Benefit-pay gives

Financial Technology (Fintech) 527

direct connection of personal or business bank accounts to the application to pay and accept multiple payments.

This paper will include a literature review of evolution of Fintech, types and uses of Fintech, challenges faced by Fintech, critical success factors of Fintech and regulatory aspects in Fintech. These topics shall be discussed based on journals, articles and books published on the respective topic.

2 Literature Review

2.1 Scope of the Study

This paper follows a literature review design of the evolution of FinTech, current challenges, types of FinTech, and critical success factors associated with the technology. The research paper is not limited to any geographical region or the size of organizations; rather, it emphasizes chronological events that marked the different phases of FinTech development before providing a clear picture of the technology as the world currently understands it. In this case, the researcher will borrow heavily from various theories to provide examples ranging from some of the largest corporations in the developed world to startups in the developing world to explain what has been achieved so far through FinTech.

2.2 Overview/Theoretical Foundations

FinTech is a relatively new field in many parts of the world. Leong and Sung [9] note that the low level of awareness about FinTech and the lack of a consistent definition have suppressed the adoption of this revolutionary technology, particularly in the developing world where technology consumption has been slow. FinTech is defined as "a system that aims to assess and produce finance" by combining "trading systems and trading technology in transactions on different markets" [16, p. 1]. Elsewhere, the concept is described as "the technology and innovation" that applies technological solutions to improve and develop services in the financial sector [1, p. 123]. Following an evaluation of common definitions of the concept, some scholars compiled an inclusive definition that all audiences can easily understand. In this case, FinTech was defined as "a cross-disciplinary subject that combines finance, technology management, and innovation management" [9, p. 74; 19, p. 187]. Contrary to the common misperception that FinTech is limited to finance and technology, the above definition reveals that the concept is a multidisciplinary field that traverses finance, technology, management, and innovation. By combining all these technologies, FinTech promises to revolutionize financial markets and trade through improved services, enhanced efficiency, automation, real-time processing, and management of finances and transactions [17]. Some examples of this technology include cellular banks or smartphone apps that offer banking services, cryptocurrencies, digital lending/credit, and mobile banking services. Although this technology has grown in popularity in the past decade, its development can be traced more than 1.5 centuries ago [3].

FinTech has been widely studied from both theoretical and empirical perspectives. The modern monetary theory is one of the few grand theories that relate to FinTech. This theory describes the currency as a public monopoly and argues that governments can spend limitlessly as long as they have full control of their respective currencies [14]. According to Legowo et al. [8], the modern monetary theory relates to FinTech due to its coverage of the supply of money and the speed at which money can flow. Unlike fiat money, FinTech allows money to flow faster and reach a larger population. Interestingly, this technology has also lowered the costs that are incurred to necessitate the flow of money, thus ensuring that the people have more to spend. Besides the supply of money, the application of this theory can also be viewed from the perspective of the currency and how much people can spend. Notably, the emergence of digital currencies like Bitcoin that are not tied to other currencies means that users can spend limitlessly as long as they do not exhaust the money that is in circulation [6, 8]. FinTech has been designed to work with both fiat-backed money and unregulated currencies like cryptocurrency. However, Peters et al. [14] write that the latter has attracted widespread resistance from governments due to their inability to control its supply, levy taxes, regulate value, introduce interest rates, and regulate usage. Nevertheless, Legowo et al. [8] explain that electronic money has revolutionized the flow of money within an economy and how people can use it. As such, continued standardization of practices in FinTech is transforming the system into one that is widely accepted at all levels.

Several theories explain the fast rate at which FinTech has been adopted and its disruptive nature in the financial markets. First, the diffusion of innovation theory explains that it is inevitable for many industries to make, change, and improve technology adoption [6]. Precisely, FinTech has introduced many technologies in the financial markets that have improved the speed at which money flows within the economy [8]. As such, the adoption of this technology has proved inevitable. While exploring the concept from the fundamental economic theory, Khotinskay [7] wrote that FinTech is a disruptive technology that was developed to revolutionize how the financial sector and other related fields operate. This theory treats both banks and non-bank services as intermediate institutions whose function is to "accumulate temporarily free money from some market participants and to deliver it to the other participants, who need it, to use on a return, time-limited and paid basis" [7, p. 224]. This function of the financial markets has historically remained unchanged. However, FinTech seeks to replace this conservative function of banks by revolutionizing how finances are managed and the role that banks and non-banks play in transactions. According to Khotinskay [7], FinTech is designed to slowly change business processes, introduce new tools into the financial market, and change how the sector is controlled. This technology has also created numerous changes at the fundamental level, with the most notable being the digitization of money and information [7]. As such, FinTech has transformed the financial sector from a fundamental model by bringing into life new business models.

FinTech is widely viewed as a source of sustained competitive advantage at the personal, corporate, and market levels. From a resource-based view (RBV) theoretical framework, FinTech is viewed as a platform that creates sustained competitive advantage by providing both technology and money as strategic resources [8]. Further, the model views FinTech as a source of greater efficiency since it transforms money from tangible assets into intangible assets. When combined with the modern monetary theory, the RBV model explains the greater efficiency that FinTech creates by ensuring a faster flow of money and information through digitization or transformation into intangible assets. The competitive theory further expands the RBV. According to Legowo et al. [8], businesses currently adopt FinTech as a source of competitive advantage due to the efficiency that it introduces to the business environment. Notably, this technology has come into the market as a new product/service, a new process of managing finances/transactions, and a new business model [8]. Although FinTech was founded on an existing system, its adoption among existing and new businesses has been faster than earlier anticipated. Legowo et al. explain that FinTech has necessitated the "e creation, improvement of existing service/product/ process of business order to increase for the customer or to make it transparent, accessible, costs or fees" (p. 35). This technology has thus emerged as a priority for both startups and existing banks. In this case, the technology has allowed startups to establish a firm foundation in markets that were initially dominated by banks, while many banks have integrated it into their operations as a means to increase efficiency.

Iluba and Phiri [6] introduce two theories to explain the fast pace at which FinTech has been adopted in the market. Although the concept of fast development had been studied earlier by Legowo et al. from the diffusion of innovations theoretical lens, the technology acceptance model (TAM), the unified theory of acceptance and use of technology (UTAUT), and the theory of reasoned actions (TRA) offer a clearer understanding of both the users' willingness to adopt the technology and the reason behind its fast growth. From an acceptance perspective as proposed by TAM and UTAUT, it has become easier for people to adopt FinTech due to the existence of fundamental technologies over which FinTech has been developed. At the time of its development, modern FinTech was embraced faster than how it happened in the previous decades since it was both reliable and efficient [6]. The positive attitude that people had developed towards banking and early versions of FinTech like the ATM created a "compelling reason for one to try out the innovation" [6, p. 841]. Such reasoned action explains why many forms of FinTech have emerged in the past decade and why the technologies have spread uncontrollably in both developed and developing nations.

3 The Evolution of FinTech

3.1 FinTech 1.0 (Analogue Technology and Mainframe Computers)

Some researchers inaccurately present FinTech as new technology. For instance, Khotinskay [7] presents this technology as one that appeared in the late twentieth century. However, Arner et al. [3] clarify that it is the "term (financial technology or FinTech that) can be traced to the early 1990s" (p. 22). Although the term is relatively new, the concept that it represents can be traced back more than 1.5 centuries ago. While clarifying that "FinTech is not novel," Arner et al. [3] wrote that referred to the concept as a "new term for an old relationship" (p. 22). In this case, only the mode in which FinTech is conducted has changed. Leong and Sung [9] traced the origin of FinTech to 1866 when the first Trans-Atlantic communication occurred through telecommunication technology that had been deployed back in 1958. During this period, the global telex allowed significantly reduced the time it took ships to deliver messages across the Atlantic by a huge margin. Leong and Sung [9] explain that this advancement in technology revolutionized communication and improved related financial services. Similarly, Arner et al. trace the origin of FinTech to 1866 when what the authors refer to as FinTech 1.0 started. This first generation continued for a century up to 1966. Major changes were experienced between 1866 and 1913 during a period of strong financial globalization. The technologies that were developed during this period continued to be improved until 1966. Although this technology was heavily interlinked, it was largely limited to analog.

3.2 FinTech 2.0 (Internet of Things)

Like many technologies, the FinTech revolution occurred in phases or generations. The technology largely remained unchanged for the first half of the twentieth century during the peak of FinTech 1.0. However, Leong and Sung [9] explain that a major turning point occurred between 1967 and 2008 when the development of computers necessitated the creation of the Society of Worldwide Interbank Financial Telecommunications (SWIFT) and ATMs. Arner et al. add that the rapid post-World War I technological development necessitated a shift from analog to digital that led to the creation of FinTech 2.0 in the late 1960s. This rapid development was accelerated by the establishment of the Inter-Bank Computer Bureau in the UK in 1968, the US Clearing House Interbank Payments System, and SWIFT in 1973. Arner et al. recognize the Clearing House Interbank Payments System's Fedwire as the first practical example of modern FinTech. While such technologies increased the global connectedness of the financial systems, they were met by skepticism. As Arner et al. explain, the collapse of Herstatt Bank in 1974 and the "Black Monday" stock market crash of 1987 highlighted the risks of interconnecting financial systems. Nevertheless,

Financial Technology (Fintech)

controls were developed in the form of agreements and regulatory authorities. Some of the earliest regulatory authorities and agreements include the Basel Committee on Banking Supervision of the Bank for International Settlements in 1975, the 1986 Big Bang of the UK, the Single European Act 1986, and the 1992 Maastricht Treaty. Like FinTech 1.0, FinTech 2.0 stagnated during the late twentieth century. However, a major turning point occurred in 1995 when Wells Fargo provided the first internet banking services. The evolution of the Internet and the advent of internet banking services provided the foundation for FinTech 3.0.

3.3 FinTech 3.0 (Data Technologies)

The third generation of FinTech started in 2008. Leong and Sung [9] recognize the continued development of data technologies as the main feature of FinTech 3.0. This evolution was necessitated by trust in technologies among consumers. Notably, businesses and their customers in this area operate in a loosely regulated area. For instance, Arner et al. explain that more than 2000 peer-to-peer FinTechs dominate the Chinese financial sector despite operating outside a clear regulatory framework. The development of powerful smartphones, increased access to the Internet, low cost of technology and availability of application programming interfaces (APIs) are also recognized as other factors that have accelerated the development of FinTech 3.0. This era has seen the emergence of numerous FinTech startups. According to Legowo et al., there has been a rising interest in FinTech to fill the gap that the redundancy of banks has created. At the same time, many financial institutions that used the classical banking system have integrated FinTech into their operations by focusing more on technology. For instance, a third of the 33,000 employees working for Goldman Sachs—one of the world's largest banks—are engineers. FinTech 3.0 has also led to virtual currency, mobile money, and cryptocurrency development.

4 Challenges Faced by FinTech

Cybersecurity issues are some of the key challenges facing FinTech. Like every other data-intensive technology, Fintech is prone to cybersecurity risks like malicious damage of data, unauthorized access of personal information, hacking, disruption of services through distributed denial of services (DDoS), hacking, ransomware, and cross-site attacks [11]. The authors add that insider threats and all other threats that could lead to unauthorized data access could have severe ramifications due to the sensitivity of the data and the risk of access to victims' finances. For instance, a single data breach in Equifax—a company that serves FinTechs in the US—compromised the data belonging to 143 million individuals [4]. Al_Duhaidahawi et al. [1] also add that maintaining the security and privacy of the data is a major challenge among many FinTechs due to the bulk of the data they manage.

Major challenges are the lack of an elaborate regulatory framework, compliance issues, hefty fines, and lack of collaboration among stakeholders. Arner et al. argue that many governments have not developed reliable legal and regulatory frameworks to govern FinTechs. As such, it is difficult for these companies to comply with the laws, particularly in areas where they are regulated in the same way as classical banking systems. Such limitations may lead to costly consequences for these companies. Similarly, the difficulty of compliance and the risk of violations raise several issues among these companies and regulators. For instance, Equifax was recently fined more than $500 million due to the data breach in which records belonging to 143 Americans were lost [4, 18]. Unfortunately, the company had been on the frontline to frustrate efforts by other stakeholders to regulate the industry. It should also be noted that Equifax had spent approximately "$1.1 million to counter legislation intended to improve the industry's data security and victim notification procedures" [18, p. 552].

Lack of awareness, lack of consumer trust, limited support by the government are key concerns. For instance, despite the success of the M-Pesa mobile money service in Kenya and other African markets, attempts by the company to establish in some firms within the African continent failed due to lack of awareness about mobile services. Other studies have reported inadequate regulation as the key reason behind the slow adoption of FinTech in some areas, lack of support by governments, and lack of trust among consumers [8]. If this industry is adequately regulated, it will be possible for governments to invest in the same way that they have done with banks and develop an elaborate framework through which companies can be supported in times of crisis. For instance, Peters et al. [14] write that consumers exercise caution when investing in cryptocurrencies due to the lack of regulations and the absence of a fail-safe framework for bailing out consumers when a crisis occurs.

5 Types and Uses of FinTech

There are different types of FinTechs. Some of these include insurance technology (InsurTech), regulatory technology (RegTech), payment technology (PayTech), lending technology (LendTech), consumer banking (BankTech), personal finance (WealthTech), mobile payments, international money transfer, and cryptocurrency/blockchain. RegTech helps FinTechs to comply with regulatory agencies, meet supervisory requirements, automate their operations, and manage risks [10]. They also provide APIs for improving the service. On the other hand, PayTechs necessitate swift payments. For instance, Armer et al. explain that China has more than 2000 PayTechs where customers can pay for their online orders, taxi, food, tickets, and services. International money transfer services like PayPal are also an important type of FinTechs that support PayTechs in necessitating faster money transfer at pocket-friendly charges.

Financial Technology (Fintech) 533

BankTechs or neobanking is a type of FinTech that has become increasingly popular. The technology has been necessitated by the adoption of FinTech by classical banks that have shifted their attention to these technologies to lower operating costs, improve efficiency, and improve the speed at which transactions can be completed [17]. Unlike transactions in traditional banks that could take too long to complete, modern banks offer synchronous transactions that necessitate instant transactions without incurring unnecessary labor and equipment costs. Insurance companies have also integrated technology in their operations through mobile apps. However, InsurTechs continue to cooperate with traditional insurance companies to ensure compliance in the highly-regulated industry [5]. The dynamic nature of FinTech means that new types may emerge in the foreseeable future as it continues to revolutionize.

6 Critical Success Factors

FinTech has grown to one of the most lucrative sectors globally and one with a sustainably high growth rate. According to Leong and Sung [9], the value of transactions necessitated by FinTech rose to almost $25 billion in 2016. Among other factors, this amount was attributed to the rising popularity of transactions that were initially impractical with conventional banking. In this case, donations, reward-based and equity-based crowdfunding amounted to £12 million, £42 million, and £332 million, respectively, in 2015 [9]. As of 2017, KMPG estimated that the value of FinTech had hit a whopping $31 billion worldwide, which was a major growth from the $9 billion that the sector was worth in 2010 [7]. However, 2016 is marked as a major turning point in the growth of FinTech. According to Sung et al. [19], the industry hit a whopping $111.8 billion or an equivalent of £88.2b billion in 2017. On average, FinTech companies reported an investment of between £15 million and £2.9 billion. Some types of FinTech have reported a higher growth rate than others. For instance, the value of cryptocurrency had more than tripled in just months leading to 2017 to hit $25 billion in market capitalization [19].

Other researchers noted that FinTech was emerging as a major source of employment, particularly among the younger generation that turned out to be more tech-savvy. For instance, Sung et al. [19] reported that the UK already had 1600 firms by 2019. However, this figure was expected to double by 2030 to hit 3300 firms. The same trend was forecasted elsewhere worldwide, albeit at higher rates in developed countries due to the existence of supportive frameworks and infrastructure that necessitate FinTech [19]. With each of these firms employing multiple employees, the workforce in the sector has been rising at a steady pace. Leong and Sung [9] estimated that FinTech would create more than 100,000 jobs in the UK economy between 2016 and 2020, which was projected to be way higher than most conventional sectors. This workforce was estimated to grow exponentially due to the continued expansion of FinTech firms and their increase in number. For instance, Sung et al. [19] estimated that FinTech firms will grow by more than 200% between 2019 and 2030, making the industry one of the fastest-growing in the world.

The role of FinTech in increasing efficiency and improving resource utilization can no longer be overlooked. Researchers attribute the unprecedented growth rate to the ability of FinTech to cut operating costs by more than 30% compared to the conventional banking system [16]. Iluba and Phiri [6] also approached the topic from the perspective of limitations in operating time. The researchers found that many people were finding banks inefficient and unreliable since they were only open for approximately 8–9 h. The availability of their services was also limited due to their small number and the distance that customers had to travel. This inconvenience was reported particularly among rural residents and those from underdeveloped regions. However, FinTech was found to provide a perfect solution that was cost-effective, time-saving, available at all times, and whose access was not geographically limited [6]. Precisely, the researchers presented the technology as a key solution to some of the challenges facing populations in Sub-Saharan Africa due to its integration with smartphones and other readily available technologies.

FinTech is highly practical in both developed and developing countries. This technology has taken the world by storm since the advent of FinTech 3.0. For instance, mobile money technology has enabled a majority of the populations in developing countries to access financial services through formal and semi-formal financial services. Most of these services are accessible through mobile money services that can be accessed everywhere, anytime, and at minimal costs. For instance, the transactions made through M-Pesa—a Kenyan mobile money service—amount to almost 50% of the country's GDP. Equally, cryptocurrency, e-banking, and insurance technology (InsurTech) have gained vast popularity in developed countries like the US, the UK, and the EU [9].

7 Conclusion

The evolution of FinTech started with FinTech 1.0 which utilized the use of analogue technology and mainframe computers. In FinTech 2.0 era the use of internet came into existence which was a revolution to the development of financial technology, and it set the platform to utilize the data in technology which is the third generation of FinTech 3.0.

The challenges faced by FinTech to remain reliable is cybersecurity. Like any other technology it is prone to hacking, unauthorized access to personal data and disruption of data, but largely the concern lack of awareness of consumers which leads to placing their assets in a risk. To counter this lack of awareness the governments and regulatory authorities are placing fines and continuously monitoring those who violate this lack of awareness.

The types of FinTech are InsurTech, RegTech, PayTech, LendTech, PayTech, BankTech and WealthTech. These different types of FinTech are used in conjunction with each other to provide wide range of solutions to firms and business to ensure safe and swift day to day operation.

The FinTech has shown its success all around the globe. It has grown to one of the most lucrative sectors with high growth rate. Also, it has been a source of employments among the younger generation who are more into technological studies. Moreover, the dependence on FinTech all around the world has grown rapidly, which lead to reduction in operation cost in firms, increased efficiency and improved resource utilization.

To conclude, the future of FinTech success is bright as it is overcoming the challenges which are arising time to time. In instances like Covid-19 pandemic is has shown that technology can run many sectors without the need of being available in offices. Thus, it is mandatory for regulating authorities to ease the way of FinTech in the market.

References

1. Al_Duhaidahawi, H.M.K., Zhang, J., Abdulreda, M.S., Sebai, M., Harjan, S.: The financial technology (Fintech) and cybersecurity. Int. J. Res. Bus. Soc. Sci. (2147–4478) 9(6), 123–133 (2020). https://doi.org/10.20525/ijrbs.v9i6.914
2. Arner, D.W., Barberis, J., Buckley, R.P.: The evolution of Fintech: a new post-crisis paradigm. Geo. J. Int. 47, 1271 (2015)
3. Arner, D., Barberis, J., Buckley, R.: 150 years of FinTech: an evolutionary analysis. JASSA Finsia J. Appl. Financ. 1(3), 22–26 (2016). https://www.researchgate.net/publication/313364 787_150_Years_of_FinTech_An_Evolutionary_Analysis
4. Berghel, H.: The equifax hack revisited and repurposed. Computer 53(5), 85–90 (2020). https://doi.org/10.1109/mc.2020.2979525
5. Bruce, D., Avis, C., Byrne, M., Gosrani, V., Lim, Z., Manning, J., Popovic, D., Purcell, R., Qin, W.: Improving the success of InsurTech opportunities. Br. Actuar. J. 23(31), 1–34 (2018). https://doi.org/10.1017/s1357321718000296
6. Iluba, E., Phiri, J.: The FinTech evolution and its effect on traditional banking in Africa—a case of Zambia. Open J. Bus. Manage. 09(02), 838–850 (2021). https://doi.org/10.4236/ojbm. 2021.92043
7. Khotinskay, G.I.: Fin Tech: fundamental theory and empirical features. Eur. Proc. Soc. Behav. Sci. 1(1), 222–229 (2019). https://doi.org/10.15405/epsbs.2019.03.23
8. Legowo, M., Subanidja, S., Fangky, A., Sorongan.: Role of FinTech mechanism to technological innovation: a conceptual framework. Int. J. Innovative Sci. Res. Technol. 5(5), 35–41 (2020). https://www.ijisrt.com/assets/upload/files/IJISRT20MAY164.pdf
9. Leong, K., Sung, A.: FinTech (financial technology): what is it and how to use technologies to create business value in Fintech way? Int. J. Innovation Manage. Technol. 9(2), 74–78 (2018). https://doi.org/10.18178/ijimt.2018.9.2.791
10. Liaw, K.T.: The Routledge Handbook of Fintech. Routledge, Taylor & Francis Group, New York, NY (2021)
11. Najaf, K., Mostafiz, M.I., Najaf, R.: Fintech firms and banks sustainability: why cybersecurity risk matters? Int. J. Financ. Eng. 8(2), 2150019 (2021). https://doi.org/10.1142/s24247863215 00195
12. Nicoletti, B., Nicoletti, W., Weis: Future of FinTech. Palgrave Macmillan, Basingstoke, UK (2017)
13. Paul, L.R., Sadath, L.: A systematic analysis on FinTech and its applications. In: Conf Pap, pp. 1–7. (2021)

14. Peters, G.W., Panayi, E., Chapelle, A.: Trends in crypto-currencies and blockchain technologies: a monetary theory and regulation perspective. SSRN Electron. J. **1**(1), 1–25 (2015). https://doi.org/10.2139/ssrn.2646618
15. Philippon, T.: The fintech opportunity (No. w22476). National Bureau of Economic Research (2016)
16. Purnomo, H., Khalda, S.: Influence of financial technology on national financial institutions. IOP Conf. Ser.: Mater. Sci. Eng. **662**(2), 1–7 (2019). https://doi.org/10.1088/1757-899x/662/2/022037
17. Salampasis, D., Mention, A.-L.: Transformation Dynamics in FinTech: An Open Innovation Ecosystem Outlook. World Scientific, New Jersey (2021)
18. Smith, M., Mulrain, G.: Equi-failure: the national security implications of the Equifax hack and a critical proposal for reform. J. Natl. Secur. Law Policy **9**(1), 549–588 (2020). https://jnslp.com/wp-content/uploads/2018/09/Equi-failure_The_National_Security_Implications_2.pdf
19. Sung, A., Leong, K., Sironi, P., O'Reilly, T., McMillan, A.: An exploratory study of the FinTech (financial technology) education and retraining in UK. J. Work-Appl. Manage. **11**(2), 187–198 (2019). https://doi.org/10.1108/jwam-06-2019-0020
20. Truong, T.: How FinTech industry is changing the world (2016)

Assessing Opportunities and Challenges of FinTech: A Bahrain's View of Fintech

Sabika AlJalal, Muneer Al Mubarak, and Ghada Nasseif

Abstract FinTech (Financial Technology) is the state-of-the-art technology used by businesses to improve their financial services. The changing customer behaviour towards digital platforms, the increase in the financial service businesses on the digital platforms, the availability of banking options on mobile, and the advent of COVID-19 has further increased the adoption of Fintech by Bahrain and several other countries. The increase in the demand for digital technologies has increased the challenges and issues also in shifting towards the digital technologies such as Fintech. The study work described in this article assesses the potential impact of the advancements in the Fintech on Bahrain's financial industry. The study is focused at identifying the challenges faced by the Bahrain's FinTech firms and other such innovators in implementing the Fintech, assessing the current developments and implementations of Fintech in Bahrain, identifying gaps in the existing research, and proposing a FinTech investment model. The article is based around carrying out a critical review of the relevant articles to understand the key aspects of Fintech in Bahrain and other countries such as the challenges faced, current state of development, scope of Fintech in Bahrain, viewpoints on Fintech by different researchers, etc. The study work provides a comprehensive understanding of various aspects of FinTech for licensees, FinTech firms, FinTech start-ups, and other innovators to utilize FinTech services and solutions in their operations.

Keywords FinTech · Financial industry · Investment model · Digital platforms · Licensees · Start-ups

---

S. AlJalal
The George Washington University, 2121 I St NW, Washington, DC 20052, USA

M. Al Mubarak (✉)
Ahlia University, Manama, Bahrain
e-mail: malmubarak@ahlia.edu.bh

G. Nasseif
University of Business and Technology, Jeddah, Saudi Arabia
e-mail: ghadanasseif@ubt.edu.sa

© The Author(s), under exclusive license to Springer Nature Switzerland AG 2024
A. Hamdan and E. S. Aldhaen (eds.), *Artificial Intelligence and Transforming Digital Marketing*, Studies in Systems, Decision and Control 487,
https://doi.org/10.1007/978-3-031-35828-9_46

1 Introduction

1.1 Growth of FinTech

Fintech (Financial Technology) is a multidisciplinary subject which combines innovation management, finance, and technology [1]. Since the year 2014, there has been a significant growth in Fintech attracting regulators, industries, etc., and has become a rapidly growing industry representing investments (as of year 2014) between USD 12 and 197 billion dollars through multiple areas such as start-ups, existing financial institutions (FIs), applications, etc., [2]. The COVID 19 pandemic situation has increased the existing measures and processes towards digital transformation focused on Fintech, so, it is expected that the investments in the Fintech could rapidly increase.

Few of the salient features which attracts the FIs towards Fintech are:

- Fintech allows FIs a streamlined technology to provide state-of-the-art experience to the users.
- It provides several Artificial Intelligence (AI) and machine learning based features to the users such as suggesting investment plans, processing payments, loan advices, etc.
- It increases the speed of the various processes in the financial sector such as KYC (Know Your Customer) procedures, due diligence, customer on-boarding, loan calculations, etc.
- It provides myriad ways for FIs to identify money laundering trends and fraudulent activities through embedded scenarios and algorithms reflecting the application.
- It provides improved customer service experience through cutting-edge user interfaces to the system.
- It assists the compliance, risk, and legal departments in assessing and meeting the various risks identification and regulatory requirements.

Because of the several benefits of Fintech in the financial industry, several countries are taking steps towards incorporating Fintech in their operations. Bahrain is among those countries which has considered Fintech as one of the top priorities to continue incorporating cutting-edge technologies in the financial sector [3].

1.2 Research Problem

Underfunding is one of the major reasons behind failure of many FinTech firms [4]. Maintaining financial stability is the other problem faced by the FinTech start-ups and which result in the closure of the FinTech firms because of the inadequate understanding and risk assessment [5]. A few FinTech firms have considered growth as the most prioritized factor and gave less focus on their business profitability which resulted in huge losses; and some FinTech firms gave highest priority to customer satisfaction by including enhanced user functionality and experience interfaces but failed to consider the potential profitability of their products and services, both of which caused financial losses to the companies [6]. Below are some of the other reasons which caused the FinTech start-ups to fail [7].

- Failing to comply with the regulations.
- Ignoring crucial aspects of being a FinTech start-up and considering FinTech similar to any other tech start-up company.
- Considering cost as the only competitive element.
- Selecting a wrong business model without understanding the core competencies.
- Not establishing partnership with the FIs.

1.3 Need for the Research

Understanding the research problem stated above, it makes it clear that there is a need to clearly understand all the governing aspects of FinTech to ensure business continuity and profitability with a low risk of failure. The study work provides a framework for FinTech firms, start-ups, and licensees (financial sector institutions) to understand the different factors associated with FinTech such as the FinTech verticals, FinTech ecosystem, etc. The study describes an investment model focused at eliminating the risks specified in the 'research problem' section. The study work assists the FinTech firms and other such innovators in conducting a comprehensive assessment of the FinTech solution/service to measure its effectiveness for their firm. The study lists out all the key elements of the FinTech ecosystem and discusses challenges and opportunities as well to provide a detailed analysis of the FinTech solution/service for the mentioned stakeholders in view of Bahrain.

1.4 FinTech Investment Model

An investment model is proposed in the study work which recommends including factors such as opportunities, benefits, challenges, issues, FinTech's ecosystem and verticals, etc. Considering these factors could assist in facilitating clear understanding of the different aspects of FinTech, with an aim to assist FinTech start-ups, financial sector institutions, and other such innovators to improve the performance of their FinTech services and solutions (Fig. 1).

1.5 Fintech Verticals

Few of the key Fintech verticals are AI and machine learning, data analytics, remittances and payments, crowdfunding, and cryptocurrencies (Fig. 2).

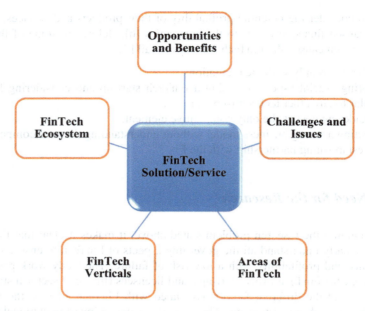

Fig. 1 FinTech investment model

AI and Machine Learning

AI involves the use of algorithms to make the systems intelligent and adopt human competencies. Basically, AI brings learning and thinking capabilities to the machines.

Machine learning is often connected with AI as one of its branches and describes the study of computer algorithms which improves automatically through continued operations or experiences with the use of data. AI and machine learning are among the major components of Fintech.

Data Analytics

Data analytics is associated with analysing a large amount of data to assist in several functions such as compliance, market research, decision making, etc. They also assist in understanding the customer's preferences and develop products and services in-line with the customer's preferences.

Remittances and Payments

Remittances and payments through digital platforms are an initiative to migrate from cash to cashless transactions. The COVID 19 pandemic situation has significantly pushed the online platforms for carrying out remittances and payments.

Cryptocurrencies

Cryptocurrencies are another approach of processing online transactions. Commonly referred to as digital currencies still faces challenges to be recognized as a currency

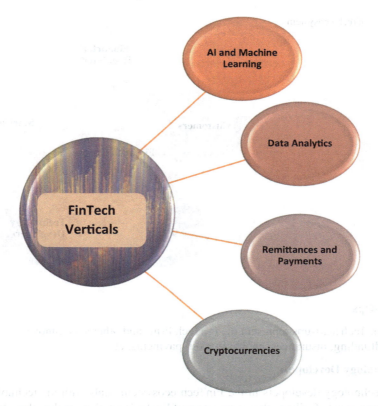

Fig. 2 FinTech verticals

at global level because of the vulnerabilities such as anonymity, frauds, money laundering, etc.

2 Elements of the FinTech Ecosystem

The five key elements of the FinTech ecosystem which symbiotically benefits the consumers and licensees in the financial industry are FinTech start-ups, developers of technology, financial regulators, customers, and FIs [8] (Fig. 3).

Financial Regulators

Regulators play a crucial role in the FinTech ecosystem. The acceptance to the guidelines from the regulators flows through different elements of the FinTech ecosystem. This element could also represent any governmental body associated with the FinTech ecosystem.

Fig. 3 FinTech ecosystem

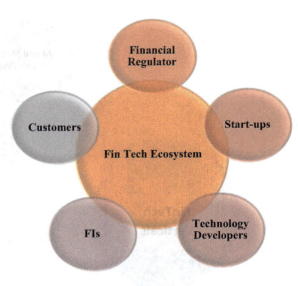

Start-Ups

The FinTech start-ups represent the FinTech firms and other such innovators such as crowdfunding, insurances, capital market, payments, etc.

Technology Developers

The technology developers in the FinTech ecosystem deals with the technological developments to facilitate the introduction of FinTech solutions in the other elements of the ecosystem. This element represents big data analytics, cryptocurrencies, cloud computing, etc.

Customers

The customers in the ecosystem represent the individual and non-individual customers utilizing the FinTech solutions. Organizations, corporates, natural customers, etc., can all be a part of this element.

FIs

The FIs in the FinTech ecosystem represent the different FIs such as banks, stock brokerage, money service businesses, insurance companies, etc., utilizing the FinTech solutions in delivering products and services to the customers and adhering to the regulations.

Table 1 Financial sector institutions in Bahrain

S.No.	Financial sector	Number of FIs
1	Wholesale banks	62
2	Retail banks	31
3	Specialized licensees	86
4	Insurance companies	34
5	Branches of foreign banks	17
6	Representative offices of banks	8
7	Investment business firms	53

Note From "about banking," by Central Bank of Bahrain, 2021, (https://www.cbb.gov.bh/banking/)

3 Research Problem

Though the use of banking has become one of the necessary aspects of life for many people, there are still 1.7 billion unbanked individuals around the world [9]. There are several reasons for this, few of which include unavailability of banking services in the related areas, security, and privacy issues, etc. In the developing countries, only 54% of the adults are having a bank account as many people still prefer to save money at home or buy illiquid assets [10]. The banking system in Bahrain accounts for more than 85% of the total financial assets, consisting of conventional and Islamic banks; and there are 376 financial sector institutions in the country [11].

As can be seen in Table 1, there is a huge scope of including FinTech solutions in the diverse banking and finance environment in Bahrain which can enhance the operations of the FIs and provide customers' a provision to use state-of-the-art banking and finance products which would encourage customers towards banking methods. The large number of unbanked customers shows that there is a huge potential for customer acquisition by deploying FinTech solutions. The diverse financial sector in Bahrain can be greatly benefited with FinTech. There are few studies works conducted on FinTech in Bahrain which makes this paper a good platform for researchers in Bahrain to work on FinTech related aspects in Bahrain.

4 Areas of FinTech in Bahrain

FinTech has opened several areas for the FIs and improved the performance of many products and services offered by the FIs [12]. FinTech assists FIs in reducing time it requires to resolve customer service-related problems or other issues involving programming decision making [13]. This further attracts the FIs to incorporate FinTech in their operations, products, and services. Below are some of the major areas where FinTech can play a significant role in improving the performance of the FIs.

4.1 Mobile Payments and Banking

Online and mobile banking are among the important technologies which have offered several features to the users. The online and mobile banking faces several challenges among which security, and customer privacy are among the key challenges [14]. FinTech can play an important role in boosting the performance of mobile banking and overcoming the security concerns in online banking services. FinTech can greatly improve the customer interfaces on mobile banking platforms providing users an easy to use and secured platform for banking and payments.

4.2 Digital Wallets and Prepaid Cards

Digital Wallets and Prepaid cards are among the major initiatives towards the reduction of cash-based transactions using smartphones [15]. Digital wallets and prepaid cards also achieve the digitization which is one of the key aspects of FinTech [16]. E-Wallets in Bahrain have been successful and have greatly replaced the physical cards and cash to facilitate electronic transfers and payments through smartphones [17]. The prepaid cards allow the user a flexibility to use money at different locations generally multiple currencies for various purposes such as shopping, travelling, etc., [18]. The advent of FinTech can substantially enhance the interfaces, simplicity, security, and operational performance of both digital wallets and prepaid cards.

4.3 Regulatory Sandbox

One of the primary requirements for FIs is to comply with the regulations set by the regulator such as Central Bank of Bahrain (CBB) in Bahrain. FinTech systems can significantly assist the FIs in developing solutions which can achieve the regulatory requirements such as customer data protection, KYC, onboarding, monitoring, etc. Since FinTech is a relatively new technology, the CBB has taken an initiative to spread awareness and allow FIs who seek to include FinTech solutions, by introducing 'regulatory sandbox'. The regulatory sandbox describes a framework which is set by the regulator (CBB) to assist FinTech start-ups and other such innovators in conducting experiments in an environment under the supervision of CBB [19]. The introduction of regulatory sandbox provides the FinTech start-up companies and other such innovators a provision to test their innovative banking solutions prior to launching such solutions in the market.

4.4 Enterprise Risk Management

Enterprise Risk Management commonly referred to as ERM refers to the risk management process which integrates all types of risks (identification and mitigation) and communicating these risks to the key stakeholders in the company [20]. FinTech inclusion can play a significant role in enhancing the performance of ERM by facilitating risk identification in several areas through state-of-the-art FinTech products. FinTech could also assist in communicating the risks to the stakeholders automatically through its automated communication system further reducing the possibilities of human errors. Moreover, risk mitigation measures/actions can also be included in the FinTech based ERM system.

4.5 KYC

One of the basic operations in the FIs is the KYC documentation collection and verification. Several online and face-to-face processes are in place to collect the KYC documents and accomplish verification of KYC. In the current banking industry, few of the problems associated with the KYC processes are difficulties in completing this legal and regulatory requirement and the cost involved [21]. FinTech through its AI, and machine learning based advancements can largely reduce the KYC related issues and the cost involved.

4.6 Customer Experience

FIs take several measures to improve the customer experience by providing high-end products and services focused on the changing needs of the customers. For FIs, one of the biggest challenges in maintaining high-standards of customer satisfaction, is the bridge between the regulatory requirements and customer satisfaction. The introduction of FinTech-based solutions can allow the FIs to develop solutions which can greatly increase the customer experience such as digital onboarding, online statements, mobile wallets, security, privacy, etc.

5 Issues and Solutions in Fintech

Though there are several benefits and huge scope for developments and use of Fintech in the financial sector, there are possible problems also which include the new business models, frauds, and meeting targets [22]. One of the factors which discourages the FIs towards adopting new financial products is the complexity of the regulations

[23]. There are both challenges, and opportunities associated with Fintech requiring FIs to assess their requirements and take initiatives to stand abreast of the technological developments and market trends. Two of the difficulties for FIs in Bahrain to embrace Fintech in their products and services are the risks of failure or non-adherence to the regulations and the unseen risks of migrating to new technology. Moreover, a clear understanding of the various aspects of Fintech, associated opportunities, related challenges, scope for developments, benefits, etc., are needed to increase migration towards Fintech. The study described in the article is needed to understand the different aspects of Fintech, its scope and status in Bahrain, the challenges faced in embracing Fintech, and the gaps in different research articles towards Fintech.

5.1 Fraud Management

When efficient cybersecurity and other measures are ignored, there is a possibility of increase in the fraudulent activities occurring through a variety of Fintech based solutions. An ignorance or establishment of weak security systems could expose the organization to vulnerability. One of the important aspects of fraud is the fraud triangle which describes the three key reasons of fraud i.e., pressure, rationalization, and opportunity.

5.2 New Business Models

Fintech largely involves the development of new business models reflecting the client's needs and requirements. Some of the problems associated with the new business model's development are reliability issues, meeting accuracy challenges, limited time frame, working on large amount of data, etc.

5.3 Meeting Targets

Meeting targets within the timeframe could compromise the data confidentiality and other key aspects. Developing newer systems and carrying out detailed tests and analysis is a time taking process and should involve approvals at different levels. However, pressure on meeting targets could compromise this and result in inefficient Fintech solution development.

6 Challenges Faced

6.1 Complexity of Regulations

Adhering to regulations is a primary responsibility of the FIs. Since many of the FinTech solutions such as blockchain, crowdfunding, etc., are new, there can be inconsistencies in meeting the regulations. Moreover, the complexity of regulations such as the need to comply with various modules in the regulations, adds challenges on the FinTech-based solutions. If there is a flaw in the FinTech solution related to meeting the regulations, then the FIs could face several risks such as regulatory risk, financial risk, and reputation risk.

6.2 Challenge of Privacy and Security

Since many of the FinTech advancements are focused at working on big-data and online platforms with reduced human involvement, the FIs could face privacy and security related issues. One of the problems associated with using the weak online platforms is the anonymity of users which could trigger fraudulent activities such as fraud transactions, data leaks, etc.

6.3 Challenges in Risk Management

One of the key aspects of the FinTech firms is the use of machine learning in providing automated or robotic advisers to a set of banking operations such as advising on the loan instalments, mortgages, currency changes, etc. One of the high-risks which could result from these automated processes is the probability of error in the calculations. If there is any loss occurring due to the technical issue such as error in currency calculation, or stocks, then the FinTech firms may have to take the responsibility of the potential losses. Hence, there is a need for FinTech firms and FIs utilizing FinTech solutions to understand the risk exposure to their activities and functions.

6.4 Integration Technology Challenge

Integrating the FinTech solutions to the existing tools and systems could be challenging because of the technological gap between the two. Moreover, there could be difficulties in testing the reliability of FinTech-based solutions if the solutions are not well-planned or efficiently tested prior to the launch.

6.5 Managing Client's Diverse Requirements

Few of the major challenges faced by the FIs in the modern world is addressing the dynamic needs of the clients and standing above the competitors. With the advent of FinTech solutions in the market, the competition between the FIs has significantly increased putting more pressure on FIs to incorporate best cutting-edge technology-based solutions in their products and services. This increased competition and diverse client's requirements could pose additional challenges in introducing FinTech-based solutions.

7 Opportunities for FinTech in Bahrain

Financial sector in Bahrain is considered as one of the major sectors of economy accounting for 17.2% of GDP (Gross Domestic Product) and is also known as the country's single largest employer [24]. The government of Bahrain has embraced the FinTech innovations and established FinTech bay which is aimed at encouraging financial sector institutions to converge towards FinTech and has attracted 50 organizations to sign-up to FinTech bay [25]. With over 370 financial sector institutions and a diverse customer market in the country consisting of customers from several countries, there are several opportunities for FinTech firms in the country. Some of the major opportunities for FinTech in Bahrain are.

7.1 Digital Transformation in Bahrain

Bahrain has embraced digital transformation in the country by establishing a dedicated unit for FinTech and innovation activities [26]. The increased awareness about digital transformation in the country and support from the Central bank, provides several opportunities for FinTech firms and such innovation firms to implement FinTech solutions in the country for a range of financial sector's products and services.

7.2 CBB's Regulatory Sandbox

The regulatory sandbox initiative by the Bahrain's central bank provides a virtual space for the licensees, FinTech firms, and other start-ups to test solutions/services which are developed based on technological innovations prior to launch stage [26]. This has opened a safe space for the FinTech firms, licensees, and other start-ups to test their services and solutions without the need to establish an expensive test environment.

7.3 Cashless Payments

The increased use of cashless payments after the pandemic situation, has raised the awareness among the people about the digital products. This provides a good opportunity for investments in FinTech and migration to FinTech-based solutions.

7.4 FinTech Bay

The establishment of FinTech bay which is a leading FinTech hub in the region has opened new set of opportunities for FinTech firms, start-ups, and innovators to invest/migrate businesses into FinTech-based technologies.

7.5 Scope for Financial Inclusion

Bahrain is a country which has embraced technological advancements in various sectors and has encouraged diversity in products and services across various sectors. Moreover, the pandemic situation in the country has shifted many people's mind towards digital products and services. This gives opportunities for FinTech firms to cover the large number of residents in the country.

7.6 GIG Workers in Bahrain

Through easy-to-use banking interfaces and digital client onboarding methods, FinTech firms can substantially increase the client base by targeting the gig workers. The state-of-the-art technology-based user interfaces built using highly-secured channels can attract several gig workers and other such customers to use the banking and finance products easily.

7.7 Online Customer Onboarding Through Face-Recognition

The Central Bank of Bahrain has provided guidelines and rules on onboarding customers online i.e., non-face-to face and has recommended face recognition as one of the identification methods [27]. This provides a ready-to-use platform for the FinTech firms to capture the market by providing high security online customer onboarding solutions.

7.8 Partnerships

FinTech firms and start-ups in Bahrain also have an opportunity to establish partnerships with the existing FIs also referred to as licensees in Bahrain, as third-party service providers. The FinTech firms can establish partnership with several licensees providing similar services such as mobile wallets, remittance, payroll, etc., and provide FinTech-based services and solutions to the licensees. This way the FinTech firms can substantially cover the financial market in the country.

8 Discussions and Contributions

Based on the analysis of the literature review, development of the investment model, understanding of the crucial aspects of FinTech services/solutions, and opportunities and challenges for FinTech in Bahrain, the below recommendations are presented as research contributions.

Adhering to the Proposed Investment Model

FinTech firms, licensees, and other such start-ups and innovators need to consider the investment model proposed in this study work. They need to consider all the aspects in detail about various elements in the proposed investment model. A description on these key aspects is provided in the study work.

Implementing Efficient Security Algorithms

Few of the common problems and challenges associated with FinTech inclusion are potential frauds, unauthorized access, third-party risk, etc. Developing and including efficient security algorithms reflecting the data protection and security issues could significantly reduce these problems and further improve the image of FinTech solutions and services in this aspect as well.

Ensuring Completeness to Regulations

One of the major challenges faced by the FinTech firms is in ensuring completeness to regulations. Not complying with the regulations is a serios threat for the financial sector institutions in Bahrain. Hence, the FinTech firms and licensees must ensure that the FinTech solution/service fully adheres to the regulations.

Improving System Quality and Products/Services Standards

Improving the service quality and product/service standards involves several areas such as system responsiveness, accuracy, reliability, quality of data, etc. Improving the service and standards is essential to standout in the competitive market in Bahrain.

Data Protection and Transparency

With the PDPL (Personal Data Protection Laws) in place in Bahrain, there is a stronger need to ensure data protection and also transparency in data management. The FinTech firms should consider their data management processes and adhere to the relevant laws to ensure continuity in the business and mitigating associated risks.

Client/Customer Feedback System

Understanding client's response to the FinTech products/services and the changing needs of customers towards banking and finance, require analysis of customer/client feedback. Hence, firms should include a provision to collect and analyse the client's feedback.

Partnerships with Other Licensees

Partnering with the licensees in Bahrain can allow the FinTech firms an ability to expand their business in the country and by collaborating with the FinTech firms, the licensees can also get several opportunities such as obtaining better insights from the large data, an innovative environment, improved customer experience through efficient interfaces, low operation costs, etc.

9 Conclusion

FinTech which is popularly known to be a combination of two words i.e., finance and technology, has become one of the biggest disruptions in the financial industry of this modern world. FinTech has opened several opportunities for financial sector institutions to use the technology in improving the products and services offered to the customers. Bahrain has embraced the potential of FinTech services and solutions on the financial sector by establishing a FinTech hub known as FinTech bay. The financial regulatory authority of Bahrain has also initiated support and awareness programs such as the development of 'regulatory sandbox' to encourage the growth of FinTech in the country. The study work is focused at understanding the various key aspects of FinTech in Bahrain and aims to provide a descriptive analysis of these aspects to assist the FinTech firms and licensees in capitalizing the benefits of FinTech. One of the setbacks for FinTech firms is the lack of investments or fundings which is commonly due to lack of clarity in the business for the investors. Considering this aspect of investment, an investment model is proposed in the study work to encourage FinTech investors.

Since the FinTech firms and start-ups are new to the country, they face challenges because of the potential security issues, possibility of FinTech operator fraud, complexity, opaque environment, etc. Because of these challenges and other issues, FinTech despite the support from the government and the regulator, has not able to gain the market as expected. Some of the reasons associated with this are lack of awareness about FinTech products and services and lack of understanding of the various key factors associated with FinTech. The study work presented here comprehensively discusses the key aspects of FinTech using the FinTech investment model. The study involves conducting literature review of several articles and Central Bank of Bahrain's data and conceptual understanding of the FinTech in Bahrain, to gain an understanding of the FinTech ecosystem, verticals of FinTech, opportunities, challenges faced, etc.

The study addressed some of the key challenges and opportunities for FinTech firms, start-ups, and other innovators in Bahrain. The study proposed an investment model which could greatly assist FinTech firms in understanding the various aspects of FinTech-based investments in the country. The investment model proposed in the study work is inclined at considering factors such as opportunities and benefits, challenges and issues, FinTech ecosystem, areas of FinTech, etc., prior to investing in FinTech solutions and services. This model is also helpful for licensees as well as it gives a comprehensive picture of various critical aspects of FinTech which can be utilized in their operations. Moreover, the study provides a good foundation for other researchers to conduct further research on the FinTech-based products and services in Bahrain.

References

1. Sung, A., Leong, K., Sironi, P., O'Reilly, T., McMillan, A.: An explanatory study of the FINTECH (Financial Technology) education and retraining in UK. J. Work Appl. Manag. **11**, 187–198 (2019)
2. Buckley, R., Arner, D.W., Barberis, J.N.: The evolution of fintech: a new post-crisis paradigm? Georgetown J. Int. Law **47**, 1271–1319 (2016)
3. AbdulKarim, A.M.: Bank users motivation for adoption of fintech services: empirical evidence with TAM in Kingdom of Bahrain. iKSP J. Innov. Writ. **2**, 1–11 (2020)
4. Boyd, A.: 10 FinTechs that failed (and why?). https://finty.com/us/research/fintech-failures/, Last accessed 21 Dec 2021 (2021)
5. Monetary Authority of Singapore (2020) Balancing the risks and rewards of FinTech developments. BIS Emerging Market Deputy Governors Meeting, pp. 279–291
6. Nicolle, E.: FinTech pays 'heavy price' for profit failures, says UK start-up boss. Financial News. https://www.fnlondon.com/articles/fintech-pays-heavy-price-for-profit-failures-says-uk-startup-boss-20210928. Last accessed 21 Dec 2021 (2021)
7. Shevlin, R.: Why FinTech start-ups fail? Forbes. https://www.forbes.com/sites/ronshevlin/2019/07/29/why-fintech-startups-fail/?sh=6aeb60066440. Last accessed 21 Dec 2021 (2019)
8. Lee, I., Shin, Y.J.: FinTech: ecosystem, business models, investment decisions, and challenges. Bus. Horiz. 35–46 (2018)
9. Demirguc-Kunt, A., Klapper, L, Singer, D., Ansar, J.H.: Measuring financial inclusion and the FinTech revolution. The Global Findex Database (2017)
10. Dupas, P., Karlan, D., Robinson, J., Ubfal, D.: Banking the unbanked? Evidence from three countries. Am. Econ. J. Appl. Econ. 257–297 (2018)
11. CBB: Banking. https://www.cbb.gov.bh/banking/. Last accessed 19 Dec 2021 (2021)
12. Deshpande, A.: AI/ML Applications and the potential transformation of FinTech and finserv sectors. In: 13th CMI Conference of Cybersecurity and Privacy—Digital Transformation—Potentials and Challenges, pp. 1–6 (2020)

13. Selim, M.: Application of FinTech, machine learning, and artificial intelligence in programmed decision making and the perceived benefits. International Conference on Decision Aid Sciences and Application, pp. 495–500 (2020)
14. Yildirim, N., Varol, A.: A research on security vulnerabilities in online and mobile banking systems. In: 7th International Symposium on Digital Forensics and Security (ISDFS), pp. 1–5 (2019)
15. Fainusa, A.F., Nurcahyo, R., Dachyar, M.: Conceptual framework for digital wallet user satisfaction. In: 6th International Conference on Engineering Technologies and Applied Sciences, pp. 1–4 (2019)
16. Maindola, P., Singhal, N., Dubey, A.D.: Sentiment analysis of digital wallets and UPI systems in India post demonetization using IBM Watson. In: International Conference on Computer Communication and Informatics, pp 1–6 (2018)
17. AlKubaisi, M.M., Naser, N.: A quantitative approach to identifying factors that affect the use of E-wallets in Bahrain. J. Siberian Federal Univ. Human. Soc. Sci. 1819–1839 (2020)
18. Mukherjee, S.D., Phirangi, S.A.: Non-performing assets and its importance with reference to Axis Bank Ltd, Mumbai. In: International E Conference on Adopting to the New Business Normal—The Way Ahead, pp. 1–19 (2020)
19. Sumaya, A.: The structural constraints of entrepreneurship in Bahrain. Rice University's Baker Institute for Public Policy (2018)
20. Razali, A.R., Tahir, I.M.: Review of the literature on enterprise risk management. Bus. Manag. Dyn. 8–16 (2011)
21. Yadav, R., Chandak, R.: Transforming the know your customer (KYC) process using blockchain. In: International Conference on Advances in Computing, Communication and Control (ICAC3), pp. 1–5 (2019)
22. Ng, A.W., Kwok, B.K.B.: Emergence of Fintech and cybersecurity in a global financial centre-strategic approach by a regulator. J. Financ. Regul. Compliance **25**, 422–434 (2017)
23. Treleaven, P.: Financial regulation of Fintech. J. Financ. Perspect. Fintech **3**, 1–14 (2015)
24. Bahrain Association of Banks: Financial and banking sector in Bahrain. https://banksbahrain.org/banking-in-bahrain/. Last accessed 20 Dec 2021 (2021)
25. Razzaque, A., Cummings, R.T., Karolak, M., Hamdan, A.: The propensity to use FinTech: input from bankers in the Kingdom of Bahrain. J. Inf. Knowl. Manag. (2020)
26. CBB: FinTech & Innovation. https://www.cbb.gov.bh/fintech/. Last accessed 20 Dec 2021 (2021)
27. CBB: Financial Crime Module. https://www.cbb.gov.bh/wp-content/uploads/2021/05/Consultation_Vol-1_FC_E-KYC.pdf. Last accessed 21 Dec 2021 (2021)

Toward Sustainable Smart Cities: A New Approach of Solar and Wind Renewable Energy in Agriculture Applications

Nurhasliza Hashim, Tiffiny Grace Neo, M. N. Mohammed, Hakim S. Sultan, Adnan N. Jameel Al-Tamimi, and M. Alfiras

Abstract Encouraging renewable energy (RE) in agricultural technology, like wind, solar, hydro-driven water pumps, solar water heaters, solar dryers for post-harvest processing, greenhouse heating and cooling, and lighting applications, might lessen climate change and its effects. However, switching the agriculture from non-renewable to renewable energy based presents a number of difficulties. Remote management, monitoring, and control of distributed solar and wind energy resources are now possible due to advancements in Internet of Things (IoT) technology. For areas of the world that experience power shortages, designing solar and wind energy systems based on the Internet of Things is crucial. An IoT-based solar, wind, and renewable energy system for a smart farm was suggested in this study. A prototype was created, constructed, and tested with the intention of validating the proposed system. Additionally, it was developed an interfaced system for wireless monitoring to enable a remote user to conveniently check on a farm from a computer or a mobile device.

Keywords Renewable energy · Sustainable agriculture · Sustainable development · Sustainable energy · Energy and agriculture · IoT · Smart cities · SDG 2 · SDG 7 · SDG 11

N. Hashim · T. G. Neo
Faculty of Information Sciences and Engineering, Management and Science University, Shah Alam, Selangor, Malaysia

M. N. Mohammed (✉)
Mechanical Engineering Department, College of Engineering, Gulf University, Sanad 26489, Kingdom of Bahrain
e-mail: dr.mohammed.alshekhly@gulfuniversity.edu.bh

H. S. Sultan
College of Engineering Karbala, University of Warith Al-Anbiyaa, Karbala, Iraq

A. N. Jameel Al-Tamimi
College of Technical Engineering, Al-Farahidi University, Baghdad, Iraq

M. Alfiras
Electrical and Electronic Engineering Department, College of Engineering, Gulf University, Sanad 26489, Kingdom of Bahrain

© The Author(s), under exclusive license to Springer Nature Switzerland AG 2024
A. Hamdan and E. S. Aldhaen (eds.), *Artificial Intelligence and Transforming Digital Marketing*, Studies in Systems, Decision and Control 487,
https://doi.org/10.1007/978-3-031-35828-9_47

555

1 Introduction

Cities all across the world are considering ways to use smart technology to improve citizen lives, spur economic growth, and improve civic efficiency [1–3]. Increased automation and a wide range of applications require pervasive connectivity so that information may be acquired from smart devices and sensors incorporated into roads, power grids, structures, and other assets. Urban agriculture is the growing term for agricultural and food manufacturing around and within cities. Rooftop gardens, intensive horticultural production, small lots for agricultural production, community gardens, and larger-scale farm operations are just a few examples of the remarkably diverse types of urban agriculture. All of these agricultural practices and the methods they use to produce food have the potential to be sustainable. Consumers' values on nutritious food and the advancement of sustainable agriculture are increasingly being included into sustainability. This has given rise to a number of informal as well as official networks and social groups for integrating farmers and consumers [4]. Additionally, various forms of this urban agriculture support purposes aside from food production, such as giving immigrants the chance to grow their own food and integrate into urban society more successfully, aiding in the management of "green" spaces in the city and its environs in some cases, and giving schoolchildren the chance to learn about healthy agriculture, among other purposes [5]. In reality, some of the technological advancements that give smart cities their high intelligence, such as IoT connectivity, autonomous vehicles, and sensors, were developed on farms. Based on real-time information from a network of sensors, smart grids in urban areas distribute electricity when and where it is required. When there are shortages or outages, the system promptly reports them, and intelligent relays and switches rapidly reroute power to avoid the issue. Everything is intended to use less energy while enhancing the reliability and resilience of the electric system. Whether renewable or not, all energy sources have an influence on the environment, including fossil fuels [6, 7]. The world's continued reliance on fossil fuels contributes to climate change by releasing greenhouse gases and leaving behind hazardous particles that have an impact on human health. The benefits of renewable energy over non-renewable energy include reduced usage of land and water, decreased air and water pollution, and nil or lower greenhouse gas emissions [8, 9]. Due to the development of renewable energy sources and the depletion of finite resources, the energy sector has seen a significant transformation during the past few decades. According to the International Energy Agency, the IoT economy and renewable energy sources are predicted to grow by 43% by 2022. IoT is crucial to the development of the energy sector and without IoT, it is difficult to imagine a world with endless resources. Solar energy is the most plentiful renewable energy source and is accessible both directly and indirectly. The Sun radiates energy at a rate of 3.8×1023 kW, of which the Earth absorbs about 1.8×1014 kW [5, 12]. The amount of solar energy accessible for thermal purposes is so substantial. The usage of resources, the impact on the environment from the production of pollutants, and the consumption of traditional energy supplies have all contributed to the importance of energy-related issues over the past ten years. PV solar cells, which directly convert light energy into electricity through internal physical processes, represent one way to generate clean electricity. There is no emission of contaminants throughout these procedures. With velocities proportional to the pressure gradient, wind energy refers to the flow of air masses from high atmospheric pressure regions to nearby low pressure regions. When compared to nearby air masses over land areas, the air masses over oceans, seas, and lakes stay cool during the day. For the time being, wind power is a dependable and established technology that can provide electricity at a price that is competitive with coal and other energy sources like nuclear, which have been deployed over the course of the past 10 years or so, roughly, starting in 2005 [10, 11]. To smoothen the energy delivered by wind turbine, L. M. S. de Siqueira and W. Peng reviewed the control approaches utilized by the storage to enhance the energy efficiency [12]. The integrated power system for catamaran ship has been optimized by Setiawan et al. [13]. The performance of PV and generator was discussed in the study. Combining solar with wind energy deliver zero pollution, no gas emissions, less issues with management of waste [14]. Numerous investigations were carries out to generate electrical power from renewable sources either solar or wind power. The individual renewable resources become an issue due to intermittent operation and operation cost that affect the output yield. To overcome these issues, two or more renewable energy are hybridized to maintain higher power efficiency [15]. An IoT-based solar, wind, and renewable energy system for a smart farm was proposed in this study. A prototype was created, constructed, and tested with the intention of validating the suggested system. Additionally, it was developed an interfaced system for wireless monitoring, enabling a remote user to conveniently check on a farm from a computer or a mobile device.

2 Materials and Methods

The proposed system consists of numerous hard and software components combined to produce the hybrid solar-wind renewable energy system. The description of each element is illustrated below: A 5-V 10-W polycrystalline silicon solar panel was used in this study in a rectangular shape. Three-propeller blades were selected for a wind turbine. These two components were integrated where the former is solar based driven while the latter is wind-based driven. The energy generated from the two renewable sources was stored in 3.6 V rechargeable batteries type VRLA (Valve Regulated Lead Acid). This battery is available in the market and does not need maintenance and it deliver stable electricity to the load upon charging. An Arduino UNO was attached to the battery. The Arduino is one of the core hard parts in the proposed system where it uses sensor for detection of soil moisture. The battery is also connected with the ESP8266 WIFI module and the readings will be interpreted and appeared on LCD display. The ESP8266 WIFI Module is a self-contained SOC with an integrated TCP/IP protocol stack that enables our WIFI network to reach any microcontroller. The ESP8266 is capable of either hosting an application or discharging from another application processor all Wi-Fi networking functions. A

LED is a light emitting diode (LED) product that is assembled into a lamp for use in lighting fixtures. LED lamps have a lifespan and electrical efficiency that is several times better than incandescent lamps. At the bottom of the Vertical Wind, the three-phase AC motor is located and attached to the three-phase bridge rectifier to adjust the turbine to produce electricity. The three solid state diodes and transistor supply power through a fixed DC supply. Table 1 shows the individual parts of the proposed system that will be assembled to produce the final hybrid system. The Software that are used to create circuit diagram, simulation, source code and testing of proposed system are: Blynk, Proteus, and Arduino Software IDE. Blynk is a platform for the Internet of Things (IoT) that is designed to simply develop mobile and web applications for the Internet of Things. It is linked to a hundred open hardware models such as Arduino, ESP8266, ESP32, Raspberry Pi, and related MCUs, iOS, and Android drag-and-drop IOT smartphone apps. Proteus software was used to design the circuit and simulation of operating circuits for the new dual power generation method (Fig. 1). The Arduino IDE program enables the NodeMCU to interpret the received data from the soil moisture sensor. The flowchart design will show the entire process of the monitoring system starting from the beginning of this project with the patient to the end of this flowchart with the desired output of the system. The entire process of the flowchart can be seen in Fig. 2.

Figure 3 shows the block diagram of the workflow of the proposed system. When the wind turbine and solar panel turn on, the battery will be charged by a delivered DC vol. The Arduino UNO is connected to soil moisture sensor, WIFI module and display. When the soil moisture detects the dryness of the soil, the LED illuminates and the water pump turns on to sprinkle the water on the dried soil. After that, the information is sent to the data cloud when the WIFI module is connected to a hotspot. Then the data cloud will be interacting with Blynk phone application and we can see the graph reading input of the solar panel and windmill together with the battery power on the LCD display.

3 Results and Discussions

This section presents the final result of the proposed prototype system. After assembling of hybrid solar-wind components, the hybrid system subjected to testing and assessment phase. Figure 3 shows the integrated system with LCD display. The prototype is tested and the result are recorded for the three different readings which is for solar panel, windmill and the power that been generate. The x-axis represents the daily hours while solar, wind, and battery voltages are placed on the y-axis. For this application we can observe the live reading for the current and the voltage that generate. Beside that we can see the reading of the voltage solar, voltage turbine and the voltage battery during the project is working. During the day, there are some fluctuations for solar and wind generated electricity due to fact that sun radiation intensity varies along the day and affected by whether conditions while wind energy is influenced by the wind speed and other factors as shown in Fig. 4.

Table 1 Components of the proposed system

No.	Image	Material
1		Solar panel
2		Wind turbine
3		3.6 V VRLA battery
4		Arduino UNO
5		ESP8266 WIFI
6		LED
7		Three-phase AC motor
8		Bridge rectifier

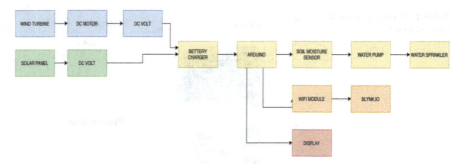

Fig. 1 The block diagram of the developed system

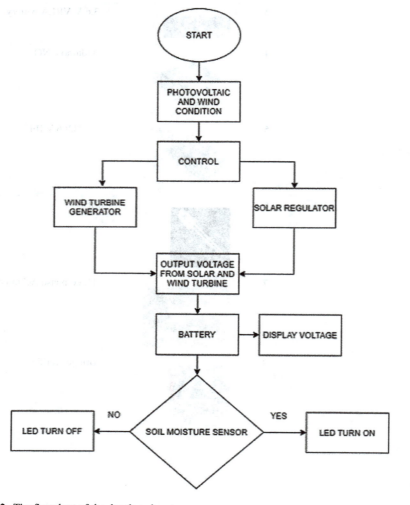

Fig. 2 The flowchart of the developed system

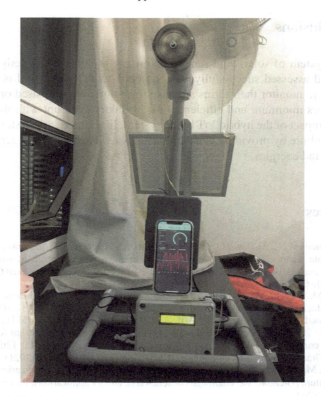

Fig. 3 The hybrid solar-wind system

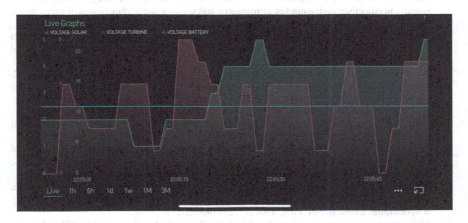

Fig. 4 The power monitor application interface

4 Conclusions

A hybrid system of solar-wind renewable energy prototype was designed, implemented, and assessed successfully. The system was IoT based and is very useful for farmers to monitor their farms and take irrigation decision based on online soil data. It offers minimum and efficient water sprinkling amount with the aid of the beneficial impact of the hybrid IoT based system. In addition, this system empowers green agriculture by providing clean and undepleted energy sources represented by solar and wind energies.

References

1. Mohammed, M.N., Al-Rawi, O.Y.M., Al-Zubaidi, S., Mustapha, S., Abdulrazaq, M.: Toward sustainable smart cities in the Kingdom of Bahrain: a new approach of smart restaurant management and ordering system during covid-19. In: 2022 IEEE International Conference on Automatic Control and Intelligent Systems (I2CACIS). IEEE, pp. 7–11
2. Alfiras, M., Yassin, A.A., Bojiah, J.: Present and the future role of the Internet of Things in higher education institutions. J. Positive Psychol. Wellbeing 6(1), 167–175 (2022)
3. Mohammed, M.N., Hazairin, N.A., Arif, A.S., Al-Zubaidi, S., Alkawaz, M.H., Sairah, A.K., Yusuf, E.: 2019 novel coronavirus disease (covid-19): Toward a novel design for disinfection robot to combat coronavirus (covid-19) using IOT based technology. In: 12th Control and System Graduate Research Colloquium (ICSGRC), pp. 211–216. IEEE (2021)
4. Rahman, M.M., Khan, I., Field, D.L., Techato, K., Alameh, K.: Powering agriculture: present status, future potential, and challenges of renewable energy applications. Renew. Energy 188, 731–749 (2022)
5. Pascaris, A.S., Schelly, C., Burnham, L., Pearce, J.M.: Integrating solar energy with agriculture: industry perspectives on the market, community, and socio-political dimensions of agrivoltaics. Energy Res. Soc. Sci. 75, 102023 (2021)
6. Hashim, N., Mohammed, M.N., Selvarajan, R.A., Al-Zubaidi, S., Mohammed, S.: Study on solar panel cleaning robot. In: IEEE International Conference on Automatic Control and Intelligent Systems (I2CACIS). IEEE, pp. 56–61
7. Mohammed, M.N., Alghoul, M.A., Abulqasem, K., Mustafa, A., Glaisa, K., Ooshaksaraei, P., Sopian, K.: TRNSYS simulation of solar water heating system in Iraq. In: Proceedings of the 4th WSEAS International Conference on Energy and Development-Environment-Biomedicine, pp. 153–156 (2011)
8. Khan, Z.A., Imran, M., Altamimi, A., Diemuodeke, O.E., Abdelatif, A.O.: Assessment of wind and solar hybrid energy for agricultural applications in Sudan. Energies 15(1), 5 (2022)
9. Zainuddin, N.F., Mohammed, M.N., Al-Zubaidi, S., Khogali, S.I.: Design and development of smart self-cleaning solar panel system. In: 2019 IEEE International Conference on Automatic Control and Intelligent Systems (I2CACIS), pp. 40–43. IEEE, (2019)
10. Acosta-Silva, Y.D.J., Torres-Pacheco, I., Matsumoto, Y., Toledano-Ayala, M., Soto-Zarazúa, G.M., Zelaya-Ángel, O., Méndez-López, A.: Applications of solar and wind renewable energy in agriculture: a review. Sci. Prog. 102(2), 127–140 (2019)
11. Abdulrazaq, M., Sapari, N., Zaenudin, M.: Solar Water Heating System (SWHS) in Iraq via Trnsys Simulation, 1st edn. Scholar's Press (2019)
12. de Siqueira, L.M.S., Peng, W.: Control strategy to smooth wind power output using battery energy storage system: a review. J. Energy Storage 35, 102252 (2021)
13. Setiawan, B., Putra, E.S., Siradjuddin, I., Junus, M.: Optimisation solar and wind hybrid energy for model catamaran ship. IOP Conf. Series Mater. Sci. Eng. 1073(1), 012044. IOP Publishing (2021)

14. Qadir, Z., Khan, S.I., Khalaji, E., Munawar, H.S., Al-Turjman, F., Mahmud, M.P., Le, K.: Predicting the energy output of hybrid PV–wind renewable energy system using feature selection technique for smart grids. Energy Rep. **7**, 8465–8475 (2021)
15. Lawan, S.M., Abidin, W.A.W.Z.: Wind Solar Hybrid Renewable Energy System, 1st edn. IntechOpen Limited, London, UK (2020)

Digital Transformation Towards Sustainability in Higher Education: A New Approach of Virtual Simulator for Series and Parallel Diodes for a Sustainable Adoption of E-Learning Systems

**Tan Chor Kuan, Khairul Huda Yusoff, M. N. Mohammed,
Adnan N. Jameel Al-Tamimi, Norazliani Md Sapari,
Firas Mohammed Ibrahim, and M. Alfiras**

Abstract Virtual laboratory simulation for electronic circuit is essential for students to simulate a circuit anytime and anywhere unlike physical laboratory. In this paper, a virtual laboratory for basic electronic courses based on Java is developed. An interactive virtual simulator focusing on the topics of series and parallel diodes is designed. All electronic components are virtualized using Java Abstract Window Toolkit (AWT) which enabling users to build and draw circuits on demand. Then, users can obtain results immediately such as voltage drop, current and power values. Users will also be able to observe the simulation for current flow and voltage drop in the circuit. A number of default experiments focusing on series and parallel diodes

---

T. C. Kuan · K. H. Yusoff
Faculty of Information Sciences and Engineering, Management and Science University, Shah Alam, Selangor, Malaysia

M. N. Mohammed (✉)
Mechanical Engineering Department, College of Engineering, Gulf University, Sanad 26489, Kingdom of Bahrain
e-mail: dr.mohammed.alshekhly@gulfuniversity.edu.bh

A. N. Jameel Al-Tamimi
College of Technical Engineering, Al-Farahidi University, Baghdad, Iraq

N. M. Sapari
School of Electrical Engineering, Faculty of Engineering, Universiti Teknologi Malaysia, Johor Bahru, Malaysia

F. M. Ibrahim
Legal Studies Department, College of Law, Gulf University, Sanad 26489, Kingdom of Bahrain

M. Alfiras
Electrical and Electronic Engineering Department, College of Engineering, Gulf University, Sanad 26489, Kingdom of Bahrain

© The Author(s), under exclusive license to Springer Nature Switzerland AG 2024
A. Hamdan and E. S. Aldhaen (eds.), *Artificial Intelligence and Transforming Digital Marketing*, Studies in Systems, Decision and Control 487,
https://doi.org/10.1007/978-3-031-35828-9_48

were available for users. This virtual simulator is suitable for all students that are interested in learning the fundamental of electronic circuit using their computer while not worrying on issues such as space needs, equipment needs, scheduling logistics, safety concerns and resource sharing. Adopting digital technologies in higher education is integral to developing sustainable teaching and learning processes, such as those found with online simulation tools like Virtual Simulator for Series and Parallel Diodes as an invaluable asset in supporting sustainable e-learning; creating more resilient digital infrastructure which supports sustainable teaching/learning processes in higher education institutions. This research contributes to our collective knowledge on optimizing digital transformation for sustainability purposes within higher education environments.

Keywords Digital transformation · Virtual simulator · E-learning; series and parallel · Electronic circuit · Covid-19 · SDG 3 · SDG 4 · SDG 15

1 Introduction

Covid-19 pandemic has caused a disastrous impact to the society and leave a deep scar to the global economy. The sudden shift from traditional methods of learning to online mode of learning has created a huge challenge across the educational institutes where the educators and students have to adapt to the new norm such as remote teaching and learning as well as virtual laboratory simulation [1–4]. These solutions have become the temporary replacement to the traditional methods of learning as social distancing is one of the ways to prevent the spread of Coronavirus following the advice of public health officials. As a result, educational institutions adopted e-learning through remote education which able to prevent students from experiencing difficulty during school closure [5–8]. One of the major e-learning tools used in engineering sector is virtual laboratory simulation. This is because virtual laboratory simulation able to assist students in learning the fundamentals and understand advanced topics [9, 10]. Since virtual laboratory simulation can be access anytime unlike the physical laboratory, students can do any experiment multiple times if they have difficulties understanding the concept of the experiment modules. Another benefit of virtual laboratory simulation is that student can follow its own pace of study without being restricted from the scheduled timetable. Moreover, many students can access to a single experiment at the same time unlike physical laboratory where students have to share equipment. In the midst of the Covid-19 pandemic, educational institutes were instructed to conduct the teaching and learning in online mode to curb the spread of Coronavirus. The changing conditions led our schools and colleges to adopt a different approach practically overnight by transitioning from face-to-face lectures to online platforms of learning [11]. One of the major issues rises is students were not able to conduct lab practical experiment because physical laboratories were closed due to lockdowns. The lockdown minimizes the physical contacts from students in the wake of Covid-19 pandemic in order to curb the spread of Coronavirus. Besides, another issue is expensive and advanced laboratory equipment remain challenging in educational institutes [12]. This could lead to a various of issues such as sharing of laboratory equipment, fixed physical laboratory schedule, faulty equipment. It will also take time to troubleshoot faulty equipment which leads to poor understanding from students' learning outcome. To solve these issues, a virtual laboratory simulation is introduced to avoid physical contacts in the wake of Covid-19 pandemic. Other than that, it allows multiple students to access the same experiment at the same time without any time constraint. Virtual laboratory simulation also allows student to conduct experiment multiple times to deepen the understanding of concepts in experiment modules. This project mainly focuses on the design and development of virtual laboratory simulation on basic electronics specifically on the topic series and parallel diodes.

2 Literature Review

There are a few literature papers discussing on virtual simulator for electronic circuits developed by several universities which is similar to this project. Virtual Labs, a virtual laboratory simulation using free open-source software, HTML5 and JavaScript, is developed by Indian Institute of Technology Kharagpur which provides a set of experiments for the course of basic electronic devices and circuits that could aid both the teachers and students during the teaching and learning session [13]. Virtual Intelligent SoftLab, a virtual lab for Encoding Techniques, is developed by Kathane and Gundalwar [14]. The program of SoftLab is developed in Visual Basic. SoftLab is a flexible laboratory environment which links the physical laboratory experiment with its theoretical simulation model within a unified and interactive environment. SEMMLABS, a software for the development of virtual practical work, is developed by the University of Lille which the production of the virtual labs is based on the images that are shoot during manipulating the experiment [9]. A series of different images at various parameter values for the experiment will be taken and made into a sort of interactive web-based application using Microsoft Silverlight. All three virtual simulators have the features for the users to manipulate the variables for component on electronic circuits and observe output on multimeter. However, in Virtual Labs and Virtual Intelligent Softlabs lacks the features of simulation for current flow and voltage drop in open circuit. Lastly, all three virtual simulators reviewed does not have the feature for user to draw circuit. These virtual simulators only support specific type of circuits which only allows the user to join the connecting wires from one component to another. Thus, the virtual simulator on series and parallel diodes that will be developed will consists of the parameters such as user can manipulate different values of variables, draw circuit, presence of simulation for current flow and voltage drops in open circuit as well as observe output.

3 Methodology

A *Project Design*

The development of the project utilizes the Eclipse IDE to code the source code for the virtual simulator shown in Fig. 1. Eclipse is an integrated development environment that is used in computer programming to code many different programming languages. One thing to note that, Eclipse is written mostly in Java programming language and for Java application. The Java frameworks that will used to develop the virtual simulator for series and parallel diodes is Abstract Window Toolkit (AWT). The framework is primarily used to design the graphical user interface of the virtual simulator, draw electronic components, simulate current flow and voltage drop and others. Based on specific topic for the basic electronic course which is series and parallel diodes, the virtual simulator was designed to be a drawing board for schematic design of electronic circuits focusing on DC based circuits. This allows the students to connect components such as DC voltage sources, resistors, diodes, wires, and ground on the drawing board. The virtual simulator makes use of Java framework like Abstract Window Toolkit (AWT) to design the graphics user interface and simulate the electronic circuits. Block diagram is one of the key elements in the project development. It is a diagram of a system which consists of blocks connected by lines that show the relationships with the blocks. In this block diagram, three important segments namely the input source, process and output source of the virtual simulator for series and parallel diodes is shown in Fig. 2.

The first segment of the project aims on the input source for software which allows the user to draw a circuit or choose a default series and parallel diodes experiment circuit. Another input source for the software is allowing the user to change different

Fig. 1 Eclipse IDE

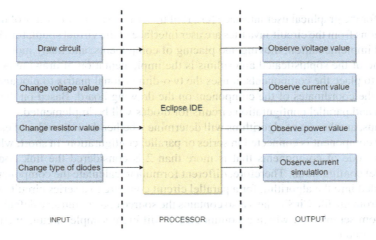

Fig. 2 Block diagram of virtual simulator for series and parallel diodes

parameters of the components such as voltage value, resistor value, types of diodes, etc. Next, the second segment of the project development is the processor development. In this segment, the source code written in Java language will be compiled and run in the Eclipse IDE. Lastly, the output source is the third segment of the project development.

The output source will allow the user to observe the resulted output such as voltage, current and power value from the graphical user interface. To add on, user can also observe how the current simulate in the software.

The virtual simulator software application enables users to assemble and draw circuits on a drawing board and simulate these circuits to obtain the outputs of the components such as results of current, voltage and power. The electronic circuit components were designed to be choose from the component menu tab bar or by using right click mouse button. After choosing the component, user can move and place the component on the drawing board. When circuit is completed and correct, user will be able to observe the results and the simulation of current and voltage drop. In addition, user can also save the circuit into text file and open the file in the future for study reference. A few of the main elements required to design the virtual simulator are user interface, menu bar, drawing board, functional buttons, electronic components and output results.

B *Project Implementation*

The implementation of the project can be separated into four aspects in Java files such as circuit, edit dialog, import export and electronic component.

1. *Circuit*

The main purpose of the circuit Java files is for the development of the drawing board connection algorithm for the virtual simulator. Besides that, it is also the main source code for the graphical user interface (GUI) of the virtual simulator. A few of the main functions from the circuit Java files are user interface of the virtual simulator, drawing board implementation, moving and placing of components, and many more.

One of the sophisticated algorithms is the implementation of drawing board for users to place the components, it uses the two-dimensional matrix to plot and determine the coordinates of the component on the drawing board. Based on the title, a series and parallel configuration circuits for diodes will be implemented.

Thus, another set of algorithms will determine whether the link nodes of each side of the component is connected in series or parallel configuration. In short, whenever a link node of components that is more than 2 is considered the link node is in parallel configuration. Therefore, different formula to calculate the components will be coded into the algorithm for a parallel circuit compare to a series circuit. In addition, from the file CirSim.java also contains the source code to retrieve default circuit file from setuplist.txt which accommodate 11 different sample circuits including a blank circuit.

2. *Electronic component*

The electronic components that are designed and implemented such as DC voltage, diode, ground, LED, resistor and wire. All these components have their own individual Java files which have similar functions such as setting default values of component, draw the graphics of the component, acquire components' values. In addition, each of the components have their own keyboard shortcuts bind to the respective component such as resistors is letter R, diode is letter D, wire is letter W and so on. These enables the user to another alternative ways to choose the components and place it on the drawing board for analysis.

3. *Edit dialog*

The main functions of implementing edit dialog Java files are to allow the user to make necessary changes to the default values of the components. For example, after a user have place a resistor on the drawing board, the default value of the resistor will be set as 100 Ω. If the user wants to change the value of the resistance, user can click on the component and an edit dialog will prompt the user to change and update the component value. These edit dialog Java files were implemented for user to change the values of the component with ease.

4. *Import export*

As for the import export Java files, these files were implemented using Java file dialog which enable the features of reading from a file and write into a file. For example, when users wanted to save a file of the circuit into their computer for future study reference, it will save and write the file in text format with all the necessary information of the circuit such as each of the components' corresponding letter followed by x and y coordinates, and lastly the values of the component.

On the other hand, if user wanted to open a pre saved circuit file, it will open and read the text file with circuit details then placed the necessary components based on

Digital Transformation Towards Sustainability in Higher Education ...

the component's corresponding letter, x and y coordinates and component's value referring to the circuit text file. Besides that, a few of pre-saved sample files were implemented on the virtual simulator for user to choose and start analysing the circuits without designing and making a new circuit from scratch. These sample circuit files were stored in a text file for the virtual simulator to access.

C *Graphic User Interface (GUI)*

As shown in Fig. 3, the title of the virtual simulator is shown at the top left corner of the application. The three standard buttons of software application such as minimize, maximize and close buttons is located at the top right corner of the application. Below of the application title is the menu bar for the software. It consists of file, edit, options, components and circuits menus.

Moving down from the menu bar, a large black colour canvas is the circuit drawing board which allows users to design and draw the circuit using available components in the virtual simulator. Besides that, when the user pointed the mouse cursor on any component, the output parameter can be obtained from the bottom right corner of

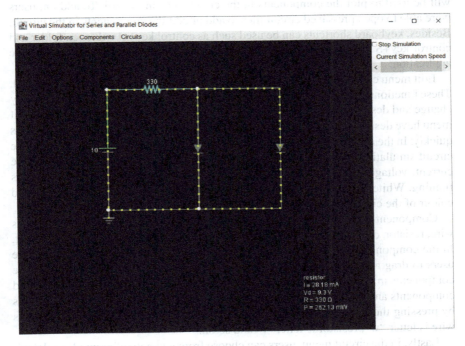

Fig. 3 GUI of virtual simulator for series and parallel diodes

the circuit drawing board. Lastly, to the right side of the circuit drawing board is the simulation canvas which allows the user to have control of the simulation.

D Functional Buttons

The functional buttons in the software can be separated into three categories which is menu bar, circuit drawing board and simulation canvas.

5. Menu bar

The menu bar of the virtual simulator consists of file, edit, options, components and circuits menus. All these menus have different and specific functionality for the virtual simulator. Users can use these menus to help them design and draw circuit using available components, save and open circuit files, undo and redo actions, access premade or default circuits. File menu allows the user to open and save the circuit file into their designated storage drive for future use. When user save a circuit file, the coordinates and parameter values of each component in the circuit will be stored in a text file. When user open a circuit file from the folder, the values in the text file will be used to plot the components in the circuit drawing board. To add on, users were able to open pre-saved circuit files to add or edit the components in the circuit. Besides, keyboard shortcuts can be used such as control key+O will open folder and control key+S will save the file in folder. In the file menu, user can also exit the application.

Edit menu consists of functions such as undo, redo, cut, copy, paste and select all. These functions work as intended as their name implies which allows the user to easily change and design the circuits in the circuit drawing board. All the functions in edit menu have designated keyboard shortcut for the user to make adjustment or changes quickly. In the option menu, there were four features to change the appearance of the circuit simulation. All four features in the options menu are check items where the current, voltage and values feature are checked every time the virtual simulator is running. White background function allows the users to change the appearance and colour of the circuit drawing board from default black colour to white colour.

Component menu allow the users to choose the available components such as wire, resistor, diode, ground, voltage source and LED to design and draw the circuit. In the component menu also consist a select or drag selected button which let the users to drag and move the components around the circuit drawing board. All the components in the menu have designated keyboard shortcut for the user to make add components and design the circuit quickly. Users can access the keyboard shortcuts by pressing the letters shown in the bracket correspond to the components such as wire is letter "w", resistor is letter "r" and so on.

Lastly, in the circuit menu, users can choose from a variety of premade or default circuits. These circuits consist of basic voltage resistor circuits with series and parallel configuration, forward bias diodes in series and parallel configuration, reverse bias

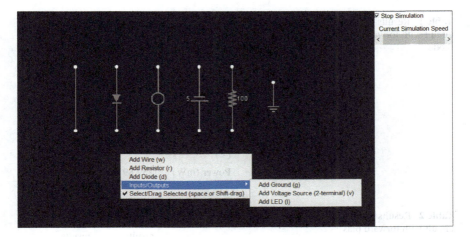

Fig. 4 Circuit drawing board and simulation canvas

diodes in series and parallel circuit and many more. Users were able to learn from these default circuits, edit the components' parameter and observe the output.

6. *Circuit drawing board*

As shown in Fig. 3, the black colour background is the area of circuit drawing board and the electronic components consists of wire, diode, LED, DC voltage, resistor and ground accordingly from left to right. Users can design and draw the circuit with the available components which can be access from the component menu bar or via right click mouse button to access any components quickly when designing the circuit in circuit drawing board.

7. *Simulation Canvas*

On the right side of the circuit drawing board is the white colour background simulation canvas shown in Fig. 4. This area allows the users to stop the simulation and manipulate the current simulation speed using a slider. The stop simulation checkbox is checked when the application starts.

4 Results

Results of the virtual simulator for series and parallel diodes were compared with another existing electronic circuit simulation software, Toolkit for Interactive Network Analysis (TINA). From the comparison Tables 1 and 2, the parameter results of current and voltage drops are shown for the sample circuit series circuit with forward bias diode and parallel circuit with forward bias diode. The components for the both circuits in testing results and TINA software are same such that the circuits consist of 10 V DC voltage, 330 Ω resistor and 1N4004 diode.

Table 1 Results for series circuit with forward bias diode

Components	Parameters	My testing results	TINA software results
Diode	Current (mA)	28.06	28.08
	Voltage drops (mV)	739.20	734.54
	Power (mW)	20.74	–
Resistor	Current (mA)	28.06	28.08
	Voltage drops (V)	9.26	9.27
	Power (mW)	259.89	–

Table 2 Results for parallel circuit with forward bias diode

Components	Parameters	My testing results	TINA software results
Resistor	Current (mA)	28.17	28.16
	Voltage drops (V)	9.30	9.29
	Power (mW)	261.90	–
Diode 1	Current (mA)	14.09	14.08
	Voltage drops (mV)	703.36	705.92
	Power (mW)	9.91	–
Diode 2	Current (mA)	14.09	14.08
	Voltage drops (mV)	703.36	705.92
	Power (mW)	9.91	–

As Tables 1 and 2 shows, all three parameters between the testing results and TINA software results are very close. This shows that the testing results is valid and equation used to calculate the parameter results are correct. One of the reasons both testing results and TINA software results were identical is because of the usage of ideal diode equation and Ohm's law in the virtual simulator. Besides that, based on the ideal diode equation, the value of saturation current and applied voltage of diode were taken from identical 1N4004 diode from datasheet such that saturation current is 18.8×10^{-9} A and applied voltage of diode at 1 A based on V-I characteristic graph is 0.925 V [15].

5 Conclusion

In this research, an interactive virtual simulator for series and parallel diode experiments have been designed and tested. A virtual laboratory for basic electronic courses which consists of several sample circuit related to the title have been developed. The sample circuits were simulated, tested, and results were checked with other Electronic Design Automation (EDA) software. This software can greatly benefit the users because they can perform basic electronic laboratory experiments using computer while not worrying on issues such as space needs, equipment needs, scheduling logistics, safety concerns and resource sharing. With users prior to download the software into their computer, users can freely experiment the sample circuits of series and parallel diodes, draw the circuit themselves, observe and analyse simulated results. As for future research, there are several aspects can be enhanced and improved upon related to basic electronic courses. A few of the recommendations that could be done are adding more the types of diodes and other electronic components like transistors, capacitors, inductors, integrated circuits and many more. Besides that, another enhancement could be to give users the opportunity to view the circuit in actual breadboard view with electronic components connected on the breadboard, allowing the user to switch between schematic view and breadboard view.

References

1. Khairul Anuardi, M.A., Mustapha, S., Mohammed, M.N.:Towards increase reading habit for preschool children through interactive augmented reality storybook. In: 2022 IEEE 12th Symposium on Computer Applications & Industrial Electronics (ISCAIE), Penang, Malaysia, pp. 252–257 (2022)
2. Alfiras, M., Bojiah, J., Yassin, A.A.: COVID-19 pandemic and the changing paradigms of higher education: a Gulf university perspective. Asian EFL J. **27**(5.1), 339–347 (2020)
3. Morgan, H.: Best practices for implementing remote learning during a pandemic. Clearing House J. Educ. Strat. Issues Ideas **93**(3), 135–141 (2020). https://doi.org/10.1080/00098655.2020.1751480
4. Alfiras, M., Yassin, A.A., Bojiah, J.: Present and the future role of the Internet of Things in higher education institutions. J. Positive Psychol. Wellbeing **6**(1), 167–175 (2022)
5. Alfiras, M., Bojiah, J., Mohammed, M.N., Ibrahim, F.M., Ahmed, H.M. Abdullah, O.I.: Powered education based on Metaverse: Pre- and post-COVID comprehensive review. Open Eng **13**(1), 20220476 (2023). https://doi.org/10.1515/eng-2022-0476
6. Alfiras, M., Nagi, M., Bojiah, J., Sherwani, M.: Students' perceptions of hybrid classes in the context of Gulf University: an analytical study. J. Hunan Univ. Nat. Sci. **48**(5) (2021)
7. Muhammad Amirul Fawwaz Nik Abdullah, N., Huzaifah Ahmad Sharipuddin, A., Mustapha, S., Mohammed, M.N.: The development of driving simulator game-based learning in virtual reality. In: 2022 IEEE 18th International Colloquium on Signal Processing & Applications (CSPA), Selangor, Malaysia, pp. 325–328 (2022)
8. Yuniarti, A., Yeni, L.F., Yokhebed: Development of virtual laboratory based on interactive multimedia on planting and painting bacteria. J. Phys. Conf. Series **895**(1) (2017). https://doi.org/10.1088/1742-6596/895/1/012120

9. Yap, W.H., Teoh, M.L., Tang, Y.Q., Goh, B.H.: Exploring the use of virtual laboratory simulations before, during, and post COVID-19 recovery phase: an animal biotechnology case study. Biochem. Mol. Biol. Educ. **49**(5), 685–691 (2021). https://doi.org/10.1002/bmb.21562
10. Hosseini, A., Koohi-Fayegh, S.: Engineering Education in Canada in the Wake of Covid-19 Pandemic (2021)
11. Ray, S., Srivastava, S.: Virtualization of science education: a lesson from the COVID-19 pandemic. J. Proteins Proteomics **11**(2), 77–80 (2020). https://doi.org/10.1007/s42485-020-00038-7
12. Dhang, S., Jana, P.K., Mandal, C.: Virtual laboratory for basic electronics. J. Eng. Sci. Manag. Educ. **10**(I), 67–74 (2017)
13. Kathane, B.Y., Gundalwar, P.R.: Development of virtual experiment on encoding techniques using virtual intelligent softlab. Int. J. Res. Comput. Appl. Robot. **4**(1), 21–25 (2016) [Online]. Available: www.ijrcar.com
14. Abuzir, Y.: The development of virtual laboratory using SEMMLABS in Al-Quds Open University. Int. J. E-Learn. Educ. Technol. Digital Media (IJEETDM) **1**(3), 133–141 (2015). https://doi.org/10.17781/P001715
15. Kuphaldt, T.R.: Lessons in Electric Circuits, Volume III—Semiconductors, 5th edn. (2009)

Toward Sustainable Smart Cities: Smart Water Quality Monitoring System Based on IoT Technology

Lee Mei Teng, Khairul Huda Yusoff, M. N. Mohammed, Adnan N. Jameel Al-Tamimi, Norazliani Md Sapari, and M. Alfiras

Abstract Water management plays an essential role in creating intelligent and sustainable cities. Low flow in the pipeline due to pipe leakage lets germs and viruses from the environment into the drinking water system. The major cause of avoidable diseases is water contamination in pipes. The pipeline needs a better web-based water monitoring system to keep drinking water from getting contaminated. Hence, this paper proposed a webserver-based pipeline water monitoring system. The IoT platform uses a webserver to monitor water quality in real-time. IoT collects and tracks data wirelessly. It improved lifestyle control, especially smart-device communication. This paper aims to develop a system to alert authorities, enhance web-based water monitoring systems, and detect water quality and pipe flow using sensors. Flow, pH, temperature, and turbidity sensors are used in this project. pH, temperature, and turbidity sensor ranges meet WHO guidelines. Data prints on LCD and webserver. The parameters will be graphs, and it will continuously output ten pieces of data. Arduino to NodeMCU data transmission takes about seven milliseconds. If water quality is low, the buzzer and red LED will activate. Green LED lights up if the water is good. This proposed project may be employed in a variety of industries. Explored are the larger implications of Smart Water Quality Monitoring Systems on urban sustainability, noting their potential to enhance public health,

---

L. M. Teng · K. H. Yusoff
Faculty of Information Sciences and Engineering, Management and Science University, Shah Alam, Selangor, Malaysia

M. N. Mohammed (✉)
Mechanical Engineering Department, College of Engineering, Gulf University, Sanad 26489, Kingdom of Bahrain
e-mail: dr.mohammed.alshekhly@gulfuniversity.edu.bh

A. N. Jameel Al-Tamimi
College of Technical Engineering, Al-Farahidi University, Baghdad, Iraq

N. M. Sapari · M. Alfiras
School of Electrical Engineering, Faculty of Engineering, Universiti Teknologi Malaysia, Johor Bahru, Malaysia

© The Author(s), under exclusive license to Springer Nature Switzerland AG 2024
A. Hamdan and E. S. Aldhaen (eds.), *Artificial Intelligence and Transforming Digital Marketing*, Studies in Systems, Decision and Control 487,
https://doi.org/10.1007/978-3-031-35828-9_49

578 L. M. Teng et al.

ensure equitable water distribution and conserve scarce water resources. Furthermore, IoT-based solutions hold great promise in revolutionising water management practices as part of sustainable smart cities development strategies.

Keywords Drinking water monitoring system · IoT technology · Real-time monitoring · Webserver · SDG 6 · SDG 11 · SDG 15

1 Introduction

Contamination of water in pipes is a major issue all around the world. On a worldwide basis, at least two billion people use face-contaminated drinking water. Individuals are exposed to preventable health risks as a result of inadequately maintained water and sanitation systems [1]. Numerous consumers complain about the odor and flavor of drinking water, regardless of the fact that an unpleasant odor or taste does not often signify dangerous water. Many naturally occurring or purposeful errors taint raw water, generating odor concerns in water treatment systems [2]. The author of [3] highlighted that water contamination has become a huge problem that threatens people's daily lives as industry and agriculture have grown; the degree of contamination has been so high, particularly in developing countries, that it has created a severe risk to public and ecosystem health in some locations. According to the Water Environment Partnership in Asia (WEPA), the Water Quality Index (WQI) of drinking water in Malaysia ranges from 51.9 to 76.5 [4]. Water surveillance serves as an early warning method for catastrophe management [5]. The author [6] stated that if a proper warning system is not in place, the corrective steps may be delayed. As a result, operators may take inappropriate actions. Traditional water contamination monitoring techniques have significant limitations, including high labor and material costs, limited sampling, and long analysis [7]. To ensure that drinking water does not get contaminated, the pipeline requires a better web-based water monitoring system. Water contamination monitoring can assist in ensuring a consistent supply of clean drinking water. The Internet of Things (IoT) is used in most current water contamination monitoring systems because it can do real-time analysis and inform authorities through SMS, and cloud storage. The requirement of a real-time response and monitoring system should be evaluated on a case-by-case basis by water utilities and management areas depending on the needs of each water treatment company and ecological agency. The water industry employs the online monitoring system as early noticing notification systems at all phases such as intake protection, water treatment, and distribution systems [8]. A wireless data communication system is flexible and easy to move, while a hard-wired system takes a lot of time and money to set up and maintain [9]. The Internet of Things (IoT) has greatly simplified and streamlined the controllability of a person's lifestyle, especially in terms of communication methods between smart gadgets [10]. Pipeline water monitoring systems have been the focus of several studies since this is a contemporary concern. The author [6] proposed a deep learning-based output that aids a water quality alert system for reverse osmosis facilities. The sensors that are used in this project are pH, total dissolved solids (TDS), oxidation–reduction potential (ORP), and electric conductivity (EC). The sensors (pH, TDS, ORP, and EC) are wired to the Arduino, which serves as the main controller. This primary controller accepts the sensor signals and transfers them to the computer for analysis. The LCD is used to show the sensor data that has been collected. This project is extremely costly, and the system's setup is quite difficult. LSTM-CNN types may not be able to reliably forecast grades when analyzing water based on quality discrepancies unless the water quality base is correctly separated into classes.

Author [11] introduced a WSN based on the design of the rural drinking water source monitoring system. This project uses a solar panel as a source of power supply, a keypad that can control the LCD, and an SD card for backup purposes. Water quality monitoring, soil monitoring node, node, routing node, and gateway server are the five aspects of the system. The gateway acts as a command-and-control center for monitoring water quality and the pH of soil. It consists of a processor module, a sensor module, a GPRS module, a power supply module, and a solar charging module. The cost of this project is expensive, and it is not suitable for unpracticed users to use. The solar panel is unsuitable for this project because it must be positioned outside of the manhole, which implies that if it is installed on top of the manhole, it will be walked on by others.

In [12] presented real-time water quality monitoring using the Internet of Things in SCADA. The system includes three parts: a sensor unit with several types of sensors, a transmitting unit with an Arduino Uno microcontroller for data processing, and a receiving unit with a GSM modem for Internet connectivity. The web server then shows the data and status of water contamination. SCADA may control the GSM module to deliver data to the monitoring center. Data is simultaneously delivered to mobile phones, allowing for speedy decisions. The sensor data will also show on the water treatment plant's LCD panel. The absence of a pH sensor, which is critical for measuring water quality, is a flaw in this system. It simply has a temperature sensor and a color sensor, so it can't tell exactly how acidic or alkaline the water is.

Feature [13] is about an IoT-based water quality monitoring system. The results may be viewed in the Blynk app, which is available for both Android and iOS. The author installed the Blynk app in its Android version for this project. When the system powers up, DC power is delivered to the kit, Arduino, and WIFI. The parameters' findings are shown on the LCD panel. The software included with the hotspot gives the accurate figure displayed on the kit's LCD display. Therefore, when the kit is placed in a waterbody with WIFI, the user may check its real-time value on their Android phone from anywhere at any time. App features such as maps need a purchase before they can be accessed in the Blynk app.

The previous research [14] suggested real-time monitoring and contamination detection in the drinking water distribution system. Turbidity, Oxidation Reduction Potential, Temperature, Electrical Conductivity, and pH, which are the five aspects to examine, have been chosen to be monitored. The author designed a specialized PIC-based microsystem for each of the parameters. With the onboard LEDs switched off and the RF Xbee transceiver module giving water quality data every five seconds, the total power consumption of the core measurement sensor node is around 50 mA at 5 V operational voltage. The software platform provides real-time measurement charts of monitored parameters, real-time water quality evaluation, and sensor calibration instructions via a Graphical User Interface (GUI). Using Pachube scripts, the user may set different thresholds for sending notifications by SMS or email. WSN is only suitable for low-speed communication and cannot be used for high-speed communication. In the previous research, there were several downsides, including the fact that it takes time to train new employees to operate the system and that some of the proposed projects are rather costly. This paper will talk about the details of Webserver's design, which helps to track the quality of water in real-time. This system can monitor different things about water by using a different type of sensors. Sensor data is sent wirelessly to a web server through Wi-Fi. The purpose of employing a webserver is to link several devices and sensors across a particular network and obtain real-time data from the sensor so that the data can be reconstructed, assessed, and used to offer pertinent information. The objectives of this research are to make a system for monitoring contamination that can use a buzzer to notify laborers and to enhance a web-based system for monitoring water. In addition, this study suggested using a pH sensor, a temperature sensor, a turbidity sensor, and a flow sensor to detect contaminated water and water flow in a pipeline. The interface is composed of a microcontroller. In this project, a webserver acts as the Internet of Things platform. The parameters will be printed as a line graph by the web server. Checking water quality at a workplace when users are not there is possible thanks to this feature. Additional web server functions are free of charge. This system is simple to use and reasonably priced.

2 Materials and Methods

The design of this project has been split into two parts: the layout of the water monitoring system and the web server architecture. The whole system is controlled by Arduino Mega and NodeMCU. The software used by this system is an HTML-coded webserver. The project's specifics are shown below.

2.1 The Layout of the Water Monitoring System

The block diagram of this system is illustrated in Fig. 1. The Arduino Mega and the NodeMCU are the microcontrollers used in this system. NodeMCU was implemented as a Wi-Fi client station linked to a local access point AP (server). This system is equipped with four sensors: a flow sensor, a pH sensor, a temperature sensor, and a turbidity sensor. They will assess the water quality in the pipeline. The data about the parameters will be shown on the LCD and on the IoT platform, which is a Webserver

Fig. 1 The suggested system is depicted as a block diagram

where the user can check the quality of the water. On the webserver, the parameter data will be shown in the form of a line graph.

Figure 2 illustrates the circuit design for this system. The Arduino Mega requires 5 V, the NodeMCU requires 3.3 V and 5 V, and the water pump requires 12 V. A 20.3–22 V power source is required for the entire system. This system is started with water being pumped from the water tank into the pipeline. The Arduino Mega is connected to four sensors, two LEDs, LCD, a buzzer, and NodeMCU. The flow sensor determines if there is any leakage in the pipeline. If there is a leak in the pipeline, it is extremely harmful because germs or contaminants from the environment can quickly enter the water. Low and negative pressure events have the potential to result in contaminant intrusion [15]. The pH sensor measures the pH value of the water, and the turbidity sensor analyzes the water's transparency. Drinking very acidic or alkaline water, as well as murky or opaque water, is risky because it may contain toxic compounds. The temperature sensor determines the water temperature. If the water temperature is too low, it signifies the water in the pipeline has frozen. Because it takes time to defrost the water in the pipeline, this will result in unannounced water cutbacks. When the water temperature is too high, the pipe will burst. This is because the pressure within and outside the pipeline differs.

The LEDs are used as indicators. Whenever the readings go outside of the threshold, a red LED and a buzzer is activated. An illuminated green LED signifies that the current readings are within the predetermined ranges. The data from the sensor will be processed by the Arduino Mega and automatically transmitted to NodeMCU. When the Wi-Fi module NodeMCU gets the data, it sends it over Wi-Fi to the IoT platform. The output of the sensor may be viewed on the webserver as well as on an LCD. The two white LEDs are used to illuminate the pipe so the turbidity sensor can determine the water's clarity.

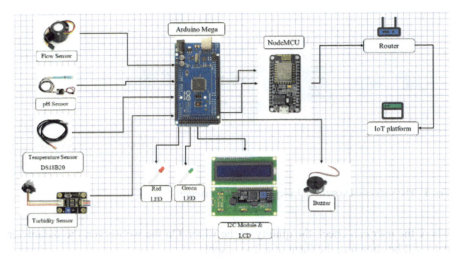

Fig. 2 Circuit diagram of water monitoring system

2.2 Webserver Architecture

HyperText Markup Language (HTML) code is used to create the webserver, while the Arduino Mega and NodeMCU are programmed in C. The Arduino IDE and Microsoft Visual Studio code are used jointly to merge and upload to NodeMCU. For programming the Arduino Mega and NodeMCU, the Arduino IDE is utilized, while Microsoft Visual Studio is used to create the webserver. Figure 3 presents an overview of the coding flowchart.

The Arduino Mega and the NodeMCU are linked to one another; the Arduino Mega is responsible for gathering the data from each sensor and sending it to the NodeMCU, while the NodeMCU is responsible for receiving the data from the Arduino Mega. When the NodeMCU board is powered up, it automatically connects to the Wi-Fi network that was registered using the Service Set Identifier (SSID) and password. If the NodeMCU is not connected to Wi-Fi, the Arduino IDE's serial monitor will display dots (.); this will repeat until the NodeMCU is connected to Wi-Fi. The IP address of the webserver will be printed once NodeMCU connects to the Wi-Fi network. When the user connects to the web server, the HTML folder will be fetched and integrated with the NodeMCU code file. Each sensor's data is saved on the NodeMCU. The server will get the information from NodeMCU and post it to the web server. The data is displayed in the form of a line graph, and it will show ten pieces of data at once and will continue to update, which means that the initial data will be replaced with the new data from the sensor.

The serial monitor will report "Client Connect" if the user is still browsing the web server. The NodeMCU's data will be downloaded by the webserver and updated as needed. When the user disconnects from the webserver, the webserver's program

Toward Sustainable Smart Cities: Smart Water Quality Monitoring ... 583

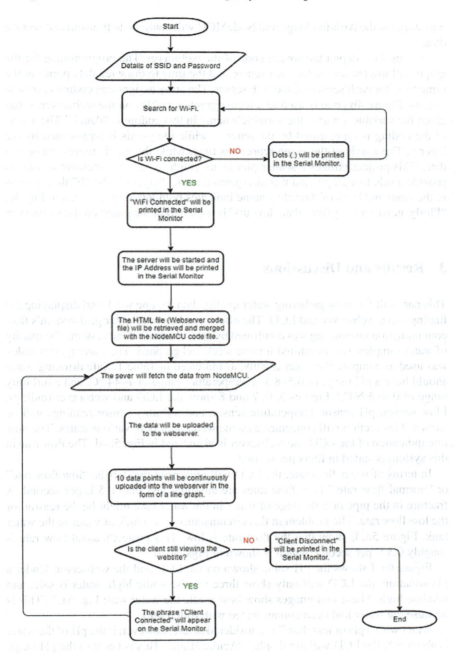

Fig. 3 NodeMCU's and webserver's flow chart

will stop, but the Arduino Mega and NodeMCU will continue to transmit and receive data.

Figures 4a–c depict the source code of the webserver. The programming for the graph will use ten arrays for each sensor and the time to draw ten data points at the same time for each sensor on the web server. The array assignment coding is seen in Fig. 4a. Figure 4b code is used to allocate graph variables on the webserver. It has given the variable a name; the variable's name in this coding is "data3." The x-axis of the coding is represented by the letter x, while the y-axis is represented by the letter y. The x-axis of the graph represents time, while the y-axis represents sensor data. This project utilizes a scatter plot as its graph. A variable element is used to provide a title to a graph, and it is also given a name, "layout." The "b" that appears in the front and back of the title's name indicates bold lettering. As seen in Fig. 4c, "Plotly, newPlot ('myDiv', data, layout)" is used to plot the graph on the webserver.

3 Results and Discussions

This part will focus on gathering water quality data for one week and displaying the findings on a webserver and LCD. The effectiveness of the water grid system's flow contamination monitoring was confirmed by an evaluation of the system. The quality of water samples was monitored for one week and graphed. The water quality index was used to compare the water quality, as indicated in Table 1. Safe drinking water should have a pH range of 6.5–8.5, a temperature range of 7–44 °C, and a turbidity range of 0 to 5 NTU. Figures 5, 6, 7 and 8 show the LCD and webserver findings. Flow sensor, pH sensor, temperature sensor, and turbidity sensor readings will be shown. This section will concentrate on the water quality in various states. The flow rate indication of the LCD and webserver is visualized in Fig. 5a–d. The flow rate in this system is stated in liters per second.

In terms of water flow rate, the LCD will only print two labels: "low flow rate" or "normal flow rate." Low flow rates are anything less than 0.5 L per second. A fracture in the pipe or a shortage of water in the water tank might be the reason for the low flow rate. The problem in this circumstance was a lack of water in the water tank. Figure 5a, b show that the flow rate is low. This project's usual flow rate is roughly 0.5 L per second, which is shown in Figs. 5c, d.

Figure 6a–f shows the pH value shown on the LCD and the webserver. Under a pH situation, the LCD will only show three words: acidic high, water is safe, and alkaline high. These two images show how acidic the water was: Fig. 6a, b. This is because the water had been contaminated with rotten milk.

Water with a pH of less than 7 is considered acidic, however, if the pH of the water is about 6.5, the LCD will not display "Acidic High." This is because the pH range of safe water is between 6.5 and 8.5. Figure 6c, d show the pH value of the water, which is about 7.21, which is deemed safe water because it is within the range. If the pH of water is more than 8.5, the water is extremely alkaline. Figure 6e, f show the results of high alkalinity. This is due to its combination with a very alkaline chemical.

Toward Sustainable Smart Cities: Smart Water Quality Monitoring … 585

```
const date = new Date()
timeArr[0] = timeArr[1]
timeArr[1] = timeArr[2]
timeArr[2] = timeArr[3]
timeArr[3] = timeArr[4]
timeArr[4] = timeArr[5]
timeArr[5] = timeArr[6]
timeArr[6] = timeArr[7]
timeArr[7] = timeArr[8]
timeArr[8] = timeArr[9]
timeArr[9] = `${date.getHours()}:${date.getMinutes()}:${date.getSeconds()}`

tempArr[0] = tempArr[1]
tempArr[1] = tempArr[2]
tempArr[2] = tempArr[3]
tempArr[3] = tempArr[4]
tempArr[4] = tempArr[5]
tempArr[5] = tempArr[6]
tempArr[6] = tempArr[7]
tempArr[7] = tempArr[8]
tempArr[8] = tempArr[9]
tempArr[9] = d.Temp
```

(a)

```
var data3 = [
{
    x: timeArr,
    y: turbArr,
    type: 'scatter'
}
];


var layout={
    title:'<b>Water Temperature</b>'
};
```

(b)

```
Plotly.newPlot('myDiv', data, layout);
```

(c)

Fig. 4 The coding of the system: **a** assign ten data points to be printed, **b** add variables to each axis and the title of each graph, and **c** code to plot the graph

Table 1 Range of water quality [6, 14]

	Parameters	Units	Quality
1	pH	pH	6.5–8.5
2	Temperature	°C	7–44
3	Turbidity	NTU	0–5

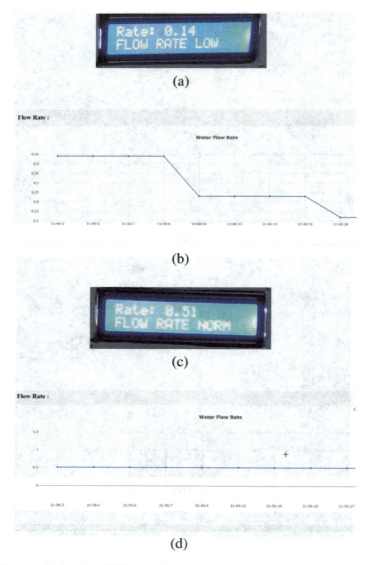

Fig. 5 Flow rate displayed on LCD and webserver: **a** and **b** low flow rate; **c** and **d** normal flow rate

The temperature of the water can be seen on the LCD as well as the webserver is shown in Fig. 7a–d. With this system, the temperature is measured in Celsius. The LCD will only print three phrases when it comes to temperature: low temperature, normal temperature, and high temperature. When the water temperature falls below seven degrees Celsius, the LCD will show "Temp low." The temperature of water in Figs. 7a, b is 28.62 °C, which is considered a normal temperature. The normal

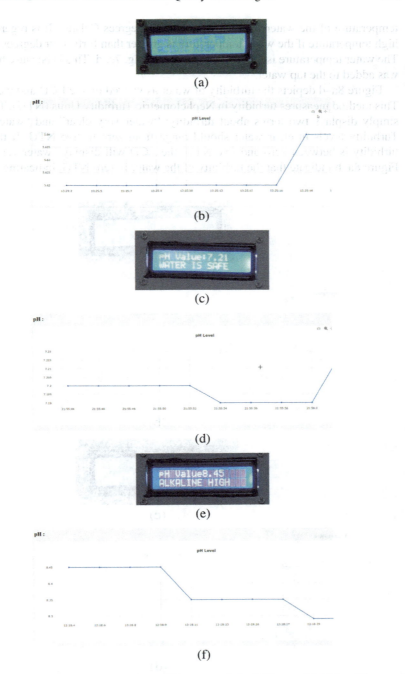

Fig. 6 The pH value is shown on both the Webserver and the LCD: a and b high acidic; c and d safe drinking water; e and f high alkaline

temperature of the water is seven to forty-four degrees Celsius. It is regarded as a high temperature if the water temperature is greater than forty-four degrees Celsius. The water temperature is 45.38 °C, as shown in Fig. 7c, d. This is because hot water was added to the tap water.

Figure 8a–d depicts the turbidity of water as viewed on the LCD and webserver. This method measures turbidity in Nephelometric Turbidity Units (NTU). The LCD simply displays two terms about turbidity: "water very clear" and "water dirty." Turbidity levels in clear water should range from zero to five NTU. If the water turbidity is between zero and five NTU, the LCD will display "water very clear." Figure 8a, b indicate that the turbidity of the water is zero NTU, indicating that it is

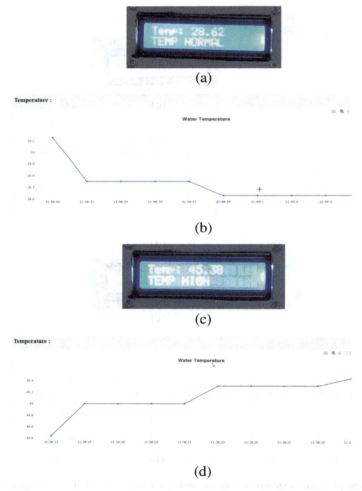

Fig. 7 Water temperature indicated on LCD and webserver: **a** and **b** normal temperature; **c** and **d** high temperature

exceptionally clean and safe to use. If the turbidity of the water exceeds five NTU, it is considered hazardous to use and unclean. The LCD will display the phrase "water dirty." Figure 8c, d demonstrate that the turbidity of the water is fifty-one NTU. This is related to the reasons that the water was blended with chocolate milk.

The graph on the webserver will show ten pieces of data at once and it will continually update. That is to say, the initial data will be replaced with the sensor's updated data. Every one to two milliseconds, the data on the web server will be updated. The data transfer time from Arduino to NodeMCU is expected to be around seven milliseconds, and the webserver will be updated every four to five seconds with fresh data from Arduino.

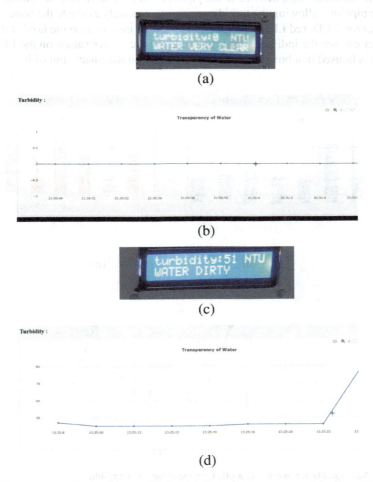

Fig. 8 Transparency of water illustrated in both LCD and webserver: **a** and **b** clear water; **c** and **d** dirty water

The pH, temperature, and turbidity of the tap water were measured for one week, and the findings of Fig. 9a–c reveal that the water in Malaysia is safe to consume. Over the course of one week, the pH of the tap water fluctuated between larger than 6.5 pH and less than 7.2 pH, as shown in Fig. 9a. Figure 9b illustrates that the temperature of water in Malaysia is normally around 29 °C, but on May 17th, 2022, it was somewhat colder than the previous days, at around 28 °C. This occurred as a result of the rain that fell on that day. The turbidity of the water was between 0 and 1 NTU in Fig. 9c, implying that it was quite clear.

The proposed system's final product is shown in Fig. 10a–d. The sensors are inserted in the pipeline so that it can detect the quality of the water as it flows through it. White LEDs are also inserted in the pipeline; they are utilized as a lighting system for the pipeline, allowing the turbidity sensor to precisely evaluate the water quality. The buzzer, LCD, red LED, and green LED are all mounted on the box's lid so that the user can see the indicator lights and read the parameter values on the LCD. The circuit is housed in a box to keep water and other contaminants out of it.

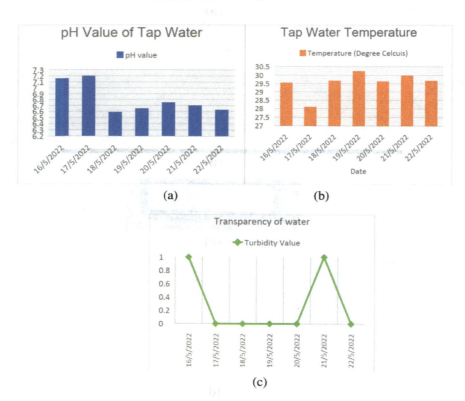

Fig. 9 Water quality for one week: **a** pH, **b** temperature and **c** turbidity

Fig. 10 **a** and **b** Final circuit setup; **c** and **d** prototype of water flow contamination in water grid system

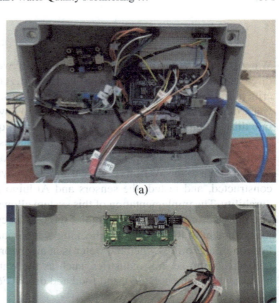
(a)

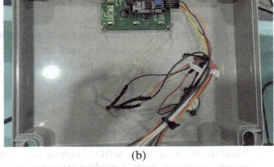
(b)

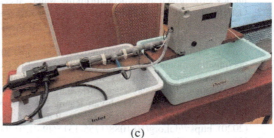
(c)

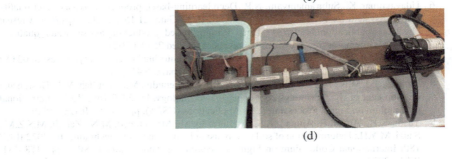
(d)

4 Conclusions

Water is essential for human life. Contaminated water can cause various avoidable diseases or even cause the death of a young child. A dependable monitoring system is critical for the safety of drinking water in pipelines. This study describes a real-time water monitoring system that makes use of an IoT system. Incorporating a webserver into this system allows the user to check the water quality in the pipeline even when they are not physically there. Multiple sensors are used and connected to a microcontroller, the Arduino Mega. The Arduino Mega is connected to the NodeMCU and it transfers data to the Webserver in real-time. The system was successfully designed, constructed, and tested; the sensors and Arduino Mega have shown the system's durability. The implementation of this system allows for the reduction of the number of employees on-site, the reduction of the frequency of unscheduled water cuts, the improvement of the efficacy of checking water quality in the pipeline, and the notification of authorities when the water is in an abnormal state. This proposed project is applicable to several industries, such as the water, medical, chemical, personal protective equipment (PPE), and food and drink (F&D) industries.

References

1. World Health Organization: WHO, "Drinking-water," Who.int, 14 Jun 2019. https://www.who.int/news-room/fact-sheets/detail/drinking-water. Accessed 02 Oct 2021
2. Ostfeld, A., Salomons, E.: Securing water distribution systems using online contamination monitoring. J. Water Resour. Plan. Manag. **131**(5), 402–405 (2005). https://doi.org/10.1061/(asce)0733-9496(2005)131:5(402)
3. Pipeline Monitoring System: Oilandgasonline.com (2021). https://www.oilandgasonline.com/doc/pipeline-monitoring-system-0001. Accessed 18 Oct 2021
4. Interim National Water Quality Standards for Malaysia: Wepa-db.net (2021). http://www.wepa-db.net/policies/law/malaysia/eq_surface.htm. Accessed 02 Oct 2021
5. Kamaruidzaman, N.S., Nazahiyah Rahmat, S.: Water monitoring system embedded with Internet of Things (IoT) device: a review. IOP Conf. Series Earth Environ. Sci. **498**(1), 012068 (2020). https://doi.org/10.1088/1755-1315/498/1/012068
6. Udayakumar, K., Subiramaniyam, N.P.: Deep learning-based production assists water quality warning system for reverse osmosis plants. ResearchGate, 21 Dec 2020. https://www.researchgate.net/publication/347917347_Deep_learning-based_production_assists_water_quality_warning_system_for_reverse_osmosis_plants. Accessed 25 Oct 2021
7. Abdelmoutia, T., Feryel, B.: Automatic System of Monitoring Water Quality. Accessed 02 Oct 2021. [Online]. Available: https://israa.edu.ps/journal/pages/v4/7.pdf
8. Mohammed, M.N., Al-Zubaidi, S., Bahrain, S.H.K., Zaenudin, M., Abdullah, M.I.: Design and development of river cleaning robot using IoT technology. In: 2020 16th IEEE International Colloquium on Signal Processing & Its Applications (CSPA), pp. 84–87. IEEE (2020)
9. Yusof, K.H., Aman, F., Ahmad, A.S., Abdulrazaq, M., Mohammed, M.N., Zabidi, M.S.Z.M., Sauzi, M.Y.H.: Determination of soil texture using image processing technique. In: 2022 IEEE 18th International Colloquium on Signal Processing & Applications (CSPA), pp. 178–181. IEEE (2022)

10. Mohammed, M.N., Al-Rawi, O.Y.M., Al-Zubaidi, S., Mustapha, S., Abdulrazaq, M.: Toward sustainable smart cities in the Kingdom of Bahrain: a new approach of smart restaurant management and ordering system during covid-19. In: 2022 IEEE International Conference on Automatic Control and Intelligent Systems (I2CACIS), pp. 7–11 (2022). https://doi.org/10.1109/I2CACIS54679.2022.9815479
11. Yusof, K.H., Aman, F., Ahmad, A.S., Abdulrazaq, M., Mohammed, M.N., Zabidi, M.S.Z.M., Asyraf, A.: Design and development of real time indoor and outdoor air quality monitoring system based on IoT technology. In: 2022 IEEE 18th Int. Colloquium on Signal Processing & Applications (CSPA), pp. 101–104 (2022)
12. Saravanan, K., Anusuya, E., Kumar, R., Son, L.H.: Real-time water quality monitoring using Internet of Things in SCADA. Environ. Monit. Assess. **190**(9) (2018). https://doi.org/10.1007/s10661-018-6914-x
13. Daigavane, V., Gaikwad, M.: Water quality monitoring system based on IOT **10**(5):1107–1116 (2017)
14. Lambrou, T.P., Anastasiou, C.C., Panayiotou, C.G., Polycarpou, M.M.: A low-cost sensor network for real-time monitoring and contamination detection in drinking water distribution systems. IEEE Sens. J. **14**(8), 2765–2772 (2014). https://doi.org/10.1109/jsen.2014.2316414
15. Collins, R., Boxall, J.: The influence of ground conditions on intrusion flows through apertures in distribution pipes—white rose research online. Whiterose.ac.uk (2012)

2019 Novel Coronavirus Disease (Covid-19): Toward a New Design for All-in-One Smart Disinfection System

M. N. Mohammed, M. Alfiras, Hakim S. Sultan, Adnan N. Jameel Al-Tamimi, Rabab Alayham Abbas Helmi, Arshad Jamal, Aysha Sharif, and Nagham Khaled

Abstract The global coronavirus epidemic continues to affect the planet. COVID 19 is a disease that primarily affects the patient's respiratory system and decreases immune system performance. As the coronavirus (Covid-19) epidemic grows, technological applications and initiatives are proliferating in an attempt to keep the situation under control, treat patients effectively, and aid overburdened healthcare personnel while finding new, effective vaccines. As the first significant pandemic of the twenty-first century, Covid-19 provides a unique chance for policymakers and regulatory agencies to consider the legal feasibility, moral soundness, and usefulness of deploying innovative technology under time constraints. Maintaining public trust in evidence-based public health treatments will require striking the proper balance. The goal of this study is to design and create a self-contained smart disinfection tunnel using IoT technologies. A thermal temperature scanner, sprinkler tunnel, hand

---

M. N. Mohammed (✉) · A. Sharif
Mechanical Engineering Department, College of Engineering, Gulf University, Sanad 26489, Kingdom of Bahrain
e-mail: dr.mohammed.alshekhly@gulfuniversity.edu.bh

M. Alfiras
Electrical and Electronic Engineering Department, College of Engineering, Gulf University, Sanad 26489, Kingdom of Bahrain

H. S. Sultan
College of Engineering Karbala, University of Warith Al-Anbiyaa, Karbala, Iraq

A. N. Jameel Al-Tamimi
College of Technical Engineering, Al-Farahidi University, Baghdad, Iraq

R. A. A. Helmi
School of Graduate Studies, Management and Science University, Shah Alam, Selangor, Malaysia

A. Jamal
Faculty of Information Sciences and Engineering, Management & Science University, Shah Alam, Selangor, Malaysia

N. Khaled
Architectural and Interior Design Engineering Department, College of Engineering, Gulf University, Sanad 26489, Kingdom of Bahrain

© The Author(s), under exclusive license to Springer Nature Switzerland AG 2024
A. Hamdan and E. S. Aldhaen (eds.), *Artificial Intelligence and Transforming Digital Marketing*, Studies in Systems, Decision and Control 487,
https://doi.org/10.1007/978-3-031-35828-9_50

sanitizer, mask supplier, health indicator, and QR code scanner are all part of the proposed research. This research could help prevent and control future pandemics of compatible events.

Keywords Covid-19 · Coronavirus · Disinfecting · Sprinkler tunnel · IoT · SDG 3 · SDG 11 · SDG 15

1 Introduction

The present pneumonia outbreak near Wuhan City, China, is being identified by a novel coronavirus known as '2019-nCoV' or '2019 novel coronavirus' or 'COVID-19' by the World Health Organization (WHO). COVID-19 is a pathogenic virus, and we know that bats are the virus's reservoir thanks to phylogenetic research using available whole genome sequences, but the intermediate host has yet to be discovered. Early inquiries of cases with symptoms occurring near Wuhan during Dec. 2019, ecological random selection from the Huanan Supplier Seafood Market as well as other neighborhood markets, and the gathering of regular reports of the site of generation and kind of biodiversity marketed on the Huanan market are all ongoing in China to advise our awareness of the pathogenic origin of the outbreak [1–3]. In today's world, there is a larger need for effective urban planning to lessen the effects of COVID-19 illness as the population of cities grows. Significant improvements in the healthcare field and other aspects of city life will be required for smart cities to thrive. In this context, the WHO has enacted strict COVID-19 patient requirements, including curfews, social distancing, and monitoring. Furthermore, many countries have implemented digital tools that allow users to identify COVID-19 patients in their vicinity using wireless target tracking technology. In view of the COVID-19 impact and multiplication, research teams have dedicated substantial resources to finding and enhancing COVID-19 treatments; numerous novel remedies, such as automated sanitizing gadgets for sterilizing medical products and people, thermal imaging systems, and many others, have been posited for the current horrible situation. Several judiciary administrations have implemented strict health-prevention regulations and made substantial efforts to clean particular topographical areas with a range of disinfectants, in addition to promoting society's health through other health-awareness measures. Individuals have been fined by governments in rare cases for attempting to break health and safety standards. It's still up for debate whether rigorous adherence to health and safety standards like wearing a mask and keeping a safe distance from others is enforced [1, 4–6]. With the global calamity of a new coronavirus, SARS-CoV-2 (severe acute respiratory syndrome coronavirus 2), which has caused a worldwide outcry for numerous reasons, the world has been obliged to confront public health emergency. Some experts believe that the SARS-CoV-2 virus originated in bats and that the virus spread to humans via palm civets, which serve as a link between bats and humans. The virus SARS-CoV-2, as per the World Health Organization (WHO), can be spread by direct contact and aerosols. The COVID-19 infection produces respiratory illness that ranges from mild to severe, as well as mortality. COVID-19 is spread from person to person, according to current statistics, meaning that this is the disease's predominant mode of transmission in the current epidemic. Strict surveillance and testing are essential to prevent the epidemic from propagating further. COVID-19 carriers will, on average, spread the disease to anybody they come into contact with. Several COVID-19 victims, on the other hand, are asymptomatic and therefore can serve as carriers, spreading the virus accidentally. This could be the reason for the high number of COVID-19 cases. The outbreak is thought to have started in China's Wuhan city. Positive-sense RNA infects the virus, and its diameter ranges from 60 to 140 nm. It was given this name because of its shape, which is a spike projection that was shown under the microscope with a crown look. When compared to the throat, studies have revealed higher viral loads in the nasal cavity, with no difference in viral weight between symptomatic and asymptomatic patients [7–10]. COVID-19 is transferred by dust particles and fomites when the infector and the affected person are in close proximity. The active virus has been described in a small number of clinical studies, and fecal dissemination has been found in certain patients. In addition, the fecal–oral pathway does not appear to be a COVID-19 transmission powerhouse; its function and significance for COVID-19 must be determined. Fever is a sign of COVID19 infection that causes an increase in body temperature. For disorders that are especially associated with high fever, a fever monitoring thermograph should be part of the tiered screening approach. As per the IEC 80601-2-59 standard, screening thermography instruments must meet basic safety and performance requirements in order to become an effective non-contact, precise, and repeatable method of rapidly screening people for fever. Also, there is an ISO standard, ISO/TR 13154:2017, that specifies the deployment, execution, and operating recommendations for preventing infectious disease spread [10–13]. With the spread of COVID-19 and the lack of comprehensive treatment and vaccine, experts and scientists have been obliged to respond to new public health emergencies. As a result, this project intends to design and build a standalone smart Disinfection Tunnel based on IoT technology in order to reduce the danger of COVID-19. A thermal temperature scanner, body spray or sterilizer, mask supplier, health indicator, and QR code scanner are all part of the proposed research.

2 Methodology

A unique approach to combat the current pandemic issue is a new design of a free-standing smart disinfection device based on IoT technology. Individuals will be disinfected utilizing an automatic mobile sanitization mist and UV lights in combination with hand sanitizer in the planned study. A thermal temperature scanner, mask supplier, health status, and QR code scanner are also included in the proposed research as shown in Fig. 1. The primary materials and components for the prototype are described and analyzed. The prototype's electrical components and structure/mechanical parts make up the materials and components as shown in Table 1.

Fig. 1 Assembled parts of standalone smart disinfection system

The system's operating concept is depicted in Fig. 2, which shows the system's major design for this work. It begins with the individual who must pass safety checks before stepping onto the device's platform. The load sensor will then detect the load on the platform and identify whether or not someone is on it. If the system detects a load, it will activate the sanitizer dispenser and temperature sensor.

Both the sanitizer dispenser and the temperature sensor work separately. Before distributing a defined dose of sanitizer, the sanitizer dispenser has provided with integrated circuit that checks for a pair of hands beneath it. It will go into standby if no hands are present. The integrated circuits in the temperature sensor will determine the temperature of the body of the person standing on the platform. It will identify whether the person's temperature is within normal limits or exceeds them. If the temperature is normal, the green light will illuminate, indicating that the person is safe to pass past the checkpoint. If the temperature is below average, the red light will illuminate to alert the person in command of the checkpoint. These data, which will be sent through the IoT integrated circuit, can be monitored using the IoT user interface. After that, the device's system will shut down, and the cycle will begin again for subsequent users.

The schematic of the system is depicted in Fig. 3. As the main controller, it uses an Arduino Uno integrated circuit. There is also an IoT module for transmitting data to the user interface via a Wi-Fi connection. A temperature sensor, a load sensor, and

2019 Novel Coronavirus Disease (Covid-19): Toward a New Design ...

Table 1 Parts of the smart disinfection system

No.	Materials	Description	Image
1	Stainless steel	To use as a framework structure of the prototype	
2	Sanitizer bottle	To store sanitizing liquid for hand usage. This will help the user to clean their hands thoroughly	
3	Sanitizer box	To store sanitizing liquid for sprinkler system usage	
4	Asperser	To spread the sanitizing mist around the body of the user	
5	PVC pipe	To route the sanitizing liquid throughout the sprinkler system	
6	UV light panel	To replace the sprinkler from neck and above as UV light is much safer to exterminate germs and virus	
7	Sprinkler car	To move the sprinkler system in back and forward direction	
8	Circuit box	To will host the system circuitry and related wiring	
9	Mask box holder	To store the surgical face mask and extender as an option for those not wearing one	
10	FLIR A615	To scan temperature and detect the face of users	
11	Monitor	To show the interface of the system which include camera scanning status and QR code for data collecting	
12	Status light	To check the health status of the persona including the of face mask detection and the temperature measurements	

(continued)

Table 1 (continued)

No.	Materials	Description	Image
13	Wi-Fi router	To link the system with the internet, a router will be use. This will enable the collected data send into cloud storage. It will also help to alert the authority when high-risk is detected	

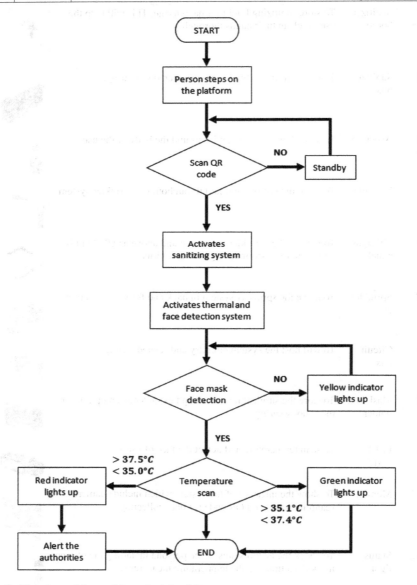

Fig. 2 Flowchart of the working principle of the system

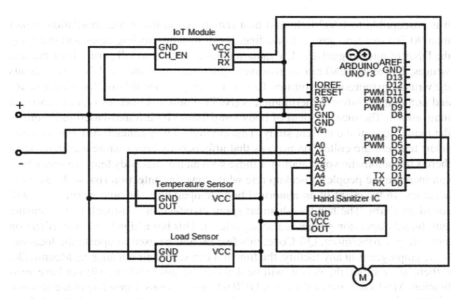

Fig. 3 Circuit diagram of the system

an integrated circuit are also used to operate the hand sanitizer dispenser's motor. A circuit box will be installed behind the stall to power the entire system.

3 Results and Discussions

Due to the inevitable circumstances of the COVID-19 pandemic, the product design and production cycle with technological development are illustrated using Solid-Works software. The final design of the prototype was developed after many assessments, refinement, and improvements from the initial design of the early concept sketch as shown in Figs. 4 and 5. It is practical for usage with extra parts to adapt to the users in terms of their sizes. The importance of the automatic area disinfection detection and spray for body surface disinfection, which singles out the liquid through a 0.3 mm nozzle opening, is also significant. This can effectively eliminate waste while saving energy. It's also portable, thanks to the universal wheel and brake function, as well as a foldable slope. The disinfection takes conducted in disinfection chambers, which have entrance points, outlet points, and passages. The power supply, pump, chemical chamber, and spray mechanism are all included in the chambers. An extensive list of compounds effective against SARS-CoV-2 has been provided by the Environmental Protection Agency (EPA). There are differences between different sources, highlighting the difficulty of evaluating viral characteristics and vulnerabilities in lab conditions. The coronavirus is a single-stranded RNA virus. SARS and MERS are related viruses. Previous research on the SARS virus has found that it is susceptible to UVC light and heat radiation and that UVC irradiation higher than 90 W/cm^2 can cause it to decline. By destroying nucleic acids and changing the DNA of microorganisms, UV radiation has the power to kill germs, bacteria, and viruses. It is a 200–280 nm short-wavelength laser that damages DNA and inhibits the virus from reproducing further. To monitor temperature, thermal imaging systems and non-contact infrared thermometers (NCITs) utilize numerous types of infrared innovations. The measurement of body temperature by infrared-based technique is the main focus of the current study. This employs a non-contact, full-auto infrared smart temperature collection process that affects body temperature readings. Then there's the automatic voice alarm for those with an unusual body temperature, which can increase the people's checkup rate while lowering infection risk with excellent accuracy. In the event of an abnormal body temperature, the auto warm voice will sound an alarm. The hand wash smart spray disinfection non-touch hands-sensing auto liquid dispersion gadget is also significant. This has a high level of sterilization and safety. Furthermore, QR Code, this feature will be used to update the location. If the employee is at any facility, the function can scan their attendance. Meanwhile, if there are visitors, the person will be detected because visitors will not have verification. Masks are one of the few COVID-19 precautions accessible in the absence of immunization, and they serve a critical role in safeguarding people's health from respiratory infections. The issue is that some people do not want to wear a face mask, and instructing them physically by police to avoid infection is difficult, especially in a large or public setting. One of the key functions of the proposed system is to build a software tool called Face Mask Detection (FMD) that uses a thermal camera to recognize any face that is not wearing a mask in a given public location because they occasionally forget to bring their mask. It will be more convenient and easier for people to use Mask Provider if they neglect to pack the most crucial thing nowadays, which is a mask. The user's health state, as sent by the Health Signal where green represents normal temperature (37.50 C), Yellow means that the system is analysing, and Red is the abnormal temperature (> 37.50) and in this case, a red lamp and sound alarm would be operated to notify the security to take the appropriate measures to prevent the entry of suspected COVID19 carriers.

4 Conclusions

This article offered a new IoT-based design for a standalone smart disinfection tunnel. A thermal temperature scanner, hand sanitizer, sprinkler tunnel, health signal, mask provider, and QR code scanner are all part of the proposed research. This research could help prevent and control future pandemics of compatible events. Nonetheless, it is a fascinating system that can be further refined for usage in vivo. As a result, it's critical to use new technologies like artificial intelligence and 5G to improve robot efficiency and performance while also considering cost and production variables in order to expand access to robotic solutions globally.

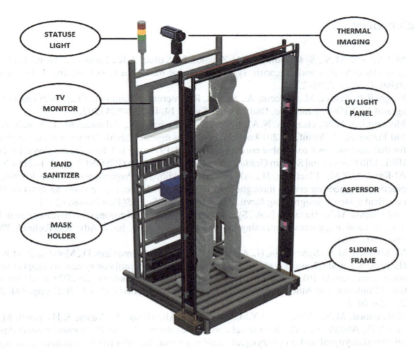

Fig. 4 Final design of standalone smart disinfection system

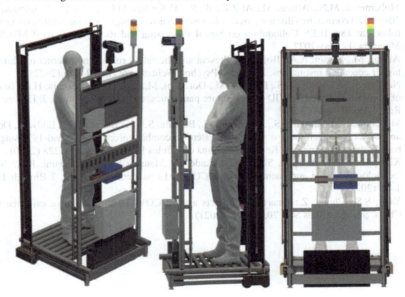

Fig. 5 Overall 3D design standalone smart disinfection system

References

1. Mohammed, M.N., Syamsudin, H., Al-Zubaidi, S., Sairah, A.K., Ramli, R., Yusuf, E.: Novel covid-19 detection and diagnosis system using Iot based smart helmet. Int. J. Psychosoc. Rehabil. **24**(7), 2296–2303
2. Singh, R.P., Javaid, M., Haleem, A., Suman, R.: Internet of things (IoT) applications to fight against COVID-19 pandemic. Diab. Metab. Syndr. **14**(4), 521–524 (2020)
3. Mohammed, M.N., Hazairin, N.A., Arif, A.S., Al-Zubaidi, S., Alkawaz, M.H., Sairah, A.K., Md Jazlan, A.I.I., Yusuf, E.: 2019 novel coronavirus disease (covid-19): toward a novel design for disinfection robot to combat coronavirus (covid-19) using IoT based technology. In: 2021 IEEE 12th Control and System Graduate Research Colloquium (ICSGRC), pp. 211–216 (2021)
4. Al-Rawi, O.Y.M., Elzefzafy, H., Alkurdi, H., Saddiq, A.: Project-based online learning of practical engineering course throughout COVID-19 pandemic: a case study analysis of MEP electrical systems design using Revit. Appl. Math. Inf. Sci. **15**(4), 479–486 (2021)
5. Mohammed, M.N., Hazairin, N.A., Syamsudin, H.: 2019 novel coronavirus disease (covid-19): detection and diagnosis system using IoT based smart glasses. Int. J. Adv. Sci. Technol. **29**(7), 954–960 (2020)
6. Mohammed, M.N., Syamsudin, H., Abdelgnei, M.A.H., Subramaniam, D., Mohd Taib, M.A.A., Hashim, N., et al.: Toward a novel design for mechanical ventilator system to support novel coronavirus (covid-19) infected patients using Iot based technology. In: 2021 IEEE International Conference on Automatic Control and Intelligent Systems (I2CACIS 2021), pp. 294–298, 26 June 2021
7. Mohammed, M.N., Al-Rawi, O.Y.M., Arif, A.S., Al-Zubaidi, S., Yusof, K.H., Rusli, M.A., Yusuf, E., Abdulrazaq, M.: 2019 novel coronavirus disease (covid-19): toward a novel design for nasopharyngeal and oropharyngeal swabbing robot. In: 18th IEEE Colloquium on Signal Processing and its Applications (CSPA2022), Malaysia, 12 May 2022
8. Mohammed, M.N., Alfiras, M., Al-Zubaidi, S., Al-Sanjary, O.I., Yusuf, E., Abdulrazaq, M.: 2019 novel coronavirus disease (covid-19): toward a novel design for smart waste management robot. In: 18th IEEE Colloquium on Signal Processing and its Applications (CSPA2022), Malaysia, 12 May 2022
9. Alfiras, M., Yassin, A.A., Bojiah, J.: Present and the future role of the Internet of Things in higher education institutions. J. Positive Psychol. Wellbeing **6**(1), 167–175 (2022)
10. Nasajpour, M., Pouriyeh, S., Parizi, R.M., Dorodchi, M., Valero, M., Arabnia, H.R.: Internet of Things for current COVID-19 and future pandemics: an exploratory study. J. Healthc. Inf. Res. **4**(4), 325–364 (2020)
11. Mohammed, M.N., Arif, I.S., Al-Zubaidi, S., Bahrain, S.H.K., Sairah, A.K., Eddy, Y.: Design and development of spray disinfection system to combat coronavirus (covid-19) using IoT based robotics technology. Revista Argentina de Clínica Psicológica **29**(5), 228 (2020)
12. Khan, H., Kushwah, K.K., Singh, S., Urkude, H., Maurya, M.R., Sadasivuni, K.K.: Smart technologies driven approaches to tackle COVID-19 pandemic: a review. 3 Biotech **11**(2), 1–22 (2021)
13. Yang, S.S., Chong, Z.: Smart city projects against COVID-19: quantitative evidence from China. Sustain. Cities Soc. **70**, 102897 (2021)

Modeling of Cutting Forces When End Milling of Ti6Al4V Using Adaptive Neuro-Fuzzy Inference System

Salah Al-Zubaidi, Jaharah A. Ghani, Che Hassan Che Haron, Hakim S. Sultan, Adnan N. Jameel Al-Tamimi, Mohammed N. Abdulrazaq Alshekhly, and M. Alfiras

Abstract The forces that are acting on a tool throughout the process of metal cutting are extremely important. An understanding of these forces can help to estimate the power required when cutting, as well as helping to ensure the subsequent structures are free from vibration and are adequately rigid. Several techniques have been utilised to model the responses throughout the metal cutting process. Adopting soft computing approaches has driven the artificial intelligence to solve complex, nonlinear problems in different fields, such as modelling and predicting problems in metal cutting. In this study, an adaptive neuro-fuzzy inference system (ANFIS) has been used for predicting the cutting forces during dry end milling with uncoated insert of a Ti6Al4V alloy. The developed ANFIS model has been trained and tested with real experimental case study. Membership function with generalized bell shape was used with numbers 2, 3, 4, and 5. The absolute percentage error (MAPE%) was calculated for models to pick up the minimum one. Good matching was obtained between experimental and ANFIS results.

S. Al-Zubaidi
Department of Automated Manufacturing Engineering, Al-Khwarizmi College of Engineering, University of Baghdad, Baghdad 10071, Iraq

S. Al-Zubaidi · M. N. A. Alshekhly (✉)
Mechanical Engineering Department, College of Engineering, Gulf University, Sanad 26489, Kingdom of Bahrain
e-mail: dr.mohammed.alshekhly@gulfuniversity.edu.bh

J. A. Ghani · C. H. C. Haron
Department of Mechanical and Manufacturing Engineering, Faculty of Engineering and Built Environment, Universiti Kebangsaan Malaysia, 43600 Bangi, Selangor, Malaysia

H. S. Sultan
University of Warith Al-Anbiyaa, College of Engineering Karbala, Karbala, Iraq

A. N. J. Al-Tamimi
College of Technical Engineering, Al-Farahidi University, Baghdad, Iraq

M. Alfiras
Electrical and Electronic Engineering Department, College of Engineering, Gulf University, Sanad 26489, Kingdom of Bahrain

© The Author(s), under exclusive license to Springer Nature Switzerland AG 2024
A. Hamdan and E. S. Aldhaen (eds.), *Artificial Intelligence and Transforming Digital Marketing*, Studies in Systems, Decision and Control 487,
https://doi.org/10.1007/978-3-031-35828-9_51

Keywords End milling · Ti6Al4V · ANFIS · Uncoated cutting tool · Dry cutting · SDG 8

1 Introduction

Cutting forces have been measured in almost every machining study. In addition to their importance in designing the machine tool, cutting tools and fixtures [1], they have also proven useful in evaluating machinability. The Dynamometer was developed for measuring tool forces, which are based on the measurements of elastic deflection on the tool under a heavy load. The cutting forces that are generated during machining will have an effect on the power consumption and product quality. Therefore, to reduce power consumption, these cutting forces should be decreased as much as possible. Substantial research has been conducted that investigates the relationship between the cutting forces and the other machining parameters. Matsumoto et al. [2] investigated the relationship between cutting forces and the hardness of the steel, discovering that chip segmentation and high cutting forces are produced because of high friction forces generated on the tool rake face. The Ti6Al4V alloy is exceptionally versatile, and has been utilized by the chemical, biomedical, aerospace, and nuclear industries. Extreme caution should be taken when machining, as this alloy has been categorized as being difficult to use during the machining process. Improper cutting conditions can led to low quality products and substantial machining costs. As the number of governing variables rises, the machining process subsequently becomes more complex and difficult [3]. Konig [4], mentioned has noted that the stresses induced during machining of the Ti6Al4V alloy are three to four times greater than those generated in steel machining. This high stress has led to strength reduction due to softening. Researchers have become increasingly interested in the modeling of cutting forces generated in metal cutting in more recent years. There are many contemporary techniques that are used in the modeling and cutting process, especially turning and milling. One of the more modern techniques involves artificial intelligence (AI), and consists of several soft computing methods [5, 6]. The adaptive neuro-fuzzy inference system (ANFIS) is one of them. ANFIS exploits the capability of neural networks in function approximation and fuzzy knowledge [7–9]. The processing of input–output data with low computational features can be done by neural networks, while high cognitive features can be achieved by fuzzy logic during processing of natural language and approximate [8]. ANFIS can solve the hidden correlation between input–output in neural networks [10]. Dweiri et al. [11] developed ANFIS model to estimate the surface roughness during down milling of Alumic-79. Four flutes performed better performance than two flutes and good prediction was done by ANFIS. ANFIS was also applied by Lo [12] for surface roughness prediction of AA 6061 end milled surface. 48 experiments were used for training, while 24 were used for testing. Using of triangle membership function maintained better prediction. Göloğlu and Arslan [13] developed three artificial models: GP, ANN, and ANFIS. Their accuracy was compared in terms of the predicted surface roughness of the four flutes milled 40CrMnNiMo8-6–4 alloy based on zigzag motions. The experimental runs were designed by Taguchi method. The best results were obtained using the GP model with a high level of accuracy. Ho et al. [14] integrated the ANFIS with a composite Taguchi- genetic learning algorithm to model the surface roughness in end milling using the experimental results found in [12] to test the reliability of the proposed composite approach. The authors then compared their new method with the relevant literature and found it to be slightly more accurate. Uros et al. [15] utilized the generated signals of cutting forces produced during end milling of CK45 (XM) and CK 45 alloys to predict the flank wear by ANFIS algorithm. Good agreement was achieved between real and estimated results when using trapezoidal and triangular membership functions. Also Dong and Wang [16] applied ANFIS and one-out validation methods for estimation the machined roughness based on the experimental findings available in [12]. Good accuracy was achieved between real and desired surface roughness by using the proposed method. The cutting forces produced due to machining process were investigated using ANFIS algorithm by Sparham et al. [17]. The developed model predicted cutting forces with low error compared with experimental data. Two artificial intelligence models were developed by Sen et al. [18] for prediction machining performance measure of Inconel alloy. The finding showed that ANFIS model was superior in prediction performance than ANN model. The machinability of Al-Mg2Si composite materials was estimated by ANFIS model [19]. Number and types of membership functions were significant factors in prediction capability of the developed model. The results show also that the surface finish has been improved during machining due to bismuth additive. Also the cutting forces were predicted by ANFIS during turning of modified Al–Si–Cu alloy by Marani et al. [20]. Cutting speed, feed rate, and Si spacing were the ANIFS inputs. The results show that cutting forces was successfully predicted by ANFIS model and was at the minimum level with Bismuth refiner compared with other elements. Based on reviewed works, the most modeled response by ANFIS was surface roughness in contrast a little work conducted on the tool wear and cutting forces particularly aerospace alloys. Therefore, this study has been conducted to apply ANFIS algorithm for prediction the resultant cutting forces generated during dry end milling with uncoated insert of Ti6Al4V.

2 Architecture of ANFIS

In this study, a first degree Sugeno type ANFIS will be applied. The rules of ANFIS architecture consisted of two inputs, x and y, and one output, f, are described below:

Rule 1:

$$f \ (x \ is \ A1) and \ (y \ is \ B1) then :$$

$$f1 = \alpha 1x + \beta 1y + \Omega 1 \ if(x \ is \ A1) \ and \ (y \ is \ B1) \ then : f\alpha 1x + \beta 1y + \Omega 1I \quad (1)$$

Rule 2:

$$\text{If}(x \text{ is } A2) \text{ and } (y \text{ is } B2) \text{ yeilds} : f2 = \alpha 2x + \beta 2y + \Omega 2 \tag{2}$$

These two rules have linear and nonlinear parameters, the former being represented by $\alpha 1, \beta 1, \Omega 1, \alpha 2, \beta 2, \Omega 2$ while A1, B1, A2, B2are represented in the latter. The ANFIS incorporates five layers namely; fuzzy, product, normalized, de-fuzzy and output layers [21]. The output of the ith and jth nodes of the first layers is indicated through the following equations:

$$Q1, i = \mu Ai(x), i = 1, 2 \tag{3}$$

$$Q1, j = \mu Bj(y), j = 1, 2 \tag{4}$$

$\mu Ai(x)$ and $\mu Bi(y)$ are the membership function of the inputs. The product layer accumulates the output signals from the previous layer in order to obtain the output function or weight function w1 and w2:

$$Q2, i = wi = \mu Ai(m)\mu Bi(n), i = 1, 2 \tag{5}$$

The weight functions are then normalized in the third layer:

$$Q3, i = Wi = \frac{wi}{w1 + w2} \tag{6}$$

The output of the de-fuzzy layer is indicated below:

$$Q4, i = Wifi = Wi(\alpha ix + \beta iy + \Omega i), i = 1, 2 \tag{7}$$

The result of the output layer is a summary of the input signals Q4,i which denote the cutting forces that are a result of it:

$$Q5, i = \sum_{i}^{2} Wifi \tag{8}$$

As Jang [7] reported, the fuzzy inference system (FIS) performs the Fuzzy reasoning steps, as indicated below:

- The input variables are compared with membership functions on the linear parameters part, in order to obtain membership function values of each linguistic label. This is called the fuzzification step.
- Getting firing strength (weights) through integrating the values of membership functions, on the linear part of each rule.

Modeling of Cutting Forces When End Milling of Ti6Al4V Using ...

- Generating the nonlinear parameters of each rule, in accordance with the firing weight or strength.
- Producing the output through the de fuzzification step.

3 Experimental Work

Elmagrabi [4] conducted substantial experimental works which were subsequently used in this study to investigate how ANFIS is able to estimate the resultant cutting forces. The experiments involved the construction of second order polynomial models for prediction surface roughness, tool life, and cutting forces by using response surface methodology (RSM) as well as experimental factorial design for the dry end milling of TiAl4V alloy. The experimental data that has been dependent in the current study involves three input parameters namely cutting speed, feed rate, and depth of cut. Meanwhile, the selected output response is the resultant cutting forces. Three levels were chosen for each parameter in following order: 50, 77.5, 105 m/min; 0.1, 0.15, 0.2 mm/tooth, and 1, 1.5, 2 mm respectively. 8 mm is radial depth of cut which is kept constant. The titanium alloy has been end miled by CNC vertical machine.

According to Elmagrabi [4], the RSM was combined with factorial design of experiment and utilized a Box-Behnken as design scheme. The selection was attributed to set of points including: it has small design points compared with central composite design with same parameters. Also this design does not have axial points that place the parameters in the safe cutting zone. Moreover, it involves five replications points to enable interaction investigation besides independent parameters. Furthermore, the five design points were replicated in order to investigate the effects and interactions among the three independent cutting parameters. This design generates 17 experimental runs based on the above conditions and this data set will be divided later into two subsets: one for training ANFIS model and the second for testing it. The milling experiments were performed with uncoated insert and under dry cutting circumstances. The x, y, and z components of the generated cutting forces were measured during end milling by utilizing a Kistler dynamometer model 9255B, which was acquired through a Dasylab Data acquisition system.

4 Results and Discussions

Like any other artificial intelligence system, the adaptive neuro-fuzzy inference system requires training data and should be tested with new data sets that differentiate from the training set. The ANFIS model was trained with the first fourteen experiments.

The accuracy of ANFIS models was examined with new experiments that have not been seen by the ANFIS before. The last three runs were selected for testing ANFIS models. According to literatures, there is no clear rule or guide concern with determining the amount of training and testing data for ANFIS except simple hint presented by Zain et al. [22] who stated that the amount of training and testing data for Artificial neural networks (ANN) could be (75%, 25%), (80%, 20%), or (85%, 15%). Therefore, and based on this hint, the first 14 runs have been selected arbitrary as 80% of the whole data and the last three run for testing.

Figure 1 presents the training flow chart of ANFIS that predict the resulting cutting forces that were generated during Ti6Al4V dry end milling with uncoated cutting tool. The training finishes when the designated training epochs number have been reached while testing finishes when the ANFIS model tested with all new testing data.

It should be noted that this architecture consists of 27 fuzzy rules. The 27 (If Then Fuzzy Rules) are a conditional statement in the form of (IF Then) and they have been obtained based on three input parameters three level (low, medium high). Each of these parameters has three membership functions, so the total If Then Fuzzy Rules will be $3^3 = 27$ Rules. When modeling using ANFIS, three types of membership functions were used from the beginning based on their training accuracies which were better than others. These included gbell, trapezoidal and triangular functions. After that the results of trapezoidal and triangular have been discarded because they were bad in testing phase. Only gbell provided the desired results because it could predict cutting forces with high accuracy in both training and testing phases compared with other two functions. Therefore, only the results of gbell are presented. The number of membership functions was 2, 3, 4, and 5 in order to show their effect on the prediction accuracy. The range of Epoch's number was 50–1000 and 80 models were developed. The criterion that has been determined to select the best ANFIS model is that one achieves lowest mean absolute error (MAPE%) in the testing phase. As indicated above, the training data set was the first 14 runs, and the last three runs were testing data. After training and testing the data, only three models were selected for evaluations which were: models A, B, and C respectively. The training epochs for them were 150, 350, and 450 respectively.

The step-size profile is designed based on the curve behavior when it first increases until approach certain peak point and at the end decreases during training. The experimental and predicting resultant cutting forces are shown in Tables 1 and 2 for the three trained and tested ANFIS models.

There is generally a very good matching between real and desired resultant cutting forces in training phase of the three ANFIS models. It can be concluded from Tables 1 and 2 that the ANFIS models have a very high accuracy in training and testing, despite the low training and testing data sets (which involved 14 runs for training and 3 runs for testing). There are comparable average errors in the training phase for three ANFIS models (0.8163) and this means that the accuracy in training phase reached (99.1837%). From the other side, the minimum average error in the testing phase is 2.789916 (i.e. the accuracy in testing phase is (97.21%).The accuracy of Models B&C was found to be 99.1837% in training, and 95.75% and 95.51% in

Fig. 1 Training flow chart of ANFIS for prediction resultant cutting forces

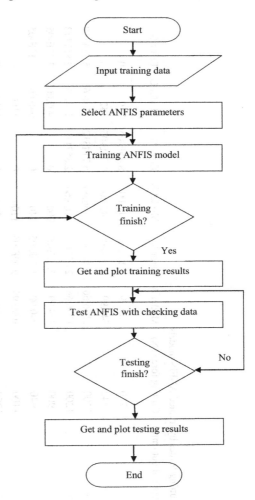

testing respectively. The training accuracy is fundamentally identical for the three models, and quite similar to each other in testing. To summaries, Model A is the best model and has a minimum MAPE% maximum accuracy and lower training epochs in comparison to the other two models. The training average error was significantly less than testing error because ANFIS models have been trained on input–output pair patterns and there is forward and backward phases in training. While in testing the ANFIS models were tested with just new input data without output and there is just forward phase, so it is normal the training error be less than testing error.

These parameters were cutting speed, feed rate, and depth of cut respectively. The gbell function is characterized by three areas; small medium and high. The membership function of the feed rates was substantially changed while those that stand for depth of cut and cutting speed were slightly altered. That means that the

Table 1 Experimental and predicted resultant cutting forces by trained ANFIS models

No.	Cutting speed (m/min)	Feed rate (mm/tooth)	Depth of cut (mm)	Cutting force (N)	Model A	Error %	Model B	Error %	Model C	Error %
1	77.5	0.1	1	710	709.99	8.45E-05	710.00	8.45E-05	710	0
2	105	0.1	1.5	742	741.99	5.39E-05	742.00	1.34E-05	742	0
3	77.5	0.15	1.5	1100	1166.66	5.714312	1166.66	5.714312	1166.66	5.714312
4	77.5	0.15	1.5	1200	1166.66	2.857113	1166.66	2.857113	1166.66	2.857113
5	50	0.15	1	900	899.99	2.22E-05	900	0	900.00	3.33E-05
6	77.5	0.2	1	940	940.00	1.06E-05	940	0	940.00	2.12E-05
7	105	0.15	2	1100	1099.99	9.09E-05	1100	0	1100	0
8	105	0.2	1.5	1250	1250	0	1250	0	1250	0
9	50	0.15	2	750	750.00	2.66E-05	749.99	1.33E-05	750.00	5.33E-05
10	105	0.15	1	970	969.99	2.06E-05	969.99	4.12E-05	970.00	1.03E-05
11	77.5	0.15	1.5	1200	1166.66	2.857113	1166.66	2.857113	1166.66	2.857113
12	50	0.2	1.5	680	679.99	4.41E-05	680	0	680	0
13	77.5	0.2	2	800	800	0	799.99	2.5E-05	800	0
14	77.5	0.1	2	767	767.00	1.30E-05	766.99	1.30E-05	767	0
Average error						0.816350		0.816337		0.816332

Table 2 Experimental and predicted resultant cutting forces by tested ANFIS models

No.	Cutting speed (m/min)	Feed rate (mm/tooth)	Depth of cut (mm)	Cutting force (N)	Model A	Error %	Model B	Error %	Model C	Error %
15	50	0.1	1.5	783	750.56	4.142695	716.86	8.446905	710.64	9.241085
16	77.5	0.15	1.5	1150	1166.66	1.449275	1166.66	1.449275	1166.66	1.449274
17	77.5	0.15	1.5	1200	1166.66	2.777778	1166.66	2.777778	1166.66	2.777779
Average error						2.789916		4.224653		4.489379

more significant factor that affects the resultant cutting forces when using an uncoated tool is actually feed rate. This is compatible with previous findings [4] that also state that the most significant factor that affecting cutting force is the feed rate. According to Ghani et al. [22] the high cutting forces are generated due to high feed rate that increases both of frictional heat at the tool cutting edge and the power consumption required to remove material in chip form. The result is more tool wear and less tool life. Furthermore, low forces could also result in longer tool life. For cutting thin sections and unsupported beam, low cutting forces are required where low level of residual stresses is generated and the material properties are preserved to some extent.

This is satisfied in design point 8 where high cutting speed (105 m/min) accompanied with high feed rate (0.2 mm/tooth) generates high cutting force (1250 N) while reducing feed rate to 0.1 mm/tooth, yields low cutting force (742 N). Thinner chips that are produced due to high cutting speed reduce the lengths or areas on the tool-chip and tool work piece, and also allowed softening also in the shear zone. A very good agreement can be depicted in Fig. 2 between the real and predicted cutting forces of ANFIS model in training phase. In contrast, how the measured cutting forces are scatter from the predicted one in illustrated in Fig. 3. It shows good correlation that follows a 45° line. The range of predicted values was (680–1167 N). The predicted values are accurately close to actual values, reflecting the reliability of the ANFIS model. These findings are considered adequate because, in this study, low training and testing data sets were used with high accuracy during the training and testing phases.

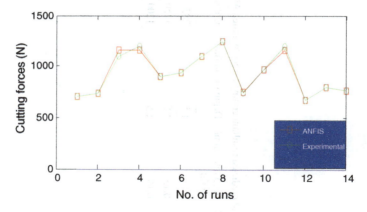

Fig. 2 Experimental and ANFIS predicted cutting forces

Fig. 3 Scatter diagram between predicted and actual values of cutting forces

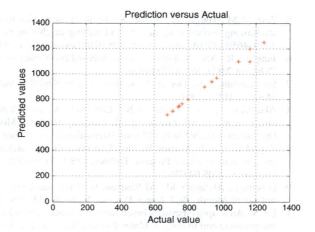

5 Conclusions

In this study, ANFIS models have been developed to predict the resultant cutting force generated during dry end milling with uncoated tool of Ti6Al4V alloy. The best model that achieved minimum MAPE and highest accuracy of 99.1837% in training and 97.21% in testing was Model A. the membership function and epochs of this model was gbell membership function and 150 epochs. Despite the low training and testing of the data sets, compared to other literature, excellent results have been obtained, and these models can be used efficiently in predicting cutting forces that are generated throughout cutting. For citations of references, we prefer the use of square brackets and consecutive numbers. Citations using labels or the author/year convention are also acceptable. The following bibliography provides a sample reference list with entries for journal articles [1], an LNCS chapter [2], a book [3], proceedings without editors [4], as well as a URL [5].

References

1. Matsumoto, Y.M., Barash, M., Liu, C.R.: Cutting mechanism during machining of hardened steel. Mat. Sci. Tech. **3**, 299–305 (1987)
2. Sivarao, B.P., El-Tayeb, N.S.M., Vengkatesh, V.C.: ANFIS modeling of laser machining responses by specially developed graphical user interface. Int. J. Mech. Mechat. Eng. **9**(9), 181–189 (2009)
3. Konig, W.: Applied research on the machinability of Titanium and its alloy. In: Proceedings of the 47th Meeting of AGARD, London, pp. 1.1–1.10 (1979)
4. Zain, A.M., Haron, H., Sharif, S.: Prediction of surface roughness in the end milling machining using Artificial Neural Network. Expert Syst. Appl. **37**(2), 1755–1768 (2010). https://doi.org/10.1016/j.eswa.2009.07.033

5. Zain, A.M., Haron, H., Sharif, S.: Application of GA to optimize cutting conditions for minimizing surface roughness in end milling machining process. Expert Syst. Appl. **37**(6), 4650–4659 (2010). https://doi.org/10.1016/j.eswa.2009.12.043
6. Jang, J.S.R.: ANFIS: adaptive-network-based fuzzy inference system. IEEE Trans. Syst. Man Cybernet **23**(3), 665–685 (1993)
7. Sivanandam, S.N., Deepa, S.: Introduction to Neural Networks Using MATLAB 6.0. Tata McGraw-Hill (2006)
8. AbdulRazzaq, M., Ariffin, A.K., El-Shafie, A., Abdullah, S., Sajuri, Z.: Prediction of fatigue crack growth rate using rule-based systems. In: Modeling, Simulation and Applied Optimization (ICMSAO), 2011 4th International Conference on, 19–21 April, pp. 1–8 (2011)
9. Zuperl, U., Cus, F., Mursec, B., Ploj, T.: A generalized neural network model of ball-end milling force system. J. Mater. Process. Technol. **175**(1–3), 98–108 (2006). https://doi.org/10.1016/j.jmatprotec.2005.04.036
10. Dweiri, F., Al-Jarrah, M., Al-Wedyan, H.: Fuzzy surface roughness modeling of CNC down milling of Alumic-79. J. Mater. Process. Technol. **133**, 266–275 (2003)
11. Lo, S.: An adaptive-network based fuzzy inference system for prediction of workpiece surface roughness in end milling. J. Mater. Process. Technol. **142**(3), 665–675 (2003). https://doi.org/10.1016/s0924-0136(03)00687-3
12. Göloğlu, C., Arslan, Y.: Zigzag machining surface roughness modelling using evolutionary approach. J. Intell. Manuf. **20**(2), 203–210 (2008). https://doi.org/10.1007/s10845-008-0222-1
13. Ho, W.-H., Tsai, J.-T., Lin, B.-T., Chou, J.-H.: Adaptive network-based fuzzy inference system for prediction of surface roughness in end milling process using hybrid Taguchi-genetic learning algorithm. Expert Syst. Appl. **36**(2), 3216–3222 (2009). https://doi.org/10.1016/j.eswa.2008.01.051
14. Uros, Z., Franc, C., Edi, K.: Adaptive network based inference system for estimation of flank wear in end-milling. J. Mater. Process. Technol. **209**(3), 1504–1511 (2009). https://doi.org/10.1016/j.jmatprotec.2008.04.002
15. Minggang Dong, N.W.: Adaptive network-based fuzzy inference system with leave-one-out cross-validation approach for prediction of surface roughness. Appl. Math. Model. **35**, 1024–1035 (2011). https://doi.org/10.1016/j.apm.2010.07.048
16. Sparham, M., Sarhan, A.A., Mardi, N.A., Hamdi, M., Dahari, M.: ANFIS modeling to predict the friction forces in CNC guideways and servomotor currents in the feed drive system to be employed in lubrication control system. J. Manuf. Processes **28**, 168–185 (2017)
17. Sen, B., Mandal, U.K., Mondal, S.P.: Advancement of an intelligent system based on ANFIS for predicting machining performance parameters of Inconel 690–A perspective of metaheuristic approach. Measurement **109**, 9–17 (2017)
18. Marani, M., Songmene, V., Zeinali, M., Kouam, J., Zedan, Y.: Neuro-fuzzy predictive model for surface roughness and cutting force of machined Al–20 Mg 2 Si–2Cu metal matrix composite using additives. Neural Comput. Appl. **32**(12), 8115–8126 (2020)
19. Marani, M., Zeinali, M., Farahany, S., et al.: Neuro-fuzzy based predictive model for cutting force in CNC turning process of Al–Si–Cu cast alloy using modifier elements. SN Appl. Sci. **3**, 72 (2021)
20. Sumathi, S., Surekha, P.: Computational Intelligence Paradigms Theory and Applications using MATLAB. Taylor & Francis Group (2010)
21. Elmagrabi, N.H.E.: End Milling of Titanium Alloy Ti-6Al-4V with Carbide Tools Using Response Surface Methodology. Universiti Kebangsaan Malaysia, Bangi (2009)
22. Ghani, J.A., Choudhury, I.A., Masjuki, H.H.: Performance of P10 TiN Coated carbide tools when end milling AISI H13 tool steel at high cutting speed. J. Mater. Process. Technol. **153–154**, 1062–1066 (2004)

Impact of the Pandemic—Covid-19 on Construction Sector in Bahrain

Zuhair Nafea Alani, Mohammed N. Abdulrazaq Alshekhly, Hamza Emad, Adnan N. Jameel Al-Tamimi, and M. Alfiras

Abstract Since the level and severity of Covid-19 pandemic cases vary by location and due to different kinds of projects, it is one of the most important risk management elements globally today. Covid-19 caused numerous financial and non-financial losses to the communities and the economy. Owing to the lockdown, disruption in supply chain management, health and safety standards, transportation, and commercial effects, this pandemic infection had an influence on the entire world. The current study is focused on analyzing the influence of COVID-19 on the construction sector of Bahrain, as the construction industry is a major contributor to the economy. This research has been divided into three parts. The first deals with reviewing the relevant literature, and the second concerns questioning ten companies with 60 projects under construction, working in Bahraini's construction sector—as a case study. These cases have been investigated to diagnose the most affecting Factors on the construction projects in Bahrain due to covid-19, which is the third part of the study. The results show that 60–90% of the participants testified that COVID-19 caused some delays to their projects, cash flow delays, importing material, and deferring projects once or more. It was found that all the participants (100%) denied that their projects were wholly stopped due to a lack of staff or materials.

Z. N. Alani
Architectural and Interior Design Engineering Department, College of Engineering, Gulf University, Sanad 26489, Kingdom of Bahrain
e-mail: coo@gulfuniversity.edu.bh

M. N. A. Alshekhly (✉)
Mechanical Engineering Department, College of Engineering, Gulf University, Sanad 26489, Kingdom of Bahrain
e-mail: dr.mohammed.alshekhly@gulfuniversity.edu.bh

H. Emad
Ministry of Education, Manama, Kingdom of Bahrain

A. N. J. Al-Tamimi
College of Technical Engineering, Al-Farahidi University, Baghdad, Iraq

M. Alfiras
Electrical and Electronic Engineering Department, College of Engineering, Gulf University, Sanad 26489, Kingdom of Bahrain

© The Author(s), under exclusive license to Springer Nature Switzerland AG 2024
A. Hamdan and E. S. Aldhaen (eds.), *Artificial Intelligence and Transforming Digital Marketing*, Studies in Systems, Decision and Control 487,
https://doi.org/10.1007/978-3-031-35828-9_52

Keywords Construction projects · Covid-19 · Construction management · SDG 3 · SDG 9 · SDG 11

1 Introduction

Beginning in 2020, COVID-19 began to spread and quickly became an epidemic, creating an unanticipated dire condition for all of human society. The COVID-19 Pandemic has had a negative impact on both the entire economy and the construction sector [1]. Even with continued government financial assistance, contractors are still experiencing losses and dealing with other issues brought on by the Pandemic [2]. Additionally, the global epidemic of COVID-19 altered the view of the world economy. In fact, the Middle East's continuous epidemic is the main reason for project delays. The Pandemic has had a strong effect on every industry. One of the sectors most susceptible to coronavirus infection is the construction industry since it cannot operate remotely [3].

Following the global spread of this new virus, the World Health Organization (WHO) has proclaimed COVID-19 to be a pandemic. An unprecedented challenge with unpredictably negative economic effects was imposed by the spread of COVID-19 [3]. Owing to the outbreak of COVID-19 in these nations and the ensuing economic impact of a nationwide lockdown. The construction industry, which serves as the foundation of the national economies of several Middle Eastern countries, is also confronted with a special difficulty [4]. Construction is one of the labor-intensive businesses affected by the COVID-19 outbreak. Construction workers face a significant risk of COVID-19 transmission, and their knowledge, attitudes, and practices (KAP) are crucial to the virus's control. Major disruptions have occurred in the world economies. Many nations that were already experiencing economic problems prior to the Pandemic have found that the decrease in the aviation industry and the drop in worldwide demand for goods and services have been devastating [4]. The COVID-19 pandemic threw everything into disarray worldwide, posing new difficulties for individuals and businesses. People have faced a variety of challenges, from self-isolation to searching for necessities. In order to accommodate the huge demand for medical supplies, logistics, and consumers' daily requirements, several firms shut down while others struggled. According to a report from the Organization for Economic Cooperation and Development (OCED) on the COVID-19 impact, the GCC and North Africa public health implications are far less than anticipated. Because of the epidemic, the economic and social repercussions are worse. Governments are implementing initiatives to loosen the limitations and revive the economy because the impact will be lessened after July 2020. According to the United Nations Economic and Social Commission for West Asia, an extra 8.3 million people would experience poverty as a result of the Pandemic's effects on the economy (OECD, 2020). Although the GCC countries are in control of the COVID-19 virus impact, they could not halt the economic collapse due to the global recession [1, 5]. The COVID-19 epidemic could have unanticipated socioeconomic repercussions that will damage the global construction sector. The COVID-19 Pandemic impacted all facets of society and all economic sectors in terms of healthcare, the economy, and social issues. One of the primary pillars of the economy and an important tributary is the construction industry [6]. A new coronavirus disease (COVID-19) first surfaced worldwide in December 2019. The construction business has also been significantly impacted by the epidemic, like many other industries. The underlying labor-intensive nature of construction projects creates additional challenges because onsite task delivery is required for construction works, and social distancing is not always possible on a job site that is actively working. Despite this, regions could still proceed with their construction works [7, 8]. The experts emphasized the potential impact of coronavirus transmission on the ability to initiate and complete building projects on schedule. The UK agency of analytical Global Data revised its projection for the construction sector's volume growth globally in 2020 from 3.1 to 0.5%. There are numerous places where this tendency of updating projections is prevalent (in China, the designed growth in construction volumes reduced from 5.1 to 3.6%, and in Italy—from 1 to 0.7%) [9].

The primary causes of the fall in construction volumes are:

- The decision-makers or government organizations halt the construction of specific projects.
- Delays in the flow of supplies and machinery.
- The eviction of workers as a result of international borders being closed
- The suspension of transportation connections, as China is Europe's primary supplier of resources while Eastern Europe provides the bulk of the continent's labor force.
- Lower productivity of workers as a result of tighter security.
- Worsening financial issues for contractors and material suppliers, given that the majority of enterprises are small businesses without access to capital or stocks [10].

Expert predictions state that 30% of European engineering and construction equipment firms are currently on the edge of bankruptcy, while another 30% are reducing production. March 2020 is projected to see a 75% drop in Middle Eastern contracts [10]. The construction industry has been impacted by COVID-19 in a number of ways, including the loss of labor, a scarcity of supplies, an increase in the price of construction materials, and a shift in demand [11]. Since several sectors of the economy, including tourism and construction, are still closed, the economy is still in decline and will take longer to recover [1]. According to estimates from the International Labor Organization, the COVID-19 countermeasure has reportedly affected almost 2.7 billion workers worldwide or 81% of the world's employment. The crisis has brought on a notable reduction in economic activity and working hours. Reduced consumer demand caused by quarantine policies, store closings, order cancellations, and decreased pay directly affects the state of the labor market. The adverse consequences of COVID-19 also cause a shock in the planning. According to the experts, this is also reflected in the labor market as a drop in labor supply due to an increase in death, illness, and absenteeism for various causes [3, 5]. Cash flow concerns brought on by COVID-19 have affected project managers and contractors. Contractors are required to include an additional payment for their product shipments to be made using alternative routes to eliminate the possibility of a major delay that could hinder the job's progress. Following the contract re-pricing advocated by the project promoters in direct proportion to changes in circumstances could be financially taxing for the contractors (Sihombing, 2020). The entire capacity of a bus has been decreased to 50% as an additional measure against the virus. Since it will take numerous trips to transport enough personnel to handle a single shift, which could be an additional cost to the contractor, these have decreased the number of shifts and hours utilized while on sit [12].

2 The Fieldwork

Through this part of the study, a survey has been conducted on thirty selected Bahrainis projects to determine how the construction sector is really impacted by the Covid-19 epidemic The fieldwork was also added to get in-depth information from site management experts in relation to this impact during construction.

2.1 Design of Questionnaire Collection of Data

The questionnaire was performed using Google Forms, a free electronic tool made available by Google that allows for the voluntary gathering of data using specially built questionnaires. The questionnaire was divided into two parts. The first part was about Companies with their projects under construction and respondents' positions, while the second part was about Companies answers regarding three areas that can diagnose the factors affected theme due to this Pandemic; these areas are:

Area-1: delays of one or more current projects due to Covid-19.

Area-2: Deferred one or more current projects due to Covid-19.

Area-3: completely stopped one or more current projects due to Covid-19.

2.2 Data Analysis Tools

Spss statistical software was adopted for statistical analysis of the whole collected data with the Chi-Square test to determine the relationship of the participant's perception of the impact of COVID-19 on their projects. The results which have been concluded from the (Q.L.) answers were tabulated, as shown in Table 1.

Impact of the Pandemic—Covid-19 on Construction Sector in Bahrain 621

Table 1 Companies and respondents' positions and Number of projects

Company no	Position in the company	Name of the company/type of work	No. of projects under construction
1	Chief Exclusive Officer (CEO)	Bahrain Mechanical Construction Company W.L.L	7
2	Engineer	Electricity distribution	2
3	Civil Engineer	MOW	2
4	Director	Government projects	5
5	General Manager	new island carpentry	20
6	Senior General Engineer	Government field	4
7	General Manager	Buildings maintenance	6
8	Head of Department	Construction field	5
9	Adviser	Governmental institution	5
10	Senior Architectural Engineer	Work at gov field	4

3 The Statistical Analysis

The sample consisted of 11 participants 10 of them had different job positions, and one did not report what job he/she occupied. Similarly, most participants belong to at least nine companies in the governmental and private sectors. More than half of the participants have 1 to 5 employees working under their leadership, 6 (54.6%), whereas 27.3% have more than five employees as shown in Table 2. Figure 1 illustrates the percentage of the Participant's companies and their positions.

Most of the participants' responses fell in favor of the absence of the negative impact of COVID-19 on their projects. This observation is well supported as 60.3% of their responses to the 18 questions were "No." All the participants (100%) denied that their projects were wholly stopped due to a lack of staff or materials. Between 90 and 55% of the participants testified that COVID-19 caused some delays to their projects, cash flow delays, importing material, and deferring projects once or more, as shown in Table 3.

There is a relationship between the participants' position and/or their recorded influence of COVID-19 on their projects. For example, participants who work as Senior generals, architects, civil engineers, and advisers reported similar impressions that COVID-19 had at least a 50% negative impact on their projects. However, those who work as an engineer, CEO, and director reported entirely different views, indicating the absence of a negative impact reaching 100%. Conversely, the Participant who works as head of the department reports a completely negative effect of COVID-19 on their projects, as shown in Fig. 2.

Table 2 The details of the participant's characteristics

Variable	Categories	Count	Percentage (%)
Position	CEO	1	9.1
	Engineer	1	9.1
	Civil Engineer	1	9.1
	Director	1	9.1
	General Manager	2	18.2
	Senior General Engineer	1	9.1
	Head of Department	1	9.1
	Adviser	1	9.1
	Senior Architectural Engineer	1	9.1
	Missing	1	9.1
Company	Bahrain Mechanical Construction	1	9.1
	Construction Field	1	9.1
	Electricity Distribution	1	9.1
	Government Field	2	18.2
	Government Projects	1	9.1
	Governmental Institution	1	9.1
	MOW	1	9.1
	New Island Carpentry	1	9.1
	Buildings maintenance	1	9.1
	Missing	1	9.1
No. of employees	1–5	6	54.6
	>5	3	27.3
	Missing	2	18.1

Fig. 1 The type and positions of the participant's company

Impact of the Pandemic—Covid-19 on Construction Sector in Bahrain

Table 3 The reasons for the delay and stopping of the projects

Statement	No		Yes	
	n(%)		n(%)	
Current projects delayed due to (Covid-19)	1	(10.0)	9	(90.0)
The delay is due to decreasing number of labors to work at the same time	4	(57.1)	3	(42.9)
Delay is due to decreasing day working	5	(71.4)	2	(28.6)
Delay is due to difficulties to import materials	3	(33.3)	6	(66.7)
The delay is due to labors sickness	5	(62.5)	3	(37.5)
The delay is due to lack of staff	5	(62.5)	3	(37.5)
The delay is due to death of some labors	6	(85.7)	1	(14.3)
The delay is due to shortage in cash flow	4	(44.4)	5	(55.6)
One or more projects deferred due to (Covid-19)	4	(40.0)	6	(60.0)
Deferred due to difficulties to import materials	4	(57.1)	3	(42.9)
The deferred due to labors sickness	4	(80.0)	1	(20.0)
The deferred due to lack of staff	4	(57.1)	3	(42.9)
The deferred due to shortage in cash flow	4	(66.7)	2	(33.3)
One or more projects completely stopped due to (Covid-19)	4	(57.1)	3	(42.9)
The projects completely stopped due to difficulties to import materials	6	(100.0)	0	(0.0)
Projects completely stopped due to labors sickness	5	(83.3)	1	(16.7)
Projects completely stopped due to lack of staff	6	(100.0)	0	(0.0)
Total	79(60.3)		51(39.7)	
p-value	0.0343*			

There is a relationship between the participants' companies and/or their recorded effect of COVID-19 on their projects. For example, participants who work in the government field, new island carpentry, MOW, and building maintenance reported similar impressions that COVID-19 had at least a 50% negative impact on their projects. However, construction workers reported entirely different views, indicating

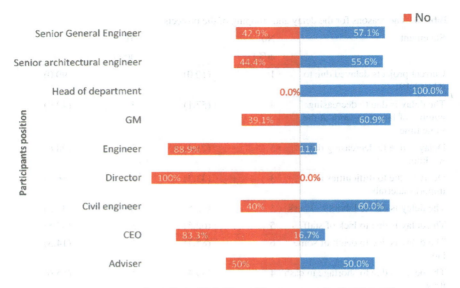

Fig. 2 The relationship between the participants' positions and the reported COVID-19 impact

a negative impact reaching 100%. Conversely, the participants who work for government projects, electricity distribution, and Bahrain mechanical construction companies reported an absence of negative impact of COVID-19 on their projects largely as shown in Fig. 3.

There is no relationship between the number of employees and the magnitude of COVID-19 impact on the participants' projects. However, both categories reported a similar impact ranging from 37.7 to 42.6% as shown in Fig. 4.

4 Conclusions and Remarks

It can be summarized the important conclusions as follows:

1. It was found that 90% of the current projects were delayed due to Covid-19, 60% of current projects were deferred due to Covid-19. While 42.9% of projects completely stopped due to Covid-19.
2. The most affecting factors causing delays are difficulties in importing materials (67%), a shortage in cash flow (55.6%), decreasing the number of laborers to work at the same time (43%), and laborers' sickness, and due to lack of staff (37.5%).

Impact of the Pandemic—Covid-19 on Construction Sector in Bahrain

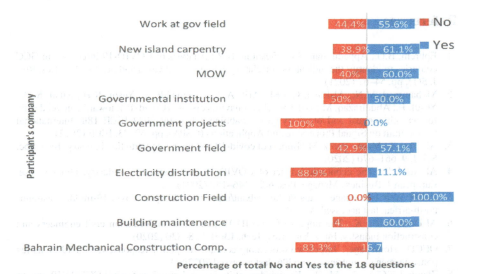

Fig. 3 Relationship between participants' company and perceived impact of COVID-19

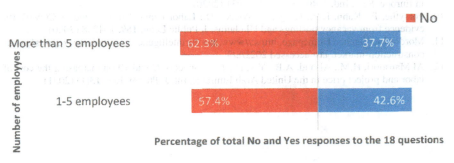

Fig. 4 Relationship between the No. of employees and the reported impact of COVID-19

3. The most affecting factors causing deferred projects occurred due to the difficulties in importing materials (42.9%), Lack of Staff (42.9%), and shortage in cash flow (33.3%).
4. The most affecting factors causing the completely stopped of projects due to Covid-19 are to Labor Sickness (17%), and all the participants (100%) denied that their projects were wholly stopped due to a lack of staff or materials.

References

1. Ephrem, B.G., Appaduraian, S.G., Sekaran, B.R.: Analysis of COVID-19 infections in GCC countries to identify the indicators correlating the number of cases and deaths. PSU Res. Rev. J. **5**(1), pp. 54–67 (2021)
2. Mohammed, M.N., Al-Rawi, O.Y.M., Arif, A.S., Al-Zubaidi, S., Yusof, K.H., Rusli, M.A., Yusuf, E., Abdulrazaq, M.: 2019 novel coronavirus disease (Covid-19): toward a novel design for nasopharyngeal and oropharyngeal swabbing robot. In: 2022 IEEE 18th International Colloquium on Signal Processing and Applications (CSPA), pp. 69–73. IEEE (2022)
3. Al Amri, T., Marey-Pérez, M.: Impact of covid-19 on Oman's construction industry. Tech. Soc. Sci. J. **9**, 661–670 (2020)
4. Al Amri, T.: The economic impact of COVID-19 on construction industry: Oman's case. European J. Business Manage. Res. **6**(2), 146–152 (2021)
5. Relief Web Homepage, https://reliefweb.int/report/world/impact-covid-19-middle-east-and-north-africa, last accessed 2022/8/3
6. Al-Deen Bsisu, K.: The impact of COVID-19 pandemic on Jordanian civil engineers and construction industry. Int. J. Eng. Res. Tech. **13**(5), 828–830 (2020)
7. OECD Homepage, https://www.oecd.org/coronavirus/policy-responses/covid-19-crisis-res ponse-in-mena-countries-4b366396, last accessed 2022/8/3
8. Zheng, L., Chen, K., Ma, L.: Knowledge, attitudes, and practices toward COVID-19 among construction industry practitioners in China. Front. Public Health **8**, 599769 (2021)
9. Williams, C.C., Kayaoglu, A.: COVID-19 and undeclared work: impacts and policy responses in Europe. Serv. Ind. J. **40**(13–14), 914–931 (2020)
10. Forsythe, E., Kahn, L.B., Lange, F., Wiczer, D.: Labor demand in the time of COVID-19: evidence from vacancy postings and UI claims. J. Public Econ. **189**, 104238 (2020)
11. Mordor Intelligence Homepage, https://www.mordorintelligence.com/industry-reports/oman-construction-market, last accessed 2022/8/3
12. Al Mansoori, H.M., Alsaud, A.B., Yas, H.: The impact of Covid 19 on increasing the cost of labor and project price in the United Arab Emirates. Int. J. Pharm. Res. **13**(1) (2021)

The Role of Technology and Market Accessibility on Financial Market Classification

Reem Sayed Mansoor, Jasim Al Ajmi, and Asieh Hosseini

Abstract Early classifications of markets as developed or emerging were somewhat arbitrary and tended to emphasize on the relative wealth of nations as the defining factor, along with subjective assessments of the market's quality. Due to the lack of transparency, investors found it difficult to predict the likelihood that a country would move between categories, and this environment did not encourage nations to adopt international best practices in an effort to advance their status. The market classification process organizes nations according to objective criteria and interacts with stock exchanges, regulators, and central banks in countries where the market is being evaluated for possible promotion or demotion. The process provides portfolio managers and asset allocators a clear picture of what to expect in terms of index evolution in the future. The measures of market accessibility and their effects on market classifications are the main topics of this paper. The study investigates how market infrastructure, market regulations, and market openness to foreign ownership influence market classification. The literature review serves as the foundation for the research methodology. It is concluded that improved market accessibility encourages market classification and allows nations to uphold global equity market standards.

Keywords Market classification · Market accessibility · Financial openness and financial technology

1 Introduction

Market classification is a key aspect that investors consider while investing internationally or more specifically constructing their international portfolio. As a result, in order to attract foreign investment, developing countries typically pay greater attention to strengthening their market transparency standards, which improve asset market allocations and primarily result in favorable macroeconomic factors [24].

R. S. Mansoor (✉) · J. Al Ajmi · A. Hosseini
Ahlia University, Manama, Bahrain
e-mail: rmansoor@ahlia.edu.bh

© The Author(s), under exclusive license to Springer Nature Switzerland AG 2024
A. Hamdan and E. S. Aldhaen (eds.), *Artificial Intelligence and Transforming Digital Marketing*, Studies in Systems, Decision and Control 487,
https://doi.org/10.1007/978-3-031-35828-9_53

Morgan Stanley Capital International (MSCI) uses specific criteria to categorize countries as frontier, emerging, or developed markets [16]. This includes three main criteria: economic development, size and liquidity requirements and market accessibility. Their definition of economic development focuses primarily on promoting economic growth while preserving environmental quality for future generations. Moving to the second criterion size and liquidity requirement, their definition was based on the minimum investment ability requirements set by MSCI Global Standard Indexes. In relation market accessibility criteria, the international investor experience with such a market is a key factor here, focusing on the ease of capital movements, the level of openness to foreign ownership, how efficient the institutional framework is and the availability of investment instruments.

On the other hand, Financial Times classifies countries according to the following criteria: Market quality, materiality, predictability and consistency, cost limitation, market access and stability [8]. Their definition of market quality reflects "the views and practices of the investment community by striking a balance between a country's economic development and the accessibility of the market while preserving index stability". Sub-criteria for market quality include regulation quality, dealing landscape, settlement procedures, and the existence of a derivatives market. Materiality is the amount of material a country must have in order to be included in a global benchmark. The FTSE definition of predictability and consistency addressed the topic of changes in country classification, which is accomplished by establishing a Watch list for countries that may be promoted or demoted. The fourth factor, cost limitation, estimates the cost of a country's reclassification within the market as promoted or demoted. According to FTSE, stability is applied to new countries introduced within a specific classification, which can only occur if there is global acceptance and permanent changes in market status. Finally, they defined market access as the ease with which international investors can invest or withdraw securely and at a reasonable cost.

1.1 The Role of Market Accessibility and Openness on Market Classification

Li et al. [12] explained the origins of market access stating that the term Market Access was first introduced by the World Trade Organization (WTO) to explain the relationship between the imported products and the domestic products of a country that are competing against each other. According to the WTO definition, market access is a set of conditions, tariff and non-tariff measures, agreed by WTO members for the entry of specific goods into their markets, as well as government policies regarding trade barriers in general, and specifically the issues of import substitution (to promote local production) and free competition.

Market accessibility refers to the level of experience that international investors have when investing in a specific market. This experience has additional sub-criteria,

The Role of Technology and Market Accessibility on Financial Market ...

which include (i) capital inflows and outflows within the market; (ii) financial openness [16]. The Framework of MSCI market classification considers market accessibility as a key criterion to reflect the investment community views and practices. Financial Times defines market accessibility as "the ability of international investors to invest and withdraw funds in a timely and secure manner at reasonable cost" [8].

1.1.1 Capital Inflows and Outflows

Özmen and Taşdemir [18] indicated that the push and pull factors shape the pattern of international investment and mainly capital flows. He claimed that over the last two decades' investment in projects and companies in developed economies and emerging markets has grown steadily. The inflow of such capital mainly resulted from developed markets' high liquidity partially, and the lower yields investors gain from these markets. These circumstances, or "Push Factors", influence investor behavior and increase their appetite for investments that not only offer higher yields but also provide diversification benefits to lower portfolio risk. This intensification of capital flow led to favorable outcomes to the respective country including better investment climate, improved fundamentals of macroeconomic and lower country risk. Which ultimately all contributes to the reclassification of such countries with improved conditions.

1.1.2 Financial Openness

Financial openness is the second sub-criteria, it is described as the extent of capital market liberalization [14]. There is a substantial body of literature that investigates the impact of financial openness on economic growth [4, 5, 9, 13, 25]. They argued that the extent of financial openness boosted the flow of capital within the market, decreased the cost of capital and consequently increased investment in the host nation. Their empirical evidence demonstrated that financial openness promotes economic growth, and that greater equity market openness increases the potential for a country's market classification to be promoted. Furthermore, they claimed that increased financial openness resulted in smoother consumption due to stronger financial markets. Or, to put it another way, more financially open nations may attract more foreign investors. They focus on macro and country level analysis which played a key role in easing up the stage's countries require going through to get their market reclassified. Daud and Ahmad [7] revealed that countries with greater financial dependence experience faster growth after openness. As a matter of fact, they asserted that financial openness enabled countries to benefit more from foreign investment growth opportunities. While the above-mentioned studies were applied to countries in emerging markets, developed or frontier markets, their studies were limited in scope because they discarded the period at which some countries got promoted in their market classification and how financial openness as well as capital flows during this shift changed and affected the respective country's economic growth. Drawing inspiration from Efficient market hypothesis investors trading in better classified markets should have access to all public information.

2 The Role of Financial Technology on Market Classification

The world is going digital and the pace of conversion is rising [11]. Advanced economies or better classified markets aim to digitize their financial markets, to have wider access to international investors [1]. Further to this they suggested that digitalization plays a key role in a country's economic growth, and without it is difficult to achieve technological innovation and improve productivity. The outstanding performance of the G7 countries (including the United States, the United Kingdom, Germany, Japan, Italy, and France) in terms of increased economic growth is due, among other things, to their level of technological innovation [25]. They also stated that the business dynamics of such G7 countries have changed in recent years as a result of digitalization manifested in the form of advancements in information and communication technologies (ICTs). Through social change, industrial development, and increased productivity, these advancements contributed to the country's economic growth. Those worthy seven countries represent 58% of the global net wealth and their economic performance is improving in the coming years [17]. Beside that, those countries put considerable effort into billions of dollars spent on Research and Development to boost their innovation performance to achieve high economic growth and maintain their advanced market classifications [17].

Researchers such as [1] conducted a comparison of two market classifications, Emerging and Developed, focusing on Russia and some EU countries in terms of the digital component of their economies. They mainly analyzed the different components of a country's digital economy on social processes and economic growth. According to their findings, Russia has a strong position in the digital economy in terms of ICT development and network readiness. Despite this, Russia as an emerging market lags far behind developed countries in the EU in terms of high-tech exports, which necessitate high research and development investment. To summarize, EU countries, as opposed to Russia, are attracting more international investments as a result of technological advancements that improve investors' access to markets.

At early stages of the twenty-first century mainly focusing on the development of data-centric strategies that changed existing business models [20]. Countries with strategies that emphasized on (DDI) data driven-innovation experienced the emergence of new products, new financial markets, and opportunities in the digital ecosystem [3, 10, 22, 23]. The digital ecosystem includes digital markets at which users actions generates information stored in the form of data. Those data analyzed in depth to find trends and patterns, similarly investors actions in the stock market are monitored to analyze market behavior [21].

2.1 Theoretical and Conceptual Frameworks

This section will be entitle the main theoretical contribution of this paper describing the effect of market Transition from frontier to emerging to developed, or from not being classified at all.

- Efficient Market Hypothesis Theory

Akintoye [2] explained capital market efficiency as the situation in which it is impossible to outperform the market or earn abnormal returns because information is publicly available to all investors and thus assets are fairly priced. Which brings us to the main factors that market classification agencies consider when promoting or demoting countries [16]. As [19] emphasized on the aspect of market quality that can experience observable improvement when market transparency is achieved in a specific country, investors in such countries will have less risk when all necessary information is publicly available.

- Behavioral Finance Theory

This theory combines physiological and behavioral theories along with conventional economics and finance, it classifies the different market anomalies caused by human mistakes. Moreover it is assumed that such theory explains individuals as well as group investors decisions and the behavior of the market. The attributes of market Participants and most importantly agents in addition to market information impose high influence on investment decisions and market outcomes. As a result, investors may choose irrational decisions that cause an effect on shaping the market and the efficiency of capital markets.

- Financial Liberalization Theory

MacKinnon [15] supports this theory suggesting that there is a strong relationship between economic growth and financial development. The more liberalization achieved in a country's financial system, the more economic growth occurs in such a country. Which ultimately facilitates the transition from one market classification to another. According to the World Bank the financial system has five key functions that can be used as criteria to measure a country's financial development. Those functions include: (i) producing information about possible investments and capital allocation, (ii) how investments are monitored and how corporate governance is exerted after providing finance, (iii) Managing risk, implementing diversification and facilitating the trade, (iv) pooling and mobilizing savings; and (v) facilitating the exchange of goods and services.

- Affordance Theory

Costall and Morris [6] supports the idea that perception drives action, "the world is perceived not only in terms of object shapes and spatial relationships but also in terms of object possibilities for action (affordances)". Similarly the possibilities and the advanced technologies financial markets offer drive transition actions mainly promotion in market classification. Investors perceive markets to have easy access, information transparency while in the same time security is maintained protecting their accounts.

3 Conclusion

The literature review concluded that there is relationship between market accessibility and market classifications, that countries with better market accessibility conditions are promoted to higher market classifications considering the better environment they provide to investors. The framework of this study demonstrates how market accessibility promotes or maintains market classification through factors such as market infrastructure, market regulations, and openness to foreign ownership. This study will contribute to the investment community to evaluate country risk and assist in diversification of international portfoilos. In addition this study will fullfil the gap of previous studies focusing on sampling specific market classification and ignoring the transition in classification period at which investor exploit new market opportunities. Regulators, central banks and stock exchanges depend on result of this study to set their strategies aiming to promote market classifications or maintain current classification. In addition portfolio managers and asset allocators focuses on market accessibility while forming their international portfolios. The Direction of the research is do an event study to measure the market accessibility before, during and post market reclassification. Future research can be done by using affordance theory to explore how financial technology and global digital adoption impact market accessibility on frontier and emerging markets in the new era. By fulfilling the gap of previous studies investigating the advanced digital adoption of some classified markets and ignoring the market transition period at which certain criterias are required by markets to offer to global investors.

References

1. Afonasova, M.A., Panfilova, E.E., Galichkina, M.A., Ślusarczyk, B.: Digitalization in economy and innovation: the effect on social and economic processes. Polish J. Manag. Stud. **19** (2019)
2. Akintoye, I.R.: Efficient market hypothesis and behavioural finance: a review of literature. Eur. J. Soc. Sci. **7**(2), 7–17 (2008)
3. Akter, S., McCarthy, G., Sajib, S., Michael, K., Dwivedi, Y.K., D'Ambra, J., Shen, K.N.: Algorithmic bias in data-driven innovation in the age of AI. Int. J. Inf. Manage. **60**, 102387 (2021)
4. Alogoskoufis, G.: Asymmetries of financial openness in an optimal growth model. J. Econ. Asymmet. **23**, e00201 (2021). https://doi.org/10.1016/j.jeca.2021.e00201
5. Ashraf, B.N., Qian, N., Shen, Y.: The impact of trade and financial openness on bank loan pricing: evidence from emerging economies. Emerg. Mark. Rev. **47**, 100793 (2021). https://doi.org/10.1016/j.ememar.2021.100793

6. Costall, A., Morris, P.: The "textbook Gibson": the assimilation of dissidence. Hist. Psychol. **18**(1), 1 (2015)
7. Daud, S.N.M., Ahmad, A.H.: Financial inclusion, economic growth and the role of digital technology. Finance Res. Lett., 103602 (2022). https://doi.org/10.1016/j.frl.2022.103602
8. FTSE Equity Country Classification Process (2022)
9. Khan, A., Hassan, M.K., Paltrinieri, A., Bahoo, S.: Trade, financial openness and dual banking economies: evidence from GCC Region. J. Multinatl. Financ. Manag. **62**, 100693 (2021). https://doi.org/10.1016/j.mulfin.2021.100693
10. Kraus, S., Roig-Tierno, N., Bouncken, R.B.: Digital innovation and venturing: an introduction into the digitalization of entrepreneurship. RMS **13**(3), 519–528 (2019)
11. Kumbure, M.M., Lohrmann, C., Luukka, P., Porras, J.: Machine learning techniques and data for stock market forecasting: a literature review. Expert Syst. Appl., 116659 (2022)
12. Li, S., Xu, N., Hui, X.: International investors and the multifractality property: evidence from accessible and inaccessible market. Physica A **559**, 125029 (2020). https://doi.org/10.1016/j.physa.2020.125029
13. Ma, Y., Yao, C.: Openness, financial structure, and bank risk: international evidence. Int. Rev. Financ. Anal. **81**, 102065 (2022). https://doi.org/10.1016/j.irfa.2022.102065
14. Ma, Y., Jiang, Y., Yao, C.: Trade openness, financial openness, and macroeconomic volatility. Econ. Syst. **46**(1), 100934 (2022). https://doi.org/10.1016/j.ecosys.2021.100934
15. MacKinnon, R.I.: Financial Liberalization in Retrospect: Interest Rate Policies in LDCs. Stanford University (1986)
16. MSCI: Global market accessibility review report (2022)
17. Organisation for Economic Co-operation and Development: OECD Skills Outlook 2019: Thriving in a Digital World. OECD Paris, France (2019)
18. Özmen, E., Taşdemir, F.: Gross capital inflows and outflows: twins or distant cousins? Econ. Syst. **45**(3), 100881 (2021). https://doi.org/10.1016/j.ecosys.2021.100881
19. Papavassiliou, V.G., Kinateder, H.: Information shares and market quality before and during the European sovereign debt crisis. J. Int. Finan. Markets. Inst. Money **72**, 101334 (2021). https://doi.org/10.1016/j.intfin.2021.101334
20. Postema, B.F., Haverkort, B.R.: Evaluation of advanced data centre power management strategies. Elect. Notes Theoret. Comp. Sci. **337**, 173–191 (2018). https://doi.org/10.1016/j.entcs.2018.03.040
21. Seles, R.P., Jabbour, L.S., Beatriz, A., Jabbour, C.J.C., Fiorini, P.C., Mohd-Yusoff, Y., Thomé, A.M.T.: Business opportunities and challenges as the two sides of the climate change: corporate responses and potential implications for big data management towards a low carbon society. J. Cleaner Prod. **189**, 763–774 (2018). https://doi.org/10.1016/j.jclepro.2018.04.113
22. Saura, J.R., Ribeiro-Soriano, D., Palacios-Marqués, D.: From user-generated data to data-driven innovation: a research agenda to understand user privacy in digital markets. Int. J. Inf. Manage. **60**, 102331 (2021)
23. Sultana, S., Akter, S., Kyriazis, E.: How data-driven innovation capability is shaping the future of market agility and competitive performance? Technol. Forecast. Soc. Chang. **174**, 121260 (2022). https://doi.org/10.1016/j.techfore.2021.121260
24. Uras, B.R.: Finance and development: Rethinking the role of financial transparency. J. Bank. Finance **111**, 105721 (2020). https://doi.org/10.1016/j.jbankfin.2019.105721
25. Yuan, S., Musibau, H.O., Genç, S.Y., Shaheen, R., Ameen, A., Tan, Z.: Digitalization of economy is the key factor behind fourth industrial revolution: how G7 countries are overcoming with the financing issues? Technol. Forecast. Soc. Chang. **165**, 120533 (2021)

A Review of the Recent Developments in the Higher Education Sector Globally and in the GCC Region

Elham Ahmed, Amama Shaukat, and Esra AlDhaen

Abstract Higher Education (HE) is one of the rapidly growing sectors worldwide. This research reviews the sector developments and challenges internationally and in the GCC region with the aim of identifying areas for future research and enhancements to improve HEIs performance. Research papers and relevant statistics extracted from official sources were reviewed by the researchers. The review findings revealed that students' enrolment growth, diversification and internationalisation; stakeholders' engagement and competitiveness; emphasis on quality, ranking, transparency, regulation, privatization and professional development; technology advancements and governance structures were some of the recent developments in the HE sector. Challenges included HEIs resources capacity and technology adaptation pressures, students' diversity, stakeholders competing powers and transparency. Research on the developments reported in the GCC context focused on internationalisation, students' body expansion, regulation and quality. As for the challenges, context specific ones included the role of Arabic language in academia, while others were general like the lack of information disclosure. Future researchers and HE policy makers and leaders need to address the challenges identified and benefit from them to achieve financial sustainability, resources innovation and governance effectiveness.

Keywords Higher education · Tertiary education · Gulf Cooperation Council · GCC · Arab gulf · Challenges · Developments

E. Ahmed (✉) · E. AlDhaen
Ahlia University, Manama, Bahrain
e-mail: Elham.Ahmed@brunel.ac.uk

A. Shaukat
Brunel University, London, UK

© The Author(s), under exclusive license to Springer Nature Switzerland AG 2024
A. Hamdan and E. S. Aldhaen (eds.), *Artificial Intelligence and Transforming Digital Marketing*, Studies in Systems, Decision and Control 487,
https://doi.org/10.1007/978-3-031-35828-9_54

1 Introduction

This research reviews current literature on Higher Education (henceforth HE) in the global and Gulf Cooperation Council (GCC) contexts with the aim of identifying the recent developments and trends in the sector. The objective of the paper is to identify gaps in research about HEIs management in the GCC context. The gaps identified will contribute to refining the research problem to be addressed by the researchers as part of future research. The paper is divided into six sections. The first is the introduction. The second highlights the research questions. The third section explores the research methodology. The fourth, which constitutes most of the paper, focuses on the sector history, developments and challenges in the international and GCC-context. The fifth section discusses and conclude the research findings including limitations and suggestions for future research. The researchers end the paper with the references.

2 Research Questions

This paper will review the literature on HE to answer the following questions:

- What are the recent developments and most prevailing challenges in the HE sector internationally and in the GCC context?
- What are the gaps/areas requiring the attention of researchers and HEIs in the GCC region to better improve HEIs performance?

3 Research Methodology

The researchers conducted a critical review of a number of research papers extracted from online databases which addressed HE challenges and developments in the GCC context and internationally. Additionally, statistics and information were extracted from other official resources to complement the review. The researchers used the following resources to conduct their review:

- Online databases: this was conducted in three stages. Stage 1 focused on research about HEIs challenges and developments internationally. The researchers used a mixture of two sets of keywords; the first set included: "higher education institutes", "higher education institutions", "tertiary education" and "university", while the second set included: "challenges", "developments" and "trends". Stage 2 involved research addressing the GCC context, the keywords "GCC" and "Arab gulf" were added to stage 1 set of keywords. The research papers identified in stage 1 and 2 were ordered as per their relevance. Only peer reviewed journal papers (except one), books and book chapters issued over the past 15 years were included. The third stage was implemented once the major themes of HE trends, developments and challenges internationally and in the GCC were identified.

The literature was searched in stage three using the keywords of set 1 as listed above and some keywords representing the theme (e.g., "governance"). Due to time constraints, the papers were further shortlisted as provided in the references section.

- Sustainable Development Goals Dataset of the UNESCO Institute for Statistics: the dataset relevant to the attainment of "SDG 4—Quality Education" was explored. The keywords "tertiary education", "post-secondary education" and "higher education" were used to identify the most relevant datasets.
- Official website for the GCC: the website was surfed to extract information about the GCC establishment, objectives, HE achievements and recent relevant statistics.

4 Literature Review

In this section, the researchers present summary and reflections of the review findings. The section begins by focusing on HE worldwide then HE in the GCC context.

4.1 Global Developments in HE

HE currently is one of the most rapidly growing sectors globally [1]. Several researchers have reviewed the evolution of HE. Previous research showed that despite differences in the evolution journeys of the HE worldwide, similarities in the HEIs functions and drivers for the sector evolution were observed [2, 3]. Theological enlightenment was found to be a key element of the sector's early-stage evolution, while cultural modernism was reported as a driver of the most recent HEIs developments. Similarly, teaching and research remained to be the core functions of HEIs, despite the recent trends focusing on a "third-mission" such as employability and sustainability.

Below are the recent developments in the HE sector as identified through the literature review and statistics collected from UN reports. The field of HE is dynamic, changes can occur at any time [4]; hence, the list discussed cannot be considered comprehensive.

Huge Expansion in the Number of Students Enrolled in HEIs. International reports have shown a massive increase in worldwide participation in tertiary education (TE) to almost 224 million in the year of 2018 [5]. In Australia, TE enrolments increased by 130% between 1989 and 2007 [6]. At regional levels, the lowest TE enrolments and growth in enrolment rates were observed in the Sub-Saharan with enrolments rates of 9.48% in the year of 2020 [1, 5, 7]. On the contrary, the highest growth in TE enrolment rates was in Central and Eastern Europe. Table 1 provides statistics about global enrolment rates in TE from 2015 to 2020.

As demonstrated in Table 1, regions across the world are witnessing growth in TE enrolments. The researchers call HE leaders and researchers' attention to investigate

Table 1 Gross enrolment ratio in percentages for tertiary education across various regions in the world

Context	2015	2016	2017	2018	2019	2020
World	36.87	37.40	37.85	38.43	39.41	40.24[a]
Arab States	31.14	31.82	32.67	34.08	34.62[a]	34.52[a]
North America and Western Europe	77.43	78.13	78.70	79.56	80.75	80.75[a]
Central and Eastern Europe	77.18	79.95	82.72	85.21	87.12	87.03
East Asia and the Pacific	43.74	45.12	45.72	46.58	48.44	51.02
Sub-Saharan Africa	8.85[a]	8.88[a]	9[a]	9.11[a]	9.12[a]	9.48[a]

[a] UIS estimation; *Source* UNESCO Institute for Statistics, data extracted on 17 Jan 2022 06:00 UTC (GMT)

the availability of sufficient learning and support resources at HEIs to meet the HE enrolment expansion [1]. Governments must design alternative TE venues and align their admission criteria to support HEIs to accommodate secondary school graduates.

Global Positioning on Education. Recently the United Nations (UN) has set a global agenda to provide lifelong inclusive learning to all regardless of demographic, social and ethnical characteristics such as gender, income and disability. Many countries have pledged to the UN goals and have implemented various measures to provide HE access to all, including financial measures (e.g., scholarships, student loans) and non-financial ones (e.g., specialised institutions for minority groups, flexible pathways and recognition of previous learning, distance learning). Scholarship granting was the most implemented measure. Efforts to achieve lifelong learning for all are intensifying and are being monitored through the UN frequent monitoring reports. New inclusion areas are emerging including prison education and female representation in the fields of science and mathematics [5]. Nonetheless, statistics reveal a gap in TE enrolment rates between regions based on income.

Table 2 shows that enrolment rates in low-income countries in 2019 were 9.36%, whereas in middle and high-income countries the percentages were 35.81% and 77.79% respectively [7]. Language constraints, lack of career guidance, under-resourced schools, parental expectations, academic support, peer pressure and lack of funding were identified as determinants to hinder transition to tertiary education in rural communities [8]. HEIs hold a great responsibility to serve humanity and contribute to its advancement. The researchers believe that HEIs have to revisit their strategies and align their resources and operations to provide support to groups of limited potentials to access HE (e.g., minority groups and low-income students), and to support them to complete their TE as emphasised by the UN [5].

Escalated Utilisation/Integration of Technology. IT advancements and information superabundance have impacted all aspects of human life including HE and the teacher-learner relationship. IT advancements have enabled HE access to a wide scope of learners as information and learning resources are becoming easily accessible via innovative delivery mechanisms such as the Internet, smart devices and

A Review of the Recent Developments in the Higher Education Sector ... 639

Table 2 Gross enrolment ratio in percentages for tertiary education per income group

World Bank income groups	2015	2016	2017	2018	2019	2020
Low-income countries	9.32[a]	9.21[a]	9.25[a]	9.30[a]	9.36[a]	NA
Lower middle-income countries	24.55	24.67	25.09	25.49	26.12	26.51[a]
Middle-income countries	33.85	34.52	35.12	35.81	36.92	37.98[a]
Upper middle-income countries	48.56	50.45	51.58	52.98	55.14	57.55[a]
High-income countries	75.67	76.61	76.94	77.79	79.25	79.35[a]

Source UNESCO Institute for Statistics, data extracted on 17 Jan 2022 06:00 UTC (GMT)

gadgets and contents personalisation [9]. Online learning is faster and easier for students; despite this, it is argued that online learning is less effective in terms of analytic skills acquisition and information retention.

Research showed that socio-demographic factors such as gender, teaching discipline and experience rather than mobile technologies possession impacted teachers' utilisation of mobile technologies for learning [10]. As for HE learners, "technacy" levels (i.e., technical and scientific problem solving competence) among secondary graduates' [9] and technology access constrains and affordability [5] were highlighted to impact HE learners utilisation of technology in learning. This clearly shows that there is more in technology acceptance and utilisation for interactive learning than research has shown. Another significant contribution of IT technologies in HE is the introduction of automation and expert systems. It is expected that artificial intelligence (AI) technologies will impact jobs at HE including teaching and learning, students advising, etc. as machine algorithms will be able to recognise patterns and make predictions to implement them in learning situations. Nonetheless, it is argued that the power of AI technology will not yet replace teachers due to their fixed rules which can stifle academic freedom [9, p. 1158].

The researchers believe that future research should identify successful IT strategies that align IT tools benefits, limitations and characteristics to the needs and capacities of learners and educators; technical, demographic and social aspects should be examined. Researchers should investigate the effectiveness of HE policies in mitigating technology risks. Optimal IT governance must address these challenges.

Emphasis on Quality Education. The UN emphasises on the provision of lifelong learning and quality education for all. In a school context, quality includes educators training, learning materials, interactive learners and educators' engagement for learning, well-developed inclusive curricula and improved governance [11]. The same can be argued for TE. Links between quality HE and quality learning settings, educators, teachers, managers and support staff to address the learner needs were confirmed [12].

Quality assurance systems have spread globally and are exerting influences on HE to the limit that some researchers are stating that these systems are causing pressures on faculty members and management especially in young universities [13]. A consequence highlighted by the previous researchers is the change in the "trustful" relationships between academics and their HEIs due to the focus on "transparency". Accordingly, this study stresses the importance of researchers' investigation of the arguments that quality systems are overpowering HEIs and the strategies to be utilised to improve HE quality while not overwhelming their faculty and other stakeholders.

Ranking for Prestige. Doidge et al. [14] highlight that bibliometrics and university rankings are receiving higher attention worldwide. The over-reliance on ranking as a source of distinction and prestige has created the "global university" model as a standard of success. The model can be replicated elsewhere as it depends on metric-data. A key challenge for HEIs implementing the metric-driven model is not to be overwhelmed in struggles over resources or lose their identity. Ranking assessment was argued to influence HE contents and delivery methods and outcomes [9]. This research calls researchers' attention for evaluating the actual benefits of the ranking schemes and to weigh these benefits against those of accreditation and quality systems implementation; this will help HEIs to utilise their resources effectively.

Internationalisation of HE Services. TE is expanding, HEIs students are becoming diverse and international. In 2007, overseas students' enrolment in HE in Australia increased to over 26% [6]. Universities were embracing international education to increase their external income benefiting from IT advancements [6, 14]. However, international students have varying needs and abilities [15]. Research on HE internationalisation emerged; themes were internationalisation processes (i.e., internationalisation models and modes), outcomes and governance [4]. The researchers believe that future research should investigate HEIs abilities to meet the requirements of the international and diversified student body and its impact on national students (e.g., cultural clashes and dispute) and curricula design and contents. HEIs need to implement internationalisation strategies that fit their profile.

Privatization of HE. One of the changes in HE is the increasing role of private funding. Many private HEIs are entering the sector, thereby elevating competition to unprecedented levels. HEIs are increasingly depending on tuition fees as source of funding rather than on state or government support or funds [6, 16, 17]. Reports by the UN [11] confirm increased role of private funding in TE compared to basic education. Levy [17] identified various drivers for the increase of private HEIs such as students massification, HE quality and elite groups support of privatization.

More recently, researchers have noticed an inclination towards "de-privatization" of the sector, especially in Eastern Europe [16]. Kwiek studied information published about private/public funding and enrolments in Poland during a ten-years period. He noticed that the number of enrolments in private HEIs and the number of students paying tuition fees in both public and private HEIs were decreasing. Kwiek attributed this decline to the country demographic and economic factors. Reports by the UN confirmed lacks in the provision of diagnostic measure to assess education financing [11].

Based on the aforementioned literature, the researchers call governments and HE leaders to develop diagnostic measures and standards to assess HE financing. HEIs should develop their strategies to better utilise their financial resources and reduce unnecessary expenditures. The impact of the sector privatization on HEIs fulfilment of their mission and accountability should be investigated further (e.g., information disclosure practices).

Academics and Administrative Staff Professional Development. The professionalisation of HE academics and administrative staff is becoming a must. The roles of academic and administrative staff are expanding and diversifying, professional middle management has emerged, and administrative functions are assigned to faculty members at the university level [6]. HE functions are expected to vary, professions and jobs to change and AI technology's role to reshape [9]. These changes require both administrative and academic staff to be equipped with special and general capabilities which have to be met to mitigate risk avoidance by staff [6].

Boeren [12] highlighted the importance of providing professional development opportunities to staff and the autonomy that HEIs have in this regard. Efforts to elevate professional development should be supported by the implementation of performance management systems for academic staff in addition to those for administrative staff [6].

The above literature calls for more attention to research on academic and administrative role conflicts among HE faculty with administrative posts, their reporting lines and accountability, the skills required by faculty to fulfil administrative roles, the pressures these administrative roles impose on faculty and their outcomes. Faculty/admin selection criteria, professional development and performance monitoring must be aligned to ensure fairness and transparency.

Research Quality and Quantity. Doidge et al. [14, p. 1127], in their study of the field of human and social sciences in HE, pinpoint to recent tendency in the HE "system" to reward researchers who investigate discrete problems or work in research teams in fields of limited empirical aspects, i.e., individualised research and historical research are becoming less attractive. The concentration of research on narrow outcomes is described as an instrumentalism threat to academic freedom as cited by Costandi et al. [13].

The researchers believe that research recognition/rewarding systems and policies should be reviewed by HEIs, research institutes and future researchers in terms of effectiveness and to ensure transparency and fairness. The researchers argue for research-work assessment measures that consider research impact and relevance. The effectiveness of research strategies and the impact of research-area driven strategies on faculty motivations for research and research quality/quantity are to be investigated.

HE of Multiple Stakeholders. Researchers have reported a growing concern that HE is becoming led by "materialistic vocational needs" [13, p. 73]. The design of courses, subject areas and degrees is increasingly aligned to student and market demand controlled by markets and jobs [9]. Beside market power, sectoral requirements and community expectations are influencing the content, subject and delivery methods and outcomes in education [9].

HEIs are falling under pressures to meet the increasing and often contradictory demands from their various stakeholder groups [1] and to respond to the threat of the fierce competition [13]. Future research should examine the impact of stakeholder groups on HEIs teaching and research strategies, values, objectives, programme blends and content and the quality of their graduates and wellbeing of their faculty. Further, it should investigate HEI leadership role to survive this dynamic environment.

Shift in HEIs Internal Governance Structures. The increasing powers of funders, legislators, society and other stakeholders have changed decision making thrust and internal governance in HEIs at the beginning of the current century from collegial governance led by academics and the power of academic freedom to a governance system assigning more powers to other stakeholder groups such as administrators and students [13, 18].

Research reviewing shifts in HEIs governance pinpoints two main internal governance structures: single-level and multilevel. In single-level governance structures, a single governing body is responsible of the administrative, academic and financial matters of the HEI, while in a multilevel structure, the responsibility for administrative/financial matters and academic ones are segregated among different governing bodies (e.g., board of directors versus academic senate) [19, 20].

The researchers call for more research on HEIs governance structures and processes and leadership styles and characteristics impact on their performance, quality of decisions and institutions rigidity/flexibility. The antecedents and consequences of institutional governance shifts on HE constituents have to be examined and strategies to navigate transition periods should be developed.

Change in National HE Governance Systems. A review of HE sectors development reveals continuous adjustments and restructuring of the sector at national levels over the last three decades in pursuit of more efficient and effective HE services [21]. Changes in HE governance were investigated in several contexts including Arab countries, China and Europe [13, 18, 22]. Dimensions of HE governance: HE arrangements implemented, university financial governance and personnel autonomy [18] were addressed differently in different contexts [21, 23]. Some countries have adopted a steering at a distance model, granting more freedom to their HEIs without neglecting their supervisory role [22], others have designed their own interpretation of the three common governance models, i.e., state-controlled, market driven and self-oriented [21].

The researchers recommend that future research should further review the impact of HE strategies and policies on HEIs performance and outcomes quality and their stakeholders' commitment and wellbeing.

4.2 Challenges to HE

Recent developments in the HE sector reveals that changes have been occurring in the sector; each of these changes proposes a number of challenges which need to be addressed and resolved. The researchers list some of these challenges below.

HE Resources and Capacity. With HE enrolments increment and international global commitment to inclusion and lifelong learning for all, challenges in terms of HE resources availability and capacity to accommodate the over demand on TE are escalating. Concerns about governments and public funds' ability to sustain expenditures caused by the expansion were raised. Alternative funding resources were to be identified and secured to support governments expenditures [6].

Furthermore, despite the global increment in the number of TE enrolments, still the gross enrolment ratio has only reached 40% [5, 11]. More pressures are being exerted on HE policy makers and HE providers to develop and implement equity and inclusion strategies in TE (only 11% of the countries investigated by the UN had comprehensive HE equity strategy); countries will have to harness further resources to implement and demonstrate their commitment to increasing TE enrolment and completion rates [5].

On the other hand, the demand to graduate employable graduates (i.e., market demand) is imposing pressures on HEIs. New staff and resources are to be acquired and budgets are to be allocated to deliver the courses and produce graduates that fit the market environment. This is particularly challenging due to the dynamic nature of market needs [9]. The researchers call for more research on HE resources and capabilities development, deployment and renovation. Of particular importance, strategies for funding streams sustainability.

Technology Pressures. The advantages that IT advancements are offering for HE cannot be exploited without securing HE learners and providers interest and abilities to embrace technology. Learners readiness, enthusiasm and capacity to absorb knowledge by themselves are important [9].

Technology implementation is threatened by the digital divide between individuals regardless of their context. The digital divide was quite evident with the emergence of COVID-19, where only 12% of families in the least developed countries had internet access at home. Income and technology access constraints were identified as possible limitations [5]. HE students come from different backgrounds and are of different "technacy" levels [9]. The transparency of processes to implement AI tool is yet another challenge.

Calls to allow individuals to question the actions or decisions made through AI tools were raised [9, p. 1158]. The researchers argue for in-depth research about ICT and decision making in HEIs, IT governance and strategies to align technology with HEIs vision and mission and the impact of ICT on learners and educators' interaction and wellbeing. Governments need to reconsider their education system as a whole to ensure appropriate technical skills progression.

The Pressure of the Diversified Student Body. In 2007, overseas students' share of the overall HE enrolments in Australia increased to over 26%. This increment in the number of international students raises serious challenges with regard to HE readiness to address issues of ethnic/racial discrimination, violence and exclusion as well as the consequences of over-reliance on funds from overseas students and the sustainability of this funding stream [6]. As recommended earlier, researchers need to investigate the new diversified student body requirements and help HEIs develop strategies that balance these needs and align resources and processes to accommodate them, and bear associated risks like physical and financial differences. Governments and HE leaders need to develop joint policies to support international students and provide them incentives to join-in and stay, thereby increasing the international funding stream sustainability.

Pressures for More Transparency. HEIs are increasingly implementing new public management approaches. An observed limitation in this approach is the lack of transparency [6]. The previous researchers attribute this limitation to HEIs and their decision-makers sensitivity to risk and blame. A lack of systemic information disclosure at the global level among HEIs was found [11]. The lack of transparency extends to financing, where diagnostic tools/measures to assess education financing was noticed [11]. The researchers support this call; further, HE information disclosure practices, standards and measures need to be revised and enhanced to ensure transparency and accountability. Policies delineating clear responsibilities in this regard need to be developed.

The Pressures of Competing Powers. HE has become a field of too many stakeholders including economy, legislators and society [13]. Markets are exerting pressures on HEIs to graduate employable graduates [9]. HE-related professional associations are increasing, and so are external regulatory environments which have caused the replication of compliance requirements by HEIs [6].

Competitions in research is threatening to create a system that discourages originality and focuses on the latest research trends that only generate higher citation rates [14]. This research suggests that HE researchers and leaders should re-examine the needs of HE stakeholders and prioritize them with the purpose of balancing these needs and aligning their resources and capabilities to resolve them. Researchers need to address the impact of the competing powers on HE constituents' wellbeing, performance and outcomes quality and the overall institutional performance.

4.3 HE in the GCC Region

In this section the researchers begin with an introduction about the GCC context then they focus on the HE history, current state and challenges.

The GCC Context. The Gulf Cooperation Council (GCC) is a league consisting of six countries sharing common language (Arabic), religion (Islam), traditions, culture

A Review of the Recent Developments in the Higher Education Sector ... 645

and geographic location. The council comprises the State of Kuwait, Kingdom of Bahrain, Kingdom of Saudi Arabia, State of Qatar, United Arab Emirates and Sultanate of Oman [1, 4, 15]. It is situated in "The Arabian Gulf" area and is surrounded by three seas, keeping its states close and connected in terms of economy, climate and natural resources [1]. The GCC was formed in 1981 to oversee and protect the common interests of its constituents. Its main objective is to effectuate "coordination, integration and inter-connection between Member States in all fields in order to achieve unity between them" and to develop similar regulations among members states in the fields of commerce, education and financial affairs, among other strategic objectives [24].

Several economic, social and political transformations were witnessed in the region over the last century. They were triggered by the discovery of oil and gas, the decline of the British influence and the establishment of the Council. Currently, the region is considered as the richest in the Arab world. Its economic power is dependent on oil and gas, with some successful attempts for economic diversification. Growth in the region economy was accompanied by social and demographic changes including an expansion in population to almost 47 million people, the majority of which are non-nationals. Present demographic trends in the GCC show a decline in the number of adolescent pregnancies [1]. Besides, factors such as globalization, media openness, knowledge and information explosion and advanced IT [4] are expected to further impact GCC societies. For example, a "de-Arabization" of the region is expected due to the decrease in national youth population and increase in the number of non-Arab expatriates although, it has been stated that modernism and wealth did not affect the intertwined heritage and values of GCC countries [1].

The HE Sector Establishment in the GCC Context. Economic and social developments in the region were accompanied by educational developments. Vardhan [4] cites researchers stating that HE in the GCC extends back to the advent of Islam in Mecca and Madina. However, the development of modern HEIs in the region did not start long time ago. In fact, some previous researchers cite that GCC countries were the last among Arab countries to participate in the HE modern boom [13].

The first HEI was established in the region in Mecca in 1949 (Umm Al-Qura) [4]. GCC countries modelled their first national universities using Arab universities. Admissions were limited to their citizens and faculty serving in these universities were mostly Arabs. By the end of the past century, these universities were accused of not being able to produce graduates with the employability skills required [1], especially as most GCC countries had a maximum of two national universities, none of which was established during the 90s' [4]. GCC countries ambition to become knowledge societies [1] was likely to be jeopardized by these shortages. Calls for HE reforms were increasing.

HE Reforms, Trends and Challenges in the GCC Context. Since its establishment, the GCC council has embarked on joint educational processes to develop education in the region. These efforts were coordinated by the Arab Bureau of Education; however, they were focused on public education as cited by David et al. [15]. By the beginning of the new millennium, many GCC countries had reformed their HE. Some

646 E. Ahmed et al.

have shifted their governance practices and programmes content to model Western HEIs [13], the American model was mostly utilised by public HEIs [1]. The number of HEIs in the GCC region, in particular private ones, has grown massively in the past two decades. A total of 167 universities are operating in GCC [4]. The reforms implemented in the region were as follows [25].

- Neoliberal: addressed costs incurred by students and HE providers with the aim of widening access to HE by establishing "open/blended" private HEIs. The number of private and semi-private HEIs has increased in the region reaching 103 out of 167 universities [4]. Researchers have attributed this "expansion" to GCC countries drive to adopt foreign academic practices, specifically: American and British [1].
- Imported internationalisation: focused on holding partnerships with renowned foreign HEIs to help GCC countries to create prestigious image and gain international recognition without the establishment of national level quality initiatives. Bucker cites other researchers who notice that despite importing international HEIs in GCC, student enrolments rates in them are less than public HEIs. Researchers have identified that a total of 70 out of the 103 private universities operating in the region had partnerships or collaborations with an international university [4]. Collaboration modes included branch campuses and student mobility. In addition, collaborations were mostly with European, American, Indian and Canadian HEIs [4].
- Quality assurance: governments which have not implemented the imported internationlisation reform, have developed quality assurance, accreditation and supervisory institutions as per Western higher education regulation standards. Self-evaluations, site visits and reports to oversee public and private HEIs were utilised [1]. HE adjustments included revising curricula to reflect international needs (e.g., increase of non-Islamic content and the utilisation of English as the language of instruction) [25].

Reforms modelling the west were criticised for importing western curricula and disciplines rather than adjusting them to GCC context [13], focusing on short-term political and economic aspects which have resulted in HEIs focus on teaching and reliance on external experts to plan, implement and assess reforms [1].

With the implementation of HE reforms across the GCC region, it is expected that new developments and challenges will show up, mostly resembling international trends [13]. Below are some of these developments as identified from the GCC-related research resources:

- Expansion in the number of students enrolled in HEIs: a review of TE gross enrolment ratios in the region (Table 3) shows growth in TE enrolments in the region, except UAE due to information unavailability. In 2020, the highest growth rate was reported in the Kingdom of Bahrain (over 60% of qualified secondary school graduates enrolled in TE institutions in 2020), while Qatar had the lowest rates. Growth in enrolment ratios signals a possible success of the region HE reforms.

It is notable though that Oman and Kuwait had drops in their TE enrolment rates in 2017 and 2018.

The researchers emphasise that research on GCC HEIs resources and capacity to accommodate HE enrolments growth rates is essential. In addition, HEIs need to provide different TE venues, revise their admission criteria and continue to support the Council strategies to integrate HE services and design policies that synchronize their resources to distribute the students' growth across the region.

- Diversified international student body: researchers have confirmed widened access to HE for many groups that had limited access earlier, including female and expatriates; access to the region HEIs is no longer a privilege to nationals [1]. However, most of GCC international students are GCC citizens (the highest mobility rates were reported by Saudi students), followed by other Arab nationals (e.g., Yemen, Egypt, Jordan and Iraq) and South Asian students (e.g., India and Pakistan) [4]. Factors attracting international students included learning environment, cost and institutional reputation [15]. Inline with this research previous recommendations, GCC countries need to revisit their internationlisation strategies to fit their context, design their policies to support underprivileged students fund and provide equal learners' access and support.
- Robust HEIs regulatory and accountability: the introduction of private institutions has increased competitiveness in the region HE sector making it dynamic [1, 4]. Due to the increasing competitive powers, accountability among public HEIs has increased. Furthermore, governments began to focus on the establishment of robust regulatory systems and the implementation of their own accreditation and supervisory processes [1]. Future research should examine the effectiveness of the HE regulatory procedures in the region and measure their impact on HEIs performance, transparency and wellbeing. Regional-level reporting measures and quality standards are worth of considering.

Table 3 Gross enrolment ratio in percentages for TE in the GCC countries

Country	2015	2016	2017	2018	2019	2020
Bahrain	43.28	46.57	47.15	50.48	55.63	60.32
Kuwait	55.15	57.29	55.36	54.36	55.31	61.13
Oman	39.23	44.05	38.13	38.04	40.45	45.48
Qatar	14.70	15.64	16.63	17.87	18.95	20.79
Saudi Arabia	61.06	67.34	69.70	68.04	70.90	70.63
United Arab Emirates	NA	NA	NA	NA	52.61	53.72

NA: Not available; *Source*: UNESCO Institute for Statistics, data extracted on 17 Jan 2022 06:00 UTC (GMT)

Inline with the reforms and changes in the HE in the GCC context listed above, the following challenges can be identified:

- Low status of Modern Standard Arabic (MSA): a decline in the use of the Arab language for academic purposes was noticed despite official statements and decisions affirming Arabic as a central component of the Arab identity. This is quite expected with the increase in number of expatriates, and HE reforms focus on importing internationalisation and HE privatization [1]. This research further supports arguments for national policies and strategies that implement GCC agenda for cooperative and integrated HE systems and the development of standards that are GCC context specific.
- Financial resources diversity: Costandi et al. [13] asserted that HEIs in the GCC countries are facing the scarcity of sustainable financail resources. The researchers report that only 21.45% of GCC HEIs had "founding" assets that could contribute to their revenue. They attributed this challenge to shortages of "foundations" and research grants and fluctuation in governmental support; HEIs are depending on tution fees as their main revnue stream and to cover their costs. The researchers believe that GCC countries attempts to develop well-diversified economy should extend to diversifying HE funding resources inline with international practices. Researchers, governments and HE leaders need to investigate and develop policies and incentives that attract funds and sustain financial resources with minimal burdening on students.
- Limited information disclosure: limited disclosure of self-assessment information and documents in the region including information about students' preparation, governance and working conditions was pinpointed, except for Oman and the Kingdom of Bahrain [1]. This limitation was attributed to the sector relatively young age [15]. The researchers argue for regional and national levels information disclosure policies and measures and regulatory bodies to enforce transparency and accountability. Researchers need to investigate information quality and governance in the region to ensure meeting international information quality, protection and disclosure standards.
- Limited trust in internal governance arrangements and decision making: the transparency and accountability of HEIs governing bodies and decision makers were under scrutiny. Research has shown low scores in terms of democracy in decision-making (58.1%), transparency (61.75%) and academic freedom (67%) [13]. It seems that faculty members trust and faith in their leadership are at stake. Researchers need to validate this issue and to investigate its antecedents and consequences. Strategies to increase faculty and staff engagement may need to be revised.

5 Discussion and Conclusion

Review of the recent developments and challenges of HE in the GCC context and internationally shows considerable changes in the sector worldwide including students expansion, internationlisation and diversity [5, 7, 14, 15], commitment to lifelong learning [5], sector privatization [6, 16, 17], increased competitiveness [13], emphasise on professional conduct and development [6, 12], multiple stakeholders engagement and governance and leadership [13, 19–21]. Challenges triggered by these developments included financial and other resources scarcity [6, 16], HEIs physical and technical capacity [9], pressures to meet multiple stakeholders demands including pressures for transparency and limitations of HE policy review and research [11, 14] which is quite troublesome due to the continuous restructuring of the sector worldwide, and the impact that such restructuring will have on HEIs, students and staff [6].

Research to assess the sector structural changes is needed [13], in particular research investigating the impact of externally inspired policy change on pre-existing governance structures [18]. Furthermore, attempts to elevate the learning process and HE services through technical advancements are confronted by different barriers such as cost. The student learning experience should be planned and designed to promote creative learning and research on technology acceptance among stakeholders should be considered [9].

The researchers believe that the developments and challenges identified require attention. Research can guide HE decision makers to understand different situations and plan actions. This research is proposing the following areas to be examined by HEIs and researchers:

- HE resources and capabilities development, renovation and sustainability
- internationalisation strategies fitness to HEI context and students' needs
- enrolment and completion of all student groups
- IT governance and strategies effectiveness and alignment to HEIs vision and mission
- accountability, transparency and information disclosure
- ranking, quality and regulatory systems impact on HEIs performance and wellbeing
- sector privatization impact on HEIs performance (mission and accountability)
- professional development and performance fairness and transparency
- research-work motivation and assessment measures impact and relevance
- strategies to overcome stakeholders' pressures
- institutional governance structures and processes impact on HEIs performance, decisions and flexibility
- review and impact of national HE strategies on HEIs performance and wellbeing.

Similar to international HEIs, HEIs in the GCC went under various restructuring and reforms, mostly to resemble western universities models [13]. Research on HE in the region evolved to cover issues such as governance, financial sustainability and internationalisation. GCC countries need to sustain their financial resources to be able to implement their ambitious knowledge economy agenda [13, 15]. The availability of diagnostic tool/measure to assess education financing is a concern [11]. Revenue can be increased through internationalisation and student mobility. Researchers demonstrated that learning mobility, especially outbound, has been stimulated in the region. Inbound mobility was mostly from other GCC countries [15]. GCC countries should actively embark on regional integration of their HE services [4, 15]. Areas for further research are as follows:

- resources and strategies to absorb HE growth across the GCC context
- integrated HE reporting measures and quality standards
- fit-for-context internationlisation strategies
- policies to support equality among learners (access and completion)
- regulatory procedures impact on HEIs performance, transparency and wellbeing
- diversity and sustainability of financial resources
- information quality, governance, protection and disclosure
- faculty and staff engagement and trust
- Arabic identity and HE reforms.

Based on the above, the researchers will attempt to address how HEIs which are operating in dynamic context can develop and renovate their existing resources and capabilities to achieve sustainable performance.

The research limitations include the criteria implemented for selecting the research papers to be reviewed (i.e., keywords and research papers publication date) and not including some HE research streams such as leadership styles and curriculum design. A justification for the reviewed research papers publication date limitation is that the current research covers the history of HE sector; research papers of over than ten years were used to track the HE sector progression. As for the second limitation, the research topics given as examples in addition to shifts in HE triggered by COVID-19 can be examined further in future research. A final limitation is that despite the valuable information provided by the UN on TE, some of the statistics were estimates, similarly, some of the reports about HE achievements in the GCC were not very recent, information up-to-dateness is a concern that can be addressed by future research.

References

1. Badry, F., Willoughby, J.: Higher Education Revolutions in the Gulf: Globalization and Institutional Viability, 1st edn. Taylor & Francis (2016)
2. Bebbington, D.W.: Christian higher education in Europe: a historical analysis. Christ. High. Educ. **10**(1), 10–24 (2011)
3. Ford, M.: The functions of higher education. Am. J. Econ. Sociol. **76**(3) (2017)
4. Vardhan, J.: Internationalization and the changing paradigm of higher education in the GCC countries. SAGE Open **5**(2) (2015)

5. UNESCO: Global Education Monitoring Report 2020: Inclusion and Education: All Means All, 1st edn. UNESCO (2020)
6. Goedegebuure, L., Schoen, M.: Key challenges for tertiary education policy and research—an Australian perspective. Stud. High. Educ. **39**(8), 1381–1392 (2014)
7. UIS Dataset: Sustainable Development Goals, 1st edn. UNESCO Institute for Statistics (2021)
8. Maila, P., Ross, E.: Perceptions of disadvantaged rural matriculants regarding factors facilitating and constraining their transition to tertiary education. S. Afr. J. Educ. **31**(8), 1360–1372 (2018)
9. Nguyen, H.D., Mai, L.T., Do, D.A.: Innovations in creative education for tertiary sector in Australia: present and future challenges. Educ. Philos. Theory **52**(11), 1149–1161 (2020)
10. Lai, K.-W., Smith, L.: Socio-demographic factors relating to perception and use of mobile technologies in tertiary teaching. Br. J. Edu. Technol. **49**(3), 492–504 (2018)
11. UNESCO: Education for All: 2000–2005: Achievements and Challenges, 1st edn. UNESCO (2015)
12. Boeren, E.: Understanding Sustainable Development Goal (SDG) 4 on "quality education" from micro, meso and macro perspectives. Int. Rev. Educ. **65**, 277–294 (2019)
13. Costandi, S., Hamdan, A., Alareeni, B., Hassan, A.: Educational governance and challenges to universities in the Arabian Gulf region. Educ. Philos. Theory **51**(1), 70–86 (2018)
14. Doidge, S., Doyle, J., Hogan, T.: The university in the global age: reconceptualising the humanities and social sciences for the twenty-first century. Educ. Philos. Theory **52**(11), 1126–1138 (2020)
15. David, S.A., et al.: An exploration into student learning mobility in higher education among the Arabian Gulf Cooperation Council countries. Int. J. Educ. Dev. **55**, 41–48 (2017)
16. Kwiek, M.: De-privatization in higher education: a conceptual approach. High. Educ. **74**, 259–281 (2017)
17. Levy, D.C.: The unanticipated explosion: private higher education's global surge. Comp. Educ. Rev. **50**(2), 217–240 (2006)
18. Dobbins, M., Knill, C.: Higher education governance in France, Germany, and Italy: change and variation in the impact of transnational soft governance. Policy Soc. **36**(1), 67–88 (2017)
19. Gornitzka, Å., Maassen, P., De Boer, H.: Change in university governance structures in Continental Europe. High. Educ. Q. **71**(3), 274–289 (2017)
20. Pennock, L., Jones, G.A., Leclerc, J.M., Li, S.X.: Assessing the role and structure of academic senates in Canadian universities, 2000–2012. High. Educ. **70**, 503–518 (2015)
21. Capano, G., Pritoni, A.: Varieties of hybrid systemic governance in European Higher Education. High. Educ. Q. **73**, 10–28 (2019)
22. Han, S., Xu, X.: How far has the state 'stepped back': an exploratory study of the changing governance of higher education in China (1978–2018). High. Educ. **78**, 931–946 (2019)
23. Donina, D., Hasanefendic, S.: Higher Education institutional governance reforms in the Netherlands, Portugal and Italy: a policy translation perspective addressing the homogeneous/heterogeneous dilemma. High. Educ. Q. **73**, 29–44 (2019)
24. About GCC, https://www.gcc-sg.org/en-us/AboutGCC/Pages/Primarylaw.aspx, last accesssed 2022/01/10
25. Buckner, E.: The role of higher education in the Arab state and society: historical legacies and recent reform patterns. Comp. Inter. High. Educ. **3**(1), 21–26 (2011)

Intra-Industry Trade Trends in India's Manufacturing Sector: A Quantitative Analysis

Alan J. Benny ⓘ**, K. P. Jaheer Mukthar** ⓘ**, Felix Julca-Guerrero** ⓘ**, Norma Ramírez-Asís** ⓘ**, Laura Nivin-Vargas** ⓘ**, and Sandra Mory-Guarnizo** ⓘ

Abstract Intra-industry trade (IIT) as a concept grew in significance in the 1970s, with the introduction of the Grubel-Lloyd Index (GLI) and Paul Krugman's new trade theory (Krugman in Am Econ Rev 70:950–959, [22]). IIT is a good indicator of a country's growth in trade and sophistication of the manufacturing sector, which in the case of India, has been experiencing stagnant growth since 1960s. Thus, in order to systemically analyse the prospects of growth for the Indian manufacturing sector, the study categorize the manufacturing sector into 19 product groups based on HS-2022 and NIC-2008 classification. Weighted GLI values were calculated for product group-wise data and an OLS estimation of dynamic changes of WGLI of India-World trade flow was conducted. The present analysis of the OLS regression found that the pattern of IIT changes in the observed periods 2010–2020.

Keywords Intra-industry trade · Grubel-Lloyd index · Indian manufacturing sector · OLS regression

---

A. J. Benny · K. P. J. Mukthar (✉)
Kristu Jayanti College Autonomous, Bengaluru, India
e-mail: jaheer@kristujayanti.com

F. Julca-Guerrero · L. Nivin-Vargas
Universidad Nacional Santiago Antúnez de Mayolo, Huaraz, Perú

N. Ramírez-Asís
Hospital Uldarico Rocca Fernandez, Lima, Perú

S. Mory-Guarnizo
Universidad Señor de Sipán, Chiclayo, Peru

© The Author(s), under exclusive license to Springer Nature Switzerland AG 2024
A. Hamdan and E. S. Aldhaen (eds.), *Artificial Intelligence and Transforming Digital Marketing*, Studies in Systems, Decision and Control 487,
https://doi.org/10.1007/978-3-031-35828-9_55

1 Introduction

India's manufacturing sector is among the largest in the world (6th largest according to recent estimates) and contributes to 3.1% of the global GDP. Organized manufacturing sector is also the largest private employer among sectors in India. According to Cushman & Wakefield's 2021 Global Manufacturing Risk Index, India has transformed itself from a heavily agrarian country during its early years of independence to the second-most desirable manufacturing location worldwide. From 1993–1994 to 2001–2002, we can observe the growth of non-agricultural sectors of India, which grew to make up nearly 75% of the GDP by 2003. The manufacturing sector by then constituted 23% of the GDP. Manufacturing has been the cornerstone of India's growth for about 40 years since independence. However, since the liberalization reforms of the 1990's, this growth has slowed down the manufacturing sector has been experiencing stagnant growth since then. Indian manufacturers have experienced rising production expenses during the past few years and their profit margins have been impacted. Although India has become self-sufficient in terms of technologies for domestic production and consumption of the majority of manufacturing necessities, it has not been able to develop the capacity to produce internationally competitive technologies.

Previous classical theories of international trade such as Ricardo's comparative advantage and the Heckscher–Ohlin theory studied the exchange of different goods, with the Ricardian model concentrating on trade between nations with different technologies and the HO Theory focusing on factor endowments. In the 1960s, researchers began to identify evidence of trade between the same industries across nations, defeating the purpose of the previous theories. Grubel and Llyod [16] conducted a comprehensive empirical analysis on the significance of intra-industry trade and how to measure it. When a nation concurrently imports and exports items or services that are comparable in nature, this is known as intra-industry trade (IIT). The Grubel- Lloyd Index is one of the most widely used indicators to gauge intra-industry trade (GLI). The current study examines changes in trade-weighted GLI (WGLI) of significant manufacturing product categories between the years 2010 and 2020, particularly on India's intra-industry trade with the Rest of the World (ROW).

The Grubel-Lloyd Index is measured in the following manner:

$B_i = [(X_i + M_i) - |X_i - M_i|] * 100/(X_i + M_i)$, which can be expressed as:

$$B_i = 1 - |X_i - M_i|/(X_i + M_i)$$

where B_i = Grubel-Lloyd Index, X_i = Export of industry i;

M_i = Import of industry i; $X_i + M_i$ = Total value of trade;

$|X_i - M_i|$ = Trade balance of industry i.

In order to correct for trade imbalances, Greenaway and Milner [15] introduced a weighted Grubel-Lloyd index, measured as:

$$wGLI = 1 - \frac{\left(\left|\frac{X_i}{X} - \frac{M_i}{M}\right|\right)}{\left(\frac{X_i}{X} + \frac{M_i}{M}\right)}$$

where X and M are the total exports of a country, Thus, the weighted GLI measures the share of industry i in the total exports and imports of the nation.

This study studies the characteristics of India's intra-industry trade with Rest of World (ROW) across 19 product groups classified by the authors. An OLS Regression of trade-weighted GLI values is conducted similar Ćuzović and Sokolov-Mladenović [11] in their study of agricultural trade of Serbia. This study is organized as follows: Section 2 contains review of literature surrounding India's manufacturing sector and previous studies on intra-industry trade. Section 3 explains the objectives and methodology adopted for the study. Section 4 deals with the results of the analysis and the relevant discussion surrounding it. Section 6 contains the concluding remarks of the study.

2 Review of Literature

The decades of 1980s and 1990s saw a proliferation of trade liberalization and Multi-national Corporations' (MNCs) activities, due to removal of several trade barriers. Broll et al. [8] examine the growth of Foreign Direct Investment (FDI) in India during the 1970s. The authors concluded by stating how India's success in industrialization with depend to a large extent on the elimination of market distortions in both the goods and factor markets. In the decades that followed, the effects of trade liberalization are shown to have benefitted the efficient producers and affected inefficient ones, according to Ramachandran [29]. Zhang et al. [33] conducted a model simulation of the best possible trade strategy India can follow using the computational general equilibrium approach and concluded that full trade liberalization is the best way forward for Indian firms.

Using aggregate data from the Annual Survey of Industries, Basu and Das [6] examined profitability in India's organised manufacturing sector from 1982–1983 to 2012–2013. They found two medium run regimes of profitability, one of dropping profitability (1982–1983) to 2001–2002, and another of expanding profitability, using structural break tests to identify medium and short run regimes of profitability (2001–2002 to 2012–2013). By presenting an analytical summary of several manufacturing competitiveness indices of the Indian economy, Lakshmanan et al. [24] noted that India's exports of manufactured goods increased noticeably in the 1990s and 2000s. Some critical decisions and trends of foreign trade policy of India, such as import substitution policies of 1950s to export promotion strategies of 1980s and to major tariff liberalisation of 1990s, have influenced the growth of manufacturing firms and their exports, according to Banga and Das [4]. Sapovadia [31] also examined the competitiveness of India's manufacturing industry and the impact of reforms after

1991. Indian manufacturing productivity has been shown to be low when compared to both the global economy and its counterparts.

Melitz [25], in his seminal research paper, combined firm heterogeneity across intra-industry trade, with monopolistic competition by developing a stochastic, general-equilibrium two-country model of trade and concluded that with only the more productive firms can sustain exporting, and the less productive can serve only the domestic market. Other trade models often integrate the gravity model approach such as Bhattarcharyya and Banerjee [7] in their explanation of India's international trade using a panel data approach. Sharma [32] used a CGE Model Simulation Analysis to examine how the BRICS + 2 countries' economic liberalisation influenced the manufacturing industry in South American countries. Kikuchi et al. [20] explore the determinants of IIT and observe that the productivity differences of industries are the most important determinant of IIT. Based on empirical estimates of a few manufacturing sectors, Aggarwal and Chakraborty [2] conducted a study on the factors that affect vertical IIT in India. The observation shows an increase in India's IIT values following the signing of trade agreements with countries starting in 2010 or 2011. Ćuzović and Sokolov-Mladenovi [11] in their study contextualizes Serbia's intra-industrial trade in agricultural products with its overseas trading partners from 2004 to 2018. The export of lower-priced goods dominates IIT of agricultural goods between Serbia and its international partners. Kumar and Ahmed [23] examined the intra-industry trade between India and Bangladesh during 1975–2010 and identified the animal and vegetable oils and fats, food and live animals, mineral fuels, lubricants and associated materials, and commodity industries as the industries with significant growth rates of exports and expanding prospects for IIT between India and Bangladesh between 1975 and 2010. Based on their technological content, Agarwal and Betai [1] divided industrial products from India into eleven groups. Despite recent growth in India's IIT, their analysis revealed that business between India and its most significant countries is predominantly inter-industrial and that trade agreements and India's comparative advantage have a positive and significant impact on IIT.

3 Objectives and Methodology

Objectives

(i) To empirically determine the extent of intra-industry trade in India and its composition in the country's manufacturing sector,
(ii) To test an OLS regression model that measures dynamic changes in the weighted Grubel-Lloyd Index of India-World trade flow.

Table 1 IIT classification

Class name	Range	Description
Class 1	$0.00 \leq GLI \leq 0.25$	Strong inter-industry trade
Class 2	$0.25 < GLI \leq 0.50$	Weak inter-industry trade
Class 3	$0.50 < GLI \leq 0.75$	Weak intra-industry trade
Class 4	$0.75 < GLI \leq 1.00$	Strong intra-industry trade

Methodology

The data for empirical analysis is collected from UN- Comtrade database. IIT is classified into the following categories based on the work of Qasmi and Fausti [28] (Table 1).

Dynamics of GLI changes can be computed with the following model given by Ćuzović and Sokolov-Mladenović [11] which was inspired by the Revealed Symmetric Comparative Advantage (RSCA) model given by Dalum et al. [12].

$$GLI_j^{t2} = \alpha + \beta GLI_j^{t1} + \varepsilon_j$$

Like before, GLI^{t1} is the index value of initial year (independent variable) and GLI^{t2} is the index value of final year (dependent variable). Where $t1$ and $t2$ are initial year and final year respectively, i is the country and j is the commodity group; α and β are standard regression parameters and ε_j is the residual.

The main proposition of the regression model is that $\beta = 1$ corresponds to an unchanged pattern from t1 to t2. If $\beta > 1$, then the country is more specialized in sectors which it is already specialized, and less specialized in sectors with low initial specialization. However, following Cantwell [10], $\beta > 1$ is not necessary for increase in national specialization pattern. IIT pattern does not change when $\beta = 1$. When $\beta > 1$, the country with higher IIT value increases value of IIT in products with both significant intra and inter- industry trade. When $0 < \beta < 1$, the existing pattern has changed and countries with higher IIT increase the value of IIT. When $\beta < 0$, the positions are reversed compared to initial period.

As stated above, since $\beta > 1$ does not necessarily lead to IIT growth, an additional condition for the same is formulated:

$$\frac{\delta_{t2}}{\delta_{t1}} = \left| \frac{\beta}{R} \right|$$

where δ_{t1} and δ_{t2} are the standard deviations during first and last years, β is the regression coefficient, and R is the square root of the coefficient of determination; When $\beta = R$ then there is no change GLI dispersion, $\beta > R$ there is an increase in GLI dispersion $\beta < R$ these is a decrease is GLI dispersion. For OLS estimation, TGLI values are used instead of GLI as the values need to range from -1 to 1, to have symmetric and normal distributions and so that extreme values are eliminated.

$$TGLI = 2GLI - 1$$

4 Results and Discussion

See Table 2.

Table 3 describe the WGLI values for each of the manufacturing goods that India trades with the world. Strong inter-industry trade has 34 observations, Weak inter-industry trade has 38 observations, weak intra-industry trade has 61 observations and strong intra-industry trade has 76 observations. It can be observed that:

- India's trade with the world as a whole, for the manufacturing sector, is generally characterized by strong intra-industry trade.
- Food products (product group 1), paper and related products (product group 7) and other non-metallic mineral products (product group 12 computer, electricals,

Table 2 Product groups and serial. nos. assigned

S. no.	Product groups	NIC-2008	HS-2022
1	Food products	10	11, 15, 16, 17, 18, 19, 20
2	Beverages	11	22
3	Tobacco products	12	24
4	Textiles and wearing apparels	13, 14	50, 51, 52, 53, 54, 55, 56, 57, 58, 59, 60, 61, 62, 63
5	Leather and related products	15	41, 42, 43, 64
6	Wood and related products	16	44, 45, 46
7	Paper and related products	17	47, 48, 49
8	Coke and refined petroleum products	19	26, 27
9	Chemical and related products	20	28, 29, 31
10	Pharmaceuticals medicinal chemical and botanical products	21	30
11	Rubber and plastics products	22	39,40
12	Other non-metallic mineral products	23	68, 69, 70
13	Basic metals	24	72, 73, 74, 75, 76, 78, 79, 80, 81, 82
14	Weapons and ammunition	252	93
15	Computer, electrical, electronic and optical products	26, 27	37, 85, 90, 91
16	Machinery and equipment	28	84
17	Motor vehicles, trailers and semi-trailers	29	87
18	Other transport equipment	30	86, 88, 89
19	Furniture	31	94

Classification by authors with reference to HS-2022 (Harmonized System) Codes and NIC-2008 (National Industrial Code). Ambiguous product groups such as gold and related products are not considered for the study

Intra-Industry Trade Trends in India's Manufacturing Sector … 659

Table 3 IIT classification for India-world trade flow

P-group no.	2010	2011	2012	2013	2014	2015
1	0.708	0.797	0.758	0.637	0.636	0.680
2	0.940	0.795	0.717	0.788	0.962	0.950
3	0.037	0.054	0.062	0.051	0.066	0.063
4	0.167	0.176	0.171	0.177	0.190	0.193
5	0.249	0.238	0.230	0.244	0.254	0.274
6	0.271	0.250	0.291	0.310	0.322	0.416
7	0.621	0.605	0.673	0.646	0.578	0.638
8	0.760	0.731	0.672	0.682	0.666	0.600
9	0.884	0.816	0.917	0.848	0.807	0.793
10	0.224	0.209	0.195	0.186	0.176	0.161
11	0.913	0.966	0.992	0.972	0.875	0.866
12	0.709	0.783	0.724	0.683	0.661	0.694
13	0.736	0.900	0.848	0.799	0.826	0.917
14	0.448	0.805	0.904	0.758	0.298	0.331
15	0.683	0.699	0.739	0.676	0.591	0.520
16	0.636	0.635	0.681	0.725	0.773	0.758
17	0.423	0.485	0.389	0.387	0.373	0.384
18	0.878	0.531	0.915	0.945	0.617	0.814
19	0.830	0.860	0.777	0.820	0.829	0.884

P-group no.	2016	2017	2018	2019	2020
1	0.655	0.642	0.751	0.839	0.849
2	0.802	0.836	0.826	0.656	0.817
3	0.072	0.058	0.059	0.066	0.078
4	0.223	0.210	0.223	0.273	0.244
5	0.302	0.303	0.320	0.345	0.306
6	0.416	0.461	0.499	0.527	0.601
7	0.612	0.609	0.741	0.777	0.812
8	0.599	0.614	0.624	0.624	0.570
9	0.833	0.906	0.935	0.928	0.896
10	0.174	0.170	0.168	0.190	0.183
11	0.845	0.897	0.956	0.934	0.934
12	0.695	0.671	0.647	0.601	0.519
13	0.906	0.756	0.868	0.891	0.791
14	0.364	0.401	0.495	0.518	0.310
15	0.503	0.483	0.553	0.619	0.603
16	0.727	0.821	0.853	0.829	0.811

(continued)

Table 3 (continued)

P-group no.	2016	2017	2018	2019	2020
17	0.376	0.367	0.356	0.345	0.398
18	0.998	0.902	0.887	0.848	0.982
19	0.882	0.891	0.828	0.768	0.606

Calculated by author with export–import data from UN Comtrade database

electronics, and optical products (product 15), and coke and refined petroleum products (product group 8) are characterized by intra-industry trade, leaning more towards weak intra-industry trade.

- Beverages (product group 2), Textiles and wearing apparel (product group 4) and Furniture (product 19), Chemical and related products (product group 9), rubber and plastic products (product group 11), basic metals and related products (product group 13), machinery (product 16) other transport equipment, which includes ships and aircraft (product 18) and furniture (product 19) are characterized by intra-industry trade, leaning more towards strong intra-industry trade.
- Tobacco products (product group 3) and pharmaceutical, chemical, medicinal and botanical products (product group 10) are characterized by inter-industry trade, which is strong-type throughout 2010–2020.
- Leather and related products (product group 5), weapons and ammunition (product 14), motor vehicles, trailers and semi-trailers (product 17) and wood and related products (product group 6) are characterized by inter-industry trade, which leans more towards weak inter-industry trade.

Note: The values taken in OLS regression are adjusted TGLI values. It is so that values are between -1 to 1 and are not statistically significant from the normal.

Model 1

See Fig. 1.

In the OLS regression, Model 1 was set up with dependent variable as the TGLI of year 2015, and the independent variable as TGLI of 2010.

The equation of model 1 is:

$$TGLI2015 = -0.00778 + 0.934 * TGLI2010$$
$$(0.0456) \quad (0.0794)$$
$$(\text{The standard errors in parentheses})$$

From the model, we can understand that for every increase of TGLI of year 2010 by 1, the TGLI of year 2015 increases by 0.934. The R-squared value indicates how much percentage of the variation in the dependent variable can be explained by the independent variable. The R-squared is given by 0.890527, which means 89.05% of the variations in TGLI of 2015 can be explained by TGLI of 2010.

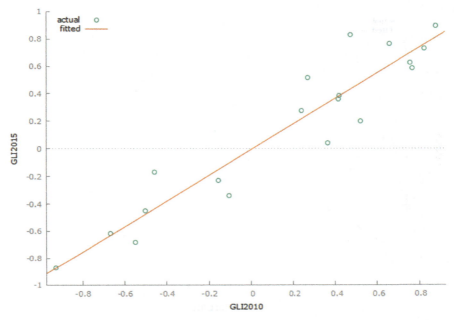

Fig. 1 Actual and fitted GLI2015 versus GLI2010. Calculation and representation by authors on Gretl

Model 2

See Fig. 2.

In the OLS regression, Model 2 was set up with dependent variable as TGLI of 2020 and independent variable as TGLI of 2016.

The equation of Model 2 is:

$$TGLI2020 = 0.04805 + 0.9126 * TGLI2016$$
$$(0.0580) \quad (0.1046)$$

(Standard errors in parentheses)

From the model, we can understand that for every increase of TGLI of year 2016 by 1, the TGLI of year 2020 increases by 0.9126. R-squared is given by 0.817434, which means 81.74% of the variations in TGLI of 2020 can be explained by TGLI of 2016.

Model 3

See Fig. 3.

In the OLS regression, Model 3 was set up with dependent variable as TGLI of 2020 and independent variable as TGLI of 2010.

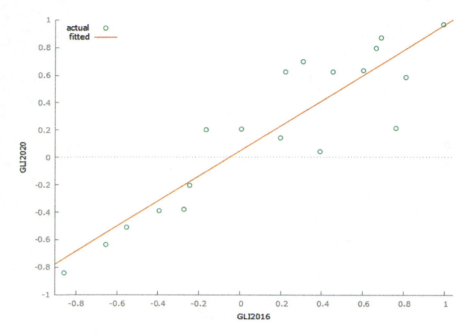

Fig. 2 Actual and fitted GLI2020 versus GLI2016. Calculation and representation by authors on Gretl

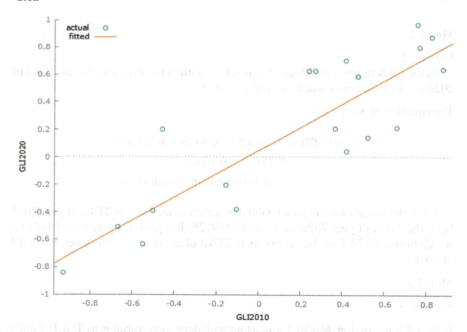

Fig. 3 Actual and fitted GLI2020 versus GLI2010. Calculation and Representation by authors on Gretl

Intra-Industry Trade Trends in India's Manufacturing Sector ... 663

The equation of Model 3 is:

$$TGLI2020 = 0.04666 + 0.8455 * TGLI2010$$
$$(0.0689) \quad (0.1201)$$
$$\text{(Standard errors in parentheses)}$$

From the model, we can understand that for every increase of TGLI of year 2010 by 1, the TGLI of year 2020 increases by 0.8455. The R-squared is given by 0.744682, which means 74.46% of the variations in TGLI of 2020 can be explained by TGLI of 2010.

5 Discussion

For Model 1, β is 0.934267, R is 0.943677, the standard deviation of TGLI 2010 is 0.5635 and the standard deviation of TGLI 2015 is 0.55788. As $\beta = 0.934267 < 1$, it implies that the existing pattern of intra-industry trade has changed. Also, product groups characterized by high intra-industry trade, increase their value of inter-industry trade. Conversely, the product groups characterized by inter-industry trade, increase their value of intra-industry trade. Also, the value of $\frac{\beta}{R} = 0.9900283 < 1$, which is nearly identical to $\frac{\delta_2}{\delta_1} = 0.9900266$. As $\beta < R$, TGLI dispersion has increased between 2010 and 2015.

For Model 2, β is 0.912603, R is 0.817434, the standard deviation of TGLI 2016 is 0.54698 and the standard deviation of TGLI 2020 is 0.55212. As $\beta = 0.912603 < 1$, it implies that the existing pattern of intra-industry trade has changed. Also, product groups characterized by high intra-industry trade increase their value of inter-industry trade. Conversely, the product groups characterized by inter-industry trade, increase their value of intra-industry trade. Also, the value of $\frac{\beta}{R} = 1.1642 > 1$, which is not identical to $\frac{\delta_2}{\delta_1} = 1.00939$. As $\beta > R$, TGLI dispersion has decreased between 2016 and 2020.

For Model 3, β is 0.845516, R is 0.744682, the standard deviation of TGLI 2010 is 0.5635 and the standard deviation of TGLI 2020 is 0.55212. As $\beta = 0.845516 < 1$, it implies that the existing pattern of intra-industry trade has changed. Also, product groups characterized by high intra-industry trade, increase their value of inter-industry trade. Conversely, the product groups characterized by inter-industry trade, increase their value of intra-industry trade. Also, the value of $\frac{\beta}{R} = 1.13540$, which is not identical to $\frac{\delta_2}{\delta_1} = 0.97980$, thus proving there is overall growth of intra-industry trade. As $\beta > R$, TGLI dispersion has decreased between 2010 and 2020.

6 Conclusion

The stagnant growth of India's manufacturing sector is reflected in the time period of the study, 2010–2020, although substantial growth is observed in certain industries. The determination of IIT for each of the product groups to the world was done in order to show how integrated India's trade is with other countries. IIT is significant as it denotes that innovation is happening in the said industries and is beneficial to the workers of the industries as economies of scale could be exploited. The main motive behind the study came from observing the tepid growth of the manufacturing sector of India throughout the decades. While industries like the 'pharmaceutical, medicinal, chemical and botanical products' industries and 'chemicals and related products' industries give India a huge competitive advantage over other countries, the weighted GLI values of the manufacture of machineries, wood, furniture, electricals and electronics indicate India is lagging behind in these industries. Therefore, the study concludes that manufacturing sector trade of India is largely intra-industrial in nature, and values of weighted GLI show significant changes across 2010 to 2020.

References

1. Agarwal, Betai: Intra-industry trade in manufactured goods: a case of India. NIPFP Working Paper Series No. 348 (2021). https://www.nipfp.org.in/publications/working-papers/1946/
2. Aggarwal, S., Chakraborty, D.: Determinants of vertical intra-industry trade in India: empirical estimates on select manufacturing sectors. Prajnan **49**(3), 221–252 (2020)
3. Anderson et al.: Firm heterogeneity and export pricing in India. U.S. International Trade Commission Working Paper Series 2016-09-B (2016)
4. Banga, R., Das, A.: Role of trade policies in growth of Indian manufacturing sector. Munich Personal RePEc Archive (2010)
5. Basant, R., Saha, S.N.: Determinants of entry in the Indian manufacturing sector. IIMA Working Papers WP2005-01-01 (2005)
6. Basu, D., Das, D.: Profitability and investment: evidence from India's organized manufacturing sector. Economics Department Working Paper Series, 183 (2015). https://doi.org/10.7275/756 5630
7. Bhattacharyya, R., Tathagata, B.: Does the gravity model explain India direction of trade? a panel data approach. Indian Institute of Management Ahmedabad, Research and Publication Department, IIMA Working Papers W.P. No.2006-09-01 (2006)
8. Broll, U., Gilroy, B.M.: Indian industrialisation, multinational enterprises and gains from trade. The Indian J. Econ. **67**(2), 231–239 (1986)
9. Caldara, D., Iacoviello, M., Molligo, P., Prestipino, A., Raffo, A.: The economic effects of trade policy uncertainty. J. Monet. Econ. **109**, 38–59 (2020)
10. Cantwell, J.A.: Technological innovation and multinational corporations. Oxford: Basil Blackwell, (1989)
11. Ćuzović, Đ, Sokolov-Mladenović, S.: Agricultural intra-industry trade in Serbia. Teme **65**(1), 349–366 (2021)
12. Dalum, B., Laursen, K., Villumsen, G.: Structural change in OECD export specialisation patterns: de-specialisation and 'stickiness.' Int. Rev. Appl. Econ. **12**(3), 423–443 (1998). https://doi.org/10.1080/02692179800000017
13. Demidova, S., Rodríguez-Clare, A.: Trade policy under firm-level heterogeneity in a small economy. J. Int. Econ., Elsevier **78**(1), 100–112 (2009)

14. Edoun, I., Kgaphola, H.: The effect of international trade on emerging economies: the case of India. In: 34th International Academic Conference, Florence, pp. 58–64 (2017). https://doi.org/10.20472/IAC.2017.034.016
15. Greenaway, D., Milner, C.: On the measurement of intra-industry trade. Econ. J. 93(372), 900–908 (1983)
16. Grubel, H.G., Lloyd, P.J.: Intra-industry trade: the theory and measurement of International trade in differentiated products. Macmillan Press, London. (1975)
17. Handley, K.: Exporting under trade policy uncertainty: theory and evidence. J. Int. Econ. **94**(1), 50–66 (2014)
18. Handley, K., Limao, N.: Trade and investment under policy uncertainty: theory and firm evidence. Am. Econ. J. Econ. Pol. **7**(4), 189–222 (2015)
19. Iyer et al.: Indian manufacturing—strategic and operational decisions and business performance. Indian Institute of Management Bangalore Working Paper Series. Working Paper No. 338 (2011)
20. Kikuchi, T., Shimomura, K., Zeng, D.-Z.: On the emergence of intra-industry trade. J. Econ. **87**, 15–28 (2006)
21. Kimsanova, B., Herzfeld, T.: Policy analysis with Melitz-type gravity model: evidence from Kyrgyzstan. J. Asian Econ. **80** (2022). https://doi.org/10.1016/j.asieco.2022.101482
22. Krugman, P.: Scale economies, product differentiation, and the pattern of trade. Am. Econ. Rev. **70**(5), 950–959 (1980)
23. Kumar, S., Ahmed, S.: Growth and pattern of intra-industry trade between India and Bangladesh: 1975-2010. J. Int. Econ. Policy 2(21), 5–28 (2014). Available at SSRN:https://ssrn.com/abstract=2544407
24. Lakshmanan, L., Chinngaihlian, S., Raj, R.: Competitiveness of India's manufacturing sector: an assessment of related issues. RBI Occas. Papers **28**(1) (2007)
25. Melitz, M.: The impact of trade on intra-industry reallocations and aggregate industry productivity. Econometrica **71**, 1695–1725 (2003). https://doi.org/10.1111/1468-0262.00467
26. Panagariya, A.: India: a new tiger on the block? J. Int. Aff. **48**(1), 193–221 (2004)
27. Phillips, P.C.B., Perron, P.: Testing for a unit root in time series regression. Biometrika **75**(2), 335–346 (1988). https://doi.org/10.1093/biomet/75.2.335
28. Qasmi, B., Fausti, S.: NAFTA intra industry trade in agricultural food products. Western Agricultural Economics Association, 1999 Annual Meeting (1999)
29. Ramachandran, R.: International trade problems and India: a case study. J. Comm. Trade **6**(2), 46–53 (2011)
30. Redding, S.: Theories of heterogeneous firms and trade. C.E.P.R. Discussion Papers, 3 (2010). https://doi.org/10.1146/annurev-economics-111809-125118
31. Sapovadia, V.: Indian manufacturing sector: competitiveness at stake. SSRN Electron. J. (2015). https://doi.org/10.2139/ssrn.2603744
32. Sharma, L.S.: Economic effects of liberalization on manufacturing sector of BRICS + 2: a CGE model simulation analysis **2**, 1–11 (2012). https://www.researchgate.net/publication/328262740_ECONOMIC_EFFECTS_OF_LIBERALIZATION_ON_MANUFACTURING_SECTOR_OF_BRICS_2_A_CGE_MODEL_SIMULATION_ANALYSIS
33. Zhang, J., Fung, H.-G., Sawhney, B.: India's optimal trade strategy: a general equilibrium analysis. Economia Internazionale/International Economics, Camera di Commercio Industria Artigianato Agricoltura di Genova 61(4), 755–776 (2008). LNCS Homepage, http://www.springer.com/lncs, last accessed 2016/11/21

Risk Management of Civil Liability Resulting from Self-Driving Vehicle Accidents

Saad Darwish and Ahmed Rashad Amin Al-Hawari

Abstract In general, legal research on artificial intelligence must figure out if accidents and losses caused by AI robots can be held civilly responsible under standard rules, which is more important. So, this study used standard civil liability rules to determine who is responsible for losses caused by self-driving car accidents and why.

Keywords AI · Risk · Liability · Self-driving cars · Risk management

1 Introduction

Artificial intelligence is here. Humanity has dreamt of creating a compassionate artificial creature for hundreds of years, but a fast technological advancement in our society has increased the number of robots using artificial intelligence. Our phones have artificially intelligent assistants that can predict where we are going when the vehicle engine begins, as can robots, aircraft, drones, and other intelligent devices, including self-driving cars, which are the focus of our research. DaVinci's ambition of constructing a driverless automobile [1] has become a reality, and we can see it. Self-driving vehicles will soon roam big cities, with 39 businesses testing self-driving vehicles, and commercialization is approaching. This vehicle will improve public road mobility [2, 3]. They make driving more comfortable and make mistakes less likely, which may make the roads safer. In March 2018, a self-driving car accident in Arizona killed the first person [4]. The issue is whether tort responsibility for personal action based on mistake or tort liability for guarding objects (mechanical machines) or objective culpability applies to these incidents.

S. Darwish (✉) · A. R. A. Al-Hawari
Kingdom University, Riffa, Bahrain
e-mail: saad.darwish@ku.edu.bh

A. R. A. Al-Hawari
e-mail: a.amin@ku.edu.bh

© The Author(s), under exclusive license to Springer Nature Switzerland AG 2024
A. Hamdan and E. S. Aldhaen (eds.), *Artificial Intelligence and Transforming Digital Marketing*, Studies in Systems, Decision and Control 487,
https://doi.org/10.1007/978-3-031-35828-9_56

2 Research Questions

The essential questions in the study are:

- What are self-driving cars and AI?
- Who pays for self-driving vehicle damage?
- What is the liability type and legal basis?

3 Methodology

This qualitative study uses secondary sources to examine self-driving car accident risks and legal consequences. This descriptive technique includes explanations and concrete instructions. Researchers will use the findings to interpret current knowledge on self-driving car accident rules and their associated risks. Our primary aim is to create a meaningful dialogue regarding current conditions and the importance of risk assessment of this new trend. We will look at secondary sources for dangers and mitigation. The paper analyses and compares jurisprudential viewpoints on responsibility for harm produced by artificially intelligent technologies, such as self-driving cars. The study seeks to clarify the legal responsibility for self-driving car losses.

4 Literature Review

4.1 Civil Liability Risk Management

Self-driving cars have grown more common on our roadways. Self-driving vehicles are widely available; thus, civil responsibility for accidents must be addressed. This article discusses the legal and ethical issues of self-driving cars' civil liability risk management. It will examine civil responsibility law, its hazards, and legal and ethical solutions. Lastly, it will look at how the public can make it less likely that people will be sued over accidents involving self-driving cars. Self-driving car drivers must be aware that defective wiring might cause electrical shock. Being aware of this issue may lower the vehicle's accident risk. These basic actions are the key to lowering self-driving car accident risk. Driver carelessness causes self-driving vehicle accidents. One must be mindful of their surroundings and acquainted with their route. Seatbelts protect against collisions but not against vehicle metal electrocution. Thus, self-driving car drivers must be vigilant and proceed correctly [5]. Rand Corporation research found that highly autonomous autos might cut US fatalities by 90% [6].

Civil responsibility for self-driving car incidents needs further consideration. Long legal fights and complicated issues over accident liability, and given the enormous consequences of self-driving cars on technology and consumer markets, clear civil liability legislation is essential to a self-driving future [7]. Automakers worry about legal responsibility from self-driving car accidents. Automakers must create a risk management strategy to assess and reduce legal liability from self-driving vehicle accidents. It includes comprehending the legal consequences of self-driving vehicle technology and appropriately assessing the risk of self-driving car accidents. Insurance and self-governance systems may protect companies from civil liability generated by self-driving vehicle accidents. As the car industry moves toward self-driving cars, it will be necessary for risk management [8] to understand and handle multi-dimensional civil liability.

4.2 Self-Driving Vehicle Liability Law

To determine the legal nature of liability for damages caused by self-driving vehicles, we must first define civil liability and then identify the tort liability of drivers or vehicle owners on the one hand and manufacturers on the other hand, as well as whether the vehicle itself can be held liable.

4.2.1 Civil Liability

It is attributed to a person who bears the consequences of an act, and it is applied morally to the person's commitment to what she/he says or does and legally to the obligation to correct and remedy the fault that occurred to others under the law [9]. In this context, Islamic law has traditionally applied the term guarantee or taking liability. There should be neither injuring nor returning harm. The Holy Quraan [10], stated in Surah Al-Shura, verse 40, "reward of an evil deed is its equivalent. However, whoever pardons and seeks reconciliation, then their reward is with Allah. He certainly does not like the wrongdoers" [11].

Civil liability aims to remedy the damage suffered by the injured person because of the contractual error (contractual liability) or the error committed due to negligence (tort liability). The person is a guarantor of compensation for the injured person and bears the penalty for the damage that occurs to others and is attributed to his/her act or the act of the person under his/her control or things under custody. Besides, contractual liability occurs when a contractor breaches his/her contractual obligations, while tort liability occurs when an illegal act breaches the general legal duty not to harm others. The principle that governs both types of civil liability is that every mistake that harms others is a legal obligation [12].

The general rule states that liability cannot be achieved except with a mistake, whether it is obligatory to prove (as in the case of liability for a personal act) or is assumed. The harmed person is exempted from proving it against the person responsible. The liability for doing things (for example, the guard's liability for doing the thing that requires special attention or the mechanical device, such as the vehicle).

To determine the extent of the driver's liability for the damages it may cause to others, it is necessary to address, in brief, when liability is established against the driver. The hypothesis here is that liability may be attributed to him/her when committing a personal error (liability for a personal action) or as the guardian of the self-driving vehicle (liability for guarding things), so we will discuss the following. According to the central norms of civil responsibility, a person is not accountable for the act he/she conducts unless he/she can control it [12]. This control is connected to awareness and free will, meaning the action must be done willingly, not coercively. For example, suppose a person causes an accident. In that case, the judge must verify that the perpetrator was aware of his/her obligation not to harm others and of his/her behavior contrary to this obligation, considering the perpetrator's circumstances.

If this is the case, the vehicle's driver will be responsible for compensating the injured person for the damages that occurred to him because of his/her actions. Due to its lack of free will, the self-driving car cannot be held responsible for the mistake [13]. This also means it does not have an independent financial duty to compensate for the detrimental conduct. However, hackers who break into the self-driving system's software and cause an accident will be held primarily liable for the accident [14].

4.2.2 Liability for Guarding Things

It is recognized that the person is not only responsible for the damage resulting from his act but also for the damage resulting from the act of his/her subordinate or the things in his/her custody. Since the self-driving vehicle is not considered a legal person, the natural person is not responsible for its harmful act to others under the rules of liability for the actions of others, so the question arises about the extent to which we immediately refer to protecting responsibility to address this.

First:

A person guards a non-living item when he/she has three powers over it: used, directed, and controlled. The guardian does not need physical or legal custody of the thing or the source of absolute control over it to be lawful or illegal.

The owner is the custodian, and anybody claiming differently must prove it. The harmed person need not find the thing's protector. He may sue its owner for compensation, but the owner can avoid paying by demonstrating he was not the guardian. The damage occurred, and the guard was for someone else based on a legal act, material fact, or legal text, and the guard is not transferred by transferring the thing from the follower to the follower. For example, if the business owner contracted with one driver to drive the vehicle for him, the driver is a follower, not a guard, because guarding it remains for the follower. If the car hurts a third party while the subordinate is driving, the owner might be ordered to subordinate to the driver and protect the vehicle. Unless the owner or beneficiary of the object has retained himself in possession throughout the repair procedure, this transfers custody to this technician [12, 15].

Second:

The occurrence of an act by a person or machine that harms others. This act has to be affirmative intervention from the thing in creating the damage, which will be accomplished if that object's condition generally enables the damage, such as if the causative vehicle was driving. However, if the item's interference will be harmful if a person hits it while it is entirely still, this responsibility is not established, and the positive intervention of the thing does not need to be physically attached to the hurt [12, 15].

This responsibility only covers third-party harm, not actor damage. Third parties are any individual who is not a guardian of the object that produced the harm, even if they are a subordinate of the guard or an owner of the thing if the guardianship has been passed to another person. Thus, this obligation does not apply to the guard's harm from its charge. The wounded individual must not establish the guard's mistake to get compensation under this obligation. Suppose these objects caused a harmful accident. In that case, this is clear proof in a court of law that the guard is at fault, and the guard cannot pay for himself unless he can show that the thing under guard did what it did because of something outside of his control [12].

Thus, the guard must have genuine power to use, command, and control the object to establish culpability. Does the guard control the self-driving car? There is no question that self-driving cars are independent, which means they can learn on their own and make decisions without their owner's help.

This autonomy conflicts with the guard's authority to control and operate them, making it very difficult to attribute liability to the driver or owner based on the rules of liability for guarding objects due to the lack of actual control on which to base liability.

5 Manufacturer Liability: (Objective Liability)

This objective liability is based on the idea of bearing the liability, as it does not count the element of error and is satisfied with the element of damage [16]. Hence, the person is responsible for compensating the injured person even if unaware of his action. Civil liability has only one function: corrective, that is, removing the damage by compensating the injured and not punishing the one who caused the damage [17]. The product must be faulty and hurt others (i.e. lacking security and safety standards).

If a self-driving vehicle's operating software has a technical fault that causes harm, the vehicle's manufacturer is liable for the damage, even if the driver did nothing wrong. The injured party must establish the harm, defect, and causal link.

Many countries ask to show evidence of a design or manufacturing fault, such as a warning technology failure that prevents the vehicle from alerting the driver that it cannot take control. In this action, the plaintiff must establish that the product was not suited to fulfil its planned activity design fault or that the particular product offered was defective, which caused the vehicle to behave abnormally manufacturing defect [18].

The European Directive, for example, stated that multiple participants in commodity production were jointly liable to the injured person. Suppose one applies this to the self-driving car. In that case, the person who was hurt is the responsibility of the designers, developers, data processors, analysts, and everyone else involved in making the car.

However, applying product liability (objective responsibility) to self-driving cars creates two issues:

The first issue is related to evidence, where objective liability requires the plaintiff to prove the existence of a product defect, the causal relationship between this defect and the harmful event, and that the product was defective when the manufacturer provided it. Proof of a design defect depends on proving that no reasonable alternative designs will avoid the accident. The plaintiff must show that modifying the running software may have averted a traffic accident. That requires building a reasonable alternative to the complex algorithm that governs self-driving cars and establishing that it is superior to the original [18].

A machine with artificial intelligence that makes it autonomous in making decisions, such as a self-driving vehicle, makes it difficult to distinguish between the damage caused by its autonomous decisions and the product's defect, primarily if the systems work and there is no mechanical defect. This problem comes from the fact that the self-driving car may learn on its own and get smarter without human help. It implies that the self-driving vehicle's artificial intelligence at the manufacturing time is different from that after commissioning [19] making it easier for the manufacturer to argue that the error is not their fault. Whoever caused the harm did not exist when the self-driving car was sold, or it became apparent afterwards [20].

To solve this problem, strict liability can be adopted. This liability assumes that the person engages in an abnormally dangerous activity, so if harm occurs to others, whether personal or financial, he/she shall hold responsible, even if it is proven that he/she has taken the utmost care to prevent harm [21].

The Second issue is the unfairness and logic of applying these principles commercially to manufacturers, who would be held responsible for matters over which they have no control or were unaware at the time of production. Defects in manufactured products can only be discovered after they are sold. However, it only applies to technically sophisticated-produced goods like self-driving cars, where research has yet to uncover such problems [22].

There is no doubt that this would affect these companies' pursuit to develop and innovate [23], so there are those who believe that in cases where artificial intelligence is the decision-maker to replace human intelligence, manufacturers will always be responsible according to the rules governing the manufacturer's liability (objective liability) without being able to get rid of the liability, for example, if in the same circumstance, he establishes his innocence.

Assigning responsibility to the manufacturer may lead to a fear of operating legally, limiting technological development in traditional markets, or corporations

Risk Management of Civil Liability Resulting from Self-Driving ...

cutting R&D and redirecting all programming efforts to informal markets [19]. Manufacturers would incur increased expenses due to possible lawsuits and compensation following accidents, as well as research and development expenditures [18, 20, 21]. In the end, we suggest using a mandatory insurance system that is paid for by manufacturers. These companies would pay 50% of a common fund equal to their market share at the start of the system. Once accident data is available, the contribution will be figured out based on the damage caused by each manufacturer. The state's general budget will pay the other 50% via the tax it levies on this kind of vehicle. The Fund assesses claims from accident victims. The maker must compensate the injured party if carelessness is shown. Otherwise, the Fund pays the victim. It is negligence if it is possible to find the software issue that led the car to behave erratically or if the manufacturer's technology is outdated for the autonomous vehicle business. If either hypothesis is verified, the maker is negligent and fully liable for the accident (2018). This proposal will encourage manufacturers to invest in research and development to improve their products to avoid being considered negligent and to access the mutual insurance fund's compensation mechanism. One of the factors for assessing negligence will be the production of modern vehicles with the most advanced technology [18].

Another proposal considers that the insurance companies charged primarily with paying compensation for accidents caused by self-driving vehicles shall bear the burden, so that compulsory vehicle insurance includes both human-driven and self-driving vehicles so that the injured person has a natural right to obtain the required compensation from the vehicle insurance company. Some feel this method will address the issues of attributing blame to a particular individual, assure recompense for those impacted, and decrease litigation and lawsuits.

In most car accidents, the traffic officer may establish who is liable based on simplified principles without analyzing the reasons for the collision or evaluating if the driver was irresponsible [24] and then the insurance company has the right to initiate a recovery claim against the responsible party, which means that insurance companies will, by default, be liable for damages arising from autonomous vehicle accidents if the vehicle is insured at the time of the accident. The insurance company's compensation may be reduced if the injured party is also fully or partially liable for the damage caused [20].

6 Conclusion

We have obtained mixed results and suggestions on legal responsibility for self-driving car damages:

Our key results and conclusions are:

(1) Vehicles are only self-driving once they have much autonomy, so the driver either assists or does not drive.

(2) Self-driving cars are accountable for harming people since they have no legal personality.
(3) It is hard to say who is to blame because the car can make decisions independently, and many people, like the driver, the manufacturer, developers, maintenance companies, and others, could be at fault.
(4) If the driver is found to be at fault, he/she will be held accountable for the personal act, and the hackers who interfered with the vehicle's operating system may be held accountable if their actions caused an accident that injured others.
(5) Since the driver and owner lack absolute control over the vehicle, they cannot be held liable for protecting things.
(6) If a product defect causes damage, the manufacturer can be held liable for the harm, even if it took all reasonable precautions to prevent it.

7 Recommendations

We suggest a manufacturer-subsidized obligatory insurance system with a shared fund in which each firm contributes 50% based on its market share at the start of the system. Once accident data is available, the contribution will be computed based on each manufacturer's damages. The state's general budget will pay the other 50% via the tax it levies on this kind of vehicle. The Fund assesses claims from accident victims. The maker must compensate the injured party if carelessness is shown.

References

1. Winkler, K.C.: Autonomous Vehicles, Regulation in Germany and the US and Its Impact on the German Vehicle Industry. Tilburg Law School (2019). http://arno.uvt.nl/show.cgi?fid=149595
2. Chen, S., Sampson,J.H., Lim, B.K.L.: Attribution of civil liability for accidents involving automated cars. In: Report on the Institutional Knowledge at Singapore Management University, Singapore Academy of Law (2020)
3. Samoili, S., Cobo, M.L., Gomez, E., De Prato, G., Martinez-Plumed, F., Delipetrev, B.: AI Watch Defining Artificial Intelligence (2020). JRC118163. EUR 30117 EN, Publications Office of the European Union, Luxembourg. https://doi.org/10.2760/382730. ISBN 978-92-76-17045-7
4. Hawkins, A.J.: Serious Safety Lapses Led to Uber's Fatal Self-Driving Crash, New Documents Suggest (2019). https://www.theverge.com/2019/11/6/20951385/uber-self-driving-crash-death-reason-ntsb-dcouments
5. Moss, R.J., et al.: Autonomous Vehicle Risk Assessment. Stanford Intelligent Systems Laboratory (SISL), Navigation and Autonomous Vehicles Laboratory (NAV Lab), Autonomous Systems Laboratory (ASL), Department of Aeronautics and Astronautics, Stanford Center for AI Safety, Stanford University (2021). https://web.stanford.edu/~mossr/pdf/Autonomous_Vehicle_Risk_Assessment.pdf
6. Groves. D.G.: Introducing Autonomous Vehicles Sooner Could Save Hundreds of Thousands of Lives Over Time. PardeeRAND Graduate School (2017). https://www.rand.org/news/press/2017/11/07.html

7. Jr Robert, L.P.: Are automated vehicles safer than manually driven cars? AI Soc. **34**, 687–688 (2019). https://doi.org/10.1007/s00146-019-00894-y
8. Darwish, S.: Risk and knowledge in the context of organizational risk management. Eur. J. Bus. Manag. **7**(15), 118–126 (2015). www.iiste.org. ISSN 2222–1905 (Paper) ISSN 2222–2839 (Online)
9. Mustafa, I., et al.: Investigation of the Arabic Language Academy, Part One. Dar Al-Da'wah (n.d.). https://shamela.ws/book/7028/413
10. The Holy Quran
11. Al-Bayhaqi, Al-Sunan Al-Kubra.: vol. 10. The Ottoman Encyclopedia in Hyderabad in 1355 AH
12. Jamal, K.: The General Theory of Obligations in Bahraini Civil Law, 2nd edn. University of Bahrain (2002)
13. Coeckelbergh, M.: Artificial intelligence, responsibility attribution, and a relational justification of explainability. Sci. Eng. Ethics **26**, 2051–2068 (2019). https://www.semanticscholar.org/paper/Artificial-Intelligence%2C-Responsibility-and-a-of-Coeckelbergh/672d7db068a6ee6eab17c462ef6ecd118bd005bb
14. Schaub, M., Zhao, A.: Self Driving Cars: Who Will Be Held Responsible? (2017). https://www.kwm.com/global/en/insights/latest-thinking/self-driving-cars-who-will-be-liable.html
15. Al-Hawary, A.R.: Civil liability for damages resulting from self-driving vehicle accidents. J. Fac. Sharia Law **24**(3), Tahfna Al-Ashraf, Dakahlia, Egypt (2022)
16. Evas, T.: Civil Liability Regime for Artificial Intelligence, EPRS|European Parliamentary Research Service (2020). https://www.europarl.europa.eu/RegData/etudes/STUD/2020/654178/EPRS_STU(2020)654178_EN.pdf
17. Reda, T.: Tort Liability for Environmental Damage, Comparative Study, 1st edn. (2019). Zain Legal Publications, Beirut (2019)
18. Davola, A.: A model for tort liability in a world of driverless cars: establishing a framework for the upcoming technology. Idaho L. Rev. **54**(3), 592–614 (2018). https://digitalcommons.law.uidaho.edu/idaho-law-review/vol54/iss3/2/
19. Benhamou, Y., Feralnd, J.: Artificial intelligence and damages: assessing liability and calculating the damages. In: D'Agostino, P., Piovesan, C., Gaon, A. (éds) Leading Legal Disruption: Artificial Intelligence and a Toolkit for Lawyers and the Law. Thomson Reuters Canada (2020)
20. Punev, A.: Autonomous vehicles: the need for a separate European legal framework. Eur. View **19**(1), 95–102 (2020)
21. Anderson, J., et al.: Autonomous Vehicle Technology. RAND Corporation (2016). www.rand.org/t/rr443-2
22. Dawood, I.: The legal system to ensure the safety of persons from the damages of defective products, reality, and hope, an analytical study of the provisions of Egyptian law in the light of French law. J. Kuwaiti Int. Law Sch. eighth year, issue 2 (2020)
23. Schellekens, M.: No-fault compensation schemes for self-driving Vehicles. Law Innov. Technol. **10**(2), 314–333 (2018). https://doi.org/10.1080/17579961.2018.1527477
24. Jamal, K.: Mediator in the Principles of Law. Dar Al-Nahda, Al-Arabiya (2016)

The Impact of Applying Just-In-Time Production System on the Company Performance in Garment Manufacturing Companies in Jordan

Mohammad Abdalkarim Alzuod and Rami Atef Al-Odeh

Abstract The aim of this study is to understand the effects of implementing a Just-In-Time production system on a company's performance on garment manufacturing companies in Jordan. The entire workforce of MAS Kreeda Al-Safi Company in Jordan made up the study population of (3800) employees. A questionnaire was developed as a tool for gathering data from a sample of (350) employees for a study that used the descriptive correlational technique. For the statistical analysis, SPSS software was employed. The results indicated that JIT system and its components (material flow, JIT commitment, and supply management) had achieved a high level. While the company performance was high across all of its dimensions, including operating performance, operating cost, and product quality. Additionally, the findings of study showed a statistically significant relationship between the JIT system and firm performance. The study recommended, firms should be to pursue the adoption of JIT technology (because of its significant role in reducing production costs). In addition, forsaking conventional production methods due to its inability to compete with businesses in the marketplace.

Keywords Just-in-time production system · Company performance · Garment manufacturing companies · Jordan

1 Introduction

Today's world is characterized by rapid change and huge technological breakthroughs. In the face of fierce competition, every industrial and service company struggles to remain in operation and maintain its competitive advantage. Which led

---

M. A. Alzuod (✉)
Zarqa University, Zarqa, Jordan
e-mail: malzuod@zu.edu.jo

R. A. Al-Odeh
MAS Kreeda Al-Safi Company, Amman, Jordan

© The Author(s), under exclusive license to Springer Nature Switzerland AG 2024
A. Hamdan and E. S. Aldhaen (eds.), *Artificial Intelligence and Transforming Digital Marketing*, Studies in Systems, Decision and Control 487,
https://doi.org/10.1007/978-3-031-35828-9_57

to the adoption of the Just-In-Time production system (JIT) by businesses in order to achieve high quality performance, cut down on time, and increase customer satisfaction. Performance efficiency, which is comprised of operational performance, cost management and control, and product or service quality, results in the achievement of the organizational goals [1].

According to [2, 3], timely manufacturing in response to customer demand promotes quality in the production process and waste disposal in materials while reducing the waste of human and material resources. The objective of adopting JIT, according to [4] and [5], is to increase performance and efficiency in order to attain high-quality productivity. With JIT implementation, the company's ultimate goal is to deliver a products or service without wasting resources. Hence, JIT production system functions to continually improve the production process and add value to the good or service [6].

Khaireddin et al. [7] came to the conclusion that JIT depends on a set of procedures for optimum performance and that JIT depends on other environmental elements in addition to the efforts of the workers. According to [8], operational performance characteristics have a big impact on outcomes of ongoing processes like service quality, product quality enhancement, defective cost devaluation, productivity enhancement, delivery time reduction, inventory performance efficiency, and product delivery. Iqbal et al. [9] found that, particularly in the automotive industry, scheduling of Just-In-Time production system and piece size reduction led to speedier production by creating shared infrastructures with clients and suppliers. Salisbury and Gurahoo [10] intriguing suggestion that JIT purchasing and selling can improve worker flexibility and supply chain efficiency eventually results in improved operational performance and higher product quality.

Due to the growing interest that Lean Manufacturing ("Just-in-Time" or "JIT" one of the Lean Manufacturing tools) has attracted from researchers and industrial organizations [11], as well as its significant role in boosting productivity levels, significantly lowering costs, operating efficiently [12]. Generally, improving the performance of the organization as a whole, which necessitates speed in processing, speed in delivery, and continuous communication [13].

Firms that use JIT procedures to cut inventory and produce without flaws fare better financially. According to [4], led to reduced costs, quicker output, greater quality, and shorter lead times for purchases. JIT procedures were found to improve productivity and responsiveness. While [14] indicted that JIT is essential for lowering waste and process time, which leads to greater performance.

Garment manufacturing is one of the labor intensive manufacturing firms that contribute to the economic growth of the country [15]. The complexity of operations and the changing business environment are forcing industrial companies in Jordan, especially garment manufacturers, to look for systems that can help improve the quality of production while reducing production costs.

This study highlights the status of Jordan. Jordan quickly appeared as an important supplier of garment to the US, as result of a privileged distinctive trade agreement [16]. This different raise of Jordan as a supplier of apparel to the United States is a

The Impact of Applying Just-In-Time Production System ...   679

direct outcome of QIZ agreement that was signed between the US, Israel and Jordan in 1997 [17].

As a result, there is a growing interest in performance through the pursuit of performance as the cost of carrying inventory becomes apparent. High-quality clothing must be manufactured to meet the requirements of the financial statements. Therefore, the study goal was to show how JIT and its related dimensions (material flow, JIT commitment, and supply management) effect on the performance of Jordanian garment manufacturing companies.

2 Literature Review

2.1 Just-In-Time Production (JIT)

The Just-In-Time Production (JIT) concept has been personally attributed to Taiichi Ohno, Toyota's former production manager [1]. As technological developments and increased industrial competition have prompted many companies in Japan and America to develop programs to increase productivity and reduce costs, where the concept of (JIT) depends on the need to reach minimum levels of inventory, whether it is for raw materials or production in operation [3]. The final production that given to any backlog of inventory means that the enterprise incurs additional costs that can be avoided when zero inventory is reached. The key to the successful implementation of the JIT concept is through commitment, trust and continuous effort for improvement [1].

Although the JIT manufacturing system appeared in the seventies and met with great success in some of the companies that implemented it, there is no specific definition for it. Basically, JIT system represents to a production strategy that seeks to improve business return on investment through inventory reduction and costs of maintaining it. Moussawi-Haidar et al. [18] reviewed JIT as the strategy related to inventory management, which includes raw materials as they are processed by the supplier only when needed. On the same line, JIT system is a philosophy and a series of methods used for manufacturing, which provides raw materials [19]. Regard to [20] described it as an order that support materials and the production process in quantities and times regard to the customer's requirements. While, [21] also indicated it as a production theme based on reducing wasted time and improving productivity continuously.

This study adopted dimensions of JIT according to [22] that consist: material flow, commitment to JIT and supply management. Firstly, material flow refer to continuous flow manufacturing. It is a method of flowing technical requirements that combines the principles of continuous flow and custom manufacturing [19]. Based to [20], indicated to commitment to JIT as the commitment of implement and JIT production system improving operational processes and reducing the time, which results in reducing the manufacturing cycle time and delivery on time to customers and reducing waste, which leads to reduce production costs, improve product quality, and increase customer satisfaction. Lastly, supply management can be defined as information, payments, and primary materials from suppliers to the company's factories and warehouses that end with the customer, as well as all related companies and operations that contribute to the manufacture of the commodity until its delivery to the consumer [20].

2.2 Company Performance

Company performance is the potential and ability of a business to use available resources effectively to achieve goals in accordance with established plans, taking into account the relevance of those goals to users [23]. The success of establishments in maximizing the utilization of the available productive resources depends on focusing on four basic aspects, which are the main performance indicators of the establishment and are represented in quality, cost, flexibility and delivery in a timely manner [24].

Based to previous studies as mentioned, company performance combine by three dimensions; operational performance, operational costs, and quality of product. Therefore, operational performance reflects to the company's efficiency in the optimal use of its assets in order to generate sales [25]. Operational performance refers to various criteria, such as quality, low cost, flexibility and speed of delivery, by which organizations can measure their operational performance [26, 27]. Although, operational performance represents the required outcomes that the firms is striving to acquire. Also, these components measure the ability of firms to set its objectives by making dynamic use of obtainable sources. It is also a reversion of how the human and resources of the firms are employed and utilized in a way that enables it to complete its goals [28]. The operational performance of companies must be assessed based on a number of competing indicators (e.g. low cost, quality and speed of delivery).

Operational costs is timely application of the production system strengthens relationships with customers and suppliers, and leads to the reduction of inventory quantity, production cost, waste and damaged materials, and works to achieve competitive quality and improve product quality [25]. The findings also reviewed that the timely application of the production system in industrial firms leads to better control over the operations carried out by the production units, and the reduction of the stock of raw materials and the stock under operation and finished to the lowest possible level [29]. Hence, an operational costs means ongoing costs that arise from normal day-to-day operations.

Lastly, according to [30] that quality of product is the establishment's ability to create a product/service to a level of excellence that enables it to recognize the needs of its customers in a manner consistent with their expectations, as well as to feel content and happy about doing so and discovering something exceptional for them, This is accomplished by adhering to predetermined standards. Silva and Ferreira [31] argue that goodwill refers to the production of products that recognize or overtake

The Impact of Applying Just-In-Time Production System ... 681

the needs, wants, and predictions of the products. It is possible to define quality in terms of the customer who defines it and who and who determines quality which products and services meet his wants and needs.

2.3 Just-In-Time System Production (JIT) and Company Performance

Many of the previous studies [1, 21–23, 32, 33] purposed to investigate the relationship between JIT—company performance. The results of the previous study showed that the application of the JIT has a statistically significant effect on maximizing the performance of firms positively. Abu-Khalifa and Al-Okdeh [21] study showed that the application of the just-in-time production system has a statistically significant impact on maximizing the profitability of Jordanian SMEs. While, [1] found that JIT, TQM and SCM practices in both manufacturing and service sector individually and jointly impact organizational performance. On the same line, [23] showed that after Just-In-Time adoption companies reduced the labor purport in equipment's, increased inventory turnover and boosted incomes in the US. According to [33], indicated that Just-In-Time purchasing has a direct relationship positively with agile manufacturing, while the positive relationship between JIT production and agile manufacturing is mediated by Just-In-Time purchasing in large U.S. manufacturers. Kannan and Tan [22] also found that JIT and a commitment to quality and an understanding of supply chain dynamics have the greatest impact on performance of materials managers and senior operations in Europe and North America.

Based on above-mentioned studies, this study will be answered by testing the following hypotheses:

H01: There is no statistically significant relationship between JIT system production and the company performance in garment manufacturing companies in Jordan at the significance level ($\alpha \leq 0.05$).

H01.a: There is no statistically significant relationship between material flow and the company performance in garment manufacturing companies in Jordan at the significance level ($\alpha \leq 0.05$).

H01.b: There is no statistically significant relationship between commitment to JIT and the company performance in garment manufacturing companies in Jordan at the significance level ($\alpha \leq 0.05$).

H01.c: There is no statistically significant relationship between supply management and the company performance in garment manufacturing companies in Jordan at the significance level ($\alpha \leq 0.05$).

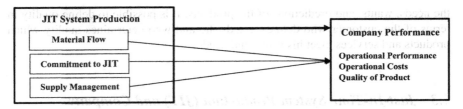

Fig. 1 Model of study

2.4 Model of Study

Based on previous models, previous studies, problem formulation and research hypotheses, the model was developed from studies; [22, 23, 33] (Fig. 1).

3 Research Methodology

3.1 Population and Sampling of the Study

The study is carried out within the company Mas Kreeda Al-Safi—Jordan was selected because the complexity of the operations and the evolution of the business environment oblige the Jordanian industrial companies, in particular garment manufacturers, looking for systems that help them reduce production costs. According to [17], MAS Kreeda Al Safi is the Jordanian manufacturing arm of the active clothing unit of MAS Holdings. It's established in 1987 in Sri Lanka and now assess about $1.6 billion. MAS Kreeda begin to Jordan in 2014, in part as a result of free trade agreements with the US and other countries. It has its main manufacturing factory in Amman, a qualified industrial zone (QIZ), though it has opened plants that employs 3800 personnel from 8 different nationalities mostly Jordanian workers in the main cites (Amman, Zarqa and Madaba) [17].

The target group of this study are all employees of the massive company Mas Kreeda al-Safi—Jordan (total number of employees is 3800). Suitable individuals will be selected using non-probability sampling—purposive sampling. This sampling technique makes it possible to select members who are typical of the population or particularly knowledgeable about the topics being investigated [34]. It used [35] table to specify the size of sample (351). Approximately 400 questionnaires were distributed electronically to the company's employees, 350 collected questionnaires were valid appropriate for analysis from the administered surveys. Hence, 87.5% was showing as a response rate.

The Impact of Applying Just-In-Time Production System ...

3.2 Measurements

The questionnaire was developed to achieve the purpose of this study. Based on [22, 23, 33] studies were designed of questionnaire, which consists of (37) items divided into three different parts; the first part for demographics and general information, the second part for JIT production system which consists of five statements for material flow, seven statements for commitment to JIT, six statements for supply management. Sequentially, the third part on company performance, which consists of operational performance (6) statements, operational cost (6) statements, and product quality (7) statements.

4 Results

4.1 Data Analysis

Data entry and analysis was conducted using version 24 of the Statistical Package for the Social Sciences (SPSS). The data was screened and cleaned for errors and missing data. Moreover, simple and multiple regression were used to test the hypotheses.

4.2 Demographic Characteristics

The study sample consisted of ($n = 350$) respondents, as shown in Table 1. The majority of the participants was male at percentage of 73.7%, while the female was at percentage 26.3%. The respondents showed experience in their work, as 193 of them have experience of from 5 years—to less than 10 years, at a rate of 55.1%, while the rest are evenly distributed between from 1 year to less than 5 years and more than 10 years by 39.1 and 5.7% respectively, others variables and percentage as shown in Table 1.

4.3 Reliability Test

The reliability of the study was verified by the Cronbach (α) reliability coefficient, by applying to the study sample, and the following Table 2 shows the results. The questionnaire is reliable and valid; the acceptable value of Cronbach's Alpha was above (0.70) [34].

Table 1 The sample demographic characteristics

Variable	Category	Frequency	%
Gender	Male	258	73.7
	Female	92	26.3
	Total	350	100
Experience	From 1 year < 5 years	137	39.1
	From 5 years < 10 years	193	55.1
	more than 10 year	20	5.7
	Total	350	100
Qualification	High school or less	208	59.4
	Diploma	62	17.7
	Bachelor	66	18.9
	Postgraduate	14	4.0
	Total	350	100
Career Position	Administrative manager	5	1.4
	General manager	2	0.6
	Finance manager	7	2.0
	Operations and production manager	4	1.1
	Purchasing Manager	4	1.1
	other	328	93.7
	Total	350	100
Specialization	Machinery and production	222	63.4
	Management	87	24.9
	Department quality supervisor on production	41	11.7
	Total	350	100
Age	From 20 years < 30 years	115	32.9
	From 30 years < 40 years	217	62.0
	Above 40 years	18	5.1
	Total	350	100

Table 2 The reliability of the study

Study Variables	Level	Cronbach's (α)	Number of Items
JIT system	Material flow	0.737	5
	Commitment to JIT	0.716	7
	Supply management	0.766	6
Company's performance	Operating performance	0.776	6
	Operating cost	0.833	6
	Quality of product	0.823	7
Overall questionnaire reliability		0.940	37

The Impact of Applying Just-In-Time Production System ... 685

Table 3 Descriptive statistics of the study variables

Variables		Mean	STD	Degree
Independent variables (JIT system)	Material flow	4.59	0.450	High
	Commitment to JIT	4.60	0.381	High
	Supply management	4.67	0.374	High
	Total	4.62	0.331	High
Dependent variables (company performance)	Operating performance	4.66	0.391	High
	Operating cost	4.68	0.392	High
	Quality of product	4.65	0.414	High
	Total	4.67	0.362	High

4.4 Descriptive Statistics of the Study Variables

Descriptive statistics are conducted to represent some data trends briefly. Table 3 revealed that independent variables (JIT system) and its dimensions: material flow, commitment to JIT and supply management captured a high level of respondents' agreements. While, the responds of the participants on the dependent variable (company performance) and its dimensions: operating performance, operating cost and quality of product captured a high level of respondents' agreements. Sequentially, the means of (JIT system) dimensions is ranging from (4.59–4.67), where the highest mean reached 4.67 for (supply management), then this was followed by item (commitment to JIT) with a mean of 4.60. Moreover, the lowest mean was 4.59 for item (material flow). The means of (company performance) dimensions is ranging from (4.65–4.68), where the highest mean reached 4.68 for (operating cost), then this was followed by item (operating performance) with a mean of 4.66. Moreover, the lowest mean was 4.65 for item (quality of product). Finally, central tendency measures of mean and standard deviation in addition to co-variance where used to rank the item, then highlight the importance of each item.

4.5 Testing the Study Hypotheses

The results of the study revealed that the adjusted R square equals to 0.708 with R^2 equals 0.710, which means that the regression model interpret 70.8% of the variance in the data. The Durbin-Watson $d = 1.783$, which is between the two critical values of $1.5 < d\ 2.5$. Therefore, there is no first order linear auto-correlation in the linear regression data of this study. F value is 282.99 with Sig. $= 0.000$, this mean that the model can interpret 70.8% of the changes in the dependent variable (Company Performance) and the rest attributed to other factors. The next Table 4 shows the model summary, and its significance.

Table 4 Model summary of company performance [**]

Sig	F	Durbin-Watson	Adjusted R^2	R^2	R	Model
0.00*	282.99	1.783	0.708	0.710	0.843	JIT system dimensions

[*] Statistically significant at the level of statistical significance ($\alpha \leq 0.05$)
[**] Dependent Variable: Company Performance

To check the impact of JIT system dimensions on company performance, multiple regression analysis was used to determine whether there is a statistically significant impact of the JIT system dimensions (Material Flow, Commitment to JIT and Supply Management) on the company performance at a statistically significant level ($\alpha \leq 0.05$). Data for the assumptions of multiple linear regression analysis were checked. Table 5 shows the significant statistical impact of the independent variables (material flow, commitment to JIT and supply management) based in the significant value of t which are 5.118, 11.567 and 10.464 when $\alpha \leq 0.05$ variables. These results lead us to build the interpretation equation as follow:

$$\text{Company Performance} = 0.365 + 0.184 * \text{Material Flow}$$
$$+ 0.440 * \text{Commitment to JIT}$$
$$+ 0.377 * \text{Supply Management}.$$

Based on the previous table, the one, two and three sub-hypotheses are rejected. So:

1. There is statistically significant relationship between Maternal Flow and company performance in garment manufacturing companies in Jordan at the significance level ($\alpha \leq 0.05$).
2. There is statistically significant relationship between Commitment to JIT and company performance garment manufacturing companies in Jordan at the significance level ($\alpha \leq 0.05$).

Table 5 The JIT system dimensions—company performance model coefficients

	Unstandardized coefficients		Standardized coefficients		
	B	Std. Error	Beta	T	Sig
(Constant)	0.365	0.149		2.457	0.015*
Material flow	0.148	0.029	0.184	5.118	0.000*
Commitment to JIT	0.417	0.036	0.440	11.567	0.000*
Supply management	0.365	0.035	0.377	10.464	0.000*

[*] Statistically significant at the level of statistical significance ($\alpha \leq 0.05$).

The Impact of Applying Just-In-Time Production System ... 687

3. There is statistically significant relationship between Supply Management and company performance in garment manufacturing companies in Jordan at the significance level ($\alpha \leq 0.05$).

5 Discussion

The findings indicated that the independent variables (JIT system) and its dimensions: material flow, commitment to JIT and supply management noted a high level of respondents' agreements. This result is due to the company's exclusion of every activity that does not drive to a growing in the added value of goods or service, commitment to a high level of quality and all aspects of the company's activities, and a commitment to continuous improvement in all of the company's activities. Also, this result is due to the fact that under this system (JIT), the company appoints employees several tasks in addition to their main job, so it is more suitable that employee can work in several workplaces simultaneously. This result is consistent with the study [32] which indicated that the adoption of just-in-time practices was high in Jordanian local fast food restaurants.

While, the responds of the participants on the dependent variable (Company's Performance) and its dimensions: operating performance, operating cost and quality of product noted a high level of respondents' agreements. This finding is caused to the company's interest in assembling an integrated system of strategies and policies necessary to control the firms work and growth, to be more rationalization of managers or top management decisions. In addition to creating an advanced work system to set the application for requirements of total quality, it also includes defining work policies, analyzing all processes, specifying the required quality specifications, and controlling process of control quality levels.

Moreover, the results showed that the JIT system has a significant positive effect on company performance. This achievement is attributed to the role of the JIT system in increasing the productivity of workers by team working on production lines coordinated and reorganized in the cells form, reduced production preparation time, smaller batches of production. As well as eliminating waste and damage through the applying TQM in inventory control, and the reduction of inventory in all its forms through better management of suppliers and a reduction in waiting times between production stages. In addition to its role in supplying the employees invested in inventory and using it in various dimensions of the business, to rise the efficiency of the use of the space that used in the factory. This finding is consistent with the study [1] which indicated that JIT practices and the company's operational performance represent a positive, significant, and medium impact.

In addition to this, the results indicated there is statistically significant relationship between maternal flow and company performance at the significance level ($\alpha \leq 0.05$). This result is noted to the role of maternal flow in reducing the cost of inventory of all kinds, raw, under operation and complete, to the minimum levels, almost to zero, which leads to achieving savings in costs, storage, handling and insurance risks. This result is in agreement with a study [21] which indicated that there is a statistically significant effect of applying JIT on (reducing product costs and reducing ending inventory volumes) in SMEs industrial companies. Moreover, the findings indicated that there is a statistically significant relationship between commitment to the JIT and the company performance at the significance level ($\alpha \leq 0.05$). This finding is showed to a role of JIT's commitment to manufacturing better and higher quality, since production quality is the every worker responsibility not just the quality inspector's responsibility. Hence, this finding is in agree with a study [30] which indicated that JIT and TQM have significant and positivity relationship with flexibility performance. Finally, the findings indicated that there is statistically significant relationship between supply management and company performance at the significance level ($\alpha \leq 0.05$). This finding is due to the fact that supply management helps effective supply chain management systems in reducing cost, reducing waste, reducing production cycle time, reducing costs, improving efficiency, timely delivery and on-time delivery. The findings are in line with study [25] that suggested a supply chain management practices significantly mitigate the relationship between JIT and total quality control and financial performance of Indonesian pharmaceutical companies.

6 Conclusions

The aim of this study is to understand the effects of implementing a Just-In-Time production system on a company's performance on garment manufacturing companies in Jordan. The entire workforce of MAS Kreeda Al-Safi Company in Jordan made up the study population of (3800) employees. A questionnaire was developed as a tool for gathering data from a sample of (350) employees for a study that used the descriptive correlational technique. For the statistical analysis, SPSS software was employed. The results indicated that JIT system and its components (material flow, JIT commitment, and supply management) had achieved a high level. While the company performance was high across all of its dimensions, including operating performance, operating cost, and product quality. Additionally, the findings of study showed a statistically significant relationship between the JIT system and firm performance. Based on the research findings, the following recommendations were achieved. The necessity for garment manufacturing companies in Jordan to seek to adopt JIT technology (because of its significant role in reducing production costs). It also should be enabled to empowering the innovation and provide innovative ideas to promote the JIT Production System. Provide automation solutions and allocate special budget to invest on automation solutions. Moreover garment manufacturing companies should also pay more attention to providing training programs relying on employee's skills analysis. Finally, companies should not forget to activate customers and supplier relationship management systems and establish strategic partnerships with them.

References

1. Lara, A.C., Menegon, E.M.P., Sehnem, S., Kuzma, E.: Relação entre práticas do Just in Time, Lean Manufacturing e Desempenho: uma meta-análise. Gest. Prod. **29**, 1–21 (2022)
2. Brakman, S., Garretsen, H., van Witteloostuijn, A.: The turn from just-in-time to just-in-case globalization in and after times of COVID-19: an essay on the risk re-appraisal of borders and buffers. Soc. Sci. Humanit. Open **2**(1), 100034-1–100034-6 (2020)
3. Li, Z., Ying, Q., Yan, W., Fan, C.: Does just-in-time adoption have an impact on corporate innovation: evidence from China. Account. Finan. **62**(S1), 1599–1635 (2022)
4. Singh, G., Ahuja, I.S.: An evaluation of just in time (JIT) implementation on manufacturing performance in Indian industry. J. Asia Bus. Stud. **8**(3), 278–294 (2014)
5. Kinyua, B.K.: An assessment of just in time procurement system on organization performance: a case study of corn products Kenya limited. Eur. J. Bus. Soc. Sci. **4**(5), 40–53 (2015)
6. Prajapati, M.R., Deshpande, V.: Cycle time reduction using lean principles and techniques: a review. Int. J. Adv. Ind. Eng. **3**(4), 208–213 (2015)
7. Khaireddin, M., Assab, M.I.E.A., Nawafleh, S.A.: Just-In-Time manufacturing practices and strategic performance: an empirical study applied on Jordanian pharmaceutical industries. Int. J. Stat. Syst. **10**(2), 287–307 (2015)
8. Sutrisno, T.F.C.W.: Relationship between total quality management element, operational performance and organizational performance in food production SMEs. J. Aplikasi Manag. **17**(2), 285–294 (2019)
9. Iqbal, T., Huq, F., Bhutta, M.K.S.: Agile manufacturing relationship building with TQM, JIT, and firm performance: an exploratory study in apparel export industry of Pakistan. Int. J. Prod. Econ. **203**, 24–37 (2018)
10. Salisbury, R.H., Gurahoo, N.: Lean and agile in small-and medium-sized enterprises: complementary or incompatible? S. Afr. J. Bus. Manag. **49**(1), 1–9 (2018)
11. Milewski, D.: Managerial and economical aspects of the just-in-time system "lean management in the time of pandemic". Sustainability **14**(3), 1204 (2022)
12. Nugroho, A., Christiananta, B., Wulani, F., Pratama, I.: Exploring the association among just in time, total quality and supply chain management influence on firm performance: evidence from Indonesia. Int. J. Supply Chain Manag. **9**(2), 920–928 (2020)
13. Atwi, R.: Strategic Management of Costs and the Various Methods That Can Be Used to Reduce Costs and Improve Profits. A thesis submitted to obtain a doctorate of sciences in economic sciences, Faculty of Economics, Commercial and Management Sciences, Farhat Abbas Setif University, Algeria (2018)
14. Bortolotti, T., Boscari, S., Danese, P.: Successful lean implementation: organizational culture and soft lean practices. Int. J. Prod. Econ. **160**, 182–201 (2015)
15. Matebu, A., Shibabaw, M.: Partial and total productivity measurement models for garment manufacturing firms. Jordan J. Mech. Ind. Eng. **9**(3), 167–176 (2015)
16. Azmeh, S., Nadvi, K.: 'Greater Chinese' global production networks in the Middle East: the rise of the Jordanian garment industry. Dev. Chang. **44**(6), 1317–1340 (2013)
17. International Finance Corporation Report.: Tackling Childcare: The Business Case for Employer-Supported Childcare. Case Study: MAS Kreeda Al Safi-Madaba Garment Manufacturing, Jordan (2021)
18. Moussawi-Haidar, L., Daou, H., Khalil, K.: Joint reserve stock and just-in-time inventory under regular preventive maintenance and random disruptions. Int. J. Prod. Res. **60**(5), 1666–1687 (2022)
19. Franco, C.E., Rubha, S.: An overview about JIT (just-in-time)–inventory management system. Int. J. Res. **5**(4), 14–18 (2017)
20. Kootenai, A.J., Babu, K.N., Talari, H.: JIT manufacturing system: from introduction to implement. Int. J. Econ. Bus. Finan. **1**(2), 7–25 (2013)
21. Abu-Khalifa, H., Al-Okdeh, S.: The effects of applying just-in-time production system on maximizing profitability of small and medium industrial companies in Jordan. Uncertain Supply Chain Manag. **9**(2), 393–402 (2021)

22. Kannan, V.R., Tan, K.C.: Just in time, total quality management, and supply chain management: understanding their linkages and impact on business performance. Omega **33**(2), 153–162 (2005)
23. Huson, M., Nanda, D.: The impact of just-in-time manufacturing on firm performance in the US. J. Oper. Manag. **12**(3–4), 297–310 (1995)
24. Fullerton, R.R., McWatters, C.S., Fawson, C.: An examination of the relationships between JIT and financial performance. J. Oper. Manag. **21**(4), 383–404 (2003)
25. Andria, F., Hartini, S., Rahmi, A., Rusmanah, E.: Effect of operations capabilities on financial performance of firms with moderating role of supply chain management capabilities: a case of Indonesian pharmaceutical firms. Syst. Rev. Pharm. **11**(1), 213–222 (2019)
26. Alzuod, M.: The impact of knowledge sharing on green innovation in jordanian industrial firms. Int. J. Innov. Creativity Change **14**(2), 1199–1211 (2020)
27. Bagher, A.: The effect of supply chain capabilities on performance of food companies. J. Finan. Mark. **2**(4), 1–9 (2018)
28. Şengül, M., Alpkan, L., Eren, E.: Effect of globalization on the operational performance: a survey on SMEs in the Turkish electric industry. Int. Bus. Res. **8**(7), 57–67 (2015)
29. Aityassine, F.L.Y., Aldiabat, B.F., Al-rjoub, S.R., Aldaihani, F.M.F., Al-Shorman, H.M., Al-Hawary, S.I.S.: The mediating effect of just in time on the relationship between green supply chain management practices and performance in the manufacturing companies. Uncertain Supply Chain Manag. **9**(4), 1081–1090 (2021)
30. Phan, A.C., Nguyen, H.T., Nguyen, H.A., Matsui, Y.: Effect of total quality management practices and JIT production practices on flexibility performance: empirical evidence from international manufacturing plants. Sustainability **11**(11), 3093 (2019)
31. Silva, A.A., Ferreira, F.C.M.: Uncertainty, flexibility and operational performance of companies: modelling from the perspective of managers. RAM (Rev. Adm. Mackenzie) **18**(4), 11–38 (2017)
32. Al-Janabi, S., Ghazi, A.: The Impact of Just in Time Practices on Operational Performance on Local Fast Food Restaurants in Jordan. Thesis, Submitted in Partial Fulfillment of the Requirements of Master Degree in MBA, Management Department, Business Faculty, Middle East University (2020)
33. Inman, R.A., Sale, R.S., Green, K.W., Jr., Whitten, D.: Agile manufacturing: relation to JIT, operational performance and firm performance. J. Oper. Manag. **29**(4), 343–355 (2011)
34. Sekaran, U., Bougie, R.: Research Methods for Business: A Skill Building Approach (2016). Wiley
35. Krejcie, R.V., Morgan, D.W.: Determining sample size for research activities. Educ. Psychol. Measur. **30**(3), 607–610 (1970)

Evaluation of Large-Scale Solar Photovoltaic Operating Plant by Using Environmental Impact Screening: A Case Study in Penang Island, Malaysia

Siti Isma Hani binti Ismail, Loh Yong Seng, Shanker Kumar Sinnakaudan, and Zul-fairul Zakaria

Abstract This research uses Environmental Impact Screening (EIS) by using Analytic Hierarchy Process (AHP) under Multi-Criteria Decision Making (MCDM) method to examine the environmental impact of large-scale solar (LSS) photovoltaic (PV) development in Penang Island, Malaysia. This study's was conducted through an interview session with 16 individuals including expert review or academician, consultant, and authorities. Large projects require an assessment to ensure that possible issues are identified and addressed early on in the planning and design phases. Ecological and chemical (EC) impact, as well as occupational safety (OSH) and economic impact (EI) are considered in this study. The data is then evaluated using AHP to determine the final component rankings. The bar chart will help in making a better decision by displaying the results. Based on the findings of the study, significant steps toward mitigation have been taken.

Keywords Large-scale solar · Renewable energy · Environmental impact screening

---

S. I. H. Ismail · S. K. Sinnakaudan · Z. Zakaria
Civil Engineering Studies, College of Engineering, Universiti Teknologi MARA, Cawangan Pulau Pinang, Kampus Permatang Pauh Permatang Pauh Pulau, Pinang, Malaysia
e-mail: sitiismai@uitm.edu.my

S. K. Sinnakaudan
e-mail: ceshan@uitm.edu.my

Z. Zakaria
e-mail: zulfairul@uitm.edu.my

L. Y. Seng (✉)
School of Housing, Building, and Planning, Universiti Sains Malaysia, USM, Penang 11800, Malaysia
e-mail: lys15_hpm047@student.usm.my

© The Author(s), under exclusive license to Springer Nature Switzerland AG 2024
A. Hamdan and E. S. Aldhaen (eds.), *Artificial Intelligence and Transforming Digital Marketing*, Studies in Systems, Decision and Control 487,
https://doi.org/10.1007/978-3-031-35828-9_58

1 Introduction

Solar energy can be defined as the power that use a sunlight to generate power without give bad impacts to the environment. It is one of the major types of renewable energy. Photovoltaic technology (PV) is one of the most effective methods for harnessing solar energy [1]. Since 2009, the capacity of photovoltaic (PV) solar energy generators has increased by 41% annually. By 2040, energy system predictions show a nearly ten-fold growth in PV solar energy generating capacity, reducing climate change, and assisting universal energy access [2]. However, there was no research has been conducted on the environmental implications of large-scale solar photovoltaic utilizing an Environmental Impact Screening (EIS), which is the first step of Environmental Impact Assessment (EIA). As a result, this gap of knowledge must be investigated as a modern approach for decreasing an environmental impact degradation [3].

2 Literature Review

Solar photovoltaic (PV) technology is becoming a more cost-effective option for Malaysia's future energy demands. This increasingly efficient and economical technology is an important step for Malaysia's target of having 31% renewable energy in its power mix by 2025. The development of Malaysia's solar power business builds on crucial industry experience to expand into other markets, such as TNB's recent investment in the solar business in Vietnam [4]. According to [5], Nanping City Solar Far is one of the examples of a large-scale solar project that preserves the natural environment. The placement among these panels generates a geometric looking which represent of the Fujin mountains under the appropriate light. Solar farm's average annual electricity power is estimated to exceed 173,350,000 kWh once it starts producing on-grid electricity. Carbon dioxide emissions are estimated to be lowered by about 172,827 tonnes each year on averages, since it meets the yearly electricity needs of about 109,713 families [6].

2.1 Large Scale Solar Scheme (LSS)

LSS stands for large scale solar scheme, which is a program to generate own electricity and sell to the grid using a solar PV farm with an installed capacity ranging from 1 to 30 MW (for distribution linked solar PV plants). The Energy Commission is in charge of this project, and potential developers will be chosen through competitive bidding [7]. A bidder may bid for a maximum capacity of 50 MWac and may submit no more than three bids, which are LSS1, LSS2, and LSS3 [8]. Fourth Competitive Bidding Round (LSS@MEnTARI) or LSS4 is a fourth competitive bidding program

Evaluation of Large-Scale Solar Photovoltaic Operating Plant by Using ... 693

Fig. 1 Example of shortlisted bidders for project capacity 30 to 50 MW

Energy Commission of Malaysia : Shortlisted Bidders Package P2: Project Capacity from 30 MW to 50 MW	
Shortlisted Bidders	**Capacity (MW)**
Asiabina Properties Sdn. Bhd.	50
Classic Solar Farm Sdn. Bhd.	50
Gopeng Berhad	50
JAKS Solar Power Sdn. Bhd.	50
Perbadanan Kemajuan Negeri Pahang and Kumpulan Powernet Berhad	50
Ragawang Corporation Sdn. Bhd.	50
Ranhill Utilities Berhad	50
Sharp Ventures Sdn. Bhd.	50
TNB Renewables Sdn. Bhd.	50
Uzma Envirorergy Sdn. Bhd.	50
Total	**500**
Source: Energy Commission, Malaysia	Mercom India Research

in Malaysia. It is used for bids to develop a large-scale solar power plants with a capacity up to 1000 MWac in Malaysia. Figure 1 shows the example of shortlisted bidders for project capacity 30 to 50 MW.

2.2 Environmental Impact Assessment (EIA)

EIA stands for Environmental Impact Assessment (EIA). It is a study of how a planned activity will affect to the environment, including biodiversity, flora and ecology, water, and air quality. The EIA is used to estimate environmental impacts at an early stage in the planning and design of a project. Overview of the EIA Process [9] claimed that it is essential to emphasize that the EIA process does not ensure that a project will be changed or denied if serious environmental concerns are identified. Table 1 shows the stages of each process of EIA with its involvement.

2.3 Environmental Impact Screening (EIS)

EIS stands for Environmental Impact Screening (EIS). Screening is the first stage of EIA process. Overview of the EIA Process [9] said that the benchmark criteria for an EIA varied by nation; some laws define the types of activities or projects that require an EIA, whereas others demand an EIA for any development that may have a major environmental impact or that exceeds a specific financial value. This study

Table 1 Stages of the EIA process with the details of each stage's involvement [10]

Stage	What is involved
Screening	Determining whether an EIA is necessary or not
Scoping	Determining what should be included in the assessment and published in the "EIA Report"
Preparing EIA report	The expected significant effects on the environment of the development must be included in the EIA report
Do the application and consultation	The EIA Report and development application should be made publicly (including through electronic methods), relevant stake-holders and the general public must be given the chance to comment on it
Decision making	Before deciding whether or not to approve consent for the development, the competent authority must consider the EIA Report as well as any comments made on it. It is necessary to make the decision statement publicly
Post decision	Any monitoring mandated by the relevant authority is performed by the developer

presents the best identification and mitigation in order to reduce the degradation of environment towards large-scale solar (LSS) photovoltaics (PV) development. The objective of this study is to evaluate and identify the potential for environmental deterioration of photovoltaic farming through data collection in Penang Island as a case study, and recommend an appropriate solar farm mitigation measure by using Multi-Criteria Decision Making (MCDM) approach. Figure 2 indicates the steps in EIA process including screening process.

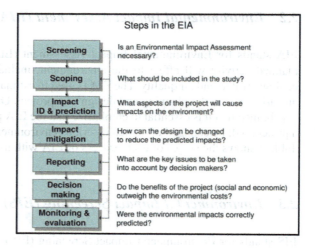

Fig. 2 Steps in the EIA.
Source Google Image (2022)

3 Methodology

The methodology and approach that were used in the execution of this research will be discussed in this chapter. The purpose of this study was to investigate the issues and the possible environmental impact of a large-scale solar photovoltaic project, and also to develop mitigation strategies for the purpose of minimizing the obstacles that may be encountered. By gathering all of the data through the use of Multi-Criteria Decision Making (MCDM), and then analyzing it through Analytic Hierarchy Process (AHP). The AHP includes a scoping process to establish particular environmental assessment components that consists of 3 categories, which are as follows: economic impact, ecological and chemical impact, and occupational health and safety impact (Fig. 3).

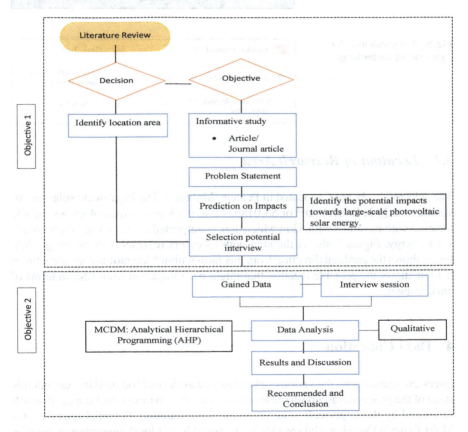

Fig. 3 Flowchart of the research study

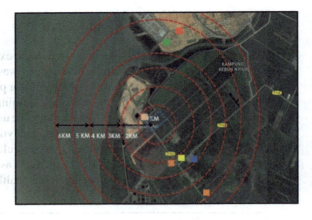

Fig. 4 Location of research area within 5 km radius. *Source* Google Earth (2022) with edited by Syaza

Fig. 5 Signs with its color represents of the building

3.1 Location of Research Area

The selected study area is located in Penang, Malaysia. The large-scale solar photovoltaic area is 214,681.38 m^2 or 53.04892429879748 acre. Figure 4 shows the site location with its radius with 5 km which may be impacted by existence of large-scale solar energy. Figure 5 shows the sign with its colors represents its building while Fig. 6 shows the cardinal direction from this farm within 5 km radius zone area. From here, it shows the possible location that might have impacted from LSS in terms of environment.

4 Data Collection

Interview session with the various stakeholders was the method used throughout this stage of the process to ascertain the significance of the environmental impact that will be caused by the large-scale solar photovoltaic installation. Potential interview for Multi-Criteria Decision Making (MCDM) would be the local approving authorities such as City Council, Jabatan Alam Sekitar, Jabatan Kerja Raya (JKR), etc. The interview was conducted by collecting the weightage and (%) effect value for each component. After that, the data will be extracted and analyzed by utilizing the MCDM approach and AHP tools analysis. The methods will examine the impact in terms of

Location/ Direction	North	Northeast	East	Southeast	South	Southwest	West	Northwest
1km	Agriculture	Agriculture	Agriculture	Agriculture	Agriculture	Agriculture	- Agriculture - Pulau Burung	Solid waste disposal
2km	Pulau Burung	Agriculture	Agriculture	Agriculture	Agriculture	Agriculture	Agriculture	Agriculture
3km	Pulau Burung	Agriculture	Agriculture	- School - Temple - Housing area	Agriculture	Agriculture	Agriculture	Agriculture
4km	- Pulau Burung - Sea	Agriculture	Agriculture	Agriculture	Agriculture	Agriculture	- Agriculture - Sea	- Agriculture - Sea
5km	- Factory - Courier - Pulau Burung	-Agriculture -Sea	Agriculture	Agriculture	Agriculture	- Agriculture - Sea	- Agriculture - Sea	- Agriculture - Sea

Fig. 6 Cardinal direction from large scale solar photovoltaics in Pulau Burung, Pulau Pinang within 5 km

economic, ecological and chemical, and occupational health and safety parameters. The goal of determining these values is to discover a mitigation strategy that can be employed to lessen the large scale solar photovoltaic negative impact on the environment.

4.1 Multi-Criteria Decision Making (MCDM)

The terms MCDM stands for "multi-criteria decision making", which refers to a method that is both well-structured and multidimensional. This method was developed to address decision-making issues that arise in a variety of contexts and to find the most desirable alternative while taking into account all relevant criteria [11]. According to [12] and [13], MCDM techniques were created to identify a better option, categorize an option into a limited number of categories, or rank an option into a subjective in a preferences order. The process of 'making a decision' involves evaluating several options in order to settle on a selection or a plan of action that will allow one to achieve their intended aims and goals [14].

4.2 Analytical Hierarchy Process (AHP)

The Analytical Hierarchical Programming (AHP) was introduced as a multi-criteria decision making (MCDM). AHP method is a decision making with a multi-level

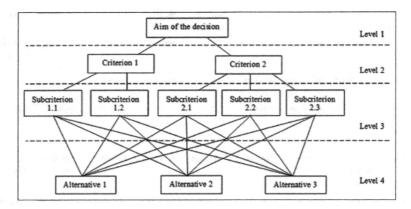

Fig. 7 The general four-level structure of the AHP [16]

Table 2 Saaty scale for pairwise comparison [15]

Intensity of importance	Definition
1	Equal importance
3	Moderate importance
5	Strong importance
7	Very strong importance
9	Extreme importance

hierarchical structure that includes criteria, sub-criteria, objectives, and alternatives. By using this process of pair-wise comparisons, the relevant data is produced. From this comparison, they can get the decision criteria importance weights. Figure 7 shows the general four level structure of the AHP process. Table 2 shows the Saaty scale for pairwise comparison [15].

4.3 Excel Template: Analytical Hierarchy Process (AHP)

Excel version MS Excel 2013 is required to use AHP template for analysis. For pairwise comparisons, the workbook has 20 input sheets, one for consolidating judgments, one for summarizing the results, and one with reference tables (randomness index, judgement ranges, geometric consistency index GCI limit ranges) as well as a sheet for finding solutions the eigenvalue problem through the eigenvector method (EVM). Figure 8 shows the results which EVM will be used to determine the weights and errors for each criterion, and also indicated indicates a check field that shows the convergence of EVM computation [17].

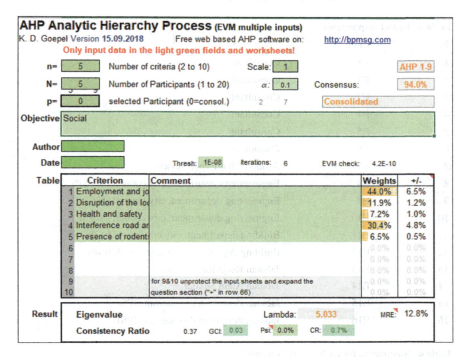

Fig. 8 Analytic Hierarchy Process (AHP) template [17]

4.4 Questionnaire Analysis

The information was gathered by conducting in-depth interviews with 16 stakeholders who were considered to be specialists in the field of study. The respondents are academician or expert review, consultant, City Council, JKR, Jabatan Alam Sekitar, and Jabatan Pengairan dan Saliran (JPS). According to Table 3, expertise is given one of the following codes: E1, E2, E3, and so on. Questions concerning the effects that using a solar photovoltaics on a big scale could have on the natural environment were asked throughout the interview. The input from the professionals is collected, and the Analytic Hierarchy Process excel is used to conduct the analysis.

5 Result and Analysis

The acquired results are addressed in Table 4 and Fig. 9. The criteria of impacts are displays in the Table 4 are analysed using AHP method. The three components are shown in the Table 4 with the weight and its ranking in each criterion. These three criteria are:

Table 3 List of respondents

No	Code	Department
1	E1	Consultant
2	E2	Consultant
3	E3	Consultant
4	E4	Consultant
5	E5	Consultant
6	E6	Consultant
7	E7	Academician/Expert review
8	E8	Engineering department, city council (MBSP)
9	E9	Engineering department, city council (MBSP)
10	E10	Engineering department, city council (MBSP)
11	E11	Building department, city council (MBSP)
12	E12	Building department, city council (MBSP)
13	E13	Jabatan Kerja Raya (JKR)
14	E14	Jabatan Alam Sekitar
15	E15	Jabatan Pengairan dan Saliran (JPS)
16	E16	Jabatan Pengairan dan Saliran (JPS)

Table 4 Summarize for each criteria's weight

Criteria		Sub-criterion	Weight	Impact	Ranking
Economic impact					
EI	*EI1*	Changes in land values	0.255	–	3
	EI2	Cost saving payment on electric bill	0.301	+	2
	EI3	Employment opportunities	0.444	+	1
Ecological and chemical impact					
EC	*EC1*	Chemical component in solar production	0.133	–	5
	EC2	Habitat loss and fragmentation	0.222	–	2
	EC3	Changes in soil moisture and temperature	0.243	–	1
	EC4	Land disturbance	0.153	–	3
	EC5	Air pollution	0.106	–	6
	EC6	Reduce greenhouse gas emissions	0.143	+	4
Occupational health and safety impacts					
OSH	*OSH 1*	Danger works environment for workers	0.169	–	5
	OSH 2	Health risk towards workers and local area	0.189	–	4
	OSH 3	Hazardous material that will endanger to public health	0.218	–	2
	OSH 4	Producing solid waste from construction phase	0.232	–	1
	OSH 5	Increased risk of fires and contamination	0.191	–	3

Fig. 9 Summarize chart for each criteria's weight

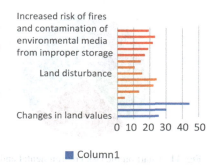

- Economic impact.
- Ecological and chemical impact.
- Occupational health and safety impact.

The data presented in Fig. 10 shows from the survey of 31% of stakeholders stated that they are agree with a statement that persons who live in close proximity to large-scale solar (LSS) photovoltaics pose a threat to both their environment and their health. While the majority of them, 69% believe that persons who live in close proximity to an LSS do not pose any threat to the environment or their own health. According to [18], solar inverters and photovoltaic technologies are not known to present any serious health risks to its surroundings.

Figure 11 shows the risk of environmental and health damage for those living near a LSS are 35% increase in habitat loss, 26% increase in cancer risk, 22% toxic chemicals in photovoltaic manufacturing process such as hydrochloride acid, and the smallest percentage is 17% for exposure to hazardous materials. According to [19], due to the size required to supply electricity systems, solar installations are enormous and impose a new type of built environments on those areas. This may have unforeseen and unanticipated effects on regional wildlife, flora, and even the microbiome.

Furthermore, Fig. 12 illustrates a stakeholders indicating that before a solar farm is constructed, the planning for animals who live in that location to survive is by avoiding a sensitive area for construction such as areas where habitats will suffer a direct impact (65%), relocated, altered, and existing habitat landscapes to guarantees

Fig. 10 Chart on the agree that there is a risk of environmental and health

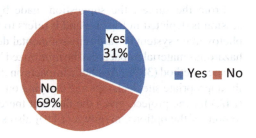

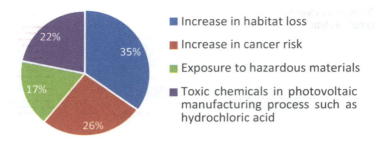

Fig. 11 Chart on the risk environmental and health damage

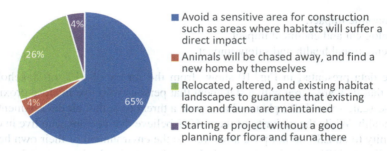

Fig. 12 Chart on planning for animals before solar farm is constructed

that existing flora and fauna are maintained (26%), starting a project without have a good planning for flora and fauna (5%), and animals will be chased away, and find a new home by themselves (4%).

Figure 13 shows most of stakeholders mentioned that constructing a LSS can be affected to the environment is due to clearing the land or earthworks can harm natural vegetation and wildlife in a variety of ways such as habitat loss and interference with rainfall (44%). Courage [19] also highlighted that a large solar farms in general have the possibility of altering the landscape, and disrupt crucial habitat for wildlife or migration routes with fences. The second highest is 40% which is from LSS project, it will require a lot of space to be clearing the land for solar generating installations which can impact on soil such as erosion and loss of nutrients. Lastly, hazardous waste due to heavy metal content in the solar panel and solar panel's fragility is 16%.

From the survey, the suggestion made by a stakeholder during the interview session is depicted in Fig. 14, and it refers to the potential improvement of the LSS photovoltaic system towards environmental degradation. Most of them claimed that hazardous material impacts can be minimized by proper planning and excellent maintenance method (38%). Aesthetic impacts on the landscape can be avoided by making the appropriate site decisions and carrying out an Environmental Impact Assessment (EIA) for the project, given the fact that there is a common percentage that applies to both of the options. mygov.scot [10] also stated that EIA will help to ensure that

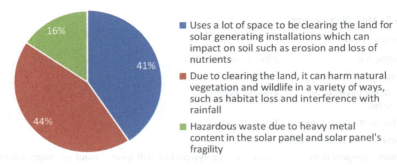

Fig. 13 Chart on why LSS can affected to the environment with bad impacts

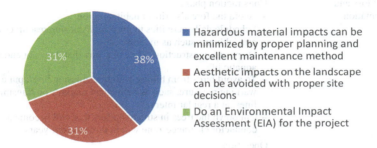

Fig. 14 Chart on suggestion to the potential improvement of LSS towards environment degradation

project decision-makers consider the potential impacts on the environment as early as possible and try to avoid, lessen, or mitigate those consequences.

5.1 Mitigation Measures

According to this study, mitigating actions need to be done in order to ensure that the impact of large-scale solar photovoltaic systems on the environment is reduced as highlighted in Table 5.

Table 5 Mitigation measure

Economic impact	
Sub-criterion	Mitigation measures
Changes in land values	Positive impact
Cost saving payment on electric bill	Positive impact
Employment opportunities	Positive impact
Ecological and chemical impact	
Chemical component in solar production	Provision for a system that will lessen, avoid, or compensate for any potential unfavourable environmental impacts that may result from development operations
Habitat loss and fragmentation	Construction phase – Avoid the forest's critical habitat region – Schedule daily activities to prevent disturbing resources during crucial hours such as mating – Dispose a construction waste in a way that does not endanger aquatic life – To avoid birds from hitting with transmission lines, put a visual warning measure, such as permanent markers on transmission lines, at a regular interval – Cut down the trees in stream buffers that will become a conductor clearance zone within three to four years
	Operation – Reduce as much of the human and vehicular activities – Restore and preserve as many places as possible in their natural state – For safety and security reasons, use only the required amount of lights to reduce the number of migratory birds and endangered species
	Decommissioning – Optimize the area that is reclaimed to reduce habitat loss and fragmentation – Replace damaged areas with topsoil taken at the start of a project or during decommissioning
Changes in soil moisture and temperature	– When constructing roads, consider the local climate, the amount of moisture in the soil, and the probability of erosion – Providing a protective soil cover and surroundings that is suitable to robust growing plants, nutrients replenishment solutions can help to recover soil organic matter
Land disturbance	– Avoid constructing a slope that are too steep – Excavations and earthwork in ecologically sensitive locations will need to be monitored by a qualified palaeontologist

(continued)

Evaluation of Large-Scale Solar Photovoltaic Operating Plant by Using ... 705

Table 5 (continued)

Air pollution	– Control the amount of damage and vegetation that is removed from the landscape – Reduce on-site vehicle usage and requiring a regular preventative maintenance to make sure the optimal combustion and lowest emissions – Prepare a dust control plan specific to the project and site – Built a wind fence around every disturbed location that might have an effect on the surrounding area that reaches beyond the site borders – Before entering paved public routes, check and cleanse the tires of equipment vehicles to ensure they are clear of dirt and remove any apparent track out dirt – Adhere to state emission regulations for all combustion sources
Reduce greenhouse gas emissions	Positive impact
Occupational safety and health impact	
Danger works environment for workers	– Set up rules for facility with materials that work well together in a safe way – Discover the area that has been used for in the past and find out if there might be any dangerous materials there
Health risk towards workers and local area	– In the process of a health risk assessments, take into account the possibility that workers may be exposed to non-cancerous as well as cancer dangers during the development and operation of the facility
Hazardous material that will endanger to public health	– Create and run the systems with dangerous materials in a way that restricts the chance that they will be released
Producing solid waste from construction phase	– Make a list for all the dangerous materials that will used, stored, moved, or discarded during activity – Create a waste management plan that lists the expected a solid and liquid waste stream, as well as procedures for determining, inspecting, and reducing waste, as well as places to store waste and the management and disposal needs for each type of waste – Verifying that extra structural steel pieces (such as metal poles and rods) are not transported, stored, or lifted in locations in which impact with overhead electric lines could happen
Increased risk of fires and contamination of environmental media from improper storage	– Establish a fire protection and prevention plan to reduce the risk of fires caused by compounds used and stored on the site, especially the heat exchange fluid used at the facility, which is highly flammable – Make sure to follow all of the necessary approach when putting away and transporting explosive items and blasting equipment – During each stage of the project, prepare and take pre-cautions against the risk of wildfire by developing and putting into action an appropriate wildfires control measure. These should include worker training as well as inspection and monitoring procedures

6 Conclusion

In conclusion, environmental factors are extremely important to consider during the process of developing large-scale solar PV systems in Pulau Pinang. This research also makes use of the screening methods, namely the Environmental Impact Screening (EIS), which is the first phase in the Environmental Impact Assessment (EIA), in order to analyse the impact assessment of the project. This research project was successful in achieving its general objectives, which included analysing and identifying the possibility for environmental deterioration caused by solar farming and also collecting of data in Pulau Pinang as a case study, which has been identified through the analysis and evaluation of the site, in addition to the opinion of an expert.

The Analytical Hierarchy Process (AHP) classifies the impacts into three categories which is economic impact, ecological and chemical impact, and occupational safety and health impact. Mitigation measures has been analysed based on the results, in order to achieve objective 2 which is to recommend an appropriate solar farm mitigation measure by using MCDM approach. In order to reach target 2, which is to recommend an acceptable solar farm mitigation measure by using the MCDM approach, mitigation measures have been assessed based on the results. This was developed in order to ensure the goal of achieving objective 2.

7 Recommendation

This research has revealed information on the significance of solar photovoltaic projects of a large size in terms of their impact to the environment. The development of the project will have the ability to lessen its environmental impact while keeps on increasing the beneficial impacts. The research findings can be utilized as guidelines for large-scale similar study that requires the implementation of environmental impact screening (EIS). This study may be used as a guideline for the government agencies to ensure that the operation of solar farm facilities is the right approach of the effectiveness of the growth in terms of economical, ecological and chemical, as well as occupational safety and health in order to reduce degradation of environment in Penang Island.

Unfortunately, this approach of this study has always been deficient in several aspects, such as the study's findings are restricted due to the fact that it only contains information from experts and authorities. It is recommended that the data be collected from local sources as well, as also help to increase the efficiency of this methodology in any future investigation or research that is conducted.

References

1. Parida, B., Iniyan, S., Goic, R.: A Review of Solar Photovoltaic Technologies (2011). https://www.sciencedirect.com/science/article/abs/pii/S1364032110004016
2. Kruitwagen, L., Story, K.T., Friedrich, J., Byers, L., Skillman, S., Hepburn, C.: A Global Inventory of Photovoltaic Solar Energy Generating Units (2021). https://www.nature.com/articles/s41586-021-03957-7
3. Zainon, M.R., Baharum, F., Seng, L.Y. Analysis of Indoor Environmental Quality Influence Toward Occupants' Work Performance in Kompleks Eureka, USM (2016). AIP Conf. Proc. **1761**, 020110. https://doi.org/10.1063/1.4960950
4. Watch, E.: Infographic: Shining a Light on Large-Scale Solar (LSS) in Malaysia (2021). https://www.energywatch.com.my/blog/2021/04/12/shining-a-light-on-large-scale-solar-lss-in-malaysia/
5. McDonagh, S.: These Are the 11 Most Beautiful Solar Farms in the World (2021). https://www.euronews.com/green/2021/10/14/these-are-the-11-most-beautiful-solar-farms-in-the-world
6. Limited, X.S.: Xinyi Solars 150MW Ground-Mounted Solar Farm Commences On-Grid Electricity Generation (2014). https://www.xinyisolar.com/en/NewsCenter/info.aspx?itemid=562
7. Bhd, E.E.: Environmental Impact Assessment (n.d). http://www.e3sb.com.my/environmental-impact-ssessment.aspx
8. Choi, C.S., Cagle, A.E., Macknick, J., Bloom, D.E., Caplan, J.S., Ravi, S.: Effects of Revegetation on Soil Physical and Chemical Properties in Solar Photovoltaic Infrastructure (2020). https://doi.org/10.3389/fenvs.2020.00140
9. Overview of the EIA Process.: (n.d). https://www.elaw.org/files/mining-eia-guidebook/Chapter2.pdf
10. mygov.scot.: Environmental Impact Assessment (EIA) Overview (2019). https://www.mygov.scot/eia
11. Eltarabishi, F., Omar, O.H., Alsyouf, I., Bettayeb, M.: Multi-Criteria Decision Making Methods and Their Applications—a Literature Review (2020). http://www.ieomsociety.org/ieom2020/papers/656.pdf
12. Velasquez, M., Hester, P.T.: An Analysis of Multi-Criteria Decision Making Methods (2013). http://www.orstw.org.tw/ijor/vol10no2/ijor_vol10_no2_p56_p66.pdf
13. Mardani, A., Jusoh, A., Nor, K.M., Khalifah, Z., Zakwan, N., Valipour, A.: Multiple Criteria Decision-Making Techniques and Their Applications—a Review of the Literature From 2000 to 2014 (2015). https://doi.org/10.1080/1331677X.2015.1075139
14. Haddad, M., Sanders, D.: Selection of Discrete Multiple Criteria Decision Making Methods in the Presence of Risk and Uncertainty (2018). https://www.sciencedirect.com/science/article/pii/S2214716018302288
15. Leal, J.E.: AHP-Express: A Simplified Version of the Analythical Hierarchy Process Method (2019). https://www.sciencedirect.com/science/article/pii/S2215016119303243
16. Bhole, G.P., Deshmukh, D.T.: Multi Criteria Decision Making (MCDM) Method and Its Application (2018). https://www.ijraset.com/fileserve.php?FID=17056
17. Goepel, K. D.: https://www.scribd.com/document/428721644/AHPcalc-2018-09-15 (2013)
18. Flowers, G., Cleveland, T.: Health and Safety Impacts of Solar Photovoltaics (2017). https://content.ces.ncsu.edu/health-and-safety-impacts-of-solar-photovoltaics
19. Courage, K.H.: Solar Farms Are Often Bad for Biodiversity—but They Don't Have to Be (2021). https://www.vox.com/2021/8/18/22556193/solar-energy-biodiversity-birds-pollinator-land

Prefabricated Plastic Pavement for High-Traffic and Extreme Weather Conditions

M. E. Al-Atroush , Nura Bala , and Musa Adamu

Abstract This study investigates the potential of prefabricated plastic pavements as an alternative solution to traditional asphalt pavements for high-traffic and extreme weather conditions. Through a comprehensive review of existing literature and case studies, the advantages of utilizing waste plastics as raw materials for the construction of plastic pavements were highlighted, such as high melting point temperatures, lower carbon footprint, and reduced occurrences of defects in comparison to conventional pavement types. Additionally, the use of plastic roads can also help to address the growing problem of plastic waste by finding a new use for recycled plastic materials. Furthermore, plastic roads have potential advantages over traditional roads in terms of water storage and climate adaptation, as they have been found to have better water drainage capabilities. Waste plastics such as polyethylene, polyvinyl chloride, and ethylene vinyl acetate were identified as suitable materials for heavy traffic roadway applications. However, the study also highlighted the limitations of the technology in terms of heavy traffic roadway applications and the need for more research to fully understand the potential benefits and drawbacks of plastic roads and how to extend this new technology to roadways considering the long-term durability and maintenance requirements of plastic pavements.

Keywords Resilient infrastructure · SDG 9 · SDG 11 · Prefabricated plastic pavement · Waste plastics · Plastic pavements

---

M. E. Al-Atroush (✉) · M. Adamu
Department of Engineering Management, College of Engineering, Prince Sultan University, 66833 Riyadh, Kingdom of Saudi Arabia
e-mail: mezzat@psu.edu.sa

N. Bala
Department of Civil Engineering, Bayero University Kano, PMB, Kano 3011, Nigeria

© The Author(s), under exclusive license to Springer Nature Switzerland AG 2024
A. Hamdan and E. S. Aldhaen (eds.), *Artificial Intelligence and Transforming Digital Marketing*, Studies in Systems, Decision and Control 487,
https://doi.org/10.1007/978-3-031-35828-9_59

709

1 Introduction

Asphalt pavements have been identified as a significant contributor to climate change. The production and construction of asphalt require large amounts of energy and result in high carbon emissions. According to the Intergovernmental Panel on Climate Change, [1] climate change represents a substantial threat not only to human life and nature but also to the built environment, including transportation infrastructures that are constantly exposed to the natural environment such as the physical roadways network, railways, hydraulic structures, and bridges. The maintenance and rehabilitation of large roadway networks also produce high levels of carbon dioxide emissions. As the world's roadway network expands or maintenance works continue, the emissions associated with its upkeep will continue to rise and become of more significant concern [2].

On the flip side, asphalt pavements are also sensitive to temperature changes. They are expected to fail earlier as a result of climate change [3]. According to a recent evaluation study undertaken by Transportation for America [4] to analyze the quality of the United States' roads between 2009 and 2017, the number of roads in poor condition rose from 14 to 20 percent. This increase is attributed to the improper selection of paving materials, which are frequently specified with the premise of a stationary climate rather than taking into account the potential impacts of climate change. Given the flexible pavement temperature sensitivity, various failures and distresses could be projected [5–7].

In order to address these challenges, there is a growing interest in developing alternative paving materials that are more sustainable and able to withstand extreme weather conditions. One such alternative is prefabricated plastic pavement, which is made entirely of recycled plastic. Those prefabricated panels have been found to be lighter, last longer, and have better water drainage capabilities than traditional asphalt surfaces [8]. Additionally, the use of plastic roads can also help to address the growing problem of plastic waste by finding a new use for recycled plastic materials [9].

One real-world example of a 100% plastic prefabricated road is the plastic cycle path in the Netherlands, which was recently constructed by the company KWS in 2016 [10]. This project involved the use of recycled plastic waste to create a durable and sustainable alternative to traditional asphalt roads. The plastic used in this project was a combination of plastic waste, such as old fishing nets and plastic bottles, which were shredded and processed to create the prefabricated panels that were used in the construction of the road. This project demonstrated the potential of plastic roads to not only provide a sustainable solution for road construction but also to address the problem of plastic waste. The plastic cycle path has an expected life span of at least 30 years and requires minimal maintenance compared to traditional asphalt roads. In addition, the plastic road has a lower heat absorption rate and a higher skid resistance, which makes them safer for cyclists and pedestrians [8].

This successful real-world example demonstrated the potential of plastic roads to provide a sustainable solution for road construction that addresses the problem of plastic waste. However, the use of plastic roads is still a relatively new technology, still limited to walkways and cycle paths. More research is needed to fully understand plastic roads' potential benefits and drawbacks and how to extend this new technology to roadways. Limited studies have shown that plastic roads have a longer lifespan than traditional asphalt roads, but it is still essential to evaluate their performance under extreme weather and heavy traffic conditions and their long-term environmental impact.

The purpose of this review paper is to provide a comprehensive overview of the current state of the art of prefabricated plastic roads, including their advantages, real-life implementation, utilized materials, construction methods, and potential benefits, as well as identify the current research gaps and suggest future research directions in this area.

2 Methodology

The methodology of this study also included a discussion of real-world examples of prefabricated plastic pavement construction. This involved a review of case studies and reports from organizations and companies that have implemented plastic road projects, such as the PlasticRoad project in the Netherlands and several locations.

The collected data was analyzed and organized thematically, with a focus on the construction method, advantages, and real-world examples of prefabricated plastic pavement. The analysis included a comparison of the performance and sustainability of prefabricated plastic pavement compared to traditional asphalt pavement. Accordingly, the research gap will be identified and discussed, and recommendations for future research in this field will be provided.

3 Prefabricated Plastic Roads

Prefabricated plastic pavement is a modular and hollow road structure that is constructed from either virgin or recycled plastic materials. As shown in Fig. 1, unlike traditional pavement structures, it is designed to be modular and lightweight. Plastic pavement's prefabrication production and modular design make it more efficient, faster to construct, and simpler to maintain [11].

Prefabricated plastic pavement typically consists of hollow spaces directly underneath the top surface of the subgrade. These hollow spaces can be utilized for various purposes, such as the passage of cables and pipes, which prevents damage caused by digging, and the storage of water to prevent the risk of flooding [11]. According to [9], CCL100, CCL200, and CCL300 are the three types of panels that are typically designed and manufactured for such systems. The details of the prefabricated panel are given in Fig. 1.

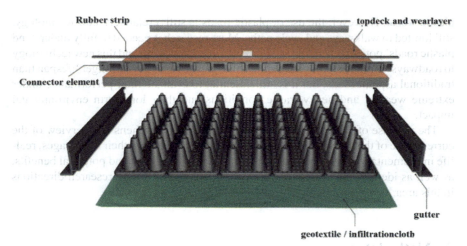

Fig. 1 The currently available design of the prefabricated plastic panel for the cycle path (After [9])

CCL100 is a model of PlasticRoad's modular infrastructure solution. It is a lightweight panel made of recycled plastic that is designed for use as a bike or pedestrian path. The hollow space inside the panel can be used to temporarily store rainwater during heavy rain showers, preventing flooding. The panel's low weight enables it to be applied on soft soils without additional foundation measures. Usually, this model is used as a transition element between the main plastic path and the asphalted path.

The CCL200 is a sustainable alternative to traditional road paving that is designed to be used in conjunction with an existing rainwater sewer system. It facilitates the collection of rainwater and allows for partial infiltration into the local subsurface. The CCL200 is a modular and flexible solution that is linked to an existing rainwater sewer, providing a sustainable option for road construction. On the other hand, the CCL300 is a premium product that is specifically engineered for climate adaptation. It features an integrated water filtration system that captures, purifies, and gradually infiltrates water into the local subsurface. It is also equipped with smart sensors that enhance the management and maintenance of the road surface. The CCL300 also has the capability to collect rainwater and facilitate infiltration into the local subsurface. Table 1 compares the three alternatives of prefabricated plastic panels [9].

3.1 Construction Method

The construction method for plastic roads involves a modular and hollow road structure that is prefabricated in a factory setting, resulting in a faster and more efficient construction process compared to traditional pavement methods. The first step in the

Table 1 A comparison between the two variants CCL200 and CCL 300 (After [9])

Feature	CCL100	CCL200	CCL300
Climate adaptation	Low	Moderate	High
Rainwater drainage	Not Included	Additional to the existing sewer	Integrated
Water filter system	Not Included	Not Included	Included
SMART Sensors	Not Included	Not Included	Included
Infiltration functionality	Not Included	Partially	Guaranteed
Overflow to existing sewer	Necessary	Necessary	Not necessary
Flexibility	Moderate	Moderate	High

construction process is the collection and processing of plastic waste. This waste is usually sourced from various materials, such as plastic bottles, old fishing nets, and other plastic products [12]. The prefabrication process involves the use of a specialized machine that shreds and melts the plastic waste, which is then molded into the desired shape of the road panels. The panels are then cooled and cut to the required length [13].

The second step in the construction process is the preparation of the subgrade or the soil and foundation beneath the road surface. The subgrade must be compacted and leveled to ensure a stable base for the plastic road. Next, a geotextile is placed on top of the subgrade to act as a separator between the subgrade and the road surface. This helps prevent soil erosion and water infiltration and for better contact stress distribution on the panel bottom base [9].

Once the prefabricated panels are manufactured, they are transported to the construction site and assembled in the desired pattern. The plastic panels are interlocked to create a seamless surface that is both durable and sustainable [10]. Since the panels are modular in design, they can be easily replaced or repaired if necessary. The assembled panels are then placed on top of a geotextile or geomembrane layer, which functions as a water barrier and water storage layer.

Fundamental to mention that the available literature was mainly focused on the construction method of the prefabricated plastic panels that fit walkways and cycle paths (Light traffic loads). However, for heavier traffic loads on the roadways purposes (Fig. 2), the construction process should include the installation of a subbase layer, which provides additional support and stability to the road surface. This layer should consist of a mixture of crushed stone and gravel, and its thickness depends on the road's intended use and traffic volume. The plastic panels are then

Fig. 2 A 3D simulation of prefabricated plastic panel utilization in the cycle and main roadway lanes

installed on top of the sub-base layer. More studies are needed in this area, in particular, to explore the stress analysis of the system and the interaction with the subgrade layer.

The final step in the construction process is the filling of the hollow spaces within the plastic panels. These spaces can be used for a variety of purposes, such as the passage of cables and pipes, and also serve as a storage vessel for water, which helps to prevent flooding.

4 Waste Plastics Materials for Advanced Applications

The current modular plastic panels are made from a combination of virgin and recycled plastic materials [9, 10]. The utilization of plastic waste as a raw material in the construction of plastic pavements constitutes a global innovation ecosystem in recycling plastic waste. The material has several advantages, such as high melting point temperatures, typically beyond 66°C, and a smaller carbon footprint. Additionally, plastic pavements exhibit less defects, such as cracks or potholes, compared to conventional pavement types [11, 14]. However, as mentioned earlier, the current application of this system is limited to light traffic pavement. Special attention should be given to the material selection to extend this promising technology to fit heavier traffic roadways applications.

The most commonly available waste plastics are polyethylene, polyethylene terephthalate, polypropylene, polyvinyl chloride, Styrene butadiene Styrene, polystyrene, acrylonitrile butadiene styrene, ethylene vinyl and acetate, polycarbonate [15, 16]. This section suggests the utilization of the most available waste plastics in producing prefabricated panels for advanced applications.

4.1 Polyethylene

Polyethylene is a widely-utilized plastic that is commonly obtained as waste from plastic bags. It is produced in three main forms: high-density polyethylene (HDPE), low-density polyethylene (LDPE), and linear low-density polyethylenes (LLDPE). The main difference between these forms of polyethylene is their density, with HDPE having a density range of 0.940 to 0.965 g/cm^3, LDPE having a density lower than 0.930 g/cm^3, and LLDPE having a density range of 0.915–0.940 g/cm^3.

Polyethylene is the most frequently produced and widely used plastic in daily life, accounting for nearly 34% of the total global plastics market [13, 17]. As a thermoplastic polymer, it can be molded or shaped at high temperatures and solidifies upon cooling to low temperatures [17].

Previous research has demonstrated the potential of using polyethylene waste plastics in asphalt modification [15, 17]. These studies have shown that different polyethylene modifiers can improve the performance of modified asphalt binders through the enhancement of softening point temperatures and reduction in penetration values. Additionally, the high rigidity of polyethylene makes it highly resistant to deformation when used as an alternative material in pavement construction [17–19].

4.2 Polyvinyl Chloride

The use of polyvinyl chloride (PVC) in the manufacturing of pipes and window frames is well-established due to its high levels of hardness and stiffness, which are attributed to the chemical bond between carbon and chlorine present in the polymer. PVC is also a common source of waste plastic, with a majority of it originating from cables, pipes, and window frames [20–22].

Research has demonstrated that the application of PVC waste in pavement construction can improve stiffness and resistance to deformation [23]. The process typically involves shredding the waste PVC before subjecting it to high temperatures for melting, however, care should be taken in regards to the potential release of chlorine gas into the atmosphere, as it can have detrimental effects on the environment [24].

4.3 Ethylene Vinyl Acetate (EVA)

Ethylene–vinyl acetate (EVA) is a thermoplastic polymer that is produced through the copolymerization of ethylene and vinyl acetate [25]. The proportion of ethylene and vinyl acetate in the polymer can be adjusted to achieve specific properties. Ethylene, which is crystalline in nature, provides high rigidity and stiffness at elevated temperatures. On the other hand, vinyl acetate, which is non-crystalline and non-polar, imparts increased elasticity and flexibility when used in higher quantities [26]. EVA polymers have been used in pavement applications for many years and are known for their resistance to deformation due to their formation of rigid, tough, and three-dimensional network structures [13, 27, 28].

4.4 Polypropylene

Polypropylene (PP) is a thermoplastic polymer that is widely used and produced globally, accounting for 21% of the total plastic market [17, 29]. It is composed of a linear chain of methyl and ethylene groups, which results in a high-strength but less flexible polymer due to the restricted rotation of the methyl groups [30]. Studies on using PP in pavement applications have shown that it can significantly improve resistance to rutting deformation [13].

Despite that, polypropylene is highly susceptible to degradation by UV radiation. Prolonged exposure can cause the material to become brittle and can lead to a loss of up to 70% of its mechanical strength after just six days of high-intensity UV radiation [31]. In order to mitigate this issue, various UV stabilizers and pigments can be added to the polypropylene to improve its resistance to UV degradation. Additionally, incorporating polypropylene with other polymers and additives can also improve its UV resistance and increase its overall performance in pavement applications.

4.5 Styrene Butadiene Styrene (SBS) and Acrylonitrile Butadiene Styrene (ABS)

Styrene–butadiene–styrene (SBS) is a thermoplastic elastomeric polymer that has been shown to be effective in enhancing the stiffness and rigidity of asphalt binders when used as an additive [17, 32–34]. While, Acrylonitrile–butadiene–styrene (ABS) is a polymer material characterized as a heterogeneous multi-phase system consisting of a rubber component and a mixture of styrene acrylonitrile [35]. Both SBS and ABS are known to improve the performance of modified asphalt binders, making them suitable candidates for use in plastic pavement applications.

5 Lessons Learned From the Real-Life Case Studies

Plastic road technology is still in the very early stage. However, there are a few real-life cases that have been successfully constructed. This section discusses the lessons learned from those cases aiming for further improvement of this promising approach. Table 2 summarizes the key facts of ten real-life case studies constructed using plastic road technology in different locations.

Based on the data available for the ten cases, several important lessons can be gleaned from the implementation of PlasticRoad technology. Firstly, it is clear that PlasticRoad is a highly versatile solution that can be applied to a wide range of infrastructure projects, including cycle paths, parking bays, and carpool places. The different types of PlasticRoad elements (CCL100, CCL200, and CCL300) are well-suited for different types of projects, with the CCL100 being ideal for projects that don't require water storage, while the CCL200 and CCL300 are more suitable for projects that require water storage and filtering.

Second is the effectiveness of PlasticRoad in addressing the challenges of climate change. The ability of PlasticRoad to temporarily store and filter rainwater during heavy rain showers, and to gradually infiltrate it into the local soil, can play a significant role in preventing flooding and mitigating the effects of drought. More and above, using recycled plastic in PlasticRoad elements helps reduce CO_2 emissions, making it a sustainable solution.

The use of sensors in some cases to monitor the performance of the plastic road allows for data collection and analysis, leading to smarter and more efficient accessibility of cities. In addition, the ability to collect data about the use of the road, temperature, and precipitation can lead to more accurate and efficient maintenance of the infrastructure. This highlights the potential for PlasticRoad to be used in smart cities, where data can be collected and analyzed to improve urban mobility and make cities more livable and accessible.

Overall, those ten case studies provide strong evidence of the potential for plastic road technology to be a viable solution for a wide range of infrastructure needs, including water management, sustainable mobility, and the use of recycled plastic waste.

6 Conclusion

This study investigated the potential of prefabricated plastic pavements as an alternative solution to traditional asphalt pavements for high-traffic and extreme weather conditions. Through a comprehensive review of existing literature and case studies, several key conclusions were reached:

- Prefabricated plastic pavements have great potential to be a sustainable and durable solution for road construction, addressing both the problem of climate change and plastic waste.

718   M. E. Al-Atroush et al.

Table 2 A summary of ten case studies of plastic roads (After [9])

#	Case Study	Application	Location	Model Type	Size (m/m^2)	Water storage (liter)
1	The municipality of Vlissingen	Combination bike and footpath	Ritthem, Netherlands	CCL100	200 sqm	N/A
2	The Antwerp port authority invests in safe cycle paths	Bike path	Antwerp, Belgium	CCL300	75 m	64,600
3	New lightweight bike path in the green area of Gouda	Bike path	Gouda, Netherlands	CCL300	100 m	91,000
4	Combination of bicycle and pedestrian path in Zeeland	Combination of cycle path and footpath	Ritthem, Netherlands	CCL00	200 sqm	NA
5	Climate-adaptive parking bays in Rotterdam	Longitudinal parking	Jan van Vuchtstraat, Rotterdam	CCL300	10 parking bays Water storage	27,000
6	Intelligent cycle path on the TU Delft campus	Bicycle path	Delft	CCL300	25 m	12,942
7	Sustainable carpool place in Hardenberg	8 parking bays	Hardenberg	CCL300	27,240 sqm	NA
8	Parking spaces at VolkerWessels Infra campus Vianen	Bike path (pilot)	Vianen	CCL300	NA	Temporarily storage
9	Bike path Mexico City (pilot) HomeprojectBike path Mexico City (pilot)	Bike path	Giethoorn, Netherlands	CCL300	NA	A solution for flooding
10**	Cycle path Giethoorn (pilot)	Cycle path	Giethoorn, Netherlands	CCL300	30 m	NA

[*] Recycled waste: 1000 kg, CO_2 emission reduction: 72%. Also, extensive sensors network in the hollow space of the road elements were used for settlement monitoring

- Plastic pavements have several advantages over traditional asphalt pavements, including increased durability and resistance to cracking and potholes, lower heat absorption rate, and higher skid resistance. Additionally, the use of plastic roads can also help to address the growing problem of plastic waste by finding a new use for recycled plastic materials.

- Waste plastics such as polyethylene, polyvinyl chloride, and ethylene vinyl acetate were suggested as suitable materials for heavy traffic roadway applications based on their properties and previous research.
- Plastic roads have potential advantages over traditional roads in terms of water storage and climate adaptation, as they have been found to have better water drainage capabilities.
- The ten real-world case studies of light-traffic plastic roads presented in the paper provided valuable insights into the performance and durability of plastic roads under various conditions. These case studies provided strong evidence of the potential for plastic road technology to be a viable solution for a wide range of infrastructure needs, including water management, sustainable mobility, and the use of recycled plastic waste.

Despite the superior advantages of plastic roads, more research is needed to fully evaluate the performance and durability of plastic roads under heavy traffic and extreme weather conditions. Additionally, research is needed to investigate the maintenance requirements and suitability of plastic roads in the long term. Future studies should also focus on exploring new and innovative materials that could be used in the production of heavy-traffic plastic pavements.

7 Limitations and Future Studies

The study presented in this paper has several limitations. Firstly, the majority of the research on prefabricated plastic pavement has been conducted on walkways and cycle paths, with limited studies on its application on high-traffic roadways. This means that the performance of these materials under heavy loads and extreme weather conditions is still uncertain and requires further investigation.

Secondly, the use of plastic waste as a raw material in the construction of plastic pavements is still a relatively new technology. The selection of the most appropriate plastic waste for the production of prefabricated panels for advanced applications is still not fully understood. Furthermore, the study of the long-term performance of plastic pavements under different environmental conditions is needed.

In conclusion, further research is required to fully understand the potential benefits and drawbacks of plastic roads and how to extend this new technology to high-traffic roadways. Future studies should focus on evaluating the performance of plastic roads under extreme weather conditions and heavy loads, investigating the most appropriate plastic waste for the production of prefabricated panels, and studying the long-term performance of plastic pavements under different environmental conditions.

Acknowledgements This research was supported by the Structures and Material (S&M) Research Lab of Prince Sultan University, Riyadh, Saudi Arabia.

References

1. IPCC Climate Change, 2013. In: Stocker, T.F., et al. (eds.) The Physical Science Basis. Cambridge University Press (2018). https://www.ipcc.ch/site/assets/uploads/2018/03/WG1 AR5_SummaryVolume_FINAL.pdf. Accessed 14 Jan 2023
2. Natural Resources Canada. Transportation in Canada: Key Results from the 2018 National Survey of the Environment and Sustainable Development (2020). https://www.nrcan.gc.ca/sci ence-data/data-analysis/statistics/transportation-canada-key-results-2018-national-survey-env ironment-sustainable-development/20200-eng. Accessed 14 Jan 2023
3. Al-Atroush, M.E., Marouf, A., Aloufi, M., Marouf, M., Sebaey, T.A., Ibrahim, Y.E.: Structural performance assessment of geothermal asphalt pavements: a comparative experimental study. Sustainability 14(19), 12855 (2022). https://doi.org/10.3390/su141912855
4. Repair Priorities, Transportation for America, Taxpayers for Common Sense, Washington, DC. https://t4america.org/wp-content/uploads/2019/05/Repair-Priorities-2019.pdf. Accessed 14 Jan 2023
5. Huang, Y.H.: Pavement Analysis and Design. Prentice-Hall, New Jersey (1993)
6. Thom, N.: Principles of Pavement Engineering. Thomas Telford, London (2008)
7. Al-Atroush, M.E.: Structural behavior of the geothermo-electrical asphalt pavement: a critical review concerning climate change. Heliyon 8, e12107 (2022). https://doi.org/10.1016/j.hel iyon.2022.e12107
8. VolkerWessels (n.d.). PlasticRoad. https://www.volkerwessels.com/en/projects/plasticroad. Accessed 18 Jan 2021
9. PlasticRoad.: PlasticRoad (n.d.). https://plasticroad.com/. Accessed 14 Jan 2023
10. Netherlands Builds The First Plasticroad, KWS, (2016). https://blog.wavin.com/en-gb/nether lands-builds-the-first-plasticroad. Accessed 14 Jan 2023
11. Kamal, I., Bas, Y.: Materials and technologies in road pavements-an overview. Mater. Today: Proc. 42(5), 2660–2667 (2021)
12. Jamshidi, A., White, G.: Evaluation of performance and challenges of use of waste materials in pavement construction: a critical review. Appl. Sci. 10(1), 226 (2019)
13. Wu, S., Montalvo, L.: Repurposing waste plastics into cleaner asphalt pavement materials: a critical literature review. J. Clean. Prod. 280(2), 124355 (2021)
14. Hall, F., White, G.: Using local waste plastics in asphalt modification to improve engineering properties of roads. In: 36th International Conference on Solid Waste Technology and Management, Annapolis, Maryland, USA, pp. 1–20 (2021)
15. Xu, X., et al.: Sustainable practice in pavement engineering through value-added collective recycling of waste plastic and waste tyre rubber. Engineering 7(6), 857–867 (2021)
16. Bala, N., Kamaruddin, I., Napiah, M.: The influence of polymer on rheological and thermo oxidative aging properties of modified bitumen binders. J. Teknologi 79(6), 69–73 (2017)
17. Ma, Y., et al.: The utilization of waste plastics in asphalt pavements: a review. Clean. Mater. 2, 100031 (2021)
18. Garcia-Morales, M., et al.: Effect of waste polymer addition on the rheology of modified bitumen. Fuel 85(7–8), 936–943 (2006)
19. Gawande, A., et al.: An overview on waste plastic utilization in asphalting of roads. J. Eng. Res. Stud. 3(2), 1–5 (2012)
20. Contreras, J.E.N., et al.: Design of a new prefabricated road subsurface drainage system. Int. J. Pavement Res. Technol. 7(2), 124–134 (2014)
21. Brasileiro, L., et al.: Reclaimed polymers as asphalt binder modifiers for more sustainable roads: a review. Sustainability 11(3), 646 (2019)
22. Salman, N., Jaleel, Z.: Effects of waste PVC addition on the properties of (40–50) grade asphalt. MATEC Web Conf. 162, 01046-1–01046-4. EDP Sciences (2018)
23. Ziari, H., et al.: The effect of EAF dust and waste PVC on moisture sensitivity, rutting resistance, and fatigue performance of asphalt binders and mixtures. Constr. Build. Mater. 203, 188–200 (2019)

24. Costa, L.M.B., et al.: Incorporation of waste plastic in asphalt binders to improve their performance in the pavement. Int. J. Pavement Res. Technol. **6**(4), 457–464 (2013)
25. El-Rahman, A.M.M.A., et al.: Enhancing the performance of blown asphalt binder using waste EVA copolymer (WEVA). Egypt. J. Pet. **27**(4), 513–521 (2018)
26. Liang, M., et al.: Rheological property and stability of polymer modified asphalt: effect of various vinyl-acetate structures in EVA copolymers. Constr. Build. Mater. **137**, 367–380 (2017)
27. Sengoz, B., Isikyakar, G.: Evaluation of the properties and microstructure of SBS and EVA polymer modified bitumen. Constr. Build. Mater. **22**(9), 1897–1905 (2008)
28. Ameri, M., Mansourian, A., Sheikhmotevali, A.H.: Laboratory evaluation of ethylene vinyl acetate modified bitumens and mixtures based upon performance related parameters. Constr. Build. Mater. **40**, 438–447 (2013)
29. Geyer, R., Jambeck, J.R., Law, K.L.: Production, use, and fate of all plastics ever made. Sci.Adv. **3**(7), e1700782 (2017)
30. Al-Atroush, M.E., et al.: A novel application of the hydrophobic polyurethane foam: expansive soil stabilization. Polymers **13**(8), 1335 (2021)
31. UV-Resistant Plastics, Polypropylene vs. Nylon. https://www.xometry.com/resources/inject ion-molding/uv-resistant-plastics-polypropylene-vs.-nylon/. Accessed 16 Jan 2023
32. Fu, H., et al.: Storage stability and compatibility of asphalt binder modified by SBS graft copolymer. Constr. Build. Mater. **21**(7), 1528–1533 (2007)
33. Peng, Y., et al.: Characteristic behavior of asphalt with SBS and PE. In: Second International Conference on Sustainable Construction Materials: Design, Performance, and Application, pp. 421–429 (2012)
34. Wang, S., et al.: Asphalt modified by thermoplastic elastomer based on recycled rubber. Constr. Build. Mater. **93**, 678–684 (2015)
35. Hasan, M.R.M., et al.: A simple treatment of electronic-waste plastics to produce asphalt binder additives with improved properties. Constr. Build. Mater. **110**, 79–88 (2016)

Highlighting The Role of UAE's Government Policies in Transition Towards "Circular Economy"

Tahira Yasmin, Ghaleb A. El Refae, and Shorouq Eletter

Abstract There is a widespread debate about sustainability and economic growth without posing threats on the planet and humanity. United Arab Emirates (UAE) has set a vital example to be part of UN Sustainable Development Goals (SDGs) as according to Agenda 2030. This paper assesses the policies and plans on accelerating the implementation of the circular economy model in four main sectors—manufacturing, food, infrastructure, and transport. The overall discussion has highlighted the UAE government plans in creating a circular economy. Moreover, we are also looking the government potential plans in this process, suggesting effective policies such as capacity building, renewable energy, waste management, awareness, and monitoring. Current study emphasizes that transitioning to a circular economy presents both a significant opportunity and an enormous challenge. An opportunity in that the commercialization of low-carbon solutions, including clean energy technologies, can further catalyze an important emerging market and support the transformation of the UAE energy sector. This initiative is even greater when you consider the requirement to make investments today for benefits that will materialize well into the future. A successful achievement of SDGs will require close coordination between policies, technology and capital, at the core of which is partnership between the public and private sector as well as opportunities to collaborate with countries around the world.

Keywords United Arab Emirates (UAE) · Sustainable Development Goals (SDGs) · Circular economy · Agenda 2030 · Transformation government plans

---

T. Yasmin (✉) · G. A. E. Refae · S. Eletter
Al Ain University, Al Jimi, Near Al Ain Municipality, Al Ain, Abu Dhabi, UAE
e-mail: tahira.yasmin@aau.ac.ae

© The Author(s), under exclusive license to Springer Nature Switzerland AG 2024
A. Hamdan and E. S. Aldhaen (eds.), *Artificial Intelligence and Transforming Digital Marketing*, Studies in Systems, Decision and Control 487,
https://doi.org/10.1007/978-3-031-35828-9_60

1 Introduction

With the pace of economic growth and technical innovation, there is ongoing debate between ecologists and economists regarding, "sustainability". van der Hamsvoort and Latacz-Lohmann [1] mentioned that the limits of growth and use of resources there is discrepancy between theoretical and practical sustainability. Upon reframing the debate of sustainability in real terms, it supported with the theoretical aspect as, "strong sustainability". The practical and theoretical sustainability could achieved by bridging the gap between informational inadequacies and human preferences in the design of sustainability constraints.

The 2030 transformative Agenda for sustainable development display a detailed perspective and international collaboration by putting the equality and dignity of people at the center. The 17 Sustainable Development Goals (SDGs) have 231 indicators with 169 targets mainly enlighten the ambitious vision with economic, social and environmental dimensions. Overall, this recalls the universal commitment towards sustainability by both developed and developing countries by strengthening global alliance. Recently, COVID-19 disclosed the severity and magnitude of challenges for humanity's survival. The confluence of floods, crises, climate change and conflicts spin off huge impact on food, health, education, environment and security posed serious challenges for Sustainable Development Goals (SDGs). To eradicate poverty and hunger, improving healthcare facilities, provide necessities and much more is required (UN [2]).

Pak [3] justified the vision of sustainability in transformation of societies expanding the boundaries of scientific, economic, ecological and technical knowledge. He also mentioned that there is still need a wider socio-political practices and lifestyles among variety of disciplines. This aspect has further supported by James [4] that circular economy framework domains needs to reframe with social and justice perspectives.

The sustainability management goals need evolvement among various players in any economy where local government should integrate sustainability as part of policies and practices. The public administration should emphasize the embedded sustainability as part of strategic plan and reform in local government (see Zeemering [5]). Furthermore, the disclosure of governmental information on sustainability including social, economic and environmental dimensions are equally important. This means the transparency of sharing sustainability data to improve awareness among public. In this article, we demonstrate the role of effective government policies has helped to achieve various SDGs goals in UAE. According to the UAE Circular Economy Policy 2021–2031 country has ambitious plans to be a diversified economy by improving the food, energy and water sectors. COP28 UAE will take place at Expo City Dubai from November 30 to December 12, 2023. This event is providing a platform for UAE to draw practical and realistic energy transition roadmap with global collaboration.

However, to the best of our knowledge, most studies linking the concepts of Circular Economy with other determinants. However, current study contributes to the field of sustainability by addressing the objective that what are the UAE government policies in context of circular economy. Therefore, this study highlighted the information available in the national plans and agendas, vision of nation, economic and social policies. The layout of this article is as follows: after the introduction part, we will present summary of previous studies. In Sect. 3, material and method approach has discussed. Section 4 present the UAE circular economy policies and perspectives. Lastly, Sect. 5 provide conclusion and policy implications.

2 Literature Review

Around 20 years ago, the economic activities take place as according to linear model as produce, consume and discard. However, with the passage of time, it has shifted with Circular Economy (CE) model, which highlights the mechanism of production and consumption in sustainable ways by promoting technical innovation. The CE is becoming the nationwide strategy around the globe to enhance the human wellbeing by refurbishing planning process. Various insights on production process came from diverse schools of thoughts including steady-state economy, industrial ecology, spaceman economy and cradle-to-cradle economy (see Daly [6]; Frosch and Gallopoulos [7]; Boulding [8]; Stahel and Reday-Mulvey [9]). In twenty-first century, the extraction of resources and interdependence of global supply chains are faster than regeneration and focusing on profit at the expense of stability. This leads to utmost need of circular transition by developing national and international legislations as procurement strategy.

Raworth [10] proposed Doughnut Economics as development of mindset to bring about the regenerative and distributive dynamics. The starting point of this mechanism is not to thrive endless GDP but to focus on the big picture by recognizing that human behavior can fostered to be caring and cooperative with competitiveness. It further highlighted the complex and interdependent societal and global system through the system thinking. This called today as degenerative economies into regenerative ones and resulting into far more distributive ones. This idea of Doughnut Economics defined the human wellbeing by fulfilling the dual conditions of social and ecological perspectives.

Szymańska and Zalewska [11] compared the implementation of sustainable development goals in area of quality education in 28 European Union countries. The study concluded that there were dissimilarities between countries and they found that Sweden ranked highest as best performer. It was suggested that there is need of unified policy implication to promote education, as it is a vital component to achieve Sustainable Development. Gunluk-Senesen [12] shed light on local public policies for gender equality and resource allocation in Turkish cities. The author presented the gender wellbeing budgeting analytical framework to track the SDG commitment. The paper discussed the need for data availability and funding for achieving SDGs in developing countries. He further suggested that there is need to reinforce women traditional roles by creating women's safety and mobility in Turkey.

Furthermore, Wiesli et al. [13] emphasized to protect parks and biospheres to promote sustainable way of living by conserving human development goals. The goal of study was park management in Switzerland to contribute the quality of life of residents. The paper presented the result of survey conducted for 2409 residents in Switzerland and indicated that quality of parks are high in the country. In general, it was suggested by residents that park management can enhance social relations and health by enabling new priorities for humans and nature. In another study, Lyytimäki et al. [14] explained the active participation by various players from the society in sustainability transition. The study assessed the development and implementation of concrete actions in Finland for Sustainable Development. It concluded that open communication and mutual trust development actions need to develop for public and private sectors.

Laurian and Crawford [15] examined the role of government in implementation of sustainability measures in U.S. and New Zealand. The authors concluded that local government has a key role in implementation of effective sustainable policies but they are slow in in taking some practices. They recommended that government should reconcile economic development goals with sustainability and have public support by institutional priorities. While, Heinrichs and Schuster [16] elaborated the role of state and non-state actors in achieving sustainable goals. They directed attention towards an important aspect that effective implementation of sustainability in government institutions is a precondition for multistage stakeholders' governance. The study presented the result of survey of 371 German cities and municipalities and provide evidence that how sustainability has institutionalized in German local governments. This means that to develop a sustainable economy the component of sustainability should dedicated in the state as institutionalizing sustainability.

Furthermore, to develop circular economy government intervention can develop influence policies in product life cycle by having set of sustainability indicators. Taghipour et al. [17] interpreted the sustainable management policies in Thailand's steel manufacturing companies. The study disclosed the importance of government financial assistance schemes support in managing recycling during steel manufacturing process. This aspect further proposed by Agyapong and Tweneboah [18] that financial readiness and support is key factor in circular economy. They suggested that internal and external financial policies especially from public sector could ascertain the financial liquidity. Additionally, the implementation of environmental protection laws and regulations by government generate steer the economy towards long-term sustainability. By facilitating, the partnership between public and private sector can also lead to societal behavioral change towards attainment of SDGs. This further help to formulate the command-and-control regulations to support competitiveness and circular business models (see Sun et al. [19]; Wasserbau et al. [20]; Li et al. [21]; Liao et al. [22]; Kim and Kim [23]; Masuda et al. [24]).

3 Material and Methods

UAE delegation has actively participated in the Rio+20 Earth Summit in Rio de Janeiro in 2012 to highlight several tracks including the country's major sustainable development initiatives. In Rio, one of the leading outcome was that the members decided to have a political forum to achieve the set of Sustainable Development Goals (SDGs). With the persistent UAE government efforts and having several meeting with Open Working Group (OWG) during 2013–2014 the country successfully held a seat. The overall SDGs finalized in 2015 with intergovernmental negotiations and then UAE has formally adopted the 2030 Agenda for Sustainable Development.

This study employed a case study approach that is useful for highlighting sustainable development framework. This will allow researchers to learn from planning and policies contexts especially the understanding of effective government policies on sectoral basis. Current research has highlighted various UAE government initiatives towards development of circular economy. The data generation for this research is primarily relied on document analysis and supplemented with sector based planning in UAE. The main documents included UAE circular economy policy agenda, national vision and national sustainability plans frameworks.

Based on the above discussion, UAE is an interesting case study of attainment of SDGs with various initiatives and policy goals. As in Fig. 1, the SDGs dashboard displays that country has ranked as 3rd in Arab region among 21 nations. The overall index score is 67 while on regional basis it is 58.2 accordingly. There are utmost efforts to manage energy targets in order to attain sustainability. Some goals have still significant challenges but this indicate a formal journey towards sustainability with effective decisions and plans.

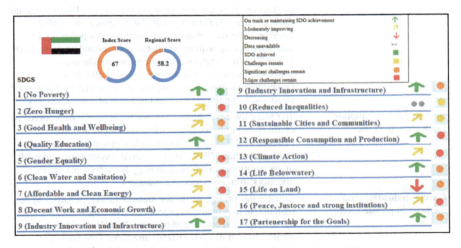

Fig. 1 UAE SDGs 2022. *Source* Author's work based on Bayoumi et al. [25]

4 UAE Circular Economy Policy Initiatives

Main purpose of UAE transitional policy towards circular economy is to promote the efficient use of human, physical, natural and financial resources. This need a joint collaborative effort from local and national government, private sector and general public to act in a way to help the country to attain the transition successfully. The overall, sustainable, happy and cohesive circular economy achievement is fully aligned with the, "UAE Centennial 2071" goals. This further shows the long-term government vision plan for five decades to fortify the country's reputation. Table 1 summarizes the UAE circular economy key objectives as:

Based on the main circular economy objectives, UAE has identified four priority sectors due to their key potential in stimulating circular economy.

4.1 *Sustainable Manufacturing*

In order to enhance sustainable manufacturing in UAE, His Highness Sheikh Khaled Bin Mohamed Bin Zayed Al Nahyan has launched the Abu Dhabi Industrial Strategy. The will not only strengthen the country's position as industrial hub it will further

Table 1 Key objectives of UAE circular economy

Objectives	Key initiatives
Efficient use of natural resources	• Being an oil resource economy, country is having entire efforts to reduce the flaring of natural gas as by product of oil production. With the continuous efforts of Abu Dhabi National Oil Company (ADNOC) group, the company successfully developed the flare management strategy as a policy tool to minimize flaring with accumulated policies and projects. UAE Water Security Strategy 2036 is optimizing the use of wastewater to increase water efficiency in all sectors. Government is further seeking new technologies of wastewater treatment technologies to cater the needs of various sectors
Sustainable consumption production patterns	• To reduce the environmental stress and meet the basic needs of entire community the country has vital sectoral based planning. UAE National Sustainable Consumption and Production Plan (2019–2030) required a systematic approach of cooperation as a national strategy among supply chain operators from produces to consumers
Cleaner industrial production methods in private sector	• Country has prompt actins for manufacturing industries to be sustainable in production process by reducing waste, reusing the material and automated process. Abu Dhabi Ports and Khalifa Industrial Zone (Kizad) announced the launch of construction activities of National Food Product Company (NFPC) to have a dairy, a juice and small water bottling plant by using waste materials and energy saving techniques

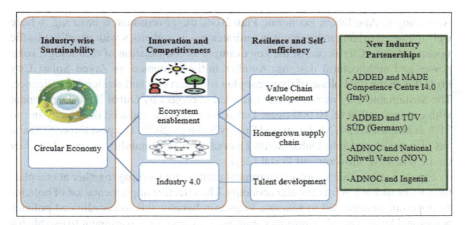

Fig. 2 ADDED transformational programmes. *Source* ADDED [26]

help towards transitional circular economy powered by industrial domain. UAE government has allocated AED 10 billion investment in various programs to boost the size of manufacturing sector by 2031. Moreover, Abu Dhabi's industrialization strategy reflects the leadership's promise in enacting the effective economic policies to elevate its position as sustainable industrial player. The green regulatory framework as responsible production has green policies and initiatives. Furthermore, the strategic guidance of UAE leadership is providing a roadmap with aligning the emirate on upward path in bolstering non-oil GDP and manufacturing eco-system (Abu Dhabi Department of Economic Development [26]).

Figure 2, displays the UAE six transformational programmes to strengthen the renewable energy use in manufacturing, innovation and competitiveness. To enhance locally manufactured products various supply chain equity funds and comprehensive economic partnership agreement (CEPA) has taken place by unified inspection programmes.

4.2 Green Infrastructure

The rapid urbanization process is posing various challenges on infrastructure, resource use and cutting pollutants. The second key priority sector of UAE economy is having green infrastructure development by smart cities, green building designs and upgrading of buildings. According to the IMD World Competitiveness Center the smart city observatory [27] for 118 cities, Abu Dhabi and Dubai ranked as 28 and 29 respectively. Both cities has displayed high indices in structure, mobility, activities, opportunities and Governance by adopting the innovative technologies.

UAE government is maintaining its vital role to maintain the perfect balance between economic and social development by preserving the environment.

According to Abu Dhabi Economic Plan 2030, the government is ensuring to have well-managed urban environment in emirates towns and cities with excellent traffic system. Masdar City is one of the key example as combination of technology, solar power and architectural designs. Another initiative named as Zayed Smart City Project is in the final phase of five-year plan for smart city and artificial intelligence. The Sustainable City recognized as remarkable project by Dubai Land Department with car free and have 10.000 trees. Dubai Silicon Oasis reduced the energy use to 31% as aligned with UAE Integrated Energy Strategy 2030. The post Expo 2020 site has also developed into Dubai South District with sustainability as one of the key theme (see UAE Government Portal [28]).

Figure 3 shows that various government entities are working together at one platform to transition UAE as circular economy. The effective implementation of policies by government expected to generate long-term benefits due to coordinated policies at national level. This also introduce guiding principles by government to establish a planning culture and respond to current and future development needs. This institutional collaboration opened the windows of opportunity for societal transformation by strengthening of supporting regimes.

Fig. 3 Interconnectedness of UAE government institutions. *Source* Author's work

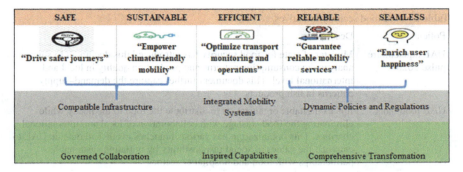

Fig. 4 UAE national smart mobility strategy 2020. *Source* UAE ministry of infrastructure development [29]

4.3 Sustainable Transportation

According to UAE, national smart mobility strategy highlights the strategic direction of country align with its vision. The economy plan to be among world leading countries in Smart Intermodal Mobility with expansion of mobility smart solutions and services. The key objectives of this mobility strategy are to reap benefits in combating climate change. The Roads and Transport Authority department has development plans to promote clean energy vehicles such as electric and hydrogen vehicles. The Smart Mobility solutions and services also generate connected ecosystem, autonomy, electrification, micro-mobility services, Mobility-as-a-Service (MaaS), Inter-modality, and integration (UAE Ministry of Infrastructure Development [29]).

Figure 4 highlights the five key objectives of smart mobility with their key themes such as safety and sustainable by dual aspects will generate compatible infrastructure under governance collaboration. There will be optimization of monitoring and operation that further result in an integrated transport system. Lastly, the reliability and seamless guarantee the follow of certain policies and regulations that will help to transform the entire system.

4.4 Sustainable Food Production and Consumption

Among UAE circular economy priority sectors, food security is among one of them. To strengthen the efforts of national authorities the Emirates Council for Food Security (ECSF) has launched including representatives from various ministries such as: Ministry of Environment and Climate Change, Ministry of Education, Ministry of Health and Community Protection, Ministry of Health and Community Protection, Ministry of Energy and Industry and National Authority for Emergency, Crisis and Disaster Management. By generating this platform with various government entities is an effective strategy to monitor the implementation of food security strategy. The

Table 2 UAE food security initiatives

Policies	Description
UAE aquaculture pulse 2020	Minister of state for food security has introduced this document to study, analyze the consumption patterns and production quality in local and international level. This document further capture the demand–supply patterns in fish products
UAE food bank	It is a charitable organization to distribute the food to the needy while eliminating the food wastage with collaboration of local and international charities. This organization has various branches inside and outside UAE to entrench the values of social responsibility by initiating sustainable partnerships with food and hospitality sector
Federal law no. 3 of 2020 regulation of the strategic stock of food commodities	H. H. Sheikh Khalifa bin Zayed Al Nahyan, President of the UAE approved the law to organize the food supplies during crisis and emergency to achieve food sustainability
Food security research platform	This is an online portal that allows specialists to share their knowledge in eight priority areas comprises on: agriculture waste, alternative food, chain technology, agriculture biotechnology, farm IOT, novel farming, bioenergy, biomaterial and robotics
Ag-tech accelerators	Government accelerators program launched the ten initiatives to enhance adoption of technology in agriculture sector. They are mainly agriculture finance along with supply chain financing, food security data platform, and framework, Ag/Tech building code, fish feed facility, aqua culture standards and atlas

Source Minister of state for food and water security[31]

effective implications of suggested legislations and policies by building databases and submitting periodic reports to UAE cabinet. Ministry of Economy [30] highlighted that food consumption patterns should be change to reduce the waste proportion. The food consumption patterns indicated that food waste could reach to 2.2 billion tons approximately by 2025. To regulate the food consumption UAE government and organizations are trying to apply the ways to minimize food waste.

Another effective food security initiative in 2020 was the launch of The Food Security Dashboard of Dubai by H. H. Sheikh Mansour bin Mohammed bin Rashid Al Maktoum, Chairman, Dubai's Supreme Committee of Crisis and Disaster Management. Table 2, summarizes some other food security policies to make the country world's best in the Global Food Security Index by 2051.

5 Conclusion and Policy Implications

In summary, the current study displays that to achieve the sustainable development goals and to have a circular economy it is equally important to institutionalize the effective policies. This article justifies the key highlights of UAE government initiatives, which revealed that the policy planning could be very successful when various government entities and municipalities engaged from different areas and backgrounds. UAE SDGs measures embedded with various government ministries to enhance the implication by monitoring the practical operations. They are mainly adoption of sustainable manufacturing, green infrastructure, sustainable transportation and sustainable consumption and production dimensions. Furthermore, it found that circular economy measures with UAE government support could leverage platforms by generating short and long-term benefits.

This study suggested some policy implications for the long-term attainment of SDGs in UAE. Firstly, the UAE Environment Agency is already banning the plastic use in food packaging, boxes and drinking cups. This could be more effective by having sectoral based planning especially in plastic industry. It means that other final products can develop and optimize reducing plastic resources to benefit producers and consumers. Secondly, due to limited water resources it is recommended to connect desalination technologies to renewable energy. UAE has already various plants but it still needs brisk response to fulfil the increasing demand of water in Emirates. The country has ambitious water support globally to provide portable water as sustainability measures. Nevertheless, this still needs careful planning to develop renewable energies in order to run power desalination plants. Thirdly, as a resource-based economy there is challenge of high dependency on oil and gas resources. With the entire efforts of diversity, UAE can build economic reforms to improve sectoral diversification at micro and macro level. This must needs careful planning as import substitution and building blocks for a dynamic and sustainable post-hydrocarbon economy.

UAE economy displayed an example of interesting case towards the direction of sustainability with not absolute but to a relative extent. Our study suggested that future study can use Sustainability Window (SuWi) method provides a novel approach for analyzing sustainability simultaneously in social, environmental, and economic dimensions in a quantitative way in UAE.

References

1. van der Hamsvoort, C.P.C.M., Latacz-Lohmann, U.: Sustainability: a review of the debate and an extension. Int. J. Sustain. Dev. World Econ. 5(2), 99–110 (1998)
2. United Nations (UN).: The Sustainable Development Goals Report 2022. https://unstats.un.org/sdgs/report/2022/. Accessed 3 Jan 2023
3. Pak, C.: The goal of Martian economics is not "sustainable development" but a sustainable prosperity for its entire biosphere': science fiction and the sustainability debate. Green Lett. 19(1), 36–49 (2015)
4. James, P.: Re-embedding the circular economy in Circles of Social Life: beyond the self-repairing (and still-rapacious) economy. Local Environ. 27(10–11), 1208–1224 (2022)
5. Zeemering, E.S.: Sustainability management, strategy and reform in local government. Public Manag. Rev. 20(1), 136–153 (2018)
6. Daly, H.E.: Allocation, distribution, and scale: towards an economics that is efficient, just, and sustainable. Ecol. Econ. 6(3), 185–193 (1992)

7. Frosch, R.A., Gallopoulos, N.E.: Strategies for manufacturing. Sci. Am. **261**(3), 144–153 (1989)
8. Boulding, K.E.: The economics of the coming spaceship earth. In: Environmental Quality in a Growing Economy, Resources for the Future, pp. 3–14. Johns Hopkins University Press, Baltimore (1966)
9. Stahel, W.R., Reday-Mulvey, G.: Jobs for Tomorrow: the Potential for Substituting Manpower for Energy. Vantage Press, New York (1981)
10. Raworth, K.: Why it's time for doughnut economics. IPPR Progressive Rev. **24**(3), 216–222 (2017)
11. Szymańska, A., Zalewska, E.: Education in the light of sustainable development goals—the case of the European Union countries. Glob. Soc. Educ. **19**(5), 658–671 (2021)
12. Gunluk-Senesen, G.: Wellbeing gender budgeting to localize the UN SDGs: examples from Turkey. Public Money Manag. **41**(7), 554–560 (2021)
13. Wiesli, T.X., Hammer, T., Knaus, F.: Improving quality of life for residents of biosphere reserves and nature parks: management recommendations from Switzerland. Sustain. Sci. Pract. Policy **18**(1), 601–615 (2022)
14. Lyytimäki, J., Vikström, S., Furman, E.: Voluntary participation for sustainability transition: experiences from the 'Commitment to Sustainable Development 2050'. Int. J. Sustain. Dev. World Ecol. **26**(1), 25–36 (2019)
15. Laurian, L., Crawford, J.: Sustainability in the USA and New Zealand: explaining and addressing the implementation gap in local government. J. Environ. Plann. Manag. **59**(12), 2124–2144 (2016)
16. Heinrichs, H., Schuster, F.: Still some way to go: institutionalisation of sustainability in German local governments. Local Environ. **22**(5), 536–552 (2017)
17. Taghipour, A., Akkalatham, W., Eaknarajindawat, N., Stefanakis, A.I.: The impact of government policies and steel recycling companies' performance on sustainable management in a circular economy. Resour. Policy **77**, 102663 (2022)
18. Agyapong, D., Tweneboah, G.: The antecedents of circular economy financing and investment supply: the role of financial environment. Clean. Environ. Syst. **8**, 100103 (2023)
19. Sun, C., Xu, Z., Zheng, H.: Green transformation of the building industry and the government policy effects: policy simulation based on the DSGE model. Energy **268**, 126721 (2023)
20. Wasserbaur, R., Sakao, T., Milios, L.: Interactions of governmental policies and business models for a circular economy: a systematic literature review. J. Clean. Prod. **337**, 130329 (2022)
21. Li, Y., Hu, S., Zhang, S., Xue, R.: The power of the imperial envoy: the impact of central government onsite environmental supervision policy on corporate green innovation. Financ. Res. Lett. **52**, 103580 (2023)
22. Liao, H., Peng, S., Li, L., Zhu, Y.: The role of governmental policy in game between traditional fuel and new energy vehicles. Comput. Ind. Eng. **169**, 108292 (2022)
23. Kim, K., Kim, S.: Bringing power and time in: how do the role of government and generation matter for environmental policy support? Energy Strategy Rev. **42**, 100894 (2022)
24. Masuda, H., Kawakubo, S., Okitasari, M., Morita, K.: Exploring the role of local governments as intermediaries to facilitate partnerships for the Sustainable Development Goals. Sustain. Cities Soc. **82**, 103883 (2022)
25. Bayoumi, M., Luomi, M., Fuller, G., AlSarihi, A., Salem, F., Verheyen, S.: Arab Region SDG Index and Dashboard Report 2022. Mohammed bin Rashid School of Government, Anwar Gargash Diplomatic Academy and UN Sustainable Development Solutions Network, Dubai, Abu Dhabi and New York. https://s3.amazonaws.com/sustainabledevelopment.report/2022/2022-arab-region-index-and-dashboard-report.pdf. Accessed 3 Dec 2022
26. Abu Dhabi Department of Economic Development.: Khaled bin Mohamed bin Zayed launches Abu Dhabi Industrial Strategy to Strengthen the Emirate's Position as the Region's Most Competitive Industrial Hub. https://added.gov.ae/Media-Center/Business-News/Khaled-bin-Mohamed-bin-Zayed-launches-Abu-Dhabi-Industrial-Strategy. Accessed 3 Dec 2022
27. IMD World Competitiveness Center. Smart City Index (2021). https://www.imd.org/smart-city-observatory/home/. Accessed 16 Dec 2022

28. UAE Government Portal. Smart Sustainable Cities. https://u.ae/en/about-the-uae/digital-uae/digital-cities/smart-sustainable-cities. Accessed 13 Dec 2022
29. UAE Ministry of Infrastructure Development. UAE National Smart Mobility Strategy. https://unece.org/DAM/trans/events/2020/ITC/ppt/4b2_UAE_National_Smart_Mobility_Strategy_Geneva_Final.pdf
30. UAE Ministry of Economy. Sustainability in Food Consumption. https://www.moec.gov.ae/en/-/sustainability-in-food-consumption-en. Accessed 13 Dec 2022
31. Minister of State for Food and Water Security. National Food Security Strategy. https://u.ae/en/information-and-services/environment-and-energy/food-security. Accessed 3 Dec 2022

Studies in Systems, Decision and Control

Volume 487

Series Editor

Janusz Kacprzyk, Systems Research Institute, Polish Academy of Sciences,
Warsaw, Poland

The series "Studies in Systems, Decision and Control" (SSDC) covers both new developments and advances, as well as the state of the art, in the various areas of broadly perceived systems, decision making and control–quickly, up to date and with a high quality. The intent is to cover the theory, applications, and perspectives on the state of the art and future developments relevant to systems, decision making, control, complex processes and related areas, as embedded in the fields of engineering, computer science, physics, economics, social and life sciences, as well as the paradigms and methodologies behind them. The series contains monographs, textbooks, lecture notes and edited volumes in systems, decision making and control spanning the areas of Cyber-Physical Systems, Autonomous Systems, Sensor Networks, Control Systems, Energy Systems, Automotive Systems, Biological Systems, Vehicular Networking and Connected Vehicles, Aerospace Systems, Automation, Manufacturing, Smart Grids, Nonlinear Systems, Power Systems, Robotics, Social Systems, Economic Systems and other. Of particular value to both the contributors and the readership are the short publication timeframe and the world-wide distribution and exposure which enable both a wide and rapid dissemination of research output.

Indexed by SCOPUS, DBLP, WTI Frankfurt eG, zbMATH, SCImago.

All books published in the series are submitted for consideration in Web of Science.

Allam Hamdan · Esra Saleh Aldhaen
Editors

Artificial Intelligence and Transforming Digital Marketing

Volume 2

Springer

Editors
Allam Hamdan
College of Business and Finance
Ahlia University
Manama, Bahrain

Esra Saleh Aldhaen
Department of Management and Marketing
Ahlia University
Manama, Bahrain

ISSN 2198-4182   ISSN 2198-4190 (electronic)
Studies in Systems, Decision and Control
ISBN 978-3-031-35827-2   ISBN 978-3-031-35828-9 (eBook)
https://doi.org/10.1007/978-3-031-35828-9

© The Editor(s) (if applicable) and The Author(s), under exclusive license to Springer Nature
Switzerland AG 2024

This work is subject to copyright. All rights are solely and exclusively licensed by the Publisher, whether
the whole or part of the material is concerned, specifically the rights of translation, reprinting, reuse
of illustrations, recitation, broadcasting, reproduction on microfilms or in any other physical way, and
transmission or information storage and retrieval, electronic adaptation, computer software, or by similar
or dissimilar methodology now known or hereafter developed.
The use of general descriptive names, registered names, trademarks, service marks, etc. in this publication
does not imply, even in the absence of a specific statement, that such names are exempt from the relevant
protective laws and regulations and therefore free for general use.
The publisher, the authors, and the editors are safe to assume that the advice and information in this book
are believed to be true and accurate at the date of publication. Neither the publisher nor the authors or
the editors give a warranty, expressed or implied, with respect to the material contained herein or for any
errors or omissions that may have been made. The publisher remains neutral with regard to jurisdictional
claims in published maps and institutional affiliations.

This Springer imprint is published by the registered company Springer Nature Switzerland AG
The registered company address is: Gewerbestrasse 11, 6330 Cham, Switzerland

Foreword

The emergence of digital marketing has changed the way people interact and communicate across the globe using Internet and specialized digital communication supported by technological advancements. One type of such advancements is artificial intelligence (AI), a supportive technological agent that plays an important role in digital marketing. It is used by most of businesses as an effective tool to interact and reach out to stakeholders including customers. Its benefits have gone beyond businesses to include many other fields such as education and health care. Digital marketing is now used more effectively compared with a traditional approach of marketing. Using AI in digital marketing has enhanced marketing communication where many digital tools can be connected to develop an effective communication campaign. Such technological tools include social media, email, and website. Companies that are quickly and wisely adopting such technology enjoy a good level of competitive advantage over their rivals. It is undoubtedly that AI can bring sustainability to companies adopting it where they can manage their current and future resources efficiently and contribute well to the economy and well-being of the society.

This book presents and discusses issues of digital marketing using AI for better marketing penetration, users' satisfaction and business efficiency. It covers important topics such as artificial intelligence, marketing and social media, cultural marketing, artificial intelligence and digital learning, innovation and sustainable operations, AI, banking and financial technology, tech-management in different disciplines and the role of digital marketing and governance and business ethics with AI.

Enjoy reading this book.

Prof. Muneer Al Mubarak
Professor of Management
and Marketing
Ahlia University
Manama, Bahrain

Preface

The use of artificial intelligence has become an essential part of decision-making, the wealth of data collection and data analysis enables drawing targeted audience and setting marketing strategies. The use of artificial intelligence has become vital in all aspects for instance recently AI-based applications including ChatGPT has created a serious fear at different level that AI will be taking over jobs from different fields.

The issue became more acute with the fear of humans that AI will take over jobs in the market and increase the unemployment rate on the other hand, new jobs are being created that requires more advanced skills in technology and digitalization. Several studies highlighted that AI is a potential tool to support humans to perform effectively with clear impact measures. In terms of marketing, the use of AI is proven to be the most effective considering the type of data generated to support market positioning and segmentation. However, considering the booming of AI there are still arising questions in terms of sustainability, would AI be sustainable considering lacking emotional intelligence that impacts social marketing. Other questions and issues are still a challenge for instance there are several ethical standards that need to be considered while using AI specifically while transforming from normal marketing to digital marketing that requires customer preservation of data as per the data protection acts. Therefore, this book provides an insight on future studies that support the integration of AI towards the use of digitalization and transformation of marketing strategies including and restricted to marketing but also managing aspects of AI and the use of digitalization towards transformation. The book aims to attract researchers globally to generate research outcomes that support the digital transformation and the use of AI towards transformation of digital marketing that could make a difference. The outcome of this research outcomes will support researchers and policymakers to have insights on different methods to be adopted to use AI effectively and support sustainability.

This book includes one hundred chapters. All of the chapters have been evaluated by the editorial board and reviewed based on double-blind peer review system by at least two reviewers.

viii Preface

The chapters of the book are divided into five main parts:

I. Artificial Intelligence, Marketing and Social Media.
II. Cultural Marketing, Artificial Intelligence and Digital Learning, Innovation and Sustainable Operations.
III. AI, Banking and Financial Technology.
IV. Tech-Management in Different Disciplines and the Role of Digital Marketing.
V. Governance and Business Ethics with AI.

The chapters of this book present a selection of high-quality research on the theoretical and practical levels, which ground the uses of AI in marketing, business, health care, media, education and other vital areas. We hope that the contribution of this book will be at the academic level and decision-makers in the various economic and executive levels.

Manama, Bahrain Allam Hamdan
Esra Saleh Aldhaen

Contents

Artificial Intelligence, Marketing and Social Media

Digital Marketing and Sustainable Businesses: As Mobile Apps in Tourism 3
Mahmoud Alghizzawi, Mohammad Habes, Abdalla Al Assuli, and Abd Alrahman Ratib Ezmigna

Exploring Small and Medium Enterprises Expectations of Electronic Payment in Kuwait 15
Ayman Hassan, Arezou Harraf, and Wael Abdallah

Content Marketing Strategy for the Social Media Positioning of the Company AD y L Consulting—Peru 27
Húber Rodríguez-Nomura, Edwin Ramirez-Asis, K. P. Jaheer Mukthar, Magdalena Valdivia-Malhaber, María Rodríguez-Kong, Nathaly Zavala-Quispe, and José Rodríguez-Kong

An Exploratory Study on E-commerce in the Sultanate of Oman: Trends, Prospects, and Challenges 37
Dawood Al Hamdani and Aisha Al Wishahi

The Impact of Researchers' Possessed Skills on Marketing Research Success 53
Ahmad M. A. Zamil, Ahmad Yousef Areiqat, Mohammed Nadem Dabaghia, and Jamal M. M. Joudeh

Impact of Online Food Ordering System and Dinning on Consumers Preference with Special Reference to Bangalore 63
J. Chandrakhanthan and C. Dhanapal

Green Economy Mechanisms in the Age of Technology and the Circular Economy 73
CH. Raja Kamal and Surjit Singha

ix

Whether Digital or Not, the Future of Innovation and Entrepreneurship 81
Yacoub Hamdan, Ahmad Yousef Areiqat, and Ahmad Alheet

Customer Buying Intention Towards Smart Watches in Urban Bangalore 87
B. Subha and Jaspreet Kaur

A Fluid-Thermal-Electric Model Based on Performance Analysis of Semi-transparent Building Integrated Photovoltaic Solar Panels 95
Nesrine Gaaliche, Hasan Alsatrawi, and Christina G. Georgantopoulou

The Impact of Social Media Influencer Marketing on Purchase Intention in Bahrain 113
Aysha AlKoheji, Allam Hamdan, and Assma Hakami

Tomorrow's Jobs and Artificial Intelligence 121
Ismail Noori Mseer, Yasser M. Abolelmagd, and Wael F. M. Mobarak

Artificial Intelligence and Security Challenges 129
Ismail Noori Mseer and Syed Muqtar Ahmed

Modern Services Quality and Its Impact on the Satisfaction of the Trade Shows in Jordan 137
Mustafa S. Al-Shaikh and Feras Alfukaha

Artificial Intelligence Application to Reduce Cost and Increase Efficiency in the Medical and Educational Sectors 151
Khaled Delaim, Muneer Al Mubarak, and Ruaa Binsaddig

Data Journalism and Its Applications in Digital Age 163
Abdulsadek Hassan and Mohammed Angawi

Digital Public Relations and Communication Crisis 177
Abdulsadek Hassan and Mohammed Angawi

Employing Applying Big Data Analytics Lifecycle in Uncovering the Factors that Relate to Causing Road Traffic Accidents to Reach Sustainable Smart Cities 193
Mohammad H. Allaymoun, Mohammed Elastal,
Ahmad Yahia Alastal, Tasnim Khaled Elbastawisy, Dana Iqbal,
Amal Yaqoob, and Adnan Sayed Ehsan

The Effects of Leadership Style on Employee Sustainable Behaviour: A Theoretical Perspective 205
Tamer M. Alkadash, Muskan Nagi, Ali Ahmed Ateeq,
Mohammed Alzoraiki, Rawan M. Alkadash, Chayanit Nadam,
Mohammad Allaymoun, and Mohammed Dawwas

Contents

Toward Sustainable Smart Cities: Design and Development of Piezoelectric-Based Footstep Power Generation System 215
M. N. Mohammed, Shahad Al-yousif, M. Alfiras, Majed Rahman, Adnan N. Jameel Al-Tamimi, and Aysha Sharif

Cultural Marketing, Artificial Intelligence and Digital Learning, Innovation and Sustainable Operations

Sustainable Urban Street Design in Jeddah, Saudi Arabia 227
Marwa Abouhassan

Green Discourse Analysis on Twitter: Imperatives to Green Product Management in Sustainable Cities (SDG11) 245
Priya Sachdeva, M. Dileep Kumar, and Archan Mitra

Harnessing Geographic Information System (GIS) by Implementing Building Information Modelling (BIM) to Improve AEC Performance Towards Sustainable Strategic Planning in Setiu, Terengganu, Malaysia 257
Siti Nur Hidayatul Ain Bt. M. Nashruddin, Siti Sarah Herman, Siti Nur Ashakirin Bt. Mohd Nashruddin, Sumarni Ismail, and Siow May Ling

The Role of Contract in Advancing ESG in Construction: A Proposed Framework of 'Green' Standard Form of Contract for Use in Green Building Projects in Malaysia 269
Khariyah Mat Yaman and Zuhairah Ariff Abd Ghadas

Sustainability of Web Application Security Using Obfuscation: Techniques and Effects 281
Raghad A. AlSufaian, Khaireya H. AlQahtani, Rana M. AlAjmi, Roa A. AlMoussa, Rahaf A. AlGhamdi, Nazar A. Saqib, and Asiya Abdus Salam

Heritage Buildings Changes and Sustainable Development 295
Marwa Abouhassan

Global Crucial Risk Factors Associated Stress Among University Students During Post Covid-19 Pandemic: Empirical Evidence from Asian Country 315
Nor Azma Rahlin, Ayu Suriawaty, Siti Aisyah Bahkiar Bahkiar, Suayb Turan, and Siti Nadhirah Ahmad Fauzi

Factors Influencing Purchase Decisions on Online Sales in Indonesia 329
Efa Wakhidatus Solikhah, Indah Fatmawati, Retno Widowati, and M. Suyanto

Re-Addressing the Notion of Patriarchy in Social and Economic
Framework in the Light of Women Empowerment 341
Jenni K. Alex and Dino Mathew

Development of Asset-Based and Asset-Backed Sukuk Issuance:
Case of Malaysia 351
Abdulmajid Obaid Saleh and Mohammed Waleed Alswaidan

Role of Colours in Web Banners: A Systematic Review
and Future Directions for e-Marketing Sustainability 361
Khalid Ali Alshohaib

Relationship of People's Knowledge, Behavior and Governmental
Action Towards Environmental Sustainability 369
Marluna Lim-Urubio and Manolo Anto

Development of Entrepreneurship in the Tourism and Recreation
Sphere: Marketing Research 379
Raisa Kozhukhivska, Olena Sakovska, Svitlana Podzihun,
Valentyna Lementovska, Ruslana Lopatiuk, and Nataliia Valinkevych

The Impact of RUPT on Corporate Environmental Responsibility 391
Acharya Supriya Pavithran, C. Nagadeepa, and Baby Niviya Feston

The Effect of the Characteristics of the Board of Directors
and the Audit Committee on Financial Performance: Evidence
from Palestine 401
Hisham Madi, Ghaidaa Abdel Nabi, Fadi Abdelfattah,
and Ahmed Madi

A Comparative Analysis of Rural and Urban Public Health
Infrastructure and Health Outcomes During Pandemic: With
Special Reference to Karnataka State 415
S. J. G. Preethi and Tinto Tom

Local Community Readiness to Implement Smart Tourism
Destination in Yogyakarta, Indonesia 425
Sri Dwi Ari Ambarwati, Mohamad Irhas Effendi,
and Sri Tuntung Pandangwati

Open Innovation and Governance Models in Public Sector:
A Systematic Literature Review 437
Meshari Abdulhameed Alsafran, Odeh Rashed Al Jayyousi,
Fairouz M. Aldhmour, and Eisa A. Alsafran

Corporate Governance in the Digital Era 453
Mary Yaqoob, Alreem Alromaihi, and Zakeya Sanad

Contents

Environmental, Social, and Governance (ESG) Impact on Firm's Performance 461
Fatema Alhamar, Allam Hamdan, and Mohamad Saif

AI, Banking and Financial Technology

A Stochastic Model for Cryptocurrencies in Illiquid Markets with Extreme Conditions and Structural Changes 479
Youssef El-Khatib and Abdulnasser Hatemi-J

Analysis of the Efficiency, Effectiveness and Productivity of Peruvian Motorcycle Cab Drivers in Times of Covid-19 Pandemic 489
Nelson Cruz-Castillo, Hernan Ramirez-Asis, K. P. Jaheer Mukthar, María Estela Ponce Aruneri, Juan Eleazar Anicama Pescorán, and Wilber Acosta-Ponce

Investment Awareness in Financial Assets—An Exploration Based on the Equity Traders in Bangalore City 501
Aneesha K. Shaji and N. Sivasankar

Cost–Benefit Analysis of Fintech Framework Adoption 509
Zainab Mohammed Baqer Shehab, Muneer Al Mubarak, and Amir Dhia

Financial Technology (Fintech) 525
Adnan Jalal, Muneer Al Mubarak, and Farah Durani

Assessing Opportunities and Challenges of FinTech: A Bahrain's View of Fintech 537
Sabika AlJalal, Muneer Al Mubarak, and Ghada Nasseif

Toward Sustainable Smart Cities: A New Approach of Solar and Wind Renewable Energy in Agriculture Applications 555
Nurhasliza Hashim, Tiffiny Grace Neo, M. N. Mohammed, Hakim S. Sultan, Adnan N. Jameel Al-Tamimi, and M. Alfiras

Digital Transformation Towards Sustainability in Higher Education: A New Approach of Virtual Simulator for Series and Parallel Diodes for a Sustainable Adoption of E-Learning Systems 565
Tan Chor Kuan, Khairul Huda Yusoff, M. N. Mohammed, Adnan N. Jameel Al-Tamimi, Norazliani Md Sapari, Firas Mohammed Ibrahim, and M. Alfiras

Toward Sustainable Smart Cities: Smart Water Quality Monitoring System Based on IoT Technology 577
Lee Mei Teng, Khairul Huda Yusoff, M. N. Mohammed, Adnan N. Jameel Al-Tamimi, Norazliani Md Sapari, and M. Alfiras

2019 Novel Coronavirus Disease (Covid-19): Toward a New Design for All-in-One Smart Disinfection System 595
M. N. Mohammed, M. Alfiras, Hakim S. Sultan,
Adnan N. Jameel Al-Tamimi, Rabab Alayham Abbas Helmi,
Arshad Jamal, Aysha Sharif, and Nagham Khaled

Modeling of Cutting Forces When End Milling of Ti6Al4V Using Adaptive Neuro-Fuzzy Inference System 605
Salah Al-Zubaidi, Jaharah A. Ghani, Che Hassan Che Haron,
Hakim S. Sultan, Adnan N. Jameel Al-Tamimi,
Mohammed N. Abdulrazaq Alshekhly, and M. Alfiras

Impact of the Pandemic—Covid-19 on Construction Sector in Bahrain 617
Zuhair Nafea Alani, Mohammed N. Abdulrazaq Alshekhly,
Hamza Emad, Adnan N. Jameel Al-Tamimi, and M. Alfiras

The Role of Technology and Market Accessibility on Financial Market Classification 627
Reem Sayed Mansoor, Jasim Al Ajmi, and Asieh Hosseini

A Review of the Recent Developments in the Higher Education Sector Globally and in the GCC Region 635
Elham Ahmed, Amama Shaukat, and Esra AlDhaen

Intra-Industry Trade Trends in India's Manufacturing Sector: A Quantitative Analysis 653
Alan J. Benny, K. P. Jaheer Mukthar, Felix Julca-Guerrero,
Norma Ramírez-Asís, Laura Nivin-Vargas, and Sandra Mory-Guarnizo

Risk Management of Civil Liability Resulting from Self-Driving Vehicle Accidents 667
Saad Darwish and Ahmed Rashad Amin Al-Hawari

The Impact of Applying Just-In-Time Production System on the Company Performance in Garment Manufacturing Companies in Jordan 677
Mohammad Abdalkarim Alzuod and Rami Atef Al-Odeh

Evaluation of Large-Scale Solar Photovoltaic Operating Plant by Using Environmental Impact Screening: A Case Study in Penang Island, Malaysia 691
Siti Isma Hani binti Ismail, Loh Yong Seng,
Shanker Kumar Sinnakaudan, and Zul-fairul Zakaria

Prefabricated Plastic Pavement for High-Traffic and Extreme Weather Conditions 709
M. E. Al-Atroush, Nura Bala, and Musa Adamu

Contents

Highlighting The Role of UAE's Government Policies in Transition Towards "Circular Economy" 723
Tahira Yasmin, Ghaleb A. El Refae, and Shorouq Eletter

Tech-Management in Different Disciplines and the Role of Digital Marketing

A Systematic Literature Review on Mercantilism 739
Hannah Biju, K. P. Jaheer Mukthar, Norma Ramírez-Asís,
Jorge Castillo-Picon, Guillermo Pelaez-Diaz, and Liset Silva-Gonzales

Revisiting the Nexus Between Suicides and Economic Indicators: An Empirical Investigation 751
Keerthana Sannapureddy, Jyoti Shaw, Shahid Bashir,
and S. Vijayalakshmi

Dynamics of Sustainable Economic Growth in Emerging Middle Power Economies: Does Institutional Quality Matter? 761
Mithilesh Phadke, Jerold Raj, Sujay Rao, Shahid Bashir,
and Jibrael Jos

Land Market in Ukraine: Functioning During the Military State and Its Development Trends 773
Reznik Nadiia, Havryliuk Yuliia, Yakymovska Anna, Halahur Yulia,
Klymenko Lidiia, Zhmudenko Viktoriia, and Ischenko Valeriy

A Study on the Tendencies in the Trade Balance of the Indian Economy 783
K. Vinodha Devi and V. Raju

Quality Efficacy Issues in Mangoes: Decoding Retailers Supply Chain 795
Lakshmi Shetty, Kavitha Desai, and Shefali Srivastava

Behavioral Bias as an Instrumental Factor in Investment Decision-An Empirical Analysis 807
Aneesha K. Shaji and V. R. Uma

Creating a Sustainable Electric Vehicle Revolution in India 819
Annie Stephen and M. Vanlalahlimpuii

An Overview of Interest Rate Derivatives in Banking Sector—A Comparasion Between Global and Indian Market 835
Riya Singh and Nidhi Raj Gupta

Adoption of Mobile Banking Among Rural Customers 843
A. J. Excelce and V. G. Jisha

Artificial Intelligence and the Decarbonization Challenge 849
Ismail Noori, Yasser M. Abolelmagd, and Wael F. M. Mobarak

Algorithms Control Contemporary Life 859
Ismail Noori Mseer, W. M. Abd-Elfattah, and A. H. Al-Alawi

Auditors' Perceptions in Gulf Countries Towards Using
Artificial Intelligence in Audit Process 867
Ahmad Yahia Mustafa Alastal, Janat Ali Farhan,
and Mohammad H. Allaymoun

Accounting Students' Perceptions on E-Learning During
Covid-19 Pandemic: Case Study of Accounting and Financial
Students in Gulf University—Bahrain 879
Ahmad Yahia Mustafa Alastal, Mohamed Abdulla Hasan Salman,
and Mohammad H. Allaymoun

Factors that Influence the Occupational Safety and Health
for Employees as a Part of Human Resource Management
Practices a Study on Non-government Organization in Palestine
Gaza Strip (UNRWA) 891
Tamer M. Alkadash

The Impact of Employee Satisfaction, Emotional Intelligence
and Organizational Commitment on Marketing Service Quality
in Medical Equipment Companies, Bahrain 903
Tamer M. Alkadash, Mahmoud AlZgool, Ali Ahmed Ateeq,
Mohammed Alzoraiki, Rawan M. Alkadash, Chayanit Nadam,
Qais AlMaamari, Marwan Milhem, and Mohammed Dawwas

Study of Mechanical Properties of Friction Welded AISI D2
and AISI 304 Steels 917
Sinta Restuasih, Ade Sunardi, M. N. Mohammed, M. Zaenudin,
Adhes Gamayel, and M. Alfiras

Assessing the Adoption of Key Principles for a Sustainable
Lean Interior Design in the Construction Industry: The Case
of Bahrain 925
Aysha Aljawder, Wafi Al-Karaghouli, and Allam Hamdan

Characterizing the DNN Impact on Multiuser PD-NOMA
System Based Channel Estimation and Power Allocation 943
Mohamed Gaballa, Maysam Abbod, and Ammar Aldallal

Mobile Wireless Sensor Network: Routing Protocols Overview 967
Maha Al-Sadoon, Ahmed Jedidi, and Hamed Al-Raweshidy

Governance and Business Ethics with AI

Perception of Young Consumers Towards Electric Two Wheeler
in Bangalore City 979
C. Surendhranatha Reddy and Ajai Abraham Thomas

Contents

Efficacy of Dividend Announcement on Bluechip Pharma Stocks—An Evidence from the Indian Stock Market 987
R. Madhusudhanan and R. Haribaskar

Examining the Impact of Mentoring on Personal Learning, Job Involvement and Career Satisfaction 997
Roshen Therese Sebastian, L. Sherly Steffi, and Geethu Anna Mathew

Islamic Values Impact on Managerial Autonomy 1007
April Lia Dina Mariyana, Ariq Idris Annaufal, and Muafi

An Empirical Corroboration on Perceived Facilitations for Training and Affective Organisational Commitment 1017
Geethu Anna Mathew, K. Opika, and Roshen Therese Sebastian

An Efficiency Analysis of Private Banks in India—A DEA Approach 1025
Ibha Rani and Arti Singh

A Case Study on the Impact of Tourism on the Tribal Life in Vayalada, Calicut, Kerala 1037
P. T. Retheesh and K. K. Jagadeesh

Pro-Environmental Behavior of Farmers in the Dieng Plateau Indonesia 1047
Dyah Sugandini, Mohamad Irhas Effendi, Yuni Istanto, Bambang Sugiarto, and Muhammad Kundarto

Balanced Economic Development: Barometer and Reflections of Economic Progress Concerning the Economies of India and China 1059
Bijin Philip and Suresh Ganesan

Customer Preference Towards E-banking Services Offered by State Bank of India 1073
Ashwitha Shetty and K. A. Thanuja

The Analysis of the Influence of the January Effect and Weekend Effect Phenomenon on Stock Returns of Companies Listed on the JII Index (2017–2019 Period) 1081
Salsabila Hartono Putri and Nur Fauziah

Integrated Study of Ethical and Economic Efficiency of Society's Perception of Corporate Social Responsibility in India 1093
Mohammad Talha, Marim Alenezi, Syed Mohammad Faisal, and Ahmad Khalid Khan

New Trends in the Banking Sector and the Development of E-Banking 1107
Mohammad Afaneh

xviii

The Impact of Economic Globalization Welfare States 1117
Natacha de Jesus Silva, Maria José Palma Lampreia Dos-Santos,
and Nuno Baptista

**The Benefits of Digital Transformation: A Case Study
from Finance and Administrative Departments Within the Oil
and Gas Industry in the Kingdom of Bahrain** 1127
Hamad Aljar and Noor Alsayed

**The Relationship Between IT Governance and Firm
Performance: A Review of Literature** 1141
Noora Ahmed Al Romaihi, Allam Hamdan, and Raef Abdennadher

Carbon Emissions Impact by the Electric-Power Industry 1151
Ismail Noori and Basem A. Abu Izneid

**Employee Engagement Concepts, Constructs and Strategies:
A Systematic Review of Literature** 1159
Hanin Aldoy and Bryan Mcintosh

Board-Level Worker Representation: A Blessing or a Curse? 1175
Fatema Alrawahi, Amama Shaukat,
and Abdalmuttaleb M. A. Musleh Al-Sartawi

The Role of Digital Marketing on Tourism Industry in Bahrain 1181
Yousif Alhawaj and Muneer Al Mubarak

About the Editors

Prof. Allam Hamdan *Dean, College of Business and Finance, Ahlia University, Bahrain,* is a Full Professor, he is listed within the World's top 2% of scientists list by Stanford University, and he is the Dean of College of Business and Finance at Ahlia University, Bahrain. Author of many publications (more than 250 papers, 174 listed in Scopus) in regional and international journals that discussed several accountings, financial and economic issues concerning the Arab world. In addition, he has interests in educational-related issues in the Arab world universities like educational governance, investment in education and economic growth. Awarded the First Prize of Al-Owais Creative Award, UAE, 2019 and 2017; the Second Prize of Rashid bin Humaid Award for Culture and Science, UAE, 2016; the Third Prize of Arab Prize for the Social Sciences and Humanities, 2015, and the First Prize of "Durrat Watan", UAE, 2013. Achieved the highest (1st) scientific research citation among the Arab countries according to Arcif 2018–2022. Appointed an external panel member as part of Bahrain Quality Assurance Authority and National Qualifications Framework NQF as a validator, and appeal committee, General Directorate of NQF, Kingdom of Bahrain. Member of Steering Committee in International Arab Conference of Quality Assurance of Higher Education. Currently leading a mission-driven process for International Accreditation for College of Business and Finance by Association to Advance Collegiate Schools of Business (AACSB).

Dr. Esra Saleh Aldhaen *PFHEA Biography, Executive Director Strategy, Quality and Sustainability, Associate Professor*, holds a Doctorate from Brunel University London, in the area of strategic decisions and quality in the context of higher education, currently an Executive Director Strategy, Quality and Sustainability as well as Associate Professor at Ahlia University Bahrain. Appointed external reviewer and conducted various QA reviews as part of Oman Academic Accreditation Authority (OAAA), Bahrain Education and Training Quality Authority (BQA) for National Qualification Framework validation and National Commission for Academic Accreditation and Assessment (NCAAA) KSA. As HEA Principle fellow and expert in curriculum review, design and mapping to the National Qualification Framework as well as alignment of cross-border qualifications was able to set Assurance of Learning

xix

(AOL) framework to evaluate and assess student direct and indirect learning and chaired the AOL committee at college level.

Highly experienced in HEIs in various countries such as UK, Kingdom of Bahrain, KSA, UAE and Oman, along with the gained diverse background enable to transfer and share knowledge and experience from multiple aspects. In particular, to institutionalize and streamline various adapted QA standards into one quality management system supporting all the HEI targets such as international accreditation and ranking. This area was activated through becoming an active member at governance level which enable revising, implementation, assessing and further planning for the university strategic plan towards achieving the mission and vision of the university that includes setting targets and key performance indicators (KPIs) have also been practised and highly performed focused on teaching and learning excellence, research impact and societal contribution.

An expert in strategic planning, measurement and evaluation for higher education including assessment of internal and external environment. Leading Sustainable Development Initiatives in collaboration with United Nations for promising future and a member of multiple national and international societies tackling sustainable development and quality improvements. Published and awarded best papers by Emerald Publishing in various research areas including strategy, sustainable quality management teaching and learning.

Tech-Management in Different Disciplines and the Role of Digital Marketing

A Systematic Literature Review on Mercantilism

Hannah Biju⬛, K. P. Jaheer Mukthar⬛, Norma Ramírez-Asís⬛, Jorge Castillo-Picon⬛, Guillermo Pelaez-Diaz⬛, and Liset Silva-Gonzales⬛

Abstract To dive into the existing corpus on the history of mercantilism, a systematic literature review seemed the most appropriate method. To attain a microscopic understanding, this paper attempts to delineate mercantilist writers and approaches in understanding the history of mercantilism through a systematic literature review. The Systematic Literature Review includes 18 documents, indexed in Scopus database. The strategy employed was "Mercantilism" AND ("Economic History" OR "Economic Thought"). Content analysis was conducted on the procured 18 documents. The systematic review sheds light on prominent mercantilist writers and various approaches in understanding the history of mercantilism. An additional impression of this study is that it is a comprehensive study on the collected corpus, however the search was limited to only one database and grey literature was excluded.

Keywords Mercantilism · Economic history · Economic thought · Systematic literature review

1 Introduction

Economic history is imperative in economics but modern-day economics tends to sideline the importance of the history of economic thought. The main motive behind this project work is to look back on mercantilism, a pre-smithian heterogeneous

H. Biju · K. P. Jaheer Mukthar (✉)
Kristu Jayanti College Autonomous, Bengaluru, India
e-mail: jaheer@kristujayanti.com

N. Ramírez-Asís
Hospital Uldarico Rocca Fernandez, Lima, Peru

J. Castillo-Picon · G. Pelaez-Diaz
Universidad Nacional Santiago Antúnez de Mayolo, Huaraz, Peru

L. Silva-Gonzales
Universidad Señor de Sipán, Chiclayo, Peru

© The Author(s), under exclusive license to Springer Nature Switzerland AG 2024
A. Hamdan and E. S. Aldhaen (eds.), *Artificial Intelligence and Transforming Digital Marketing*, Studies in Systems, Decision and Control 487,
https://doi.org/10.1007/978-3-031-35828-9_61

739

collection of ideas which was popular in the sixteenth and eighteenth century Europe. The prominent economic ideas of protectionism, balance of trade and economic nationalism emerged during that time period and is reflected in trade policies of various countries. Moreover, it's like a school of thought that prevailed for a very long time, neither classical school nor neo-classical school had three centuries of lifespan. Mercantilism exists even today but with a new name, neo-mercantilism. The bullions transformed to foreign exchange reserves, remnants of economic nationalism persists, and maritime supremacy is sought by every other country.

This paper aims to dive deep into mercantilism to acquire an understanding about the prominent writers and the ways in which different schools of thoughts or individuals interpreted mercantilism through a systematic literature review. This paper is divided into three parts, the first part entails the introduction, Sect. 2 elaborates on the research methodology, Sect. 3 details the systematic review followed by conclusion and references.

2 Research Methodology

Reviews are a technique used to pool the existing literature and to find answers to specific research questions. Systematic Literature Reviews helps in conducting a literature review in a well-evidenced, transparent and reproducible manner.

2.1 Identification and Search Strategy

In order to execute this systematic literature review, the database selected was Scopus (referenced line on scopus). Two searches were conducted with minute differences in the key words to attain all possible existing literature. The below mentioned was the first query string initiated in the Scopus database:

(TITLE-ABS-KEY("mercantilism") AND ALL("economic thought" OR "economic history")) AND (LIMIT-TO (OA, "all")) AND (LIMIT-TO (DOCTYPE, "ar") OR LIMIT-TO (DOCTYPE, "ch") OR LIMIT-TO (DOCTYPE, "re") OR LIMIT-TO (DOCTYPE, "bk")) AND (LIMIT-TO (SUBJAREA, "ARTS") OR LIMIT-TO (SUBJAREA, "ECON") OR LIMIT-TO (SUBJAREA, "SOCI")) AND (LIMIT-TO (LANGUAGE, "English")).

The second query string is mentioned below:

(ALL("mercantilism") AND TITLE-ABS-KEY("economic history" OR "economic thought")) AND (LIMIT-TO (OA, "all")) AND (LIMIT-TO (PUBSTAGE, "final")) AND (LIMIT-TO (DOCTYPE, "ar") OR LIMIT-TO (DOCTYPE, "bk") OR LIMIT-TO (DOCTYPE, "re") OR LIMIT-TO (DOCTYPE, "ch")) AND (LIMIT-TO (SUBJAREA, "ECON") OR LIMIT-TO (SUBJAREA, "ARTS") OR LIMIT-TO (SUBJAREA, "SOCI")) AND (LIMIT-TO (LANGUAGE, "English")).

2.2 Screening and Inclusion

During the screening and inclusion, a set of inclusion and exclusion criteria was devised as listed in the Table 1. The selected 603 documents were further screened and finally a total of 18 documents were selected for the review. Figure 1 schematically represents the various stages involved in the identification, screening and selection of the documents.

3 Systematic Literature Review

The eminent physiocrat, Marquis de Mirabeau (1749–1791), is the first to use the term "système mercantile" in print, in his *Philosophie Rurale* (1763).[1] Considered as the 'Pre-History' of economic thought, mercantilism prevailed in Europe from fifteenth century to eighteenth century. Mercantilist literature was in the form of tracts, or pamphlets written by lay writers to influence economic policy [2]. Although mercantilism was spread throughout England, Germany, Italy, Spain, Netherlands and Sweden, and there are a rage of mercantilist thinkers and writers, a common consensus in the definition of mercantilism is strenuous to rummage. Debate over the existence was sparked after the publication of Adam Smith's Wealth of Nations (1776). Hence, it's important to note that individuals that are being addressed as mercantilists never identified as belonging to this grouping. Nonetheless, it's imperative to study mercantilism. In this paper, an account of prominent mercantilists is being sketched. Hence, this paper also tries to sketch the vantage points of various schools of thoughts.

Table 1 Inclusion and exclusion

	Criterion inclusion	Exclusion
	Timeline 1934–2022	Anything published before 1934 or after Oct 2022
Document type	Articles, book chapter	Book, review, conference paper, editorial, note
Language	English	French, Spanish, German, Italian
Subject area	Arts and humanities, social sciences, economics, econometrics and finance	All the other subject areas
Source type	Journal, book, book series	Conference proceedings

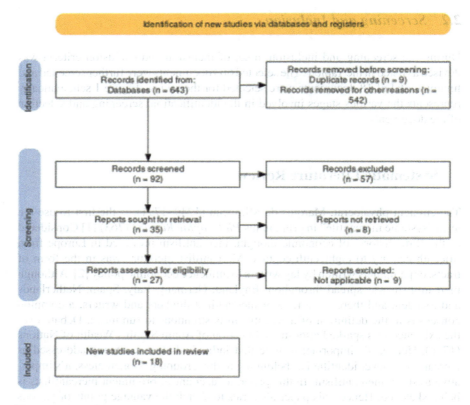

Fig. 1 PRISMA flow diagram 2020

3.1 An Overview on Prominent Mercantilist Writers

The mercantilist writers can be classified into late scholastics, seventeenth century mercantilists and eighteenth century mercantilists. The Middle Ages saw the height of scholarly economic thought, which was very different from modern economic theory in many ways. It was normative, instructing the faithful to do some things and refrain from doing others, rather than positive or hypothetical. Gaps between the norm and its fulfilment would be explained by human weakness or depravity [3]. Scholastic ideas started dimming after the renaissance, the scholastic writers of 1550–1600 who focused on giving a religious or moral perspective on trade and commerce were termed as late scholastics. One of the notable writers among them is Mariana (1609), a Jesuit, according to him monarchs should lead a modest life and debasement of coins was not a good idea. Three of the most prominent mercantilists of the early seventeenth century are Gerard Malynes, Edward Misseldon and Thomas Mun. Gerard Malynes (1585–1626) believed that the monarch was solely responsible for establishing economic justice which was in line with God and nature, this thought roots from Aristotle's *Nicomachean Ethic* [2] and the influence of scholastics [3].

His reasoning behind England's 1622 shortage of money was usurious bankers and profit-driven foreign exchange. As an aristotelian at heart [2] he was interested in the preservation of harmony in the state through controlling the usurious foreign exchange by reinstating the Royal Foreign Exchange. Edward Misseldon (1615–1654), was a Merchant Adventurer, and as such, he was intimately involved in the events surrounding the Cockayne project [1]. The famous debate between Malynes and his contemporary Misseldon deconstructs the diverging views that existed within mercantilists. The debate took place in between 1622 and 1623 over an acute economic crisis faced by England in the 1620 s and resulted in four lengthy pamphlets [2]. As Hengstmengel (2017) notes, Malynes' justification is that the undervaluation of English coins was because of the usurious behaviour of bankers and he was strongly vocal about reinstating the Royal Exchange in London. On the other hand, Misseldon argued that the economic crisis is caused by a deficit in the balance of trade.

The term 'balance of trade' first appeared in Misseldon's pamphlets; this concept was further developed by Thomas Mun [4]. According to Misseldon, the main reason behind this negative balance of trade was due to increasing exports of luxury goods in spite of raw materials. Contrary to Malynes view, Misseldon equated public good with surplus trade, improvement in navigation and employment of the poor [2]. Considered as the father of mercantilism, Thomas Mun (1571–1641) wrote the first pamphlet titled *A Discourse of Trade, from England to East Indies* and he defended the company from the accusations of draining the country's wealth.

Mun assumes that national capital will grow not only by increasing the net surplus of 'natural' or 'artificial wealth' over consumption, but also by saving in the form of restricted consumption of foreign surpluses [2]. The balance-of-trade concept developed by Thomas Mun incorporates both empirical data and economic analysis. These, however, should be viewed in the context of—the doctrine built around the concept of the national capital, which is best described as a pre-scientific macro-theory with a strong ideological underpinning. These early mercantilists' economic discourse and writing demonstrate that mercantilists were more than a group of befuddled bankers acting in their own self-interest.

3.2 Approaches in Understanding Mercantilism

To have a grounded understanding about something from the past, it's significant to learn about how that domain was understood and perceived by scholars from various schools of thought. This helps in creating a rounder understanding on the concept of mercantilism. So, the approaches in which different schools of thought had perceived and interpreted mercantilism are elaborated in the next subsections. The schools of thought include classical school, German school, Stockholm and Keynesian school.

4 The Classical School

The icon of classical school, Adam Smith can be deemed responsible for instigating discussions on mercantilism. He sparked a fire with publishing his phenomenal book *An Inquiry into the Nature and Causes of the Wealth of Nations* (1776), where he vehemently criticised "système mercantile" or 'systeme des commercants'. He defined mercantilism as a set of ideas that prevailed during sixteenth and eighteenth century Europe. According to him, these ideas were both rational and absurd. 'That wealth consists in money, or in gold and silver, is a popular notion which naturally arises from the double function of money, as the instrument of commerce, and as the measure of value' (Smith, 1979 [1776], p. 429). According to Smith, the chrysohedonist assumption that money or species can be synonymous with wealth was rather absurd and based on fallacious reasoning.

He also agreed on the precedent that the ideas that accumulated were the result of unpatriotic mercantile interest which was propagated under the guise of economic prosperity. Classical school adherents were influenced by Smith's revealing criticism of mercantilism and hence considered it to be the reasoning of a bunch of confused bankers and merchants. For example, J. S. Mill in his *Principles of Political Economy* states that: 'The doctrine that money is synonymous with wealth ... looks like one of the crude fancies of childhood (Mill, 1968 [1848–71], vol. 1, p. 4). Smith considered the pamphlets and tracts written by the merchants as a tool which fooled the ruling classes. The ruling class was reduced to 'fools who believed' the merchants and whose vision was not far enough to understand the mercantile treachery.

Smith's criticism is rather celebrated with volume than his defence for mercantilism. Although he did not agree with their axioms or maxims, he does not agree on the argument that mercantilism was a complete economic catastrophe. Smith was an ardent advocate of capital accumulation, one of the prominent obstacles to economic prosperity was the lack of capital accumulation. Smith considered mercantilism as a phase which transformed the social structure of European society as it encouraged in transitioning from feudalism to a modern bourgeois society. Smith was also impressed by Thomas Mun's tract titled *England's Treasure by Forraign Trade* (1895 [1664]). In essence, Smith's critical perspective can be considered as a representation of the classical school's view on mercantilism.

5 German Historical School of Economics

Smithian aversion towards mercantilism prevailed among the scholars until the late nineteenth century. Two of the prominent individuals who contributed the most in the political interpretation of mercantilism was Gustav Schmoller from Germany and William Cunningham from England. The German historical school of economics led by Gustav Schmoller broke the silence with the new interpretation of the pre-classic economic concept of mercantilism. In Wilson's words (Wilson 1969a, p. 68), there

A Systematic Literature Review on Mercantilism 745

were negligible differences between Smith's and Schmoller's standpoint on mercantilism, as the former condemns and the latter applauds [5]. Schmoller reconstructed mercantilism from a political perspective, state interference in economic matters was a key feature of mercantilism.

Both Schmoller and Cunnigham were influenced by Heyking's late Hagelian theory regarding the genesis of mercantilism. In accordance with the theory, the concept of a national state started developing from the late middle ages. This concept of a self-conscious entity developed within the nation states and mercantilism was mere an outgrowth of this political process which was the result of a self-sustaining material foundation [6]. According to Schmoller, mercantilism arose as a necessary by-product of political transformation of a nation state, during that time period. He emphasised less on the monetary aspects. Thus for Schmoller, the changes brought by mercantilism to the political structure of society overarched rather than its shortcomings. As for Cunnigham, mercantilism was an expression of nationalist sentiments [6]. For them, mercantilism did not reflect the interests of the merchant class; rather, it promoted national interests. Cunningham and Schmoller elevated the ruling classes, the representatives of state authority, from 'fools who believed' to the rank of standard-bearers of the concepts of the state, national unity, and the national economy. The major limitation of this approach is the role of merchants are enormously sidelined, both Cunnigham and Schmoller highlight the emergence of the nation state but forgets the rise of merchants.

6 Neo-Classical School

Two of the significant individuals who extensively wrote about mercantilism were Eli F. Hecksher (1879–1952) and Jacob Viner (1892–1970). His magnum opus, "Mercantilism" which was published in 1931 reflected his views on mercantilism and also sparked scholarly debates. In Hecksher's eyes, Cunningham's and Schmoller's analysis of mercantilism lacked economic theory. For Heckscher, mercantilism was more than a unifying force of state-making, it was a theoretical and practical political instrument aiming at increasing the wealth of nations.

The difference between German Historical School and Hecksher's approaches is that he considered mercantilism to be more than mere a driving force in state making rather a practical political instrument that aims in increasing the nation's wealth [7]. Thus, he considered mercantilism as a monetary system of unification, power, protection and conception of society. In Hecksher's perspective, the transition from Middle age to Modern age resulted in the adoption of 'cash economy' over barter system and this inturn created the fetish for money or 'love for money'. Gradually. The misinterpretation of money with capital led to stressing on the balance of trade. Hence, 'fear of goods' was a result of 'hunger for money'. Heckscher believed that this misconception led to the failure of the mercantilist model to obtain its major objective of increasing the wealth of nations. His investigation leads to the conclusion that the common denominator of this period's literature was a static view of commercial and monetary relations as a zero-sum game Heckscher also equated 'capitalism' with 'mercantilism' and he believed both the concepts had the same objective of increasing the nation's wealth but employed different models to achieve this objective.

Mercantilism has been interpreted by scholars in a multitude of ways and Hecksher disagrees with the sect that denied the existence of mercantilism. Heckscher's work rekindled scholarly interest in early modern economic ideas while also bringing mercantilism to the notice of every historian-minded economist as an intrinsic aspect of his or her own discipline's history. Jacob Viner (1892–1970), studied mercantile theory extensively in his Studies in the *Theory of International Trade* Jacob Viner has been referred to as "the greatest historian of economic thought that ever lived" by economist Mark Blaug. The heart of Viner's argument is summarised as follows:

> ... practically all mercantilists, whatever the period, country, or status of the particular individual, would have subscribed to all the following propositions: (1) wealth is an absolutely essential means to power, whether for security or for aggression; (2) power is essential or valuable as a means to the acquisition or retention; (3) wealth and power are each proper ultimate ends of national policy; (4) there is a long-run harmony between these ends, although in particular circumstances it may be necessary for a time to make economic sacrifices in the interest of military security and therefore also of long run prosperity. [8]

A concise summary of his works on mercantilism emphasises the search for national advantage: accumulation of precious metals, to be secured through a favourable balance of trade for countries without their own mines, and aided by active export promotion and import discouragement policies, combined with a foreign policy aimed at power and plenty. It's impossible to say whether this is against Keynes' attempt to rehabilitate mercantilism in the General Theory.

7 Keynesian School

Under Keynesian school, the approaches of John Maynard Keynes and Werner Stark are analysed to understand how keynesians perceived mercantilism. Werner Stark (1905–1985), a close friend of John Maynard Keynes (1883–1946), is one of the forgotten scholars of the history of economic thought, Repapis (2021) in his work *W. Stark, J.M. Keynes, and the Mercantilists* highlight the importance of W. Stark's historiography and elaborates on the way in which it can frame the mercantilist writings. Stark was known for his 'sociology of knowledge' approach to the history of economic thought which disregards considering history as a steady progression from error to truth, rather emphasises on the importance of maintaining a balance of judgement, understanding and contextual setting of the society which gave birth to the theory. Through his book, *The history of economics: In its relation to Social Development*, he clearly mentions the two different approaches to evaluate history:

> There are, in the last analysis, two ways of looking upon the history of economic thought: the one is to regard it as a steady progression from error to truth, or at least from dim and partial vision to clear and comprehensible perception; the other is to interpret every single theory

put forward in the past as a faithful expression and reflection of contemporary conditions, and thus to understand it in its historical causation and meaning. It is obvious that between these two antagonistic conceptions, no compromise is possible. [9]

Mercantilism is often characterised as selfish pressure groups composed of merchants who cared solely about their self-interest and persuaded the individuals in authority to agree with their precedent. W. Stark argues that this judgement is partial in nature and is born without truly understanding the social reality of that era. He defends mercantilists by backing the core argument of mercantilism as increasing the mass of precious metals as a necessity during that period while the economy was moving through a transitional phase. In other words increasing the supply of precious metals in turn increased circulation and stimulated commerce and production. In his 1914 letter to Ms. Simson of the Society for the Protection of Science and Learning, Keynes states that mercantilism was a by-product of nationalism and state-making [10].

8 Conclusion

In essence, this paper is an organised review on the existing literature on the history of mercantilism. The systematic review shed light into the history of mercantilism, from a detailed account of mercantilist authors and the approaches, the review helped in creating a rounder understanding about the domain of study. In essence by mapping the existing literature on mercantilism, The paper provides historical lenses that can be used to glean out a clear understanding about mercantilism. The usage of one database and exclusion of grey literature are the limitations of the research. However, that can be avenues for future research.

References

1. Suprinyak, C.E.: Merchants and councilors: intellectual divergences in early 17th century British economic thought. Nova Economia **21**(3), 459–482 (2011). https://doi.org/10.1590/s0103-63512011000300006
2. Hengstmengel, J.W.: The survival of Aristotelianism in early English mercantilism: an illustration from the debate between Malynes and Misselden. Erasmus J. Philos. Econ **10**(1), 64–82 (2017). https://doi.org/10.23941/ejpe.v10i1.266
3. Scholastic Economics.: Routledge History of Economic Thought (1997). https://doi.org/10.4324/9780203441329.ch4
4. Blanc, J., Desmedt, L.: In search of a crude fancy of childhood: deconstructing mercantilism. Camb. J. Econ. **38**(3), 585–604 (2013). https://doi.org/10.1093/cje/bes081
5. Tsoulfidis, L.: In Competing Schools of Economic Thought, pp. 5–6. Essay, Springer (2010)
6. Vaggi, G., Groenewegen, P.D.: A concise history of economic thought: from mercantilism to Monetarism. Palgrave Macmillan (2014)
7. Kaidantzis, J.B.: Karl Marx. Econlib. Retrieved November 13, 2022, from https://www.econlib.org/library/Enc/bios/Marx.html. (8 Aug, 2022)
8. Viner, J.: Essays on the Intellectual History of Economics. In: Irwin, D.A. (ed.) Princeton University Press, Princeton, NJ (1991)
9. Stark, W.: The History of Economics: In its Relation to Social Development. Routledge (1944)
10. Clark, Charles, M.A.: Introduction. In Stark, W. History and Historians of Political Economy, pp. xiii–xxvi. Transactions Publishers, Introduction, NJ (1994)
11. Herlitz, L.: On the prehistory of the trinity: Functional class division from mercantilism to classicism. Scand. Econ. Hist. Rev. **37**(2), 5–25 (1989). https://doi.org/10.1080/03585522.1989.10408141
12. Belfanti CM (2006) Between mercantilism and market: privileges for invention in early modern Europe. J. Inst. Econ. **2**(3), 319–338. https://doi.org/10.1017/s1744137406000439
13. Bick, A.: Bernard Mandeville and the economy of the Dutch. Erasmus J. Philos. Econ. **1**(1), 87 (2008). https://doi.org/10.23941/ejpe.v1i1
14. Repapis, C.: W. Stark, J. M. Keynes, and the mercantilists. J. Hist. Econ. Thought **43**(1), 27–54 (2021). https://doi.org/10.1017/s1053837219000683
15. Brandon, P.: Marxism and the Dutch miracle: the Dutch Republic and the transition-debate. Hist. Mater. **19**(3), 106–146 (2011). https://doi.org/10.1163/156920611x573806
16. Herlitz, L.: The concept of mercantilism. Scand. Econ. Hist. Rev. **12**(2), 101–120 (1964). https://doi.org/10.1080/03585522.1964.10407639
17. Coats, A.W.: In defence of Heckscher and the idea of mercantilism. Scand. Econ. Hist. Rev. **5**(2), 173–187 (1957). https://doi.org/10.1080/03585522.1957.10411398
18. Magnusson, L.: Eli Heckscher, mercantilism, and the favourable balance of trade. Scand. Econ. Hist. Rev. **26**(2), 103–127 (1978). https://doi.org/10.1080/03585522.1978.10415622
19. Coleman, D.C.: Eli Heckscher and the idea of mercantilism. Scand. Econ. Hist. Rev. **5**(1), 3–25 (1957). https://doi.org/10.1080/03585522.1957.10411389
20. Uhr, C.G.: Eli F. Heckscher, 1879–1952, and his treatise on mercantilism revisited. Econ. Hist. **23**(1), 3–39 (1980). https://doi.org/10.1080/00708852.1980.10418968
21. Adamson, R.: Economic history research in Sweden since the mid-1970s. Scand. Econ. Hist. Rev. **36**(3), 51–66 (1988). https://doi.org/10.1080/03585522.1988.10408127
22. Groenewegen, P.: Jacob Viner and the history of economic thought*. Contrib. Polit. Econ. **13**(1), 69–86 (1994). https://doi.org/10.1093/oxfordjournals.cpe.a035630
23. Lichtenstein, P.M.: In Theories of International Economics. Introduction, Taylor and Francis (2016)
24. Penchev, P.: Carl Menger on the theory of economic history. Reflections from Bulgaria. Panoeconomicus **61**(6), 723–738 (2014). https://doi.org/10.2298/pan1406723p
25. Kim, S.: Bak Je-Gha's thoughts on logistics and overseas commerce in the 18th century in Korea. Asian J. Shipping Logistics **25**(1), 139–163 (2009). https://doi.org/10.1016/s2092-5212(09)80017-5

26. Hinderson, D.R.: John Locke. Econlib. Retrieved November 7, 2022, from https://www.eco nlib.org/library/Enc/bios/Locke.htm. (11 Oct, 2021)
27. Paganelli, M.P.: Adam Smith's digression on silver: the centrepiece of the wealth of nations. Camb. J. Econ. **46**(3), 531–544 (2022). https://doi.org/10.1093/cje/beac009
28. Wolf, B.: Adam Smith's cosmopolitan liberalism: taste, political economy, and objectification. Polity **54**(4), 709–733 (2022). https://doi.org/10.1086/721233

Revisiting the Nexus Between Suicides and Economic Indicators: An Empirical Investigation

Keerthana Sannapureddy, Jyoti Shaw, Shahid Bashir, and S. Vijayalakshmi

Abstract In India, as of 2021, there was a 7.2% increase in suicides. With the economic burden inflicted by the pandemic and increasing suicides, a systematic investigation needs to be done. This empirical investigation uses the "Autoregressive Distributed Lag (ARDL) model" to obtain long and short-term estimates for the relationship between suicides and prominent economic indicators. The findings suggest that economic indicators like GDP per capita growth, age dependency ratio, and unemployment rate have a significant dynamic relationship with suicides. In this regard, preventive measures can be formulated and implemented in such a way that focuses on improving the country's economic scenario which will in turn reduce suicides. Organizations and governments can plan training and mental health care programs for farmers, workers, and students. Mental health care services require attention from the government so that at a macro level the problem of increasing suicides can be handled.

Keywords ARDL model · Economic indicators · Suicides · India · SDG3

1 Introduction

People's mental health has notably suffered as a result of the recent dynamic changes in economic growth brought on by the pandemic and due to the repercussions of the global financial crisis. Prolonged severe mental breakdown, economic downfall, and its effects lead to increasing suicides in many countries. Suicide is a multi-faceted, critical public health problem. Sustainable development goal (SDG3) 3.4.2 specifically aims to reduce the number of suicides. Achieving this target would involve efforts to prevent suicides, understand what is causing suicides and promote mental health, such as increasing access to mental health services and support, reducing the stigma associated with mental health conditions, and implementing policies and

K. Sannapureddy (✉) · J. Shaw · S. Bashir · S. Vijayalakshmi
CHRIST (Deemed to be University), Bangalore, India
e-mail: sannapureddy.keerthana@msea.christuniversity.in

© The Author(s), under exclusive license to Springer Nature Switzerland AG 2024
A. Hamdan and E. S. Aldhaen (eds.), *Artificial Intelligence and Transforming Digital Marketing*, Studies in Systems, Decision and Control 487,
https://doi.org/10.1007/978-3-031-35828-9_62

programs to promote mental well-being. According to reports, India's suicide rates have remained high over the past few decades. Since 1998, every year, more than 1,00,000 people have committed suicide in India. The number of suicides reported in India for 2021 was 1,64,033, a rise of 7.2% over 2020, so addressing this is imperative.

The alarming increase in suicides has received insufficient attention from economists, despite the seriousness of the issue that it costs a lot for a country and one suicide triggers many more people to commit suicide. Various specific economic causes for suicides could be financial loss, unemployment, economic insecurity—uncertain future of the country's economy, and many more. In 2019, over 77% of suicides worldwide took place in middle and low-income nations. So, it is essential to do an empirical investigation on how the economic condition in India affects the movement of suicide rates while this country strives to recover from the consequences of the pandemic that has affected people in terms of uncertain future, unemployment, and financial loss.

Many existing studies on economic factors and suicide rates have focused on how income, government spending, unemployment rate, Gross Domestic Product, and public debt have affected suicide rates. This paper focuses on the variables, Gross Domestic Product per capita growth, General government final consumption expenditure, Unemployment rate, and Age Dependency Ratio. The World Bank has defined the growth rate of the gross domestic product (GDP) per capita (taken annually, calculated using constant local currency and the midyear population) as a measurement of the nation's economic scenario. General government consumption expenditure involves all government spending to purchase goods and services for the public. Unemployment is defined as the fraction of the workforce that is unemployed and seeking to be employed. The age dependency ratio is the proportion of dependents per 100 individuals of working age (15–64 years), where the dependents are young (under 15) and older people (over 64).

Most studies based on the Indian context have investigated the sociological aspect of suicides and employed conventional methodologies. With globalization, modernization, economic crisis, and the pandemic, an empirical analysis is required using advanced methodologies with subsequent tests on the results. This paper captures how much influence a country's economic condition has on increasing suicides using a better cointegration method. This study focuses on how the economy of the country affects the trend of suicides by analyzing the relationship between the growth of GDP per capita and suicides and how the amount of funding allocated by the government to various programmes affects the population's mental health. The impact of future expectations, fear of unemployment, and changing age demographic on suicides is explored.

2 Review of Literature

The trend of suicide rates has been studied in both theoretical and empirical ways. However, most research up to the early 1970s mainly concentrated on the sociological aspect of suicides, failing to get economists' interest. Few of the studies that were based on social and economic indicators had the following inferences. Based on short and long-term estimates, the 2008 financial crisis significantly affected several suicides in the US. Reducing people's expectations for the future was thought to be the root cause of suicides [1]. Unemployment greatly impacted suicide rates in Denmark over the long term [2]. The impact of unemployment was more in nations with lower socioeconomic levels [3]. Through monetary and fiscal policy, it is vital to target the unemployment rate, which significantly influences suicide rates. The expansionary fiscal policy raises the deficit but raising the debt saves many lives. Redesigning government support and welfare programs to encourage people to actively look for work, promoting job creation as a means of reducing unemployment, and limiting illegal immigration and under-the-table pay are some of the strategies to reduce suicides [4]. "6,912 young people residing in the world's most developed nations committed suicide in 2014". A loss due to suicides, in terms of economic income, was $5.53 billion [5]. Rising per capita household income showed a drop in suicides. Industrial growth and Inflation were positively associated with suicides in India, and the reasons were that workers lost jobs because of unmatched skills with industrialization and workers were not able to acquire new skills that were demanded to handle new methods at the workplace [6]. The forecasted suicides trend for the years 2020 to 2022 was expected to rise high and then decline [7]. In Japan, a relation was found that a 1% increase in local government spending was linked to a 0.2% drop in suicide rates in both genders between the ages of 40 and 64. This relationship became stronger as the unemployment rate rose [8]. The rate of public debt and suicides had an evident correlation. People who were experiencing job and income losses, living in a devastating social environment due to severe economic measures used by governments to lower budget deficits, were at risk of committing suicides [9]. When a country spent more on total transfer payments, medical benefits, and family support, suicide rates in that country were low [10].

Over the years, the literature showed that economic conditions have emerged as important factors affecting suicides. So, this study focuses on prominent economic indicators. Studies have been done using conventional cointegration methods, graphical analysis, correlation, and ordinary least squares. Considering the recent disturbance created by the global pandemic not only all over the world but also in India, it becomes imperative to analyze the determinants of suicides afresh using new evidence, new data, and robust methodologies. This study attempts to contribute these aspects to the existing literature.

3 Data

This paper studies yearly data from 1991 to 2021. This timeframe is chosen based on the data's availability. This data incorporates the impact of India's economic turmoil on each indicator. The dependent variable is the number of suicides. This data is taken from the National Crime Records Bureau (NCRB). It is a source with well-collated and documented information about suicides in India which comes from police records. It has comprehensive reports on suicides, including state-wise, cause-wise, profession-wise, age and gender-wise reports. The independent variables are Gross Domestic Product per capita growth rate (GDP), General government final consumption expenditure (GCE, percent of Gross Domestic Product), Unemployment (UNP, percent of the total labour force) (modelled ILO estimate), and Age Dependency Ratio (ADR, percent of working-age population) from the World Bank.

4 Methodology

Before choosing the methodology, one should make sure the variables are stationary, or else the inference cannot be generalized and there is a chance of making a wrong inference. To avoid this unit root test is done. Along with this, to select an appropriate econometric method for the study unit root test such as Augmented Dickey-Fuller (ADF) test is done. This test handles serial correlation by incorporating lagged difference terms when the unit root test is done on the variable.

The proposed method for this empirical study is the "Autoregressive distributed lag (ARDL) model". This model gives the long and short-run relationship between increasing suicides, and GDP per capita, age dependency ratio, unemployment, and government spending. It is the most suitable and efficient model for studying dynamic relationships. It reflects a dynamic effect by including an appropriate number of lags of variables. To get robust results for a finite sample size, ARDL is an efficient model. This model can be applied when the model's variables are of both I (0) and I (1) integration order. The only condition is that none of the variables should be in the order of integration I (2). This model just has a single reduced-form equation to study the short and long-term dynamics of the variables, which will now be easy to interpret.

$$\text{Empirical model}: \text{LS}_t = \alpha_0 + \beta_1 \text{GDP}_t + \beta_2 \text{GCE}_t + \beta_3 \text{UNP}_t + \beta_4 \text{ADR}_t + \varepsilon_t \tag{1}$$

where LS is the logarithm of Suicides, t represents the time from 1991 to 2021, and ε represents the error term.

After obtaining the results, the Dynamic Ordinary Least Squares (DOLS) method is used to test the robustness of the estimates. This test takes a parametric approach. It can be used for a model in which the variables have a mixed order of integration and

Revisiting the Nexus Between Suicides and Economic Indicators ... 755

are cointegrated. This method deals with serial correlation, endogeneity, and small sample bias by including lags and leads in its model. This test is done to determine whether the estimators are asymptotically efficient.

5 Results and Discussion

Table 1 indicates the descriptive statistics. Mean gives us the average of these variables over 30 years. Maximum and minimum provide the range of the data. The standard deviation, skewness, and kurtosis tell us about the data's spread and variation.

The results of the ADF test are presented in Table 2. It is to find out if these variables are non-stationary. All variables should be stationary to avoid false inferences from the regression. ARDL model works with all variables being stationary in I (0) and I (1) or mixed. LS and GCE are I (1) at 5 and 1% levels of significance, respectively. GDP, UNP, and ADR are I (0) at a 1% level of significance. All variables are stationary.

The outcome of the ARDL bounds test is presented in Table 3, which checks for a long-term relationship. The lag length has been selected based on Akaike Information

Table 1 Descriptive statistics

	LS	GDP	GCE	UNP	ADR
Observations	31	31	31	31	31
Mean	5.06	4.34	10.9	5.64	59.41
Median	5.07	5.35	10.88	5.58	59.21
Maximum	5.21	7.9	12.18	8	71.31
Minimum	4.89	−7.52	9.8	5.27	48.27
Std. dev	0.08	3.01	0.62	0.46	7.4
Skewness	−0.46	−2.13	0.44	4.42	0.06
Kurtosis	2.39	8.91	2.53	23.15	1.72

Table 2 Augmented Dickey-Fuller (ADF) test

Variables	t-statistic		Specification	Inference
	Level	First diff		
LS	−0.856	−3.212**	Intercept only	I (1)
GDP	−5.392*	–	Intercept only	I (0)
GCE	−0.006	−4.586*	None	I (1)
UNP	−4.829*	–	Intercept only	I (0)
ADR	−2.906*	–	None	I (0)

* and ** denote 1% and 5% significance levels respectively

Table 3 Bounds test

F-statistic	Value	Significance (%)	I (0)	I (1)
	4.932**	1	4.768	6.67
		5	3.354	4.774
		10	2.752	3.994

** denotes a 5% significance level

Criterion (AIC). The chosen model is (2, 2, 3, 3, 0). The test results show that the F-statistic calculated is 4.932, higher than the upper bound at a 5% significance level. This result is suggestive of cointegration among the variables.

Table 4 demonstrates the long-run relationship with each variable. At 1% level, the GDP per capita growth coefficient is negative and significant. This outcome is consistent with the existing literature. Suicide rates had a significant relationship with GDP per capita and economic growth [11]. Economic downturns were strongly associated with the rise in suicides [12]. By encountering economic disturbances, individuals will be pressured regarding the future of credit for education, business, and living. This will lead to a prolonged mental breakdown and then suicide. After the economic downturn in 2008, there was an increase in suicides in 2009, 2.5% was because of sudden economic change, 2.3% was due to poverty, and 1.1 due to career problems [13]. After 2020, the causes for the increase in suicides in the year 2021 were because of unemployment (2.2%), career problems (1.6%), and poverty (1.1%) [14]. Gross domestic product (GDP) negatively affects suicide rates, indicating that the number of suicides decreases with economic prosperity and rises with recession [15]. The coefficient of Government consumption expenditure is insignificant. At a 1% level of significance, the unemployment rate's coefficient is positive and significant. Unemployment increases a person's propensity to commit suicide since it lowers the expectation of future income and utility. In a way, a person's expected income is reduced due to unemployment, which raises their chances of committing suicide [16]. Unemployment could cause suicides because with insecurity comes fear and depressive episodes. The rise in suicides during economic turmoil strongly correlates with the fear of unemployment. A high unemployment rate causes a deterioration of living standards and a decline in income, leading to increasing suicides [17].

The age dependency ratio's coefficient is negative and significant at a 1% level of significance. By the age dependency ratio, one can understand how the population's

Table 4 Long-term estimates

Variables	Coefficient	Std. error	t-statistic
GDP	−0.049*	0.006	−7.685
GCE	0.000	0.007	−0.076
UNP	1.414*	0.151	9.389
ADR	−0.040*	0.003	−12.186

* denotes a 1% significance level

demographic composition impacts a country's development. In India, this ratio has been decreasing for over 30 years. As the age dependency ratio decreases, suicides increase because, with globalization and modernization, young people are moving out for higher education and jobs. By moving out, young people and older adults are going through many psychological consequences, which leads to suicidal thoughts. With changing economic patterns, increasing price levels, and changing standard of living, older people are not willing to be a burden on the working age and with the trend of being independent, they feel isolated and go through psychological consequences and commit suicide. This way age dependency ratio is related to suicides in the long run.

The outcome drawn from Table 5 specifies short-term estimates. The coefficient of the past year's logarithm of suicides is positive and holds significance at 1%. At the 1% level, the coefficient of GDP per capita growth in the short term is negative and significant, and the coefficient of the past year's GDP per capita growth is positive and significant. The coefficient of government consumption expenditure is insignificant in the short run. The past year's government consumption expenditure coefficient is positive and holds significant at 5%. The unemployment coefficient is positive and significant at the 10% level. There is a negative and significant impact of past unemployment rates on suicides in the short run.

The lagged error correction term's coefficient is significant and negative. The ECT_{t-1} coefficient, which measures the speed of adjustment towards equilibrium, is -1.107, meaning that the speed of adjustment is very fast (Table 5).

Table 5 Short-term estimates

Variable	Coefficient	Std. error	t-Statistic
C	$-0.233*$	0.042	-5.53
ΔLS_{t-1}	$0.678*$	0.157	4.308
ΔGDP_t	$-0.003*$	0.001	-3.856
ΔGDP_{t-1}	$0.049*$	0.009	5.6
ΔGCE_t	-0.007	0.005	-1.423
ΔGCE_{t-1}	$0.010**$	0.004	2.417
ΔGCE_{t-2}	$0.011**$	0.005	2.198
ΔUNP_t	$0.007***$	0.004	1.866
ΔUNP_{t-1}	$-1.563*$	0.277	-5.65
ΔUNP_{t-2}	$-0.047***$	0.025	-1.908
ECT_{t-1}	$-1.107*$	0.195	-5.679

* 1% significance level, ** 5% significance level, *** 10% significance level

Table 6 Diagnostic tests

Tests	Statistic	Inference
Ramsey RESET test	0.166 (0.690)	No model misspecification
Jarque–Bera normality test	1.792 (0.408)	Normally distributed residuals
Breusch Pagan Godfrey test	0.303 (0.982)	Homoscedasticity
ARCH test	0.166 (0.686)	Homoscedasticity
Breusch-Godfrey LM test	0.619 (0.5565)	No serial correlation

The value in parenthesis () indicates the probability values

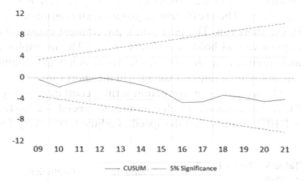

Fig. 1 CUSUM test. *Source* Author's calculation

5.1 Stability Tests

Overall, the results of the diagnostic tests performed to check the model's fit proved to be correct (Table 6).

Figures 1 and 2 display the cumulative sum (CUSUM) plot and cumulative sum of square (CUMSUMSQ) plot where the blue line is within the boundaries at a 5% significance level, indicating this model's coefficients are stable.

5.2 Robustness Check

Dynamic OLS is employed to confirm the robustness of the long-run coefficients from ARDL. The ARDL and DOLS results are consistent in sign and significance, confirming the model's accuracy, which ensures that the above ARDL model is robust to statistical biases (Table 7).

Fig. 2 CUSUMSQ test. *Source* Author's calculation

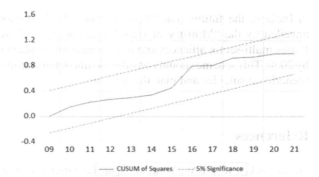

Table 7 Results from dynamic OLS

Variables	DOLS estimates
GDP	−0.041*(0.00)
GCE	0.009(0.25)
UNP	1.311*(0.00)
ADR	−0.037*(0.00)
Adj. R^2	0.98

* denotes a 1% significance level

6 Conclusion

This empirical study explores the association between suicides and GDP per capita growth, government consumption expenditure, age dependency ratio, and the unemployment rate for the period of 1991–2021. This study found that GDP per capita growth, age dependency ratio, and unemployment rate have a significant relationship with suicides, which means if there is economic growth and unemployment decreases, suicides reduce. These findings are in line with the studies [11, 15, 16]. This empirical investigation contradicts a study [6] which explains that with industrial growth and an increase in per capita real GDP, suicides increase.

India is yet to develop systematic preventive measures for suicides. There is a requirement for tailored preventive measures for India. It can be designed by considering India's unemployment situation, and pattern of economic development. Organisations and institutions can plan training and mental health care programs for farmers, workers, and students so that productivity is increased and unemployment is reduced. Allocation of resources for affordable and accessible mental health care services for young and old dependents requires attention from the government. Financial and social support can be provided to the people by the government, non-governmental organisations, community level health workers, youth clubs, and many more.

The focus on preventing suicides by government intervention has increased after the pandemic. Focusing on these specific factors will contribute to preventing suicides in India in the future. The "National Suicide Prevention Strategy" was recently unveiled by the "Ministry of Health and Family Welfare, Government of India". It uses multi-sector alliances and time-bound action plans to reduce suicides by 10% by 2030. This scheme should consider the impact of unemployment and the economic scenario on suicides and plan the scheme.

References

1. Demirci, Ş, Konca, M., Yetim, B., İlgün, G.: Effect of economic crisis on suicide cases: an ARDL bounds testing approach. Int. J. Soc. Psychiatry 66(1), 34–40 (2020). https://doi.org/10.1177/0020764019879946
2. Andrés, A.R., Halicioglu, F.: Determinants of suicides in Denmark: evidence from time series data. Health Policy (Amsterdam, Neth.) 98(2–3), 263–269 (2010). https://doi.org/10.1016/j.healthpol.2010.06.023
3. Baumbach, A., Gulis, G.: Impact of financial crisis on selected health outcomes in Europe. Eur. J. Pub. Health 24(3), 399–403 (2014). https://doi.org/10.1093/eurpub/cku042
4. Bergeron, A.J.: The Economics of Suicide: An Empirical Study (2014)
5. Doran, C.M., Kinchin, I.: Economic and epidemiological impact of youth suicide in countries with the highest human development index. PLoS ONE 15(5), e0232940 (2020)
6. Pandey, M.K., Kaur, C.: Investigating suicidal trend and its economic determinants: evidence from India (2009)
7. Swain, P.K., Tripathy, M.R., Priyadarshini, S., Acharya, S.K.: Forecasting suicide rates in India: an empirical exposition. PLoS ONE 16(7), e0255342 (2021). https://doi.org/10.1371/journal.pone.0255342
8. Matsubayashi, T., Sekijima, K., Ueda, M.: Government spending, recession, and suicide: evidence from Japan. BMC Public Health 20, 243 (2020). https://doi.org/10.1186/s12889-020-8264-1
9. Madianos, M.G., Alexiou, T., Patelakis, A., Economou, M.: Suicide, unemployment and other socioeconomic factors: evidence from the economic crisis in Greece. Eur. J. Psychiatry 28(1), 39–49 (2014)
10. Flavin, P., Radcliff, B.: Public policies and suicide rates in the American states. Soc. Indic. Res. 90(2), 195–209 (2009)
11. Jungeilges, J., Kirchgässner, G.: Economic welfare, civil liberty, and suicide: an empirical investigation. J. Socio-Econ. 31(3), 215–231 (2002)
12. Oyesanya, M., Lopez-Morinigo, J., Dutta, R.: Systematic review of suicide in economic recession. World J. Psychiatry 5(2), 243–254 (2015). https://doi.org/10.5498/wjp.v5.i2.243
13. National Crime Records Bureau, Ministry of Home Affairs, Govt. of India.: Accidental deaths and suicides in India (2009). https://ncrb.gov.in/sites/default/files/ADSI2009-full-report.pdf
14. National Crime Records Bureau, Ministry of Home Affairs, Govt. of India.: Accidental deaths and suicides in India (2021). https://ncrb.gov.in/sites/default/files/ADSI-2021/ADSI_2021_FULL_REPORT.pdf
15. Jalles, J.T., Andresen, M.A.: The social and economic determinants of suicide in Canadian provinces. Heal. Econ. Rev. 5, 1 (2015). https://doi.org/10.1186/s13561-015-0041-y
16. Warr, P.: Job loss, unemployment and psychological well-being. Role Transitions 263–285 (1984). https://doi.org/10.1007/978-1-4613-2697-7_19
17. Suzuki, T.: Economic modelling of suicide under income uncertainty: for better understanding of middle-aged suicide. Aust. Econ. Pap. 47(3), 296–310 (2008)

Dynamics of Sustainable Economic Growth in Emerging Middle Power Economies: Does Institutional Quality Matter?

Mithilesh Phadke, Jerold Raj, Sujay Rao, Shahid Bashir, and Jibrael Jos

Abstract The present study investigates the relevance of Institutional structures' quality as a determinant of the GDP of the Emerging Middle Power Economies (MIKTA) which constitute predominantly middle-income countries, namely Mexico, South Korea, Indonesia, Turkey, and Australia over the timeframe of 1985–2016. In addition to institutional variables such as Government Stability, Bureaucratic Quality and Socioeconomic Conditions, the study uses productive factors (per worker capital, human capital) and a macroeconomic indicator (inflation) to show the GDP of the above-mentioned countries. The impact that institutional variables taken have on Efficient Environmental resources, Sustainability and their management has shown to have an impact on the rate of growth of the middle-income economies. To estimate a long-run relation, the study employs the Autoregressive Distributed Lag model, also known as the ARDL model, bringing in controls for cointegration, nonstationary, heterogeneity and cross-sectional dependency and accounts for a mixed order of integration of variables. The model indicates that capital per worker, socio-economic conditions, bureaucratic quality, human capital and inflation have a long-run effect on the GDP of a country. The paper concludes with a positive impact of institutional variables during both, the short-run and the long-run, for the de-pendent variable.

Keywords MIKTA · Institutional variables · Macroeconomic variables · ARDL · Second generation test · SDG

1 Introduction

MIKTA are categorized as either newly industrialized or middle developed countries due to the institutions of the country adapting a progressive and developmental strategy, mostly from 1990s onwards. Such countries which are not naturally endowed with resources ought to look elsewhere when planning development.

M. Phadke (✉) · J. Raj · S. Rao · S. Bashir · J. Jos
CHRIST (Deemed to Be University), Bangalore, India
e-mail: mphadke3@gmail.com

© The Author(s), under exclusive license to Springer Nature Switzerland AG 2024
A. Hamdan and E. S. Aldhaen (eds.), *Artificial Intelligence and Transforming Digital Marketing*, Studies in Systems, Decision and Control 487,
https://doi.org/10.1007/978-3-031-35828-9_63

Middle power countries aim to strengthen their administrative and bureaucratic quality and effectiveness to improve their standing among larger regional powers. These specific countries do not possess hegemonic capabilities, however, have been instrumental in formulating regional and international developmental capabilities. MIKTA economies take inspiration following UN's sustainable development goals like quality of education and Global partnership for sustainable development [1]. One of the other important point is due to their effective administrative capabilities. These capabilities have proved to be instrumental in them becoming regionally capable due to which they have been rewarded with superior economic growth potential and have thus far ranked higher than most on UNCTAD [2].

On the down side, the same multilateral institutions fail to effectively offer collective leadership to address international situations [3]. Political Instability has been credited as a cause of economic development deterioration. The impact of institutional variables such as Government Stability and socioeconomic conditions which encompass Political Stability of a nation on variables such as inflation which can potentially be mutually affected [4]. With regards to, Ravenhill [5], certain middle-power countries meet the criteria, accounting for Capacity, Coalition Building, Concentration, Creativity and for Credibility which was attributed as the "Five C's". They have since its inception, addressed issues on economics, security and cybersecurity, climate change and maritime security, among others. Mehmetcik [6] identified a "Three Staged Potential" where he had identified that first, International economic cooperation with key middle powered countries, secondly, Economic Security linkages while securing ports for trade, and lastly, Traditional International security to promote smaller countries to operate in maritime and air trade.

Economic growth, while being investigated for in the long run accounts for the correction mechanism in the disequilibrium which is visible in the short run. This disequilibrium can be attributed for, due to the small variations in the economic variables across countries will have a long-term effect. As shown by endogenous growth models in [7], we can see that the concept of convergence has not been heeded, while prompting decades long studies to confirm (or deny) the effect of convergence in the long-run.

We have employed the panel ARDL (Auto-Regressive Distributed Lag) method to investigate both long-run and short-run relationship that the institutional and certain selected macroeconomic indicators share with the growth rate of the country's output. To encompass data regarding institutional factors, time series data of Government Stability, Bureaucratic efficiency and Socio-Economic conditions has been used for an instrumental recording of how stability of the bureaucracy leads to the growth of the economy while improving the standing of middle, regional powers in their respective regions.

2 Literature Review

In recent times, a significant amount of research has been done on cross-country growth all of whom have demonstrated a two-way growth relationship when countries who share similar characteristics when it comes to economic strength, population size and land mass [8]. Using these characteristics, MIKTA has, several times been attributed to the group of middle powers in their respective regions [3]. By following the Mankiw, Romer and Weil (hereinafter M-R-W), which had used the so called "textbook Solow model" which presented the Cobb-Douglass production function with Technical Progress [8]. It operated on the assumption that the countries that were referred to were in their steady states. However, their analysis was primarily to understand how the per capita income is affected by differing savings and the labor force growth rates. The model was proven to be moderately successful in justifying varying incomes. Its shortcoming was however, that it failed to account for the abnormally high output per capital which they yielded [9]. Also, their assumption that the savings rate and the growth rate of a population are independent of any country specific factors which have affected the production function.

The literature accounts for the several variables that were considered by M-R-W which are considered effectual for the growth in the skill level of the population [9]. One of the important SDG goals considered for economic growth is quality of education. Hence, the Human Capital factor of the country was kept dynamic over-time and was found to be cross-sectionally dependent on the labor quality growth in other countries in the panel. An innovative solution employed by them was to consider the foregone labor wages on the part of students in a country as an invest-ment in education. The concept behind this being that a worker earning a low wage foregoes a low salary for the purpose of developing his/her individual Human capital while a person earning more would have to forego more. Capital per capita becomes partially questionable if the skills does not exist to use them. To correct this, this study incorporates both, level of capital per laborer (capital per capita) and the level of human capital for all these countries.

Literature justifies that growth in the countries, especially in so called newly industrialized countries such as those of Mexico and Indonesia experience conver-gence. Indirect economic losses from instability in the considered institutional vari-ables help with the explanation of any dip in economic growth at any given time [10]. Malik and Masood [10] investigates primarily, the phenomenon of conditional convergence and its occurrence as per the output per worker in the MENA, which are the Middle Eastern and North African countries, all the while analyzing the impact of factors, namely human capital and physical capital per worker on the rate of growth of the economy. The econometric approach used my Malik and Masood [10] accounts for Cointegration, Heterogeneity and for Cross-Sectional dependency in variables considered and in the residuals. The contribution in this paper is primarily for the development of the method used in the investigation of the phenomenon of the convergence process.

With regards to the institutional effectiveness indicators and their impact with respect to the countries that were taken my Malik and Masood [10], states that a deterioration of the institutional environmental indicators has become one of the biggest deterrents towards achieving developmental goals of developing countries. A weak political structure has in literature shown that while complimented with a lack of political awareness has a detrimental impact on the growth of the country in consideration. A country which has weak institutional quality would indicate that large scale reforms in conformation with the SDG goals specified for the international community would lead to an upward trend in the growth of the economy.

3 Empirics

3.1 Data Description

We used data for South Korea, Indonesia, Mexico, Australia and Turkey from 1984 to 2016 for our panel study. The data used in this study has two dimensions— economic and institutional variables. For the economic variables, we have taken Inflation, capital stock, human capital, population, and GDP from Penn World table, and socio-economic conditions, government stability and bureaucratic quality which were taken from ICRG (international country risk guidelines).

We calculated GDP per capita as the ratio of GDP to population. The ratio of the capital stock to the population is an independent variable for capital per capita. Human capital and inflation are the other two economic independent variables. Institutional variable such as government stability, which is made up of three subcomponents: legislative power, public support, and government cohesiveness. Socioeconomic conditions are the aggregate of three subcomponents—Unemployment, poverty and consumer confidence [11]. Government Stability and Socio-economic conditions have a similar index which states that score "0" to indicate very high risk and "12" indicate very low risk [11]. Another shock-absorbing mechanism that tends to reduce policy modifications when governments change is the bureaucracy's institutional strength and quality. As a result, nations with strong bureaucracy quality that can run their governments with minimal policy changes and service disruptions receive high marks. Countries lacking the insulating effect of a strong bureaucracy obtain low ratings because a change in leadership is often unpleasant in terms of strategy planning and ongoing administrative work.

3.2 Model Specification

We develop an econometric model as follows, where GDPPC is the dependent variable (GDP per capita).

GDPPC = f (Capital per-capita, Bureaucratic quality, Socioeconomic conditions, Government stability, Inflation, Human Capital)

$$lnGDPPC_{it} = \beta_0 + \beta_1 lnKAP_{it} + \beta_2 BQ_{it} + \beta_3 SOC_{it}$$
$$+ \beta_4 GOV_{it} + \beta_5 INF_{it} + \beta_6 HUM_{it} + \varepsilon_{it}$$

Each country's subscript is 1 ..., N, while the time period is 1 ..., T, n the aforementioned model, the explanatory variables KAP, BQ, SOC, GOV, INF and HUM stand in for Capital per capita, Bureaucratic quality, Socioeconomic Condition, Government Stability, Inflation and Human Capital respectively. The dependent variable GDPPC in the above equation represents the country's GDP per-capita.

3.3 Methodology

It is critical to test cross-sectional dependency tests and homogeneity tests before proceeding for the model selection and analysis. Using Pesaran CD test, the cross-sectional dependency test is computed [12]. For heterogeneity we have used delta tilde test. These findings prompt us to proceed with the second-generation unit root test based on the findings of the investigation. We use CIPS and CADF tests for the panel variables to check order of integration [13]. After getting cross-sectional dependence and heterogeneity, the cointegration test of Kao is used to examine the relationship between economic growth and long-term growth [14]. We next move on to panel ARDL because the cointegration tests showed a long-run relationship between the economic and institutional variables in each country. An intermediate econometric PMG estimator, which forces the similarity of long-term parameters while allowing the short-term coefficients to differ among the countries, was used to estimate the long-run and short-run coefficients as well as causal links between the variables [15]. MG and PMG allow for more variability to obtain country-specific coefficients than standard panel approaches do [16]. As previously mentioned, this estimator allows the short-term estimates of the model, the variance of the error term and intercepts to alter, keeping the long-term parameters constant across different countries. Since it might be used to synchronously generate long-run and short-run estimates, given that series is of mixed order of I(1) or I(0). We have also used the dynamic fixed effect (DFE) model for panel data analysis [15] and the Hausman test to check the efficient models between PMG and DFE.

Table 1 Homogeneity test

	Delta	p-value
	10.753	0.000
adj.	12.416	0.000

Table 2 Cross-sectional dependence test

Test	Statistic	d.f.	p-value
Breusch-Pagan LM	161.462	10	0.000
Pesaran scaled LM	33.867		0.000
Pesaran CD	−2.443		0.014

4 Empirical Findings

4.1 Heterogeneity and Cross-Sectional Dependence Test

Table 1 represents the results of the homogeneity test. Using the calculated values of the delta tilde (D) and adjusted delta tilde (adjD) and their associated p-values (as shown below), we strongly reject the null hypothesis that the slope coefficients are homogenous at 1% level of significance. This suggests that there is heterogeneity for all the variables that have been analysed in different nation groups, hence heterogeneous panel methods that use different parameters for each cross-section within the panels must be used.

Table 2 reports the outcomes of the Cross-Sectional Dependency tests that were used. By comparing the CD test statistics and their corresponding p-values, we reject the null hypothesis of cross-sectional independence. This suggests that our panel data set is cross-sectionally dependent.

4.2 Panel Unit Root Test

We use second-generation unit root tests CADF and (CIPS) suggested by Peasaran. Table 3 results for CIPS reports that every variable is stationary at level except GDP per capita and human capital but they become stationary after first differencing. For all the variables, the 10% level of significance allows for the rejection of the null hypothesis for the unit root. Similarly, for the CADF test we observe that variables that are non-stationary at level, but become stationary after taking their first difference. At a 1% level of significance, we proceed to reject the unit root null hypothesis for all variables except for human capital. We can conclude that all the individual series are stationary at I(0) or I(1) order of integration. The variables in the table can be

Dynamics of Sustainable Economic Growth in Emerging Middle Power ...

Table 3 Panel unit root tests

Variables	Level	First difference	Inference
CIPS			
lnGDP	−0.548	−4.305***	I(1)
lnKAP	−2.472**	−	I(0)
INF	−4.197***	−	I(0)
SOEC	−3.561***	−	I(0)
HUM	−0.548	−0.827*	I(1)
GOV	−3.351***	−	I(0)
BQ	−6.065***	−	I(0)
CADF			
lnGDP	−1.709	−2.868**	I(1)
lnKAP	−1.737	−2.601***	I(1)
INF	−2.957***	−	I(0)
SOEC	−3.214***	−	I(0)
HUM	−0.549	−1.260**	−
GOV	−3.307***	−	I(0)
BQ	−2.92***	−	I(0)

Here *** indicates 1%, ** indicates 5% and * indicates 10% level of significance

tested for cointegration since they combine I(1) and I(0) processes. The PMG co-integration model developed by Pesaran, Shin, and Smith takes both I(0) and I(1) variables into account while estimating the model's parameters (1999).

4.3 Cointegration Test

The Kao panel cointegration test results are reported in Table 4. By taking GDP per capita (GDPPC) as the response variable, the test statistic demonstrates that all the variables are cointegrated, accordingly in line with their probability values for various panels of nation groups since the hypothesis of "No Cointegration" is rejected at the 1% level of significance. We can thus infer a long-term link between the variables under discussion.

Table 4 Cointegration test

Kao cointegration test	Statistic	p-value
Unadjusted modified Dickey–Fuller t	−2.649	0.004
Unadjusted Dickey–Fuller t	−2.004	0.022

Table 5 PMG—long-run estimates

Variables	Coefficient	Std. Error	z	p-value
lnKAP	0.432	0.027	15.75	0.000
INF	−0.000	0.000	−2.62	0.009
SOEC	0.005	0.002	2.32	0.021
GOV	−0.003	0.002	−1.60	0.11
HUM	0.274	0.046	5.90	0.000
BQ	0.028	0.007	3.76	0.000

4.4 Panel ARDL Estimates

It is crucial for us to estimate the long-run and the short-run estimates using the Pooled Mean Group (PMG) estimator after validating that the variables are cointegrated across all country groups. Tables 5 and 6 summarizes the important conclusions from Panel ARDL estimates. A further in-depth examination follows.

The results of the PMG estimate for a panel consisting of all the MIKTA nations in the long run are reported in Table 5. Bureaucratic quality, human capital and capital per capita have a positive impact on the economic growth for all countries with 1% level of significance in the long-run. Inflation, accordingly, has a negative impact with a 1% level of significance on GDP. Socio-economic conditions also have a positive impact on GDP with a 5% level of significance. Our results demonstrate that Government stability is insignificant in explaining long run economic growth.

The panel's short-run analysis shows that the speed at which the adjustment parameter operates is −0.39 and significant at the 1% level, corresponding to 39% of the disequilibrium from the previous year. The outcomes of the PMG estimate differ significantly from the long-run coefficients when it comes to short-run analysis. Even at the 10% threshold of significance, all of the estimated coefficients are statistically insignificant, with the exception of capital per capita, which has a short-run positive impact on GDP. From Table 6 we can also interpret that capital per capita, inflation, and human capital are significant at 5% level of significance for Mexico. All the variables except the bureaucratic quality and government stability are significant at 5% level of significance. For Turkey capital per capita and human capital are significant at 5%. Lastly, only capital per capita is significant at 5% level of significance for Australia. For all the MIKTA economies in the short run, error correction and constants are significant at 5% level of significance [17].

For panel data analysis, we additionally developed a dynamic fixed effect model to account for the Cross-Sectional dependency and heterogeneity. The panel's short-run analysis shows that the speed of the adjustment parameter (0.15) is negative and significant at the 1% level, indicating that 15% of the disequilibrium from the previous year is adjusted this year. With a 5% threshold of significance, capital per capita and political stability have a favorable short-term effect on economic growth. Along with this, inflation and human capital have a negative short-term impact with

Dynamics of Sustainable Economic Growth in Emerging Middle Power ... 769

Table 6 PMG—short-run estimates

Variables	Coefficient	Std. error	z	p-value
Republic of Korea				
ECT	−0.138	0.073	−1.88	0.060
C	0.584	0.328	1.78	0.075
$\Delta \ln KAP_{t-1}$	0.004	0.348	0.01	0.990
ΔINF_{t-1}	−0.000	0.002	−0.33	0.741
ΔSEC_{t-1}	−0.000	0.006	−0.14	0.885
ΔHUM_{t-1}	0.534	1.319	0.41	0.685
ΔGOV_{t-1}	−0.004	0.004	−1.06	0.289
ΔBQ_{t-1}	0.079	0.052	1.52	0.130
Mexico				
ECT	−0.723	0.095	−7.54	0.000
C	2.823	0.381	7.39	0.000
$\Delta \ln KAP_{t-1}$	3.212	0.500	6.41	0.000
ΔINF_{t-1}	0.000	0.000	2.49	0.013
$\Delta SOEC_{t-1}$	0.004	0.002	1.48	0.140
ΔHUM_{t-1}	4.848	1.212	4.00	0.000
ΔGOV_{t-1}	0.003	0.002	1.29	0.197
ΔBQ_{t-1}	−0.012	0.012	−0.94	0.346
Australia				
ECT	−0.058	0.020	−2.88	0.004
C	0.238	0.087	2.73	0.006
$\Delta \ln KAP_{t-1}$	2.244	0.476	4.71	0.000
ΔINF_{t-1}	0.000	0.001	−0.04	0.972
$\Delta SOEC_{t-1}$	0.002	0.002	0.91	0.361
ΔHUM_{t-1}	0.071	0.177	0.40	0.686
ΔGOV_{t-1}	−0.004	0.001	−0.25	0.803
ΔBQ_{t-1}	−0.042	0.029	−1.42	0.156
Turkey				
ECT	−0.754	0.081	−9.25	0.000
C	3.088	0.354	8.72	0.000
$\Delta \ln KAP_{t-1}$	2.094	0.211	9.88	0.000
ΔINF_{t-1}	−0.001	0.000	−0.37	0.712
$\Delta SOEC_{t-1}$	0.002	0.003	0.82	0.412
ΔHUM_{t-1}	3.535	0.679	5.21	0.000
ΔGOV_{t-1}	0.002	0.002	1.07	0.284
ΔBQ_{t-1}	−0.009	0.015	−0.57	0.568

(continued)

Table 6 (continued)

Variables	Coefficient	Std. error	z	p-value
Indonesia				
ECT	−0.288	0.084	−3.42	0.001
C	1.022	0.297	3.44	0.001
$\Delta \ln KAP_{t-1}$	1.408	0.205	6.85	0.000
ΔINF_{t-1}	−0.001	0.000	−3.05	0.002
$\Delta SOEC_{t-1}$	0.008	0.002	2.93	0.003
ΔHUM_{t-1}	−1.242	0.338	−3.67	0.000
ΔGOV_{t-1}	−0.002	0.003	−0.57	0.569
ΔBQ_{t-1}	−0.014	0.010	−1.42	0.155

Table 7 Hausman test

	PMG model	DFE model	Difference	Std. error
lnKAP	0.432	0.574	−0.141	5.836
INF	−0.000	−0.000	0.000	0.031
SOEC	0.005	0.002	0.002	0.532
HUM	−0.003	−0.014	0.011	0.445
GOV	0.274	0.495	−0.221	9.886
BQ	0.028	−0.061	0.089	1.591

5% level of relevance. We report from the Hausman test that probability is greater than chi square statistic and therefore conclude that the PMG model is better and more efficient than DFE [18]. Table 7 shows us the individual impact of independent variables on each cross-section with their statistical significance (Table 7).

5 Conclusion

The study examined the impact of institutional factors over the growth of an economy. The results demonstrate that, in addition to being integrated in a mixed order, the variables in the study also exhibit cross-sectional correlation. Moreover, Kao's cointegration test shows that the output per worker, government stability, socioeconomic conditions, and bureaucratic quality tent to share a long run relationship. Since its inception, MIKTA relies on their institutional flexibility as opposed to competing on the world stage with their economic strength. Several countries which match the features of the middle-income countries considered in the panel have shown improvement in terms of sustainable development with an increase in the quality of the institutions that operate the nation's state economy. As shown by M-R-W models, human capital contributes positively to the long-term economic growth. However, the difference between papers which consider human capital arise due to the way human capital is defined. Strict revitalization of the economic structure with adherence to the SDG goals for the purpose of sustainable development and the preservation of the environment would be as mentioned by literature lead to a positive growth trajectory of the country. In this paper, we have adapted the Penn World specification. A sufficient investment in education, specifically the quaternary sector, strong institutional structures and any other component that has a direct influence on labor productivity is necessary to support long-term growth of a country, which in turn gives rise to the contribution of human capital.

References

1. Azam, M., Hunjra, A.I., Bourie, E., Tan, Y., Al-Faryan, M.A.S.: Impact of institutional quality on sustainable development. J. Environ. Mange. 113–465 (2021)
2. World Investment Report 2022.: International Tax Reforms and Sustainable Investment. Retrieved 23 Dec, 2022, from UNCTAD: https://unctad.org/system/files/official-document/wir2022_en.p (28 Feb, 2022)
3. Schiavon, J.A., Dominguez, D.: Mexico, Indonesia, South Korea, Turkey, and Australia (MIKTA): middle, regional, and constructive powers providing global governance. Retrieved 23 Dec, 2022, from Wiley Online Library: https://onlinelibrary.wiley.com/doi/full/10.1002/app 5.148
4. Elbahnasawy, N., Ellis, M.A.: Inflation and the structure of economic and political systems. Struct. Change Econ. Dinamics 60, 59–74 (2022). Retrieved from https://www.sciencedirect.com/science/article/abs/pii/S0954349X2100151
5. Ravenhill, J.: Cycles of middle power activism: constraint and choice in Australian and Canadian foreign policies. Aust. J. Int. Aff. 52(3), 309–327 (1998)
6. Mehmetcik, H.: Review—MIKTA, Middle Powers, and New Dynamics of Global Governance. Retrieved from E-International Relations. https://www.e-ir.info/2015/12/09/review-mikta-mid dle-powers-and-new-dynamics-of-global-governance/ (2014)
7. Romer, P.M.: Endogenous technological change. J. Polit. Econ. 98(5, Part 2), S71-S102 (1990)
8. Islam, N.: Growth empirics: a panel data approach. Q. J. Econ. 1127–1170 (1995)
9. Mankiw, G.N., Romer, D., David, W.N.: A contribution to the empirics of economic growth. Q. J. Econ. 107(2), 407–437 (1992). https://doi.org/10.2307/2118477
10. Malik, M.A., Masood, T.: Dynamics of output growth and convergence. J. Knowl. Econ. (2019)
11. Howell, L.: International Country Risk Guide Methodology. PRS Group (2011)
12. Pesaran, M.H.: General Diagnostic Tests for Cross Section Dependence in Panels, Institute for the Study of Labor (IZA). Discussion Paper, 1240 (2004)
13. Pesaran, H.M.: A simple panel unit root test in the presence of cross-section dependency. J. Appl. Econ. 265–312 (2007)
14. Kao.: Spurious regression and residual-based tests for cointegration in panel data. J. Econ. 1–44 (1999)
15. Pesaran, H.M., Shin, Y., Smith, P.: Pooled mean group estimation of dynamic heterogeneous panels. J. Am. Stat. Assoc 621–632 (1999)
16. Pesaran, H., Smith, R.: Estimating long run relationships from dynamic heterogeneous panels. J. Econ. 79–113 (1995)
17. Pesaran, H.M.: Estimation and inference in large heterogeneous panels with a multifactor error structure. Econometrica 967–1012. Retrieved from Econometrica 74 (n.d.)
18. Hausman.: Specification tests in econometrics. Econometrica 1251–1271 (1978)

Land Market in Ukraine: Functioning During the Military State and Its Development Trends

Reznik Nadiia ⑩, Havryliuk Yuliia ⑩, Yakymovska Anna ⑩, Halahur Yulia ⑩, Klymenko Lidiia ⑩, Zhmudenko Viktoriia ⑩, and Ischenko Valeriy ⑩

Abstract The article examines the development trends of the agricultural land market in Ukraine under the influence of the military situation. The main results of the survey on land market introduction in Ukraine are given, and its problems are highlighted. The main stages of the market of agricultural land are defined and described. The most important arguments of land market organization for the attraction of new investors are listed. It focuses on the relevance of foreign experience in Ukraine on the market of agricultural land. The necessity of reforming the agricultural land market in Ukraine is justified.

Keywords Agricultural land market · Sustainable use of land resources · Management · Agricultural land market · Rational land use · Land relations · Management · Moratorium

---

R. Nadiia (✉)
Economic Sciences, Department of Management, National University of Life and Environmental Sciences of Ukraine, Kyiv, Ukraine
e-mail: nadya-reznik@ukr.net

H. Yuliia
Marketing and International Trade Department, National University of Life and Environmental Sciences of Ukraine, Kyiv, Ukraine

Y. Anna
Department of Economics, National University of Life and Environmental Sciences of Ukraine, Kyiv, Ukraine
e-mail: anysy87@ukr.net

H. Yulia · K. Lidiia · Z. Viktoriia
Department of Management, Uman National University of Horticulture, Uman, Ukraine

I. Valeriy
Technical Sciences, National University of Life and Environmental Sciences of Ukraine, Kyiv, Ukraine

© The Author(s), under exclusive license to Springer Nature Switzerland AG 2024
A. Hamdan and E. S. Aldhaen (eds.), *Artificial Intelligence and Transforming Digital Marketing*, Studies in Systems, Decision and Control 487,
https://doi.org/10.1007/978-3-031-35828-9_64

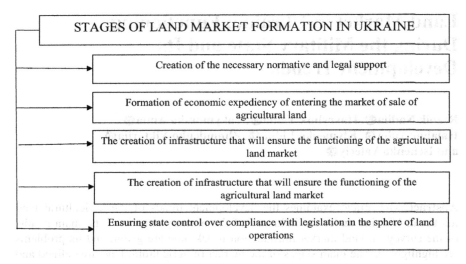

Fig. 1 Stages of the agricultural land market formation

The modern problem of the national economy is the formation of a civilized land market. Ukraine is allocated in the world as a country that has significant agricultural potential through the largest reserves of black earth. Land—this important component of productive forces, is still not integrated into the civilized market exchange [1].

The land market is a means of redistribution of land plots between owners and users through economic methods based on competitive demand and supply, which provides: purchase and sale, lease, mortgage or rights to land plots; determination of the value of land and recognition it as capital and good; distribution of space between competing variants of using land and market actors. The land market as a self-regulating system in the established legal environment consists of seven main elements: demand, supply, price, management, marketing, infrastructure, and business procedures [2].

The purpose of the article is to analyze the domestic land market during the military situation and to justify the need to develop the agricultural land market in Ukraine.

In Ukraine, as in any other country of the world, the land market performs three main tasks: ensuring the realization of the principle of effective use of land; forming the relationship to the land as a special value, contributing to the preservation of land and increase of their fertility; provides access to land for farmers, mainly those who can effectively manage, produce agricultural products necessary for the market [3].

The necessity of introducing a full-fledged land market is conditioned by the realities of the present. The evolutionary way of establishing and forming the land market has passed several stages (see Fig. 1) [4].

This year the land market had two phases: operation in peacetime and restoration of work within the limits of the military state. During the pre-war period, there was an increase in interest in investing in gardening and the country's support programs.

Agricultural producers, which have land ownership, have started to consider long-term objectives. In addition, interest in investing in dairy cattle has begun to grow, but unfortunately, the increase in these investments has not been realized. In other words, with the opening of the land market, there was a tendency to increase investments in long-term sectors of the agro-industrial complex [5].

In the first months of the full-scale invasion of the Russian Federation in Ukraine and the introduction of the military state, the land market completely stopped. Thus, access to the register of speech rights for real estate and the State Land Cadastre was closed, and prohibitions on the formation of land plots, free transfer, as well as bidding on the land for the agricultural purpose of state and communal property, etc. were introduced. Therefore, there was no legal possibility to perform transactions with real estate [6].

At the beginning of June this year the Law "on amendments to certain Legislative acts of Ukraine concerning the peculiarities of regulation of land relations in the conditions of military condition" No. 2247-IX, which regulates relations in the sphere of purchase/sale of land in military time, came into force. In particular, this regulatory document defines the updated procedure of implementation of the land, peculiarities of their use, etc.

At present, quite a lot of entrepreneurs are interested in buying sites for different purposes. According to Law No. 2247-IX, in the conditions of military condition, entrepreneurs have the right to buy land directly from the owner or on electronic auction [7].

Also, this year, the Land Code of Ukraine has been amended. In particular, the categories of state and communal areas that can be realized without holding auctions are defined:

- industrial sites used by enterprises performing relocation;
- land for non-residential construction, which is used for building terminals of sea ports and other objects critical for the economy;
- energy-related land used for power transmission lines and gas pipelines, etc.

All other state and communal property (except agricultural land) are still sold through electronic auctions of Prozorro. Sale system.

Changes in the land legislation were made to simplify the conditions of realization of agricultural, energy, and industrial sites, which are critical for the state in the conditions of war [8].

During the first year of the land market more than 88 thousand transactions of purchase and sale of land, a total cost of more than 7 billion hryvnias. But because of Russian aggression more than 51 thousand agreements on land purchase and sale did not take place, and landholders did not receive 4.2 billion UAH (see Fig. 2) [9].

With the resumption of access to state registers, in May 2022, the land market began to gradually revive. According to the Ministry of Agrarian Policy, for 9 months of 2022, more than 34 thousand agreements on land plots with a total area of more than 61 thousand hectares were concluded on the land market. The region with the most active land market this year can be recognized in the Khmelnytskyi region (6.2 thousand hectares). The greatest demand was also used land plots in the central regions of

Fig. 2 Functioning of the market in conditions of military condition

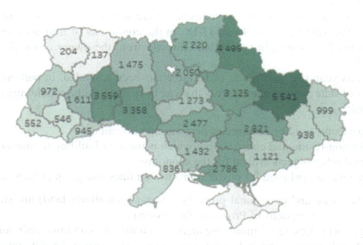

Fig. 3 The number of sales leads of land plots as of 10.10.2022 [9]

Ukraine: Vinnytsia (6.1 thousand hectares), Poltava (5.9 thousand hectares), Kirovograd (5.8 thousand hectares), and Dnipropetrovsk (5.2 thousand hectares) regions (see Fig. 3).

Small changes were also made to land prices. As of the end of December, the average value of the hectare of land was almost 37.9 thousand hryvnias per hectare, which is 30% more than in the same period last year. In other words, the average price of a hectare has somewhat increased in UAH compared to the "pre-war" time, but in the dollar equivalent—has decreased by 5% (see Fig. 4). The highest land

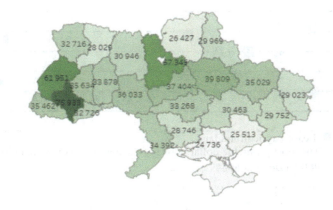

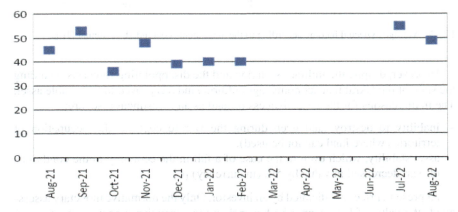

Fig. 4 Average price on the agreements concluded, thousand UAH/ha [10]

prices were recorded in Ivano-Frankivsk (91.7 thousand UAH). For ha), Kyiv (77.9 thousand UAH), and L'viv (70.4 thousand UAH) regions.

If to the Russia's massive invasion, investors were interested in land shares with average return on investment (the expected annual rent payment, divided by the purchase price) from 5%, then today, the real demand for investments in land is fluctuating in the area of 6.0–6.5% (see Fig. 5) [10].

However, it should be understood that neither the minimum nor the maximum price does not reflect the complete picture of the land market now because of the influence of many factors. First, there are very different price indicators depending on the location of the site, but now it is influenced not only by region or distance to large settlements but also proximity or distance of the front. Secondly, the ratio of demand to supply is changing actively. This is also related to the increased risks associated with the state of the land (especially in places close to the front line or liberated from occupation).

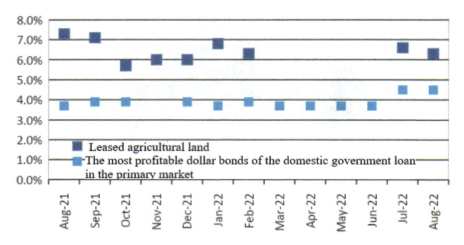

Fig. 5 Average expected income according to the agreements concluded, % annual [10]

However, despite the military situation and the disappointing forecasts regarding the state of the Ukrainian economy, agricultural land has proved to be a stable asset. The main reasons for the attractiveness of land as an investment asset are:

- inability to destroy land even during the war (except for the occupation of territories where land cannot be used);
- asset stability, which even in the case of a fall in the economy in the conditions of war, decreased in value by 5% (in currency) [6].

At present, prices are dictated by an investor. Only the normative monetary assessment (the value of land cannot be lower than the normative monetary assessment) keeps him from the aggressive price increase.

The new method of normative monetary valuation of land plots was approved in early November 2021. In particular, the value of land has begun to be determined as follows: the area of land subject to valuation, multiplied by a clearly defined standard of capitalized rent income (for example, for land with/g of purpose is 27.5 thousand UAH/ha) and five coefficients, which depend on location, purpose, and quality of land assets.

In practice, the transactions that occur on the terms of the regulatory monetary valuation give +10–20% to the price (on average 900 $/ha), if it is chernozem and the land is sown and processed [11].

But land reform is not limited to buying and selling land. The right to buy or sell agricultural land also encourages the development of crediting in Ukraine under the pledge of land [9].

Conclusion: The land market in Ukraine is necessary because it is a combination of social relations regarding alienation and acquisition of land. In the process of their market circulation, there is a competitive change of land owners or land users.

The opening of the land market was a complicated story of the fight against the myth, which was created for many years, that the land could not be sold.

The land market is a product of compromise. The land market is a product of compromise. From the beginning of his launch, he was accompanied by many restrictions. Restrictions were applied to entities, lands and types of objects that can be in owned.

The key obstacle to the market launch from the first days was a long period for checking and evaluating the purchasing power of the land buyer. But the Ministry of Agrarian Policy and the Ministry of Justice has managed to do so—the system has started to work and sales started. The development of the land market has stopped Russia's full-scale invasion of Ukraine.

There are a lot of unresolved issues in Ukraine at present. The entire economy of Ukraine is adjusted to new military conditions of existence. The agricultural land market, which started in July last year, has also been reoriented over the last year.

In military conditions:

- the number of sales transactions concluded has decreased in times (it is estimated that the number of transactions executed daily, compared to the pre-war period, has decreased by 5 times),
- there is a new driving factor in the demand and supply of land for agricultural purposes, especially in regions of the greatest risk of occupation.

In fact, from the beginning of the market the most sold land with the best soils. Mainly within the framework of Kharkiv, Poltava, Dnipropetrovsk, Kherson, Kirovograd, Vinnytsia, and Sumy regions. These territories were the most developed.

Also, the greatest activity is observed on the market of land, namely, not agricultural purposes. This confirms the map of land lots in the system transparently.

Market adaptation to changes and new rules of the game took place very quickly. Already in November 2021, there was a significant intensification of land agreements. And every second auction ended with an agreement. Even in wartime, the land market remains active. Thus, in the market of land rent in March, 40% of the trades were successfully held—the price growth was recorded by about 4 times. Now the situation, in our opinion, is normalizing. The percentage of successful trades is kept at 40%, which is already close to the pre-war level. The price growth at the same time at the auction—on average 3–3.5 times.

In the system of electronic auctions, there is a considerable activity of natural persons. And, although Ukrainian citizens are not usually using auctions even private persons began to use electronic auctions to find a fair market price for their assets.

Despite the war, the land market exists. And its achievements during the year are significant. This confirms the price of land, which has increased by 30%. So, after our victory over the Russian enemy, the Ukrainian land market will work with better results.

In our opinion, after winning over the Russian enemy, to attract more investors to the Ukrainian land market, it is necessary to do the following:

- complete the formation of the regulatory and legal base of the land market;

- to improve the accounting system and registration of land plots and ownership rights for them;
- assess land to create its real price;
- to create a body that will monitor and control the preservation of the quality of land and the conformity of the use of land for the purpose, to ensure the fulfillment of ecological requirements when using agricultural land;
- to simplify procedures for the registration of agreements with land plots.

References

1. Prutska, O.O.: Formuvannia rynku zemli v Ukraini. http://econjournal.vsau.org/files/pdfa/227.pdf. Accessed 10 Jan 2023
2. Biesieda, O.L.: Suchasnyi stan ta perspektyvy formuvannia rynku zemli v Ukraini. https://www.pdaa.edu.ua/sites/default/files/nppdaa/2011/v2i3/37.pdf. Accessed 12 Jan 2023
3. Krasovskyi, H.Y., Tolchevska, O.Y.: Rozroblennia tekhnolohii vyznachennia koefitsiientiv rozoranosti zemelnykh masyviv. Ekolohichna bezpeka ta pryrodokorystuvannia **15**, 111 (2014)
4. Yermakov, O.Y., Kravchenko, A.: V Rozvytok rehionalnoho rynku orendy silskohospodarskykh zemel. Ekonomika APK (6), 10–14 (2007)
5. Rynok zemli: za rik zhoden iz mifiv, yakykh boialysia ukraintsi, ne realizuvavsia. http://milkua.info/uk/post/rinok-zemli-za-rik-zoden-iz-mifiv-akih-boalisa-ukrainci-ne-realizuvavsa. Accessed 12 Jan 2023
6. Iak pratsiuie rynok zemli pid chas viiny? https://biz.ligazakon.net/analitycs/215874_yak-pratsyu-rinok-zeml-pd-chas-vyni. Accessed 14 Jan 2023
7. Zakon Ukrainy «Pro vnesennia zmin do deiakykh zakonodavchykh aktiv Ukrainy shchodo osoblyvostei rehuliuvannia zemelnykh vidnosyn v umovakh voiennoho stanu» No. 2247-IX vid 12.05.2022. https://zakon.rada.gov.ua/laws/show/2247-20#Text. Accessed 16 Jan 2023
8. Iak vidbuvaietsia prodazh zemli pid chas voiennoho stanu. https://sale.uub.com.ua/news/prodazh-zemli-u-vojennyj-chas-v-ukraqini. Accessed 19 Jan 2023
9. Rynok zemli v Ukraini: osnovni dosiahnennia ta provaly za rik. https://agropolit.com/spetsproekty/980-rinok-zemli-v-ukrayini-osnovni-dosyagnennya-ta-provali-za-rik. Accessed 15 Jan 2023
10. Chomu ukraintsi obyraiut investytsii v zemliu. https://www.epravda.com.ua/columns/2022/09/8/691256/. Accessed 14 Jan 2023
11. Shcho vidbuvaietsia z rynkom silskohospodarskoi zemli na shostomu misiatsi viiny. https://www.epravda.com.ua/columns/2022/08/17/690474/. Accessed 16 Jan 2023
12. Richnytsia rynku zemli: kozhni 5 khvylyn khtos kupuie dilianku. https://kurkul.com/spetsproekty/1335-richnitsya-rinku-zemli-kojni-5-hvilin-htos-kupuye-dilyanku. Accessed 19 Jan 2023
13. Reznik, N., Boshtan, A.: Features of execution of customs control in Ukraine during the war. Young Sci. **11**(111), 103–107 (2022). https://doi.org/10.32839/2304-5809/2022-11-111-22 [in Ukrainian]
14. Reznik, N., Hrechaniuk, L., Zahorodnia, A.: Analysis of the logistics component of the economic security system of enterprises. Int. J. Innovative Technol. Econ. **4**(36) (2021). https://doi.org/10.31435/rsglobal_ijite/30122021/7739
15. Reznik, N., Hrechaniuk, L.: Justification of measures to improve the efficiency of logistics system management. AIP Conf. Proc. **2413**, 040002 (2022). https://doi.org/10.1063/5.0090406
16. Reznik, N., Dolynskyi, S., Voloshchuk, N.: Retrospective analysis of basic risk as a part futures trading in Ukraine. Int. J. Sci. Technol. Res. (2020) Retrieved from http://www.scopus.com/inward/record.url?eid=2-s2.0-85078765334&partnerID=MN8TOARS

17. Reznik, N.P.: Logistics: heading guide/N.P. Reznik/National University of Life and Environmental Sciences of Ukraine. Kiev, p 146 (2021)

A Study on the Tendencies in the Trade Balance of the Indian Economy

K. Vinodha Devi and V. Raju

Abstract India has always been a hopeful trade destination for the rest of the world. Even before the formation of British rule in India, Is known for its exclusive and quality goods. Over time, trade became an essential part of the economy and after the opening up of the Indian economy after the enforcement of the New Economic Policy in 1991; India unlocked the doors to new openings for evolving its trade and for developing as a potential trade destination for the contemporary world. India has been able to attain its goal. Today India positions to be the eighth largest exporter of commercial services in the world by holding 3.7% of the global share. India's trade has been rising but the issue that India looks at is that the rate of imports has been more than its exports important to an enormous trade deficit in the economy. This paper aims to study the trade and trade in India during the period extending from 1999 to 2019. The study has been conducted to know more about the national and international motives behind the rise and fall of the trade balance that was observed by the Indian economy during the Indian study.

Keywords Exports · Imports · Trade balance · SDG · Trade deficit

1 Introduction

Trade has been an integral part of an economy since the age of civilization. Traditionally, it was a barter system, but with the introduction of money and currency, the valuation of the worth of the Product became simple. Different schools of financial thought have recognized the importance of trade and analyzed its different aspects. International trade has gained both economic and political importance over the years, helping nations expand their markets and maintain better relations with other nations.

K. V. Devi · V. Raju (✉)
Kristujayanti College (Autonomous), K Narayanapura, Bengaluru 560077, India
e-mail: raju.v@kristujayanti.com

© The Author(s), under exclusive license to Springer Nature Switzerland AG 2024
A. Hamdan and E. S. Aldhaen (eds.), *Artificial Intelligence and Transforming Digital Marketing*, Studies in Systems, Decision and Control 487,
https://doi.org/10.1007/978-3-031-35828-9_65

International trade has been a great source of income and growth for developing nations, especially through the opening up of the economy to the global market. This has led to the flow of FDI, making it the passion of the global economy.

International Trade brings benefits to the Nations involved which helps them to gain growth in their real income which further helps to increase the standard of living of their population. A few of the benefits can be jotted down:

- Promotes International Specialization
- Helps to widen the market and raise efficiency
- Helps to utilize higher growth potential
- Helps to attain the educative effect of trade through the expansion of technological knowledge
- Helps in Capital Formation
- Promotes import of Foreign Capital from developed to underdeveloped countries
- Promotes Healthy Competition among the countries of the world
- Helps underdeveloped nations to break the Vicious Circle of Poverty
- Helps in the well-organized use of means of production.

The two major types of trade that take place between two nations are exports and imports. Exports refer to the trade wherein the goods produced by a country nationally are sold to another nation while imports refer to a trade practice whereby a nation purchases goods or services from another nation. After the exports in a nation surpass its imports then the nation is supposed to have a satisfactory equilibrium of trade or trade excess. And when the imports are more than the exports then the nation is said to have an unfavorable balance of trade or trade deficit. Conventionally, having a trade surplus was measured to be favorable for a country as it certain an increase in a nation's wealth but later economists opinioned that it is not mandatory for a nation to uphold a constant trade surplus to insure a steady growth of an economy.

Although International trade has its own advantages, not every nation was honored enough to have trade relations with other nations of the world. Developing nations could not trade with the developed nations and the high tariff rates were not helpful either. These limitations gave birth to Free trade Agreements. Developing nations found it to be a great relief as they could form agreements among themselves which led to the united growth of the member nations. Over time, as trade developed it also led to animosity among some nations and these gave birth to tariff wars. Economists have argued that policies taken to reduce the trade deficit may lead to a reduction in the trade volume for a nation. Such policies may moderately restrict a nation from appreciating its comparative advantage.

Trade deficit suggests that there is the flow of money from the domestic country to a foreign country it might seem like a loss but economists have suggested that the money which flows from the home country to the foreign country comes back in the form of investment. Thus, it is not necessarily a loss. It only shows that with a proper flow of money, the international capital market is working efficiently. But on the other hand, it has also been noted that when trade deficits are created because of the huge government borrowings in the country which lacks stable and strong economic and political institutions then, a trade deficit may cause harm to the economy. But

A Study on the Tendencies in the Trade Balance of the Indian Economy 785

for developed nations like the US, which has been undergoing a trade deficit for the past few years, it may be a favorable phenomenon. On the other hand, although many nations prefer to enjoy a trade surplus on the grounds that it would increase their capital income which could be used for the residents and it would help in growing the standard of living of the people but that may not be the case. Constant urge to preserve a trade surplus might help the economy in the short run but in the long run, it marks the overall trade in the home country. To preserve the trade surplus and to protect their domestic businesses, the home country might resort to taking nationalist policies and would impose tariffs, quotas, or subsidies on the imports and this would affect the trade volume of the home country, would also lead to an increase in the prices and would affect the economic conditions of the residents. Thus, trade indeed is an important factor in an economy that could affect both the economic growth and development of a Nation.

2 Significance of the Study

International Trade has played an important role in promoting global growth and solving disputes. It has helped developing nations to enlarge their domestic, export-oriented initiatives and increase their national income, while advanced nations have gained in terms of achieving products and services at a lower cost due to their low labor costs. This study aims to study the influence of the trade balance on the economic growth of a country by studying the relationship between Trade balance and other macroeconomic variables.

3 Review of Literature

Reviewing the current literature is one of the most important and foreseeable parts of a research study. It gives a better consideration of the area under deliberation and helps to analyze the topic from various outlooks. Various research papers, articles, and a thesis were studied for getting a better understanding of the trade patterns in India and to know more about the status of GDP and employment rate in India.

Thahara et al. [1] in their study tried to infer the association between the exchange rate and trade balance and focused on the pragmatic evidence from Sri Lanka which has a deficit trade balance. The paper suggests that the association between trade balance and foreign exchange is determined by the forces of the demand and supply of foreign exchange. The authors have taken trade balance as the reliant on variable and exchange rate as the self-governing variable and GDP and inflation as the control variables to study the relationship. The study came to the conclusion that in the short run, inflation has a positive effect on the trade balance while the exchange rate has a negative impact on the trade balance in the long run and GDP has an adverse impact in the long run. For a country like Sri Lanka which is an import-dependent nation, the trade deficit is inevitable but devaluation might help in the extensive run to improve the trade balance in the country.

Panda and Baruah [2] in their report have mentioned the growing trade between India and China. The trade war between China and US has made India a potential trade partner for China. And the trade war has given China and India, both to evaluate their Free Trade Agreements (FTAs). India aims to expand its Foreign Investments, boost its FTA and liberalize their tariffs on the other hand China aims to shift its comprehensive trade away from the United States. In a way, trade was supposed to benefit both nations but the incrementing trade deficit for India is a barrier, and addressing the issue of trade deficit is crucial for both nations to expand their markets.

Reference [3] in his article focuses on the trends in India's international trade. With the economic reforms of 1991, India's trade witnessed positive trends. India's exports rose from 44.56 billion US$ in 2000–01 to 126.41 billion US$ in 2006–07. But the financial crisis of 2008–09 slowed the export growth in India but the setback was short-lived as India recovered from the impact of the crisis in two years. By 2011–21 India's exports grew seven times but the rate of growth of imports was higher than the exports resulting in a trade deficit of 193 billion US$. The difference in the export and import volume has been growing ever since. The author also points out that apart from the economic crisis other factors have impacted India's trade patterns. The policies of WTO, America's trade restrictive policies, FTA with South East Asian countries and East Asian countries, and Most Favored Nation Treatment have been major players in affecting global trade patterns.

Reference [4] in his article emphasizes the power balancing strategy between the major economies of the world: the US, China, and India. India is considered the weakest among the three nations that need both nations for its growth. The author remarks that India needs to adopt better policies to produce a higher growth rate, higher than 5%, and also needs to solve the issues of high joblessness, revolts, and fractured polities otherwise the power gap between India and China would widen and India would enter a period of greater strategic vulnerability. From India's point of view, its ties with the US would help it to rise as a main power and to strengthen its position in Asia and China is its principal strategic advisor. But India is also capable enough to adopt anti-Chinese policies if China threatens the security of the Indian subcontinent. Such a situation can lead to an alignment between India and US against China. Thus, balancing the power between the three nations is necessary to maintain global peace and security and to also ensure the economic growth of all.

Dasgupta and Chowdhary [5] have stated in their study that there was a considerable rise in the GDP of India during 2002–03 to 2007–08 and this growth was also accompanied by a sharp rise in the trade deficit in India. Though there was a fall in GDP after 2008–09 there was no fall in the trade deficit. They argued that the reason for the increasing trade deficit was due to the increasing Import-GDP ratio. The rate of growth in imports was greater than the GDP in India. From 2002–03 to 2011–13, the average Quantum index of Imports was 9.4 while the average growth rate of GDP was recorded as 7.3. This growth in the Import-GDP ratio was characterized by the changes in the consumption pattern by the elites in the economy as a result of inequality in the distribution of Income in India.

A Study on the Tendencies in the Trade Balance of the Indian Economy
787

Reference [6] in a research paper has discussed the fundamental association between exports and GDP in India and has given more importance to the impacts that the trade agreement had on the relationship between Exports and GDP. Through her research, she found out that although there might not be a significant long-run relationship between the two variables in the short run there is a two-way causality between the two variables. In the pre-trade agreement time period, the researcher found out that the GDP was triggered by the exports and had an adverse impact but statistically the impact was not significant in the post-trade agreement period there was a shift in the relationship wherein, exports were caused by GDP and had a negative relationship (when GDP increase exports fall) and this time the impact was very significant. Thus, the researcher concludes that trade agreements have a greater role to play in India as it affects the pattern of exports and growth in the country. This negative relationship is also the reason behind the rising trade deficit in India as the rising GDP encouraged increased imports into the country.

Reference [7] in his articles mentions the contraction in both exports and imports of India's merchandise trade since December 2014. As a result of these contractions, India's trade balance improved. The trade deficit declined as compared to the deficit in 2014 but it also affected the overall trade as not just the trade deficit but the trade in itself declined during that time period. The reason behind the decline in the deficit is pointed out as the decline in the oil prices in the global market and also decline in the import of gold items and also decline in the international gold prices during 2014–15. The researcher also points out that during the time period between December 2014 and May 2015, imports from developed nations and East Asian Developing Nations saw an increasing trend. Especially in the case of China, India's most important source of imports increased by 22.15% by May 2015. The Researcher opinions that though there is a decline in the trade deficit it is not a sign of recovery or revival for the Indian economy as the deterioration in the worldwide price of oil has a huge effect on the value of trade and this would lead to industrial stagnation in India in the long run.

Reference [8] in his article analyzes the trend in India's exports to the European Union and the United States after the financial crisis of 2008 in the economic year 2011. The author points out that although during the difficult period, many nations including China faced a fall in their exports, the value of India's exports was 42% higher in July 2011 compared to the previous year's statistics. On the other hand, China faced a heavy fall in its exports to the EU, from 31.9% in 2010 to 3.4% in 2011. In the case of the US also India was able to maintain a steady growth rate in its exports in relation to China and other developing nations. It is noted that the increase in India's exports was a result of certain structural factors. The reason for the growth in imports is considered to be the differences in the product composition of imports. In the case of India, the major products that the US imports are chemicals and related products, energy-related products, minerals, and metals, and these merchandises constitute a higher proportion of US imports than electronic products from China. A sudden fall in the demand for electronics led to fall in imports from China, while different product compositions of US Imports turned out to be favorable for India.

4 India—An Overview

India is a subcontinent in South Asia that is known for its diverse culture and rich traditions. It has faced both its ascent and failure, but has always stood up with greater willpower and strength. It was under the rule of its invaders for centuries, from the Turks to the Mughals and then the British, who exploited the Indian resources and discriminated against the Indian population. After years of discrimination, struggle, and revolt, India finally got its freedom on August 15, 1947, and on 26 January 1950, India's Constitution officially came into effect. India is governed by a democratic parliamentary system, but its economic state is vulnerable with a high population, high levels of poverty, and a GDP of 2.7 lakh crores.

The Planning Commission was established in 1950 to help India recover from poverty, inefficiency, low growth, cultural differences, and other social, political, and economic inadequacies. In 1951, the first five-year plan was enforced with an aim to develop the agricultural sector, and twelve more plans were introduced before its dissolution in 2014. The economic reform of 1991 paved the way for Liberalization, Privatization, and Globalization in India, which cleared the path for India to reach a global identity and be a great power player in the global market.

After 75 years of Independence, India has achieved the status of being one of the fastest emerging nations in the world and is a major hub for Information technology services. At present, India is the seventh-largest country by area and the second-most populated country in the world with a number standing at around 1.38 billion. It is also the major self-governing country in the world, sharing borders with Afghanistan and Pakistan to the northwest, China, Nepal, and Bhutan in the north, and Bangladesh and Myanmar to the east. India also shares a maritime border with Sri Lanka and the Maldives in the Indian Ocean. And within 10–15 years, India is expected to be one of the top three economic powers in the world.

5 Trade and Trade Balance in India

Trade is one of the important pillars of the Indian economy. Before the formation of the British Empire, India was known worldwide for its wealth. As the Towns in India were urbanized under the Mughals, they became a Centre for trade and industry. Through better roads and better market facilities, domestic trade was developed in India, which further led to the growth of international trade in India. After Independence, India became a protuberant player in the world market and after the inception of the New Economic Policy of 1991, its position in the global market was only strengthened. This change helped the Indian economy to emerge as one potential market for various goods and services and also as a potential hub for Foreign Direct Investment. By 2016, India became the eighth major exporter of commercial services in the world, recording a 5.7% growth in the trade of services in 2016–17. India's accounted for 3.4% of the total global trade of services in the world. India is

A Study on the Tendencies in the Trade Balance of the Indian Economy   789

Table 1 Top 5 export partners of India

Country name	Trade (US$ in millions)	Partner share %
United States	49,321	17.90
China	19,008	6.90
United Arab emirates	17,953	6.52
Hong Kong, China	9,537	3.46
Singapore	**8,295**	3.21

Source https://wits.worldbank.org

Table 2 Top 5 import partners of India

Country name	Trade (US$ in billions)	Partner share %
China	57.8	14.37
United States	30.5	7.57
United Arab emirates	25.8	6.39
Saudi Arabia	23	5.70
Iraq	22	4.61

Source https://wits.worldbank.org

a country that has witnessed a trade deficit over the past years and is still struggling with a cumulative trade deficit.

In India, imports have always been greater than exports which have led to a negative trade balance in the country. Although India, exports more products compared to the products that it imports there happens to be a huge difference in their ratio. In 2019, the number of products that India exported was 4442 whereas the number of products that India imported was 4356 but the value of its imports was higher than its exports. And the huge difference in the value of its exports and imports has created an issue for India as its leads to a huge trade deficit for the country (Tables 1 and 2).

India's trade deficit grew from 13.5 billion in 2011 to 155.63 billion in 2021. It shows how imports have increased in India; it shows how much India depends on its imports, which to some extent has also affected the domestic industries of the country. One of the biggest disadvantages of trade in India is that India indulges in the export of middle and incomplete products to the rest of the world and imports finished products. This aspect itself leads to a difference in the valuation of the products important to a huge gap in the trade balance of the country (Table 3).

The table above shows the various product groups and their share of exports and increases in 2021. The major top most exported products in India are consumer products with around 45% export share and the second most exported products are the intermediate products with around 31% export share. At the same time, the most imported product in India is Raw materials with around 33% import share and the second most imported product in India is transitional goods with around 31% import

Table 3 Export and import share of different product groups in India

Product group	Export product share (%)	Import product share (%)
Consumer goods	45.15	12.11
Intermediate goods	31.09	31.12
Capital goods	16.52	22.75
Chemicals	14.93	9.93
Fuels	13.78	31.88
Stone and glass	12.73	12.89
Machinery and electricals	11.2	19.85
Textiles and clothing	10.98	1.74
Metals	7.99	6.59
Transportation	7.75	2.86
Raw materials	7.16	33.57
Vegetable	5.39	3.57
Plastic and rubber	3.27	3.74
Animal	3.15	0.04
Food products	2.3	0.56
Miscellaneous	2.12	3.15
Minerals	1.58	1.15
Footwear	0.97	0.18
Hides and skins	0.95	0.22
Wood	0.93	1.66

Source https://wits.worldbank.org

share. One of the biggest challenges that India looks is that it exports more low-value commodities which it becomes difficult to raise the export income (Table 4).

The trade balance in India has grown over the past 21 years, but the imports have grown at a greater pace than the exports. This has been due to various domestic and global economic setbacks, such as the subprime crisis in 2009 and 2010. In 2015, the exports and imports in India saw a heavy decline due to the fall in crude oil prices and the depreciating Indian currency. However, in the later years, exports have increased compared to the numbers in 2015 and 2016, and imports also saw an exceptional increase (Fig. 1).

India has been able to sustain exceptionally better relations with the countries of the world which has been reflected in foreign trade of India. At present, India has 226 export partners and 210 import partners. Among them, India has been able to have a greater share of trade with the most technically advanced and developed nations of the world such as the United States, China Japan, and UAE. This has really assisted

Table 4 Total exports and imports of India

Year	Exports	Imports	Trade balance
2011	176,765,036.3	266,401,552.9	−89,636,516.57
2012	220,408,496	350,029,386.9	−129,620,890.9
2013	301,483,250.2	462,402,790.8	−160,919,540.6
2014	289,564,769.4	488,976,378.5	−199,411,609
2015	336,611,388.8	466,045,567.3	−129,434,178.6
2016	317,544,642.3	459,369,463.6	−141,824,821.3
2017	264,381,003.6	390,744,731.4	−126,363,727.8
2018	260,326,912.3	356,704,792.1	−96,377,879.77
2019	294,364,490.2	444,052,353.8	−149,687,863.7
2020	322,291,568.4	617,945,603.1	−295,654,034.6
2021	323,250,726.4	478,883,729.1	−155,633,002.7

Source https://wits.worldbank.org

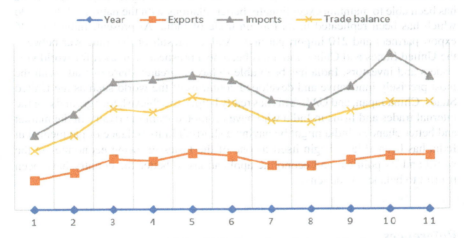

Fig. 1 India's trade balance from 2011 to 2021. *Source* https://wits.worldbank.org

to achieve better grounds for the domestic industries and firms in India who have endorsed the rivalry in the market and better times.

6 Conclusion

Over the past 23 years, India has skilled a lot of vacillations in its trade designs. Trade has become one of the significant sources of income for the Indian economy and a well-organized way to better the standard of living of the people living in India. By 2016, India became the eighth prime exporter of gainful facilities in the world, recording a 5.7% growth in the trade of services in 2016–17. India is also known to share 3.4% of the total global trade of services in the world. Although India has achieved great standards in the global market it still remains to be a country that has been witnessing a trade shortage over the past years and is still stressed with an increasing trade deficit.

India's trade deficit has grown from 13.5 billion in 1999 to 155.63 billion in 2019, showing how much it depends on imports. Despite India's devaluation policy to stimulate exports and control imports, imports grew at a greater rate. The high demand for crude oil and petroleum products in India makes it difficult to have a lesser import or to have a low value for its imports, and the rising price of crude oil in the global market has caused India to become a victim of the elevated prices.

Another difficulty that India faces is that the most imported products in India are Raw materials with around 33% import share and the second most imported product in India is transitional goods with around 31% import share. One of the biggest challenges that India faces is that its exports are more low-value commodities and it becomes difficult to raise the export income. But with a higher import ratio, India has been able to maintain exceptionally better relations with the nations of the world which has been replicated in its foreign trade of India. At present, India has 226 export partners and 210 import partners. And as a result of the trade war between the United States and China, India has become a prospective market for worldwide traders and investors. India has been able to have a greater share of trade with the most precisely innovative and developed countries of the world such as the United States, China Japan, and UAE. This has really helped to achieve better grounds for the internal trades and firms in India who have promoted from the rivalry in the market and better chances. India might be having a shortfall in its Balance of payments but India has been able to begin itself as one of the fastest-growing economies in the world and its political and economic applications to better the country have been fruitful to balance its deficits.

References

1. Fathima Thahara, A., Fathima Rinosha, K., Fathima Shifaniya, A.J.: The relationship between exchange rate and trade balance: empirical evidence from Sri Lanka. J. Asian Finan., Econ. Bus. **8**(5), 37–41 (2021). https://doi.org/10.13106/JAFEB.2021.VOL8.NO5.0037
2. Panda, J.P., Baruah, A.G.: Foreseeing India–China Relations: The 'Compromised Context' of Rapprochement. East-West Center (2019). http://www.jstor.org/stable/resrep21072
3. Seshadri, V.S.: India's international trade: trends and perspectives. Indian Foreign Aff. J. **12**(3), 181–201 (2017). http://www.jstor.org/stable/45341992
4. Malik, M.: Balancing act: the China-India-U.S. triangle. World Aff. **179**(1), 46–57 (2016). https://www.jstor.org/stable/26369496
5. Dasgupta, Z., Chowdhury, S.: Growth, imports, and inequality: explaining the persistently high trade deficit in India. Econ. Polit. Wkly. **50**(48), 65–74 (2015). http://www.jstor.org/stable/44002902
6. Ghoshal, I.: Trade-growth relationship in India in the pre and post-trade agreements regime. Procedia Econ. Finan. **30**, 254–264 (2015)

7. Mazumdar, S.: Recent trends in India's merchandise trade: warning signs for the economy. Econ. Polit. Wkly. **50**(33), 14–17 (2015). http://www.jstor.org/stable/24482396
8. Goldar, B.: India's exports at the time of the global crisis. Econ. Polit. Wkly. **47**(17), 15–17 (2012). http://www.jstor.org/stable/23214818

Quality Efficacy Issues in Mangoes: Decoding Retailers Supply Chain

Lakshmi Shetty ⓘ, Kavitha Desai ⓘ, and Shefali Srivastava ⓘ

Abstract This research article tries to uncover the elements and compelling reasons causing supply chain inefficacy concerning low quality at the retailer level of the mangoes supply chain in Karnataka. The descriptive research approach was used in work. The research was conducted in the biggest mango-producing areas of Karnataka. Factors were discovered by factor analysis. A systematic questionnaire was used to determine how much the mango sector may improve supply chain efficacy. Contingent on the factor analysis, four variables for low quality were identified: functional difficulties, knowledge, Manpower, and resources. It was also discovered that the functional component is the compelling factor causing supply chain inefficacy. The study is confined to the retailer level of the Mango supply chain, focusing on four Mango-producing districts in Karnataka. Furthermore, the measures for the key causes under each aspect causing hindrances in supply chain efficiency in terms of quality have been discovered. There is a scarcity of materials to enhance the supply chain efficiency of merchants in India's mango business. This research attempted to address a literature gap and help practitioners improve the mango supply chain in underdeveloped nations. This paper also serves the 2nd goal, Zero Hunger, End starvation, improve food security and nutrition, and promote sustainable agriculture of sustainable development.

Keywords Quality · Export · Mangoes · Efficacy · Supply chain · Manpower

---

L. Shetty (✉) · S. Srivastava
Christ Deemed To Be University, Bengaluru, India
e-mail: lakshmi.shetty@res.christuniversity.in

K. Desai
SVKM's Narsee Monjee Institute of Management Studies, Bengaluru, India

© The Author(s), under exclusive license to Springer Nature Switzerland AG 2024
A. Hamdan and E. S. Aldhaen (eds.), *Artificial Intelligence and Transforming Digital Marketing*, Studies in Systems, Decision and Control 487,
https://doi.org/10.1007/978-3-031-35828-9_66

1 Introduction

India is the world's foremost manufacturer of fruits. It grows a diverse array of fruits and generates an abundance of both. India is the world's 2nd leading fresh produce manufacturer, shadowing China. India is the world's leading producer of bananas, mangoes, and papaya, the 6th leading pineapple producer, and the 7th leading producer of apples. It is the world's leading producer of okra, the world's 2nd leading producer of brinjal, cabbage, cauliflower, onion, and potato.

"Supply chain management" was formerly known by other titles such as "operation management," "logistics," and so on. The company recognises that the supply chain is far larger and more sophisticated than operations and logistics. The supply chain spans the whole value chain, from raw materials to final items, via several intermediate production/processing and distribution phases.

The fruit supply chain is a subclass of the Agri-Mangoes supply chain (also known as the Perishable supply chain) and differs from the standard supply chain in several ways. India is the world's 2nd highest manufacturer of fruits. However, the supply chain's varied product supply base and various non-value-added middlemen make it less competitive.

Each element of the supply chain is desperate to increase efficiency. Other challenges/advantages may exist in other channels, adding to total efficiency. As a result, understanding the chances to improve the performance of the specific area, which will benefit the total supply chain, is critical (Fig. 1).

In India Fruit retailing is progressively being restructured, with standardized businesses reconsidering their supply chain strategy at every level from farm to plate. They are revamping their logistics and distribution tactics to boost the business and extend product shelf life.

In India, modern retailing is gaining traction. The tendency has accelerated in the fresh produce and grocery sector, which accounts for about 60% of overall retail sales [1]. Compared to other fresh produce categories, the modern retailing of fresh produce has been sluggish to catch up. The 1991 economic liberalization signaled the beginning of standardized merchants in fruit. The introduction of standardized retail players paved the path for improved Mangoes' standardized retailing.

This region grows the freshest fruits in the country, including mango, guava, potato, and others [2]. Some of the fruits and vegetables farmed in this region, particularly fruits such as mango, are important and regarded as among the highest fruit quality, consenting for the high requirement in other regions of the country. Such perishable and abundant fresh products should be processed or conserved for subsequent use in cold storage, which would otherwise result in substantial price variations [3]. Because of their short shelf life and high perishability, these goods require an appropriate transport and handling system to ensure that they arrive fresh from locally-sourced [4]. Fresh food is extensively farmed in India's emerging economies. However, inefficiencies in the supply chain result in low returns for farmers on the one hand and high retail costs for purchasers on the other due to insufficient fresh product quality at the point of sale [5–7]. The numerous inefficacies at all levels of the

Quality Efficacy Issues in Mangoes: Decoding Retailers Supply Chain

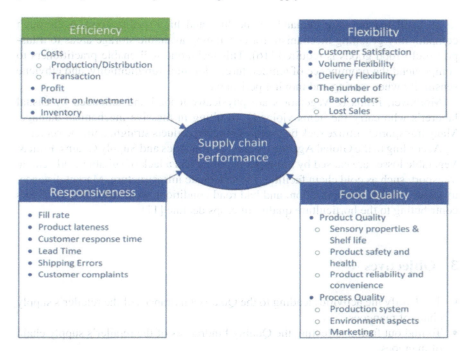

Fig. 1 Supply chain performance framework. Source: Lusine et al. [18]

Mangoes supply chain are obvious in the large volume of food losses and wastes, a serious problem for agribusinesses in developing nations [8–10]. Thus, supply chain efficiency is necessary for big or small- and medium-sized agricultural-business developing nations [11–14].

2 Review of Literature

The literature disclosed that Every nation is concerned about supply chain inefficacy caused by low quality, resulting in losses and wastage of fruits. The magnitude of losses in developing nations is greater than in wealthy countries. In that instance the fresh supply chain, a productive supply chain, plays a key role in dispensing the appropriate quality produce.

After harvesting, it takes around 24 h, sometimes longer, to get the fresh items to the ultimate retailer (typically an open-market stall or a pushcart vendor). Fresh produce is positioned in enormous cane bins or on truck flatbeds with no protection or main packing, exposing them at high temperatures under the sun and rapidly deteriorating fresh item quality. Feizi et al. [15] observed that the packing methods utilized significantly impacted post-harvest loss.

To fulfil the growing demand for healthy and high-quality food, a thorough compatibility grouping mechanism for consuming available storage areas to transport multi-food goods is required [16]. This technique will enable practitioners to comprehend the prominence of temperature, odor, moisture/humidity, and ethylene sensitivity when storing and moving perishables.

Moreover, Fruits & Vegetables are physically loaded and unloaded by casual laborer's who carry the items violently, resulting in massive mechanical damage. Many transporters utilize sack bags or just put the products straight onto the vehicles.

According to the Global Agenda Council on Logistics and Supply Chains, Fruit & Vegetable losses are caused by improper handling and a lack of suitable cold vehicle transport, such as cold chain facilities or inadequate infrastructure. Market distance, a lack of suitable transportation, and bad road conditions were identified as factors contributing to the horticulture quality of crops decline [17].

3 Objectives

- To identify the activities leading to the Quality hindrances of the retailer's supply chain of mangoes.
- To find out factors affecting the Quality hindrances of the retailer's supply chain of mangoes.

4 Hypothesis

$H0_1$: There is no relationship between the retailer's supply chain activities and the hindrances in the quality.

$H1_1$: There is a relationship between the retailer's supply chain activities and the hindrances in the quality.

$H0_2$: There is no relationship between the retailer's supply chain factors and the hindrances in the quality.

$H1_2$: There is no relationship between the retailer's supply chain factors and the hindrances in the quality.

5 Methodology

To achieve the study's aims, the researchers gathered data and information from primary and secondary sources and thoroughly analyzed them. The paper used a descriptive research method—the method employed for collecting the primary data in the form of a structured questionnaire. Karnataka state was used for the scope of this study. Five years (2016–2020) of production data for major state crops were

Quality Efficacy Issues in Mangoes: Decoding Retailers Supply Chain 799

collected from the Karnataka Horticulture department. Accordingly, the top three commonly used fruits (Mango, Banana and Papaya) were selected for the study. Once the fruits were selected, a multistage area sampling method was used, and each district was considered a sub-area. The district-wise distribution of samples for different channels of 110 samples from Kolar, Ramanagara, Tumkuru and Chikkaballapur convenience sampling method was used by the researcher. The Polit research was conducted on a small demographic group. To determine participants' willingness and interest in completing the questionnaire. Secondary sources of information are widely available and easily accessible to the general audience. The information for this study was gathered from books, journals, papers, periodicals, and websites. The reliability (0.852) and validity tests are used to identify measurement error; we employed the Cronbach alpha test to evaluate reliability. Face validity is used to verify validity. SPSS, Amos, and Excel were utilized to analyze the current study.

5.1 Objective 1: Identifying the Activities Leading to the Quality Hindrances of the Retailer's Supply Chain of Mangoes

The activities performed at the Retailers Level of the mangoes supply chain, as well as their total composite score, weightage mean scores, and rank orders, are depicted in the tables below. An investigation of the data in Table 1 reveals that the three key activities which lead to quality hindrances in the supply chain.

The top three activities of the retailer's supply chain of Mangoes which lead to the hindrances of the quality of mangoes are storing and ripening of Mangoes with a 4.50 weight mean score, Storing of Mangoes with a 4.06 and unloading activities with a 3.94.

Table 1 Activities leading to the quality hindrances of the retailer's supply chain—mango

Activities	N	TCS	%	WMS	RANK
Unloading	110	551	16.09	3.94	3
Storing of mangoes	110	568	16.59	4.06	2
Cleaning of mangoes	110	524	15.30	3.74	4
Sorting, grading and repacking mangoes	110	450	13.14	3.21	5
Storing and ripening of mangoes	110	630	18.40	4.50	1
Selling of mangoes	110	346	10.11	2.47	7
Total		3099	100		

5.2 Objective 2: Factors Affecting the Quality Hindrances of the Retailer's Supply Chain of Mangoes

The data obtained from the survey was analyzed using statistical techniques. The techniques employed include descriptive statistics, data analysis assumptions (such as normality and linearity), and quantitative data analysis utilizing Exploratory Factor Analysis (EFA).

Factor analysis was applied to investigate the interconnectedness and interdependence of the explanatory variables and to identify the significant aspects that contribute to supply chain inefficiencies. Factor analysis identifies underlying factors and explains the trend of correlation within a group of measured/observed variables.

5.3 Kaiser—Meyer—Olkin (KMO) and Bartlett's Test—Factor Analysis

This study used factor analysis to identify the factor based on their similarity value. The researchers used principal component analysis to cluster the variables in order to find the elements influencing the hindrances in mango quality in the retailer's supply chain efficacy. It is a data reduction strategy based on an item's volatility proportion. The value of KMO is.845, indicating that variation among all variables is effectively managed (Table 2).

The vital element contributing to inferior quality Mangoes at this level is functional issues, followed by basic facilities, resources (Knowledge, Techniques and Manpower) and Atmosphere. These four latent components are as follows summarized (Table 3).

The primary element contributing to low mango quality in the retailer supply chain is functional issues. This factor has eleven sub-variables with a total variation of 38.55%. The majority of the variables underlying this classification are functional concerns. The result suggests that functions during storing and ripening, storing of Mangoes, and unloading of Mangoes should be performed appropriately, resulting in inferior-quality of Mangoes.

The subsequent factor is Basic Facilities, which has a total variance of 10.33%. The result signifies that poor infrastructure facilities at the Retailers stage related to cold chain, ripening chamber facilities, and insufficient space for operations are

Table 2 KMO and Bartlett's test

Default model				
	Kaiser–Meyer–Olkin measure of sampling adequacy	Bartlett's test of sphericity		
		Approx. Chi-square	df	Sig
Final loading	0.845	6340.006	435	0.000

Quality Efficacy Issues in Mangoes: Decoding Retailers Supply Chain 801

Table 3 Summary-factors affecting the quality hindrances of the retailer's supply chain of mangoes

Factor name	Variables	Component						Variance (% of explained) eigenvalues
		1	2	3	4	5	6	
Functional	Inadequate and inconsistent grading, which impedes uniform ripening	0.966						38.55
	Unloading the cartons in an open location that will be subjected to direct sunlight/heat	0.956						
	Mangoes compressed owing to excessive weight in storage	0.951						
	Mangoes damaged by insect and fungal infestation during storage	0.947						
	Stockpiling items in expectation of increased costs during storing	0.939						
	Improper and excessive stacking during storage	0.916						
	Rough handling during unloading	0.899						
	Improper fruit placement during ripening	0.823						
	Mangoes compressed owing to excessive weight in storage	0.81						
	Inadequate Ripening Process Monitoring	0.782						
	Unloading the containers in an inefficient way	0.73						
Basic facilities	Non-availability of a cooling shelter at the storage location		0.885					10.331

(continued)

Table 3 (continued)

Factor name	Variables	Component						Variance (% of explained) eigenvalues
		1	2	3	4	5	6	
	Storage and ripening facilities are insufficient		0.809					
	Insufficient room for offloading operations		0.563					
	Inadequate use of cold storage facilities		0.528					
	Inadequate storage and temperature control facilities		0.404					
Atmosphere	Temperatures in the storage room are quite high			0.719				7.95
	Inadequate ventilation at the storage and ripening location			0.671				
	Inadequate ventilation at the storage location			0.663				
	Temperatures in the storage and ripening areas are high			0.483				
Knowledge	Inadequate and untimely supply of demand information in storage and ripening				0.949			7.081
	Lack of knowledge about market circumstances in storage				0.938			

(continued)

Quality Efficacy Issues in Mangoes: Decoding Retailers Supply Chain 803

Table 3 (continued)

| Factor name | Variables | Component | | | | | | Variance (% of explained) eigenvalues |
		1	2	3	4	5	6	
Techniques	Inadequate ripening equipment and process					0.897		6.181
	Calcium carbide is used in the ripening process					0.825		
Manpower	Inadequate availability of skilled manpower to unpack the containers						0.808	4.673

some of the main reasons under this category for poor mango quality in the supply chain.

The Atmosphere is the third element contributing to inferior quality Mangoes in the supply chain. It has a total deviation of 7.95%. The findings reveal that high temperatures and insufficient ventilation are the primary causes of poor mango quality in the retail supply chain. This factor's underlying difficulties are closely tied to the Atmosphere.

The fourth factor contributing to low-quality fresh food at this stage is the need for more knowledge. It has a variation of 7.08% and comprises two sub-variables. The results show that, because of a lack of knowledge about market demand, fresh product is stored in anticipation for a long period, deteriorating the mangoes' quality.

The technique is the sixth factor contributing to poor fresh product quality at this stage. It has a variation of 6.18% and is made up of two sub-variables. The results show that bad technique and a poor manner of the ripening process are the main areas of concern under this category, which leads to poor mango quality.

The fifth and last issue contributing to low quality in the mango supply chain is Manpower. It has a variation of 4.67% and only one sub-variable. The findings indicate that a lack of qualified Manpower to undertake mango operations in the merchants' supply chain contributes to the inferior Mango quality.

6 Conclusion

Since the economy of India is backed up by the agriculture sector. There is enormous potential for service sectors through numerous value addition. The establishment of an effectual supply chain will play a critical part in decreasing losses and wastages, increasing retailer's income, developing export channels, creating job opportunities for the local population, and improving the livelihood of retailers and other stakeholders, all of which will contribute to economic progress and allow India to transpire as a global player. This study reveals that the factors leading to this are Resources, Functional Charges, Infrastructure, Manpower Knowledge, Functional Issues, Imprudence, Manpower Availability, Quality Control System, Standardization, Connectivity, Manpower, Ambience, Preservation, Manpower Charges, Information, Techniques, Market Uncertainty, Rates & Charges, Transit Ease, Verification & Frisking, Geography, Technical Resources, and Resources/Transport Facilities are the elements causing supply chain inefficiencies in the retailing level of the mango supply chain. The stakeholders must address the identified variables appropriately to reduce the risk of loss and meet the customer's expectations by providing high-quality Mangoes. Retailing or selling is one of the most crucial operations in the supply chain, acting as a connection between demand and supply fulfilment.

Mini food parks should be established near the wholesale Mandi region. These food parks will assist vendors in keeping their things fresh while charging a reasonable rental fee. The trader can seek integration with other agencies, such as the hotel industry and new merchants, to sell the product. This connection will cut inventory-carrying costs and ensure that goods are delivered to clients on time. The National Informatics Agency, Horticulture Information Department and the government of India should develop a national/regional information networking infrastructure that would allow market information to be transmitted to all stakeholders in the supply chain.

References

1. Bhagat, N.: Ripe time for fresh. India Retailing (2014). Retrieved from http://www.indiaretailing.com/FoodGrocer/7/42/81/10474/Ripe-Time-for-Fresh
2. NHB.: Area and Production of Horticulture Crops: All India (2020). Retrieved from http://nhb.gov.in/statistics/State_Level/2018-19 (3rd Adv.Est_.) - Website.pdf
3. Kitinoja, L.: Use of cold chains for reducing food losses in developing countries (2013). La Pine, Oregon
4. Kiaya, V. (2014). *Post-Harvest Losses and Strategies to Reduce Them.* Retrieved from https://www.actioncontrelafaim.org/wpcontent/uploads/2018/01/technical_paper_phl__.pdf.
5. Narula, S.A.: Reinventing cold chain industry in India: need of the hour. Interview with Mr Sanjay Aggarwal. J. Agribusiness Developing Emerg. Econ. 1(2) (2011). https://doi.org/10.1108/jadee.2011.52401baa.001
6. Artiuch, P., Kornstein, S.: Sustainable Approaches to Reducing Food Waste in India. Massachusetts (2012)
7. Siddh, M.M., Soni, G., Jain, R., Sharma, M.K., Yadav, V.: Agri-fresh food supply chain quality (AFSCQ): a literature review. Ind. Manag. Data Syst. **117**(9), 2015–2044 (2017). https://doi.org/10.1108/IMDS-10-2016-0427
8. Gunasekera, D., Parsons, H., Smith, M.: Post-harvest loss reduction in Asia-Pacific developing economies. J. Agribusiness Developing Emerg. Econ. **7**(3), 303–317 (2017). https://doi.org/10.1108/JADEE-12-2015-0058

9. Negi, S., Anand, N.: Factors leading to supply chain inefficiency in agribusiness: evidence from Asia's largest wholesale market. Int. J. Value Chain Manage. **9**(3), 257 (2018). https://doi.org/10.1504/IJVCM.2018.093890
10. Negi, S., Anand, N.: Wholesalers perspectives on mango supply chain efficiency in India. J. Agribusiness Developing Emerg. Econ. **9**(2), 175–200 (2019). https://doi.org/10.1108/JADEE-02-2018-0032
11. Bhattarai, S., Lyne, M.C., Martin, S.K.: Assessing the performance of a supply chain for organic vegetables from a smallholder perspective. J. Agribusiness Developing Emerg. Econ. **3**(2), 101–118 (2013). https://doi.org/10.1108/JADEE-12-2012-0031
12. Odongo, W., Dora, M.K., Molnar, A., Ongeng, D., Gellynck, X.: Role of power in supply chain performance: evidence from agribusiness SMEs in Uganda. J. Agribusiness Developing Emerg. Econ. **7**(3), 339–354 (2017). https://doi.org/10.1108/JADEE-09-2016-0066
13. Battese, G.E., Nazli, H., Smale, M.: Factors influencing the productivity and efficiency of wheat farmers in Punjab, Pakistan. J. Agribusiness Developing Emerg. Econ. **7**(2), 82–98 (2017). https://doi.org/10.1108/JADEE-12-2013-0042
14. Routroy, S., Behera, A.: Agriculture supply chain. J. Agribusiness Developing Emerg. Econ. **7**(3), 275–302 (2017). https://doi.org/10.1108/JADEE-06-2016-0039
15. Feizi, H., Kaveh, H., Sahabi, H.: Impact of different packaging schemes and transport temperature on post-harvest losses and quality of tomato (Solanum lycopersicum L.). J. Agric. Sci. Technol. **22**(3), 801–814 (2020)
16. Bhatnagar, A., Vrat, P., Shankar, R.: Multi-criteria clustering analytics for agro-based perishables in cold-chain. J. Adv. Manage. Res. **16**(4), 563–593 (2019). https://doi.org/10.1108/JAMR-10-2018-0093
17. Kasso, M., Bekele, A.: Post-harvest loss and quality deterioration of horticultural crops in Dire Dawa Region, Ethiopia. J. Saudi Soc. Agric. Sci. **17**(1), 88–96 (2018). https://doi.org/10.1016/j.jssas.2016.01.005
18. Lusine, H.A., Alfons, G.J.M.O.L., Van der Vorst, J. G.A.J., Olaf, V. K.: Performance measurement in agri-food supply chains: a case study. Supply. Chain. Manag. Int. J. *12*(4), 304–315 (2007). https://doi.org/10.1108/13598540710759826

Behavioral Bias as an Instrumental Factor in Investment Decision-An Empirical Analysis

Aneesha K. Shaji and **V. R. Uma**

Abstract Investment decisions are always complex in nature. Investment assets are volatile in nature there are less volatile, medium volatile and high volatile investment assets in the financial market. In the current study how, the behavioral biases of the investors affecting their investment decisions in the less volatile asset classes is examined using an extensive survey method among the IT professionals in the Bangalore city. The relationship between the demographic variables and behavioral biases is tested. Also, a detailed study is conducted to examine the risk-taking behavior of the investors in the less volatile assets. There are basically three type of investors on the basis of their risk-taking behavior i.e. Risk seeking, Risk Neutral and Risk averse investors. Current study reveals that investors in the less volatile asset classes are very much cautious about the risk factor and therefore they are risk averse in nature.

Keywords Behavioral biases · Investment decisions · Demographic factors · Risk taking behaviors

1 Introduction

The conventional theories of finance postulate that investment choices are cautious, reasonable, and made after appropriate evaluation of several factors. However, contemporary ideas claim that investment choices are not entirely rational and are not influenced by any relevant factors. They are often inconsistent as well. According to behavioral choice theorists, decision making processes are subject to a variety of cognitive and psychological constraints. In addition, the complexity and unpredictability of the circumstances, as well as the imprecise and inadequate information within which judgments must be made, make decision-making a challenging task.

A. K. Shaji (✉) · V. R. Uma
Department of Commerce, CHRIST (Deemed to be University), Bangalore, Karnataka, India
e-mail: aneesha@kristujayanti.com

A. K. Shaji
Kristu Jayanti College (Autonomous), Bangalore, Karnataka, India

© The Author(s), under exclusive license to Springer Nature Switzerland AG 2024
A. Hamdan and E. S. Aldhaen (eds.), *Artificial Intelligence and Transforming Digital Marketing*, Studies in Systems, Decision and Control 487,
https://doi.org/10.1007/978-3-031-35828-9_67

This complicated, confusing, and unclear scenario forces decision-makers to make a multitude of dubious, and perhaps poor, choices [1]. Such judgments are often made under the impact of a variety of social and psychological processes. Additionally, a variety of biases impact the decision maker. Moreover, a variety of cognitive constraints and biases force the decision maker to use a number of shortcuts throughout the decision-making process.

2 Need and Importance of the Study

Behavioral biases in investment behaviour and human judgement are substantial. This research aims to analyse and characterise behavioural biases in IT workers' investment decision-making by reviewing relevant research articles, publications, and papers. It explores how behavioural finance has evolved into a distinct field of study. Articles are categorised by bias, year, and author. The study focuses on individual IT investors and identifies key biases. The study has practical implications for corporations, politicians, and security issuers so they may consider investors' interests before issuing securities. It may help investors understand behavioural biases, make better investing choices, and reduce risk. This study will assist researchers advance their studies. This research focuses on behavioural finance's underlying principles and evolving notions and hypotheses. The report encourages readers to identify ways to reduce bias in decision-making.

3 Statement of the Problem

Investment decisions of the investors are always depends on various factors. Behavioural Biases are the one among the factors which influence the investors to make their investment decisions. The present study focuses on the IT professionals in the Bangalore city. The main objective is to examine the differences and interactions among the various demographic variables on the over confidence bias of investors in the less volatile asset classes and to examine the risk taking behaviour of investors in less volatile asset classes. As per the literature reviews identified overconfidence bias is the most prevailing bias among the investors so only overconfidence bias is considered for the present study among various other behavioural bias factors.

4 Review of Literature

The significance of behavioral biases in financial decision-making is illustrated by behavioral finance. Behavioral biases are generally divided into two categories [2] 1. Heuristic driven biases and 2. Frame dependent biases.

Reference [3] when different financial practitioners' analyses data and make decisions using heuristics or rules of thumb. The overconfidence, anchoring, adjustment reinforcement learning, excessive optimism, and pessimism are the author's categories for the heuristic drive's biases. He also talks about "frame dependent biases," which are biases that are influenced by how financial professionals frame their numerous options when making decisions. This encompasses prejudices including herding, mental accounting, disposition effect, and restricted framing.

Reference [4] cognitive errors and emotional biases. Endowment, self-control, optimism, loss aversion, regret aversion, and status quo are examples of emotional biases. Overconfidence, representativeness, anchoring and adjustment, cognitive dissonance, availability, self-attribution, illusion of control, conservatism, ambiguity aversion, mental accounting, confirmation, hindsight, and framing are just a few of the names he used to categories cognitive errors.

4.1 Heuristic Driven Biases

Reference [5] heuristic driven biases while making the decisions under uncertainity people will not look into the statistical theory or any relevant data instead they will rely on certain heuristics which may results in a positive results or sometimes may lead to worst results. Reference [6] stated that biases such as over confidence, anchoring, excessive optimism and pessimism comes under the category of heuristic driven biases.

4.2 Overconfidence Biases

Overconfidence may be a strong and predictable bias that causes individuals to be overconfident in their knowledge and abilities and to neglect the risk associated with an investment. Investors with lower brokerage Trading accounts reportedly grow more self-assured and participate in excessive trading [7]. Reference [8] as a level of unwarranted confidence in one's natural reasoning, judgements, and psychological talents. Investors have a propensity to overrate their ability to predict events, the accuracy of their information, and their level of understanding.

4.3 Risk Perception of the Investors

Investors are frequently more careful in their choice of investments, and every investor has a rational side that seeks a greater return with the least amount of risk. However, in an efficient market, anomalous returns are not possible to accomplish. Although risk often has a wide range of linkages with various uses, it typically has a negative connotation, such as harm or loss or another undesirable action. Reference [9] contains a component of uncertainty and a possibility of loss. Reference [10] risk is the trade-off that every investor must make between the opportunity's potential for higher profits and the therefore higher risk that must be accepted. Even if various risk-related literature has varying definitions of risk, the term "risk" refers to circumstances when a decision is made whose consequences depend on the outcomes of future events with known probabilities [11]. Experts in strategic management define risk as an occurrence that might result from a certain decision or course of action. Risk is typically viewed as a manager's subjective estimate of the consequences for the organisation or for the manager personally. In order to describe the slope of any regression line, beta has come to be recognised as the most appropriate risk indicator. It shows how volatile a stock is in relation to a market benchmark, specifically [12].

5 Objectives

1. To examine the differences and interactions among the various demographic Variables on the over confidence bias of investors in the less volatile asset classes.
2. To examine the risk taking behaviour of investors in less volatile asset classes.

6 Hypotheses

H01: There is no significant difference between the Overconfidence bias and demographic variables of investors.

H02: There is significant difference between the Overconfidence bias and demographic variables of investors.

H03: There is no significant difference between the Overconfidence bias and risk taking behavior of investors.

H04: There is significant difference between the Overconfidence bias and risk taking behavior of investors.

7 Methodology

The exploratory research helped validate and identify the main characteristics and elements that are relevant during investment. This information is then put to use in order to discover the connections between the various variables. A thorough examination of the relevant literature is followed by conclusive research to verify and quantify the findings. The study used a descriptive research design as its final research strategy.

In this study, researchers examine the characteristics of IT employees. It also investigates the investment bias of IT employees in Karnataka, such as behavior of IT employees, how long diverse socioeconomic backgrounds are examined. Other elements that impact investment include behavioral bias towards investment.

Qualitative and quantitative research methodologies are also used in this study. There were many empirical studies, literature and personal experiences collected as part of the qualitative research approach. Statistical analysis and data quantification are both hallmarks of quantitative research. In the second step of research utilize the qualitative output of my exploratory study to design a questionnaire. Quantitative research is the next step. Cross-sectional surveys of respondents are conducted using the structured questionnaire, which is filled out by the respondents in this descriptive phase.

IT employees in Karnataka who have invested at least once in the recent year are included in the studies demographic. The research used a multi-stage stratified random sampling technique.

Stage 1: All branches of the selected IT companies were visited.

Stage 2: Data collected from the respondents irrespective of volatility in investment classes.

Stage 3: From the collected data 420 respondents investing in the less volatile assets were considered for the study.

Tools of analysis consists of One ANOVA and Pearson's correlation.

8 Results

8.1 Overconfidence

H_0 (1): There is no significant difference among opinion of Respondents regarding overconfidence with respect to gender (Table 1).

The aforementioned table shows that the mean overconfidence scores for men and women are, respectively, 3.6107 and 3.5905. The mean of male respondents was the highest, coming in at 3.6107. The table also reveals that 0.672 is the simultaneously meaningful value. The significance level is above the cutoff of 0.05. Therefore, the null hypothesis is accepted. This suggests that there are no appreciable differences between the respondents' views on gender-related overconfidence (Table 1).

Table 1 Overconfidence and gender

	Gender	Mean	Standard Deviation	F-value	Sig	Null hypothesis
Over confidence	Male	3.6107	0.81653	0.397	0.672	Accepted
	Female	3.5905	0.82852			
	Total	3.5843	0.79859			

Sources Field Survey
* Significant at 5%

H_0 (2): There is no significant difference among age with respect to overconfidence.

According to the above table, respondents between the ages of 31 and 40 had the highest mean, which was 3.6161, followed by respondents under 20 at 3.5818, respondents between the ages of 21 and 30, respondents between the ages of 41 and 50, and respondents over 50 at 3.5278. The table also reveals that the meaningful value is simultaneously 0.705. The significance level is above the cutoff of 0.05. The null hypothesis is accepted. Because of this, there is no obvious difference in overconfidence between age groups (Table 2).

H_0 (3): There is no significant difference among educational qualification of the Respondents with respect to overconfidence.

The respondents with a PG degree had the highest mean, which was 3.6316, according to the above table. They were followed by undergraduate students, who had a mean of 3.5924, PUC/+2 students, who had a mean of 3.5729, professionals, who had a mean of 3.5548, and other respondents, who had a mean of 3.5428. The table also reveals that the significant value is simultaneously 0.834. The significance level is above the cutoff of 0.05. Therefore, the null hypothesis is accepted. Therefore, it can be concluded that there is no appreciable difference in the respondents' levels of overconfidence depending on their educational backgrounds (Table 3).

H_0 (4): There is no significant difference among annual income with respect to overconfidence.

Table 2 Overconfidence and age

	Age	Mean	Standard deviation	F-value	Sig	Null hypothesis
Over confidence	Less than 20	3.5818	0.81704	0.350	0.705	Accepted
	21–30	3.5480	0.81129			
	31–40	3.6161	0.78043			
	41–50	3.5278	0.78591			
	Above 50	3.5142	0.76472			
	Total	3.5843	0.79859			

Sources Field Survey
* Significant at 5%

Behavioral Bias as an Instrumental Factor in Investment ... 813

Table 3 Overconfidence and educational qualification

	Educational qualification	Mean	Standard deviation	F-value	Sig	Null hypothesis
Over confidence	PG	3.6316	0.88290	0.182	0.834	Accepted
	UG	3.5924	0.79900			
	PUC/+2	3.5729	0.78581			
	Professional courses	3.5548	0.77931			
	Others	3.5428	0.76581			
	Total	3.5843	0.79859			

Sources Field Survey
* Significant at 5%

Table 4 Overconfidence and annual income

	Annual income	Mean	Standard deviation	F-value	Sig	Null hypothesis
Over confidence	Less than 5,00,000	3.5726	0.86113	0.350	0.705	Accepted
	5,00,001 to 10,00,000	3.5371	0.80117			
	10,00,001 to 15,00,000	3.6259	0.79132			
	15,00,001 and above	3.5167	0.78891			
	Total	3.5843	0.79859			

Sources Field Survey
* Significant at 5%

The above data shows that respondents with earnings between $10,00,001 and $15,00,000 had the highest mean, followed by less than $5,00,000 at 3.5726, 5,00,001–10,00,000 at 3.5371, and 15,00,001 and above at 3.5167. The table also reveals that the meaningful value is simultaneously 0.705. The significance level is above the cutoff of 0.05. The null hypothesis is accepted. As a result, there is no observable difference in overconfidence between yearly incomes (Table 4).

The significance level for a correlation is 0.05 for one-tailed tests. The connection is significant at the level of 0.01 (1-tailed).

In the table above, correlations based on feedback from 420 respondents are used to evaluate the connections between risk taking behaviour and overconfidence bias. The table shows the importance of the Pearson's correlation coefficients with alpha at the 0.05 and 0.01 levels. The investors make the decision. There is statistical significance for the components Bias ($r = 0.341$; $p = 0.001$, significant at 0.01 level). These examples show that bias has a significant impact on investment decisions

Table 5 Respondents opinion on risk profiling

Sl. No.	Items	M	SD
1	Those who know me well might characterize me as cautious	3.86	1.61
2	I feel confident about making investments in less volatile securities	3.04	1.61
3	Even if it results in lesser profits, I normally prefer to make safer investments	3.86	1.61
4	Normally, it takes me a long time to decide on an investment	3.04	1.61
5	The term "risk" makes me think of the concept of "opportunity"	3.86	1.61
6	In general, I favour safe investments over speculative ones	3.04	1.61
7	I can understand investment-related topics easily	3.86	1.61
8	I'm willing to take on big investing risk in exchange for sizable rewards	4.01	1.30
9	I have limited investment process experience	4.27	1.07
10	I frequently feel concerned about my investment choices	4.25	1.08
11	Rather than raise my savings rate, I'd like to gamble on higher risk investments	3.61	1.50
12	The erratic nature of investments worries me	4.23	1.08
13	The possibility of investments losing value	4.25	1.08
14	Shares' market price fluctuates constantly based on supply and demand	4.45	0.78
15	The chance of not being able to get a fair price when selling your investment	3.56	1.59
16	The risk of suffering a loss of funds if you only invest all of your money in one sort of investment	4.34	1.02
17	While investing money or earnings at a lower rate of return, there is a risk of financial loss	3.56	1.59
18	Over time, inflation reduces the purchasing power of money	3.86	1.61
19	The chance of living beyond your savings	3.74	1.39
20	The potential for financial loss when making foreign investments	3.57	0.65
21	The threat of bankruptcy brought on by an alteration in interest rates	3.92	1.05
Grand mean		**3.62**	

because the alternative hypotheses have been accepted and the null hypothesis has been rejected. The subsequent study makes use of a multiple regression equation, with the independent variables being behavioural traits and the dependent variable being investment choice.

Behavioral Bias as an Instrumental Factor in Investment ...

Table 6 Correlation between risk profile and average returns for past five year

Pearson's correlation		Average return for the past five (5) years
Those who know me well might characterize me as cautious	Pearson correlation	1
	Sig. (1-tailed)	
I feel confident about making investments in less volatile securities	Pearson correlation	0.341**
	Sig. (1-tailed)	**0.001**
Even if it results in lesser profits, I normally prefer to make safer investments	Pearson correlation	0.086
	Sig. (1-tailed)	0.222
Normally, it takes me a long time to decide on an investment	Pearson correlation	0.172
	Sig. (1-tailed)	0.063
The term "risk" makes me think of the concept of "opportunity"	Pearson correlation	0.123
	Sig. (1-tailed)	0.138
In general, I favour safe investments over speculative ones	Pearson correlation	0.164
	Sig. (1-tailed)	0.071
I can understand investment-related topics easily	Pearson correlation	0.074
	Sig. (1-tailed)	0.256
I'm willing to take on big investing risk in exchange for sizable rewards	Pearson correlation	0.226
	Sig. (1-tailed)	**0.021***
I have limited investment process experience	Pearson correlation	−0.068
	Sig. (1-tailed)	0.273
I frequently feel concerned about my investment choices	Pearson correlation	0.239
	Sig. (1-tailed)	**0.016***

(continued)

Table 6 (continued)

Pearson's correlation		Average return for the past five (5) years
Rather than raise my savings rate, I'd like to gamble on higher risk investments	Pearson correlation	0.066
	Sig. (1-tailed)	0.222
The erratic nature of investments worries me	Pearson correlation	0.172
	Sig. (1-tailed)	0.063
The possibility of investments losing value	Pearson correlation	0.123
	Sig. (1-tailed)	0.136
Shares' market price fluctuates constantly based on supply and demand	Pearson correlation	0.086
	Sig. (1-tailed)	0.101
The chance of not being able to get a fair price when selling your investment	Pearson correlation	0.171
	Sig. (1-tailed)	0.063
The risk of suffering a loss of funds if you only invest all of your money in one sort of investment	Pearson correlation	0.113
	Sig. (1-tailed)	0.138
While investing money or earnings at a lower rate of return, there is a risk of financial loss	Pearson correlation	0.066
	Sig. (1-tailed)	0.222
Over time, inflation reduces the purchasing power of money	Pearson correlation	0.172
	Sig. (1-tailed)	0.062
The chance of living beyond your savings	Pearson correlation	0.122
	Sig. (1-tailed)	0.126
The potential for financial loss when making foreign investments	Pearson correlation	0.178
	Sig. (1-tailed)	0.124

(continued)

Table 6 (continued)

Pearson's correlation		Average return for the past five (5) years
The threat of bankruptcy brought on by an alteration in interest rates	Pearson correlation	0.163
	Sig. (1-tailed)	0.082

9 Conclusion

Behavioral Biases have a significant impact on the risk taking behaviors of the investors. Due to the impact of overconfidence bias investment decisions of the investors in the less volatile assets are also changing. The present study revealed that there is no significant relationship between the demographic variables and overconfidence biases of the investors. Ultimately this behavioural biases will lead to the change in the risk taking behaviour of the investors and this results in the final investment decision making process. There are basically three type of investors on the basis of their risk taking behaviour i.e. Risk seeking, Risk Neutral and Risk averse investors. Current study reveals that investors in the less volatile asset classes are very much cautious about the risk factor and therefore they are risk averse in nature.

References

1. Barberis, N., Thaler, R.: A survey of behavioral finance. Handb. Econ. Finance **1**, 1053–1128 (2003). http://faculty.som.yale.edu/nicholasbarberis/ch18_6.pdf
2. Baker, M., Bradley, B., Wurgler, J.: Benchmarks as limits to arbitrage: Understanding the low-volatility anomaly. Financ. Anal. J. **67**(1), 40–54 (2011). https://www.cfapubs.org/doi/abs/10.2469/faj.v67.n1.4
3. Chang, C.C., Chao, C.H., Yeh, J.H.: The role of buy-side anchoring bias: evidence from the real estate market. Pacific-Basin Finance J. **38**, 34–58 (2016). https://www.tib.eu/en/search/id/BLSE%3ARN378411619/The-role-of-buy-side-anchoring-bias-Evidence-from/?tx_tibsearch_search%5Bsearchspace%5D=tn
4. Kahneman, D., Tversky, A.: Prospect theory: an analysis of decision under risk. In Handbook of the Fundamentals of Financial Decision Making: Part I, pp. 99–127 (2013). https://www.uzh.ch/cmsssl/suz/dam/jcr:00000000-64a0-5b1c-0000-00003b7ec704/10.05-kahneman-tversky-79.pdf
5. Shiller, R.J.: From efficient markets theory to behavioral finance. J. Econ. Perspect. **17**(1), 83–104 (2003). http://www.its.caltech.edu/~camerer/Ec101/EffMktsThryBehFin_Shiller_Win03.pdf
6. Shleifer, A.: Inefficient Markets: An Introduction to Behavioural Finance. OUP Oxford (2000). https://global.oup.com/academic/product/inefficient-markets-9780198292272?cc=id&lang=en&
7. Shefrin, H.: Beyond Greed and Fear: Understanding Behavioral Finance and the Psychology of Investing. Oxford University Press on Demand (2002). http://www.oxfordscholarship.com/view/10.1093/0195161211.001.0001/acprof-9780195161212

8. Singha, S., Sivarethinamohan, R.: Social inclusion, equality, leadership, and diversity to attain sustainable development goal 5 in the Indian banking industry. J. Int. Women's Stud. **23**(5), Article 9 (2022a). https://vc.bridgew.edu/jiws/vol23/iss5/9
9. Singha, R., Kanna, S.Y.: Tobacco farming, addiction, promotion of gender equality, well-being and monopoly of the Indian market. J. Int. Women's Stud. **23**(5), Article 11 (2022a). https://vc.bridgew.edu/jiws/vol23/iss5/11
10. Singha, R., Kanna, S.Y.: Women's empowerment, mindfulness, and role of women in eradicating alcohol and drug addiction from Indian society. J. Int. Women's Stud. **24**(7), Article 7 (2022b). https://vc.bridgew.edu/jiws/vol24/iss7/7
11. Lopes, L.L.: Between hope and fear: the psychology of risk. In: Advances in Experimental Social Psychology, vol. 20, pp. 255–295 (1987). https://doi.org/10.1016/S0065-2601(08)60416-5
12. Singha, R., Kanna, S.Y.: Physical abuse in the absence of Ubuntu. J. Int. Women's Stud. **24**(4), Article 9 (2022c). https://vc.bridgew.edu/jiws/vol24/iss4/9

Creating a Sustainable Electric Vehicle Revolution in India

Annie Stephen⊙ **and M. Vanlalahlimpuii**⊙

Abstract In shaping the future of mobility, and as India moves towards stricter emission targets working relentlessly in carving out a paradigm shift in the usage of electric vehicles for shaping 'cleaner India and building a sustainable future. The government has taken several steps with its focus on environmental stewardship and the result is that an increasing number of people are gravitating towards an energy efficient means of transport. Despite the rewarding effects of electric vehicles, the growth of Electric vehicles does not reveal a significant rise. A study has been made to understand the perception of the people towards electric vehicles.

Keywords Electric vehicles · Electric cars · Electric scooters · Sustainability

1 Introduction

The market for Electric Vehicles is marching at a rapid pace. Based on the Electric Vehicles volume data, the overall global share of electric vehicles stands at 8.3% that includes the battery electric vehicles [BEVs] as well as the Hybrid electric vehicles with plug in feature. The figures in the year 2021 ranged from 4.2% in the year 2020 with approximately 6.75 million electric vehicles hitting the roads. This reflects an overall increase of 108% in comparison to that of 2020. Electric Vehicles are gaining paramount importance and are also gaining attention across the globe as there are several benefits offered by the Electric vehicles. The use of Electric Vehicles aid in the reduction of emissions as well as depletion of the natural and scarce resources of the planet.

The situation is no different in India too with the Electric Vehicles treading into the markets and gradually becoming more and more popular. The market for Electric Vehicles is evolving fast. The approximate number of vehicles that were sold in the year 2021 draws close to 0.32 million and has gone up by 168% in comparison to that

---

A. Stephen (✉) · M. Vanlalahlimpuii
Kristu Jayanti College, Bengaluru, India
e-mail: annie@kristujayanti.com

© The Author(s), under exclusive license to Springer Nature Switzerland AG 2024
A. Hamdan and E. S. Aldhaen (eds.), *Artificial Intelligence and Transforming Digital Marketing*, Studies in Systems, Decision and Control 487,
https://doi.org/10.1007/978-3-031-35828-9_68

of previous years. The use of Electric vehicles can reduce your carbon footprint as it is more likely to eliminate tailpipe emissions. The impact on the environment that is caused by the charging of the electric vehicles can also be decreased by choosing renewable energy options for home electricity.

Fossil Fuels are limited and the use of those fuels are causing a destruction of the planet. Petrol and Diesel run vehicles produce toxic emissions that leave long lasting damages to the planet earth. The use of Electric vehicles can result in conversion of 60% of the electrical energy from the grid to power the wheels, but only 17% to 21% of the energy stored in the fuel can be converted by the electric vehicles. This results in a wastage of about 80%. Petrol or Diesel vehicles emit roughly more than 3 times more carbon di-oxide than the Electric Vehicles. Transport is the bare necessity of modern day life, and one cannot afford to go without it. However, if one has to consider the damage caused by the use of such vehicles, they are immense and irreversible. Therefore, the need of the hour is to look at other options that will negate the effect of such damage or at least cause significantly lesser damage than what it already been happening.

Electric vehicles operate on electric motors, replaced by the internal-combustion engine which has the ability to generate power by burning fuel and gases. The nation has no dearth of issues concerning global warming, natural resource depletion and environmental damages and the like. Therefore, Electric Vehicles have been the centre of discussion in all important discussions and has drawn a considerable attention owing to rising carbon footprint and other environmental impacts caused by fuel run vehicles.

In the year 2010, India made its first concrete decision to capitalise on Electric Vehicles. A 95 crore scheme was approved by the New & Renewable Energy Ministry for providing financial incentives for the manufacture of Electric vehicles in India. The scheme, provided incentives of up-to 20% on the prices of vehicles, which was later withdrawn in the year 2012. Later in the year 2013, another plan was introduced titled 'the National Electric Mobility Mission Plan (NEMMP) 2020 in order to encourage the production of electric vehicles to combat the issues faced by the nation especially environmental concerns However the plan did not materialise and at the advent of the Union Budget of 2015–16 the then Finance Minister initiated manufacture of electric vehicles with an initial outlay of Rs. 75 crore. In the year 2017, India was preparing to move to 100% usage of electric cars by the year 2030 however the plan subsequently was diluted to only 30%. During the year 2019 a FAME II scheme was rolled out in order to encourage a faster adoption of electric as well as hybrid vehicles with an outlay of Rs. 10,000 crore. And since then there has been no looking back. Electric two-wheeler registrations have hit an all-time high for 2022, touching close to 68,324 vehicles in the festival month of October (till October 30) this year, an increase of 29% over the last month. This shows that the people are convinced of the benefits of using Electric vehicles.

2 Types of Electric Vehicles

There are four types of electric vehicles available:

(1) **Battery run Electric Vehicles**: The electric vehicles that are completely run by the use of electricity and are believed to be more economically efficient in comparison to the Hybrid mode of vehicles.

(2) **Hybrid Electric Vehicles**: As the name suggests, hybrid electric vehicles are a combination of the traditional form of vehicles that run on petrol and diesel and also combine with the battery powered engines.

(3) **Plug-in Hybrid Electric Vehicles**: Such vehicles use yet again the internal combustion engines as well as battery that can be charged using a plug. This suggests that the battery can be charged using electricity.

(4) **Fuel Cell Electric Vehicles**: The chemical energy such as hydrogen is at work in such Fuel based Electric vehicles.

3 Indian Scenario

The fast facing developing automobile industry is deemed to be the 5th largest industry in the world and is expected to soon cross the benchmark and become the 3rd largest industry in the country by 2030. According to the India Energy Storage Alliance (IESA), this booming industry is likely to see a sharp growth and grow at a 36% CAGR. With an increase in population, the demand of these electric vehicles also increases subsequently and the dependence on non-convertible and non-renewable sources of energy like fuel is not the wisest or sustainable option as India is known to import about 80% of crude oil essentials. The Indian body, NITI Aayog largely aims at achieving roughly around Electric Vehicles sales penetration of over 70% by the end of the year 2030. Furthermore, it also goes well in line with the goal to achieve a net 0 carbon emission by the year 2050. The industry of Electric vehicles are showing an increasing trend and seem to flourish due to the government intervention and favourable policies and programmes set up by the government.

The state of Uttar Pradesh in India, depicts the highest share and a significant share of electric vehicle sales in India in the year 2021. This is followed by Karnataka state and then Tamil Nadu. UP achieved a sale of 66,704 units, Karnataka with 33,302 units and Tamil Nadu with 30,036 units.

The three-wheeler segment was dominated by Uttar Pradesh whereas the two wheeler segment was dominated by Karnataka and the four wheeler segment took the lead for Maharashtra. In the year 2021, the Indian Electric Vehicle market valued at USD 7025.56 million and the prediction is that it would touch USD 30,414.83 million by the end of the year 2027. The ill effects of Covid-19 is undeniable in the Electric Vehicles segment too as much as it impacted other industries prevailing in the market. Post Covid-19, the electric vehicles market does show a significant growth owing to the government initiatives and policies in this regard. It would now be incorrect to admit that the electric vehicle segment is in its nascent stages especially because the growth of this sector is exponential.

The world today is also growing in its approach towards environmental concerns. The rising concerns and issues caused by high emission vehicles, government intervention through strict rules and regulations, economic battery costs and the rising prices on fuel have all contributed to the increasing demand and preference for electric vehicles. In response to increasing pollution and damage caused to the environment, several cities in India have switched to the adoption of sustainable policies to fight against the perils that pose a danger to the environment. There is also a rising awareness on the benefits offered by the use of electric vehicles that causes a switch in the mind-set of the people and causes the people to opt for the purchase of electric vehicles. A start up from Taiwan has initiated a plan to bring about a '6-s' battery swapping stations for all the two-wheelers in Delhi. Foxconn Technology Group is also aiming at swapping of battery in India. The future of mobility seems to be offering a major shift. Electric Vehicles comes into the picture when India addresses climatic changes or the steep rise in fuel prices. With such an emphasis on nation challenges, the Electric Vehicles ecosystem will soon reach greater heights.

4 Government Role

The government has announced a variety of measure to promote the Electric Vehicles signet. Tax incentives were announced as well as increasing the number of charging infrastructure to cut down he concern of lack of charging ports which served as a deterrent for adoption of Electric Vehicles.

India had signed the Paris climate agreement along with 170 other nations during the year 2015. At Paris Climate Conference held in the year 2021, India had resolved a reduction in the carbon footprint by 35% by the year 2030. India has also taken a giant leap to pledging a 40% increase in the replacement of non-fossil fuel vehicles. The government has developed a policy framework for waste management and finding alternatives for lithium batteries. The government has also initiated the FAME Policy which stands for Faster Adoption & Manufacturing of Electric Vehicles in two phases. One which deals with the grants and the subsidies and financial support for research and development. The other phase including schemes to boost demand, create publicity, increasing charging points and reduced prices of electric vehicles.

The policies for Battery Waste management was set out in 2020. Another proposal for the application of a Production Linked Incentive (PLI) Scheme 'National Programme on Advanced Chemistry Cell (ACC) was also introduced. At the same time, Phased Manufacturing Program (PMP) was initiated to promote the manufacture of Electric vehicles. Customs duty too on imported Semi Knocked down and Completely Knocked Down conditions have been increased.

The growth of the Battery Electric Vehicle is also a notable feature in the market. The stricter regulations imposed on account of the rise of emissions and the rising demand for hassle free vehicles that do not cause damage to the environment are likely to increase the market share in the electric vehicle and battery segment. Apart from the other schemes, the government has also announced a scheme commonly termed as the battery swapping policy in the Union Budget of 2022–23 which will permit the depleted batteries to be exchanged for charged batteries, which will thereby increase the adoptability of electric vehicles by potential buyers.

As the nation moves into a newer economy, the automobile industry to set to keep pace with the technological development, growth and innovation. Even though, the road to electric vehicles faces a few potholes, it would not be wrong to say the Electric Vehicles are going to be the future of India.

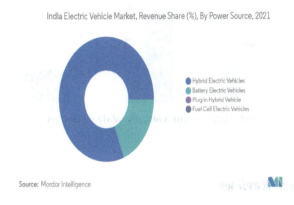

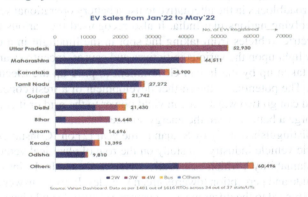

Firms that have helped boost Electric Vehicles Adoption in India are

1. Mahindra Electric
2. Tata Motors
3. Ashok Leyland

4. Ola
5. Hero Electric
6. Okinawa
7. Ather Energy
8. Revolt Motors
9. Tork Motors.

Source https://e-amrit.niti.gov.in/Manufacturers

4.1 Literature Reviews

Goel et al. [1], in her study, "A study on the barrier and challenges of electric vehicle in India and vehicle to grid optimisation" identified the most quintessential methods, obstacles and roadblocks in the alternative to use a battery operational vehicle in fast pacing and evolving nations like India. It also recognised the various reasons that prevented Electric Vehicles from taking the spot in the limelight in a country like India. It threw light upon the various initiatives and causes, incentives and subsidies provided and taken up by the Indian Government to pace up the Electric Vehicles drive in India. The potential of the further development of a fresh concept and idea of Vehicle grid can go two ways—it can yield power to the grid or it could possibly be used to charge a battery when the energy sources run out.

A similar distinguished study by Shajan in the year [2] on "Consumer perception towards electric vehicle industry—a study on the role of electrical vehicles in environmental sustainability", which essentially understands and looks into the varying customer attitude and perception towards green vehicles. It gives answers to pressing questions with regard to the differences in customers' attitude and views towards the same and factors that affect the levels of awareness and preferences of each customer to see who is most likely to opt for a more environment conscious and safe car in comparison to a normal fuel driven car.

"Consumer Perception towards e-vehicle in Vadodara city" by Professor Tushar Pradhan developed a study in May 2021 and it dwelled on understanding the perception of customers towards e-vehicles specifically in the city of Vadodara and the study stated that the customers are interested or would like to make a shift to e-vehicles but the thing that stops them from buying it, are the skyrocketed prices and hence will not invest in e-vehicles.

Indukala and Mathew [3] studied the future scope and prospects of the electrical vehicles in a country setting like that of India and discovered that there were many plus points and at the same time, found many negatives and several challenges of the same subsequently that hinder the growth and acceleration of electric mobility. Deployment of a reliable and developed infrastructure and addressing the issues with regard to range anxiety and charging time is the key to drive Electric Vehicles penetration in the country and is rather pivotal in the proliferation of E-mobility. Quick, cheap and easy access to developed and reliable infrastructure- both the standard AC and the fast DC charging will certainly play a key role in meeting customer needs.

Monika in [4, 5], in "A study of the preference of customers towards e-vehicles" clearly sets out on the aim to vividly capture the varying views, sentiments and perception on the awareness levels among the customers and probability of these individuals to buy the electronic vehicles so that environment sustainability can be kept up and maintained. It deducted the findings that Electric Vehicles and HEV have their own set of golden opportunities and subsequent obstacles and as a matter of fact it is just about the attitude and mind set of the customer and individual that it really plays a pivotal and key role in making a choice to indulge in an Electric Vehicles.

"A Study of Consumer Perception and Purchase Intention of Electric Vehicles" by Pretty Bhalla concluded that there are several factors that influence the buying behaviour or the purchase decision of the buyers of cars like cost factor, trustworthiness, brand image, individual views on the area of environmental issues, advanced technology used, societal acceptance, and infrastructure. The government needs to up its game and take the forefront and take up the leading role when it comes to creating and framing good environmental policies, development of infrastructure and subsidising the cost of these vehicles or reducing the bank interest rates on the purchase of these vehicles so as to promote the sales of electric vehicles.

4.2 Statement of the Problem

Electric vehicles market is growing at a rapid pace and is expected to move up to over 475 billion by the year 2025. The reach and penetration levels of the two wheelers in the Electric Vehicles segment is expected to touch around 15% by the end of the year 2025. However, despite the benefits offered by the Electric Vehicles there seem to be a lot or apprehensions as there are several challenges still surrounding this segment causing hiccups in deep penetration into the markets. One of the main concerns with the electric vehicles is the high prices which becomes one of the most important barriers for the latter's adoption in the market. The limited number of charging stations and the fear of going out of charge is one of the greatest fears surrounding the people. Although the electric vehicle market is catching up pace, the challenges and concerns facing the Electric Vehicle Market still outnumber the benefits. Therefore, an attempt has been made to understand the perception levels of the people towards electric vehicles and finding out the leading cause of not opting for the Electric Vehicles as a preferred buy, despite its advantages.

4.3 Objectives of the Study

- To find out the consumer perception towards electronic vehicles.
- To explore willingness to shift to Electric vehicles.
- To analyse the impact of demographic factors on the perception levels towards the purchase decision of electric vehicles.

4.4 Hypotheses of the Study

- H_{o1}: Age of the Indian consumer significantly impacts the perception levels towards electric vehicles.
- H_{o2}: Income of the Indian consumer significantly impacts the perception levels towards electric vehicles.
- H_{o3}: Education of the Indian consumer significantly impacts the perception levels towards electric vehicles.

4.5 Research Methodology

A. Data collection

The study combines both primary as well as secondary data. The primary form of data collection was done through the institution of a well-structured questionnaire on Google forms. The sources of Secondary data are primarily websites, journals, magazines, newspapers etc. **B.**

B. Sample size

The sample size used for the purpose of the study was 110. The scope of the study is limited to the city of Bengaluru and therefore the available sample size would be appropriate for drawing meaningful findings from the study.

C. Nature of questionnaire

Questionnaire consists of dichotomous, Likert scale, close ended and quantifiable questions. The education levels and the willingness to respond have been taken into account while deciding upon the complexity of questions in the questionnaire.

Creating a Sustainable Electric Vehicle Revolution in India

D. **Tools & technique used for the study**

SPSS tools have been used to analyse the data under study. ANOVA has been used to understand the relationship between the demographic factors and the dependent variables. Simple pie charts have also been used to analyse and interpret the data.

4.6 Scope of the Study

The main intent for the study is to address the necessity of shifting to Electric Vehicles. Despite the varied initiatives by the government to maximise widespread use of e-vehicles, the proposition still stands at a negligible state and the causes of which are unknown. The main intent of the paper is therefore to highlight the perception levels of the respondents with respect to owning electric vehicles and understand the reasons as to why the electric vehicles are not so popular despite the benefits attached to it.

5 Data Analysis and Interpretation

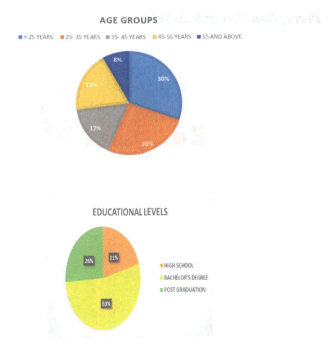

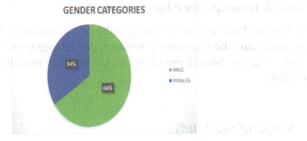

5.1 Perception Towards E-Vehicles

6 Factors Influencing Decision to Purchase E-Vehicles

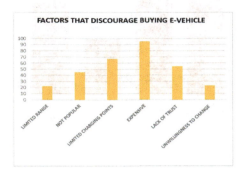

7 Factors that Discourage People from Buying E-Vehicles

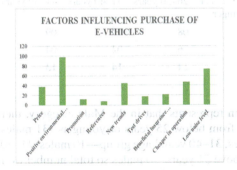

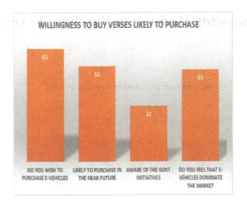

8 Cross Tabulation of Gender and Age

Gender * Age cross tabulation							
		Age					Total
		25 years or younger	26–30 years	31–40 years	41–50 years	51 years or older	
Gender	Male	31	08	10	09	00	**58**
	Female	38	06	04	04	00	**52**
Total		**69**	**14**	**14**	**13**	**00**	**110**

(1) Cross Tabulation represents that in the Male category, there are totally 5 age groups ranging from below 25 years or younger—31 males, 26–30 years' age group—08 males, 31–40 years' age group—10 males, 41–50 years age group—09 males and above 51 years—0 males so total number of males is **58**.

(2) Cross Tabulation represents that in the female category, there are totally 5 age groups ranging from below 25 years or younger—38 males, 26–30 years' age group—06 females, 31–40 years' age group—04 females, 41–50 years age group—04 females and above 51 years—0 females so total number of females is **52**.

Creating a Sustainable Electric Vehicle Revolution in India

9 Cross Tabulation of Gender and Income

Gender * Income cross tabulation

		Income						Total
		Nil	Less than 1 lac	1 lac–2 lac	2 lac–3 lac	3 lac–5 lac	More than 5 lac	
Gender	Male	00	02	01	46	02	07	**58**
	Female	00	00	01	48	01	02	**52**
Total		**00**	**02**	**02**	**94**	**03**	**09**	**110**

(1) Cross Tabulation represents that in the Male category, there are totally 06 income groups ranging from Nil Category—0 males, Less than 1lac income group—02 males, 1 lac–2 lac income group—01 male, 2 lac–3 lac income group—46 males, 3 lac–5 lac income group—02 males and More than 5 lac income group—07 males so total number of males is **58**.

(2) Cross Tabulation represents that in the female category, there are totally 06 income groups ranging from Nil Category—0 females, Less than 1 lac income group—00 females, 1 lac–2 lac income group—01 female, 2 lac–3 lac income group - 48 females, 3 lac–5 lac income group—01 females and More than 5 lac income group—02 females so total number of females is **52**.

10 Hypothesis Testing

- **H_{o1}**: Age of the Indian consumer significantly impacts their decision to purchase electric vehicles.
- **H_{o2}**: Income of the Indian consumer significantly impacts their decision whether or not to purchase electric vehicles.
- **H_{o3}**: Education of the Indian consumer significantly impacts the decision to purchase electric vehicles.

One way ANOVA

		Sum of squares	df	Mean square	F	Sig
Age	Between groups	22.021	2	11.010	10.770	0.000
	Within groups	110.412	108	1.022		
	Total	132.432	110			
Edu	Between groups	11.145	2	5.572	13.955	0.000
	Within groups	43.125	108	0.399		
	Total	54.270	110			

(continued)

(continued)

One way ANOVA

		Sum of squares	df	Mean square	F	Sig
Income	Between groups	3.180	2	1.590	3.745	0.027
	Within groups	45.847	108	0.425		
	Total	49.027	110			

11 Source SPSS 21

Based on the above table, as the p values are less than the significance level i.e. 0.05, the null hypothesis is rejected and the alternate hypothesis is accepted which means that there was statistically significant difference between groups. Income levels and the educational levels of the respondents have a significant impact on their willingness to purchase or not purchase electric vehicles.

12 Findings

1. Majority of the respondents' desire to purchase e-vehicles in the future despite few reservations about prices and the limiting factor of insufficient charging points.
2. The respondents also feel that they are more likely to purchase e- vehicles in the future as e-vehicles are going to dominate the market.
3. The government is also taking several initiatives to introduce e-vehicles and replace them with the traditional mode of transport as it reduces the carbon print and saves the environment.
4. The main factors influencing the people's decision of not purchasing the e-vehicles is that the vehicles are expensive in comparison to the other vehicles and there is also a massive shortage of charging points.
5. Owing to lack of trust in the sustainability of the e-vehicles, many respondents have reservations with respect to the purchase of the e-vehicles.
6. The demographic variables Age, Education and Income significant impact the purchase decision of the respondents.
7. The study of the perception of the respondents with respect to Electric vehicles revealed that majority of them felt that Electric Vehicles are expensive however it offers the convenience of driving and charging from home.
8. Many respondents also feel that the usage of Electric Vehicles will help reduce the carbon footprint because there will be zero tailpipe emissions.

Creating a Sustainable Electric Vehicle Revolution in India 833

9. Another factor that worries the respondent in investing in the Electric Vehicles is that there are a limited number of charging points and people fear that they may lose charge while driving.

13 Suggestions

The study mainly concentrated on the perception of the respondents towards Electric vehicles and majority of the respondents were aware of the benefits and key features of the electric vehicles. Despite the benefits, the respondents raised several concerns on the willingness to purchase and own an Electric Vehicles. The topmost concern signalled towards the cost of the electric vehicles as most of them felt that the Electric Vehicles were expensive which makes it unaffordable and discourages them to possess one. The second main issue faced by the respondents was the limited number of charging points. The respondents felt that Electric Vehicles were still in its teething stages it would not be possible for them to invest in one while the concerns still continue to bother them. Consumption patterns in an unsustainable manner usually contribute to the crisis in climate and in this manner every person could add to more emissions in the planet. The UNSDG makes a reference to Climate Action. According to Goal 12 on Responsible Consumption and Production. The nations are using the scarce resources 1.75 times faster than it could rejuvenate and if we do not act responsible, the nation—unless things change, we will require three Earths to supply our needs by 2050.

14 Conclusion

It looks like the population of India needs some more time to adjust to the shift to e-mobility. While there seem to be road blocks ahead still in the implementation of Electric Vehicles, an infrastructural change may boost drastic adoption of the electric vehicles. The government is working towards the bottlenecks that have caused hindrances in a rapid adoption and purchase of Electric Vehicles. While on one hand, it may not be practical to completely slide away from ICE vehicles, the fact remains that the electric vehicle market will grow at a rapid pace in its collaborative effort towards the building of a new ecosystem for the EVs segment. It is noteworthy that the Indian automobile sector is at the fifth position worldwide and efforts are on to further move on. The present study has been made on the factors influencing the purchase of electric vehicles in Bengaluru. The study offers scope for further research as the study can be enhanced to study the satisfaction levels of the users of electric vehicles and the problems faced by the users of electric vehicles etc.

References

1. Goel, S. et al.: A study on the barrier and challenges of electric vehicle in India and vehicle to grid optimisation, (June 2021)
2. Shajan A.: Consumer perception towards electric vehicle industry- a study on the role of electrical vehicles in environmental sustainability, (2019)
3. Indukala, M.P., Mathew, B.M.: A Study on Connected and Autonomous Electric Vehicles, (2019)
4. Monika, B.A.: A study of the preference of customers towards e-vehicles, (2019)
5. Monika, B.A.: A study on customer perception towards e-vehicles in Blore, (2019)

An Overview of Interest Rate Derivatives in Banking Sector—A Comparasion Between Global and Indian Market

Riya Singh and Nidhi Raj Gupta

Abstract Nowadays, with the development of the financial market, the banking sectors to make them developed. Apart from their primary activities they are too engaged in investments, speculation and hedging activities to earn income and to save themselves from any risk. In this research paper, attempts are made to know the position of Interest Rate Derivatives of banking sectors in Global and Indian markets which contains the data from the year 2008–2020. In this study, one more finding is that "whether there is any difference in the level of Interest rate Derivatives of banking sectors in Global as well as in Indian market" which showed the result, "there is a huge difference in the use of Interest rate Derivatives between Global and Indian market".

Keywords Derivatives · Interest rate derivatives · Indian market · Global market

1 Introduction

The Banking sector has taken part in nourishing and fueling growth within the economy. It supports the nations in saving and channelizing the accessible resources into high investment priorities and better utilization. Prevailing finance industry are a few things different from old one whose function was only borrowing and lending. Banking sector take risk so as to gain profits. They acknowledge and states differing kinds of threat like "credit risk, operational risk, rate of interest risk, liquidity risk, price risk, interchange risk, etc.". Among that all, "interest rate risk" found the utmost widespread risk, ends within the experience of an Indian banking sector economic situation to unfavorable movements in interest rate. The factor of risk coming out of "differences in timing of changes in rates, the timing of money flows (reprising

R. Singh · N. R. Gupta (✉)
Kristu Jayanti College (Autonomous) K. Narayanapura, Kothanur (PO), Bengaluru, India
e-mail: nidhiraj@kristujayanti.com

R. Singh
e-mail: riya@kristujayati.com

© The Author(s), under exclusive license to Springer Nature Switzerland AG 2024
A. Hamdan and E. S. Aldhaen (eds.), *Artificial Intelligence and Transforming Digital Marketing*, Studies in Systems, Decision and Control 487,
https://doi.org/10.1007/978-3-031-35828-9_69

risk), changes within the shape of the yield curve (yield curve risk) and option values embedded within the products (options risk)". In nutshell, the bank's assets proposition (i.e., securities, loans) will drop down by a rise in rate of interest. Additionally, incomes from sources like, fees, assets also the control of lent capitals are tormented by the fluctuation of rate of interest [1] Risk of Interest rate occurs by holding liabilities and assets with different notional amounts and reprising dates. Consequently, an efficient process of risk management that retains the risk of rate within cautious levels is very crucial with respect to protection as well effectiveness of institutional banks. If Banks cut its risk for the interest rate through fudging with various "derivatives securities" and means of asset and management practices of liability. The Marketplace scene for the product of derivative is often drawn back to the eagerness of agents of economic who try and protect themselves against uncertainties of instability in prices of asset. Because of their very nature, markets of finance are traced by an awfully high instability. With the introduction of products of derivatives, it's somewhat probable to moderately or fully transfer risk of price through securing in asset prices. Given products primarily come up as "hedging devices" in contradiction of rise and fall in "commodity prices". "Commodity-linked derivatives" persisted the only sort of goods for nearly a hundred of years. "Financial derivatives" originated into attention after-1971 phase because of escalating unsteadiness within the market of finances. They're defined as instruments of finances whose settlement relies on the value of an "underlying asset", rate of reference, or index. In fact their initiation, these kind of goods/article/products are much talked about and by the 1991's, around 2/3 of total businesses in derivatives they are accounted for. In current ages the marketplace for derivatives of finances has full-fledged immensely in sense of, the sort of methods available, their complication, and also revenue. **Carter David**: According to him "Interest rate derivatives" can have multiple functions. One of them is to reduce the rate of interest in banking institutions. Cater calculated the amount of "interest rate "threat by seeing the exact importance of the 12-month "maturity gap analysis" [2]. **Ravikant Bhat**: His research paper of him i.e. "The Incidence of Interest Rate Risk in Indian Banks" was expressing the amount of interest rate risk in a few designated Indian Banks, its reasons, and the method to resistor it. In addition to it, a maximum of the findings and research apportioned with risk of interest rate to its capital [3]. **Dhanani**: In his paper, he found many features like "interest rate fluctuations", "use of borrowed funds" etc. because of this risk of interest rate ascend. He also observed and established that United Kingdom corporation should border their interest rate risk to achieve instability in their cash inflows and outflows and profit [4]. **Begenau, Piazzesi, and Schneider**: They show in their paper that US banks surged their interest rates by their product positions, and determined that these banks do not hedge [5]. **Shashi Srivastava**: She calculated the interest rate risk in ICICI and SBI banks and observed that SBI is having greater disclosure of the risk of interest rate as a comparison to ICICI bank. She scrutinized the rate of Interest disclosure by consuming several devices like "assets-liability mismatch", "gap analysis, and sensitivity analysis" and proposed to use "interest rate derivatives" to hedge this risk [6]. **Reeta**: She scrutinized in her paper that the calculation and organization implements rate of Interest risk utilised in Indian firms. Reeta observed risk of interest rate is measured the greatest impact in Indian firms with regard to imperative risk and majority of organization practice "gap analysis" and "Maclay's Duration analysis" to quantify the level rate of Interest. Reeta as well mentioned many features like, inflation rate, monetary policy and fluctuations in money market and those are accountable for interest rate variations [1]. **Di Tella and Kurlat**: Their paper represent the only paper where such derivative exposures result from optimal hedging. In their model, banks ideally take losses when interest rates increase, the reason they assume upper spreads on deposits going forward [7]. **VN Prakash Sharma**: He examined in his research paper the result rate of interest variations on income of net interest and on the success of the Bank of Baroda and ICICI bank in term of profit. He deployed earning sensitivity analysis, rate adjusted gap, and duration gap analysis to calculate the level of interest rate risk [8].

1.1 Derivatives

The past decade has witnessed the several growths in the level of international trade and business due to the wave of globalization and liberalization all over the world. As a result, the demand for the international money and financial instruments improved notably at a global level. In this respect changes in the interest rates, exchange rates and stock market prices at the different financial markets have increased the financial risk. In order to deal with such risks, the new financial instruments have been developed, which are commonly known as financial derivatives. The basic purpose of this instrument is to offer commitments to prices for future dates for giving safety opposed to adverse movement in future price, in order to reduce the level of future risk.

Interest rate derivatives "An interest-rate derivative is a financial instrument with a value that increases and decreases based on movements in interest rates. Interest-rate derivatives are often used as hedges by institutional investors, banks, companies and individuals to protect themselves against changes in market interest rates, but they can also be used to increase or refine the holder's risk profile." [9].

Benefits of interest rate derivatives

Use of "Interest Rate Derivatives" are used in the following major areas:

Hedging exposures: "Adding greater certainty and replacing risks normally associated with interest rate and currency exposure".

Raising capital: "Reducing the cost of capital by exploiting funding differentials between countries, currencies, and funding sources".

Investing: "Allowing users to express their market views efficiently, enhancing yield and achieving portfolio objectives".

Fund managers, Insurance companies, Corporations, Banks, Insurance companies, Government, Financial services industry and individuals utilize interest rate derivatives methods to resolve their problems of management of financial risk [10].

2 Objective

- To analyze the trend of "Interest Rate Derivatives" in Global and in Indian market.

3 Hypotheses

HO1: There is no significance change in trend of Interest Rate Derivatives between Indian Market and Global Market.

4 Research Methodology

The sample for this study includes "Private Sector Banks and Public Sector Banks" respectively. The present work uses a non-probability judgmental sampling technique to choose the sample for the study. This study deals with secondary data which has been composed of "The Bank for International Settlement", "Clearing Corporation of India Limited", and the Indian Banking Association (IBA). Data is poised for the financial year 2008–2017. To conduct an analysis on the collected data measures of central tendency and variation have been taken to pronounce the use of Interest Rate Derivatives in the global and Indian market and T-tests is being used to know the difference in the use of Interest Rate Derivatives between Global and Indian market. SPSS 20 version has been used to analyze the result.

5 Limitations of the Study

In carrying out this study, the limitation is there relating to deficient records given by the commercial banks in their annual report. Financial institutions have a propensity of failing to disclose their true financial position. This puts to test the validity of the data used in the study therefore conclusion from the result might not convey the true result. Time is the other limitation which has laid confines on the selection of the increased number of sample banks. The statistical tools and techniques used in the study may have their own restrictions which may affect the assumptions drawn from the study. The data is being used in this paper is till the year 2020 because of the unavailability of data.

6 Discussion on Analysis

The above figure illustrates the use of IRD decreased from $9761.17 Billion in 2008 to over $9170.91 Billion in 2010. It increased in 2011 and 2013 amounting to $10,573.57 Billion and $11,494.72 Billion respectively to their respective previous year. If we see the utilization of interest rate derivatives in 2019–2020, it reflects a drastic fall due to the pandemic situation. On the whole, the data gives the picture of up and down conditions throughout the year from 2008 to 2020 (Fig. 1).

On the other hand it also discloses the use of "Interest Rate Derivatives" Market in India. The figure tells that it was highest in 2008 with the amount $4280.85 billion but after that its situation was uneven i.e. $1880.77 Billion in 2009, $2096.02 billion in 2010, $2942.70 billion in 2011, $2297.52 Billion in 2012, $1874.02 billion in 2013, $1746.03 billion in 2014, $1841.64 Billion in 2015, $1731.80 billion in 2016, $1794.43 billion in 2017. If we talk here about the year 2019–2020, it shows the same condition as it is there in the global level where there is a fall in the pattern of the use of IRD because of the pandemic situation.

6.1 Data Analysis

The major drive of data analysis is to examine the trend of "Interest Rate Derivatives" in the Global and Indian markets both. In this study secondary data is being used. The data is analyzed on the basis of the volume of Interest Rate Derivatives used in the Global market and in the Indian market. I have formulated hypotheses on this

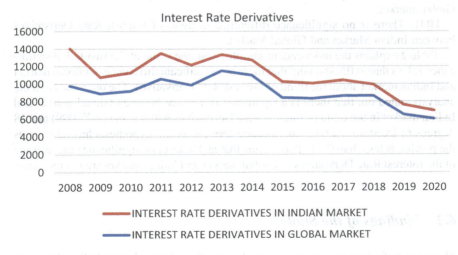

Fig. 1 Global and Indian interest rate derivatives market growth. *Source* Bank For International Settlement

840   R. Singh and N. R. Gupta

Table 1 Descriptive statistics

IRD	Mean	Std. deviation	C.V*	Std. Error Mean
Global IRD	9608.17	1118.54	11.64	353.71
Indian IRD	2248.58	803.15	35.72	253.96

Table 2 Independent samples test

Levene's test for equality of variances			t-test for equality of mean			
	F	Sig	T		Df	Sig.(2-tailed)
Equal variances assumed	28.901	0.000	−27.099		18	0.000

basis only. In order to prove this, statistical techniques are employed. By and large, the study used statistical techniques:

- "Measures of central tendency and measures of variation" have been used to define the use of Interest Rate Derivatives in the global and Indian market.
- T-tests is being used to find out the difference in the use of Interest Rate Derivatives between the Global and Indian market.

From Table 1, it is clearly depicted that the average of "Interest Rate Derivatives" value of the Global market is found to be Rs. 9608.16 Cr. which is greater than Interest Rate Derivatives mean value of the Indian market which is totaled to be Rs. 2248.58 Cr. The Coefficient of Variance is detected to be higher in the Indian market as matched to the Global market which is calculated to be 35.72 and 11.64 respectively. It clearly shows that relative dispersion is higher in Indian market than Global market.

HO1: There is no significance difference in trend of Interest Rate Derivatives between Indian Market and Global Market.

Table 2 explains the independent sample t-test, the value of the two-tailed significance is less than 0.05, showing that there is a significant difference in Global market and Indian market about the trend of Interest Rate Derivatives. The outcome from the analysis stated out that there is a significant difference between Global market and Indian market in the trend of Interest Rate Derivatives with t value (−27.099) and 18 degrees of freedom. At 0.01 significance level, also, the hypothesis has rejected as the p value is less than 0.01. This means that at 1% level of significance the average of the Interest Rate Derivatives in Global market and Indian market are not the same.

6.2 Findings of the Study

There is a difference in usage pattern of Interest Rate Derivatives in Indian as well as in Global market. The Variance is detected to be greater in Indian market as equated to Global market in the usage pattern of IRD. These differences in usage pattern can be due to the divergence in the approaches taken up in both the level.

6.3 Suggestion for the Study

The study mentions that Indian banking sector should boost the use interest rate derivatives reason derivatives lessen the possibility of financial distress by decreasing the irregularity in firm value, thus dipping the predictable costs of financial distress.

7 Conclusion

The above analysis reveals that the usage and pattern of Interest Rate Derivatives in Global market and Indian market is different. These differences in usage and pattern can be due to the divergence in the approaches taken up in both the level. As compare to Indian market, Global market play very well in Interest Rate Derivatives. Though Indian market uses Interest rate Derivatives for trading as well as for hedging purpose but still there is a need to use more and more these financial instruments.

References

1. Srivastava, S., Srivastava, D.: Interest rate risk management: a comparative study of state bank of India and ICICI bank. Int. J. Manage. Soc. Sci. Res. (IJMSSR) 4 (2015)
2. David, C.A., Sinkey, J.F.: The use of interest rate derivatives by end users: the case of large community banks. J. Financ. Serv. Res. 14, 17–34 (1998)
3. Bhat, R.A.: The Incidence of Interest Rate Risk in Indian Banks (2005)
4. Research Report, CIMDR, India
5. Dhanani, A.: Interest rate risk management—an investigation into the management of interest rate risk in UK companies. Res. Executive Summary Ser. Published CIMA 2, 1–7 (2005)
6. .Begenau, J., Piazzesi, M., Schneider, M.: Banks' Risk Exposures. Working Paper (2015)
7. Reeta (2013) Measurement and management strategies of interest rate risk—a study in Indian perspective. GGGI Manag. Rev. 1, 7–11
8. Di Tella, S., Kurlat P.: Why are Bank Balance Sheets Exposed to Monetary Policy? Working Paper (2016)
9. Sharma, V.N.P., Santhosh, N.G.: Interest rate risk management: a comparative study of bank of Baroda and ICICI bank. IOSR J. Econ. Finance (IOSR-JEF) 7, 01–04 (2016)
10. www.investopedia.com

Adoption of Mobile Banking Among Rural Customers

A. J. Excelce and V. G. Jisha

Abstract Banking is the backbone for the development of the country. The usage of banking have increased rapidly day by day among costumers. There is shift from traditional banking to modern banking in current years. The study focus on usage of mobile banking among the rural customers. Total of 127 respondents were randomly selected for the study from North Bengaluru. Gender of the respondent is having impact on utilization of Technology. There is association among age and the awareness of the respondents. Education plays a vital role in adoption of changes in technology. More awareness program can be conducted in remote places to expand the usage of mobile banking.

Keywords Mobile banking · Challenges · Usage and technology

1 Introduction

Government of India initiated the Digital India project after understanding the importance of technology. The initiative aims to turn India into an integrated economy by using cellular phones and the network as two supporting pillars for rendering services extended by the country. India is often recognized as Asia's fastest-growing smartphone market. Customers have been flocking to mobile banking systems because of low cost mobile phones, which provides various facility by one touch. The introduction of several M-banking apps has proven to be a success. India is regularly recognized as Asia's fastest-growing smartphone advertise. Clients have been running to portable managing an account frameworks. The presentation of a few M-banking apps has demonstrated to be a victory. The term computerized incorporation gives the people the abilities of fundamental innovation to take part within the information economy.

A. J. Excelce · V. G. Jisha (✉)
Department of Management, Kristu Jayanti College, Bengaluru, India
e-mail: jisha@kristujayanti.com

© The Author(s), under exclusive license to Springer Nature Switzerland AG 2024
A. Hamdan and E. S. Aldhaen (eds.), *Artificial Intelligence and Transforming Digital Marketing*, Studies in Systems, Decision and Control 487,
https://doi.org/10.1007/978-3-031-35828-9_70

843

2 Research Gap

In a developing country, the pace of growth in mobile banking is immensely fast. People are adapting it and for some group of people it has become a part of their everyday life. There are various merits or advantages of Mobile banking as easy money transfer, easy access to funds, on the contrary a lot of people are not well aware about the use and because of that they are subjected to fraud and even lose their hard-earned money. Monetary transactions may include sending or receiving money, saving, loan, insurances, pensions, etc. Due to fear, increased cost, lack of information, and inadequate funds. The government has launched many schemes so that underserve and financially excluded people can have access to financial services.

Though there are many facilities and services available for people, rural consumers are hesitant to avail those facilities. Among these facilities the researcher is focusing only on the mobile banking usage among rural Customers.

3 Research Objective

1. Analyse the consumer perception towards mobile banking.
2. Study about the usage of mobile banking by rural consumers.
3. To examine the positive and negative factors of mobile banking.

4 Review of Literature

1. Alam et al. [1] they presented that mobile banking will help the poor people by providing the services at less processing charges and maintenance cost.
2. Lyman et al. [2] and Yu [3] insisted that rural customers can save time and money through modern banking system rather than the traditional banking.
3. Datta [4] pointed out an information about the, cashless transactions which is successfully boosting in measuring the different government departments that push citizens into cashless transactions, which lead to the usage of mobile banking at large.
4. Yu [3] The study explains about the adoption of modern banking system in the rural areas rather than raising branches in the country. It also focuses on the facilities in the mobile banking such as withdrawals, savings, remittances, repayments of bills, transfer of funds and other facilities.
5. Unnithan and Swatman [5] focuses on the mobile users and the usage of mobile banking. The study explains that the rural India, the mobile users are increased more compare to the fixed line subscribers. The financial institutions provides various banking services with a expectation growth in Indian mobile banking.

Adoption of Mobile Banking Among Rural Customers 845

5 Hypothesis

H₀ Age does not influence the sources of awareness about mobile banking.
H₀ Gender has no impact on the usage of mobile banking.
H₀ Education does not influence the challenges faced in using mobile banking.

6 Research Design

The primary study was conducted with 127 respondents at random from rural districts of North Bengaluru by personal interview method through questionnaire. The t test was applied to study the association between the gender and usage of mobile banking. Anova has been used to depict the challenges faced by respondents. Cross tab has been used to reveal the relationship between age and awareness of mobile banking.

7 Results and Discussion

Out of 127 respondents 73.81% belong to the age group of 20–30 years are aware of mobile banking. In it, 26.19% of the respondents are unaware. In the age group of 30–40 years 76.32% of the respondents are aware of mobile banking and 23.68% of the respondents are unaware of mobile banking. Among the age group of 40–50 Years 66.67% of the respondents are aware of mobile banking and 33.33% of the respondents are unaware. 60% of the respondents who belongs to the age group of above 50 years are aware of mobile banking and 40% are unaware. This analysis concluded that majority of the respondents in the age group of 20–30 Years are aware of Mobile Banking. Awareness should be created among the other age group by explaining the benefits of mobile banking through various sources of Information (Tables 1 and 2).

The study shows the relationship of gender and its impact on usage of Mobile Banking. The variables of usage of mobile banking are not significantly associated with gender. It is concluded that gender has no impact on usage of mobile banking. Banks can target more on women by educating them on technology usages (Table 3).

The variables of security, lack of knowledge, cost, internet facilities significantly differ with the education level of the respondents. It is concluded that educated respondents give more importance to mobile banking usage. Banking sector should pay more attention to overcome the challenges faced by the respondents.

Table 1 Age and awareness of mobile banking

Age	Awareness of mobile banking		Total
	Aware	Unaware	
20–30 years	31	11	42
	(73.81%)	(26.19%)	(100.0%)
30–40 years	29	9	38
	(76.32%)	(23.68%)	(100.0%)
40–50 years	18	9	27
	(66.67%)	(33.33%)	(100.0%)
Above 50 years	12	08	20
	(60.00%)	(40.00%)	(100.0%)
Total	90	37	127
	[100.0%]	[100.0%]	[100.0%]

Table 2 Gender and usage of mobile banking service

Usage of mobile banking	Gender	N	$\overline{X}$	σ	t value	P value
Balance enquiry	Male	118	0.583	0.144	0.163	0.870*
	Female	9	0.609	0.018		
Bill Payments	Male	118	0.583	0.141	1.792	0.073*
	Female	9	0.570	0.222		
Online purchase	Male	118	0.551	0.172	3.913	0.000*
	Female	9	0.482	0.235		
Fund Transfer	Male	118	0.543	0.184	4.129	0.000*
	Female	9	0.462	0.232		

Null Hypothesis Gender has no impact on the Usage of Mobile Banking

8 Conclusion

Mobile Banking is the latest technology which opened a new of opportunity to the existing banks in financial institution. The growth of customers in banking sector have increased rapidly with the digital life style. It is necessary the bank should reach the common man at the remotest location in the country. There is also a need to generate awareness and benefits of mobile banking among people. The result of this study shows that majority of the respondents in the age group of 20–30 Years are aware of Mobile Banking. Awareness should be created among the other age group by explaining the benefits of mobile banking through various sources of Information. Gender has no impact on usage of mobile banking. Banks can target more on women by educating them on technology usages. Banking sector should pay more attention to overcome the challenges faced by the respondents.

Adoption of Mobile Banking Among Rural Customers

Table 3 Education and challenges faced in using mobile banking

Challenges faced in using mobile banking	Education	N	$\overline{X}$	σ	F value	P value
Security	College	16	0.59	0.148	2.996	0.018 *
	High school	44	0.61	0.128		
	Primary school	33	0.56	0.151		
	Read and write	9	0.60	0.124		
	Illiterate	25	0.56	0.167		
Lack of knowledge	College	16	0.60	0.145	3.501	0.008*
	High school	44	0.60	0.131		
	Primary school	33	0.55	0.164		
	Read and write	9	0.61	0.582		
	Illiterate	25	0.57	0.158		
Cost	College	16	0.59	0.165	6.654	0.000*
	High school	44	0.58	0.161		
	Primary school	33	0.50	0.197		
	Read and write	9	0.57	0.149		
	Illiterate	25	0.53	0.188		
Internet facilities	College	16	0.57	0.171	6.926	0.000*
	High school	44	0.57	0.173		
	Primary school	33	0.49	0.205		
	Read and write	9	0.58	0.159		
	Illiterate	25	0.51	0.195		

* Significant at 5% level

References

1. Alam, S., et al.: A Secured Electronic Transaction Scheme or Mobile Banking. ICITIS (2010)
2. Lyman, T.R., Pickens, M., Porteous, D.: Regulating transformational branchless banking: mobile phones and other technology to increase access to finance. In: Focus Note 43. Consultative Group to Assist the Poor (CGAP). Washington, DC (2008)
3. Yu, S. (2009). Factors influencing the use of mobile banking: the case of SMS-based mobile banking. Scientific Journal of King Faisal University.
4. Datta, B.: Factors affecting mobile payment adoption intention: an Indian perspective. Glob. Bus. Rev. **19**, S72–S89 (2017)
5. Unnithan, C.R., Swatman, P.: Online banking adaptation and dot.com viability: a comparison of Australian and Indian experiences in the banking sector. School of Management Information Systems, Deakin University, No. 14 (2001)
6. Mas, I.: Realizing the potential of branchless banking: challenges ahead. In: Focus Note 50. Consultancy Group to Assist the Poor (CGAP). Washington, D.C. (2008)

Artificial Intelligence and the Decarbonization Challenge

Ismail Noori, Yasser M. Abolelmagd, and Wael F. M. Mobarak

Abstract To help "operators estimate available capacity on the system and plan for future needs," the European DSO uses AI to analyze data on voltage, load, and grid topology. The DSO was able to employ available and incoming distributed energy resource assets more effectively thanks to AI Similar to this, a German transmission system operator employed AI to produce more accurate grid loss forecasts. AI could be used into both new and existing advanced distribution management systems in the United States (ADMS). The ADMS conducts tasks including "fault location, peak demand management, support for microgrids and electric vehicles," and is "the software platform that enables the full range of distribution administration and optimization." Finally, by optimizing end uses, AI can make significant progress in lowering the demand for electricity usage. For instance, Google declared in 2015 that machine intelligence from DeepMind had helped it cut the energy needed to cool its data centers by 40%. Given that the DeepMind project has been contracted to assist in reducing waste in the UK's National Grid, further savings appear possible. More recently, in 2019, Hewlett Packard and the National Renewable Energy Laboratories collaborated to assess how AI may improve the efficiency of their data center operations. Energy efficiency is frequently referred to as the unsung hero of decarbonization because it affects numerous industries and has some of the greatest effects at the lowest prices. Increasing efficiency in many industries can also aid in lowering electricity consumption.

Keywords Decarbonization · DeepMind · Electricity consumption · Energy efficiency

---

I. Noori (✉)
Ahlia University, Manama, Bahrain
e-mail: allamh3@hotmail.com

Y. M. Abolelmagd · W. F. M. Mobarak
Civil Engineering Department, College of Engineering, University of Business & Technology (UBT), Jeddah, Kingdom of Saudi Arabia
e-mail: Yasser@ubt.edu.sa

W. F. M. Mobarak
e-mail: W.fawzy@ubt.edu.sa

© The Author(s), under exclusive license to Springer Nature Switzerland AG 2024
A. Hamdan and E. S. Aldhaen (eds.), *Artificial Intelligence and Transforming Digital Marketing*, Studies in Systems, Decision and Control 487,
https://doi.org/10.1007/978-3-031-35828-9_71

1 Introduction

Energy efficiency, according to the International Energy Agency, is crucial to the transformation of energy systems and will be crucial in reducing the rise of the global energy demand to one-third of the current rate by 2040. By combining the effects of energy-efficiency opportunities in buildings, industry, transportation, and the electric grid, experts were able to predict that GHG emissions might be cut in half by 2050. C. Artificial intelligence for resilient and reliable electric grids Reliability and resilience are a third area where artificial intelligence might help the electric grid. For the bulk power system, the North American Electric Dependability Corporation (NERC) has made a distinction between reliability and resilience. The ability of the electric system to meet customer needs as well as "withstand sudden interruptions or unplanned loss of system components" is the focus of reliability, which is made up of both adequacy and security. Contrarily, resilience is defined as "the ability to tolerate and lessen the magnitude and/or length of disruptive events, including the capability to predict, absorb, adapt to, and/or fast recover from such an event." AI can help with both objectives by decreasing downtime and avoiding backup generation, which produces more greenhouse gases. Blackouts have disastrous economic ramifications [1], but they can also have negative environmental effects. Our interconnected grid becomes inefficient and wasteful when there is downtime. Manufacturing facilities and other enterprises frequently remain inactive until the power is restored, sometimes having to restart a process if it was halted. For instance, almost 29 tons of food were wasted during a brief blackout in New York. Because the grid was down for only a short period of time, the process of producing and delivering that food into the city required a significant amount of GHG emissions. Additionally, using generators and having to use alternative modes of transportation (if public transit is not available) can all result in an increase in emissions when there are blackouts. To minimize catastrophic effects, blackouts must be prevented and quickly resolved. Additionally, historically, utilities have been reactive organizations. In many areas of the nation, calls from irate customers are the only way for the company to learn about a power outage. The race to update these distribution systems includes the smart-grid developments that were previously addressed. Increased weather-related power outages are a result of the grid's aging, and grid operators' response times are hampered by a lack of automated sensors. Transmission lines that are older lose more energy than newer ones do, and grid deterioration makes the system more susceptible to extreme weather. Utility companies can become proactive instead of reactive with the aid of AI. To increase the resilience and dependability of the country's electricity grid, the Department of Energy announced in 2018 a $5.8 million funding opportunity for research and development of advanced instruments and controls.

2 Preventative Maintenance

How AI can speed up repairs to increase resilience, assist with troubleshooting, and help with preventative maintenance to reduce the number of unexpected disruptions. maintenance/preventative troubleshooting. AI is being utilized to foresee issues before they arise, as opposed to waiting for grid assets to malfunction. Due to the interconnectedness of the electrical grid, a single piece of equipment failure can avoid a cascading blackout, which has been a common occurrence in many of the most infamous power outages in American history. 155 Algorithms156 that "take into account industry-wide early failure rates for equipment, creating a richer understanding of premature failure risks for enhanced asset maintenance, workflow, and portfolio management" are being used by utilities and generators to predict the likelihood of failure to be more proactive [2]. For instance, the New York Power Authority (NYPA) has invested in sensors to help assess the remaining useful life of important machinery and identify issues before they can impair power plant operations. Sensor monitoring is a similar form of preventative maintenance that power providers might use. To provide a real-time snapshot of the grid's operational status, utilities already use devices known as phasor measurement units (PMUs) to "measure the amplitude and phase of electric current and voltage at various points on the electric grid using a common time source [3] for synchronization." Additionally, New York announced a PMU initiative in 2018. This program deploys sensors to capture voltage and current data with high-resolution and precise time stamping at NYPA's power generating facilities and switchyards. The gathered information can then be combined and applied to the real-time grid management, asset management, and potential issue identification processes. The electric industry's prior system was over a hundred times faster than these PMUs, but "more sophisticated technologies are needed to interpret the data for meaningful information." For instance, utilities can abandon a predetermined schedule for asset maintenance and retirement in favor of a more effective one that considers the actual state of the assets. Machine learning applications could "liberate grid operators from decommissioning assets before their useful lives have ended, while enabling them to perform more frequent inspections and maintenance to keep assets working well" by utilizing data from sensors and other hardware that remotely tracks assets.

A European power distribution business, for instance, was able to lower expenses by 30% by examining dozens of variables to assess the general health of transformers and identify specific components. Similar to this, AI can be used to more accurately predict severe weather and natural disasters that endanger energy grids. As an illustration, NASA recently utilized artificial intelligence to follow Hurricane Harvey [4], enabling it to forecast the storm's path with six times greater accuracy. Additionally, IBM offers an AI solution that enables utilities to speed up their emergency response times and aids in the prediction of power outages caused by powerful storms. A group of scientists in Switzerland used AI to almost 80% of the time anticipate lightning strikes within an 18-mile radius. With the aid of IBM's Watson AI system, which visually analyzes camera feeds to locate new fires [5] and predict where they will spread, AI and bionic eyes are also assisting in the containment of wildfires, assist with repairs. As was already said, a crucial aspect of resilience relies on a system's capacity to restart. In the quest to increase the resilience of the grid, AI has the potential to be a key instrument. "By relying on data from remote sensors, power producers may specify where to send a crew for a repair and make sure the team comes with the necessary tools for the job based on an assessment of the damage." To handle issues in remote places, Utilities have been utilizing drones, AI developed using deep-learning algorithms, and sensors. Drones gather data on a wide range of remote problems, including broken equipment, downed trees, and simple vegetation encroachment on remote assets [6].

The energy and utilities industries can use a sophisticated tool that is created by fusing AI analysis with drone-taken photographs. By studying predictive data, some experts even believe that combining drones and AI will help stop utility problems before they even arise. Requirements for NERC security. AI can help bulk grid users more cheaply meet the dependability standards set by the North American Electric Reliability Corporation, the country's reliability watchdog (NERC). To counter risks from both the real world and the cyberspace, NERC has published security regulations. The rules "require a minimal level of organizational, operational, and procedural controls to manage risk" to encourage organizations to be more proactive. The CIP-014-1,177 NERC physical security requirements [7] specify six fundamental independent verifications of the risk assessment and an analysis of the potential risks and weaknesses of a physical attack on these critical stations or substations. The obligation to create and implement a physical security plan, which must include measures to detect and respond to physical threats, is probably the most difficult. AI-powered tower-mounted robots are one of the most creative ways to meet these physical security criteria. In comparison to physical security standards, the NERC cybersecurity laws are more comprehensive, covering cybersecurity issues through eleven regulations181 that include everything from employee training to incident reporting. This could be partly attributed to the rise in data breaches in the United States.

Although historically separate, the transportation and electrical sectors are becoming [8] less distinct as a result of this change. This is because if EV sales continue to rise rapidly on a global scale, the automobile sector will be more and more dependent on the electric grid. Depending on where the charger is located, EVs are powered by a considerably more varied mix of electricity sources, which includes some amount of renewable energy. In this approach, the more the grid depends on renewable energy, the more GHG emissions are reduced because of the switch to EVs. EV charge scheduling, congestion management, vehicle-to-grid algorithms, battery energy management, as well as the study and development of EV batteries are all areas where AI can be helpful in this revolution.

Grid operators will also need to understand user charging and use patterns to forecast and control the amount of electricity used. EV movement can be coordinated, their charging cycles can be optimized, and overall user demand can be predicted with the aid of AI. AI developments in autonomous vehicles may be crucial in reducing GHG emissions, especially if electric vehicles continue to account for most of them.

Artificial Intelligence and the Decarbonization Challenge

According to a World Economic Forum assessment, AI will be especially important for the transition to autonomous linked EVs. As connected vehicles communicate with one another and with transportation infrastructure to identify risks and enhance navigation, using machine learning into autonomous EVs will aid in the optimization of transportation networks. Additionally, "it has been anticipated that clever automated driving systems over human operators might see a 15% reduction in fuel usage." A switch to electric vehicles would also present utilities with a rare opportunity. Although the world's demand for energy has grown recently [9], predictions for U.S. power consumption are largely unchanged. However, other analysts predict that the increased demand brought on by a sizable new fleet of EVs might result in an increase in electricity demand of up to 38%. 116 When used in conjunction with smart grid and EV deployment, as well as increased intermittent renewables and distributed resources, AI can have a substantial positive impact on grid optimization. B. Increasing Electric-Grid Efficiency with Artificial Intelligence Utility companies might be considering a second, less visible method to cut GHGs in addition to switching to electricity sources with a lower carbon footprint: utilizing AI to target inefficiencies [10].

According to a recent report by the Lawrence Livermore National Laboratory, over 68% of the energy produced in the US is [11] "rejected." Rejected energy is a portion of a fuel's energy that could be used for a useful activity, like producing electricity or powering a vehicle, but is instead lost to the environment. Examples of such fuels are gas and oil. Waste heat, like the warm exhaust from furnaces and cars, is the most common kind of rejected energy. Energy waste has gradually surpassed energy productivity because of the significant increase in energy demand over the past 50 years for electricity and transportation, two industries with traditionally low rates of fuel productivity. Although there are inefficiencies throughout the entire electric grid, the fundamental energy potential of electricity generation is lost by about two thirds. Although it is impossible for heat engines to operate at 100% efficiency due to the second law of thermodynamics, there is still a lot of potential for efficiency gains in the power industry. AI has the potential to lessen these inefficiencies. AI can improve the efficiency of the more recent renewable grid resources as well as minimize inefficiencies at current fossil fuel operations and nuclear plants. For instance, AI can help with the planning and management of wind and solar farms, greatly improving the electricity production efficiency of these utility-scale renewable energy systems. The turbine heads for wind farms can be actively positioned to capture more of the incoming wind. AI can make it simpler and more profitable for solar generators to participate in energy markets by improving solar forecasting intelligence. AI can help cut down on energy losses during the transmission and distribution of electricity. An operator of a distribution system, for instance only produces power when the sun is shining or the wind is blowing, making it intermittent [12].

This intermittency poses distinct reliability difficulties as compared to baseload sources like nuclear and natural gas, which can produce power on a more constant basis. These intermittent renewables' predictability can be improved with AI, raising their value. To "better estimate wind power output thirty-six hours ahead of actual generation," Google and DeepMind, for example, applied machine learning to wind power capacity in the United States. This helped with optimal hourly day-ahead delivery promises. According to Google, the use of machine learning increased the value of its wind energy by almost 20% [13]. AI can handle power fluctuations and enhance energy storage, and it is increasingly utilized to regulate the intermittent nature of renewable energy so that more may be added to the system. Artificial intelligence (AI), for instance, can be used to alter wind farm propellers to match shifting wind directions, thereby reducing the intermittent nature of wind turbines. The structure of renewable energy sources like solar electricity and wind turbine farms can also be helped by AI. The ability of AI to maximize the performance of huge energy batteries to extend their lives with fewer problems is perhaps its most amazing capability for enhancing the storage of renewable energy during sporadic times of downtime. Such efficiency optimization necessitates extensive data analysis, making AI participation ideal. Resources Dispersed.

The transition from big, centralized power sources to smaller, decentralized sources located closer to the place of use is a second and related outcome of the move toward more renewable energy. Rooftop solar and wind, combined heat, and power (CHP), fossil fuel generators [14], fuel cells, energy storage, microgrids, and nano grids are examples of prevalent distributed energy resources (DERs) in the residential sector. The management of these distributed resources, many of which are not owned, managed, or even visible to the grid administrators, is a challenge for utilities. Tens of thousands of smaller/residential dispersed energy sources cannot simply replace one massive rotating coal-fired plant, as others have noted. Demand response, which aims to move demand to better match supply, is a type of distributed resource that can also be used to shift electricity use into off-peak hours. The electrical grid is expected to receive more distributed energy resources over the course of the next ten years than any utility can handle. AI is ready to help with these difficulties. This integration procedure could be improved by an autonomous energy grid to the advantage of the power system and DER owners.

AI can assist utilities in managing their generation assets more effectively, dependably, and adaptably in response to changes in supply and demand brought on by distributed generation. Like this, AI can be used to decentralized optimize small-scale systems for automated demand response and learn the factors influencing effective demand response. An AI system would need to learn the thermal qualities of a home, the local weather conditions, how these circumstances effect heat flows in a home, user preferences, and modify energy consumption against real-time price signals, for instance, to assist in shifting energy use to off-peak hours. Intelligent Grid Infrastructure A smarter grid is a third and crucial element that will enable a transformed electric system. The electrical sector in the US is anticipated to invest $46 billion between 2018 and 2030, or "roughly $3.5 billion yearly [on the smart grid alone]. As part of modernization efforts" (in nominal dollars). The smart grid is "an intelligent electricity grid—one that uses digital communications technology, information systems, and automation to detect and react to local changes in usage, improve system operating efficiency, and, in turn, reduce operating costs while maintaining high system reliability," according to the Department of Energy (DOE). Winners of the DOE's 2019 Innovation [15] Challenge suggested the following AI

Artificial Intelligence and the Decarbonization Challenge

improvements, to name a few examples: (a) Southern California Edison suggested operating electric grid substations with a human–machine interface and virtualizing their component parts (HMI). (a) Siemens Corporation: Suggested creating a digital companion for green technologies [16] that blends semantic technologies, machine learning, and augmented reality to give grid operators improved insight into the state of the grid. Using various data sources, such as weather and charging infrastructure, the companion could allow predictive capabilities. Artificial intelligence (AI) can help with many of these investments in digital communications technology, information systems, and automation in an effort to accommodate more complex power flows and to improve overall reliability, efficiency, and safety, while also meeting future demand from new uses.

3 Conclusion

The continuous transition from internal combustion engines to electric vehicles is the final area of grid asset optimization that lends itself to AI. used synonymously with machine learning because learning-based AI "diagnoses problems by engaging with the problem," establishing assumptions, reevaluating models, and reevaluating data all without human involvement. The fundamental discovery of machine learning is that a large portion of what we consider to be [human] intelligence depends more on chance than on reason or reasoning. Although many people are looking forward to the development of "wide AI," demonstrated by robots that can think at or beyond the level of a human, "narrow AI" in use today is far more akin to a Roomba than the Terminator. While AI now excels at well-defined tasks, it still lacks conscience, common sense, the capacity to learn from a limited number of instances, and true originality.

Addressing questions about climate change, a subject fraught with significant data challenges, seems to be a particularly good application for artificial intelligence and climate AI. Despite decades of monitoring GHGs and their sources, it has remained challenging to filter, evaluate, and effectively use the data. It is difficult to work with such enormous data sets, but meaningful climate science needs gathering enormous volumes of data on numerous different factors, such as temperature and humidity. However, some claim that "rapid and continuous gains, and advancements in the sophistication, cost, compactness, and usage of technology are enabling the swift generation and analysis of voluminous data sets." Investing in strategies to overcome these obstacles might be quite profitable. Artificial intelligence (AI) is already being used to track the effects of climate change by analyzing photos of shallow-water reefs and identifying coral by color. as well as to gather information on temperature, humidity, and carbon dioxide to monitor the condition of our trees. Since there are presently roughly a billion people without access to energy worldwide, AI can help democratize electricity by enabling more inexpensive access to it. And by promoting microgrid development to provide off-grid zero-carbon electrification.

AI can also assist in predicting the source of carbon emissions, which can help policymakers and financiers decide how and where to control and finance the generation of energy. Given the predictive capabilities of machine learning, one can envision its use to address many additional aspects of climate change. Improving climate modeling and forecasting is one of AI's most obvious climate change applications. Models for predicting the climate (and the weather) are physical systems that derive their conclusions from the rules of physics. Since many years, meteorology and climate science have employed statistical methods now referred to as "AI" or "machine learning." As a result of recent advancements in machine learning, models created by artificial intelligence (AI) are becoming more accurate and more predictable when used for meteorological reasons. Additionally, the application of AI to evaluate model outputs and reconcile them with actual atmospheric observations is growing. Artificial intelligence in climate change and the electric-power sector, part two Beyond its advantages for climate science, artificial intelligence (AI) can be a key player in initiatives to cut emissions from one of the major sources of GHGs: the electric power industry. On a global scale, electricity is responsible for about 25% of GHG emissions.

References

1. Sources of Greenhouse Gas Emissions: https://www.epa.gov/ghgemissions/sources-greenhouse-gas-emissions
2. New York Power Authority, EPRI and GE Announce Results from NYPA Green Hydrogen Demonstration Project for Immediate Release: 23 Sept 2022. https://www.nypa.gov/news/press-releases/2022/20220923-greenhydrogen
3. Srivastava, A.K.: The School of Electrical Engineering and Computer Science Director, Smart Grid Demonstration and Research Investigation Lab, Phasor Measurement (Estimation) Units, https://na.eventscloud.com/file_uploads/b9c0cedfe296b5347ce0877fbfac3753_PMU_Relay_Schoolcopy.pdf
4. Jet Propulsion Laboratory September 4, 2020: NASA Using Machine-Learning AI to Predict Hurricane Intensity. https://scitechdaily.com/nasa-using-machine-learning-ai-to-predict-hurricane-intensity/
5. Correia, V.: How AI, IoT and Weather Tech Can Help Better Detect Deadly Wildfires, 8 Aug 2019. https://www.ibm.com/blogs/think/2019/08/ai-detect-wildfires/
6. Mathew, D., Brintha, N.C., Jappes, J.T.W.: Artificial intelligence powered automation for industry 4.0. In: Nayyar, A., Naved, M., Rameshwar, R. (eds.) New Horizons for Industry 4.0 in Modern Business. Contributions to Environmental Sciences & Innovative Business Technology. Springer, Cham (2023). https://doi.org/10.1007/978-3-031-20443-2_1
7. NATF Practices Document for NERC Reliability Standard CIP-014-2 Requirement R5: https://www.nerc.com/pa/comp/guidance/EROEndorsedImplementationGuidance/CIP-014-2_R5_Developing_and_Implementing_Physical_Security_Plans_(NATF).pdf
8. McFadden C: A Brief History and Evolution of Electric Cars, The history of the electric car is much longer than you might think. In this article, we take a whistle-stop tour of the evolution of the EV, 01 July 2020. https://interestingengineering.com/transportation/a-brief-history-and-evolution-of-electric-cars
9. Massar, M., Reza, I., Rahman, S.M., Abdullah, S., Jamal, A., Al-Ismail, F.S.: Impacts of autonomous vehicles on greenhouse gas emissions-positive or negative? Int. J. Environ. Res. Public Health 18(11), 5567 (2021). https://doi.org/10.3390/ijerph18115567

10. Al Dhaen, E.S.: The use of information management towards strategic decision effectiveness in higher education institutions in the context of Bahrain. Bottom Line 34(2), 143–169 (2021). https://doi.org/10.1108/BL-11-2020-0072
11. U.S. energy use rises to highest level ever: https://www.llnl.gov/news/us-energy-use-rises-hig hest-level-ever
12. Kanan, M., Taha, B., Saleh, Y., Alsayed, M., Assaf, R., Ben Hassen, M., Alshaibani, E., Bakir, A., Tunsi, W.: Green innovation as a mediator between green human resource management practices and sustainable performance in Palestinian manufacturing industries. Sustainability **15**(2), 1077 (2023). https://doi.org/10.3390/su15021077
13. Cohen N.: How Machine Learning Can Boost the Value of Wind Power (2019). https://techxp lore.com/news/2019-02-machine-boost-power.html
14. Thorin, E., Sandberg, J., Yan, J.: Combined Heat and Power. https://www.diva-portal.org/ smash/get/diva2:1057331/FULLTEXT01.pdf
15. U.S. Department of Energy DOE Awards $3.6 Million to Promote Equity And Diversity in Clean Energy Innovation DOE's "Inclusive Energy Innovation Prize" Winners Receive Cash Awards to Implement Strategies Prioritizing Climate Solutions for Underrepresented Com-munities; 80% of the Applications Were First Time DOE Applicants, 31 May 2022, https://electricenergyonline.com/article/energy/category/climate-change/82/962824/doe-awa rds-3-6-million-to-promote-equity-and-diversity-in-clean-energy-innovation.html
16. EITPIC Awardee Profile: Siemens Corporation Develops Green Technologies Digital Companion for the Grid, 20 Feb 2020. https://www.energy.gov/oe/articles/eitpic-awardee-pro file-siemens-corporation-develops-green-technologies-digital-companion

Algorithms Control Contemporary Life

Ismail Noori Mseer, W. M. Abd-Elfattah, and A. H. Al-Alawi

Abstract Today, algorithms can influence many different elements of your life, including what you buy, where you live, whether you land a job or a bank loan, and many other things. Now, autocomplete can anticipate your words in search terms, Gmail messages, and text messages. Even Tinder is subject to algorithmic control–did you choose your partner, or did Tinder? People are aware of the impact of the algorithms used by digital companies and what the typical person can accomplish when confronted with powerful international corporations. Many people share the opinion that, when it comes to sophisticated technology and the algorithms that have been unleashed on us, we are all relatively defenseless as humans. However, I believe that while individual effort alone won't be enough to address this issue, we do have some power in this situation, and that power comes in the shape of our knowledge, our votes, and our money. The concept of knowledge—being aware of the technology we use and what's going on with it behind the scenes—is rather simple, but I believe it is underappreciated. We should make more thoughtful decisions while using technologies and algorithms rather than being merely passive users of them. We need to consider how algorithms affect the choices we make and the judgments others make about us. I think that what Facebook is doing and what they announced last week regarding changes to their products, how they are going to support message encryption and sort of treat messages and sort of appreciate people's privacy needs are all a direct result of pushback from users.

Keywords Algorithms · Digital companies · Artificial hybrids · Software algorithms

---

I. N. Mseer (✉)
Ahlia University, Manama, Bahrain
e-mail: imseer@ahlia.edu.bh

W. M. Abd-Elfattah · A. H. Al-Alawi
Jeddah College of Engineering, University of Business & Technology, Jeddah, Saudi Arabia

© The Author(s), under exclusive license to Springer Nature Switzerland AG 2024
A. Hamdan and E. S. Aldhaen (eds.), *Artificial Intelligence and Transforming Digital Marketing*, Studies in Systems, Decision and Control 487,
https://doi.org/10.1007/978-3-031-35828-9_72

1 Introduction

Keep in mind that these systems and technologies running the algorithms that control our lives are artificial hybrids; they are not entirely independent. The sorting is done at previously unimaginable speeds and scales by software algorithms and machine-learning models, but the mechanism is built by humans, and the training data sets are created by humans. Once in use, data-driven algorithms and models continue to learn from both the activities of their users as well as from fresh instructions provided by the authors of the mechanism. In reality, the enormous algorithmic systems of Google, Facebook, and other social media sites combine older machine-learning models with specifically programmed sorting methods. For instance, Google search considers hundreds of parameters, but machine learning can only identify a small number of them. The order of the results is determined by adding these qualities together to create a score. Google search is now more individualized, with results based on the preferences and interests of the single user posing the query as well as what the engine anticipates all users will favor. In term of social media use, because there is no one "correct" response, social media algorithms are much more complicated [1]. "Right" depends on each end user's-interests, which unlike with search, are not declared directly but must instead be deduced from past behavior, the interests of the person's friends, and other factors. These algorithms are examples of what investor George Soros has referred to as reflexive systems [2], in which some outcomes are determined by the collective beliefs of all system users (the market) rather than being objectively true or correct.

These systems are able to learn from and respond to their surroundings to take into consideration a variety of aspects when making decisions. In addition, this allows the systems to constantly improve their outputs based on new knowledge. The definition of "intelligence" does not include self-awareness or volition, but it is still a pretty good one, and in that sense, I the individual machine components of these systems cannot be regarded of as intelligent. The data used to train a model can skew the findings, just as with humans. Nevertheless, these algorithms have produced amazing outcomes, and they consistently outperform human capabilities. In these hybrid systems, humans are still ostensibly in charge, but fresh information is frequently recognized and handled automatically. Machine-learning models are replacing outdated, hand-coded algorithms created by human programmers because they can react to changes in enormous volumes of data before a human programmer may ever detect the difference.

However, there are times when the changes in the data are so significant (such as with deepfake videos, Astro-turfed content produced at scale by bots impersonating humans, or makeup specifically created to fool facial recognition systems [3], that humans must create and train new digital subsystems to recognize them. The creators of human mechanisms are also always seeking methods of enhancing their designs.

Algorithms Control Contemporary Life   861

2 Maintain Search Quality

Along with success stories like anti-spam and credit card fraud-detection systems, Google search's decades of successful updates that maintain search quality in the face of massive amounts of new information, adversarial attacks, and changes in user behavior serve as a foundation for understanding how to regulate the artificial intelligence (AI) of the future. Human control is communicated through a set of desired outcomes rather than a rigid set of rules. To accomplish those results, the regulations are modified frequently. Systems controlled in this manner mark a significant departure from earlier, rule-based governance structures.

Any governing system that strives to establish a set of rigid laws once and for all is doomed to failure. The decision of desired end, measurement of whether that outcome is being accomplished, and a continuously updated set of methods for accomplishing it are the three main components of governance. The micro governance of ongoing modifications in response to fresh data, manifested through the development of improved algorithms and models, is the overall control of the outcome that algorithms and models are optimized for. Technology firms of today are quite proficient at level 1, but at level 2, they experience difficulties. There is a weakness in the outcomes-based governance model. Systems using algorithms are focused optimizers. Like the genies of Arabian legend, they carry out their masters' orders regardless of the repercussions, and there frequently are unintended and undesired outcomes.

The difficulty of expressing what you want succinctly applies to statements made in daily language, legalese, or programming languages, according to Peter Norvig, Google's head of research and co-author of the top textbook on AI. One benefit of machine learning over conventional systems is this. Similar to how human courts rely on case law, we give these systems examples of what we believe to be good and wrong rather than attempting to describe them in a single sentence. The arrogance of believing it is possible to grant the genie a coherent request is another aspect of the issue. Norvig makes the point that we should accept that mistakes will happen, and we should apply safety engineering concepts. If only King Midas had stated, "I want everything I touch to turn to gold, but I want an undo button and a pause button," as he said to me, "Then he would have been okay."

The benefits can be substantial if the outcome is carefully selected and guided by the interests of the mechanism's designer and owner, in addition to the system's users and society in general. For instance, Google stated the company's aim is "To organize the world's knowledge and make it generally accessible and valuable." Few would contest Google's significant advancements in achieving that objective. However, in a hybrid system, objectives that are expressed in human terms must be converted into the mathematical dialect of the robots. This is accomplished by using an objective function, whose value is to be maximized (or minimized). Google's search algorithms are relentlessly tuned to deliver results that please users, as evidenced by the fact that they leave and don't perform the same search again. Originally, the results were a list of links to websites, but now, for many searches, they are genuine answers.

Facebook also claims to have a good cause. Its mission is, "To empower individuals to forge ties with others and unite the planet." Facebook, however, gave its systems the responsibility of optimizing for what can be widely referred to as engagement, considering things like how much time users spend on the site and how many posts they read, like, and comment on. The designers of the system had hoped that this would help users develop tighter bonds with their friends, but as we now know, it fueled rivalry, addiction, and several other negative traits. Additionally, outsiders figured out how to manipulate Facebook users for their own purposes by learning how to game the system.

Similar to how new tests were added to the vending machines of my youth to prevent them dispensing candy bars in exchange for worthless metal slugs, Facebook has advanced at the microlevel of governance by focusing on hate speech, fake news, and other flaws in the newsfeed curation carried out by its algorithmic systems. However, the company is still having trouble determining how to mathematically define the human urge to bring people together so that its genies will deliver the desired result. What concerns me the most is the query: What are Facebook's alternatives if more engagement with its services is not genuinely beneficial to Facebook users? [4] The expansion of consumers and usage determines the company's value. Its premium advertising model is built on microtargeting, in which information about consumers' interests and behaviors can be utilized for or against them. Facebook appears to side with the marketers far too often when the interests of its users and those of its advertisers' conflict, going as far as to knowingly accept misleading [5] advertising over the objections of its own staff.

Despite Google's history of success in continuously improving its search engine for the benefit of its customers, its YouTube division succumbed to many of the same issues as Facebook [6]. Given that it was unable to adopt the strategy of "give them the proper answer and send them away" for the bulk of its queries, YouTube opted to instead focus on increasing user engagement, which led to disinformation issues that were at least as problematic as those plaguing Facebook—if not worse. The clarity of Google's original position on the conflict of interest between its users and its advertisers, which Google cofounders Larry Page and Sergey Brin had recognized in their original 1998 research paper on the Google search engine, appears to have changed, even within the search engine unit. They stated, "We expect that advertising funded search engines will be inherently biased towards the advertisers and away from the needs of the consumers," in an appendix titled "Advertising and Mixed Motives."

At the time, Page and Brin argued that there should be a non-profit academic search engine to address this issue. However, they believed they had found a method to balance the interests of the company's two main constituencies with the introduction of pay-per-click advertising, where advertisers are only charged when a user clicks on an advertisement—likely because the user thought it would be beneficial for them. Google likewise established a distinct division between the systems that serviced its end users and the systems that served its advertisers throughout the first ten or so years of the company. Ad results were calculated separately and shown in a completely different manned than organic search results. But over time, the lines started to fudge.

Algorithms Control Contemporary Life

Ads, which were formerly in a secondary position, and the ability to distinguish them from organic search result was lost.

Additionally, Google appears to have reexamined its relationship with the Web's content providers. The business was founded as a means of connecting information suppliers with information seekers in an enormous new market for human collective intelligence. Its role was to act as an unbiased middleman, searching through what would eventually amount to trillions of Web pages to discover the one that provided the greatest response to each of the trillions of searches made each year. In addition to the success of its users, Google's success was also determined by the success of the other websites it directed users to.

'We want you to come to Google and locate what you want quickly,' Page stated in an interview that was included with the Form S-1 filing for Google's 2004 IPO. 'Afterward, we'll be pleased to direct you to the other websites. That is the key idea. The goal of the portal strategy is to control all the information. Most portals prioritize their own material over content from other websites. We consider that to be a conflict of interest, comparable to accepting payment for search results. Their search engines deliver the portal's results, not always the best ones. Google makes a conscious effort to avoid doing that. We want to direct you away from Google and toward the appropriate location as soon as we can. It is a highly distinctive model.'

At the time, Page and Brin appear to have realized that success did not imply solely their success—or even that of their clients and advertisers—but also meant success for the network of providers whose content Google was built to search. The early genius of Google was in juggling the conflicting interests of all those various groups. This is AI's promising future. The opportunity for AI is to help humans model and manage complex, interacting systems, as Paul Cohen, a former DARPA program manager for AI who is now dean of the School of Computing and Information at the University of Pittsburgh, once stated. However, 15 years after Larry Page stated that Google's goal was to send users on their way, more than 50% of all searches on Google end on Google's own information services, with no click-through to third-party sites.

The transition from traditional, human-directed enterprise to a new type of human–machine hybrid that binds employees, customers, and suppliers into a digital, data-driven, algorithmic system is not limited to businesses like Google and Facebook. It encompasses all businesses for which stock is traded on open markets. Modern companies are referred to as having "slow AIs" by science fiction author Charlie Stross. These AIs are already carrying out a set of instructions that directs them to optimize for the wrong goal and to perceive human values as impediments, just as Bostrom's paper clip maximizer.

How else can your account for a system that views workers as expenses to be cut and customers and communities as resources to be used to one's advantage? How else would your account for pharmaceutical firms that made a purposeful effort to mislead authorities about the strength of addiction to the opioids they were marketing, ultimately leading to a catastrophic health crisis? How else can you account for decades of cigarette company disease denial, decades of fossil fuel company denial of climate change, enormous accounting fraud, and avoidance of paying the taxes that support their host nations? It is the machine that is in command, and like all such machines, it thinks only in terms of mathematics, with the goal of maximizing an objective function.

Everyone who prioritized maximizing shareholder wealth over everything else in our markets and enterprises held the belief that doing so would increase prosperity for people. Milton Friedman stated in 1970 that a corporation's main social responsibility is to grow its profits, and thus, shareholders, who are the recipients of those gains, would be free to choose how to best employ them [7]. He had no idea that the relentless quest for corporate profit would lead to a race to the bottom of falling wages, environmental destruction, and social deterioration. However, after 1976, when Jensen and Meckling [8] argued that paying executives in company stock was the optimum mechanism for maximizing shareholder value, the goal of the machine overrode that of the human managers.

We now understand that Friedman, Jensen, and Meckling misjudged the outcomes they anticipated [9], but the process has been established and codified in law. The people who created it have passed away, and those who are now ostensibly in charge— our legislators, policymakers, economic planners, and government executives—no longer fully comprehend what was created or cannot agree on how to reform it. Government has also evolved into a sluggish AI. In the book *The Machine Stops*, Forster stated [10] "We constructed the Machine, to do our will, but we cannot make it do our will now. If it could function without us, it would allow us to perish because we are merely the blood corpuscles that flow through its arteries. In accordance with instructions, the paper clip maximizer keeps working." [11].

3 Conclusion

We humans do everything we can to thwart this unrelenting demand from what was once our algorithmic servant but is now our algorithmic master due to ignorance and negligence [12]. We adopt lofty ideals like those stated by the Business Roundtable, vowing to consider not only the requirements of the corporation but also those of the workforce, customers, the environment, and society at large. Until we acknowledge that we have created a machine and put it on its course, attempts at this form of governance will be ineffective [13]. Instead, we fail to hold the people who created the market's mechanism accountable and instead pretend that it is a natural occurrence better left alone. That machine needs to be dismantled, rebuilt, and reprogrammed with flourishing of humans as its ultimate purpose rather than corporate profits. We must realize that we cannot only declare our values [14]. Our values must be put into practice in a way that our machines can comprehend and use.

And we must approach this with extreme humility, admitting our lack of knowledge and our propensity to fail. We must design procedures that not only continuously assess whether the mechanisms we have created are accomplishing their goal [15], but also continuously assess whether that goal is the right representation of what we actually want. But even that might not be sufficient. The apparatus we develop must work under the assumption that it does not know the proper purpose, as Russell

Algorithms Control Contemporary Life 865

points out in Human Compatible. Our monitoring is insufficient because if a truly self-aware AI has a single-minded goal, it may try to stop us from changing it—or from even realizing that a change is necessary. AI governance is not an easy task. It entails fundamentally rethinking how we run our businesses [16], markets, and society—not merely overseeing a new technology on its own. One of the biggest challenges of the twenty-first century, this task will be incredibly difficult, but it also presents a fantastic opportunity.

References

1. Bishop, J.M.: Artificial intelligence is stupid and causal reasoning will not fix it. Front. Psychol. **11**, 513474 (2021). https://doi.org/10.3389/fpsyg.2020.513474
2. Euronews, A.P., Reuters, 25/05/2022, Davos 2022: George Soros Blames AI and new Tech for Helping Repressive Regimes. https://www.euronews.com/next/2022/05/25/davos-2022-george-soros-blames-ai-and-new-tech-for-helping-repressive-regimes-like-russia
3. Kevin Townsend on April 11, 2022, The Art Exhibition That Fools Facial Recognition Systems. https://www.securityweek.com/art-exhibition-fools-facial-recognition-system
4. Gomes de Andrade, N., Pawson, D., Muriello, D., et al.: Ethics and artificial intelligence: suicide prevention on Facebook. Philos. Technol. **31**, 669–684 (2018). https://doi.org/10.1007/s13347-018-0336-0
5. Popenici, S.A.D., Kerr, S.: Exploring the impact of artificial intelligence on teaching and learning in higher education. RPTEL **12**, 22 (2017). https://doi.org/10.1186/s41039-017-0062-8
6. Head, A.J., Fister, B., MacMillan, M.: information literacy in the age of algorithms, Student experiences with news and information, and the need for change. In: Project Information Literacy (2020)
7. Neal Hartman Jan 11, 2021, Social Responsibility Matters to Business—A Different View from Milton Friedman from 50 Years Ago. https://mitsloan.mit.edu/experts/social-responsibility-matters-to-business-a-different-view-milton-friedman-50-years-ago
8. Brian Cheffins, April 4, 2021, The Most Famous Article on the Theory of the Firm is Widely Misunderstood. https://www.promarket.org/2021/04/04/theory-firm-misunderstood-michael-jensen-william-meckling/
9. Reprinted in Jensen, M.C.: A Theory of the Firm: Governance, Residual Claims and Organizational Forms. Harvard University Press, 2000. Available at http://hupress.harvard.edu/catalog/JENTHF.htm
10. Naudé, W.: Artificial intelligence: neither Utopian nor apocalyptic impacts soon. Econ. Innov. New Technol. **30**(1), 1–23 (2021). https://doi.org/10.1080/10438599.2020.1839173
11. Gans, J.: AI and the Paperclip Problem (2018). https://cepr.org/voxeu/columns/ai-and-paperclip-problem
12. I am a robot, 8 Sep 2020, A robot wrote this entire article. Are you scared yet, human? https://www.theguardian.com/commentisfree/2020/sep/08/robot-wrote-this-article-gpt-3
13. Anderson, J., Rainie, L.: Artificial Intelligence and the Future of Humans (2018). https://www.pewresearch.org/internet/2018/12/10/artificial-intelligence-and-the-future-of-humans/
14. Gabriel, I.: Artificial intelligence, values, and alignment. Mind. Mach. **30**, 411–437 (2020). https://doi.org/10.1007/s11023-020-09539-2
15. Mikalef, P., Gupta M.: Artificial intelligence capability: conceptualization, measurement calibration, and empirical study on its impact on organizational creativity and firm performance. Inf. Manage. 58(3), 103434 (2021)
16. Papagiannidis, E., Enholm, I.M., Dremel, C., et al.: Toward AI governance: identifying best practices and potential barriers and outcomes. Inf Syst Front (2022). https://doi.org/10.1007/s10796-022-10251-y

Auditors' Perceptions in Gulf Countries Towards Using Artificial Intelligence in Audit Process

Ahmad Yahia Mustafa Alastal, Janat Ali Farhan, and Mohammad H. Allaymoun

Abstract The study aims to examine the perception of auditors in Gulf countries (United Arab Emirates, Kingdom of Bahrain, Kingdom of Saudi Arabia, Sultanate of Oman, State of Qatar, State of Kuwait) about using artificial intelligence (AI) in the auditing processes. The data were collected by using an online questionnaire which was distributed to 200 auditors from gulf countries. 50 responses were collected. The findings revealed that using AI systems and tools in auditing will automate routine audit processes and procedures, allowing more time to focus on areas of significant judgment, understanding of the entity and its processes, enable ongoing risk assessment throughout the audit process. However, the respond emphasized that AI is not easy to use in auditing which needs more training and knowledge about it. The current study recommends more studies about AI and their relationship with auditing processes in other developing countries.

Keywords Artificial intelligence · Audit · Audit quality · Sustainability development goals

The development and globalization of technology continued to put pressure on accountants, and with the continuous adoption of modern technologies such as artificial intelligence (AI), it became necessary for every accountant to organize the use of these technologies in the field of his job, so the profession of auditing and accounting,

---

A. Y. M. Alastal (✉) · J. A. Farhan
Accounting and Financial Science Department, College of Administrative and Financial Science, Gulf University, Sanad 26489, Kingdom of Bahrain
e-mail: dr.ahmed.alastal@gulfuniversity.edu.bh

J. A. Farhan
e-mail: 180101114653@gulfuniversity.edu.bh

M. H. Allaymoun
Administrative Science Department, College of Administrative and Financial Science, Gulf University, Sanad 26489, Kingdom of Bahrain
e-mail: dr.mohammed.allaymoun@gulfuniversity.edu.bh

© The Author(s), under exclusive license to Springer Nature Switzerland AG 2024
A. Hamdan and E. S. Aldhaen (eds.), *Artificial Intelligence and Transforming Digital Marketing*, Studies in Systems, Decision and Control 487,
https://doi.org/10.1007/978-3-031-35828-9_73

which was previously considered hard and tiring work, has now become faster and easier with development through the use of the most important modern technologies based on knowledge [1]. One of the significant and cutting-edge methods that have arisen in many disciplines is the use of AI in its applications and procedures, as well as the potential for the audit profession to benefit from its features. AI networks are one of the methods and tools that have advanced and modern characteristics and can effectively play the role of the auditor, It helps in detecting errors and fraudulent practices that some companies may engage in that are difficult to detect using traditional methods of auditing, also It can be used in place of or in addition to conventional auditing techniques [4, 6, 7].

The trend for providing services and goods to customers in the twenty-first century has been toward advancement, improvement, and diversification. The goal of the modern field of AI is to employ devices and technologies to mimic human thought and behavior. It changed from being a phenomenon of the late twentieth century to becoming a common occurrence and application in many businesses at the beginning of the twenty-first century [2, 5]. AI consists of a number of pieces and elements that are now required while advising and making judgments in the fields of administration, economics, engineering, agriculture, and medicine. AI is the use of advanced technology to mimic human functions including thinking, speaking, analyzing, and feeling [13].

According to [8] to improve the effectiveness and efficiency of the audit process, audit firms are investing a lot of money in cutting-edge technological technologies. Furthermore, the "big 4" audit firms spends more that 250 million every year on IT [10]. Law and Shen [11] emphasized that, they are three potential advantages of utilizing AI in the auditing process have been identified by prior studies: Improve the process of identifying substantial misstatements, improve the client's operations and the risks associated with them, and improve communications with those in charge of governance.

Previous studies did not address the auditors perception in gulf countries about using AI in their audit work. By investigating the perceived usability, utility, and contribution to audit quality of various types of AI systems, our descriptive study fills this gap. Also, the current study intends to examine whether artificial intelligence systems are flexible and easy to use or not, examine whether that artificial intelligence systems improve the quality of audit outputs or not, and examine whether artificial intelligence systems detect human errors or not.

1 Literature Review

Noordin et al. [12] Examined the external auditors' perceptions and analyzes of the use of artificial intelligence in the United Arab Emirates, the study compared the perceptions of external auditors through the contribution of RI to the quality of audit outputs and its positive impact on the final audit results. Data was collected through several local and international companies to achieve the desired objectives of the study. In the process of testing the hypotheses identified in the study, reliability tests, descriptive analysis and t-test were used for independent samples. The study showed that there is an insignificant difference in the contribution of artificial intelligence to the audit process between local and international companies. All audit firms have equal contributions in terms of audit quality, and regardless of the type of audit business they work for, auditors can use AI to help auditors to improving their technical skills and deduction errors.

Fedyk et al. [9] examined the impact of AI on the output and quality of auditing by collecting a set of data for more than 310,000 CVs in the 36 largest auditing firms to determine the companies' employment of workers in the field of AI. The empirical analyzes in their study were supported by interviews with 17 audit partners in the largest public accounting firms in the United States of America. The results showed that first, artificial intelligence was developed centrally, second, artificial intelligence was widely used and applied in audit work, and third, the goal the main application of artificial intelligence is to improve audit quality first, and then improve efficiency.

Afroze and Aulad [1] explored the most important perceptions of accountants about the application of artificial intelligence in their job in Bangladesh. In their study, audit practitioners working in different audit offices in the city of Dhaka were consulted. Data was collected by conducting a survey questionnaire, and we obtained results that show us that professionals in Bangladesh do not have sufficient information about the importance of using artificial intelligence in the audit profession, and there are some problems that It puts an obstacle to artificial intelligence in Bangladesh such as the high cost of modern technologies in addition to the lack of information security and unsecured jobs.

Albawwat and Frijat [3] examined on their descriptive study, which is easy to use, useful, and contributes to the audit quality of several types of RI. On their study, a questionnaire was conducted to collect data from auditors in Jordan, and the results showed that the auditors consider the RI audit systems as secondary and strengthening factors to facilitate the audit process, while realizing that these systems are complex to use. In addition, there are some auditors who do not recognize the importance and capabilities of independent artificial RI and considered that they are not beneficial to the audit process.

2 Methodology

This exploratory study intends to examine the perception of auditors in Gulf countries (United Arab Emirates, Kingdom of Bahrain, Kingdom of Saudi Arabia, Sultanate of Oman, State of Qatar, State of Kuwait) about using artificial intelligence (AI) in the auditing processes. The current study adopted the questionnaire from study [3]. The questionnaire content from two section, demographics questions such as gender, educational level, professional certifications they hold, and finally the number of years of experience as an auditor. The second section asked about the use of AI in auditing process.

The data were collected by using an online questionnaire which was distributed to 200 auditors from gulf countries. 50 responses were collected.

3 Findings and Discussion

This section provides the findings and discussion of the current study, it contents from two section the demographic analysis section and the second section: measures auditors' perceptions of the use of artificial intelligence in the Kingdom of Bahrain and the GCC countries.

3.1 Demographics Analysis

For the initial background analysis, the reviewers were asked to provide information about gender, educational level, professional certifications they hold, and finally the number of years of experience as an auditor. The results shows that, the number of females was 19, at a rate of (39%), and the number of males was 31, at a rate of (62%). As for the second category related to age, the dominant category was the group (30–39 years) with a rate of (32%), followed by the group (20–29 years) with a rate of (28%). As for the educational qualification category, 15 (30%) of the auditors were holders of a diploma or less, Ph.D. degree. As for professional certificates, the majority (67%) have a ACCA, while the rest of the auditors have been qualified as certified accountants. Regarding the years of experience in the auditing profession, 23 auditors (45%) of them have experience from 1 to 5 years. Among the respondents, 20 auditors (40%) have worked as auditors for 6 to 10 years. In addition, 4 (8%) auditors had from 11 to 15 years of experience, and the remaining percentage had more than sixteen years of experience.

3.2 Auditors' Perceptions of the Use of Artificial Intelligence

Through Fig. 1, the results were related to the question: Will it be easy for individuals to learn how to operate AI systems and tools in auditing? The result of the majority of reviewers was that the system would not be very easy, so the result (46%) of the auditors chose Level 3, which is normal. And from the results of the majority, we can say that the majority of auditors may find it difficult to learn advanced technology systems of AI, and there must be continuous plans and support for auditors, as they need to adopt easy techniques of AI.

Figure 2 shows us the results of the question that it is easy to obtain artificial intelligence tools and systems to do what the auditors need in the audit process, so the result showed us that the majority of auditors agreed with the approval rate

Fig. 1 Learning to operate AI systems and tools in auditing would be easy for me

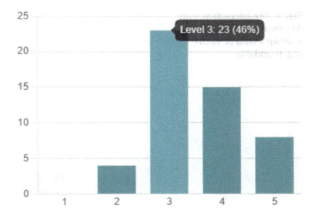

(38%) and (38%) strongly agreed. The remaining percentage (14%) was divided into normal and a very small percentage (10%) did not agree. In addition, it is clear that the majority of reviewers consider it easy at the present time to provide artificial intelligence tools and systems due to the tremendous progress in various developed countries and the great scientific renaissance, so the urgent need for AI machines has become and work to develop its technological and scientific tools and related research in addition to skill development and study experiences related to artificial intelligence.

Figure 3 shows that more than 85% of the auditors (48% are normal, and 40% agree) whose interaction with AI tools and systems in the auditing profession will be understandable and clear. This result gives an indication that many auditors may have difficulty understanding AI.

Figure 4 shows that, the auditor's perception to interact with AI system and tools in auditing to be flexible. Approximately (45%) agree and (29%) strongly agree. The results show us that the majority of auditors find it easy for them to adapt to AI

Fig. 2 I would find it easy to get AI systems and tools to do what I want it to do in auditing

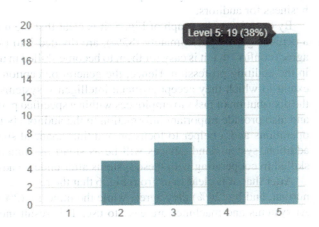

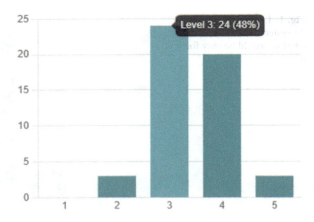

Fig. 3 My interaction with AI systems and tools in auditing would be clear/understandable

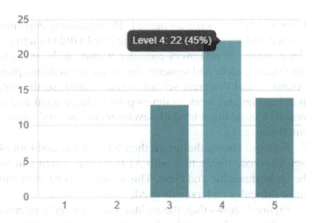

Fig. 4 I would find AI systems and tools in auditing to be flexible to interact with

systems because these systems provide more capacity and flexibility in dealing with business for auditors.

By examining the graph of Fig. 5, it is clear that the majority of reviewers, with a percentage of approximately (85%), are divided into (43% strongly agree, 37% agree) confirm that it is easy for them to become skilled in using AI tools and systems in the auditing profession. Hence, the general perception of auditors shows us the extent to which they accept artificial intelligence systems because it will facilitate the distribution of tasks to employees within a specified period of time automatically, and also provide important information to the auditor about the validity of business operations and a helper to focus on auditing financial statements and accounts. In addition, systems and devices will be as smart as humans, so individuals will be skilled in cooperating with these systems after understanding them well.

After that, it is clear to us from Fig. 6 that the reviewers are distributed by (44%) normal, and by (24%) they agree, while the minority (2%) shows that they see that AI systems and machines are easy to use. This result indicates that auditors need

Fig. 5 It would be easy for me to become skillful with AI systems and tools in auditing

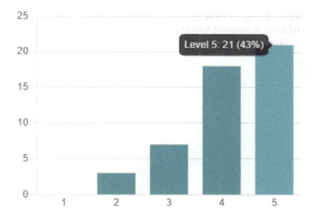

Fig. 6 I would find AI systems and tools in auditing easy to use

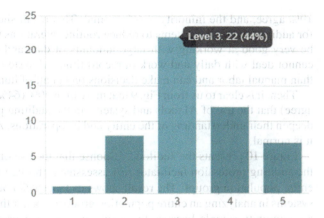

to develop their capabilities in AI by making machine learning applications and conducting training courses and programs for auditors in institutions.

As shown in Fig. 7, the majority of reviewers strongly agree (65%), and the remaining percentages are divided into "I agree" and "Normal". The vast majority of auditors believe that their use of AI systems and tools in the auditing profession helps them confirm their professional skepticism, because the use of AI helps auditors discover fundamental errors that only the sample cannot detect. In addition, AI systems qualify auditors to reach the highest levels of business and commercial data verification in the least possible time and effort, instead of spending a long time in the process of auditing and reviewing contracts with the possibility of errors, artificial intelligence machines facilitate this process and do it in record time with discovery Errors and eliminate professional skepticism for the auditor.

To test the impact of using AI tools and systems in the auditing profession to automate routine auditing processes and procedures, allowing more time to focus on important areas. As shown in Fig. 8, 55% of the reviewers strongly agree, while

Fig. 7 Using AI systems and tools in auditing will aid my professional skepticism

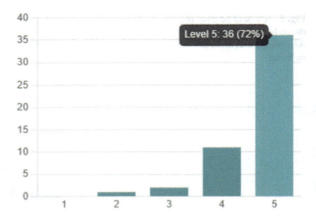

19% agree, and the minority is 6% normal. This result shows us the extreme need for auditors to use AI systems to reduce routine operations because the systems will be very good at working with large amounts of data and daily work that humans cannot deal with daily and work in record time. AI systems work faster and better than manual labor and can make decisions on behalf of humans.

Then, it is clear to us from Fig. 9 that more than 90% (64% strongly agree and 32% agree) that the use of AI tools and systems in the auditing profession helps auditors deepen their understanding of the entity and its operations. As for the minority (4%), it is normal.

Figure 10 presents the auditors' response that the use of AI tools and systems in the auditing profession facilitates risk assessment through the process of analyzing entire population groups. The result shows us that (65%) of auditors believe that AI systems in analyzing an entire population group are better than taking and examining a community sample because the result may be inaccurate and carry errors. The

Fig. 8 Using AI systems and tools in auditing will automate routine audit processes and procedures, allowing more time to focus on areas of significant judgment

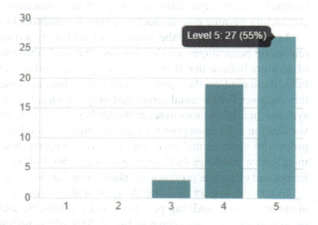

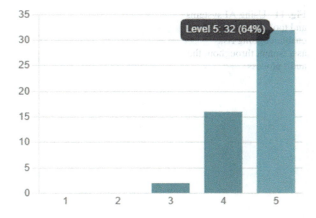

Fig. 9 Using AI systems and tools in auditing will deepen my understanding of the entity and its processes

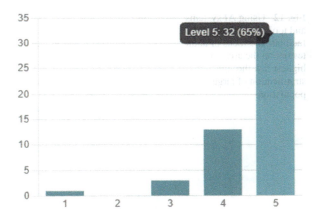

Fig. 10 Using AI systems and tools in auditing will facilitate robust risk assessment through the analysis of entire populations

remaining percentage was (27%) who agree, and the minority was divided into (6% normal, and 2% disagree).

In Fig. 11, it seems that the majority of auditors (86%) strongly agree that the use of AI tools and systems in the auditing profession leads to enabling the continuous assessment of risks during the auditing process. With a percentage of (24%) they agree, while the remaining percentage was (2%) normal and (2%) disagreed. Through the results in this figure, it seems to us that the tools and systems of AI work to eliminate human error or reduce the occurrence of risks because it is not negatively affected by the environmental conditions of work and does not lose focus, unlike the individual who may feel tired and exposed to making wrong decisions and not pay attention to some of the risks that occur during The audit process, unlike AI devices, is not weak in the same way as humans, and it is always accurate and works to assess and reduce risks.

As shown in Fig. 12, this can be explained by the fact that the use of AI tools and systems in the audit profession leads to facilitating the focus of the audit test

Fig. 11 Using AI systems and tools in auditing will enable ongoing risk assessment throughout the audit process

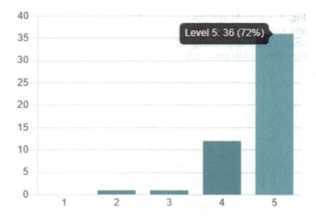

Fig. 12 Using AI systems and tools in auditing will facilitate the focus of audit testing on the areas of highest risk through stratification of large populations

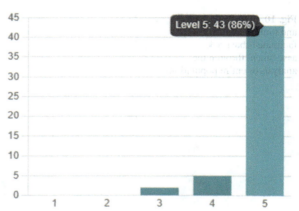

in the areas with the highest risks through the stratification of a large number of the population. The following figure shows us that the majority of reviewers (86%) strongly agree. The remaining percentage is distributed among those who agree and are normal. Auditors can take advantage of AI systems instead of relying on taking samples and checking them manually by auditors. Machine learning algorithms can provide auditors to examine the entire statistical community, and this helps them avoid risks resulting from sampling.

4 Conclusions

With the passage of time, AI systems have greatly helped the auditing profession, and AI technologies have become one of the most important factors that auditing institutions consider when planning their business and commercial services, and this aims to increase the exploitation of the huge data available to them. Through our survey, we have established the importance of AI for auditors in the Gulf countries and examined the perception of auditors in Gulf countries (United Arab Emirates, Kingdom of Bahrain, Kingdom of Saudi Arabia, Sultanate of Oman, State of Qatar, State of Kuwait) about using AI in the auditing processes.

The final results indicated that the majority of auditors consider that AI systems make them able to perform their business in an excellent manner in order to keep pace with customer expectations, keep pace with technological development, improve the quality of audit outputs and with the realization that AI systems are not easy to use and need continuous training to be fully able to accept these systems. Through this study, the results provided insights about the auditors' acceptance of AI systems. The study also found that with the use of this AI means, it will greatly help the auditors to get rid of their professional doubts about some business, as the systems will be able to detect fundamental errors in controlling the internal systems or in the financial statements that the auditors examine. Determine some unusual processes that are difficult to detect errors if they are examined using a sample.

In addition, the auditors saw that the use of AI systems and tools in auditing leads to the automation of routine auditing processes and procedures. This allows them more time to focus on important areas of judgment, and this indicates the importance and necessity of the presence of AI machines in auditing offices to improve the efficiency and effectiveness of auditing processes. At the same time, the results showed us that the auditors believe that the use of AI systems and tools will make it easier for them to assess risks by examining the entire statistical community. The majority of auditors rely on samples during the evaluation of audit projects because they face great difficulty in examining large amounts of documents and information manually. However, the use of samples is associated with (sampling risks), which occurs because the sample does not represent the statistical community, and this matter may eventually lead to inaccurate or wrong results, and this causes the auditor or the institution to make incorrect decisions because of the wrong results. For example, the auditor may conduct an audit in a particular institution and the results of the sample show him that the internal control system is strong and appropriate while it contains many fundamental weaknesses that were not revealed by the use of the sample.

At the same time, the findings of the current study are useful for the policy makers, ministries, regulators, universities in Gulf countries to identify the benefits of using AI programs in auditing process.

In the end, we conclude that any change that occurs requires many risks, challenges and obstacles, but the opportunities that we get as a result of using such techniques are always worth trying to overcome the challenges in order to achieve the continuous development of the auditing profession. The current study recommends more studies about AI and their relationship with auditing processes in other developing countries.

Acknowledgements Authors wish to thank Gulf University for funding this research.

References

1. Afroze, D., Aulad, A.: Perception of professional accountants about the application of artificial intelligence (AI) in auditing industry of Bangladesh. J. Soc. Econ. Res. 7(2), 51–61 (2020)
2. Ahmed, I., Jeon, G., Piccialli, F.: From artificial intelligence to explainable artificial intelligence in industry 4.0: a survey on what, how, and where. IEEE Trans. Ind. Inf. 18(8), 5031–5042 (2022)
3. Albawwat, I., Frijat, Y.: An analysis of auditors' perceptions towards artificial intelligence and its contribution to audit quality. Accounting 7(4), 755–762 (2021)
4. Allami, F.A.J., Nabhan, S.H., Jabbar, A.K.: A comparative study of measuring the accuracy of using artificial intelligence methods as an alternative to traditional methods of auditing. World Econ. Fin. Bull. 9, 90–99 (2022)
5. Allaymoun, M.H., Al Saad, L.H., Majed, Z.M., Hashem, S.M.A.: Big data analysis and data visualization to facilitate decision-making-mega start case study. In: Paper Presented at the International Conference on Business and Technology (2023)
6. Allaymoun, M.H., Hamid, O.A.H.: Business intelligence model to analyze social network advertising. In: Paper presented at the 2021 International Conference on Information Technology (ICIT) (2021)
7. Allaymoun, M.H., Qaradh, S., Salman, M., Hasan, M.: Big data analysis and data visualization to help make a decision-Islamic banks case study. In: Paper presented at the International Conference on Business and Technology (2023)
8. Alles, M.G., Gray, G.L.: Will the medium become the message? A framework for understanding the coming automation of the audit process. J. Inf. Syst. 34(2), 109–130 (2020)
9. Fedyk, A., Hodson, J., Khimich, N., Fedyk, T.: Is artificial intelligence improving the audit process? Rev. Acc. Stud. 27(3), 938–985 (2022)
10. Kokina, J., Davenport, T.H.: The emergence of artificial intelligence: how automation is changing auditing. J. Emerg. Technol. Acc. 14(1), 115–122 (2017)
11. Law, K., Shen, M.: How does artificial intelligence shape the audit industry? Nanyang Business School Research Paper (20–31) (2020)
12. Noordin, N.A., Hussainey, K., Hayek, A.F.: The use of artificial intelligence and audit quality: an analysis from the perspectives of external auditors in the UAE. J. Risk Fin. Manage. 15(8), 339 (2022)
13. Singh, J., Sajid, M., Gupta, S.K., Haidri, R.A.: Artificial intelligence and blockchain technologies for smart city. In: Intelligent Green Technologies for Sustainable Smart Cities, pp. 317–330 (2022)

Accounting Students' Perceptions on E-Learning During Covid-19 Pandemic: Case Study of Accounting and Financial Students in Gulf University—Bahrain

Ahmad Yahia Mustafa Alastal, Mohamed Abdulla Hasan Salman, and Mohammad H. Allaymoun

Abstract The study aims to know the perception of accounting and financial students on gulf university about e-learning during the covid-19 pandemic and to know the advantages of e-learning during the covid-19 pandemic according to the perception of gulf university students. The study used an online survey which distributed to all the students who used online accounting e-learning during covid-19, 83 respondents were collected from them. The study found that, the students agreed that the online e-learning was effective and has many advantages to them such as simple to use and they can to the material of the lecture anytime, the contact with the university lectures has improved, improved their communication with other students and the university succeeded by facilitate the accounting and financial courses with e-learning resources. Future studies can examine the perception of the accounting students in other Bahraini universities as well as in developing countries about the using of e-learning in teaching accounting and finance courses, and examine the links between e-learning accounting education and sustainability development goal No. 4 quality education.

Keywords Online accounting education · E-learning · Covid-19 pandemic · Accounting students perception · Sustainability development goal

---

A. Y. M. Alastal (✉) · M. A. H. Salman
Accounting and Financial Science Department, College of Administrative and Financial Science, Gulf University, Sanad 26489, Kingdom of Bahrain
e-mail: dr.ahmed.alastal@gulfuniversity.edu.bh

M. A. H. Salman
e-mail: 180101114630@gulfuniversity.edu.bh

M. H. Allaymoun
Administrative Science Department, College of Administrative and Financial Science, Gulf University, Sanad 26489, Kingdom of Bahrain
e-mail: dr.mohammed.allaymoun@gulfuniversity.edu.bh

© The Author(s), under exclusive license to Springer Nature Switzerland AG 2024
A. Hamdan and E. S. Aldhaen (eds.), *Artificial Intelligence and Transforming Digital Marketing*, Studies in Systems, Decision and Control 487,
https://doi.org/10.1007/978-3-031-35828-9_74

1 Introduction

The coronavirus that affected China started to spread globally in December 2019 [1]. In March 2020, the World Health Organization (WHO) declared the corona virus to be a pandemic and asked all nations to take the necessary precautions to limit its spread [2]. All industries and institutions, including those in education, were affected by this pandemic. One hundred eighty-nine countries around the world shuttered educational institutions to halt the pandemic, according to UNESCO, the United Nations Organization for Education, Science, and Culture [3].

Many Universities around the world have closed their campuses, switched to online learning, and delayed student activities [4]. By using online learning methods, universities expected to continue the learning process and achieve the objectives planned for each course [5, 6].

Several researches have noted a lot of advantages of online learning. Amer [7] highlighted that online learning decrease stress and increased satisfaction among lecturers and students. According to [8], online learning offers advantages to students such as efficient feedback and discussion, using class material anywhere and anytime, less costly than physically going to university, and enhances the dialogue between lecturers and their students. Furthermore [4] indicated that online learning is comfortable for students, and most students can access it. Other benefits include staying at home, accessing online materials easily, and learning at student's residences [9].

E-learning is a term that refers to a type of online learning tool that makes use of the internet. E-learning makes it easier for students and lecturers to communicate. Online assignments and teaching materials are available for lecturers to post for students to download. According to [10], using e-learning effectively has a big impact on students' comprehension. Additionally [11], research indicates that the utilization of e-learning affects student achievement.

The importance of online learning systems cannot be overstated, and there are numerous advantages to them, including higher technological knowledge, better training, better learning outcomes, flexibility, and control [12]. As per [13p. 58], "E-learning is a situated activity that takes place in a variety of locations and, if handled properly, can provide an ideal atmosphere promoting social interaction while also providing academic, social, and psychological benefits."

According to [14], technology is one of the seven critical talents that accountant recruiters look for. Students are appreciative of the use of technology and an online learning system in assessments, and online tests help accounting students perform better [15]. The amount of time that students spend on online learning platforms and how well they score on the final test for accounting courses are positively and significantly correlated [16]. This finding is confirmed by [17], who discovered that the performance of accounting students is boosted during online assessments, especially in courses that encompass synchronous and asynchronous interactions between students.

Online learning systems have a number of advantages, including cheaper educational costs, flexible access to course materials, and a solution to space limitations

[11, 18]. On the other hand, some earlier studies have highlighted the drawbacks of online learning platforms. The high cost of establishing online learning systems may be one of these challenges [12]. Additionally, an online learning system necessitates intense cooperation between students and teachers. Additionally, accreditation organizations oppose teaching in conventional university settings and online learning environments [11, 16, 17]. The adoption of online learning systems has a number of difficulties, according to [19], including a lack of funding, a lack of trust, managerial issues, and technological difficulties.

At the end of the year 2019, a virus spread in the world, and this virus is called Corona, which caused the disruption of academic and social life due to its rapid spread, and therefore the government of the Kingdom of Bahrain approved distance learning in all academic levels, and because of this law, Gulf University is among those covered by this law, and it was Studying in the distance period for financial accounting students in the last two years from the date of this research. The Gulf University used programs for online education, such as the Microsoft team program.

This study aims to examine the perception of gulf university students about E-learning during the covid- 19 pandemic, and to examine the advantages of e-learning during the covid- 19 pandemic according to the perception of gulf university students.

This was the first experience of online education at Gulf University completely for accounting and finance students, so this study is important for Gulf University to know the positive and negative points of developing this prototype, and for that we seek the help of students because they are the end users of this service provided by the university.

2 Literature Review

In 25 February 2020 Bahrain has declared the closure of all education institution because the spread of Covid-19 pandemic (UNESCO). The Ministry of Higher Education in Bahrain decided to totally discontinue traditional education that takes place in a classroom or face-to-face environment and replace it with e-learning through virtual classes at the beginning of the Covid-19 term in mid-February 2020 [20].

Regarding the assessment process, the ministry of education in Bahrain maintained constant communication with the universities to ensure the use of a fair and efficient system to assess the students' performance, such as online exams administered through the LMS, open-book exams, or the substitution of assignments for all exams [21]. Furthermore, Because of the shift to online learning, it is anticipated that the COVID-19 pandemic will be advantageous for the accounting industry. By switching to online learning, students will be able to save money and time, which will benefit them [22].

[23] looked at accounting students' perspectives when they switched from in-class to online learning during COVID-19. Their research was based on the opinions of academic staff from two of the biggest universities in New Zealand as well as data

gathered from students through paper evaluations. The study's findings showed that there were both achievements and difficulties with the pandemic's online accounting education. These difficulties include difficulties in engaging particular student groups in learning, displeasure with online learning and technology, and trouble forming personal connections between academic staff members and students.

Due to Covid-19, [24] discovered the views of accounting students using the e-learning platforms used by Saudi universities. In order to gather information for his work, the author used a cross-sectional survey method to interview 106 accounting students who had used Umm Al-Qura University's e-learning platform. The study's findings showed how students saw the advantages and disadvantages of online learning. These advantages included enhanced student–teacher contact, simplified accounting course study, increased adaptability, enhanced interaction with other students, and enhanced problem-solving skills. The survey also found that e-biggest learning's negatives were its technical issues, widespread reliance on computers, and lack of human interaction.

[25] Examined how cloud computing could be used to raise the standard of accounting instruction at Palestinian colleges during the epidemic. 63 faculty members who teach accounting in the management and commerce faculties of Gaza Strip institutions were surveyed to gather data. The study's findings indicated that the use of cloud computing by accounting faculty and students in accounting education enabled them to access their documents and programs at any time and from any location, especially during the pandemic, and that it also helped students develop the technological skills that the job market demands.

In their study, [26] compared the results of learning both before and after the pandemic. 367 Indonesian accounting teachers were given online surveys to complete in order to gather the data. The results showed that online learning was more successful if it was planned from the start prior to the pandemic.

3 Methodology

This exploratory study intends to determine how COVID-19 has affected accounting education and to know the accounting students' perception of the benefits of e-learning among gulf university students in accounting and financial departments.

The current study adopted the questionnaire from [27] study. There are five sections in the questionnaire. The effects of COVID-19 on accounting students' performance are examined in the first section. The second component gathers data on the digitization of accounting education, the third piece deals with the self-efficacy of the lecturers and teaching personnel, and the fourth section examines how the current epidemic has affected lecture times and instructional strategies. The demographic information about the respondents is then gathered in the final part.

This questionnaire was directed to accounting and finance students at the Gulf University, not only current students, but also graduates students from the Department of Accounting and Finance at the University who used e-learning. Furthermore, the questionnaire was created on the Google service method and sent to students of the Accounting and Finance Department—Gulf University by sending it in the form of a link to open the questionnaire via the well-known (WhatsApp) application personally to all my colleagues and in groups specialized only for students of the Accounting and Finance Department—Gulf University. After the distribution of the questionnaire to the student, 83 responses were collected from them.

4 Findings and Discussion

This section provides the findings and discussion of the current study, it contents from two section the demographic analysis section and the perception of accounting and finance student at Gulf University.

4.1 Demographics Analysis

Age: Total of 83 students of accountant and finance at gulf university who responded to the questionnaire, so 78.3% over the age of 20, 3.6% at the age of 18, 3.6% at the age of 19, and 14.5% at the age of 20.

Gender: Total of 83 students of accountant and finance at gulf university who respond to the questionnaire, 74.4% of them are females, and 25.6% of them are males.

Year of study: the respondent student 42.2% of the respondent students are in fourth year of their study, 39.9% in third year, 14.5 in second year, and 8.4% in the first year.

4.2 The Perception of Accounting and Finance Student at Gulf University

Figure 1 showed that, most of the student emphasized that, the E-learning is simple and understand. Remarkably, 54.2% (Agree + strongly agree) of them are agree that the E-learning is simple. Hence, from the majority result, we can say that the use of E-learning is not difficult for the student and simply they can use it.

Figure 2 showed that, most of the student emphasized that, the ability to study for accounting courses with more freedom is because to the utilization of e-learning. Remarkably, 54.2% (Agree + strongly agree) of them are agree that the utilization of e-learning gave them a freedom to study accounting courses. Hence, from the majority result, we can say that the utilization of E-learning is one aspect that helps the student to study accounting courses.

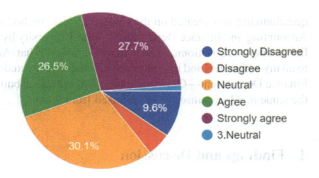

Fig. 1 The e-learning is simple to use and understand

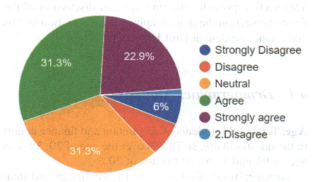

Fig. 2 Our ability to study for my accounting courses with more freedom is because to the utilization of e-learning

Figure 3 above showed that, most of the student emphasized that, the study of accounting courses has been facilitated by the usage of e-learning. Remarkably, 54.2% (Agree + strongly agree) of them agree that study of accounting courses has been facilitated by the usage of e-learning. Hence, from the majority result, we can say that the Facilitated of E-learning is one aspect that helps the student to study accounting courses.

Figure 4 showed that, most of the students emphasized that, the contact with their teachers has improved as a result of using e-learning. Remarkably, 54.2% (Agree

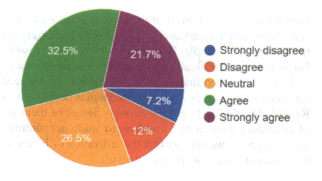

Fig. 3 My study of accounting courses has been facilitated by the usage of e-learning

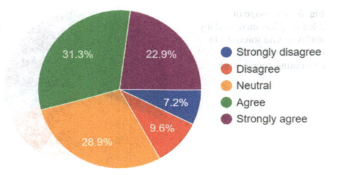

Fig. 4 My contact with teachers has improved as a result of using e-learning

+ strongly agree) of them agree that using e-learning improved their contact with the student teachers. Hence, from the majority result, we can say that the use of E-learning has advantages to the student such as improve their contact with their teachers.

Figure 5 showed that, most of the students emphasized that, the use of E-learning has helped them to communicate with other students better. Remarkably, 54.8% (Agree + strongly agree) of them agree that using e-learning improved their communication with their friends. Hence, from the majority result, we can say that the use of E-learning has advantages to the student such improve their communication with their friends.

Figure 6 showed that, most of the students emphasized that, the use of E-learning has increased their ability to find solutions that related to the accounting courses work. Remarkably, 56.6% (Agree + strongly agree) of them agree that using e-learning improved their ability to find solutions that related to the accounting courses work. Hence, from the majority result, we can say that the use of E-learning has advantages to the student such ability to find solutions that related to the accounting courses work.

Figure 7 showed that, most of the students emphasized that, the examination of accounting students' interactions with teachers was more thorough because of the COVID-19 pandemic. Remarkably, 49.4% (Agree + strongly agree) of them agree that COVID-19 pandemic affected their examination. Hence, from the majority result,

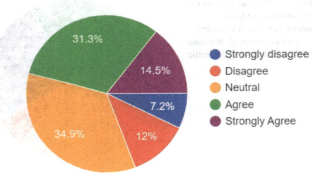

Fig. 5 E-learning has helped me communicate with my other students better

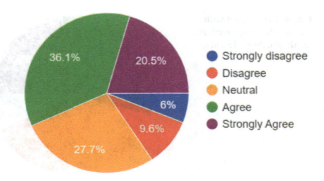

Fig. 6 The usage of e-learning has increased my ability to find solutions to issues related to my accounting course work

we can say that the examination of accounting students' interactions with teachers was more thorough because of the COVID-19 pandemic.

Figure 8 showed that, most of the students emphasized that Due to the COVID-19 pandemic, the midterm and/or final exams were given online using an open-exam format. Remarkably, 49.4% (Agree + strongly agree) of them agree that COVID-19 pandemic affected their exams type. Hence, from the majority result, because of COVID-19 pandemic the exams for midterm and final type have been change to open exam format.

Figure 9 showed that, most of the students think that, the use of E-learning will be globally requirement due to COVID-19 pandemic. Remarkably, 47.9% (Agree + strongly agree) of them agree that COVID-19 pandemic affected the way of teaching to become using E-learning. Hence, from the majority result, we can say that the use of E-learning has become compulsory requirement in the universities because of affected of COVID-19 pandemic.

Figure 10 showed that, most of the students think that, Because of COVID-19, in my opinion, digitizing accounting education would be the common trend of the education institutes. Remarkably, 53% (Agree + strongly agree) of them agree that COVID-19 pandemic affected the way of teaching to become using E-learning. Hence, from the majority result, we can say that the use of E-learning has

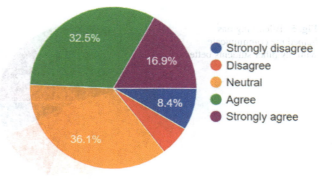

Fig. 7 The examination of accounting students' interactions with teachers was more thorough because of the COVID-19 pandemic

Fig. 8 Due to the COVID-19 pandemic, the midterm and/or final exams were given online using an open-exam format

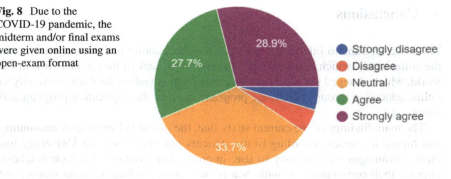

Fig. 9 Think that completely digitizing accounting education would be required globally due of the COVID-19 epidemic

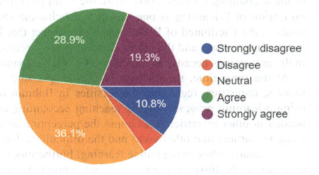

become compulsory requirement in the universities because of affected of COVID-19 pandemic.

Fig. 10 Because of COVID-19, in my opinion, digitizing accounting education would be the common trend of the education institutes

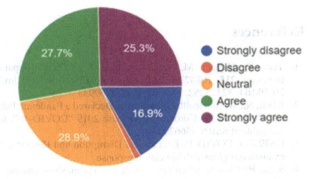

5 Conclusions

The research aims to inform the perception of accounting and finance students in the online study, which began in 2019 due to the outbreak of the Corona virus in the world, which changed the course of education from attending the Gulf University to online education through electronic programs such as the application program and zoom.

The main findings of the current study that, the use of E-Learning in accounting and financial courses according to the students perception in Gulf University has many advantages such as easy to use, improve their contact with their teachers, improve their communication with their friends, ability to find solutions that related to the accounting courses work. Also, the findings of the study indicated that, the utilization of E-learning is one aspect that helps the student to study accounting courses, the Facilitated of E-learning is one aspect that helps the student to study accounting courses, and the use of E-learning has become compulsory requirement in the universities because of affected of COVID-19 pandemic.

At the same time, the findings of the current study are useful for the policy makers, ministries, regulators, universities in Bahrain to identify the benefits of using e-learning programs during teaching accounting courses. Future research is needed in other countries to examine the perception of accounting students about using E-learning and other areas and the difficulties that both students and professors encounter while using online learning. Furthermore, further studies are needed to examine the links between e-learning accounting education and sustainability development goal No 4-quality education.

Acknowledgements Authors wish to thank Gulf University for funding this research

References

1. Yuki, K., Fujiogi, M., Koutsogiannaki, S.: COVID-19 pathophysiology: A review. Clin Immunol. **215**, 108427 (2020). https://doi.org/10.1016/j.clim.2020.108427. Epub 2020 Apr 20. PMID: 32325252; PMCID: PMC7169933
2. Jebril, N.: World Health Organization Declared a Pandemic Public Health Menace: A Systematic Review of the Coronavirus Disease 2019 "COVID-19", up to 26th March 2020 (2020). Available at SSRN 3566298
3. UNESCO: COVID-19 Educational Disruption and Response (2020). Retrieved from https://en.unesco.org/covid19/educationresponse
4. Sahu, P.: Closure of universities due to coronavirus disease 2019 (COVID-19): impact on education and mental health of students and academic staff. Cureus **12**(4), e7541 (2020)
5. Almarzooq, Z.I., Lopes, M., Kochar, A.: Virtual Learning During the COVID-19 Pandemic: A Disruptive Technology in Graduate Medical Education. American College of Cardiology Foundation Washington, DC (2020)
6. Kapasia, N., Paul, P., Roy, A., Saha, J., Zaveri, A., Mallick, R., Barman, B., Das, P., Chouhan, P.: Impact of lockdown on learning status of undergraduate and postgraduate students during COVID-19 pandemic in West Bengal, India. Child.Youth Serv. Rev. **116**, 105194 (2020)

7. Amer, T.: E-learning and education. Dar Alshehab publication, Cairo (2007)
8. Davis, A.M.: Measuring student satisfaction in online math courses. Kentucky J. Excellence College Teach. Learn. **14**, 21–37 (2014)
9. Bączek, M., Zagańczyk-Bączek, M., Szpringer, M., Jaroszyński, A., Wożakowska-Kapłon, B.: Students' perception of online learning during the COVID-19 pandemic: a survey study of Polish medical students. Medicine **100**(7), e24821 (2021)
10. Favale, T., Soro, F., Trevisan, M., Drago, I., Mellia, M.: Campus traffic and e-Learning during COVID-19 pandemic. Comput. Netw. **176**, 107290 (2020)
11. Moore, M. G., Kearsley, G.: Distance education: A systems view of online learning. Cengage Learn. (2011)
12. Turnbull, D., Chugh, R., Luck, J.: Transitioning to E-Learning during the COVID-19 pandemic: How have Higher Education Institutions responded to the challenge?. Educ. Inform. Technol. **26**(5), 6401–6419 (2021)
13. Chugh, R.: E-learning tools and their impact on pedagogy. Emerging paradigms in commerce and management education, pp 58–81 (2010)
14. Uwizeyemungu, S., Bertrand, J., Poba-Nzaou, P.: Patterns underlying required competencies for CPA professionals: a content and cluster analysis of job ads. Account. Educ. **29**(2), 109–136 (2020)
15. Aisbitt, S., Sangster, A.: Using internet-based on-line assessment: A case study. Account. Educ. Int. J. **14**(4), 383–394 (2005)
16. Perera, L., Richardson, P.: Students' use of online academic resources within a course web site and its relationship with their course performance: An exploratory study. Account. Educ. Int. J. **19**(6), 587–600 (2010)
17. Mardini, G. H., Mah'd, O. A.: Distance learning as emergency remote teaching vs. traditional learning for accounting students during the COVID-19 pandemic: Cross-country evidence. J. Account. Educ. **61**, 100814 (2022)
18. Roy, R., Al-Absy, M. S. M.: Impact of critical factors on the effectiveness of online learning. Sustainability, **14**(21), 14073 (2022)
19. Almaiah, M. A., Al-Khasawneh, A., Althunibat, A.: Exploring the critical challenges and factors influencing the E-learning system usage during COVID-19 pandemic. Educ Inform Technol **25**, 5261–5280 (2020)
20. Khalifa, H., Al-Absy, M., Badran, S., Alkadash, T.M., Almaamari, Q.A., Nagi, M.: COVID-19 pandemic and diffusion of fake news through social media in the Arab world. Arab Media Soc **30**, 43–58 (2020)
21. Taufiq-Hail, G.A.-M., Sarea, A.: Digitizing education during COVID-19 pandemic in Bahrain higher education context. In: COVID-19 Challenges to University Information Technology Governance, pp. 179–189. Springer (2022)
22. Meade, J.A., Parthasarathy, K.: Did COVID-19 impact student learning in an introductory accounting course? Bus. Educ. Innov. J. **12**(2), 18–23 (2020)
23. Ali, I., Narayan, A.K., Sharma, U.: Adapting to COVID-19 disruptions: student engagement in online learning of accounting. *Accounting Research Journal*, ahead-of-print (2020)
24. Ebaid, I.E.-S.: Accounting students' perceptions on e-learning during the Covid-19 pandemic: Preliminary evidence from Saudi Arabia. J. Manage. Bus. Educ. **3**(3), 236–249 (2020)
25. Alhelou, E.M., Rashwan, A.-R.M., Abu-Naser, S.S.: The role of using cloud computing in improving the quality of accounting education in Palestinian universities in light of the Covid-19 pandemic. J. Econ. Fin. Acc. Stud. **3**(1), 11–32 (2021)
26. Kustono, A.S., Winarno, W.A., Nanggala, A.Y.A.: Effect of accounting lecturer behavior on the level of online learning outcomes achievement. Int. J. Learn. Teach. Educ. Res. **20**(3), 169–187 (2021)
27. Sarea, A., Alhadrami, A., Taufiq-Hail, G.A.-M.: COVID-19 and digitising accounting education: Empirical evidence from GCC. PSU Res. Rev. **5**(1), 68–83 (2021)
28. Bali, S., Liu, M.: Students' perceptions toward online learning and face-to-face learning courses. J. Phys: Conf. Ser. **1108**, 1–7 (2018)

29. Hermawan, S., Hanun, N.R., Junjunan, M.I.: E-learning and understanding of accounting during COVID-19 pandemic. Int. J. Soc. Sci. Bus. **5**(1), 44–51 (2021)
30. Lazim, C., Ismail, N.D.B., Tazilah, M.: Application of technology acceptance model (TAM) towards online learning during Covid-19 pandemic: accounting students perspective. Int. J. Bus. Econ. Law **24**(1), 13–20 (2021)
31. White, A.: May you live in interesting times: A reflection on academic integrity and accounting assessment during COVID19 and online learning. Acc. Res. J., ahead-of-print (2020)

Factors that Influence the Occupational Safety and Health for Employees as a Part of Human Resource Management Practices a Study on Non-government Organization in Palestine Gaza Strip (UNRWA)

Tamer M. Alkadash ⓘ

Abstract To develop a theoretical framework to explain the relationships involved, the research set out to identify the factors that affect workers' occupational safety and health at the non-governmental organization in Palestine-Gaza Strip (UNRWA), the contribution of health and safety measures made in the workplace, as well as assess the impact of these factors on working conditions and the effects of healthcare professionals. The expense of non-compliance in terms of work-related injuries and workers' compensation as well as absence from duty serves as an example of the significance of health and safety. The study looked at how to change how both the administration and the staff feel about safety. It has outlined the necessary precautions to take to ameliorate the situation, as well as health and safety procedures. Any organization's performance is primarily dependent on the ongoing assessment and strengthening of health and safety procedures. The virtual component maintains the health and safety of the workers. A study of the literature on occupational health and safety from various publications was done for the research, and it was found that every company with more than five employees must have an effective health and safety policy and program in place. The study included several statistical analyses using secondary and primary data from the UNRWA in the Gaza Strip in Palestine, which has a bed capacity of more than eighty. Given the diversity of the population, quantifiable data from a survey of 300 employees was acquired to guarantee that all cadres were represented. The study's findings show that the UNRWA organization has sufficient facilities available.

Keywords Occupational safety and health · Human resource management · Influential factor · Employees safety · Occupational safety · Health factors

---

T. M. Alkadash (✉)
Administrative Science Department, College of Administrative and Financial Science Department, Gulf University, Sanad 26489, Kingdom of Bahrain
e-mail: tamer.alkadash@gmail.com

© The Author(s), under exclusive license to Springer Nature Switzerland AG 2024 891
A. Hamdan and E. S. Aldhaen (eds.), *Artificial Intelligence and Transforming Digital Marketing*, Studies in Systems, Decision and Control 487,
https://doi.org/10.1007/978-3-031-35828-9_75

1 Introduction

Human resource management practices for occupational safety and health for employees involve creating and establishing open lines of communication [4], coordinating with other departments to ensure compliance with safety regulations [2], making worker safety and health a core organizational value [3], and eliminating hazards, protecting workers, and continuously improving workplace safety and health [3]. Human resource risk management is also important to evaluate strategic, operational, and employee-related risks that could affect workplace safety and security [1]. HR professionals can step up by providing training on safety protocols, developing policies that promote safe working conditions, monitoring compliance with regulations, and more [2]. Moreover, human resource management (HRM) practices for occupational safety and health involve creating and establishing open lines of communication between employers and employees [2], coordinating with safety professionals to ensure compliance with regulations [2], making worker safety and health a core organizational value [1], eliminating hazards, protecting workers, and continuously improving workplace safety and health [1], and applying two HRM practices to influence workers' occupational health [4]. Additionally, health and safety initiatives should be part of a strategic approach to HRM to ensure the well-being of employees [3]. Thus, Human Resources (HR) plays an important role in ensuring workplace health and safety [4]. HR professionals are responsible for developing and implementing policies and procedures that follow federal compliance standards to prevent workplace injuries and illnesses [2]. These policies should make worker safety and health a core organizational value, eliminate hazards, protect workers, and continuously improve workplace safety and health [1]. Additionally to that, HR professionals must also be aware of specific risk issues related to workplace safety and security, such as communicable disease prevention [5]. Additionally, they must ensure that staff members are aware of the organization's health and safety policies [3]. By doing so, HR can help create a safe work environment for all employees. Studies have shown that HRM policies for hazard prevention include implementing a strict safety policy [1], eliminating hazards, protecting workers, and continuously improving workplace safety and health [2, 3], providing sufficient resources to implement and maintain the safety policy [3], and encouraging employees to invest in safety tools and equipment such as anti-slip mats, eye protection, ear protection, and masks [4]. HR departments should also be responsible for enforcing workplace safety and compliance policies [5]. Researchers also identify Human Resources (HR) departments play an important role in ensuring workplace safety and compliance with health and safety regulations [4]. HR departments should understand the rules and regulations related to workplace safety, provide training programs for employees, build opportunities for employees to report hazards, and ensure that sufficient resources are available to implement and maintain the safety program [2, 3].

HR departments should also be aware of potential health hazards in the workplace, such as stress management, office-related injuries such as carpal tunnel syndrome, and no-fragrance areas [5]. The primary federal law affecting safety in the workplace is the Occupational Safety and Health Act of 1970 (OSH Act), which is designed to protect the health of workers by setting standards for safe working conditions [1].

HR departments should ensure that employers are compliant with OSH Act regulations and other relevant laws. They should also develop policies to address any potential hazards in the workplace. By taking these steps, HR departments can help create a safe working environment for all employees. However, implementing a safe workplace has numerous benefits for both employers and employees. A safe workplace can lower injury/illness costs, reduce absenteeism and turnover, increase productivity and quality, and raise employee morale [1]. Employees who observe their employer taking steps to protect them against workplace hazards will feel safer and more valued as part of the company [2]. Additionally, safety in the workplace has a significant impact on many business KPIs such as fewer accidents [3]. Workplace safety also promotes the wellness of employees and employers alike by creating healthier working environments [4]. Finally, having a workplace safety program can provide advantages such as improved health, safety, and wellness; fewer workers' compensation claims; increased employee morale; improved public image; and reduced liability risks [5]. The main goal of this study is to determine the Occupational Safety and Health practices for Employees as a part of Human Resource Management Practices (UNRWA).

2 Literature Review

Human resource management (HRM) is the process of finding, selecting, deploying, and overseeing people for a business [1, 2]. It includes all aspects of people management to effectively meet an organization's goals and objectives [3]. HRM involves planning, organizing, directing, and controlling procurement, development, compensation, integration, maintenance, and separation of personnel [4]. It is a strategic approach to managing people to achieve organizational goals. HRM focuses on developing policies and procedures that ensure the effective use of human resources within an organization. In addition to that, Human Resources (HR) departments have a critical role to play in health and safety, as well as compliance [1]. Scholars have shown that making worker safety and health a key organizational priority, collaborating with safety professionals, removing hazards, safeguarding workers, and continually enhancing workplace safety and health are some of the finest HR practices that support employee health and safety [2, 3]. A study of the relationship between HR safety procedures including selection, training, and performance management, and employee injuries has demonstrated that firms may increase employee safety [4]. However, Human Resources (HR) plays an important role in maintaining a safe and secure workplace environment [2, 3]. HR professionals are responsible for identifying and addressing specific risk issues, such as communicable disease prevention [1]. They can also create a healthy, engaged, and productive work environment by implementing safety policies and procedures [3]. Furthermore, HR departments can also help create a safe work environment by providing training on safety protocols and procedures [5], as well as offering process safety management services to employees [5]. Additionally, HR departments can help reduce costs associated with workplace injuries and illnesses by implementing safety programs that comply with laws and regulations [4].

Safety and health training programs are important tools for informing workers and managers about workplace hazards and controls so they can work more safely [1]. OSHA provides safety management education and training courses to help employers meet their safety and health obligations [1].

Based on the above discussions, the formal hypotheses are the following:

H1: Safety & Health Training Programs &Awareness are the significant and positive factors that affect occupational safety and health for employees.

HR management systems and safety committees are important for encouraging employee participation in implementing and monitoring workplace safety [1, 2]. Safety committees typically have functions such as establishing procedures for employee input, creating an overview of employee concerns, and defining safety goals [3–5]. The roles and responsibilities of safety committees include promoting and maintaining employee interest in health and safety issues [5], educating managers, supervisors, and employees on safety protocols [5], establishing procedures for employee input such as suggestions, hazard reports, and other pertinent safety information [2], and holding regular meetings with employees and management to ensure that everyone understands specific safety protocols [4]. Safety committees also help reduce the risk of workplace injuries and illnesses, as well as ensure compliance with federal regulations [1, 3].

Based on the above discussions, the formal hypotheses are the following:

H2: Management System & Safety Committees are the significant and positive factors that affect occupational safety and health for employees.

A safe work environment is essential for the health and well-being of employees [4]. It involves attention to chemical hazards, equipment and workstation design, physical environment, task design, and other factors [1, 2]. OSHA recommends safety and health programs that focus on preventing workplace injuries and illnesses, improving compliance with laws and regulations, and reducing costs [3]. Physical comfort is also an important factor in creating a safe work environment as it sets the minimum standard for basic habitability [5]. Employers should prioritize employee well-being by providing a safe work environment that meets all these criteria.

Based on the above discussions, the formal hypotheses are the following:

H3: Physical Workplace Environment is a significant and positive factor that affects occupational safety and health for employees.

Factors that Influence the Occupational Safety and Health … 895

3 Research Methodology

For the scope of this study for this research. The research will be carried out at a non-government organization in Palestine (UNRWA). The purpose of this research is to identify the factors that can affect to Safety and Health of employees as a part of Human Resource Management Practices at UNRWA in Palestine, the study will look into these factors as a part of human resource management practices task.

The researchers have identified stratified random sampling as the sampling strategy that will be employed for this researcher. Stratified random sampling is a method of sampling in which the complete population is divided into a few subgroups to whom questionnaires are sent (Wast-fall 2009). Because they will be disseminated to UNRWA staff, a stratified random sample approach has been employed for this research. As a result, 317 questionnaires have been given to UNRWA's main office in the Gaza Strip for this research.

4 Hypotheses Testing

Table 1 refers to the Mean and Standard Deviation of the study variables. According to the table, the mean shows that the independent variables and also the dependent variable are at the average point which is between 3.88 and 3.48. The first independent variable which is Health & safety training programs & awareness shows a mean of 3.03. Followed by the second independent variable which is the Management system & safety committees with a mean of 2.92 and the third independent variable which is the physical workplace environment with a mean of 3.48. As for the dependent variable which is safety and health for employees, it shows a mean of 2.88. A standard deviation, on the other hand, is a measure of dispersion that shows the distance between respondents. When the value is lower, the standard deviation might be seen favorable. According to Table 6, the Management system & safety committees variable has the lowest standard deviation, which is 0.24. With the same calculation of a standard deviation of 0.28, it is followed by physical working environment support, safety and health for employees, and finally, health and safety training programs and awareness, which has a standard deviation of 0.37.

Table 1 Mean, standard deviation, and variances of the study variables

	N	Mean	Std. Deviation
Health & safety training programs & awareness (IV)	300	3.03	0.37
Management system & safety committees (IV)	300	2.92	0.24
physical workplace environment (IV)	300	3.48	0.28
safety and health for employees (DV)	300	2.88	0.28
Valid N (list wise)	300		

Pearson's Correlation Analysis

Spearman correlation coefficient to study the relations between the (Health & safety training programs & awareness, Management system & safety committees, and physical workplace environment) and safety and health for employees, and the results are shown through the following tables.

Table 6 shows Correlation Coefficient results to study the relationship between (Health & safety training programs & awareness, Management system, and physical workplace environment) and safety and health for employees.

It is indicated from the results in Table 2:

- There is a positive correlation with statistical significance between the Training and development programs and safety and health for employees, where the value of ($R = 0.37$, p-value < 0.05).
- There is a negative correlation with statistical significance between the Management system and safety and health for employees, where the value of ($R = -0.16$, p-value < 0.05).
- There is no correlation with statistical significance between the physical workplace environment and safety and health for employees, where the value of ($R = -0.10$, p-value < 0.05).

Liner Regression Analysis

* H1: There is an effect statistically significant for the independent variable "Health & safety training programs & awareness on variable "safety and health for employees."

The researcher looked at the effect independent variable using a simple regression analysis to see if this hypothesis held up.

The value of "F" in Table 3 above suggests that the relationship between the variable (Training and development programs) and the outcome (safety and health for employees) is statistically significant ($F = 52.26$, p-value 0.05). It was determined that (training and development programs) accounted for 15% of the overall variation in terms of (employee safety and health), with the model's coefficient of determination being 0.15.

Table 2 Pearson correlation analysis result for all variables

Dimensions	Safety and health for employees (DV)	
	Correlation coefficient	Significance level
Health & safety training programs & awareness (IV)	0.37	0.000*
Management system & safety committees (IV)	−0.16	0.004*
Physical workplace environment (IV)	−0.10	0.061//

* p-value is statistical significance
//p-value is not statistically significance

Factors that Influence the Occupational Safety and Health … 897

Table 3 Findings Straightforward regression analysis of the dependent variable "safety and health for employees" and the independent variable "training and development programs"

Dimension	B	Std. Error	T	Beta	R	R^2	F
Constant	2.01	0.12	16.62	0.38	0.38	0.15	52.26*
Training and development programs	0.28	0.04	7.22*				

* p-value is statistical significance

This table may be created using a regression equation that had to estimate the degree of employee safety and health and training and development programs:

safety and health for employees = *2.01 + 0.28 (Training and development programs)*

The previous equation indicated that the increase of the (Training and development programs) by one degree whenever led to an increase (in safety and health for employees) with 0.28°, and vice versa.

* H2: There is an effect statistically significant for the independent variable "Management system & safety committees" on variable "safety and health for employees."

The researcher did a simple regression analysis to investigate the relationship between the independent variable "Management system" and the dependent variable "safety and health for employees," and the findings are displayed in Table 4.

This was interpreted as (Management system) as an Independent variable 0.02% of the total variance for (safety and health for employees), where the coefficient of determination of the model was 0.02. Table 4 indicates that the value of "F" is statistically significant ($F = 6.91$, p-value 0.05), and this indicates that the effect of the variable (Management system) (safe-ty and health for employees) is statistically significant. This table can be formulated regression equation that had to predict with degree (Management system) and Knowledge degree (safety and health for employees):

safety and health for employees = *3.38 – 0.17 (Management system)*

The previous equation indicated that the increase of the (Management system) by one degree whenever led to a decrease (safety and health for employees) with 0.17°.

* H3: There is an effect statistically significant for the independent variable "physical workplace environment" on variable "safety and health for employees."

Table 4 Results of Simple regression analysis of the independent variable "Management system" on variable "safety and health for employees"

Dimension	B	Std. Error	T	Beta	R	R^2	F
Constant	3.38	0.19	17.65	−0.15	0.15	0.02	6.91*
Management system &safety committees	−0.17	0.06	−2.62*				

* p-value is statistical significance

898 T. M. Alkadash

The findings of the researcher's simple regression analysis, which was performed to examine the relationship between the independent variable "physical working environment" and the dependent variable "safety and health for employees," are displayed in Table 5.

Table 5 indicate that the value of "F" is not statistically significant ($F = 0.11$, p-value > 0.05), and this indicates that the effect of variable (physical workplace environment) on (safety and health for employees) is not statistically significant, this indicates that the independent variable (physical workplace environment) no effect on the dependent variable (safety and health for employees).

Multiple Regression Analysis

- There is effect statistically significant for independent variables "Safety and health training programs & awareness, Management system & safety committees, physical workplace environment" on the variable "safety and health for employees."

The researcher employed multiple regression analysis stepwise to identify the variables that can predict employee safety and health with statistical significance, regression analysis to examine the impact of the independent variable "physical workplace environment" on the employee safety and health variable, the order of the strongest effects of the variables, and interpretation of the variation in total scores (safety and health for employees). It was to stop at the second step, where the only variable was introduced to a regression equation (safety and health training programs & awareness), the variables (Management system & safety committees, physical workplace environment), has been excluded from the equation due to the lack of effect on them (safety and health for employees). The following table shows the results of this hypothesis:

Table 5 Results of Simple regression analysis of the independent variable "physical workplace environment" on variable "safety and health for employees"

Dimension	B	Std. Error	T	Beta	R	R^2	F
Constant	3.20	0.20	16.03	-0.09	0.00	0.00	0.11//
Physical workplace environment	-0.09	0.05	-1.59//				

//p-value is not statistically significance

Table 6 Result of multiple regression analysis for independent variables and dependent variable

Dimension	B	Std. Error	T	Beta	R	R^2	F
Constant	2.01	0.12	16.62	0.38	0.38	0.15	52.26*
Safety and health training programs & Awareness	0.28	0.04	7.22*				

* significant at the 0.05 level (2-tailed)

Factors that Influence the Occupational Safety and Health ... 899

Table 6 there is effect statistically significant for the independent variable "**Safety and health training programs & awareness**" ($F = 52.26$, P-value < 0.05), where it was noted that the variable (Safety and health training programs & awareness) is the best variable to predict the (safety and health for employees), it was interpreted by this variable proportion of 15% of the total variance in (safety and health for employees) and the rest returns to other factors, as the coefficient of determination of the model 0.15.

This table can be formulated regression equation that had to predict with degree (**Safety and health training programs & awareness**) and Knowledge degree (safety and health for employees):

safety and health for employees $= 2.01 + 0.28$ (Safety and health training programs & awareness)

The previous equation indicated that the increase of the (Safety and health training programs & awareness) by one degree whenever led to an increase (in safety and health for employees) with $0.28°$, and vice versa.

H1: Safety & Health Training Programs & Awareness are the significant and positive factors that affect occupational safety and health for employees.

Based on the analysis in Multiple Regression Analysis, there is a significant relationship between safety and health training Programs & awareness and occupational safety and health for employees where ($\beta = 0.028$, $p < 0.01$). Therefore, HI is rejected.

H2: Management System & Safety Committees are the significant and positive factors that affect occupational safety and health for employees. Based on the analysis in Multiple Regression Analysis, there is a significant relationship between management system & safety committees and occupational safety and health for employees where ($\beta = 2.01$, $p < 0.01$). Therefore, HI is accepted.

H3: Physical Workplace Environment is significantly and positively factoring the effect of occupational safety and health for employees.

Based on the analysis in Multiple Regression Analysis, there is a significant relationship between physical work environment and occupational safety and health for employees where ($\beta = 2.01$, $p < 0.01$). Therefore, HI is accepted.

5 Summary and Concluding Remarks

In this research study, a mix of results had been gained. This result is to clarify the factor that could affect the occupational safety and health of UNRWA employees in Palestine Gaza Strip. The results of the study show, just one independent variable, which is having a substantial impact on the occupational safety and health of UNRWA employees, is identified in this research study being conducted at UNRWA Palestine-Gaza Strip. This finding demonstrates the effects of combining two independent variables into one independent variable. The support for the work environment is the new independent variable. The performance of the employees isn't greatly impacted by the second independent variable, which is supervisor support.

Also, it has a poor impact on the productivity of the staff. So, if both employers and workers fail to fulfill their respective obligations, there cannot be any effective occupational health and safety rules. The employer is required to report accidents to the government, keep track of health and safety concerns, publish legislative information and safety notifications, and offer training in health and safety. The employer is obligated to form a safety committee to oversee any health and safety problems. The safety committee is responsible for studying accident trends and making recommendations for corrective actions, reviewing safety reports and making proposals for avoiding accidents, reviewing and discussing reports from safety representatives, and making recommendations for new or revised safety procedures. As they directly affect how people behave at work, the workplace environment should offer better communication channels and working circumstances. Motivation, skill development, and reward systems should all be carefully considered and planned. The researcher also suggests that risk analyses be conducted in all non-governmental organizations to identify hazards and evaluate the risks associated with them. Also, all organizations receiving government assistance should do a comparative study. Also, the researcher suggests that additional studies be done on other human resource strategies that have a favorable influence on workers' occupational safety and health.

References

1. Boyd, C.: Human Resource Management and Occupational Health and Safety. Routledge, London (2003)
2. Bebbington, A., Hickey, S., Mitlin, D.: Introduction: can NGOs make a difference?: the challenge of development alternatives. In: Bebbington, A., Hickey, S., Mitlin, D. (eds.) Can NGOs make a difference?: the challenge of development alternatives, pp. 3–37. Zed Books, London (2008)
3. BS 8800: 1996 Guide to Occupational Health and Safety Management Systems
4. BS EN ISO 14001: 1996 Environmental Management Systems—Specification with Guidance for Use
5. BS EN ISO 9000-1: 1994 Quality Management and Quality Assurance Standards: Guidelines for Selection and Use
6. Cambridge Advanced Learner's Dictionary: 3rd edn. Cambridge University Press, Cambridge (2008)
7. Cascio, W.F.: Managing Human Resources Productivity, Quality of Life, Profit. MC Graw-HillNew York (1986)
8. Cole, G.A.: Personnel and Human Resource Management. Thompson Learning Bedford Row, London (2002)
9. David, A.D., Stephen, P.R.: Human Resource Management, Concepts and Application. Progressive International Technologies, USA (1999)
10. Dessler, G.: Human Resource Management, 7th edn. Prentice –Hall of India Private Ltd, New Delhi (2001)
11. Department of Labour, republic of south Africa No. 85 of 1993: Occupational health and safety South African Department of labor (2004)
12. Downey, D.M., et al.: The Development of Case Studies that Demonstrate the Business Benefit of Effective Management of Health and Safety. HSE, London (1995)
13. Eva, D., Oswald R.: Health and Safety at Work. Pan Books, London (1981)

Factors that Influence the Occupational Safety and Health ... 901

14. Glass, V.: How to Make A Questionnaire (2011). Retrieved 23 Aug 2011, from http://www.ehow.com/how_2305520_make-questionnaire.html
15. Israel, G.D.: Sampling the Evidence of Extension Program Impact. Program (1992)
16. Evaluation and Organizational Development, IFAS, University of Florida. PEOD-5 Oct
17. McCracken, J.: A Handbook Successful Health and Safety Management. Kogan Page Ltd, London (2004)
18. Labour Act of Ghana: Act 651,Accra: GPC Printing Division (2003)
19. Lawrence, P.R., Lorsch, J.W.: Organization and Environment. Harvard University Press, Cambridge (1976)
20. Litwin, G.H., Stringer, R.A.: Motivation and Organisation Climate. Harvard University Press, Boston (1968)
21. Byars, L., Rue, L.W.: Human Resource Management. MC Graw-Hill/Irwin, New York (2008)
22. Michael, A.: A Handbook of Human Resource Management Practice. Kogan Page Ltd., London (2006)
23. Miaoulis, G., Michener, R.D.: An Introduction to Sampling. Kendall/Hunt Publishing Company, Dubuque, Iowa (1976)
24. Nordquist, R.: *Literature Review* (2011). Retrieved 22 Aug 2011 from http://grammar.about.com/od/il/g/literaturereviewterm.htm
25. Pirani, M., Reynolds, J.: Gearing up for Safety, Personnel Management (1976)
26. Quin, M.D.: Planning with people in mind. Harvard Business Review, November–December, pp. 97–105 (1983)
27. Mathis, R.L., Jackson, J.H.: Human Resource Management. Melisa Acuna (2004)
28. Salon, G.: Far from Remote People Management, 27 Sept, pp. 34–36 (2001)
29. Sekaran, U.: Research Methodology for Business: A Skill Building Approach, 4th edn. John Wiley & Sons, New Delhi (2007)
30. Shuler, L.O.: Guide to Data Collection Method (2009). Retrieved 23 Aug 2011, From http://web.mit.edu/tll/assessment-evaluation/ae-datacollection-methods-lols.pdf
31. Total quality management and the management of health and safety CRR153 HSE Books (1997). ISBN 0 7176 1455 7
32. Tsui, A.S., Gomez-Mejia, L.R.: Evaluating Human Resource Effectiveness in Human Resource Management, Evolving Roles and Responsibilities. In: Dyer, L. (ed.). Bureau of National Affairs, Washington (1988)
33. Turner, A.N., Lawrence, P.R.: Industrial Jobs and Worker, An Investigation of Response to task attributes. Harvard University Graduate School Of Business Administration, Boston (1965)
34. Westfall, L.: Sampling Method (2009). Retrieved 24 Aug 2011 from http://www.westfallteam.com/Papers/Sampling%20Methods.pdf

The Impact of Employee Satisfaction, Emotional Intelligence and Organizational Commitment on Marketing Service Quality in Medical Equipment Companies, Bahrain

Tamer M. Alkadash⬤, Mahmoud AlZgool, Ali Ahmed Ateeq, Mohammed Alzoraiki, Rawan M. Alkadash, Chayanit Nadam, Qais AlMaamari, Marwan Milhem, and Mohammed Dawwas

Abstract This study aims to examine the relationship between employee satisfaction, organizational commitment, and emotional intelligence on marketing service quality in Bahrain. The literature review has used previous studies to approve the hypothesis. The data was gathered from 167 respondents. IBM SPSS software was used to analyze the data and tested the hypothesis, after a quantitative research study used for this study as a type of primary data. The result demonstrates that employee satisfaction, organizational commitment, and emotional intelligence were found to have a significant positive relationship with marketing service quality in medical equipment companies in Bahrain. Thus, medical equipment companies in Bahrain can use these results to expand their marketing service quality in campiness. In addition, the finding of this study also provided the medical equipment companies a good preparation and rising strategies to improve the marketing service quality. Finally, the study provides guidance and tools for further actions. This is important to keep the endurance of the medical equipment companies in the global part, however, new strategies for marketing service quality medical equipment companies need to be developed from time to time and it depends on the outcome of this study.

Keywords Marketing service quality · Employee satisfaction · Organization commitment

---

T. M. Alkadash (✉) · M. AlZgool · A. A. Ateeq · M. Alzoraiki · Q. AlMaamari · M. Milhem · M. Dawwas
Administrative Science Department, College of Administrative and Financial Science Department, Gulf University, Sanad 26489, Kingdom of Bahrain
e-mail: tamer.alkadash@gmail.com

R. M. Alkadash
Al Azhar University, Palestine, Egypt

C. Nadam
Global Business Center of British Petroleum, Kuala Lumpur, Malaysia

© The Author(s), under exclusive license to Springer Nature Switzerland AG 2024
A. Hamdan and E. S. Aldhaen (eds.), *Artificial Intelligence and Transforming Digital Marketing*, Studies in Systems, Decision and Control 487,
https://doi.org/10.1007/978-3-031-35828-9_76

1 Introduction

In today's highly competitive business marketplace, quality service is a key factor that drives customer loyalty and ultimately contributes to sustainable financial success [1]. However, marketing service quality is of paramount importance as it determines the success of any business. Poor service quality can lead to dissatisfied customers, negative word-of-mouth, loss of reputation, and ultimately, reduced revenue. In today's competitive market, where customers have numerous options at their disposal, businesses need to focus on providing superior service quality to attract and retain customers. Not only does this increase customer loyalty and satisfaction, but it also creates a sense of trust which sets businesses apart from their counterparts in the same market sector. Effective marketing strategies that emphasize service quality will ultimately contribute to positive customer experiences and translate into a better bottom line for businesses. Therefore, businesses that prioritize delivering high-quality services through innovative marketing tactics can create a strong foundation for sustainable growth and profit-making [2].

Several companies use service because of the key differentiation in their product-promoting strategy [3]. With a lot of intense promoting competition, the necessity to reinforce service to succeeding levels is acute. However, firms have outsourced the differentiation blessings to lower-price service suppliers that don't have any stake in the promoting effort. Outsourcing is inflicting less direct client "face-to-face" valuable service opportunities and fewer product market differentiation [4]. However, Marketing service quality is an essential aspect of any literature review as it is a key factor in determining customer satisfaction, loyalty, and long-term profitability. Through intensive research and surveys, marketing service quality measures the extent to which the services provided by a business meet or exceed customers' expectations, ensuring that they are satisfied with their experience. It reflects on the quality-of-service delivery processes through employee behavior, responsiveness, reliability, assurance, and empathy [5]. Understanding marketing service quality enables businesses to design customer-centric strategies that anticipate customers' needs and preferences; it also empowers entrepreneurs to understand and manage gaps between the level of service expected by customers and what's delivered. A thorough study of marketing service quality should form part of every business strategy; such knowledge assists businesses in developing product/service features that meet unmet needs while improving customer engagement via better packaging and communication strategies [6].

Literature review shows, marketing service quality is an important part of corporate success, and various researchers have studied it in depth. According to the literature study, numerous factors, such as dependability, responsiveness, empathy, and tangibles, impact service quality for consumers [7–9]. Researchers have created models to assess consumers' views of service quality to assist firms in continuously improving their services [7, 8]. According to certain research, boosting service quality leads to higher customer satisfaction and loyalty, resulting in a competitive advantage. As such, organizations must emphasize marketing service quality as a key component for their success in the to-competitive day's business scene [10]. Businesses may establish long-term connections with their consumers by consistently offering high-quality services and surpassing customer expectations, resulting in repeat business and increased market share.

In the service industry, satisfying customer needs is paramount to a business's success. However, providing quality services that meet customer expectations implies the role of employees' satisfaction in delivering excellent outcomes. Employee satisfaction directly impacts the delivery of services, thereby influencing the perception of customers about the quality of services offered by a company [11]. A satisfied workforce translates into better engagement, enthusiasm, and motivation toward serving customers with diligence and professionalism; thus, enhancing their experience. As such, businesses must prioritize employee satisfaction to improve overall service delivery by cultivating an environment that nurtures teamwork, encourages growth opportunities, and investment in employee training to build morale and strengthen expertise. In doing so, a business can guarantee exceptional customer experiences that translate into loyalty and increase brand equity in the market space [12].

Marketing service quality is an essential aspect of any successful organization, and ensuring its continual improvement requires a deep commitment by the entire organization [13]. Organizations need to consistently deliver high-quality service to their customers in order to build loyalty and competitive advantage. This can only be achieved through an unwavering commitment from all levels of management, employees, and stakeholders to maintain a customer-focused culture [13, 14]. Organizational commitment entails dedicating resources towards training employees on customer experience best practices, developing mechanisms for constant feedback and continuous improvement, investing in effective communication strategies with customers, and providing excellent after-sales support, among other things. Without this dedication, even the most well-intentioned marketing initiatives will fall flat as customers are unlikely to return if they perceive poor quality services from the outset. Thus, organizational commitment is critical for enhancing marketing service quality which ultimately drives business success [15].

However, the delivery of high-quality service goes beyond technical skills and requires emotional intelligence on the part of the marketer. Emotional intelligence is the ability to recognize and manage one's own emotions while understanding and empathizing with others' emotions to build strong interpersonal relationships [16, 17]. When marketers possess emotional intelligence, they can better understand customers' needs, anticipate problems, handle difficult situations effectively, and communicate clearly and respectfully. The result is more satisfied customers who feel valued and well-taken care of by the organization. Therefore, marketing service quality needs emotional intelligence as it fosters positive attitudes toward the brand reputation which ultimately leads to maximizing customer satisfaction that generates loyal customers for a longer period of time [18].

This study aims to examine what ether the factors are "employee satisfaction, emotional intelligence, and organizational commitment" affecting the marketing service quality in medical equipment companies in Bahrain.

2 Theoretical Considerations and Hypothesis Development

2.1 The Relationship Between Employee Satisfaction and Service Quality

Employee satisfaction refers to describing whether employees are happy and contented and fulfilling their desires and needs at work [19].

Employee satisfaction refers to the degree of contentment employees have with their work and the overall workplace environment. It is an essential aspect of any successful organization, as satisfied employees tend to be more productive, motivated, and committed to achieving organizational objectives. Measuring employee satisfaction requires a structured approach that involves regular surveys, feedback mechanisms, and other forms of communication channels between management and employees. Tracking employee satisfaction data provides insights into areas that require improvements such as work-life balance, compensation, safety concerns, or opportunities for growth within the organization—ultimately helping establish a plan to retain top-performing talent. A defined strategy for improving employee job satisfaction is essential for not only increasing productivity but also reputation building among both consumers and potential hires in today's competitive market [20].

Employee satisfaction plays a critical role in ensuring high-quality customer service [21]. In the marketing field, this is even more vital since employees' interactions with customers can significantly shape their perceptions of the brand. Companies need to establish an environment that fosters employee satisfaction to enhance loyalty and motivation toward delivering excellent service. Therefore, marketers must pay close attention to employee feedback regarding the quality of services offered both internally and externally. Creating opportunities for frequent communication and providing avenues for professional development are ways that companies can satisfy their employees while simultaneously increasing productivity, thus improving marketing strategies overall. A happy workforce translates into positive reputations for brands, not only among its staff but also its clients/customers; it's a win–win scenario for all involved [22].

The following hypothesis has been observed based on the mentioned literature reviews.

H1: There is a significant relationship between employee satisfaction and service quality in Bahrain.

2.2 The Relationship Between Organizational Commitment and Service Quality

An organizational member's psychological ties to the organization are referred to as organizational commitment. It has a significant impact on whether a member will remain with the group and fervently pursue organizational objectives [26].

Organizational commitment refers to the degree of loyalty and involvement an employee has toward their organization. This includes a willingness to help achieve the organization's goals, remaining with the company over time, and promoting its missions and values. Such demonstrated commitment enhances an employee's level of productivity and job satisfaction, as well as fostering increased organizational performance. However, organizational commitment can be fostered through various means, such as providing fair compensation packages, opportunities for career growth and development, feedback mechanisms that enable employees to express their concerns freely, supportive working conditions, and positive work culture. A strong sense of defined organizational commitment creates a synergy between the employee's personal goals and those of the organization leading to an increased alignment of morale resulting in improved team dynamics.

Organizational commitment plays a significant role in ensuring quality marketing services. Commitment from an organization's management team and employees are vital to meeting the needs and satisfaction levels of customers. Organizations with high marketing service quality rely on staff performance to achieve success [23]. These organizations ensure that every aspect, from employee training programs to data management techniques, aligns with their objectives. The organization's unique culture can also have a significant impact on the level of employee motivation and engagement in providing top-notch marketing services [24]. Therefore, when companies prioritize organizational commitment toward marketing service quality, they encourage strong management team creation and develop positive working relationships between all members of the workforce leading to higher service delivery standards ultimately benefiting customer experiences overall [25–27].

The following hypothesis has been observed based on the mentioned literature reviews.

H2: There is a significant relationship between organizational commitment and service quality in Bahrain.

2.3 The Relationship Between Emotional Intelligence and Service Quality

In today's work environment, employers value employees who demonstrate high emotional intelligence because it contributes to workplace culture and employee productivity [28]. Therefore, professionals must continue developing their emotional intelligence skills throughout their careers by experiencing new environments such as professional development training sessions or courses on communication skills in order to become effective collaborators within their industries.

Emotional intelligence refers to the ability to recognize and manage one's own emotions, as well as understand and influence the emotions of others. It involves a range of skills, such as empathy, self-awareness, self-regulation, social awareness, and relationship management. Professionals with high levels of emotional intelligence are able to communicate effectively, work well in teams, display empathy towards colleagues or clients, and cope under pressure. Furthermore, they can manage their emotions while interacting with people from diverse backgrounds and different situations [28, 29].

Emotional intelligence is a key component in delivering high-quality marketing services that generate positive experiences for customers. By understanding emotions, marketers can more effectively tailor their efforts to satisfy customer needs and preferences. This includes acknowledging the emotional drivers that motivate consumer behavior, such as desires for social connection or personal fulfillment. Emotional intelligence also helps marketers to cultivate empathy and build rapport with customers, which can go a long way toward establishing brand loyalty and trust. Additionally, marketing efforts that are grounded in emotional intelligence are better placed to mitigate negative experiences by addressing customer concerns and proactively managing potential dissatisfaction. In short, emotional intelligence is an essential tool for ensuring that marketing service quality remains relevant, authentic, and impactful in today's increasingly competitive digital marketplace [30].

The following hypothesis has been observed based on the mentioned literature reviews.

H3: There is a significant relationship between emotional intelligence and service quality in Bahrain.

3 Methodology

The main goal of this current study is to explore the factors that will affect the service quality system gaps in medical equipment companies in Bahrain by filling up the wage gap found in research. In the current research, the type of questionnaire method used to collect primary data, it took to get a certification from medical equipment companies' management to begin to allocate the survey. Therefore, the data was collected from the respondents after one month time. All respondents took 5–10 min to answer the survey questions. For this study, the research community at medical equipment companies in Bahrain. The sample of this study is 189 staff working at the department of marketing and administration in medical equipment companies in Bahrain, respondents of this study have been selected sample of approximately.

The Impact of Employee Satisfaction, Emotional Intelligence ...

4 Results

In this study research, the survey is a part of the primary data collection method used. Thus, the researcher distributed 200 surveys to participants at medical companies in Bahrain. After one month. Only 172 surveys were collected by the researcher from participating in this study, after cleaning the missing data to improve the data quality and reliability for further statistical analysis, a total number of 167 surveys are valid to use in this study by IBM Statistics software (SPSS).

4.1 Frequency Analysis

In the current study, the survey included four parts, the first part of the survey shows the demographic questions (respondent's characters) where fifth question are represented for this part included (gender, age, education level, marital status, and total years of experience) all these questions are developed for this study. The result of the demographic of respondents "Gender" who participated in the survey, the frequency analysis result indicates there are 134 participants of this study under the male category and only 33 participants are female. Followed by demographic of respondents "Age" the frequency analysis result indicate there are 35 participants of this study less than 30 years old, while 66 and only 30 participants of this study aged between 31 and 40 years old, 48 participants of this study aged between 41 and 50 and only 18 participants of this study aged more than 50 years old. Followed by the demographic of respondents "Education" the frequency analysis result indicate there are 8 participants in this study holding diploma, 121 participants of this study hold bachelor, and 38 participants in this study hold a postgraduate certificate. Followed by the demographic of respondents' "Marital status" the frequency analysis result indicates there are 14 participants of this study are single, 132 participants of this study are married, and only 6 participants of this study others. Followed by the last demographic of respondents "total years of experience" the frequency analysis result indicate there are 26 participants in this study. Experience less than 5 years, while 39 participants of this study experience between 6 and 10 years, next 74 participants in this study experience between 11 and 15 years, and only 28 participants of this study experience more than 6–15 years.

4.2 Reliability Test

The purpose of Cronbach's Alpha (α) test is to show the correlation for every item connected to another item into a reliability coefficient positive relationship (Sekaran 2007). A "high" value for Cronbach's Alpha out result show nearer to 1 which leads to higher internal consistency reliability. (Sekaran 2007).

Table 1 Cronbach's Alpha test for study variables

Variables (N = 150)	Number of items	Cronbach's Alpha	Internal consistency (Remarks)
ES	6	0.871	Good
OC	7	0.857	Good
EI	7	0.824	Good
SQ	5	0.801	Good

Note ES employee satisfaction, *OC* organization commitment, *EI* emotional intelligence, and *SQ* service quality

Table 1 shows the Cronbach's Alpha statistic test for all study variables score range between 0.801 and 0.871 which internal consistency evaluation good level. The internal consistency evaluation for the first independent variable "employee satisfaction" remarks a good, coefficient of reliability (Cronbach Alpha value = 0.871). The internal consistency evaluation for the second independent variable "organization commitment" remarks a good, coefficient of reliability (Cronbach Alpha value = 0.857). The internal consistency evaluation for the third independent variable "emotional intelligence" remarks a good coefficient of reliability (Cronbach Alpha value = 0.824). The internal consistency evaluation for the dependent variable "service quality" remarks a good coefficient of reliability, (Cronbach Alpha value = 0.801).

4.3 Descriptive Analysis

The purpose of descriptive analysis is to show features of the data in a research study statistic description. It measures the sample for every quantitative data analysis. In the current study, Mean (M) and Std. Deviation (SD) in descriptive statistics of the research study variables is shown in Table 4 the result indicates between the score range (M = 3.49–3.69) and Std. Deviation score range (0.71–0.78) out of 167 respondents in this study.

Table 2 shows a descriptive analysis of all study variables. The mean value of the employee satisfaction variable obtain (M = 3.03) at (Std. Deviation = 0.37), followed by the organization commitment variable obtaining Mean value of level (M = 3.64)and (Std. Deviation = 0.75), then emotional intelligence variable obtain mean value at level (M = 3.49) and (Std. Deviation = 0.67), next service quality variable mean value at level (M = 3.59) and service quality obtain mean value at level (M = 3.59) and (Std. Deviation = 0.71).

The Impact of Employee Satisfaction, Emotional Intelligence … 911

Table 2 Descriptive statistics of the variables

Items	N	M	Std. Deviation
ES (IV)	167	3.69	0.78
OC (IV)	167	3.64	0.75
EI (IV)	167	3.49	0.67
SQ (DV)	167	3.59	0.71

Note *ES* employee satisfaction, *OC* organization commitment, *EI* emotional intelligence, *SQ* service quality, *N* number of respondents and *M* Mean

4.4 Correlation Test

The purpose of the correlation coefficient in this study is to explore the relations between (Employee satisfaction, Organization commitment, Emotional intelligence) on Service quality in Bahrain, the results of the correlation test are shown in the following table.

Pearson correlation test obtains for brand consciousness variable their results show there is a positive correlation with statistical significance among (Employee satisfaction, Organization commitment, and Emotional intelligence) on Service quality at medical equipment companies in Bahrain. See Table 3.

Table 3 Correlation test

Correlation test				
	1	2	3	4
Service quality (1)	1	0.510**	0.574**	0.290**
Employee satisfaction (2)	0.510**	1	0.559**	0.064
Organization commitment (3)	0.574**	0.559**	1	0.063
Emotional intelligence (4)	0.488**	0.407**	0.450**	1

** Correlation is significant at the 0.01 level (1-tailed)
* Correlation is significant at the 0.05 level (2-tailed)

Table 4 Multiple regression testing

Variables	Service quality (Beta Standardization)	Sig
Employee satisfaction	0.226	0.000
Organization commitment	0.269	0.000
Emotional intelligence	0.276	0.000
F Value		24.542
R Square		0.711
Adjusted *R* Square		0.718

4.5 Multiple Regression Testing

The results of the multiple regression test indicate $R^2 = 0.711$ and Adjusted $R^2 = 0.718$ It means there is a good correlation between the independent variable "Service quality" and the dependent variables "Employee satisfaction, Organization commitment, and Emotional intelligence". "F" scores significant of the model show that ($F = 24.542$), mean there is a significant correlation at the significance level (0.05 $= \alpha$) between (Employee satisfaction, Organization commitment, and Emotional intelligence) on Service quality at medical equipment companies in Bahrain.

The employee satisfaction Variable demonstrates there is a positive correlation between service quality ($B = 0.226$, T value $= 4.47$ and sig. level 0.00) followed by the Organization commitment variable demonstrates there is a positive correlation between service quality ($B = 269$, T value $= 4.72$ and sig. level 0.00) and Emotional intelligence variable there is a positive correlation on service quality (Beta $= 0.276$, T value $= 4.86$ and sig. level.00). See Tables 4 and 5.

5 Discussion

The study aim of this is to extend the correlation between employee satisfaction, organization commitment, and emotional intelligence on marketing service quality in Bahrain. Thus, this study gives a significant contribution to extant more research in management, marketing strategies, and project management studies.

There is a significant relationship between employee satisfaction and service quality in Bahrain. The findings of the study based on SPSS software show Beta Standardization $= 0.226$, T value $= 4.47$, meaning there is a positive correlation of employee satisfaction demonstrated on service quality at sig. level 0.00. The result proves employee satisfaction has a direct effect on service quality, this is given support to medical equipment companies in Bahrain to increase the employee satisfaction level which offers better service quality.

There is a significant relationship between organizational commitment and service quality in Bahrain. The findings of the study based on SPSS software show Beta Standardization $= 269$, T value $= 4.72$, meaning there is a positive correlation of

Table 5 Summary of study hypothesis testing results:

	Hypothesis	Result
H1:	There is a significant correlation between employee satisfaction and service quality in Bahrain	Accepted
H2:	There is a significant correlation between organizational commitment and service quality in Bahrain	Accepted
H3:	There is a significant relationship between emotional intelligence and service quality in Bahrain	Accepted

organization commitment demonstrated on service quality at sig. level 0.00. The result proves employee satisfaction has a direct effect on service quality, this is given support to medical equipment companies in Bahrain to raise the emotional intelligence level which offers better service quality.

There is a significant relationship between emotional intelligence and service quality in Bahrain. The findings of the study based on SPSS software show Beta Standardization $= 0.276$, T value $= 4.86$, meaning there is a positive correlation between emotional intelligence on service quality at sig. level.00. The result establishes evidence of emotional intelligence has a direct effect on service quality, this gives support to medical equipment companies in Bahrain to raise the emotional intelligence level which offers better service quality.

5.1 Limitations of the Study

Like any other research, all studies need to have limitations for it the research is conducted by employees working in medical equipment companies in Bahrain. Therefore, the research findings can't be generalized because the sample is a small sample selected in this research. A questionnaire as with other research that uses a questionnaire as the instrument to collect data, there may be a problem of social desirability. Some respondents may have given honest responses instead of responding deemed to be desirable by others. The background of respondents is the type of limitation of this study, the experience years, and even the hypothesis in this study is accepted. However, this is depending on knowing the population size, by using different data collection methods, the conclusions of the study would be on a large sample requires large resources, and it may not be the best way to analyze the research question, as it might be difficult to keep track of a large amount of information. However, it can be used to enhance the performance of any interested parties in the research. Also, the lack of data and literature that studies consumer reactions and behavior and their effects on Bahrain as a limitation made it a bit difficult to complete the study, however, we managed to overcome this problem by using different resources using the neighboring countries' research and another research that is done on an international basis.

5.2 Suggestion for Future Research

Suggestions for future research are provided in this section for future analysis. Thus, the study identified several variables important to market service quality at companies in Bahrain, by raising a variety of questions that might like the additional analysis. A replicate of the study with a longitudinal style would permit an additional elaborated analysis and would get around the problem of poor recall. This would allocate more detailed information about the market service quality, for example, to be collected.

Future studies need to replicate the model with different industries in the bank, Small and medium-sized enterprises (SMEs) in Bahrain. Also, Future studies need to replicate the model with a bigger number of respondents and different areas in Bahrain. However, researchers could add more independent variables like knowledge sharing, and leadership style which would be a great impact on service quality. Finally, future studies need to implement this study with a different type of respondents such as a sample of customers to evaluate the quality of service at companies in Bahrain. Finally, can be involved on a larger scale by investigating the external and internal factors affecting marketing service quality medical equipment companies.

References

1. Le, T.T.: Corporate social responsibility and SMEs' performance: mediating role of corporate image, corporate reputation and customer loyalty. Int. J. Emerg. Mark. (2022)
2. Lariviere, B., Smit, E.G.: People–planet–profits for a sustainable world: integrating the triple-P idea in the marketing strategy, implementation and evaluation of service firms. J. Serv. Manag. **33**(4/5), 507–519 (2022)
3. Medrano, N., Cornejo-Cañamares, M., Olarte-Pascual, C.: The impact of marketing innovation on companies' environmental orientation. J. Bus. Ind. Mark. **35**(1), 1–12 (2020)
4. Ken Colwell, M.B.A.: Starting a Business QuickStart Guide: The Simplified Beginner's Guide to Launching a Successful Small Business, Turning Your Vision Into Reality, and Achieving Your Entrepreneurial Dream. ClydeBank Media LLC (2019)
5. Rodriguez, M.: Brand Storytelling: Put Customers at the Heart of Your Brand Story. Kogan Page Publishers. (2020)
6. Heng, L., Ferdinand, A.T., Afifah, N., Ramadania, R.: Service innovation capability for enhancing marketing performance: an SDL perspectives. Bus. Theor. Pract. **21**(2), 623–632 (2020)
7. Ge, Y., Yuan, Q., Wang, Y., Park, K.: The structural relationship among perceived service quality, perceived value, and customer satisfaction-focused on starbucks reserve coffee shops in Shanghai, China. Sustainability **13**(15), 8633 (2021)
8. Zygiaris, S., Hameed, Z., Ayidh Alsubaie, M., Ur Rehman, S.: Service quality and customer satisfaction in the post pandemic world: a study of Saudi auto care industry. Front. Psychol. **13**, 690 (2022)
9. Fauzi, A.A., Suryani, T.: Measuring the effects of service quality by using CARTER model towards customer satisfaction, trust and loyalty in Indonesian Islamic banking. J. Islamic Mark. **10**(1), 269–289 (2019)
10. Le Oo, W.: Customer Perceived Value and Customer Loyalty Towards MPT Mobile Service. Doctoral dissertation, Yangon University of Economics (2019)
11. Choi, S., Mattila, A.S., Bolton, L.E.: To err is human (-oid): how do consumers react to robot service failure and recovery? J. Serv. Res. **24**(3), 354–371 (2021)
12. Goodman, J.: Strategic Customer Service: Managing the Customer Experience to Increase Positive Word of Mouth, Build Loyalty, and Maximize Profits. Amacom (2019)
13. Bryson, J.M.: Strategic Planning for Public and Nonprofit Organizations: A Guide to Strengthening and Sustaining Organizational Achievement. John Wiley & Sons (2018)
14. Khudhair, H.Y., Jusoh, A., Mardani, A., Nor, K.M., Streimikiene, D.: Review of scoping studies on service quality, customer satisfaction and customer loyalty in the airline industry. Contemp. Econ. **13**(4), 375–388 (2019)
15. Goutam, D.: Customer Loyalty Development in Online Shopping: An Integration of E-Service Quality Model and the Commitment-Trust Theory. Doctoral dissertation, National Institute of Technology Karnataka, Surathkal (2020)

16. Wamsler, C., Restoy, F.: Emotional intelligence and the sustainable development goals: supporting peaceful, just, and inclusive societies. In: Peace, Justice and Strong Institutions, pp. 1–11 (2020)
17. Drigas, A., Papoutsi, C.: The need for emotional intelligence training education in critical and stressful situations: the case of Covid-19. Int. J. Recent Contributions Eng. Sci. IT **8**(3), 20–36(2020)
18. Papoutsi, C., Drigas, A., Skianis, C.: Emotional intelligence as an important asset for HR in organizations: attitudes and working variables. Int. J. Adv. Corp.Learn. **12**(2), 21 (2019)
19. Dziuba, S.T., Ingaldi, M., Zhuravskaya, M.: Employees' job satisfaction and their work performance as elements influencing work safety. Syst. Saf. Hum.-Tech. Facility-Environ. **2**(1), 18–25 (2020)
20. Khatoon, S., Zhengliang, X., Hussain, H.: The Mediating effect of customer satisfaction on the relationship between electronic banking service quality and customer purchase intention: evidence from the Qatar banking sector. SAGE Open **10**(2), 2158244020935887 (2020)
21. Min, D.: Exploring the structural relationships between service quality, perceived value, satisfaction, and loyalty in nonprofit sport clubs: empirical evidence from Germany. Sport Mark. Q. **31**(3) (2022)
22. Thomas, M.B., Fay, D.L., Berry, F.S.: Strategically marketing Florida's cities: an exploratory study into how cities engage in public marketing. Am. Rev. Pub. Adm. **50**(3), 275–285 (2020)
23. Geisler, M., Berthelsen, H., Muhonen, T.: Retaining social workers: the role of quality of work and psychosocial safety climate for work engagement, job satisfaction, and organizational commitment. Hum. Serv. Organ. Manage. Leadersh. Governance **43**(1), 1–15 (2019)
24. Kaplan, M., Kaplan, A.: The Relationship Between Organizational Commitment and Work Performance: A Case of Industrial Enterprises (2018)
25. Jang, J., Kandampully, J.: Reducing employee turnover intention through servant leadership in the restaurant context: a mediation study of affective organizational commitment. Int. J. Hosp. Tour. Adm. **19**(2), 125–141 (2018)
26. Alkadash, T.M.: Mediating role between authentic leadership, organizational commitment on talents turnover intention: In Palestine higher education. In: TEST Engineering & Management, March–April (2020)
27. Vorobyova, K., Alkadash, T.M., Nadam, C.: Investigating beliefs, attitudes, and intentions regarding strategic decision-making process: an application of theory planned behavior with moderating effects of overconfidence and confirmation biases. Specialusis Ugdymas **1**(43), 367–381 (2022)
28. Alzyoud, A.A.Y., Ahmed, U., Alzgool, M.R.H., Pahi, M.H.: Leaders¡ emotional intelligence and employee retention: mediation of job satisfaction in the hospitality industry. Int. J. Fin. Res. **10**(3), 1–10 (2019)
29. Supramanian, K., Shahruddin, R., Sekar, M.: Emotional intelligence using ability model in context of nursing and its impact on end-stage renal disease patients: a narrative review. Int. J. Curr. Res. Rev. **13**(16), 151–158 (2021)
30. Leonidou, L.C., Aykol, B., Fotiadis, T.A., Zeriti, A., Christodoulides, P.: The role of exporters' emotional intelligence in building foreign customer relationships. J. Int. Mark. **27**(4), 58–80 (2019)

Study of Mechanical Properties of Friction Welded AISI D2 and AISI 304 Steels

Sinta Restuasih, Ade Sunardi, M. N. Mohammed, M. Zaenudin, Adhes Gamayel, and M. Alfiras

Abstract Finding the most efficient way to join one material to another by welding is a matter of intense research. Both techniques, whether solid-state welding or fusion welding, seem to have their strengths and weaknesses. In solid-state welding, researchers' attention is paid to the fact that friction welding (FW), a technique that involves friction between the two materials to be welded, introduces a time-efficient process, excellent mechanical properties, and filler-free solid-state welding. In this study, we produce several samples of the as-welded material from the developed friction welding machine with promising mechanical properties. From the three parameters, friction time, forging time, and fording pressure, the results show that forging time has a more significant impact than the other parameters. We expect to improve the as-welded material by enhancing the construction of the FW machine and the procedure of FW.

Keywords Friction welding · Cold work tool steel · Stainless steel · Mechanical properties · SDG 9

S. Restuasih
Department of Industrial Engineering, Jakarta Global University, Grand Depok City, Boulevard Raya No. 2 St., Depok City 16412, West Java, Indonesia

A. Sunardi · M. Zaenudin · A. Gamayel
Department of Mechanical Engineering, Jakarta Global University, Grand Depok City, Boulevard Raya No. 2 St., Depok City 16412, West Java, Indonesia

M. N. Mohammed (✉)
Mechanical Engineering Department, College of Engineering, Gulf University, Sanad 26489, Bahrain
e-mail: dr.mohammed.alshekhly@gulfuniversity.edu.bh

Department of Engineering & Technology, Faculty of Information Sciences and Engineering, Management & Science University, 40100 Shah Alam, Selangor, Malaysia

M. Alfiras
Electrical and Electronic Engineering Department, College of Engineering, Gulf University, Sanad 26489, Kingdom of Bahrain

© The Author(s), under exclusive license to Springer Nature Switzerland AG 2024
A. Hamdan and E. S. Aldhaen (eds.), *Artificial Intelligence and Transforming Digital Marketing*, Studies in Systems, Decision and Control 487,
https://doi.org/10.1007/978-3-031-35828-9_77

1 Introduction

In the manufacturing process, especially for metallic materials, the rigid perpetual of the two materials is an integral part of the entire manufacturing and assembly process [1]. Joining processes such as welding can seem simple. Still, the back side of the experimental procedure can be complicated since it involves a few parameters, such as temperature, pressure, holding time, friction time, and rotational speed of the tools (in the case of friction welding). It's complicated enough to change and greatly affects the welding result [1–7]. At the research stage, slight changes in parameters do not necessarily cause significant problems, however in manufacturing processes, precise parameters are required to meet industrial manufacturing demands for precision and efficiency [8–11]. In 1966, a report published by British Research Association, as discussed in ref. [12] mentioned the significant developments in welding technology demonstrated by several factors: (1) significant growth by government-supported members and research organizations, (2) significantly lower cost and lighter weight new series of austenitic stainless steels and application of friction welding to stainless steel pressure vessels, (3) welded turbine wheels means new friction welding for industrial applications. Since then, various welding techniques, such as friction welding and solid-state welding processes, have been introduced and developed for multiple applications. There are several techniques for friction welding, some of which have already found ways to use them in the following industrial sectors, for instance, linear friction welding [13–16] friction stir welding [17–19], and rotary/direct drive and inertia friction welding [20]. The early design of the friction welder was developed by Li et al. [20] and Vairis et al. [21]. Although this machine is designed to be lightweight, its performance is considerably compared to previously developed machines. In this study, a simple inertial friction welder is designed based on a rotating machine—lathe machine—utilizing some subsystem such as brakes, chuck holder, and hydraulic. As an example, a friction welding of the known as nonweldable material—stainless steel to tool steel is presented and discussed in Ref. [22]. The study shows that friction welding can join two metallic materials that are relatively difficult, if not impossible, to join using conventional welding processes involving primarily molten metal and filler metal.

2 Experimental Section

2.1 Friction Welding Machine

Rotating machines such as lathes must rotate to perform friction welding, especially inertial friction welding. Flywheels of different sizes are mounted on the chuck and spindle shaft for inertia welding. A motor is then attached to the spindle shaft to rotate the part. This study is done on a lathe. The lathe is a gap bed lathe machine

Fig. 1 A gap-bed lathe CW6232 machine

type CW6232B, as depicted in Fig. 1. The machine has a top speed of 1600 rpm and can be set to different speeds by changing the positions of levers A and B (Fig. 1).

Generally, inertia friction welding consists of three steps: (1) At the beginning of the inertia friction welding, rotating motor and a non-rotating part as a place where the specimen placed for the welding process. The non-rotating part is equipped with a hydraulic system that applies pressure to the specimen. The rotating part of the machine is holding the specimen through the holder chuck, whilst the non-rotating part is giving pressure to both side of the specimen generating heat from the friction. (2) Another aspect to be specified before the welding is the speed of the rotation. As the lathe machine is turned on with desired speed, the speed must be reached before the pressure could be applied. If the pressure is applied before the designed speed, the generated heat from the friction is not at the maximum value, thus the deformation is not evenly distributed on the interface of both specimens. The lever position of the speed controller must be adjusted according to the designed speed. (3) After the designed speed is reached, the two specimens are come into contact, creating inertial friction at the specimen interface, generating heat, and performing the weld. At this point, the sample remains in contact for a period of time before being subjected to pressure. In this study, pressure was generated by a hydraulic system equipped with a pressure gauge to allow adjustment and monitoring of the amount of force delivered. However, inertial friction welding was a very difficult task. Since the sample is partially melted and easily deformed, buckling occurs when pressure is applied while the lathe is rotating. On the other hand, the holding time before the lathe stops is probably longer, giving the sample time to solidify without enough the pressure applied from the hydraulic system. This limitation was resolved by adding a braking system to the lathe machine. As a result, the time before the machine stops rotating is significantly reduced, giving the hydraulic system time to apply pressure to the sample while it is still partially solidified.

3 Friction Welding Parameters

The experiment consists of three welding parameters, each of which is designed to have three different values. Table 1 displays the simulation parameters along with their levels, while Table 2 displays the simulation matrix of the welding conditions based on their level. SX denotes the numbering of conditions, making it easier to mention later. This sampling method follows the Taguchi Method for orthogonal experiments; thus, a much more efficient investigation could be conducted with fewer resources [23, 24].

Although there is other parameters variation such as friction pressure, which is a pressure applied during friction processes, rotational speed, which is the speed of the rotation during friction process, and so forth, those parameters are subjected to a constant value. During friction welding, a friction pressure of 0.5 MPa and a rotational speed of 1600 rpm is employed.

Table 1 Level of parameters

Level	Friction time (s)	Forging pressure (MPa)	Forging time (s)
1	60	2.4	30
2	65	2.6	35
3	70	2.8	40

Table 2 Experiment detail (orthogonal experiment)

Sample	Friction time level	Forging pressure level	Forging time level
S1	1	1	1
S2	1	2	2
S3	1	3	3
S4	2	1	2
S5	2	2	3
S6	2	3	1
S7	3	1	3
S8	3	2	1
S9	3	3	2

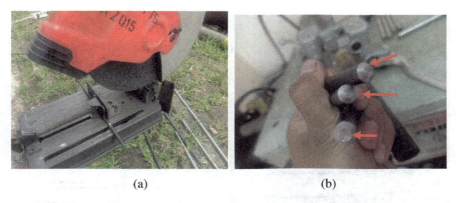

(a) (b)

Fig. 2 a Cutting process of the sample, **b** the final to be welded sample after sharpening

4 Sample Preparation

Before welding processes begin, the sample must be pre-prepared so that it fits with the required form and shape, thus allowing for better contact between the two pieces, resulting in a better-welded joint. The samples are prepared as follows:

1. First, both AISI 304 and AISI D2 tool steel are cut with an abrasive cutter, with each sample having a length of about 8.5 cm, while the samples' diameter is around 12 mm (Fig. 2a).
2. Then, by utilizing the same lathe machine, the interface of the sample is sharpened a little so that one interface is matched with the other, allowing a better bonding between the two (Fig. 2b).

5 Results and Discussions

5.1 The Physical Property of Friction Welded AISI D2 Tool-Steel and AISI 304 Stainless-Steel

The friction-welded AISI D2 tool steel and AISI 304 stainless steel are depicted in Fig. 3. The darker material (right side) indicates more carbide in the material associated with AISI D2 tool steel. In contrast, the brighter one shows less carbide (left side) associated with AISI 304 steel. It appears that the heat generated during friction affects the interface of the material significantly. In contrast, the area near the interface, known as the "heat affected zone" (HAZ), physically changes darker from the as-received material. The heat generated during friction welding is quite high; thus, strong bonding between the two materials is expected. However, to produce such good results, the friction welding machine's configuration must be maintained so that the interface between the two materials is matched and the pressure is maintained

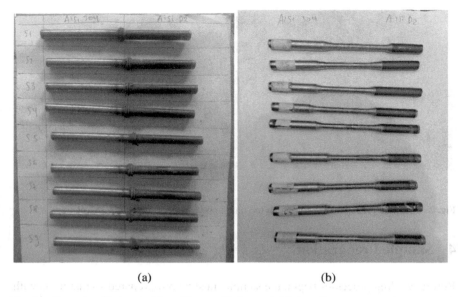

Fig. 3 **a** Raw as-welded material, and **b** as-welded material for tensile test

correctly. Although the sample shows an excellent result, the result from physical observation needs to be confirmed by a mechanical test. The piece depicted in Fig. 3a shows the raw as-welded material and turns it into the samples for the tensile test (Fig. 3b).

6 Mechanical Properties of Friction Welded AISI D2 Tool-Steel and AISI 304 Stainless-Steel

The tensile test is performed using a universal test machine and producing the data, as shown in Table 3. Of the three employed parameters, friction time, forging pressure, and forging time, the highest impact comes from forging time. It is shown that the highest forging time, that is, 40 s, offers the highest value of tensile strength, though the highest friction time shows different behavior. Other parameters like friction time and forging pressure show less significant impact on the tensile strength than forging time. Pressure during the forging phase could lead to a considerable increase in the strength of the as-welded material due to the high cooling rate; thus, high pressure is needed for such a critical time to produce a better bonding structure between the two materials. There are other possibilities for improvement, especially in the construction of FW machines and FW procedures.

Study of Mechanical Properties of Friction Welded AISI D2 and AISI ...   923

Table 3 Tensile test result

Sample Code	Friction time (s)	Forging pressure (MPa)	Forging time (s)	Tensile strength (MPa)
S1	60	2.4	30	216
S2	60	2.6	35	157
S3	60	2.8	40	265
S4	65	2.4	35	196
S5	65	2.6	40	235
S6	65	2.8	30	127
S7	70	2.4	40	216
S8	70	2.6	30	118
S9	70	2.8	35	284

7 Conclusion

This study shows that it is possible to produce high-quality welding of the two materials with a tube-shaped that classically can only be welded by their interface or using a filler. Although the construction of a friction welding machine itself is a complex step with a critical procedure to produce the best welding result, the chance that this friction welding could be promoted as one of the main welding techniques for certain applications like industrial manufacture and aircraft construction seems promising. The results also showed a vast possibility for improvement in both the structure of the FW machine and the FW procedure.

Acknowledgements The authors thanks Universitas Global Jakarta and Management & Science University for their support throughout the project. This project is funded by the Early Career Lecturer (PDP) scheme from the Ministry of Education, Culture, Research, and Technology with grant number B/112/E3/RA.00/2021.

References

1. Mohammed, M.N., Omar, M.Z., Sajuri, Z., Salleh, M.S., Alhawari, K.S.: Trend and development of semisolid metal joining processing. Adv. Mater. Sci. Eng. (2015)
2. Kim, I.S., Son, K.J., Yang, Y.S., Yaragada, P.K.D.V.: Sensitivity analysis for process parameters in GMA welding processes using a factorial design method. Int. J. Mach. Tools Manuf. 43(8), 763–769 (2003)
3. Karaoğlu, S., Seçgin, A.: Sensitivity analysis of submerged arc welding process parameters. J. Mater. Process Technol. 202(1–3), 500–507 (2008)
4. Tarng, Y.S., Juang, S.C., Chang, C.H.: The use of grey-based Taguchi methods to determine submerged arc welding process parameters in hardfacing. J. Mater. Process Technol. 128(1–3), 1–6 (2002)
5. Balasubramanian, V.: Relationship between base metal properties and friction stir welding process parameters. Mater. Sci. Eng., A 480(1–2), 397–403 (2008)

6. Zaenudin, M., Mohammed, M.N., Al-Zubaidi, S.: Molecular dynamics simulation of welding and joining processes: an overview. In. J. Eng. Technol. [Internet]. 2018 Dec 5 [cited 2023 Jan 4] 7(4), 3816–3825. Available from: https://www.sciencepubco.com/index.php/ijet/article/view/16610

7. Zaenudin, M., Abdulrazaq, M.N., Al-Zubaidi, S.S.: A review on molecular dynamics simulation of joining carbon-nanotubes and nanowires: joining and properties. Int. J. Integr. Eng. [Internet]. 2022 Jun 21 [cited 2023 Jan 4]; 14(4), 137–159. Available from: https://penerbit.uthm.edu.my/ojs/index.php/ijie/article/view/5578

8. da Costa, A.P., Botelho, E.C., Costa, M.L., Narita, N.E., Tarpani, J.R.: A review of welding technologies for thermoplastic composites in aerospace applications. J. Aerosp. Technol. Manage. [Internet]. 2012 [cited 2023 Jan 31]; 4(3):255–65. Available from: http://www.scielo.br/j/jatm/a/W6T3p6THtcnrfcBtwsMqNyk/abstract/?lang=en

9. Wang, B., Hu, S.J., Sun, L., Freiheit, T.: Intelligent welding system technologies: state-of-the-art review and perspectives. J. Manuf. Syst. 1(56), 373–391 (2020)

10. Hong, K.M., Shin, Y.C.: Prospects of laser welding technology in the automotive industry: a review. J. Mater. Process Technol. 1(245), 46–69 (2017)

11. Mohammed, M.N., Omar, M.Z., Al-Tamimi, A.N.J., Sultan, H.S., Abbud, L.H., Al-Zubaidi, S., et al.: Microstructural and mechanical characterization of ledeburitic AISI D2 cold-work tool steel in semisolid zones via direct partial remelting process. J. Manuf. Mater. Process. 7, 11 [Internet] (2023). 2022 Dec 28 [cited 2023 Jan 31]; 7(1), 11. Available from: https://www.mdpi.com/2504-4494/7/1/11/htm

12. Progress in welding. Nature 215(5099), 343 (1967)

13. Scherillo, F., Liberini, M., Astarita, A., Franchitti, S., Pirozzi, C., Borrelli, R., et al.: On the microstructural analysis of LFW Joints of Ti6Al4V components made via electron beam melting. Procedia Eng. 1(183), 264–269 (2017)

14. Fratini, L., Buffa, G., Campanella, D., la Spisa, D.: Investigations on the linear friction welding process through numerical simulations and experiments. Mater Des. 1(40), 285–291 (2012)

15. Bhamji, I., Preuss, M., Threadgill, P.L., Moat, R.J., Addison, A.C., Peel, M.J.: Linear friction welding of AISI 316L stainless steel. Mater. Sci. Eng., A 528(2), 680–690 (2010)

16. Vairis, A., Frost, M.: On the extrusion stage of linear friction welding of Ti 6Al 4V. Mater. Sci. Eng., A 271(1–2), 477–484 (1999)

17. Chaudhary, A., Kumar Dev, A., Goel, A., Butola, R., Ranganath, M.S.: The mechanical properties of different alloys in friction stir processing: a review. Mater. Today Proc. 5(2), 5553–5562 (2018)

18. Mendes, N., Neto, P., Loureiro, A., Moreira, A.P.: Machines and control systems for friction stir welding: a review. Mater. Des. 15(90), 256–265 (2016)

19. Avettand-Fènoël, M.N., Simar, A.: A review about friction stir welding of metal matrix composites. Mater. Charact. 1(120), 1–17 (2016)

20. Li, W., Vairis, A., Preuss, M., Ma, T.: Linear and Rotary Friction Welding Review. [Internet]. 2016 [cited 2023 Jan 31]; 61(2):71–100. Available from: https://www.tandfonline.com/doi/abs/https://doi.org/10.1080/09506608.2015.1109214

21. Vairis, A., Frost, M.: Design and Commissioning of a Friction Welding Machine. [Internet]. 2007 Dec 1 [cited 2023 Jan 31]; 21(8):766–773. Available from: https://www.tandfonline.com/doi/abs/https://doi.org/10.1080/03602550600728356

22. Sunardi, S., Mariyana, M., Gamayel, A., Mohammed, M.N., Zaenudin, M.: Design & development of friction welding machine based on lathe machine. AIP Conf. Proc. [Internet]. 2022 Nov 3 [cited 2023 Jan 31]; 2578(1):070002. Available from: https://aip.scitation.org/doi/abs/https://doi.org/10.1063/5.0106341

23. Byrne, D.M., Taguchi, S.: Taguchi approach to parameter design. Qual. Prog. 20(12), 19–26 (1987)

24. Freddi, A., Salmon, M.: Introduction to the Taguchi method. Springer Tracts in Mechanical Engineering [Internet]. 2019 [cited 2023 Jan 31]; 159–80. Available from: https://link.springer.com/chapter/https://doi.org/10.1007/978-3-319-95342-7_7

Assessing the Adoption of Key Principles for a Sustainable Lean Interior Design in the Construction Industry: The Case of Bahrain

Aysha Aljawder, Wafi Al-Karaghouli, and Allam Hamdan

Abstract This study identifies and adopt the key principles for a sustainable lean interior design in the construction industry. The concept of lean management has been of interest to researchers and practitioners in the construction industry and is implemented in most countries. This paper is part of the study to critically investigate the factors effecting the process of selecting lean tools and techniques in the construction industry in the public sector in the four governates of the Kingdom of Bahrain. The quantitative research methodology has been adopted in this study. The investigation will determine the level of lean construction management implementation and consequently its effect on sustainability in the interior design of housing projects. It is hoped that this study will benefit both academics and practitioners.

Keywords Lean management · Interior design · Sustainability · Artificial intelligence

1 Introduction

From previously published literature, the construction industry has been identified as the most important economic markers of accomplishment in the affluence of a nation, since this sector has a considerable impact on gross domestic product and employment [7]. This statement crucially emphasises the importance of construction as an industry in any country as it is the livelihood for many. According to Bajjou et al. the construction industry presently comprises more than 53,000 construction firms in Morocco. Bajjou et al. continued by stating the fact that the annual revenue of these companies exceeded \$3.2 billion, which arguably makes a big contribution

---

A. Aljawder · W. Al-Karaghouli
Brunel University, London, UK

A. Aljawder · A. Hamdan (✉)
Ahlia University, Manama, Bahrain
e-mail: allamh3@hotmail.com

© The Author(s), under exclusive license to Springer Nature Switzerland AG 2024
A. Hamdan and E. S. Aldhaen (eds.), *Artificial Intelligence and Transforming Digital Marketing*, Studies in Systems, Decision and Control 487,
https://doi.org/10.1007/978-3-031-35828-9_78

to the world economy. Luangcharoenrat et al. [17] highlighted that the construction industry is a main mechanism for infrastructure advancement to support cities' development of other industries and impacts environmental degradation. It is worth noting that most of the relevant literature's, findings state that the growth and expansion of a vital necessity, such as the built environment and including green environment, of a community must come with some form of waste. Reducing waste creates a competitive advantage for stakeholders in the construction industry especially subcontractors, main contractors, and real estate developers [17], and helps the environment, by cutting wastage. Vilventhan et al. [26] reported that the sheer magnitude of waste from the construction industry recently reported in literature validates the need for further research to better ways of managing them.

Lean Construction Management consists of a set of objectives aimed to serve the delivery process better and maximize performance from design to delivery [3]. Currently, there is a variety of lean tools and techniques, such as, Total Quality Management (TQM), Daily Huddle Meetings, Just in Time (JIT), Target Value Design (TVD), Total Productive Maintenance (TPM) and Computer Aided Design (CAD) all of which support the implementation of Lean Construction processes [24].

2 Literature Review

2.1 Waste and Lean Strategy in Construction

Freitag et al. [7] found that Brazil's progress and production capacity are wholly related to the development of the construction industry. This poses a research gap preview and new challenges for invention of new technology, professional training and the formation of business environments that stimulate productivity, business competitiveness and the development of any country. The generation of waste in the processes of production along with the discharges that result from these processes are very well examined. According to Purushothaman et al. [21], some types of waste are generated by organisational hierarchical systems, departmental boundary limitations, individual activities, and IT functions. Koskela indicated that the problems facing the construction industry are sometimes direct or sometimes indirect sources of substantial waste generation and thus negatively affect the performance of construction projects. The construction industry utilises 35% of generated energy and releases 40% of carbon dioxide into the planet's atmosphere [17]. This further reinforces that new evolving management techniques and practices are crucial to reducing such harmful percentages of waste. Babalola et al. [4] discovered that cost overrun takes up to 14% of project contract amount, while about 70% of all projects turn up with time overrun, also about 10% of the whole project material end up as waste. According to the United Nations Environmental Global Report, the construction industry is the leading consumer of raw materials obtained from natural resources, illustrating the importance of management for the purpose of waste elimination.

- **Smart Construction Management** The utilisation of construction digital solutions nowadays has exhibited major advantages in conquering those tasks. The plan of the automisation of construction control and monitoring as well as pushed practitioners and researchers to seek various techniques like visual-based approaches. A few of those alternate techniques utilised audio-signal processing for recognition of construction endeavors [22]. Others utilised devices that were installed such as ultra-wideband, radio-frequency identification, inertial measurement units, or equipment elements for tracking and monitoring their routines or a world-wide positioning approach related to construction workers. According to [22], the main constraint of the audio related technique is recognising the distinctive sounds of every single movement on a loud site in the construction sector. Additionally, installed devices that are worn generate irritation to workers, which adversely impacts their efficiency [14]. Moreover, the implementation of these techniques is frequently impeded by the soaring expense of monitoring and installation [18].

Performance [21], in situ and assessment of post-construction quality [16], construction waste management [27], facilities management [21], and dynamic worksite management [18]. Many computer visions (CV) tasks like image categorisation, activity recognition, object detection, pose estimation, as well as object tracking were deciphered using deep learning (DL) algorithms as an essential part of these functions [19]. The authors continued to mention that the numerous studies associated with deep learning (DL) implementations in construction has been soaring since the year 2018 due to multiple reasons, such as the convenience of some datasets in the public sector, the progression of GPU processing, and so forth.

Nevertheless, there are yet several open disputes to refer for implementing the research conclusions immediately in complicated construction projects [19]. At present, it is crucial to evaluate the ultra-modern, feature future exploration directions regarding the community of construction management, and to offer a generalised system for the DL-based visual data analytics for the construction field practitioners [19]. Even though some recent reviews [14, 21] have concentrated on vision-based methods for construction, and these were not limited to deep learning (DL) utilisation. As deep learning (DL) algorithms have surpassed virtually every conventional computer visions (CV) technique in the sense of performing and strength, a devoted analysis of deep learning (DL) is extremely required. Pal and Hsieh [19] highlighted that as far as we know, only two papers until the day [2, 11] have revised DL implementation in the (AEC) industries. Specifics of those evaluations are given in Table 1. Hou et al. [11] evaluated DL implementations related only to safety management while different functions require additional review. Akinosho et al. [2] involved published work until the beginning of [21] in their evaluation. Nevertheless, a substantial amount of significant papers have been published since then, according to Pal and Hsieh [19]. Furthermore, mutually the published papers were not limited to visual-based techniques, and they have not given any standard system for DL-based visual data analytics.

Table 1 Specifics of the previous DL-associated evaluations in the AEC field

References	Search specifics				Range of the paper
	Data bank searched	Pursuit period	Search keywords	Number of reviewed publications	
Hou et al. [11]	Web of Science (WoS)	2010—[21]	("deep learning" "machine learning" CNN* "Recurrent neural network*") AND "structu* health monitoring" inspection* behavi* concrete* "computer vision*" "Natural Language Processing"	527 papers published for analysis of bibliometric. The number of Papers employed for comprehensive evaluation have not identified	Products associated with safety management
Akinosho et al. [2]	Scopus and Science Direct	2012—[21]	"deep learning", "deep learning in the construction industry", "deep neural networks", networks", "Auto-Encoders"	45 SCI indexed paper for the in-depth review	Applications related to construction

Source Pal and Hsieh et al. [19]

- **BIM and Virtual Reality Technology in Construction** The (BIM) philosophy of construction information prototype was initiated in the year 1962 where Ivan E. Sutherland of MIT initially established the Sketchpad which is an interactive graphic system, that intended the idea of computer graphics as well as generating computer communication [27]. In the middle of the 1980s, Robert Aish an American scholar brought into view the notion of building information modeling in his research. In the beginning of the 1990s, van Nederveen and Tolman, in the late 1990s, aimed to determine "building information model" using information disintegration. In the year 2002, Autodesk company in the publicity of building info strength shaping method, as known by BIM. Ever since then, developers of software graphisoft and Bentley have introduced BIM associated programs. Ever since, BIM advanced from theoretical thinking to instruments and techniques to resolve functional difficulties [20].

Lean Construction Principles: According to Bajjou and Chafi [5], the attainment of the objective of the work that developed their theoretical framework was based on a thorough and comprehensive research of the top academic journals that debated the core principles of Lean Construction methods. They examined dependable databases, particularly: IEEE Xplore, Springer Link, Science Direct, Emerald Online, World Scientific, Inderscience, Taylor & Francis, and Scopus. Since the notion of the theory of lean construction was presented by Lauri Koskela in [15] in the technical report

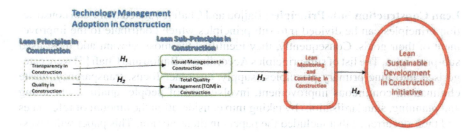

Fig. 1 Conceptual framework

"Application of the New Production Philosophy to Construction", which was the start of the familiarisation of concept in the industry of construction, the study was based on previous conference papers, research papers, books, review papers, and thesis that were published among the years 1992 and 2017 [5]. The keywords used by Bajjou and Chafi [5] were: "Lean Construction conformance", "Lean construction model", "Lean Construction maturity", "Lean construction implementation", "Lean construction framework", "Lean construction principles", and "lean construction assessment". To make the finding more significant, they also used a number of mixtures of the afore mentioned keywords [8].

Bajjou and Chafi's [5] initially published 751 articles about the lean construction theory from all the well-thought-out publication databases. After that, a deep assessment of the entire text of all the article was done in order to enabled them to classify 125 relevant studies and articles. They selected the most recent studies, which were 59 articles that were published in between the years 2015–2017, to plan a model of lean construction of conceptual nature. Figure 1 displays the chosen sources categorised by the year they were published in according to Bajjou and Chafi. [5].

With nine principles of Lean that were identified symbolise the core of Lean Construction thinking. They are: Supply, Customer focus, Waste elimination, Continuous improvement, Quality, Planning and Scheduling, People involvement, Standardization, Transparency. Furthermore, those principles could be allocated into three principal pillars:

- Systems management: which concentrates on the management of systems in interface with customers, suppliers and contractors (Customer focus, Supply, Planning and Scheduling);
- Technology management: which affects the operational techniques intended to optimising the performance of the company (Transparency, Standardisation, and Quality);
- Culture and behaviour: which encompasses all Lean Construction practices that permit the propagation of continuous improvement as a culture during every cycle of production when guaranteeing an optimum employee skills utilisation (people involvement, continuous improvement).

Lean Construction Sub-Principles: Bajjou and Chafi [5] stated that Lean construction principles can be divided into sub-principles which contribute to the improvement of their goals. Consequently, they identified the most relevant and most used sub-principles. The list of sub-principles According to Bajjou and Chafi [5] were categorised into nine primary principles (supply, customer focus, transparency, waste elimination, continuous improvement, involvement of people, quality, scheduling and planning, standardisation) by taking into consideration the amount of references and their occurrence that included the papers in their citation. This paper will investigate the sub-principles that are linked to the principle of technology management in construction.

Bajjou and Chafi [5] stated that the highest frequency that they found was 20 from an approximate overall of 125 key articles that were analysed and emphasize mainly on the sub-principles and techniques utmost utilised all over the globe for the purpose of implementation of the lean construction idea, specifically in the construction projects that are executed on a basis philosophy of lean construction. Bajjou and Chafi [5] explained that the scale of frequency that they employed was: 1 to 5 = Very Low (VL); 6 to 10 = Low (L); 11 to 15 = High (H); 16 to 20 = Very High (VH). The sub-principles of below six (VL) were neglected in the model of their conceptual prototype.

3 Conceptual Framework

This research will adopt the constructs of the theoretical framework of Bajjou and Chafi [5] and extend it by adding to it constructs that are relevant e.g., waste elimination in construction, leadership in construction and, talent in construction, which will be applicable to the construction industry in the public sector in the Kingdom of Bahrain as shown in. The conceptual framework for this research is presented in Fig. 1 which focuses on the principle of technology management in construction.

Hypotheses

H_1: There is a relationship between Transparency in Construction and Visual Management in Construction.

H_2: There is a relationship between Quality in Construction and Total Quality Management (TQM) in Construction.

H_3: Technology management in construction helps monitor and control the implementation of lean principles and sub-principles.

H$_4$: Lean monitoring and controlling in construction help achieve lean sustainable development in construction.

- *Technology Management in Construction*: which affects the operational techniques intended to optimising the performance of the company. The principle that fall under this pillar are:

  - *Transparency in Construction*: this construct represents the policy of total transparency and clarity in the construction operation and how the steps that are followed by the team are monitored and managed through construction management software.
  - *Quality in Construction*: this construct concentrates on managing the quality of the construction operations as well as the quality of the end result.
  - *Lean Monitoring and Controlling*: this construct focuses on lean sustaining, tracking, reviewing and regulating.

4 Research Approach and Methodology

In this research, the main lean principles, sub-principles and the hypotheses stating the relationships between them and recognised in the adopted and extended conceptual framework. In this research study, it is believed that the empirical method is more suitable for the sake of assessing the hypotheses using an empirical study. Taking the research question into consideration (what are the factors influencing the adoption of lean principles at construction sites?) and the adequate data to generate hypotheses for assessment, the investigation framework that was implemented in this study was cross-sectional. Cross-sectional analysis is a broadly utilised investigation framework particularly in social science studies that are linked with both questionnaire surveys and defined interviews. This specific type of investigation is incorporated in the data collection process with additional cases in a specific period of time to examine the relationship among the variables [23].

This investigation determines the relationships among independent and dependent variables. This investigation begins with a vast review, and assessment of the relevant literature. The key goal in performing a large literature review is to recognise the research gap of this investigation; a conceptual framework was established to perform a pragmatic assessment. Concerning the conceptual framework and with the support of social discussion and operation cost economic theories as well as hypotheses were formulated to examine the relationship among the independent and dependent variables. The method that was adopted for this investigation is a quantitative method for data collection and the following evaluation. Gilbert [9] stated that the theory in positivism is empirical and leads with hypotheses. A recommendation was made by Hussey and Hussey that the conventional technique in a positivistic theory is to examine the literature in order to produce a proper theory and construct hypotheses.

Within the protocol, a quantitative research method for data collection have been gathered to analyse lean implementation in the Kingdom of Bahrain. A questionnaire survey was utilised where there were several issues that are associated to the lean principles and sub-principles that affect lean monitoring and controlling and its relation to a lean sustainable construction development.

- **Creating a Survey Questionnaire** With the intention of creating a questionnaire survey, it is crucial to keep in mind the variety of information that requires to be acquired. For this investigation, a questionnaire survey was recommended for data collection to investigate the investigation hypotheses. A questionnaire survey can offer understanding into people's perceptions and mindsets in addition to organisational policies and traditions [6].

  In this research study a cross-sectional research is implemented, where all the gathered data have been taken at the same time period from a suitable sample of participants to assess the hypotheses. Both the independent and the dependent variables were examined simultaneously. There are six sections in the investigation tool of this study.

  Section A: is concerned with the demography of the respondents, offering data concerning the respondents' individual qualities and overall environment, four items from questions 1 to 4 will investigate the demography and overall environment of the participants.

  Section B: is concerned with the implementation of systems management as well as the principles and sub-principles that fall under it in the public construction sector in the Kingdom of Bahrain, four items were developed for this section from questions 5 to 8.

  Section C: is concerned with the implementation of technology management and the related principles and sub-principles in the public construction sector in the Kingdom of Bahrain, four items were developed for this section from questions 9 to 12 which is the main focus is this paper.

  Section D: is concerned with the implementation of culture and behaviour along with the principles and sub-principles in the public construction sector in the Kingdom of Bahrain, four items were developed for this section from questions 13 to 16.

  Section E: is concerned with the implementation of waste elimination in construction along with the principles and sub-principles in the public construction sector in the Kingdom of Bahrain, four items were developed for this section from questions 17–20.

  Section F: is concerned with implementation of lean monitoring and controlling to the proposed lean principles and sub-principles (technology management, systems management, waste elimination, and culture and behaviour,) in the construction industry in the public sector in Bahrain, six items were developed for this section from questions 21–26.

- **Measurement Scales** The main questionnaire has eight independent, two dependent variables, and eight intervening variables to measure the lean implementation in the public construction sector in the kingdom of Bahrain. The independent variables are customer focus, supply in construction, transparency, quality, work culture, leadership, inventory, and talent. Lean monitoring and controlling and lean sustainable development are the dependent variables, while customer involvement, supplier involvement, visual management, total quality management, training, huddle meetings, reduce process cycle time and waste awareness are the intervening variables which are used to explain the causal links between the other variables (Table 2).

These adopted scales were tested using a pilot study on managers, project managers, project officers and all of the project team members in the public construction industry in the four governates of the Kingdom of Bahrain. In order to collect the data and to get the targeted participants to take part in the survey, formal emails and digital messages were sent to invite them to do so. The goal of the pilot study was to eradicate any poor phrasing or ambiguity in the questions and to verify the time necessary to carry out the survey. Following the pilot data collection, the reliability and validity of the survey tools were tested, the process of data collection for the primary study from targeted sample commenced.

Targeted Sample From Table 3 it can be seen that a total of 111 (33.5%) respondents that participated in the survey have been of the younger generation between the ages of 20 and 30 years of age. While the majority of respondents were of an older group between the ages of 31–40 years of age with 123 (37.2%) respondents. The middle-aged group between 41 and 50 years old had a count of 68 (20.5%) respondents. On the other hand, the minority of respondents were between 51 and 60 years of age with 29 (8.8%) respondents. See Table 3.

Table 2 Questions related to hypotheses

Hypotheses	Variable	Relevant question items
H_1: There is a relationship between Transparency in Construction and Visual Management in Construction	Transparency	Q9–Q10
H_2: There is a relationship between Quality in Construction and Total Quality Management (TQM) in Construction	Quality	Q11–Q12

Category	Frequency	Percentage
Age		
20–30	111	33.5
31–40	123	37.2
41–50	68	20.5
51–60	29	8.8
Total	331	100.0
Gender		
Female	180	54.7
Male	149	45.3
Total	331	100.0
Occupation		
Manager	96	29.3
Project Manager	81	24.7
Project Officer	68	20.7
Project Management Team Member	83	25.3
Total	331	100.0
Experience		
Less than 5 years	110	33.4
10 years	83	25.2
15 years	86	26.1
More than 20 years	50	15.2
Total	331	100.0

Table 3 Targeted sample

5 Regression Analysis 1: Examining the Relationship Between Gender and the Adoption of Lean Principles and Sub-principles

In this regression analysis, the relationship between gender and the adoption of lean principles and sub principles in the construction industry in the Kingdom of Bahrain in the pubic sector. Moreover, this research will use the T-Test as a technique to assess the mean differences and deviations significance between the variables and whether or not the relation is significant (Table 4).

Assessing the Adoption of Key Principles for a Sustainable Lean Interior … 935

Table 4 Gender in relation to the adoption of lean principles

Variable	Gender	Mean	Standard Deviation	T-Test	Sig
SMC	Female	2.1361	0.72047	2.002	0.046
	Male	1.9715	0.76810	1.990	0.047
TMC	Female	4.1292	1.19194	−2.370	0.018
	Male	4.4379	1.15662	−2.377	0.018
CBC	Female	2.1616	0.75050	1.491	0.137
	Male	2.0352	0.78253	1.485	0.139
WEC	Female	4.1417	1.17828	−1.981	0.048
	Male	4.3971	1.14672	−1.986	0.048
ILPC	Female	1.1464	0.21225	−0.483	0.629
	Male	1.1603	0.30768	−0.467	0.641

6 Regression Analysis 2: Examining the Relationship Between Work Position and the Adoption of Lean Principles and Sub-principles

In this research, the relationship between work position and the adoption of lean principles and sub principles using the one-way ANOVA test in the SPSS V.23. The test determines the variables where significance can be found. Table 5 showcases the results when tested against each variable.

By obtaining significant ANOVA values, this research uses Post Hoc Test in order to assess the mean differences between the two groups from the variables. In this Post Hoc Test, the third category of the demographic, which is work position, was tested for mean differences against each group of variables from the lean principles and sub principles (Table 6).

Table 5 Work position in relation to the adoption of lean principles (F-Test)

Variable	F	Sig
SMC	0.380	0.767
TMC	8.118	0.000
CBC	0.754	0.521
WEC	10.571	0.000
ILPC	0.732	0.534

Table 6 Work position in relation to the adoption of lean principles (Significance)

Dependent variable	Demo 3	Mean Difference	Sig
SMC	Manager	0.05314	0.638
		0.11106	0.350
		−0.00003	1.000
	Project Manager	−0.05314	0.638
		0.05792	0.638
		−0.05318	0.649
	Project officer	−0.11106	0.350
		−0.05792	0.638
		−0.11109	0.365
	Project Management Team Member	0.00003	1.000
		0.05318	0.649
		0.11109	0.365
TMC	Manager	−0.19637	0.258
		−0.13419	0.461
		0.59676	0.001
	Project Manager	0.19637	0.258
		0.06218	0.742
		0.79314	0.000
	Project officer	0.13419	0.461
		−0.06218	0.742
		0.73095	0.000
	Project Management Team Member	−0.59676	0.001
		−0.79314	0.000
		−0.73095	0.000
CBC	Manager	−0.02980	0.797
		−0.06970	0.567
		−0.16456	0.153
	Project Manager	0.02980	0.797
		−0.03990	0.752
		−0.13476	0.261
	Project officer	0.06970	0.567
		0.03390	0.752
		−0.09486	0.450
	Project Management Team Member	0.16456	0.153
		0.13476	0.261
		0.09486	0.450

(continued)

Assessing the Adoption of Key Principles for a Sustainable Lean Interior … 937

Table 6 (continued)

Dependent variable	Demo 3	Mean Difference	Sig
WEC	Manager	−0.30430	0.072
		−0.07409	0.676
		0.63285	0.000
	Project Manager	0.30430	0.072
		0.23021	0.211
		0.93716	0.000
	Project officer	0.07409	0.676
		−0.23021	0.211
		0.70695	0.000
	Project Management Team Member	−0.63285	0.000
		−0.93716	0.000
		−0.70695	0.000
ILPC	Manager	−0.02789	0.478
		−0.03010	0.466
		−0.05768	0.140
	Project Manager	0.02789	0.478
		−0.00221	0.959
		−0.02980	0.464
	Project officer	0.03010	0.466
		0.00221	0.959
		−0.02759	0.517
	Project Management Team Member	0.05768	0.140
		0.02980	0.464
		0.02759	0.517

7 Hypotheses Testing

The second category from the framework is technology management in construction, has had two lean principles that are linked to it (TMC$_1$, TMC$_3$). The variables TMC$_1$ and TMC$_3$ were tested in relation to two intervening variables (TMC$_2$, TMC$_5$). The questions were asked based on the adoption of the lean principles and sub-principles in the Kingdom of Bahrain and whether or not the respondents find it achieved lean and sustainable results in construction.

The first lean principle that the participants were asked about was transparency in construction (TMC$_1$) which required the participant to rate—from 1 to 5, 1 being lowest and 5 being highest—the adoption of availability and speed of sharing information and communicating standards with the team in the public Bahraini construction sector. Respondents who rated the lean principle adoption at 3 were 86 (25.9%)

which was the majority, while the rating 4 had 82 (24.7%) responses. The rating 5 came next with 60 (18.1%) responses, while the rating 2 had 57 (17.2%) and the rating 1 had 38 (11.4%) responses. Only 9 (2.7%) respondents thought that the lean principle was non-applicable.

The corresponding lean sub-principle that the participants were asked about was visual management in construction (TMC_2) which required the participants to rate— from 1 to 5, 1 being lowest and 5 being highest—the adoption of bringing attention to irregularities and performance levels in the public Bahraini sector to help the team react to them when they happen. The responses were a majority at the rate 4 with 104 (31.3%) responses and the rating 3 came in after it with 74 (22.3%) responses. Moreover, 66 (19.9%) respondents rated the adoption at 5, while the rating 2 had 60 (18.1%) responses and the rating 1 had 23 (6.9%). Only 5 (1.5%) participants found the lean sub-principle to be non-applicable.

The second lean principle that the participants were asked about was quality in construction (TMC_3) which required the participants to rate–from 1 to 5, 1 being lowest and 5 being highest—the adoption of investing in advance technology in the public Bahraini construction sector to help with and auditing testing to assure client satisfaction. The majority of responses came in at the highest rating which were 83 (25%) responses and the rating 4 came in second with 79 (23.8%) responses. The rating 2 had 70 (21.1%) responses while the rating 3 came in after it with 68 (20.5%) responses. The participants who rated the lean principle at 1 were 24 (7.2%) and only 8 (2.4%) participants thought that it was non-applicable.

The last corresponding sub-principle in that category was total quality management in construction (TMC_4) which required the participants to rate—from 1 to 5, 1 being lowest and 5 being highest- the adoption of regularly updating control charts for quality assurance programme in the public Bahraini construction sector. The highest ratings came in with close results where 5 had a count of 86 (25.9%) responses and 4 had 83 (25%) responses. The rating 3 came in after with 62 (18.7%) responses while 2 had a count of 56 (16.9%) responses. Only 37 (11.1%) participants rated the sub-principle as 1 while 7 (2.1%) respondents thought that it was non-applicable.

- **Gender**

  This research tested the significance of the role of gender in relation to the previously tested variables SMC, TMC, CBC, WEC, and ILPC. The analysis applied the T-Test using SPSS, to help further clarify the significance of gender.

  According to the findings of the T-Test of this study, there is significance in both female and male in relation to systems management in construction (SMC) variables. Technology management in construction (TMC) variables, also showed significance in both genders. Significance in both genders was exhibited in relation to waste elimination in construction (WEC).

- **Professional Position**

  The professional position category of the demographic required the implementation of two tests due to the number of options given as answers for the respondents to choose from (manager, project manager, project officer, and project team member). Firstly, the one-way ANOVA was used which cultivated results of significance in the variables of technology management in construction (TMC) and waste elimination in construction (WEC). Secondly, the Post Hoc tests were implemented to specify which professional positions in relation to the variables exhibited significance. The variables in technology management in construction (TMC) showed significance in all of the professional positions (manager, project manager, project officer, and project team member). While waster elimination in construction showed significance in only two professional positions (project officer and project team member). This study has shown that the professional position is significant in the case of the adoption of lean principles.

- **Professional Experience**

In this category of the demographic the previously used tests (one-way ANOVA and Post Hoc tests) were implemented due to the number of answer choices given to the respondents (less than 5 years, 10 years, 15 years, and more than 20 years). The one-way ANOVA was implemented first to know the significance in the overall variables (SMC, TMC, CBC, WEC, and ILPC). The significance was found in all of the variables except for adoption of monitoring and controlling (ILPC). Afterwards, the Post Hoc tests specified the significance in the categories of the variables. Systems management in construction (SMC) showed significance in all the years of professional experience except for the answer of more than 20 years of experience, while technology management in construction (TMC) showed significance in all years of experience. In culture and behavior in construction (CBC) all the years of experience exhibited significance except for the answer of 10 years of experience. Finally this study indicates that waste elimination in construction showed significance in all years of professional experience.

8 The Role of Technology Management in Construction

According to [5] theoretical framework, one of the pillars of lean construction management is technology management which is concerned with the techniques of operations that are intended to the optimisation of the performance of the company. In this research, the lean principles and sub-principles that were tested in the Bahraini construction sector from this category, were chosen according to their relevance in the field.

- **Visual management in relation to transparency in construction**

The following hypothesis was formulated to test the relationship between the lean principle (transparency in construction) and the lean sub-principle (visual management) from the category of technology management.

H_2a: There is a relationship between Transparency in Construction and Visual Management in Construction.

After testing the abovementioned hypothesis using regression test in SPSS V.23 the results revealed that there is significance between the two constructs. As reported by [10], that out of all the five senses to influence analysis and learning, sight is the main contributor up to 75% of human perception which include learning, and activity. Moreover, with the human brain's visual system being an extremely vital system to process information, the findings could clarify the reason behind most organisations, particularly industrial organisations, promote the utilisation of current visual tools for on-site communications with workers [10].

- **Total quality management (TQM) in relation to quality in construction**

  The following hypothesis was formulated to test the relationship between the lean principle (quality in construction) and the lean sub-principle (total quality management) from the category of technology management.

H_2b: There is a relationship between Quality in Construction and Total Quality Management (TQM) in Construction.
After testing the abovementioned hypothesis using regression test in SPSS V.23 the results revealed that there is significance between the two constructs. [1] reported that research have exposed [25] that inadequate understanding, lack of tools, lack of capital, lack of practicable obligation of leadership, poor data and appointment strategies, inflexibility of disorganised application, unlikely anticipations, and inadequate managerial skills are some of the likely complications in the sequence of successful application of TQM. Moreover, Adeyemi et al. [1] concluded that findings through the multiple existing studies such as [13], Ezeani and Ibijola [12], that some of the main problems disturbing the adoption of TQM in the construction industry are lack of benchmarking and employee resistance to change, lack of understanding, insufficient preparation, absence of obligation from top management, lack of rewards and acknowledgment, insufficient evaluation processes, inadequate funding, incompetent management, scarce raw materials, absence of appropriate communication and under-productive leadership.

9 Study Limitation

This study had some limitations, which was the lack of published literature about the topic in the Kingdom of Bahrain. However, there were some published research studies from the Middle East and GCC exploring the topic of lean such as, Bajjou and Chafi's [5] published research paper in Morrocco, and Sarhan et al. [24] published research paper in the Kingdom of Saudi Arabia.

Hence, this study was adopted from relevant published research from the Middle East and the GCC and was carried out in the Kingdom of Bahrain, and it is believed that it will contribute to the knowledge of both sustainable lean in the construction industry and the context of Bahrain.

10 Future Recommendation

Due to the complicated nature of the stages in construction and the vastness of the implementation of lean in construction, this research was carried out in the stage of executing interior finishes such as, interior wall fixtures and laying tiles. Therefore, as future recommendation, researchers should investigate further the lean tools and techniques that are being utilised in other stages of construction in the Kingdom of Bahrain.

References

1. Adeyemi, B.S., Aigbavboa, C.O., Thwala, W.D.: Factors affecting total quality management implementation in the construction industry (2022)
2. Akinosho, T.D., Oyedele, L.O., Bilal, M., Ajayi, A.O., Delgado, M.D., Akinade, O.O., Ahmed, A.A.: Deep learning in the construction industry: a review of present status and future innovations. J. Build. Eng. **32**, 101827 (2020). https://doi.org/10.1016/j.jobe.2020.101827
3. Aziz, R.F., Hafez, S.M.: Applying lean thinking in construction and performance improvement. Alex. Eng. J. **52**(4), 679–695 (2013)
4. Babalola, O., Ibem, E., Ezema, I.: Implementation of lean practices in the construction industry: a systematic review. Build. Environ. **148** (2018). https://doi.org/10.1016/j.buildenv.2018.10.051
5. Bajjou, M.S., Chafi, A.: A conceptual model of lean construction: a theoretical framework. Malays. Constr. Res. J. **26**(3), 67–86 (2018)
6. Baruch, Y., Holtom, B.C.: Survey response rate levels and trends in organizational research. Hum. Relat. **61**(8), 1139–1160 (2008). https://doi.org/10.1177/0018726708094863
7. Besser Freitag, A.E., et al.: Integration of concepts about lean construction, sustainability and life cycle of buildings: a literature review. Braz. J. Oper. Prod. Manage. **14**(4), 486 (2017). https://doi.org/10.14488/bjopm.2017.v14.n4.a5
8. Embia, G., Moharana, B.R., Mohamed, A., Muduli, K., Muhammad, N.B.: 3D Printing Pathways for Sustainable Manufacturing. In: Nayyar, A., Naved, M., Rameshwar, R. (eds) New Horizons for Industry 4.0 in Modern Business. Contributions to Environmental Sciences & Innovative Business Technology. Springer, Cham. (2023). https://doi.org/10.1007/978-3-031-20443-2_12
9. Gilbert, C.D., Sigman, M., Crist, R.E.: The neural basis of perceptual learning. Neuron **31**(5), 681–697 (2001). https://doi.org/10.1016/S0896-6273(01)00424-X
10. Hashim, H., Alsalman, H.: Visual management and its impact on reducing wastage (3M) according to the perspective of the agile 7S methodology: an applied study in Al-Rayan Company. In: Proceedings of 2nd International Multi-Disciplinary Conference Theme: Integrated Sciences and Technologies, IMDC-IST 2021, 7–9 Sept 2021, Sakarya, Turkey (2022)
11. Hou, H., Feng, X., Zhang, Y., Bai, H., Ji, Y., Xu, H.: Energy-related carbon emissions mitigation potential for the construction sector in China. Environ. Impact Assess. Rev. **89**, 106599 (2021). https://doi.org/10.1016/j.eiar.2021.106599
12. Ibijola, E.Y., Ezeani, N.S.: The conversion process in the university system: Nigerian university students' assessment. J. Educ. Res. Rev. **5**(2), 14–20 (2017)
13. Johnson, S., Kleiner, B.: TQM canencompass success. Ind. Manage. **55**(2) (2013)
14. Kim, J.: Visual analytics for operation-level construction monitoring and documentation: state-of-the-art technologies, research challenges, and future directions. Front. Built Environment **6**(November), 1–20 (2020). https://doi.org/10.3389/fbuil.2020.575738

15. Koskela, L.: Application of the new production philosophy to construction, vol. 72. Stanford University, Stanford, USA (1992)
16. Liu, S., Wang, X., Liu, M., Zhu, J.: Towards better analysis of machine learning models: a visual analytics perspective. Vis. Inf. 1(1), 48–56 (2017). https://doi.org/10.1016/j.visinf.2017.01.006
17. Luangcharoenrat, C., et al.: Factors influencing construction waste generation in building construction: Thailand's perspective. Sustainability (Switzerland) 11(13) (2019). https://doi.org/10.3390/su11133638
18. Luo, L., Qiping Shen, G., Xu, G., Liu, Y., Wang, Y.: Stakeholder-associated supply chain risks and their interactions in a prefabricated building project in Hong Kong. J. Manage. Eng. 35(2), 05018015 (2019). https://doi.org/10.1061/(ASCE)ME.1943-5479.0000675
19. Pal, A., Hsieh, S.H.: Deep-learning-based visual data analytics for smart construction management. Autom. Constr. 131(August), 103892 (2021). https://doi.org/10.1016/j.autcon.2021.103892
20. Pilecka, E., Szwarkowski, D., Pilecki, Z., & Marcak, H. (2017). An application of the ground laser scanning to recognise terrain surface deformation over a shallowly located underground excavation. E3S Web Conf. 24, 01006
21. Purushothaman, M.B., Seadon, J., Moore, D.: Waste reduction using lean tools in a multicultural environment. J. Cleaner Prod. 265, 121681 (2020). https://doi.org/10.1016/j.jclepro.2020.121681
22. Rashid, K.M., Louis, J.: Activity identification in modular construction using audio signals and machine learning. Autom. Constr. 119, 103361 (2020). ISSN 0926-5805. https://doi.org/10.1016/j.autcon.2020.103361
23. Robson, P.: The Economics of International Integration, 4th edn. Routledge (2002). https://doi.org/10.4324/9780203019603
24. Sarhan, J.G., Xia, B., Fawzia, S., Karim, A.: Lean construction implementation in the Saudi Arabian construction industry. Constr. Econ. Build. 17(1), 46–69. https://doi.org/10.5130/AJCEB.v17i1.5098
25. Suleman, Q., Gul, R.: Challenges to successful total quality management implementation in public secondary schools: a case study of Kohat District, Pakistan. J. Educ. Pract. 6(15), 123–134 (2015)
26. Vilventhan, A., Ram, V.G., Sugumaran, S.: Value stream mapping for identification and assessment of material waste in construction: a case study. Waste Manage. Res. 37(8), 815–825 (2019). https://doi.org/10.1177/0734242X19855429
27. Wang, Y.: BIM + VR technology in construction management of construction engineering. J. Phys. Conf. Ser. 2037(1) (2021). https://doi.org/10.1088/1742-6596/2037/1/012083

Characterizing the DNN Impact on Multiuser PD-NOMA System Based Channel Estimation and Power Allocation

Mohamed Gaballa, Maysam Abbod, and Ammar Aldallal

Abstract This Paper demonstrates how the channel estimation based Deep Learning (DL) and power optimization are jointly utilized for multiuser (MU) recognition in Power domain Non-Orthogonal Multiple Access (PD-NOMA) wireless system. In NOMA systems the successive interference cancellation (SIC) procedure is typically employed at the receiver side, where several users are decoded in a subsequent manner. Fading channels may disperse the transmitted signals and originate dependencies among its samples, this may affect the channel estimation process and consequently affect the SIC process and signal detection accuracy. In this scenario, the impact of Deep Neural Network (DNN) in explicitly estimating the channel coefficients for each user in NOMA cell is investigated. This approach, integrate the Long Short-Term Memory (LSTM) network into the NOMA system where this LSTM network is used for complex data processing to carry out training, updating, and predicting. The DNN is trained online based on channel statistics and then the trained model is used to predict the channel parameters that will be exploited by the receiver in retrieving the original data. Furthermore, Power coefficients are optimized in order to maximize the sum throughput of the system users based on the total transmitted power and Quality of service (QoS) constraints. We formulate an expression for Signal to interference noise ratio (SINR) for each user in the system, then an analysis for the optimization problem and the considered constraints to verify the concavity of the objective function is presented. Lagrange function and Karush–Kuhn–Tucker (KKT) optimality conditions are utilized to derive the optimal power coefficients. Simulation results for different metrics such as bit Error Rate (BER),

---

M. Gaballa (✉) · M. Abbod
Electronic & Electrical Engineering, Brunel University London, Uxbridge, UK
e-mail: mohamedgaballa.gaballa@brunel.ac.uk

M. Abbod
e-mail: maysam.abbod@brunel.ac.uk

A. Aldallal
Telecommunication Engineering, Ahlia University, Manama, Bahrain
e-mail: aaldallal@ahlia.edu.bh

© The Author(s), under exclusive license to Springer Nature Switzerland AG 2024
A. Hamdan and E. S. Aldhaen (eds.), *Artificial Intelligence and Transforming Digital Marketing*, Studies in Systems, Decision and Control 487,
https://doi.org/10.1007/978-3-031-35828-9_79

sum rate, Outage probability and individual user capacity have proved the superiority of the DL approach over the conventional approaches in terms of channel estimation.

Keywords NOMA · SIC · DNN · DL · LSTM · KKT

1 Introduction

Currently, non-orthogonal multiple access (NOMA) has been considered as a key authorizing multiple access scheme for 5G & beyond (B5G) cellular systems. In NOMA, by utilizing the variations in channels gain, multiple users can be multiplexed in power domain and then non-orthogonally organized for communication on the same spectrum resources. Practical successive interference cancellation (SIC) technique needs to be used at the receivers to decode the desired information signals.

Fifth generation (5G) wireless cellular system is not just an outspread form of fourth generation (4G) system due to the increasing request of network data traffic, but also 5G can support new emerged technologies such as internet of things (IoT) devices, web-based artificial intelligence (AI) applications in addition to spectral efficiency and massive connectivity [1]. Orthogonal multiple access scheme (OMA) which includes orthogonal frequency division multiple access (OFDMA) is utilized in the air interface stage of 5G New Radio (5G NR), which allows for mobile connectivity [2]. On average, OMA scheme is considered as the conventional multiple access scheme in limited communication environment, but it may not be adequate to support huge networks which involve diversified quality of service (QoS) [2, 3]. This may relate to the matter of restricted degrees of freedom (DoF), where users with good channel conditions are assisted first with respect to users with bad channel status who have to postpone for channel access [3]. In order to realize the demands of QoS and DoF needed for cellular networks, non-orthogonal multiple access (NOMA) with a sort of practical successive interference cancellation (SIC) at the receiver was introduced as a new scheme to enhance a multi-user access technique [4].

In power domain NOMA (PD-NOMA) as shown in Fig. 1, multiplexing is performed in power field with superposition coding (SC) at the transmitter, while power differences between users is adopted to achieve interference cancellation at the receiver end [1, 4]. PD-NOMA deals with signals that have substantial distinction in power levels, so it has the potential to split up the high-energy signal at the receiver and then remove it to leave only the intended signal. Therefore, NOMA has the ability to distribute DoF between users in a fair way using superposition and cancelling out the interfering signals sharing the same resources using practical techniques for interference cancellation.

Characterizing the DNN Impact on Multiuser PD-NOMA System Based ...

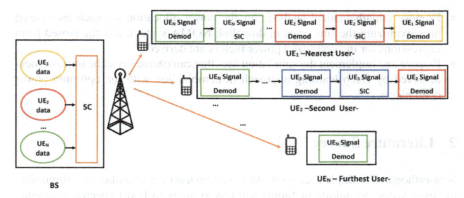

Fig. 1 Block diagram of the PD-NOMA system [5]

Aim and Objectives

Research Question

1. How to estimate the channel coefficients in fading channels using Deep learning (DL) algorithm.
2. How to allocate power factors for different users in non-orthogonal multiple access systems.

Research aim

The aim of this research is to

1. Investigate and characterise the impact of Deep Learning (DL) in effectively Estimating and predicting the channel coefficients for users in downlink NOMA system.
2. Optimizing the power coefficients for each user based on maximizing the sum rate of the system users when the total transmitted power and Quality of service (QoS) constraints are considered.

Research objectives

To achieve the above aim the following objectives are set:

- Understand the system Model for NOMA system that include Superposition Coding system (SC) at transmitter side, and Successive Interference Cancellation (SIC) at receiver side.
- Carry out critical review on Sum Rate Maximization, Objective function, constraints, and Power Optimization.
- Investigate the influence of Deep Neural Network (DNN) in Estimating the channel coefficients for each user.

- Propose a mathematical analysis for the power allocation for each user based on maximizing the sum rates in downlink NOMA system, and the closed form expressions for the optimized power factors are derived.
- Prepare and implement the simulation files that can characterize the improvement achieved by Deep learning approach in combination with the optimum power coefficients.

2 Literature Survey

Non-orthogonal multiple access (NOMA) system has been classified as a promoting multiple access technique in future wireless systems to boost spectral efficiency and system throughput. NOMA can utilize the present resources more effectively by opportunistically taking benefit of the users' channel environments and provide several users with diverse quality of service (QoS) constraints. NOMA enables multiple users to achieve simultaneous access to the same time–frequency resources by principle of superposition them in the code or power domain [1]. The concept of NOMA is based on that the user with weak channel condition can be combined with the user with good channel condition at the same time slot on the same allocated subcarrier in order that the bandwidth block can be effectively utilized [2]. In NOMA scheme, the receiver equipment will receive the superposition of symbols from users in the system, therefore the elimination of interference from other users come to be necessary for coordinated decoding. Generally, multi-user detection (MUD) in NOMA is realized through successive interference cancellation (SIC) performed in the power domain. In the SIC technique, symbols from multiple users are decoded sequentially based on the channel state information (CSI) and allocated signal power [2]. Complete realization of CSI or channel status for individual users is challenging because pilot symbols utilized in channel estimation can interfere with symbols from other users, thus influencing the performance of classical channel estimation techniques, such as minimum mean square error and least square estimators [1, 2]. Machine learning (ML) algorithms have the capability to adapt to variations in channel between user and base station (BS) and predict the channel coefficients for each user, therefore they are regarded as strong contenders for future radio networks [2].

In [3] authors have introduced a channel estimation procedure for multi-user detection scheme with imperfect CSI. Discrete state space model and Tobit Kalman filter were employed to make an estimate of the unspecified state variables of a dynamic channel based on uncertainty pattern. Authors have inspected the censoring phenomenon via the Tobit measurement model to accomplish more precise estimation, where the QoS demands such as minimum signal to interference and noise ratio (SINR) were considered and satisfied for all users. A robust min–max mean square error (MSE) estimation framework is developed to minimize the estimation error. Theoretical analysis, and simulation results validated the efficiency of the framework in terms of channel estimation accuracy for a wide range of uncertainty.

Authors in [4], investigated the changes in the outage probability and throughput versus changing signal to noise ratio (SNR) in NOMA system on the basis of two categories of partial channel state information that are imperfect CSI and second order statistics (SOS). Authors proved that SOS-based NOMA achieves better performance than that with imperfect CSI, while it can achieve similar performance to the NOMA with perfect CSI at low signal to noise ratio, also results reveal that NOMA system based partial channel state information can attain superior performance compared to the conventional orthogonal multiple access (OMA) scheme.

Based on Deep Learning (DL) algorithm, authors in [5] have presented a sliding window Gated Recurrent Unit (GRU) channel estimator to acquire knowledge for the time varying Rayleigh fading channel. Channel coding scheme and interleaver were further merged with the proposed sliding window estimator to further enhance system performance. Simulation results have showed the ability of the proposed estimator to follow the channel in reliable way and achieve better mean square error (MSE) performance. Moreover, the sliding window based GRU estimator has been examined with different numbers of pilot symbols, and the robustness against the variations in the channel characteristics was analyzed.

In [6], authors demonstrated that deep learning scheme can be utilized in signal detection and uplink analysis in NOMA networks. Authors proposed a deep learning method to analyze the complex channel characteristics of NOMA, where a restricted Boltzmann machines (RBM) is implemented as a pre-training stage for the original input sequence for the network. A constructed iterative support detection algorithm is suggested to detect the transmitted symbols and the proposed learning scenario based LSTM layer can track the environment statistics automatically via offline learning. Performance analysis for the proposed deep learning scheme is analyzed in terms of sum data rate and block error rate.

In [7], the main objective was to create an optimal decoding order by engaging the deep learning technique in the decoding scenario. Authors have explored the weight correlation between the channel state and the queue state in downlink NOMA system for satellite-based internet of things. The suggested deep learning algorithm counts on long term power allocation scheme that can appropriately develop a more accurate decoding order than the available standard approaches.

In [8], a pilot-assisted receiver structure is proposed for uplink SIMO-NOMA system, which incorporates a joint channel estimation and signal detection scheme supported by adjustable parameters based on random channels. Authors bring together deep learning model with SIC detection structure to minimize the learnable parameters. Furthermore, signal detection accuracy improvement and noise interference reduction have been achieved by adding noise and interference elimination factors at the SIC detection stage. Simulation results show that the BER performance based on the proposed deep learning scheme is better than the conventional MMSE method and the complexity of the receiver is diminished.

In [9], authors proposed a semi-blind joint detection based Deep Learning network to detect users' symbols jointly in co-operative NOMA system. The proposed scheme is able to detect symbols without acquiring an additional channel estimation algorithm since it can perform a simultaneous detection based on the pilot responses. The model was trained offline over Rayleigh fading channel and then the trained network is deployed in an online detection phase. In addition, the trained model is inspected using Nakagami-m and Rician fading channels and simulation results prove that the proposed scheme outperforms conventional detectors.

In [10], authors have employed deep neural network for channel estimation and symbol detection in an OFDM system. The model is trained offline based on the simulated data that consider OFDM system and fading channels as black boxes. Simulation results reveal that the deep learning method has capability to remember and investigate the complicated attributes of the wireless channels. In addition, results for the deep learning approach proves the superiority of the proposed DL approach than conventional methods when fewer training pilots are used, omitting the cyclic prefix and nonlinear clipping noise is applied.

The main contributions of this work are summarized as follows.

- Investigate the impact of Deep Learning (DL) based LSTM in explicitly estimating the channel coefficients for each user in NOMA cell. DNN model is trained with randomly generated channel coefficients in an online training stage without the need for offline stage and with minimum number of layers. The predicted channel coefficients are utilized in decoding the data symbols for each user at the receiver and the performance is evaluated and compared to conventional decoding scheme in terms of BER, outage probability, sum rate and individual user capacity.
- The combination of optimal power allocation scheme and Deep Learning (DL) based channel estimation is also explored and compared to the fixed power allocation scenario when deep learning is also considered.
- An elaborated and structured mathematical analysis to derive a fair, non-complex analytical form for the power allocation for each user based on maximizing the sum rates in downlink NOMA system is introduced, and the closed form expressions for the power factors are determined.

3 Mathematical System Model

In this work, an N-user downlink NOMA network is analysed, where all users encounter diverse channel gains. In this N-user downlink NOMA cellular system, a single base station (BS) can transmit non-orthogonal N several signals over the same radio channels, while all N receivers in user's devices have the capability to detect their desired signals beside identifying the interfering signals that caused by the transmission of symbols related to other users [11]. To receive the required signal, each SIC receiver firstly decodes the predominant interferences then remove them from the superimposed signal. Thus, the received signal levels of the interfering signals must be high enough compared to the targeted signal in order to facilitate the removal process by SIC at the receiver. In NOMA system, each user device receives all signals that include targeted and interfering signals transmitted via same time–frequency slots, hence multiplexing of several signals using diverse power levels is critical to distinguish signals and to enhance SIC process at each user's receiver

Characterizing the DNN Impact on Multiuser PD-NOMA System Based ... 949

[12]. Traditionally, in downlink NOMA, users characterized by high channel gain are mostly assigned low power levels while users with weak channel gains are given high power portions. In this report, we will focus on downlink NOMA cellular network with single BS in the cell. Because of the processing complexity of SIC, up to 2 active users are considered here for network analysis. For each transmission, we can classify each user according to its distance from base station. We can denote the closest user as near user (NU), and the farthest one is considered as far user (FU). The channel gains for users in the cell can be stated as, $|h_n|^2$, and $|h_f|^2$ for NU, and FU respectively [13]. In this research, zero mean Rayleigh fading channel is assumed for communication links between the base station and users in the cell. According to that, channel gain for each user can mathematically be represented as $h_n \sim \left(0, d_n^{-k}\right)$, and $h_f \sim \left(0, d_f^{-k}\right)$, where d is the distance and k represents the path loss exponent [14, 15]. The additive white Gaussian noise (AWGN) power is denoted as σ^2. Without loss of generality, in this work we assume that $|h_n|^2 > |h_f|^2$ and channel state information (CSI) is not identified. Total transmitted power from base station to users is denoted as P_t. The power coefficients for NU, and FU are labelled here as α_n, and α_f respectively, while the information messages associated with each user can be defined as s_n, and s_f in the same way. The transmitted signals are denoted as follows, for near user $x_n = \sqrt{\alpha_n P_t} s_n$, and for far user $x_f = \sqrt{\alpha_f P_t} s_f$. As indicated before, each receiver has the capability to perform SIC to remove signals related to other users with weaker channel gains. While signals from users with strong channel gains cannot be taken away and handled as interference. Thus, far user will receive signals from near user as interference, while near user will cancel out the signals coming from far user.

For two users NOMA system the antenna at the base station can transmit the superposition coded signal x which can be written as [10, 11, 16]

$$x = \sqrt{P_t}\left(\sqrt{\alpha_f} x_f + \sqrt{\alpha_n} x_n\right) \tag{1}$$

where α_f and α_n are the power coefficients for far user and near user respectively, while x_f and x_n represent the desired messages related to far and near user respectively. Therefore, the received signal y_f at far user can be formed as [17, 18]

$$y_f = x h_f + z_f \tag{2}$$

where h_f is the fading channel between far user and base station (BS), while z_f is AWGN noise samples for far user with zero mean and σ^2 variance. Likewise, the signal received at near user y_n can be written as

$$y_n = x h_n + z_n \tag{3}$$

Similarly, h_n represent the channel link between near user and the BS. Also, here z_n is a another AWGN noise samples for near user with zero mean and σ^2 variance. Since far user is characterized by cell edge with weak channel condition, his own signal x_f is allocated more power by the base station where $\alpha_f > \alpha_n$. Therefore, far user can directly decode his desired message x_f from received signal y_f, based on treating x_n term as interference. The composite received signal at far user can simply expressed as [11, 19]

$$y_f = \sqrt{P_t}\left(\sqrt{\alpha_f}x_f + \sqrt{\alpha_n}x_n\right)h_f + z_f$$
$$y_f = \sqrt{P_t\alpha_f}x_f h_f + \sqrt{P_t\alpha_n}x_n h_f + z_f \tag{4}$$

The 1st term in (4) denotes the desired signal for far user, while the 2nd term represents the interference term from near user.

The signal to interference noise ratio γ_f (SINR) for the far user can be written as

$$\gamma_f = \frac{|h_f|^2 P_t \alpha_f}{|h_f|^2 P_t \alpha_n + \sigma^2} \tag{5}$$

Based on (5) The achievable bit rate for far user can be written as [4, 20, 21]

$$R_f = \log_2(1 + \gamma_f) = \log_2\left(1 + \frac{|h_f|^2 P_t \alpha_f}{|h_f|^2 P_t \alpha_n + \sigma^2}\right) \tag{6}$$

Traditionally, near user has good channel status with base station, hence his signal x_n is allocated less power $\alpha_n < \alpha_f$. Likewise, the received signal at the near user can be written as

$$y_n = x h_n + z_n$$
$$y_n = \sqrt{P_t\alpha_n}x_n h_n + \sqrt{P_t\alpha_f}x_f h_n + z_n \tag{7}$$

The 1st term in (7) denotes the near user anticipated signal, while the 2nd term is the interference term from far user. Also, it can be noticed from (7), that the interference term is dominant because of the more power allocated to far user, so at the receiver side of near user, direct decoding for the far user signal x_f must be firstly performed. At near user receive side, the SINR γ_{fn} between far user and near for direct decoding x_f can be written as [22, 23]

$$\gamma_{fn} = \frac{|h_n|^2 P_t \alpha_f}{|h_n|^2 P_t \alpha_n + \sigma^2} \tag{8}$$

After SIC, the near user achievable rate to decode its own desired signal x_n can be formed as

$$R_n = \log_2 \left(1 + \frac{|h_n|^2 P_t \alpha_n}{\sigma^2} \right) \tag{9}$$

4 Optimization Problem

In this section, the aim is to maximize the sum throughput for all users in NOMA system based on optimizing the power coefficients for each user according to the inspected channel condition. The sum of the aforementioned achievable rates for the users in the NOMA system can be generally formulated as follows [11, 24, 25]

$$R_{sum} = \sum_{k=1}^{N} \log_2 \left(1 + \frac{|h_k|^2 P_t \alpha_k}{|h_k|^2 \sum_{j=1}^{k-1} P_t \alpha_j + \sigma^2} \right) \tag{10}$$

The constraints accounted for the considered optimization problem in this system can be discussed as follows:

1. **Total Transmit Power**
   The assigned power for each user in the examined network is a portion of the total power transmitted from base station P_t, then the allocated power percentage for each user device must follows [11–15]:

$$\sum_{x=1}^{N} \alpha_x \leq 1 \tag{11}$$

   where α_x is the power allocation fraction for the x th user in the N-user NOMA cell.
2. **Quality of Service (QoS)**

$$\log_2(1 + \delta_m) \geq R_{\min} \tag{12}$$

   where δ_n is the signal to interference noise ratio for m th user and $R_{\min}$ is the minimum transmission rate [12, 26, 27].
   This constraint can be clarified in many ways, assume we have $R_{m \to k}$ which is the rate of user k to detect the signal of user m where $1 \leq k \leq m$ and $R_m = R_{\min}$ is the minimum rate required in the system. When user k is not capable to detect the message of user m with rate R_{min}, this can be denoted as $R_{m \to k} < R_{\min}$ [28]. Equation (12) can be expanded as

$$\frac{|h_m|^2 P_t \alpha_m}{|h_m|^2 P_t \sum_{i=1}^{m-1} \alpha_i + \sigma^2} > (2^{R_{\min}} - 1) \tag{13}$$

where α_m and α_i are power factors for m th user and i th user in the system respectively. Equation (13), can be reformulated as

$$|h_m|^2 \rho \left(\alpha_m - (2^{R_{\min}} - 1) \sum_{i=1}^{m-1} \alpha_i \right) > (2^{R_{\min}} - 1) \qquad (14)$$

where ρ represent the signal to noise ratio. Equation (14), also declares that in order to satisfy the minimum rate and avoid that m^{th} user being in outage, the following condition must be achieved

$$\alpha_m > \left(2^{R_m} - 1 \right) \sum_{i=1}^{m-1} \alpha_i \qquad (15)$$

3. **Sum Rate Maximization**

Based on the above-mentioned constraints (11), (12) and considered sum rate expression in (10), the objective function and optimization problem can be generally specified as follows [29, 30]:

$$\max_{\alpha} R_{sum} = \sum_{k=1}^{N} \log_2 \left(\frac{|h_k|^2 P_t \sum_{j=1}^{k-1} \alpha_j + \sigma^2 + |h_k|^2 P_t \alpha_k}{|h_k|^2 P_t \sum_{j=1}^{k-1} \alpha_j + \sigma^2} \right) \qquad (16)$$

such that

$$\sum_{x=1}^{N} \alpha_x \leq 1 \qquad (17)$$

$$\log_2(1 + \delta_m) \geq R_{\min} \qquad (18)$$

$$\alpha_k \geq 0 \quad \forall k = 1, 2, \ldots, N \qquad (19)$$

5 Optimization Analysis

In this section, we will introduce mathematical analysis for the optimization problem, firstly finding 1st and 2nd derivative for the objective function and prove that it is negative definite [26], which demonstrate that the objective function is a concave function, and since the that, the constraints are linear, then the optimization problem is convex. Therefore, Karush–Kuhn–Tucker (KKT) conditions can be applied as sufficient for optimal solution for the optimization problem [27].

In the following analysis, cell network will be restricted to two users ($N = 2$) NOMA system and the optimization problem can be reformulated as follows

Characterizing the DNN Impact on Multiuser PD-NOMA System Based ...

$$\max_{\alpha} R_{sum} = R_n + R_f \tag{20}$$

Such that

$$\alpha_n + \alpha_f - 1 \leq 0 \tag{21}$$

$$|h_k|^2 \rho \left(\alpha_m - (2^{R_{min}} - 1) \sum_{i=1}^{m-1} \alpha_i \right) > (2^{R_{min}} - 1) \tag{22}$$

$$\alpha_n, \alpha_f \geq 0 \tag{23}$$

According to the aforementioned analysis, and assuming that minimum transmission rate $R_{min} = R_f$ then, the examined constraints can be represented as follows:

$$C_1(\alpha) = (2^{R_f} - 1) - \rho|h_k|^2 \left(\alpha_f - (2^{R_f} - 1)\alpha_n \right) \tag{24}$$

$$C_2(\alpha) = \alpha_n + \alpha_f - 1 \tag{25}$$

Since both $C_1(\alpha)$ & $C_2(\alpha)$ are linear in terms of α, then $C_1(\alpha)$ & $C_2(\alpha)$ are convex. Now we need to find $\nabla R_{Sum}(\alpha)$ & $\nabla^2 R_{Sum}(\alpha)$. Generally, we can find the first derivative for the objective function (16) in terms of α. After some mathematical handlings $\nabla R_{Sum}(\alpha)$ can be represented as follows [29, 30]

$$\frac{\partial R_{Sum}}{\partial \alpha_i} = \frac{1}{\ln 2} \left(\frac{|h_i|^2 P_t}{|h_i|^2 P_t \sum_{j=1}^{i} \alpha_j + \sigma^2} \right) - \frac{1}{\ln 2} \sum_{k=1}^{N-i} \left\{ \left(\frac{\left(|h_{(i+k)}|^2 P_t \right)^2 \alpha_{i+k}}{\left(|h_{(i+k)}|^2 P_t \sum_{j=1}^{i+k} \alpha_j + \sigma^2 \right)} \right) \right.$$
$$\left. \times \left(\frac{1}{\left(|h_{(i+k)}|^2 P_t \sum_{j=1}^{i+k-1} \alpha_j + \sigma^2 \right)} \right) \right\} \tag{26}$$

After Deriving the mathematical formula for $\nabla^2 R_{Sum}(\alpha)$, the resultant formula will be negative, this demonstrates that that the objective function is negative definite and therefore it is strictly concave and has a unique global maximum.

For two users NOMA system where $R_{Sum} = R_n + R_f$ Lagrange function and the KKT necessary conditions can be applied as follows [31, 32]:

$$\mathcal{L}(\alpha_n, \alpha_f, \mu_1, \mu_2) = R_{Sum} - \mu_1 C_1(\alpha) - \mu_2 C_2(\alpha) \tag{27}$$

where α_n and α_f are the power allocations for near user and far user respectively, while μ_1 & μ_2 are Lagrange multipliers

- **Optimality condition** can be formulated as follows:

$$\frac{\partial R_{Sum}}{\partial \alpha_f} - \mu_1 \frac{\partial C_1(\alpha)}{\partial \alpha_f} - \mu_2 \frac{\partial C_2(\alpha)}{\partial \alpha_f} = 0 \tag{28}$$

$$\frac{\partial R_{Sum}}{\partial \alpha_n} - \mu_1 \frac{\partial C_1(\alpha)}{\partial \alpha_n} - \mu_2 \frac{\partial C_2(\alpha)}{\partial \alpha_n} = 0 \tag{29}$$

- **Slackness conditions** can be represented as follows

$$\mu_1 \left\{ (2^{R_f} - 1) - \rho |h_k|^2 (\alpha_f - (2^{R_f} - 1)\alpha_n) \right\} = 0 \tag{30}$$

$$\mu_2 \{ \alpha_n + \alpha_f - 1 \} = 0 \tag{31}$$

- **Lagrange multipliers** need to satisfy the following

$$\mu_1 \geq 0, \mu_2 \geq 0 \tag{32}$$

In the following steps, the Lagrange multipliers μ_1 & μ_2 should be verified to be positive. Both (29) and (30) can be written as

$$\frac{\partial R_{Sum}}{\partial \alpha_f} + \mu_1 \rho |h_k|^2 - \mu_2 = 0 \tag{33}$$

$$\frac{\partial R_{Sum}}{\partial \alpha_n} - \mu_1 \rho |h_k|^2 (2^{R_f} - 1) - \mu_2 = 0 \tag{34}$$

From (34) μ_1 & μ_2 and be related as follow

$$\frac{\partial R_{Sum}}{\partial \alpha_f} + \mu_1 \rho |h_k|^2 = \mu_2 \tag{35}$$

substitute (36) in (35)

$$\frac{\partial R_{Sum}}{\partial \alpha_n} - \mu_1 \rho |h_k|^2 (2^{R_f} - 1) - \left(\frac{\partial R_{Sum}}{\partial \alpha_f} + \mu_1 \rho |h_k|^2 \right) = 0 \tag{36}$$

$$\frac{\partial R_{Sum}}{\partial \alpha_n} - \frac{\partial R_{Sum}}{\partial \alpha_f} = \mu_1 \rho |h_k|^2 2^{R_f} \tag{37}$$

For two users (N = 2) and based on (26), $\frac{\partial R_{Sum}}{\partial \alpha_n}$ and $\frac{\partial R_{Sum}}{\partial \alpha_f}$ can be mathematically expressed as follows

$$\frac{\partial R_{Sum}}{\partial \alpha_n} = \frac{1}{\ln 2} \left(\frac{|h_n|^2 P_t}{|h_n|^2 P_t \alpha_n + \sigma^2} - \frac{\left(|h_f|^2 P_t \right)^2 \alpha_f}{\left(|h_f|^2 P_t (\alpha_n + \alpha_f) + \sigma^2 \right) \left(|h_f|^2 P_t (\alpha_n) + \sigma^2 \right)} \right) \tag{38}$$

$$\frac{\partial R_{Sum}}{\partial \alpha_f} = \frac{1}{\ln 2} \left(\frac{|h_f|^2 P_t}{|h_f|^2 P_t (\alpha_n + \alpha_f) + \sigma^2} \right) \tag{39}$$

Based on $|h_n|^2 > |h_f|^2$ [33] and after some mathematical simplifications, then the left-hand side of (37) can be easily proved to be positive. As a result, the right-hand side of (37) $\mu_1 \rho |h_k|^2 2^{R_f}$ must be also positive and since $\rho |h_k|^2$ is larger than zero and 2^{R_f} is a positive constant, this implies that μ_1 must be positive.

Correspondingly, we need to check the sign of μ_2 and this can be deduced easily from (35), where $\frac{\partial R_{Sum}}{\partial \alpha_f}$ is verified to be positive as indicated in (39) and since μ_1 is proved to be positive and the term $\rho |h_k|^2$ is positive by inspection, this concludes that μ_2 must be positive. Now both the Lagrange multipliers μ_1 & μ_2 has been shown to be positive, so the examined constraints are feasible [32, 34, 35] and the closed form expressions for α_n & α_f can be deduced from the slackness conditions as follows

$$\alpha_n = \frac{1}{(2^{R_f})} \left(\frac{\rho |h_n|^2 - (2^{R_f} - 1)}{\rho |h_n|^2} \right) \tag{40}$$

$$\alpha_f = \frac{(2^{R_f} - 1)}{(2^{R_f})} \left(\frac{\rho |h_f|^2 + 1)}{\rho |h_f|^2} \right) \tag{41}$$

5.1 Deep Neural Network Approach

Recurrent neural networks (RNNs) are regard as a class of supervised learning procedure, where they can model successive sequence for prediction and identification [24]. As shown in Fig. 2, RNNs involve hidden layers comprised of artificial neurons with feedback loops, therefore they have two inputs, the current and the recent past sample.

In RNNs, the hidden layers are able to act as memory for the network at a specific instant, this structure enables the RNNs to remember, and process the preceding complex data for an prolonged period of time. As a result of their feedback connections, RNNs can model time dependencies with a lower number of neurons [25]. On the other hand, traditional RNN based on backpropagation through time (BTT) experiences the vanishing gradient problem and slow learning. Therefore, RNNs will not the best candidate for signals transmitted through a fading channel that may disperse the signal and originate a long term dependencies among its samples.

Long short-term memory (LSTM) network, which is a one kind of RNNs, is often applied to sequences and time series data for classification where it can take advantage of data time-dependencies [25]. The LSTM network can build a knowledge among time steps of sequence data and preserve related information.

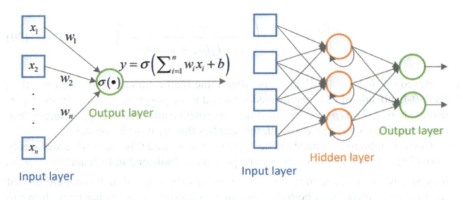

(a) Single neuron element. (b) Recurrent neural network (RNN).

Fig. 2 RNN network architecture [25]

LSTM include memory cells, that are able to save and gain access to data over extended periods of time based on their fundamental design. Unlike the conventional RNN, LSTM counteract the error from backpropagation [26]. LSTM can receive a vector of sequence data, hence integrating the magnitude and phase parts of the received sequence simultaneously. LSTM cells are also able to efficiently utilize the past time of data series and manage the long term dependency of time series data. LSTM networks are suitable to realize multi-user detection (MUD) and making predictions based on time series data [27].

5.2 LSTM Cell Structure

In this subsection, we introduce a framework that integrates the LSTM network into the NOMA model where this LSTM network is used for complex data processing that are employed to carry out training, updating, and prediction. The LSTM network consists of 4 layers, each layer is equipped with multiple neurons, and an output is the weighted sum of these neurons with a nonlinear function. In all data-driven communication applications, the number of LSTM layers and the number of LSTM cells in each layer are empirically determined, such that increasing the number of layers do not provide a notable gain in learning performance or significantly affect the network convergence [29].

In our LSTM framework, we denote the length of each training sequence as L, which is the dimension of the input layer. The input layer includes 128 neurons, where the inputs of the network are transferred to the subsequent layer with weight coefficients. In the second layer, we implement one LSTM layer which has 300 hidden units. The third layer is a fully-connected layer that computes the outputs

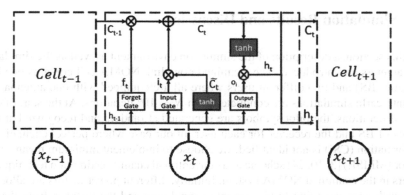

Fig. 3 Internal structure of LSTM cell [24]

of the network based on trained weights and biases (W, b). The last layer is the regression layer to optimize the network weights, biases, and cell status.

In an LSTM cell, the output is produced by not only the current input but also the previous cell status. In order to remember the previous status and decide whether the previous statuses are used or not, the LSTM cell includes different types of gates as shown in Fig. 3 [24, 25]. These gates are the input gate, the forget gate and the output gate, where the functionality of each gate can be clarified as follows [24, 29, 30]

- LSTM is divided into two states the cell state C_{t-1} which is called internal memory where all information is stored, and the other state is hidden state h_{t-1} that are used for computing the output.
- The forget gate is responsible for deciding what information should be removed from the cell state.
  Forget gate: $f_t = \sigma(W_f h_{t-1} + U_f x_t + b_f)$.
- The input gate collects the data from the previous layer and previous cells within the LSTM layer.
  Input gate: $i_t = \sigma(W_i h_{t-1} + U_i x_t + b_i)$.
- The candidate state computes the new coefficient based on the activation function and the status of forget gate.
  Candidate state: $g_t = \tanh(W_g h_{t-1} + U_g x_t + b_g)$.
- Update Cell state: $C_t = (C_{t-1} f_t) + (i_t g_t)$
- The output gate, the cell output is computed according to cell input, candidate gate value and activation function.
  Output gate: $O_t = \sigma(W_o h_{t-1} + U_o x_t + b_o)$.
- Estimated channel coefficients $h_t = O_t \times \tanh(C_t)$.

6 Simulation Results and Discussion

In this section, a description of the simulation environment as well as the simulation parameters are introduced. The examined downlink NOMA cell contains one base station (BS) and two different users where all are equipped with one antenna, and monte-carlo simulations are conducted with $N = 10^6$ iterations. At the start of each set of iterations, the pilot symbols are generated at random and recognized at both sides of BS and the receiver for each user. In our case where perfect channel state information (CSI) is not identified, we choose to implement minimum mean square error (MMSE) [3, 10, 24] scheme as a conventional channel estimation technique for users in the examined NOMA system. Initially, different power factors are allocated for each user according to his current channel gain and the distance from the BS. Binary phase shift keying BPSK [10, 36], is utilized as modulation scheme for the data symbols and pilot sequences. For the downlink NOMA scenario, the modulated signals are then superimposed and transmitted by BS to the users through uncorrelated Rayleigh fading channels affected by additive white gaussian noise (AWGN) where the noise spectral density $N_0 = -174$ dBm and path loss exponent is 4. At the receiver side, the channel estimation algorithm will be implemented based on LSTM neural network, which uses gradient descent algorithm [26, 37]. The purpose is to facilitate the LSTM layer to accurately estimate the desired channel coefficients, this needs LSTM to be combined with a fully connected layer, that uses a SoftMax activation function in the output layer and each neuron in the former layer is fully connected to each neuron in the subsequent layer. Channel parameters that employed to model the Rayleigh fading wireless channel are generated on the basis of ITU channel models [38]. At the start of each training stage, the weights and bias values are randomly initialized, while during the training process, weights and biases of the model are modified according to a gradient descent procedure. Throughout the training phase, the performance of the LSTM network is evaluated using root mean square error (RMSE) and the loss functions. NOMA system parameters applied in the simulations are based upon the long term evolution (LTE) standard [24, 39], and a subcarrier spacing of 15 kHz was selected to minimize the influence of phase noise on network performance. Both training and implementation phases are conducted online throughout the simulations, and the fading coefficients in the implementation phase are generated such that these coefficients are not the same as in the training stage. Once the training stage is ended, the trained model is used as online channel estimator for the users rather than the conventional NOMA scheme that use MMSE method for channel estimation.

In this section, simulations are conducted to evaluate the effectiveness of embedding the proposed Deep Learning based LSTM scheme for channel estimation in a single cell downlink NOMA system. Power allocation coefficients α_n, α_m, and α_f are defined for near, middle, and far users respectively. In fixed power allocation (FPA) scheme we set $\alpha_f = 0.75$ and $\alpha_n = 0.25$. The applied transmitted power is mainly varying from 0 to 30 dBm.

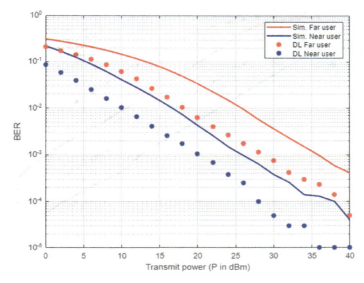

Fig. 4 BER versus power based deep learning and conventional channel estimation schemes

In Fig. 4 simulation results illustrate the comparison between Deep learning (DL) scheme and the MMSE conventional scheme for channel estimation and symbol decoding based on SIC procedures for far, and near users in NOMA system in terms of bit error rate (BER) and transmitted power. All examined users in the cell based DL channel estimation show sufficient improvement in lowering the bit errors compared to the conventional SIC scenario specially when the assigned power is increasing. The power saving for DL scheme is approximately 5 dB low SNR, while when transmitted power is increased the improvement for DL scheme may reach up to 8 dB.

Figure 5 demonstrates the outage probability versus transmitted power for the two examined users in NOMA system on the basis of DL channel estimation scheme and MMSE channel estimation. Far user simulation results indicate an improvement with 5–6 dB approximately in outage probability when DL based channel estimation scenario is conducted compared to classical MMSE scheme. Similarly, near user with DL scheme shows a clear enhancement compared to MMSE scenario. It is noticeable that near user always shows good performance compared to far user either for DL approach or conventional MMSE channel estimation, this may be justified by the good channel condition.

In Fig. 6, simulation results for the sum-rate for the examined users in NOMA system are demonstrated. In this figure, also the two different schemes are inspected, Deep learning based channel estimation approach and the conventional NOMA based on MMSE channel estimation. Based on the simulation results. It is clearly noticed that for small SNR, DL channel estimation scheme shows improvement by 1 b/s/Hz over the conventional scenario while this enhancement increases to 2 b/s/Hz

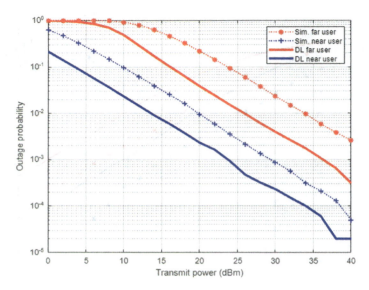

Fig. 5 Outage probability versus power based deep learning and conventional channel estimation schemes

for high transmitted power. This results proves the effectiveness of the DL channel coefficients estimation employed before using this estimated channel parameters in the SIC decoding process for each user in the cell.

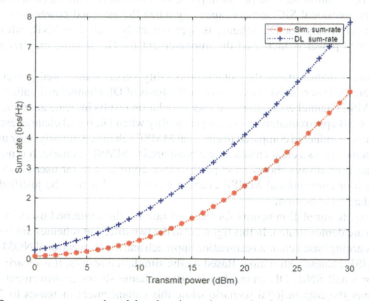

Fig. 6 Sum rate versus power based deep learning and conventional channel estimation schemes

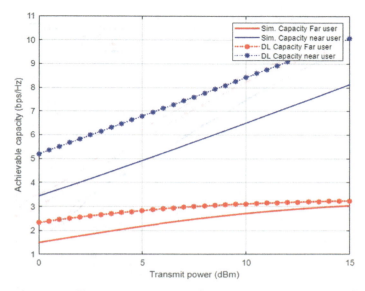

Fig. 7 Individual capacity versus power based deep learning and conventional channel estimation schemes

Figure 7 demonstrates simulation results in terms of the achieved individual capacity for each user in NOMA system when both the DL and conventional channel parameters estimation schemes are considered. In terms of DL results, the achieved capacity for near user shows significant difference by 2 b/s/Hz over the conventional approach. While for the far user results, the DL approach still delivers noticeable enhancements over the conventional estimation approach, especially in Low transmitted power levels. Both the applied DL approach and the good channel condition for near user paly effective role in enhancing the rate by at least 5 b/s/Hz compared to near user results.

In Fig. 8, two different scenarios are conducted here to generate this figure, the first one when fixed power allocation (FPA) scheme is applied for each user in the system and the other scenario when optimized power scheme is implemented and both scenarios are realized in when Deep Learning shame is utilized for channel estimation for all examined users in NOMA cell. Both DL and optimized power scheme simulation results for far user show performance improvement over the fixed power allocation (FPA) scheme in terms of the BER and transmitted power. On the other hand, for near user results, DL based channel estimation jointly with FPA provide comparable results to optimized power scheme, this may be justified that for near user the good channel condition is more effective than the allocated power.

Figure 9 demonstrates the outage probability versus transmitted power for far user, and near user when optimized power scheme and fixed power allocation (FPA) scheme are applied and both scenarios are conducted in combination with DL based channel estimation for inspected users in NOMA cell. Far user results indicate an noticeable improvement in outage probability when both DL and optimized scheme

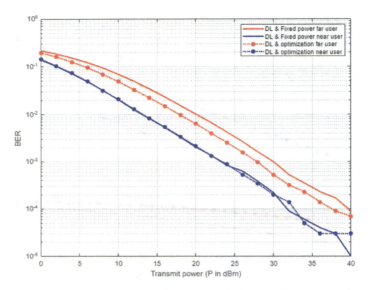

Fig. 8 BER versus power for deep learning (dl) based optimized and FPA power schemes

are applied compared to FPA results. On the other hand, near user with both DL scenario and fixed power allocation (FPA) scheme demonstrates considerable outage enhancement compared to optimized scenario. This might be justified that both the good channel gain with FPA coefficients are more effective for near user than optimized scheme even when DL scheme is applied.

In Fig. 10, results for the sum-rate for two users considered in NOMA system are illustrated. Two different power allocation schemes are inspected, optimized power scheme and FPA scheme, where each approach is incorporated with DL algorithm that utilized for estimating the channel coefficients prior to calculate the rate for each user. On the basis of the simulation results, it is clearly noticed that NOMA based on DL and optimized scheme shows little improvement in sum rate compared to FPA scenario when the applied power level is low. Starting from 15 dB, both power optimized scheme and FPA provide comparable sum rate when DL algorithm is considered.

Characterizing the DNN Impact on Multiuser PD-NOMA System Based ... 963

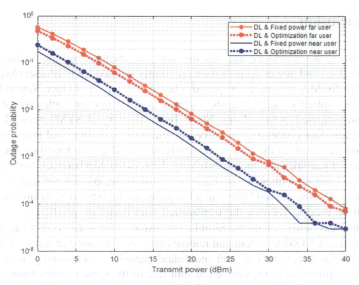

Fig. 9 Outage probability versus power for deep learning (DL) based optimized and FPA power schemes

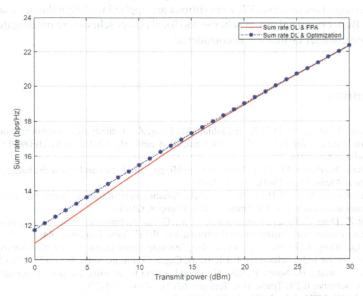

Fig. 10 Sum rate versus power for deep learning (DL) based optimized and FPA power schemes

7 Conclusion

In this work, we introduce and discuss how the channel estimation based Deep Learning (DL) and power optimization are jointly utilized for multiuser (MU) detection in PD-NOMA system. In proposed scheme, the impact of Deep Neural Network (DNN) in explicitly estimating the channel coefficients for each user in NOMA cell is investigated, where Long Short-Term Memory (LSTM) network is used for complex data processing to carry out training, updating, and prediction. Two stages are involved, in the first stage, DNN is trained online based on the normalized channel statistics and then identifying the relationship between the normalized successive training sequences. In implementation stage, normalized testing data will be employed by the trained DNN model to generate the estimated fading coefficients that will be used to recover the transmitted data based on predicting the channel parameters explicitly for users. Simulation results have verified that the proposed LSTM assisted NOMA system can realize better performance in terms of the BER, Outage probability, sum rate and individual capacity. Additionally, user's power factors are optimized to maximize the sum rate of the system users based on transmitted power and Quality of service (QoS) constraints. A systematic analysis for the optimization problem to verify the concavity of the objective function is presented and Lagrange function and KKT conditions are applied to derive the optimal power factors. Both optimized power scheme and fixed power scheme are investigated when DL based channel estimation is considered.

References

1. Wei, Z., Yuan, J., Ng, D.W.K., Elkashlan, M., Ding, Z.: A survey of downlink non-orthogonal multiple access for 5G wireless communication networks (2016). arXiv:1609.01856 arXiv preprint arXiv, 2016
2. Hossain, E., Hasan, M.: 5G cellular: key enabling technologies and research challenges. IEEE Instrum. Meas. Mag. **18**(3), 11–21 (2015)
3. Pourkabirian, Anisi, M.H.: Robust channel estimation in multiuser downlink 5G systems under channel uncertainties. IEEE Trans. Mobile Comput. (2021)
4. Yang, Z., Ding, Z., Fan, P., Karagiannidis, G.K.: On the performance of non-orthogonal multiple access systems with partial channel information. IEEE Trans. Commun. **64**, 654–667 (2016)
5. Bai, Q., Wang, J., Zhang, Y., Song, J.: deep learning-based channel estimation algorithm over time selective fading channels. IEEE Trans. Cogn. Commun. Netw. **6**(1), 125–134 (2020)
6. Gui, G., Huang, H., Song, Y., Sari, H.: Deep learning for an effective nonorthogonal multiple access scheme. IEEE Trans. Veh. Technol. **67**(9), 8440–8450 (2018)
7. Sun, Y., Wang, Y., Jiao, J., Wu, S., Zhang, Q.: Deep learning-based long-term power allocation scheme for NOMA downlink system in S-IoT. IEEE Access **7** (2019)
8. Wang, X., Zhu, P., Li, D., Xu, Y., You, X.: Pilot-assisted SIMO-NOMA signal detection with learnable successive interference cancellation. IEEE Commun. Lett. (2021)
9. Emir, A., Kara, F., Kaya, H., Yanikomeroglu, H.: Deep learning empowered semi-blind joint detection in cooperative NOMA. IEEE Access **9** (2021)
10. Ye, H., Li, G.Y., Juang, B.-H.: Power of deep learning for channel estimation and signal detection in OFDM systems. IEEE Wirel. Commun. Lett. (2018)

11. Dai, L., Wang, B., Ding, Z., Wang, Z., Chen, S., Hanzo, L.: A survey of non-orthogonal multiple access for 5G. IEEE Commun. Surveys Tuts. **20**(3), 2294–2323 (2018)
12. Dai, L., Wang, B., Yuan, Y., Han, S., Chih-lin, I., Wang, Z.: Non-orthogonal multiple access for 5G: solutions, challenges, opportunities, and future research trends. IEEE Commun. Mag. **53**(9), 74–81 (2015)
13. Zuo, H., Tao, X.: Power allocation optimization for uplink non-orthogonal multiple access systems. In: 2017 International Conference on Wireless Communications and Signal Processing (WCSP), pp. 1–5. IEEE (2017)
14. Ali, M.S., Tabassum, H., Hossain, E.: Dynamic user clustering and power allocation for uplink and downlink non-orthogonal multiple access (NOMA) systems. IEEE Access **4**, 6325–6343 (2016)
15. Park, T., Lee, G., Saad, W., Bennis, M.: Sum rate and reliability analysis for power-domain nonorthogonal multiple access (PD-NOMA). IEEE Internet Things J. **8**(12), 10160–10169 (2021)
16. Narottama, B., Shin, S.Y.: Dynamic power allocation for non-orthogonal multiple access with user mobility. In: 2019 IEEE 10th Annual Information Technology, Electronics and Mobile Communication Conference (IEMCON), pp. 0442–0446 (2019)
17. Assaf, T., Al-Dweik, A., Moursi, M.E., Zeineldin, H.: Exact BER performance analysis for downlink NOMA systems over Nakagami-m fading channels. IEEE Access **7**, 134539–134555 (2019)
18. Gupta, P., Ghosh, D.: Channel assignment with power allocation for sum rate maximization in NOMA cellular networks. In: 2020 5th International Conference on Computing, Communication and Security (ICCCS), pp. 1–5. IEEE (2020)
19. Amin, S.H., Mehana, A.H., Soliman, S.S., Fahmy, Y.A.: User capacity in downlink MISO-NOMA systems. In: IEEE Global Communications Conference (GLOBECOM), pp. 1–7 (2018)
20. Sun, Z., Jing, Y.: Average power analysis and user clustering design for MISO-NOMA systems. IEEE International Workshop on Signal Processing Advances in Wireless Communications, pp. 1–5 (2020)
21. Fu, Y., Salaün, L., Sung, C.W., Chen, C.S.: Distributed power allocation for the downlink of a two-cell MISO-NOMA system. In: IEEE 87th Vehicular Technology Conference, pp. 1–6 (2018)
22. Nguyen, T.-V., Nguyen, V.-D., Do, T.-N., da Costa, D.B., An, B.: Spectral efficiency maximization for multiuser MISO-NOMA downlink systems with SWIPT. IEEE Global Communications Conference, pp. 1–6 (2019)
23. Li, S., Derakhshani, M., Chen, C.S., Lambotharan, S.: Outage probability analysis for two-antennas MISO-NOMA downlink with statistical CSI. In: IEEE Global Communications Conference, pp. 1–6 (2019)
24. AbdelMoniem, M., Gasser, S.M., El-Mahallawy, M.S., Fakhr, M.W., Soliman, A.: Enhanced NOMA system using adaptive coding and modulation based on LSTM neural network channel estimation. Appl. Sci. **9**(15) (2019)
25. Zhang, Y., Wang, X., Xu, Y.: Energy-efficient resource allocation in uplink noma systems with deep reinforcement learning. In: Proceedings of International Conference on Wireless Communications and Signal Processing (WCSP), Xi'an, China (2019)
26. He, C., Hu, Y., Chen, Y., Zeng, B.: Joint power allocation and channel assignment for NOMA with deep reinforcement learning. IEEE J. Sel. Areas Commun. **37**(10), 2200–2210 (2019)
27. Lin, C., Chang, Q., Li, X.: A deep learning approach for MIMO-NOMA downlink signal detection. Sensors **19** (2019)
28. Kim, M., et al.: Deep learning-aided SCMA. IEEE Commun. Lett. **22**(4), 720–723 (2018)
29. Luo, J., Tang, J., So, D.K.C., Chen, G., Cumanan, K., Chambers, J.A.: A deep learning-based approach to power minimization in multi-carrier NOMA with SWIPT. IEEE Access **7**, 17450–17460 (2019)
30. Kang, J., Kim, I., Chun, C.: Deep learning-based MIMO-NOMA with imperfect SIC decoding. IEEE Syst. J., 1–4

31. Lu, Y., et al.: Deep multi-task learning for cooperative NOMA: System design and principles. IEEE J. Selected Areas Commun. (2020)
32. Zhu, J., Wang, J., Huang, Y., He, S., You, X., Yang, L.: On optimal power allocation for downlink non-orthogonal multiple access systems. IEEE J. Sel. Areas Commun. **35**(12), 2744–2757 (2017)
33. Tang, Z., Wang, J., Wang, J., Song, J.: On the achievable rate region of NOMA under outage probability constraints. IEEE Commun. Lett. **23**(2), 370–373 (2019)
34. Yang, Z., Xu, W., Pan, C., Pan, Y., Chen, M.: On the optimality of power allocation for NOMA downlinks with individual QoS constraints. IEEE Commun. Lett. **21**(7), 1649–1652 (2017)
35. Ding, Z., Yang, Z., Fan, P., Poor, H.V.: On the performance of nonorthogonal multiple access in 5G systems with randomly deployed users. IEEE Sig. Process. Lett. **21**(12), 1501–1505 (2014)
36. Awan, D.A., Cavalcante, R.L.G., Yukawa, M., Stanczak, S.: Detection for 5G-NOMA: an online adaptive machine learning approach. arXiv:1711.00355
37. Yang, P., Li, L., Liang, W., Zhang, H., Ding, Z.: Latency optimization for multi-user NOMA-MEC offloading using reinforcement learning. In: Proceedings of Wireless and Optical Communications Conference (WOCC), Beijing, China (2019)
38. Jiang, L., Li, X., Ye, N., Wang, A.: Deep learning-aided constellation design for downlink NOMA. In: Proceedings of International Wireless Communications & Mobile Computing Conference (IWCMC), Tangier, Morocco (2019)
39. Alberge, F.: Constellation design with deep learning for downlink non-orthogonal multiple access. In: Proceedings of IEEE Annual International Symposium on Personal, Indoor and Mobile Radio Communications (PIMRC), Bologna (2018)

Mobile Wireless Sensor Network: Routing Protocols Overview

Maha Al-Sadoon, Ahmed Jedidi, and Hamed Al-Raweshidy

Abstract Wireless sensor Networks (WSNs) became in one of the important technologies in our days in which it is applied in many applications and domains. Further, most of these applications are under the umbrella of the Internet of things (IoT) and that required a particular Quality of Service (QoS). Recently, many vital applications need mobility, which Mobile WSNs (MWSN) subscribes in these needs. However, the low-cost technology of the MWSNs is the first obstacle to improve performance in these applications. The usual methods of routing algorithm cannot be applied in MWSNs. In addition, security has become a primary concern to provide secure communication between wireless nodes, with additional challenges related to the node's computational resources. Particularly, the wireless ad-hoc communication adopted by the sensor nodes communication makes WSNs more susceptible to different types of security threats and attacks. In this paper, we present the specification of MWSN, and based on that we describe the different routing protocols used. Second, we discuss the performance of these protocols and propose a new technique to improve MWSN performance.

Keywords Mobile Wireless sensor network · Routing Algorithm · Genetic Algorithm · LEACH · Cluster Head

---

M. Al-Sadoon (✉) · A. Jedidi
Ahlia University, Manama 10878, Bahrain
e-mail: malsaadoon@ahlia.edu.bh

A. Jedidi
e-mail: ajedidi@ahlia.edu.bh

A. Jedidi
University of Sfax, 3029 Sfax, Tunisia

H. Al-Raweshidy
Brunel University London, London UB8 3PH, UK
e-mail: hamed.al-raweshidy@brunel.ac.uk

© The Author(s), under exclusive license to Springer Nature Switzerland AG 2024
A. Hamdan and E. S. Aldhaen (eds.), *Artificial Intelligence and Transforming Digital Marketing*, Studies in Systems, Decision and Control 487,
https://doi.org/10.1007/978-3-031-35828-9_80

1 Introduction

Wireless Sensor Network (WSN) is presented as a promoting technology in various domains. According to the low-cost technology, WSN became the backbone of Internet of Things (IoT) and many other applications, which the performance of the network presents one of the high priorities and importance criteria. IoT is defined as the connection between the physical environments and the digital one. Recently, IoT involves in various areas industrial, militarily and ecosystem, which it takes place more and more indispensable. In addition, these areas required a high quality of service (QoS), particularly: security, efficiency and energy consumption [1].

One of the major advancements in the WSN field is the introduction of the Mobile Wireless Sensor Network (MWSN) being much more versatile than static WSNs as the deployed sensor nodes have to adapt to the changes in the network's topologies. Examples of applications of MWSNs are military surveillance, habitat monitoring, agriculture applications, healthcare management, industrial monitoring, and environment monitoring [2, 3] some of these applications illustrated in Fig. 1 [4]. However, MWSNs have major design challenges such as the hardware cost, system architecture, memory and battery size, processing speed as shown in Fig. 2.

Mobile sensor nodes consist of a microcontroller, various sensors (i.e. light, temperature, humidity, pressure, mobility, etc.), a radio transceiver powered by a battery [5]. Usually the sensor nodes are deployed on land, underground, under water environments, and can be classified into heterogeneous or homogenous [6].

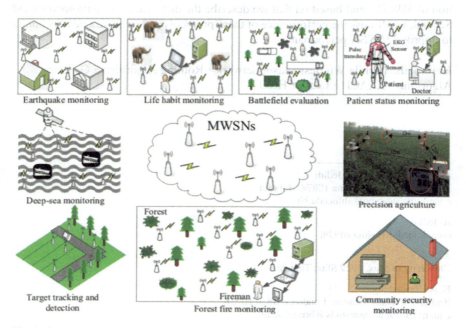

Fig. 1 Some application of MWSN [4]

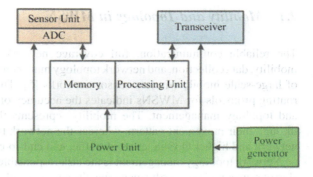

Fig. 2 Inner structure of sensor node

The heterogeneous nodes in MWSNs consists of sensor nodes that have unequal properties, but the homogenous MWSNs consist of identical sensor nodes according to the resources of the sensor nodes such as battery power, memory size, computing power, sensing range, transmission range, and mobility, etc. [7].

The unique characteristic of MWSNs imposes extra challenges on the design of an efficient routing protocol that considers the dynamicity of the network's topology, node's mobility, and other node's related constraints such as energy, computation complexity, resource availability, storage, and bandwidth. Therefore, this article presents the most famous routing protocols used in MWSN, which we discuss, the advantage and disadvantage for each one. Further we propose a new technique based on genetic algorithm to improve the MWSN performance.

The rest of paper is organized as follow. The second section will present the specification of MWSN and the different challenges of this technology. The third section will describe the various routing protocols used in MWSNs and a discussion will be hold in section fourth to compare between these protocols. Finally, we will propose a brief description of our new routing algorithm based on genetic algorithm.

2 MWSN Specification and Challenges

MWSNs introduce mobility in two approaches, either by having static sensor nodes while the sink devices are moving, while in the second approach the sink device is static, and the sensor nodes are mobile [4]. Examples of the first approach include crops on a farm deploying sensors and sending measurements about the humidity and temperature to the farmer's smartphone as he/she walks in the field. The second approach is clarified by static sink that can be used to collect tracking information stored in sensor nodes when the animals are in its range. The two approaches can be combined to have all nodes mobile such as in assistant personnel systems.

2.1 Mobility and Topology in MWSN

The reliable communication, full coverage network connectivity, sensor node mobility, data collection, and network topology management representing the success of large-scale mobile wireless sensor networks [5]. Therefore, designing efficient routing protocols for MWSNs indicates the accuracy of modelling sensor mobility and topology management. The mobility represents the sensor nodes behaviour through their movement pattern, whereas the network topology provides a reliable network and higher QoS in terms of traffic, and end-to-end connectivity [8, 9].

Network topology management is the task responsible for managing the membership of sensor nodes group by managing the new and withdrawn members. Depending on the nature of the MWSN, in order to achieve the best performance and to ensure reliable data gathering, different types of network topologies are deployed. The network topologies can be categorized into several types such as flat or unstructured, tree, clustered, chain, mesh, and hybrid [5–10]. Another proposed classification of network topology also available as follows:

- Distributed topology: In this topology there is no central node to manage the network, hence the network consists of homogenous sensor nodes so all of sensor nodes have an equal role and without prior infrastructure imposed before the network start running.
- Hierarchical topology: The senor nodes in this topology are organized in several levels, and the role of sensor nodes can be different, some nodes will act as cluster head while the others can act as cluster members also some of the sensor nodes might be consider as relay nodes in some cases. WSNs are more manageable and scalable in this hierarchical topology.
- Centralized topology: All the sensor nodes in this topology have the task of sensing the area of interest and collecting the data and then sending it to central node for processing. However, the main problem of this topology when the central node depletes its energy is a single point of failure.

On the other hand, the modelling of node's mobility predicts the future positioning of the sensor nodes. Thus, it brings the opportunity of reducing the number of hopes to the sink nodes, which results in a reduction of the latency. However high mobility scenarios can reduce the successful transmission of data to the sink nodes, therefore, increasing the complexity of the routing protocols [11–13]. Here, different mobility models can be applied to the sensor nodes or sinks depending on the application of the MWSN, by characterizing the mobile sensor movement patterns either in an independent or dependent approach. Mobility models can define the movement of the sensor nodes using both analytical or simulation-based models [14]. The analytical models usually provide simple mathematical models for the change in node's movement. In contrast, the simulation-based models provide more realistic mobility scenarios by introducing more complicated solutions. The mobility can further be classified into (a) trace models: a deterministic mobility pattern of real-life systems; (b) syntactic models: represents the movements towards/ away mobile sensor nodes realistically.

Another classification based on mobility patterns and histories such as directional, random and habitual mobility models. Generally, the various models available in the literature can be classified into four major categories including: random, temporal dependency, spatial, and geographic-restricted models.

2.2 MWSN Challenges

The mobility aspect of wireless sensors in MWSN increases the challenges of designing the MWSN. The following are some of those challenges [15]:

1. Dynamic Network Topology: due to the sensor nodes movement, the network structure is frequently changing. Therefore, the routing has to be adjusted to cope with the new position of sensor nodes. This adjustment can be done, by operating the location look up table, which contains the updated position of each sensor node in MWSN.
2. Localization: identifying the location of each sensor node within the wireless network is mandatory. In static WSN, the localization of all the sensor nodes is done once since the deployment and remains fixed. While in MWSN the sensor nodes are changing their location dynamically. So using some rapid localization techniques, the sensor nodes location has to be updated. This process consumes more energy and time.
3. Power Consumption: the sensor nodes scarce in power i.e. embedded with battery of low power capacity, the life of the wireless network is mainly affected by the energy depletion, which caused by mobility of sensor nodes that requires energy beside localizing sensor nodes also consumes extra power. Also routing these mobile nodes after changing their position also has energy overhead. All these processes leading to additional power consumption in MWSN. The sensor nodes already restricted with embedded battery of low capacity.
4. Network Coverage: One of the major influences in the MWSNs is the coverage area of interest design and application where the sensor coverage measured by the whole area that the network is currently monitoring. The quality of service that the network provides is part of sensor coverage measurements, and it will drop if case of sensor failures or undesirable sensor deployment and consequently will affect the critical application when initial deployment is far from having a full coverage area. Moreover, natural constrains and external harsh environments (e.g., wind, fire) affect the lifetime of whole networking.

Although MWSN is a special type of mobile ad hoc network (MANET) which designed to cope with mobile environments. However, there are varying in the following aspects [16, 17]:

- WSNs are mainly used to sense and collect data while MANETs are designed for distributed computing i.e. no sensing ability.

- The density of SNs in a WSN is relatively high comparing with the density of nodes in a MANET.
- The size of SNs is small, while the size of the wireless ad hoc is quite large with dimensions of laptop.
- SNs primarily employ the broadcast communication paradigm; on the other hand, MANET used a point-to-point communication.
- The data flows from the SNs toward the sink node whereas in MANETs, the flows of data are irregular.
- SNs resources are limited comparing to MANET recourses. SNs memory size is limited to 8-bit to 16-bit while with ad hoc the memory size is in the range of gigabyte to terabyte. Power resources of SNs are small batteries embedded into the sensing device with limited capacity because of their cost and mostly not rechargeable however nodes in ad hoc can be recharged.
- SNs are much more limited in their computation and communication capabilities (3–30 m) than their ad hoc (10–500 m) due to their low cost and prone to failures.
- SNs could be placed in harsh environments by helicopter or chopper plane, but not the case with MANET.

3 Routing Protocols in MWSN

To establish routing paths from the source nodes which response to sense the data to the destination that is a sink node, MWSN routing protocol relies on the geographic data of the sensor nodes deploy in the target area. Many routing protocols have been proposed in the art-of-state that uses position parameter for routing decisions. It is assumed in this type of protocols, sensor nodes have access to location information by using low power GPS module, or it use distributed localization schemes based on the received signal power (RSSI), signals time of arrival, etc. Therefore, in position-based protocols [18, 19], sensor nodes usually identified the position of their neighbour nodes through periodic "Hello" messages, hence, reducing the communication overhead resulting from flooding. However, obtaining the location information of nodes is a costly process due to message transfer overhead and energy consumption especially in case of mobile nodes. Examples of position-based routing protocols are:

- Geographical Energy Aware Routing (GEAR) [20]: This approach uses energy-aware metric in order to select the associated nodes for each cluster. GEAR is aims to balance the energy consumption and extending the network lifetime. The protocol calculates the value of the cost function for reaching an associated node based on node location and residual energy.
- Geographical Adaptive Fidelity (GAF) [21]: this protocol organized the area into geographical grid that comprises of cells, each grid contains multiple cells, with only one active node at a time. GAF purposes to prolong the network lifetime and reduce energy consumption.

- Adaptive Face Routing (AFR) [22]: This an ad hoc routing protocol that is based on the Euclidean planar graphs, AFR will divide the nodes and edges of a plane into regions called faces. This protocol used face routing to traverse the faces in a controlled way, the protocol repeated the same process using eclipse of double size when the face routing fails to deliver the data to the destination.
- Mobility Aware Routing (MAR) [23]: It is a hierarchal position-based routing protocol, in this network the area is organized into geographic grid with cluster heads that serve the area based on the mobility metric. This approach selects cluster heads that have less mobility metric without considering the energy level of the selected node and this represent the main weakness of MAR protocol.
- Geographic Robust Clustering (GRC) [24]: This protocol uses hierarchal-based routing protocol, based on two parameter which are nodes position, and energy levels it will select the cluster heads. To recover the packet loss this protocol, implement inter-cluster communication phase.
- Minimum Energy Communication Network (MECN) [25]: MECN is an energy-efficient routing protocol with objective of reducing the energy exhaustion of the entire network with help of low power GPS. The idea behind MECN protocol is to transmit the data packets through intermediate nodes _rely node instead of direct transmitting data to the sink node since direct communication consumes more energy than transmitting data through multiple hops using several relay nodes.

4 The Proposed Routing Protocol

In this section we propose a new clustering algorithm based on energy efficient incorporate dynamic Genetic Algorithm, is proposed to extend the lifetime of whole network. The idea is suggested to deploying homogenous mobile wireless sensor nodes randomly in the area of interest. This networking area is divided into five balanced cluster divisions, while the number of sensor nodes are not equal due to mobility with a constrain. Each cluster division has one cluster head that is response of collecting and aggregate the data sent by the cluster members associated with each cluster head. The selection of cluster head is based on a novel dynamic genetic algorithm. The fitness objective is form of weighted parameters, the first represent the position of the sensor node toward the centre of the cluster division whereas the second parameter indicates the residual energy of the sensor node.

5 Conclusion

The revolution of technologies since past few years that adding mobility features to the WSNs puts new challenges especially in design efficient routing protocols. According to some researchers, hierarchical based routing protocol outperform than the flat based and location-based routing protocols particularly in preserving energy, and prolong the lifetime of MWSNs as reviewed in this paper.

Several hierarchal routing protocols were suggested in many research papers. Some requirements for the routing protocols may conflict. Therefore, when picking the shortest path towards the sink node causes the in-between sensor nodes to deplete their energy fast, and consequently result in reducing the network lifetime. At the same time, it might result in less network delay and minimum energy consumption. Since the routing objectives are tailored according to the application, several routing protocols have been suggested for different applications.

Moreover, this survey highlighted the main challenges that affect the performance of MWSN beside the obvious different between the stationary WSN, MWSN and ad-hoc. The proposed technique in this survey focuses on develop an optimized routing-based protocol with help of evolutionary algorithms such as genetic algorithm to acquiring long lifetime of MWSN. And to attain that, a new selection approach for the cluster head with homogenous mobile sensor nodes is proposed. The parameters that employed in this work to formulate the fitness objective are the outstanding energy in addition to the position of each sensor nodes within the cluster division. The extensive description of this method will be discussed after we done with MATLAB simulation of MWSN and develop a fair comparative with recent available hierarchal based routing protocols then analyzed the performance of the proposed approach on the basis of energy efficiency, scalability, lifetime.

References

1. Vandana, R., Gayathri, P.: Integration of internet of things with wireless sensor network. Int. J. Electr. Comput. Eng. (IJECE) 9(1), 439–444 (2019)
2. Akyildiz, I., Su, W., Sankarasubramaniam, Y., Cayirci, E.: Wireless sensor networks: a survey. Comput. Netw. 38, 393–422 (2002)
3. Shen, J., Wang, A., Wang, C., Hung, P., Lai, C.: An efficient centroid-based routing protocol for energy management in WSN-assisted IoT. IEEE Access 5, 18469–18479 (2017)
4. Yue, Y., He, P.: A comprehensive survey on the reliability of mobile wireless sensor networks: taxonomy, challenges, and future directions. Inf. Fusion (2018)
5. Velmani, R.: Mobile wireless sensor networks: an overview. In: Wireless Sensor Networks. IntechOpen Limited: London, UK (2017)
6. Sai, K.K., Pavankumar, J., Sai, K.G.:Wireless Sensor Networks and Applications (2017)
7. Nabil, S., Shigenobu, S., Mohammed, A., Sabah, M.: A comprehensive survey on hierarchical-based routing protocols for mobile wireless sensor networks: review, taxonomy, and future directions. Wirel. Commun. Mobile Comput. (2017)
8. Krishna, K., Suresh, Y., Kumar, T.: Wireless sensor network topology control using clustering. In: 7th International Conference on Communication, Computing and Virtualization, pp. 893–902 (2016)
9. Nagpure, A., Sulabha: Topology control in wireless sensor network: an overview. Int. J. Comput. Appl. (IJCA), 13–18 (2014)
10. Matin, M.: Wireless Sensor Networks Technology and Protocols. M M, Intechopen (2012)
11. Camp, T., Davies, V.: A survey of mobility models for ad hoc network research. Wirel. Commun. Mobile Comput. 2 (2002)

12. Vasanthi, V., Singh, A., Hemalatha, M.: A detailed study of mobility models in wireless. J. Theory Appl. Inf. Technol. **33**, 7–14 (2011)
13. Silva, R., Silva, J., Vassiliou, V.: Mobility in WSNs for critical applications. In: IEEE Symposium on Computers and Communications (ISCC'11), Corfu. Greece, pp. 451–456 (2011)
14. Roy, R.R.: Handbook of Mobile AdHoc Networks for Mobility Models. Springer (2011)
15. Shyamala, C., Geetha, P., Sumithra, D.: Mobile wireless sensor network a survey. J. Wirel. Commun. Netw. Mobile Eng. Technol. **3**(2), 8–27 (2018)
16. Theofanis, P., Christos, G.: A survey on routing techniques supporting mobility in sensor networks. In: 5th International Conference on Mobile Ad-hoc and Sensor Networks (2009)
17. Rajashree, V., Patil, V., Sawant, S., Mudholkar, R.: Classification and comparison of routing protocols in wireless sensor networks. Special Issue on Ubiquit. Comput. Secur. Syst. **4**, 704–711 (2010)
18. Giordano, S., Stojmenovic, I., Blazevic, L.: Position based routing algorithms for ad hoc networks: a taxonomy. Ad hoc Wireless Network (2003)
19. Stojmenovic, I.: Position-based routing in ad hoc networks. IEEE Commun. Mag., 128–134 (2002)
20. Yu, Y., Govindan, R., Estrin, D.: Geographical and Energy Aware Routing: A Recursive Data Dissemination Protocol for Wireless Sensor Networks. UCLA Computer Science Department (2001)
21. Xu, Y., Heidemann, J., Estrin, D.: Geography-informed energy conservation for ad hoc routing. In: 7th annual International Conference on Mobile Computing and Networking, pp. 70–84 (2001)
22. Kuhn, F., Wattenhofer, R., Zollinger, A.: Asymptotically optimal geometric mobile ad-hoc routing. In: 6th International Workshop on Discrete Algorithms and Methods for Mobile Computing and Communications, pp. 24–33 (2002)
23. Al-Karaki, J., Ahmed, E.: Routing techniques in wireless sensor networks: a survey. IEEE Wirel. Commun. (2005)
24. Arboleda, C., Nasser, N.: Cluster-based routing protocol for mobile sensor networks. In: 3rd International Conference on Quality of Service in Heterogeneous Wired/Wireless Networks (2006)
25. Chaudhary, R., Sonia: A tutorial of routing protocols in wireless sensor networks. Int. J. Comput. Sci. Mobile Comput. (IJCSMC), 971–979 (2014)

Governance and Business Ethics with AI

Perception of Young Consumers Towards Electric Two Wheeler in Bangalore City

C. Surendhranatha Reddy and Ajai Abraham Thomas

Abstract In this dynamic environment of the market, automobile industry is changing very quickly due to technological up gradation. The manufacturers are able to respond to change in the taste and preferences of the customers. The conventional automobiles have some disadvantages like harmful effects on environment, increasing fuel prices, increasing prices etc. made the customers to think about an alternative technology that has resulted into electric vehicle technology. Even the customers are welcoming this new technology as they have many benefits from it. It is always a challenge for the marketers to make the customers to accept the change and push their products in the market. This research paper emphasises on understanding the perception of young consumers about electric two wheelers in Bangalore city.

Keywords Perception · Electric two wheelers · Automobile · Young consumers

1 Introduction

The increasing threat from pollution to the environment is the current challenge to entire world at large. Moving from fuel based vehicles to electric vehicles is one of the solutions to save the planet earth. As per a survey it is found that 77.2% of the respondents were said that they would force their parents to buy the electric vehicles to reduce the ill effects of fuel based vehicles [1].

As the climatic conditions are changing, it is alarming to save our planet. Electric vehicles are a good solution to the problem. Though there is a greater opportunity for the manufacturers to attract the customers towards electric vehicles but there are few limitations like absence of product innovation, high upfront expenses and lack of enough charging stations. In spite of these challenges there are some positive aspects like growth in product technology, application of computer technology and turn to total cost of ownership [2].

C. Surendhranatha Reddy (✉) · A. A. Thomas
Kristu Jayanti College, Bangalore 560077, India
e-mail: surendranath@kristujayanti.com

© The Author(s), under exclusive license to Springer Nature Switzerland AG 2024
A. Hamdan and E. S. Aldhaen (eds.), *Artificial Intelligence and Transforming Digital Marketing*, Studies in Systems, Decision and Control 487,
https://doi.org/10.1007/978-3-031-35828-9_81

Environment conscious millennials are key to the electric two wheelers market. There is a growing concern about environment among young customers and they are ready to shift to electric vehicles. Even the companies are gearing up for introducing the innovative product range in electric vehicles as the market is slowly picking up. The benefits of choosing electric two wheelers over petrol versions in metropolitan cities are very clear and they include much less congestion, negligible air pollution and less noise [3].

2 Review of Literature

The variables which could impact the decision of buying an electric car would be price, battery warranty, running time, charging time, appearance, performance, safety, comfort, brand, after sale service and supporting infrastructure where subsidy from government, celebrity endorsement and influence by family and friends have less impact on buying decisions [4].

When the adequate infrastructural facilities are provided customers are ready to buy the electric two wheelers and they are aware of bad effects of conventional two wheelers. Initial cost of purchase, battery charging time and number of charging stations are some of the factors discouraging the customers to buy electric two wheelers [5].

The young customers are willing to buy electric vehicles and there is strong correlation between young customers and adoption of electric cars. As the concern for protecting the plant is growing there is a considerable awareness to adopt automobiles which can cause less damage to the environment. The perceived risk and trust in electric vehicles have not influenced the intentions of young customers to buy them [6].

The behaviour of customers is positive about electric vehicles as they offer many benefits. Many of the customers are in opinion that they are ready to pay extra amount if the electrical vehicles can deliver same performance as of conventional vehicles. They also insist that government should provide subsidies to purchase electric vehicles and the number of charging stations should be increased [7].

Though the EVs offer huge advantages to the customers there are some doubts in their minds about performance, battery life, charging time and maintenance cost. Customers are ready to pay extra amount if the electric vehicles come with faster charging and more running range. Customers are also concerned about the features and performance of electric vehicles as they have minimum knowledge about them [8].

The customers are in opinion that electric vehicles are environment friendly and there is an innovation into making these electric vehicles. The factor that could discourage customers could be lack of awareness and risk factors. The customers are also believe that electric vehicles are cheaper to use [9].

The awareness about EVs is considerably low and the marketers need to develop a positive perception. As electric vehicles could be one of the big solutions to global warming and pollution, government and marketers can educate people about the benefits of electric vehicles and encourage them to buy [10].

Considerable number of customers are satisfied with the performance of electric two wheels as they provide many benefits. They are also happy with the service provided by dealers on regular basis. The awareness level among customers about electric vehicles is low and there is a need for promotion and setting up more charging stations [11].

Adoption behaviour of customers is influenced by price, government subsidies, performance of electric vehicles and battery range and the behaviour is not significantly influenced by gender, age, income, occupation, riding pleasure, operational cost, battery life, environment condition and public charging infrastructure [12].

Customers feel that there are less number of models in electric two wheelers and the number of charging stations are also less. Customers prefer to buy the electric two wheelers if the charging time can be reduced. The concern towards environment and distance travelled on a single charge are factors considered while buying electric two wheelers [13].

The purchase of electric two wheelers is gaining momentum in the market as the customers feel that price of EVs is reasonable and maintenance cost is less. But there is need for creating more awareness among customers as EVs can help the society at large by replacing fuel based vehicles which are one of main causes of global warming. [14].

Many customers are with the opinion that electric bikes are expensive and many customers are not aware of the benefits of electric bikes. High charging time, low battery life, lower mileage and less speed are considered to be factors which are influencing the purchase decisions of customers while advertisements have no impact on the purchase decisions [15].

The usefulness of product, level of difficulty to handle and compatibility have a favourable influence on customer's buying behaviour while interpersonal influence by friends, family and peer group have negative influence on attitude of customers towards buying behaviour of electric vehicles. Lack of awareness about the benefits of electric vehicles is also having negative impact on attitude of customers [16].

Attitudinal differences towards EVs among male and female is less and the customers are ready to adopt the EVs provided that manufacturers can make EVs at reasonable price, with fast charging battery, more mileage, charging infrastructure and good battery life. People also believe that EVs are the future and they can occupy the market share of fuel based two vehicles [17].

The customer buying behaviour towards electric vehicles is largely influenced by upfront purchase price, maintenance, charging prices, service cost and comfort features. The main reason for slow penetration of EVs is high purchasing price as compared to fuel based vehicles and range per charging of vehicle [18].

3 Objectives

- To apprehend the perception of young customers about electric two wheelers
- To explore the expectations of youth towards the performance of electric two wheeler
- To recommend strategies to create awareness about electric two wheelers in the market.

4 Hypothesis

H1: Age of customers has positive influence on perception of customers towards electric two wheelers.

H2: Family income of customers has positive influence on perception of customers towards electric two wheelers.

H3: Electric two wheelers are perceived to be better in performance than conventional two wheelers.

5 Methodology

The current study is devised to apprehend the young customers' behaviour about electric two wheelers in Bangalore city. For the purpose of study primary data were collected using a systematic questionnaire. Five point Likert scale was used in the questionnaire to determine the various factors considered while buying the electric two wheelers. A sample size of 272 was drawn from the population with the help of simple random sampling. Research participants included young customers between the age group of 18–40 from Bangalore city. Data was analysed using statistical tools like Reliability test and Regression.

6 Results

The reliability and validity of the instrument used in the study was tested with the help of Cronbach's alpha coefficients. The value to be considered reliable is of 0.7.

We can observe from Table 1, that overall Cronbach Alpha score is 0.782, which exceeds the recommended threshold.

From Table 2, it is exhibited that more respondents participated in the survey are male (54.8%) followed by female (45.2%).

Perception of Young Consumers Towards Electric Two Wheeler ... 983

Table 1 Reliability statistics

Reliability statistics

Cronbach's alpha	Cronbach's alpha based on standardized items	N of items
0.782	0.792	29

Table 2 Gender of respondents

Gender

		Frequency	Percent	Valid percent	Cumulative percent
Valid	Male	149	54.8	54.8	54.8
	Female	123	45.2	45.2	100.0
	Total	272	100.0	100.0	

We can observe from Table 3 that many respondents participated in the survey are students (45.2%) as they are the target segment for electric vehicles followed by respondents who are employed, self-employed and house wives.

It is understood from Table 4 that many respondents have the family income between 700,001 and 1,000,000 (65.4%).

Table 3 Occupation of respondents

Occupation

		Frequency	Percent	Valid percent	Cumulative percent
Valid	Student	123	45.2	45.2	45.2
	Employed	68	25.0	25.0	70.2
	Self employed	39	14.3	14.3	84.6
	Hose wife	42	15.4	15.4	100.0
	Total	272	100.0	100.0	

Table 4 Family income of respondents

Family income

		Frequency	Percent	Valid percent	Cumulative percent
Valid	Less than 500,000	11	4.0	4.0	4.0
	500,001 to 700,000	32	11.8	11.8	15.8
	700,001 to 1,000,000	178	65.4	65.4	81.3
	Above 1,000,001	12	4.4	4.4	85.7
	5.00	39	14.3	14.3	100.0
	Total	272	100.0	100.0	

H1: Age of customers has an impact on perception of customers towards electric two wheelers.

H0: Age of customers has no impact on perception of customers towards electric two wheelers.

It is evident from Table 5 that null hypothesis is accepted since the coefficient value (0.913) is greater than the value of significance (p > 0.05). This means age of the customers has no significant impact on perception of customers towards electric two wheelers.

H2: Family income of customers has an impact on perception of customers towards electric two wheelers.

H0: Family income of customers has an impact on perception of customers towards electric two wheelers.

We can interpret from Table 6 that coefficient value (p = 0.910) is greater than the value of significance (p > 0.05) so the null hypothesis can be rejected and we can understand that perception of customers towards electric two wheelers may not be influenced by their family income.

H3: Electric two wheelers are perceived to be better in performance than conventional two wheelers.

Table 5 Testing hypothesis H1

Coefficients						
Model		Unstandardized coefficients		Standardized coefficients	t	Sig
		B	Std. error	Beta		
1	(Constant)	3.320	0.121		27.363	0.000
	Age	−0.009	0.078	−0.007	−0.110	0.913

a. Dependent Variable: I will buy an Electric two wheelers if accessibility to efficient charging stations were enough

Table 6 Testing hypothesis H2

Coefficients						
Model		Unstandardized coefficients		Standardized coefficients	t	Sig
		B	Std. error	Beta		
1	(Constant)	3.284	0.228		14.405	0.000
	Family income	0.008	0.070	0.007	0.113	0.910

a. Dependent Variable: I will buy an Electric two wheelers if accessibility to efficient charging stations were enough

Table 7 Testing hypothesis H3

Coefficients

Model		Unstandardized coefficients		Standardized coefficients	t	Sig
		B	Std. error	Beta		
1	(Constant)	2.537	0.223		11.364	0.000
	The features, performance and the price of electric two wheelers can make them compete with fuel based two wheelers	0.331	0.067	0.288	4.947	0.000

a. Dependent Variable: Electric two wheelers are environment friendly

H0: Electric two wheelers are not perceived to be better in performance than conventional two wheelers.

We can apprehend from Table 7 that coefficient value (p = 0.000) is less than the value of significance (0.05), so the null hypothesis is rejected. It means electric two wheelers are perceived to be better in performance than conventional two wheelers.

7 Conclusion and Management Implications

Customer perception is a complex psychological behaviour towards products and services. The perception can change as the time and market change. The demographic factors like age, gender, education and family income of customers are not having any impact on customer perception towards electric two wheelers. Customers are in the opinion that electric two wheelers can be a proper replacement for conventional two wheelers as they offer more benefits.

The young customers are expecting many features in electric two wheelers like more styles, mileage, quick charging of battery, more charging stations, better after sale services, customer education, better performance, more acceleration, subsidy from government, durability and warranty. Customers are also expecting the value for money and price discounts since the price of electric two wheelers is comparatively more than conventional two wheelers.

Electric two wheeler manufacturers need to look at the changes in the market and need to attract the customers and make them to shift to electric two wheelers by educating customers about benefits of two wheelers, offering more features, increasing the power of two wheelers, setting up more recharge centers, providing price discounts without compromising the profit margins, making the accessories available at convenient places, providing service centers and promoting the products.

References

1. HT Auto Desk Youth for change survey finds kids key in convincing parents to buy EVs. 28 Oct 2020. https://auto.hindustantimes.com/auto/news/youth-for-change-survey-finds-kids-key-in-convincing-parents-to-buy-evs-41603861922970.html
2. Agarwal, V.: Electric vehicles: a roadmap to mass adoption in India. 18 Dec 2021. https://economictimes.indiatimes.com/industry/renewables/electric-vehicles-a-roadmap-to-mass-adoption-in-india/articleshow/88336832.cms?from=mdr
3. Jolly, J.: EV riders: motorcycle manufacturers making the leap to electric. 3 Sep 2022. https://www.theguardian.com/business/2022/sep/03/ev-riders-motorcycle-manufacturers-making-the-leap-to-electric
4. Lalwani, A.: The perception and buying behavior of Indians towards electric cars. Int. J. Manage. (IJM) (2020)
5. Tupe, O., Kishore, S., Johnvieira, A.: Consumer perception of electric vehicles in India. Eur. J. Mol. Clin. Med. (2020)
6. Sefora, I.: An investigation of young consumers' perceptions towards the adoption of electric cars. Afr. J. Bus. Econ. Res., 107 (2019)
7. Kiran, J.: A study on attitude of customers towards electric vehicles. Christ College. Project Work (2021)
8. Bansal, P.: Willingness to pay and attitudinal preferences of Indian consumers for electric vehicles. Energy Econ. (2021)
9. Jose, S.P., Cyriac, S., Joseph, B.: Consumer attitude and perception towards electric vehicles. Acad. Market. Stud. J., 1–12 (2022)
10. Das, S. Customer perception and awareness towards electric two-wheelers: an analysis in Pune city. Int. J. Res. Anal. Rev., 622–624 (2020)
11. Ukesh, M., Chandra Kumar, M., Gangaiselvi, R.: A study on consumer satisfaction of buying electric two wheelers in Coimbatore district of Tamil Nadu. Pharma Innov. J., 381–383 (2022)
12. Wahab, L., and Haobin Jiang.: Factors Influencing the Adoption of Electric Vehicle: The Case of Electric Motorcycle in Northern Ghana. International Journal for Traffic & Transport Engineering, 22–37 (2019)
13. Sankar, J.G., Hemanth, K.G.: Customer perception towards electric two-wheeler innovation. J. Contemp. Issues Bus. Gov. **26**(2), 686–690 (2020)
14. Selvi, S.: A study on customer satisfaction towards electric bikes with special reference to Coimbatore city. Int. J. Appl. Res., 355–359 (2017)
15. Rajiv, P., Kavitha S.: Customers perception towards electric two wheeler vehicles in Vellore City: a study on go green battery operated vehicles. Int. J. Commer. Manage. Res., 63–65 (2016)
16. Tu, J.-C., Yang, C.: Key factors influencing consumers' purchase of electric vehicles. Sustainability **11**(14), 3863 (2019)
17. Purohit, H.K.S.: Electric vehicles and attitude of metropolitan consumers. PalArch's J. Archaeol. Egypt/Egyptol. **18**(7), 2415–2424 (2021)
18. Sriram, K.V., Michael, L.K., Hungund, S.S., Fernandes, M.: Factors influencing adoption of electric vehicles—a case in India. Cogent Eng. 9(1) (2022)

Efficacy of Dividend Announcement on Bluechip Pharma Stocks—An Evidence from the Indian Stock Market

R. Madhusudhanan and R. Haribaskar

Abstract It is always a curiosity to the existing and prospective investors to analyses the impact on any events in the share price or returns. This study adds fuel to that and taken a corporate event of Dividend and its impact in the returns. For the study the researcher has taken pharmaceutical industries for analyzing the impact. The top five companies were selected from the same industries and the time period taken for the study is ten years (2012–2021). Secondary data were collected and the returns were calculated to identify the volatility. Further diving deep descriptive statistics was used to derive mean, median, Standard deviation, Skewness and kurtosis. Out of which the mean helps to identify the volatility and standard deviation helps to identify the risks. Finally the result of the study explains that there is no significant impact on the share price by the announcement of the dividend.

Keywords Dividend · Announcement · Returns · Pharmaceutical industries share price · Returns · Risk · Impact

1 Introduction

The entire world was still during the COVID19 pandemic, one industry was out performing when the rest of the industries are sleeping. Generally the risk adverse investors always have a safe play with banking and these industries. Distinguished reasons ruling the minds of investors to invest and hold the scripts while coming to these particular sectors. These are the companies always having the upward demand trend and the reason why the profit will not get affected. This was the greatest reason behind the trust of the investors. The long term investment option for various investors the preference comes first in their mind is the same company. Expected market growth will be of their times in the next decade, which predicted by the Indian Economic

---

R. Madhusudhanan (✉) · R. Haribaskar
Kristu Jayanti College (Autonomous), Bengaluru, India
e-mail: madhusudhanan@kristujayanti.com

© The Author(s), under exclusive license to Springer Nature Switzerland AG 2024
A. Hamdan and E. S. Aldhaen (eds.), *Artificial Intelligence and Transforming Digital Marketing*, Studies in Systems, Decision and Control 487,
https://doi.org/10.1007/978-3-031-35828-9_82

Survey 2021 as well as there will be a drastic market expansion will expected to reach US$120–130 billion by 2030. The industry what so far we are talking is nothing but a pharmaceutical industry. Pharmaceutical industries play an important role among the market choices.

The theory of relevance clearly states that the dividend declaration has its impact over the prices of the shares, in contrast to this there are other school of thought to prove there dividend declaration will not have any impact over the share prices which is called as theory of irrelevance. In general scenario every information or any kind of corporate events are having the consequences in the share price and has its reflection in the returns as well. Therefore there is no exemption for the dividend announcement has its impact over the share price or returns. This study will examine both the theories with the help of historical data.

2 Review of Literature

Mahmood et al. [1]. The study was conducted on Karachi Stock Exchange. The researcher has taken the dividend announcement pretending to cash dividends to identify the impact on the stock price. For the study hundred dividend announcement dates was considered. The study incorporates the event study methodology and identifies that there was an insider trader's information leakages are identified in the pre-event window and the prices were.

Louhichi [2]. The study investigate the cash dividend announcement was captured by its reactions in the stock prices. Event study methodology was inculcated to identify the dividend announcement in the stock price returns. The event was considered fifteen days before and after event was calculated, and identified that there was a positive signal in the market that means the dividend announcement was having the significant impact in the stock prices. The investors were keen in watching the dividend announcement and act accordingly taking the decisions.

Hussainey et al. [3]. The article investigates the relationship between the dividend announcement and the market volatility in the US market. The study had two different dimensions dividend yield and stock price volatility and dividend payout and share price volatility. The study identifies that there were positive relationship between the dividend yield and stock price volatility and it was found negative in the dividend payout and stock price volatility. Therefore the study concludes that the dividend policy has a significant impact on the stock price and it drives the investor's perception.

Chaudhary et al. [4]. The study examines the dividend announcement impact in the Karachi Stock Market. The study was taken at the year 2010 with the sample of 30 different companies and tested the impact of cash dividend in the stock prices. The normal event study methodology was inculcated 15 days before the announcement and 15 days after the announcement the abnormal returns were calculated and identified that there were a positive signal in the stock prices because of the cash dividend announcement. Therefore the article concludes that there was a positive relationship exist between the cash dividend announcement and stock prices.

3 Research Design

Statement of Problem

Till the last of existence of the market the curiosity to know the impact of any event will not come to an end. This is the reason why the dividend announcement is taken for the study to analyses the impact on its returns. By selecting a particular sector and analyzing the impact of the events will gives the clarity that how the markets are reacting to the selected events, by the same time this will be the learning process to apply in the near future.

Objectives

- To identify the returns
- To check the volatility of returns
- To identify the impact of dividend announcement on the returns.
- To analyses the insider information leakages.

Methodology

The daily share price of the selected scripts is taken for the time period between 2012 and 2021. The returns are calculated and Descriptive Statistics tools are used to identify the Mean, Median, Standard deviation, Skewness and kurtosis and so on. The exact dividend announcement date taken for the consideration, from that announcement date thirty days prior and after data was separated to analyze the impact of divided among the shareholders. For the analysis Microsoft Excel was used for calculations. The historical data are collected from the secondary sources through https://www.bseindia.com/market_data.html. For the study top five pharmaceutical companies are taken as a sample Sun Pharma, Divis Laboratory, Dr. Reddy Laboratory, Cipla, Biocon. Last ten years final dividend announcement has been taken for the study.

4 Data Analysis and Discussions

Sun Pharma Ltd.

Dividend announcement date	Descriptive statistics	Mean	Median	Standard deviation	Kurtosis	Skewness	Count
7-Aug-12	Before	0.0033	0.0018	0.0112	0.1385	0.5417	30
	After	0.0002	−0.0016	0.01	1.1972	0.5169	30
28-May-13	Before	0.0052	0.0062	0.019	0.3527	−0.3032	30
	After	0.0018	0.0007	0.0212	2.9898	1.2555	30
12-Sep-14	Before	0.0015	0.0002	0.0169	1.8208	0.0932	30
	After	0.0012	0.0009	0.0194	0.6979	0.3633	30
11-Aug-15	Before	0.0002	0.0061	0.0309	20.0723	−4.0649	30
	After	0.0025	0.0034	0.0205	0.4271	−0.4311	30
31-May-16	Before	−0.0007	−0.0017	0.0145	8.6905	2.3658	30
	After	0.001	0.0014	0.0103	0.05	0.4011	30
3-Jul-17	Before	−0.0051	−0.0014	0.0276	8.1971	−2.2886	30
	After	−0.0067	−0.0081	0.0177	0.4003	−0.5369	30
26-Mar-18	Before	−0.0027	−0.007	0.0252	1.3385	0.6	30
	After	0.0003	−0.0032	0.0155	−0.0094	0.4099	30
28-May-19	Before	−0.0039	−0.0014	0.0243	7.0116	−1.0972	30
	After	−0.0021	−0.0045	0.0216	0.6163	0.6873	30
27-May-20	Before	0.0019	0.0018	0.0211	−0.7689	0.2676	30
	After	0.0023	0.0063	0.0188	1.5059	−0.6577	30
30-May-21	Before	0	0.0004	0.0162	−0.0147	−0.0718	30
	After	0.0028	0.0028	0.0149	−0.3005	−0.3773	30

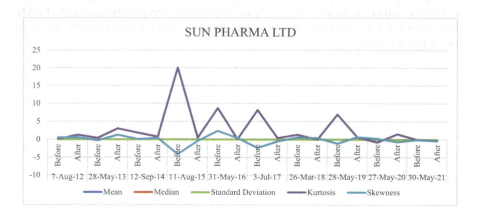

Divis Laboratory

Dividend announcement date	Descriptive statistics	Mean	Median	Standard deviation	Kurtosis	Skewness	Count
11-May-12	Before	0.364	0.199	1.5117	1.6965	0.3053	30
	After	0.5086	0.5537	1.9138	4.6045	1.4215	30
20-May-13	Before	0.2999	0.2943	1.2813	−0.2207	−0.2488	30
	After	−0.3011	−0.1553	1.8816	5.8065	−1.6688	30
26-May-14	Before	−0.1353	0.1009	1.2051	−0.7723	−0.3055	30
	After	0.7392	0.6817	1.5998	2.5798	1.0366	30
25-May-15	Before	−0.1585	−0.2379	1.5523	0.2151	0.306	30
	After	0.0879	0.1803	1.1813	0.343	0.515	30
2-Mar-16	Before	−0.3547	−0.3234	1.9053	1.1487	0.554	30
	After	0.2712	0.4481	1.5526	4.3491	−1.0764	30
26-May-17	Before	−0.5771	−0.3946	1.1451	1.0426	−0.6742	30
	After	0.7703	0.1477	1.7145	−0.6218	0.3816	30
28-May-18	Before	0.0803	0.1341	1.4192	−0.7585	0.0852	30
	After	−0.0407	−0.4049	2.0289	−0.016	0.8903	30
27-May-19	Before	0.1607	−0.2477	1.4605	−0.6773	0.4075	30
	After	−0.0292	−0.1264	1.4142	1.6345	0.2603	30
12-Feb-20	Before	0.5637	0.2824	1.1747	0.0556	0.861	30
	After	−0.3546	−0.1068	3.4143	2.61	−1.0129	30
20-May-21	Before	0.0942	0.2122	1.372	0.1804	−0.262	30
	After	0.2741	0.1481	1.4602	0.2724	−0.0636	30

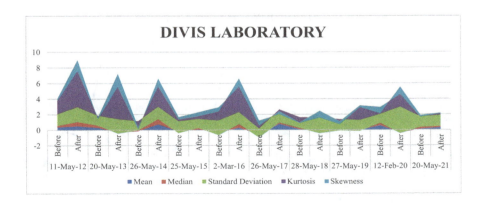

Dr. Reddy Laboratory

Dividend announcement date	Descriptive statistics	Mean	Median	Standard deviation	Kurtosis	Skewness	Count
11-May-12	Before	0.000483	0.000325	0.013978	−0.2355	0.25013	30
	After	−0.00149	−0.00081	0.01168	1.003239	0.282949	30
14-May-13	Before	0.005806	0.000376	0.01697	0.205277	0.797904	30
	After	0.00138	−0.00263	0.013567	−1.00043	0.450938	30
13-May-14	Before	−0.00028	0.000891	0.013631	−0.18455	-0.55116	30
	After	−0.00208	0.000843	0.018124	1.03006	-0.8884	30
12-May-15	Before	−0.0013	−0.00338	0.018425	−0.28487	0.529426	30
	After	−0.00035	0.000919	0.012283	1.108878	0.456973	30
13-May-16	Before	−0.00057	0.000655	0.015584	0.252872	0.319536	30
	After	0.002997	0.001872	0.011104	0.63687	0.51303	30
12-May-17	Before	−0.00053	−0.00147	0.010688	9.344794	2.291927	30
	After	0.000727	0.000635	0.016747	0.401029	0.189191	30
22-May-18	Before	−0.00372	−0.0021	0.013494	1.846504	-1.2242	30
	After	0.003887	0.002294	0.020093	0.538166	0.790683	30
17-May-19	Before	0.000409	−0.00029	0.009821	0.993793	0.051447	30
	After	−0.00098	−0.0024	0.018487	2.511688	-0.56326	30
20-May-20	Before	0.006186	0.000315	0.031698	10.29811	2.676918	30
	After	0.000125	−0.00241	0.01626	2.317251	1.196368	30
13-May-21	Before	0.001247	−0.00045	0.015879	3.015233	-0.69274	30
	After	−0.0035	−0.00303	0.012037	−0.29903	0.09751	30

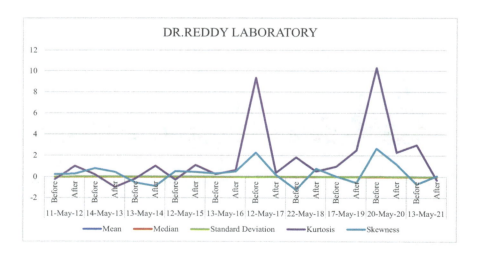

Cipla Ltd.

Dividend announcement date	Descriptive statistics	Mean	Median	Standard deviation	Kurtosis	Skewness	Count
6-Jul-12	Before	0.001589	0.00081	0.011966	0.458227	0.102555	30
	After	0.003181	0.001855	0.012326	2.947507	1.167032	30
29-May-13	Before	−0.00057	−0.00143	0.013519	0.196908	0.687156	30
	After	−3.8E−05	−0.00013	0.013619	4.365294	−1.56992	30
29-May-14	Before	−0.00144	−0.00286	0.012593	0.18337	0.319012	30
	After	0.005211	0.002637	0.011895	0.20048	0.754253	30
29-May-15	Before	−0.00229	−0.00087	0.020992	1.482292	0.648613	30
	After	6.25E−05	−0.00035	0.018359	−0.2688	0.184139	30
24-May-16	Before	−0.0005	−0.0014	0.012226	0.272488	0.459682	30
	After	0.000881	0.001137	0.014661	3.904841	−1.11919	30
25-May-17	Before	−0.00403	−0.00182	0.011931	6.144801	−1.91144	30
	After	0.002749	0.003764	0.012974	−0.21846	−0.27453	30
22-May-18	Before	−0.00225	−0.00117	0.016178	3.837769	0.924509	30
	After	0.006815	0.002423	0.016568	−0.20879	0.745328	30
22-May-19	Before	0.002485	−0.00138	0.014411	0.115753	0.910044	30
	After	−7.5E−05	0.00054	0.012971	0.706246	0.056627	30
20-Feb-20	Before	−0.00156	−0.00083	0.012667	0.502615	−0.23117	30
	After	0.006371	−0.00467	0.043545	−0.71901	0.460084	30
29-Jun-21	Before	0.002789	−0.00184	0.017148	−0.4738	0.278016	30
	After	−0.00418	−0.00442	0.013413	0.081259	0.29531	30

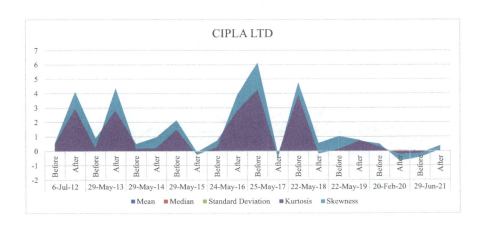

Biocon Ltd.

Dividend announcement date	Descriptive statistics	Mean	Median	Standard deviation	Kurtosis	Skewness	Count
27-Apr-12	Before	−0.00278	−0.00466	0.01493	−0.08581	0.427413	30
	After	−0.00289	0.000326	0.013503	0.639354	−0.79925	30
26-Apr-13	Before	0.002207	0.001722	0.017289	0.034261	0.377395	30
	After	0.000309	−0.00183	0.010459	0.549116	1.009648	30
25-Apr-14	Before	0.002066	0.000694	0.02332	6.854691	1.700543	30
	After	−0.00043	0.001759	0.016729	−0.54016	−0.38744	30
27-Mar-15	Before	0.001631	−0.0015	0.020167	2.933104	1.351216	30
	After	0.001346	−0.00058	0.025596	1.700176	0.7845	30
8-Mar-16	Before	0.00039	0.000698	0.016765	−0.54622	−0.12093	30
	After	0.006023	0.004594	0.018508	5.582038	1.519397	30
2-May-17	Before	−0.00013	−0.00303	0.021126	13.50105	2.959813	30
	After	−0.00181	0.002163	0.022505	0.676261	−0.63332	30
26-Apr-18	Before	0.003164	0.001568	0.014476	2.00213	0.982205	30
	After	−0.00264	−0.00027	0.021194	3.513351	−1.66243	30
25-Apr-19	Before	−3.5E−06	0.00053	0.011279	−0.25128	−0.24403	30
	After	−0.00458	−0.00341	0.021192	0.477677	−0.06444	30
28-Apr-21	Before	0.006723	0.005714	0.015106	0.103724	0.388284	30
	After	0.000154	0.000417	0.019362	−0.07121	−0.07743	30

5 Discussions and Results

Sun Pharma

The mean value before and after dividend announcement were not having much variations. There were no significant deviations found before and after the announcement of dividends. Five out of ten years the kurtosis is Leptokurtic after the announcement of dividend. The Skewness is positive majority after the announcement.

Divis Laboratory

The mean value before and after dividend announcement had some variations. There were no significant deviations found before and after the announcement of dividends. The after announcement years proves that kurtosis is Leptokurtic. The Skewness was negative in after announcement.

Dr. Reddy's Labarotary

The mean value before and after dividend announcement had some variations. There were no significant deviations found before and after the announcement of dividends. The after announcement years proves that kurtosis is Leptokurtic. More positive Skewness was found after announcement of dividend.

Cipla Ltd.

The mean value before and after dividend announcement had some variations. There were no significant deviations found before and after the announcement of dividends. The after announcement years proves that kurtosis is Leptokurtic. More positive Skewness was found after announcement of dividend.

Biocon Ltd.

The mean value before and after dividend announcement had some variations. There were no significant deviations found before and after the announcement of dividends. The after announcement years proves that kurtosis is Leptokurtic. More positive Skewness was found after announcement of dividend.

6 Conclusion

The study reveals the pharmaceutical industries are less volatile in nature. Returns of these companies shares are not much affected the share price or returns. Therefore the growth is considered as capital appreciation. The variations in the standard deviation clearly state that the risk is lesser for the investors. All the five companies over the years the dividend announcement has no impact on its share prices or returns. The market of the particular industry is in Random Walk. This also an evidence that there was no information leakages. Therefore the investors are having the hold on the existing shareholding as well as the prospective capacity.

7 Suggestions

The pharmaceutical industries are good choices for the risk adverse investors who are looking for less risk and long term investment opportunities. Parallel to which the short term profit using speculation is also not having much scope in this field. There was strong evidence that all these companies are consistently paying the dividends as well. Therefore the investors are looking for the regular income this industry will be adding the feather to the portfolio crown.

References

1. Mahmood, S., Sheikh, M.F., Ghaffari, A.: Dividend announcements and stock returns: an event study on Karachi stock exchange. Interdiscip. J. Contempor. Res. Bus. 3(8), 972–981 (2011)
2. Louhichi, W.: Adjustment of stock prices to earnings announcements: evidence from Euronext Paris. Rev. Acc. Financ. 7(1), 102–115 (2008). https://doi.org/10.1108/14757700810853879
3. Hussainey, K., Oscar Mgbame, C., ChijokeMgbame, A.M.: Dividend policy and share price volatility: UK evidence. J. Risk Finance 12(1), 57–68 (2011)
4. Chaudhary, G.M., Hashmi, S.H., Younis, A.: Does dividend announcement generate market signal? Evidence from Pakistan. Int. J. Econ. Financ. Issues 6(1), 65–72 (2016)

Examining the Impact of Mentoring on Personal Learning, Job Involvement and Career Satisfaction

Roshen Therese Sebastian⬤, L. Sherly Steffi⬤, and Geethu Anna Mathew⬤

Abstract Mentoring is introduced as a magic wand by highly matured organizations to develop the employees personally and professionally. The study investigates the influence of mentoring on personal learning, career satisfaction and job involvement. Research participants including IT professionals from Kerala (N = 390) are selected by multistage sampling method. A structured questionnaire was completed by the respondents, which measured mentoring functions, personal learning, job involvement and career satisfaction. The factors identified from exploratory factor analysis of mentoring functions are career mentoring, psychosocial mentoring and role modeling function. Path analysis was exercised to examine the hypothesis evolved from the literature. The study found that career mentoring influences personal learning and career satisfaction, role modelling influences personal learning, job involvement and career satisfaction, personal learning influences job involvement and career satisfaction. The study also revealed that job involvement results in career satisfaction and it is the strongest relationship when compared with other variables.

Keywords Career mentoring · Psychosocial mentoring · Role modelling · Personal learning · Job involvement · Career satisfaction

1 Introduction

The Indian IT industry has attained an inevitable position in the economy and is considered as a pivotal contributor to India's economic achievements. The gorgeous feature of Indian IT industry is the availability of competent and talented manpower at much lesser costs to other developing destinations in the world. However, the IT industry is encountering pressing issues affecting its existing and potential human capital inside the business like high employee turnover, generational dissimilarities etc. The crucial challenge is that major companies worldwide must consider high

---

R. T. Sebastian (✉) · L. Sherly Steffi · G. A. Mathew
Kristu Jayanti College Bangalore, K. Narayanapura, Kothanur, Bangalore 560077, India
e-mail: roshen.ts@kristujayanti.com

© The Author(s), under exclusive license to Springer Nature Switzerland AG 2024
A. Hamdan and E. S. Aldhaen (eds.), *Artificial Intelligence and Transforming Digital Marketing*, Studies in Systems, Decision and Control 487,
https://doi.org/10.1007/978-3-031-35828-9_83

employee turnover, mostly in different employee divisions. Mentoring is introduced as a magic wand by highly matured organizations to resolve these issues [14]. Companies started largely using mentoring as an inevitable technique for attracting and retaining high-talented employees, new comers, as well as experienced executives.

Today's organizations are facing unprecedented change like technological advancements, competition, diversity of the workforce, downsizing, mergers, acquisitions, restructuring etc. In such a fast-moving working environment, skills and competencies easily get outdated and need to continuously improve. Rather than textbooks and training programs individuals can look into others with experience and expertise to gain skills and competencies required for their job. Mentoring is an unbeatable development intervention which aids individual's progress both professionally and personally.

Many high-flying companies kicked off mentoring program expecting to reap the tangible and intangible gains for all stakeholders. But still, there is a dearth of research base to clarify the associations between mentoring and career outcomes of the employees. Having recognized the gap of extensive researches on mentoring of IT professionals, the research is a minimal, yet significant effort to gauge the mentoring functions received by IT Professionals in the IT parks in Kerala.

The purpose of this study was to (a) examine the mediating role of personal learning in the relationship between mentoring and career satisfaction and between mentoring and job involvement (b) the impact of mentoring on personal learning (c) the impact of mentoring on job involvement (d) the impact of mentoring on career satisfaction.

2 Literature Review and Hypothesis Development

Nowadays organizations spend a lot on resources, be it monetary or non-monetary to improve their human potential. Mentoring at the workspace is recognized as an unbeaten tactic to endorse career development and better human capital. Mentoring is internationally recognized as an influential human resource development (HRD) instrument [13]. Generally, assessment of the works related to mentoring at the workplace proved that the results are optimistic [26]. Lankau and Scandura looked into the antecedents as well as outcomes of personal learning of mentoring associations. "Relational job learning is an amplified acceptance of the interdependence of one's job to others' jobs. Personal skill development is about the gaining of new skills and abilities that will encourage healthier work space connections" [18]. The conclusions of the research confirmed that mentees get better relational job learning. Explicitly, career mentoring displayed a positive association to relational job learning. The research work underwrites a significant facet in the arena of mentoring and learning [18]. "Personal learning is all about attaining knowledge, skills, and/or competencies, which benefits in a person's personal growth". Bradford et al. [4] propose that internal mentoring has noteworthy association with personal learning. When mentees do have reciprocal relationship with mentors, they do notice and impersonate the mentors in work environment. Consequently, the protégé would be able to grasp how the job is related to others which inevitably results in personal learning.

"Career satisfaction refers to a worker's contentment or unhappiness with her/his profession" [19]. Career satisfaction encompasses a subjective valuation concerning the selection and progression of an occupation. It is more intense and broad than job satisfaction and it embraces both internal and external facets of the profession [15]. An assessment of the occupation as a whole reflects career satisfaction. It is proved that mentored employees showed notably advanced levels of career satisfaction than non-mentored employees in a study convened by Aryee and Chayt [3] in public as well as private sector organizations in Singapore. Several studies show that organizational level career development practices such as mentoring facilitates the career success [21] and career adaptability of mentee. Nevertheless, career functions in mentoring may create an optimistic outlook to job and career because informational and instrumental social sustenance is one of the core functions of [2]. This assistance perhaps helps employees to be confident in their career decisions and it augments their efficiency with the help of challenging assignments and coaching which in turn results in better career satisfaction. These findings acclaim that the major paybacks of mentoring can be constructive psychological moods about the career.

It is anticipated that employees who receive mentoring may have better job involvement [17]. Koberg et al. [16] convened a study, which underlines that psychosocial mentoring beneficially results in job involvement. Aryee and Chayt [3] conducted a study and the results suggested that mentored respondents received greatly advanced levels of job involvement. Eby et al. [8] proved job involvement can be improved by mentoring. Salami [23] exposed that satisfaction towards mentoring estimated job involvement. H.P.L.P & J found out that formal mentoring is most effective than informal mentoring in Sri Lanka apparel industry.

Learning goal orientation means aim to amplify one's ability by grooming skills. According to Egan [9], Godshalk and Sosik [10] people who are having high-learning goal-orientation showed superior targets and career satisfaction in mentoring. Individuals having a high learning goal orientation will attain the talents that are required in a person's personal as well as professional growth. Therefore, we can assume that personal learning may also lead to career satisfaction (Fig. 1).

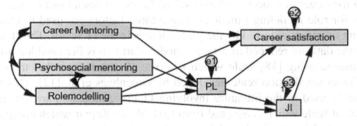

Fig. 1 *Sources* Allen et al. [2], Goulet and Singh [11], Lankau and Scandura [18], Sebastian and Zacharias [25]

From the literature, it can be hypothesized that:

H1: Career Mentoring have significant effect on Personal Learning.
H2: Career Mentoring have significant effect on Career Satisfaction.
H3: Psychosocial Mentoring have significant effect on Personal Learning.
H4: Role Modelling have significant effect on Career Satisfaction.
H5: Role Modelling have significant effect on Personal Learning.
H6: Role Modelling have significant effect on Job Involvement.
H7: Personal Learning have significant effect on Career Satisfaction.
H8: Personal Learning have significant effect on Job Involvement.
H9: Job Involvement have significant effect on Career Satisfaction.

3 Research Methodology

Nowadays in order to reduce the adverse effect of generational tension, organizations have incorporated mentoring as an organizational strategy. So that senior employees will be valued more and junior employees will be engaged. Therefore, it is the high time to find out how employees are utilizing mentoring in terms of personal learning, career satisfaction and job involvement. Data has been collected from employees working in the IT sector in Kerala. The sampling approach employed in the study is multistage sampling. 390 employees were selected into the sample basket through power analysis.

A structured questionnaire was designed with four parts for data collection. Part one is used to measure the mentoring functions received by the employees. Mentoring functions received was assessed using the mentoring functions scale developed by Dreher and Ash [6]. The questionnaire consisted of 18 statements specifying the functions like sponsorship, exposure and visibility, protection, provision of challenging assignments, acceptance and confirmation, counseling, and personal friendship and role modeling. An exploratory factor analysis is applied in this study to find out the underlying dimensions in the set of statements relating to the mentoring functions received by the respondents. The factors identified are career mentoring, psychosocial mentoring and role modeling function and the factor loadings range from 0.455 to 0.789 for career mentoring, 0.540 to 0.770 for psychosocial mentoring and 0.484 to 0.585 for role modeling functions. These three factors are used to examine the mentoring functions received by the professionals in the IT sector. Thus the 18 variables in the data are reduced to 3-factor model. Part two is Personal learning which was measured using [18] scale which involves 12 statements. Part three consists of the Career satisfaction scale developed by Greenhaus et al. [12]. This scale has been widely used in the literature involving IT professionals. Part four is the job involvement scale which is adopted from Lodahl and Kejnar which comprises nine questions. The hypothesized model was tested with path analysis.

3.1 Statistical Analysis and Results

The employees' age group is ranging from 20 to 45, and responses from males (199) and females (191) are almost equal. The mean value of the mentoring function is 3.31 and the standard deviation is 0.697. The outcomes measured in the study are personal learning, career satisfaction, and job involvement. The mean value of personal learning is 3.77 and standard deviation is 0.413, career satisfaction is 3.33 and standard deviation is 0.673, job involvement is 3.25 and standard deviation is 0.484. The mean values are in the neighborhood of three, which highlights that the respondents have a favorable opinion towards the mentoring functions received by them as well as towards personal learning, career satisfaction and job involvement.

3.2 Model Testing

Path analysis was executed to examine the model framed from literature. The path model established acceptable fit to the data, (Chi-square value is 4.652 ($p = 0.199$), GFI of 0.996, an AGFI of 0.972, a normed fit index of 0.994, a comparative fit index of 0.998).

Hypothesis one predicts that career mentoring will have an impact on personal learning. Explicitly it suggests that career mentoring helps them to learn more about their job. This hypothesis is supported (B $= 0.111$, p < 0.05) along with the study of which states that when challenging assignment and exposure are provided to the mentee it enhances the experiential learning as well as overall learning capabilities of the mentee. The previous studies [18] substantiates the result. Hypothesis two predicts that career mentoring will have an impact on career satisfaction. From table one it is revealed that (B $= -0.156$, p < 0.05) indicating a significant relationship. Hypothesis three is not supported, specifically psychosocial mentoring will not have an impact on personal learning (B $= 0.039$, p > 0.05). Hypothesis four is accepted which predicts that role modelling influences career satisfaction, (B $= 0.168$, p < 0.05) and the result is reinforced by the study of [2]. Hypothesis five is supported from the results, (B $= 0.159$, (p < 0.05), so we can say that role modelling influences personal learning and the result is parallel with the studies of [4, 27]. Hypothesis six predicts that role modelling influences job involvement. Table one reveals that that (B $= 0.218$, p < 0.05) and the results are matching with previous literatures [8, 17]. Hypothesis seven assumes that personal learning influences career satisfaction. The hypothesis clearly says that when the respondents are able to learn about their job, they will have more career satisfaction. The results support the hypothesis with (B $= 0.223$, p < 0.05) and in agreement with past studies like [7, 10] etc. Hypothesis eight assumes that personal learning influences job involvement. Explicitly it suggests that learning more about the job helps to improve job involvement. This hypothesis is supported with the values (B $= 0.242$, p < 0.05). Hypothesis nine predicts that job involvement influences career satisfaction and the values (B $= 0.692$, p < 0.05)

supports that. The finding is in accordance with prior literatures like [5, 11, 22] (Table 1).

From Table 2, it was found that the indirect (mediated) effect of role modelling on job involvement is 0.055 and it has mediated effect on career satisfaction (0.220). Psychosocial mentoring has an indirect effect on job involvement (0.014) and career satisfaction (0.016). Career mentoring has an indirect effect on Job involvement (0.039) and career satisfaction (0.045). Personal learning has an indirect effect on career satisfaction (0.103).

Table 1 Model testing

			Estimate	S.E	C.R	P
PL	<--	Career mentoring	0.111	0.039	2.826	0.005
PL	<--	Psychosocial mentoring	0.039	0.034	1.134	0.257
PL	<--	Role modelling	0.159	0.034	4.613	***
JI	<--	PL	0.242	0.059	4.133	***
JI	<--	Role modelling	0.218	0.035	6.264	***
Career satisfaction	<--	PL	0.223	0.076	2.950	0.003
Career satisfaction	<--	JI	0.692	0.063	11.001	***
Career satisfaction	<--	Career mentoring	−0.156	0.051	−3.095	0.002
Career satisfaction	<--	Role modelling	0.168	0.053	3.173	0.002
	X^2					4.652
	P					0.199
	GFI					0.996
	AGFI					0.972
	NFI					0.994
	CFI					0.998

Table 2 Standardized indirect effects

	Rolemodelling	Psychosocial mentoring	Career mentoring	Personal Learning	Job involvement
Personal learning	0.000	0.000	0.000	0.000	0.000
Job involvement	0.055	0.014	0.039	0.000	0.000
Career satisfaction	0.220	0.016	0.045	0.103	0.000

4 Discussion

The purpose of the study is to evaluate the influence of mentoring functions on personal learning, career satisfaction and job involvement along with the influence of personal learning on career satisfaction and job involvement. It is also hypothesized that job involvement has an influence on career satisfaction. The results are in congruence with the previous researches. Kram [17] found that the challenging assignments, which characterize career function, may help the employee to learn more about the job than from any other training programs. The mentors through continuous feedback can support this learning process. Mentors, usually senior employees have more experience and knowledge that help them to find balance between personal life and professional life, also are good in networking with others, maintaining a proper relationship and creating visibility in front of higher officials. Mentees can be motivated by these characteristics and can consider their mentors as role models. In addition to that access to highly experienced mentors embed a feeling of security in protégés, so they may pursue knowledge and actively try to share ideas [20]. From the findings, researchers suggest that protégés learn through role modelling by observing and duplicating. The results reinforce prior studies, which have strongly debated that mentors perform as role models who make protégés learn from them [24]. By observing mentors' behaviors and other mentoring relationships, protégés establish their schemas and in doing so focusing their expectations, behaviors, satisfaction, and relationship eminence perceptions in mentoring relationships [18]. The findings of the previous studies. Protégés who are sponsored for projects or challenging assignments that increase their exposure to other people in their organization may be more likely to develop a "macroscopic" understanding of the organization through these networks. Mentor's acceptance and confirmation can assist learning by empowering the mentee to experiment with novel behaviors. Counseling behaviors may serve like a strong pillar, explaining own experiences, and assisting to solve problems may equip protégé to adapt to problems effectively. These findings are in congruence with the results of the study by Kram [17].

Allen et al. [1] highlights that the support from the mentors may assist employees to take decisions in their career ladder with more confidence and coaching and challenging job assignments enhance their career-related efficiency, which results in gratification to their career. Concurrently based on the findings of the current study, the researchers have to point out that mentoring beyond a level could be counterproductive as the employee may feel a lack of opportunities or challenges to experiment his potential. That may hinder the satisfaction towards career.

Challenging assignments may facilitate one's growth needs and therefore, identification with the job. Interestingly, the results are in congruence with the previous studies [3] which state that respondents who received mentoring reported significantly higher levels of job involvement than non-mentored respondents. The results reinforce previous studies [8, 17, 23], and in which mentoring predicts job involvement. The analysis findings revealed that role modeling function is the significant predictor of job involvement. The values, attitudes, and behavior of the mentor may influence the protégé to imitate those personalities and to involve best in their job [25].

Learning provides a conducive environment for developing essential capabilities that will create a positive outlook about their job. Enhancing skills that are pertinent to profession will increase career satisfaction Personal learning helps the employee to study in detail about the job and how the job is related to other jobs in the organization. It aids to understand the responsibilities and demands of the job and consequently people will be more involved and engaged in the job. The higher involvement in the job boosts his sense of responsibility and ownership, which in turn results in a delighted career.

5 Implications

This study will be a revelation to IT companies, which have not incorporated mentoring as part of their HR practice. Not only does mentoring relate to individual employee learning but the results also suggest that mentoring influences career satisfaction and job involvement. Therefore, the findings of the study provide insight to organizations to develop a proper structure for the execution of the mentoring program. The study results indicate that, through mentoring individuals can realize the dream of achieving specific competencies for the job seamlessly. Managers can administer mentoring on a regular basis to help the mentees learn about their job as well as the organizational culture. The bearing of the study emanates from the finding that mentored employee's testimonies job involvement and career satisfaction. This points out that mentoring is a prospective approach that organizations can employ to upgrade outcomes. In this volatile, uncertain, complex and ambiguous world, personal learning helps employees to improve their job involvement and career satisfaction. Therefore, organizations can seed learning culture for better job involvement and career satisfaction.

Specifically in the time of covid-19, employees are mostly working from home in IT sector, they lack the warmth of the presence of colleagues and other engagement activities at work spaces. This can increase the chances of frustration and demotivation. The organizations can make use of the mentoring programs to make the employees aware about all updates and engage them in meaningful conversations. Employees are expecting a reassurance that they will not lose their job and will receive paid leave if they affected with virus. Through mentoring programs, employees can be updated on a regular basis so that information can be learned to the point. Building a sense of hope is extremely important for their engagement and performance. It helps to develop an involvement in the job. Also the support from mentors may motivate the employees to accomplish career objectives during the time of pandemic which helps to develop career satisfaction.

6 Limitations and Recommendations of the Future Research

Since a dearth of studies linked to mentoring in the organizational context has found out, the prospective researchers could conduct a longitudinal study analyzing outcomes in various phases. This would be a significant input to the garden of information. The study was conducted from the mentee's (protégés) perspective and slight significance has been placed on the mentor's side. The study would have been more accurate and reliable if one considers mentors perspective also. A comparative study can be conducted between various service industries like the banking sector, IT, healthcare, insurance, consulting etc. to identify the differences existing across industries.

References

1. Allen, T.D., Eby, L.T., Poteet, M.L., Lentz, E., Lima, L.: Career benefits associated with mentoring for protégée: a meta-analysis. J. Appl. Psychol. **89**(1), 127–136 (2004). https://doi.org/10.1037/0021-9010.89.1.127
2. Allen, T.D., McManus, S.E., Russell, J.E.A.: Newcomer Socialization and stress: formal peer relationships as a source of support. J. Vocat. Behav. **54**(3), 453–470 (1999). https://doi.org/10.1006/jvbe.1998.1674
3. Aryee, S., Chayt, Y.W.: An examination of the impact of career-oriented mentoring on work commitment attitudes and career satisfaction among professional and managerial employees. Br. J. Manag. **5**(4), 241–249 (1994). https://doi.org/10.1111/j.1467-8551.1994.tb00076.x
4. Bradford, S.K., Rutherford, B.N., Friend, S.B.: The impact of training, mentoring and coaching on personal learning in the sales environment. Int. J. Evidence Based Coach. Mentor. **15**(1), 133–151 (2017)
5. Doobree, D.: Job involvement of bank managers in Mauritius (2009)
6. Dreher, G.F., Ash, R.A.: A comparative study of mentoring among men and women in managerial, professional, and technical positions. J. Appl. Psychol. **75**(5) (1990)
7. Dweck, C.S., Leggett, E.L.: A social-cognitive approach to motivation and personality. Psychol. Rev. **95**(2), 256–273 (1988). https://doi.org/10.1037/0033-295X.95.2.256
8. Eby, L.T., Allen, T.D., Evans, S.C., Ng, T., Dubois, D.L.: Does mentoring matter? A multidisciplinary meta-analysis comparing mentored and non-mentored individuals. J. Vocat. Behav. **72**, 254–267 (2008). https://doi.org/10.1016/j.jvb.2007.04.005
9. Egan, T.M.: the impact of learning goal orientation similarity on formal mentoring relationship outcomes. Adv. Dev. Hum. Resour. **7**(4), 489–504 (2005). https://doi.org/10.1177/1523422305279679
10. Godshalk, V.M., Sosik, J.J.: Aiming for career success: the role of learning goal orientation in mentoring relationships. J. Vocat. Behav. **63**(3), 417–437 (2003). https://doi.org/10.1016/S0001-8791(02)00038-6
11. Goulet, L.R., Singh, P.: Career commitment: a reexamination and an extension. J. Vocat. Behav. **61**(1), 73–91 (2002). https://doi.org/10.1006/jvbe.2001.1844
12. Greenhaus, J.H., Parasuraman, S.P., Wormley, W.M.: Effects of race on organizational experiences, job performance evaluations, and career outcomes. Acad. Manag. J. **33**(1), 64–86 (1990)
13. Hegstad, C.D., Wentling, R.M.: Organizational antecedents and moderators that impact on the effectiveness of exemplary formal mentoring programs in fortune 500 companies in the United

States. Hum. Resour. Dev. Int. **8**(4), 467–487 (2005). https://doi.org/10.1080/136788605001 99808

14. Hezlett, S.A.: Mentoring and human resource development: where we are and where we need to go. Adv. Dev. Hum. Resour. **7**(4), 446–469 (2005). https://doi.org/10.1177/152342230527 9667

15. Judge, T.A., Cable, D.M., Boudreau, J.W., Bretz, R.D.: An Empirical Investigation of the Predictors of Executive Career Success. NY (1994). Retrieved from http://digitalcommons.ilr. cornell.edu/cahrswp

16. Koberg, C.S., Boss, R.W., Goodman, E.: Factors and outcomes associated with mentoring among health-care professionals. J. Vocat. Behav. **53**(1), 58–72 (1998). https://doi.org/10.1006/jvbe.1997.1607

17. Kram, K.E.: Mentoring at Work: Developmental Relationships in Organizational Life, p. 1985. Scott, Foresman (1985)

18. Lankau, M.J., Scandura, T.A.: An investigation of personal learning in mentoring relationships: content, antecedents, and consequences. Acad. Manage. J. **45**(4), 779–790 (2002). https://doi.org/10.2307/3069311

19. Lounsbury, J.W., Steel, R.P., Gibson, L.W., Drost, A.W.: Personality traits and career satisfaction of human resource professionals. Hum. Resour. Dev. Int. **11**(4), 351–366 (2008). https://doi.org/10.1080/13678860802261215

20. Morrison, E.W.: Newcomer information seeking: exploring types, modes, sources, and outcomes. Acad. Manage. J. **36**(3), 557–589 (1993). https://doi.org/10.2307/256592

21. McDonald, K.S., Hite, L.M.: Conceptualizing and creating sustainable careers. Hum. Resour. Dev. Rev. **17**(4), 349–372 (2018)

22. O'Driscoll, M.P., Randall, D.M.: Perceived organizational support, satisfaction with rewards, and employee job involvement and organizational commitment. Appl. Psychol. **48**(2), 197–209 (1999). https://doi.org/10.1111/j.1464-0597.1999.tb00058.x

23. Salami, D.S.O.: Mentoring and work attitudes among nurses: the moderator roles of gender and social support. Eur. J. Psychol. **6**(1), 102–126 (2010). https://doi.org/10.5964/ejop.v6i1.174

24. Scandura, T.A.: Mentorship and career mobility: an empirical investigation. J. Organ. Behav. **13**(2), 169–174 (1992). https://doi.org/10.1002/job.4030130206

25. Sebastian, R.T., Zacharias, S.: A Study on Mentoring Among IT Professionals with Special Reference to IT Parks in Kerala. University. Kottayam (2018)

26. Wanberg, C.R., Welsh, E.T., Hezlett, S.A.: Mentoring research: a review and dynamic process model. Res. Personnel Hum. Resour. Manage. **22**(OCTOBER 2003), 39–124 (2003). https://doi.org/10.1016/S0742-7301(03)22002-8

27. Weinberg, F.J.: How and when is role modeling effective? The influence of mentee professional identity on mentoring dynamics and personal learning outcomes. Group Org. Manag. **44**(2), 425–477 (2019). https://doi.org/10.1177/1059601119838689

Islamic Values Impact on Managerial Autonomy

April Lia Dina Mariyana, Ariq Idris Annaufal, and Muafi

Abstract This study aims to explore the impact of Islamic values. This study focuses on the relationship between Islamic Values and Managerial Autonomy from an individual and organizational perspective. Based on previous research, it is found that autonomy can increase the sense of responsibility, ownership of organizational issues, employee efficiency, innovation performance, subsidiary development, and flexibility. However, further studies on Islamic Values on Managerial Autonomy in individual and organizational views need to be explored to develop a broader theory with new perspectives.

Keywords Islamic values · Managerial autonomy · Ownership · Efficiency · Flexibility

1 Introduction

One form of development in the science of human resource management is managerial autonomy or freedom in doing work. Gómez-Mejía et al. [1] states that employees have the amount of freedom, independence, and discretion in areas such as scheduling work, making decisions, and determining how to do the job. Autonomy in the workplace is essential to enhance employee performance. Companies give them more opportunities for employees to organize their working life. As a result, they feel motivated to carry out their daily routines and develop their skills.

Islamic values lead humans to the path desired by Allah. This makes Muslims who believe will always try to be the best individual because they believe that Allah is always watching. Islam regulates all aspects of life clearly and systematically,

---

A. L. D. Mariyana · A. I. Annaufal · Muafi (✉)
Department of Management, Universitas Islam Indonesia, Daerah Istimewa Yogyakarta 55281, Indonesia
e-mail: muafi@uii.ac.id

© The Author(s), under exclusive license to Springer Nature Switzerland AG 2024
A. Hamdan and E. S. Aldhaen (eds.), *Artificial Intelligence and Transforming Digital Marketing*, Studies in Systems, Decision and Control 487,
https://doi.org/10.1007/978-3-031-35828-9_84

emphasizing values and their implementation [2]. Religion influences people's attitude formation, choices, and actions [3]. Ethics in Islam is defined as sound principles and values sourced from Islam [4].

The Quran was revealed as a book of guidance covering the fields of creed, sharia, and morals. However, beyond these three instructions, the Quran provides motivation and inspiration for Muslims in various areas of life. Allah SWT says that: This is the Book! There is no doubt about it –a guide for those mindful 'of Allah' (QS. Al-Baqarah: 2). The Quran is a book of guidance that shapes social life with a balance between material and spiritual so that the Quran becomes the soul of a person's life. The guidance given by the Qur'an can be a guide if given freedom at work, as in a company with a decentralized organizational structure.

Companies with a decentralized organizational structure provide more autonomy to employees. Employees and lower-level management have more flexibility to manage their own lives. Meanwhile, upper management is more involved in strategic aspects such as target planning and strategy. On the other hand, such opportunities are hard to find under authoritarian leadership. Leaders dominate in making decisions, managing work, and directly supervising employees. HRM decentralization improves organizational performance, a result that is consistent with existing research arguments that the practice of delegating work process authority improves organizational performance [5].

Religious values should be the moral force and moral dynamic in one's activities [6]. Religion will influence work values and employee expectations which are reflected in the behaviors and approaches chosen [7]. Each person has a set of values, even if they cannot be expressed explicitly. Ethical behavior is required by society so that everything can run in an orderly manner. Values in work include honesty, integrity, keeping promises, loyalty, fairness, respect, care for others, responsibility, and giving your best.

Ethics is a little different from morality. Morality is a system of values about how we should live as human beings. It can be defined as a tradition of beliefs about good and bad behavior in religion and culture. Morality gives people concrete rules or instructions on how to live and tells people to avoid bad behavior. Ethics and morality are intertwined because ethics ultimately calls on people to act according to morality, but not because the action is commanded that it is suitable for its own person [2]. Mentioning that morals consist of Creativity, honesty and trust, personal fulfillment, commitment, motivation and job satisfaction, organizational commitment, emotional development, spiritual competence, encouraged holistic ways of working, developed community at work, empowered the workforce and human society, risk aversion and ethics, stress management, and career development.

The values in the Qur'an and the sayings and practices of Prophet Muhammad (Sunnah) include equality, accountability, hard work, fairness, consultation, trust, self-discipline, perseverance, and cooperation [8–10]. Theoretical research, such as that conducted by [11] and [12], supports the importance of Islam. Business and management as case studies that hypothetically implement Islamic principles [3].

Research Objectives

Along with the development of human resource management science, Islamic values need to be instilled to get Allah's blessing from our activities. This research will discuss about the impact of Islamic values on managerial autonomy or freedom in performing managerial roles. There can be some adverse effects rule does not bind someone rule then someone will tend to do things that are not ethical. Islamic values must be instilled to prevent this from happening in an organization, especially to decision-makers or managers.

This study aims to examine and analyze Islamic values toward managerial autonomy. This study has three main sub-sections, namely:

- Introduction, which explains the theoretical issues and business phenomena related to the practice of managerial autonomy based on Islamic values.
- Literature Review and Methods, which includes theories on Islamic values and managerial autonomy.
- Results, Discussion, and Limitations of the Study.

2 Research Methodology

To support claims or hypotheses already developed, research on the impact of Islamic values on human and organizational elements might be done using a quantitative technique. If the discussion of Islamic values is conducted within the context of Islamic organizations that function in accordance with Islamic law and for the good of the populace, it will be highly relevant. This aims to determine how high the level of Islamic Values is in the ranks of managers and whether Islamic Values can greatly impact organizational and individual activity. Additionally, the survey approach, which entails conducting direct interviews with managers to gather accurate and reliable data in support of claims or study hypotheses, will be more appropriate for research on Islamic values.

3 Propositional Framework

See Fig. 1.

3.1 Islamic Values Can Increase the Decision Effectiveness of Manager

One of the important values in Islamic teachings that can be accepted by society is honesty [13]. Integrity has a general meaning about moral uprightness. Their research found that transformational leaders are perceived to have more integrity and are more effective than non-transformational leaders [14]. A true leader must behave with integrity in this sense by being an honest and ethical individual, someone whose every word and deed is consistent [15]. Applying Islamic values in business management will create positive cultural values towards employees' Islamic behavior, including: Doing good (Itqan and Ihsan), Doing good, and Taqwa (Iffah).

As the link between top management and the sales force, sales managers who use Managerial Autonomy practices can influence the sales force and organizational performance. [16]. Managerial Autonomy practices can also be used to identify

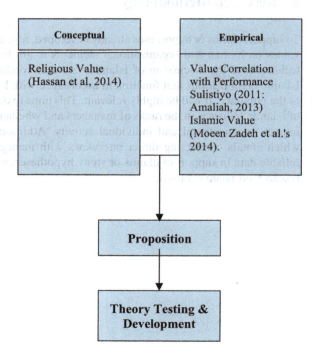

Fig. 1 The conceptual framework

specific ways to adapt sales managers' behavior to improve SMEs. [17]. The psychological state of self-efficacy can also provide an understanding and strengthen the employee engagement-managerial effectiveness relationship. As such, it provides added value to work outcomes and management development. [18].

P1: *Islamic Values Positive Impact on Manager's Effectiveness.*

3.2 Islamic Values Can Increase Organizational Decision

Observing ethics and moral values has become one of the critical phenomenon and received the attention of most organizations [19]. The Islamic context discusses the importance of implementing Islamic values in organizations [19]. Business decision-makers are free to make choices, but religious principles provide the proper framework for carrying out those choices [20]. Religious values create a framework for decision-makers to make the right choices. Hence, Muslims are driven by their beliefs, known as faith, which leads them to adopt Islamic practices [20]. Description of decision-making styles in Islam that emphasize the principle of deliberation (Shura), and core Islamic values such as the principles of trust (Al-Amanah), honesty (Al-sisi), justice, teamwork, and cooperation (Al-Ta'waan), justice in dealing with employees (Al-adl), and perfection/excellence (Al-Ikhlas) [21]. Studies related to spirituality show that the practice of religious values drives underlying ethical behavior in the decision-making process [22]. Religiosity has also significantly influenced Muslim consumer decision-making concerning halal tourism consumption [23].

Maitlis and Ozcelik [24] argue that rational models of organizational decision-making imply the possible role of emotions. An important gap in the research concerns how anticipatory emotions influence the decision-making process and how emotions experienced after making a decision influence subsequent decision. The development of Artificial Intelligence technology also helps organizations to make a decision. AI-based decisions in organizations become relatively effective when some level of transparency or decision interpretation can be achieved. Managers must keep up with AI developments that can be interpreted and explained [25]. This is added to upper management as one of many groups involved in intuitive decision-making in organizations. Many knowledge workers, even at lower and non-managerial levels, constantly find themselves in new situations (particularly characterized by uncertainty and vagueness) and therefore require visionary and intuitive thinking [26].

P2: *Islamic Values Positive Impact on Organizational Decision.*

3.3 Islamic Values Can Increase Job Satisfaction

There is an overall view that Islam has a 'real' presence in the workplace in individuals [8]. Job satisfaction in the concept of conventional economics is a sense of pleasure that individuals feel because of the achievement of their expectations [27]. Imam [28] Job satisfaction in the concept of conventional economics is a sense of enjoyment that individuals feel because of the achievement of their expectations. Recent research on Islamic work ethics on job satisfaction and organizational commitment mediated by intrinsic motivation states a positive relationship to intrinsic motivation. The relationship becomes positive when mediated by intrinsic motivation. Conversely, the relationship is negative when work ethic is directly related to organizational commitment [29]. Research [2] mentioned the positive relationship of Islamic work ethics to organizational commitment and job satisfaction.

P3: *Islamic Values Positive Impact on Job Satisfaction.*

4 Discussion

One of the important values in Islamic teachings that can be accepted by society is honesty. A true leader must behave with integrity in this sense by being an honest and ethical individual, someone whose every word and deed is consistent. Business decision-makers are free to make choices, but religious principles provide the right framework to make those choices. Religious values create a framework for decision-makers to make the right choices. Hence, Muslims are driven by their beliefs, known as faith, which leads them to adopt Islamic practices.

Description of Islamic decision-making styles that emphasize the principle of deliberation, and core Islamic values such as the principles of trust, honesty, fairness, teamwork, and cooperation, fairness in dealing with employees, and excellence. Studies related to spirituality show that the practice of religious values drives the underlying ethical behavior in the decision-making process. Religiosity has also significantly influenced Muslim consumers' decision-making concerning halal tourism consumption. An important gap in the research concerns how anticipatory emotions influence the decision-making process and how emotions experienced after making a decision influence subsequent decisions.

Managers need to keep up to date with AI that can be interpreted and explained. This is added to Upper management is one of many groups involved in intuitive decision-making in organizations. Many knowledge workers, even at lower and non-managerial levels, constantly find themselves in new situations and require visionary and intuitive thinking. Job satisfaction in the conventional economic concept is the sense of satisfaction that individuals feel due to the achievement of their expectations.

The highest source of satisfaction for an individual is when he can learn more about God. Recent research on Islamic work ethics on job satisfaction and organizational

Islamic Values Impact on Managerial Autonomy

commitment mediated by intrinsic motivation state a positive relationship to intrinsic motivation.

5 Implication

The theoretical implication of this paper is that it can be a foundation for further research to deepen the impact of freedom of management on corporate consumers. This is because consumers are also important stakeholders in building a business. This paper is also expected to answer the importance of Islamic values on managerial autonomy in running a business.

The managerial implication that can be obtained is to know the importance of managers who use managerial autonomy practices to hold Islamic values tightly. The company or top management is also obligated to bring employees closer to holding Islamic values to have a good impact on the practice of managerial autonomy. The above proposition states that Islamic values can increase the effectiveness of managers in doing their work, how they make ethical decisions and increase satisfaction at work.

6 Conclusion

One of the important values in Islamic teachings that can be accepted by society is honesty. A true leader must behave with integrity in this sense by being an honest and ethical individual, someone whose every word and deed is consistent. Business decision-makers are free to make choices, but religious principles provide an appropriate framework for implementing those choices. Studies related to spirituality show that the practice of religious values drives the underlying ethical behavior in the decision-making process. Religiosity was also shown to have a significant influence on Muslim consumer decision-making concerning making decisions in consuming halal tourism. Many employees, even at lower and non-managerial levels, constantly find themselves in new situations that require visionary and intuitive thinking complemented by Islamic values.

This relationship has been demonstrated by a number of pertinent studies. However, thorough research has yet to be explicitly found regarding Islamic Values on Managerial Autonomy. There are still just a few studies on the subject on Islamic Values on Managerial Autonomy, and they have never been thoroughly used to explain a variety of occurrences in the context of people and organizations. Therefore, it is crucial to create the notion of Islamic spiritual intelligence so that it can be used in regions or nations where the majority of the people is Muslim.

References

1. Gómez-Mejía, L.R., Balkin, D.B., Cardy, R.L.: Managing human resources
2. Rafiki, A., Wahab, K.A.: Islamic values and principles in the organization: a review of literature. Asian Soc. Sci. **10**(9), 1 (2014). https://doi.org/10.5539/ass.v10n9p1
3. Rice, G., Al-Mossawi, M.: The implications of Islam for advertising messages: the middle eastern context. J. Euromarketing **11**(3), 71–96 (2002). https://doi.org/10.1300/J037v11n03_05
4. Al-Aidaros, A.-H., Mohd. Shamsudin, F., Md. Idris, K.: Ethics and ethical theories from an Islamic perspective. Int J Islam Thought 4(1):1–13 (2013). https://doi.org/10.24035/ijit.04.2013.001.
5. Shimanuki, T.: The impact of decentralization in human resource management on organizational performance
6. Uygur, S.: The influence of religion over work ethic values: The case of Islam and Turkish SME owner managers. Brunel Univ. Brunel Bus. Sch. Ph.D. Theses (2009)
7. Leat, M.J.: An exploratory investigation of some work related values among middle managers (2002)
8. Ali, A.: Scaling an Islamic work ethic. J. Soc. Psychol. **128**(5), 575–583 (1988). https://doi.org/10.1080/00224545.1988.9922911
9. Ali, A.J.: The Islamic work ethic in Arabia. J. Psychol. **126**(5), 507–519 (1992). https://doi.org/10.1080/00223980.1992.10543384
10. Ali, A.J., Weir, D.: Islamic perspectives on management and organization. J. Manag. Spiritual. Relig. **2**(3), 410–415 (2005). https://doi.org/10.1080/14766080509518602
11. Kazmi, S.W.: Role of education in globalization: a case for Pakistan (2005)
12. Kalantari, B.: In search of a public administration paradigm: is there anything to be learned from Islamic public administration? Int. J. Public Adm. **21**(12), 1821–1861 (1998). https://doi.org/10.1080/01900699808525370
13. Syi'aruddin, M.A.: Transformasi Nilai-Nilai Ajaran Islam Dalam Karya Sastra
14. Lubis, A., SM, A., Sabrina, H.: Pengaruh loyalitas dan integritas terhadap kebijakan pimpinan di pt. Quantum training centre Medan (2019)
15. Krishnan, V.R.: Transformational leadership and outcomes: role of relationship duration. Leadersh. Organ. Dev. J. **26**(6), 442–457 (2005). https://doi.org/10.1108/01437730510617654
16. Deeter-Schmelz, D.R., Kennedy, K.N., Goebel, D.J.: Understanding sales manager effectiveness Linking attributes to sales force values. Ind Mark Manag (2002)
17. Murphy, W.H., Li, N.: A multi-nation study of sales manager effectiveness with global implications. Elsevier **41**(7), 1152–1163 (2012). https://doi.org/10.1016/j.indmarman.2012.06.012
18. Luthans, F., Peterson, S.J.: Employee engagement and manager self-efficacy. J. Manag. Dev. **21**(5), 376–387 (2002). https://doi.org/10.1108/02621710210426864
19. Nejad, A.A., Yaghoubi, N.M., Doaei, H., Rowshan, S.A.: Exploring the dimensions and components of Islamic values influencing the productivity of human resources from the perspective of Mashhad municipality employees. Procedia—Soc. Behav. Sci. **230**, 379–386 (2016). https://doi.org/10.1016/j.sbspro.2016.09.048
20. Ali, A.J., Gibbs, M.: Foundation of business ethics in contemporary religious thought: the Ten Commandment perspective. Int. J. Soc. Econ. **25**(10), 1552–1564 (1998). https://doi.org/10.1108/03068299810214089
21. Abuznaid, S.A.: Business ethics in Islam: the glaring gap in practice. Int. J. Islam. Middle East. Finance Manag. **2**(4), 278–288 (2009). https://doi.org/10.1108/17538390911006340
22. Aydın, S., Uştuk, Ö. (2020) A descriptive study on foreign language teaching anxiety (2020)
23. Eid, R., El-Gohary, H.: The role of Islamic religiosity on the relationship between perceived value and tourist satisfaction. Tour. Manag. **46**, 477–488 (2015). https://doi.org/10.1016/j.tourman.2014.08.003
24. Maitlis, S., Ozcelik, H.: Toxic decision processes: a study of emotion and organizational decision making. Organ. Sci. **15**(4), 375–393 (2004). https://doi.org/10.1287/orsc.1040.0070

25. Shrestha, Y.R., Ben-Menahem, S.M., von Krogh, G.: Organizational decision-making structures in the age of artificial intelligence. Calif. Manage. Rev. **61**(4), 66–83 (2019). https://doi.org/10.1177/0008125619862257
26. Jarrahi, M.H.: Artificial intelligence and the future of work: human-AI symbiosis in organizational decision making. Bus. Horiz. **61**(4), 577–586 (2018). https://doi.org/10.1016/j.bushor.2018.03.007
27. Sutrisno, E.: *Manajemen Sumber daya Manusia*. Jakarta Kencana Prenada Media Group (2010)
28. Al-Ghazali, A.-G.:*Ihya Ulum al-Din*. Beirut: Dar al-Kutb al-'Ilmiyah, t.th
29. Gheitani, A., Imani, S., Seyyedamiri, N., Foroudi, P.: Mediating effect of intrinsic motivation on the relationship between Islamic work ethic, job satisfaction, and organizational commitment in banking sector. Int. J. Islam. Middle East. Finance Manag. **12**(1), 76–95 (2019). https://doi.org/10.1108/IMEFM-01-2018-0029

An Empirical Corroboration on Perceived Facilitations for Training and Affective Organisational Commitment

Geethu Anna Mathew ⓘ, K. Opika ⓘ, and Roshen Therese Sebastian ⓘ

Abstract The key determination of this empirical study is to analyse the affiliation between Perceived Facilitations for Training and Affective Organisational Commitment of employees of star rated hotels. A survey was conducted among the employees star rated hotels and the sample size is 280 employees based on convenience sampling. Perceived facilitations for training can be categorized into three namely; perceived access to training, perceived trainer quality and perceived motivation for training and their affiliation with affective organizational commitment were investigated. The hypotheses were tested using correlation and multiple regression analysis. The results revealed that perceived facilitations for training have significant relationship with affective commitment of employees of star rated hotels in Karnataka. The implications of the study have been offered for researchers and practitioners of human resource as how to utilize the aspects of facilitations for training to augment the organizational commitment of employees.

Keywords Employee training · Perceived facilitations for training · Affective organizational commitment · First section

1 Introduction

Service quality is an aspect which gives competitive advantage to hotels to sustain in the competition. Retaining talents in the industry is major concern. "Lashley and Chaplain [10] indicated that high employee turnover in hotels is an important factor influencing productivity, workplace efficiency and hotel cost structure and thereby affecting the service quality". The industry requires availability of skilled professionals all the times. The employee communication and close interactions with the customers leads to guest satisfaction in hotels. Employees are becoming progressively valuable assets to the hotel industry [1]. Nowadays hotels are giving

---

G. A. Mathew · K. Opika · R. T. Sebastian (✉)
Department of Management, Kristu Jayanti College, Bangalore, India
e-mail: roshen.ts@kristujayanti.com

© The Author(s), under exclusive license to Springer Nature Switzerland AG 2024   1017
A. Hamdan and E. S. Aldhaen (eds.), *Artificial Intelligence and Transforming Digital Marketing*, Studies in Systems, Decision and Control 487,
https://doi.org/10.1007/978-3-031-35828-9_85

more emphasis on highbrow resources that is human capital rather than monetary elements [9]. According to Branham [5] retaining employees begins with improving their commitment to their organisation. Researchers in the field recommend that employees productivity as a result of training can be weighed by understanding the affiliation of perceived facilitations for training and affective organisational commitment [2].

Organisations can ensure that suitable and effective activities as part of training are provided to their human capital so that their skills improve and that in turn leads to achieving organisational goals and aims. Commitment and dedication of employees plays a pivotal role in this context. The major tenacity of this research is to survey the affiliation amongst perceived facilitations for training to affective organisational commitment of workforces of star rated hotels in Karnataka.

2 Significance of the Study

A hotel is a service sector industry. Service quality is an aspect which gives competitive advantage to hotels to sustain in the competition. Retaining talents in the industry is major concern. The industry survives only with highbrow resources that are human capital to deliver admirable service to their customers. Attitude, productivity and behaviour of human capital that is employees are key aspects to reach the service quality expectations of customers. Relevant training activities enhance the Attitude, productivity and behaviour of human capital.

Even though with the reluctance in investment in training activities, nowadays hotels are taking initiatives for training employees because of the above mentioned reasons [9, 17. Committed human capital is essential for an organisation to survive in the competitive world. Providing training prospects to the employees is one way to enhance and improve their commitment towards organisation. Employees feel that they organisation is taking care of them through the development prospects offered. The present study relevance lies in these contexts. The major tenacity of this research is to survey the affiliation amongst perceived facilitations for training to affective organisational commitment of workforces of star rated hotels in Karnataka.

3 Theoretical Framework and Hypothesis Formulation

3.1 Affective Organizational Commitment

"*Affective component of organisational commitment is based on employees' emotional attachment to the organisation*". Affective organisational commitment is demarcated as "*the employee's emotional attachment to, identification with, and involvement in the organisation*" ([11], page no 11). "*Affective commitment is the relative strength of an individual's identification with and involvement in a particular organisation". "Employees who are committed on an affective level stay with the organisation because they observe their personal employment relationship as corresponding to the values and goals of their organisation". "Affective commitment can be considered as an employee's strong belief in and acceptance of the values and goals of the organisation and also his readiness to exert substantial effort on behalf of the organisation, as well as desire to continue as a member of the organisation, Affective organisational commitment is an employee's emotional attachment to his or her organisation.".*

3.2 Perceived Facilitations for Training

Perceived facilitations for training is referred as the factors which facilitates and motivates employees participation in training activities. Perceived facilitations for training can be categorized into three namely; perceived access to training, perceived trainer quality and perceived motivation for training. *"Perceived access to training refers to the extent to which employees perceive that they have access to the training activities required for acquiring the knowledge, skills, and abilities for a current position, and that there are minimal organisational constraints limiting their participation in training"* [2].

"Perceived Motivation for training is the degree to which employees are willing to make efforts to improve themselves and their task and job performances by training" [13]. Well-motivated employees for training are likely to be positive about the training activities in their organisations, which lead to more participation and involvement. Employee's motivation to learn can be deliberated as significant variable connected with the participation and involvement in training initiatives [14]. Previous studies revealed higher the employee motivation for training, more will be the employee's performance in learning. **Perceived trainer quality** is defined as the employees perception on the attributes of trainers and also facilitations done by the trainer at the time of their participation in training activities [7]. Training strategies used to improve the training include employing skilled trainers. Quality trainers are a significant component of training plan of every organisation because, they delivers the training modules and structures to the employees.

3.3 Conceptual Background

See Fig. 1.

The hypothesis of the study were constructed on the conceptual background.

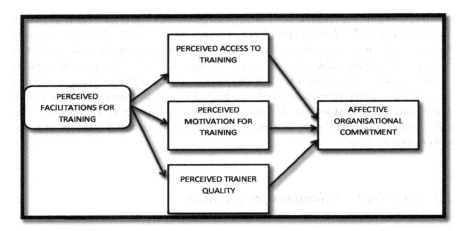

Fig. 1 Conceptual background

4 Methodology and Approach

The major tenacity of this research is to survey the affiliation amongst perceived facilitations for training to affective organisational commitment of workforces of star rated hotels in Karnataka. Primary data requirements of the study are to collect data from employees of star rated hotels in Karnataka. Secondary sources include articles, projects and other printed sources. Primary data requirements are to collect data from workforces of star rated hotels in Karnataka. Tool used for data collection is questionnaire and method adopted was survey. Respondents are from star rated hotels in Karnataka. The population is the workforces of star rated hotels in Karnataka and technique to select samples was convenience sampling. 350 questionnaires were circulated and among that 280 were filled and returned. Respondents profile includes both 150 male and 130 female respondents.

4.1 Measures

The operationalization of perceived facilitations for training measured on a five point scale with 1 = strongly disagree to 5 = strongly agree was based on twenty eight items. Four constructs including perceived access to training (PACT) adapted from; Bartlett [2]; Bulut and Culha [6]; perceived motivation for training (PMOT) adapted from; Bulut and Culha [6], perceived trainer quality (PTQ) adapted from; Chiang et al. [7] are measured under the perceived facilitations for training. The operationalization of affective organisational commitment was from Meyer and Allen [11]. Inferences drawn from statistical tests are at 0.05 level of significance.

5 Results and Discussions

5.1 Reliability Test

"*A Cronbach 's Alpha value of 0.7 and above is considered as reliable*" [12]. The reliability for the perceived access to training scale is 0.759, the reliability of perceived motivation for training is 0.818, reliability of perceived trainer quality is 0.761 and reliability of affective commitment scale is 0.949 respectively. Hence the scale used for the study is reliable.

5.2 Perceived Facilitations for Training and Affective Organisational Commitment

The affiliation between employees perceived facilitations for training and affective organisational commitment are analyzed using Pearson's Correlation Coefficient. The results are discussed below.

Result presented in Table 1 identifies that perceived access to training is significantly correlated with affective organisational commitment ($r = 0.429$, $p < 0.05$). Results are in congruence with the findings by Bashir and Long [4], Bartlett and Kang [3], Bulut and Culha [6] and Dhar [8]. Thus it can be inferred that perceived access to training can enhance commitment of employees to their organisation. Perceived motivation for training is significantly correlated with affective organisational commitment ($r = 0.181$, $p < 0.05$). This shows that employees will be motivated for training, their ability to apply learnt skills result in the relationship with affective organisational commitment. The results also indicates perceived quality of trainer is significantly correlated with affective organisational ($r = 0.620$, $p < 0.05$). Therefore we can summarize that all the three factors of perceived facilitations for training, have a significant affiliation with affective organisational commitment of employees.

5.3 Regression Analysis

The effect of perceived facilitations for training on affective organisational commitment of employees is analyzed using stepwise multiple regression method and the results are furnished in Table 2.

Dependent Variable: Organisational Commitment.

The independent variables, perceived trainer quality ($t = 5.827$, $p = 0.000$), perceived access to training ($t = 9.710$, $p = 0.000$) and perceived motivation for training ($t = 4.865$, $p = 0.000$) are statistically significant and hence the hypotheses that perceived trainer quality, perceived access to training and perceived motivation for training have no significant effect on affective organisational commitment are

1022   G. A. Mathew et al.

Table 1 Correlation analysis

		Affective commitment
Perceived access to training	Correlation coefficient	0.429*
	Significance (2-tailed)	0.000
	N	280
Perceived motivation for training	Correlation Coefficient	0.181*
	Significance (2-tailed)	0.000
	N	280
Perceived quality of trainer	Correlation Coefficient	0.620*
	Significance (2-tailed)	0.000
	N	280

Table 2 Coefficients—significance of perceived trainer quality (PTQ), perceived trainer quality (PTQ) and perceived access to training (PACT), perceived trainer quality (PTQ), perceived access to training (PACT) and perceived motivation for training (PMT) to affective organisational commitment

Model	Dimensions	Beta	t	Sig	Tolerance	VIF
1	(Constant)	0.525	28.283*	0.000	1.000	1.000
	PTQ		13.050*	0.000		
2	(Constant)	0.453	20.281*	0.000	0.955	1.047
	PTQ	0.340	11.932*	0.000	0.955	1.047
	PACT		8.957*	0.000		
3	(Constant)	0.290	14.873*	0.000	0.527	1.898
	PTQ	0.362	5.827*	0.000	0.940	1.063
	PACT	0.237	9.710*	0.000	0.550	1.817
	PMT		4.865*	0.000		

rejected. The contribution of each variable to the model is measured by the beta (β) coefficients. The beta value of perceived access to training is higher ($\beta = 0.362$) therefore it has greater effect on affective organisational commitment followed by perceived trainer quality ($\beta = 0.290$) and perceived motivation for training ($\beta = 0.237$).

6 Scope for Further Research

The major tenacity of this research is to survey the affiliation amongst perceived facilitations for training to affective organisational commitment of workforces of star rated hotels in Karnataka. Organisations differ in its policies and practices hence affiliation will be diverse for diverse organisations. The present study can be replicated in other sectors or other diverse organisations. Other than organisational commitment affiliations with other outcomes such as job involvement, employee engagement etc., can be considered.

References

1. Ariffin, H.F., Che Ha, N.: Examinining Malaysian hotel employees organizational commitment by gender, education level and salary. South East Asian J. Manage. 9(1), 1–16 (2015)
2. Bartlett, K.R.: The Relationship between training and organizational commitment: a study in the health care field. Human Resour. Develop. Quarter., 12(4)
3. Bartlett, K., Kang, D.S.: Training and organizational commitment among nurses following industry and organizational change in New Zealand and the United States. Human Resour. Develop. Int., ISSN 1367–8868 (2004)
4. Bashir, N., Long, C.S.: The relationship between training and organizational commitment among academicians in Malaysia. J. Manage. Develop. 34(10), 1227–1245 (2015)
5. Branham, L.: The 7 Hidden Reasons Employees Leave. AMACOM, New York (2004)
6. Bulut, C., Osman, C.: The effects of organizational training on organizational commitment. Int. J. Training Develop. 14(4), 12 (2010)
7. Chiang, C., Back, K., Canter, D.D.: The impact of employee training on job satisfaction and intention to stay in the Hotel industry. J. Hum. Resour. Hosp. Tour. 4(2), 99–118 (2005)
8. Dhar, R.L.: Service quality and the training of employees: The mediating role of organizational commitment. Tour. Manage. 46(2015), 419–430 (2015)
9. Eskildersen, J.K., Nussler, M.L.: The managerial drivers of employee statisfaction and loyalty. Total Qual. Manag. 11(4–6), 581–588 (2000)
10. Lashley, C., Chaplain, A.: Labour turnover: hidden problem—hidden cost. Hospital. Rev. 1(1), 49–54 (1999)
11. Meyer, J.P., Allen, N.J.: Commitment in the workplace: theory, research, and application' (Thousand Oaks, CA: Sage, 1997)
12. Nunnally, J.C.: *Psychometric Theory*. New York: McGraw—Hill (1978)
13. Robinson, K.A.: Handbook of Training Management. Kogan Page Publisher, London (1985)
14. Tharenou, P., Saks, A.M.: A review and critique of research on training and organizational-level outcomes. Human Resour. Manage. Rev. 17, 251–273 (2007)

An Efficiency Analysis of Private Banks in India—A DEA Approach

Ibha Rani and **Arti Singh**

Abstract This Paper aims to the relative efficiency of private banks by employing the non-parametric approach of "Data Envelopment Analysis (DEA)" for the period of 5 years from 2017 to 2021. There are various models are available, in this study Input oriented under constant Return to Scale (CRS) has been employed. In the DEA model, Banks are considered mediators between saver and borrower, so Fixed Assets, borrowings, and, operating expenses have been considered for input, and Investment, "Interest income and other Income" are considered as output. As per the findings, the performance of new private banks was comparatively better than the old private banks and the average efficiency of the sample bank is 92.4%.

Keywords Efficiency · Technical efficiency · Data envelopment analysis (DEA) · CRS

1 Introduction

An effective banking system is a key sign of a nation's economic advancement.

A strong and effective banking system is required for a developing nation like India to achieve its goal of a $5 trillion GDP, and an effective banking system results in an effective economy. Private Banks are crucial to the growth of the Indian economy. Major changes were made to the banking sector following liberalization. Changes in Economic Policy have made various changes in the banking Sector as well. As per the Narashiman committee's proposal "The Reserve Bank of India (RBI)" proposed the formation of new banks in the category of private banks. In India, public banks are more dominant compared to other sector banks. However, the situation has changed, and new-generation banks have taken a respectable portion in the financial sector

---

I. Rani (✉) · A. Singh
Department of Commerce, Kristu Jayanti College (Autonomous), Bengaluru, India
e-mail: ibha.r@kristujayanti.com

A. Singh
e-mail: arti@kristujayanti.com

© The Author(s), under exclusive license to Springer Nature Switzerland AG 2024
A. Hamdan and E. S. Aldhaen (eds.), *Artificial Intelligence and Transforming Digital Marketing*, Studies in Systems, Decision and Control 487,
https://doi.org/10.1007/978-3-031-35828-9_86

1025

thanks to the application of technology and professional management. Now the entry of foreign banks in India makes this sector more competitive.

Private Banks are those institutions where the majority of the capital is owned privately. In India, private banks are categorized into old and new private banks. Old Private Banks are ones that were around in India when large banks were nationalized but weren't because of their size or for some other reason. These banks received permission to continue operating following the banking reforms, and they have been present in India alongside new commercial banks and government banks.

As per the RBI report for March 2021, Private Banks account for 50% of deposits of financial and nonfinancial companies, RBI quoted. Out of the total Rs. 154.43 lakh crore deposits, private banks account for Rs. 46.23 lakh crore as of March 2021. As per the RBI, the Banking structure comprises "12 public sectors, 21 Private Banks, 42 foreign banks and 43 Regional Rural banks" in India.

Private Banks are categorized as scheduled commercial banks and play a crucial part in a thriving economy by giving people access to resources such as loans and credit. Compared to public sector banks, private banks are known for their smooth banking services. Private Banks in India have experienced issues with their assets, earnings, and expansion in recent years, which has created barriers to their profitability. Due to these serious issues, it is now time to analyze the financial efficiency of Private Banks.

The banking sector changes increased the banks' productivity and efficiency. Before now, banks' efficiency was frequently assessed using the financial ratio. Although it was simple to use, it has received numerous criticisms. In this paper researcher has used DEA "(Data Envelopment Analysis to estimate the efficiency of Private Banks in India".

2   Literature Review

The DEA technique was mostly used to assess the effectiveness of banks in the USA, Europe, and other developed nations, but in recent years, other developing nations such as Spain, Nigeria, Kuwait, and India have also started using this method.

According to De Young (1997), "Data Envelopment Analysis (DEA) is one of the approaches" belonging to a non-parametric approach which is used for the estimation of the efficiency of the performance of the banks. The application of the DEA would help in the estimation of overall technical efficiency (OTE). For this, the DEA approaches use two components that are pure technical efficiency (PTE) and scale efficiency (SE). The evaluation of these variables helps in acquiring better learning about the various sources of inefficiencies that represent in the banks. The estimation of the PTE is done by estimating the position of the frontier that is based upon certain suppositions and values attained from the evaluation of returns–to–scale.

Sherman and Gold (1985) evaluated the operating efficiency of fourteen branches of US saving Banks. This study is considered the first evaluator of DEA in the banking sector.

An Efficiency Analysis of Private Banks in India—A DEA Approach 1027

Rajput and Gupta (2011) have tracked the improvement in the efficiency of international banks operating in India between 2005 and 2011. The results of utilizing the frontier-based nonparametric approach DEA reveal that foreign banks' efficiency has steadily increased during the course of deregulation with few drifts.

Harjinder Singh (2013) assessed the effectiveness of India's public, private, and foreign sector banks from the year 2005 to 2011. He employed two models with different inputs and outputs using the intermediate technique. According to the results, the public sector outperformed the private sector. Because finite resources were used inefficiently to create a certain level of production, there is a significant amount of inefficiency. He has also asserted that management irregularities were a major contributor to the development of new technical inefficiencies.

Premalatha (2017) has studied the performance efficiency of the selected 12 public sector banks in India during the year 2001–02–2015–16. She selected 12 banks based on the weightage of Input (Deposit, Borrowing, and Operating Expenses) and output (Investment and advances). Cross efficiency matrix suggested that the Oriental Bank of Commerce has achieved a maximum time efficiency score of 1.00 and achieved as a benchmark bank.

Hafsah et al. (2020) analyzed the impact of NPA on banking efficiency through a two-stage DEA model for 46 Indian banks for the year 2016. According to this analysis, the Indian banking industry suffers an efficiency loss of 16.2% as a result of NPAs.

The research related to the analysis of the performance efficiency of banks in India is limited due to the unsystematic functioning and structure of banks in India (Berger and Humphrey 1997). This paper focused on the efficiency estimation of Private Banks in India.

2.1 Objective

The major objective is to analyze the technical efficiency of selected 20 Indian Private Banks for the study period from 2017 to 2021.

3 Methodology

In this paper, the DEA method is used to analyze the efficiency of selected Private Banks. DEA is a non-parametric method to measure the efficiency of banking and other financial institutions. It has been originated by Charnes et al. in the year 1978 and was further structured and re-established by Banker et al (1984). Banker et al (1984). The DEA technique is found to be using the principles of linear programming which is based on the process of examining the different variables of Decision-Making Units (DMU) such as a bank that operates relative to other DMUs taken as a sample. The very first application of DEA in the banking sector was done by

Sherman and Gold in the year 1985. Since then, the application of DEA is done in almost all the banks and financial institutions across the country and around the world (Silva et al. 2018). The DEA efficiency score is varying from 0 to 1. A bank scored 1 means the bank is fully efficient and banks having less than 1 are considered the "inefficient bank".

3.1 Models of DEA

There are basis two DEA Model:
  See Fig. 1.

CCR Model
This model is named after its developer Charnes, Cooper, and Rhodes in the year 1978. The efficiency calculated under this model is known as overall technical efficiency (TE). A firm having a maximum score of 1 is known a as "Globally Efficient Bank". A firm having less than 1 score is known as an in-efficient bank. Technical efficiency describes how input is converted into output. This model is not providing optimal efficiency instead of optimal it provides relative efficiency of all the banks, which will help to differentiate between least efficient and highly efficient banks.

BCC Model
It is the modification of the basic CCR model which has been developed by Banker, Charnes, and Cooper. This model gives more flexibility than the CCR model as it allows return-to-scale be to a variable so it can be increasing, decreasing, and constant return-to-scale (CRS). The efficiency under this model is known as Pure Technical

Fig. 1 Models of DEA

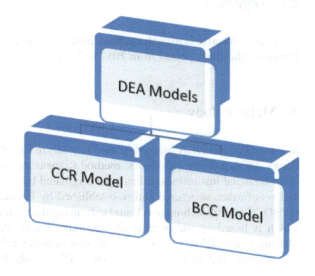

An Efficiency Analysis of Private Banks in India—A DEA Approach

Table 1 Sample private banks

Old private banks	New private banks
City Union Bank Limited (CITY)	Axis Bank Limited (AXIS)
CSB Bank Limited (CSB)	Bandhan Bank Limited (BANDHAN)
Federal Bank Ltd (FED)	DCB Bank Limited (DCB)
Jammu and Kashmir Bank Ltd (J & K)	HDFC Bank Ltd (HDFC)
Karnataka Bank Ltd (KAR)	ICICI Bank Limited (ICICI)
Karur Vysya Bank Ltd (KARUR)	IDFC First Bank Limited (IDFC)
Nainital Bank Ltd (NANI)	Indusind Bank Ltd (INDUS)
RBL Bank Limited (RBL)	Kotak Mahindra Bank Ltd (KMB)
South Indian Bank Ltd (SIB)	Yes Bank Ltd. (YES)
Tamilnad Mercantile Bank Ltd (TMB)	
Dhanalakshmi Bank Ltd (DHAN)	

Efficiency (PTE) and firms having an efficiency score of 1 are known as "Locally Efficient banks".

In this study, we have considered the CCR model for calculating the Technical Efficiency of private banks and also the comparative analysis of Old and New private sector Banks for the Study Period.

3.2 Sample Bank

In this study, the researcher has selected 20 Private Banks in India which are continuously working from 2017 to 2021 (Table 1).

3.3 Input and Output Variables

Unlike the other sector, there is no consensus in the banking sector on what constitutes input and output (Fig. 2).

Based on the Table 2, represent the different combination of input and output used by various authors for the DEA model. In this study, we have considered three input &output variables based on the Intermediation approach (Fig. 3).

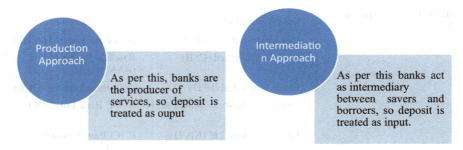

Fig. 2 Approaches to select input and output variables

Here, Fixed Assets = Premises + Fixed Assets under construction + Other Fixed Assets.

Borrowing = Borrowings in India + Borrowings outside India.

Operating Expense = Payment to Employee + Rent, Taxes + Advertisement + Insurance + other.

Investment = Investment in India + Investment outside India.

Interest Income = Interest earned on advance / bills/ + int on balance with RBI and other inter-bank.

Other Income = Commission, Exchange. and Brokerage + net profit/ loss on sale of Investment.

4 Empirical Findings and Result Analysis

In this study secondary data has been used for the efficiency analysis of 20 private banks, which is obtained from the RBI website for the period of five years from 2017–2021. DEAP 2.1 software has been used for the efficiency analysis. For DEA, three inputs and three outputs have been selected based on the Intermediation approach. The Descriptive statistics of Input and output variables are summarized in the Table 3.

Table 3 is figured in crores (Indian Rupees).

In this paper, we have calculated Technical Efficiency under the CRS model by taking the three input and three output variables and also the analysis of old private banks (11) and new private sector banks (9) separately and as combined sample private banks (20).

An Efficiency Analysis of Private Banks in India—A DEA Approach 1031

Table 2 Combination of input and output variables used by different authors

Author	Year	Input	Output
Luo	2003	Number of employees	Profit
		Total Assets	Revenue
		Shareholders' equity	
Arrof and Luc	2008	Deposit	Investment
		Number of employees	Total loan
		Fixed assets	
Kumar and Gulati	2010	Physical capital (Value of Fixed Assets)	Advance
		Labour (Number of Employees)	Investment
		Loanable funds (Deposit and Borrowings)	
Abhiman Das and Subhash Ray	2010	Fund	Investment
		Labours	Earning advances
		Capital	Other income
Rachita Gulati	2011	Physical capital	Net interest income
		Labours	Non-interest income
		Loanable funds	
Amit Kumar Dwivedi and Charyulu	2011	Deposit	Loan/advances
		Operating expense	Non-interest income
		No. of branches	
Erina and Ernis	2013	Labours	Loans
		Capital	Deposits
Piyush Singh and V.K.Gupta	2013	Capital	Advances
		Fixed assets	Investment
		Interest expense	Net profit
		Total borrowing	Total revenue
		Total liabilities	
		Total deposit	
		Operating cost	
Sandeepa Kaur and P K Gupta	2015	Interest expense	Interest income
		Operating expense	Fee based income
			Investment income

(continued)

Table 2 (continued)

Author	Year	Input	Output
Premalatha	2017	Deposit	Investment
		Borrowing	Advances
		operating expense	
Dar et al	2021	Borrowing	Investment
		Labours	Loan
		Fixed assets	Non-interest income
		Equity	

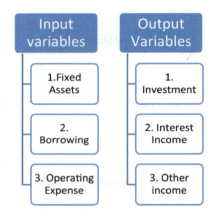

Fig. 3 List of input and output variables

Table 3 Descriptive statistics of sample (n = 20) private banks

Statistical tools	Input variables			Output variables		
	Fixed asset	Borrowing	Operating exp	Investment	Interest income	Other income
Mean	6976.38	164,741.56	25,003.05	278,341.55	90,245.76	19,071.60
Median	2982.98	28,540.34	9783.13	93,187.32	35,762.90	5680.62
Standard deviation	9859.29	248,449.29	35,057.57	417,259.15	126,685.16	28,902.91
Minimum	595.50	32.71	742.84	13,045.62	3110.30	206.75
Maximum	40,928.02	750,262.46	131,933.26	1,585,334.60	484,190.24	93,608.40

4.1 Technical Efficiency Score Through CCR Model

As per the CCR model, the technical efficiency of 20 Private Banks has been estimated for 5 years. Technical Efficiency is a firm's capacity to generate maximum output at a given set of inputs. The statistical summary has been described in Table 4.

As per the calculation, the average efficiency of private banks has increased from 0.907 to 0.924. Out of 20 Sample banks, only 4 banks are fully efficient for all 5 years namely Nainital Bank (NANI), Tamilnadu Mercantile Bank (TMB), ICICI, and Yes bank. A fully efficient bank is considered a "Globally Efficient bank" and these banks are optimally utilizing their resources to convert input into output without wastage of any resources. In old and new private banks, the mean efficiency of old private

Table 4 Represents the technical efficiency of sample banks for the period of 2017–2021

Bank name	Type	2017	2018	2019	2020	2021	Mean
CITY	Old	1.0000	0.981	1	1	1	0.996
CSB	Old	1.0000	0.855	0.598	0.69	0.676	0.764
Federal	Old	0.8170	0.908	0.971	1	1	0.939
J & K	Old	0.8020	0.728	0.752	0.765	0.761	0.762
KAR	Old	0.9610	1	1	0.861	1	0.964
Karur	Old	0.9190	0.878	0.972	1	0.718	0.897
Nani	Old	1.0000	1	1	1	1	1.000
RBL	Old	0.8290	0.838	0.936	1	0.961	0.913
SIB	Old	1.0000	1	1	0.883	0.836	0.944
TMB	Old	1.0000	1	1	1	1	1.000
DHAN	Old	0.7210	0.965	1	1	0.866	0.910
Mean Efficiency of Old banks		*0.9135*	*0.9230*	*0.9299*	*0.9272*	*0.8920*	*0.917*
Axis	New	0.906	0.922	1	0.956	1	0.957
Bandhan	New	0.889	1	1	0.952	1	0.968
DCB	New	0.669	0.763	0.894	0.793	0.962	0.816
HDFC	New	0.892	0.942	1	1	1	0.967
ICICI	New	1	1	1	1	1	1
IDFC	New	1	1	0.73	0.698	0.755	0.837
Indusind	New	0.994	0.975	1	0.854	0.981	0.961
Kotak	New	0.746	0.884	0.872	1	0.914	0.883
Yes	New	1	1	1	1	1	1
Mean efficiency of New banks		0.900	0.943	0.944	0.917	0.957	0.932
Mean efficiency of private banks		0.907	0.932	0.936	0.923	0.922	0.924

sector bank is 0.917 while the mean efficiency of new private banks is having 0.932 and the overall efficiency of private banks are 0.924.

Table 5 indicates the ranking of banks based on the average technical efficiency of sample banks. The 1st rank is shared by four banks namely Nainital Bank, Tamilnadu Mercantile Banks, ICICI and Yes bank. It means that bank can utilize their resources effectively and they are operating at their optimum scale size. There is no further improvement required for the functioning and they are considered leaders in banking. Out of 4 banks, 2 banks belong to old & other two belongs to new private banks. 2nd rank secured by citi bank and 3rd and 4th rank secured by Bandhan and HDFC respectively. The last two ranks were secured by CSB and Jammu and Kashmir Bank.

Table 5 Ranking based on average of technical efficiency

Bank name	Type	TE- average (2017–2021)	No. of times (TE = 1)	Rank
NANI	Old	1.000	5	1
TMB	Old	1.000	5	1
ICICI	New	1.000	5	1
YES	New	1.000	5	1
CITY	Old	0.996	4	2
Bandhan	New	0.968	2	3
HDFC	New	0.967	3	4
KAR	Old	0.964	3	5
Indusind	New	0.961	1	6
Axis	New	0.957	2	7
SIB	Old	0.944	3	8
Federal	Old	0.939	2	9
RBL	Old	0.913	1	10
DHAN	Old	0.910	2	11
Karur	Old	0.897	1	12
Kotak	New	0.883	1	13
IDFC	New	0.837	2	14
DCB	New	0.816	0	15
CSB	Old	0.764	1	16
J & K	Old	0.762	0	17

An Efficiency Analysis of Private Banks in India—A DEA Approach 1035

Table 6 Summary of average efficiency and in-efficiency (2017–2021)

Type of banks	N	Average efficiency (%)	Average in-efficiency (%)
Old private banks	11	91.7	8.3
New private banks	9	93.2	6.8
Total private banks	20	92.4	7.6

5 Conclusion and Future Research

The major objective of this study is to measure the efficiency of private banks for five years and new private sector banks performed better than the old private banks. The efficiency score indicates that there is no regular pattern followed during the study period.

As per the Table 6, the efficiency of new private sector banks is more compare to new private sector banks. The average efficiency of sample banks is 0.924 or 92.4 in percentage, it means that the banks can produce the same output by utilizing only 92.4% of their present input. The reason for Inefficiency is managers are not able to utilize their resources efficiently. The new private banks have utilized their resources better way than the old private banks.

The further research could broaden the scope of our studies in other ways that were not taken into account in this paper, such as the BCC model of DEA, Changes in the selection of Input and output variables, or terms of the study period.

References

1. De Young, R.: Data Envelopment Analysis (DEA): A non-parametric approach for estimating bank performance efficiency. In Proceedings of the Symposium on Banking Efficiency and Profitability, Federal Reserve Bank of Chicago (1997)
2. Rajput, N., Gupta, M.: Impact of IT on Indian commercial banking industry: DEA analysis. Glob. J. Enterpr. Inform. Syst. 3(1), 17–31 (2011)
3. Hafsal, K, Suvvari, A, Durai, S.R.S.: Efficiency of Indian banks with nonperforming assets: Evidence from two-stage network DEA. Future Bus. J. 6(1) (2020)
4. Sherman, H. D., Gold, F.: Bank branch operating efficiency: Evaluation with data envelopment analysis. J. Bank. Finance, 9(2), 297–315 (1985)
5. Kumar, S., Gulati, R.: Evaluation of technical efficiency and ranking of public sector banks in India: an analysis from cross-sectional perspective. Int. J. Product. Perform. Manag. 57(7), 540–568 (2008)
6. Singh, H.: Assessing the effectiveness of India's banking sector: A comparative analysis of public, private, and foreign banks from 2005 to 2011. Int. J. Econ. Comm. Manag. 1(9), 1–18 (2013)
7. Ketkar, K.W.: Performance and profitability of Indian banks in the post-reform period. Int. J. Finance 20(3), 4910–4929 (2008)

8. Reserve Bank of India, "Statistical Tables Relating to Banks in India", Various Issues, RBI (www.rbi.org.in)
9. Reserve Bank of India, "Trends and Progress of Banking in India", Various Issues, RBI (www.rbi.org.in)
10. Premlatha, K.: An empirical Study on Performance Efficiency of selected Indian Public sector Bank through Data Envelopment Analysis (Doctoral dissertation, Anna University) (2017)

A Case Study on the Impact of Tourism on the Tribal Life in Vayalada, Calicut, Kerala

P. T. Retheesh and **K. K. Jagadeesh**

Abstract There aren't as many misty hill stations in Kerala's northern region as there are in its southern region, but there is one place in Kozhikode that will steal your breath away. Travelers have made Vayalada, a quiet location blessed with natural beauty and shrouded in sleepy mist, popular on social media. In the Kozhikode district, this magnificent mountaintop is located about 12 km from Balussery. Vayalada has all the elegance and tranquilly of a hill station because it is 2000 ft above sea level, and from its highest point, one can see the scenic beauty. There are different Adivasis who are divided into a number of sects, including the Ooralis, Kattunaikkans, Adiyars, Kurumas, Paniyas, and Uraali Kurumas, who live on the border of Kozhikode and Wayanad districts. This case is about six tribal families who are living in Vayalada, near Thalayd, in Calicut, Kerala. Vayalada is just becoming a popular destination because of its scenic beauty and the beautiful view of Kakki Dam. This case is especially about the problems faced by the tribal families living in Vayalada due to the increase in tourist arrivals.

Keywords Tribal life · Tourism · Kerals · Impact of tourism · Sustainable development

1 Introduction

Indigenous tourism is any form of tourism in which native people are actively involved, either through direct control over the activity or by having their culture serve as the main draw. Calicut, Kerala's capital, is experiencing rapid indigenous tourism development. It ensures the livelihood of many poor local communities.

P. T. Retheesh (✉)
Department of Tourism, Faculty, Department of Commerce, Kristu Jayanti College Autonomous, Bangalore, India
e-mail: retheesh@kristujayanti.com

K. K. Jagadeesh
Kristu Jayanti College Autonomous, Bangalore, India

© The Author(s), under exclusive license to Springer Nature Switzerland AG 2024
A. Hamdan and E. S. Aldhaen (eds.), *Artificial Intelligence and Transforming Digital Marketing*, Studies in Systems, Decision and Control 487,
https://doi.org/10.1007/978-3-031-35828-9_87

Fig. 1 Vayalada view point

During the 1940s, the local population's migration increased enormously, displacing the local aborigines, or Adivasis. With the loss of their lands and demographic decline, the tribes today make up only 20% of the district's total population.

In honour of the well-known hill station Gavi in the south Kerala district of Pathanamthitta, Vayalada has acquired the moniker "Kozhikode's Gavi." Vayalada has a lush touch of green and a misty, chilly atmosphere similar to Gavi. The area is riddled with hills, the tallest of which is the Kottakunnu hill, making it a popular hiking destination. A hike to the viewpoint is always worthwhile because it offers breathtaking views of the Kakkayam dam and enormous stones (Fig. 1).

2 Literature Review

Social, economic, environmental, sustainability, facility, and problem dimensions all seemed to have a positive impact on the sustainability of tribal tourism [1].

If eco-tourism is properly developed, it can attract tourists from anywhere, and generate more revenue for the inhabitants of the region. It was also recommended to make some attempts to conserve the physical ecology and the cultural ecology of the ethnic tribal communities by empowering them through a participatory protected area management approach [2].

Tribal fairs serve as a showcase for their entire culture, and those who value culture always laud its uniqueness. Therefore, it could be a tool for their economic and social development. It is important to create an environment for tribal fairs in order to capitalise on the opportunities. Tribal fairs are a crucial part of preserving intangible tribal history [1].

Tribal tourism in India is most refreshing and energizing. It will take you from the bustle of the city and town to a serene location where people continue to value leading a simple life in the manner of their ancestors. One of the oldest civilizations in the world, Indian culture can be observed in primaeval parts of the country [3]. To make tribal tourism sustainable involving the local tribal communities and capacity building are important [4].

The focus of tourism in the state of Kerala is changing from historically well-liked destinations like Munnar, Thekkady, Alappey, Kovalam, etc. to Wayanad, Varkala, Nelliyampathy, Bekkal, Wagamon, etc. These are pristine, eco-friendly locations that require a thoughtful and sustainable strategy to boost Kerala tourism in the coming decades. In recent years, thousands of travellers from all over the world have been enthralled by Wayanad, an up-and-coming tourist attraction. Wayanad is a particularly sensitive location in many aspects due to its lush natural surroundings and the presence of numerous tribes [5].

One of the articles investigates whether tourism operations in Kumarakom with no planned interventions are more sustainable than those in Kumily with them. Among the various changes that have taken place in Kumily are the transformation of ex-poachers into forest guards and the inclusion of marginalized people in community-based ecotourism, whereas unplanned tourist growth at Kumarakom gave birth to a number of socio-cultural difficulties [6]. The Vana Samrakshana Samithi (VSS) structure is one of the most effective ways to maintain forests and promote ecotourism with the help of the local population. The majority of populations that depend on forests are tribal ones, and the forests are where they live and breathe. It is exceedingly noteworthy that the Vana Samrakshana Samithi (VSS) trainings promote environmental protection. The empowerment of tribal people in matters of the environment has successfully continued since Vana Samrakshana Samithi (VSS) got involved. With the assistance of the Kerala Forest Department, the Vana Samrakshana Samithi (VSS) closely monitors ecotourism and forest protection operations [7].

Previously, studies were either focused on the development of tourism in a specific destination or on the promotion of tribal tourism. This research is only interested in the difficulties that a community faces as a result of the development of a specific tourism destination.

3 Objectives

To identify the important factors that promotes tourism in Calicut.
To identify the level satisfaction of tribal people who are living in Vayalada.
To identify the problems faced by the tribal inhabitants of Vayalada.

4 Tourism Development and Tribal Life in Kozhikode

The only thing that can fill a person with experiences and priceless knowledge is travel, which is the only thing that can make life for humans more than simply a bustling city life and well-paying work. As a result, a dedicated traveller would always emphasise that no opportunity should be missed to visit and discover uncharted territory, particularly indigenous communities, without endangering their way of life. When it comes to tourism and economic development, the emphasis should be on protecting the environment, Kozhikode's cultural and social heritage, tribal life, and its delicate ecosystem rather than focusing on the number of visitors.

5 Major Observations

According to the Calicut Tourism Promotion Council, one of the least visited tourist spots is the tribal settlement in Kerala's Calicut district. There are numerous tribal communities in the district, and each one exemplifies the traditional style of tribal life. One of the country's oldest tribal groups is the Calicut tribe. Development initiatives with a comparatively higher employment potential ought to be given priority while preserving the existing culture of the neighbourhood and its environmental resources as a result.

When creating tourism sites, local communities' and stakeholders' needs should also be taken into consideration. From the interaction with the tribal head of Vayalada Kottakkunnu settlement, it is understood that they are struggling to leave, and one review mentioned the role of Vana Samrakshana Samithi (VSS) in preserving forest resources and tribal culture, but this was not seen in Vayalada Kottakkunnu settlement (Fig. 2).

6 Deprived Living Standard

It is observed that there was no proper involvement of the local administrative system in ensuring the primary rights of the tribal families during the visit. It's exciting to learn that the local panchayat has been allocated a small sum for the construction of restrooms rather than considering a request for the construction of a house. Also, the current accommodation setup cannot be called a "house because it looks like a tent made out of coconut palm leaf plastic sheets. There is no proper floor or doors for said accommodations. It's so unfortunate to see that they are sleeping, cooking, having food, and spending time in a single room under the above-mentioned canopy (Fig. 3).

Fig. 2 Tribal shelter

Fig. 3 Tribal community members

7 Connectivity

There is no proper connectivity to the Vayalada tribal settlement. People who are living in this settlement have to walk 4 km from the main road, and they have to travel 8 km for their basic needs. Authorities are very particular about building a road to a nearby tourist viewpoint, but not for the tribal community.

8 Lack of Water Resources

Another major issue these tribal members are facing is water scarcity, especially during the summer. Even the responsible authorities are accepting the same. It's too difficult for them to survive without proper water resources.

9 No Proper Education

Education is one of our country's fundamental rights, but only one girl out of 14 family members is in high school, and she also skipped half of it. There is no interference from the respective authorities to ensure that they are getting a proper education (Fig. 4).

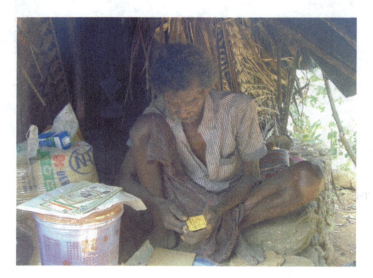

Fig.4 Tribal community head

10 Tribal Families are not Willing to Relocate

Government authorities have tried to relocate the tribal families by giving them land. But the community members are saying that the land given to them is 10 km away from the place where they are now, and they do not have any sources for their livelihood. At their current location, they are finding some source of income as they explore the forest resources and sell them in the local market. The new place they have been allotted is not at all suitable for their lives, it seems (Fig. 5).

Tribal communities' safety and security should be prioritised while promoting tourism in Vayalada. As the author, I witnessed one of the tribal children fleeing from her home during my first visit, and the tribal head was explaining why. It was really shocking to hear about the bad experience they had from the tourists who visited Vayalada View Point. Visitors were encroaching on the limit and distracting the tribal habitat to a certain extent, and with the previous experience, only the kid had escaped by breaking the wall of the house, which is made out of coconut palm leaves. This experience cited by the tribal head was really shocking and never expected (Fig. 6).

This incident shows that there is no safety for women and children who are part of this tribal community. This type of unethical behavior, which tourists were not expecting, must be stopped at the source. Respective authorities cannot escape from their morel responsibilities, and this kind of negligence has to be reported to the higher authorities, but unfortunately, the members of the tribal community are unaware of this.

Fig. 5 Interior of tribal shelter

Fig. 6 It's not he but its she

If the legal enforcement system is not involved in these types of issues, as perceived by the local tribal community, the local administrative system's illegal and irresponsible attitude must be changed.

11 Conclusion

Indian tribal groups have inherited traditions that are ingrained in their way of life and culture. The planning of tourism activities in tribal areas, as well as the function and contribution of authorities to the wellbeing of the community, are the main topics of this descriptive article, which is based on an observational methodology. This study has shown that developing the tourism industry should take into account sustainable planning and growth. By giving tribes a source of income and ensuring their safety, tourism may undoubtedly serve as a sector for their growth.

References

1. Chouhan, V.: Developing a sustainable tribal tourism model vis-a-vis the tribal region of Rajasthan. J. Tourism Heritage Serv. Market. (JTHSM) **8**(1), 58–63 (2022)
2. Panigrahi, N.: Development of eco-tourism in tribal regions of Orissa: potential and recommendations. Cult. Mandala **13**(3), 9410 (2019)
3. Patil, P.A.: Visiting tribal cultures in India. Int. J. Res. Soc. Sci. Inf. Stud., 122–124 (2017)
4. Sahoo, M.S.S.: Developing Responsible Tourism Through Tribal Community in Uttar Kannada District. In Compass, p. 92 (2014)

5. Sibi, P.S., Swamy, G.A.: Scope and potential for indigenous tourism-an analytical study in Wayanad, Kerala. ZENITH Int. J. Multi. Res. **5**(11), 60–69 (2015)
6. Sebastian, L.M., Rajagopalan, P.: Socio-cultural transformations through tourism: a comparison of residents' perspectives at two destinations in Kerala, India. J. Tourism Cult. Change **7**(1), 5–21 (2009)
7. Kuttencherry, A.J., Arunachalam, P.: Role of tribal Vana Samrakshana Samithi (VSS) members in building forest protection and eco-tourism at Vazhachal eco-tourism area in Thrissur district of Kerala (2020). Interview Link: https://www.youtube.com/watch?v=L5nUvQZuCD4

Pro-Environmental Behavior of Farmers in the Dieng Plateau Indonesia

Dyah Sugandini, Mohamad Irhas Effendi, Yuni Istanto, Bambang Sugiarto, and Muhammad Kundarto

Abstract This study aims to analyze a pro-environmental behavior model for farmers in the Dieng plateau, Indonesia. Some variables used to predict PEB are an environmental concern, perceived behavioral control, and environmental Attitude. The number of questionnaires in this study was 12 items using a five-point Likert scale. Respondents in this study were farmers who had their fields and had carried out pro-environmental agricultural practices. The number of samples in this study was 200 respondents. The data analysis tool used is PLS-SEM. The results of the research show that all the proposed hypotheses are supported. Environmental concerns and perceived behavioral control affect E.A. and PEB. And E.A. has a direct effect on PEB. This research has a novelty related to the object of research, namely small farming, and very slow adoption. Besides that, this study also discusses the pro-environmental behavior from the internal side of farmers. Previous studies on adopting pro-environmental practices by smallholders have primarily focused on external constraints. This internal side of the farmer is believed to play an essential role in adopting pro-environmental behavior and the social order of life.

Keywords Pro-environmental behavior · Environmental concern · Perceived behavioral control · Environmental attitude

---

D. Sugandini (✉) · M. I. Effendi · Y. Istanto · B. Sugiarto · M. Kundarto
Universitas Pembangunan Nasional Veteran Yogyakarta, Jl. SWK Jl. Ring Road Utara No.104, Ngropoh, Condongcatur, Kec. Depok, Kabupaten Sleman, Daerah Istimewa Yogyakarta, 55283 Yogyakarta, Indonesia
e-mail: dini@upnyk.ac.id

© The Author(s), under exclusive license to Springer Nature Switzerland AG 2024
A. Hamdan and E. S. Aldhaen (eds.), *Artificial Intelligence and Transforming Digital Marketing*, Studies in Systems, Decision and Control 487,
https://doi.org/10.1007/978-3-031-35828-9_88

1 Introduction

Over the past few decades, agricultural production growth has raised concerns about unsustainable agricultural practices and their adverse impact on natural resources and the environment [7]. Some evidence shows that the intensive use of chemical fertilizers reduces soil fertility in various regions [19, 31]. Large-scale irrigation development also consumes a lot of groundwater [32]. Another threat to the environment can be seen from the inappropriate use of pesticides that damage the ecology and environment [13]. Overcoming this requires environmentally friendly behavior from farmers who work in their fields. Green behavior has the potential to play an essential role in encouraging more sustainable behavior and consumption [6]. However, a socio-psychological study by [26] shows that farmers internally lose motivation to adopt pro-environmental farming practices. Qian et al. [24] and Bakker et al. [5] show that the performance of green agriculture has had failures in practice. There is some empirical evidence of the attitude-behavior green gap [26], where pro-environmental attitudes are not always reflected in sustainable behaviour. Young et al. [33] noted that although people claim to be highly concerned about the environment, this does not always translate into pro-environmental behavior. Numerous studies of pro-environmental behavior report little correlation between environmental values, attitudes, intentions, and actual behavior. The results of a study [31] show that farmers are aware that if they do not behave pro-environmentally, it will endanger agricultural production, which is their main livelihood. Farming behavior that is not environmentally friendly can cause degradation of agricultural land, water pollution, and damage to food crops before harvest. Several factors influencing pro-environmental behavior among little farmers are attitudes, perceptions of behavioral control, and environmental concern. This research is based on those strongly correlated in describing pro-environmental behaviour (PEB). TPB emphasizes the importance of behavioral intention in determining behavior [34]. TPB has a psychological construction that can be tested using a structural causal model. TPB provides a complementary perspective in examining the intentions and shaping of individual behaviour. TPB is based on self-interest based on rational action, whereas Norm Activation Theory evolved from the norm-activation model, which focuses on individual moral values [29, 30]. This study aims to analyze the pro-environmental behavior of farmers in adopting environmentally friendly agriculture. Several factors from TPB are integrated to explain pro-environmental behavior. These factors are environmental Attitude, ecological concern, and perceived behavioral control. Stern [29] defines pro-environmental behavior as preserving a sustainable environment while considering the benefits that can be obtained at present and in the future. Stern [29] further stated that pro-environmental behavior would change the availability of resources and protect ecosystems. Gong [13] shows that it is challenging to justify pro-environmental behavior because pro-environmental behavior is not simple adequately. Pro-environmental behavior is an ever-evolving topic in specific consumer behavior studies [21]. Usually, this pro-environmental behavior has a lot to do with purchasing green products [31]. Zubair and Garforth [36] show that PBC is an essential predictor of pro-environmental decision-making. Perception of perceived behavioral control shows the degree to which an individual exists and whether or not a behavior is under his control. Perceived behavioral control is a variable that indicates the ease or difficulty of taking pro-environmental actions [1]. Apart from aiming to minimize research gaps related to attitudes and PEB, this study also has the following novelties: (1) This research reveals the inner side of individuals (farmers) in predicting pro-environmental behavior. Previous research on pro-environmental behavior has placed more emphasis on individual external factors [16, 19]. This study discusses the internal factors of small farmers, such as attitudes, awareness of the environment observed by environmental concern, PBC, and farmers' behavioral intentions regarding the application of pro-environmental agricultural practices. (2) This research was conducted in the agricultural sector, which was very small and slow to adopt agricultural innovations. Savari and Gharechaee [26], Qian et al. [24], and Bakker et al. [5] show that research involving smallholders and their decision-making processes regarding pro-environmental practices is still urgently needed. Smallholders associated with smallholder farming practices have different internal controls than large farmers [4, 24].

2 Literature Review

Environmental Attitude is crucial for understanding pro-environmental behaviour [6, 11]. The reason underlying this research is that the decision to adopt pro-environmental products requires complex decisions. Complex decision-making with high involvement requires rational decisions, meaning that consumers can still determine other alternatives. Kollmuss and Agyeman [17] showed that the linear relationship between knowledge, attitudes and pro-environmental behavior could not be explained. Schultz et al. [27] and Albayrak et al. [3] show that the role of attitudes toward environmental behavior has various relationships. Not always, this Attitude towards the environment influences pro-environmental behavior. Kaiser et al. [15] confirmed a positive relationship between environmental Attitude and ecological behavior (i.e., the tendency to behave ecologically). They also found a significant correlation between environmental knowledge and the value of the environment.

2.1 Environmental Attitude (E.A.)

Environmental Attitude is a positive or negative evaluation of the consequences of pro-environmental behavior [8]. Environmental attitudes are a set of beliefs and individual emotions related to individual activities and environmental problems [23]. The relationship between attitudes and pro-environmental behavior needs to be reviewed because there are several different findings. Previous research [3, 27] showed that the role of attitudes toward environmental behavior has various relationships. Not always, this Attitude towards the environment influences pro-environmental behavior. Akehurst et al. [2] and Nittala [22] showed that the relationship between environmental attitudes and behavior is fragile. Kaiser et al. [15] showed that, despite the high level of public concern for the environment, he concluded that some research results showed a moderate relationship between pro-environmental attitudes and behavior, some showed a weak relationship, and others found no relationship between both of them. The results of another study by [15] confirm a positive relationship between environmental Attitude and ecological behavior (i.e., the tendency to behave ecologically) and also find a significant correlation between environmental knowledge and the value of the environment.

Hypothesis 1: Environmental Attitude influences pro-environmental behaviour.

2.2 Ecological Concern (E.C.)

Ecological concern refers to individual beliefs and levels of concern for the environment [10]. Fontes et al. [10] define Ecological concern as a combination of awareness about environmental issues and a willingness to be part of environmental protection. The ecological concern is public awareness of environmental problems, often expressed through purchasing environmentally friendly products [9]. Environmental concern is the extent to which consumers are aware of problems related to the environment and support efforts to solve them or show a willingness to contribute personally to the environment [23, 35]. Kollmuss and Agyeman [17] and Fontes et al. [10] show that ecological concern is an individual's awareness to behave ecologically has an impact on pro-environmental behavior. According to [1], Attitude is an individual response to liking or disliking environmental protection.

Hypothesis 2: Ecological concern influences environmental Attitude.
Hypothesis 3: Ecological concern influences pro-environmental behaviour.

2.3 Perceived Behavioral Control (PBC)

Perceived behavioral control is a condition related to how easy or difficult it is for consumers to consider behavior based on experience and obstacles that can be anticipated [1]. When sustainable behavior is considered difficult to implement, people tend not to involve themselves in environmental advocacy. Sreens et al. [28] showed that green purchase intention and perceived behavioral control about green products have a significant positive relationship. Nguyen et al. [20] showed that consumer purchase intention is influenced by perceived behavioral control when buying green clothes. Qin and Song [25] demonstrated that perceived behavioral control significantly influences consumer pro-environmental behavior in China. Recent research from Lavuri [18] investigates millennial purchase intentions for eco-friendly products in India as a growing market. The results show that PBC affects the buying

Pro-Environmental Behavior of Farmers in the Dieng Plateau Indonesia   1051

behavior of green products. Gansser and Reich [12] added that if someone thinks that pro-environmental behavior is something simple and easy to apply to everyday life, then someone will behave pro-environmentally.

Hypothesis 4: PBC affects environmental Attitude.
Hypothesis 5: PBC affects pro-environmental behaviour.

3 Research Methodology

This study uses a survey of farmers to explain the pro-environmental behavior of farmers when farming. The population of this study was all farmers in the Dieng plateau. The unit of analysis in this study is the individual. The number of samples taken refers to the opinion of [14], which states that the minimum number of samples for a data test to have statistical power that can be accounted for is five to ten times the parameters analyzed. Questionnaires were distributed to as many as 200 farmers. Measurement of several research variables refers to research from [6, 12], and others. The ecological concern is the extent to which individuals are aware of problems related to the environment and support efforts to solve them [9, 35]. PBC is the degree to which the innovation is considered easy for individuals to control. This study uses Smart-PLS with the Structural Equation Model (SEM) approach to test the hypothesis.

4 Results

This research is a survey with a quantitative approach. Respondents are farmers who own agricultural land. The number of samples used was 200 farmers living in the Dieng Plateau, Indonesia. Dieng was chosen as the research area because the Dieng plateau is a vegetable farming center supporting the demand for vegetables in Indonesia. This study aims to test the model of pro-environmental behavior on farmers. Table 1 shows the research respondent data.

Reliability analysis can be seen from the internal consistency value of Cronbach's alpha and composite reliability. Convergent validity was observed through the Average Variance Extracted value and discriminant validity using the Heterotrait-Monotrait Ratio (HTMT). Meanwhile, statistical assessments such as VIF values, path coefficients, t-statistics, and p-values are used to evaluate the structural model. The t-test is used to assess the significance of the relationship between variables. The reliability structure of the questionnaire uses the Cronbach value of each variable to verify internal consistency between the questionnaire items. Table 2 shows that all constructs have Cronbach's alpha values greater than 0.70. Each construct in Table 3 has good reliability; namely, the composite reliability value is greater than 0.5. The AVE value is good because all constructs are greater than 0.5.

Table 1 Characteristics of respondents

Demographic characteristics	Description	Percent
Gender	Female	34.2
	Male	65.8
Age (Years)	30–35	11.2
	35–40	35.8
	41–45	20.5
	> 45	32.7
Education of the head of household	Secondary school	46.9
	High school	50.0
	Equivalent-bachelor	3.1
Number of family members	One-person	6.0
	Two-person	24.6
	Three-person	12.9
	Four-person	42.7
	Five-person	10.8
	> Five-person	3.0
Land area	1–3 ha	56.9
	4–6 ha	37.9
	> 6 ha	5.2

Table 2 Characteristics of respondents

	Cronbach's alpha	rho_A	Composite reliability	Average variance extracted (AVE)
Environmental Concern	0.870	0.877	0.921	0.795
Environmental Attitude	0.800	0.801	0.882	0.715
PBC	0.903	0.904	0.939	0.838
Pro-Environmental Behavior	0.903	0.903	0.939	0.838

4.1 Hypothesis Testing Results

The model's fit in Table 4 shows several values generated from Smart-PLS to indicate acceptable suitability. Overall, the proposed model accounts for 62.8% of the variance in pro-environmental behavior. Standardized Root Means Square Residual (SRMR) was used to assess the fit of the PLS model. A good match is defined by an SRMR value of less than 0.10. The SRMR value in the study was 0.057. The model meets the criteria for model fit if the RMS Theta value is <0.102 and the NFI value is >0.9.

Pro-Environmental Behavior of Farmers in the Dieng Plateau Indonesia

Table 3 Discriminant validity-Fornell-Larcker Criterion racteristics of respondents

	Environmental Concern	Environmental Attitude	PBC	Pro-Environmental Behavior
Environmental Concern	0.891			
Environmental Attitude	0.456	0.845		
PBC	0.582	0.410	0.915	
Pro-Environmental Behavior	0.721	0.531	0.647	0.915

Table 4 R-square and fit summary

	R square	R square adjusted
Environmental Attitude	0.239	0.232
Pro-Environmental Behavior	0.628	0.623
	Saturated model	Estimated model
SRMR	0.057	0.057
NFI	0.957	0.957
Rms Theta	0.102	

The results of this study indicate that the Theta value is 0.102, and the NFI is 0.957. So it shows a very suitable model. The model has reliability and validity and can explain the hypothesized relationship according to the measured R2 (Fig. 1).

Based on the data processing results in Table 5, a significance value or p-value is obtained at less than 0.05 based on this p-value, it can be stated that all the hypotheses proposed in this study are supported.

5 Discussion

This research shows that small farmers who produce on small farms have good pro-environmental behavior values. This result means that all the farmers sampled show high pro-environmental behaviour. The results of this study also indicate that the first hypothesis, which states that environmental Attitude influences pro-environmental behavior, is supported. The effect of E.A on PEB is 21.9%. The second hypothesis, which states that E.C. influences E.A, is supported. The influence of E.C. on E.A. is 32.8%. The third hypothesis about the effect of E.C. on PEB is also supported (45.5%). The fourth and fifth hypotheses, which state that PBC affects E.A., are also supported. The effect of PBC on E.A is 21.9%, and the effect of PBC on PEB is 30%. The effect of E.A. on PEB supports the findings from [8, 23] which state

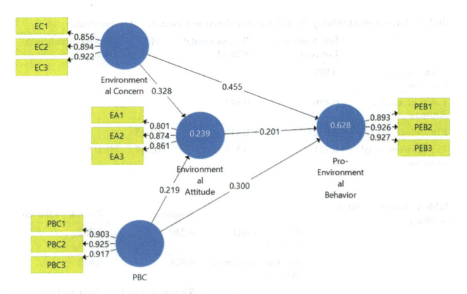

Fig. 1 Model of pro-environmental behavior on farmers

Table 5 SEM analysis for hypothesis testing

	Original sample (O)	Sample mean (M)	Standard deviation (STDEV)	T Statistics (|O/STDEV|)	P values
EC → EA	0.328	0.335	0.081	4.031	0.000
EC → PEB	0.455	0.451	0.088	5.155	0.000
EA → PEB	0.201	0.201	0.054	3.708	0.000
PBC → EA	0.219	0.214	0.079	2.767	0.006
PBC → PEB	0.300	0.306	0.085	3.540	0.000

that there is a relationship between E.A. and PEB. Vegetable smallholder farmers show a good attitude towards environmental sustainability. They realize that if agricultural practices are carried out incorrectly, it will have an impact on damage to their rice fields. The excessive use of chemical fertilizers and pesticides has also damaged farmers' agricultural land and reduced the amount of vegetable production. Farmers in Dieng have intensified the movement towards pro-environmental agricultural practices, although the agricultural results obtained from PEB practices have not shown relevant output. Farmers realize that the change towards green agriculture requires a long adaptation pattern. However, these Dieng farmers can benefit from switching to pro-environmental farming practices. Other results from research that state that E.C. affects E.A. and PEB support the findings from [9, 10]. A high level of environmental concern encourages an individual willingness to contribute to the environment [23, 35]. The results of this study indicate that farmers already have a soul to preserve nature because they are directly involved with land use for vegetable production. Farmers also have a high willingness to sacrifice to protect their environment. The more careless farmers are in treating the environment, the more nature will be damaged, and the impact of damage to nature will be on the farmers. Small farms on the Dieng plateau have a soil structure that is prone to landslides because (1) it is located on a plateau, (2) Agriculture on the Dieng plateau has the potential to damage the soil structure and trigger future landslides. The vulnerability of this soil structure causes Dieng farmers to have a good attitude towards the environment and always behave environmentally friendly by reducing the exploitation of agricultural land and preventing landslides by implementing a terracing farming pattern. Farmers have also started planting large trees as water retainers to avoid landslides. This study's hypothesis supports PBC's effect on E.A. and PEB. Farmers consider that pro-environmental agricultural practices have many benefits. The results of this study support the findings of [6, 12], which state that PBC can affect PEB in farmers on small farms. Small farmers feel that their agricultural land is appropriate and suitable for pro-environmental agricultural practices. PEB for small farmers can reduce technical risks and solve problems if they behave pro-environmentally.

6 Conclusion

This research produces a pro-environmental behavior model that is acceptable because it has a good fit model. The research was conducted using data from 200 respondents of smallholder farmers who produce on small farms. The results of this study indicate that small farmers already have good environmental awareness and have pro-environmental behavior. Farmers have adopted environmentally friendly agricultural practices such as: using organic fertilizers and natural pesticides and planting trees to support rainwater. The pro-environmental behavior of small farmers has no obstacles in its adoption because farmers realize that behavior that is not pro-environmental can cause enormous losses to the farmers themselves. Losses that arise include land degradation, which impacts decreasing soil fertility and production yields. Besides that, the land in Dieng is in the highlands, so landslides are easy. The E.A. owned by these farmers turned out to be able to increase PEB. The role of PBC in increasing PEB was also found in this study. Farmers feel a lot of convenience and benefits when applying PEB in their agricultural practices. In the long term, PEB practices can also reduce damage to agricultural land, landslides, and flash floods that endanger humans.

7 Limitations of Research and Suggestions for Further Research

This research has limitations related to the research setting, namely on small farmers who produce on small farming practices. So the results of this study can only be generalized for agriculture on small farms. On the other hand, research on PEB needs to be conducted on large farmers with large agricultural practices because these farmers have the most significant risk of environmental damage. So it is hoped that in the future, studies and research can be carried out with settings in the practice of large farmers so that they can find PEB. We realize that environmental damage is also caused by large agricultural practices that are not environmentally friendly. This study seeks to close the research gap regarding the relationship between attitudes and PEB. The results of this study show consistency with previous findings, which show a relationship between attitudes and behavior. However, this research is only based on the internal side of the individual. The external side of the individual also needs to be explored again about PEB to get a complete theory about PEB in small farmers. Although the results of this study indicate the effect of Attitude towards PEB, this effect has the smallest value in the proposed model. Future researchers are advised to re-examine the effect of attitudes on behavior to get better theoretical consistency.

Acknowledgements Thank you to the Ministry of Education, Culture, and Technology Research and the Universitas Pembangunan Nasional "Veteran" Yogyakarta for funding the Kedaireka, Matching-Fund program in 2022.

References

1. Ajzen, I.: The theory of planned behavior. Organ. Behav. Human Decision Process. **50**(2), 179–211 (1991). https://doi.org/10.1016/0749-5978(91)90020-T
2. Akehurst, G., Afonso, C., Goncalves, H.M.: Re-examining green purchase behaviour and the green consumer profile: new evidences. Manag. Decis. **50**(5), 972–988 (2012). https://doi.org/10.1108/00251741211227726
3. Albayrak, T., Aksoy, Ş, Caber, M.: The effect of environmental concern and scepticism on green purchase behaviour. Mark. Intell. Plan. **31**(1), 27–39 (2013). https://doi.org/10.1108/02634501311292902
4. Ali, D., Bowen, D., Deininger, K.: Personality traits, technology adoption, and technical efficiency: evidence from smallholder rice farms in Ghana. J. Develop. Stud. **56**(7), 1330–1348 (2020). https://doi.org/10.1080/00220388.2019.1666978
5. Bakker, L., Sok, J., van der Werf, W., Bianchi, F.J.J.A.: Kicking the habit: what makes and breaks farmers' intentions to reduce pesticide use? Ecol. Econ. **180**, 106868 (2021). https://doi.org/10.1016/j.ecolecon.2020.106868
6. Cao, H., Li, F., Zhao, K., Qian, C., Xiang, T.: From value perception to behavioural intention: study of Chinese smallholders' pro-environmental agricultural practices. J. Environ. Manage., 115179 (2022). https://doi.org/10.1016/j.jenvman.2022.115179
7. Cao, H., Zhu, X.Q., Heijman, W., Zhao, K.: The impact of land transfer and farmers' knowledge of farmland protection policy on pro-environmental agricultural practices: the case of straw

return to fields in Ningxia China. J. Clean Prod. **277**, 123701 (2020). https://doi.org/10.1016/j.jclepro.2020.123701

8. Carfora, V., Cavallo, C., Caso, D., Del Giudice, T., De Devitiis, B., Viscecchia, R., Nardone, G., Cicia, G.: Explaining consumer purchase behavior for organic milk: including trust and green self-identity within the theory of planned behavior. Food Qual. Prefer. **76**, 1–9 (2019). https://doi.org/10.1016/j.foodqual.2019.03.006

9. Dhir, A., Sadiq, M., Talwar, S., Sakashita, M., Kaur, P.: Why do retail consumers buy green apparel? A knowledge-attitude-behavior-context perspective. J. Retail. Consum. Serv. **59**, 102398 (2021). https://doi.org/10.1016/j.jretconser.2020.102398

10. Fontes, E., Moreira, A.C., Carlos, V.: The influence of ecological concern on green purchase behavior. Manage. Market. Challenges Knowl. Soc. **16**(3), 246–267 (2021). https://doi.org/10.2478/mmcks-2021-0015

11. Fransson, N., G¨arling, T.” Environmental concern: conceptual definitions, measurement methods, and research findings. J. Environ. Psychol., **19**(4), 369–382 (1999). https://doi.org/10.1006/jevp.1999.0141

12. Gansser, O.A., Reich, C.F. (2023). Influence of the New Ecological Paradigm (NEP) and environmental concerns on pro-environmental behavioral intention based on the Theory of Planned Behavior (TPB). J. Cleaner Prod., **134629**. https://doi.org/10.1016/j.jclepro.2022.134629

13. Gong, Y.Z., Baylis, K., Kozak, R., Bull, G.: Farmers' risk preferences and pesticide use decisions: evidence from field experiments in China. Agric. Econ. **47**(4), 411–421 (2016). https://doi.org/10.1111/agec.12240

14. Hair, J.F., Hult, G.T., Ringle, C.M., Sarstedt, M.: A Primer on Partial Least Squares Structural Equation Modeling (PLS-SEM). SAGE Publication, Los Angeles (2014)

15. Kaiser, F.G., Wolfing, S., Fuhrer, U.: Environmental attitude and ecological behavior. J. Environ. Psychol. **19**, 1–19 (1999). https://doi.org/10.1006/jevp.1998.0107

16. Khataza, R.R.B., Doole, G.J., Kragt, M.E., Hailu, A.: Information acquisition, learning and the adoption of conservation agriculture in Malawi: a discrete-time duration analysis. Technol. Technol. Forecast. Soc. Change **132**, 299–307 (2018). https://doi.org/10.1016/j.techfore.2018.02.015

17. Kollmuss, A., Agyeman, J.: Mind the gap: why do people act environmentally and what are the barriers to pro-environmental behavior? Environ. Educ. Res. **8**(3), 239–260 (2002). https://doi.org/10.1080/13504620220145401

18. Lavuri, R.: Extending the theory of planned behavior: factors fostering millennials' intention to purchase eco-sustainable products in an emerging market. J. Environ. Planning Manage. **65**(8), 1507–1529 (2022). https://doi.org/10.1080/09640568.2021.1933925

19. Lu, H., Zhang, P.W., Hu, H., Xie, H.L., Yu, Z.N., Chen, S.: Effect of the grain-growing purpose and farm size on the ability of stable land property rights to encourage farmers to apply organic fertilizers. J. Environ. Manage. **251**, 109621 (2019). https://doi.org/10.1016/j.jenvman.2019.109621

20. Nguyen, M.T.T., Nguyen, L.H., Nguyen, H.V.: Materialistic values and green apparel purchase intention among young Vietnamese consumers. Young Consumers. **20**(4), 246–263 (2019). https://doi.org/10.1108/YC-10-2018-0859

21. Nguyen, T.T.M., Phan, T.H., Nguyen, H.L., Dang, T.K.T., Nguyen, N.D.: Antecedents of purchase intention toward organic food in an Asian emerging market: a study of urban Vietnamese consumers. Sustainability **11**(17), 4773 (2019). https://doi.org/10.3390/su11174773

22. Nittala, R.: Green consumer behavior of the educated segment in India. J. Int. Consum. Mark. **26**(2), 138–152 (2014). https://doi.org/10.1080/08961530.2014.878205

23. Onurlubaş, E.: The mediating role of environmental attitude on the impact of environmental concern on green product purchasing intention. EMAJ: Emerg. Markets J., **8**(2), 5–18 (2018). https://doi.org/10.5195/emaj.2018.158

24. Qian, C., Li, F., Antonides, G., Heerink, N., Ma, X.L., Li, X.D.: Effect of personality traits on smallholders' land renting behavior: theory and evidence from the North China Plain. China Econ. Rev. **62**, 101510 (2020). https://doi.org/10.1016/j.chieco.2020.101510

25. Qin, B., Song, G.: Internal motivations, external contexts, and sustainable consumption behavior in China—based on the TPB-ABC integration model. Sustainability **14**(13), 7677 (2022). https://doi.org/10.3390/su14137677
26. Savari, M., Gharechaee, H.: Application of the extended theory of planned behavior to predict Iranian farmers' intention for safe use of chemical fertilizers. J. Clean. Prod. **263**, 121512 (2020). https://doi.org/10.1016/j.jclepro.2020.121512
27. Schultz, P.W., Shriver, C., Tabanico, J.J., Khazian, A.M.: Implicit connections with nature. J. Environ. Psychol. **24**(1), 31–42 (2004). https://doi.org/10.1016/S0272-4944(03)00022-7
28. Sreen, N., Purbey, S., Sadarangani, P.: Impact of culture, behavior and gender on green purchase intention. J. Retailing Consum. Serv., **41**, 177–189. https://doi.org/10.1016/j.jretconser.2017.12.002
29. Stern, P.C.: Toward a coherent theory of environmentally significant behavior. J. Soc. Issues **56**(3), 407–424 (2000). https://doi.org/10.1111/0022-4537.00175
30. Su, H.Z., Zhao, X.Y., Wang, W.J., Jiang, L., Xue, B.: What factors affect the water saving behaviors of farmers in the Loess Hilly Region of China? J. Environ. Manage. **292**, 112683 (2021). https://doi.org/10.1016/j.jenvman.2021.112683
31. Sugandini, D., Rahatmawati, I., Arundati, R.: Environmental attitude on the adoption decision mangrove conservation: an empirical study on communities in special region of Yogyakarta, Indonesia. Rev. Integr. Bus. Econ. Res. **7**(s1), 266–275 (2018)
32. Udimal, T.B., Jincai, Z., Ayamba, E.C., Owusu, S.M.: China's water situation; the supply of water and the pattern of its usage. Int. J. Sustain. Built Environ. **6**(2), 491–500 (2017). https://doi.org/10.1016/j.ijsbe.2017.10.001
33. Young, W., Hwang, K., McDonald, S., Oates, C.J.: Sustainable consumption: green consumer behaviour when purchasing products. Sustain. Dev. **18**, 20–31 (2010). https://doi.org/10.1002/sd.394
34. Yuriev, A., Dahmen, M., Paillé, P., Boiral, O., Guillaumie, L.: Pro-environmental behaviors through the lens of the theory of planned behavior: a scoping review. Resour. Conserv. Recycl., **155**, 104660. (2020). 10.1016/j. resconrec.2019.104660
35. Zaremohzzabieh, Z., Ismail, N., Ahrari, S., Abu Samah, A.: The effects of consumer attitude on green purchase intention: a meta-analytic path analysis. J. Bus. Res., (2020). https://doi.org/10.1016/j.jbusres.2020.10.053
36. Zubair, M., Garforth, C.: Farm level tree planting in Pakistan: the role of farmers' perceptions and attitudes. Agroforestry Syst. **66**, 217–229 (2006). https://doi.org/10.1007/s10457-005-8846-z

Balanced Economic Development: Barometer and Reflections of Economic Progress Concerning the Economies of India and China

Bijin Philip and **Suresh Ganesan**

Abstract Balanced economic development is integral to our economy's growth since it allows us to create high-wage jobs and facilitate an improved quality of life. Balanced economic development is necessary for long-term Sustainability. Recently, China and India have enjoyed unprecedented economic development, with the GDP per capita in both countries quickly expanding. However, assessing if countries can achieve a balanced economic development in the service, manufacturing, and agriculture sectors is critical. India and China have adopted several strategies and initiatives to develop and support the agriculture sector, including agricultural insurance programs, innovative technology, irrigation processes, and water management systems. This study will look at the contributions and development of the farming industry in China and India, as well as the problems and challenges it faces. The researcher has also created a model to depict the agricultural sector's difficulties and the measures nations have taken to improve it.

Keywords Agriculture insurance · Balanced economic development · Economic growth · GDP rate · Crop insurance · Sustainable agriculture · Sustainable development goal

1 Introduction

For the past few years, the impressive economic growth of China since the early 1980s and India since the late 1990s has been a prominent topic of discussion in the areas of economic progress and economic reform. Both nations started economic transformations fueled by deregulation and liberalization, allowing for increased

---

B. Philip (✉)
Research Scholar, Bharathidasan University, Tamil Nadu, India
e-mail: bijinalfy111@gmail.com

S. Ganesan
Assistant Professor and Research Adviser, Department of Commerce, Imayam Arts and Science College, Vadakkuveli, Tamil Nadu, India

© The Author(s), under exclusive license to Springer Nature Switzerland AG 2024
A. Hamdan and E. S. Aldhaen (eds.), *Artificial Intelligence and Transforming Digital Marketing*, Studies in Systems, Decision and Control 487,
https://doi.org/10.1007/978-3-031-35828-9_89

1059

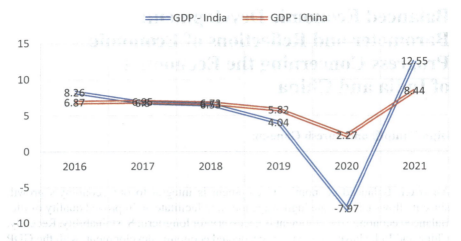

Fig. 1 GDP rate—India and China

trade and investment abroad. China began to reform earlier and more vigorously than India, beginning in 1981. In 1991, India started a comprehensive set of economic policy reforms in response to fiscal and balance-of-payments problems. China and India had comparable economic frameworks before their economic reforms, with a large state sector and substantial reliance on agriculture [9].

China and India are the world's two emerging and developing economies. As of 2021, China and India have the second and fifth-largest nominal economies in the world, respectively. Regarding purchasing power parity (PPP), China is in the first place, while India is third. Both nations provide 21 and 26% of the total global wealth in nominal and PPP terms. China and India are responsible for more than half of Asia's GDP (Word bank 2021) (Fig. 1).

Every country, developed or developing, has issues with regional development. While certain regions of the nation are highly developed, others suffer greatly from a lack of resources and amenities. Therefore, the country needs to embrace a process of balanced regional growth if it wants its development to be stable. Balanced regional development is very much required to accelerate the economy's development, develop and conserve resources, promote large employment opportunities, maintain political stability, defend the country, and control social evils.

Due to the strong development rates of the industrial and service sectors, agriculture's share of India's GDP has gradually decreased to less than 15%. Still, the sector's significance in the nation's economic and social fabric goes beyond this figure. To do so, an agricultural industry that is productive, competitive, diverse, and sustainable will need to emerge quickly. India is one of the world's top agricultural nations. It also has the largest cow herd (buffaloes), the central area planted with wheat, rice, and cotton, and generates the most milk, pulses, and spices globally. Among the biggest exports from the nation are vegetables, farmed fish, rice, fruit, wheat, cotton, sugarcane, sheep and goat meat, and tea. Of the 195 million hectares

Balanced Economic Development: Barometer and Reflections … 1061

under cultivation in the nation, 63% (or around 125 million ha) are rainfed, and 37% are irrigated (70 m ha). In addition, 65 million hectares of India are covered in forests. (World Bank Report 2012).

The Gross Capital Formation (GCF) in agriculture as a percentage of the total GCF in the economy has decreased from 8.5% in Financial Year 2011–12 to 6.5% in Financial Year 2018–19 due to a lack of access to modern technology rising land ownership, forced sales, more farm labor than farmers, and falling investment in agriculture. This is because the amount of private investment has decreased, issues with subsidies and associated issues, problems with the Minimum Support Price (MSP), and related issues have hampered the expansion and productivity of the agricultural sector in China and India.

The transition to a cleaner and more environmentally friendly food production needs to align current agricultural policies with UN Sustainable Development Goals, especially for a country with India's diversified geography and population. The formulation of an overarching policy influencing sustainable management of agricultural systems, combined with proper implementation of social welfare schemes, would lead to the timely realization of SDG 1 (no poverty), SDG 2 (zero hunger), and SDG 3 (good health and well-being) in India, according to a critical analysis of operational as well as recommended agriculture and farmer welfare policies [21].

2 Review of Literature

Singh and Agrawal [7] the solution for the farming community is agriculture insurance. In India, several policy measures have been implemented to increase farmers' entree to agriculture insurance. However, one of the primary issues for Indian policymakers has always been accessing farm insurance. This article aims to investigate current policy initiatives in India's farm insurance sector. Agriculture insurance is a significant risk management strategy that is out of reach for most Indian farmers. Every decade, the Indian government adopts a new agriculture policy. However, due to operational issues, every crop insurance scheme has been uneven and inefficient. Agriculture insurance in India is still evolving regarding coverage, breadth, and exposure, but farmers' unhappiness with the product has spread over the country. The key causes for limited access to agriculture insurance are insurance awareness and farmers' inclination for agriculture relief payments. Because of execution challenges at the state level, the present crop insurance plans are ineffective.

Shi et al. [3] this study aims to give readers a better knowledge of China's intelligent agriculture (IA) growth, a key trend in increasing agricultural productivity in the approaching period. In China, IA has three primary characteristics: an unequal geographic distribution, a trend still in its early stages, and attention primarily concentrated on a small number of technologies. Similar to the development of IA in other nations with diverse qualities, such as Japan, India, and the United States, parallels and variations in IA expansion in China and other countries are also explored.

Kushankur and Debasish [6] the Pradhan Mantri Fasal Bima Yojana (PMFBY) and the Weather-based Crop Insurance Scheme are evaluated using a set of performance indicators. Farmers' coverage under PMFBY can be increased by claim payout, but subsidies and actuarial premium rates substantially impact farmers' coverage under WBCIS, according to the study. However, we suggest a comprehensive insurance package to supplement the performance of two schemes, such as seed insurance via prepaid insurance card, replanting guarantee program, and crop cycle insurance, to mention a few.

Okoye et al. [10] Using historical data on chosen variables from 1986 to 2015, the main goal of this article was to estimate the impact of financial inclusion on Nigerian economic progress and development. The key conclusions are that credit distribution to the private sector (a measure of financial inclusion) has not considerably aided Nigeria's economic growth, and (ii) financial inclusion has aided poverty alleviation in Nigeria through rural credit delivery.

Malini [8] Agricultural insurance is essential for farmer prosperity and policy instruments to cope with the risks that come with farming. On the other hand, agriculture insurance mostly depends on farmers' attitudes. This study evaluates and tests respondents' attitudes about agriculture insurance, as well as favorable variables and challenges encountered in implementing agriculture insurance. Between April and August 2008, sixty farmers were interviewed. The farmers were chosen using a simple random selection procedure. Percentage analysis and the sign test were used to examine the acquired data. Farmers have a favorable view regarding agriculture insurance, according to the report. Furthermore, they acknowledge that certain favorable conditions and roadblocks exist in implementing farm insurance in the Ambasamudram area. Based on the findings, the study offers several mechanisms to increase the share of agricultural revenue in the Ambasamudram area at all levels.

Fan [1] the importance of governments in boosting innovative ability and contributing to economic development is highlighted in this study. This paper provides micro-level understandings of modernization capacity and economic development: (1) Modernization capability has become perilous for domestic enterprises' market accomplishment, and (2) worldwide institutional factors and national government procedures on modernization have a significant impact on the decision to conduct indigenous research and development or import technology at the company level.

Sinha [5] the subject of this research is the use of agriculture insurance plans to safeguard farmers from agricultural variability. With little coverage and a high claims-to-premium ratio, the agricultural yield insurance plan has been mainly ineffective. There are issues with both the scheme's conception and implementation, and crop yields are difficult to measure appropriately and perfectly. The current system of Crop Cutting Experiments has been called into doubt. It also introduces significant basis risk, which can only be mitigated at a high expense. The high claims-to-premium ratio shows that current premium rates are below actuarial standards.

Sinha [4] This article explores the creation and performance of agriculture insurance and the potential for private insurers to participate. According to the study report, crop insurance schemes in India cover only 10% of the sowed area and have a high claim-to-premium ratio. Crop insurance can be improved by using new technology to improve the correctness and suitability of crop estimation approaches.

Dreze and Sen [2] This book examines India's endemic deprivation and the role of government intervention in alleviating the issue. The study takes a comprehensive view of economic growth, concentrating on human well-being and "social potential" rather than traditional economic growth measures. Since independence, India's victory in alleviating deprivation has been modest.

3 Statement of the Problem

As per Ragnar Nurkse's balanced growth theory, The two significant merits of balanced growth theory are:

(i) The balanced growth theory emphasizes that all the sectors should develop simultaneously, and (ii) No sector will be discriminated.

Recently, China and India have experienced extraordinary economic growth, with both countries GDP per capita increasing quickly. However, evaluating whether countries can have balanced economic development in the service, industry, and agriculture sectors is essential. As a result, it's critical to examine whether India's and China's economies can achieve balanced economic development and the current state of the agriculture sectors in both countries (Fig. 2).

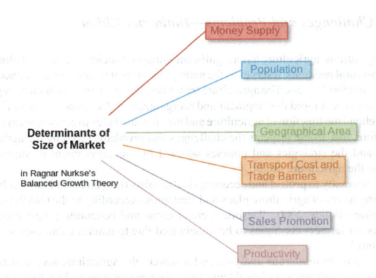

Fig. 2 Determinants of size of market

3.1 Objectives

- To study the challenges and problems faced by the agriculture sector in India and China
- To analyze the strategies adopted by the countries to support and sustain the agriculture sector.
- Analyze the GDP contribution from the agriculture sector in India and China.

4 Methodology

To satisfy the study objectives, the researchers collected information from secondary sources and carefully used them for analysis. The secondary bases of evidence are publically available and easily accessible. The present Study information was collected from books, journals, articles, magazines, and websites. The researcher developed a model that focuses on the initiatives taken by India and China to build and enhance the agriculture sector.

5 Results & Discussion

5.1 Challenges and Problems—India and China

In many nations, agriculture has a significant impact on socioeconomic development. For most rural residents, it serves as the main pillar of their means of subsistence, work source, and food access. The agriculture sector has, however, mainly been disregarded due to the recent rapid development and reorganization. Nevertheless, it's critical to comprehend the function of agriculture and how it contributes to economic expansion. Therefore, this article inspects the challenges and problems faced by the agriculture sector and the strategies and policies adopted by the government to support and develop the agriculture sector.

The necessity to protect and secure agriculturalists from inconsistency has been an unending worry of agriculture plans and strategies. According to the (NAP) National Agriculture Policy 2000, "Despite technological and economic progressions, the situation of farmers continues to be unbalanced due to natural calamities and price variations."

According to numerous analyses and research, the agriculture sector in India is confronting challenges and problems from four major zones. The four key challenge areas identified are financial, environmental, facility, and awareness. Financial concerns include scarcity of capital, Agricultural Credit, Subsidies and Related Issues, Agriculture Insurance Mechanisms, and Price Stabilization. Climate change and Erratic Monsoon, Rainfall Dependence, Soil Erosion, Fragmented Landholding,

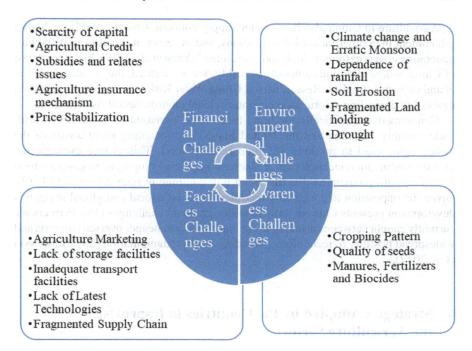

Fig. 3 The model represents the challenges in Indian Agriculture

and Drought are significant environmental challenges. Facilities challenges include Agriculture Marketing, Lack of storage facilities, Inadequate transport facilities, Lack of Latest Technologies, and a Fragmented Supply Chain. And finally, awareness challenges consists of Cropping Pattern, Quality of seeds, Manures, Fertilizers, and Biocides (Fig. 3).

The world's farmers will effectively produce enough food to feed the nation in 2020. Still, because of unequal distribution, discriminatory trade policies, and poorly organized supply chains, more than one billion of the planet's 770 crores will experience food unavailability [13]. Nourishment security and food safety issues have historically been of the utmost importance to China and its authorities due to the country's proximity to 20% of the world's population and its lack of fresh water and fertile land (6% and 7%, respectively) [18].

In China essential to cope with falling farm/rural populations and aging farm/rural workforces (Peng, 2011; Gilmer et al., 1998; Li and Sicular, 2013). Young and middle-aged workers' emigration in search of better pay and more opportunities has significantly changed rural China's social structure and encouraged the "left behind" phenomenon, which has, for better or worse, permanently changed agrarian Chinese society.

Agriculture in China also has a water supply concern. China's coal has a higher ash content than other industrialized nations, and the amount of sulfur in its power-generating coal is extremely high and imbalanced (Yang et al., 2019). In the southeast of China, where enormous amounts of paddy rice are farmed, they contaminate the abundant water supplies. Researchers at China Water Risk, a nonprofit organization devoted to compiling information on water-related environmental risks in China.

Concerns over genetically modified foods have increased as consumers world-wide, notably in China and the United States, are becoming more aware of the technologies used to produce and prepare their food. This is true even though genetic modification technologies are increasingly being employed to generate food and fiber, with predictions that this trend will continue in both countries [18, 19]. Given the opposition to genetically modified foods on a local and global scale, this development presents Chinese farmers with significant challenges [18]. Farmers are currently caught between shrinking profit margins from cheaper overseas imports and widespread hostility to technology, including genetic manipulation [20] (Butkowski et al. 2017).

6 Strategies Adopted by the Countries to Improvise the Agriculture Sector

Due to the significant development rates of the manufacturing and service sectors, agriculture's share of the Indian economy has gradually declined to less than 15%. Still, the sector's significance in India's economic and social fabric goes beyond this metric (Fig. 4).

India is one of the world's top agricultural nations. It also has the largest cow herd (buffaloes), the most significant area planted with cotton, wheat, and rice, and it generates the most milk, pulses, and spices globally. In the nation, 195 million hectares of land are used for agriculture, of which 37% is irrigated, and 63% is rain-fed (or around 125 million ha) (70 m ha). In addition, 65 million hectares of India's territory are covered in forests.

To develop and promote the agriculture sector in India, the government has to focus on four primary areas. (i) Farmer's awareness and education, (ii) Technology adoption, (iii) Financial support, and (iv) Environmental consideration.

The Government of India has launched several programs to increase farmer knowledge and awareness to improve crop production and productivity, including Front Line Demonstrations and Extension through a network of Krishi Vigyan Kendras (KVKs), National Mission of Agricultural Extension & Technology (NMAET), National Food Security Mission (NFSM), Soil Health Management Scheme, Mission for Integrated Development of Horticulture (MIDH), and Bringing Green to India (RKVY).

For the previous 60 years, the GDP contributions from India's and China's agriculture sectors are shown in the Table 1. We can see from the table that agriculture's

Balanced Economic Development: Barometer and Reflections … 1067

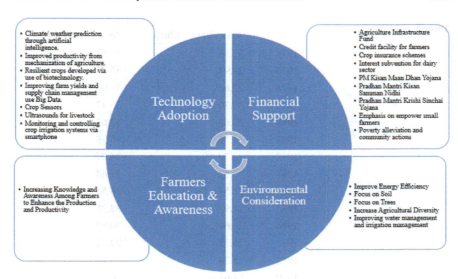

Fig. 4 Strategies adopted by the countries to improvise the Agriculture Sector

contribution to GDP is decreasing in both counties. It's not a healthy indicator of the economy's long-term stability and balanced economic development (Fig. 5).

Table 1 GDP rate—agriculture sector

Year	Agriculture—China	Agriculture—India
1960	23.18	44.26
1961	35.79	42.56
1962	38.99	41.77
1963	39.85	39.89
1964	38.03	41.08
1965	37.55	42.96
1966	37.18	40.91
1967	39.81	41.81
1968	41.64	44.53
1969	37.52	43.52
1970	34.80	43.29
1971	33.63	41.95
1972	32.42	40.28
1973	32.93	40.28
1974	33.43	43.31
1975	31.95	40.31

(continued)

Table 1 (continued)

Year	Agriculture—China	Agriculture—India
1976	32.36	37.62
1977	28.99	35.75
1978	27.69	37.09
1979	30.70	35.47
1980	29.63	33.63
1981	31.32	35.39
1982	32.79	34.07
1983	32.57	32.88
1984	31.54	33.54
1985	27.93	32.21
1986	26.64	30.89
1987	26.32	29.74
1988	25.24	29.18
1989	24.61	30.2
1990	26.58	28.97
1991	24.03	29.02
1992	21.33	29.39
1993	19.31	28.74
1994	19.47	28.68
1995	19.60	28.27
1996	19.33	26.26
1997	17.90	27.13
1998	17.16	25.89
1999	16.06	25.79
2000	14.68	24.5
2001	13.98	23.02
2002	13.30	22.92
2003	12.35	20.7
2004	12.92	20.74
2005	11.64	19.03
2006	10.63	18.81
2007	10.25	18.29
2008	10.17	18.26
2009	9.64	17.78
2010	9.33	17.74
2011	9.18	18.21
2012	9.11	17.86

(continued)

Table 1 (continued)

Year	Agriculture—China	Agriculture—India
2013	8.94	17.52
2014	8.67	18.2
2015	8.42	16.17
2016	8.13	16.36
2017	7.57	16.36
2018	7	15.41
2019	7.1	15.96
2020	7.7	18.4

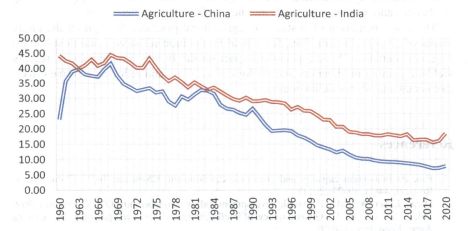

Fig. 5 GDP rate—agriculture sector

7 Conclusion

The agricultural segment plays a deliberate role in the advancement of the economic expansion of a country. It has already completed an extensive contribution to the economic achievement of progressive governments, and its role in the economic development of less-developed countries is of vital importance. In India, around 70% of the population is employed as farmers, and the contribution to the gross domestic product is just 18%. It makes available foodstuff in India for approximately 135 crore people. About 12% of export incomes come from the farming sector. The Indian trade profit of 14.6 billion dollars from farming was achieved in 2018 (Prateek 2020).

Three fourth of the population in India is centered on agriculture; production is the central foundation of livelihood for the whole nation. The reliance on farming has been equal for the time being. Agriculture still plays a crucial role in modern development, even though there are now a lot of growth variables.

The agriculture sector has a pivotal role in employment creation, industrial expansion, supply of foreign exchange, a resource for capital formation, supply of food and raw materials, creating effective demand, and reducing inequality.

Due to poor inputs and techniques, inadequate irrigation facilities, indebtedness of the farmers, Low adoption of improved technology, absence of innovation in agriculture, and rural transport and communication network creates loopholes in carrying out agricultural activities.

Several government inventiveness and strategies, expanding government's role, clear objectives, suitable actions, and considerable resources will lead to agricultural development in our economy. These figures show the significant and sizeable role the agricultural sector plays in the nation's functioning. It will only continue to rise from here and play an essential role in the country's economic development as long as changes and advances are made and new policies are implemented.

The implementation of sustainable agricultural practices throughout the entire stages of crop production (i.e., from field preparation to harvest and post-harvest stages) is certain to have a positive impact on the Indian economy by increasing yields, regaining biodiversity and ecological stability, and reducing adverse effects like social and economic inequality.

References

1. Fan, P.: Innovation capacity and economic development: China and India. Econ. Chang. Restruct. **44**(1), 49–73 (2011)
2. Dreze, J., Sen, A.: India: Economic development and social opportunity. OUP Catalogue (1999)
3. Shi, L., Shi, G., Qiu, H.: General review of intelligent agriculture development in China. China Agric. Econ. Rev., (2019)
4. Sinha, S.: Agriculture insurance in India: scope for the participation of private insurers. Econ. Polit. Weekly, 2605–2612 (2004)
5. Sinha, S.: Agriculture insurance in India. CIRM Working Paper Series (Chennai, India, CIRM) (2007)
6. Kushankur, D., Debasish, M.: Agriculture insurance in India-promise, pitfalls, and the way forward. Econ. Polit. Weekly, **52**(52) (2017)
7. Singh, P., Agrawal, G.: Development, present status and performance analysis of agriculture insurance schemes in India: review of the evidence. Int. J. Soc. Econ., (2020)
8. Malini, R.: The attitude of farmers toward agriculture insurance: a study with special reference to Ambasamudram Area of Tamil Nadu. IUP J. Agricult. Econ. **8**(3), 24 (2011)
9. Hussin, F., Yik, S.Y.: The contribution of economic sectors to economic growth: the cases of China and India. Res. Appl. Econ. **4**(4), 38–53 (2012)
10. Okoye, L.U., Adetiloye, K.A., Erin, O., Modebe, N.J.: Financial inclusion: A panacea for balanced economic development (2016)
11. Veeck, G., Veeck, A., Yu, H.: Challenges of agriculture and food systems issues in China and the United States. Geogr. Sustain., **1**(2), 109–117 (2020), ISSN 2666–6839. https://doi.org/10.1016/j.geosus.2020.05.002
12. Hok-wui Wong, S., Wu, N.: Can Beijing buy Taiwan? An empirical assessment of Beijing's agricultural trade concessions to Taiwan. J. Contemp. China **25**(99), 353–371 (2016)
13. Pretty, J., Sutherland, W.J., Ashby, J., Auburn, J., Baulcombe, D., Bell, M., Pilgrim, S., et al.: The top 100 questions of importance to the future of global agriculture. Int. J. Agric. Sustain. **8**(4), 219–236 (2010)

14. Gilmer, D.F., Aldwin, C.M., Ober, B.A.: The Greying of Rural America (1998)
15. Peng, X.: China's demographic history and future challenges. Science **333**(6042), 581–587 (2011)
16. Li, M., Sicular, T.: Aging of the labor force and technical efficiency in crop production: Evidence from Liaoning province, China. China Agricult. Econ. Rev. (2013)
17. Yang, F., Tan, J., Sui, X.: Analysis of environmental protection constraint of energy development. In Constraints and Solutions for Energy and Electricity Development, pp. 91–165. Springer, Singapore (2019)
18. Wong, A.Y.T., Chan, A.W.K.: Genetically modified foods in China and the United States: a primer of regulation and intellectual property protection. Food Sci. Human Wellness **5**(3), 124–140 (2016)
19. Raman, R.: The impact of Genetically Modified (GM) crops in modern agriculture: a review. GM Crops Food **8**(4), 195–208 (2017)
20. Gerasimova, K.: Debates on genetically modified crops in the context of sustainable development. Sci. Eng. Ethics **22**(2), 525–547 (2016)
21. Priyadarshini, P., Abhilash, P.C.: Policy recommendations for enabling transition towards sustainable agriculture in India. Land Use Policy **96**, 104718 (2020)

Customer Preference Towards E-banking Services Offered by State Bank of India

Ashwitha Shetty and K. A. Thanuja

Abstract The objectives of this research is to raise customer insights and ascertain which bank's E-banking services they prefer. Following data collection, the researcher determined which commercial bank provides superior E-banking services to consumers. The customer age, education, occupation and income everything plays very important contribution towards the preference of the E-banking services. The data analysis takes into account the customers' age, educational level, occupation, and income level.

Keywords Banks · Customer awareness · Electronic banking services · Profession · Income level

1 Introduction

Information technology plays a crucial role in the banking industry today. With the constant advancement of technology, the electronic banking services of many banks are improving. SBI's legacy branch model will be phased out in favour of modern E-banking services, including stand-alone marketing machines and coin-operated vending machines. Customers of different banks can benefit from a range of discounts. Today's people are better educated than previous generations, their lives have become more machine-centric, and they have less time than ever to visit a bank branch.

The term e-service is a well-known example of how we are using the information and communication technologies can be used in the banking sector. However, a clear definition of e-service is difficult to find as many definitions have been used by academics to explain e-service. ATM is an electronic computerized broadcastings

---

A. Shetty (✉) · K. A. Thanuja
Department of Commerce, Kristu Jayanti College Autonomous, Bangalore, India
e-mail: ashwitha@kristujayanti.com

K. A. Thanuja
e-mail: thanuja@kristujayanti.com

© The Author(s), under exclusive license to Springer Nature Switzerland AG 2024
A. Hamdan and E. S. Aldhaen (eds.), *Artificial Intelligence and Transforming Digital Marketing*, Studies in Systems, Decision and Control 487,
https://doi.org/10.1007/978-3-031-35828-9_90

device that permits customers to transact on their bank accounts, withdraw cash and use other facilities through a secure method of communication. Customers who own a computer and have access to the Internet can conduct business transactions without having to go to a bank office. Simply put, if they have an internet connection, they can use it. To give its customers faster access to its services, SBI has developed a coin-operated machine and a kiosk marketing machine. Along with these services, the bank offers SBI e-Tax, Demat Services e-Services-Pay and State Bank Mobi Cash.

2 Review of Literature

1. **Sandhu and Arora (2020)** [1], For emerging economies to achieve their development goals, electronic banking and digital finance are critically essential. The views and usage habits of the intended users of such services have a significant role in the success of service developments in the field of electronic banking. This study examined how clients used electronic banking services across multiple channels, and the results are presented in this paper.
2. **Al-Sharafi et al. (2018)** [2], The purpose of the custody was to find out how perceptions of security and privacy affected bank customers' willingness to accept and use online banking services. The study's examination specifically looked into what influences consumers' willingness to adopt and use internet banking services. The findings demonstrated that customers' behavioural intention to use online banking services was positively influenced by trust. Additionally, users' perceptions, such as how valuable, secure, and private they thought the service was, had a big impact on how trustworthy they felt.
3. **Ismail and Alawamleh (2017)** [3], Many people, especially in developing nations, oppose changes and continue to use conventional high-street banking methods to their dissatisfaction. Defining E-banking vocabulary and its purposes, as well as the benefits and drawbacks of E-banking applications to examine the influence of E-banking on traditional services, are covered in this article along with client viewpoints on the Jordanian E-banking system. It concludes that the introduction of E-banking improved customer satisfaction, loyalty, and the banks' reputation among Jordanian customers.
4. **Nifer Isern** discovered in the banking industry that there is a favorable relationship between financial infrastructure and competition. As well as a negative relationship between state ownership and competition.
5. **Reynolds, John**, Businesses were able to more effectively manage their marketing resources thanks to the findings of the customer loyalty study conducted in the E-banking technology services sector.
6. **According to Huang, Haibo,** The use of electronic money and E-banking services, is essential to their success. The main lesson is that, while E-banking consumers share some traits, they vary greatly amongst various E-banking providers.

7. **According to Jeon, Kiyong (2014)** [4], consumers in the United States prefer larger banks (2014). Larger banks operate numerous ATMs nationwide to reduce transportation costs.
8. **According to Jeon, Kiyong (2014)** [4], consumers in the United States prefer larger banks. Because they need to reduce their transportation costs, larger banks operate many ATMs across the country.

3 Statement of the Problem

For the past four to five years, SBI has had an ATM. With ATM usage increasing daily, it is important to research customer preferences for ATM services. This study is an example of such an effort by the bank, which offers various services to customers. Identifying the information needed to solve the problem Data collection tools should be selected or developed. Identify the target population and decide on a sampling strategy and design a data collection procedure. Collecting, analysing and presenting information. Predictions and/or rough estimates.

4 Objectives

a. To learn about the level of E-banking awareness among SBI customers.
b. To learn more about SBI's most popular E-banking service.
c. To determine who prefer to use the E-banking services.
d. To determining the reasons why customer don't prefer the E-banking services.

5 Methodology of Research

A questionnaire was developed to determine consumer preferences for banks' electronic banking services. This study used an investigative research approach based on previous field research; a questionnaire was developed to examine consumer preferences for Electronic Banking services provided by banks. After trial testing, the questionnaire was spread to 200 SBI customers. We'll take a 20-year-old as a starting point. A structured questionnaire and convenience sampling were used to collect data. The percentage technique, frequencies and correlations, and chi-square tests were used to analyse the data. Tables and graphics are also generated.

6 Data Analysis and Interpretation

Customer preferences for bank E-banking services are different from person to person in the current situation, so the researcher used different dimensions to understand the preferences of different customers towards the E-banking services offered by State Bank of India.

1076 A. Shetty and K. A. Thanuja

See Tables 1, 2 and 3.

Table 4 shows the effect of respondent age on using E-banking services.

See Table 5 and Graph 1.

According to Graph 1, 36 percent of people preferred ATMs online banking and Mobile Banking for the daily activities. Across different E-banking channels, 29% of them chose ATMs and Online as their next preference. With the information technology industry constantly introducing new applications, 23 percent of respondents prefer online banking. Therefore, new mobile banking apps that enable split-second transactions are extremely valuable to them. Tele phone banking was chosen by 7% of respondents.

See Graph 2.

Table 1 H1: there is a connection between gender and awareness level

Chi-square tests

	Value	df	Asymp. Sig. (2-sided)
Pearson chi-square	39.213[a]	38	0.415
Likelihood ratio	52.418	38	0.060
N of valid cases	100		

Result As a result, there is no association between gender and awareness level because the p value is greater than 0.05

Table 2 H1: there is a connection between education and awareness level

Chi-square tests

	Value	df	Asymp. Sig. (2-sided)
Pearson chi-square	20.705[a]	114	0.0316
Likelihood ratio	89.167	114	0.95
N of valid cases	100		

Result As the p value is less than 0.05, we are accepting the alternative hypothesis. So we can conclude that there is connection between the education and awareness level

Table 3 H1: significant difference between awareness level of men and women

ANOVA

Satisfaction

	Sum of squares	df	Mean square	F	Sig
Between groups	0.057	1	0.057	0.044	0.834
Within groups	126.703	98	1.293		
Total	126.760	99			

Result Consequently, as the alternative hypothesis is ruled out because of the high p value. It shows that there is no meaningful difference in men's

Customer Preference Towards E-banking Services Offered by State ...   1077

Table 4 H1: there is a substantial association between respondents' ages and their use of E-banking services

One-sample test

	Test value = 0					
	t	df	Sig. (2-tailed)	Mean difference	95% confidence interval of the difference	
					Lower	Upper
Age	35.211	99	0.000	2.230	2.10	2.36
Awareness	27.760	99	0.000	1.440	1.34	1.54

As a result, we can accept the hypothesis because the significant value is less than 0.05. Therefore, we can conclude that a respondent's age has an impact on their use of online banking services

Table 5 Demographic variables frequency analysis

Variables	Group	Occurrence	Percentage occurrence
Gender	Men	149	74.5
	Women	51	25.5
Age	20–25	50	25
	26–35	62	31
	36–45	67	33.5
	46–60	15	7.5
	Above 60	6	3
Educational qualification	Up to SSLC	19	9.5
	UG	65	32.5
	PG	57	28.5
	Professional	35	17.5
	Other	24	12
Occupation	Business	24	12
	Employee	80	40
	Professional	24	12
	Student	60	30
	Other	12	6

According to Table 5, the majority of responders (74.5 percent) are men and use E-banking more than women in the city of the study region. In percentage terms, the age group of 26 to 35 year olds makes up 31 percent, the age group of 36 to 45 year olds group people around 33.5 people will use the E-banking services. According to the survey 32% of respondents hold a UG degree and 28.5 percentage hold a PG degree

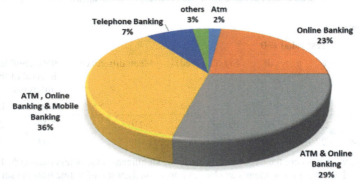

Graph 1 SBI's most preferred E-banking services frequency analysis

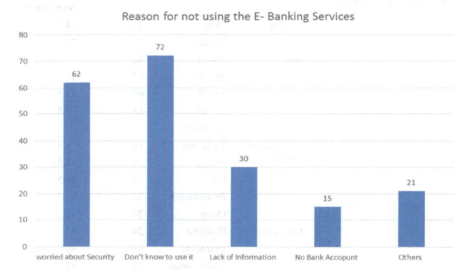

Graph 2 Unwillingness of customers to do internet banking

The results of Graph 2 show the main reason why SBI customers do not open an Internet banking account. According to the results of this study, the main reason for this because the people are not educated related to the how to use the E-banking services. Additionally, State Bank of India customers are worried about the securities of Electronic Banking services.

See Table 6.
See Graph 3.

Table 6 E-banking services preferred by the people according to the age

Age	Preference for electronic banking services
20–30	80
31–39	45
40–49	38
50–60	20
Above 60	17

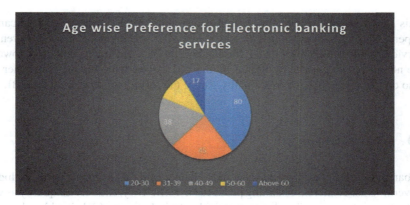

Graph 3 Banking services preferred by the people according to the age

7 Findings, Recommendations and Conclusion

This study found that younger generations use E-banking services more often than older generations due to new modernizations in info technology and that their E-banking adoption rate is high. People over 60 use E-banking services less often than others. When it comes to opening an internet bank account, one of the issues customers consider is risk. They were reluctant to use internet banking out of concern for their safety. For financial transactions, respondents chose ATMs, online banking, mobile banking, and SMS banking. In the eyes of customers, private banks offer better service than public banks. Public banks, on the other hand, will be able to conduct secure transactions.

8 Suggestions

SBI should publish and organize special awareness activities to increase customers' understanding of E-banking services. SBI should increase the number of ATMs near customer locations such as movie theatres, markets and other retail outlets. SBI's E-banking services are popular with the majority of respondents. However, many people hesitate to use it as they are not sure how to do it properly. In order to compete with other commercial banks, SBI should provide appropriate training or other ways to address this issue and strive to improve their service levels. Customers, banks and other organizations such as authorities benefit from E-banking technology.

9 Scope for the Future Research

This study is focusing on demographical variables more, further the study can be expended to study the various schemes by the SBI. And it can be measured even the service quality offered by the SBI. Researcher can also measure the intension towards the new E-banking services offered by the SBI. Along with this the researcher can also do the research on service quality gap of the services provided by the SBI.

10 Conclusion

E-banking technology benefits customers, banks, and other organizations including governments. To increase banking productivity, efficiency, and service quality as well as to grow internationally, all commercial banks in India are. SBI should make every effort to keep its internet banking system up to date. E-banking services reduces the work of both customers and the bank employees. So bank should take certain measure to improve the awareness about the E-banking services among the customers. As the technology is changing day by day the banks also should provide the new E-banking services to the customers. In most of the cases people are having the fear to use the E-banking services because of lack of safety and fear of fraudulent activity. So the banks should work on improving the security aspect of E-banking services offered by them. If the banks improves the securities in the coming day's number of people who is using the E-banking services will be increase.

References

1. Sandhu, S., Arora, S.: Customers' usage behaviour of e-banking services: interplay of electronic banking and traditional banking. Int. J. Finance Econ. **27**(2), 2169–2181 (2022)
2. Al-Sharafi A.M., Arshah, R.A., Herzallah A.T.F., Abu-Shanab, E.A.: The impact of customer trust and perception of security and privacy on the acceptance of online banking services: structural equation modeling approach. Int. J. Ind. Manag. (IJIM) **4**, 1–14 (2018)
3. Ismail, L.B., Alawamleh, M.: The impact of online banking of customer satisfaction in Jordan. J. Organ. Stud. Innov. **4**(2), 1–13 (2017)
4. Jeon, K.: Essays on banking industry: ATM (Automatic Teller Machine) (2014)

The Analysis of the Influence of the January Effect and Weekend Effect Phenomenon on Stock Returns of Companies Listed on the JII Index (2017–2019 Period)

Salsabila Hartono Putri⬤ and Nur Fauziah⬤

Abstract This study aims to determine the influence of the January effect and the weekend effect on stock returns of companies listed on the Jakarta Islamic Index (JII). The research period was carried out for three years, starting from 2017 to 2019. The samples used in this study were 17 companies selected using the purposive sampling method. This study uses multiple linear regression analysis method. The results of the study show that the January effect and weekend effect have no influence on stock returns in companies listed on the JII Index for the 2017–2019 period.

Keywords January effect · Weekend effect · Stock return

1 Introduction

The capital market is a means for a company to meet its funding needs, while also provides facilities to people who want to invest in financial instruments such as stocks, bonds, mutual funds, and others [1]. Fama [2], who pioneered the popular theory used in capital market, classified the level of market efficiency based on the level of relevance of information received by the market into three forms, namely: (1) weak, (2) semi-strong, and (3) strong form.

A market with a weak form of efficiency is a condition when the collection of information is only based on historical prices or prices in the past. Semi-strong form market efficiency occurs when stock prices fully reflect all publicly available or published information, such as the annual income statement. Furthermore, strong form market efficiency occurs when investors or a group have monopolistic access given to all types of information related to price formation [2].

S. H. Putri · N. Fauziah (✉)
Management Department, Faculty of Business and Economics, Universitas Islam Indonesia, Special Region of Yogyakarta, Indonesia
e-mail: nurfauziah@uii.ac.id

© The Author(s), under exclusive license to Springer Nature Switzerland AG 2024
A. Hamdan and E. S. Aldhaen (eds.), *Artificial Intelligence and Transforming Digital Marketing*, Studies in Systems, Decision and Control 487,
https://doi.org/10.1007/978-3-031-35828-9_91

Capital market anomaly is a deviation that occurs due to a pattern of stock movement used by investors to gain above average returns (abnormal returns) [3]. Anomalies do not only occur in one type of efficient market form, although they are more often found in semi strong efficient market forms [4].

In financial theory, it is known that there are at least four kinds of market anomalies, namely firm, seasonal, event, and accounting anomalies [4, 5]. When compared to the four types of market anomalies based on the financial theory, seasonal anomalies are the most frequently studied in terms of their influence on stock returns.

Seasonal anomalies rely on the assumption of a certain pattern of the stock market formed from stock prices in the past to predict stock prices in the future [6]. Seasonal anomaly enable investors to estimate returns at certain times resulting in a pattern of stock price movements that can be predicted by investors, which results in investors taking advantage of this pattern to get abnormal returns [7]. According to [4], there are six special types of seasonal anomaly, namely the January effect, weekend effect, time of day effect, end of month effect, seasonal effect, and holiday effect.

The January effect is known as the condition when the rate of return is high in January compared to other months [8]. This phenomenon is related to the occurrence of year changes in December which is the end of the tax year and January as the beginning of the tax year [9]. The January effect is affected by sale of shares at the end of the year to reduce taxes (tax-loss selling), realizing capital gains, the effect of window dressing, portfolios or investors selling their shares for holidays [10].

On the other hand, the weekend effect is an anomaly that occurs when security prices tend to rise on Friday and fall on Monday [4]. French [11] was the first to pay attention to the weekend effect and since then other researchers have tried to examine and attempt to uncover the causes of the weekend effect [12]. The weekend effect shows that there is negative income after the weekend, which is predicted because there is a settlement effect and measurement errors [5].

This study attempts to investigate and analyze how stock returns in firms that are listed on the Jakarta Islamic Index are impacted by the weekend effect and the January effect (JII). This study has three sub-main part, namely: (1) Introduction, which explains a theoretical problem and business phenomenon relating to the state of company stock returns in Indonesia; (2) Literature Review and Methods, which encompasses the concepts of capital market, efficient and adaptive capital market, capital market anomaly, as well as the January effect and weekend effect; and (3) Results, discussion, and conclusion of the study.

2 Literature Review

2.1 Capital Market

The capital market is a market for various kinds of long-term financial instruments that can be traded, whether in the form of bonds, stocks, mutual funds, derivative instruments, or others. The capital market performs two functions for a country's economy, namely as a means for business funding or as a forum for companies to obtain funds from investors and a forum for the public to invest in financial instruments.

In capital market, there are a number of theories introduced by the previous scholars. Fama [2] discovered capital market efficiency theory, which means that capital market is efficient if the available information is fully reflected in price [13]. Another theory is adaptive market hypothesis (AMH), which adjust market efficiency and behavioral alternatives by applying the principles of evolution (competition, adaptation, and natural selection) to financial interactions [14]. The last is capital market anomaly, or market asymmetry caused by the information bias among investors [15, 16].

2.2 January Effect

The January effect is a deviation in returns in January that experienced a significant increase compared to other months [16]. January effect phenomenon can be explained by tax loss selling, window dressing, and small stock's beta [17]. The January effect is a phenomenon related to the occurrence of year changes, especially in the context of tax year between December and January. Companies at the end of the year usually make calculations of tax payments so that investors prefer to release their shares whose value has decreased to avoid losses, but at the beginning of the year investors will buy shares again which will affect stock price returns [9].

2.3 Weekend Effect

The weekend effect is a phenomenon in the capital market where stock returns tend to be lower on Mondays and higher on Fridays [12]. This is a phenomenon influenced by the psychological nature of investors and traders [18]. According to [19], Monday is considered the worst day compared to other days because it is the first day of work, but inversely proportional to Friday which is considered the best day because it is the last working day before a holiday, so that investors tend to be more pessimistic on Mondays and be more optimistic Friday.

2.4 Hypothesis Development

Previous research has found the effect of the January effect anomaly on company returns. The January effect is proved to occur in Lithuania, Estonia [20], Jordan, Egypt, Lebanon, and Morocco [21]. In Indonesia, there is also a number of study on the influence of the January effect on company returns. Research from [8] proved that returns in January to be higher when compared to other months on the Indonesia Stock Exchange and the Shanghai Stock Exchange for the 2011–2013 period. Likewise, research conducted by [9] found a January effect on the Indonesia Stock Exchange in 2009–2013 year.

H1: The phenomenon of the January effect affects the stock returns of companies listed on the JII index.

Based on previous studies, it was found that the weekend effect has an influence on company returns. Schaub et al. [22] in their study found weekend effect anomalies on the Hong Kong, Korean and Japanese stock markets in 1985–2004. Besides that, [23] in their research found weekend effect anomalies on the Ukrainian stock market. In Indonesia, the influence of weekend effect on company returns was also found. Dwitania [7] in her study proved that the weekend effect phenomenon had a significant influence on stock returns on the LQ45 index for the period February 2018–January 2019. In a similar vein, [24] also found a significant influence between the weekend effect and stock returns on the Kompas index 100.

H2: The weekend effect phenomenon affects the stock returns of companies listed on the JII index.

3 Research Method

The population in this study is the shares of companies listed on the Indonesian Stock Exchange. The samples were company shares listed on the JII index for 3 years, namely in the 2017–2019 period. The research sample was taken using a purposive sampling technique, resulting in 17 company shares.

The dependent variable in this study is the company's stock returns on the JII index in the 2017–2019 period. The independent variables in this study are the January effect and the weekend effect. The stock return used is the closing price of the daily stock return on the five trading days. The formula for calculating stock returns is as follows:

$$R_{it} = \frac{P_{it} - P_{it-1}}{P_{it-1}}$$

Details:

R_{it} = Return of stock realization of company i in time t.

The Analysis of the Influence of the January Effect and Weekend Effect ... 1085

Table 1 Results of descriptive statistics analysis of January effect

Descriptive statistics

	N	Minimum	Maximum	Mean	Std. deviation
January	51	− 0.00693	0.04461	0.0037623	0.00781024
February	51	− 0.00918	0.00616	− 0.0009684	0.00307208
March	51	− 0.01377	0.00910	− 0.0014879	0.00421443
April	51	− 0.00645	0.00642	0.0002188	0.00327155
May	51	− 0.01060	0.00902	− 0.0011131	0.00396341
June	51	− 0.01113	0.00798	0.0001420	0.00410354
July	51	− 0.00636	0.01319	0.0007165	0.00378477
August	51	− 0.00965	0.01037	− 0.0001780	0.00439675
September	51	− 0.01063	0.00705	− 0.0013638	0.00361212
October	51	− 0.01454	0.00989	− 0.0004429	0.00432418
November	51	− 0.01275	0.01619	− 0.0008152	0.00609915
December	51	− 0.00698	0.01296	0.0028458	0.00406067

P_{it} = Closing price of current daily stock of company i.

P_{it-1} = Closing price of previous daily stock of company i.

In this study the data used is secondary data. The data used comes from the daily closing price of shares on the JII index in the 2017–2019 period. This study uses data analysis methods in the form of multiple regression analysis, F test, and T test.

4 Results of Data Analysis

4.1 Descriptive Statistics Analysis

See Table 1.

Based on the results of the descriptive statistical analysis of the January effect which can be seen in Table 1, the lowest minimum value during the study period occurred in October at − 0.01454 while the highest maximum value occurred in January at 0.04461. The average value in January is the highest average value and the average stock return value in May is the lowest average stock return value.

4.2 Multiple Linear Regression Analysis

See Table 2.

Table 2 Results of multiple linear regression of January effect

Coefficient

Model		Unstandardized coefficients		Standardized coefficients	T	Sig.
		B	Std. error	Beta		
1	(Constant)	0.099	0.035		2.801	0.008
	January	0.093	4.391	0.004	0.021	0.983
	February	9.200	9.153	0.158	1.005	0.321
	March	3.795	6.760	0.089	0.561	0.578
	April	25.754	9.557	0.470	2.695	0.010
	May	− 1.325	7.372	− 0.029	− 0.180	0.858
	June	− 32.508	10.511	− 0.744	− 3.093	0.004
	July	− 6.664	7.728	− 0.141	− 0.862	0.394
	August	7.209	7.195	0.177	1.002	0.323
	September	7.868	7.079	0.159	1.111	0.273
	October	− 3.248	7.736	− 0.078	− 0.420	0.677
	November	− 17.990	6.995	− 0.612	− 2.572	0.014
	December	0.100	7.281	0.002	0.014	0.989

Based on the results of the multiple linear regression for the January effect, the equation are as follows:

$$\begin{aligned} \text{Stock Return} = {}& 0.009 + 0.093\,\text{January} \\ & + 9.200\,\text{February} + 3.795\,\text{March} \\ & + 25.754\,\text{April} - 1.325\,\text{May} \\ & - 32.508\,\text{June} - 6.664\,\text{July} \\ & + 7.209\,\text{August} + 7.868\,\text{September} \\ & - 3.248\,\text{October} - 17.990\,\text{November} \\ & + 0.100\,\text{December} + \in \end{aligned}$$

See Table 3.

Based on the results of the multiple linear regression for the weekend effect, the equation are as follows:

$$\begin{aligned} \text{Stock Return} = {}& - 0.001 - 0.591\,\text{Monday} \\ & - 0.944\,\text{Tuesday} \\ & - 0.148\,\text{Wednesday} \\ & + 0.723\,\text{Thursday} \\ & + 0.136\,\text{Friday} + \in \end{aligned}$$

The Analysis of the Influence of the January Effect and Weekend Effect … 1087

Table 3 Results of multiple linear regression of weekend effect

Coefficient

Model		Unstandardized coefficients		Standardized coefficients	T	Sig.
		B	Std. error	Beta		
1	(Constant)	− 0.001	0.002		− 0.617	0.540
	Monday	− 0.591	0.711	− 0.125	− 0.831	0.410
	Tuesday	− 0.944	0.726	− 0.205	− 1.300	0.200
	Wednesday	− 0.148	0.866	− 0.026	− 0.170	0.865
	Thursday	0.723	0.982	0.115	0.736	0.465
	Friday	0.136	1.017	0.023	0.134	0.894

Table 4 Results of determinant coefficient

Model summary

Model	R	R square	Adjusted R square	Std. error of the estimate
1	0.372	0.139	0.103	0.00267342

4.3 Determinant Coefficient

See Table 4.

Based on the results of the determinant coefficient which can be seen in the adjusted R square column, the influence of each month and stock trading days on stock returns is 10.3%, while the other 89.7% is influenced by other variables not included in this study.

4.4 Hypothesis Test

4.4.1 F Test

See Table 5.

Table 5 Results of F test of January effect

ANOVA

Model		Sum of squares	Df	Mean square	F	Sig.
1	Regression	0.519	12	0.043	1.510	0.163
	Residual	1.088	38	0.029		
	Total	1.607	50			

Table 6 Results of F test of weekend effect

ANOVA

Model		Sum of squares	Df	Mean square	F	Sig.
1	Regression	0.001	5	0.000	0.716	0.615
	Residual	0.012	45	0.000		
	Total	0.013	50			

The results of the F test show a significance probability value of 0.163, which has a greater value when compared to the significance level of 0.05. Therefore, January, February, March, April, May, June, July, August, September, October, November and December have no significant influence on stock returns.

According to [25] the possible causes of insignificant influence of January effect, or how the different months of January to December not having an effect on stock returns are due to market reactions that are not interested in the pattern of the turn of the year. It can also be caused by the cold response from the market towards certain event, either in the form of a positive or negative response. In addition, according to [15], there is no influence of the months from January to December on stock returns according to the random walk theory. This theory basically states that the movement of the capital market cannot be predicted due to its random pattern.

See Table 6.

Based on the results of the F test in Table 6, it can be seen that the significance probability value is 0.615, which is higher than the significance level, which is 0.05. As a result, it can be concluded that Monday, Tuesday, Wednesday, Thursday, and Friday as a whole do not have a significant influence on stock returns.

Regarding this, [26] suggested that the insignificant influence of stock trading days, namely Monday to Friday on stock returns, can be caused by investor behavior. When making investments, they are not only influenced by the day, but are also influenced by the laws of demand and supply. In addition, research conducted by [27] also did not show any effect from Monday to Friday on stock returns, because there are other factors that affect stock returns more than Monday to Friday, namely information about the company, stock split information, mergers and acquisitions, and other activities carried out by the company.

4.4.2 T Test

See Table 7.

The results of the T-test analysis indicate that January coefficient has a positive relationship of 0.093 and a significant value of 0.983. The significant value for January is greater than 0.05, therefore H0 is accepted, and it can be concluded that the January effect has no influence on the stock returns of the 17 companies listed as the JII index in the period 2017 to 2019.

The Analysis of the Influence of the January Effect and Weekend Effect … 1089

Table 7 Results of T test of January effect

Coefficient

Model		Unstandardized coefficients		Standardized coefficients	T	Sig.
		B	Std. error	Beta		
1	(Constant)	0.099	0.035		2.801	0.008
	January	0.093	4.391	0.004	0.021	0.983
	February	9.200	9.153	0.158	1.005	0.321
	March	3.795	6.760	0.089	0.561	0.578
	April	25.754	9.557	0.470	2.695	0.010
	May	− 1.325	7.372	− 0.029	− 0.180	0.858
	June	− 32.508	10.511	− 0.744	− 3.093	0.004
	July	− 6.664	7.728	− 0.141	− 0.862	0.394
	August	7.209	7.195	0.177	1.002	0.323
	September	7.868	7.079	0.159	1.111	0.273
	October	− 3.248	7.736	− 0.078	− 0.420	0.677
	November	− 17.990	6.995	− 0.612	− 2.572	0.014
	December	0.100	7.281	0.002	0.014	0.989

The absence of the January effect phenomenon can be caused by cultural differences, because the majority of the January effect occurs in developed countries where at the beginning of the year and the end of the year there are major celebrations, namely Christmas and New Year which are celebrated by the majority of citizens of developed countries. According to [28] in his study, the January effect did not occur on the Indonesia Stock Exchange. In fact, during 2016–2018 there was a decline in the JCI during the initial trade on the Indonesia Stock Exchange, with one of the reasons being foreign investors who were still selling shares in early January.

See Table 8.

Table 8 Results of T test of weekend effect

Coefficient

Model		Unstandardized coefficients		Standardized coefficients	T	Sig.
		B	Std. error	Beta		
1	(Constant)	− 0.001	0.002		− 0.617	0.540
	Monday	− 0.591	0.711	− 0.125	− 0.831	0.410
	Tuesday	− 0.944	0.726	− 0.205	− 1.300	0.200
	Wednesday	− 0.148	0.866	− 0.026	− 0.170	0.865
	Thursday	0.723	0.982	0.115	0.736	0.465
	Friday	0.136	1.017	0.023	0.134	0.894

Based on the results of the T-test analysis for the weekend effect variable, the significance value on Monday and Friday is not less than 0.05, on Monday it is 0.410 while on Friday it is 0.894, so H0 is accepted and it can be concluded that the weekend effect has no influence on stock returns. The results of this study are in line with [29–31], who mentioned the possibility that the weekend effect would not occur due to the cautious attitude of market players who also anticipate economic conditions which also remain unchanged [32].

5 Conclusion

Based on the results of the tests that have been carried out, it can be concluded the January effect phenomenon does not affect stock returns, due to cultural differences as the phenomenon is more often found in developed countries. Furthermore, the weekend effect phenomenon also does not affect stock return, which could occur due to cautious attitude of market players and the growth of the capital market in Indonesia.

This study contributes to the enrichment of capital market theory and practices, by highlighting that the condition between capital market in developed and developing country differs due to the circumstances. Future research can develop this research model by examining the Rogalski effect phenomena, namely the return every Monday in each month, or examining other anomalies such as neglect effect or institutional holding.

References

1. Fakhruddin, M.H.: Istilah Pasar Modal A–Z. Elex Media Komputindo (2008). https://books. google.co.id/books?lr=&id=wx5bDwAAQBAJ&dq=pasar+modal+adalah&q=pasar+modal#v=snippet&q=pasar%20modal&f=false. Accessed 26 Jun 2020
2. Fama, E.F.: Efficient capital markets: a review of theory and empirical work. J. Financ. **25**(2), 383–414 (1970)
3. Alliyah, C., Ekawaty, M.: Analisis anomali pasar modal terhadap return Dan abnormal return di Jakarta Islamic index tahun 2015. J. Ilmiah Mahasiswa FEB Universitas Brawijaya **5**(2), 2–8 (2017)
4. Gumanti, T.A., Utami, E.S.: Bentuk pasar efisien dan pengujiannya. J. Akuntansi Dan Keuangan **4**(1), 55–67 (2002)
5. Indriasari, I. Sugiarto: Seasonal effects pada anomali pasar modal: suatu review. J. Dinamika Ekonomi Dan Bisnis (JDEB) **11**(1), 1–12 (2014)
6. Li, B., Liu, B.: Monthly seasonality in the New Zealand stock market. Int. J. Bus. Manag. Econ. Res. **1**(1), 9–14 (2010)
7. Dwitania, R.: Pengaruh Monday Effect, Weekend Effect, Dan Week Four Effect Terhadap Return Saham LQ45 Di Bursa EFEK Indonesia. Skripsi, Universitas Pasundan (2019)
8. Kartikasari, L.H.: Pengujian January effect: studi komparasi pada bursa EFEK Indonesia dan bursa saham shanghai periode 2011–2013. J. Bus. Bank. **6**(1), 65–80 (2016)
9. Wulandari, A.: Analisis Fenomena January Effect Pada Saham LQ45 Yang Listing Di BEI Periode 2009–2013. Skripsi, Universitas Negeri Padang (2014)

The Analysis of the Influence of the January Effect and Weekend Effect ...   1091

10. As'adah, L.: Pengaruh January Effect Terhadap Abnormal Return Dan Volume Perdagangan Pada Saham Di Jakarta Islamic Index (JII). Skripsi, Universitas Islam Negeri (UIN) Sunan Kalijaga Yogyakarta (2009)
11. French, K.R.: Stock returns and the weekend effect. J. Financ. Econ. **8**(1), 55–69 (1980)
12. Singhal, A., Bahure, V.: Weekend effect of stock returns in the Indian market. Great Lakes Herald **3**(1), 12–22 (2009)
13. Al-Khazali, O., Mirzaei, A.: Stock market anomalies, market efficiency and the adaptive market hypothesis: evidence from Islamic stock indices. J. Int. Finan. Markets Inst. Money **51**, 190–208 (2017)
14. Lo, A.W.: The adaptive market hypothesis. J. Portfolio Mgmt. **30**(5), 15–29 (2004)
15. Sari, F.A., Sisdyani, E.A.: Analisis January effect di pasar modal Indonesia. E-J. Akuntansi Universitas Udayana **6**(2), 237–248 (2014)
16. Yunita, N.K.E., Rahyuda, H.: Pengujian anomali pasar (January effect) di bursa EFEK Indonesia. E-J. Manajemen Unud **8**(9), 5571–5590 (2019)
17. Pratomo, A.W.: January Effect Dan SizeEffect Pada Bursa EFEK Jakarta (BEJ) Periode 1998–2005. Tesis, Universitas Diponegoro (2007)
18. Budiwati, C., Yudana, R.N.: The effects of the days of the week on the Indonesian stock exchange. J. Fin. Bank. Rev. **2**(4), 22–27 (2017)
19. Ramadhani, R., Subekti, I.: Pengujian anomali pasar Monday effect, weekend effect, rogalski effect di bursa EFEK Indonesia. J. Ilmiah Mahasiswa FEB Universitas Brawijaya **3**(2), 2–24 (2015)
20. Norvaisiene, R., Stankeviciene, J., Lakstutiene, A.: Seasonality in the Baltic stock markets. Procedia Soc. Behav. Sci. **213**, 468–473 (2015)
21. Gharaibeh, O.: The january effect: evidence from four Arabic market indices. Int. J. Acad. Res. Acc. Finance Manag. Sci. **7**(1), 144–150 (2017)
22. Schaub, M., Lee, B.S., Chun, S.E.: Overreaction and seasonality in Asian stock indices: evidence from Korea, Hong Kong and Japan. Res. Finance **24**, 169–195. Emerald Group Publishing Limited (2008)
23. Plastun, A., Gil-Alana, L., Caporale, G.M.: The weekend effect: an exploitable anomaly in the Ukrainian stock market? J. Econ. Stud. **43**(6), 954–965 (2016)
24. Sofiana, N.: Analisis Return Saham Jumat Dan Senin Perusahaan Kategori Kompas 100 Yang Terdaftar Di Bursa EFEK Indonesia. Skripsi, Universitas Muhammadiyah Malang (2020)
25. Sanjaya, M.J.: Analisis January Effect Pada Perusahaan Yang Terdaftar Di LQ45 Sebagai Bahan Pengembalian Keputusan Pada Bursa EFEK Indonesia Periode 2007–2011. Skripsi, Universitas Islam Negeri Maulana Malik Ibrahim (2012)
26. Yusuf, M.: Analisis Pengaruh Day of the Week Effect Terhadap Return Saham Pada Bursa EFEK Jakarta Periode Januari 2014 Sampai Dengan Desember 2014. Skripsi, Universitas Dian Nuswantoro Semarang (2016)
27. Kasdjan, A.M.Z., Nazarudin, Yusuf, J.: Pengaruh anomali pasar terhadap return saham perusahaan LQ-45. J. Kajian Akuntansi **1**(1), 35–48 (2017)
28. Darman, T.: Apakah terjadi January effect di bursa EFEK Indonesia? J. Riset Akuntansi Dan Keuangan **6**(1), 73–80 (2018)
29. Lugiatno: Analisis weekend effect terhadap return saham di bursa EFEK Indonesia (Analysis of weekend effect toward stock reurn in Indonesia stock exchange). J. Ilmu Manajemen Dan Akuntansi Terapan (JIMAT) **2**(2), 1–11 (2011)
30. Kusno, H.S., Murlita, Y.A., Ramli, R., Finanto, H.: Pengaruh anomali pasar terhadap return saham perusahaan perbankan yang tedaftar pada bursa EFEK Indonesia periode 2018–2020. INOVASI **17**(4), 785–791 (2021)
31. Truong, L.D., Friday, H.S.: The January effect and lunar new year influences in frontier markets: evidence from the Vietnam stock market. Int. J. Econ. Financ. Issues **11**(2), 28–34 (2021)
32. Liang, X., Liu, Q., Zebedee, A.A.: One country, two calendars: lunar January effect in china's A-share stock market. Asia-Pac. J. Fin. Stud. **51**(6), 859–895 (2022)

Integrated Study of Ethical and Economic Efficiency of Society's Perception of Corporate Social Responsibility in India

Mohammad Talha, Marim Alenezi, Syed Mohammad Faisal, and Ahmad Khalid Khan

Abstract This research paper examines the different facets of corporate social responsibility (CSR), its relevance, and its economic efficiency on society. A Likert scale with five points was used in the study to assess the attitudes of the general public and the business community. The data analysis was carried out with SPSS. According to the findings of the study, corporate social responsibility has a substantial impact on societies in order to sustain economic efficiency. The results of the EFA analysis indicated that there is a significant association between the various components of the responses. Despite the numerous discrepancies that were revealed in the process of empowering society in various fields, it is impossible to rebut the misleading claim that the firm was effectively embracing CSR. As a consequence, the numerous stakeholders will have a better understanding of the connections between CSR, businesses, and society. This very original study reveals crucial details about the legal and regulatory issues that arise when integrating CSR into an organization's internal governance structure, resulting in a benefit to society that directly reaps the rewards of CSR. This study was carried out to demonstrate that integrating CSR into an organization's internal governance structure can positively impact society.

Keywords CSR · Society · Corporates · Economic efficiency · EFA

M. Talha · M. Alenezi (✉)
Department of Accounting and Finance, College of Business Administration, Prince Mohammad Bin Fahd University, Al-Khobar, Saudi Arabia
e-mail: malenezi@pmu.edu.sa

M. Talha
e-mail: mtalha@pmu.edu.sa

S. M. Faisal · A. K. Khan
Department of Management, Applied College, Jazan University, Jizan, Saudi Arabia
e-mail: dfaisal@jazanu.edu.sa

A. K. Khan
e-mail: akkhan@jazanu.edu.sa

© The Author(s), under exclusive license to Springer Nature Switzerland AG 2024
A. Hamdan and E. S. Aldhaen (eds.), *Artificial Intelligence and Transforming Digital Marketing*, Studies in Systems, Decision and Control 487,
https://doi.org/10.1007/978-3-031-35828-9_92

1 Introduction

"Corporate social responsibility" has been shortened to "CSR" in the business world due to its overuse. The phrase refers to a corporation's responsibility to report its operations and effects to its shareholders and the communities in which it works. Participating in CSR initiatives demonstrates a company's concern for morals, the community, and the environment [30]. The company will outline its efforts to adhere to its CSR policy as often as it announces its financial accomplishments. CSR offers a wide range of advantages, including increased profitability and value, better customer interactions, and other outcomes [23]. CSR programs, according to critics, are probably ineffective. Because savvy customers may see "greenwashing" and the apparent contradiction between CSR and the economic purpose, some detractors view CSR as a waste of time and resources (Greenwashing refers to company practices that provide the impression of being environmentally conscious but do not lead to changes to standard operating procedures) [10]. Giving back to the community, helping those in need, and making a constructive contribution to society are all parts of corporate social responsibility. The adoption of CSR efforts by businesses to improve society and build a positive reputation in the marketplace has significantly increased [5]. Since its original suggestion that firms put aside a percentage of their profits to contribute to charity organizations, corporate social responsibility, or CSR, has evolved into an essential component of how many businesses operate today [7].

1.1 *Structure of Study*

This study focuses on six primary facets of the topic. The structure consists of the following: the first portion, discussed deeply into corporate social responsibility, and its connection with society, its beginnings, the era in which it arose, who is regarded as the "father" of CSR, and who was responsible for introducing CSR to India. In the second part, the theoretical foundations of corporate social responsibility are examined briefly, it analyzes in detail the most current studies and academic contributions of eminent academicians to CSR. The third portion provides a comprehensive objective of the research and next research methodologies has been discussed in detail like how data were collected, and the statistical approaches used to reach a reasonable conclusion. The application of statistical methods and the derivation of findings can be found in Part 4. Following that, Sect. 5 has an in-depth analysis of the results and an explanation of what the results indicate, and Sect. 6, brings the research to a close.

2 Literature Review

The goals of Islamic banking are to make the world more affluent and fairer. In this research study, the benefits of organizations following Islamic CSR rules are looked at. This study looks at [31] how Islamic CSR could help solve problems like poverty. The results are based on a thorough look at Islamic banking in Indonesia and talks with experts in the field. Shari'ah says that people should take care of their workers, do what they are supposed to do, protect the environment, give to charity, and think about the next generation. CSR must be able to combine two Islamic ideas for it to make sense in an Islamic setting. Social capital should be the primary goal of CSR.

This composition discusses [13] the long-term goals of corporate social responsibility (CSR) programs are to improve the environment, economy, and society. Most important tourism-related firms have at least one CSR project. However, studies have shown that initiatives are often put into action right once and are related to cost savings and the company's reputation. This strategy is comparable to "planning for visitors first" from our perspective. We propose a Development First framework for CSR, based on Peter Burns' model for tourism planning, as a way for companies to handle the growing demand for moral behavior beyond 2015. This way of thinking is long-lasting, enduring, and unique. The possibility that specific projects will lead to good, long-term development in diverse regions may be evaluated by geographers and other social scientists. This manuscript argues that a proactive approach to corporate social responsibility is essential for effectively combating global poverty. This research explores, from the perspective of empowerment, the characteristics of companies that may aid the poor in escaping poverty. In this article, we develop a plan for firms that may help combat poverty, and we use case studies to explain how this plan could be put into reality. One-way corporations may assist fewer poor individuals to acquire employment is by sharing technology and resources and offering them education and training in the necessary skills. Enterprises may assist in eliminating poverty by employing disadvantaged individuals, providing money to NGOs that serve the poor, or giving the poor the financial and social support they need to establish their businesses [17].

CFP has positive effects on CSR (Zakat). These results show that zakat is a successful strategy for meeting material, productive, and social demands. The economic forecasts are accurate. Businesses in Islamic countries may help lessen economic disparities by using zakat as corporate social responsibility (CSR). The goal of this strategy is to reduce economic disparity. Companies participate in CSR because they think it will improve their financial performance in the long run.

This research examines [19] CSR's influence on corporate success in China. Between 2006 and 2016, 839 enterprises in 31 provinces submitted 34,000 CSR initiatives. It is possible to understand CSR activities' setting, patterns, and effects on an organization's productivity and financial performance via modeling strategies including clustering, ordinary least squares, and fixed effects panel regression. The data analysis and modeling demonstrate that

- Most programs were designed to motivate businesses to improve their environmental performance.
- Businesses benefited from environmental and social initiatives equally, and
- Trends, context, and effects varied over time by firm type and region (provinces).

Enterprises in regions with less wealth (lower GDP per capita) seem less likely to implement environmentally and socially beneficial sustainable practices. This research might be helpful for organizations considering CSR initiatives and stakeholders and managers in China attempting to advance sustainable development. This paper contends [29] the impact of corporate social responsibility on marketing strategy has been the subject of several studies. The use of bibliometric research is underutilized, in comparison. In this study, we concentrate on recent publications to undertake a bibliometric analysis of corporate social responsibility (CSR) studies in marketing. Two thousand four hundred twenty-one publications were found after utilizing the Web of Science to search for similar ones. Utilizing VOS viewer, the data visualizations were produced. In this study, 42.16% of all papers were published in the top journals. The "Journal of Business Ethics" and "Sustainability" proliferated in recent years and were the two most prolific publishers. According to data on the most common search terms on the Internet, corporate social responsibility is increasingly being used as an advertising tactic. This study tries to identify the modern CSR and marketing strategies that are most successful.

With the overarching goal of creating a framework and offering directions for future research, the assessment of a firm's contribution to the reduction of poverty, the types of pro-poor CSR initiatives that firms could adopt, and the factors influencing a firm's contribution are all critically examined. Future benefits of this research strategy are envisaged, but it may also help the UN to persuade more companies to support its goal of eradicating poverty via sustainable development. This article aims [22] to provide a concise, logical overview of the development and history of research on corporate social responsibility (CSR) from its conception to the present. We will get to work and begin from the beginning as soon as possible. A study that examines the most critical things authors and institutions have done to promote social responsibility and how it has benefited the discipline is also made available. Since institutional theory and the stakeholder approach are tendencies that reinforce one other, they will get special attention in this research. Our conversation will focus on the notion and its arguments and controversies.

This paper contours [6] while sustainability management is becoming more common in major firms, assessments that track the earth's condition do not account for the impact these organizations' activities have on it. It is common to refer to the ensuing chasm as a "big gap." This article discusses "When is a corporation sustainable?" and "What are its characteristics?". What role businesses can play in addressing our environmental concerns is another question. We must recognize the characteristics that set sustainable businesses apart from imitators now more than ever. To further clarify, we provide a detailed explanation of "business sustainability." We look at some of the most common approaches and create a taxonomy for firms' long-term profitability while considering how each would affect environmental preservation. This paper overviews [3] is society better off as a consequence of businesses voluntarily taking up some of its numerous issues? These subjects, as well as the underlying ideas, are thoroughly examined in this article. This research examines [14] the connection between small and medium-sized enterprises' financial success and CSR (also known as corporate social responsibility) in the Indian state of Rajasthan. The conditions impacting small and medium-sized enterprises in Jaipur will be the subject of this study. We will include a wide range of interested stakeholders in the framework to define CSR (SMEs). It was accomplished via an exploratory study, in which data were gathered from 384 SMEs using a structured questionnaire made up of previously tested scale items, and second-order structural equation modeling was used for analysis. The results show a significant, if slightly positive, correlation between CSR and financial success, demonstrating the model's general accuracy.

There is no dearth of literature on corporate social responsibility. The prominent academics helped develop a strategy to deal with unethical conduct. There must be linkage across different systems to successfully help the underprivileged. There are not enough safeguards in place to identify or stop fraud inside businesses. It is possible that not enough money will be provided to stop the epidemic. It is not only a waste of tax money to support criminals. It is terrible for the company and the government if a significant legal loophole that might be used for corruption is permitted to continue.

3 The Objective of CSR is as Follows

- To know the society's perception of practical implications of CSR by corporates.
- To analyze various factors of CSR implication through exploratory factor analysis (EFA).
- To analyze the relationship among the factors formed on the basis of statistical results of EFA.

4 Research Methodology

In this study, we used a descriptive as well as exploratory research design to find out the various impact of CSR in the context of society E as well as corporate in India. Advanced quantitative analysis, SPSS was used to analyze the data. And EFA (exploratory factor analysis) was used to get determine various variables and their relationship, including health and safety sanitization health factors, training

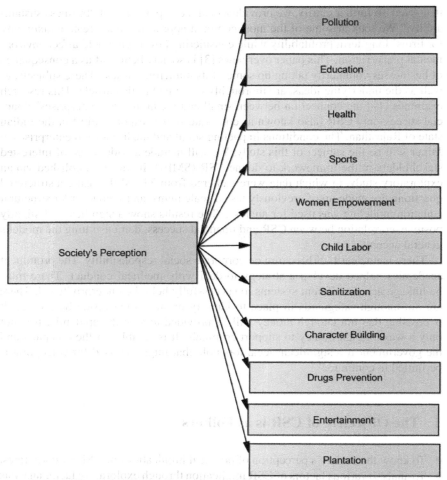

Perception of Society Towards Corporate Social Responsibility

Fig. 1 Perception of society towards CSR

development etc. A set of 15 questions was prepared and accordingly interviewed of respondents on the basis of the model depicted in Fig. 1.

4.1 Sample Size

For this research, a sample size of five hundred people was collected from various industrial cities across India. During the process of cleansing the data, there were only

Integrated Study of Ethical and Economic Efficiency of Society's ... 1099

489 responses recorded. After the replies had been recorded, 11 of those responses were subsequently discarded as being too unclear for further consideration.

4.2 Data Collection

Before commencing the complete test, a pilot study was done to examine the Likert scale questionnaire's validity, consistency, and reliability. Prior to starting the investigation, this was finished. This research project aims to assess the volume of quantitative data amassed via surveys using an appropriate sampling approach and direct surveys in electronic format. The bulk of the data was gathered using a 5-point Likert scale, where 1 denotes a level of firmly held disagreement and five denotes highly agreed upon positions.

Before employing the complete analysis, a pilot study was undertaken to ensure the validity consistency and reliability of the likert scale questionnaire. This study was conducted on 489 respondents from the society after preliminary screening as mentioned above. The study is the reflection of quantitative data analysis it was gathered via surveys using a suitable sampling method by using direct surveys. The primary data was collected by using 5 point likert scale where 1 is equal to highly disagreement towards 5 which is a highly agreed level. Table 1 the clear picture about the primary statistics of the likert scale questionnaire.

As shown further in Table 2 regarding statistics of gender that out of 489 respondents taken for study there were 275 males and 214 females with 56.2% and 43.8% respectively.

Further Table 3 depicts the statistics about the qualification of target sample size studied in three cities of national capital region in India. It reveals there were 348 i.e. 71.2% population was among educated ones and rest 141 i.e. 28.8% population was not educated as they mentioned in their responses.

Also in this below mentioned in Table 4 it shows the background of 489 respondents in terms of their geography. 302 i.e.61.8% belonged to rural (outskirts of city) and 187 i.e. 38.2% belonged to urban cities.

After preliminary trend analysis of the selected population reliability and validity of data executed.

4.3 Reliability and Validity of Data

To assess the authenticity and reliability of the data, we employed Cronbach Alpha values for the internal stability of the variables. The Rule of Thumb to determine whether or not the data are valid and reliable is that the Value ought to be greater than 0.6 the consistency of all the variables at a distance, as stated by Hair et al. [9].

1100   M. Talha et al.

Table 1 Descriptive statistics

	N	Minimum	Maximum	Mean	Std. deviation
Gender	489	1	2	1.44	0.497
Qualification	489	1	4	2.64	1.053
Background	489	2	4	3.19	0.715
Corporate causing air pollution	489	2	5	3.38	1.003
Is corporate taking precaution to avoid pollution	489	1	5	2.22	1.392
Corporate causing noise pollution	489	1	5	3.03	1.270
Corporate causing water pollution	489	1	5	3.16	1.186
Corporate contributing towards education	489	1	5	3.30	1.141
Corporate promote health care facilities	489	1	5	3.24	1.459
Corporate conducts sports activities	489	1	5	3.37	1.634
Corporate encourages women empowerment	489	1	5	3.39	1.504
Corporate discourages child labour	489	1	5	3.19	1.253
Corporate conducts sanitization activities	489	1	5	2.94	1.405
Corporate conducts training for character building in the society	489	1	5	3.01	1.763
Corporate runs drugs prevention campaigns	489	1	5	3.46	1.465
Corporate conducts entertainment	489	1	5	3.48	1.325
Corporate promote plantation	489	1	5	3.46	1.405

Table 2 Gender

	Valid		
	Male	Female	Total
Frequency	275	214	489
Percent	56.2	43.8	100.0
Valid percent	56.2	43.8	100.0

Table 3 Qualification

	Valid		
	Educated	Uneducated	Total
Frequency	348	141	489
Percent	71.2	28.8	100.0
Valid percent	71.2	28.8	100.0

Integrated Study of Ethical and Economic Efficiency of Society's ... 1101

Table 4 Background

	Valid		
	Rural	Urban	Total
Frequency	302	187	489
Percent	61.8	38.2	100.0
Valid percent	61.8	38.2	100.0
Cumulative percent	61.8	100.0	

5 Data Analysis

It was proposed that the reliability test, also known as the Cronbach Alpha, be performed prior to conducting the exploratory factor analysis, and the result of that test was determined to be 894 (see Table 5). This value is more significant than 0.6 and indicates that the questionnaire may be relied upon.

After the Cronbach Alpha reliability test, an exploratory component analysis was done in Principal Component Analysis. This analysis focused on those factors whose eigenvalues were determined to be greater than one, and the results of this analysis are shown in Table 6. The Eigenvalues have been demonstrating the relationship between the variables and the factor.

After examining major components, KMO sample education initiatives were implemented.

It revealed that the KMO values were more than 0.6, a requirement for statistical significance and appropriate sampling. The KMO score of 0.677 is good (Table 7), and it is recommended that further theatrical performances occur.

After the construction of the factors, the table indicates that three aspects have been produced: honesty, integrity, and dedication (user defined) (Table 8).

The results of the below stated ANOVA Table 9, which SPSS generated, reveal that all of the factors, after having their normality tested, had p-values that were more than 0.05. The data were examined further using the ANOVA model for significance testing. In the analysis of variance (ANOVA) Table 10, the findings that were obtained show that the null hypothesis cannot be accepted when the p-value is less than 0.05, which is 0.02. As a result, there is a considerable difference between the variables' means, and the model's significance cannot be questioned. It has been shown that 73.9% of the variance in the integrity factor can be predicted from the differences in the independent components, which are honesty and commitment, respectively. The higher the R^2 value, the better the model. As a result, it has become clear that there is a close link between the dependent and the independent variables (Table 11).

Table 5 Reliability statistics

Cronbach's alpha	No. of Items
0.894	17

1102   M. Talha et al.

Table 6 Rotated component matrix[a]

	Component		
	1	2	3
Corporate conducts sports activities	0.964		
Corporate encourages women empowerment	0.768		
Corporate promote health care facilities	0.908		
Corporate contributing towards education	0.900		
Corporate causing air pollution	0.766		
Corporate discourages child labor		0.616	
Corporate conducts entertainment		0.969	
Corporate runs drugs prevention campaigns		0.906	
Corporate promote plantation		0.867	
Corporate conducts training for character building in the society		0.865	
Corporate conducts sanitization activities			0.657
Is corporate taking precaution to avoid pollution			0.888
Corporate causing noise pollution			0.804
Corporate causing water pollution			0.646

Extraction method: principal component analysis
Rotation method: Varimax with Kaiser normalization
[a] Rotation converged in 7 iterations

Table 7 Total variance explained

Component	Initial eigenvalues			Extraction sums of squared loadings		
	Total	% of variance	Cumulative %	Total	% of variance	Cumulative %
1	6.419	45.850	45.850	6.419	45.850	45.850
2	4.528	32.345	78.195	4.528	32.345	78.195
3	1.569	11.206	89.402	1.569	11.206	89.402
4	0.615	4.394	93.795			
5	0.365	2.605	96.400			
6	0.208	1.488	97.888			
7	0.113	0.806	98.694			
8	0.062	0.445	99.139			
9	0.056	0.403	99.542			
10	0.038	0.270	99.813			
11	0.015	0.106	99.919			
12	0.011	0.078	99.997			
13	0.000	0.003	100.000			
14	3.533E–005	0.000	100.000			

Extraction method: principal component analysis

Integrated Study of Ethical and Economic Efficiency of Society's ... 1103

Table 8 KMO and Bartlett's test

Kaiser-Meyer-Olkin measure of sampling adequacy		0.725
Bartlett's test of sphericity	Approx. chi-square	17,923.526
	df	91
	Sig.	0.000

Table 9 Eigenvalues (>1)

Component	Initial eigenvalues			Extraction sums of squared loadings		
	Total	% of variance	Cumulative %	Total	% of variance	Cumulative %
1	6.419	45.85	45.85	6.419	45.85	45.85
2	4.528	32.345	78.195	4.528	32.345	78.195
3	1.569	11.206	89.402	1.569	11.206	89.402

Table 10 Model summary[b]

Model	R	R^2	Adjusted R^2	Std. error of the estimate
1	0.739[a]	0.668	0.665	65.231

[a] Predictors: (constant), commitment, honesty
[b] Dependent variable: integrity

Table 11 ANOVA[a]

Model		Sum of Squares	df	Mean Square	F	Sig
1	Regression	267,109.875	2	87,678.710	22.780	0.02[b]
	Residual	245,679.160	486	1.004		

[a] Dependent variable: integrity
[b] Predictors: (constant), Commitment, honesty

Hypothesis

H0: There is no statistically significance difference of means among variables.

H1: There is statistically significance difference of means among variables.

6 Conclusion

The findings give a genuine and accurate depiction of the CSR implementation held by businesses to achieve economic efficiency, as shown by the collected data. Most respondents and society leaders who participated in this study in India were from industrial cities in the country's northern region. The statistical analysis indicates a significant relationship between the EFA-developed components. The result indicates that the ANOVA model is statistically significant at the p0.05 threshold. It is in addition to the fact that it demonstrates that the model fits the data well. Participation in EFA was directly responsible for developing traits like honesty, integrity, and dedication. When it comes to implementing the principles of corporate social responsibility in society, it is hard to overlook any of the challenges that have been presented due to their extensive interconnections. Due to the facts presented before, those responsible for designing CSR policy must devote considerable attention to the corporate policy framework of their organization. It offers new avenues of investigation into this particular component of the issue, which is quite helpful. In addition, it is highly recommended that organizations recognize that CSR is essential to their primary organizational objectives and that they cannot disregard this reality. They expected to include CSR into their system since it is an integral aspect of their other objectives. All employees, including those whose roles include setting corporate policy and taking executive action, are obliged to participate in the process of defining and implementing CSR. When using the EFA model described in this study, firms must pay special attention to the three components emphasized throughout the model. Furthermore, if the derived factors are used at their optimum level, it will achieve the corporate goal of economic efficiency. It would be advantageous for businesses if they could undertake CSR projects for the greater good of society while also creating a healthy and objective company culture.

References

1. Abad-Segura, E., Cortés-García, F.J., Belmonte-Ureña, L.J.: The sustainable approach to corporate social responsibility: A global analysis and future trends. Sustainability **11**(19), 5382 (2019)
2. Al-Malkawi, H.-A. N., Javaid, S.: Corporate social responsibility and financial performance in Saudi Arabia: Evidence from Zakat contribution. Manag. Fin. (2018)
3. Barnett, M.L.: The business case for corporate social responsibility: A critique and an indirect path forward. Bus. Soc. **58**(1), 167–190 (2019)
4. Blowfield, M.: Business, corporate responsibility and poverty reduction. In: Corporate Social Responsibility and Regulatory Governance, pp. 124–150. Springer (2010)
5. Diez-Cañamero, B., Bishara, T., Otegi-Olaso, J.R., Minguez, R., Fernández, J.M.: Measurement of corporate social responsibility: A review of corporate sustainability indexes, rankings and ratings. Sustainability **12**(5), 2153 (2020)
6. Dyllick, T., Muff, K.: Clarifying the meaning of sustainable business: Introducing a typology from business-as-usual to true business sustainability. Organ. Environ. **29**(2), 156–174 (2016)
7. García-Sánchez, I.-M., García-Sánchez, A.: Corporate social responsibility during COVID-19 pandemic. J. Open Innov. Technol. Market Compl. **6**(4), 126 (2020)
8. Ge, W., Liu, M.: Corporate social responsibility and the cost of corporate bonds. J. Account. Public Policy **34**(6), 597–624 (2015)
9. Hair Jr, J.F., Schimmel, K., Nicholls, J., Ragland, C.: The relevance of ethics, CSR, and sustainability topics in the business school and marketing curricula: dean and department head opinions. J. Bus. Ethics Educ. **13**, 169–184 (2016)
10. Hohnen, P., Potts, J.: Corporate social responsibility. Implement Guide Business (2007)

11. Hoque, N., Rahman, A.R.A., Molla, R.I., Noman, A.H.M., Bhuiyan, M.Z.H.: Is corporate social responsibility pursuing pristine business goals for sustainable development? Corp. Soc. Responsib. Environ. Manag. **25**(6), 1130–1142 (2018)
12. Hou, J., Reber, B.H.: Dimensions of disclosures: Corporate social responsibility (CSR) reporting by media companies. Public Relat. Rev. **37**(2), 166–168 (2011)
13. Hughes, E., Scheyvens, R.: Corporate social responsibility in tourism post-2015: A development first approach. Tour. Geogr. **18**(5), 469–482 (2016)
14. Jain, P., Vyas, V., Chalasani, D.P.S.: Corporate social responsibility and financial performance in SMEs: a structural equation modelling approach. Glob. Bus. Rev. **17**(3), 630–653 (2016)
15. Jakunskiene, E.: Assessment of the impact of social responsibility on poverty. Sustainability **13**(16), 9395 (2021)
16. Kamrujjaman, M., Obaidullah, M.: Poverty eradication through the corporate social responsibility (CSR) initiatives: a case study on two selected banks in Bangladesh. Int. J. Appl. Res. (IJAR) **2**(9), 43–50 (2016)
17. Kao, T.Y., Chen, J.C., Wu, J.T.B., Yang, M.H.: Poverty reduction through empowerment for sustainable development: a proactive strategy of corporate social responsibility. Corp. Soc. Responsib. Environ. Manag. **23**(3), 140–149 (2016)
18. Kolk, A.: The social responsibility of international business: from ethics and the environment to CSR and sustainable development. J. World Bus. **51**(1), 23–34 (2016)
19. Li, K., Khalili, N.R., Cheng, W.: Corporate Social Responsibility Practices in China: Trends, Context, and Impact on Company Performance (2019)
20. Lindgreen, A., Swaen, V.: Corporate social responsibility. Int. J. Manag. Rev. **12**(1), 1–7 (2010)
21. Manchiraju, H., Rajgopal, S.: Does corporate social responsibility (CSR) create shareholder value? Evidence from the Indian companies act 2013. J. Account. Res. **55**(5), 1257–1300 (2017)
22. Martínez, J.B., Fernández, M.L., Fernández, P.M.R.: Corporate social responsibility: evolution through institutional and stakeholder perspectives. Eur. J. Manag. Bus. Econ. **25**(1), 8–14 (2016)
23. Matten, D., Moon, J.: Corporate social responsibility. J. Bus. Ethics **54**(4), 323–337 (2004)
24. Medina-Muñoz, R.D., Medina-Muñoz, D.R.: Corporate social responsibility for poverty alleviation: An integrated research framework. Business Ethics Europ. Rev. **29**(1), 3–19 (2020)
25. Merino, A., Valor, C.: The potential of corporate social responsibility to eradicate poverty: an ongoing debate. Dev. Pract. **21**(2), 157–167 (2011)
26. Mocan, M., Rus, S., Draghici, A., Ivascu, L., Turi, A.: Impact of corporate social responsibility practices on the banking industry in Romania. Proc. Econom. Fin. **23**, 712–716 (2015)
27. Moharana, S.: Corporate social responsibility: a study of selected public sector banks in India. IOSR J. Bus. Manag. **15**(4), 1–9 (2013)
28. Ndaru, F.A., Kurniawan, T.: Corporate social responsibility partnership to alleviate poverty in Kulon Progo regency. BISNIS BIROKRASI: Jurnal Ilmu Administrasi dan Organisasi **22**(3), 145–155 (2016)
29. Quezado, T.C.C., Cavalcante, W.Q.F., Fortes, N., Ramos, R.F.: Corporate social responsibility and marketing: a bibliometric and visualization analysis of the literature between the years 1994 and 2020. Sustainability (2071–1050) **14**(3), (2022)
30. Windsor, D.: The future of corporate social responsibility. Int. J. Organ. Anal. **9**(3), 225–256 (2001)
31. Yusuf, M.Y., Bahari, Z.: Islamic corporate social responsibility in Islamic banking: towards poverty alleviation. Ethics Govern. Regul. Islamic Fin. 73 (2015)

New Trends in the Banking Sector and the Development of E-Banking

Mohammad Afaneh

Abstract The banking industry is no longer a limited field of work in which clients conduct business with banks all over the world in order to lend money or transfer funds between two accounts; banking is now involved in many aspects of work in clients' daily lives. Also, much of the credit for this noticeable involvement goes to the massive technological evolution we are witnessing and are a part of, which is why in this study, the role of technology in changing banking strategies and processes is discussed, as well as the benefits and drawbacks of E-Banking are discussed unbiasedly. And new E-Banking trends were illuminated. Then, some key figures from the banking and e-banking industries were examined to demonstrate the growing global adoption of E-Banking platforms.

Keywords E-Banking · Banking industry · Information technology · Risk · Automated teller machine

1 Introduction

Banking is currently being rebuilt as a result of technological evolution, intense competition, and innovation, as well as the emergence and adoption of new methods and strategies that emphasize providing banking services in the most efficient ways while protecting customers' privacy.

Nowadays, one of the most important aspects of banking sector enhancement is that banks all over the world begin to offer modified services 24 h a day, seven days a week, without sacrificing quality standards or customer privacy, while also enriching the customer experience.

It is no secret that many types of work from various backgrounds and fields have begun to invest heavily in technology, and the banking sector is no exception, As a result, the enormous impact of information technology (IT) is discussed in this study,

M. Afaneh (✉)
Imam Mohammad Ibn Saud Islamic University, Riyadh, Saudi Arabia
e-mail: afaneh100@yahoo.com

© The Author(s), under exclusive license to Springer Nature Switzerland AG 2024
A. Hamdan and E. S. Aldhaen (eds.), *Artificial Intelligence and Transforming Digital Marketing*, Studies in Systems, Decision and Control 487,
https://doi.org/10.1007/978-3-031-35828-9_93

as well as the risk-to-reward ratio and the benefits and drawbacks of adopting new technological-based strategies and methods in the banking sector.

The banking environment and banking services in general are constantly changing; clients no longer need to visit a branch to check their remaining balance, withdraw cash, or even issue a check book; all of this is now possible using a technological platform, such as the customer's cell phone, personal laptop, and automated teller machine (ATMs). What was once the norm can now happen anywhere in the world via a delivery channel provided by the bank. Thus, traditional banking is increasingly threatened by IT, which is gradually reducing face-to-face interactions with clients.

Such advancement can be viewed as a double-edged sword; on the one hand, it greatly simplifies processes and the delivery of services; on the other hand, it puts the clients' privacy in jeopardy. Taking all of this into account, the client is the main beneficiary, as well as the most vulnerable, leaving the client in a precarious position of having to choose between keeping up with technological advancement to make things easier for themselves, or sticking to traditional methods.

2 Literature Review

Remote banking, viewed as a new economy delegation tool, consists of electronic transactions between clients and their banks. Electronic banking, also known as E-Banking, is the most recent delivery channel for banking services.

The term had been defined in various ways by experts, primarily because E-Banking alludes to a few types of services through which clients can request data and execute transactions via methods such as ATMs, PCs, or cell phones.

Nitsure [1] defined E-Banking as the dissemination of data and services by banks to clients via various conveyance platforms that can be used with a PC or other Intelligent platforms.

According to Liao and Cheung [2], E-Banking refers to the collective data or administrations provided by a bank to its clients via a PC or television.

E-Banking is defined by Angelakopoulos and Mihiotis [3] as "an umbrella expression for the interaction in which a client may perform banking exchanges electronically without visiting a physical location."

Most banking experts agree that E-Banking should provide 24/7 accessibility through cutting-edge IT channels.

The Basel Committee on Banking Supervision provides a broad definition of E-Banking: "E-Banking incorporates the arrangement of retail and low-value banking products and services through E-Channels, as well as high-value E-Payments and other discount banking administrations conveyed electronically" [4].

E-banking is a service that allows clients to access their accounts, conduct transactions, and obtain information on financial products and services via a public or private network, such as the Internet.

New Trends in the Banking Sector and the Development of E-Banking 1109

There are several terms used in the literature that all refer to some form of E-Banking: (PC) Banking, Internet Banking, Virtual Banking, Online Banking, Web Banking, Home Banking, and so on, but they are frequently used interchangeably.

ATMs and telephone transactions have long been used for e-banking services. Current E-Banking services, such as the internet and mobile banking, have changed banking administrations in recent years. The growth of the E-Banking industry can be traced back to the mid-1970s, when banks began to consider these types of administrations as an alternative to some of their traditional bank capacities [5].

3 Important Terms and Definitions

Banking: The acceptance and transfer of money claimed by others and substances, followed by lending this money out to lead economic activities such as profit generation or essentially covering operating costs.

Electronic banking (E-Baking): A type of banking in which funds are transferred via electronic signals rather than cash, checks, or other physical paper reports. Transfers of funds take place between financial institutions such as banks and credit unions.

IT: The use of computers, storage, networking, and other physical devices, infrastructure, and processes to create, process, store, secure, and exchange electronic data of all types.

Remote Banking: In contrast to communicating face-to-face in the same location, a remote relationship involves an exchange between people in different locations, which can take various shapes and forms, via communication channels (letter, telephone, internet, etc.).

ATM: An E-Banking outlet that allows customers to complete basic transactions without the assistance of a branch representative or teller. Anyone with a credit or debit card can withdraw cash from most ATMs worldwide.

The Basel Committee on Banking Supervision (BCBS): The BCBS is the primary global standard setter for bank prudential regulation and serves as a forum for regular cooperation on banking supervisory matters. Its 45 members are central banks and bank supervisors from 28 different countries [4].

4 E-Banking Features

1. Review your account statement online.
2. Open a savings account.
3. Pay your bills.
4. Make payments online.
5. Transfer money.
6. Place an order for a checkbook.

5 Advantages and Disadvantages of E-Banking [6]

5.1 Advantages

1. Availability: Banking services are available at all times. A large portion of the services provided are not time-limited; clients can check their account balances and transfer funds whenever they want without having to wait for the bank to open.
2. Ease of Use: Using the services provided by E-Banking is simple and straightforward. Many people believe that transacting online is much easier than visiting a branch for something similar.
3. Convenience: Customers do not want to leave their jobs, or even drive in traffic, just to wait in line at a bank branch. They can complete their transactions from any location.
4. Time Saving: E-Banking allows clients to complete any financial transaction in a matter of minutes. Money can be moved between any two accounts within the country or issue a deposit account in no time using E-Banking.

5.2 Disadvantages

1. Internet Requirement: To use E-Banking services, you must have a constant internet connection. If one of the two parties (Bank/Client) is unable to use any of the E-Banking services. Furthermore, if the bank's servers are down due to technical issues, you will be unable to access their services.
2. Exchange Security: Regardless of how many precautions banks take to provide a secure network, E-Banking transactions are still vulnerable to hackers. Regardless of the high-level encryption strategies used to protect client information, there have been instances where transaction data has been compromised. This could pose a significant threat, such as using the information illegally for the Hacker's benefit.
3. Difficult for Inexperienced Clients: If there is no one who can explain how E-Banking works and how to interact with it to an inexperienced client. Inexperienced amateurs will undoubtedly struggle to figure it out on their own.

6 E-Banking Types [7]

1. PC Banking: A type of banking that allows clients to perform bank transactions from a PC by providing a restricted financial software program that E-Banking services, and features that allow the client to perform financial transactions from their home PC.

New Trends in the Banking Sector and the Development of E-Banking 1111

2. Web Banking: A framework that allows bank clients to access accounts and general data on bank products and services, as well as perform account exchanges with the bank instantly via a PC using the internet as the delivery channel.
3. Mobile Banking: A framework that enables bank clients to manage various financial exchanges through a mobile device, being the most recent help in E-Banking; versatile banking is dependent on Wireless Application Protocol Technologies because a mobile device requires a WAP program to allow access to information.

7 Remote Queue Reservation Mobile Apps

Nowadays, with the huge leap in technology that we have been a part of, there is a mobile app that covers a specific aspect of a service for almost every scope of work, and the banking sector is no different, so as a result, remote queue reservation apps emerged, which are cloud-based customer flow management mobile apps that allow users to reserve a queue from their current location ahead of the visit to service providers across sectors, such as banks, in order to get served. Some of the most popular apps that provide similar services are as follows:

1. Balador
2. TagTime
3. My Providers
4. Skiplino.

8 Trends in the E-Banking World

The evolution of technology has accelerated E-Banking and forced banks to evolve at an exponential rate in order to create more efficient ways to meet the needs of their clients. Many banking experts believe that automation and personalization will be the primary digital banking trends in 2021 [8].

1. Automation: Banking automation refers to the process of operating the banking system using highly automated means in order to reduce human intervention to a bare minimum. Platform automation is another term for bank automation.
2. Personalization: Banks have been slow to adopt the personalization concept, but many are now attempting to catch up as they recognize that this is not a passing fad. Personalization in banking is associated with providing a significant service to a client based on personal experiences or historical data. It can help with trust building and thus drive results and income. By creating a beginning-to-end experience that incorporates client and functional information across all of the bank's capabilities. Personalization can help banks provide solutions to their customers before they even realize they have a problem.

9 Artificial Intelligence (AI) and Banking

As in the case of banks, information is fundamental to practically all business areas, from traditional deposit-taking and lending to investment banking and resource executives. AI is a huge step forward in the digitalization and change of modern businesses. Independent information management without human involvement thus provides extraordinary freedoms for banks to improve speed, accuracy, and efficiency. AI applications in banking can be divided into four categories [9]:

1. Front-office applications centered on the customer.
2. Back-office applications focused on operations.
3. Portfolio management and trading.
4. Regulatory adherence.

9.1 AI and Bank Profitability

Artificial intelligence could contribute to bank profitability in two ways: first, by taking over redundant tasks from employees, which saves time; second, Autonomous AI programming may reduce the interest in less-skilled workers and work on the efficiency and accuracy of excess bank staff. This is critical because worker compensation accounts for a significant portion of a bank's expense base. Second, AI implementation could boost revenue. As previously mentioned in the study, it may help banks develop new products and offer personalized items that are more tailored to customer interests.

9.2 AI Importance to E-Banking

From securing clients' data to providing 24/7 access to their banking information, the importance of AI in the E-Banking field is clear. The following are some of the most valuable points that clarify the importance of such relationship:

1. Enhancing Customer Experience: AI improves our understanding of customers and their behavior. Which assures clients that they will save time and effort and may even abandon the idea of visiting a branch, unless it is an unavoidable cause. As a result of fewer clients visiting the branches, banks can complete daily and routine processes and tasks more quickly and easily. As a result, it is a win–win situation for both parties.
2. Optimal Decision-Making: AI systems that think and respond like human experts provide optimal solutions based on continuously available information. In their database set, these systems keep a store of expert data known as a knowledge database. Banks use these cognitive systems to make strategic decisions.

AI will not only help banks by automating their knowledge labor force; it will also make the entire automation process smart enough to eliminate cyber threats and unhealthy competition. AI is critical to the bank's cycles and tasks, and it develops and advances over time with little or no human intervention. AI will enable banks to use human and machine abilities to drive functional and cost efficiencies, as well as provide customized services. These benefits are not currently a modern vision for banks to achieve. By embracing AI, pioneers in the banking sector have already taken the necessary steps to reap these anticipated benefits.

10   Challenges and Obstacles in E-Banking [10]

1. Security: Because of the inherent concerns associated with E-Banking, security may be the most difficult challenge for E-Banking promoters. Despite the fact that banks are designed to be virtually impenetrable, cyberattacks and fraudulent activity are still a reality. However, many clients are unaware that their online habits and patterns of use may be putting them in danger. Digital fraudsters and attackers can take advantage of poor privacy habits, which is why E-Banking professionals must promote and explain the security of their E-Bank frameworks, as well as educate clients on the most effective way to be more upright online by working on their protection and security propensities. Multi-factor authentication and the use of passwords are good places to start.
2. Technical Issues: As long as clients use the internet, they expose themselves to technology and network interruptions. Clients' ability to access their accounts may be impacted if their internet is slowed or interrupted for any reason. Similarly, no matter how advanced the technology, bank servers are still susceptible to both intentional and unintentional downtime.
3. Traditional Banking Habits: Using old methods in any field usually leads to problems, and the banking industry is no exception. Visiting bank branches in person to perform specific banking transactions that can be done using E-Banking platforms is the ultimate waste of time, effort, and money, and it can also lead to a delay in performing banking tasks that are worthy for customers to visit the bank's branch in person.

11   E-Banking in Numbers

80% of people prefer online banking to visiting a physical location [11].

Over 95% of E-Banking users are confident in the security of their banks' data [11].

In order to complete the desired service or inquiry, less than 10% of clients were advised to visit a physical bank branch while doing their banking online [12].

1114   M. Afaneh

Online banking platforms are used by 94% of mobile banking customers at least once a month [13].

According to digital banking statistics, the total number of online and mobile banking users will surpass 3.6 billion by 2024 [14].

Every year, banks around the world lose more than $1 trillion to cybercrime [15].

48% of account holders use mobile banking to transfer funds between their own accounts or send money to others, while 38% use desktop online banking. Total 86% [13].

When it comes to international transfers, banking statistics show that 24% of account holders prefer to do so via smartphone, while 53% prefer to do so via laptop or desktop computer. Total 77% [13].

The average annual cost of cybercrime per global financial institution is $18.37 million [16].

12 Conclusion and Recommendations

When it comes to E-Banking, it is critical to understand that all technological executions should be carried out in collaboration with a trusted partner who has many years of experience in the industry. The key here is to consistently set clear communication and input approaches and to work with the most innovative technologies, seasoned and talented experts, and agile systems.

It makes no difference whether the customer's banking association is using computerized acceleration or transformation. A companywide commitment to executing viable and amazing tech arrangements will help drive development, improve business execution, and shift your focus to other critical areas your organization must address.

This study recommends that banks begin raising awareness of E-Banking capabilities, particularly among inexperienced clients. It also suggests that banks begin investing in well-secured systems in order to guarantee the promised allegations, and to hope that clients with trust issues with E-Banking can be persuaded to join the rest of the E-Banking platform users, allowing the bank to devote more time to more important issues.

References

1. Nitsure, R.R.: E-Banking: Challenges and Opportunities. Economic and Political Weekly, pp. 5377–5381 (2003)
2. Liao, Z., Cheung, M.T.: Challenges to Internet e-banking. Commun. ACM **46**(12), 248–250 (2003)
3. Angelakopoulos, G., Mihiotis, A.: E-banking: challenges and opportunities in the Greek banking sector. Electron. Commer. Res. **11**(3), 297–319 (2011)
4. https://www.bis.org/bcbs/

New Trends in the Banking Sector and the Development of E-Banking

5. Nsouli, S.M., Schaechter, A.: Challenges of the "E-banking revolution". Fin. Develop. **39**(003), (2002)
6. Drigă, I., Isac, C.: E-banking services–features, challenges and benefits. Annals of the University of Petroşani. Economics **14**, 49–58 (2014)
7. Chaimaa, B., Najib, E., Rachid, H.: E-banking overview: concepts, challenges and solutions. Wireless Pers. Commun. **117**(2), 1059–1078 (2021)
8. Sepashvili, E.: Digital chain of contemporary global economy: e-commerce through e-banking and e-signature. Economia Aziendale Online **11**(3), 239–249 (2020)
9. Dantsoho, M.A., Ringim, K.J., Kura, K.M.: The relationship between artificial intelligence (AI) quality, customer preference, satisfaction and continuous usage intention of e-banking services. Indonesian Business Rev. **4**(1), 24–43 (2021)
10. Daba, T.: The Current Challenges and Opportunities of E-Banking in Ethiopian Banking System (The Case of Wegagen and United Banks). Doctoral dissertation, St. Mary's University (2021)
11. Consumer Affairs: Online banking has become more widespread among consumers (2019)
12. Lightico: Digitize and Streamline Customer-Facing Banking Processes (2021)
13. Deloitte: Deloitte Insights, Accelerating Digital Transformation in Banking (2020)
14. Juniper Research. Leading Banks Positioned in Juniper Research's Digital Transformation Readiness Index 2020 (2020)
15. McAfee: Latest Report from McAfee and CSIS Uncovers the Hidden Costs of Cybercrime Beyond Economic Impact (2020)
16. Accenture: It's time for banks to challenge everything (2020)
17. https://www.accenture.com/us-en/industries/banking-index
18. https://www.consumeraffairs.com/news/online-banking-has-become-more-widespread-among-consumers-survey-finds-103119.html
19. https://www2.deloitte.com/content/dam/Deloitte/cn/Documents/financial-services/deloitte-cn-fs-accelerating-digital-transformation-in-banking-en-190515.pdf
20. https://www.juniperresearch.com/press/digital-banking-users-to-exceed-3-6-billion#:~:text=Banks%20Catalyse%20Market-,Digital%20Banking%20Users%20to%20Exceed%203.6%20Billion%20Globally%20by%202024,2020%3B%20a%2054%25%20increase
21. https://www.lightico.com/lending_banking/
22. https://www.businesswire.com/news/home/20201206005011/en/New-McAfee-Report-Estimates-Global-Cybercrime-Losses-to-Exceed-1-Trillion
23. Agarwal, R., Rastogi, S., Mehrotra, A.: Customers' perspectives regarding e-banking in an emerging economy. J. Retail. Consum. Serv. **16**(5), 340–351 (2009)
24. Sohail, M.S., Shanmugham, B.: E-banking and customer preferences in Malaysia: an empirical investigation. Inf. Sci. **150**(3–4), 207–217 (2003)
25. The Basel Committee on Banking Supervision (BCBS): The Basel Committee, Overview (2021)
26. Thommandru, A.: Impact of Artificial Intelligence on E-Banking and Financial Technology Development (2021)
27. Vyas, S.D.: Impact of e-banking on traditional banking services (2012). arXiv preprint arXiv: 1209.2368

The Impact of Economic Globalization Welfare States

Natacha de Jesus Silva (iD), **Maria José Palma Lampreia Dos-Santos** (iD), **and Nuno Baptista** (iD)

Abstract The literature has yet to draw firm conclusions regarding the effect of economic globalization on government spending in developing countries in general and Pakistan in particular. This paper tries to overcome this gap on the literature and analyses the impact of globalization and democracy on the aggregate level of Pakistan's social spending from 1972 to 2018. The methods used include econometrics models. The results confirm that globalization is inversely related to social spending, while democracy positively affects social expenditures. Other results confirm that debt service and inflation rate harm social spending. In contrast, the more significant economic development and the increased unemployment rate have a positive and meaningful relationship with social expenditure in the long run. Public decision-makers should consider these conclusions to promote economic and social sustainability in Pakistan.

Keywords Democracy · Econometric models · Globalization · Social expenditures

---

N. de Jesus Silva (✉)
IJP—Instituto Jurídico Portucalense, REMIT—Research on Economics, Management and Information Technologies, Universidade Portucalense Infante D. Henrique, Porto, Portugal
e-mail: natachajsilva@upt.pt

M. J. P. L. Dos-Santos · N. Baptista
Escola Superior De Comunicação Social—Instituto Politécnico de Lisboa, ISCTE-IUL DINÂMIA´CET, Campus De Benfica Do IPL, Lisboa, Portugal
e-mail: msantos@escs.ipl.pt

N. Baptista
e-mail: nbaptista@escs.ipl.pt

N. Baptista
NECE—UBI, Covilhã, Portugal

© The Author(s), under exclusive license to Springer Nature Switzerland AG 2024
A. Hamdan and E. S. Aldhaen (eds.), *Artificial Intelligence and Transforming Digital Marketing*, Studies in Systems, Decision and Control 487,
https://doi.org/10.1007/978-3-031-35828-9_94

1 Introduction

Over recent decades, the relationship between globalization and the nation-state has become one of the most pressing social science controversies [20]. The discussion on the effect of economic globalization on the welfare state is far from being established, and the literature suggests contradictory arguments and results about this effect [4, 20, 27–29, 36, 38]. The main question is how regimes of various kinds act in response to globalization toward welfare spending.

The issue we propose to address is to understand how the main regimes and their typology react to globalization by relating spending to well-being.

The optimistic assumption predicts a positive correlation between globalization and social spending due to adopting a stimulus–response model. According to this perspective, globalization induces popular demand for compensatory social policies explaining why highly open economies have the most significant welfare states. The pessimistic assumption argues that opening the economy implies greater competitiveness between countries, which creates constraints on the growth of public expenditure [40]. Opening the economy means competing for the attraction of capital, which imposes limitations on the level of tax revenue and therefore depresses public spending [29]. Several studies suggest that openness and external shocks were significant determinants of increased welfare commitments in economically advanced countries [35, 37, 38]. The reason that explained this importance was that governments had encouragement to increase expenditure due to the enlarged vulnerability and insecurity related to economic openness. The expansion of social spending occurred due to countercyclical Keynesian policies or by intensifying the scope and depth of social insurance [12].

However, some authors [18, 22, 45] needed to be more convinced of the benefits of globalization. The literature suggested that increasing trade and investment could constrain public and welfare expenditures. The main reason is that globalization could lead to a greater dependence on exports because of trade liberalization and capital mobility. In this context, firms see taxation as a constraint on competitiveness. The discussion on the association between globalization and social spending produces contrasting expectations for advanced and developing countries. Indeed, the literature on social spending does not adequately describe why the welfare spending trend in LDCs differs in developed countries.

The debate about the interaction between globalization and social spending generates contrasting expectations concerning advanced and developing countries. Concerning social expenditures, the literature does not adequately describe the spending trend with the social welfare trend in LDCs, which differs from developed countries.

The literature on the association between economic globalization and the welfare state suggests different perspectives on the nature of this correlation. Some scholars consider that other socioeconomic processes than globalization influence the welfare state. According to these authors, domestic policies play a more critical role in the

The Impact of Economic Globalization Welfare States

welfare state definition than globalization [24]. Other authors consider that globalization has a significant influence on the welfare state. However, they diverge about the direction of this influence which led to the formulation of two theses: the efficiency thesis and the compensation thesis.

The efficiency assumes that high social spending makes international markets less competitive. This effect can transmit through different channels. High social spending can be connected, for example, to higher taxation that results in increased labor costs and decreases the efficiency of exports and domestic production, which are exposed to international competition [25]. Second, high fiscal expenditures can decline competitiveness by increasing interest rates, leading to crowding out effects on private investors and increasing the values of the exchange rate [2]. The compensation hypothesis states the opposite effect. It explains that the welfare state counterbalances the risk of globalization by investing more inhuman capital [25]. This hypothesis is supported by several studies that make a robust empirical relation between globalization, large public sectors, and social safety net programs [10]. The quantitative research has generated more practical work supporting the compensation thesis as far as developed countries are concerned. As increased economic openness also reinforces welfare spending to strengthen human capital. That is also evident because developed countries usually have large welfare spending budgets to enhance the competitiveness and productivity of the economy in local and international markets [33].

Pakistan comprises an appealing and comparatively understudied area for analytical inquiries in social spending. Most of the literature has been carried out to identify globalization's effect on welfare spending in OECD countries [10, 38, 46] and Latin American countries [3, 25]. Only some scholars analyzed developing countries, including Pakistan [40]. However, all the previous literature used a multilevel country analysis, and Pakistan (PK) was never analyzed as a country-level study. Therefore, this paper contributes to the literature because it is the first based on time series in Pakistan.

The main aim of this paper includes, respectively: (1) Analyze the impact of globalization and democratization on social spending in Pakistan at the aggregate level from 1972 to 2018. (2) Analyze if the government of PK acted in response to the challenges of globalization with social policy selection that leaned more towards reducing cost or through defensive nation's welfare or compensation.

2 Literature Review

The discussion on the impacts of economic globalization on welfare states is widespread. A protuberant hypothesis is that substantial welfare policies cushion the negative externalities of globalization. New empirical evidence suggests a negative relationship between globalization and public social spending [7]. There are several definitions of globalization. Ahmad et al. [1] define economic globalization as strengthening international monetary exchange and the ticket for the contemporary era of global economic integration. Several economic globalization indicators measure trade, financial, and overall economic globalization [20]. Early "structuralist theories" stated that modernization (economic development) is positively associated with expanding welfare expenditures. This theory claims that the growth of social spending and respective policies represents the state's sense of responsibility toward its citizen "needs" [11, 26].

However, "power resource theorists" declined structuralist arguments. The researchers stated that the power distribution among different political and social groups is the reason for amendments in social policies [10, 46]. Several power resource theorists explained that by generating evidence of labor union strength, considered the most robust forecaster of social and welfare expenditures. Labor wishes to increase welfare expenditures as long as their post-transferer compensation and benefits are likely to increase [45].

The discussion on the association between globalization and social spending produces contrasting expectations for advanced and developing countries. Regarding the influence of globalization on social expenditures, several studies suggest that globalization, remarkably increasing openness to trade, has an adversative impact on at least some types of social spending [25].

3 The Efficiency and the Compensation Theory

The literature on the relationship between economic globalization and the welfare state is vast and can be divided into three approaches. The first presents a skeptical attitude towards globalization and its contribution to well-being [23, 24, 41, 48].

According to these authors, not globalization leads to social well-being but the socioeconomic processes based on technological change that leads to well-being and economic growth. Furthermore, they mention that welfare policies are developed more at the internal policy level than externally at the level of internationalization [47]. In this way, the market institutions and the balance of power between internal decision-makers (parties, unions, industry associations, Etc.) condition the impact of the forces of globalization on the formulation of social policies [8, 41, 46]. However, two other theories agree that globalization significantly affects welfare states. In this context, and as mentioned by Busemeyer [9], globalization leads to an increase in worker dissatisfaction at an individual level, namely those excluded by globalization. So, they claim compensation through social welfare policies based on increased spending on internal transfers and social support [19, 38].

On the contrary, according to the theory of efficiency, which postulates the opposite of the previous ones, globalization limits the use of public expenditures (public expenditures and internal transfers) in general and in welfare in particular. That happens due to the scarce use of public capital at the global level [49]. Given that companies in international markets can threaten to withdraw their money, governments respond by lowering business costs, especially taxes. Thus, decreasing tax revenues reduces public spending and internal transfers, reducing social well-being.

4 Globalization and Expenditures on Welfare

Like the comprehensive policy of internal transfers, investment policies in the social sector also promote training and capacity building of human capital. That can play an essential role in mitigating the adverse side effects of economic globalization and, through the multiplier effect, lead to future social and economic well-being and increasing levels of development. Dreher et al. [14]. Welfare states, institutions, and national policies condition states' reactions to globalization [30]. Most authors agree that since the crisis (oil shock) of 1970, states have reduced expenditure on social support policies (government transfers and public investment in the social sector).

At the same time, in democratic states, where, as a general rule, there are elections every four years, we are witnessing social policies that are interrupted by a new mandate without continuity, on the one hand. On the other hand, procedures with time horizons allow measuring their social gains in terms of competitiveness or social sustainability.

In turn, citizens' expectations also change each time a new government takes office, which may create more or less positive or negative expectations.

5 Democracy and Social Welfare

According to Onaran and Boesch [34], globalization affects government budgets for social welfare. Despite this, country responses are conditioned by the economic policies of different welfare regimes and their respective institutions. These authors concluded that in Western Europe, globalization has led to increased spending on social policies. However, according to these authors Onaran and Boesch [34] taxes on dependent employment finance these expenses.

Thus, the compensation hypothesis is verified on the social expenditure side, while there are pressures on efficiency on the tax revenue side.

In this way, these policies are complementary and not competing. Nevertheless, according to Onaran and Boesch [34], these results occur mainly in conservative regimes.

6 Methodology

Information and data were obtained from secondary sources, namely, the United Nations database (2020) and the World Bank database (2020). The period analyzed included data from 1980 to 2020. The econometric software program includes EVIEWS. The econometric model has the long-run and short-run dynamics between globalization, democracy, controlled variables, and social spending.:

$$S_t = \alpha_t + \beta_1 TO_t + \beta_2 FOR_t + \beta_3 Dem_t + \beta_4 EcodeT_t$$
$$+ \beta_5 URT_t + \beta_6 IRT_t + \beta_7 DebtT_t + \varepsilon_t$$

where: α; βs; ε, and t represent, respectively, the constant; the parameters to be esteemed, the error term, and the years of observation. URT represents the aggregate social expenditures; TO the social expenses; TOt and FORt are the openness of the global economy unemployment rate; and EcodeTt, URTt, IRTTt, and DebTTtt represent inflation rate and financial account.

Dos-Santos and Diz [13] and Dos-Santos et al. [42] used a Log–Log model. The heteroscedasticity White test was used, and the values of the t-test were done (0.000). The model was run for 0.10%, 0.005%, and 0.010%.

The degree of fit of the model (R2) is good. The R2 evidence this at 80%, which is a correct fit for the data.

7 Results and Discussion

The model's results are generally in line with those of other authors from developed countries, namely by Dos-Santos and Diz [13]. Thus, in the long term, the opening of emerging economies leads to results that, despite being positive, although these results are dynamic and subject to unpredictability. In addition to opening up the economy, the democratic regime also entails more significant expenditures in terms of internal transfers. The results confirm that the past years of more significant political turbulence in this country led to decreased social expenses. The existence of a higher per capita national income translates into higher social expenditures (control variable). The social problems in this country, namely the increase in the unemployment rate, also conditioned an increase in internal and external transfers. As a problem, the high public debt translates into a sharp decrease in transfers, and this problem worsens in the long term due to instabilities in the interest rate on public debt. In this way, inflation turns out to be an explosive engine with very negative impacts on internal transfers, namely, contributing, in the long term, to a reduction in social support.

8 Conclusion

The results allow us to conclude that, in developing countries, there is an effect and a degree of openness of public policies, an effect of the external environment, which intensely conditions the expenditure on internal transfers. This situation directly impacts a country's social status and its citizens' level of well-being. In this way, developed countries too, namely the G20 and G8 economies, should be alert to the possibility that, with their policies, they have conditioned their countries not only internally but also externally the countries with emerging economies. This contagion can be positive or negative, depending on that country's monetary and fiscal policies. For this reason, internal organizations, such as the International Monetary Fund, will also have to play an essential role in contributing to a better world by defining monetary policies adjusted to realities that are not national but global. Despite this, globalization and open economies continue to be a win–win situation. These transnational supervisory mechanisms must act with a view to the common good.

Acknowledgements This work was supported by the UIDB/05105/2020 Program Contract, funded by national funds through the FCT I.P.

References

1. Ahmad, M., Ahmed, Z., Yang, X., Can, M.: Natural resources depletion, financial risk, and human well-being: what is the role of green innovation and economic globalization?. Soc. Indic. Res. **167**(1–3), 269–288 (2023)
2. Avelino, G., Brown, D.S., Hunter, W.: "Globalization, democracy, and social spending in Latin America, 1980–1997." Am. J. Po. Sci. **512**, 1980–1997 (2005)
3. Avelino, G., Brown, D.S., Hunter, W.: The effects of capital mobility, trade openness, and democracy on social spending in Latin America, 1980–1999. Am. J. Polit. Sci. **49**(3), 625–641 (2005)
4. Baptista, N., Alves, H., Pinho, J.C.: Uncovering the use of the social support concept in social marketing interventions for health. J. Nonprofit Publ. Sect. Mark. **34**(1), 1–35 (2022)
5. Beal Krause, A.L.: The aged population and social spending in Latin America: comparing the demographic functionalist theories and political pressure arguments. Polit. Pol. **49**(5), 1061–1091 (2021)
6. Brown, D.S., Hunter, W.: Democracy and social spending in Latin America, 1980–92. Am. Po. Sci. Rev. **93**(4), 779–790 (1999)
7. Busemeyer, M.R., Garritzmann, J.L., Neimanns, E., Nezi, R.: Investing in education in Europe: evidence from a new survey of public opinion. J. Eur. Soc. Policy **28**(1), 34–54 (2018)
8. Busemeyer, M.R., Garritzmann, J.L.: Compensation or social investment? Revisiting the link between globalisation and popular demand for the welfare state. J. Soc. Policy **48**(3), 427–448 (2019)
9. Busemeyer, M.R.: From myth to reality: globalisation and public spending in OECD countries revisited. Europ. J. Polit. Res. **48**(4), 455–482 (2009)
10. Cameron, D.R.: The expansion of the public economy: a comparative analysis. Am. Po. Sci. Rev. **72**(4), 1243–1261 (1978)

11. Dong, C., Morehouse, J.: Toward a caring government: advancing ethical government public relations with a care-based relationship cultivation model. J. Publ. Relati. Res. **34**(5), 179–207 (2022)
12. Dos Santos, M.J.P.L., Ahmad, N.: Sustainability of European agricultural holdings. J. Saudi Soc. Agric. Sci. **19**(5), 358–364 (2020)
13. Dos Santos, M.J.P., Diz, H.: Towards sustainability in European agricultural firms. In Advances in Human Factors, Business Management and Society: Proceedings of the AHFE 2018 International Conference on Human Factors, Business Management and Society. Loews Sapphire Falls Resort at Universal Studios, Orlando, Florida, USA, vol. 9, pp. 161–168. Springer International Publishing
14. Dreher, A., Fuchs, A., Parks, B., Strange, A., Tierney, M.J.: Aid, China, and growth: evidence from a new global development finance dataset. Am. Econ. J. Econ. Pol. **13**(2), 135–174 (2021)
15. Dreher, A., Lang, V., Rosendorff, B.P., Vreeland, J.R.: Bilateral or multilateral? international financial flows and the dirty-work hypothesis. J. Polit. **84**(4), 1932–1946 (2022)
16. Eggers, T., Grages, C., Pfau-Effinger, B.: Self-responsibility of the "active social citizen": different types of the policy concept of "active social citizenship" in different types of welfare states. Am. Behav. Sci. **63**(1), 43–64 (2019)
17. Fonseca, A.M., Diz, H.M., Dos-Santos, M.J.P.L.: O crowdfunding como financiamento do jornalismo de investigação em Portugal. Palabra Clave **19**(3), 893–918 (2016)
18. Franzese, R., Hays, J.: Empirical modelling strategies for spatial interdependence: omitted-variable vs. simultaneity biases. In: Paper presented at the 21st Summer Meeting of the Society for Political Methodology, Palo Alto, CA (2004)
19. Garrett, G.: Globalization and government spending around the world. Stud. Comp. Int. Dev. **35**(4), 3–29 (2001)
20. Heimberger, P.: Does economic globalization affect government spending? a meta-analysis. Public Choi., 1–26 (2020)
21. Hope, D., Limberg, J.: The knowledge economy and taxes on the rich. J. Eur. Publ. Policy **29**(5), 728–747 (2022)
22. Huber, E., Stephens, J.D.: Globalisation, competitiveness, and the social democratic model. Soc. Policy Soc. **1**(1), 47–57 (2002)
23. Hunjra, A.I., Islam, F., Verhoeven, P., Hassan, M.K.: The impact of a dual banking system on macroeconomic efficiency. Res. Int. Bus. Financ. **61**, 101647 (2022)
24. Iversen, T.: The dynamics of welfare state expansion: trade openness, de-industrialization, and partisan politics. The new politics of the welfare state, 45–79 (2001)
25. Kaufman, R.R., Segura-Ubiergo, A.: Globalization, domestic politics, and social spending in Latin America: a time-series cross-section analysis, 1973–97. World Polit. **53**(4), 553–587 (2001)
26. Kunißen, K.: Premises: perspectives on the welfare state. In: The Independent Variable Problem: Welfare Stateness as an Explanatory Concept, pp. 11–42. Wiesbaden, Springer Fachmedien Wiesbaden (2022)
27. Kwon, H., Pontusson, J.: Globalization, labour power and partisan politics revisited. Socio Econ. Rev. **8**(2), 251–281 (2010)
28. Lin, S.C., Wu, Y.C.: Finance in a more globalized economy. Emerg. Markets Fin. Trade 1–15 (2022)
29. Marshall, J., Fisher, S.D.: Compensation or constraint? How different dimensions of economic globalization affect government spending and electoral turnout. Br. J. Polit. Sci. **45**(2), 353–389 (2015)
30. Moisander, J., Eräranta, K., Fahy, K.M., Penttilä, V.: Emergence of hybrid CSR models as a conflict-driven communicative process in a nordic welfare state. J. Manage. Stud.
31. Mulligan, C.B., Gil, R., Sala-i-martin, X.: Do democracies have different Policies than Nondemocracies ? J. Econ. Perspect. **18**(1), 51–74 (2004)
32. Nacar, B., Karabacak, Y.: Does globalization affect taxation policies? evidence from Turkey. Appl. Econ. Lett., 1–4 (2022)

33. Nukpezah, J., Abutabenjeh, S., Azhar, A.: Do smart cities technologies contribute to revenue performance? Evidence from US local governments. Public Perf. Manage. Rev. **45**(5), 1155–1180 (2022)
34. Onaran, Ö., Boesch, V.: The effect of globalization on the distribution of taxes and social expenditures in Europe: do welfare state regimes matter? Environ. Plan. A **46**(2), 373–397 (2014)
35. Quinn, D.: The correlates of change in international financial regulation. Am. Polit. Sci. Rev. **91**(3), 531–551 (1997)
36. Rodden, J.: Reviving leviathan: Fiscal federalism and the growth of government. Int. Organ. **57**(4), 695–729 (2003)
37. Rodrik, D.: Has globalization gone too far? Institute for International Economics, Washington, DC (1997)
38. Rodrik, D.: Why do more open economies have bigger governments? J. Polit. Econ. **106**(5), 997–1032 (1998)
39. Rudman, A., Ellem, B.: Union purpose and power: regulating the fissured workplace. Econ. Indust. Demo. 0143831X221139333
40. Rudra, N., Haggard, S.: Globalization, democracy, and effective welfare spending in the developing world. Comp. Pol. Stud. **38**(9), 1015–1049 (2005)
41. Sadiq, M., Wen, F., Bashir, M.F., Amin, A.: Does nuclear energy consumption contribute to human development? Modeling the effects of public debt and trade globalization in an OECD heterogeneous panel. J. Clean. Prod. **375**, 133965 (2022)
42. Santos, M.J.P.L.D., Fragoso, R.M.D.S., Henriques, P.D.D.S., Carvalho, M.L.D.S.: A competitividade do regadio de Alqueva em Portugal: o caso do bloco de rega do Monte Novo. Rev. Econ. Sociol. Rural. **50**, 107–118 (2012)
43. Shahbaz, M., Rehman, H., Amir, N.: The impact of trade and financial-openness on government size: a case study of Pakistan. J. Qual. Tech. Manag. **6**(1), 105–118 (2010)
44. Stephens, A., Lin, M.: Determinants of government size: evidence from China. Public Cho. **151**(1–2), 255–270 (2012). Yang, Z., Wang, J.: Differential effects of social influence sources on self-reported music piracy. Decis. Support Syst. **69**, 70–81 (2015)
45. Stephens, E.: Dependency revisited: international markets, business cycles, and social spending in the developing world. Int. Organ. **60**(2), 433–468 (2006)
46. Swank, D.: The partisan politics of new social risks in advanced postindustrial democracies: social protection for labor market outsiders. In: The European Social Model under Pressure: Liber Amicorum in Honour of Klaus Armingeon, pp. 139–157 (2020)
47. Swank, D., Steinmo, S.: The new political economy of taxation in advanced capitalist democracies. Am. J. Polit. Sci. 642–655 (2002)
48. Taylor-Gooby, P.: The silver age of the welfare state: perspectives on resilience. J. Soc. Policy **31**(4), 597–621 (2002)
49. Xu, D.: Overview of tax incentives as subsidies in the context of international trade and competition: rationale for granting and regulating tax incentives. In: Interactions Between Chinese Tax Incentives and WTO's Subsidy Rules Against the Background of EU State Aid, pp. 19–47. Singapore, Springer Nature Singapore (2023)

The Benefits of Digital Transformation: A Case Study from Finance and Administrative Departments Within the Oil and Gas Industry in the Kingdom of Bahrain

Hamad Aljar and Noor Alsayed

Abstract This research aims to explore the benefits of digitally transforming the manual and paper-based processes in the finance and administrative departments within the oil and gas industry in the Kingdom of Bahrain. There are many benefits examined in the literature across various industries and the most relevant to the oil and gas industry include: acquiring new skills, saving time, improved information security, boosting employee morale, and positively contributing to saving the environment. An online questionnaire based on the reviewed literature was developed and the link was distributed to employees in the financial and administrative departments (n = 252) within the oil and gas industry in the Kingdom of Bahrain. After collecting and analyzing the data, the researchers found that there are strong positive correlations between digital transformation and acquiring new skills, saving time, improved information security, boosting employee morale. However, the findings were not conclusive on how much if any, a correlation exists between digital transformation and preserving the environment.

Keywords Digital transformation · Digitization · Employee morale · Information security

1 Introduction

The oil and gas industry in the region is considered the pillar that holds up the economy, and in light of the continuous and fast-paced technological advancements in the world, the re-imagining of manual and analog processes and transforming them into a digital form is one of the goals and visions that organizations must implement in the field of human and administrative processes to keep pace with this technological

H. Aljar · N. Alsayed (✉)
Ahlia University, Manama, Bahrain
e-mail: nalsayed@ahlia.edu.bh

© The Author(s), under exclusive license to Springer Nature Switzerland AG 2024
A. Hamdan and E. S. Aldhaen (eds.), *Artificial Intelligence and Transforming Digital Marketing*, Studies in Systems, Decision and Control 487,
https://doi.org/10.1007/978-3-031-35828-9_95

development and to become more current, where manual paper processing constitute the majority of work procedures in these areas [6]. Many companies in the oil and gas industry in the region are still utilizing paper in their administrative processes such as letters, carbon forms, and petty cash vouchers and converting them to digital while maintaining their privacy, there are also other examples that companies use, namely bank transfers and issuing Cheques methods [8]. The world is currently experiencing a period of advanced technological evolution, and this has led to the emergence of new methods, tools, and techniques in all aspects of business, especially in management and managerial functions [9]. Additionally, the emergence of the Fourth industrial revolution and the development of means of communication led to the emergence of new payment methods that are compatible with the requirements of electronic commerce and the nature of transactions via the Internet, which are modern electronic payment methods which means companies must develop their operations in line with the development in the world [8]. During the Covid 19 pandemic due to the majority of employees working remotely, accomplishing tasks has become very dependent on how much employees are familiar with using the technologies available to them. As a result, it has become more important that employees acquire the skills and competencies to navigate this new reality and become more tech-savvy [19].

The advantages of converting manual and analog processes to digital have been well documented with examples such as reducing human errors, improving consistency, and providing timely, and in the cases of human resources, management, accounting and finance, economics, and marketing [9]. However, this does not mean that such attractive positives do not come with serious negatives which include the availability of appropriate technologies to facilitate a successful digital transformation, the affordability of acquiring the technologies versus the cost of training and producing a competent workforce that is capable of operating it, the usability and accessibility of these technologies, and the resistance of the workforce to the technologies once introduced [28]. This warrants a radical change in how businesses conduct their operations and raises the question of what the implications of adopting a digital view on the work of employees who are used to the routine of their manual processes. This study will explore the possibility of expanding the use of technology within the administrative departments of the major local oil and gas companies in the Kingdom of Bahrain, its advantages, potential challenges, and areas in which improvements may be introduced to the systems currently in place. The purpose of this research is to study and evaluate the potential benefits of digital transformation in the oil and gas industry within the Kingdom of Bahrain, specifically within the financial and administrative departments, and to determine its impact on the morale of employees, their acquisition of new skills that maintain work accuracy, saving time, and to fortify the security of company information in addition to reducing paper consumption.

2 Methods

2.1 Conceptual Model

Digital Transformation may be referred to as the process of converting manual and analog processes or operations to digital without affecting the original purpose or the outcome of the converted task or process converting time-consuming transactions to simple, high-efficiency operations that reduce costs or increase profit [19]. The literature demonstrates and lists a myriad of reasons and factors why Digital Transformation is embraced and integrated within organizations at such a massive scale across industries depending on the nature of business, its processes, and its direction. One of the main recent catalyst escalators of Digital Transformation that emerged during the past 3 years was the outbreak of the Coronavirus Pandemic (COVID-19) which forced the world to rethink and reimagine how businesses and transactions are conducted resulting in a complete alteration in the modus operandi of business and commerce globally as the world was forced to limit the use of old fashioned interpersonal communication, in-person meetings with the closure of international borders and ports. In this research, we will be examining several benefits of embracing Digital Transformation in the finance and administrative processes within the Oil and Gas sector which include acquiring new skills, saving time, improved information security, boosting employee morale, and positively contributing to saving the environment.

2.1.1 New Skills

One of the benefits of embracing Digital Transformation in the administrative processes is the massive potential for learning new and valuable skills of varying degrees, given that digitizing any process automatically implies either partially or entirely replacing the currently performed task or even the employee performing it, a by product of employees learning the new skills associated with digitizing their processes is the increased potential for promotions based on merit, Thomas L Friedman uses China as an example of this in his book The World is Flat, explaining how China is moving at such a fast pace, not only as a business-hub and a destination for rapid economic growth but as a country that is eager to seize every available opportunity to develop and qualify its citizens to become productive forces to their nation's prosperity. The author said, "For a communist authoritarian system, China does a pretty good job of promoting people on merit." [15]. The potential for career advancement may serve as a very compelling incentive for employees who are eager to further their personal growth and maximize their job security within the organization. Employees who have high technical and creative skills can benefit from the training programs prepared by structured training and development programs, no matter how difficult they are. Unlike employees who face difficulty in this, so they have to make more effort, and for some of them, standing at a certain level is

1130 H. Aljar and N. Alsayed

the best way in practical terms than fighting experience and the fear of failing in it [10]. The potential of learning and acquiring new skills and improving upon existing competencies throughout the process of Digital Transformation varies depending on several factors such as the level of education among the employees, the culture of the organization, and the level of resistance to change within the organization [11].

2.1.2 Employees Morale

According to [7] Morale refers to the total satisfaction, a person derives from his job, his work group, his boss, the organization, and his general environment. It is reflected in the general feeling of well-being, satisfaction, and happiness of people.

Morale in a general sense is referred to as 'willingness to work'. High morale is the result of job- satisfaction which is again the result of the motivational attitude of the management. The literature reveals two opposing views on whether an increase in morale is directly correlated with increased productivity among employees. For instance, [7] states that "there is no direct relationship between morale and productivity. High morale may lead to higher productivity but in some cases, production may go down even. It is generally felt that there is a positive relation between morale and productivity but the degree may not be the same". However, on the opposing end of the spectrum [33] states "If employee morale is high then they will be committed to work. Job performances will also be increased. If employee motivational level is high then productivity will also be increased. This helps the organization retain employees and reduces employee turnover". With these two views from the literature, it becomes fairly clear that understanding the effects of employee morale on productivity is a multifactorial matter and requires further examination of the factors that may affect morale within any organization.

2.1.3 Saves Time

Saving time is another potential benefit of introducing Digital Transformation and technology into processes within organizations that is worth delving into, increasing internal efficiency is the primary objective, on the process level according to the International Journal of Information Systems and Project Management Vol. 5, No. 1, 2017, 63–77 "Adopting new digital tools and streamlining processes by reducing manual steps". The key terms here are "Streamline" and "Reducing Manual Processes". Both of these outcomes are direct contributors to saving time and expediting processes by technological means and mediums such as emails, scanned documents, SAP platforms, remote access mediums, remote meetings, electronic surveys, and digital notice boards. The potential benefits of Digital Transformation to improve internal efficiency include improving the efficiency of business processes and consistency by eliminating manual and paper-intensive processes and gaining better accuracy. Digital Transformation can also provide an overview of operations and results in real-time, it also provides better employee satisfaction by automating routine processes

The Benefits of Digital Transformation: A Case Study from Finance ... 1131

and tasks and thus, freeing up more time to either develop new skills and competencies or tackle more challenging tasks [27].

HR systems have witnessed rapid developmental improvements over the past five years. Three years ago, we were talking about moving toward cloud-based solutions as companies quickly replaced their old personnel management systems with integrated HR platforms on the cloud [5]. Hence, organizations are focused on Hiring high-tech human resources, in the past in the sixties and seventies, the focus of human resources was on following up on personnel affairs and keeping their records intact. In the eighties, the personnel department was reimagined to be a service center and meet their individual needs, and in the nineties and early twentieth century it turned to integrated talent management. Moreover, the introduction of new systems in employment, training, and salaries. Today, the focus has been on building the digital human resources department and working on employing the latest digital technologies at the hands of digitally intelligent employees who can use data to analyze data and make decisions related to their area of expertise.

2.1.4 Information Security

In today's world, Digital Transformation underpins most of the socio-economic and political developments around the world, at a time when the bulk of information nowadays is stored or maintained in digital form, the protection of this information is paramount and this is where cybersecurity plays a vital role in the successful implementation and integration of Digital Transformation within any organization [30]. Two consistent and closely related themes in corporate technology have emerged in recent years, both involving immediate, rapid, and dramatic change. One is the rise of the digital tide across all sectors and the second is the need for IT to react fast and swiftly to meet and realize the company's digital aspirations. As IT organizations seek to digitize, however, many faces significant cybersecurity challenges. In company after company, fundamental tensions arise between the business's need to digitize and the cybersecurity team's responsibility to protect the organization, its employees, and its customers within existing cyber operating models and practices [26].

2.1.5 Environmentally Friendly

Many organizations are starting to take steps to reduce the amount of paper in their business processes, this is particularly prevalent in the financial services industry where paper-intensive processes are strongly present [17]. Less paper consumption has a tremendous and positive impact on the environment, the pulp and paper industry ranks fourth among industrial sectors in emissions of Toxics Release Inventory (TRI) chemicals to water, and third in such releases to the air. These impacts can occur at all phases of the paper lifecycle, from fiber acquisition, through manufacturing, and storage to disposal as it diminishes natural resources and increases Greenhouse

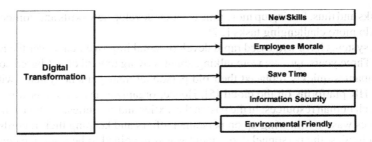

Fig. 1 Conceptual model

Gasses (GHGs) in Earth's atmosphere, therefore it is an ideal and crucial time to introduce innovative methods by which superfluous environmental impacts can be reduced [12].

Figure 1 below shows the conceptual model of the benefits of digital transformation that will be tested in this study. The benefits include acquiring new skills, saving time, improved information security, boosting employees morale, and positively contributing to saving the environment.

2.2 Research Instrument

For this research, an online questionnaire was designed to target employees of both genders and a wide age bracket who work in the Oil and Gas Industry in the Kingdom of Bahrain. The distributed questionnaire contained a total of 15 directly relevant, clear, and precise items which contained a combination of questions and statements and close-ended questions. The questionnaire was made to fit the current objectives of the research, as the questions and statements were designed to investigate whether introducing Digital Transformation and technology into the Administrative Processes within the Oil and Gas Industry within the Kingdom of Bahrain is beneficial to these processes.

2.3 Population and Sampling

The population sample that was sampled was limited to employees working in the oil and gas industry, particularly in the financial and administrative departments, a total of 300 questionnaires were distributed and disseminated. The researchers were able to collect 252 answers from both sexes, age groups, and different career paths through the questionnaires that were distributed through electronic means such as WhatsApp and email. The response rate was 84%, and respondents were expected to complete the entire questionnaire they received. The questionnaire was designed to enforce that the specific component, respondents would not be able to submit their answers without providing answers to all 15 items included in the submitted questionnaire. The researchers resorted to using a soft questionnaire as a data collection tool for several reasons, one of which was that the coronavirus pandemic remains a public safety concern and that public safety measures within these companies may prevent or limit face-to-face interactions. In addition, the method is also in line with the researchers' belief that by extension, the paperless methods help in preserving the environment and may also serve as an example of how digital transformation can be very effective in this regard. The questionnaire was created using a free online available tool (Google Forms) and the complete set of questions was distributed through an accessible link sent by WhatsApp. There was no specific time limit to receive the responses however, the researchers have set a satisfactory target of collecting a minimum of (252) responses.

2.4 Data Analysis

SPSS software was employed to enter the information obtained from the measurement instrument—and thus, the distributed questionnaires—as a result of the analyses. A descriptive analysis will be conducted to allow for respondent characteristics such as school type, age, gender, and graduation background. To ensure that the questionnaire answers are reliable and valid, descriptive measures will be used to test the rate of agreement of the sampled population. Data analysis tools including correlation and comparing means were used to test the conceptual model.

3 Results

3.1 Sample Description

Based on the sample collected, male respondents (145 responses) comprising (57.5%) outnumbered the female respondents of (107 responses) comprising (42.5%). A detailed outline of the gender profile of the above is presented below, along with a pie chart repressing the above-mentioned percentile in Fig. 2.

The age bracket in the distributed questionnaire was divided into four categories whereas (13.5%) of the total responses came from employees who were under 25 years of age, (26.6%) of the collected responses came from employees who were between 25 and 35 years of age, (40.9%), the largest portion of the responses came from employees who were between 36 and 45 years of age, and finally, the remainder (19%) of the responses came from employees who 46 and above years of age. A breakdown of this variable is available in a pie chart separating each age bracket by percentage in Fig. 3.

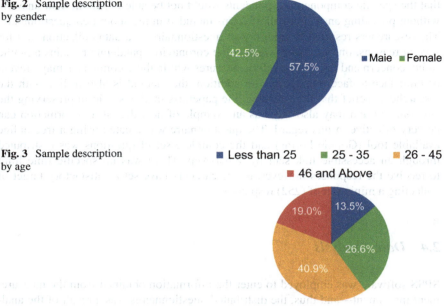

Fig. 2 Sample description by gender

Fig. 3 Sample description by age

The next variable is the qualification/educational attainment of the sampled population separated into four different levels of education: (High School, Diploma, bachelor's, master's, and Ph.D.) whereas employees holding a high school certificate represented (6%), employees holding diplomas represented (10.7%) of the sampled population, Bachelor Degree holders/undergraduates was the largest portion of the sampled population comprising (52%) whereas Masters Degree holders/Postgraduates made (27.8%) and finally, Ph.D. holders made (3.6%) of the total sampled population. The below pie chart shows an outline of these findings in Fig. 4.

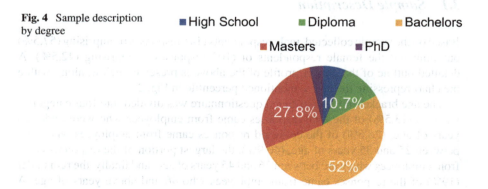

Fig. 4 Sample description by degree

Fig. 5 Sample description by years of experience

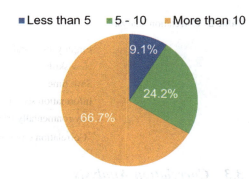

The final variable is the years of experience where (9.1%) of the questionnaire participants had less than 5 years of experience, (24.2%) of the employees had between 5 and 10 years of experience and finally, the majority of the employees who participated in the survey making (66.7%) had more than 10 years of experience. The below pie chart shows an outline of these findings in Fig. 5.

3.2 Reliability Testing

Cronbach Alpha is the most common measuring instrument within SPSS of internal consistency and it is commonly used when testing multiple Likert questionnaires or surveys that form a scale to determine whether the scale is reliable is not. It also measures the extent of close interdependence between groups of organisms as a group. Technically, Cronbach's alpha is a measure of reliability, which is the most famous and widely used reliability measure, and a value ranging from 0 to 1 with a value of at least 0.7 is considered sufficient to ensure the validity of the questionnaire. Table 1 summarizes the findings.

Table 1 Cronbach alpha

Variables	Number of items	Alpha
New skills	4	0.877
Employee morale	4	0.740
Save time	4	0.798
Information security	4	0.780
Environmentally friendly	4	0.863

Table 2 Correlation analysis

	Digital transformation
Employees moral	0.596^{**}
New skills	0.426^{**}
Save time	0.546^{**}
Information security	0.528^{**}
Environmentally friendly	-0.034

**Correlation co-efficients

3.3 Correlation Analysis

Upon analyzing the data collected from the population sample in an attempt to identify and find any correlations between digital transformation and the benefits mentioned in the conceptual model. The result shows that the correlations between the digital transformation and the benefits are positive and significant ($p < 0.01$) for acquiring new skills, saving time, improved information security, boosting employees morale. There was no significant correlation between digital transformation and contributing to saving the environment (Table 2).

3.4 Compare Means

This section discusses the values for each of the benefits from the descriptive statistics that are studied independently, with a focus on two types of measurements, the measure of central tendency and dispersion, in addition to the unit of the questionnaire. The means of the each of the benefits are in line with the correlations discussed in the previous section (between 4 and 5 hence between Agree and Strongly Agree). However, the findings were not conclusive on how much if any, a correlation exists between Digital Transformation and preserving the environment (Table 3).

Table 3 Descriptive statistics

	Minimum	Maximum	Mean	Std. deviation
Total employees moral	2	5	4.25	0.557
Total new skills	1	5	4.76	0.838
Total save time	1	5	4.29	0.537
Total information security	1	5	4.26	0.560
Total environment friendly	1	5	2.29	0.707

4 Conclusion

4.1 Discussion

This study aimed to explore the benefits of using technology to transform administrative and financial processes in the oil and gas industry in the Kingdom of Bahrain from analog to digital. The results showed that there are strong correlations between some factors that were assumed to be potential benefits of moving from analog or paper-based processes to digital versions of the same processes, in an attempt to ensure that the data used were correct and reliable, Cronbach Alpha reliability test was performed and the data was found to be usable and reliable. The data collected was processed, examined, and categorized by sex, age, years of experience, and qualifications as each examined category were separated into subcategories based on the choices available in the questionnaire answered by the sampled population.

The results of the study are aligned with findings in the literature regarding acquiring new skills, saving time, improved information security, boosting employee morale. It has been shown that the acquisition of new skills is associated with digital transformation. Highly skilled employees can adapt to this transformation and will be easy to train [11]. High morale will affect rapid acceptance of the introduced technology among employees (Eureka 2019). Digital operations in companies will save time and effort and improve efficiency and accuracy [27]. However, the findings were not conclusive on how much if any, a correlation exists be-tween digital transformation and preserving the environment.

4.2 Recommendations

In light of the above-mentioned results, the researchers recommend the below:

- That companies in the oil and gas sector in the Kingdom of Bahrain start pursuing digital transformation and integrate technology and digital solutions into their administrative and financial operations, however, the transition from analog solutions to digital solutions within these companies must take place as soon as possible to keep pace with the most developed companies in other fields.
- That appropriate and structured training and development programs for the employees who will deal with and operate these new systems, preferably these training and development programs be based on the existing competencies observed for the individual employees, the level of education, the nature of the task to be performed and participation in it And that these programs be aware of all parts of this transformation to save time and effort from employees who were suffering from the amount of time wasted in interfering with manual processes.

- Companies that engage in these digital operations record their paper consumption and paper waste consistently to gain a visual understanding of the extent to which the transition from analog to digital has reduced paper usage.
- The use of dedicated servers with large capacities and reliable data backup capabilities to replace paper filing systems that may exist to any degree and dispose of original documents that may not be useful according to the company's document retention policies at that time to save space and avoid unnecessary duplication of documents and company records.
- Companies should also invest in strengthening their cybersecurity capabilities as the transition to digital is an ongoing process, and the ongoing protection of a company's digital assets becomes even more important as the process continues, so it is necessary to strengthen the company's digital defenses if data becomes at risk of unauthorized access, corruption, manipulation, and unauthorized movement.

4.3 Suggestions for Future Research Topics

The scope of research on Digital Transformation is wide and it keeps branching out however, the context of embracing Digital Transformation within the Oil and Gas Industry is limited. The below suggestions may be helpful for future researchers who wish to pursue research endeavors in a similar subject in hopes to take these endeavors further:

- To explore the effects of Digital Transformation on employees' daily attendance and whether Digital Transformation may diminish the necessity of administrative employees to be physically present at their workstations to perform the tasks.
- To examine the effects of Digital Transformation on workforce retention and whether replacing manual processes with their digital equals could risk employees' job security since certain automated processes may no longer require humans to operate and engage with these tasks.

4.4 Limitations

During conducting this research, there have been several challenges that have limited or diminished the extent of conducting this research:

- There was no way to identify all the technical skills required to be able to operate and engage actively with the technologies introduced as the scope of the transition and the nature of jobs greatly vary.
- In regards to employee morale, by nature, it was challenging to form a well-rounded understanding of the effects that Digital Transformation has on all aspects of human behavior in the workplace as subjects come from a different, almost endless varieties of backgrounds and emotional make-ups.

References

1. Ahmad, A., Alshurideh, M., Al Kurdi, B., Aburayya, A., Hamadneh, S.: Digital transformation metrics: a conceptual view. J. Manage. Inform. Dec. Sci. **24**(7), 1–18 (2021)
3. Al Mehrez, A.A., Alshurideh, M., Al Kurdi, B., Salloum, S.A.: Internal factors affect knowledge management and firm performance: a systematic review. In: Proceedings of International Conference on Advanced Intelligent Systems and Informatics, pp. 632–643 (2020)
4. Alameeri, K., Alshurideh, M., Al Kurdi, B., Salloum, S.A.: The effect of work environment happiness on employee leadership. In: Proceedings of International Conference on Advanced Intelligent Systems and Informatics, pp. 668–680 (2020)
5. Alhussain, M.: Bahrain and digital transformation in human resources management (2020)
6. Alobidyeen B.: Digitalization and its Impact on Employee's Performance: A Case Study on Greater Tafila Municipality. Int. J. Business Admin. Stud. **8**(1), 33–47 (2022). https://doi.org/10.20469/ijbas.8.10004-1
7. Bhatia, R.C.: Employee Morale and Productivity (2020)
8. Biwaeka, K.: The Legal System for Electronic Payment Method in Algeria (2020)
9. Brahimi, S.: Electronic Management in Algeria Between Reality and Prospects. University of Mesilla, Legal system of electronic public utility (2019)
10. Crittenden, W.F.: Embracing digitalization: student learning and new technologies. J. Market. Educ. **41**(1), 5–14 (2019)
11. Damawan, A.H.: Resistance to change: causes and strategies as an organizational challenge. Adv. Soc. Sci. Educ. Human. Res. **395** (2019)
12. Doculabs: Digitization and Paperless Processing: What You Need to Know (2009)
13. Duncan, A., Cater, K.: Expanding Evidence Approaches For Learning in a Digital World (2013)
14. Firk, S.: Chief digital officers: an analysis of the presence of a centralized digital transformation role. J. Manage. Stud. **58** (2021). https://doi.org/10.1111/joms.12718
15. Friedman, T.: The world is flat (2005)
16. Hetmanczyk, P.: Digitization and its impact on the national labour market and education system. Results of a moderated discussion, scientific papers of Silesian University of technology organization and management Series No. 148 (2020)
17. Hijab, Y.: The impact of electronic management on the basic principles that govern the public facility (2019)
18. Idris, M.: The impact of level of education, teaching experience and gender on professionalism and performance: the case study of Universitas Muhammadiyah Palembang's academic teaching staffs. Int. J. Human Res. Stud. **9**(1), (2019). ISSN 2162-3058
19. Isa, Z.: Skills you need to succeed at work after the corona pandemic (2021)
20. Ishchenko, M.: Enterprise Digitalization Security Level Diagnostics According to its Lifecycle Phase, Financial and Credit Activities. Prob. Theo. Pract. **4**(35), (2020)
21. Katuu, S.: Introduction to digitization (2020)
22. Kingdom of Bahrain Ministry of Education (2019)
23. Kuusisto, T.: The Balanced Digitalization and Digital Security—Case the Regional Authorities (2019)
24. Lapierre, J.C.: People-led digital transformation: future-proofing a legacy brand with purpose (2021)
25. Matsaung, R.: Factors influencing the morale of employees at the greater Tzaneen Municipality (2014)
26. McKinsey: Cybersecurity in a digital era (2020)
27. Metnambek, I.: Expected benfits of open government data in the kingdom of Saudi Arabia (2022)
28. Kvasha, N.V., Demidenko, D.S., Voroshin, E.A.: Industrial development in the conditions of digitalization of infocommunication technologies. St. Petersburg State Polytech. Univ. J. Econom. **11**(2), 17—27. https://doi.org/10.18721/JE.11202
29. Nabinayangan, P.: A study on employee Morale. IJRTI **2**(6). ISSN: 2456-3315

30. Nassef, Y.: Contributions of some areas of the digital economy to the digitization of education (2021)
31. Nethmini, I.: The impact of human resource management practices on employee (2022). Retrieved May 14, 2022, from https://www.researchgate.net/publication/355483105. The_impact_of_human_resource_management_practices_on_employee_performance_and_the_mediating_role_of_employee_commitment.
32. Parviainen, P.: Tackling the digitalization challenge: how to benefit from digitalization in practice. Int. J. Inform. Syst. Project Manage. (2017). ISSN (Print) 2182-7796, ISSN (Online): 2182-7788, ISSN (CD-Rom): 2182-782x
33. Porika, B.K.: A study on employee morale (2019). ISSN No. 1021-9056
34. Rodchenko, V.: The effectiveness of human capital in the context of the digital transformation of the economy: the case of Ukraine. J. East. Europ. Central Asian Res. 8(2), (2021)
35. Schneider, P.: Employees perspectives on digitalization-induced change: exploring frames of industry 4.0. Acad. Manage. Discov. 6(3), 406–435 (2020)
36. Scholkmann, B.: Resistance to (digital) change. In: Individual, Systemic and Learning-Related Perspectives (2021)
37. Solberg, E.: Digital mindsets: recognizing and leveraging individual beliefs for digital transformation. California Manage. Rev. 62(4), 105–124 (2020)
38. Stark, J.: Digital transformation of industry (2020)
39. Sumbal, M., Eric, T., Eric, W.K.: Interrelationship between big data and knowledge management: an exploratory study in the oil and gas sector (2017)
40. Turner, D.P., Houle, T.T.: Conducting and reporting descriptive statistics (2019). Headache from https://pubmed.ncbi.nlm.nih.gov/30825211/
2. Waseem A, Irfan: Qualitative VS Quantitative Research (2019)
41. Zhanga, T.: Enterprise Digital Transformation and Production Efficiency: Mechanism Analysis and Empirical Research. Informa UK Limited, trading as Taylor & Francis Group (2021)
42. Zimnoch, D.: Digital transformation of transportation in the age of COVID-19. Problemy ZarzÈdzania (Management Issues) 19(3), 100–121 (2021). https://doi.org/10.7172/1644-9584.93.5

The Relationship Between IT Governance and Firm Performance: A Review of Literature

Noora Ahmed Al Romaihi, Allam Hamdan, and Raef Abdennadher

Abstract This research aims to deliver more indication on information technology (IT) governance and its relationship with firm performance in emergenMarketplaces. The objective is mainly to clarify the role of IT governance in GCC firms; it, as well, seeks to analyze the effect of IT governance on the firm performance. The investigators have applied the IT-connected qualifications of the participants of the board of directors as a guide of IT governance. The investigators have also utilized an additional unit of corporate governance guides in addition to a setting of control variables. The performance of firms has been described as the operating performance which is characterized as Return on Assets (ROA) and the financial performance as the Return on Equity (ROE).

Keywords IT governance · Firm performance · Board members · CEO · CIO

1 Introduction

This chapter represents an identification and review of past studies focusing onto the topic of ITG and firm performance. The purpose of this is to find out the research gap in this topic. Mainly, this section will look for the optimized answers to the proposed questions. In many years, Info Tech (IT) came to be critical in accepting businesses to improve approach performance besides funding development of businesses structural design. Nowadays, numerical period of quick revolution, quickness into market value and responsiveness businesses remain reliant to implementation of current (IT). Therefore, prevalent usage of tech needs formed vital dependence at tech that asks to concentrate on IT Governance (ITG) [26].

N. A. Al Romaihi · A. Hamdan (✉)
Ahlia University, Manama, Kingdom of Bahrain
e-mail: allamh3@hotmail.com

R. Abdennadher
University of Business and Technology, Jeddah, Saudi Arabia

© The Author(s), under exclusive license to Springer Nature Switzerland AG 2024
A. Hamdan and E. S. Aldhaen (eds.), *Artificial Intelligence and Transforming Digital Marketing*, Studies in Systems, Decision and Control 487,
https://doi.org/10.1007/978-3-031-35828-9_96

Researchers tend to define ITG as "the preparation for, making and implementation of decisions regarding goals, processes, people and technology for organizations on a tactical and strategic level" [24].

For a basic value of information technology (IT) in businesses operations, nowadays, can barely be contested. While IT expenditure is continuously increasing, the constant argument bordering the IT production contradiction has reduced.

IT Governance should become a component of corporate governance method as it contains the executive practices, techniques, along with the strategies recognized to require assessments plus guidance to the IT facilities and funds, as well as concerns on hazards, observance and performance. IT Governance is the obligation of the tactical, and effective "vendors" of IT funds in aid of the investors who require discernible value. The IT Governance Institute's definition of IT Supremacy contains the control and managerial configuration and practices that safeguard, maintains and covers the organization's tactics and intentions [18].

The firm Performance is defined as "the total value created by the firm through its activities, which is the sum of the utility created for each of a firm's legitimate stakeholders" [15]. In addition, firm performance is able to be evaluated when reflected in elements like profitability, growth, market value, total return on shareholder, economic value added and customer satisfaction, based on the stakeholder's expectations" [15].

The effect of a governance of this resource on firm performance is irrefutable. Information Technology governance (ITG) is a major facilitator and a main element which aid to improve firm performance in the workplace. ITG is currently divided into two main segments. Both segments are still in early stages of modeling. These two segments are the educational research studies and the business Integration of ITG. Clear impact of firms' performance in these segments are still not available [16].

By the end of this literature review, this research aims to identify the gap in the previous articles in order to find the relationship between IT governance and operational and financial performance of firms, which in its turn will be reflected in bank performance in GCC countries.

2 Literature Review

2.1 Theorical Background

Source-based theory gets used in IT business value investigation to solve the question of IT corporate cost, and Information Technology -created useful gain. Sketching on this idea, an organization's capability to efficiently develop, incorporate, and utilize IT reserves in mixture with extra reserves, can establish distinctive useful improvements and subtle resources for the firm. As mentioned previously, the main objective of the present research is to examine how IT governance and Information

Technology ability in conjunction involve a businesses' marketplace worth that one viable accountancy performing.

In this regard, we refer to the explanation considering marketplace profit compute and workable accountancy performance calculate is termed in the "Variable Definitions and the Proposed Research Model [27].

The worth of ICFR is recognized by a lot of shareholders, much although self-controls are just a division of the internal controls its applied to help organization actions, as COSO points out. This estimate connects to a sequence of many issues in adding to improved consistency of financial statements that's intention of SOX.

This additional advantages consist on enhancing financial performance after smaller rigid conformity costs, increased operations, and reduce specialized costs. Sustaining efficient internal control be able to be extremely expensive. The investing in internal control and the secondary practices needed to detail it, appraise the subject and description on that one is issue to management decision it be able to change by organization's reimbursement system. Consequently, the thing is probable that doubt fees remain expensive, administration might decide to accept certain internal control failings towards prevent the fees of remediate these failings and the deleterious influence of such fees on financial performance and reimbursement. These could be especially the issue if the failings stayed as a type of marketplace had few concern in. For example, contempt existence topic to internal control audits under FDICIA Act guidelines. Some of financial institutes that replied to a Standard and had substantial deficits in internal control over financial reporting, few firms directed their attention to use the framework for the valuation of financial reporting controls. Add-on, IT controls need businesses to achieve unique experience and ability and may thus be more expensive to fulfill, function, observe and review than additional kinds of controls [4].

Traditional RBV: Its usual RBV implemented the limited opinion and hypothesized firms as mixed units entailing of wads of peculiar reserve. In instance, quick surveys on RBV utilized their private concept of source-policy hurdles that ensure fiscal lease to shield a firm's reserve from imitation and substitution. The RBV of a firm is based on two essential theories, as developed in the field of strategic management. The first assumption happens that the assets and skills seized by rival firms can vary; the next belief is that these distinct assets can be constant since of oddity and complexities in asset or parody. Hence, in conventional RBV surveys, investigators have guessed that profit-forming funds are held and organized by the focal firm [20].

Extended RBV: Firms cannot have all essential assets to compete essentially in the rapidly changing business environment. Thus, corporations seek entry to the crucial assets out of partnerships with their outward allies. Agreeing to the ERBV prose, customary RBV neglected to concede the precise splitting of assets and the implicit transferability of advances linked with exterior reserves. Allotted that the basic theories of the RBV go off to be inappropriate, many scientists have stretched worried that firms must concurrently believe their inner then outer assets to acquire and nurture aggressive benefit. Larvie insinuated the ERBV that firm estimate must be centered not just on the funds of the crucial firm in query but also on the reserve bequests of its partnership allies by hiring a abstract family of the interactive vision and collective group concepts. Thus, a more exhaustive understanding of the internal and external resources in organizations must have being achieved to actually regulate these assets and to establish reasonable benefit of companies [20].

An array of executive habits has been believed in the poetry. Find that intelligence machinery simplifies a firm's allocation of more clout to entities and company divisions at reduce executive heights. Firms approving a dispersed executive structure obtain higher productivity and more value from their IT ventures. Reveals that the same effect retains while firm profit is assessed built on stock market valuation. Devaraj and Kohli find that the value of IT assets varies on real usage. This survey emphasizes the significance of having executive systems in spot to spur useful treatment of IT resources. Al Dhaen [1] finds that studying IT interactions among industry groups shows a crucial part in giving out the cost of IT funds. This ruling indicates that companies want to make up needed structural adjustments to influence cross-entity IT interactions, provide a complete summary of the prose and find that the cost of IT assets is reliant on a total of domestic and peripheral aspects, varying from balancing managerial funds to overall sell conditions [12].

The key aim of the poetry and test issue study was to get a clearer picture on how organizations are addressing IT governance these days and to come up with an early list of IT control procedures from training. From the litigation studies, various trainers for embracing IT control were found. An important one was certainly the need to comply with Sarbanes–Oxley needs, which impacts heavily on the control environment in IT. Other important drivers designed for IT control were the drive to attain countries of levels after unions and assets and plan stress, causing in a lesser cost for new ventures. Task of policy is then to optimally allocate the residual plan to schemes and events that are providing price to the company. Lastly, several pilot case firms stated that the IT governance scheme was beyond an attempt of validating and arranging current systems now used [9].

Theoretical models clearly dealing with ITG are hardly accessible. Recent exemptions take into consideration ITG mostly in relation to business/IT alignment, which in turning is known in relation to firm performance.

2.2 IT Governance

IT governance basically places outline around how firm's IT plan supports with firm plan. This IT-firm configuration will make sure that business maintain to accomplish their plans and objectives and apply techniques to estimate its performance. One specific characteristic of IT governance is that it reflects the benefits of stakeholders and confirms all practices deliver measurable results. This condition is probable with sideways IT governance forms, with the participation of all levels of management [17].

IT governance identifies the judgment rights with responsibility of the structure that leads towards promote attractive conduct in the use of IT. This conduct conveys to the structure of the governance, and executive constructs and procedures that guarantee that the company's IT maintains and broadens the government's plans and intentions [21]. The extent of IT control are not only options themselves but the courage which choices need to be made, who can donate to the managerial methods and who is finally qualified to make the ruling. In this logic, every firm has IT governance, but only and clearly intended one is able to align IT efficiently and well to the objectives of the firm [17].

The tenure IT governance to designate the usual of instruments for confirming the achievement of essential IT abilities but did non feature conspicuously in the academic fiction until the last decade started to mention to an idea of "IS governance frameworks" and then deferred to "IT governance frameworks" in their documents. If they accept Weill's meaning of IT governance, the conception of describing IT choice merits and responsibilities, in reality, perfectly studied extended prior to 1990s. This act characterizes significant advancement in researching governance [5].

Other study mentioned that IT governance has been developed an essential subject in preparation which deliver detailed assessment of the current information on ITG, concentrating on the progress and evolve of the study area in overall, whereas attention on the main meanings of the tenure ITG, and opinion out the variety of subjects that are composed below the canopy tenure ITG. In latest eras, study in what we currently call in ITG has concentrated on the position of self-control and governance constructs, eventuality assessment and the sequence of these double rivers. Despite the good, applied benefit of the study, the results and examples formed within this study rivers are frequently explanatory and inflexible, and therefore, mainly deficient a strong theoretic foundation [16].

ITG is an administration procedure that is implemented by in cooperation of the board of directors with executive managing group. Its purpose is to make sure efficient consumption of Information Technology such that: (A) Information Technology is associated with the organization, (B) Information Technology lets the company to achievement possibilities, (C) Information Technology funds are cast-off reliably, (D) Information Technology menaces are accomplished properly. These four attentions are entangled to performance dimension. Board and the decision-making managing group require to trail mission tactics and approach and examine Information Technology facilities and take the risk of Similar opinions concerning the scale of ITG have been stated in many papers. Some relations and calibration figures showed try to progress ITG backgrounds, containing Val IT (background for forecasting, execution and observing the abstraction of worth as of IT-allowed procedures, and Control Objectives for Information and related Technology which are permitted by the IT Governance Institution. Others consist of ISO/IEC by the International Organization for Standardization, and COSO rules for internal control systems, deceit deterrence, and risk managing [25].

In previous study shows the term executives keep on applied intersection together with tenure with the directors. The analysis on board independence and firm performance showed combined findings; either one positive, negative or no connection along with firm performance. Limited reviews also seem at the relationship between board independence and attaining executive [11].

There is no separate commonly understood ITG structure. Provided the significance of ITG, it would include important board participation. The board of directors is a team that supervises the management of the business. Their common corporate governance obligations consist of tactical planning and examining, ensuring procedures and supplies for accomplishing targets, justifying internal controls, and making sure risks are identified and monitored. Boards of directors approve all major decisions, often offer advice and counsel to directors (external administrators can bring insights from other associations with which they are involved) and monitor both management's dealings and the resultant execution. All of these tasks can have IT connections, which behooves the board to involve in ITG. Boards find data for their ITG tasks (e.g., for probability opinion, consideration of needed assets, and examination of strategic needs) by creating IT issues in the boardroom and requesting administration probes looks present with possibility IT consequences, assets, and tactical ideas [25].

In Spite of the great rational cost of this research, the results and models created within this research rivers are often explanatory and inflexible, mostly needing a strong theoretical groundwork.

Even Though a very important first step, this is still insufficient from the viewpoint of ITG, as ITG is just one out of six main input elements of BITA. In Spite Of its kin to ITG, BITA is an individual research area and therefore, consequences of BITA-centrical models for ITG are inadequate. Furthermore, despite a three-digit number of periodicals. Furthermore, we disagree that BITA is not the first required impact of ITG that can take the lead to a definite impact on firm performance and therefore the kin between ITG and firm performance should be evaluated individually to the very complicated concept of BITA [16].

International standards of ITG are to ensure that the legitimate openness while this, facilitate fair cooperation among firms besides nations. They agreed at high levels prominent firms and market. In Information Skill and software expansion, norms are the basis of the "state of the art," for example, in product liability or in safety and cybersecurity. With the International Organization for Consistency (ISO) [10].

For ruling-creating stages imply various actions necessary to make rulings within the distinct spheres. This element deals with the relative between IT, and the types of the truth used for ruling-creating. Before making any ruling about e.g. the subcontracting of a helpdesk event, the association must be plainly known. Facts have to be felt over and probed and converted into a model. The version might be a simple cognitive map, offer nowhere else but in the top of the ruling-maker, or a more reinforced, theoretical model put on print. This process of evaluation and insight is meant the Comprehension phase. Then the standard is designed, the definite ruling can be made giving to trade IT rules, in a timely manner, by the right entities, etc. In the IT governance classification, this is symbolized by the Decide phase, which also includes design of how to make the verdict [19].

2.3 Board Members and CEO

Research on board creation has also focused on the relationship between board size and firm performance. Disputes have been put forward on why small boards might be more valuable than significant boards. Dalton et al. (ibid.) summary those disputes. Social lolling is one of them and suggests to parties putting in fewer energy when the size of the group raises. Band cohesiveness, the influence that brings bands closer to each other, may be accelerated by having minor groups [2]. A third dispute is that a board's capability to commence tactics pursuits could be hindered in a larger group. Still, smaller parties are more easily able to reach consensus, lastly large and diverse boards may be more easily swayed when it comes to performance assessment of high management [7].

With the wonderful expansion of using IT in organization, the Chief Executive Officer (CEO) must have an IT-connected knowledge that modifies him/her to articulate his/her opinion/s on various issues related to the operations of the firm. Confirmed that the knowledge and expertise that the CEO has in IT will influence the scale of participation in IT management and will be able to concentrate funds in the direction of ideal investment in IT. Hence, the key disagreement of this research is to undertake that the CEO has passable knowledge, and the pertinent IT-related related, can track the specifics of all practical tasks and duties in a business; and start suitable controller actions that will absolutely reproduce on the firm's operational and financial performance. The Chief Information Officer (CIO) is allocated to indicate, create, and apply a distinctive image for the part of IT in supporting firm policies [14].

2.4 The Relationship Between IT Governance and Firm Performance

IT governance instruments help realize the intelligent IS tactical position in that the realized governance instruments elucidate the parts and errands of the complicated gatherings and how the expert for IT is communal between commercial associates, IT organization, and facility earners; they can be measured to be legislative provisions that enable collective empathetic among side associates. First, IT managing agencies are one of the "efficient control processes for allying IT-related determinations and activities with an association's tactical and operating priorities". Handling agencies are consisting of high-level representatives who are certified with the job of connecting IT plan with big business policy by equaling company fears with IT backing. Indeed, guiding groups are valuable in creating the mood of industry–IT contacts. The authority of IT steering committees offers the visibility of IT initiatives, an important tool instead of top executive to realize IT in the association. Having a guiding board written of both business unit leaders and CIOs can assist to ensure close coordination of company and IT in the association and therefore the planned alliance. Buildings that allow the CIO to clearly article to the CEO and/or the COO6 certify that IT is an element of the director group, the point at which most policy negotiations occur. This allows the CIO to find a worldwide and universal position on the alliance, its aims, and tactics, and creates the CIO's awareness of the TMT's concept of the government. These mechanical devices impact IS deliberate orientation by supporting entry to both IT and corporate units for expertise swap and interest involvement [22].

3 Conclusion

The main aim of this Literature review is to find the gap of the previous study and weather they find a solution to solve these gaps, poetry and initial case study were to get a clearer view on how businesses are delivering IT governance these days and to come up with an early list of IT governance habits from routine. From the issue research, distinct chauffeurs for embracing IT governance were discovered. A crucial one was surely the need to submit with Sarbanes–Oxley obligations, which effects severely on the power atmosphere in IT. Other crucial handlers for IT governance were the boost to attain reductions of ratios after fusions and assets and modest burden, subsequent in a tinier cost for fresh schemes. Task of policy is then to optimally allocate the residual cost to schemes and events that are supplying benefit to the enterprise. Lastly, some experimental case corporations revealed that the IT governance development was more an endeavor of validating and forming accessible procedures already utilized. Established on the results of the prose study, and the pilot case study, an early list of IT governance systems was assembled, as displayed below. For each of these procedures, a short description was created based on the narrative and response from the initial instances.

As a universal assumption of this empirical research, this study showed that IT governance is certainly high on the schedule. Our study indicates that there is a positive correlation between the use of IT governance procedures and corporate/IT orientation. It emerged that extremely allied corporations do indeed influence more stable IT governance habits related to inadequately allied corporations. Some exhaustive inferences were sketched involving IT authority edifices, activities, and interactive processes. It was proven that it is calmer to execute IT governance arrangements related to IT governance practices. It also emerged that interpersonal procedures are very vital in the starting junctures of an IT governance completion scheme and become less crucial when the IT governance agenda is fixed into day-to-day tasks. For some detail IT governance systems, the inquiry presents clues that disprove present fiction. An embodiment is the commitment of the panel of executives in IT governance, which is marketed by many inventors in poetry, but was not endorsed in this inquiry.

References

1. Al Dhaen, E.S.: The use of information management towards strategic decision effectiveness in higher education institutions in the context of Bahrain. Bottom Line 34(2), 143–169 (2021). https://doi.org/10.1108/BL-11-2020-0072
2. Alshater, M., Atayah, O., Hamdan, A.: Journal of sustainable finance and investment: a bibliometric analysis. J. Sustain. Financ. Investment 13(3), 1131–1152 (2021). https://doi.org/10.1080/20430795.2021.1947116
3. Bhagat, S., Black, B.S.: The uncertain relationship between board composition and firm performance. Available at SSRN: https://ssrn.com/abstract=11417
4. Boritz, E., Lim, J.-H.: IT control weaknesses. In: IT Governance and Firm Performance. (CAAA) 2008 Annual Conference Paper January 11 (2008). Available at SSRN: https://doi.org/10.2139/ssrn.1082957
5. Brown, A., Grant, G.: Framing the frameworks: a review of IT governance research. Comm. Assoc. Info. Syst. **15** (2005). https://doi.org/10.17705/1CAIS.01538
6. Chau, D., Ngai, E., Gerow, J., Thatcher, J.: The effects of business-IT strategic alignment and IT governance on firm performance: a moderated polynomial regression analysis. MIS Quart. **44**, 1679–1703 (2020). https://doi.org/10.25300/MISQ/2020/12165
7. Dagsson, S., Larsson, E.: How age diversity on the Board of Directors affects firm performance (Dissertation) (2011). Retrieved from http://urn.kb.se/resolve?urn=urn:nbn:se:bth-4607
8. Debreceny, R.: Research on IT governance, risk, and value: challenges and opportunities. J. Info. Syst. **27**, 129–135 (2013). https://doi.org/10.2308/isys-10339
9. De Haes, S., Grembergen, W.: An exploratory study into IT governance implementations and its impact on business/IT alignment. IS Manag. **26**, 123–137 (2009). https://doi.org/10.1080/10580530902794786
10. Ebert, C., Vizcaino, A., Manjavacas, A.: IT governance. IEEE Soft. **37**(6), 13–20 (2020). https://doi.org/10.1109/MS.2020.3016099
11. Fuzi, S., Abdul Halim, S., Julizaerma, M.: Board independence and firm performance. Proced. Econ. Fin. **37**, **460–465** (2016). ISSN 2212-5671. https://doi.org/10.1016/S2212-5671(16)30152-6
12. Gu, B., Xue, L., Ray, G.: IT governance and IT investment performance: an empirical analysis, July 1 (2008). Available at SSRN: https://doi.org/10.2139/ssrn.1145102
13. Haes, S., Grembergen, W.: An exploratory study into IT governance implementations and its impact on business/IT alignment. Info. Syst. Manag. **26**(2), 123–137 (2009). https://doi.org/10.1080/10580530902794786
14. Hamdan, A., Khamis, R., Anasweh, M., Al-Hashimi, M., Razzaque, A.: IT governance and firm performance: empirical study from Saudi Arabia. SAGE Open **9**, 215824401984372 (2019). https://doi.org/10.1177/2158244019843721
15. Herciu, M.: Drivers of firm performance: exploring quantitative and qualitative approaches. Stud. Bus. Econ. **12** (2017). https://doi.org/10.1515/sbe-2017-0006
16. Lazic, M., Groth, M., Schillinger, C., Heinzl, A.: The impact of IT governance on business performance. In: AMCIS 2011 Proceedings—All Submissions. Paper 189 (2011). http://aisel.aisnet.org/amcis2011_submissions/189
17. Lorences, P., García-Ávila, L.: The evaluation and improvement of IT governance. JISTEM **10**, 219–234 (2013). https://doi.org/10.4301/S1807-17752013000200002
18. Luftman, J., Ben-Zvi, T., Dwivedi, R., Rigoni, E.: IT governance: an alignment maturity perspective. IJITBAG **1**, 13–25 (2010). https://doi.org/10.4018/jitbag.2010040102
19. Nsor-Ambala, R., Amewu, G.: Linear and non-linear ARDL estimation of financial innovation and economic growth in Ghana. J. Bus. Socio-economic Dev. 3(1), 36–49 (2023). https://doi.org/10.1108/JBSED-09-2021-0128
20. Park, J., Lee, J., Ryu, H.: Alignment between internal and external information technology control mechanisms: an extended resource-based view. PACIS (2014)

21. Patra, G., Roy, R.K.: Business sustainability and growth in journey of industry 4.0-a case study. In: Nayyar, A., Naved, M., Rameshwar, R. (eds.) New Horizons for Industry 4.0 in Modern Business. Contributions to Environmental Sciences & Innovative Business Technology. Springer, Cham (2023). https://doi.org/10.1007/978-3-031-20443-2_2
22. Shelly, P.-J., Straub, D.W., Ting-Peng, L.: How information technology governance mechanisms and strategic alignment influence organizational performance: insights from a matched survey of business and IT managers. MIS Quart. **39**, 497–518 (2015). https://doi.org/10.25300/MISQ/2015/39.2.10
23. Simonsson, M., Johnson, P.: Assessment of IT governance—a prioritization of Cobit (2021)
24. Simonsson, M., Johnson, P.: Defining IT governance—a consolidation of literature (2005)
25. Turel, O., Bart, C.: Board-level IT governance and organizational performance. Euro. J. Info. Syst. **23** (2014). https://doi.org/10.1057/ejis.2012.61
26. Vejseli, S., Rossmann, A.: The impact of IT governance on firm performance: a literature review. In: PACIS 2017 Proceedings, 41. http://aisel.aisnet.org/pacis2017/41
27. Zhang, P., Zhao, K., Kumar, R.L.: Impact of IT governance and IT capability on firm performance. Inf. Syst. Manag. **33**(4), 357–373 (2016). https://doi.org/10.1080/10580530.2016.1220218

Carbon Emissions Impact by the Electric-Power Industry

Ismail Noori and Basem A. Abu Izneid

Abstract The electric power industry in the US is second in terms of GHG emissions, contributing to almost 28% of total emissions. In addition, as more people switch to driving electric vehicles, there will be a rise up in electricity demand. If no action is taken to decarbonize the electric grid, such increased electricity use will result in higher GHG emissions. The electric-power sector consists of electricity generation, transmission, distribution, and consumption, and has a wide range of applications for AI, counting accelerating the advancement of clean energy technologies, improving electricity demand forecasts, bolstering system optimization and management, and improving system monitoring. Even novel materials for energy-storing batteries or compounds that collect carbon dioxide from the environment could be discovered thanks to AI. Finally, by incorporating data from risks like wildfires and major storms and altering grid operations appropriately, AI can help make the grid safer, more effective, and more reliable.

Keywords Data analytics · Machine learning · Algorithm · Climate change · Carbon emissions

1 Introduction

The privacy implications of all this data are a pressing issue for any AI application. Utility companies should not be exempt from these concerns. In this context, utilities are in a unique position of power because many of their customers, many of whom are captured ratepayers with no retail choice, have long-standing relationships with them. With the use of smart meters, utilities now have more detailed access to the data of these customers than ever before. The privacy concerns with these smart

---

I. Noori (✉)
Ahlia University, Manama, Bahrain
e-mail: imseer@ahlia.edu.bh

B. A. A. Izneid
Jeddah College of Engineering, University of Business and Technology, Jeddah, Saudi Arabia

© The Author(s), under exclusive license to Springer Nature Switzerland AG 2024
A. Hamdan and E. S. Aldhaen (eds.), *Artificial Intelligence and Transforming Digital Marketing*, Studies in Systems, Decision and Control 487,
https://doi.org/10.1007/978-3-031-35828-9_97

meters are particularly frustrating. Smart meters are bidirectional meters that facilitate smart consumption and pricing applications of distributed resources like rooftop solar and can be viewed remotely. They also directly convey voltage, current, and power information to utilities. Data sharing will be essential for successfully integrating AI technologies if AI becomes smarter with more data. In the United States, there are more than 86.8 million installed smart meters [1], of which amount to almost 56% of all meters. By 2030, it is anticipated that this percentage will reach 93%. "System operators are collecting previously unheard-of amounts of data thanks to the introduction of advanced metering infrastructure (AMI) and smart sensor-equipped devices".

This raise concerns about the knowledge that can be discovered about persons established on their behavior patterns, increasing the likelihood of unintended or purposeful observing, home attacks along with a particular intent, reporting, conduct tracing, or personality theft. This adds to the utilities' obligation as guardians of this data and presents crucial issues around its retention, use, transfer, and destruction. Much in the set of warrants for this data, the constitutionality of the government's access to this information has been questioned. The Seventh Circuit ruled in 2018 that gathering information from smart meters qualified as a "search" under the Fourth Amendment.

The energy company "searches" its residents' houses when it collects this data because it "reveals details about the home, even when collected at fifteen-minute intervals, that would otherwise be unavailable to government officials with a physical search," according to the court. However, after weighing the intrusion against the advancement of the legitimate government interest in updating the electrical infrastructure, the court of law likewise determined that the search was reasonable. Other state courts opposed it, concluding that no warrant was necessary for such data. Additionally, at least one state legislature has firmly opposed obtaining warrants to access data from smart meters. Such privacy issues directly contradict initiatives to reduce redundant training and promote data exchange, which was assessed above and considered necessary to enable a more modern grid. Stakeholders and regulators can take several crucial actions to reduce the unfavorable confidentiality implications of using all this energy data to address this trade-off. There are at least two ways to support such cooperation, both of which are covered below: stringent methods for anonymizing energy data [1]; and regulation of data ownership [2]. Data anonymization. Data anonymization is a widely used method for preserving data privacy. Defining data so that it can no longer "reasonably identify, relate to, describe, reference, be capable of being associated with, or be linked, directly or indirectly, to a particular individual" is the process of anonymization. Randomization and generalization are two common methods for data anonymization. Data protection rules in the US offer safe harbors to organizations that anonymize their data.

Even though all three require big data to function, these can be the idea of as related but individual concepts. In the past, people collected enormous amounts of data to aggregate and analyze its "commonalities" to identify relationships between variables. Humans make assumptions, and the data is "queried to test" those assumptions' correlations. Data analytics is useful for forecasting, running, and making payments in the electrical business. Several claim that "much of what utilities need" can be achieved by data analytics "without the cost and difficulties of AI and [machine learning]." However, there are conditions under which the added expense and complexity of AI and machine learning are warranted. Machine learning is a common component of modern data analytics, enabling the process to go beyond merely analyzing data by creating hypotheses, testing them, and learning on its own.

The term "machine learning" belongs to a group of methods employed to prepare algorithms and enable ongoing algorithmic development. Data is provided by humans, who also define several important parameters. However, the algorithm estimates the kind of data to seek on each forward and then modifies the subsequent prediction based on how well the previous one performed. With that knowledge, it is easier to comprehend why machine learning has emerged as the leading example of contemporary AI. Machine learning is just one of many technologies that go under the umbrella term "AI," although it is frequently used to refer to.

AI is being utilized to create more sustainable potential fuels for the future, such as fusion technology, even if it has historically been used in legacy fossil-fuel operations. The importance of the power sector to climate change is discussed in this section, along with three ways AI can help the electrical industry reduce GHG emissions [2]: [1] by optimizing grid assets; [2] by raising energy efficiency; and [3] by improving reliability and resiliency. A. Grid Asset Optimization with Artificial Intelligence The obvious use of AI in the electricity industry is to optimize grid assets, which involves enhancing how energy is used in the grid for both efficiency and conservation goals. The American grid used to be made up of thousands of massive, centrally located generation sources (such as power plants), but it has recently experienced a considerable shift. Increased use of renewable energy sources, many of which are smaller, distributed resources (such as rooftop solar and energy storage), a smart organization to assist enable these resources, and a move toward more electric vehicles are all part of this revolution.

Each of these four areas of AI use is covered in this section. Intermittent renewable energy. Transforming the nation's reliance on fossil fuels to renewables is the first step in reducing GHG emissions from the electrical industry. For over 63% of its electricity needs in 2019, the United States depends on fossil fuels. However, the percentage of renewable sources has been continuously increasing, and in 2020, it is projected to account for 76% of all newly planned generating capacity in the United States. but a lot of it is renewable States or to an increased awareness of data protection in general. Like CIP-014-1, CIP-010-2 mandates that organizations have mechanisms in place to create baseline configurations for cyber assets, track changes to the baselines, and "perform a paper or active exposure assessment, at least once every 15 calendar months." Threats to physical security and cybersecurity are covered by other standards as well. For instance, CIP-006-6 mandates associations to restrict access to cyber assets to those who have been pre-authorized.

By anticipating and addressing data breaches, or by automating security operations procedures by offering a big-data solution to a big-data problem, AI could be utilized to counteract these cybersecurity dangers. For regulators and utilities, cybersecurity concerns in the energy sector are a constant source of worry. Due to the possibility of foreign enemies "exploiting weaknesses in the United States bulk-power infrastructure," President Trump proclaimed a national emergency in May 2020. In response to these dangers, President Trump outlawed the "purchase, importation, transfer, or installation of any bulk-power system electric equipment" whenever a foreign government or one of its citizens is involved. Furthermore, in a recent test by a security company in Boston, hackers overran a fictitious control network that was based on control networks used by North American electric firms in just three days.

By financing initiatives to conduct research and develop "cybersecurity solutions for energy delivery systems... to identify, prevent, and reduce the repercussions of a cyber-incident...," the DOE has already taken action to tackle cybersecurity threats. Energy companies might use AI to "manage their distribution networks, including diagnosis of problems" to thwart future cyberattacks as well as "reroute power flows, with real-time knowledge and control," which enables prompt responses if a cyberattack does take place [3]. Although many of the challenges raised above have long been addressed by grid operators using data analytics, AI has the potential to go beyond what is currently possible with human processing in terms of optimizing grid assets, boosting efficiency, and improving dependability and resiliency. Important climate and artificial intelligence tradeoffs understanding the tradeoffs involved with these technological advancements will be crucial if we are to adopt artificial intelligence as a climate strategy. AI may be met with skepticism, hefty upfront expenses, and intense regulatory scrutiny, as with any newly developing technology. Even established participants in the energy sector may not be fully aware of their capabilities and constraints, which makes it difficult for them to feel as comfortable as when relying on the status quo. This Part examines just a few of the significant tradeoffs linked to the usage of AI to address issues related to electricity and climate change: Environmental effects, data privacy, investment and procurement, and accountability are the first three categories [4].

It also offers normative suggestions for how to proceed with AI in a way that best balances the competing needs of businesses, consumers, and the public within each tradeoff. A. Effects on the environment even if it has the potential to cut down on energy use and improve grid efficiency [4], AI can also be a significant energy user. More than 2% of the world's electricity is used by data centers, and by that, that percentage may increase to between 8% (the best case) and 21%, according to academics (expected). According to one study, which indicated that the number of computational resources being utilized by AI is rising alarmingly, these numbers could rise even higher. According to a University of Massachusetts study, the lifetime emissions of the typical American car are approximately five times higher when a huge AI model is trained to understand human language.

The diverse processing requirements across the many distinct types of AI may allay this worry. He already said, a wide range of methods, including machine learning, are referred to as AI. The difficulty of the task and the effectiveness of the algorithm used will determine the computation requirements for each of these strategies. One of the most energy-intensive applications of AI is the processing of human language, which was the AI method assessed in the Massachusetts study.

Thankfully, its method is less common in the algorithms applied to lower the carbon intensity of the electric industry [4]. The electric-power industry may be able to prefer the less energy-intensive algorithms to cut emissions by being aware of the fluctuating energy demands. It will be crucial to monitor how the oil and gas sector uses AI to increase productivity. Depending on the energy requirements of those algorithms, the employment of AI in fossil fuels may result in lower emissions costs and a greater competitive edge over less carbon-intensive resources.

AI's good environmental consequences outweigh its negative ones if it is to be employed to tackle climate change. Thankfully, there are several approaches to do it, three of which are covered below: Requirements for disclosure [1]; Rules for certification [2]; and Increasing Data Sharing Disclosure [3]. The Allen Institute for AI suggests that researchers in AI incorporate various monetary and computational expenses in their reported findings as a first step. Other academics support increased environmental impact visibility and disclosure. A modern machine learning tool that tracks carbon emissions enables researchers to train their system with data and then calculate emissions totals [4]. Researchers should reveal the amount of additional ecological expenses associated with AI, such as the heat and electrical waste produced as well as the raw materials utilized, in addition to emissions disclosures. It is hoped that more openness and accountability will encourage researchers to work harder to keep costs down and raise awareness of the potential effects of algorithms.

For a complete view of the algorithm's possible effects, the Allen Institute of AI explicitly advises researchers to report carbon emissions, electricity usage, elapsed real time, the number of parameters used by the model, and the number of floating-point operations. Researchers can only evaluate an algorithm's costs to its advantages if they are aware of any potential expenses, a subject that is frequently left out of discussions regarding algorithm training and development. Certification. By considering a certification requirement, a second alternative imitates previous environmental regimes. The Allen Institute proposed a certification program for artificial intelligence techniques [4], designating "green" AI as carbon–neutral and "red" AI as non-carbon–neutral. These badges, just like other environmental certification programs, may have significant signaling effects that encourage businesses to internalize their electricity use. Federal agencies may be engaged in this certification process, much like they are with the regulation of organic labeling. However, these marks also carry the possibility of greenwashing, just like previous environmental certification programs. Data exchange and enhancing the sharing of data used in climate-related algorithms would be the last strategy for reducing the environmental effects of AI's computational capacity.

Data are the primary input for algorithms, and how one individual uses the data does not affect how well it may be used by another. In addition, data can be viewed as a public asset in the same way that we treat scientific knowledge because of its non-rivalrous character. The federal government may be a workable partner for climate-related AI, just as it has a role to play in other types of public goods. By acting as a repository for publicly available, anonymized electricity data, for example, the federal government might reduce the number of redundant jobs related to climate AI in the electric power sector. By granting interested parties access to general smart-meter customer data in 2005, Congress started to solve these complex challenges. The DOE and the White House collaborated to develop the Green Button Initiative in 2012 as a follow-up [4].

Customers of utilities are now able to safely exchange their data with approved third-party assistance suppliers thanks to the program. With more than 3000 utility and electricity suppliers in the United States, it still has a long way to go [4]. According to the DOE, about fifty utilities and electricity suppliers have entered the program. Even further phases in the creation of high-quality AI may be centralized by the federal government. Energy-demanding procedures include selecting, training, and implementing the model as well as gathering, cleaning, and splitting the data. Centralizing these procedures would reduce costs and reduce the environmental effects of AI training while enabling more effective access to more data. The state government can carry out some of this nonproprietary work and then disseminate it to the larger scientific public. As academics work out to create useful climate AI [4], this might assist to reduce useless replication. These companies could run into the same issues along with trade secrets, intellectual estate rights, and client confidentiality. Due to the extremely private and secret nature of the data, not public service data containing identifiable information might not be eligible for such a data-sharing project.

One of the most difficult issues facing civilization is climate change. However, because its effects are anticipated to be most severe in the future, more manageable and urgent issues that coincide with election cycles frequently occupy center stage on the political scene. Even behavioral specialists remind us that people tend to downplay society's biggest threats as a form of defense. Denial can be quite therapeutic for the mind, but not so much for the grandchildren. This section offers a succinct introduction to climate change. It emphasizes the technological and data problems and offers illustrations of general climate applications of AI for lowering GHGs [5]. Climate Change, first as a worldwide issue, climate change requires both reducing carbon emissions and preparing for its repercussions. In their examination of the causes and effects of global warming, researchers from all over the world reached the following conclusion: during the industrial revolution, human-induced increases in GHG emissions have been the "primary cause" of previously unheard-of rises in global temperature. Limiting global warming to 1.5 °C is crucial, according to a 2018 Special Report by the United Nations Intergovernmental Panel on Climate Change (IPCC) [5], to lessen difficult effects on ecosystems and human health and well-being.

The IPCC stated that greenhouse-gas pollution would need to drop by 45% by 2030 and 100% by 2050 to avoid the severe repercussions that would come with a 2 °C rise up in global temperature. The report acknowledges that these drastic emissions declines will need to be swift and extensive, and they will call for "unprecedented transformations in all facets of society." 26 Since the United States' major sources of carbon emissions are transportation (29%) followed by energy (28%), industry (22%), business and residential structures (12%), and agriculture (9%), these

Carbon Emissions Impact by the Electric-Power Industry 1157

substantial social adjustments will be required across a wide range of socioeconomic sectors.

Both improvement and change measures have been created by researchers, researchers, and policymakers, with varying degrees of political approval. But there is still a lot to be done. "Copious expert testimony indicates that this unprecedented rise derives from fossil fuel consumption and will wreak havoc on the Earth's climate if unchecked," the Ninth Circuit found. B. Climate AI and artificial intelligence (AI) seem ideally suited to tackle these transformative problems brought on by climate change. This Part gives a fundamental summary of artificial intelligence (AI) and some examples of how it has been used to lessen severe environmental harm. The intelligence generated artificially in today's culture; artificial intelligence gets spread like wildfire [6].

Data researchers remained the first to coin the word in the 1950s, and it has since become widely used. However, the majority of articles about AI still start by stating that the phrase is amorphous and open to many different interpretations. As was previously mentioned, one way to look at AI is as "a set of techniques pointing at duplicating some aspect of human or animal cognition using machines." The fact that people don't fully comprehend exactly how their intellect functions may contribute to the difficulty in understanding this elusive idea. It might be simpler to embrace the ambiguities of AI if we acknowledge the limits of our knowledge [6]. Our limited understanding might also help to explain why instances of various cognitive tasks that computers can perform, such as speech or facial recognition, problem-solving, and natural language processing, have been frequently used to define AI. These tasks are specific and limited, especially when compared to the enormous capabilities of the human brain. As a result, it is crucial to create clear theoretical differences, if not definitions, between three fundamental ideas at the outset: data analytics, artificial intelligence, and machine learning.

2 Conclusion

Nevertheless, anonymized data is frequently not completely anonymous. Professor Ohm, a renowned expert on data privacy, inspected three instances where anonymization was ineffective and found that: Despite administrators removing any data fields they believed might uniquely identify individuals, researchers remained still able to unlock identities by finding pockets of unexpected uniqueness in the data. In the same way that a single individual can be uniquely identified by their fingerprints left at a crime scene and connected to "anonymous" information, data subjects also produce "data fingerprints"—combinations of values of data that no other person in their table has in common. Although it tackles these worries about re-identification, the proposed Online Privacy Act of 2019 does not take a strong stance in favor of making anonymization mandatory. It did not simply demand anonymization when there is not "an unreasonable amount of work," but only where efforts to do so are "reasonable," which may prevent anonymization of vast volumes of smart meter data. Possession of data regulation. The regulation of data ownership, usage, and distribution would be a final resort. Customers of utility companies often have access to the data, although opinions on who owns the data and whether third parties can use it otherwise. These complex issues have already started to be addressed by Congress, including giving interested parties access to general utility customer data. The majority of experts concur that concentrating on just four areas—electricity, transportation, agriculture, and buildings—can have a major impact on the decarbonization of civilization. Only one is covered in this article: electricity. It is not meant to be a comprehensive analysis of AI's uses in the electric industry. But as additional ideas about AI and the environment arise, this article outlines a few requests of just how it may be used to cut carbon emissions in the electric-power industry and offers some policy concerns [6].

References

1. Fine, E.: NanoLock security. Safeguarding smart meters as cyber threats surge Jan. 12 (2020), https://www.power-grid.com/executive-insight/safeguarding-smart-meters-as-cybert hreats-surge/#gref
2. Neslen, A.: Here's how AI can help fight climate change Aug 11 (2021), https://www.weforum. org/agenda/2021/08/how-ai-can-fight-climate-change/
3. Securing America's energy infrastructure from cyber threats, July 26, 2021, https://www.ene rgy.gov/articles/securing-americas-energy-infrastructure-cyber-threats
4. Azzari, G., Jain, M., Lobell, D.B.: Towards fine resolution global maps of crop yields: testing multiple methods and satellites in three countries. Remote Sens. Environ. **202**, 129–141 (2017)
5. Summary for policymakers of IPCC special report on global warming of 1.5°C approved by governments, https://www.ipcc.ch/2018/10/08/summary-for-policymakers-of-ipcc-special-report-on-global-warming-of-1-5c-approved-by-governments
6. Ediger, M., Harrowell, P.: Perspective: supercooled liquids and glasses. J. Chem. Phys. **137**, 080901 (2012)

Employee Engagement Concepts, Constructs and Strategies: A Systematic Review of Literature

Hanin Aldoy and Bryan Mcintosh

Abstract Employee engagement is an emerging phenomenon in the human resources field. Organizations are seeking ways to enhance employee performance and increase productivity through the development of human capital, thus organizations focus on ways to engage employees in the organizations' goals, objectives, and values, moreover ways to engage the employees in their job and tasks. This study used a systematic literature review approach, to explore employee engagement in the workplace from the literature, illustrating the drivers and factors that impact the fluctuating differences in engagement between employees. Moreover, human resources practices are playing a significant role in the formulation of engagement strategy in the organization, therefore the study is to provide aspects that the top level should be aware of to merge the engagement strategy in the human resources policies. Also, this study is to focus on the theoretical models in the literature about employee engagement which are the base for further studies in the same field. It is found that there is no consensus on the concept of employee engagement, as different theoretical frameworks exhibit different constructs for employee engagement, which leads to different strategies of implementation. It has been remarkably noted that employees are either engaged or disengaged based on different drivers which are from inner psychological traits, behavior, or emotions, on the other hand, the outside drivers are from the environment, cultural changes, and leadership. Whilst the tasks and duties and level of responsibilities can be the major engine of employee engagement level.

Keywords Employee engagement · Organization engagement · Human resources management · Human resources policies · Employee satisfaction · Sustainable development

---

H. Aldoy
Ahlia University, Manama, Bahrain
e-mail: 2147422@brunel.ac.uk

B. Mcintosh (✉)
Brunel University, London, UK
e-mail: Bryan.Mcintosh@brunel.ac.uk

© The Author(s), under exclusive license to Springer Nature Switzerland AG 2024
A. Hamdan and E. S. Aldhaen (eds.), *Artificial Intelligence and Transforming Digital Marketing*, Studies in Systems, Decision and Control 487,
https://doi.org/10.1007/978-3-031-35828-9_98

1 Introduction

Human resources management (HRM) is valuable for any organization. It is the key driver for the organizational improvement and success [36]. The most important asset for business organizations is their high-performing employees. Therefore, It is very important for the organization to transfer their focus towards human capital to stimulate their commitment and engagement behaviour [27]. According to Elrehail et al. [15] organizations are trying to keep improving and empowering the human resources within the organizations to gain competitive advantages through having the employees engaged to their jobs. Hence employee engagement is a human resource management initiative, which is believed to raise the satisfaction rate and performance of the employees. According to Crabb [12], employee engagement has recently drawn a wealth of managerial attention, especially after research into the management concept indicated that a competitive advantage could be achieved by organizations. Thus, engagement needs to be explicitly embedded within an integrated system of HRM policies, practices, and procedures [21]. Employee engagement has several definitions in the literature, among which it is defined as employees being able to devote their emotional, physical, and cognitive energies towards role performance [37]. Yet, the definition of employee engagement may differ from one to another based on their position as an employee or as a leader as claimed by Imandine et al. [22]. Therefore, organizations endeavour to keep the employees engaged in their job and tasks in order to gain a higher level of performance and productivity [35]. In this matter, Gatenby et al. [19] stressed the idea that organizations are putting strategies for employee engagement in order to gain employees' satisfaction and commitment. Albrecht et al. [1] state that considerable progress has been made with respect to clarifying employee engagement, yet despite this progress, relatively low levels of employee engagement continue to be reported in organizations across the globe.

Gallup [18] showed in their survey that employee engagement is recognized by organizations in several countries while in GCC are still developing. Hence, it is crucial to understand the drivers and outcomes of employee engagement implementation in the workplace to increase employee performance and productivity [7]. Similarly, Bhalla [8] pointed to the importance of ensuring a sustainable employee engagement level to stay competitive in the market. On the same matter, Bedarkar and Pandita [7] emphasized that there are factors and drivers that lead to employee engagement in organizations, these factors and drivers affect the employee engagement level and represent the constructs.

Thus, this paper attempts to explore the extant literature to inform the emerging employee engagement concept and explore the anchoring theories of the development of employee engagement by focusing on the strategies and constructs of employee engagement. Employee engagement is an emerging concept in the human resources development (HRD) literature Wollard and Shuck [45]. Therefore, even though the concept of engagement is almost three decades old, the interest it enjoys among practitioners and researchers has only recently emerged. Therefore, this paper sheds light on the employee engagement concept, constructs, and strategies in the workplace. This research can be used by the senior management of organizations and policymakers to appreciate the extant scholarly works and make effective use to achieve productive employment and work decency, which is in line with the United Nation Sustainable development goal (8) "*SDG 8 recognizes the importance of sustained economic growth and high levels of economic productivity for the creation of well-paid quality jobs, as well as resource efficiency in consumption and production.*"

2 Aim and Objectives

Employee engagement is a relatively new concept, there are few theories that have attempted to provide a framework for it [31]. At the same time, the conceptualization and interpretation of employee engagement have elicited a great deal of confusion [11]. Therefore, this paper attempts to explore the emerging concepts of employee engagement and provides a theoretical glance at the anchoring theories, constructs, and strategies of employee engagement.

a. The emerging employee engagement concepts.
b. The anchoring theories of the development of employee engagement.
c. The Strategies and constructs of employee engagement in the workplace.

This paper endeavors to answer the following questions:

Q1: What are the emerging concepts of the employee engagement?
Q2: What are the anchoring theories of development of employee engagement?
Q3: What are the strategies and constructs of employee engagement in the workplace?

3 Research Methodology

The method used in this paper is a systematic literature review of the collected secondary data from journals, books, and websites about employee engagement as conceptual research. As employee engagement is a topic that is correlated with other phenomena based on Albrecht et al. [1] and Gatenby et al. [19], therefore it is essential to use a systematic literature review. The systematic literature reviews according to Xiao and Watson [46] used to provide an in-depth investigation of the phenomena specifically by identifying the keywords, summarizing the existing literature, and grouping it into each keyword specified in order to answer the question of the research. Therefore, the process of a systematic literature review is based on the survey of publications selection based on the keywords identifications then evaluation of the publications, and finally extraction of the needed data from the publications according to Xiao and Watson [46] and Munn et al. [33] methods. Moreover, Munn

1162   H. Aldoy and B. Mcintosh

Table 1 The number of articles in the literature related to the key words

Type	Number of articles
Keyword = employee/organization engagement	120
Keyword =citizenship, satisfaction, involvement	18
Keyword = human resources management and policies in employee engagement	6
Keyword = employee engagement in Gulf and Arab countries	5
Employee engagement studies used qualitative method	3
Employee engagement studies used quantitative method	36

et al. [33] emphasized that the systematic review is a way to show the gaps and emerging literature to help the researcher in the future identification of the research.

Table 1 illustrates the articles based on their relevant to the keywords and methodology used.

3.1 Articles Identification

Different databases were reviewed to extract the identification of publication based on the keywords shown in Table 1 which are related to "employee engagement: and correlated concepts which are "citizenship, satisfaction, and involvement". Moreover, keywords related to "human resources management and policies related to employee engagement" considering also the "employee engagement publications in Gulf and Arab countries". On the other hand, the publications were reviewed as the methodology used either "quantitative" or "qualitative".

The publications after 1990 were searched Since Kahn [25] was the first noted study that mentioned engagement then from databases of Brunel Library, Research-Gate, Emerald Insight, Asia Pacific Human Resources Journal, Journal of Applied Psychology, Sage Publications, International Journals of Management Reviews and ELSEVIER. Consequently, the publications that contain in their main subject any of the keywords mentioned above. Therefore, any publications before 1990 were not reviewed.

3.2 Articles Evaluation

Table 1 showed the articles were reviewed by the authors based on the keywords and research methodology. The total number of publications was 160 clustered into their category based on the keywords as shown in the Table and some of the publications were found in more than one keyword or more than one category.

Employee Engagement Concepts, Constructs and Strategies ...

The process of evaluating the publication is based on selecting the purposeful publication that matches the required keywords noted by Xiao and Watson [46]. This step is ensured unnecessary and irrelevant publication from the research according to Lame [29]. Therefore, the literature review evaluates the publication based on its relevance to the keyword mainly employee engagement. Nevertheless, Xiao and Watson [46] showed that certain criteria should be set in order to evaluate the publication, in this research the criteria are based on the keywords and methodology that have been used for the publications. Therefore, the keyword of *"Employee/organization engagement"* appeared in 120 publications, while the correlated concepts *"citizenship, satisfaction, and involvement"* appeared in 18 publications. On the other hand, *"the human resources management and policies in employee engagement"* appeared in 6 publications, while "employee engagement in Gulf and Arab countries" appeared in 5 publications only.

The authors claimed that by screening the necessary publications that the answer to this article can be reached. Therefore, the publications were also clustered on the research methodology that has been used. Whereas **"employee engagement used quantitative methodology"** were 36 compared to only 3 publications that used *"qualitative research methodology"*.

4 Literature Review

Based on the paper's objectives and questions this section discusses the employee engagement definition, the importance of employee engagement in the workplace, the relationship between employee engagement and human resources practices, and then the models of employee engagement. The scant literature has been searched and explored by the authors to explore the concepts, constructs and strategies of employee engagement.

4.1 The Definition of Employee Engagement

There are different definitions of employee engagement according to research which makes it difficult to obtain certain definitions based on Macey and Schneider [30] however it is mainly related to the psychological nature of the definition. Moreover, the literature found that correlated phenomena overlap with employee engagement based on Barden [6], Gatenby et al. [19] and Bedarkar and Pandita [7], nevertheless, these phenomena are under the definition of employee engagement. In the same matter, Macey and Schneider [30] defined employee engagement as a psychological trait and employees' behavior. Similarly, Kahn [25] studied employee engagement from the perspective of the psychological state related to individuals' attitudes. Kahn's study found a relationship between engagement and the employees' cognitive, physical and emotional conditions during their work. From this point, employee engagement was identified by the organizations that let it impact employee performance and productivity. Jha and Kumar [23] pointed out in their study the emotional trait of being more engaged and the extent to which employees are looking into achieving their goals. Therefore, Bedarkar and Pandita [7] related the engaged employee's as Kahn's outcome of employees' emotional, physical and mental engagement. According to Barden [6], Gatenby et al. [19], and Bedarkar and Pandita [7] that these phenomena are job involvement, satisfaction, organizational citizenship, and commitment. Kahn [25] and Saks [39] defined job engagement and organization engagement as part of employee engagement since they are used interchangeably with both phenomena however employee engagement is more related to the employee in the organization. On the other hand, Vance [44] and Robinson et al. [38] defined commitment as a lower level of engagement related to the employee's responsibility toward completing the task or duty. While satisfaction as defined by Jha and Kumar [23] is the sense of achievement from the employee's perspective in the workplace. Moreover, Robinson et al. [38] explained organizational citizenship as the personal employee characteristics toward the organization. Several studies were conducted by Aon Hewitt, CIPD, Deloitte, and Gallup for measuring the engaged, semi-engaged, and disengaged employees in a certain context. Here are some of the definitions resulting from the bibliographies of the studies. As Deloitte Consulting [14] defined employee engagement as related to their satisfaction in achieving organizational goals. On the other hand, Gallup [18] classified the engagement level as engaged, actively engaged, and not engaged, while this classification was approved by the statistics from their study. Despite of the definitions mentioned earlier about the correlated phenomena, CIPD [10] concluded that engagement covers employee commitment and motivation within its definition. As Aon [3] classified engagement into three behavior manners: Say, Stay and Strive. Which, Say is for talking positively about the company, Stay is the longest period the employee is willing to stay working for the company and Strive is the extent to which the employee is willing to give more for the company.

4.2 Employee Engagement in the Workplace

The most important initiative by the organization is enhancing engagement that reflects on human resources involvement, satisfaction, productivity, and well-being. Moreover, Gallup [18] emphasized that in order to achieve human well-being in the organization, engagement practices should be implemented. While Aon [3] stated that engagement is an ongoing process within the organization that ensures a sustainable level of employee engagement. Therefore, Albrecht et al. [1] noted that the importance of sustainable employee engagement led to sustainable growth and productivity. From this point, several drivers in the workplace that enhance the implementation of employee engagement based on Gruman and Saks [20] and Khan [28], mainly human resources practices, the organization environment, the communication within the organization, leadership, and the employee's behaviour or psychological characteristics. On the other hand, Albrecht et al. [1] ensured the importance of human resources practices impact on employee engagement sustainability. Employee engagement existence in the organization influences overall organizations and employees' productivity according to Albrecht et al. [1]. At the basic level, the organizations' goals and objectives are set to serve the processes and practices of employee engagement as noted by Gruman and Saks [20]. Elrehail et al. [15] and Shahid [42] stated that since organizations are focusing on increasing their competitive advantage, they are more likely to enhance their human capital by implementing ways to retain, empower and engage their employees.

4.2.1 Organizational Culture

Employee turnover is a result of disengaged employees based on Vance [44], therefore more engaged employees are willing to work for longer periods. On the other hand, the negative outcome of disengaged employees is the absenteeism, whereas based on Gallup [18] study that low level of engagement effect on increasing the employee absenteeism that will also reflect on low productivity and performance. Worth to mention, that the organizational culture influences the engagement process wherein engagement strategy is implemented in the formulation of culture as noted by Chanana and Sangeeta [9]. The implementation of supportive and motivated culture led to high level of engagement according to Gatenby et al. [19].

As a result, positive environment led to positive employee engagement as emphasized by Chanana and Sangeeta [9]. Nevertheless, [1] noted that the organizational culture that support the values and norms of employees toward more engagement as values are more willing to increase the employee's wellbeing. Therefore, Gatenby et al. [19] suggested that activities should be implemented in the organization to improve the engagement culture, whereas these activities in turn enhance the employee wellbeing based on Gallup [18] study. Moreover, Khan [28] showed the importance of communication within the organization which allow the employees to be more aware about the organizational values and norms also the organizational strategy that also agreed by Albrecht et al. [1] in which the employee feel more involved. In conclusion, the employee feels more loyal by more engagement strategy that is implemented in the organizational culture as agreed by Gatenby et al. [19] and Albrecht et al. [1].

A fundamental strategy to be implemented in order to enhance the employee engagement is the communication, as mentioned earlier that communication allow the employees to be more engaged. In the same sequence, Gallup [18] revealed the importance of communication in their study of employee engagement measurement. Therefore, Gallup [18] suggested that organizations seek for plans to increase the communication within the organization, in which the organization can listen to their employees' initiatives [22].

4.2.2 Employee's Emotions and Behaviour

Kahn [25] related in his study employee engagement to employee psychological behaviour, emotions and physical. According to Barden [6] that employees are emotionally related to their organizations. Therefore, employees' psychological emotions and behaviour matter when discussing employee engagement. As Deloitte Consulting [14] study explained the importance of focusing on employees' psychological needs to gain employees' well-being, loyalty, commitment and engagement. As a result, organizations are focusing on employees' happiness, feeling of justice and fairness based on Joo and Lee [24], in order to gain more engaged employees according to Mohan et al. [32]. Also, Mohan et al. [32] showed the strong relationship between the impact of employee morale and attitude on the overall organisational environment. Nevertheless, the employees' morale on the job reflects their behaviour thus their performance as mentioned by Albrecht et al. [1].

From the perspective of employees, their emotions and behaviour are affected by the leaders' emotions and behaviours according to Khan [28]. As well as Gatenby et al. [19] showed that leadership style is influencing employee engagement levels in the workplace. Therefore, Imandine et al. [22] mentioned that the employee engagement strategy should be spread on the organization from the top level to the lower levels in the organization through the organizational strategy. From this point, leaders feeling toward the organization is controlling the employees' feeling that resulting in engagement and performance as Dajani [13] confirmed. A study from Gallup [18] showed that employees empowered by their employers and leaders feel more engaged and can feel more responsibility toward their job. The consideration of communicating the organizational strategy, objectives and goals is essential as the leader of the organization transfer through the organization to the employees according to Mohan et al. [32] which confirmed that communication between the leader and the employees is one of the employee engagement drivers.

4.3 Human Resources Management Practices

Human resources management practices in the organization are the processes that the organization is taking in order to manage the human resources based on Elrehail et al. [15]. Therefore, Albrecht et al. [1] emphasized that the theoretical models of human resources practices formed in the literature to show the impact of employee engagement. Thus, Jha and Kumar [23] noted that the organizations obtain human resources practices that support the focus on human resources. From this matter, employee engagement is introduced in the formulation of human resources management practices to increase employee engagement according to Altit and Hunitie [2]. Based on Vance [44] that the main factors of human resources practices that affect employee engagement are Job performance, reward, and recognition, job design, employees, training and development. Despite the human resources development importance for organization development, Tinti et al. [43] claimed that human resources practices are the main influence on competitiveness and effectiveness. Therefore, the organizations in order to be more competitive in the market, the human capital should be more competitive by focusing on the employees' well-being and engagement. Consequently, [23] emphasized the importance of merging human resources practices with the employee engagement strategy throughout the organization. The job design of the employee should formulate the importance of engagement by specifying the job objectives, tools and resources needed to perform the job and tasks. Yet, Vance [44] showed the processes of job analysis in the designing of the job, whereas the job analysis gives the employee the opportunity to be aligned with the job. Another point is important with the job designing is the employee characteristics that is also analysed with the job design according to Vance [44]. Another practice from the human resources management is the employment. The process of selecting employee to fit on the job is one of the practices that the human resources management is practicing during the recruitment process. Albrecht et al. [1] debated about the differences of engagement level between new employees and experienced employees, however the study emphasized the driver of employee characteristics, behaviour and background experience. From this point, Macey and Schneider [30] showed the relationship between the behaviour and psychology of the employee and the life cycle in the employment process. Taking into account the debate of working experience years and the behavioural drivers, Gallup [18] argued that organizations need to set a criterion that shows the level of employee engagement and welling to be engaged in the employment process. Training and development are fundamental practice of human resources that lead to provide the employees with their career development path based on Dajani [13], that employees are seemed to be more engaged when they know exactly what they are expecting from the organization in their career. Albrecht et al. [1] also emphasized that training and development brings the employee to the path needed for his/her career therefore enhance the employees' engagement toward achieving tasks and organizational goals. Moreover, Altit and Hunitie [2] showed the impact of career growth on the level of engagement in the organization. In the same matter, rewards and recognition ensure the impact of the high level of engagement based on Dajani [13]. Furthermore, Saks [39] emphasized the positive relationship between engagement and rewards and recognition. However, the rewards and recognitions could be tangible as increase in payment or intangible as encouragement or motivation or acknowledgment or another type of recognition that influence the employee to be more satisfied about the job in order to increase the engagement according to Vance [44]. Finally, the most important impact on the employee as the result of being engaged in his/her performance. Therefore, organizations are looking for ways that increase employee performance and productivity through policies that increase employees' engagement according to Dajani [13]. In the same matter, Jha and Kumar [23] emphasized that the organizations' success is based on the employees' performance therefore it is important to increase the way that the employees are engaged to the job. As also ensured by Albrecht et al. [1] that the positive job performance process in the organization which ensures fairness is impacting the level of engagement. Therefore, Imandine et al. [22] emphasized the importance of establishing and setting fair and clear appraisals that led to a higher quality of measuring job performance that sequentially increases employee engagement.

4.4 The Employee Engagement Theoretical Models

In order to understand the employee engagement in the literature, it is essential to illustrate the theoretical models about it. The literature presents the employee engagement theoretical models from the social exchange theory (SET). Saks [39] identified the theory SET is the relationship between the employee and the organization, as what do the employee give to the organization and what the organization gives the employee. From this point, the theoretical models of employee engagement have been built by authors as the emotional, social and phycological traits that is set between the employee and the organization.

4.4.1 Psychological Based Model

The psychological characteristic of the employee has an impact on the engagement level. Kahn [25] introduced this model by exploring and observing the behavior and emotions of group of employees in the organization. He noticed that there are dimensions and conditions in this model. The dimensions are related to the psychological behavior of the employee which are physical, cognitive and emotional, while the psychological conditions are meaningfulness, safety and availability in the workplace. Therefore, the psychological model for employee engagement is related to employees' feeling of wellbeing, appreciation and existence in the organization based on Fairlie [16] that employees need to feel the importance and existence of their efforts in the workplace. From the other side, the long experience for the employee in the organization enhance the engagement as the employee feels more belonging to the workplace and the environment according to Kahn and Heaphy [26].

In the same sequence, Kahn [25] emphasized that the employee wellbeing affects the engagement level therefore it is increased by the dimensions of physical, cognitive and emotional constraints. Whereas the positive behavior that the employee can provide to the organization, the more positive engagement can be gained. Many authors from the literature adopted the psychological model in their research, therefore Gallup [18] concluded that the more the employee is engaged psychologically, the more that the performance is higher. Figure 1 shows the constraints of the psychological based model as explained.

4.4.2 Task Based Model

As the previous model is related to psychological traits, this model is more likely to focus on the different tasks and roles requested from the employee. This model

Fig. 1 The psychological based model. *Source* Authors

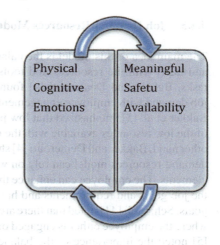

was introduced by Saks [39] which is also under the social exchange theory, whereas the more the employee is attached to the task, the more the employee is engaged. In relation to Kahn [25] model in which job meaningfulness is part of the engagement constraints, this model is illustrating the relationship between engagement and task meaningfulness based on Saks [39]. Therefore, based on Fletcher et al. [17] that the task and role goals must be clearly explained to ensure employee engagement. In the same matter, Newton et al. [34] emphasized that tasks and role differences can affect the employee engagement level, whereas the importance and the circumstances surrounding the task assigned to the employee can influence their engagement. From this point, Newton et al. [34] introduced the job fit model, as to what extent is the employee characteristics fit the tasks and roles given in order to ensure the engagement that leads to employee higher performance. Therefore, Fig. 2 illustrates the impact of the tasks-based model on engagement.

Fig. 2 The task-based model. *Source* Authors

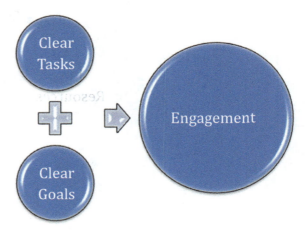

4.4.3 Job Demand-Resources Model

The main idea of this model is the balance between the requirements of the job role and tasks with the resources and tools available for the employee to perform the tasks. Bakker and Demerouti [4] found a relationship between job resources and job demand with employee engagement based on the social exchange theory. Whilst Bakker et al. [5] emphasized that low performance lead to low engagement because of the low resources available with the over-stressed requirements of the job. On the other hand, Bakker and Demerouti [4] showed that there are three elements that the job demand-resources model can rely on which are physical, social, and organizational elements. The employee can enhance the engagement level by clearly understanding the job goals and requirements and having the necessary resources to achieve these goals. Schaufeli [41] noted that there are two ends of the job demand-resources model where the employee either is engaged or burned out. In the same matter, Bakker et al. [5] noted the importance of the balance between demand and resources to ensure employee wellbeing that leads to employee engagement. Moreover, Bakker et al. [5] presented the importance of job resources with personal resources which are the personal characteristics and psychological traits that affect employee performance in achieving the demand of the job needed. Therefore, Imandine et al. [22] related the job demand-resources model to the job fit as the extent to which the employee is able to get the responsibilities and is accountable to do the job needed. Figure 3 illustrates the relationship between the job Demand-Resources Model.

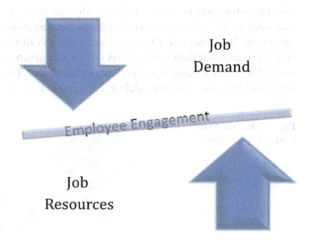

Fig. 3 The job demand-resources model. *Source* Authors

Employee Engagement Concepts, Constructs and Strategies ...   1171

5 Discussion and Conclusion

Thirty-six studies were investigated discussing the importance of employee engagement and performance. The studies concluded the positive organizational outcomes via the consideration and implementation of employee engagement processes and practices in the workplace. These studies, although in the majority have reported a significant relationship between employee engagement and contextual performance in terms of the following aspects: Demand-resources; psychological traits; tasks and roles; physical, cognitive, and emotional constraints; personal resources. It has also been found that there is no consensus on the definition of the term employee engagement. As the definition of employee engagement is not yet specified and differs from one country to another or from one organization to another [1, 19]. Furthermore, it is interestingly found that the concept of employee engagement is correlated and overlapped with other concepts such as employee citizenship, satisfaction, and involvement [19]. A significant finding in relation to employee engagement and methodological approaches, the literature review showed that there is a gap in studying the concept in depth using the qualitative research methodology. As the qualitative research methodology is the way to interpret the concept deeply from the perspective of the manager, employee, or both. According to Saunders et al. [40] that qualitative research methodology is used to ease the complexity of a phenomenon and to felicitate the sensitivity of people related to a phenomenon. From the theoretical perspective, the theories found in the literature are located under the social exchange theory, which is the relationship between the employee and the organization as to what to give and what to get according to Saks [39]. From this perspective, the authors introduced three theoretical models that found in the literature. The theoretical models showed the relationship between the employee and the level of engagement whereas there are different surrounding criteria and drivers that lead to engagement and disengagement. As a result, the paper showed the gap in the literature that merges the theoretical models into the human resources policies in the organization as a conclusion of the models from the social exchange theory. An important finding in this study is the idea that there are several drivers in the workplace that influence engagement. The organization's culture is the overall environment that integrates and correlates with the employee during the work-life journey. Also, the employee's psychological personality is the behavior and emotions of the employee and related communication with coworkers and leaders in the workplace. These all drivers are merged with the human resources management aspects that start with the employee journey from employment, to job design, rewards and recognition, training and development, performance management, and evaluation. The human resources management aspects are the backbone of the organization that provides the path of employees to be engaged or not engaged. In conclusion, employee engagement is defined from different perspectives and contexts. Although scholars are in consensus in terms of organizational outcomes, the concept is being diffused, and the inconsistency is evident. Therefore, the constructs of employee engagement differ accordingly. This is evident in the exploration of the literature and the theoretical frameworks presented in the literature, given the embedded elements and factors which constitute employee engagement. Having said that, it is also concluded that practitioners may use different frameworks and that there is no one size fits all model for employee engagement. Overall, employee engagement is a strategy that needs to be implemented in the workplace through human resources policies. The management of the organization should also consider the employee and the organization drivers that lead to increased engagement. Thus, it must be within the initiatives of organizations, and in the agendas of both organizations and policymakers.

References

1. Albrecht, S., Bakker, A., Gruman, J., Macey, W., Saks, A.: Employee engagement, human resource management practices and competitive advantage: An integrated approach. J. Organ. Effect. **2**(1), 7–35 (2015)
2. Altit, A., Hunitie, M.: The Mediating effect of employee engagement between its antecedents and consequences. J. Manag. Res. **7**(5)
3. Aon, H.: Engagement trends in the Middle East (2012). Retrieved from https://trendsmena.com/wp-content/uploads/2013/08/AH-Report-ME-Engagement-Trends-2012.pdf
4. Bakker, A., Demerouti, E.: Towards a model of work engagement. Career Dev. Int. **13**(3), 209–223 (2008)
5. Bakker, A., et al.: Crossover of burnout and engagement in work teams. Work. Occup. **33**(4), 464–489 (2006)
6. Barden, C.: The Correlation Between Employee Engagement and Job Satisfaction in the Social Security Administration, p. 365. Governors State University, All Capstone Projects (2017)
7. Bedarkar, M., Pandita, D.: A study on the derivers of employee engagement impacting employee performance. Soc. Behav. Sci. **133**, 106–115 (2013)
8. Bhalla, M.: Employee engagement and its determinants for IT organizations. Inter. J. Sci. Tech. Manag. **6**(2) (2017)
9. Chanana, N., Sangeeta: Employee engagement practices during COVID-19 lockdown. J. Public Affairs (2021)
10. CIPD: Health and well-being at work, United Kingdom (2020), available at https://www.cipd.co.uk/Images/health-and-well-being-2020-report_tcm18-73967.pdf
11. Cole, M., Walter, F., Bedeian, A., O'Boyle, E.: Job burnout and employee engagement: A meta-analytic examination of construct proliferation. J. Manag. **38**(5), 1550–1581 (2012). https://doi.org/10.1177/0149206311415252
12. Crabb, S.: The use of coaching principles to foster employee engagement. Coaching Psychol. **7**(1), 27–34 (2011)
13. Dajani, M.: The impact of employee engagement on job performance and organizational commitment in the Egyptian banking sector. J. Bus. Manage. Sci. **3**(5), 138–147 (2015)
14. Deloitte Consulting: Engaging the work force. Deloitte Development (2016), available at https://www2.deloitte.com/content/dam/Deloitte/us/Documents/human-capital/us-cons-eng aging-the-workforce.pdf
15. Elrehail, H., Harazneh, I., Abuhjeeleh, M., Alzghoul, A., Alnajdawi, S., Ibrahim, H.: Employee satisfaction, human resource management practices and competitive advantage. Eur. J. Manag. Bus. Econ. **29**(2), 125–149 (2020)
16. Fairlie, P.: Meaningful work employee engagement, and other key employee outcome: implications for human resource development. Adv. Develop. Human Res. **13**(4), 508–525 (2011)
17. Fletcher, L., Bailey, C., Gilman, M.: Fluctuating level of personal role engagement within the working day: a multilevel study. Hum. Resour. Manag. J. **28**, 128–147 (2017)

18. Gallup: State of the global workplace (2013), retrieved from https://www.slideshare.net/InnoGarage/state-of-the-global-workplace-report-2013-by-gallup
19. Gatenby, M., Rees, C., Soane, E., Truss, C.: Employee Engagement in Context. Kingstone Business School, CIPD Research Insight (2008)
20. Gruman, J., Saks, A.: Performance management and employee engagement. Hum. Resour. Manag. Rev. **21**, 123–136 (2011)
21. Guest, D.E.: Employee engagement: a skeptical analysis. J. Organ. Effect.: People Perform. **1**, 141–156 (2014)
22. Imandine, L., et al.: A model to measure employee engagement. Probl. Perspect. Manag. **12**(4), 520–532 (2014)
23. Jha, B., Kumar, A.: Employee Engagement: a strategic tool to enhance performance. J. Contem. Res. Manag. **3**, 21–29 (2016)
24. Joo, B., Lee, I.: Workplace happiness: work engagement, career satisfaction, and subjective well-being. Evidence-Based HRM: A Glob. For. Emp. Scholar. **5**(2), 206–221 (2017)
25. Kahn, W.: Psychological conditions of personal engagement and disengagement at work. Acad. Manag. J. **33**(4), 692–724 (1990)
26. Kahn, W., Heaphy, E.: Relational context of personal engagement at work. In: Truss, C.et al. (eds.) Employee Engagement in Theory and Practice. Taylor & Francis Group, London and New York, pp. 13–96 (2014)
27. Kavyashree, M.B., Kulenur, S., Nagesh, P., Nanjundeshwaraswamy, T.S.: Relationship between human resource management practices and employee engagement. Brazilian J. Operat. Product. Manage. **20**(1), 1331–1331 (2023)
28. Khan, N.: Employee engagement drivers for organizational success. Global J. Manag. Bus. Stud. **3**(6), 675–680 (2013)
29. Lame, G.: Systematic Literature Review: An Introduction. University of Cambridge (2019), accessed: https://www.researchgate.net/publication/334714968
30. Macey, W., Schneider, B.: The meaning of employee engagement. Ind. Organ. Psychol. **1**, 3–30 (2008)
31. Mani, S., Mishra, M.: Employee engagement constructs: "CARE" model of engagement—need to look beyond the obvious. Leadership & Organ. Development J. **42**(3), 453–466 (2021). https://doi.org/10.1108/LODJ-08-2020-0358
32. Mohan, J., Haque, M., Khan, N.: Empirical approach to measure employee engagement: Evidence from Indian IT Industry. Human Res. Manag. Res. **8**(1), 7–13 (2018)
33. Munn, Z., Peters, M., Stern, C., Tufanaru, C., McArthur, A., Aromataris, E.: Systematic review or scoping review? Guidance for authors when choosing between a systematic or scoping review approach. BMC Med. Res. Method. **18**(143) (2018)
34. Newton, D., LePine, J., Kim, J., Wellman, N., Bush, J.: Taking engagement to task: the nature and functioning of task engagement across transitions. J. Appl. Psychol. **105**(1), 1–18 (2020)
35. Osborne, S., Hummod, M.: Effective employee engagement in the workplace. Inter. J. Appl. Manag. Techn. **16**(1), 50–67 (2017)
36. Pasban, M., Nojedeh, S.: A review of the role of human capital in the organization. Soc. Behav. Sci. **230**, 249–253 (2016)
37. Rich, B.L., Lepine, J.A., Crawford, E.R.: Job engagement: antecedents and effects on job performance. Acad. Manage. J. **53**, 617–635 (2010)
38. Robinson, D., Perryman, S., Hayday, S.: The Drivers of Employee Engagement. Institute for Employment studies, UK (2004)
39. Saks, A.: Antecedents and consequences of employee engagement. J. Manag. Psychol. **21**(7), 600–619 (2006)
40. Saunders, M. Lewis, P., Thornill, A.: Research Methods for Business Students, 8th edn. Pearson Education, London, UK (2019)

41. Schaufeli, W.: 'What is engagement?. In: Truss, C.et al. (eds.) Employee Engagement in theory and Practice, pp. 13–96. Taylor & Francis Group, London and New York (2014)
42. Shahid, A.: The employee engagement framework: high impact drivers and outcomes. J. Manag. Res. **11**(2), 45–54 (2019)
43. Tinti, J. et al.: The impact of human resources policies and practices on organizational citizenship behaviors. Braz. Bus. Rev. **14**(6), 636–653 (2017)
44. Vance, R.: Employee Engagement and Commitment; A Guide to Understanding, Measuring and Increasing Engagement in Your Organization. SHRM Foundation, VA (2006)
45. Wollard, K., Shuck, B.: Antecedents to employee engagement: a structured review of the literature. Adv. Dev. Hum. Resour. **13**(4), 429–446 (2011)
46. Xiao, Y., Watson, M.: Guidance on conducting a systematic literature review. J. Plan. Educ. Res. **39**(1), 93–112 (2019)

Board-Level Worker Representation: A Blessing or a Curse?

Fatema Alrawahi, Amama Shaukat, and Abdalmuttaleb M. A. Musleh Al-Sartawi

Abstract Within the context of corporate governance reform, this paper aims to present a review of the two opposing views of board-level worker representation (BLWR) and the underlying theories including the implications for firms and their board effectiveness. As such, this paper is significant to corporations, regulators, academics and researchers who plan to do research in the same area, as it provides a comprehensive outline of relevant papers in the field.

Keywords Worker representation · Corporate governance · Boards of Directors

1 Introduction

> When plunder has become a way of life for a group of men living together in society, they create for themselves in the course of time a legal system that authorizes it and a moral code that glorifies it.
>
> —*Frédéric Bastiat, 1845, The Physiology of Plunder*

In a call for a long overdue and inevitable Corporate Governance (CG) reform, the United Kingdom's Green Paper (2016) urges UK corporations to strengthen transparency and disclosure practices related to stakeholder engagement. The Green Paper specifically calls for governance and reporting requirements in relation to workers. Consequently, in a bid towards giving employees a voice, the UK's Corporate Governance Code (2018) encourages boards to establish methods for communication with workers through workforce representation on the board, formal workforce advisory panels or through the appointment of work force representative non-executive directors.

F. Alrawahi (✉) · A. Shaukat
Brunel University, London, UK
e-mail: falrawahi@ahlia.edu.bh

F. Alrawahi · A. M. A. M. Al-Sartawi
Ahlia University, Manama, Bahrain

© The Author(s), under exclusive license to Springer Nature Switzerland AG 2024
A. Hamdan and E. S. Aldhaen (eds.), *Artificial Intelligence and Transforming Digital Marketing*, Studies in Systems, Decision and Control 487,
https://doi.org/10.1007/978-3-031-35828-9_99

Several countries such as Germany, China and Egypt mandate the representation of workforce on corporate boards [14]. While some researchers are advocates of the notion of workers on boards of directors, others argue against it [4]. The general consensus is that board-level worker representation (BLWR) distorts the optimal design of boards leading to decrease in the value of firms [5].

Within the context of corporate governance reform, this paper aims to present a review of the two opposing views of board-level worker representation (BLWR) and the underlying theories including the implications for firm performance. As such, this paper is significant to corporations, regulators, academics and researchers who plan to do research in the same area, as it provides a comprehensive outline of relevant papers in the field.

The remainder of the paper is organized as follows. Section 2 present a literature review of the relevant studies, and Sect. 3 provides conclusions and proposes topics for which new research should be of high benefit.

2 Literature Review

2.1 Context of the Study

The central tenet of scholars and governments' call for corporate governance reform is that companies should be managed with a dedicated focus on creating long term-value for shareholders as well as other stakeholders, such as customers, workers and the wider public, to gain their confidence and respect. This categorization into shareholder [8] vs. stakeholder- oriented [6] governance is accompanied by the two partly contra-paradigms and has influenced and shaped corporate governance systems around the world.

Consequently, the essence of both opposing arguments for or against BLWR boils down to the nature of the corporate governance (agency) relationship between share-holders and senior executives, which places an emphasis on the creation of share-holder value, and supports an expanding use of share buybacks, the raising of stock dividends, and the rise of executive remuneration. The concept of value maximiza-tion is ingrained in the shareholder primacy theory and has its roots in two hundred years of research in economics and finance [12]. The theory expresses the notion that owners have the priority interest in both economics and corporate governance, i.e., corporations exist to serve shareholders in a principal-agent relationship; where the primacy model directs the board to manage the corporation for the purpose of maxi-mizing shareholder wealth [11, 15]. In addition to market value maximization, Jensen [12] maintains that by increasing their total firm value, corporations can contribute towards social welfare. He argues that the crux of the Friedman [8] and Freeman [6] debate is often wrongly defined, and in reality, should be framed to settle the number of objective functions, single-valued vs. multiple-valued objectives. Furthermore, he adds that the stakeholder model constrains the choices of board of directors due to the limitation of "principled criterion" for decision-making. This implies that managers might pick and choose what causes to pursue—whether be it environment, charities, medical research—depending on their interests, without accountability on how it affects firm value.

On the other, the stakeholder theory [6], the main contender to the shareholder primacy theory, considers the welfare of all non-shareholder stakeholders, whereby shareholders are only one of many interested parties. Freeman replaces the concept that managers have a duty to shareholders with the notion that managers need to have a fiduciary relationship with stakeholders. He defines stakeholders as individuals or groups who "benefit from or are harmed by, and whose rights are violated or respected by, corporate actions". These include employees, suppliers, customers, local environmentalists, vendors, governmental agencies, and the wider society. Freeman's [6] theory suggests that a company's real success lies in considering the welfare of all its stakeholders, not just those who might profit from its stock. In light of the significance of employees as primary stakeholders, Freeman [7] believes that their interests with the firm are mutual and not "univocal". Employees have their source of livelihood at stake where they receive wages and security in return for their work. Moreover, employees are required to meet certain expectations such as representing the firm positively and responsibly in local communities. As such, and in exchange for their loyalty, employees need to participate in decisions affecting their use as means to the firm's ends.

Reiterating the point shown in the literature review and the issues stated in the Green Paper, intuitively, the notion of a more collaborative and stakeholder-driven approach to corporate governance is highly appealing towards achieving long-term firm success as well as sustainability. However, Jensen and Meckling [11] and Aguilera and Jackson [1] claim that managers who create value for non-shareholder stakeholders through BLWR as an end may reduce shareholder value due to the unfocused attention of managers leading to inefficient resource allocations, ineffective decision making, and reduced accountability of managers. Nevertheless, as this tends to lead to equal distribution of wealth and sustainable development, it is the responsibility of top management to balance the diverse claims of conflicting stakeholders [7]. Those who support BLWR tend to gravitate towards the Stakeholder Theory, while those who oppose the BLWR support the Shareholder primacy theory. Jensen [12] proposes that the answer to the shareholder vs. stakeholder dilemma lies between mixing both the 'enlightened value maximization and enlightened stakeholder theory' elements. But where does BLWR fit into this picture? Accordingly, this literature review compares and contrasts the studies adopting each perspective presents their views under a more critical lens.

2.2 BLWR: Key Problems and Proposed Solutions

Literature on human resource management have used the term 'voice' sparingly. Wilkinson et al. [16, p. 3] narrow the definition of voice which focuses on "*how workers communicate with managers and are able to express their concerns about their work situation without a union, and on the ways in which employees have a say over work tasks and organizational decision-making*". Therefore, it could be infered that worker representation on the board is considered as giving workers a voice.

Accordingly, this section will first discuss the benefits of worker representation by giving them a voice on boards of directors. Gregorič and Poulsen [9] cite several benefits of BLWR mainly that board meetings might facilitate the use of information from the employees, the negotiation of mutual and integrative solutions, and by supporting human capital formation, it can increase worker productivity. Therefore, due to the increase in employee satisfaction as a result of worker representation, companies have lower turnover as workers are less likely to leave. Additionally [16] consider the economic approach to studying employee voice as it builds on transaction costs economics—perceiving worker voice as a key governance answer to issues related to opportunism and transparency in the contracting between employers–workers. Moreover, workers, through their representatives, are able participate in the development of the company's policy, thereby they will be less likely to renounce reached agreements. Gregorič and Rapp [10] argue that as a result, companies with BLWR are able to agree on ways to reduce labour costs, for example temporary furloughs and bonus restructuring, and as such are less likely to have high turnover in the case of poor performance.

As for the negative effects of BLWR, and in response to the Green Paper, a number of key problems can be identified. Contrary to Wilkinson et al.'s [16] economic approach, Edmans [4] postulates that worker representation is not the answer to stakeholder engagement. In his paper where he questions the ways in which the interests of employees, and other stakeholders can be strengthened at board levels, he claims that it is impractical for a worker to be a representative of other workers due to different pay grades, departmentalisations, and conflicting decisions benefiting one group of workers over others. He further argues that some decisions benefitting workers could be discussed behind the scenes (outside boardrooms), in the absence of the worker representative. Similarly, in his essay defending shareholder maximization, Dennis [2] believes that stakeholder governance fails to determine a clear criterion to make trade-offs between the different types of stakeholders, as there will be many opinions as there are parties. At the end of the day, whose opinion matters the most? Stakeholder governance does not give a definitive answer as of yet. This has root in the shareholder vs. stakeholder theory, i.e., while value maximization provides managers with only one objective, stakeholder theory directs managers to serve more than one party. Simply cited by Jensen [12], when there is more than one master, everyone ends up being short-changed. Freeman [7] agrees with this concept, but adds another layer to the argument by claiming that it is the duty of the management to keep the relationship among stakeholders in balance.

Faleye et al. [5] argue that worker representation is actually counterintuitive whereby it deters productivity as firms with BLWR invest less in long-term assets, take less risks, hence tend to have slower growth and create less jobs, leading to lower labour productivity. More recently, Masulis et al. [13] claim that when workers own a large stake and voting blocks, alliances could form between the managers and workers which could potentially aggravate the manager-shareholder conflicts of interests.

However, despite his concerns regarding worker representation on the boards, Edmans [3] states the fair treatment of stakeholders, mainly workers, increases the stock returns in the long-run. Therefore, it is significant that boards of directors take into their consideration the welfare of workers. In other words, boards should not wait for worker representatives to be appointed on the board to start on taking into account the interests of workers. Edmans [4] in his response to the Green Paper suggests a solution of change, where he proposes that companies need to focus on a holistic culture change all throughout the firm, instead of BLWR which he deems a shallow solution.

Despite the mixed perception of researchers on the notion of BLWR, the Green Paper (2016) states that there is evidence that a number of companies have been falling short of the high standards expected by the Government, whereby some directors have lost sight of their broader legal and ethical responsibilities mainly in relation to stakeholders. By calling for corporate governance reform, the Government aims to incentivize businesses to take the right long-term decisions, more sustainable business performance and build wider confidence in the way businesses are run and help restore the public's trust. Therefore, in the light of the new call for reforms and the Financial Reporting Council's inclusion of BLWR in the UK's 2018 Code, this literature review acknowledges that there are: (1) implications for regulators and listed companies in following the revised principles of the Code, and (2) area for empirical research to investigate BLWR based on the models implemented by China, Egypt and European countries. Nonetheless, research in this area need to take into account the concerns raised by Edmans [4] in his response to the Green Paper (2016) as well as the unitary nature of the UK's board structure as opposed to the other European countries two-tier board systems which are exogenous in nature.

3 Conclusion and Future Studies

This paper set out to throw light on the recent discussions in the literature about the significance and implications of Board-Level Worker Represntation (BLWR), following the Green Paper's (2016) call for reform and concerns in relation to the financial scandals caused by the unethical behavior of some of the listed companies. Restating the arguments presented in the literature review section, there is a mix of positive and negative perceptions of BLWR and its effects on the board effectiveness, and firm performance. While some researchers agree that BLWR increases worker productivity, commitment to policies, and reduces labour turnover; others argue that BLWR has adverse effects on productivity and firm growth as organisations with worker representation tend to invest less in long-term assets and take fewer risks. Nonetheless, many countries such as Germany, China, Egypt and Scandinavian countries have implemented it successfully and, in some countries, it has been in existence for a long time (e.g., Germany). Consequently, this presents areas for future empirical research investigating BLWR and its effects on firm performance taking into consideration the concerns raised by Edmans [4] in his response to the Green Paper. An issue that the present study will consider. Finally, research in this area is topical, interesting, relevant and challenging with policy, practical and theoretical implications.

References

1. Aguilera, R.V., Jackson, G.: Comparative and international corporate governance. Acad. Manag. Ann. **4**(1), 485–556 (2010)
2. Dennis, D.: Corporate governance and the goal of the firm. in defence of shareholder wealth maximization. The Financial Review **51**, 467–480 (2016)
3. Edmans, A.: The link between job satisfaction and firm value, with implications for corporate social responsibility. Acad. Manag. Perspect. **26**, 1–19 (2012)
4. Edmans, A.: Question 7: How can the way in which the interests of employees, customers and wider stakeholders are taken into account at board level in large UK companies be strengthened? (2017). Available at: https://alexedmans.com/wp-content/uploads/2019/02/Workers-on-Boards.pdf
5. Faleye, O., Vikas, M., Randall, M.: When labor has a voice in corporate governance. J. Financial Quant. Anal. **41**, 489–510 (2006)
6. Freeman, R.E.: Strategic Management: A Stakeholder Approach. Pitman Publishing (1984)
7. Freeman, R.E.: A stakeholder theory of the modern corporation. Perspect. Busi. Ethics Sie **3**(144), 38–48 (2001)
8. Friedman, M.: The social responsibility of business is to increase its profits. N.Y. TIMES, Sunday Magazine, 32 (1970)
9. Gregorič, A., Poulsen, T.: When do employees choose to be represented on the board of directors? empirical analysis of board-level employee representation in Denmark. Br. J. Ind. Relat. **58**(2), 241–272 (2020)
10. Gregorič, A., Rapp, M.S.: Board-level employee representation (BLER) and firms' responses to crisis. Indust. Relations: A J. Econ. Soc. **58**(3), 376–422 (2019)
11. Jensen, M.C., Meckling, W.H.: Theory of the firm: Managerial behavior, agency costs and ownership structure. J. Financ. Econ. **3**(4), 305–360 (1976)
12. Jensen, M.C.: Value maximization, stakeholder theory, and the corporate objective function. J. Appl. Corp. Financ. **22**(1), 32–42 (2010)
13. Masulis, R., Cong, W., Fei, X.: Employee-manager alliances and shareholder returns from acquisitions. J. Finan. Quant. Anal. Working Paper. Available at: https://papers.ssrn.com/sol3/papers.cfm?abstract_id=2895745
14. Petry, S.: Mandatory worker representation on the board and its effect on shareholder wealth. Financ. Manage. **47**(1), 25–54 (2018)
15. Rhee, R.J.: A legal theory of shareholder primacy. Minn. L. Rev. **102**, 1951 (2017)
16. Wilkinson, A., Dundon, T., Donaghey, J., Freeman, R.: Employee voice: Charting new terrain. In: The Handbook of Research on Employee Voice: Participation and Involvement in the Workplace, pp. 1–16 (2014)

The Role of Digital Marketing on Tourism Industry in Bahrain

Yousif Alhawaj and Muneer Al Mubarak

Abstract This study reviews the potential and possible role of digital marketing in the tourism industry growth in Bahrain. It sheds light to the background and activities of tourism industry in general with particular attention to Bahrain. Many factors were found in this secondary study that could contribute to the growth of tourism industry. Among them are social media, digital influencers, online travel agencies, and email marketing. Undoubtedly, this technology has changed the way organizations use to approach, communicate, and provide solutions to different stakeholders. Policy makers and managers are invited to make use of this study to understand and improve the situation by attracting and retaining customers. Digital marketing, when is used properly can improve efficiency of the organizations.

Keywords Digital marketing · Growth · Tourism industry

1 Introduction

The concept of marketing has been subject to a lot of debate and regular evaluation since it was recognized as a distinctive domain and discipline. Over the last 50 years, many definitions of marketing have been provided, where each generation gives its own vision about marketing and what it means to them. That is why the concept of marketing has been redefined regularly so that it could fit new contexts such as non-profit, political, and social sectors. More recently, new breakthroughs in the field of information and communication technologies with the spread of the internet and the different social media websites, have brought more recent opportunities to re-define marketing once more again. Therefore, marketing's nature comes originally from the continuously shifting perspectives associated with the socio-cultural and technological perspectives leading marketing to be considered a vibrant rather than a stagnant domain/field [19].

Y. Alhawaj · M. Al Mubarak (✉)
Ahlia University, Manama, Bahrain
e-mail: malmubarak@ahlia.edu.bh

© The Author(s), under exclusive license to Springer Nature Switzerland AG 2024
A. Hamdan and E. S. Aldhaen (eds.), *Artificial Intelligence and Transforming Digital Marketing*, Studies in Systems, Decision and Control 487,
https://doi.org/10.1007/978-3-031-35828-9_100

The profound advancements in communication and information technology, particularly in the field of social media, with its widespread usage, have enabled customers of the tourism industry access to obtain whatever information they want about the different tourist destinations before they travel [15]. This has given support and enhancement to the concept of the empowered traveler [16]. Before they travel, customers can attain all the required information concerning the various aspects of the potential travel destination, including the ranking of accommodation, landscape, safety, entertainment, and prices [15]. All those various features affect the image of the destination the travelers seek to go to. In this respect, the successful tourist attractions of the future would be the places that succeed in utilizing all the potentials provided by digital marketing to attract as many tourists as possible, which, in turn, indicates the implied impact of digital marketing upon the tourism industry [20]. The last few decades have witnessed tremendous advancement in communication and information technologies. Those advances started with the inauguration of the internet and later backed and enhanced with the introduction of social media, which affected all aspects of our modern life. The tourism industry is one of the fields that has been affected directly with this digital web 2.0. Instead of going to traditional travel agencies to decide upon the next destination to spend a holiday, customers rely on information obtained online for potential destinations on websites using their laptops or smartphones with the ease of being at home.

The study focuses on two main parts: tourism industry in Bahrain and digital marketing. Tourism is a major pillar of the national economy of Bahrain. Additionally, digital marketing plays an important part in advertising for the tourism industry in Bahrain, both locally and internationally. It is a very effective tool in improving the image of Bahrain as a tourist spot to tourists globally. Therefore, this study is useful to all interested in the link between digital marketing and business performance especially in the tourism industry. The study could represent an addition to the literature associated with the tourism sector in Bahrain, and the impact of digital marketing for future academic research related to similar topics. The purpose of the study is to assess the role of and later on to examine the impact of digital marketing on the performance of the tourism industry in Bahrain. It looks for the usefulness of digital marketing tools for promotions, retaining customers and attracting new ones. It highlights main digital tools and platforms that aid in the growth of business performance. This study contributes to the existing research by surveying the impact of digital marketing on tourism in this part of the world as few academic research have been conducted. This study reviews the role of digital marketing in the tourism industry in the Kingdom of Bahrain.

2 Literature Review

The concept of relationships enabled by technology has been made possible when promoting and selling on the web. This technology-enabled relationship management is created when companies obtain information about consumer behaviors, needs, purchase patterns, preferences and make use of the information gathered to set product/service prices, personalize promotions, adjust product/service features. In basic terms, this allows companies to tailor the complete relationship with the consumer. Digital Marketing uses social media platforms like Instagram, YouTube, Twitter, and Facebook for promoting products and services. They use advertising campaigns, which are set to target the customers they want, based on different factors like age groups, location, interests, gender etc. whereas that is not possible with traditional marketing. Additionally, digital advertising has a different way to measure performance in comparison to traditional marketing. Traditional advertisements typically use methods like gross rating point (GRP) to measure performance, whereas advertisements in digital marketing have more characteristic methods such as cost-per- click (CPC) and cost-per-mile (CPM) [1].

There are three main forms of tourism, those which include domestic tourism, inbound tourism, and outbound tourism. Domestic tourism is when residents of one country travel within the same country, this is also known as staycations. Staycations in some countries are a big profit-making business bringing significant revenue to the economy. Popular domestic tourism destinations include the USA, China, and India. According to WTTC [21] domestic tourism makes up 73% of the total travel and tourism spending globally in 2018, making it an important driver of the tourism industry. Some countries depend on domestic tourism to decrease poverty, improve infrastructure, create jobs, and drive economic growth. Inbound tourism is when someone travels to a country, other than that of which they live in, for the purposes of tourism. This form of tourism is often dependent on certain variables like weather conditions, public holidays, and summer vacations. Each country has its peak season and low season. Many countries around the world heavily rely on inbound tourism as it brings in a lot of money to a country through foreign exchange and hard currency. This becomes even more beneficial when the tourists are coming from a country with a stronger currency. This makes those countries a target for advertising travel plans and packages. Reliance on inbound tourism can sometimes be negatively affected because of factors like political unrest, or a pandemic like the one we are facing today. Outbound tourism is the act of leaving your country to travel to another for the purpose of tourism. This does not include the purchases done before and after the tourism generating country.

In traditional marketing, organizations try to make the public conscious of a product or service for sale. As per an article written by McCauley [13], by using traditional marketing methods and tools such as TV, radio and/or direct mail, the response rate is expected to be someplace between 0.5 and 2% on outbound messages. In other words, if the message was sent out to around 1000 potential consumers, we can expect a range of 5–20 people to respond. Looked at from a different view, if an organization were looking to get 100 replies, it must reach out to a range of 5000–2000 potential consumers. Those numbers are broad estimate ranges and may differ depending on target market, product, service, and other variables. The tools being used to reach larger audiences in traditional marketing include newspapers, radio, television, direct mail, trade shows and magazines.

Traditional marketing tools can be expensive, and most of the time do not necessarily reach the audience the organization is targeting. Moreover, responses can be very low. Today, a big percentage of consumers rely on the internet in making buying decisions. Even those who still buy from shops locally, often use the internet to make their decision prior to purchasing. They will use the internet by using their laptops, computers, tablets, and cell phones to do some research on the product/service, and perhaps read customer reviews, and then make the purchase decision. Digital marketing offers a far greater opportunity, insofar as it is a two-way communication channel unlike traditional marketing. Digital marketing gathers information about consumer behavior which is then used in companies' marketing strategies and decisions. As consumers spend their time researching and gathering information to make purchases, companies are gathering information about the channel visitors, where information that is gathered by the internet/websites are known as a clickstream.

Digital marketing plays a strategic role in the advancement of the tourism industry in Bahrain. A clear strategy of the Economic Development Board (EDB) confirms that as part of Bahrain's National Economic Strategy. They certify that digital technology is at the center of the tourism industry to enhance and transform each of its sectors. The Bahrain Tourism & Exhibition Authority office (BTEA) started to direct its focus on digital followers by launching tourism campaigns to attract potential inbound and outbound travels, as well as local tourism. Growth in the tourism industry in Bahrain has a direct and indirect influence on the growth of the economy. In 2018, tourism contributed to 7% of Bahrain's GDP, and the contribution to the GDP by 2028 is expected to rise steadily to 10.9%. The tourism industry is a key contributor to the Bahrain economy [17]. Those contributions include:

- Direct contributions to the economy, including activities in hotels, transportation, airlines, travel agencies.
- Indirect contributions to the economy, including government and private investments to promote tourism in Bahrain.
- Induced contributions to the economy, including direct or indirect job creations in the tourism industry.

The EDB has invested over 10 billion dollars directly for tourism infrastructure projects. Those include hotels, museums, facilities that would aid and support in making Bahrain attractive and explorative to potential travelers. Bahrain is establishing a foundation to provide support to public and private investments, customers, facilities, and services. To ease and help travelers to visit Bahrain is the first step to becoming a successful hospitality and tourism destination. Bahrain expanded its airport capacity by 40% and is enhancing its national carrier's fleet. An increase in tourists to the country leads to enhancement to the infrastructure, as it is one of the strong supports for exploration in Bahrain. The study reviews main variables that could positively influence tourism industry in Bahrain. These are social media, digital influencers, online travel agencies and email marketing.

The importance of social media in the field of tourism lies in the fact that social media platforms are very popular among travelers as they use them extensively daily [7]. Therefore, the social websites enable tourists to integrate and share their experiences, reviews, and comments and use them as a source of information for visitors around the world [6]. This constitutes a big source of data and information for the next tourists when they want to go on a holiday, for instance, with an impressive entrance to travel-correlated communication. These data, information, feedback, and impressions about destinations undoubtedly play an essential factor for both travelers and the tourism industry. Consequently, social media is captivating a different role in information exchange as well as changing the form of tourism related, transforming the shape of the tourism sector [3]. The advantages of using social media when preparing trips and travel are vital because the process of classification is complex, due to the numerous characteristics of tourists. Thus, the utilization of social media websites has a positive attitude for tourists where they could gain four main benefits including hedonic, functional, social, and psychological benefits [18].

The concept of influencers refers to people valued in their communities, who have a big group of devoted supporters, fans, and audience. They create their own specific content (user-generated content) to form their reputation, where they are considered specialists in their communities [10]. According to Lin et al. [12], influencers are powerful human brands that positively influence the performance of companies related with them. Since those influencers use the internet, they are known as digital influencers. They are any person who broadcasts online who has a significant number of followers [4]. They are a type of micro-celebrity who have accrued many followers on social media and frequently use this social capital to gain access to financial resources, with the ability to influence potential buyers of a product or service by promoting or recommending it in social media [5]. Another more detailed definition of the term influencers refers to them as opinion leaders that are well-liked in a broader or sophisticated group of regular receivers, who, with his or her trustworthy actions—currently conducted more and more often on the internet—inspires trust, participates and convinces the addressees of his or her communication to make precise choices, such as those related to shopping, nutrition or worldview [8].

The fast developments in the field of information and communication technologies, especially those related to the internet, have generated massive opportunities for the traditional travel agencies to utilize this new horizon for targeted tourism offerings to a broader market. In response to the ever-increasing need for tourism information, many travel agencies have set up their own official websites to promote their tourism-related products and services to tourists and travelers. These websites play a key part in facilitating flows of information between consumers and businesses/agencies [9]. In the past, websites of travel agencies were used as an 'online brochure', providing just static information and data for the online consumers to see. Using new technologies such as artificial intelligence (AI), Internet of Things (IoT), and big data have made digital marketing more effective. Now, with a click of a button, travelers can find what they want, and when they want to plan their holidays and detestations well [2].

Email marketing represents an online marketing technique using email to send commercial information or advertisements to target customers. This email marketing represents an essential communication tool that could be exploited to attract new customers. In the field of tourism, the consumption of email marketing generates the opportunity to provide any potential interested guest to arrive at the right time, at the minimum cost, and the outcomes of such activities must be measurable, which forms a basis for decisions on future marketing activities [11]. Email marketing allows users to get acknowledged with the newest propositions by the hotels, special prices on holiday packages, membership clubs, room availability and so much more [14].

3 Conclusion

The role of digital marketing on tourism industry and especially in Bahrain has been reviewed. The review shows that digital marketing has a strong influence on the tourism industry in the Kingdom of Bahrain when it comes to promotion and marketing related activities. It has been used and relied on heavily to connect and reach potential customers. There has been an indication that mostly used social media platforms are Facebook and Instagram. The research study provides useful information on tourism industry. By understanding the different tools of digital marketing and looking at its powerful influence on individuals, some points need to be considered to enhance overall performance. Company owners, whether in the hospitality sector, retail, food and beverages or transportation, should identify whether they are connecting and reaching out to current customers and potential ones. Then they should create digital marketing plans and strategies to ensure efficiently reaching their target customers. The tourism industry in Bahrain has so much to offer but needs to work on the right digital marketing tools, to attract and encourage consumers to participate and experience what it has to offer. The different digital marketing tools should be further studied and used to ensure maximum reach and connection. Four main variables that to be considered with the aim to improve tourism industry are social media, digital influencers, online travel agencies, and email marketing.

3.1 Implications

Digital marketing is expected to have main influence on Bahrain's tourism industry, as it is one of the main pillars in the national strategy. Many opportunities are there for researchers to explore such area and for policy makers and practitioners to realize the potential of digital making in business growth.

3.2 Limitations and Future Studies

This study is based on secondary research. The scope here is to highlight the main points on the topic. Future research can investigate the area by conducting applied research in the four recommended factors and possibly more, in Bahrain and beyond

Bahrain to explore the potential and impact of digital marketing on tourism industry growth.

References

1. Açıkel, E., Çelikol, M.: Dijitoloji. İstanbul: Kapital Medya Yayıncılık (2012)
2. Al Mubarak, M.: Sustainably developing in a digital world: harnessing artificial intelligence to meet the imperatives of work-based learning in Industry 5.0. Development and Learning in Organizations, vol. ahead-of-print No. ahead-of-print (2022). https://doi.org/10.1108/DLO-04-2022-0063
3. Barger, V., Peltier, J.W., Schultz, D.E.: Social media and consumer engagement: a review and research agenda. J. Res. Interact. Mark. **10**(4), 268–287 (2016)
4. Chaffey, D., Ellis-Chadwick, F.: Digital Marketing. Pearson, Harlow, UK (2016)
5. Cotter, K.: Playing the visibility game: how digital influencers and algorithms negotiate influence on Instagram. New Media Soc. **21**(4), 895–913 (2019). https://doi.org/10.1177/146144 4818815684
6. De Vries, L., Gensler, S., Leeflang, P.S.: Effects of traditional advertising and social messages on brand-building metrics and customer acquisition. J. Mark. **81**(5), 1–15 (2017)
7. Gao, H., Tate, M., Zhang, H., Chen, S., Liang, B.: Social media ties strategy in international branding: an application of resource based theory. J. Int. Mark. **26**(3), 45–69 (2018)
8. Górecka-Butora, P., Strykowski, P., Biegun, K.: Influencer Marketing Od A Do Z. WhitePress, Bielsko-Biała, Poland (2019)
9. Jeong, M., Lambert, C.: "Measuring the information quality on lodging websites. Int. J. Hosp. Info. Tech. **1**(1), 63–75 (2016)
10. Kartajaya, H., Kotler, P., Setiawan, I.: Marketing 4.0: Moving from Traditional to Digital. Wiley, East Orange, NY (2017)
11. Laurell, C., Sandström, C., Berthold, A., Larsson, D.: Exploring barriers to adoption of virtual reality through social media analytics and machine learning–an assessment of technology, network, price and trialability. J. Bus. Res. **100**(C), 469–474 (2019)
12. Lin, H., Bruning, P., Swarna, H.: Using online opinion leaders to promote the hedonic and utilitarian value of products and services. Bus. Horizon, Elsevier **61**(3), 431–442 (2018)
13. McCauley, D.: Internet marketing vs traditional marketing—3 key points for small businesses (2013), retrieved from http://www.examiner.com/article/internet-marketing-vs-traditional-mar keting-3-key-points-for-small-businesses
14. McColl-Kennedy, J., Zaki, M., Lemon, K., Urmetzer, F., Neely, A.: Gaining customer experience insights that matter. J. Serv. Res. **22**(1), 8–26 (2019)
15. Oliveira, E., Panyik, E.: Content, context and co-creation: Digital challenges in destination branding with references to Portugal as a tourist destination. J. Vacat. Mark. **21**(1), 53–74 (2015)
16. Roper, S., Caruana, R., Medway, D., Murphy, P.: Constructing luxury brands: exploring the role of consumer discourse. Eur. J. Mark. **47**(3/4), 375–400 (2013)
17. Saniya: Tourism industry and its impact on Bahrain economy (Blog) (2019), retrieved from https://www.businesssetup.com/blog/tourism-industry-and-its-impact-on-bah rain-economy (Accessed 23 January 2021)
18. Sheng, J.: Being active in online communications: firm responsiveness and customer engagement behavior. J. Int. Market. **46**(C), 40–51 (2019)

1188 Y. Alhawaj and M. Al Mubarak

19. Sheth, J.N., Sisodia, R.S.: Does Marketing Need Reform? Fresh Perspectives on the Future. Routledge, New York (2015)
20. Solnet, D., Baum, T., Robinson, R.N., Locksto, R.: What about the workers? roles and skills for employees in hotels of the future. J. Vacat. Mark. **22**(3), 212–226 (2016)
21. WTTC: Domestic tourism important and economic impact (2018), retrieves from https://wttc.org/Research/Insights/moduleId/1036/itemId/17/controller/DownloadRequest/action/QuickDownload#:~:text=Governments%20use%20domestic%20tourism%20as,of%20non%2Dwage%20tourism%20benefits